IV International Scientific Congress "Society of Ambient Intelligence" (ISCSAI 2021)

SHS Web of Conferences Volume 100 (2021)

Kryvyi Rih, Ukraine
12 - 16 April 2021

Editors:

Serhii Hushko
Victoria Solovieva
Andrii Shaikan
Inesa Khvostina
Serhii Semerikov

ISBN: 978-1-7138-2871-6

Printed from e-media with permission by:

Curran Associates, Inc.
57 Morehouse Lane
Red Hook, NY 12571

CURRAN ASSOCIATES INC. proceedings .com

Some format issues inherent in the e-media version may also appear in this print version.

For additional information, please contact EDP Sciences – Web of Conferences at the address below.

EDP Sciences – Web of Conferences
17, Avenue du Hoggar
Parc d'Activité de Courtabœuf
BP 112
F-91944 Les Ulis Cedex A
France

Phone: +33 (0) 1 69 18 75 75
Fax: +33 (0) 1 69 28 84 91

contact-edps@webofconferences.org

Additional copies of this publication are available from:

Curran Associates, Inc.
57 Morehouse Lane
Red Hook, NY 12571 USA
Phone: 845-758-0400
Fax: 845-758-2633
Email: curran@proceedings.com
Web: www.proceedings.com

TABLE OF CONTENTS

Statement of Peer review

In submitting conference proceedings to *Web of Conferences*, the editors of the proceedings certify to the Publisher that

1. They adhere to its **Policy on Publishing Integrity** in order to safeguard good scientific practice in publishing.
2. All articles have been subjected to peer review administered by the proceedings editors.
3. Reviews have been conducted by expert referees, who have been requested to provide unbiased and constructive comments aimed, whenever possible, at improving the work.
4. Proceedings editors have taken all reasonable steps to ensure the quality of the materials they publish and their decision to accept or reject a paper for publication has been based only on the merits of the work and the relevance to the journal.

Title, date and place of the conference

IV International Scientific Congress "SOCIETY OF AMBIENT INTELLIGENCE – 2021" (ISCSAI 2021)

April 12-16, 2021 Kryvyi Rih, Ukraine

Proceedings editor(s):

Serhii Hushko
Victoria Solovieva
Andrii Shaikan
Inesa Khvostina
Serhii Semerikov

Date and editor's signature

March 10, 2021

https://doi.org/10.1051/shsconf/202110000001

On the Way to Novelty - Society of Ambient Intelligence

Serhii Hushko[1*], *Victoria* Solovieva[1], *Andrii* Shaikan[1], *Inesa* Khvostina[2], and *Serhii* Semerikov[3]

[1]State University of Economics and Technology, 50005, Kryvyi Rih, Ukraine
[2]IFNTUNG, 76000 Ivano-Frankivsk, Ukraine
[3]Kryvyi Rih State Pedagogical University, Kryvyi Rih, 50086, Ukraine

Abstract. The ability to bring benefit to the society with the help of the artificial intelligence technologies within the short time drives the developers in many spheres. Modern developments are used to confirm the economic theories, in law-making, technical developments in the field of verification, data updating, security and control.

Ambient Intelligence: Concepts and Understanding

Ambient Intelligence combines several important areas of computing which are connected through new technology available. It allows developments which change the way system help us or can potentially help us in our daily life.

While the advent of the internet was certainly a ground-breaking milestone, digital technologies have not stopped evolving. The time when we had to turn on the computer and login to browse the worldwide web is coming to an end. Now, an important strand of information processing technologies is centered on the so-called « ubiquitous computing».[1] This concept derives from the philosophical or psychological observation that the best technologies are those that fade into the background when used. [2] Together, they connect people and deliver practical information through the cheaper, more efficient and more tailored services, while weaving themselves into the fabric of everyday life.[3] This is the essence of Ambient Intelligence (AmI), a concept built around the idea that, in a not so distant future, autonomic smart environments, based on profiling technologies, embedded sensors and sophisticated autonomous software, will take an unprecedented number of decisions for us and about us, to cater to our inferred preferences.[4]

The topic of the artificial intelligence is quite urgent and needs the scientific and practical evaluation of the prospects for its application. The speed of the artificial intelligence development will be determined by the external and internal factors. The external factors mean the speed of the data exchange within the "brains" of the artificial intelligence, within its network, as well as the speed of calculations, associated with it, and the availability of the information scopes, on the basis of which it will form its assumptions. The internal factors include the availability of energy to maintain its vital activity and the presence of channels for the receipt of the new information, so that the artificial mind, figuratively speaking, would enrich itself.

ISC-SAI 2021 - current trends

The realization of IV International Scientific Congress "SOCIETY OF AMBIENT INTELLIGENCE – 2021" allows us understand what we mean by the notion "artificial intelligence"; what trends and solutions will make it possible to use it as efficiently as possible not only in the business processes of the companies, but also in all the spheres of modern society.

The Congress calls for the organization of the many - sided, open scientific platform for the cooperation of the academic community and the real sector of the economy with the aim of obtaining the new knowledge, the exchange of information and the advanced experience, the examination of modern tendencies in the economy, the key factors and peculiarities of information's technological development in a society on conditions of globalization.

In order for all the participants to understand the essence of the events, the branding of the locations and all the materials of the Congress was realized (fig.1) through the brand positioning at different stages, taking into account all the possible communications with the audience.

Fig. 1. Logotype ISC-SAI 2021.

The urgency of the information contents of the Congress is one of the main factors of its successful promotion. Many of the web resources, living on the Internet areas, do not meet the expectations of their

* Corresponding author: gushko77@gmail.com

owners, - the fact, happening quite often and being a very unpleasant circumstance - they do not return the funds that were invested into the creation and the SEO promotion of the product.

The Congress website is an online brochure of the scheduled event with widened functionality for the registration and promotion. Since there is no time for the searching optimization in this sphere, the concept of the Internet resource was realized: the page, aimed at the client's performance of the target action, consisting of ten structured blocks with the description of information. The resource provides the complete information of the events: a detailed description of the event and the format of its realization, the date and location, a list of speakers and the general information of them (fig.2).

Fig. 2. Internet-Resource ISC-SAI 2021.

One of the most important factors in the success of business development is being in contact with the customer. The attraction of each new participant means the advertising budgets, and the longer the Congress can work with the already attracted participant, the higher will be the return of the investments. Thus, it is necessary to constantly work with the participants, by measuring their level of satisfaction.

The organizing committee used indices that allow us examine the change's dynamics in the relations of the participants with the brand within some time, to compare the data obtained with the average indicators of the branch:

- NPS (Net Promoter Score) – the consumer loyalty index, reflecting the readiness to recommend the Congress to the other clients;
- CSI (Customer Satisfaction Index) - the index of participants' satisfaction.

The presented researches are aimed at the identification of the positive and negative sides of the event, as well as at the determination of the various criteria influence on the attitude of the participants towards the Congress (fig.3).

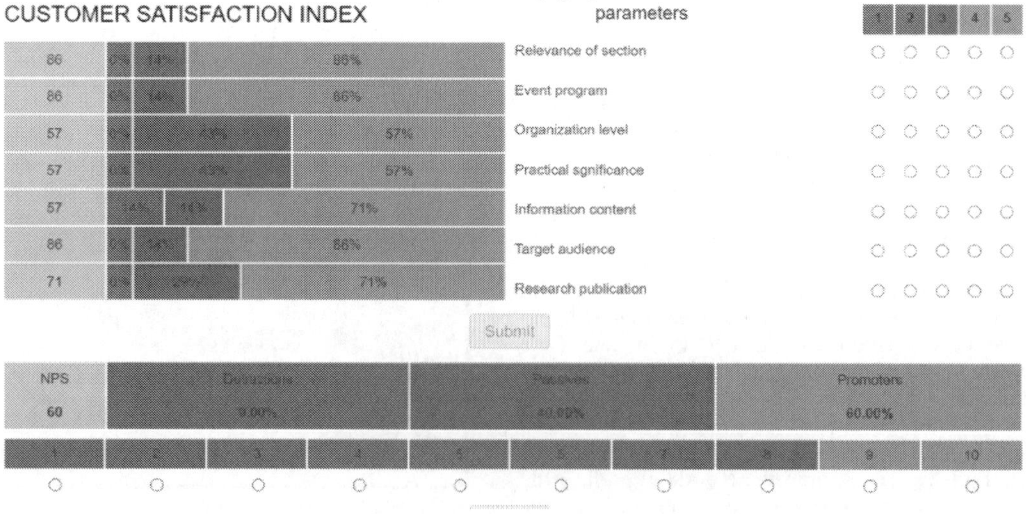

Fig. 2. Indices NPS/CSI ISC-SAI 2021.

To visualize and to analyze information about the Congress indicators, the processes and their efficiency, a dashboard of indicators was realized. The dashboards of indicators are the effective management tool that meets reasonably all the changing needs of the Congress Organizing and Programming Committees (fig.3).

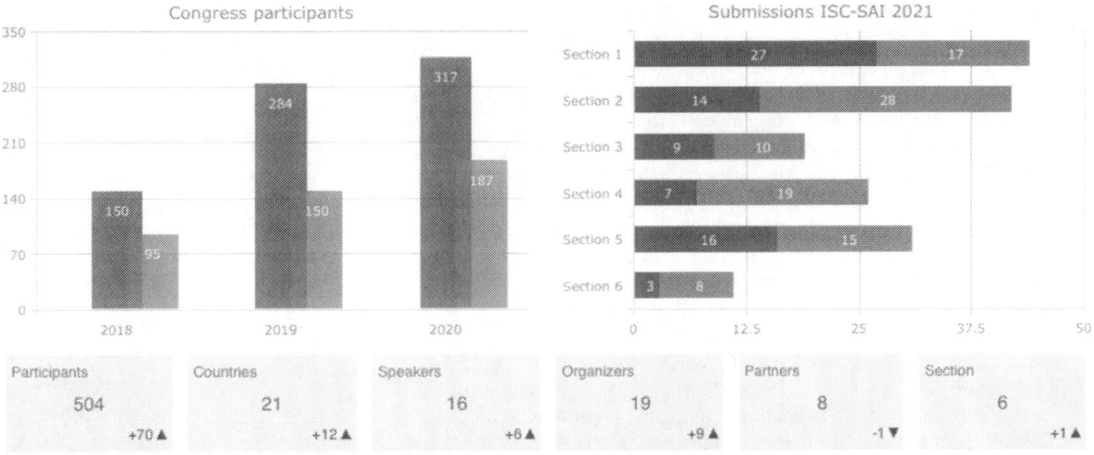

Fig. 3. Indicators Dashboard ISC-SAI 2021.

Conception of Congress:

- the work of the Congress is planned in the format of the panel discussions, according to the directions;
- the arrangement of the discussion panels according to the directions foresees the invitation of speakers from the academic and business spheres;
- the realization of the international student's competition of projects (the competition of cases' decision), according to the Congress directions. The competition will be held at two stages: the 1-st stage (national) – is the competition of the research projects/business-ideas; the 2-nd stage (international) is the decision competition of the cases, represented by the companies-partners of the Congress (the competition of projects);
- the organization of a platform of the legal clinics with the purpose of the experience exchange, the study of the best foreign and national practical trainings of the legal clinics, dealing with the different directions of the juridical education;
- the activity of the organizational and scientific committee is exercised via the procedure of the collective acceptance of organizational decisions with the division of responsibility levels of each of the organizers;
- the realization of the communications' format "business — universities – schools".

The Program Committee separated the following urgent directions for the discussion and approbation of research in the field of the Ambient Intelligence at the Congress:

- Information Technologies and Business Innovations;
- Business Communication and Linguistics;
- Digitalization of Society in National Legal Systems;
- Social Work and Artificial Intelligence;
- Ecology and Living Space;
- Mechanical Engineering and Mechatronic Systems

The investigations in these directions unite scientists, practical researchers and business representatives. This alliance foresees presenting of the solutions that may be adapted to the modern requirements, by offering the new models of distribution, the new behavior of the clients and the new business-models at that. The researchers offer the stable plans that will help to reduce the global rise in temperature, to eliminate the poverty and to protect the environment from the further destruction.

The information technologies play an important role in the creation of the new economic possibilities and business-innovations today. The aim of this direction is – to study the role of the modern information technologies in the acceleration of the economic growth and the creation of the new economic possibilities, the development and the realization of the digital economic policy and the influence of the digital economy on the national and the international investments and the creation of cost.

The digital technologies develop intensively with the growing scope of information, the increase of the vast information tracks and data bases. They are widely used in all the spheres of social life, in the activity of the state and the social institutes, - the fact, characterizing the significant development of the modern society. And due to that, the law becomes not only the tool, which provides the digitalization of the economy and the other branches, but it changes its forms, contents, structure and mechanisms. Therefore, the new digital reality presents the new demands to the science of law and to the legal practice. It allows develop the efficient tools and models of legal regulation. And the juridical investigations in this direction offer the new legal solutions for the practical researchers.

Due to the tools of the information technologies, the style of communication in the modern world has suffered the radical changes, by giving the place to the revolutionary method of communication, which ignores the traditional barriers. The new approaches to the research of the theoretical and the applied problems of the intercultural communication, the transportation of the pedagogical strategies of language training in education, the modern forms of education integration are offered today.

The artificial intellect has already become the integral part of all the spheres of life, which offers the unique methods of data processing to the modern world. And the social sphere has the new qualitative and organizational transformation today, due to the use of the information technologies. The investigations in this direction foresee the use of the artificial intellect in the social sphere and the prognostication of demand on social services.

It's important to understand that the transformation of the society and its "digitalization" helps, first of all, to promote the tools for the solution of different urgent tasks: the increase of labor productivity, the reduction of the part of the monotonous or dangerous procedures for people and, at last, for the widening of human possibilities.

The digital transformation happens slowly, unequally, it's faced with the great number of the barriers - both technological, political and the social ones. The formation of the digital society is gradually freed of the hard labor and the new professions are created, but the realization of the modern tools are spread at the very different levels and they may cause the unexpected social effects already today.

Thus, the introduction of the environmental intellect into life in all sections of the society, including the educational one too, doesn't only push a person to the life side of the road so far, but it makes him (her) to develop in the new directions.

The realization of the platform in the ISC-SAI 2021 format by the University community with the combination with the real sector of the economy will promote for the unification of the professionals and the researchers of the different districts, who discuss the modern calls of the digital society in the thematic tracks.

The Congress participants are offered to present both the theoretical and practical results and to take part in the discussion panels on the urgent problems of the present time too. The presentations of the practical works of the professionals from the different countries, the discussion of challenges, concerning the introduction and development, and the exchange of the conclusions and the corresponding results are foreseen.

References

1. This concept was firstly used by Mark Weiser of IBM. See Mark Weiser, 'The computer for the 21st century' (1991) 265 Scientific American 94.
2. N. Dijk, 'Property, Privacy and Personhood in the World of Ambient Intelligence' (2010) 12 Ethics and Information Technology 57, (n 13), p61.
3. Mary Rundle and others, At a Crossroads: 'Personhood' and Digital Identity in the Information Society (STI Working Paper 2007/7, 2008)
4. Mireille Hildebrandt and Bert-Jaap Koops, 'The Challenges of Ambient Law and Legal Protection in the Profiling Era' (2010) 73 Modern Law Review 428

Digital Audit as an Imperative for Ukraine's Way out From the COVID-crisis and a Tool to Increase the Competitveness of the State

Karina Nazarova[1], *Mariia* Nezhyva[1*], *Volodymyr* Hotsuliak[1], *Nataliia* Novikova[1], and *Oleksandra* Fedorenko[1]

[1]Kyiv National University of trade and economics, 02156, Ukraine

Abstract. The article is devoted to the issues of organization and prospects of audit development in terms of digitalization. The state and the measures needed to improve audit effectiveness in the context of automation are investigated and analyzed. The factors that negatively influence the audit digitization process and the possible directions of development of audit activity in the computerized environment are identified. The use of automation helps auditors process more data and focus on identifying risks, allowing for higher quality audits. Classification of information technology of audit, conditions for the development of audit software, factors that impede audit automation are given. Audit digitization is a new level in the audit field that is rapidly evolving and rapidly expanding in enterprises, especially those that use an automated accounting method. The software currently used by auditors needs to be refined to meet the demands of today's world. Digitization will become relevant for all audit firms seeking to gain competitive advantage and take a place in the audit services market.

1 Introduction

The rapid spread of COVID-19 is forcing many businessmen to change their priorities. Thus, companies that have traditionally provided services offline are trying to adapt their services to new realities. Despite the negative impact of the pandemic on the world economy, it significantly accelerates the development of digitalization.

Due to the digital transformation of companies there is a need to improve and optimize the functions of auditors. Large amounts of information must be used effectively to conduct a high-quality audit. Instant access to all the necessary information will allow professionals to quickly identify risks and generate business solutions.

According to the COVID-19 Global Study: Impact on Business and Counteraction, conducted by the Association of Chartered Certified Accountants (ACCA), the COVID-19 pandemic had a significant impact on auditing [1]. In our opinion, auditors in a COVID crisis need to plan their activities in accordance with current research. A survey was conducted on the impact of the pandemic on audit firms, and it was found that there were significant difficulties in working with clients. The majority of respondents (53%) said that they feel some pressure when working with clients, and more than a third (36%) said that they are not able to meet the deadlines for reporting. A quarter of respondents said that the process of gathering audit evidence had become much more complicated, and 27% said that the audit risk associated with valuing assets, meeting obligations and ensuring continuity of operations had increased.

In the context of the COVID-19 pandemic, many businesses were faced with the question of organizing further activities, as some of them already had to submit financial statements. Limited conditions for the operation of companies have a negative impact not only on themselves but also on auditors.

Under the influence of restrictions caused by the need for self-isolation, businesses faced difficulties in preparing reports and, therefore, in conducting audits from the very beginning of quarantine. Because these limitations affect not only the ability to meet and relocate, but also the auditors' ability to gather sufficient evidence to provide an audit opinion, there is a direct need for auditors to consider alternative approaches to the audit [11].

Auditors need to consider the impact of COVID-19 on the following factors:

- gathering sufficient appropriate audit evidence. At the same time, auditors must recognize that in a pandemic, it is not possible to conduct standard audit procedures in the normal manner, so the methods of gathering evidence will require adjustment or total changes. There may also be a need to develop alternative procedures;
- in the case of an audit of a group of companies, it is necessary to pay attention to how the audit of individual units within this group is planned;
- significant attention in the audit will require the auditor's ability to provide a reasonable and reasonable assessment of the continuity of activities

* Corresponding author: marijka@ukr.net

and the absence of the need or preconditions for the liquidation of the audited company;

- ability to obtain reliable and validated information from the auditor about the actual extent of the impact of COVID-19 on the activities of the audited entity and to compare it with the published;
- the auditor will need to assess the need for some of the planned procedures, as well as discuss with the auditee the possibility of providing information as quickly as possible due to the rapid change in the situation in the country [2, 19].

It should be noted that audited entities, in turn, are more profitable to work with auditors and promptly provide reliable information so that the auditor can ensure that the level of disclosure during a pandemic does not differ from the level of disclosure that auditors would expect to see in financial statements, normal operating conditions of the enterprise. In this way, auditors will have a real opportunity to communicate the existing risk to the business, industry or specifically to the company [20].

Significant growth of information, working with Big Data requires the development of new approaches to auditing. For auditors, the key point is digital audit.

Problematic and controversial issues need to be addressed in order to formulate a comprehensive approach to audit digitalization uring economic transformation and restricted business practices caused a global pandemic 2019-2020.

The economic consequences of the pandemic have been studied by Ukrainian and foreign scientists such as Kulitsky S. [9], Stavytsky A. [18], Nezhyva M. [14] and Miniailo V. [14], Naveen Donthu [6] and Anders Gustafsson [6], Alexander W. Bartik [3] and other.

S. Kulitsky studies the problems facing Ukrainian society, including economic ones [9]. He determines the impact of the pandemic on the economic and social spheres of life. Naveen Donthu [6] and Alexander W. Bartik [3] analyze consequences of the COVID-19 for different industry sectors and the economy in general. They found that a lot of businesses, especially small firms, were closed and employment had fallen.

All authors note that businesses have to use innovative strategies to meet the customer demands during the COVID-19 crisis. A. Stavytsky [18] says that the pandemic and economic crisis should form a new country in which there will be changes in the structure of the economy. M. Nezhyva and V. Miniailo [14] determine problems faced by auditors and suggest ways to solve these problems by using professional software products.

The purpose of the article is to determine the role of digitization in the financial spheres, in particular in the sphere of audit, to study the results of technology implementation in foreign countries and to identify promising areas of development for Ukraine in the conditions of COVID-19.

2 Materials and methods

The materials of the study were indicators of the spread of the COVID-19 pandemic in the world, public publications of scientists, official reports of international organizations, regulations, etc. In the study, the theoretical and methodological basis consisted of general scientific and special methods of cognition, based on a dialectical approach to the consideration of economic processes. To solve the tasks were used: methods of analysis and synthesis – to study the economic essence of the audit in terms of COVID-19; system approach – to determine the methodological foundations of the enterprise management system; methods of general analysis, etc. In the study, used general and specific methods of understanding the objective nature of economic phenomena and processes to analyze the impact of the COVID-19 pandemic on business; methods of causation – to study the impact of digitalization on the audit process; methods of grouping and compilation, sample observation – for statistical data processing. Based on the methods of systems analysis and scientific abstraction, general conclusions are formulated. In addition, a set of complementary methods of scientific research of economic processes and phenomena using statistical and analytical materials, as well as the results of their own research.

3 Statement of the main research

In Ukraine, in 2019, the Ministry and the Committee for Digital Transformation of Ukraine [13] were established to implement digital policy. The purpose of the ministry is to accelerate the digitalization of the country, which aims to bring online at least 50% of all public services.

The newly created ministry will be responsible for the formation and implementation of state policy in the field of digitalization, open data, national electronic information resources, implementation of electronic services, etc.

The primary goal of this ministry is to provide full coverage of Ukraine with high-quality high-speed Internet. Further plans are to teach about 6 million people digital skills. It is also planned to launch an online platform where everyone will be able to acquire such skills for free.

In addition, the ministry will support the IT sector, where it is planned to receive up to 10% of the country's GDP. The developers are currently working on the ACTION platform, through which government services will be provided online [7].

The International Auditing and Auditing Standards Board (IAASB) has developed a strategy for 2020-2023, taking into account the active development of technology. The main tasks for this period include: improving the internal processes of auditors through the introduction of new technologies; correct allocation of resources; compliance with the growing needs of key stakeholders.

In connection with the adoption of tax changes in Ukraine as a step in the implementation of the BEPS

Plan [10] and the prepared package of new recommendations of the Organization for Economic Cooperation and Development, tentatively called "BEPS 2.0 Plan", which aims to implement measures to address tax challenges global digitization and which consists of two parts.

The first part recognizes how the current rules of international taxation can be modified to take into account the digitalisation of the world economy and the changes it will bring. The second part deals with other issues that remain open. In particular, it is necessary to deprive multinational corporations of the opportunity to withdraw profits to low-tax jurisdictions.

Experts note that the process of digitalization will actively contribute to the growth of the world economy in the coming years and in 2025 the digital economy will cover almost a quarter of world GDP [13].

Under quarantine conditions, given the existing restrictions on movement and actual access to businesses, alternative procedures are used to obtain audit evidence. They use remote photo and video recording of the inventory process, online communication with the client's staff through secure communication channels.

It was allowed to extend and publish the financial statements and the auditor's report, which greatly simplified the audit process and allowed to postpone some of the audit procedures related to direct access to the primary documents and assets [15].

Auditors should consider how to gather sufficient audit evidence in the face of existing constraints. Given that most now work remotely, workflow control is required. This can be done with various software [12]. An example is CaseWare Cloud, which allows a company to control all workflows. And in combination with CaseWare Working Papers, the audit becomes more efficient and structured, because this software is fully automated. In addition, all stages of audit procedures comply with International Standards on Auditing.

These programs help auditors to observe the performed procedures. They allow you to view tasks online from any device, allowing auditors to stay up to date of all workflows, regardless of their location.

The following benefits of CaseWare Cloud can be highlighted:

- control work tasks by using just mobile phones or computers;
- no need for employees to be in the office;
- ability to plan an audit;
- constant updating of data, which allows you to work with current information.

In terms of privacy, the program is equipped with high-level physical security features, including compliance with SSAE 16 standarts, has PCI level 1 certification, ISO 27001 and is compatible with all security systems.

In terms of audit software, the most used programs are Word and Excel. In small firms, auditors replace specialized programs by performing simple calculations, printing standard forms of working audit documents; make queries to the electronic database; check separate calculations carried out at different areas of accounting; form accounting registers using the client's electronic database and conduct a comprehensive analysis of the financial condition of the economic entity using separate programs [4, p. 9]. In "Big 4" companies, any element of work can be done with the help of specialized programs adapted to the work of each company. Such programs include random number generators, databases that allow you to store large arrays of data, applications for automatic construction of analytical graphs, charts, and more.

Ernst & Young was able to implement a huge number of digital innovations in auditing, which improved risk identification, reduced the burden on customers and provided operational benefits and quality audits.

EY Canvas is the first online audit program, the advantage of which is the ability of the auditors to communicate with the client, regardless of their location. Audit firms have the ability to remotely perform the audit process, coordinate and manage their own employees, while providing high quality audit services [5].

With the program, auditors have access to the overall audit plan, can review the tasks on a daily basis and share their own intermediate audit results.

EY Canvas has many benefits, such as [5]:

- existence of a central plan and performing audit by teams according to this plan and a single methodology;
- quick reaction, which allows you to immediately report the findings;
- rapid adjustment of the strategy to different geographical regions and reaction to the constantly changing environment;
- constant monitoring of the audit process and timely response to inquiries;
- optimized communication with customers through an integrated online network.

In addition to EY Canvas, Ernst & Young developed a program called EY Canvas Client Portal, which allows clients to communicate directly with auditors so they can look after the progress of audit.

This program allows to [5]:

- reduce e-mail inquiries and provide better communication with customer, which saves time;
- review the status of inquiries;
- improving customer data security;
- multilingual support (supports 10 languages).

Automation and robotics processes are becoming more widespread in the world, the advantages of which are:

- security – non-interference technology, can be configured for existing IT systems in the "as is" mode, flexible configuration;
- action monitoring – recording and storing all actions in the system;
- retention and development of employees – employyees switch to painfully intellectual tasks;
- speed – reducing the time to complete tasks, increasing productivity;
- significant cost reduction;

- 24/7 mode;
- reliability – systematic work without breaks, sick leave, weekends and holidays.

Ernst & Young actively uses "smart" technologies for specific audit procedures [17]:

1) inventory - for accurate measurement of volumes of bulk materials. It is provided as a tool of visual analytics with the accuracy of measurement of 95-97% and which can be carried out by means of the smartphone camera is developed.

2) contract analysis – reduced deadlines and increased accuracy of contract verification through the use of artificial intelligence. A tool has been developed including further functions: obtaining primary rental documents with a text in the form of a picture; use of optical character and word recognition; identifies and retrieves key data; performs data verification; marks parts of the text based on their compliance with accounting rules; sends requests for the involvement of specialists to analyze deviations. As a result, auditors can focus on performing other intellectual tasks, such as preparing recommendations for the client.

3) automation of administrative processes – automatic conversion of scanned copies of documents into electronic; depersonalization of documents with the help of machine learning technology; real-time tracking of the status of performance of works within the framework of the audit task (EY Canvas).

Digitization is a new trend of the present, which requires changes in personnel, new management styles, and other systems of work organization. Recent studies indicate that most traditional companies are not ready for digitization yet. The main problem consists of people who are used to working with traditional tools and methods of interaction with each other and with clients.

The corresponding changes caused by the digitalization of financial and economic processes at enterprises led to the emergency to create conceptually new, digital economy. The digital economy is an economy based on digital computer technology. The digital economy is also sometimes referred to as the internet economy, the new economy, or the web economy. The digital economy means manufacturing, selling and delivering products through computer networks.

The digital economy and digitization processes, however, aim to use information technology in as many processes as possible.

Computer auditing should be understood as a high level of automation of auditor activity, characterized by: the use of the latest information technology as a key tool in the process of preparation and verification in the computer information system; an audit approach that assesses the reliability of the computerized information system environment as a basis for drawing conclusions about the reliability of the financial statements.

Thus, the purpose of information technology (IT) in the audit can be divided into security and functional (Fig. 1).

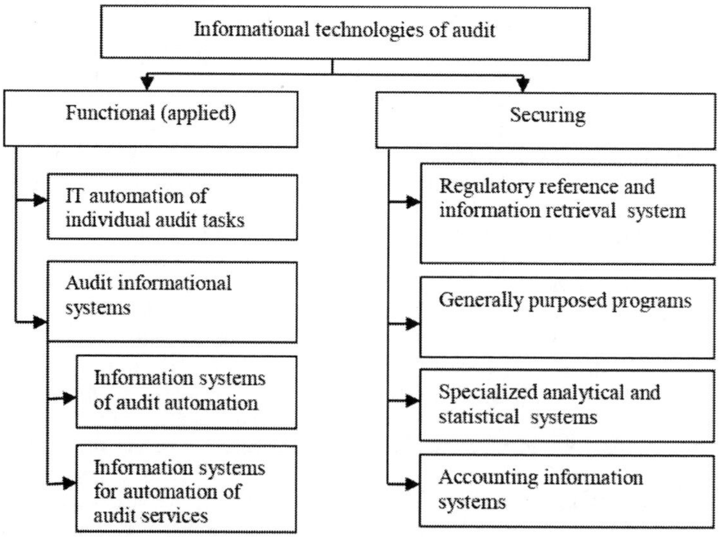

Fig. 1. Classification of information technology of audit

The use of automated software in the audit allows to obtain information faster from the enterprise-client database, gives the opportunity to process large amounts of information, accelerate the implementation of audit procedures, makes the documentation of results more convenient.

Computer-assisted audit methods can be used to perform a variety of audit procedures, including:

- testing of information processing in the client's accounting system;
- analytical review of procedures to identify uncertain cases;
- access to data files and libraries;
- tests for compliance of software and management and accounting systems.

Approximation to international auditing standards and practices requires the introduction of up-to-date technologies that would automate the financial statement verification process and help accomplish assurance and related tasks. You can formulate a number of requirements for the audit software as follows (Fig. 2).

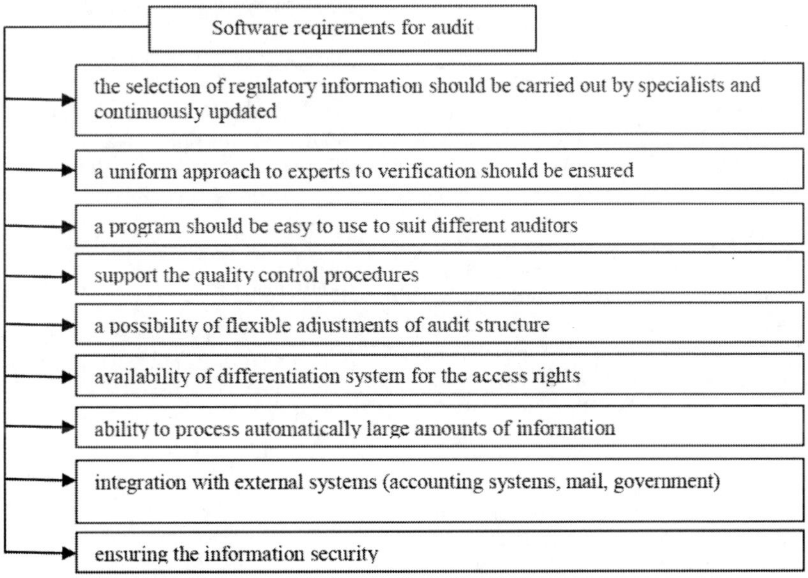

Fig. 2. Conditions for the development of audit software

However, the use of automated software in auditing is complicated by certain factors. Among them is the fact that the client companies use different software that does not allow to fully perform automated audit procedures. Thus, the main requirement for audit programs is their flexibility, ie the ability to adapt to the peculiarities of accounting in a particular enterprise.

There are certain reasons that hinder the implementation of software in the audit process (Fig. 3) [8].

The main prospects for the development of audit automation programs include:

- integration with the accounting system and accounting software;
- creation of the concept and the detailed instruction on adjustment by users of algorithms of formation of the reporting on the basis of the imported data of accounting of the enterprise;
- embedding and regular updating of audit methodology in terms of auditor's working documents, database of potential (typical) violations and distortions;
- embedding the ability to describe the customer's business processes;
- calculation of key indicators based on reporting data [16].

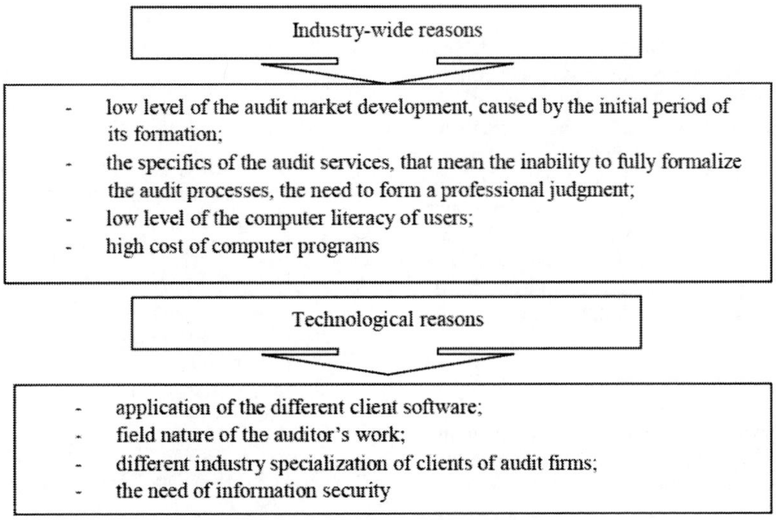

Fig. 3. Factors that prevent the digitalization of the audit

Audit information technology is a set of methods and procedures that provide the functions of collecting, accumulating, storing, processing and transmitting data using technical means to achieve the audit objective in the best possible conditions.

Digital audit provides the ability to perform audit procedures in accordance with the standards, taking into account the use of information visualization and analytical tools in order to obtain the required level of audit confidence on large data sets, as well as analytical procedures for non-financial audit

The current state and dynamics of business processes of the XXI century is accompanied by a significant increase in the amount of information that the company receives, processes and produces. This has led to the formation of large amounts of information, which in the digital economy has received its own name - big data (large arrays of information). In the end, Big data as a phenomenon has created a separate market for companies that provide a wide range of services related to such information - storage, protection, processing, visualization of large amounts of information and more.

In each market, there are companies that are audited for their own needs (so-called initiative audit) or according to the legislation of the country or supranational associations (independent external, mandatory audit). As a result of these information transformations in the digital economy there is an important, almost equal user of big data enterprises - an audit firm, the results of which are of public interest. Execution of audit procedures requires obtaining and processing large data sets not only of one business entity, but a certain number of clients-clients of the audit, which are in the portfolio of the audit firm.

In addition to the direct audit of financial statements, companies additionally order or will be required to order as mandatory non-financial audit: controls, processes, IT-audit of software and more.

In the field of audit services, there was an evolutionary transition from paper procedures to procedures in Microsoft Office software, but with the advent of Big data, such a transition to a universal mass tool for processing large databases has not yet taken place. The absence of such a single mass tool indicates a potential reduction in audit effectiveness. To maintain the required level of efficiency, companies use both extensive ways (increase in staff) and intensive In particular, the use of macro add-ons for MS Excel. Such measures are considered temporary because they do not take into account the annual growth of information; the increase in the number of employees has its limit, and the use of add-ons to MS Excel is limited by a set of tools of the software itself.

4 Conclusions

The use of information technology is a topical and important issue for the successful work of the auditor, especially in a pandemic. Digitalization of audit is a new level in the field of auditing, which is developing rapidly and is spreading rapidly in enterprises, especially those that use automated accounting.

The software currently used by auditors needs to be improved in accordance with the requirements of the modern world. In the near future, digitalization will become relevant for all audit firms that are trying to gain a competitive advantage and take a place in the market of audit services.

Problematic and controversial issues need to be addressed in order to formulate a comprehensive approach to the digitalisation of auditing in the course of economic transformation and limited business practices caused by the global pandemic of 2019-2020.

References:

1. Research by the Association of Chartered Certified Accountants (ACCA). Homepage, https://www.accaglobal.com/gb/en/news/2020/april/ACCA-Covid-19-Research-Audit.html (2020).
2. B. Aditya, R. Hartanto, L. Nugroho, *The Role of IT Audit in the Era of Digital Transformation*, IOP Conference Series: Materials Science and Engineering, **407**, 1-6. (2018).
3. A. Bartik, M. Bertrand, Z. Cullen, E. Glaeser, M. Luca, C. Stanton, *The impact of COVID-19 on small business outcomes and expectations,* Proceedings of the National Academy of Sciences, **117(30)**, 17656-17666 (2020).
4. E. Bogdanova, *Comparative analysis of functional and technical capabilities of domestic audit automation programs*, System management: electronic scientific publication, **1 (7)**, 1-17 (2010).
5. M. Delarue, M. Jeschonneck, R. Leali, *EY Canvas.* Homepage, https://www.ey.com/en_gl/audit/technology/canvas (2020).
6. N. Donthu, A. Gustafsson, *Effects of COVID-19 on business and research*, Journal of Business Research, **117**, 284-289 (2020).
7. Electronic application "ACTION". Homepage, https://plan2.diia.gov.ua/
8. O. Kudirko, *Computerization of audit in Ukraine: actual problems and real prospects,* Economy and state, **9**, 34-38 (2018).
9. S. Kulitsky, *Problems of economic development of Ukraine caused by the pandemic of coronavirus COVID-19 in the world, and finding ways to solve them.* Ukraine: events, facts, comments, **9**, 47-53 (2020).
10. Law of Ukraine On Amendments to the Tax Code of Ukraine to improve tax administration, eliminate technical and logical inconsistencies in tax legislation from 16.01.2020, № 466-IX. Homepage, https://zakon.rada.gov.ua/laws/show/466-20#Text.
11. Legal newspaper online. *Audit and coronavirus: The FRC recommends developing alternative audit procedures to gather evidence*. Homepage,

https://yur-gazeta.com/golovna/audit-i-koronavirus-frc-rekomendue-rozrobiti-alternativni-auditorski-proceduri-dlya-zboru-dokaziv.html (2020).

12. J. Merhout, D. Havelka, *Information Technology Auditing: A Value-Added IT Governance Partnership between IT Management and Audit*, Communications of the Association for Information Systems, **23** (2008).
13. Ministry and Committee for Digital Transformation of Ukraine. Homepage, https://thedigital.gov.ua/.
14. M. Nezhyva, V. Miniailo, *Digitalization of audit in the conditions of the COVID-19,* Visnik KNUTE, **3**, 123-134 (2020).
15. A. Nguyen, H. Ha, S. Nguyen, *Determinants of Information Technology Audit Quality: Evidence from Vietnam,* The Journal of Asian Finance, Economics, and Business, **7(4)**, 41-50 (2020).
16. G. Pchelyanskaya, V. Nemchenko, *Audit in terms of using computer systems*, Odessa: Phoenix (2016).
17. G. Shinin, S. Azizyan, A. Larin, *Innovation in auditing*. Homepage, https://www.ey.com/ru_ru/audit/innovation (2019).
18. A. Stavitsky, *Economic scenarios of the pandemic: how to transform Ukraine*. Economic truth. Homepage, https://www.epravda.com.ua/rus/columns/2020/04/7/659085/ (2020).
19. D. Stoel, D. Havelka, J. Merhout, *An analysis of attributes that impact information technology audit quality: A study of IT and financial audit practitioners*, International Journal of Accounting Information Systems, **13(1)**, 60-79 (2012).
20. A. Ullrich, J. Enke, M. Teichmann, A. Kreß, N. Gronau, *Audit - and then what? A roadmap for digitization of learning factories*, Procedia Manufacturing, **31**, 162-168 (2019).

Formation of the "Green Economy" in the Innovative Model of Sustainable Ukraine's Development

Natalia Zakharchenko[1*], *Dariusz* Pawliszczy[2], *Nataliia* Orlova[3], *Nataliya* Navolska[4], and *Oksana* Apostolyuk[5]

[1] Odessa I. I. Mechnikov National University, Department of Finance, Banking and Insurance, 65000 Odessa, Ukraine
[2] Wójt of gmina Gromadka, 59706 Gromadka, Poland
[3] Odessa I. I. Mechnikov National University, Management and Innovation Department, 65000 Odessa, Ukraine
[4] West Ukrainian National University, Department of Economics and Economic Theory, 46000 Ternopil, Ukraine
[5] West Ukrainian National University, Department of Management, Public Administration and Personnel, 46000 Ternopil, Ukraine

Abstract. The purpose of the study- is to substantiate the problems and prospects of forming a "green " economy and a way to implement sustainable development at the national level. Regional and global levels in the long run. The active growth of "green" enterprises, the introduction of innovative technologies and the reduction of waste and harmful emissions in various sectors of Ukraine's economy, the transition to energy-saving technologies and biofuels and "greening" of agriculture promise to ensure the country's energy security. This allows for increased incomes and formal employment in these sectors of the economy in the short and long term, while providing more environmental, economic and social benefits, which in general leads to a higher level of well-being of the population and the country. At the same time, although the transition to a "green " economy will require large-scale investments, these investments can be mobilized through sound government policies and innovative financing mechanisms.

1 Introdaction

In recent decades, the economic growth of both Ukraine and world countries has been achieved mainly through the use of natural resources; humanity did not allow reserves to recover, but allowed ecosystems to degrade and disappear. The shortage of clean drinking water on the planet, depletion of the soil, deforestation, energy overruns, carbon emissions, and other problems make it possible to create a socio-ecological and economic catastrophe. It is the transition to a "green" economy that will reduce the impact on the environment per capita without reducing the quality of life. But for the transition to a green economy, conditions conducive to this must be created. These conditions include: relevant national regulations, policies, subsidies and incentives, the global marketplace, legal infrastructure, trade and financial aid protocols. The conditions prevailing today are favorable and conducive to the continuation of the "brown" economy and the era of misallocation of capital. "Greening" of many sectors the economy can bring sustainable and positive results such as asset and output growth, decent employment and poverty reduction. In Ukraine, there is a huge potential for the development of a new innovative model of sustainable development - "green economy". This is confirmed in the article by the adoption of a number normative legal documents, the introduction of nature-saving innovative technologies at enterprises, the development of organic farms and the growth of "green" enterprises in Ukraine. However, there are many problems and obstacles on the way to the formation of a "green" economy in Ukraine. And this is also discussed in this study. The purpose of the article is to identify the prerequisites and justify the need for Ukraine's transition to a "green" economy as an innovative model of sustainable development. To achieve the goals in the article were set such following parameters as: 1) to define the basic terms and concepts; 2) substantiate the prerequisites and necessary for the development of a "green" economy in Ukraine; 3) reveal the problems and obstacles to the formation of a "green" economy in Ukraine; 4) present a SWOT- an analysis of the development of the "green" economy, which will determine the advantages and prospects for the development of the "green" economy. It is in the substantiation of the goal and the tasks set that the relevance of the study and its novelty.

A necessary condition for achieving innovative development is to change the system of social values in order to transform the established market form of management in a socialized and green environment. Thus, a "green" economy that develops on such a basis must create conditions for maintaining the health of the current generation, improving its material well-being, spiritual and social development, and all this will serve as the basis for the progress of human civilization in the future. A number of components of the "green" economy, as a rule, include the production of environmentally friendly products and the provision of adequate services, the implementation of socially responsible investments, the formation of corporate social responsibility [1].

*Corresponding autors: nvzakharchenko777@gmail.com

2 Literature review

Ukraine is currently going through difficult times in the transition to a new format of management in the world system. Against the background of deteriorating socio-economic and political situation, Ukraine did not miss COVID-19, which brought many negative consequences. The issue of ecological crisis and non-reproduction of natural resources has also become acute. The global terrestrial ecosystem has lost its capacity for natural self-regulation and reproduction, and society must now become its main regulator, and its own future depends on how it performs these functions. Mankind is faced with the dilemma "if people do not remove the contradiction between themselves and nature, the latter will solve it at the expense of humanity.

Recently, scientific research on solving economic and environmental problems has been widely deepened, which has given impetus to the formation of an innovative model of sustainable development - the "green" economy. These fundamental works are the subject of attention of domestic and foreign scientists. In particular, the following works are devoted to the development of these issues: Georgeson L. & Maslin M. [2], Yi H. & Liu Y. [3], Aldy J. [4], Carfì D. & Schilirò D. [5], Jackson T., & Victor P. [6], Vona F., Marin G., & Consoli D. [7], Iermakova O. [8], Boronos V., Plikus I. & Sokolenko L. [9], Kaletnik G. & Lutkovska S. [10] and others.

Scientists emphasize the need to use new ways to address environmental issues, in connection with which they justify the development of a "green" economy as one of the promising ways to form an innovative model of Ukraine. But the research can not be considered complete. They need further concretization and deepening.

Currently, the "green" economy is developing around the world. Its role in economic development and employment is justified in a study by Georgeson L. & Maslin M. [2] and is estimated at $ 1.3 trillion. in the United States. Compared to China, OECD members, and the G20, the United States has a larger share of the working-age population in the green economy (4%) and higher per capita sales. To remain competitive, the United States is implementing energy and environmental policies at the federal level in this sector of the economy, which are vital to combating climate change and protecting the environment.

This can be a "breakthrough" strategy for Ukraine in this direction, including the introduction of modern innovative industrial technologies based on supporting the development of high-tech industries (increasing production of domestic aerospace technologies, instrumentation, electronic technologies, nanotechnologies, medical technologies), and eco-technologies and energy-saving technologies in all sectors of the economy (Kaletnik, G. & Lutkovska S. [10]). The authors prove that the development of "green" business is impossible without the formation and use of appropriate management capacity, innovators-managers who will be able to competently "start" all other potentials of the innovation system.

Ukrainian rate to implement an "innovative model of sustainable development", which provides an increase in the quality of life of the population through innovative economic development, was announced by the Government of Ukraine, but did not implement certain measures to achieve this strategy. The authors Boronos V., Plikus, I., Sokolenko L. [9] in their study among the main reasons indicate the inefficiency of management at all levels. The authors propose measures to accelerate the integration of Ukraine's economy into the world economy by forming an innovation-oriented economy, an economy based on environmentally friendly technologies, which are the basis of innovative changes in Ukraine and provide "green growth" to which the world aspires.

That is why, Iermakova O., argues the innovation-is an integral part of the "green economy", which improves people's well-being and ensures social justice, significantly reducing risks to the environment. Most developed countries consider the "green" economy as innovative and a priority in the long run. In order to promote the implementation of the principles of the concept of "green economy" in the regions of Ukraine, the author proposes the introduction of areas of smart specialization, which should be tested in accordance with sustainable development goals in Ukraine in the context of social, economic and environmental goals. The study emphasizes that the glocalization of the innovation paradigm of the concept of "green economy" is manifested through a combination of local socio-ecological and economic environment of the country / region with integration into global networks, thus providing environmentally friendly innovation with appropriate resources (financial, human, technical, information) from local and global sources, facilitating access to them and creating opportunities for their use, which requires proper institutional support "[8].

In the works of Vona F., Marin G., Consoli D. [7] the main methodological novelty is a new measure of environmental employment, based on the content of the tasks of professions. To this end, the data of local labor markets in the United States and the reasonable consequences of "green" employment were studied. The analysis showed that one additional green job is associated with 4.2 new jobs in "non-green" activities. The authors determined that environmental employment is pro-cyclical, highly skilled and requires a 4% wage increase and is geographically linked.

Interesting are the studies of Yi H., Liu Y., which showed that "... cities located in the province with a clean energy policy have 54.3% more green jobs and 61.8% more green enterprises compared to cities located in the province without such a policy", which explains variation between cities in China [3].

The transition to a sustainable low-carbon economy is a huge problem, which is primarily related to the investment mechanism. In particular, investments should focus on resource productivity, renewable energy, clean technologies, business ecology, climate adaptation and ecosystem protection. Traditional investment models will change, say authors T. Jackson, P. Victor [6]. The function of innovations, which is focused on achieving socially significant goals of innovative sustainable development of the state, will remain important. The authors substantiate that investments in renewable energy are competitive. Other investments can only pay off in the long run. And investing in ecosystem improvement and climate adaptation may not be profitable at all, although they protect vital ecosystem services for the future and can also promote employment. That is why investing in "green" innovation can only benefit on a long-term basis.

Therefore, the definition of problems and prospects for the development of "green economy" in the formation of a model of sustainable development of Ukraine requires further in-depth study, as developed countries at the present stage form a policy of neo-industrial type.

3 Research methodology

To solve the set tasks were used such general and special methods as: morphological analysis – during clarifying the conceptual and categorical apparatus of scientific research; system-structural approach – in completing the theoretical and methodological basis of ensuring and carrying out the formation of concepts of "green" economy as a way of implementing the course of sustainable development of Ukraine's economy; economic and statistical – in the study of the dynamics organic farms and eco-based production of Ukraine; graphic – to reflect their trends; functional – in characterizing the effective tools of formation and regulation of the "green" economy in Ukraine; SWOT-an analysis and grouping – for further directions and prospects of the "green" economy, classification-analytical – to identify obstacles and drivers of the "green" economy of Ukraine; logical – to build a structure and describe the results of the study.
The statistical, factual and informational basis is compiled by the works and developments of leading domestic and foreign scientists and practitioners on research issues, current legislation, reporting data of Ukrainian enterprises, the results of their own scientific research, official publications of domestic and international organizations, Internet resources

4 Resalts

In scientific research, the concept of "green economy" is an integral attribute of innovative and sustainable development, namely meeting the needs of both present and future generations, ensuring a harmonious way of combining economic and social activities of society while preserving and properly reproducing the environment. Sustainable development is considered as a model of functioning of an integrated and innovative socio-ecological-economic system with limited parameters, which provides a balanced dynamic balance between its components for a long time. The principles of sustainable development and the model of green economy are the basis of the development strategy of the EU "Europe 2020" and countries with a high level of quality of life. Ukraine, territorially located in Europe, has been an independent state since 1991 and is currently undergoing a difficult stage of economic and social transformation.
The "green economy" according to a UNEP report - is "a low-carbon economy that uses resources efficiently and is in the interests of society as a whole." In a green economy, income and employment growth is driven by public and private investment in measures and projects that reduce carbon emissions and pollution, increase energy and resource efficiency, and work to prevent biodiversity loss and ecosystem services. These investments need to be catalyzed and supported through targeted public spending, subsidies and incentives to promote the development of "green" sectors, the development of markets for "green" technologies and innovations, improve regulatory policies and financial aid flows, the transition to "green" public procurement [11].

The rating agency Dual Citizen has developed the Global Green Economy Index (GGEI), which has established the efficiency and progress of the "green" economy in 130 countries.

The results of the expert survey showed that Ukraine ranks 120th among the world's leading countries [3].

This confirms the significant existing obstacles to the development of Ukraine's economy. But it should be noted that there are prerequisites and drivers for the development of "green" economy (table 1).

Table 1. Obstacles and drivers of development of the "green" economy of Ukraine.

Obstacles	Drivers
Insufficient awareness of green modernization of the economy, energy and resource efficiency	Informing about green modernization of the economy: local authorities; public organizations; international organizations; independent search for information
Outdated legislation and administrative barriers to green modernization	Simplification of the procedure and development of the legal framework
High cost for energy and resource efficiency measures	Calculation of expected economic efficiency from energy and resource efficiency measures and attraction of own reserves and partial attraction of credit funds
Obsolete energy and resource-consuming equipment	Application of new technological solutions for the introduction of energy and resource efficiency
Management does not understand the benefits of energy efficiency	Proving the economic attractiveness and technological feasibility of the project to increase energy and resource efficiency
Long payback period for renewable sources	Schemes of preferential business lending
Lack of experience and professional skills in implementing	Environmental management training

environmental management systems	
Disinterest and ignorance of potential buyers in environmental goods and services	Information and advertising
High costs of financing the production of environmental goods and services	Improving the company's image and expanding the market
Difficulties in obtaining loans for green modernization - a complex bureaucratic system	Deregulation
Low environmental awareness of staff	Informing staff, raising environmental awareness
Lack of processing of some types of waste in Ukraine	Promoting the introduction of new technologies for waste processing

Source: created by the authors based on [8;9].

A new strategy of the state environmental policy of Ukraine until 2030 came into force on January 1, 2020. Law 2697-VIII "On Basic Principles (Strategy) of the State Environmental Policy of Ukraine until 2030". What will be implemented in Ukraine according to the new document? The law defines 5 strategic innovative and environmental goals: 1) formation in society of ecological values and principles of sustainable consumption and production; 2) ensuring sustainable development of Ukraine's natural resource potential; 3) ensuring the integration of environmental policy in the decision-making process for innovative and socio-economic development of Ukraine; 4) reduction of environmental risks in order to minimize their impact on ecosystems, socio-economic development and public health; 5) increase liability for damage caused to the environment, in accordance with Ukraine's international obligations. Therefore, Ukrainian economy needs to prepare for change in advance - to develop business while preserving the environment [12].

Today, "green" economy offers solutions to some of society's most pressing environmental problems: design and construction of energy efficient buildings; recycling and safe waste management; development of renewable energy; wastewater treatment. "Green" economy includes three groups of companies: 1) companies whose activity is the production of environmental goods and services (traditional "Environmental goods and services sector"); 2) enterprises that have taken active and independent actions to change their products and / or processes, taking into account sustainable development ("green" companies / enterprises); 3) all other firms that take a number of measures to improve the efficiency of the process or change their image [13].

"Living Planet" and the State Ecological Academy of Postgraduate Education named the enterprises of Ukraine, which, in their opinion, are the most active in implementing environmental technologies and reducing emissions. They belong to innovative and socio-environmentally oriented enterprises, whose main mission is to preserve nature, introduce new environmentally friendly technologies and promote the development of environmental potential in the global market (table 2).

Table 2. Ukrainian companies that most actively implement environmental technologies and reduce emissions

Name of Company	Characteristic	Location
Agro-industrial holding MHP: SE "Victory Nova", PJSC "Myronivska Poultry Farm", NVF "Harvest" Katerynopil Elevator LLC [14]	Particular attention should be paid to related to environmental - corporate social responsibility, they help communities in their highest priority issues: co-financing of utilities, repair or construction of new roads, etc.	Regions: Vinnytsia, Cherkasy, Zhytomyr, Kyiv, Sumy, Ivano-Frankivsk, Ternopil, Lviv, Donetsk
NIBULON LLC [15]	is one of the largest domestic producers and exporters of agricultural products (wheat, barley, rapeseed, corn, sorghum, soybeans, sunflowers, etc.).	Regions: Mykolaiv, Zhytomyr, Luhansk, Khmelnytsky, Cherkasy, Kharkiv, Vinnytsia, Sumy, Poltava
PJSC "Dickerhoff Cement Ukraine" [16]	for 10 years provides delivery of high-quality cement to the territory of our state	Volyn region
PJSC "Zaporizhstal" [17]	The company implements resource-efficient technologies. More than 10 billion hryvnias were invested in modernization at the plant	Zaporozhye region
PJSC "ArcelorMittal Kryvyi Rih" [18]	Increases investment in the environment and continues to modernize its shops, which includes new treatment facilities	Dnipropetrovsk region

Source: created by the authors based on [14;15;16;17;18].

Ukraine has significant potential for the production of "green" products. Most Ukrainian "green" farms are located in Odesa, Kherson, Kyiv, Poltava, Vinnytsia, Zakarpattia, Lviv, Ternopil, and Zhytomyr regions. The size of Ukrainian certified "green" farms: from a few hectares (as in most European countries) - up to several thousand hectares of arable land. According to the Federation of Organic Movement of Ukraine, the area of certified agricultural land in Ukraine involved in the cultivation of various "green" products is more than 400 thousand hectares. According to this indicator, Ukraine ranks 20th among the countries of the world and first – among the countries of Eastern Europe. At the same time, the share of certified "green" areas among the total agricultural land of Ukraine is about 1%. The "green" sector in Ukraine specializes mainly in .the production of cereals, legumes and oilseeds and dairy products. If in 2002 there were 31 "green" farms registered in Ukraine, which received the status of "green", in 2015 there were 210 certified "green" farms, and in 2018 – 501 "green" farms, and the total area of agricultural land on which "green" production is carried out was 309100 ha. Studies by the Federation of the Organic Movement of Ukraine show that the modern domestic consumer market of "green" products in Ukraine began to develop in the early 2000s. In 2006, its volume amounted to 400 thousand euros, and in 2015 – up to 17.5 million euros, in 2018 - 33.0 million euros [19].

There are about 5,000 "green"enterprises in Ukraine. "Green"enterprises are branched out in all spheres of economic activity of Ukraine, namely the sphere of crop production, animal husbandry, household chemicals, dairy production, grain crops and cereals, confectionery production. The largest share of the "green" market belongs to confectionery and chocolate products (16.04%), vegetable oils (14.74%), eco-vegetables (20.22%), eco-fruits and berries (14.28%). The largest percentage of "green" enterprises is in Kyiv region (23.08%), Dnipropetrovsk (7.94%), Kharkiv (6.36%), Odessa (6.28%), Lviv (6.2%), Vinnytsia (5.08%), and the smallest - Luhansk (0.82%) and Donetsk (1.9%) (table 3).

Table 3. "Green" enterprises of Ukraine by region

Region	Number of enterprises	The most famous representatives
Vinnytsia	254	ORGANIC-D, MAK-VAR ECOPRODUCT, BIO-FERM ORGANIC PLUS
Volyn	105	OLD PORITK, TSUMANSKY LISGOSP
Dnepro	397	ELIKSIR, SENCO-UKRAINE, ECOSAD
Donetsk	95	BIOPOLE, RADIST, CHOCOLATE WORKSHOP
Zhytomyr	110	GALEKS-AGRO, FITOSUROVINA, ORGANIC MILK
Transcarpathian	161	PAN-ECO, DZERELA KARPAT, STOVCONTINENT
Zaporozhye	180	MELITOPOL MEAT COMBINAT, RODNICHOK8 GURSOY
Ivano-Frankivsk	153	ECOPRODUCT, PROMETEY (FARM), EXPORT TREYD GROUP
Kyiv	1154	Arpolcт beyk-end Milling, Agro-ros, EST-ETUAL GROUP
Kirovograd	139	EKKO-CONTINENT, UKRAGROHIM
Luhansk	41	MVL-Ukraine, Prodgroupservice, NEONOLIN-ECO
Lviv	310	Fruit, group of company, Apro
Mykolayivska	169	STODOLA, Nature's Sunshine Products (NSP), SINTA
Odessa	314	BILOCHI-etno-farm, Bolgrad sirorobniy zavod, Progresfarm, Hlibio
Poltava	139	Agnus, Agroecology, Bilocerkivska promyslova group
Sumy	89	AGROSVIT, LEV_GROUP, FIRM-ECO-SOYA
Ternopil	149	ZHIVA-ZEMLYA, Tepli gryadki, AGROZAGOTZBUT
Kharkiv	318	My Garden, Ecopolza, Mudriy dachnik, Internet – kramnitsa
Kherson	224	Amrita, Ecoferma Onuka, Euro-Service-Group
Khmelnytsky	150	ECOAGROPOSTACH, Global-Organic-Group
Cherkasy	159	Spivpratsya, Academ of organic gardening Ekogarden
Chernihiv	144	Gold Parmen, PLISKIVSKIY AGROPROMYSLOVIY COMPLEX, ECO-PRODUCT, PAN-ORGANIC
Chernivtsi	220	Lilak

Source: created by the authors based on [19].

5 Discussion

For Ukraine, "greening" of market is quite promising, especially in terms of exports to the EU, as such products are set at "zero" rates and quotas do not apply, and prices are usually on average 30% higher than conventional counterparts. In this segment, the European market is already represented by Ukrainian companies engaged in the production of birch sap, concentrated juices, sublimated, fresh and frozen fruits, cultivation and sale of cereals, etc., although the volume of their supply is clearly insufficient. Ukraine's "green" exports are amount to $ 350 million and continues to grow, while "green" imports are about $ 700 million. The global volume of "green" trade in 2020 has tripled - to $ 2-3 trillion [20].

At the same time, "greening" of agriculture involves not only the production of organic products, but also the cultivation of energy crops and their use for energy purposes - for the transition to biofuels. Thus, on the one hand, Ukraine strengthens its competitiveness in European markets by exporting products, and on the other hand, it reduces the need for traditional energy sources, contributing to the country's energy security. SWOT-analysis of the concept of "green" economy development in Ukraine is presented in table 4.

Table 4. SWOT-analysis of the concept of "green" economy development of Ukraine

Strengths
High quality products / services
Promoting stakeholder involvement and empowering local communities
Growing demand for "green" products and services that minimize environmental impact
Weaknesses
Low level of awareness about the organization of products among consumers
Higher price for "greens" than for traditional products or services
Lack of knowledge and competencies, which becomes a major obstacle to the development of "green" economy
Lack of a clear understanding of the possibilities of obtaining financial benefits from the transition to "green" economy
Opportunities
Global challenges of climate change and sustainable development
Search and generation of new ideas in various areas of business
Creating skilled jobs
Promoting energy, food and water security
Poverty reduction
Threats
Unregulated regulatory framework, which complicates the work of "green" economy
Lack or insufficiency of public and private funding for appropriate development and adaptation measures
Loss of jobs and the gradual disappearance of the "brown" economy
Lack of a single global approach to shaping green development shared by all countries and businesses, leading to controversy
Lack of correct measurement of "green" growth
Challenges to changes in consumer and production patterns and lifestyles that are commonplace

Source: created by the authors based on [20, 21].

The emergence of new solutions in the economy leads to the formation of new conceptual approaches, namely a balanced socio-ecological and economic, which covers the "green" economy, innovation and information and digital economy based on the principle of preserving the environment, "smart" and "green" production technologies, which are aimed at reducing the consumption of energy and material resources, automation of calculations, software, development of nano- and biotechnologies and creating added value through innovative intelligent solutions. The transition to a "green" economy objectively stimulates innovation effects, which must be supported by a favorable competitive environment and effective regulatory tools. It is important to emphasize that the concept of "green" economy does not replace the concept of sustainable development, but develops it and is a means of its practical implementation. "Green" development can be ensured only if the integration of environmental and economic policies so that social progress, economic growth and improving the quality of life of the population occur against the background of reducing threats to the environment [20].

In this way, the main advantages of the transition to a "green" economy are: 1) increasing wealth and ensuring the growth rate of GDP, as one of the main indicators of economic well-being; 2) the creation of new jobs to replace disappeared in the "brown" economy, which requires investments in retraining the workforce; 3) creating the value of natural capital as a source of human well-being and building it up in the interests of socio-economic progress. Among the factors influencing the "green" economy to increase the well-being of people and the welfare of the world economy are: 1) investing 0.03% of GDP by 2050 in the form of payments to forest owners intended for forest conservation can increase the added value in the forestry industry by more than 20%; 2) investing in "green" agriculture $ 100-300 billion per year until 2050 will lead to increase in soil quality and an increase in the yield of the most important crops. This will partially solve the problem of food for the population; 3) increasing efficiency in the agricultural, industrial, municipal sectors will reduce the demand for water by 20%; 4) reorientation of government spending on improving fisheries will provide for the restoration of the world's fish stocks; 5) the allocation of 1% world GDP to improve energy efficiency and increase the use of renewable energy will create additional jobs, as well as ensure the provision of competitive electricity [21]. 6) the creation of "green" cities will increase the efficiency and productivity of labor; 7) "greening" transport will reduce negative impact on the environment and social costs. All these measures can increase income and formal employment of the population in these sectors of the economy in the short and long term, while providing more environmental, economic and social benefits. To transform the economy, needed concentrated pools of assets, similar to those controlled by long-term investors, such as: state financial institutions, development banks, sovereign wealth funds, as well as some pension and insurance funds that are free from obligations on short-term payments. Slelust also rationally apply subsidies, which may facilitate Ukraine's transition to a "green" economy. Taxes and other market-based instruments can be used to stimulate the necessary investment and innovation to finance the transition.

6 Conclusion and perspectives for the research

The "green" economy puts forward new requirements for business models and their management, changing technologies and the environment of their formation and implementation on the principle of advanced development and digitalization and gives preference to information and innovation-oriented business models. This creates new challenges for new types of entrepreneurship. On the one hand, conditions are created for the emergence of new opportunities - the search and generation of new ideas in various areas of entrepreneurship, on the other - it causes a lack of knowledge and competencies, which becomes a major obstacle to the development of "green" entrepreneurs. The introduction of "green" economy in Ukraine will help solve many environmental, economic and social problems that currently constrain economic growth in the country.

References

1. Marchuk, L.P. (2014). "Zelena" ekonomika: superechnosti ta perspektyvy rozvytku ["Green" economy: contradictions and prospects for development] *Bulletin of Agrarian Science of the Black Sea Coast, 1, 34-41.* URL: file:///Users/user/Downloads/vanp_2014_1_7%20(1).pdf
2. Georgeson, L. & Maslin, M. (2019). Estimating the scale of the US green economy within the global context. *Humanities and social sciences communications, 5.* URL: https://www.nature.com/articles/s41599-019-0329-3
3. Yi H., Liu Y. (2015) Green economy in China: regional variations and policy drivers. Glob Environ Change 31, 11–19. URL: https://doi.org/10.1016/j.gloenvcha.2014.12.001
4. Aldy J. (2013) A preliminary assessment of the American Recovery and Reinvestment Act's Clean Energy Package. Rev Environ Econ Policy 7(1), 136–155. URL: https://doi.org/10.1093/reep/res014
5. Carfì, D. & Schilirò, D. (2012). A coopetitive model for the green economy. *Econimic Modelling*, 29 (4), 1215-1219. URL: https://doi.org/10.1016/j.econmod.2012.04.005
6. Jackson, T. & Victor, P. (2011). Productivity and work in the 'green economy': Some theoretical reflections and empirical tests. *Environmental Innovation and Societal Transitions* 1, 101–108. URL: https://doi.org/10.1016/j.eist.2011.04.005.
7. Vona, F., Marin, G. Consoli, D. (2019). Measures, drivers and effects of green employment: evidence from US local labor markets, 2006–2014. *Journal of Economic Geography*, 19 (5), 1021–1048, URL: https://doi.org/10.1093/jeg/lby038
8. Yermakova, O. A. (2019). Teoretyko-kontseptualnyi bazys innovatsiinoi paradyhmy «zelenoi ekonomiky» v umovakh hlokalizatsii [Theoretical and Conceptual Basis of the Innovative Paradigm of the «Green Economy» in Terms of Glocalization]. *Modern Economics*, 13, 90-94. DOI: https://doi.org/10.31521/modecon.V13(2019)-14.
9. Boronos, V., Plikus I., Sokolenko L. (2018). Formalization of The Approaches Regarding Assessing the Level of the EcoFriendly Regional Innovation Systems. *International Journal of Ecology & Development.* 33 (4), 93-102. URL: http://essuir.sumdu.edu.ua/handle/123456789/68187.
10. Kaletnik, G. & Lutkovska S. (2020). Innovative Environmental Strategy for Sustainable Development. *European Journal o f Sustainable Development.* 9 (2), 89-98. URL: https://doi.org/10.14207/ejsd.2020.v9n2p89
11. "Zeleni" investytsii u stalomu rozvytku: svitovyi dosvid ta ukrainskyi kontekst (2019) [Green Investments in Sustainable Development: Global Experience and the Ukrainian Context] URL: http://razumkov.org.ua/uploads/article/2019_ZELEN_INVEST.pdf
12. Pro Osnovni zasady (stratehiiu) derzhavnoi ekolohichnoi polityky Ukrainy na period do 2030 roku : Zakon Ukrainy 2697-VIII. [On Basic Principles (Strategy) of the State Environmental Policy of Ukraine until 2030 : Law of Ukraine 2697-VIII.] URL: https://zakon.rada.gov.ua/laws/show/2697-19#Text
13. Marushevskyi, H., Khikman, D. (2017). Zelenyi biznes dlia malykh i serednikh pidpryiemstv [Green business for small and medium enterprises] URL: http://pleddg.org.ua/wp-content/uploads/2017/11/SME-Guide-web.pdf
14. Official website. URL: https://www.mhp.com.ua/uk/partners
15. Official website. URL: https://www.nibulon.com
16. Official website. URL: https://www.dyckerhoff.com.ua
17. Official website. URL: https://www.zaporizhstal.com/ru/
18. Official website. URL: https://ukraine.arcelormittal.com/?lang=ru
19. Federation of Organic Movement of Ukraine. URL: http://organic.com.ua
20. Towards Green Growth. May 2011 URL: http://www.oecd.org/greengrowth/ 48224539.pdf
21. UNEP, 2011. Navstrechu «zelenoj» jekonomike: puti k ustojchivomu razvitiju i iskoreneniju bednosti — obobshhajushhij doklad dlja predstavitelej vlastnyh struktur [UNEP, 2011, Towards a Green Economy: Pathways to Sustainable Development and Poverty Eradication - A Synthesis Report for Government Officials]. URL: www.unep.org/greeneconomy

Digital Transformation Modelling in the Context of Slowbalization

Iluta Arbidane[1], *Hanna* Purii[2*], *Abdukhakim* Mamanazarov[3], *Serhii* Hushko[2] *and Volodymyr* Kulishov[2]

[1]Rezekne Academy of Technologies, 115 Atbrivosanas aleja, Rezekne, LV-4601, Latvia
[2]State University of Economics and Technology, 16 Medychna Str., Kryvyi Rih, 50005, Ukraine
[3]Tashkent Branch of M. V. Lomonosov Moscow State University, Tashkent 100060, Uzbekistan

Abstract. The article studies modern concepts and theories of digital transformation of society, analyzes structural changes in the economy during the fourth industrial revolution, presents different views of experts on the expected positive results of Industry 4.0 in the context of globalization. The authors suggest models and correlate increment rate dynamics with the information globalization index in the world and individual countries, and identify trends in the development of digital transformation in the context of slowbalization. The study considers possible consequences and risks of Industry 4.0 and ways to level them, outlines the priority areas of digital transformation to ensure high competitiveness of national economies in the period of slowbalization.

1 Introduction

Modern society exists in the world where information technology is integrating into all spheres of human life. The digital revolution has introduced a new way of thinking, social behaviour, building an ecosystem, and doing business, the latter changing to new standards of strategic positions in various sectoral spheres of public life and real economy and adapting to innovative conditions at an unpredictable rate.

Global labour division according to individual countries' specialization in producing certain goods and services is the determinant of developing world economic relations and one of the most important trends in the world economic progress. The countries taking advantage of opportunities and benefits of participating in global labour division experience positive rates of their economic and social growth.

In terms of digital transformation, this period of social development is characterized as the Fourth Industrial Revolution, or Industry 4.0. Transition to the Industry 4.0 concept shows that in the future such business processes as supply of raw materials, development and delivery of goods to consumers and their maintenance will be carried out through digitalization of services. That is why, the study of the world's digital transition to Industry 4.0 is of great topicality.

2 Brief analysis of the researches

In scientific circles, prospects and directions of smart technology development and digitalization have recently been paid much attention to due to the constant global transformations of economic systems. Theoretical research on the issues of digital transformation and, in particular, Industry 4.0 focuses on certain aspects of the Fourth Industrial Revolution, namely: economic problems of introducing Industry 4.0 and its impact on the development of a state economic sector; peculiarities of Industry 4.0 application to state regulations; transformation of TNC strategies in the context of the fourth industrial revolution; political consequences of Industry 4.0 introduction and others.

Regardless of the significant number of publications on this issue, the dynamic and rapid digital transformation of economic systems daily creates new challenges and risks of Industry 4.0 for society and scientists requiring the development of relevant tools to level them.

2.1 Role and extent of Industry 4.0 development in different countries

Adapting to innovations, digital transformation has reached a new level while its integration into all social spheres is becoming mandatory. Such countries as China, Great Britain, Singapore, New Zealand, Estonia, the UAE, Japan, and Israel are considered the leaders in digital technology development. When studying their experience, it is possible to identify advantages and disadvantages of social digitalization [1].

Industry 4.0 results in an increased amount of stored, processed and available knowledge associated with the array growth, namely: a set of approaches, tools and methods of processing structured and unstructured big data, a significant variety of human-perceived and effective results in the context of their continuous increase and distribution through numerous nodes of a computer network [2, 3, 4].

* Corresponding author: puriy_av@kneu.dp.ua

According to experts, the fourth technological revolution will deepen economic and social inequality leading in a short run to even greater stratification of population: technology owners and developers will receive excess profit, while low-skilled staff will earn meagre wages, and the middle class will literally disappear. However, after staff retraining, the situation will level off. Following the World Economic Forum (WEF), more than 5 million jobs will disappear as a result of the development of genetics, robotics and other technological changes. In total, about 47% of vacancies in the United States today are at risk of extinction due to changes associated with automation and transition to the digital environment. For example, 65% of today's schoolchildren will have brand new professions [5]. Robots and digital technology will form a new type of production, while technologies that combine physical, biological and digital worlds will form a new type of consumption. This will lead to even greater specialization of the service sector by creating specialized business platforms that will bring together certain groups of people in horizontal models of interaction. Goods and services will be high quality and cheap, but this can significantly reduce the cost of human labour.

New technologies will provide for implementing thoughts and ideas. However, they will increase the risks for society to be controlled by new tools via an overwhelming digital infrastructure and possible implantation of chips in human bodies to manage their consciousness. Many countries are now developing e-government, e-procurement and other blockchain systems. Most likely mass decentralization of power will be possible as the state will not be the source of wealth. Developed will be the states able to adopt a new paradigm of development and change their regulatory mechanism, which should become flexible thanks to innovative technologies and be based on principles of tripartism (close cooperation of state, business and society) [6]. A state should adjust to an individual, nature and new technologies, not the opposite. The fourth technological revolution will also affect national and international security as interstate conflicts will be largely hybrid, and the boundaries between war and peace will be blurred. The fourth wave will change not only the space where a person lives, but also his/her personality: worldview, values, behaviour, habits, lifestyle, communication, consumption, time for work and rest, approaches to education, and success criteria.

Common features of all the four technological revolutions include development of science; use of new energy sources, materials; invention of machines and technologies; work organization; expansion of labour specialization; emergence of new industries, significant changes in transportation and communications; increase of urban population; changes in the structure of society, its values, style, and manner of interaction and behaviour [7]. Concurrently, all the unresolved problems of previous revolutions are carried over to the next ones and deepened significantly.

Thus, summarizing the comprehensive scale of Industry 4.0, it is possible to identify its main consequences (Table 1) [8-12].

Table 1. Consequences of Industry 4.0.

At the global economy level	- a jump of labour productivity in processing industries; - creation of new markets and diminishment of some traditional activities; - formation of global centres of rapid industrial growth; - radical restructuring of the existing system of the world labour division by reducing obsolete elements of the technological chain of previous systems, mainly in developing countries; - reduction of the need for unskilled labour and exacerbation of global unemployment.
At the social level	- a sharp increase in the amount of scientific and technical information, emergence of fundamentally new approaches to its processing and ways to manage hardware and software tools for research and development; - blurring of disciplinary and sectoral boundaries in research and development, interdisciplinary and convergent nature of research; - higher requirements to researchers' qualifications, international competition for talented highly-qualified employees and their involvement in science, engineering, technical entrepreneurship; - compression of the innovation cycle: a significant time reduction between acquisition of new knowledge and creation of technologies, products and services, their entry into the market; - a growing role of international standards
At the individual level	- digital reality as for people's activity and communication space affects their personal level, namely: development of higher mental functions, values, motivation, interaction manner and behaviour

As a result of such digital transformation, various types of activity are transferred to the virtual dimension, this determining increased enthusiasm among the digital community leading to their worsened adaptability to the outer world and strained interaction with other types. At the same time, the manner of acting and communicating of the virtual world is transferred to the real-life one. As such, digital technologies expand and transform human abilities creating the basis for new realia. Using new digital technologies, a person creates objects of the secondary (multimedia) nature, which are easily transformed and controlled through communications, interaction, and exchange of multimedia products.

Under the Fourth Technological Revolution, information technology will be the main driver of development [13]. Nevertheless, society cannot avoid traditional manufacturing that, however, will take place in laboratories, data processing research companies, high-tech automated engineering plants, robotic shops, high-performance data centers, etc. In Ukraine, on the other hand, IT industry is traditionally contrasted with others; in particular, with mineral processing and agriculture, while in developed countries, they rely on technologies and actively implement them in the national economy. Therefore, the leaders of the "Global Innovation Index 2019" are Switzerland, Sweden, the USA, the Netherlands, Great Britain, Finland, Denmark, Singapore, Germany, Israel; in the post-Soviet space the best results are shown by Estonia (24th place), Armenia (33rd), Latvia (34th) and Lithuania (38th), then Russia (46th), Ukraine (47th place) and Moldova (58th). The worst are Belarus (72nd place) and Azerbaijan (84). Turkmenistan and Uzbekistan are not represented in the ranking. Kazakhstan is the 79th with its closest neighbours - Kyrgyzstan (90) and Tajikistan (100) that showed, according to international experts, lower results [14].

2.2 Benefits and ambiguity of multiplying effects

The development of digital transformation affects not only business efficiency, but also economic performance and preparedness for the future. Governments invest heavily in their digital economies to contribute to value creation and prosperity.

All-out digitalization has contributed to the rapid development of globalization, and global processes have led to rapid informatization of public communication. Therefore, under modern conditions, development of Industry 4.0 and globalization rate itself are significantly correlated with each other and generate a multiplying effect [15, 16].

However, in the recent decade, global economy has been experiencing a transition from total globalization to more constrained slow stabilization. Today's realia indicate growing tension between economic nationalism and globalization, which is becoming one of the determent trends at the threshold of the next decade [17].

Over the past four decades, ideas, products, money, and people have been crossing borders freely following globalization growth. This has led to rising income inequality. However, the integration rate has slowed markedly due to transition to domestic or local development and emergence of slowbalization. The close correlation between globalization and digitalization of society necessitates forecasting development of Industry 4.0 in the context of slowbalization of society.

In order to assess the degree of inclusion of individual countries in the process of globalization, four globalization indices are used to build a ranking of countries according to their degree of involvement in global flows: A.T. Keamey's globalization index, the KOF-globalization index, the CSGR globalization index and the Global Index. [1]. As such, the KOF-globalization index includes three dimensions of globalization: economic, political and social. The index is treated as a combined indicator that allows assessing the level of a country's integration in the world and comparing different countries by its components. All the countries being surveyed within the Index are assessed by 24 indicators combined into three main groups of global integration:

1) economic globalization includes the following indicators: the international trade volume; the level of international business activity; trade flows; international investment; tariff policy; international trade restrictions and taxes, and so on;

2) social globalization is expressed through such indicators as the level of cultural integration; the percentage of foreign population; international tourism; international personal contacts; the volume of telephone traffic, postal items, cross-border money transfers; information flows; development of information and communication infrastructure;

3) the third one - political globalization is characterized by the following indicators: membership in international organizations; participation in international missions (including the UN missions); ratification of international multilateral treaties; the number of embassies and other foreign missions in the country. The globalization level index is calculated as a sum of these components with the weight ratio of 36%, 38%, and 26% respectively.

All countries under this investigation are ranked according to the Globalization Index, which indicates their place among other countries studied. The final value of the index from 0 to 100 points is obtained by evaluating 24 indicators combined into three components: globalization in the economic, social and political spheres. The higher the index value of the state is, the more it is involved in world processes (Fig. 1).

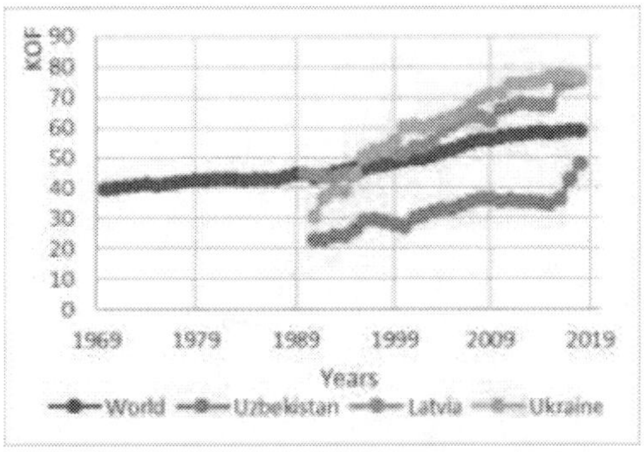

Fig. 1. Dynamics of the Globalization Index (KOF) in the world and some countries.

Globalisation has slowed from light speed to a snail's pace in the past decade [19]. This is caused by the following factors:

- first, transportation costs have been stabilized;

- second, transnational companies have found that global expansion 'burns their money', while national competitors have advantages over them;

- third, trade tends to focus on the services difficult to sell abroad;

- fourth, Chinese production has become more independent and the need in parts import has reduced.

Nowadays, society is in the fifth period of digital transformation occasionally called slow stabilization or slowbalization. Trade tension is caused by the changes continuing since the financial crisis of 2008-2009. In comparison to the world GDP, cross-border investments, trade, bank loans and delivery chains are being cut or stagnating. Globalization is giving place to a new epoch of slowdown or slowbalization, concurrently there are limited sources for Industry 4.0 development. In the context of economic egoism of leading national economies, efficient use of digital communications, development of technological infrastructure and application of new technologies can be constrained by national boundaries (Fig. 2).

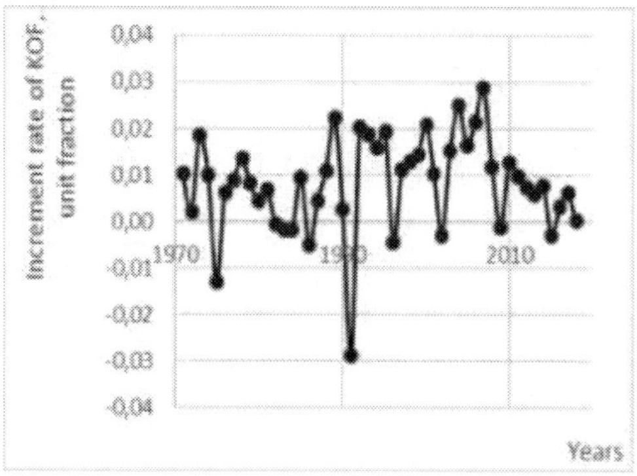

Fig. 2. Increment rate dynamics of the KOF Globalization Index, unit fraction.

Analysis indicates that the KOF Information Globalization level has also slowed down reaching its climax in 2008 (Fig. 3).

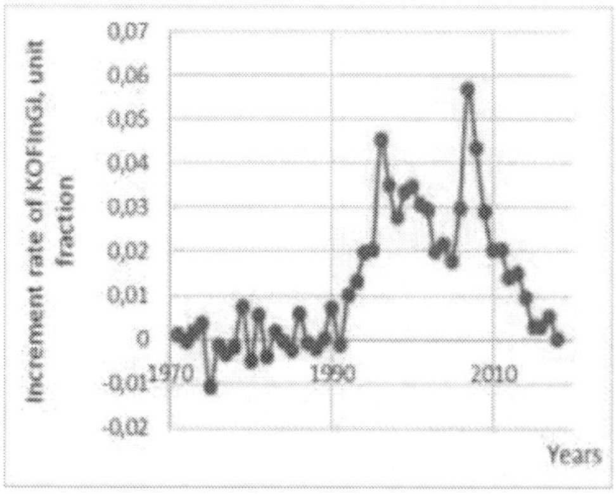

Fig. 3. Increment rate dynamics of the KOF Information Globalization Index, unit fraction.

Gradual reduction of increment rates of global information globalization reveals a current stage of decline in digitalization of society due to the law of marginal utility. That is why, after the decline stage a new wave of communication digitalization can be expected.

Analytical researches enable conclusions about stagnation both of the globalization index and the information globalization index in the world, this confirming the beginning of communicative slowbalization of national economies.

The authors studied density of interrelations between the analyzed global indices in 1970-2018 and those for some countries (Ukraine, Latvia and Uzbekistan) in 1991-2018 (Fig. 4).

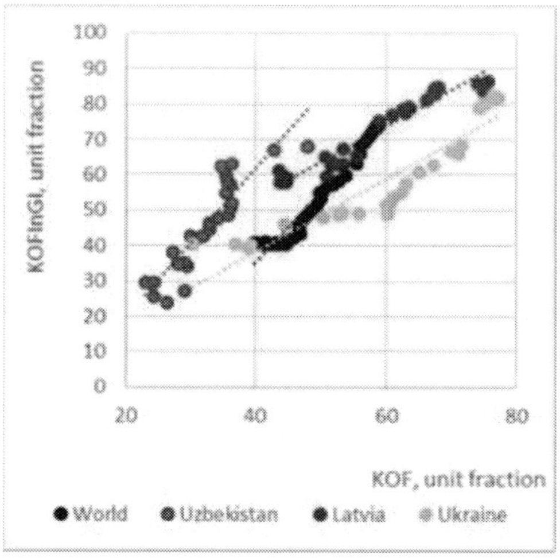

Fig. 4. Correlative dependencies between the KOF globalization index and the information globalization index (KOFInGI)

The analyzed dependencies are presented by equations with determination factors R2 (Table 2)

Table 2. Correlative dependencies between the KOF globalization index and the information globalization index in the world and in some countries.

Country	Correlation type of the information globalization and the globalization indices	R^2
World	$y_{World} = 1.9781x - 44.16$	0.9602
Latvia	$y_{Latvia} = 1.0228x + 12.446$	0.9399
Uzbekistan	$y_{Uzbekistan} = 2.1181x - 22.984$	0.8156
Ukraine	$y_{Ukraine} = 1.0292x - 2.8499$	0.864

High determination indices indicate that the obtained observations conform the models (Table 2), while the increased level of society's information communication ability is accompanied by the increased world globalization level.

Yet, further slowbalization can cause different consequences for Industry 4.0 development in different national economies. To model the determined dependencies, there are specified annual average increment rates of globalization indices in 2009-2018 (Table 3).

Table 3. The increment factor of globalization indices in 2009-2018 of the world and some countries.

Country	The increment factor, unit fraction
World	0.00485
Latvia	0.01816
Uzbekistan	0.03032
Ukraine	0.00872

As is seen from Table 3, increment levels of the globalization indices vary, which is conditioned by the analyzed countries' internal and external space. In total, low rates of globalization increment in the world provide evidence for slowbalization beginning.

The KOF increment factors (Table 3) and determined correlations between the globalization and information globalization indices (Table 2) enable modelling forecast values of the information globalization (KOFInGI) in the world and some countries in 2025-2050 (Fig. 5).

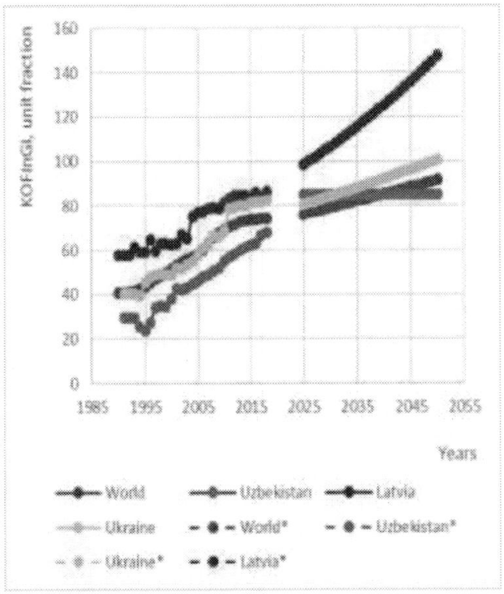

Fig. 5. Dynamics of actual and forecast values of dependency of the KOF globalization index and the information globalization index (KOFInGI) of the world and some countries.

The obtained modelling results indicate significant impacts of available digital potentials of each country's transformation at the beginning of slowbalization. Economic egoism of national economies can not only encourage Industry 4.0, but also become its digital constraint. That is why, the countries with a higher level of information globalization at the beginning of slowbalization (Latvia, Ukraine) will have rising development trends of Industry 4.0 and the economic situation will cause the rapid advance of information-communication technologies in these countries. On the contrary, In Uzbekistan, the society's digitalization level developed steadily constraining Industry 4.0 and this might cause decelerated rates of the country's information globalization.

Thus, the obtained results reveal ambiguous trends of Industry 4.0 development under slowbalization conditions.

For this reason, during the slowbalization period; on the national level, Industry 4.0 is to encourage high competitiveness of national economies by their prioritized areas of activity including:

- combining production and ICT;
- combining a client and production data;
- maximally using potential of mechatronic systems;
- managing production in an autonomous, flexible and efficient manner.

No doubt, digitalization provides a variety of advantages like numerous multiplying effects from including all the industrial chains into a single informational space. Yet, the impact of digitalization on society and economics is controversial due to some reasons:

- first, transition of economics to a new level, which is implemented by digitalization, leads to destruction of the older industrial system and welfare distribution, this characterizing digital technologies as 'undermining';
- second, introduction of new technologies is characterized by excessive optimism, i.e. a total effect from using digital dividends is much weaker than expected, in other words, there is a steady trend of reducing the world average economic rates of labour productivity with only 15% of enterprises' digitalization projects being successful;
- third, there is irregularity of distributing the positive effect from introduction of digital economy both among countries and population groups inside a country;
- fourth, digitalization is to a greater extent more profitable for trade companies and banks, while 'scattered' technologies (robotization, blockchain technology, neural networks, artificial intelligence, quantum virtual and augmented reality) used to optimize production cause reduction and elimination of jobs.

As to positive aspects, it should be noted that the digital space is becoming an environment for new generations allowing exploring the world in a new way and creating radically new research tools. Human creative potentials are also evident as when used along with digital, computer and information technologies, they make a person essentially incomplete and open to changes.

2.3 Basic risks of Industry 4.0 and their levelling

Along with numerous positive aspects of digital transformation, experts define some basic risks as well:

- vulnerability (unauthorized impacts, cyberterrorism) and illegal application of technology (video surveillance control, etc.);
- application of artificial intelligence, robotization, automation: increased social alienation caused by loss of jobs, elevated unemployment, social tension, mass surveillance, potential leakage of commercial secrets, etc.;

- application of blockchain technology associated with vulnerable security of the blockchain system itself and the service infrastructure on its basis, irremovability of information in the network (inability to correct errors and change incorrectly input data), application of tokens as a tool of 'money washing', financing of terrorism;

- application of imported microelectronics. The basic part of software (software of operating systems and database control systems) and computer equipment used, for example in Ukraine, are brought from abroad. It is not impossible that they can contain special spy chips;

- application of cloud technologies and distributed calculations – dependency on reliability of the telecommunication system; dilution of responsibility for information security and reduced control because of its distribution among companies-users, an entity and a cloud platform's owner.

- Internet functioning stability;

- impacts on social conscience. Development of big data technologies, expansion of the network space, achievements in cognitive and behavioral sciences have conditioned appearance of effective methods aimed at implicit data collection and concealed control over group behaviour;

- increased complexity of business-models and lack of highly-qualified staff.

To level all the mentioned risks and apply Industry 4.0 potential to conserving basic principles of sustainable economic development, the society's cognitive paradigm should be oriented at:

- development of the cybersecurity system and the cyberpolice structure;

- formation of the most vital employment skills in future, namely: the elevated level of emotional intelligence; communicative ability; time management; technical skills; creativity and creative approaches to problem solution;

- introduction of the 'closed educational system' without using IT;

- professionals of the future (a solar/wind power technician, a software designer, a data analyst, a digital content manager, a digital security analyst, blockchain designer, an unmanned aircraft dispatcher, a 'smart house' designer, a 3D-print designer, an augmented reality designer, an advisor on personal life issues, an urban and farm economist, a genetic engineer, a personal education guide, an end-of-life manager, an avatar designer, an extinct species revival specialist.

- acquisition of experience in building an effective system of cross-cultural communication;

- formation of creative innovative thinking based on innovating technologies;

- introduction of environmental and energy education.

Thus, digital transformation of society and software will allow exploring the world in a new way and creating radically new research tools. Being an open system and transforming the world in the autopoetic communication process, human beings infinitely expand, transforming themselves and the environment.

3 Conclusions and recommendations

Rapid development of technologies to make modern life better, potential of fast communication connections with each continent; access to information and its fast perception, processing, transmission by mobile technical devices; the community's participation in power execution, opportunities to improve life standards, accelerated development of national minorities and national regions are positive consequences of digital transformation of society.

Development of digital technologies of doing business is accompanied with risks to inviolability of personal life and private data conservation. No limitations on mass media concentration may lead to mass conscience manipulations as well as state or criminal structures control over an individual. Thus, there are negative factors affecting social conscience in the information space.

Besides, there are also significant adverse effects of digitalization including time insufficiency for personal life; application of digital technologies for entertainment industries only and not as an educational or a business resource; 24-hour surveillance of human existence; constant pressure on an individual who has to get adapted to new technologies; rapid access to information causing its distortion and elimination.

The renovated world is noted for accelerated rates of 'digital dictatorship' expansion. If the world community does not start discussing current events as problems, humanity will have all the chances of losing their own identity, culture values, morals and ethics.

Business digitalization and digital transformations of the present day are limitless opportunities for innovative advances. This very fact is both a key advantage and a threat for people. Despite of the fact that computers are able to find an answer better, it is the human who asks the question.

References

1. Official site of the platform "Industry 4.0" (Federal Ministry of Economic Affairs and Energy. Federal Ministry of Education and Research, Germany, 2020). http://www.plattform/i40.de/
2. I.V. Tarasov, Industry 4.0: Concept & Development. Business Strategies. (2018). doi:10.17747/2311-7184-2018-5-43-49
3. Big data and analytics for IBM Power Systems (IBM, 2020). http://www.ibm.com/systems/ua/power/solutions/bigdata/analytics/
4. B. Marr (Ed.). Index. In Big Data in Practice. (2016). https://doi.org/10.1002/9781119278825.index
5. S. Voitko, A. Polishchuk, O. Tkachenko, Development of the defense and industrial complex on the basis of Industry 4.0. Political Science and Security Studies Journal, (2020). doi:10.5281/zenodo.4310124

6. J.Marius Müller, Daniel Kiel and Kai-Ingo Voigt, What Drives the Implementation of Industry 4.0? The Role of Opportunities and Challenges in the Context of Sustainability. *Sustainability, (*2018). doi:10.3390/su10010247
7. R. Shevchuk, A. Melnyk, Mobile Cyberphysical System for Dynamic Display of Object Information on Digital Location. Scientific Bulletin of NLTU of Ukraine, (2017). doi:10.15421/40270436
8. Schwab K. The fourth industrial revolution. Cologny (Geneva Switzerland: World Economic Forum, 2016).
9. G. Maxton, J. Randers, In search of prosperity. Managing economic development to reduce unemployment, inequality and climate change. (Pabulum, Kyiv, Ukraine, 2017).
10. Ph. Steden, R. Kirchner. Industry 4.0 -Overview and Policy Implications. (German Advisory Group, Berlin Economics, 2018). https://www.beratergruppe-ukraine.de/wordpress/wpcontent/uploads/2018/08/PB_06_2018_en.pdf.
11. L. O. Syhyda Industry 4.0 and its Impact on the Countries of the World. World Economy and International Economic Relations. (2018). doi:10.32782/2524-0072/2018-17-9
12. N. Kraus, K. Kraus, O. Marchenko, Digital economy and innovation-entrepreneurial university in the light of competitiveness. Efektyvna ekonomika, (2020). doi: 10.32702/2307-2105-2020.3.5
13. Li Guoping, Hou Yun, Wu, Aizhi, Fourth Industrial Revolution: technological drivers, impacts and coping methods. *Chinese Geographical Science.* (2017). doi: 10.1007/s11769-017-0890-x
14. Global Innovation Index (Cornell University, INSEAD, and the World Intellectual Property Organization, 2019), https://www.wipo.int/edocs/pubdocs/en/wipo_pub_gii_2019.pdf Accessed 20 Jan 2021
15. S. Hushko, I. Mietule, K. Slyusarenko, N. Sheludiakova, Information Economy in the Context of Sustainable Development of Society and its Effect on TNC Activity. *Scientific Journal of Polonia University.* (2017). doi:10.23856/2406
16. T. Dołęgowski , S. Hushko, V. Kulishov, Problems of Instability of the World Economy Under Globalization. *Society. Integration. Educatoin Proceedings of the International Scientific Conference.* (2017). doi:10.17770/sie2017vol4.2280
17. N. Bahashova, H. Puriy, Integration of Ukraine into European Union in Globolizing World. *Scientific Journal of Polonia University.* (2017). doi:org/10.23856/2013
18. G. Savina, F. Haelg, N. Potrafke, J.-E. Sturm, The KOF Globalisation Index – Revisited. *Review of International Organizations.* (2019). doi:10.1007/s11558-019-09344-2
19. The Economist Newspaper and The Economist Group. Slowbalisation. *The Economist.* (The Economist Newspaper Ltd, 2019).

https://doi.org/10.1051/shsconf/202110001004

Digital Technologies in Finance: Modernity and Prospects

Anatoliy Suprun[1], *Tetiana* Petrishina[1*], *Marina* Sadovenko[1], *Natalya* Voloshanyuk[1] and *Serhii* Khodakevich[2]

[1] State University of Economics and Technology, Kryvyi Rih 50000, Ukraine

[2] Kyiv National Economic University named after Vadym Hetman, Kyiv 03057, Ukraine

Abstract. Digital technologies have radically changed all financial relations in society. A new financial business ecosystem based on the use of digital technologies is rapidly forming. The article examines the recent changes in the financial sector under the influence of new technologies. The benefits received by consumers of financial services and financial institutions from the introduction of digital technologies are assessed. The survey of financial services consumers conducted by the authors of the article to some extent reflects the situation with the use of digital financial services in Ukraine. However, the process of digitalization is still far from complete. The formation of a new ecosystem, along with opportunities, generates new risks. There are new, often complex and not always clear to the general public financial products and services. The low level of financial and digital literacy forms a platform for manipulation, falsification and outright fraud. The problem of cybersecurity is relevant. The situation with employment in the financial sector in the medium and long term is rather uncertain. System errors are a risk factor too which can cause significant damage to both individuals and legal entities. The article provides examples of system errors and identifies their main cause - low qualification of information technology professionals working in the public sector. The vision of the future development of the financial sphere within the framework of digitalization is formed, the necessary changes at the private and public levels are determined.

1 Introduction

The modern digital technologies provide many opportunities for the development of the financial business, but they contain many threats and dangers too. Digital technologies in finances are real, although they are far from completion. New financial instruments, such as virtual assets, they endanger the existence of a fundamental financial model that is based on credit money issued by the banking system. Many questions about the functioning of neofinance and neobanking arise at the household level. Although digital financial technologies simplify and accelerate many business processes, they require users to be financially and digitally literate. Under the new conditions, public bodies to control financial business, which is now rapidly virtualizing, has become increasingly difficult. It is necessary to improve state supervision and ensure the interests of society in obtaining reliable instruments of payment and storage of valuables. A new financial system will have to form in the next decade, and all actors in this system have to work to make it fair, secure and efficient for society as a whole.

1.1 Literature Review

The question of the functioning of the financial system in a digital economy has been raised by many scientists, public institutions and non-governmental organizations. In-depth studies in the field of digital and information technology have been made by K.Skinner [8, 10]. The impact of artificial intelligence on the near future of digital business is described in sufficient detail by Kai-fu Lee [5]. The concept of blockchain technologies application in finance was opened by M. Iansiti, K. Lakhani [2], R.Bukht, R.Heeks [15]. The scholar writings of mentioned scientists consider the impact of digitalization on socio-economic processes, including the financial sphere. The most successful examples of fintech implementation in the financial sector in Ukraine are Privatbank with the Privat24 system and Universal Bank with the monobank system. The creation and introduction of the monobank to the market is described in the book-interview with the original author by D. Dubilet [13]. Significant attention is paid to the use of digital technologies in the financial sphere by intergovernmental associations, such as the EU and the UN. Digitalization concepts of the financial sphere are presented in EU Directives and UN reports. In Ukraine, the process of digitalization of the financial and credit system led by the NBU. This is clearly reflected in its regulations, regulatory acts and legislative initiative (in particular, the preparation of legislation on the implementation of PSD2 regulations and the implementation of the open banking concept). The Verkhovna Rada of Ukraine and the Ministry of Digital Development joined the digitalization and virtualization processes by adopting in the first reading on December 2, 2020 the draft Law on Virtual Assets [21], which in

* Corresponding author: petrishina_to@kneu.dp.ua

particular legalizes cryptocurrency. Despite the significant quantity of materials, there is still space for further scientific research, because the finance digitalization process of itself is still far from complete. It should be noted that in existing studies, much more attention is paid to the benefits of digitalization and much less risk.

2 Problem statements

In view of the above, there is a task to substantiate the basic principles of functioning of the modern and future financial system in the digital economy realities and identify safe areas for its development. At the same time, attention is paid not only to the benefits but also to the risks of the neofinance functioning for modern society. Social environment in the form of researchers, private financial institutions and the state should form an understanding of the processes of digitalization in finance and a vision of opportunities and threats in the future. The purpose of the article is to form a vision of financial processes future based on an analysis of present changes that digital technology has brought to the financial industry.

3 Material presentation and results

First of all, the society of the 21st Century is characterized as informational and digital. Digital technologies are penetrating all spheres of social and economic life. The formation of a digital society has many advantages that cannot be overlooked (speed of information transfer, work with data for the benefit of business and society, a new culture of entertainment), but also has a number of obvious and hidden threats.

The debate about the new world continues at all levels: from domestic and educational to state and global.

In 2019, the UN produced a report on the Internet and technology called «The Age of Digital Interdependence» [1], which set goals until 2030 for UN member states in the use of digital technologies for the benefit of society. The basic principle announced in the document is the availability of all users to the Internet and their ability to use specialized services (public services, finance, medicine etc). One of the tasks for governments is ensuring access to financial services (through mobile money, identification / authorization and e-commerce).

The front office decisions of financial institutions are on the surface. Recent advances in remote customer service, the ability to transfer funds quickly and cheaply not only nationally but also internationally, mobile phone payment form a new ecosystem of financial institutions' relations with the client. These financial services have been developing rapidly over the past decade. Particularly noteworthy is their positive impact on the development of financial services in developing countries (in a number of African countries, mobile banking is practically the only available banking service for the general population) and in the context of the COVID-19 coronavirus pandemic on a global scale. But most digital financial decisions are hidden from consumers and concern the back offices of financial institutions

The financial institutions back office functioning models have undergone revolutionary changes over the past 3-5 years, which has provided cost optimization in internal business processes at almost every government level.

In a broad sense, the back office of a financial institution provides internal business processes for asset and liability management. In a banking institution, these are such operations as assessing the quality of the loan portfolio and individual borrowers, opening, closing and maintaining customer accounts, execution of trade and depository agreements with securities. In insurance companies, the operations and internal business processes of the back office are similar to banking, but have insurance specifics (for example, actuarial calculations).

The use of technologies related to data collection and processing, together with high-performance computing capabilities, has allowed automating decision-making processes in the field of pricing, capital allocation, investment and insurance portfolio management.

The use of cloud computing provides shared access to communication networks and data warehouses, instant access to libraries, the use of applications, the ability to use computing resources at any point where there is Internet. Management decisions based on cloud technologies optimize the costs of the back office.

Budgeting has received a new impetus, as coordination between the main back office and structural units has reached a new level through the use of a single technology platform, reporting standards and regulations, electronic document management, bringing financial responsibility centers closer to performance targets and indicators.

In 2020, financial institutions began to make more active use of blockchain technology in the management of both the front office and the back office. M. Iansiti and K. Lakhani substantiated in their scientific works that the financial industry will become a pioneer in the introduction of blockchain technologies [2].

Based on blockchain technology, the world's first cryptocurrency - Bitcoin (an asset that was incredibly successful in 2019-2020) was developed.

Except the opportunity to issue a variety of Steemcoins, the blockchain provides financial institutions with many other opportunities: namely,

- security of payments and funds transfer;
- sending controlled mobile messages;
- tracking of insurance policies issued by various companies to prevent fraud;
- fixation of various financial agreements;
- simultaneous messages of the same content without the possibility of making changes by one party (the same applies to any financial contracts);
- formation of a vicious circle in the field of reinsurance and co-insurance;
- automatic processing of claims in insurance under previously tokenized policies;

- checking specialists for their qualifications in the field of financial business (banking, insurance, brokerage); and
- fixation of financial guarantees.

Digitalization of internal business processes has opened new opportunities for banks, insurance companies and other financial institutions to develop personalized financial services and differentiate pricing. Customers' identification together with an array of data about them allows you to develop personalized financial services without significant additional operating costs. This area perhaps has the greatest potential for increasing the availability of financial services at a price and rise in the number of financial institutions clients.

The state is represented in all financial processes of society. Influence on money circulation and money supply is provided through the institutions of central banks, tax collection is performed by fiscal services, regulation and control of the financial sector is through regulators. Of course, the state cannot and has no right to stand aside from technological processes in the financial sphere.

Financial activities are subject to state supervision and control. Economic regulations of regulators are becoming stricter due to the growing volume and complexity of financial transactions and require the provision of increasing amounts of information with its complex structuring. Models based on risk assessment of financial institutions (for instance, Basel III in banking and Solvency 2 in insurance) cannot be used without appropriate information technology, processing power and software.

The large-scale participation of the state in financial circulation based on digital technologies has given rise to such terms as RegTech and SupTech.

RegTech is a new domain within the financial industry that applies technology to improve regulatory processes, especially with regards to KYC (Know Your Customer) and AML (Anti - Money Laundering) [3].

Fuzzy terminology is a fairly common phenomenon in today's digital economy. Thus, the NBU talks about decisions in the field of SupTech (supervisory technologies) and RegTech (regulatory technologies). SupTech can be defined as a set of technologies used by supervisors and market regulators to perform their functions and administrative tasks. Thus, SupTech provides a system of reporting to banks on transactions subject to financial monitoring. RegTech is a set of technologies that optimize the bank's compliance. It means control system over compliance with regulatory requirements [4].

Therefore, the front office digital technologies, first of all, are improvement of marketing activity, and back office - operational, investment and financial. The combined characteristic is the transparency of all financial institution business processes formed on a single technological platform. Each transaction leaves an «electronic footprint», each transaction is permanently tied to the center of financial responsibility and to a specific executor, and each decision can be subsequently subjected to objective analysis. Transparency of business processes builds trust in information for management decisions.

In the coming years, technologies related to the use of artificial intelligence will be introduced in the financial sector, as well as in other economy and business areas. Kai-fu Lee identifies four waves of artificial intelligence: artificial intelligence of the Internet, artificial intelligence for business, artificial intelligence of perception and autonomous artificial intelligence [5]. At the same time, in the author's opinion, the first two waves are already being realized, gradually changing the financial and digital world.

Credit analysis and monitoring using neural networks, extensive use of chatbots based on machine learning technologies, and analyzes the market and consumers with further forecasting (including forecasting of consumer preferences, their behavior in the market) have already been in use financial sector. But many factors constrain the development of artificial intelligence technology in finance and, first of all, this concerns the misunderstanding by many consumers of the technology itself, its complexity (many authors also add the possible uncontrollability factor of AI systems).

Digital technologies have radically changed and continue to change the financial sphere in a short time, a fundamentally new ecosystem of financial business has been formed, consumers of financial services have gained many advantages, and customers have become more thanks to technology.

But there is a lot of research on the threats that are emerging in the new digital society. Many of them are in plain view and this applies primarily to cybersecurity. The issue of cybersecurity is extremely important, because hacker attacks can instantly destroy both critical infrastructure and the financial systems of entire countries. At the same time, taking into account the globalization, the issue of protection should be built at the supranational level. The financial systems of different countries are interconnected. At the same time, many countries have limited opportunities to finance the necessary cybersecurity measures.

At the household level, there is a speed of change in financial services, their diversity, often incomprehensible to many users or technical limitations are threats. Gradually, a layer of people-leaders in the use of digital financial services is formed, and people-outsiders who are unable to master new financial services due to lack of education or understanding, due to lack of technical means (smartphones that support financial services). But in the meantime, financial institutions strive for the maximum possible digitalization of financial services because it significantly reduces operating costs. In the near future, there will be situations when a certain circle of people in a specific location will be cut off from the opportunity to receive financial services.

The scaling of financial services, the emergence of new types and kinds of financial assets (including virtual assets) will increase the number of frauds in the financial market. In addition to outright fraud, there are activities of financial institutions within the law, which are based on low financial literacy of the population and deliberate information misrepresentation.

The most typical example of this phenomenon is the issuance of fast loans through mobile applications (at 2-2.5% per annum for each day of credit). This problem could best be solved by proper financial and digital literacy of society and control over the financial market by the state.

The digital technologies impact on the labor market is a more complex challenge. While it had previously been assumed that automation and robotization of processes could affect the segment of unskilled jobs, now this threat has fully spread to the segment of middle and highly qualified workers, the so-called White collar workers.

It is already possible to identify a number of specialties in the financial sector, which may soon be unclaimed in the market. Such professions include insurance agents, emergency commissioners, bank clerks, call center employees, financial lawyers, and tax inspectors. Mentioned and many other professions can be fully automated using algorithms. Moreover, this list will include IT services employees of financial institutions with a certain time lag. Internal business processes on IT issues will be outsourced en masse (this is already being done now), and the algorithms themselves will create the necessary software for doing business (medium term). The rapid spread of digital technologies in the Ukrainian banking sector is accompanying by a decrease in the number of banking institutions branches (Fig.1).

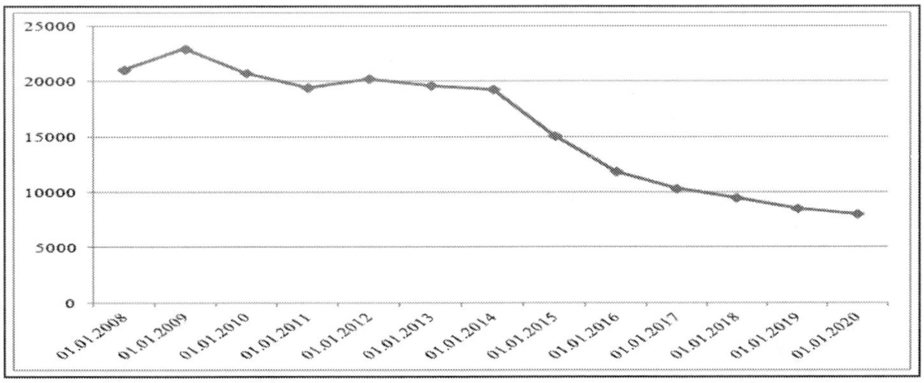

Fig. 1. Dynamics of banking institutions branches in Ukraine in 2008-2020, units
Source: generalized basing on [6]

The global financial crisis affected the market infrastructure that had already formed at that time. In particular, the number of branches in 2008-2010 decreased by 7.8%. The main reason was not the closure of banking institutions (their number rose from 175 to 182 during this period), but the development of banking technologies. The crisis has provided an impetus for the cost optimization search, and innovations and new technologies are the only pathway to do that.

The Ukrainian banking services market had been underwent the largest structural changes during the local financial crisis of 2014-2016, when 84 banks left the market [7]. The branches number was reduced by 53.5%. After the market stabilization in 2017, the figure of offices had continued to drop (until 01.10.2020 by 29%) and this reduction has already had a purely technological basis. In 2020, despite the Covid-19 pandemic, the banking system operated profitably, but the accelerating of the adoption process of remote access technologies led to the further closure of branches - their number decreased by 9% over the year.

The objectivity of the bank branches reduction was described by Chris Skinner in 2014. Historically, retail banks had an extensive branch network. In the 70's, they introduced Automated Teller Machines, in the 80's - call centres, in the 90's - the Internet, and now, in the 2000s, they are introducing mobile technologies [8]. And each of these stages has contributed to a decrease in the number of departments and staff, but there has been a significant acceleration of these processes in recent years.

The banking market of Ukraine was hit the hardest during the crisis of 2014-2016 (Fig. 2). Recovery began after 2017, when provisions for doubtful loans (NPLs) had finally formed. At the same time, there has been an increase in the share of commission income in the total income of the banking system over the past 2 years, which is always a sign of the crisis. In general, in the Ukrainian financial market, banks now do not feel competition from fintech companies, neither in the field of payments nor in the money transfers areas. The mobile Internet banking system is fully controlled by traditional banks.

Privat24 is the most popular Internet mobile banking system in Ukraine. It was created by Privatbank in 2001, but in fact this system became widespread after 2010. Then a version of the Privat24 mobile application was launched for download on smartphones that support iOS and Android operating systems. Innovations have made Ukrainian Privatbank one of the most innovative banks in the world. For instance, the bank was one of the first in the world to use one-time SMS passwords more than 10 years ago.

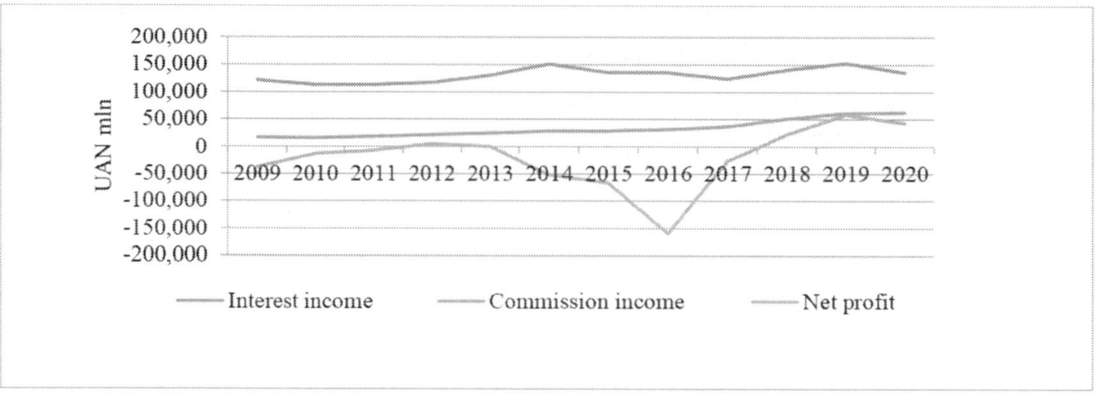

Fig. 2. Interest, commission income and net profit of Ukrainian commercial banks in 2009-2020, UAN mln
Source: generalized basing on [6]

The latest innovations, which have gained worldwide recognition, include products such as a payment mini-terminal, access to the Internet bank via QR code, online collection, as well as dozens of different mobile applications [9]. The well-known researcher of digital banking Chris Skinner also noted the innovativeness of the Ukrainian Privatbank in his work. The author noted the introduction of an open API architecture and the formation on this basis of a component-oriented banking system and the introduction of the Sender system, which allows companies to communicate with customers and share values via chat without the cost of developing applications. The Sender app received the Best of Show award from the Finovate Fall conference, which took place in New York in the fall of 2014 [10].

Talking about technologies in the banking sector of Ukraine, we cannot but mention another successful example - the Monobank project of the Ukrainian JSC «Universal Bank».

Monobank is a retail product of JSC «Universal Bank», which emerged in cooperation with the Fintech Band. Credit cards are issued for customers, it is possible to place deposits and receive other services within the Monobank. This mobile application makes financial management as convenient as possible. Monobank only works on mobile devices [11]. The Monobank project was created in 2017 by a team of specialists who once worked on the Privat24 system. The work is carried out by the Universal Bank license, which is part of the TAS group. The project was very successful and won the nomination «Best Ukrainian FinTech-startup» in the year of creation, PaySpace Magazine Awards, and in 2018 won the main award in the category «Neobank of the Year» FinAwards 2018. Now Monobank is used more than 3 million customers.

Testing was conducted to assess the public's perception of the business organization form. Testing is an experimental method of diagnosis which are often used in socio-economic research. Tests are classified on various grounds. Thus, according to the types of properties, tests are divided into tests of achievement and personality [12]. Simple personality tests to determine the priorities of using mobile banking were used in the research. Testing was conducted by the authors using Google form among State University of Economics and Technology students and teachers aged 16 to 65 years (over 650 respondents). The question was formulated as follows: What financial service do you personally use? The following possible responses were suggested:

- Privat24;
- monobank;
- other financial institutions;
- I do not use financial services.

The study results are presented in the Table 1.

Table 1. Results of mobile banking users' survey

Age of respondents / Financial institution (system)	From 16 to 25 years,%	From 26 to 45 years,%	From 45 to 66 years,%
Privat24	319 (68,6)	138 (97,9)	30 (68,2)
Monobank	138 (29,7)	33 (23,4)	6 (13,6)
Other financial institutions	6 (1,3)	2 (1,4)	-
I do not use financial services	2 (0,4)	3 (2,1)	8 (18,2)
Total	465	141	44

Source: Formed basing on the results of survey among teachers and students of the State University of Economics and Technology and University Colleges

In general, the survey corresponds to the client profile formed by Dmitry Dubilet (the original author and one of the Fintech Band founders). According to his observation, the bank's main clients belong to the age group of 20-40 years, often self-employed persons, those who choose the bank independently (monobank is not engaged in corporate lending, maintenance of pension accounts or salary projects of large enterprises). The monobank pricing model is extremely simple - Privatbank's tariffs are reduced by 10-30%. This avoids or reduces the number of offices, the utilities' cost, a large number of staff [13]. The research results also indicated that the vast majority of respondents represent Monobank not as a structure of JSC «Universal Bank», but as an independent digital bank.

There are fintech companies that provide transfer and payment services in Ukraine, but their activity are small. Monobank is more popular among young people, and Privatbank is a national leader. Few people do not use financial services at all (do not use either the elderly or for religious reasons). The situation in the banking market under the influence of technology causes two trends - banking productivity increasing and rise of return on invested capital while reducing the number of employees in the industry. Of course, this will also apply to other financial institutions. In general, this confirms the conclusion of a study conducted by Thomas Picketti that, except for a relatively short time in the twentieth century, the rate of return on investment exceeds the level of return on labor [14]. This conclusion is made in more detail, but through the prism of artificial intelligence, and in the study of Kai-Fu Lee [5].

R.Buht and R.Hicks in their research give other views, which relate, on the contrary, job growth in most areas of activity through the introduction of digital technologies [15].

In our opinion, the number of people employed in the financial sector will decline due to the introduction of digital technologies, but this is more a short-term or medium-term trend. Neither banks, nor insurance companies, nor most other financial institutions will be able to completely get rid of branches, even in the long run.

The number of branches and employees will reach a certain minimum level and stabilize. In addition, a new demand for financial services is expected to appearance, and more affluent customers will be served by people. The trend when wealthy citizens prefer to do without digital services, described by Nellie Bowles in a well-known article Human Contact Is Now a Luxury Good. The author emphasizes that in almost every type of socio-economic activity (medicine, education, business services etc.) there is a people layer who want to communicate with human professionals, rather than with their digital substitutes. The availability of digital technologies and their relatively low cost (the cost reducing process of both equipment and services is likely to be continued for a long time). Undoubtedly, the use of digital technologies no longer stresses a person status [16].

Banks and other financial companies will be sure to leave branches and individual employees for VIP services. Those who have not adapted to the new realities will also need physical care. Supervisors may establish a list of services that financial institutions will provide both online and offline to resolve potential conflicts.

Competition in the financial sector will intensify and digital technologies will play an important role in this. The adoption of the EU Directive on payment services in the domestic market (Payment Services Directive 2015/2366) introduced the concept of «open banking» [17]. The main content of this is the admission to the payment services market of non-banking financial companies, including fintech companies. «Open banking» is based on the possibility of third-party access (their list is determined by law) to the bank customers' accounts and clients' data through the API (Application Programming Interface). Such access can be used both for the implementation of payment services (they are still focused on) and for the development of new financial services by non-banking financial institutions (the same fintech companies). Banks essentially create their own competitors by providing such access.

But consumers of financial services will receive many benefits, including the improve security of payments and money transfers, the reduction of commissions, which will affect all participants in the financial system, the receipt of new financial services.

The PSD2 directive was adopted in 2015, but provided for a two-year implementation period and step-by-step entry into force and came into force only in September 2019. In Europe, banks are actively developing and implementing the Open API to not only meet PSD2 requirements, but to make money. The most popular is the launch of the freemium API (free access to accounts within the requirements of PSD2 with paid access to information and services not covered by PSD2). Most of payment institutions failed to implement enhanced customer authentication (SCA) requirements in a timely manner. This prompted the European Banking Authority (EBA) to recommend postponing the implementation of the SCA until 31 December 2020 [18].

According to the National Bank of Ukraine plans, the implementation of PSD2 in Ukraine is scheduled for completion in 2022 [19]. Today, implementation is impossible due to outdated legislation.

Technology is evolving so rapidly, and its application in business processes is becoming so profound that the financial processes forecasting becomes very vague (the idea of not being able to make objective forecasts at all deserves the attention). It seems that there is a huge field for speculation of a new type due to a combination of rather complex financial instruments (for instance, atypical derivatives with a complex structure) and digital technologies, which are currently understood only by narrow specialists (blockchain, artificial intelligence, machine learning). At the same time, the speculative component of financial markets, which is enormity today, will grow, while the real (issue of securities for project financing) will decrease. At present day, the most significant is the dynamics of the first cryptocurrency - bitcoin (Fig.3).

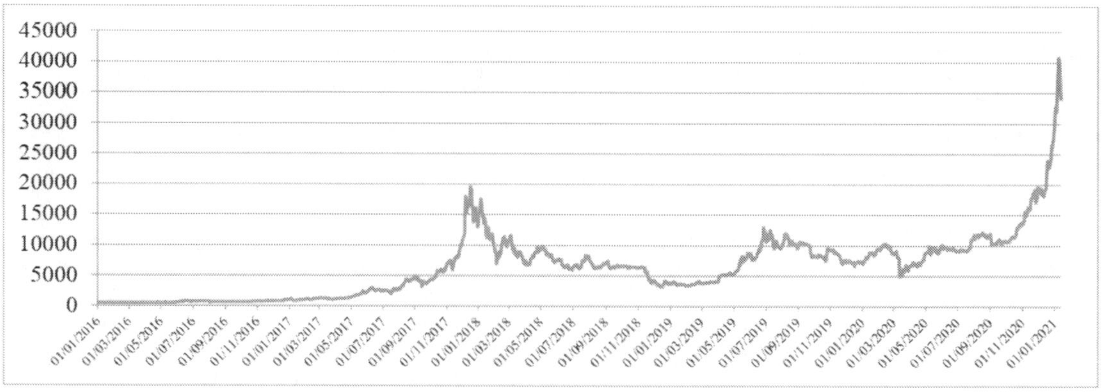

Fig. 3. Dynamics of the bitcoin exchange rate versus the US dollar in 2016-2021
Source: generalized basing on [20]

The standard deviation of the series is 5283, and the coefficient of variation of 83.56% indicates an extremely high level of risk associated with the asset.

Cryptocurrency apologists and «crypto-optimists» insist that bitcoin is money (although not controlled by the state). From our point of view, a currency with such fluctuations and complexity of application cannot perform the basic functions of money such as a measure of value and a means of payment. But it probably performs the function of accumulation. It is more correct to consider bitcoin and similar assets as virtual (digital) goods. This is exactly the view presented in the first reading of the Law of Ukraine 923-IX of September 29, 2020 «On Virtual Assets» adopted on December 2, 2020. The Ministry of Digital Transformation of Ukraine, not even the National Bank of Ukraine, has been designated as the regulator of the virtual assets market [21]. The growth of the market for virtual assets, including virtual financial assets, cryptocurrencies and other quasi-money (monetary surrogates) can be predicted with a high probability. The risk is offset by extremely high potential returns and there will always be investors who will invest in virtual assets. The situation with bitcoin is an indicator of investor interest in virtual assets.

Against the background of the COVID-19 coronavirus pandemic, there have been trends of increasing money issuance by central banks, declining interest rates and a collapse in many commodity markets, in the tourism and hotel business, and in passenger traffic. There is more «extra» money in the world, which explains the surge in interest in bitcoin.

The state should limit speculation in the market, but the main efforts need be aimed at protecting the interests of pension funds depositors, life insurance companies, time and current deposits. Investing in speculative assets ought to be prohibited for institutions licensed for such activities.

Failures of information systems related to the human factor, especially in the field of public administration, are a very important danger. The salaries of IT professionals in the public sector are much lower than in the private sector. In Ukraine, this difference is extremely high. This creates a low quality of information logistics, software and personal data protection. The authors observed the investigation of a case where two different people had one pension account (one of these people is a pensioner and the other one is of non-retirement age). The investigation was conducted by three services (Pension Fund, Fiscal Service and Employment Service) and did not reveal anything . The result was obtained by the authors themselves by simple physical verification of documents and facts. The series of passports were different (AK and AM, which is very similar in Ukrainian paper passports), and one person refused to be assigned a taxpayer identification code for religious reasons. Such situations occur periodically in the public sector and most of them are due to low qualifications of performers or low quality software.

The variety of risks posed by digital technology in the financial sector is striking in both quantity and possible consequences. The finances digitalization process is ongoing, which means that both the number of opportunities and the amount of risks will increase.

The state and representatives of the financial business should form a partnership that will aim to form a stable global system based on the following principles:

1. Accessibility - financial services should be available to absolutely all inhabitants of the planet with the help of today's information and communication technologies.
2. Fairness - discrimination in access to certain financial services depending on a person's social status is inadmissible. This will be achieved primarily through the pricing policy of financial institutions.
3. Competitiveness - digital platforms ought to be available to all participants as well as the software necessary for conducting monetary transactions.
4. Monopolization of any financial services at the national and global level is not allowed (except for the issuance of money in the short and medium term).
5. Protection of personal data - financial institutions must reach the consent of the individual to use his data when conducting financial transactions or transferring data to other financial institutions.
6. Control and regulation - only state control and regulation can keep the global financial system from complete chaos and thus protect the interests of individuals and legal entities.

4 Conclusion

Digital technologies are shaping a new financial ecosystem. There have been increasing in the efficiency and productivity of financial institutions both at the level of the front office and at the level of the back office the last 3-5 years. Customers have received many benefits and conveniences, and internal business processes have become much more optimal. But there are amount of new risks and threats which do not attract as much attention as the benefits of new digital technologies. The increase in financial services consumers through mobile and Internet technologies contributes to the growth of financial institutions income, but is the reason for the reduction of jobs in the industry. The article this is illustrated by the banking system of Ukraine example - income from commission activities and interest income grew against the background of the continuing decline in the number of banking institutions and employees in recent years.

Very serious challenges for the new financial system related to ensuring a high level of cybersecurity. The banking system demonopolization in the area of payments and remittances, provided for in the EU Directive PSD2, will raise the level of competition in the financial sector, but will lead to the emergence of a large number of new small market participants. At the same time in Ukraine in the field of funds and payments transfer is dominated by commercial banks with their payment services (Privat24, Monobank). The situation will not change, at least in the medium term. But the opening of the payment services market for non-bank financial institutions (in particular, fintech companies) may stimulate the emergence of a new type of fraud.

Use of blockchain technologies, data science, machine learning and especially artificial intelligence, show their effectiveness in many processes and operations. But according to analysis result, these technologies need further thorough research before they are ready to enter the market in full. The volume of financial speculation and outright fraud is growing (this is clearly represented by the dynamics of the bitcoin exchange rate versus the US dollar) with the development of virtual assets. The idea of the uncontrolled money functioning in the case of its implementation can plunge the world into financial chaos.

The link of state control and management in the financial sphere is problematic, because the vast majority of regulatory processes have already been digitized (which led to the emergence of such terms as SupTech and RegTech). At the same time, information technology specialists in the public sector earn much less than in the private sector. This is already affects the government regulation quality in the financial sector.

The solution to the problems of digital finance future lies only in the field of effective public-private cooperation, which should be based on the principles of financial services total accessibility, their fairness, fair competition in financial business, ensuring maximum protection of personal data and improving public supervision and control over the financial institutions.

References

1. The age of digital interdependence. Report of the UN Secretary Generals High-Level Panel on Digital Cooperation (2020). https://www.un.org/en/pdfs/DigitalCooperation-report-for%20web.pdf
2. Iansiti M. Lakhani K. *The Truth about Blockchain*. Harvard Business Review. (2017). https://hbr.org/2017/01/the-truth-about-blockchain
3. Patrick Schueffel. *The Concise Fintech Compendium* (School of Management Fribourg, 2017). https://www.researchgate.net/publication/322819310_The_Concise_Fintech_Compendium
4. Kurcev V. *SupTech, RegTech and FinTech — What is it and what is the difference* (2019) https://ain.ua/2019/05/17/suptech-regtech-ta-fintech/
5. Kai-Fu Lee. *AI Superpowers: China, Silicon Valley, and the New World Order* (translated by Punjko V.). Kyiv: Fors Ukraine, 2020, 304 p.
6. Data from supervisory statistics. Website of the National Bank of Ukraine. https://bank.gov.ua/ua/statistic/supervision-statist/data-supervision#5
7. Number of banks in Ukraine. Website of the Ministry of Finance of Ukraine. https://index.minfin.com.ua/ua/banks/stat/count/
8. Skinner C. Digital Bank: *Strategies to Launch or Become a Digital Bank* (translated by Smirnov S.). Moscow: Ivanov and Ferber, 2015, 320 p.
9. Information about the Bank. Website of PrivatBank. https://en.privatbank.ua/about
10. Skinner C. *How fintech firms are using bitcoin blockchain and mobile technologies to create the Internet of value* (translated by Jacjuk N.). Moscow: Moscow: Ivanov and Ferber, 2018, 416 p.
11. About Monobank. Universal Bank. https://www.monobank.ua/about?lang=uk
12. Azaryan A, Azaryan R. *Fundamentals of scientific research*: Kryvyi Rih: KTU Publishing Center, 2010, 230 p.
13. Vorona T., Dubilet. D. *Business on common sense. 50 ideas how to achieve yours*. Kyiv: Force Ukraine, 2020. 256 p.
14. Piketty T. *Capital in the Twenty-First Century*. Moscow: Ad Marginem, 2015, 592 p.
15. Bukht R, Heeks R. *Defining. Conceptualising and Measuring the Digital Economy*, **13**, 2, 143–172 (2018). DOI: 10.17323/1996-7845-2018-02-07
16. Bowles N. *Human Contact Is Now a Luxury Good* (New York, USA, 2019)
17. Payment Services Directive 2015/2366. https://ec.europa.eu/info/law/payment-services-psd-2-directive-eu-2015-2366_en
18. Shulyar R., Burlaka B. Fintech VS Banking. *What Will Change the New Payment Standard of the European Union*? (2020). https://marchenkopartners.com/fintehom-po-bankingu-shho-zminit-novij-platizhnij-standart-yevropejskogo-soyuzu/)

19. Payment infrastructure of Ukraine goes over to the international standard ISO 20022.
https://bank.gov.ua/ua/news/all/platijna-infrastruktura-ukrayini-perehoditime-na-mijnarodniy -standart-iso-20022
20. Bitcoin USD (BTC-USD) Stock Historical Prices & Data. https://finance.yahoo.com/quote/BTC-USD/history/
21. Draft Law of Ukraine «On Virtual Assets» from 11.06.2020 №3637

Transition from Vertical to Horizontal Industrial Policy in Ukraine: Effects on Industrial Sector Growth

Liudmyla Deineko[1], *Olena* Tsyplitska[1*], *Nadiia* Hrebeniuk[2], *and Oleksandr* Deineko[3]

[1]State Organization "Institute for Economics and Forecasting of NAS of Ukraine", Department of Industrial Policy, 01011, Kyiv, Ukraine
[2]King Danylo University, Department of Finance and Accounting, 76018, Ivano-Frankivsk, Ukraine
[3]Ukrainian-American Concordia University, Department of International Economic Relations, Business, and Management, 01054, Kyiv, Ukraine

Abstract. This paper explores the determinants of industry's value added growth related to transition of Ukrainian industrial policy from vertical to horizontal type in 2015. The method used is econometric modelling through building multiple linear regression, where the real rates of value added growth from year to year represent the dependent variable. As explanatory variables the real rates of capital investments, structural changes, institutional indicators of regulatory quality and index of economic freedom as well as dummy-variables denoting the periods associated with industrial policy transit were introduced into the model. The main finding of the study is the revealed negative effect of horizontal industrial policy on industrial sector productivity. Other results showed the significant contribution of capital investments growth rates and government support's share in the State budget expenditures. The included institutional determinants as well as structural changes indicator didn't play any significant role in providing of industry's value added growth. The results of modelling were used in recommendations for matrix (combination of vertical and horizontal instruments) policy implementation in Ukraine.

1 Introduction

Industrial policy stands as one of the most important tools in the field of creating a structurally balanced, competitive economy, within which the latest technological and institutional trends are implemented. Adjustment of conceptual approaches and targets for its development and implementation is called for, in the light of the new challenges of the Industry 4.0, "green" modernization and inclusive development in transformational economies, integration processes in Central and Eastern Europe and the growing numbers of participants in international trade agreements. If the goal of industrial policy in the XIX-XX centuries was to create a strong industrial sector, at the present stage it is adjusted to environmental, social, and institutional aspects as the prerequisites for the implementation of any technological solutions.

The signing of the Association Agreement between Ukraine and the EU was also the basis for a revision of the dominant – vertical (sectoral) – approach to the selection of industrial policy instruments. EU member states and other developed countries ensure a balance between sectoral and horizontal means of supporting industrial development, forming a so-called matrix model of industrial policy.

In Ukraine as a result of the armed conflict in the Donbas region depleting the economic potential of both the temporarily occupied territories and the country as a whole, and the introduction of the European integration vector of development, the share of sectoral support in the state budget has been reduced. Such a "transition" from the dirigiste model of industrial development management, which, to some extent, led to the oligarchizing of the Ukrainian economy, to the liberal model, created an additional "shock" in the domestic market which has led to a significant decline in conditions of COVID-19 pandemic. In mining and quarrying industry, the volumes of production fell by 4.2 (I quarter), by 8.7 (II quarter), and by 0.8% (III quarter) in 2020; in manufacturing the indicator fell respectively by 4.3, 14.7, and 5.6%.

Domestic industry, left alone with outdated fixed assets, products that do not meet international standards (in particular – European), import dependence, loss of the Russian market, critical lack of investment and credit resources etc., responded with a significant drop in production, reduced numbers of enterprises and outflow of investment and labor, negatively affecting the growth rates of the economy.

Hence, the assessment of the possible impact of state support for industrial development in Ukraine is especially relevant, to offset the negative effects of a market economy, as well as the impact of other factors – economic, institutional, and political, which will adjust the course of industrial policy to accelerate economic growth of the industrial sector. This constitutes the goal of this research.

1.1 Literature review

* Corresponding author: o.liakhovetz@gmail.com

The state industrial policy is developed for negating negative and exploiting positive externalities of a market in the conditions of certain restrictions, namely:

1) lack of information, which is manifested in the inability to accurately identify industries, sectors or activities that are significantly affected by externalities, and therefore does not allow to fully compensate for their negative impact;

2) political commitment, through which the governing bodies cannot resist lobbying and rent orientation in order to prevent the transformation of industrial policy into tools for rent transfer to related insiders.

These factors determine the ineffectiveness of institutional mechanisms for modernization of the industrial sector, including direct state support. Mathematically, this will be reflected in the feedback between sectoral aid and the growth rate of productivity of the industrial sector [1], which can be represented as:

$$g_i = \gamma s_i + \varepsilon_i , \quad (1)$$

where g_i is a variable that indicates the efficiency of the sector, for example, productivity; s_i – government intervention as a sectoral policy; γ – parameter that represent the success of industrial policy; ε – model error.

The success of an industrial policy will be manifested in the sign of the coefficient γ: if it is greater than zero, the policy is effective, if less, then vice versa. The theoretical generalization developed by D. Rodrik [1] demonstrates that, in fact, the policy will always be ineffective. Considering a simple model (2):

$$g_i = (1 - \theta_i) \cdot A, \quad (2)$$

where A – productivity growth rate; θ_i – the extent of market failure, D. Rodrik mathematically confirms that the impact of government support will always have a negative sign of the relevant parameter, so the relationship between government aid and productivity growth is negative in the model, which assumes that the government pursues only public interests – maximizing the growth of productivity of industries. Even if a state applies the optimal level of subsidies and other assistance, it cannot fully eliminate all market failures, so γ <0. This becomes the basis for legislative regulation of the extent of government intervention. Based on this, regardless of whose interests a state pursues, the policy of intervention will bring a negative result.

Determination of the effectiveness of state mechanisms in support of the industrial sector is reflected in a set of econometric models by R. Beason, D. E. Weinstein [2]; C. Criscuolo, R. Martin, H.G. Overman, J. Van Reenen [3]; A.O. Krueger, B. Tuncer [4]; R.Z. Lawrence, D.E. Weinstein [5]; J.-W. Lee [6]; R. Stöllinger et al. [7]; J. Wang [8]; H.-Y. Joo, H. Suh [9]; J. Chen, L. Xie [10]; Y. Higuchi, G. Shimada [11]. Despite the existing discourse in the economic scientific community about the essence of industrial policy, its tools and their effectiveness, empirical research using statistical and economic modeling is quite small. However, some results were obtained from the analysis of developed countries – EU member states, Japan, Turkey, and the Republic of Korea.

R.Z. Lawrence, D.E. Weinstein [5] tested a popular hypothesis that state protectionism in Japan has increased the competitiveness of manufacturing companies at the global level by building an econometric model of the impact of industrial policy on the overall productivity of the industrial sector. The model included variables such as the level of technological backwardness, cumulative growth rates in industries since 1960 as effects of "learning-by-doing", R&D expenditures, the share of imports in domestic consumption, the share of exports in total intake, the level of tariff protection and individual industrial policy measures. The simulation results refuted the hypothesis of the usefulness of state protectionism. On the contrary, the stimulating role of imports to increase the competitiveness of domestic producers was proved, especially in the period 1964-1973, which led the authors to conclude the beneficial nature of the liberal model for the Japanese economy.

A.O. Krueger, B. Tuncer [4] assessed the dependence of the growth rate of industrial sector output (emerging industries) on the growth rate of state protectionism in Turkey, in particular, in the framework of tariff policy. They've built an econometric model of the dependence of these variables, considering the cost function of the industrial sector. The main assumption was that effective government support should have resulted in lower production costs per unit of output. As a result, it turned out that in the protected industries there was no increase in productivity per unit of output to give the justification for such state protection. However, researchers suggested that the reason may lie in the inappropriate incentives structure.

Based on the correlation analysis, a negative relationship between the types of government support and the growth rate of production in Japan was found by R. Beason, D. E. Weinstein [2] in different periods from 1955 to 1990. Researchers also built a linear function of industrial policy's influence on the productivity of the industrial sector (3):

$$\Psi_i = f(L(DTAX)\ L(DJDB),\ L(DSUB),\ L(DTAR),$$
$$L(TAR),\ L(NTB)), \quad (3)$$

where $L(\cdot)$ is the lag field of policy of variables; *DTAX* – tax relief variable; *DJDB* – Japan Development Bank loans variable; *DSUB* – subsidies variable; *DTAR* – the difference between the sectoral tariff rate and the average tariff rate; *TAR* – effective rate of protection in the sector; *NTB* – the percentage of subsectors within a given sector covered by quotas.

As a result, it was concluded that state support in Japan was not directed to high-tech sectors, seemingly due to the erroneousness of government decisions in the choice of priority areas of development, and such support did not lead to productivity growth in the studied areas.

J.-W. Lee [6] studied the impact of state industrial policy on productivity growth in the Republic of Korea's industrial sector. The empirical results of econometric modeling, which evaluated the effectiveness of measures such as non-tariff barriers, customs tariffs, tax incentives and credit incentives, disprove the popular notion of the success of government intervention in the Korean economy. Empirical results have failed to confirm any positive effect of government interventions on labor productivity. Some positive effects were achieved only through tax privileges. As a result, the author believes that the Republic of Korea's success would be more significant without government intervention.

Other scholars, J. Chen & L. Xie [10] considering the predominantly sectoral approach to industrial policy in China tested the correlation between the policy measures and industrial performance from 2003 to 2015 on several regression models (including explanatory variables of capital, population, the level of informatization, transport level, human capital level, opening-up level, and the extent of government intervention). They concluded that sectoral industrial policy was the main locomotive of development in industrial sector in spite of extension of horizontal (functional) instruments in recent years.

In the context of the EU member states, researchers R. Stöllinger et al. [7] determined the impact of different types of state aid provided to EU countries on export-oriented sectors of the processing industry, using three-stage regression. Among the types of assistance were considered the following: state subsidies for environmental protection, aid for R&D, regional development, assistance to strategic sectors (shipbuilding, steel, aviation, etc.) and other types of horizontal support. To make the results more valid, government efficiency indicators (to control the quality of institutions), the share of wages in value added (as an indirect indicator of competitiveness), and the weighted import tariff rate (as an indicator of trade protection) were added to the model. The simulation results revealed that the only significant type of state support could be the expenditures on the internationalization of production, which should occur along with the improvement of governance efficiency. Thus, for some countries with a very low level of public administration efficiency, such assistance may be counterproductive. However, there is a group of countries with a high share of exports, but with a low level of government efficiency, so, depending on the institutional conditions, different strategies can be applied to increase exports. Overall, both in terms of value added and in terms of growth, sectoral processing aid in the EU has not yielded any significant results.

More recent studies allow estimating the performance and configuration of government support for industrial sector by innovation-driving, employment-increasing, and environmental-protecting measures.

The comparison between Singapore and Hong Kong industrial policies and forms of government intervention into innovative activities in the research of J. Wang [8] suggested that stronger state presence on the market leaded big companies to a significant progress in innovative development while some horizontal measures in Hong Kong became the base for some innovations emerging from small and medium business. However, in general innovative development in Hong Kong was slowing down. The author tested the hypotheses using the negative binomial regression for industries in Singapore and Hong Kong. The approach was based on the accounting of the quantity of patents issued by each country. The conclusion was to strengthen the government intervention to stimulate innovative industrial development.

The environmental issues of industrial exporting small & medium companies in Korea and China were studied by H.-Y. Joo and H. Suh [9]. The scholars used the distinction of market, environmental and export performance between the companies as dependent variables to reveal the mediating effects of different countries in the relationship between government support and company performances. As the result, the assistance for introduction of eco-innovations has become the most significant outcome of the interventions.

Y. Higuchi and G. Shimada [11] discussed the controversy of interventionist policy on the examples of Asian and African states. The have widely analyzed the employment in industrial and other economic sectors over years. The model of decomposition of gross domestic product (GDP) per capita ensured the perspective of "training-infrastructure-finance" (TIF) strategy of government intervention in industrial production sphere in African countries in order to set up the high rates of industrial productivity growth. Thus the combination of horizontal (training and infrastructure development) and vertical (direct financing) measures would provide the comparative advantages over Asian countries in the labor-intensive industries.

A team of EU researchers led by C. Criscuolo [3] has found a positive contribution between the regional selective support in European countries and the job growth, considering that only some regions of the EU are eligible for such assistance. The greatest effect was observed for small enterprises, while for the large ones it was insignificant. The following econometric model (4) was evaluated:

$$y_{r,t} = \lambda_1 NGE_{r,t} + \eta_r + \tau_r + \nu_{r,t}, \quad (4)$$

where $NGE_{r,t}$ – Net Grant Equivalent, which is the key policy variable and is defined as the maximum investment subsidy available in ward area r in year t and ranges from 0 to 35 percent; $y_{r,t}$ – employment; η_r – area fixes effect; τ_t – time dummies; $\nu_{r,t}$ – error.

The modelling revealed that the program reduced unemployment and increased employment in the manufacturing industry, but no impact on productivity was found. This gives grounds to consider the model of effectiveness of state support for industrial modernization in Ukraine, which includes both direct expenditures on sectoral development and the expenditures on R&D in relevant sectors, as well as the implementation of strategies for the development of the industrial sector.

1.2 Data and methodology

https://doi.org/10.1051/shsconf/202110001005

Since the growth of the industrial sector makes a direct contribution to the economic growth in general, it is necessary to assess whether the state budget expenditures for the restructuring of the industrial sector and its modernization are productive. The share of public expenditures to support the industrial sector in total state budget expenditures is presented in Fig. 1.

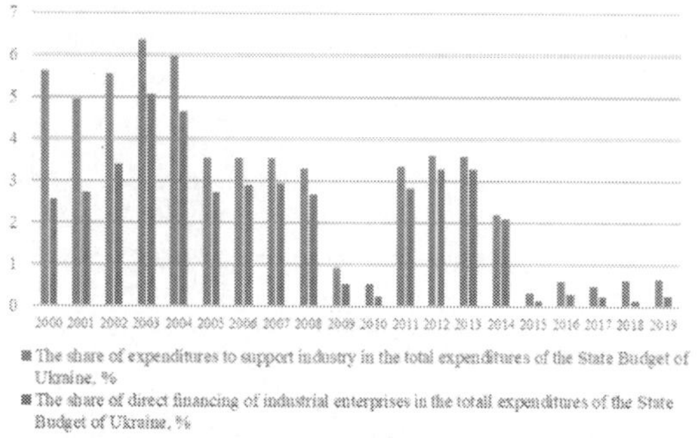

Fig. 1. The share of expenditures of the State budget of Ukraine on support for the industrial sector in 2000-2019, %

Modernization transformations and growth of gross value added in the industrial sector cannot be carried out without a corresponding inflow of capital investments. To stimulate economic growth in the sector, the dynamics of the growth of the latter must be positive. Empirical data for Ukraine in 2002-2018 indicates the connection between these two indicators (Fig. 2). The coefficient of correlation is equal to 0.77, which confirms a strong relationship between the variables.

In addition, as an important component of the impact on the productivity of the industrial sector can be considered its natural and stimulated restructuring, which, as mentioned above, should go from low to high value added activities.

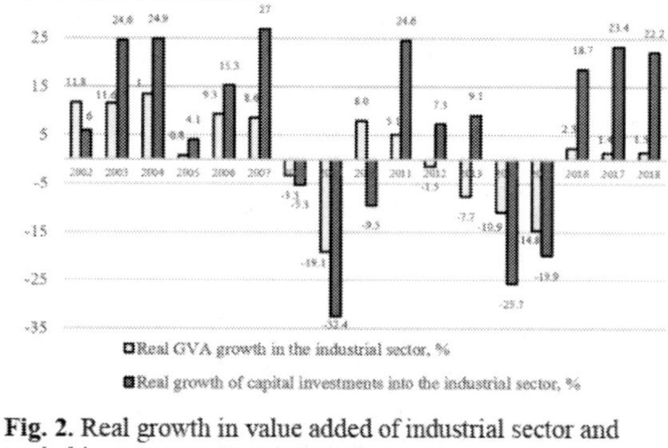

Fig. 2. Real growth in value added of industrial sector and capital investments, %, 2002-2018

Taking into account the model of endogenous economic growth proposed by B. Buhm and L.F. Punzo [12], in which structural shifts are the key factor, the inclusion of this factor in the econometric model will reveal the extent of the impact of structural adjustment of the industrial sector on its growth rate. The measurement of structural changes was carried out by determining the quadratic coefficients of structural changes (up to the previous year) (Fig. 3), calculated according to the tables "Input-output" in basic prices for the period 2001-2018.

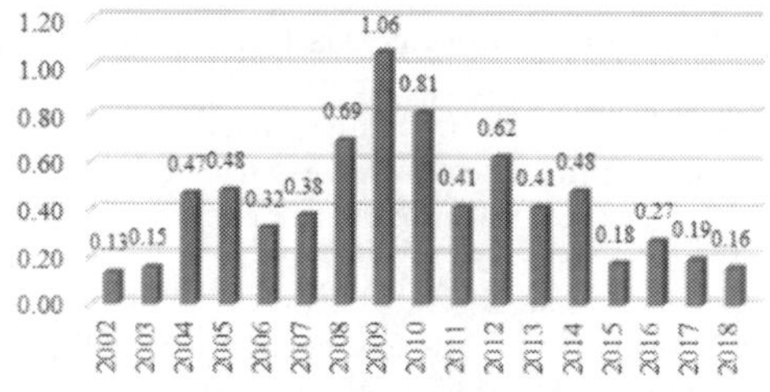

Fig. 3. Quadratic coefficient of structural changes in the industrial sector (chain), 2002-2018

The quadratic coefficient shows how many percentage points, on average, the industrial activities' shares deviate from each other. During the researched period. In order to ensure statistical homogeneity and comparability, the indicators of gross value added in different years were aggregated. Thus, in 2012-2018, the production of chemicals and chemical products, basic pharmaceutical products and pharmaceuticals and rubber and plastic products were combined into the chemical industry to meet the indicators of 2001-2011.

The dynamics of the quadratic coefficient of structural changes indicates a significant impact of the crisis of 2008-2009 on the change in the structure of industrial production. There was a reduction in the share of the extractive industry in GDP from 6.6% in 2008 to 5.1% in 2009. The share of the food industry increased (from 4.9 to 6.5%, respectively), which is a characteristic of economic stagnation, when the reduction of industrial production calls for less raw materials and investment goods, and more consumer goods part becomes more significant. In fact, the volumes of gross value added of metallurgy decreased by two times (from 5.3% to 2.5%). Significant changes also took place in 2012, when the next phase of the decline in industrial production took place. If in 2011 the mining industry reached 7.8% of GDP, in 2012 its share decreased to 6.5%, while metallurgical production increased from 1.6 to 2.3% of GDP. As for the processing industry, most of the activities producing consumer goods (food production, textile production, wood, and paper production), as well as mechanical engineering products, collectively reduced their shares in GDP. In 2016-2018, low growth rates of gross value added (GVA) in the industrial sector were accompanied by weak structural changes, as evidenced by the low values of the coefficients.

In total, for the studied period from 2002 to 2018, the coefficient of structural changes was 1.52, which indicates a significant structural adjustment of industrial production for the studied period. The share of the extractive industry in GDP increased by 2%, but, in addition to the production of furniture and other products, repair and installation of machinery and equipment, in virtually all branches of the processing industry there was a reduction in the share of GDP. Metallurgical production and production of metal products (4% of GDP), mechanical engineering (2%), food industry (1.8%), chemical industry (1.1%), textile production (0.3%), etc. suffered the most. The change in structure could not but affect the growth potential of the industrial sector.

It is expedient to assess the effectiveness of the state policy of sectoral and horizontal support of the industrial sector in the context of a specific institutional environment. As D. North pointed out, institutions provide the incentive structure of human interaction [13]. Together, such motives constitute a set of motivational levers that ensure the effective transfer of impulses from the state and the institutional environment to the market agents and encourage them to develop and, in this case, to achieve the goals of industrial modernization. To take them into account, indicators were selected that satisfy two conditions: 1) the availability of data for the researched period; 2) quantitative reflection of the most significant features of the institutional environment of modernization of the industrial sector.

In the view of the aforementioned, the Index of Economic Freedom (assessed by the Heritage Foundation) was chosen measured on the basis of 12 quantitative and qualitative factors grouped into four categories ("pillars" of economic freedom): 1) rule of law, property rights, state integrity, judicial effectiveness); 2) the amount of government intervention (government spending, tax burden, fiscal health); 3) regulatory effectiveness (freedom of enterprise, freedom of labor, monetary freedom); 4) open market (freedom of trade, freedom of investment and financial freedom) [14]. Each of the components is given the same weight in the scoring, so the deterioration of one of the indicators causes symmetrical changes in the indicator itself. The range of components of the institutional environment covered by this indicator is wide, which allows to reduce the number of input variables.

Another indicator, which is not fully reflected in the Index of Economic Freedom, but, nevertheless, has a significant impact on the activities of the industrial sector, is the Regulatory Quality measured by the World Bank [15].

Considering the characteristics of basic institutions in this model does not seem possible because they are low-dynamic and difficult to quantify.

An important step of the modeling process is to take into account the fundamental change in state policy on industrial development after the signing of the Association Agreement between Ukraine and the EU. The Agreement, first of all, requires compliance with new principles in implementing state economic policy, namely shifting emphasis from sectoral to horizontal support, affecting the structure and volume of state funding for industrial development. For this purpose, we use two dummy variables: first take into consideration the industrial policy for 2005-2014 years (equals 1 these years, and 0 in others); second represents the industrial policy for 2015-2018 years (equals 1 these years, and 0 in others). The modeling also assumes that the influence of factors is manifested in the same time period in which there is a certain rate of growth of GVA in the industrial sector.

The initial array of data for model development for Ukraine is presented in в Table 1.

Table 1. Initial data for modelling

	g_i*	PS_i	CI_i	SC_i	EF_i	QL_i
2002	11.8	5.54	6	0.13	48.2	30.1
2003	11.6	6.37	24.6	0.15	51.1	30.1
2004	13.5	5.98	24.9	0.47	53.7	39.41
2005	0.8	3.54	4.1	0.48	55.8	32.35
2006	9.3	3.54	15.3	0.32	54.4	31.86
2007	8.6	3.54	27	0.38	51.5	36.41
2008	□3.3	3.30	□5.3	0.69	51	33.01
2009	□19.1	0.93	□32.4	1.06	48.8	32.06
2010	8.0	0.54	□9.5	0.81	46.4	33.97
2011	5.1	3.34	24.6	0.41	45.8	29.86
2012	□1.5	3.60	7.3	0.62	46.1	29.86
2013	□7.7	3.60	9.1	0.41	46.3	30.33
2014	□10.9	2.21	□25.7	0.48	49.3	29.33
2015	□14.8	0.33	□19.9	0.18	46.9	29.81
2016	2.3	0.63	18.7	0.27	46.8	36.06
2017	1.4	0.51	23.4	0.19	48.1	40.38
2018	1.5	0.65	22.2	0.16	51.9	40.87

* g – real growth rate of industrial sector value added, %; *PS* – share of governmental support of industry in the state budget total expenditures, %; *CI* – real growth rate of capital investments in industrial sector, %; *SC* – structural changes in industrial sector, units; *EF* – Index of Economic Freedom, points; *QL* – quality of legislation, point; *i* – relevant year.

As the result we estimate the following multiple linear regression (5):

$$g_i = a + b_1PS_i + b_2CI_i + b_3SC_i + b_4EF_i + b_5QL_i + \\ +A_1D_{1i} + A_2D_{2i} + \varepsilon_i \,, \quad (5)$$

where D_1, D_2 – dummy variables which splitting up the industrial policy into periods of 2005-2014 and 2015-2018 years; a, A_1, A_2, $b_1 \ldots b_6$ – regression coefficients; ε – error; i – year.

Given that the factor values are different, the data were standardized, and the model was evaluated in the standardized form (6):

$$t(g_i) = \alpha + \beta_1 t(PS_i) + \beta_2 t(CI_i) + \beta_3 t(SC_i) + \beta_4 t(EF_i) + \\ + \beta_5 t(QL_i) + K_1D_{1i} + K_2D_{2i} + u_i \,, \quad (6)$$

where $t(\cdot)$ are relevant standardized variable, α, $\beta_1 \ldots \beta_5$, K_1, K_2 are coefficients of standardized regression.

Let us specify, that the dummy variables were not standardized that is why the model still has an intercept α. Standardized data are presented in Table 2.

Table 2. Standardized data for modelling

	g_i	PS_i	CI_i	SC_i	EF_i	QL_i
2002	1.1	1.4	-0.04	-1.1	-0.4	-0.8
2003	1.1	1.8	0.9	-1.1	0.5	-0.8
2004	1.3	1.6	1.0	0.2	1.3	1.6

2005	0.0	0.4	-0.1	0.2	2.0	-0.2
2006	0.9	0.4	0.4	-0.4	1.5	-0.4
2007	0.8	0.4	1.1	-0.2	0.6	0.8
2008	-0.4	0.2	-0.6	1.0	0.5	-0.1
2009	-2.1	-1.0	-2.1	2.5	-0.2	-0.3
2010	0.7	-1.1	-0.9	1.5	-1.0	0.2
2011	0.4	0.3	0.9	0.0	-1.2	-0.9
2012	-0.3	0.4	0.0	0.8	-1.1	-0.9
2013	-0.9	0.4	0.1	0.0	-1.0	-0.7
2014	-1.2	-0.3	-1.7	0.2	-0.1	-1.0
2015	-1.6	-1.3	-1.4	-1.0	-0.8	-0.9
2016	0.1	-1.1	0.6	-0.6	-0.9	0.7
2017	0.0	-1.2	0.9	-0.9	-0.5	1.8
2018	0.1	-1.1	0.8	-1.0	0.7	1.9

2 Main findings of the study

2.1. Interpretation of the model results

Using method of least squares we suggest the following estimations of standardized multiple linear regression (7):

$$t(g_i) = 2.35 - 1.33{\cdot}t(PS_i) + 1.03{\cdot}t(CI_i) - 0.28{\cdot}t(SC_i) + $$
$$+ 0.21{\cdot}t(EF_i) - 0.13{\cdot}t(QL_i) - 2.17{\cdot}D_{1i} - 4.54{\cdot}D_{2i}, \quad (7)$$

Statistical estimations of the coefficients and quality of the model are presented in Tables 3, 4, and 5.

Table 3. Regression statistics

Multiple R	0.93095
R^2	0.86667
Adjusted R^2	0.76296
Standard Error	0.48686
Observations	17

Table 4. Analysis of variance

Source of Error	df	Sum of Squares	MS	F	Significance F
Regression	7	13.87	1.98	8.36	0.0025
Residual	9	2.13	0.24	□	□
Total	16	16	□	□	□

"–" no data

Table 5. Parameters estimation

	Coefficients	Standard Error	t Stat	P-value
Intercept	2.346	0.708	3.3145	0.0090
$t(PS_i)$	□1.334	0.529	□2.5240	0.0326
$t(CIt)$	1.033	0.312	3.3110	0.0091
$t(SC_i)$	□0.280	0.292	□0.9591	0.3626
$t(EF_i)$	0.210	0.190	1.1050	0.2978
$t(QL_i)$	□0.126	0.307	□0.4119	0.6900
$t(D_1)$	□2.173	0.748	□2.9067	0.0174
$t(D_2)$	□4.536	1.287	□3.5245	0.0065

Statistical estimations of the coefficients and quality of the model give grounds to confirm the theoretical calculations of D. Rodrick on the negative impact of state support on the productivity of the industrial sector. Together with the real growth of capital investment and the type of policy, this factor turned out to be significant, and the model as a whole was adequate. The coefficients of the model (except for those near the dichotomous variables) show on how many standard deviations σ_g will the rate of GVA growth of the industrial sector g increase if the explanatory variable increases by one standard deviation σ_x. Budget expenditures for various purposes to ensure the development of the industrial sector with an increase of one sigma (2%) cause a reduction in the growth rate of GVA in the industrial sector by 1.33 sigma (or 12.8%). In turn, capital investment, increasing by one sigma (by 19.1%), causes an increase in growth by 1.03 sigma (by 9.9%).

In the model (7), the coefficients for the variable structural shifts and institutional factors were statistically insignificant.

Within the sample, structural shifts had negative effects on the growth of the industrial sector, which may indirectly indicate the destructive nature of such structural changes. The index of economic freedom, in turn, as a reflection of the institutional environment of the economy, has a positive effect on the growth of the industrial sector. And the low quality of the Ukrainian legal framework governing the economic activity of the industrial sector has forced Ukrainian businesses to adapt and develop their own mechanisms of life in the current environment, so a slight improvement in the legal framework, which reduced protectionism and decreased sectoral subsidies, together with the incompletion of deregulation and judicial reforms, had a negative impact.

Binary variables, which denote the implementation of elements of sectoral or horizontal industrial policy, demonstrate their negative impact on the growth of the sector. However, the negative impact of the 2005-2014 policy reduces the value of the standardized variable of the growth rate of GVA in the industrial sector by 2.17 units, which still leads to its positive values in general (7):

$$\alpha - K_1 = 2.35 - 2.17 = 0.18. \quad (7)$$

In turn, the policy of 2015-2018, which is characterized by a sharp reduction in government support, with the elimination of the impact of other variables, causes a significant decline in the growth rate of gross value added:

$$\alpha - K_2 = 2.35 - 4.54 = -2.19. \quad (8)$$

This is somewhat at odds with the inverse relationship between the growth of government support and the gross value added of the industrial sector. The solution to this contradiction lies in the plane of structural changes in the course of modernization: the allocation of resources to support medium- and high-tech activities, rather than traditional, stagnant industries, can change the results of modeling. Hence the thesis of the need for state support, which, although not able to eliminate all the "failures", can stimulate the modernization of the industrial sector through reasonable intervention.

2.2 Policy recommendations

The obtained results show the necessity of revision the conceptual base for industrial policy in Ukraine especially in terms of introduction of new Ministry on Issues of Strategic Industries of Ukraine. Its major designation is promoting the development of high-tech sectors of manufacturing.

The provision of optimal proportion between horizontal and vertical measures of matrix industrial policy should be ensured considering:

- historical background to neutralize market shocks while changing the paradigm of economic policy;
- business readiness and level of competitiveness to keep it in reforms;
- economic expediency, in particular, in considering old low-value-added economic activities for support;
- institutional and budget capabilities, which influence the speed of reforms;
- determination of sunrise industries of medium- and high-tech levels and their specific needs;
- careful selection of policy instruments to maintain their congruency to goals of industrial development.

3 Conclusions

Thus, the empirical analysis of the model allowed us to confirm the theoretical assumptions about the impossibility of full compensation for the distortion of market mechanisms through government support, as well as the importance of capital investment and changes in industrial policy. It was also found that the structure of state support and its scope should be revised to improve the effectiveness of horizontal instruments. In the environment of weak institutions of Ukraine, the instruments of horizontal policy most likely are less effective than sectoral policy measures. The implementation of selective instruments is limited in terms of international agreements on Association with EU or WTO. Thus, Ukraine should construct the configuration of its mixed vertical-horizontal model of industrial policy in order to achieve the maximum economic effect (increasing the growth rate of gross value added to state resources, increasing the share of exports of medium and high-tech products, and wages in the industrial sector).

References

1. D. Rodrik, Normalizing Industrial Policy (The International Bank for Reconstruction and Development, The World Bank, Washington, 2008)
2. R. Beason, D. E. Weinstein, The Rev. of Econ. and Stat. **78** (2), 286 (1996)
3. C. Criscuolo, R. Martin, H.G. Overman, J. Van Reenen, Americ. Econ. Rev. 109 (1), 48 (2019)
4. A.O. Krueger, B. Tuncer, Americ. Econ. Rev. **72** (5), 1142 (1982)
5. R.Z. Lawrence, D.E. Weinstein, NBER WP 7264 (1999)
6. J.-W. Lee, J. Econ. Gr. 1, 391 (1996)
7. R. Stöllinger, N. Foster-McGregor, M. Holzner, M. Landesmann, J. Pöschl, R. Stehrer. Res. Reports, 391 (2013)
8. J. Wang, Res. Pol. **47** (2), 399 (2017). doi:10.1016/j.respol.2017.12.008
9. H.-Y. Joo, H. Suh, Sustainability **9**, 1980 (2017). doi:10.3390/su9111980
10. J. Chen, L. Xie, Front. Bus. Res. Of China **13**, 18 (2019). doi:10.1186/s11782-019-0065-y
11. Y. Higuchi, G. Shimada, Industrial Policy, Industrial Development, and Structural Transformation in Asia and Africa, in Paths to the Emerging State in Asia and Africa. Eds. K. Otsuka, K. Sugihara, 195 (2019). doi:10.1007/978-981-13-3131-2_9
12. B. Buhm, L.F. Punzo, Productivity – investment fluctuations and structural change, in Growth and Structural Change. Eds. L.S. Punzo, L. Cycles, 47 (2006)
13. D. C. North, Institutions, Institutional Change and Economic Performance (Cambridge University Press, 1990)
14. Heritage Foundation, Index of Economic Freedom, https://www.heritage.org/index/ranking
15. Worldwide Governance Indicators: Country Data Report, The World Bank. http://info.worldbank.org/governance/wgi/index.aspx#countryReports

Internal Audit of Cybercrimes in Information Technologies of Enterprises Accounting

Ruslan Dutchak[1*], *Olha* Kondratiuk[1], *Olena* Rudenko[1], *Andrii* Shaikan[1], and *Elizaveta* Shubenko[1]

[1]State University of Economics and Technology, Department of Accounting and Taxation, Kryvyi Rih 50000, Ukraine

Abstract. The article is devoted to the problem's research of cybercrimes at the enterprise and the efficient methods of its solvation. The main trends of the cybercrime's development in the national and the global scope are defined. The analysis of the normative-legal acts on cybercrime is made. The main risks of cyberthreats before the illegal alienation of the enterprise assets are determined. The specific peculiarities of cybercrimes in the information technologies of enterprises accounting are revealed, the main ones of which are: cyberspace, the anonymity, harmful program products, the electronic (digital) track. The practicability of the internal audit use for the system opposition to cybercrimes at the enterprise is grounded. The main competences of the internal audit employees for the efficient work in the cyberspace of the enterprise are formulated. The recommendations, concerning rules of behavior for the employees in the cyberspace, are elaborated. The method of the internal audit is improved by the additional involvement of the modern ways of watching the electronic (digital) tracks of ill-intentioned persons in the cyberspace of the enterprise. The order of the juridical registration, concerning the methods use of watching the electronic (digital) track in practice of the internal audit at theenterprise, is offered.

1 Introduction

The rapid development of information technologies and their spreading practically in all spheres of the social-economic life of any country have become the main trend of the humankind development at the beginning of XXI century. The mentioned development caused the digital transformation of the processes, connected with collection, storage, processing and widening of information. The following advantages of digitalization have become undeniable for business: the increase of speed and transparency of business-processes; the growth of labor productivity; the reduction of costs on work with information; the transition from the paper documents to the electronic ones; processing of the data great volume; synchronization of information flows; acceleration of the managerial decisions-taking and others. In spite of the advantages, the new technologies are used to commit crimes, to inflame hatred, to falsify information and to interfere in private life [1]. According to the data of the World Economic Forum report, the cyberattacks belong to the five main dangers, threatening to the mankind so, as the natural disasters and the climate changes [2]. The cybercrimes, which main purpose is the expropriation of property (including the robbery of assets from the bank accounts) from their owners, are of special danger. According to the investigations data of McAfee and the Center of Strategic and International Research (CSIS), the world economy has lost more than 1 trillion of USA dollars in 2020 due to the hacker attacks [3].

The information technologies of enterprises accounting are significantly popular among the cybercriminals, namely: "SAP R3", "BAS", "OneBox", "MS Dynamics", "Perfectum", "IT-Enterprise", "1 S: Enterprise 8", "Parus 8" and the programs on the distant control of the enterprises bank account. The reason of such liking is the concentration of information on the available assets of the enterprise in the accounting, the possibility of manipulations with the accounting documents for their alienation, the distant access of employees to the accounting programs (especially actual in the period of COVID-19 pandemic) and others.

The objective necessity in protection of its assets appears for the enterprise's management with the purpose of avoiding the losses from cybercriminals. The most practicable method of such protection is the organization of the internal audit's service operation in the structure of the enterprise. However, the specificity of the struggle with the principally new cybercriminality requires the definite innovations for the traditional methods of the internal audit, which will correspond to the modern provocations of swindling in the cyberspace of the enterprise.

The problem of the internal audit is presented in the modern scientific literature by the wide interest of scientists. The analysis of the latest sources and publications on this problem allow select the main works results of the leading scientists. Thus, P. Rosati, F. Gogolin та T. Lynn [4] confirm, that the incidents with the cybersecurity affect the quality reduction of the financial reporting's audit. G. D'Onza, G. Sarens and S. DeSimone [5] determined, that the management system's maturity of risks depended on the development of the internal audit functions. J. Christopher [6] researched the negative consequences for the functioning

*Corresponding author: dutchakrr@ukr.net

of the internal audit from its role's manipulation by the enterprise's Board and higher leadership. M. N. Ivanova and A. Prencipe [7] examine the influence's (swindling's) components of the companies' higher administration on the audit remuneration. JR. Cohen, JR. Joe, JC. Thibodeau and GM. Trompeter [8] reveal the problem essence of the internal control, according to the audits of the financial reporting, being the subject of the intensive learning by the Council on Supervision of Public Company Accounting. N. Betti and G. Sarens [9] made the analysis of the internal audit's function development in the digitalized business environment. The obtained analysis showed, that the digitalized business environment influenced the functions of the internal audit from the three aspects: 1) the flexibility of the internal audit planning and the necessary knowledge will increase, but the risks of the information technologies will gain the great importance, especially the threats of cybersecurity; 2) the higher demand on the consultation measures, being performed by the internal auditors; 3) the digitalization modifies the work practice of the internal auditors in their everyday tasks. The new technologies, such as the tools of the data analysis, are gradually introduced into the departments of the internal audit, but the digital skills are considered to be the important asset. Y. Chen, B. Lin, LZ. Lu and GG. Zhou [10] proved the existence of the quality influence of the internal audit functions on the efficiency increase of the firm operation. The authors demonstrate that the competence of the internal audit makes the operational efficiency of the firm better, but the interrelation between the independence of the internal audit and the operational efficiency of the firm is insignificant. D. Weekes-Marshall [11] investigated the role of the internal audit in the process of risks management of the countries, being developed. T. Bondarenko, R. Dutchak, O. Kondratiuk, O. Rudenko, A. Shaikan and E. Shubenko [12] grounded the perspective reality of the internal audit for the solvation of conflicts in accounting of the enterprise with the help of the tools of the artificial intellect (artificial neuron networks and machine training). Yu. M. Popivnyak [13] revealed the identification methods of cyberthreats in the sphere of the accounting information's use and defined the means of its protection from stealing, damaging or loss in the cyberspace. L. P. Polovenko with S.V. Merinova [14] analyzed the instruments of the social character and offered the operating mechanisms, giving the opportunity to watch and to reveal the indications of the social engineering operatively at the early stages, warning of cyberthreats at the enterprise and opposing to the social hackering. L. Smidt, A. Ahmi, L. Steenkamp, D. P. and D. Lubbe [15] made the estimation of the generalized software's maturity level for the fulfillment of the internal audit functions. They determined that the internal audit in the whole world is still at the relatively low level of development, in spite of the intensified introduction of information technologies and the generation of the great data in the organizations.

The conducted analysis allows come to the conclusion, that the problem of the internal audit of cybercrimes in the information technologies of enterprises accounting has no any system research. Some fragments investigations of the presented topic are met in the scientific literature: the internal audit, cybersecurity, information technologies and others. The traditional methods of the internal audit do not correspond to the modern specificity of cybercrimes. That's why the absence of the internal audit's system approach to overcome the problem of cybercrimes at the enterprise is left to be the unsolved part of the lined problem.

The aim of the presented article is to research the nature of the cybercrime phenomenon in information technologies of accounting and to elaborate the theoretical-methodic principles of the internal audit for the improvement of enterprises cybersecurity.

In order to achieve the set aim, the methods of the research, which basis is the dialectical method of cognition (knowledge), is used in the article. At the same time, the following scientific methods are additionally used in the research, namely: the historical one – at studying the trends of cybercriminality; the analysis – at the research of cybercrime components, information technologies of accounting and the process of the internal audit; the synthesis - at studying the indications and the ways of cybercrime at the enterprise, new competences of the employees and new methods of the internal audit; the abstracting – at the formation of the internal audit's new tasks; the induction – at the influence's definition of some cyberthreats on the general level of the enterprise's cybersecurity; the deduction – at the research of connection between the complex of the internal audit measures with the demonstrations of cybercriminality; the explanation – at the revealing of the cybercrime essence in information technologies of accounting; the classification – at the types' definition of the fixation methods (programs) of electronic (digital) tracks; the systematization – at the regulation of the enterprise's cybersecurity components; the concretization – at the determination of the cybercrime's commitment methods; the generalization – at the formation of documents for the juridical fixation of the behavior rules for the personnel in the cyberspace of the enterprise and the fixation methods use of the electronic (digital) tracks in practice of the enterprise's internal audit.

2 Presentation of the main research material

The auditor company "PricewaterhouseCoopers" published the report on "World Research of Economic Crimes and Swindling of 2020: Questioning Results of Ukrainian Companies" in 2020. According to the data of this report, it's determined that 51% of the Ukrainian organizations had suffered from the swindling cases during the last 24 months. This indicator is higher than the average one in the world (47%) and it grew, compared with 48% in 2018. More than 1/3 of the respondents in Ukraine have suffered from 2-5 incidents of swindling for the last 24 months. The main subjects of swindling essentially in the Ukrainian organizations have become: the employee of the organization – 25%, the third side – 41%, the conspiracy between the employee and the third side – 25%. The Ukrainian respondents answered the question: "What was the approximate sum of your organization's direct losses from swindling during the last 24 months?" in the following way: 38% - less than 50 000 USA dollars, 13% - 50 000 - < 100 000 USA dollars, 13% - 100 000 - < 1 mln.USA dollars and 9% - 1 mln.USA dollars or more. However, only 59% of the Ukrainian organizations made the investigation of their worst case of swindling.

According to the data of the conducted research by "PricewaterhouseCoopers", cybercrimes occupy the fourth place among the TOP-5 ones from the most widely-spread types of economic crimes: 31% of the organizations in Ukraine suffered from them; 16% of the Ukrainian respondents are not only waiting for their cyberattacks at their organizations in the next two years, but they are convinced that the cyberattacks will be the most significant for their organization from the view-point of financial losses or the other consequences. Besides, the most part of the organizations in Ukraine is not ready enough for cyberattacks, only every third part of organizations in Ukraine has the program of cybersecurity [16].

The Article 8 of the Convention on Cybercriminality from 23.11.2001, which was ratified with the warnings and the statements by the Law of Ukraine "On Ratification of Convention on Cybercriminality" from 07.09.2005 № 2824-IV foresees that the criminal responsibility in relation to the internal legislation of the state is established for the premeditated committal, without any right for it, of the actions, resulting in the property loss of the other person by the following:

a) any introduction, change, destruction or concealment of computer data,

b) any interference into the functioning of the computer system,

gaining the economic advantages for themselves or another person, without any right for it, with the swindling or unfair purpose.

According to p. 8 of the Art.1 of the Law of Ukraine "On Main Principles of Cyber-Security Ensuring in Ukraine" from 5.10.2017 № 2163-VIII, the cybercrime (computer crime) – is the socially dangerous guilty action in the cyberspace and/or with its use, the responsibility for which is foreseen by the Law of Ukraine on Criminal Responsibility and/or which is recognized to be the crime by the international treaties of Ukraine.

The Criminal Code of Ukraine from 5.04.2001 № 2341-III (further on - CCU) determined the list of the main cybercrimes, for which the punishment is foreseen, namely:

violation of the copyright and adjoining rights (art. 176 of CCU);

taking possession of the unfamiliar property or acquiring the right for the property by the deception or the trust abuse (art. 190 of CCU);

illegal actions with the documents for remittance, payment cards and the other means of access to the bank accounts, electronic money, equipment for its production (art. 200 of CCU);

illegal use of the brand for the goods and services, the firm name, the qualified indication of the product's origin (art. 229 of CCU);

non-sanctioned interference into the operation of the electronic-computing machines (computers), the automation systems, computer networks or the systems of the electric connection (art. 361 of CCU).

Thus, you should understand that, according to art. 176 of CCU, art. 190 of CCU, art. 200 of CCU, art. 229 of CCU or art. 361 of CCU, the cybercrime is any crime, taking place with the help of the computer engineering, information technologies, computer systems, the compatible (united) communication systems and the Internet system.

According to the data of the General Public Prosecutor's Office of Ukraine, being reflected in the annual reports ("United Report on Criminal Offences") for the period of 2016-2020 [17], the number of the crimes, being committed in the cyberspace of Ukraine, is equal to the following (Table 1):

Table 1. Registered Number of Cybercrimes in Ukraine during 2016 - 2020

CCU Articles	Number of Crimes				
	2016	**2017**	**2018**	**2019**	**2020**
art. 176	157	95	126	148	118
art. 190	46019	37014	33290	32358	26830
art. 200	176	390	609	711	724
art. 229	168	121	111	62	123
art. 361	865	2573	2301	2204	2498

The data, presented in the Table 1, demonstrate the scope of the cybercrime phenomenon in the social-economic life of Ukraine. The mentioned data don't allow separate the exact indicator of cybercriminality in the very information technologies of the enterprise's accounting. The reason – the methods of filling in the mentioned report don't foresee the analytical criterion "cybercrimes at enterprises". However, the data in the Table 1 characterize successfully the environment, in which the enterprise functions and its employees live, namely:

- traditionally high scope of swindling quantity;
- the rapid quantity growth of the non-sanctioned interference into operation of the electronic-computing machines (computers) and the illegal actions with the documents for the remittance, payment cards and the other means of access to the bank accounts;
- the periodic activity of the copyright violation and the illegal use of the brand for the goods and services, the firm name.

Ukraine took the 25-th place among 160 countries in the National Cyber-Security Index (NCSI) in 2020 with the indicator of 68,83 (68,83%). The presented indicator demonstrates the level of Ukraine's readiness to prevent the cyberthreats and to control the cyber-incidents. At the same time, the digital development level (DDL) of Ukraine in 2020 was equal to 58,1 (58,1%). The difference between the estimation mark of NCSI (68,83) and DDL (58,1) is equal to 10,73. The positive result shows that the development of cybersecurity in Ukraine during 2020 anticipates its digital development by 10,73%. The countries-leaders in the National Index of Cybersecurity (NCSI), where the development of cybersecurity anticipates its digital development, are the following: Greece (NCSI – 96,1, DDL – 65,44, the difference – 30,66), the Czech Republic (NCSI – 92,21, DDL – 69,37, the difference – 22,84), Estonia (NCSI – 90,91, DDL – 79,27, the difference – 11,64), Lithuania (NCSI – 88,31, DDL – 70,95, the difference – 17,36), Spain (NCSI – 88,31, DDL – 73,24, the difference – 15,07). It's reasonably to indicate separately the countries, where the digital society is more developed than the national zone of cybersecurity, namely: Denmark (NCSI – 81,82, DDL – 83,55, the difference – (-1,73)), the Netherlands (NCSI – 81,82, DDL – 83,88, the difference – (-2,06)), Germany (NCSI – 80,52, DDL – 81,95, the difference – (-1,43)), Singapore (NCSI – 80,52, DDL – 83,11, the difference – (-2,59)), the USA (NCSI – 79,22, DDL – 82,33, the difference – (-3,11)), Great Britain (NCSI – 77,79, DDL – 83,96, the difference – (-6,04)), Switzerland (NCSI – 76,62, DDL – 85,13, the difference – (-8,51)) [18].

Ukraine took the 25-th place in the world (from 30 countries), the 10-th place among the European countries and the 3-rd one in the post-Soviet space in the rating of the National Cyber-Power Index (NCSI) for 2020. According to the estimation of the Belfer Center of Science and International Relations, the most all-embracing cyber-powers among the world countries (TOP-10) had: the USA, China, Great Britain, Russia, the Netherlands, France, Germany, Canada, Japan and Australia [19].

According to the information on the revealed cybercrimes on the territory of Ukraine in 2020, being located on the official web-site of the Cyberpolice Department of the Ukraine National Police [20], the most typical methods of cybercrimes realization in the sphere of information technologies of enterprises accounting are the following:

1) the ill-intentioned people ring up to enterprises and pretend themselves as the employees of the bank's IT-support. They try to convince the enterprise's accountant-cashier to realize some "testing" operations as if they check the payment system of the bank. They made the illegal remittances of money to the accounts, being controlled by them, as if they realized "the testing of the payment system";

2) the use of the harmful software ("virus") in order to receive the distant access to the accountant's computer and the access keys to the program of the remote control of the enterprise's current account. After that the ill-intentioned person receives the complete access to the current account of the enterprise and realizes the money remittance to the private accounts already on behalf of his/her name;

3) the evil-minded people leave the flash drives (as a rule, of the famous brands with the high price) or the CD-disk with the harmful software ("virus") in the accountant's working place intentionally, after what the accountant's curiosity makes him/her to put such an information bearer into the computer (to check the information it contains) – and realizes the non-sanctioned access to the accountant's computer and its programs in such a very way;

4) the creation of the faked copy of the enterprise's web-site page (with the famous brand) and the location of photos of the not existent product on them with the essentially reduced price (action or sale). The evil-minded people receive the complete advanced payment on their own accounts for the virtual product, after which they send the joky message of the crime's commitment;

5) the ill-intentioned person receives the access to the enterprise's letter-box, through the use of the harmful software ("virus"), which is used by its employees to post the invoices for the clients' payment for the received product or service. The ill-intentioned person changes the account's requisites of the enterprise for his/her own requisites with the help of the definite manipulations with the document. The buyer realizes the money remittance to the bank account of the ill-intentioned person, using the requisites of the false invoce;

6) the destruction of the electronic documents (bill of parcels, invoice, order), containing information on the fact of the goods consignment's loading to the certain buyer, in the information system of accounting. The information on the client's liabilities and the fact of his product's receipt are lost in the result of such actions;

7) the conspirancy of the enterprise's commercial manager with its programist, concerning the regulation of the accounting's information system in such a way, that the maximal discount is set for the concrete buyers automatically at the definite manipulations in the documents for the goods realization;

8) the employee of the enterprise's staff service appropriated illegally the copies of the documents (the passport of the Ukraine citizen and the identification number certificate) of the job seekers at the selection of the candidates for the vacant positions, due to which he/she arranged the credits for them;

9) the fabrication of the loyalty cards' duplicates for the systems of supermarkets, produced for the other citizens, in order to pay by the money, accumulated on the bonus accounts. Using the software's vulnerability of the trade network's loyalty system, the evil-minded people restored the bonus balance through the definite program manipulations, which allowed them to realize the money removal repeatedly by the unlimited number of times;

10) the forgery of the Ukraine citizen's passport (who died) and his/her financial documents for the illegal getting of credit on the personal account of the ill-intentioned person in the bank with the following making the ready cash in the bankomats;

11) the use of the electronic-computing technics to take possession of data, concerning the enterprise's employees (passports of the Ukraine citizens and the identification numbers). The swindlers arrange credits in the Internet with the help of these documents, not giving any information to their owners. The obtained assets are

transferred to the personal bank accounts and are removed as the ready cash in the bankomats;

12) the supermarket cashier of the trade system made the copies of the buyers' card accounts requisites hiddenly with the help of a special device, with the following removal of all the assets from the card. The evil-minded person received the PIN-codes of the cards by watching how the clients put them in at the POS-terminal, while paying for the product;

13) abusing the official state, the bank employee forged the bank documents for the illegal remittance of clients' money to the account under his/her control;

14) the controller-cashier in the bank institution, who had an access to the automation client system, realized the not-sanctioned issue and the activation of the bank cards independently, after which he/she created the official electronic bank documents – cheques on the cash issue;

15) the change of the information, being processed in the electronic-computing machine, with the purpose of the illegal money appropriation of the bank's clients;

16) the use of the harmful software ("virus") to break up the personal cabinets of the communal enterprise's clients and the entry of changes into the accounting indicators of the consumed services;

17) the bank employee realized the forgery of the bank documents, not giving any information to its clients, for the illegal remittance of their money to the account under his/her control;

18) the repeated replenishment of the mobile account of the evil-minded person, who used one 500-UHR denomination, to which the isothermal ribbon was fixed, which gave the possibility to use it repeatedly;

19) the entry of the untruthful news into the information system of the financial institution by the bank's employee and the forgery of the corresponding documents with the purpose of the credit contract's arrangement for the not existent person. Later on, the illegally obtained credit was transferred to the account of the evil-minded person;

20) the use of the harmful software ("virus") to break up the electronic letter-boxes, accounts, logins and passwords of enterprises' employees for the collection and spreading of the secret information, which can potentially ruin the reputation of enterprises.

In order to have the objective estimation of the scope and the variety of cybercrimes in Ukraine, you should take into account that the significant number of such crimes is left as not being included into the United Register of Pre-Court Investigations. According to the questioning results of the Ukrainian organizations, conducted by "PricewaterhouseCoopers", it was determined: 28% of the organizations in Ukraine will not probably or will never inform of such facts to the state or to the law enforcing bodies (compared with 12% of the respondents in the world); more than half (54%) of these respondents state that they are not sure in the fact that the law enforcing bodies have the necessary qualification in this field, but the other 41% - do not trust the law enforcing bodies [21]. It's reasonably to add to the mentioned reasons too, that cybercrimes are very often qualified mistakenly as the technical or the program break in the operation of information technologies of accounting; the conscious hiding of the facts, concerning cybercrimes, with the purpose of keeping the reputation of the higher leadership and the enterprise before owners, investors, business-partners, clients and controlling bodies; the complexity of the expertise procedure of the cybercrime's circumstances by the cyberpolice employees (removal of servers, computers, modems and others), paralyzing the enterprise's operation for the long time.

According to the above-mentioned state of cyberthreats in Ukraine, the main aim of the evil-minded people lies in the enterprise's money (or goods) alienation, being achieved by the illegal influence on the behavior of the employees and the influence on the functioning of the accounting's information technologies. The protection's guaranties of enterprises assets stipulate the necessity in the adequate managerial decisions-taking on cybersecurity. Such decision should take into account the principally new specificity of cybercrimes in comparison with the traditional crimes against ownership, namely: the intellectual character of crime; the place of crime – cyberspace; the unpersonification of cybercrimes; the remoteness of a cybercriminal; the means of the crime commitment – the harmful program product; the electronic form of the cybercrimes commitment's arguments; the possibility of the fast remote change or the destruction of the electronic (digital) tracks of cybercrimes commitment; the additional need in special programs and equipment for the fixation of the cybercrimes commitment's arguments and the others.

The optimal managerial decision for the system opposition to cybercrimes in the enterprises accounting's information technology is the creation of the internal audit's service. Such selection is grounded by the fact that the internal audit has the majority of the necessary knowledge (competences) for the struggle with cybercrimes in the information technologies of the accounting at the enterprise, namely:

- on the object of control – the economic operations, taking place in the cyberspace of the enterprise;
- on accounting (financial reporting, accounting calculations, methods of accounting, primary documents, the rules of operation in information systems and others);
- on the remote control programs of the enterprise's bank accounts (the order of entry into the system, the creation and transaction of payment documents, the requisites of cashless payments);
- on the means and methods of control (actual, documental, calculation-analytical and others);
- on means of electronic communications (including the information-communication technologies, the program ones, the software-hardware means, the other technical and technological means and equipment);
- on social engineering (provocation of the employees for violation of security rules or illogical actions).

However, the struggle specificity with the principally new method of crimes against the ownership requires the definite innovations into the traditional methods of the internal audit that will correspond to the modern provocations of swindling in the enterprise's cyberspace. The internal audit in this context is appealed for the improvement of the enterprise cybersecurity's existing level: to reveal, to avoid and to neutralize the real and the potential cyberthreats in time.

The methods of the internal audit requires the improvement in the part of the additional use of those methods of security and control, that will allow solve the following tasks: maintenance check-up of cybercrimes at the enterprise; the reveal of the criminal actions in the enterprise's cyberspace; the fixation of the electronic (digital) tracks in computers, systems, networks and others; the renewal of the cybercrime's commitment circumstances (the actions of the employees, the processes in the hardware and software); collection and storage of the electronic (digital) evidences of cybercrimes; the analysis of the operation's principles of the harmful program products ("virus"); the definition of the place, from which the cyberattack has started; the determination of the cybercrime's subject; the elaboration of the recommendation for the higher leadership.

The service of the internal audit should improve its own staff provision, for the complex elaboration and the use of the mentioned methods at the enterprise, in the following way: to involve the professional in cybersecurity of the enterprise's IT-infrastructure into the complement of its service; to increase the digital competence's level of the internal auditors (the second education, passing of the cources, certification, the self-education, etc.).

As the weak place in cybersecurity of information technologies of enterprises accounting is the human factor (curiosity, trust, carelessness, inability to understand), then, the methods of the internal audit should foresee the maintenance check-up of the employees' mistakes in cyberspace. It's reasonably for the internal audit's service to develop the "Protocol of Behavior Rules for Employees in Enterprise's Cyberspace". The Protocol contents should reflect the list of the rules, which the employees would be obliged to observe at their work in the cyberspace. The main recommendations of such rules are the following:

- the use of the complex access passwords (letters and numbers, not less than 8 symbols, the different decomposition of the keyboard (fingerboard);
- the logins and passwords should be kept in secret from the other persons;
- the logins and the passwords of access to the accounting and banking programs should be kept on the physically separated information bearer (flash drives) – the operation in which is possible only on the working computer and at the presence of the put on flash drive (token);
- the prohibition to use the found unknown flash drives (or CD-disks) on the enterprise's computers;
- to ensure the access to the enterprise's internal local network only through the identification of the computer MAC- address;
- the use of the multi-level authorization of access (sms dispatch with the temporary code to the employee's mobile telephone);
- the prohibition to use the social networks at the working places;
- "hygienics" of the electronic mail – not to open the letters from the incomprehensible addressers, the prohibition to set the unknown programs, the transition, according to the doubtful dispatch, the change of passwords or the other ones;
- the physical access prohibition of the strangers or their devices to the internal network of the company;
- the differentiation of access to Wi-Fi between the guest one and the system one;
- the prohibition to set the programs independently or for their renewal, not giving any information to the professional in cybersecurity;
- the limitation of the remote access to the working table from the private (home) computers of the employees;
- the prohibition to realize the dispatch of the enterprise's accounting documents with the help of the programs (Viber, Facebook, WhatsUpp, Telegram and others) on the personal smartphones of the employees.

Such a document is formaly approved by the enterprise's leader in the form of the order and is designed for the obligatory fulfillment, in the first turn, of the higher leadership, the employees of the accounting, the financial, the personnel, the warehouse and the other services of the enterprise. The internal audit service should conduct the trainings in cybersecurity systematically (not less than once a month) additionally for the increase of the digital competence of the employees and their culture of safe behavior in the enterprise's cyberspace.

The traditional methods of the internal audit require the addition by the special methods of watching the electronic (digital) track in the enterprise's cyberspace, which will allow analyze the influence on the enterprise's employees, their actions with the electronic documents, the operation of the software and the hardware, the physical connection of computer systems and others. The peculiarity of such methods is the fact that they are represented by the computer program products in the majority of cases. The recommended list of such programs and their functioning design for the internal audit are presented lower:

1) it's recommended to use the special programs-analyzers of traffic, which operation's methods allow recognize the structure of the different network protocols with the aim of catching and analyzing traffic of the enterprise's networks, - namely: to examine the caught packages and to decode them in the real time; to see IP-connection (IP-addresses, ports, sessions), MAC-addresses, the host's name and others. You may refer "Wireshark", "tcpdump", "Comm View", "Ultra Network", "RMON" and others to such programs;

2) it's reasonably for the internal audit's service to use the programs of analysis and restoration of the browsers' history for the control and the analysis of the electronic (digital) tracks, concerning the committed actions by an employee in the Internet system at the working place. The most popular among such programs are: "Hetman Internet Spy", "Starus Web Detective", "RS Browser Forensics" and others. The operation's

https://doi.org/10.1051/shsconf/202110001006

methods of the mentioned programs allow fot the internal audit's service fix the electronic (digital) evidences of the cybercrime: the history of searching, the visited pages of sites, the lists of loadings, the examined documents in the browser, the read electronic letters, correspondence in the electronic letter-boxes, revision of the correspondence in the social networks, the reviewed video-rollers and others. The value of such programs is the fact that they allow restore the very that history of the user's events, which was removed by him/her in the browser or formed on the disk with the purpose of hiding;

3) all the information technologies of enterprises accounting ("SAP R3", "BAS", "1 S: Enterprise 8" and others) practically have the standard interface mechanism of the history reflection of actions. The mentioned history contains information of all the actions of the user, being connected with the electronic document. The revision of such a history allows watch, who, when and how created or changed the existing document in the program. The fixation's arrangement of the history of changes in the electronic accounting documents and the obtaining of access for its revision are taken place at the level of the information system administration. Therefore, the methods of the information technology's functioning of the enterprise accounting are the integral component of the internal audit's methods;

4) the use of the monitoring program products at the enterprise (the hardware or the software products, containing Keylogger as a modul), which allow fix the activity of the users (or processes) and determine the identification of their participation in the definite events. It's recommended to use one of the presented programs for the tasks of the internal audit at the enterprise: "Spytech SpyAgent", "Spyrix Personal Monitor", "All In One Keylogger", "Punto Switcher", "REFOG" and others. The use of Keylogger in the internal audit's operation will allow analyze the circumstances of a cyber-incident at the enterprise efficiently: the access to information, concerning the composition of the key words and word combinations on the key-board of the employee, the selection's attempt of login and password for the access, the time of the employee's activity at the computer, catching of the mouse's clicks, catching of the exchange buffer, monitoring of the file activity, monitoring of the system register, monitoring of the tasks' turn, the documents, sent to the printer, the use of the external bearers of information (USB, CD, memory cards) informed immediately of the behavior and others. The analytical information of the mentioned type of programs allows to the internal audit determine groundly the reason-result connections in the cybercrime's circumstances.

The use of the presented programs at the enterprise should be agreed with the owner, the legal services, the system's administrator; they should be officially gained with the corresponding licences, regulated in the documents (principles, instructions) of the internal audit, etc. with the purpose of the formation legitimacy of the electronic (digital) evidences.

It's reasonably for the internal audit's service to define the circle of the officials, who are intrinsic in the risk of participation in a cybercrime, due to the character of their functional duties. The following people belong to such employees in the first turn: the higher leadership; the accountants or the financiers, who realize payments through the programs of the remote control of the bank account; the storekeepers, who realize the loading of goods; the secretaries, who receive the electronic letters on the official e-mail, and others. It's the internal audit's service priority to use the methods of cyberthreats monitoring for the very such categories of enterprise employees.

The presence of the special controlling methods (programs) of the electronic (digital) tracks in the enterprise's cyberspace allows include groundly the tasks on testing of cyberthreats (or cybercrimes) into the programs of the internal audit at the enterprise. Correspondingly, the reports of the internal audit should reveal the result of the realized control of cybercrimes and should contain the practical recommendations on their maintenance check-up at the enterprise.

3 Conclusion

The rapid development of information technologies at the beginning of XXI century caused the absolute progress of human civilization. The appearance and development of cybercriminality in the global scale have become the attendant phenomenon of such a progress. The main danger of cybercriminality lies in the illegal expropriation of property from their owners by the way of any interference into the functioning of the computer system or the manipulation with the computer data. Cybercrimes are principally different from the traditional crimes against the ownership by the fact, that the crime takes place in the cyberspace; the anonymity of cybercrimes; the harmful program product is the means of crime; the remoteness of a cybercriminal; the electronic (digital) form of evidences for the cybercrimes' commitment and others.

Cybercrimes in Ukraine take the fourth place among the most widely-spread types of economic crimes in the sphere of enterprises' economic activity. The most attractive for cybercriminality are information technologies of enterprises accounting. The real possibility of access to the information, concerning the available assets of the enterprise and the manipulations with the accounting documents for their alienation have become the reasons of such attractiveness. The actual level of cybersecurity at the Ukrainian enterprises is left to be absolutely low – only every third organization in Ukraine has the program of cybersecurity.

In order to protect the enterprise's assets from cybercriminals, it's reasonably to organize the adequate opposition at the level of the internal audit's service. However, the methods of the internal audit should take into account the specificity of struggle with the crimes against the ownership in the cyberspace of the enterprise. Therefore it's recommended to do the following for the efficient organization and functioning of the internal audit's service at the enterprise:

1) to involve the enterprise's IT-structure professional in cybersecurity into the staff complement of the internal audit's service and to increase the level of digital competence of the internal auditors;

2) to elaborate the rules of behavior for the employees in the cyberspace of the enterprise and to introduce them at the enterprise in the form of the internal document "Protocol of Behavior Rules for Employees in Enterprise's Cyberspace";

3) to add the traditional methods of the internal audit by the special methods (programs) of watching the electronic (digital) track in the cyberspace of the enterprise, namely: the programs-analyzers of traffic, the programs of the analysis and renewal of the history of browsers, the information technologies of accounting (in the part of the history analysis of the user's actions with the document), the monitoring program products (the hardware or the sofrware products, containing Keylogger as the modul); to develop the internal Principle "On Programs' Use of Watching Electronic (Digital) Track in Cyberspace of Enterprise by Internal Audit" and to introduce it at the enterprise.

Thus, the introduction of the above-mentioned recommendations into the traditional practice of the internal audit's operation will allow create the reliable system of cybersecurity at the enterprise, which will be able to oppose to the modern cybercrimes.

References

1. O. Pyschulina, Digital Economics: Trends, Risks and Social Determinants (Razumkov Center, 2020), p. 274
2. The Global Risks Report 2019. 14th Edition (World Economic Forum, 2019), p. 114
3. J. A. Lewis, Z. M. Smith, E. Lostri, The Hidden Costs of Cybercrime (CSIS, 2020), https://www.csis.org/analysis/hidden-costs-cybercrime. Accessed on 25 January 2021
4. P. Rosati, F. Gogolin, T. Lynn, European Accounting Review (2020)
5. G. D'Onza, G. Sarens, S. DeSimone, Accounting Horizons, **34 (4)**, 57-74 (2020)
6. J. Christopher, Journal of Management Inquiry, **28 (4)**, 472-483 (2019)
7. M.N. Ivanova, A. Prencipe, Journal of Accounting Auditing and Finance, 0148558X20971947 (2020)
8. JR. Cohen, JR Joe, JC. Thibodeau, GM. Trompeter, AUDITING: A Journal of Practice & Theory, **39 (4)**, 57-85 (2020)
9. N. Betti, G. Sarens, Journal of Accounting and Organizational Change (2020)
10. Y. Chen, B. Lin, LZ. Lu, GG. Zhou, Managerial Auditing Journal, **35 (8)**, 1167-1188 (2020)
11. D. Weekes-Marshall, Journal of Corporate Accounting and Finance, **31 (4)**, 154-165 (2020)
12. T. Bondarenko, R. Dutchak, O. Kondratiuk, O. Rudenko, A. Shaikan, E. Shubenko, Atlantis Press, **129**, 39-46 (2020)
13. Yu. M. Popivnyak, Business Inform, **8**, 150-157 (2019)
14. L. P. Polovenko, S. V. Merinova, Entrepreneurship and Innovations, **10**, 183-187 (2019)
15. L. Smidt, A. Ahmi, L. Steenkamp, D. P. van der Nest, D. Lubbe, Australian Accounting Review, **29**, 516-531 (2019)
16. World Research of Economic Crimes and Swindling of 2020: Questioning Results of Ukrainian Companies (PwC, 2020), p. 14
17. General Public Prosecutor's Office of Ukraine (2021), https://old.gp.gov.ua/ua/statinfo.html. Accessed on 25 January 2021
18. National Cyber Security Index (2021), https://ncsi.ega.ee/ncsi-index/. Accessed on 25 January 2021
19. Harvard Belfer National Cyber Power Index 2020 (NCPI, 2020), p. 84
20. Cyberpolice in Ukraine (2021), https://cyberpolice.gov.ua/. Accessed on 25 January 2021
21. World Research of Economic Crimes and Swindling of 2018: Questioning Results of Ukrainian Organizations (PwC, 2018), p. 19

The Use of Information Technology in Financial Management

Iryna Prikhno[1*], *Valentyna* Kuksa[2], and *Ivaylo* Mihaylov[3]

[1]Cherkasy State Technological University, Finance Department, 460 Shevchenko Blvd., Cherkasy, 18006, Ukraine
[2]Cherkasy Educational Scientific Institute University of Banking, Department of Management and Information Technologies, 164 V. Chornovil Str., Cherkasy, 18028, Ukraine
[3]Prof. Assen Zlatarov University, Department of Economics and Management, 1 Prof. Yakimov Str., 8010 Burgas, Bulgaria

Abstract. The article explores the concept of information technology and clarifies the essence of information technology in finance. The types of information technology in finance have been studied and their classification according to business entities has been carried out. A brief description of the main software products, the purpose of which is to ensure the implementation of the process of automation of financial research is given. The main tools for managing household (or person) finance using modern information technology are presented. The implementation of e-government in Ukraine is analyzed. The evaluation of the effectiveness of e-government implementation with the help of E-Government Development Index in Eastern European countries and in Ukraine is performed. A detailed analysis of the E-Government Development Index in Ukraine using a system of indicators has been studied. Digital technologies in the economy in general and in finance in particular have been studied separately. The indicators of world indexes digital economy development for Ukraine and for Eastern European countries are analyzed. The advantages and problems of the modern cryptocurrency market are clarified. The main directions of using artificial intelligence in finance are determine.

1 Introduction

The development of financial relations in the modern world requires the implementation of effective financial management in various spheres of human life.

Building an effective financial management system involves achieving this goal by performing certain tasks of organizational, informational, managerial, financial, legal, production, political, social types, and so on. This, in turn, is constantly influenced by external and internal factors, which together can have either negative or positive effects, accordingly, posing a danger (risks) for the realization of the goal or contributing to its achievement. Thus, the financial management requires the processing of a large amount of information, based on which the choice of the particular scenario of behaviour is done.

Therefore, in order to improve managerial decision-making process in finance, in order to be able to be competitive and successful in the market, to increase profitability and efficiency, to be able to respond quickly to external factors and adapt and change according to modern requirements, it is necessary to use information technology.

2 Related literature

Problems with the use of information technology in management are highlighted in the scientific works of domestic and international researchers.

In particular, the study of information technology in enterprise management is highlighted in the works of the following scientists: T. Baranovska, P. Buryak, J. Champi, V. Denysyuk, M. Hamer, A. Kozyryev, O. Kryvokon, T. Lepeyko, I. Lukasevych, K. Malachevska, L. Melnyk, J. Nabatova, A. Onopko, O. Prysyazhnyuk, O. Pushkar, A. Romanov, M. Royik, A. Shchedrin, H. Tytarenko, V. Utkin, J. Zhyhalkevych and others.

It should be noted that J. Nabatova and K. Malchevska in their scientific works explore the theoretical foundations of automation of the process of analysis and forecasting of financial results of the enterprise, analyze examples of statistical programs and tools that make it possible to perform analysis. Also, these scientists argue that "automation of analysis significantly improves the quality and accuracy of the results, frees employees from long and routine work, as well as takes the management of financial results to a new level, as it contributes to more informed and effective business decisions" [1].

Information technology in public administration is discussed in the scientific works of the following scientists: K. Andersen, L. Bakayev, O. Bakayev, A. Bersutsky, J. Bersutsky, S. Bretschnayder, P. Drucker, V. Glushkov, R. Kalyuzhny, N. Krasnozhon, M. Lepa, Z. Pavlyshyn, V. Porohnya, T. Pysarevska, V. Sytnyk, M. Tatarchuk, E. Toffler, O. Tsarenko and others.

Thus, Z. Pavlyshyn notes that "using information technologies in public administration with the ability to find alternative ways to set data, perform the functions of collecting, storing, processing, transmitting and using knowledge, really increases the efficiency of public administration, which allows to find the necessary resources". He also notes the tendency to increase the number of government agencies that use information technology, and emphasizes that without the use of information technology is now impossible to compete in the world market [2].

The problem of the application of information technology in finance has been investigated by M. Bastrikov, E. Burtseva, N. Emelyanova, A. Gavrylenko, T. Gavrylko, O. Korystin, M. Kurkov, A. Kwilinski, X. Li, L. Melnyk, I. Semakin, N. Svyrydiuk, I. Tkachenko, V. Tkachenko, O. Tomashevsky, O. Shatunova, C.A. Wang and others. In their works, they often mentioned digital finance, in particular, the impact of cryptocurrency on micro and macro indicators [3]. The following scientists have studied the problems of digitalization of the economy in more detail: O. Bilyk, R. Bukht, A. Cheko, G. Chmeruk, A. Dobrynin, R. Hicks, G. Karcheva, J. Levitska, V. Lyashenko, R. Narishman, V. Panasyuk, N. Podolchak, M. Rudenko, A. Semenog, D. Tapscott, A. Tugu, L. Zakharchenko and others. In particular, N. Podolchak, O. Bilyk and J. Levytska analyze the current state of digitalization and the main problems that hinder the intensive development of the digital economy [4].

We believe that scientists have made an invaluable contribution to the development of research in the application of information technology in financial management, but, in our opinion, taking into consideration the current processes of informatization, digitalization and globalization, some questions need more detailed and in-depth study.

3 The purpose of the research

The purpose of the article is to understand the existing information technology in finance, to systematize it, to identify its positive and negative impact on functioning finance of an enterprise (business), a household (person) and a state.

4 Methods

The methodological basis of the study were the theoretical provisions set out in the works of domestic and foreign authors. The methods of scientific abstraction and system analysis were used in the work.

5 Results

According to the Law of Ukraine "On the National Informatization Program", information technology is a purposeful organized set of information processes using computer technology that provides high speed data processing, fast information retrieval, data dissemination, access to information sources regardless of their location [5].

Note that along with the term "information technology", the term "information and communication technology" is widely used. Moreover, these two terms are often used as synonyms in scientific works. In our opinion, "information and communication technologies" is a more general term, which consists of information technology and telecommunications, media broadcasts, all types of audio and video processing and transmission of information.

Information technologies are implemented using IT tools – software (programming languages, operating systems, applications, etc.) and special equipment that allows you to process, transmit and store information (computers and computer equipment, modern communications, etc.).

Information technology of management processes refers to the methods and ways of interaction of the control and managed systems of the organization on the basis of use the modern tools for information transformation [2].

Based on the above definitions, it can be stated that information technology in financial management is the use of modern tools for the transformation of financial information or information necessary to make an effective financial decision.

Due to the availability of different entities of financial use, information technology in financial management should be divided into information technology in financial management of the enterprise (business), information technology in financial management of the household (person), information technology in state financial management of banking, budget, money and other systems. Also, given the rapid development of digital technologies and artificial intelligence, it is advisable to present separately the use of digital technologies in finance and the use of artificial intelligence in finance (Fig. 1).

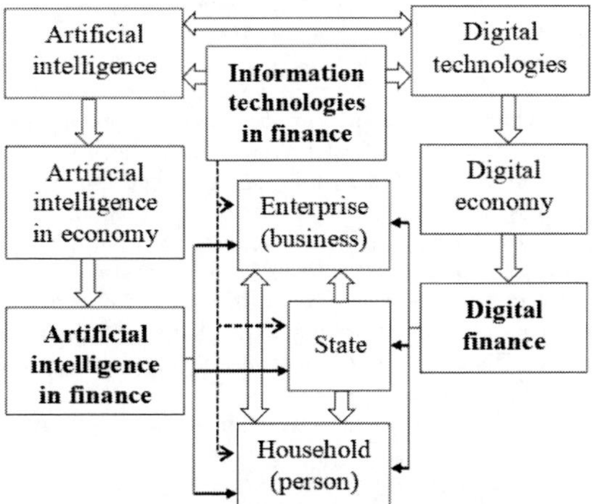

Fig. 1. Structuring the using of information technology in finance.

Source: authors' own development.

Taking into consideration the global trend of rapid development of information technology and digitalization, globalization and internationalization, the use of information technology in financial management of the enterprise (business) is one of the most important stages in the management of any enterprise, as it allows you to perform a comprehensive study of key financial indicators of the enterprise and perform effective planning and forecasting of its financial activity. The process of finding financial and analytical information and its research is quite extensive and time consuming, requires the involvement of a large number of specialists and a significant amount of various calculations, including and using methods of economic and mathematical programming and forecasting. That is why the automation of financial analysis processes has become natural and necessary, and it turned out to be possible due to the rapid development of information and digital technologies. Automation of financial analysis of the enterprise allows you to accelerate the process of evaluation of financial indicators, quickly and in the shortest possible time to identify possible risks and threats and plan further actions to minimize losses and increase profitability, allows you to identify potential domestic and foreign partners, suppliers, creditors, investors, etc. and to study the competitor activity and identify the most favorable markets in Ukraine and abroad.

The main advantages of automation of indicators of financial activity of the enterprise should include:

1) increase in productivity of financial managers: automation of calculations allows you to accelerate process of calculations and to release additional time for comprehension and evaluation of the obtained indicators and search and substantiation of alternative solutions to complex financial problems;

2) increasing the scale of the study: automation of calculations allows you to perform calculations using a large amount of information and a significant number of indicators, which indicate the possibility of a comprehensive study of the impact of all (or the vast majority) factors on enterprise financial indicators;

3) improving the quality of analysis: automation of calculations allows you to get more accurate and diagnostically reliable results of the analysis, which will help to prevent and eliminate errors and allow you to formulate conclusions and recommendations for more efficient operation of enterprise finance.

The main tool in ensuring the implementation of the process of automation of financial research is the software represented by various software products (programs, packages). The wide range of such software products presented on the market caused the need to systematize them according to the following criteria:

1. Depending on the functionality of the program for financial analysis [1]:

1.1. General purpose programs (universal programs):

STADIA is the universal statistical package, which is a set of modern methods of analysis: categorical, variance, correlation and spectral analysis, smoothing, filtering, forecasting, regression, discriminant, cluster and factor analysis, quality control methods, sequential analysis and replacement of missing data, and also contains a complete set of business and scientific graphics (functions, dependencies, maps, charts, etc.) [6];

STATGRAPHICS is a comprehensive Windows desktop product designed to perform statistical analysis, data visualization and forecasting with the help of more than 290 statistical operations, procedures and special functions [7];

IBM SPSS is a platform of statistical software that has the necessary set of functions for the analysis of its own financial, statistical, informative data. The obtained results are quite easy and clear to visualize, so the program is universal for users of different levels. SPSS offers two programs, SPSS Statistics, which is based on a top-to-bottom approach, and SPSS Modeler, which uses a bottom-up analysis [8];

STATISTICA is an universal package for comprehensive statistical analysis, which implements procedures for data analysis (data analysis), data management (data management), data mining (data mining), data visualization (data visualization). This program allows you to calculate descriptive statistics; analyze time series and predict, perform regression, discriminant, cluster, variance analysis, analysis of correspondences, etc. [9].

1.2. Professional programs:

SAS (Statistical Analysis System) is a program that provides the ability to analyze spreadsheets, perform statistical analysis, build forecasts, provides access to data, allows you to write reports. The program contains a significant number of built-in mathematical and financial functions, including depreciation, compound interest, cash flow, hyperbolic functions, factorials, combinations and mechanisms etc. [10].

1.3. Specialized programs:

MESOSAUR is a program, used for time series analysis (smoothing, filtering, autocorrelation, cross-correlation, spectral, cross-spectral analysis), allows you to perform regression analysis, solve parametric and nonparametric problems [1, 11];

DATASCOPE is a statistical package that specializes in performing multidimensional data analysis.

2. Depending on the country that uses the program [1]:

2.1. Used in foreign countries:

EViews (Econometric Views) is a statistical package that allows you to analyze econometric data of time series, perform analysis and simulate and forecast data, build regression models using a modern, innovative and easy to use interface. This program is used in financial analysis, in macroeconomic forecasting, in modeling economic processes, in forecasting market indicators, etc. [12];

Maple (Waterloo Maple) is a computer program, which allows you with the help of built-in functions of modern mathematics (more than 5000 functions) to carry out modeling, interactive visualization, gives the chance to combine results of calculations, formulas, text files, graphic objects, sound effects in one electronic document [13];

R is both a programming language and a software environment that allows you to perform statistical calculations, perform analysis and display the results of calculations in graphical form [1, 14].

2.2. Used in Ukraine:

Programs for analysis:

MS Excel: This program is one of the MS Office programs and is currently the most widespread program in Ukraine. It is used for financial analysis of domestic enterprises. The Ukrainian market presents both ready-made products developed in MS Excel, which can be used in enterprises as needed, and directly MS Excel is often used by financiers, economists, managers, accountants in enterprises for their own calculations. This program is a spreadsheet that contains ready-made built-in mathematical formulas, represented by a fairly powerful mathematical apparatus, and also allows you to create in the cells of the table needed to automate the calculation of the formula. Tables can be linked together, allowing you to process a large amount of information while using many metrics. In this program regression analysis can be performed, predictive calculations can be made, sorted, highlighted, various graphical objects can be built. MS Excel also contains a built-in programming language Visual Basic for Applications (VBA), which allows you to create algorithms for calculating certain indicators. All tables and graphical objects created in MS Excel are easily transferred to any other office program MS Office, which, of course, facilitates the process of further work with the obtained results of the analysis (writing a report, sharing files or their separate parts, etc.).

Google Sheets, Google Apps Script, Google Analytics are programs offered by Google and are free web software packages designed to perform financial analysis. Google Sheets is a spreadsheet program using powerful financial analysis tools, which is based on the use of artificial intelligence technologies, which, accordingly, allows you to get fairly accurate calculations and predict balanced directions for solving problems. Google Apps Script – with this program it is possible to create macros, which simplifies the process of financial calculations and automate the process of preparing and writing reports. Google Analytics is a program designed to perform web analytics on the Internet: it analyzes Internet sites and mobile applications [3].

Reporting Programs:

"1C: Enterprise" is a series of programs that belong to one platform and are represented by different products depending on the needs and specifics of the business. This program allows you to automate the accounting of enterprises in different areas of activity, different forms of ownership, different scales and more. "1C: Enterprise" offers the following programs:

a) "1C: Accounting" is automation of accounting and tax accounting in organizations related to any type of business. The program allows to carry out: accounting of standard operations, accounting of inventories, warehouse accounting, accounting of trade operations, accounting of commission trade, accounting of operations with container, accounting of bank and cash operations, accounting of settlements with counterparties, accounting of fixed assets, intangible and low-value assets, accounting for primary and secondary production, accounting for semi-finished products, accounting for indirect costs, VAT accounting, payroll accounting and personnel accounting. Also, the program "1C: Accounting" supports various tax schemes, allows you to carry out the final operations of the month, generate standard accounting reports, create regulated reporting, keep records of activities of several organizations, to carry out tax accounting for income tax and single tax;

b) "1C: Commercial Enterprise Management" is automation of management and accounting tasks at enterprises engaged in any type of commercial activity;

c) "1C: Manufacturing Enterprise Management" is automation of management and accounting at the manufacturing enterprise;

d) "1C: Salary and Personnel Management" is a tool for the implementation of personnel policy of the enterprise, as well as the automation of various services of the enterprise, ranging from personnel management and line managers to accounting staff. The program "1C: Salary and personnel management" allows: to keep records of activities of several organizations, to plan staffing needs, automate recruitment, to manage competencies and financial motivation of employees, to plan staffing, personnel accounting and personnel analysis, to regulate labor relations and to keep personnel records, to carry out calculation and accounting of wages and deductions of regulated taxes, to compile regulated reporting;

e) "1C: Small Business Management" is automation of operational management in small businesses;

f) "1C: CRM" is automation of customer relationship processes; helps to organize effective work of departments of sales, marketing, service at all stages of interaction with customers [15];

BAS is a line of programs that allow you to automatically perform a wide range of tasks related to the management of the enterprise in various aspects. The main programs of BAS are: Accounting, Accounting CORP, Integrated Enterprise Management, Document Management CORP, Small Business, Retail, Trade Management, Holding Management, ERP (financial management, including accounting for deposits and loans, acquiring (payment cards), flexible instruments) for maintaining a payment calendar, expanded capabilities for managing current payments, routes for reconciling applications, flexible tools for generating payment documents for future dates, inventory of cash registers and bank accounts, analytical reporting on cash flows, chart of accounts for international financial accounting, translation of transactions into the IFRS subsystem data of operative and regulated accounting, deferred display of postings in the account, creation of documents on standard operations, audit, the generator of financial reports) [15].

Note that "1C: Enterprise" and its software products are still the most widely used in Ukraine. However, it should be noted that BAS products are increasingly being developed and implemented in domestic enterprises, as they offer a number of programs that are similar to the programs offered by "1C: Enterprise", but more adapted to modern requirements, have an updated interface, which is more user-friendly, expand the user's ability to customize the program according to his needs.

Therefore, we can predict that there is a high probability that over time, new BAS products will compete and take a worthy place next to 1C or even replace 1C.

Programs for Information Exchange and Reporting:

FlyDoc is a special program, the purpose of which is the rapid exchange of documents in electronic form. This program is installed directly in the program "1C: Enterprise". The transmitted information is available only to the sender and recipient [15].

M.E.Doc is a program that allows to carry out electronic document management (reports, agreements, tax invoices, acts, invoices and other documents). With the help of this program you can create tax invoices and sign them with an electronic digital signature, register them in the Unified Register of Tax Invoices, receive receipts in return or necessary extracts from the register, exchange registered tax invoices with receipts for their registration with counterparties. The M.E.Doc program allows sending electronic reports to the State Tax Inspectorates, the State Statistics Service, the Pension Fund, the State Employment Center, and the Social Insurance Fund. In order to meet the needs of different business entities, the following software products are offered: Reporting, Reporting to the NBU for non-banking institutions, Electronic document management, VAT accounting, Excise and TTN, Salary, ARI "Cashalot" (registration of electronic checks in the State Tax Service), "SOTA" (online reporting) [15].

Information technology in household (person) financial management is associated with the emergence of modern programs, the purpose of which is to help in the efficient distribution of income and expenses. Currently, users are offered the following programs: "Home Accounting", Moneytracker, Personal Finances, "Your Money", MyBudget, IControlMyMoney, iCASH, Electronic System "All PRO Money", "DomEconom", "AbilityCash", "Buddi" and others. As you can see, in the modern world there are a significant number of tools for managing household finance (a person) using modern information technology, which is presented in the form of computer programs, mobile applications, online financial control tools.

At the state level, in the context of the development of modern information technology, the evaluation of the implementation of e-government in Ukraine deserves attention.

According to the Concept of e-government development in Ukraine, e-government is a form of public administration that promotes efficiency, openness and transparency of public authorities and local governments with the use of information and telecommunications technology to form a new type of state focused on meeting the needs of citizens [16].

This Concept defines comprehensive measures aimed at:

– modernization of public services and development of interaction between government, citizens and business through information and communications technology (development of electronic services, development of open data, development of electronic tools for involving citizens, development of electronic identification and trust services);

– modernization of public administration with the help of information and communications technology (development of electronic interaction, development of electronic document management, e-government in basic industries and support of priority reforms);

– e-government development management (formation of basic information and telecommunications infrastructure of e-government, increase of efficiency of e-government development management) [16].

There are the following levels of implementation of e-government:

G2C (government to citizens) is organization of feedback with citizens;

G2E (government to employees) is the relationship of government with officials or employees);

G2B (government to business) is the relationship between government agencies and business;

G2G (government to government) is automation of relations and document flow between departments [17].

The effectiveness of e-government implementation can be assessed using the E-Government Development Index (EGDI), which is an aggregate indicator based on three indices that reflect:

1) the level of development of online services (Online Service Index);

2) the level of development of telecommunications infrastructure (Telecommunications Infrastructure Index);

3) the level of human development (Human Capital Index).

Listed in Table 1 statistics of E-Government Development Index in Eastern European countries show that during 2014-2020 the implementation of e-government developed quite rapidly: the meaning of this Index increased on average from 0.50-0.66 to 0.71-0.85.

Table 1. Analysis of E-Government Development Index in Eastern European Countries in 2014-2020.

Country	2014	2020	Growth
Ukraine	0.5032	0.7119	+0.2087
Bulgaria	0.5421	0.7980	+0.2559
Czech Republic	0.6070	0.8135	+0.2065
Hungary	0.6637	0.7745	+0.1108
Poland	0.6482	0.8531	+0.2049
Slovak Republic	0.6148	0.7817	+0.1669

Source: developed by the authors according to the source [18, 19].

Note that the meaning of this Index in Ukraine was lower compared to other studied Indexes in Eastern European countries: 0.5032 in 2014 and 0.7119 in 2020. In Bulgaria this Index changed from 0.5421 in 2014 to 0.7119 in 2020. The highest E-Government Development Index was in Poland in 2020 – 0.8531. Although it is worth noting that the largest increase during the studied period was recorded in Bulgaria (+0.2559) and Ukraine (+0.2087).

Detailed analysis of E-Government Development Index in Ukraine and Bulgaria in 2020 (Fig. 2) showed that the level of human development in Ukraine was quite high (Human Capital Index was 0.8591 in Ukraine and 0.8401 in Bulgaria), the level of development of online services was above the middle (Online Service Index was 0.6824 in Ukraine and 0.7706 in Bulgaria), and the level of development of telecommunications infrastructure was above the middle (Telecommunication Infrastructure Index was 0.5942 in Ukraine and 0.7826 in Bulgaria). The results of the analysis showed that in Ukraine and Bulgaria it is necessary to continue working on the development of e-government.

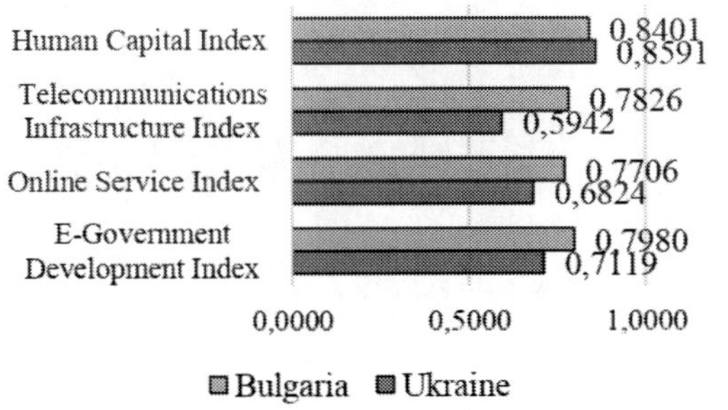

Fig. 2. Detailed Analysis of E-Government Development Index in Ukraine in 2020.

Source: developed by the authors according to the source [19].

Thus, in the context of strengthening the innovative development of information technology, the formation of an effective e-government system should continue not only with the support of public authorities, but also with the assistance of all stakeholders, as only joint efforts can ensure effective interaction of state and regional structures, as well as local governments with business (enterprises) and the person (households).

Information technology in state financial management is primarily represented by the System of Electronic Payments (SEP), which provides 96% of interbank transfers in the national currency within Ukraine. As of the end of 2019, 75 banks of Ukraine, the State Treasury Service of Ukraine and the National Bank of Ukraine were SEP participants [20].

The efficiency of the SEP in Ukraine is evidenced by the annual increase in the amount and number of payments processed in the SEP (Fig. 3).

During 2015-2019, the number of payments processed in the SEP increased by 85 million units (by 28.4%), and the total amount of payments increased by UAH 14 trillion (by 77.8%). In 2019, the SEP processed 384 million payments totaling UAH 32 trillion.

In today's world, information technology is closely linked to digital technology, which has recently become increasingly active in all areas of human life, including and the economy. The mechanism of digital technology is the encoding of information and its transmission, which contributes to a significant number of operations in a short period of time. The main tools for the application of digital technologies today are the Internet and mobile phones (smartphones), computers, laptops, tablets. Since these devices are widespread and widely used by a large number of people in all spheres of life, so the rapid development of the digital economy is natural.

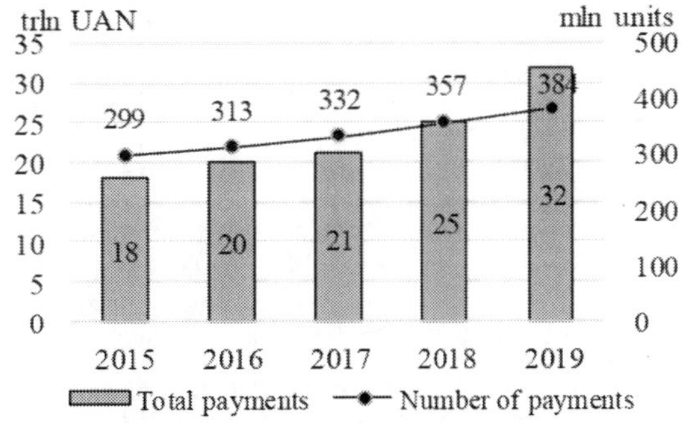

Fig. 3. Analysis of Payments processed in the SEP in Ukraine in 2015-2019.

Source: developed by the authors according to the source [20].

Today, the part of the traditional economy is declining and the part of the digital economy is growing around the world. In European countries, the process of digitization (digitalization) began in 2010, when the European Commission decided to create a Digital Single Market [21].

In Ukraine, for the first time the development of the digital economy at the state level, taking into account the European integration course of our country, was announced in 2016, when the Ministry of Economic Development and Trade of Ukraine developed a project "Digital Agenda of Ukraine – 2020", which defined the conceptual principles of digital economy implementation [22]. Also in 2018, the Cabinet of Ministers of Ukraine approved the "Concept for the development of digital economy and society of Ukraine for 2018-2020", which provided for the implementation of measures to implement appropriate incentives for digitalization of the economy, public and social spheres, awareness of existing challenges and tools for digital infrastructure development, acquisition of digital competencies by citizens, as well as identified critical areas and projects of digitalization, stimulation of the internal market of manufacturing, use and consumption of digital technologies [23].

According to this Concept, digitalization means the saturation of the physical world with electronic-digital devices, tools, systems and the establishment of electronic-communication exchange between them, which actually allows the integrated interaction of virtual and physical, that is, it creates cyberspace.

The Concept also identified the main goals of digital development, in particular: accelerating economic growth and attracting investment; transformation of economic sectors into competitive and efficient; technological and digital modernization of industry and creation of high-tech industries; accessibility of the benefits and opportunities of the digital world for citizens; realization of human resources, development of digital industries and digital entrepreneurship.

The Concept provided a plan to reach a certain level in the world rankings for the development of the digital economy. Thus, in 2020 Ukraine had to occupy:

30th place in the Networked Readiness Index (WEF) (in 2016, Ukraine ranked 64th place);

40th place in the Global Innovation Index (INSEAD, WIPO) (in 2016 – 56th place);

50th place in the ICT Development Index (ITU) (in 2016 – 79th place);

60th place in the Global Competitiveness Index (WEF) (in 2016 – 85th place) [23].

The effectiveness of the implementation of this Concept can be assessed by analyzing the global indexes of digital economy, which were officially published in 2016-2019 (Table 2) [24-27].

Table 2. Indicators of the world indexes of digital economy development in Ukraine in 2016-2019.

Index name	2016	2017	2018	2019
The IMD World Digital Competitiveness Ranking	59	60	59	54
Global Connectivity Index	55	55	54	50
Global Innovation Index	56	50	43	47
Global Competitiveness Index	85	81	83	85

Source: developed by the authors according to the source [24-27].

These indexes indicate that Ukraine occupies a rather low position in world indexes and ratings of the digital economy, and the indicators planned in the Concept as of the end of 2019 have not been achieved.

Note that in Ukraine, however, there are positive trends towards the digitalization of the economy, as almost all of the studied world indexes in our country its position has been improved in the global rankings. Note that in 2019, according to the Global Connectivity Index, Ukraine ranked 50th compared to 55th in 2016; and in the IMD World Digital Competitiveness Ranking, Ukraine ranked 54th, which is 5 positions higher than in 2016. According to the Global Innovation Index and the Global Competitiveness Index, Ukraine improved its position during 2016-2018, and in 2019 there was there letdown of position.

Comparing the world indexes of digital economy development in Eastern European countries, it is worth noting that Ukraine shows the lowest result (Fig. 4) [24-27].

Special attention needs to be paid to the study of the financial component of the digital economy (so-called "digital finance"). The point is that modern information technology can significantly change the traditional forms of money, gradually turning them into so-called cryptocurrency.

Note that cryptocurrency is a new experimental type of money that works in a distributed and decentralized system of secure exchange and transfer of digital banknotes based on cryptography [28].

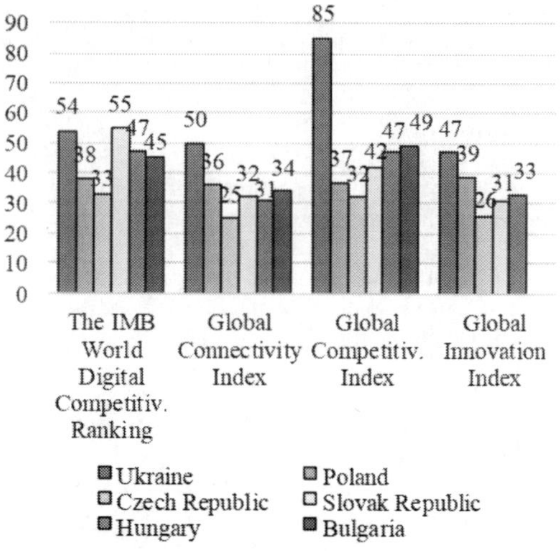

Fig. 4. Comparison of the world indexes of digital economy development in Eastern European countries in 2019.
Source: developed by the authors according to the source [24-27].

Cryptocurrency is a currency that exists in the form of digital money, has no material content and exists in circulation on the Internet, is independent of the central bank.

Cryptocurrency is a currency that exists in the form of digital money, has no material content and exists in circulation on the Internet, is independent of the central bank.

Historically, the first cryptocurrency was bitcoin (BTC), which was introduced in January 2009 [3]. In a relatively short period of time (only 12 years) the number of types of cryptocurrency has increased to 1009, but the leading position is retained by bitcoin [29].

The vast majority of cryptocurrencies have an insignificant market capitalization and a low price. Only 16 cryptocurrencies today have the market capitalization of over USD 1 billion (Table 3).

Bitcoin remains the main cryptocurrency today, as well as during the whole period of cryptocurrency operation. The number of existing bitcoins is more than 18.6 million with a total market capitalization of over USD 610.1 billion. In January 2021, the price of one bitcoin was about USD 32.8 thousand.

Ethereum ranks second in the ranking of cryptocurrencies in the world. The number of all existing ethereums now amounts to about 114.4 million, with a total capitalization of over USD 153.3 billion. In January 2021, the cost of one ethereum was USD 1.3 thousand [29].

As you can see, the use of cryptocurrencies has gained significant momentum, but still a number of issues remain open and need to be refined.

Table 3. Analysis of the top cryptocurrencies on January 25, 2021.

Cryptocurrency	Market capitalization in USD	Price in USD
Bitcoin (BTC)	610 146 914 252	32 787,74
Ethereum (ETH)	153 351 822 186	1 340,69
XRP (XRP)	12 180 662 400	0,276
Cardano (ADA)	11 441 229 292	0,368
Litecoin (LTC)	9 454 248 196	141,41
Bitcoin Cash (BCH)	8 320 149 169	446,76
Stellar Lumens (XLM)	6 229 823 631	0,282
Bitcoin SV (BSV)	3 278 703 930	176,08
EOS (EOS)	2 828 438 984	2,76
Monero (XMR)	2 534 726 108	142,23
Tezos (XTZ)	2 451 237 393	3,23
TRON (TRX)	2 186 433 125	0,031
IOTA (IOT)	1 301 877 853	0,468
Dogecoin (DOGE)	1 135 127 851	0,0089
Zcash (ZEC)	1 100 458 032	90,61
Dash (DASH)	1 095 716 842	110,24

Source: developed by the authors according to the source [29].

For example, the issue of anonymity of transactions in the bitcoin system remains controversial. The system has a public register of all transactions that have taken place throughout the history of this cryptocurrency. However, it should be noted that the real names of the participants are not displayed during transactions in the blockchain, but are indicated by user aliases. On the one hand, this allows you to protect the personal data of the e-wallet owner at the same time as an open transaction history. On the other hand, the lack of identification of a person, as in the case of traditional methods of monetary currency transactions, can have negative consequences associated with the possibility of avoiding liability in the case of illegal transactions [3].

Another problem is the legal inregulation of the status and circulation of cryptocurrencies. Thus, legal entities do not have the opportunity to use cryptocurrency, as they cannot include it in the charter and use it as an asset of the enterprise. We believe that theoretically enterprises could have a positive effect from the use of cryptocurrency, as cryptocurrency transactions save significant amount of time and labor, and there is no commission, so enterprises have the opportunity to save money and use them in production.

It should be noted that at the beginning of December 2020 the Verkhovna Rada of Ukraine adopted at the first reading the draft law "On Virtual Assets" [30]. The issue of adopting this Law is debatable. The drafters of the bill believe that the adoption of the Law will regulate legal relations regarding the circulation, storage, possession, use and conduct of transactions using cryptocurrency in Ukraine. Instead, experts, mostly agreeing with the need to legalize cryptocurrency, believe that some provisions need serious refinement. This Law does not define specific mechanisms and procedures with virtual assets, but only allows their use. In addition, although cryptocurrency acquires legal status, but it is not recognized as a means of payment, that was the main idea of creating bitcoin.

The disadvantages of using cryptocurrencies should also include the risk of significant market price fluctuations, which does not depend on the market capitalization of cryptocurrency (one of the characteristics of cryptocurrency is its high volatility). Therefore, it is difficult to predict exactly how the exchange rate of any cryptocurrency will behave in the market, and the owner of a cryptocurrency wallet has no guarantees that it will be saved.

However, the demand for cryptocurrency is appreciable, so in the future this market will continue to develop and improve.

Artificial intelligence as a separate direction of information technology development is developing more and more every year. Thus, in 2017, the market for artificial intelligence was about USD 2.42 billion, in 2021 is projected to grow to USD 16.2 billion, and in 2023 – about USD 34.4 billion [31].

Artificial intelligence is a branch of computer science that studies the processes of automating intelligent behavior.

Note that in Ukraine at the end of 2020 the Concept of Artificial Intelligence Development in Ukraine was approved, which defines that artificial intelligence is one of the priority areas in scientific and technological research, outlines the problems of state policy for artificial intelligence and identifies areas for solving them [32].

Artificial intelligence allows you to simultaneously process huge amounts of information and get the best solutions to the problem.

The financial sector is one of the most adapted to the use of artificial intelligence compared to other sectors.

Based on the analysis of domestic and foreign literature [33, 34], the following areas of application of artificial intelligence in finance were identified:

1. Trading, Wealth Management, and Investment Banking – the use of software for natural language processing (natural language processing NLP), which allows you to collect and analyze information about the state of foreign exchange markets, provide forecasts for the market conditions, provide investors with information on what shares to buy and sell for their customers, assess the various risks. The use of "robo-advisors" (or the digital advice market) allows you to effectively manage investments.

2. Digitizing Paper Documents – artificial intelligence programs help digitize paper documents. This is a necessary condition for the further use of artificial intelligence, the action of which is based on numbers.

3. Searching Through Large Databases of Documents – artificial intelligence programs help to search in the institution's own huge repositories of digital documents, sort this information, organize it.

4. Underwriting – artificial intelligence programs allow banks and insurance companies to find information about the applicant's public activities (for example, their publications on social networks), on the basis of which a conclusion is made about its reliability.

5. Credit score – such artificial intelligence programs can calculate credit scores of customers, using not only the past credit history of customers, but also other characteristics.

6. Credit risk management in portfolios – the use of artificial intelligence allows you to calculate credit risks, which take into account a significant number of macro-financial factors.

7. Chatbots – the use of chatbots allows you to inform about products and services, provide contact information, financial recommendations to the client, make payment transactions, show exchange rates and exchange currency, keep track of personal finances, transfer from card to card, send orders for trade and online acquiring and checking the counterparty.

8. Personal virtual assistant – is a program that allows you to answer customer questions on financial and banking issues like the financial advisors.

9. Training of bank employees – on the basis of artificial intelligence analysis of the conversation of the bank's specialist with the client and conclusions about the mistakes made by the bank employee and ways to correct them.

Thus, it can be noted that artificial intelligence has now established itself as a promising area of research, which is actively penetrating into various areas of human life, including and in the financial sphere.

It is worth noting that, in our opinion, the benefits of using artificial intelligence for some businesses in parallel have a negative impact for others. For example, a reduction in labor costs due to the use of artificial intelligence is a positive consequence for businesses and at the same time has a negative impact on the workforce, as jobs are reduced, unemployment rises, there is a need to improve skills or get a new education.

Another example is the increasing demand for mathematical and technical specialties in the field of artificial intelligence, which is undoubtedly positive for the workforce, but requires significant costs of the enterprise (business) to conduct research.

Thus, artificial intelligence initiates a new stage of the industrial revolution and in the future can radically change human life. It is important to understand that artificial intelligence deprives a person of the ability to think and make decisions, which can ultimately lead to a decrease in people's intellectual abilities. Therefore, introducing artificial intelligence in any field, including and financial, it is necessary to seriously weigh all the positive and negative consequences.

6 Conclusions

Summing up the results of the study, we note that today the use of information technology has reached a new stage of development, due to the rapid development of digital technologies. All countries of the world, including Ukraine and Bulgaria, faced a new challenge: the digitalization of all spheres of life. Significant steps in this direction have already been taken, but any innovation process is always accompanied by problems and shortcomings that need to be promptly eliminated.

We believe that in order to establish the effective functioning of information technology in finance, it is necessary to: 1) promote the creation of national information systems, platforms and technologies in order to reduce the share of foreign software; 2) organize and provide internships for information technology developers; 3) to carry out explanatory work and to organize trainings for the purpose of informing the population about advantages of application of information technologies; 4) stimulate enterprises to use information technology, including and finance; 5) to expand Internet coverage in Ukraine, in particular, in small settlements where there are no Internet communications, which will contribute to the expansion of information technology users.

We think that in order to establish the effective functioning of information technology in finance it is necessary to: 1) promote the creation of national information systems, platforms and technologies in order to reduce the share of foreign software; 2) organize and provide internships for information technology developers; 3) to carry out explanatory work and to organize trainings for the purpose of informing the population about advantages of application of information technologies; 4) stimulate enterprises to use information technology, including the finance; 5) to expand Internet coverage in Ukraine, in particular, in small settlements where there are no Internet communications, which will increase the number of users of information technology.

Therefore, in order to establish the effective functioning of all spheres of life with the use of digital technologies, it is necessary to continue working on finding improved rules for legislative regulation of the application of information technology in finance and create new methods of information technology management.

References

1. J. Nabatova, K. Malachevska, *Avtomatyzatsiya analizu ta prohnozuvannya finansovykh rezul'tativ pidpryyemstva* (Automation of analysis and forecasting of an enterprise's financial results). Efektyvna ekonomika. 12 (2020), http://www.economy.nayka.com.ua/?op=1&z=8405. Accessed 25 Jan 2021. doi:10.32702/2307-2105-2020.12.80
2. Z. Pavlyshyn, *Strukturno-funktsional'nyy analiz vykorystannya informatsiynykh tekhnolohiy i system u derzhavnomu upravlinni* (Structural and functional analysis of the use of information technologies and systems in public administration). Efektyvnist' Derzhavnoho Upravlinnya. 3(48), vol. 1, 162-166 (2016)
3. V. Tkachenko, A. Kwilinski, O. Korystin, N. Svyrydiuk, I. Tkachenko, Assessment of

information technologies influence on financial security of economy. Journal of Security and Sustainability. 8(3), 387-384 (2019), https://www.researchgate.net/publication/331999626_Assessment_of_information_technologies_influence_on_financial_security_of_economy. Accessed 25 Jan 2021

4. N. Podolchak, O. Bilyk, YA. Levytska, suchasnyy stan tsyfrovizatsiyi v Ukrayini (The condition of digitalization in Ukraine). Efektyvna ekonomika. 10 (2019), http://www.economy.nayka.com.ua/pdf/10_2019/6.pdf. Accessed 25 Jan 2021. doi: 10.32702/2307-2105-2019.10.4
5. Pro Natsional'nu prohramu informatyzatsiyi (On the National Informatization Program), Law of Ukraine, 4 Feb 1998 No 74/98-BP. Verkhovna Rada of Ukraine. https://zakon.rada.gov.ua/laws/show/74/98-%D0%B2%D1%80#Text. Accessed 25 Jan 2021
6. *Universal'nyy statystychnyy paket STADIA 8.0* (Universal statistical package STADIA 8.0). Allsoft Ukrayina, https://allsoft.prom.ua/p387839347-universalnyj-statisticheskij-paket.html. Accessed 25 Jan 2021
7. Statgraphics Technologies, Inc. CoreWin, https://corewin.com.ua/statgraphics. Accessed 25 Jan 2021
8. IBM SPSS software. IBM, https://www.ibm.com/analytics/spss-statistics-software. Accessed 25 Jan 2021
9. V.S. Fetisov, *Paket statystychnoho analizu danykh STATISTICA* (Package of statistical data analysis STATISTICA). (Nizhyn Mykola Gogol State University, Nizhyn, 2018), p. 114
10. Try SAS for Free. SAS, https://www.sas.com/en_us/home.html. Accessed 25 Jan 2021
11. M. Royik, O. Prysyazhnyuk, V. Denysyuk, *Ohlyad prohramnykh zasobiv statystychnoho analizu danykh* (Review of software for statistical data analysis). Efektyvna ekonomika. 7 (2017), http://www.economy.nayka.com.ua/?op=1&z=5676. Accessed 25 Jan 2021
12. What is EViews? EViews, https://www.eviews.com/home.html. Accessed 25 Jan 2021
13. Maplesoft Product Line. Maplesoft, https://www.maplesoft.com/products. Accessed 25 Jan 2021
14. What is R? The R Foundation, https://www.r-project.org/about.html. Accessed 25 Jan 2021
15. SOFTCOM, https://www.softcom.ua/ua/1c/v8. Accessed 25 Jan 2021
16. *Kontseptsiya rozvytku elektronnoho uryaduvannya v Ukrayini* (Concept development of e-government in Ukraine), Order of the Cabinet of Ministers of Ukraine, 20 Sep 2017 No 649-p. Verkhovna Rada of Ukraine. https://zakon.rada.gov.ua/laws/show/ 649-2017-%D1%80#Text. Accessed 25 Jan 2021
17. O. Tsarenko, N. Krasnozhon, *Elektronne uryaduvannya yak instrument posylennya konkurentospromozhnosti krayiny* (E-government as a tool for strengthening the country's competitiveness). Efektyvna ekonomika. 8 (2017), http://www.economy.nayka.com.ua/?op=1&z=8123. Accessed 25 Jan 2021. doi: 10.32702/2307-2105-2020.8.74
18. United Nations e-government survey 2014 (UNITED NATIONS, New York, 2014), https://publicadministration.un.org/egovkb/Portals/egovkb/Documents/un/2014-Survey/E-Gov_Complete_Survey-2014.pdf. Accessed 25 Jan 2021
19. United Nations e-government survey 2020 (UNITED NATIONS, New York, 2020), https://publicadministration.un.org/egovkb/Portals/egovkb/Documents/un/2020-Survey/2020%20UN%20E-Government%20Survey%20(Full%20Report).pdf. Accessed 25 Jan 2021
20. *Zvit z oversayta infrastruktur finansovoho rynku za 2019 rik* (Report on the oversight of financial market infrastructures for 2019). National Bank of Ukraine. 35 p. https://bank.gov.ua/admin_uploads/article/Report_oversight_2019.pdf?v=4. Accessed 25 Jan 2021
21. EU Digital Single Market. https://eufordigital.eu/discover-eu/eu-digital-single-market. Accessed 25 Jan 2021
22. *Tsyfrova adzhenda Ukrayiny – 2020* (Digital agenda of Ukraine – 2020). HITECH office. 2016. 90 p. https://ucci.org.ua/uploads/files/58e78ee3c3922.pdf. Accessed 25 Jan 2021
23. *Pro skhvalennya Kontseptsiyi rozvytku tsyfrovoyi ekonomiky ta suspil'stva Ukrayiny na 2018-2020 roky ta zatverdzhennya planu zakhodiv shchodo yiyi realizatsiyi* (On approval of the Concept of development of the digital economy and society of Ukraine for 2018-2020 and approval of the action plan for its implementation), Order of the Cabinet of Ministers of Ukraine, 17 Jan 2018 No 67-p. Verkhovna Rada of Ukraine. https://zakon.rada.gov.ua/laws/show/67-2018-p/ed20180117#n23. Accessed 25 Jan 2021
24. Digital Competitiveness Ranking 2019. IMD World Competitiveness center. 2019. 179 p. https://www.imd.org/wcc/world-competitiveness-center-rankings/world-digital-competitiveness-rankings-2019. Accessed 25 Jan 2021
25. Global Connectivity Index. Huawei Technologies CO., LTD. 2019. 53 p. https://www.huawei.com/minisite/gci/assets/files/gci_2019_whitepaper_en.pdf?v=20191217v2. Accessed 25 Jan 2021
26. Global Innovation Index 2019: Ukraine. 9 p. https://www.globalinnovationindex.org/analysis-economy. Accessed 25 Jan 2021
27. K. Schwab (ed.), The Global Competitiveness Report 2019 (World Economic Forum, Switzerland, 2019). 666 p. http://www3.weforum.org/docs/WEF_TheGlobalCompetitivenessReport2019.pdf. Accessed 25 Jan 2021
28. X. Li, C.A. Wang, The technology and economic determinants of cryptocurrency exchange rates: The case of Bitcoin. Decision Support Systems. 95, 49-60 (2016). https://www.sciencedirect.com/science/

article/pii/S0167923616302111. Accessed 25 Jan 2021

29. Cryptocurrency prices. BitInfoCharts. https://bitinfocharts.com/cryptocurrency-prices. Accessed 25 Jan 2021
30. *Pro virtual'ni aktyvy* (On virtual assets), Law of Ukraine: project, No 3637. Verkhovna Rada of Ukraine. https://w1.c1.rada.gov.ua/pls/zweb2/webproc4_1?pf3511=69110. Accessed 25 Jan 2021
31. Revenues from the artificial intelligence market worldwide from 2016 to 2025. Statista. https://www.statista.com/statistics/607716/worldwide-artificial-intelligence-market-revenues. Accessed 25 Jan 2021
32. *Pro skhvalennya Kontseptsiyi rozvytku shtuchnoho intelektu v Ukrayini* (On approval of the Concept of development of artificial intelligence in Ukraine), Order of the Cabinet of Ministers of Ukraine, 02 Dec 2020 No 1556-p. Verkhovna Rada of Ukraine. https://zakon.rada.gov.ua/laws/show/1556-2020-%D1%80#Text. Accessed 25 Jan 2021
33. S. Tsyhanov, V. Apalkova, *Rozvytok shtuchnoho intelektu: evolyutsiyni protsesy na mizhnarodnomu finansovomu rynku* (Development of artificial intelligence: evolutional processes in the international financial market). Finansy Ukrayiny. 8 (2018), http://finukr.org.ua/docs/FU_18_08_020_uk.pdf. Accessed 25 Jan 2021
34. Azulay D. Artificial intelligence in finance – a comprehensive overview. Business intelligence and analytics. Emerj. 24 Dec 2019. https://emerj.com/ai-sector-overviews/artificial-intelligence-in-finance-a-comprehensive-overview. Accessed 25 Jan 2021

Tools to Support the Development and Promotion of Innovative Projects

Dilshod Okhunov[1*], *Sergei* Semenov[2], *Saidahror* Gulyamov[3], *Dilorom* Okhunova[4], and *Mamatjon* Okhunov[5]

[1]Fergana branch of the Tashkent University of Information Technologies named after Muhammad Al-Khorazmiy, 150100, Fergana city, Republic of Uzbekistan
[2]Research Institute "CENTERPROGRAMSYSTEM", CJSC 170024, Tver city, Russian Federation
[3]Tashkent State University of Economics, 100066, Tashkent city, Republic of Uzbekistan
[4]Regional Department of Finance, 150103, Fergana city, Republic of Uzbekistan
[5]Ferghana Polytechnic Institute, 150101, Fergana city, Republic of Uzbekistan

Abstract. Competitiveness and efficiency in the modern economy are based on innovation. It is necessary to ensure the flow of well-founded innovative projects. This requires funds to support the creation and promotion of these projects. These tools include innovation methodology, mathematical models and methods, software and information tools, and other types of software. All these tools are implemented as services of a digital platform based on cloud technologies and available to various categories of users. This platform is being developed in research Institute of CPI, Tver, Russia.

1 Introduction

The innovation economy differs from the previous models of economies primarily by the need for constant changes in all elements of production and economic relations. In the economies of mass and mass production, the concept of efficiency was associated with the concept of stability for a certain period of time. In the innovation model, efficiency is primarily associated with the concept of competitiveness. The global nature of competition, the common information space, and the diversity of goods and services require any manufacturer and supplier to be flexible in making decisions and adapting to competitive conditions. Constant search and implementation of changes at different levels of production and economic relations are necessary conditions for competitiveness, regardless of the type and scale of activity.

It is the change in economic ideology that leads to the need to introduce a new concept of "innovation" as a unit for assessing the degree of changes in production and economic relations.

The key role in the formation of the philosophy of innovative economy belongs to information technologies (IT). This is due to the following important factors.

First, there is a short information link between all the elements of the production and economic system. The speed of decision-making and information exchange is a determining factor in the efficiency and competitiveness of any system.

Secondly, the safety of information. The information necessary for decision-making today has a very wide variety of forms and representations. The volume of this information is very large and is constantly growing. The ability to save the accumulated information is the most important factor in the success of any activity.

Third, the availability of information. Regardless of the scale of the task to be solved, it is necessary to be able to obtain information not only from local sources (the enterprise information system, the local database), but also from the global information space. This is due to the variety of factors that influence decision-making in the context of working in the global economic system.

Fourth, a high degree of automation of information processing. Information is often created when solving some problems, and can be used to solve others. Therefore, the ability to quickly transform information to the right form for decision-making is also a crucial factor in the need for IT.

Fifth, the ability to transform information into a form of knowledge. Knowledge differs from information in its pragmatic aspect: what to do with this information or what to do if this information is obtained. For the speed and quality of decisions, knowledge is a more valuable source of competitiveness and efficiency, as well as an object of legal protection. Modern IT allows you to effectively transform information to the form of knowledge and work successfully with them.

Ultimately, the innovation economy depends largely on the success of the use of IT capabilities, which are one of the three cornerstones of the innovation economy. IT as a type of production activity has its own characteristics, and this article is devoted to this.

* Corresponding author: mamatdilshod@rambler.ru

To understand the philosophy and ideology of the innovation economy, it is necessary to focus on its basic elements: concepts and definitions.

The term "innovation" is now one of the most widely used in the economic literature and in public speeches. Dozens (and perhaps even hundreds) of its definitions and interpretations are known. And this is quite natural for the era of the global innovation economy. However, there is still no clear understanding of this term and, most importantly, the essence of the concept itself.

While the Russian economy went its own way, not caring about competitiveness and efficiency, terminological disputes were the lot of scientists. But by joining the WTO, Russia is forced to play by the general economic rules, otherwise the domestic industry will simply disappear under the onslaught of more efficient foreign companies and technologies.

Why is it so important to understand the terms and concepts?

The fact is that there is a structural restructuring of the economic model. The role of the state in this process is crucial: it creates a new infrastructure and launches cash flows into a new economic model. Here, the negative consequences of conceptual illiteracy are fully manifested. The money is being used, and nothing positive is happening. Competitiveness, efficiency and productivity remain at the same level or even decline.

Due to the total lack of understanding of the new concept, there is a "gluing of a new label on the old concepts"everywhere. Now everywhere you can see how the doors of offices, products and technologies are labeled "innovative" and expect a miracle: how the economy will suddenly become efficient and competitive.

Let's try to sort out the confusion in terminology and concepts.

First, it is necessary to immediately determine that innovation is not the result of the evolutionary development of the previous economic model, but a revolutionary transition to a new model. And there can be no continuity in the concepts. However, it should be noted that the "product based on invention" model is not new in itself, moreover, it is already many thousands of years old. And the model of "goods based on the study of demand" is quite young (in historical terms), and it arose with the advent of mass production industry. But it was the last 150 years of industrial revolutions that made us forget about inventions as the main source of new goods. Of course, to facilitate understanding, sometimes it is necessary to use some analogies with elements of the old model, but it is analogies, and not direct inheritance of a term or concept.

The reason for the total misunderstanding of the innovation economy lies in the attempt to find direct analogies in the old model and make the old economic mechanism work in a new way. Hence the desire to use the old terminological and conceptual apparatus.

Of course, it is not only and not so much in terms, but in a lack of understanding of the essence of the new concept and new processes. This leads to both wrong decisions and wrong actions.

Secondly, the concept of "innovation" refers to the so-called indefinable concepts [1]. An indefinable concept in axiomatics is an initial, basic concept, which cannot be unambiguously defined. Any science and theory is based on some basic concepts, usually intuitive and having properties described by the axioms of this theory. So, the main undefined concepts in geometry are point, line, plane, volume, space, in mathematics-number, set, correspondence, in physics-time.

That is, an indefinable concept cannot be uniquely defined through other basic concepts of a given theory.

It follows that there can be innumerable" correct " definitions of this concept. And all of them, in a certain sense, will be both right and wrong at the same time. That is, a concept formulated quite strictly and precisely for one specific task may be completely unacceptable for another.

There is some contradiction between the generality and the usefulness of the definitions. The most general definition of a concept is usually the most useless. On the other hand, a very specific definition makes it useful only in a very narrow range of tasks. And this is constantly traced in everyday life. So, there are several hundred definitions of the concept of "system" and several dozen of the concept of "marketing". Definitions are most successful only in specific situations.

Third, the concept of "innovation" is multifaceted and multidimensional. This is a philosophy, an object, a function, an action, properties, and a method. An attempt to map a multidimensional object to a plane (or line) only gives the correct representation within that plane or line. This reasoning would be a mathematical abstraction if the tens of thousands of people and businesses trying to operate in an innovative economy did not suffer from "correct" definitions that kill innovation and innovation activity in the bud.

Fourth, there is a need for a common denominator in decision-making for all stakeholders. The fact is that each of them is incompetent in innovation in its own way: the investor understands money, the inventor understands machines, and the general definition does not help either of them. The working definition of innovation should help all parties, it should reflect the essence of the invention, the idea. The market situation is in innovation. Money matters - in innovation. Innovation is the top of the concept tree, the individual branches of which are the view of each interested party on innovation (through an idea, through money, through a market position, through production opportunities, etc.).

2 Analysis of existing definitions of the concept "innovation"

Let's see how the concept of "innovation" is defined in a number of primary sources and some modern documents. An analysis of the existing definitions reveals a wide variety of views and approaches.

Schumpeter, J. A. In [2], under innovation, he proposed to understand any possible change that occurs due to the use of new or improved solutions of a technical, technological, organizational nature in the processes of production, supply, sales of products, etc.

Foster R. defines innovation as a battle in the market between innovators (or attackers) - those who try to make money by changing the order of things - and those who defend themselves by protecting their current income [3].

Twiss B. interprets innovation as an offer on the market of something new, for which the consumer is willing to pay [4]. An invention becomes an innovation if it is successful in the market. Innovation is an application, that is, a process in which an invention or idea acquires an economic content.

According to Santo B., innovation is a social, technical or economic process that, through the practical use of ideas and inventions, allows you to create the best products and technologies that can bring additional income on the market [5]. Innovation is a technical and economic cycle in which the use of the results of research and development directly causes technical and economic changes that affect the activities of a particular field.

Drucker P. defines innovation as a special tool of entrepreneurs, a means by which a chance to open a new type of business or service is realized [6].

For Porter M., innovation is an opportunity to gain competitive advantages [7]. Innovations in a broad sense include both new technologies and new working methods. Innovation manifests itself in a new product design, in a new production process, in a new approach to marketing, or in a new methodology for improving the skills of employees. Innovations are fairly simple and small, based more on the accumulation of minor improvements and achievements than on a single major technological breakthrough.

Hamel G. and Prahalad K. believe that innovation is what brings the maximum profit, ahead of others on a global scale [8].

According to Nelson R. and Winter S., innovation is a change in routine. Viability depends on how consumers evaluate it [9].

In the OECD Frascati Position, innovation is the end result of an activity, embodied in the form of a new or improved product introduced in the market, a new or improved technological process used in practice, or in a new approach to social services.

Yakovets Yu. He believes that innovation is a qualitative change in production, which can relate to both technology and technology, as well as forms of production organization and management [10].

Morozov Yu. defines innovation in a broad sense as the profitable use of innovations in the form of new technologies, products and services, organizational, technical and socio-economic solutions of a production, financial, commercial, administrative or other nature [11].

According to Fatkhutdinov R., innovation is the final result of the introduction of new products in order to change the object of management and obtain an economic, social, environmental, scientific and technical or other type of effect [12].

According to the definition of V. Medinsky, innovation is an object introduced into production as a result of scientific research or discovery, qualitatively different from the previous analog [13].

Hipel V. defines innovation as the application of a new product or process in practice [14], and Rigs H.-as the commercial development of a new idea [15].

Freeman K., speaking about industrial innovation, defines it as technical, design, production, management and commercial activities, the production of new (improved) goods or the first commercial use of new (improved) processes or equipment [16].

Rothwell R. and Gardiner P. believe that innovation is not only the commercialization of a major improvement in a technical process (radical innovation), but also the use of small-scale changes in technological skills (improvements or minor innovations) [17].

Tidd D., Bessant D., Pavit K. define innovation as the process of turning opportunities into new ideas that are widely implemented in practice [18].

In the Law of Ukraine on Innovation Activity [19], innovations are newly created (applied) and (or) improved competitive technologies, products or services, as well as organizational and technical solutions of a production, administrative, commercial or other nature that significantly improve the structure and quality of production and (or) the social sphere. An innovative product is the result of research and (or) development work that meets the requirements established by this law. Innovative products - new competitive goods or services that meet the requirements established by this law.

Wikipedia gives the following definitions [20]. Innovation is an implemented innovation that provides a qualitative increase in the efficiency of processes or products that are in demand by the market. It is the end result of a person's intellectual activity, imagination, creative process, discoveries, inventions, and rationalization. An example of innovation is the introduction to the market of products (goods and services) with new consumer properties or a qualitative increase in the efficiency of production systems. Innovation - a new or significantly improved product (product, service) or process, a new sales method, or a new organizational method in business practice, workplace organization, or external relations. Innovation is not any innovation or innovation, but only one that seriously improves the efficiency of the current system.

Generically, this concept can also be applied to a creative idea that has been implemented.

3 Methodical section

In an innovative economy, it is very important to increase the number of innovative projects being implemented. At the same time, it is necessary to create conditions for improving the quality of project development.

These conditions are implemented in four ways (see Figure. 1)

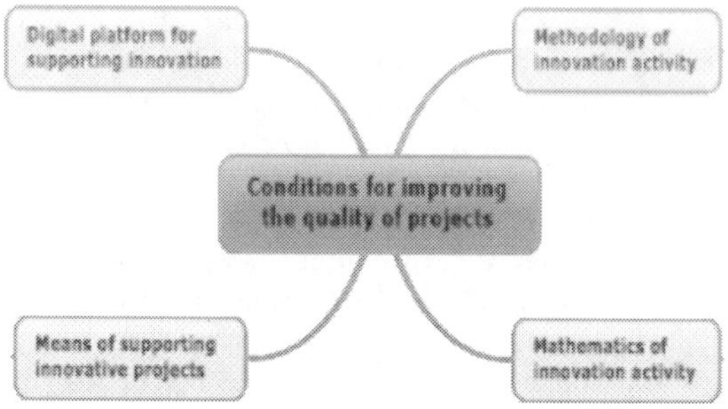

Fig. 1. Conditions for improving the quality of projects

These conditions are implemented on a digital platform for supporting innovation activities and innovative projects (next DP).

The digital platform for supporting the creation and promotion of innovative projects solves the following basic tasks:

- The methodology of innovation activity in various fields is developed and supported.
- Supports the creation and promotion of an innovative project.
- Information and methodological support for the user is provided.
- Communication between subjects interested in an innovative project is provided.
- The project is evaluated from different points of view.
- Work with intellectual property is supported.

The DP itself is an innovative project. At the initial stage of DP development, the main technical solutions were conceptualized and justified. The DP is represented as subsystems, blocks, and services. Solutions to fundamental problems in the creation and promotion of innovative projects, as well as problems of innovation management, have been worked out.

The initial step in this project is to develop an innovation Methodology. This allows, first of all, to form an optimal organization of innovation activities. Secondly, it is most effective to automate the solution of innovation tasks [21].

The methodology is developed based on the definition [22]. "Methodology is the teaching about the organization of activities."

The methodology of innovation activity elaborates the basic concepts of this activity:

<u>Characteristics of innovative activity</u>: Features, Principles [23], Conditions, Norms of this activity.

<u>Logical structure of innovative activity</u>: Subject, Object, Subject, Forms, Means, Methods, Result of activity.

Time structure of activity: Phases, Stages, and Steps of activity.

The methodology can generally be developed for any type of activity, taking into account terminology, application rules, automation conditions, etc.

Based on the basic concepts of the methodology (theoretical plan), the system and applied concepts of the applied field are worked out.

For example, an innovative project, an innovative activity, an innovative system, an innovative infrastructure, stakeholders, an innovative product (product, technology, service), efficiency, competitiveness, feasibility, investment system, stages, intellectual property, etc.

From basic concepts, on the one hand, the study is conducted in the direction of detail to elementary facts.On the other hand, we are building integrated concepts based on basic concepts and elementary facts (for example, a "well-packaged" project).

On the basis of terms of varying degrees of detail, an ontological network of concepts of innovation is constructed.

The main problems (goals) of innovation that require automated support should be recognized:

- Increasing the motivation of participants in innovation activities.
- Increasing the validity of innovative projects.
- Improving the ease of working with innovative projects.
- Reducing the subjective factor when evaluating and making decisions about supporting an innovative project.

Solving these problems will significantly increase the flow of "well-developed" innovative projects. Implementation of these projects will allow creating a wide range of competitive and effective products [24].

To solve the fundamental problems of innovation, it is necessary to solve many problems of different levels and complexity related to:

- Evaluation of the innovation project (its parts) according to several criteria;
- Justification of the novelty, feasibility and effectiveness of the innovation project;
- Preparing a decision to support (or reject) the project;
- Monitoring and managing project progress;
- Various calculations, project presentation, etc.

Various mathematical methods and models are used to solve these problems.

The DP project addresses the following issues:

- Measurement and evaluation of elements of an innovation project and elements of innovation activity.
- Decision-making on an innovative project and the possibility of its support at the stages of the life cycle.
- Analysis of data obtained from our own and third-party sources (information resources). The possibility of using information from data Centers (Big data) is considered.
- Modeling of objects and processes of innovation (including simulation of the consequences of applying the results of an innovation project).
- Management of innovation activities at different levels (from the workplace of a specialist to the level of the relevant Ministry).
- Investment mathematics that solves the problems of

substantiating the effectiveness and competitiveness of innovative projects.

The third branch of Figure 1 "means of supporting innovative projects" includes various types of support. For example, information, methodological, financial, regulatory, organizational, and other types of support.

The fourth branch of Figure 1 deals with the creation of the digital platform for supporting innovative projects and innovative activities [25].

The DP is developed as a self-organizing and self-developing system that integrates its own and third-party information resources.

Templates and scenarios for the formation of innovative projects within the use of the DP are being worked out.

The DP architecture is represented as interconnected subsystems, blocks, and services based on a cloud platform.

The DP consists of the following subsystems:

- Subsystem "formation of an innovative project".
- Subsystem "Information base of scientific and innovative activity".
- Subsystem "support for analysis and evaluation of innovative projects".
- Subsystem "support for stakeholder communications".
- Subsystem "working with intellectual property".
- Subsystem "support of experts and formation of Competence Centers".
- Subsystem "monitoring and management of innovative projects".
- Subsystem "Simulation of innovation activity".
- Subsystem "forming templates for innovative projects".
- Subsystem "automation and accounting of preliminary and scientific and technical expertise".
- Subsystem "maintaining the register of experts".
- Subsystem "system Administration and information Security" and others.

The DP provides four levels of access:

- The level of public access.
- The level of personal access.
- Certified access level.
- The level of restricted access.

Subsystems consist of blocks that combine groups of tasks. For example, the subsystem "Information base of scientific and innovative activity" includes:

- Database support block (patents, intellectual property, partners, investors, consumers of innovations, etc.).
- Block of linguistic and semantic search, etc.

The subsystem "support for analysis and evaluation of innovative projects" includes the following blocks:

- Block "Analysis of innovative projects".
- The block "Methodological support for innovation" (search for analogues, "packaging" of projects, registration of property rights, etc.).
- Block "Support for training in the design of innovative projects".
- Block "support for calculations in innovation" (investment performance indicators, quantitative, qualitative and cost assessment of the results obtained, etc.).
- Block "Support for scientific study of projects", etc.

The implementation of these tasks requires the use of a wide range of mathematical methods to solve the problems of innovation.

The DP uses mathematical methods to support decision-making in the innovation sphere. The issues of measuring and evaluating various aspects of innovative projects are considered. Without numerical values of qualitative characteristics, it is impossible to make estimates and make decisions.

The degree of innovation is the measure by which an investor decides to support a project. This measure is based on evaluation theory and measurement theory. Dimension theory is also used. In the future, it will be necessary to formulate new units of measurement related to the measurement of the degree of innovation, the degree of realizability, the degree of novelty, etc.

The methods of utility theory and value theory are used to justify project decisions. These sections are necessary for the formation of criteria for the success and feasibility of innovative projects.

Here we use mathematical methods and models that allow us to display subjective qualitative judgments of subjects on a numerical scale (of some type) in order to formalize value judgments about the project. Various types of evaluation and measurement scales are used, their applicability, ways of construction and use. One of the DP services supports evaluating the effectiveness of innovations.

To prepare a decision on project support, we use methods of generalizing information to build integrated concepts that give a General and final assessment of the project as a whole. To work with weakly structured information, the apparatus of semantics, knowledge, and ontologies in the field of innovation. Semantic methods of working with information are of paramount importance. This is due to the widespread use of the Internet, the use of cloud technologies, and great achievements in the field of web standards. Structured metainformation based on the RDF standard (and a number of others) allows you to present information in the form of knowledge. In other words, this is a step towards separating knowledge from specialists (experts) in order to be able to store and use knowledge, regardless of the presence of the source of knowledge itself. Based on this principle, virtual competence centers will be formed in the DP.

For the successful implementation of the semantic web concept, an ontological network of concepts of innovation is formed, taking into account the applied areas of activity.

The DP solves modeling problems, in particular, ontological and simulation. In fact, innovation is aimed at changing the world around us. But the consequences of these changes are not always obvious at the beginning of the project. To successfully promote the project and its implementation, it is necessary to model not only the innovative product itself, but also its impact on the design environment, production environment, distribution and use environment. Such modeling will allow, on the one hand, to prove the usefulness and profitability of an innovative product (in the case of positive modeling results). On the other hand, stop the project in time and avoid unnecessary costs (in case of negative modeling results).

It is advisable to implement the DP in three queues.

4 Conclusion

The tasks of the first stage are the"first aid" for the author (and expert) in terms of information and methodology: dictionaries, reference books, regulatory framework, standard competition documents, methodological assistance, etc. This allows you to speak the same language with each other, understand each other's problems, find compromise solutions, and so on.

The tasks of the second stage of the DP include the tasks of extracting information, creating information, converting it, storing it, presenting it to stakeholders, processing (detailing, integration), developing templates, typing, and evaluating it. As well as mechanisms for working with information that allows you to prepare a project to get a positive decision from the investor to support the project.

Tasks of the third stage are intended for the intellectualization of works implemented in the first and second stages. In fact, it should be an intelligent information and advice system using artificial intelligence methods.

References

1. URL: ru.wikipedia.org/wiki/Undetectable concept (accessed date: 10.10.13).
2. Schumpeter, J. A. Theory of economic development. Moscow: Progress, 1992. 231 p.
3. Foster R. Production update: attackers win. Moscow: Progress, 1987. 147 p.
4. Twiss B. Management of scientific and technical innovations. M.: Ekonomika, 1989. p. 37.
5. Santo B. Innovation as a means of economic development. M.: Progress, 1990. 291 p.
6. Drucker P. How to ensure success in business. K.: Ukraina, 1994. p. 41.
7. Porter M. Competition: textbook. manual; [trans. Moscow: Williams, 2001. 495 p.
8. Hamel G., Prahalad K. Competing for the future. Moscow: Olimp-Business, 2002. 216 p.
9. Nelson R., Winter S. Evolutionary theory of economic changes. Moscow: Finstatinform, 2000. p. 46.
10. Yakovets Yu. Acceleration of scientific and technological progress: theory and economic mechanism. Moscow: Ekonomika, 1988. p. 95.
11. Morozov Yu. Innovation management. M.: UNITY-DANA, 2000. p. 6.
12. Fatkhutdinov R. Competitiveness: economy, strategy management. M.: INFRA-M, 2000. P. 279.
13. Medinsky V. Innovation management. M.: Infra - M, 2002. P. 5.
14. Hipel V. U. S. high technology trade and competitiveness. Staf-freport. U.S. Department of Commerce, 1985, p. 19.
15. Rts H. Managing high technology companies Blmont. N.Y., 1983, p. 7.
16. Frimen K. The Economics of industrial innovation; 2nd edn. Frances Pinter, London, 1982, p. 38.
17. Rotvell R., Gardiner P. Invention, innovation, re-innovation and the role of the user / / Techonovation, 1985, No. 3, p. 168.
18. Tidd D., Bessant D., Pavitt K. Managing innovation. John wiley & Sons, LTD, Chichester, 2001, p. 38.
19. The law of Adornment "Pro shnovatsshnu diyalnist" / / Golos Adornment. 2002. 9 serp. p. 10.
20. URL: ru. w ikipedia.org/wiki/Innovation (date of request: 10.10.13).
21. Okhunov D.M. Theoretical and methodological aspects of the development of marketing automated information systems. – Tashkent: FAN, 2010.– 208p.
22. Novikov A.M., Novikov D. A. Metodologiya. – M.: SYNTEG, 2007. – 668 p.
23. Kupriyanov K. V., Nesterov V. V. the Principle of changes in innovation activity. Science and innovation in the XXI century: current issues, discoveries and achievements: collection of articles of the XVII International scientific and practical conference. – Penza: "Science and Education". – 2020. – 220 p.
24. Semenov S. V. Innovations. Integration of scientific knowledge into production activities. Eurasian Scientific Association. 2018. no. 6-1 (40). Pp. 43-45.
25. Kupriyanov K. V. Concept of automation tools for supporting innovation activities. Science and innovation in the XXI century: current issues, discoveries and achievements. Collection of articles of the XVII international scientific and practical conference. Penza: "Science and Education", 2020, 220 p.

The Impact of Industry 4.0 Technologies on Structural Transformation in the Manufacturing

Oksana Kushnirenko[1*], *Natalia* Gakhovich[1], and *Liliia* Venger[1]

[1]State Organization "Institute for Economics and Forecasting of NAS of Ukraine", 01011, Kyiv, Ukraine

Abstract. In the paper, we investigate the impact of global trends Industry 4.0 on the structural transformations of Ukrainian industrial sector through the prism of global technological challenges. The current structural imbalances of Ukrainian industrial development and highlights the opportunities that 4.0 technologies provide for innovative industrial renewal are considered; and the readiness for adoption of innovative technologies of Industry 4.0 in the industrial enterprises is assessed. The key conclusion of the study is that further innovative transformation of global production will lead to inevitable changes in the organization and management of production processes. Based on the research results, an effective solution to certain problems has been proposed. The key finding of the research is that representatives of Ukrainian industry though perceive the importance of this tool face many obstacles in its implementation. Considering research results the efficient solutions of identified problems were proposed. They include the establishing strengthening the strategic partnership between the state science and business institutions in the process of stimulating cluster development; improvement of innovation infrastructure, creating the innovation ecosystems for the capitalization of scientific developments and the commercialization of innovations; creation of competitive conditions for all market participants, including equal conditions for access to the raw materials market, technology transfer, protection of property rights; reforming educational infrastructure; creation of a favorable investment environment with the help of mechanisms of state guarantees for foreign investments.

1 Introduction

At the beginning of the 21st century, mankind was affected by the new industrial revolution, better known as the concept of "Industry 4.0". New technological challenges in Industry 4.0 lead to dramatic the changes taking place in consumer behavior and the management of production. Moreover, it leads to innovative transformations in many areas, breaks down barriers between them, also tended to blur the lines between suppliers, producers and consumers, and ultimately is a prerequisite for structural transformations in the economy. This has led the governments of both developed and developing countries to justify and apply new economic policy approaches to enhance the innovative development of the industrial sector. This is especially important for Ukraine, since according to Heyets V.M., the future of the country depends on the introduction of innovative development mechanisms: whether the country will move towards becoming a developed country, and potentially be left behind by emerging technologies. This is related to general patterns of social development, according to which there is a transition from a predominantly reproductive to an innovative type of development [1].

The purpose of the article is to justification of priorities for structural transformation of industrial development mindful of the increasingly important role played by the Industry's 4.0 technological challenges. The objectives of the study are: to analyze the theoretical base for the impact of Industry 4.0 on structural transformation in industrial development; to describe the methodological approaches used in the research; to assess the industrial structure changes during 2014-2019 years; to propose policy recommendations concerning the main vector and tools of stimulating adaptation of Ukrainian manufacture to the challenges of Industry 4.0.

1.1 Related Work

1.1.1 Literature Review

There are a big number of economic publications about continuous economic, social and technological changes, which are conditioning the industrial development. In this regard, special attention should be given to the studies of Ukrainian and foreign scientists. Indeed, some experts (E. Reinert, J. Stiglitz, D. Rodrik, K. Freeman, K. Perez and others) stated that bridging the technological gap between developing countries and countries with economies in transition and others is an urgent issue, and one that impedes the global development process. In this respect, the works of Ukrainian scientists occupy a special place: V. Heyts, O. Amosha, Gritsenko, V. Lyashenko, L. Deyneko, A. Chukhno, Yu. Pilipenko, Yu. Kindzersky, L. Fedulova, in which the main guidelines for reforming the

* Corresponding author: kushnksena@gmail.com1

 https://doi.org/10.1051/shsconf/202110001009

domestic economy based on leading innovative industrial development. So, Yu. Pilipenko defines technology system as a relatively autonomous element of public, based on a systematic approach to the analysis of socio-economic processes [2]. As noted by Yu. Kindzerskiy, the effectiveness of structural reforms in industry can be said only when the state creates effective institutions for this [3]. Summarizing scientific research on the study of the prerequisites for technological transformations in the industry of foreign and domestic scientists, we can observe that that their driving forces are a combination of factors: the accumulation of knowledge and the search for new methods of processing raw materials at lower costs; concentration of capital; the availability of resources (human, raw materials, financial), the formation of a unified information space and an innovative institutional system. "Industry 4.0" is considered by us as new approaches to production management and organization of production processes based on the integrated implementation of cyber-physical systems in all business processes that determine the digital transformation of the ways of human interaction, data and processes, providing new opportunities to meet the needs of consumers, using the latest bio -, nano-materials, reducing the duration of the production cycle, reducing material costs, reducing industrial injuries, introducing circular production models, optimizing the load on the environment. At the same time, for Ukraine, which is trying to ensure sustainable economic growth and the well-being of the population in difficult conditions of a military conflict, it is extremely important to determine the features of the impact of the fourth industrial revolution on the economy and society, as well as to justify the directions of adaptation of the Ukrainian industry to the latest technological challenges.

The essence and theoretical and applied aspects of the Industry 4.0 concept as a phenomenon of the Fourth Industrial Revolution were formulated and it is proved that its emergence is closely related to civilizational development and the inevitable continuation of the change of technological orders. It has been established that the unique difference between Industry 4.0 and previous industrial revolutions are: the possibility of integrating industrial automation with control systems at the global level, transforming the concept of industrial production, activating the development of global value chains.

1.1.2 Methodological Approach

The research methodology derived from the basic principles for predicting the future characteristics of useful machines, procedures or working methods, which is the basis for the development of programs, ranging from forecasting the development of products or technical capabilities to creating scenarios for predicting the impact of future technologies introduced by J. Martino [4].

We have been able to build upon the R. Kurzweil's model, where he presented an analysis of the history of technology and substantiated in the Law of Rapid Return, technological changes are exponential [5] in the further discussions on this issue.

Following the European methodology, the questionnaire is the right method to identify sentiments, trends and needs of manufacturers what becomes a ground for technological upgrading. As mentioned in OECD paper "Handbook on Constructing Composite Indicators" the assessment of complex, dynamic phenomena in broad areas, such as the environment, economy, social or technological development by a system of complex indicators (Composite indicators) in different countries and recognizes them as a useful tool for policy analysis and communication with the public [6]. For example, the manual substantiates the Technological Achievement Index (TAI), which focuses on four dimensions of technological capacity.

The main method used in the study is structural-functional and comparative analysis - to study the factual material that characterizes the current state, the practice of structural transformation of the model of economic development.

Considering the purpose of the analytical and experimental investigation the error value is satisfactory.

The data gathered during the questioning was processed using such techniques as grouping, reduction, comparison, and graph analysis.

1.2 Our Contribution

These analytical materials can be implemented in programs of industry development and seek for new effective approaches of industrial development policies.

1.3 Paper Structure

The research is organized as follows. The first section represents the theoretical base for the impact of Industry 4.0 on structural transformation in industrial development. The description of the methodological approaches used in the research is given in the second section. The third section describes main results of the assessment of industrial structure changes during 2014-2019 years. At the end of the study, we conclude the survey results and give some policy recommendations concerning the main vector and tools of stimulating adaptation of Ukrainian manufacture to the challenges of Industry 4.0.

2 Background

2.1 Status and characteristics of industrial development in Ukraine: trends, problems, needs

Industrial development could be considered as the deployment of the process of "Total Causality", where factors of supply and demand interact: on the one hand, the development of industrial production creates jobs, generates income, stimulates demand, and on the other accelerates productivity growth, which in turn stimulates the growth of the welfare of the population [7]. Continuous improvement of the production base of the dairy industry, which is part of this process, can lead to increased productivity through the development of new areas of economic activity, the use of more advanced technologies, production of more complex goods and / or inclusion in international supply chains at an increasing technological level.

The key causes of economic problems during the existence of independent Ukraine and the previous period are crisis phenomena with an interval of 5-10 years, which directly influenced the development of industry. So, in the economic literature, such crisis periods are distinguished: the formation of Ukraine as an independent state (1991-1993), the signs of which were the liberalization of prices in the absence of competition and the preservation of the command-administrative functions of the state, hyperinflation (almost 2000%, which is comparable with the indicators of Brazil and Congo), money issue, bankruptcy and liquidation of enterprises, devaluation of deposits of the population, severing economic ties with enterprises of the former USSR; the financial crisis of 1998, caused by the unreasonable debt policy of the state, due to the state budget deficit; the structural crisis of 2003-2004, which covered the food market, financial, credit and energy systems, and the like; financial crisis of 2008, caused by debt and other obligations similar to the problems of 1998; Russia's military aggression and the annexation of Crimea in 2014 and the crisis associated with the corona virus. All of the above developmental breaks required coordinated state regulation aimed at supporting domestic industrial production in the formation and implementation of the principles of the "new industrialization" policy in Ukraine.

We determine the root causes that led to the current degradation of the economic situation in Ukraine (Fig. 1).

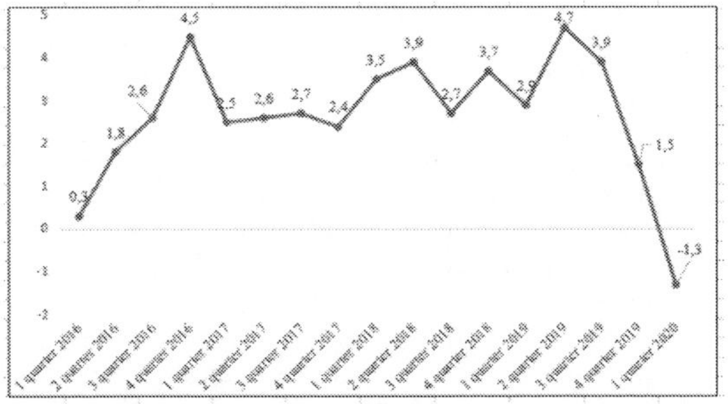

Fig. 1. Real gross domestic product (GDP): percent change from preceding quarter [8].

Once the direct causes have been identified, the next stage is to try to follow the chain of causation further.

2.2 The basic premise of researching

Modern practice of developed countries shows that in the Fourth industrial revolution only modern developed industry determines the rapid and high-quality growth of the economy. Increasing the competitiveness of the processing industry is of strategic importance. Thus, in Germany in 2011 the concept and industry development program "Industry 4.0" was developed and approved. Support for the vector of Industry 4.0 forms an appropriate industrial policy, which is reflected in the implementation of various countries programs aimed at building scientific, technological and industrial potential, improving the innovation system, updating the technological base. According to European statistics, the highest concentration of high-tech industries in the EU is observed in Italy, followed by Germany, France and Poland, together form more than 60% of the EU's high-tech industrial potential. For comparison, in Ukraine this sector is of very little importance. Thus, according to the World Bank, in Ukraine, the share of high-tech products over the past three years is gradually declining and in 2018 is 5.41% of industrial exports, and in 2015, this figure reached 8.52%. For other neighboring countries, such as Belarus, similar trends are also observed: the share of high-tech products in exports decreases from 4.85% in 2016 to 3.49% in 2018 [10]. Instead, the world average is 16.3% in 2017. Comparing the structure of Ukraine's industry in 2010 and 2019 by type of economic activity, we can note a decrease in the share of manufacturing to 64.4% and an increase in the contribution of electricity, gas, steam and air conditioning to 18.5% and mining and quarrying up to 15.9% (Fig. 2).

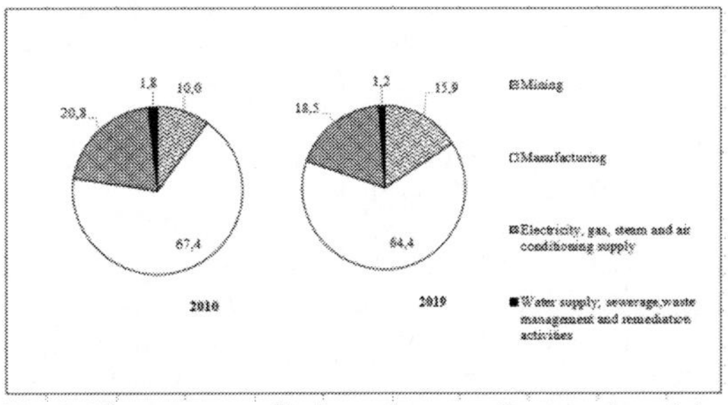

Fig. 2. The structure of industry by type of economic activity in 2010 and 2019 [9].

In turn, the processing industry is mainly (64.4% in 2019) formed by: production of food, beverages and tobacco products (33.7% of the total processing industry in 2019); metallurgical production and production of finished metal products (25.3%); production of chemicals and chemical products (14.1%) and mechanical engineering (10.3%). To date, the processing industry is dominated by production with a low level of technological processing: metallurgical production (15.6%), production of wood and its products (5.2%), production of coke and petroleum products (6.3%), 6.3 products (4.0%). Moreover, we should pay attention to the long declining trend of the share of mechanical engineering production: from 13.8% in 2010 to 10.3% in 2019.

During this period, the development of industrial production had the following features: the instability of the rate of recovery of growth after the crisis; intersectional uneven growth; annual fluctuations in certain types of production (eg food production, NACE section 10); different development trends within one type of activity (for example, beverage production: a decline in 2012-2019 half in the distillation, rectification and blending of alcoholic beverages, class 11.01 and a small increase of 1.9% in the production of grape wines, class 11.02).

Accordingly, in terms of the main types of industrial activity, the following changes took place:

- in the pharmaceutical market in 2019 there was an increase of 17.5% compared to 2012 in conditions of devaluation of the hryvnia, an increase in the cost of imported raw materials and a decline in the purchasing power of the population. Despite the inability of domestic companies to compete with global manufacturers of original drugs and the production of mostly generic drugs, leading Ukrainian firms are increasing their combined market share. In general, the annual growth of the drug market in physical terms over the past two years was 14%, and growth was supported by the reform of public procurement in medicine, the introduction of reimbursement, external demand, which contributed to obtaining GMP certificates;

- in the market of woodworking products unsustainable growth in 2016-2017, but to reach the level of production in 2012 failed (94.6%). Among the subtypes of activity there is an imbalance: sawmilling and planning production is growing faster - by an average of 4.3% annually, and production with significant processing - the manufacture of wood products, cork and others. - 1.7% per year. This market is one of the most promising, but for domestic production is determined by external demand, which formed a significant export-oriented production (86.4% in 2019). At the same time, the development of this market is characterized by significant import dependence of production in the supply of raw materials (pulp, waste paper) and components;

- in the furniture market (NACE section 31), where after losing almost 20% of sales in 2014-2015, manufacturers reoriented to new markets, and due to the growth of domestic demand exceeded the level of production in 2019 compared to 2012 by 11.2%. This was also facilitated by the progressive volume of Internet sales and cooperation of Ukrainian manufacturers with foreign ones.

Impact of Industry 4.0 technologies on transformations in important industries:

1. Since that fact that Industry 4.0 can provide customized solutions for consumer needs, even industries such as aerospace can experience the impact of advanced digital technology in manufacturing. Although the level of automation in the aerospace industry is already high enough, it will be even higher thanks to smart robotics in assembly. Some of this growth will be driven by more strict quality requirements, for example, for control that the parts have the minimum weight with the necessary strength and safety;

2. The technology of automatic control of transport will affect the reduction of logistics costs in the automotive industry;

3. The Industry 4.0 technologies implementation in the food industry begins with the identification of sources of raw materials and ends in the point of sale to the final consumer. The concept creates an "umbrella" management system that connects every step in the value chain, from food collection to final product consumption. At the same time blockchain technologies and the Internet of Things play an important role for the lowest possible costs and maximum possible output.

4. Industry 4.0 is expressed in the chemical industry mainly in two dimensions: improving the productivity of chemical production by various intelligent manufacturing technologies: predictive asset management, process control, and production modeling, including.

As well as reducing risk, it involves managing supply chains and internal operations to respond to changing customer needs and improve safety and quality.

According to the results of recent years, the industry has seen an increase in those types of industrial activities that were supported by investment, external and domestic consumer demand, as well as provided with agriculture and mining raw materials for processing (Fig. 3).

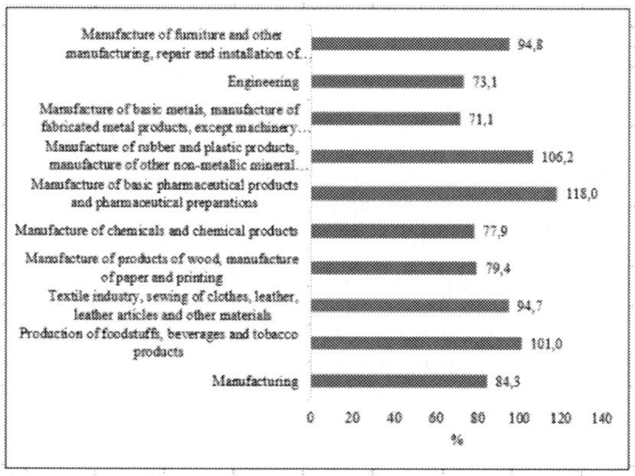

Fig. 3. Growth rates of production of processing industry 2012-2019, 2012=100% [9].

Ukrainian industry has structural problems of foreign economic relations, which have reached a critical level, in particular the predominance of exports of intermediate goods (44.7% in 2019). Share of exports of industrial goods to total exports of goods has reduced from 90.2 % to 71.7 % (at nearly 20 per cent). In addition, there are threatening growth trends in Ukrainian raw material exports - more than 70% of total industrial exports (low value - added products, including 30.7% - metallurgical products; 12% - wood, pulp and paper products, etc.) and the growth of imports of high-tech products (almost a third increased imports of engineering products), which indicates the lag of the technological level of production from world indicators. Today, the industry operates in difficult economic conditions caused by systemic problems of the industry: significant depreciation of fixed capital, lack of financial resources for its renewal, dependence on imported raw materials. Compared to 2010, in 2019 the number of enterprises engaged in innovative activities in industry decreased to 15.2%.

The most innovative manufacturing industries include (2019): production of basic pharmaceutical products and pharmaceuticals (53.8% of the total number of industrial enterprises), production of computers, electronic and optical products (34%), production of motor vehicles , trailers and semi-trailers and other vehicles (30.2%), manufacture of electrical equipment (25.2%), manufacture of chemicals and chemical products (25%), manufacture of machinery and equipment (22.6%), metallurgical production (18.9%). However, in the volume of sold products of the processing industry only 1.2% is innovative. Moreover, the share of costs for innovative activities of machine-building industries to the volume of sold products is insignificant - 1.9% [10].

2.3 The preliminary findings of the study

In the process of spreading the technologies of Industry 4.0, the place of the processing industry in the structure of the economy remains significant with a radical change in its qualitative characteristics. Today, the share of traditional industries in the industrial structure of the EU countries is at least 88%. However, the development of high-tech industries is decisive for both the national economy as a whole and for individual, predominantly industrial regions. It seems that in the context of the fourth industrial revolution, one should focus on innovative economic growth, manifested in an increase in the growth rate of macroeconomic indicators along with an improvement in the distribution of the results of such growth, in the expansion of equality of opportunities for all members of society.

Of particular importance among the technologies of Industry 4.0 is blockchain, which can be considered as an effective tool for solving the problems of industrial enterprises in big data management and analysis using artificial intelligence. This technology has three important properties: decentralization, immutability and integrity. During the COVID-19 pandemic, tracking and monitoring goods and services has become a new priority for many companies in supply chain management. And that's the kind of thing that blockchain technology does. They can be useful for companies in a variety of industries from heavy industry to fashion. These technologies enable to solve the requirement of social and environmental responsibility issues from the public and investors and help to substantiate the selection of suppliers.

Blockchain is an important cross-cutting technology in the digital economy by which developing countries can beneficially integrate into the world economy.

Expectations are especially high for blockchain in combination with other technologies, such as the Internet of Things. Blockchain is a basis for easier interaction between devices, which are autonomously administers. This facilitates various types of transactions between devices, such as registering a new device, authenticating remote users, and contacting to exchange with other devices. Blockchain would lead eventually to reduce transaction costs, supply chains and increased business competitiveness.

Intensive processes of structural transformation of the economy are taking place in the world, which are accompanied by an increase in uneven development and general uncertainty. For a number of countries seeking to cope with acute internal problems, overcome the existing serious imbalances in the structure of the economy or respond to external challenges, it is important to formulate a policy of structural transformation. In these conditions, there is an active search for new solutions for structural policy, understood as industrial policy in a broad sense (industrial policy), that is, government actions aimed at improving the business environment or the structure of economic activity in sectors or technological areas that will provide better prospects for economic growth and the creation of public goods versus no government intervention.

Considering the experience of the developed countries of the world in implementing structural transformations, we can note that most of the leading countries, through their structural policies, seek to create conditions for industrial growth and the competitiveness of the economy. For example, developing countries are trying to overcome the middle-income trap, as well as to carry out a technological "upgrade" of the economy, the Gulf countries use it to diversify their national economies, and a number of African countries see it as a means of overcoming poverty. The EU industrial policy is specifically aimed at accelerating the adaptation of industry to structural changes; creating an environment favorable for the initiative and development of enterprises in the EU, especially small and medium-sized ones; creating an environment conducive to cooperation between enterprises; promoting better use of the industrial potential of innovation, research and technological development policies. Following the best practices of developed countries, Ukraine should form an active industrial policy to ensure the prospects for economic growth and overcome the accumulated structural problems and imbalances in the development of industry caused by the incompleteness and inconsistency of reforms, deepened by world crises, political and military confrontation with the Russian Federation. The negative impact of external factors was partially mitigated after the signing of the Association Agreement with the EU by gradual adaptation of national legislation in accordance with international best practices, improvement of the business climate and deregulation of business activities, development of public-private partnerships.

Implementation of circular economy principles is particularly important for structural transformation of economy.

The introduction of the principles of a circular economy into manufacturing makes it possible to overload the economy, making it more stable and competitive. It would be beneficial for all enterprises, industries and citizens in such areas as more innovative and efficient ways of production and consumption; securing sufficient resources for business and volatile prices; opportunities for localization and social integration; optimization of waste management, promotes waste recycling and reducing the volume of landfills; energy savings as there are fewer production processes require less energy; benefits for the environment in terms of climate and biodiversity, reducing air, soil and water pollution.

The greatest benefit would result from replacing old equipment with modern, which saves energy. In addition, emphasis should be placed on the development of renewable energy sources - in the near future it would be only environmentally friendly, but also cost less than energy from traditional sources. There are also strong multiplier effect might be a circular economy development. The basis of a circular economy is implementation of circular supply chain, that maximize the added value throughout the product life cycle.

The foregoing calls for the formulation of a development and implementation the effective tools for supporting industry to increase its efficiency while reducing environmental load, namely the following directions of state incentives: improvement of legal regulation, implementation of principles sustainable development, financial support for development-oriented projects circular economy, as well as the development of regional waste management plans with focusing on the European Green Deal.

Conclusion

Given the significance of the development of high-tech industries for the competitiveness of the national economy, it is important to ensure the mechanisms of production, that are not contrary to the rules of international law. Among organizational and economic activities, which have directly or indirectly affect through the technological introduction on the markets of industrial goods and services, are the regulator and management, inspection, information support.

Annually, one of them is the establishment of a manufacturing economy as a whole, and its industrial production, as well as its digital development: technological, production and economic basis and a set of management tools. As a result of digitalization of industrial development, it is supervised by new innovations, improved productivity of practice, optimization of the structure of costs, and expanded fenestration technologies [10].

Industry 4.0 provides for the promotion of digital technologies at enterprises, attracting relevant specialists, increasing the availability of special software tools, introducing and developing information systems, databases, computing power, and spreading cloud services technology. As a result, the expected changes in the industry may be characterized by an increase in labor productivity, efficiency of production and sales of metal products, a decrease in transaction costs, an increase in innovation activity and an improvement in the quality of industrial products with a high share of added value.

It is argued that technological transformations of Industry 4.0 provide new opportunities to remote control features that allows to place production facilities in any place where there is Internet connection. Industry 4.0 technologies have been found to have a significant impact on the global transformation of the worldnamely, production organization as a set of networks linked to global value chains. Today, virtually all value-added activities, including development, design, manufacturing, testing, marketing and implementation, can be digitized. In addition, the development of the Internet of Things allows for interaction between machines and people with a high level of precision and efficiency.

Digital technologies are the driving force of structural transformation in industry due to changes in production and logistics models. Considering the above, we can highlight new technological trends for the industrial development of Ukraine, that :

- the growth of consumer requirements for the quality of products in terms of a set of properties, are mostly individual in nature;
- a new level of automation of industrial production systems;
- growing flexibility of strategic decisions in the selection of promising technologies and their resource provision, taking into account the complex of the latest materials;
- development and expansion of the functionality of SMART technologies of modern production based on the latest materials;
- anchoring digital technologies in the product value chain.

Consequently, global trends and analysis help us to propose such solutions for enhancing Industry 4.0 in Ukrainian industry:

- development of network structures; development of mini and SMART factories;
- transition to new business models corresponding to the conditions for the development of Industry 4.0;
- the formation of new development institutions that promote artificial intelligence, additive technologies, the Internet of things, BIG DATA, production robotization; supplementing basic instruments with venture financing instruments;

-development of public, educational and scientific institutions that provide an increase in the qualification level of personnel, the expansion of project innovation. The benefits derived from Industry 4.0 in industrial sector will have a multiplier effect on the whole economy.

References

1. V.M. Heyets, The fenomenon of instability as a challenge to economic development, (NAS of Ukraine, 2020)
2. Yu. I. Pilipenko, I. Yu. Guzenko, Econ. Sp., **129**, 89 (2018)
3. Yu. Kindzersky, Ukr Econ. **11**. 48. (2017).
4. J. P. Martino, Techn. For. and Soc. Ch., **70**, 719. (2003)
5. R. Kurzweil, Kurzweil Library. (2021). URL: https://www.kurzweilai.net/the-law-of-accelerating-returns
6. Handbook on constructing composite indicators: methodology and user guide. OECD. (2008). URL: http://www.oecd.org/sdd/42495745.pdf
7. United Nations conference on trade and development. Structural transformation for inclusive and sustainable growth. Trade and development report. (2016). URL: https://unctad.org/en/PublicationsLibrary/tdr2016_ru.pdf
8. State Statistics Service of Ukraine. (2021). URL: https://prm.ua/derzhstat-oprilyudniv-dani-pro-vvp-ukrayini-pislya-karantinu-ekonomika-prosila-na-11-4
9. Database of State Statistics Service of Ukraine. (2021). URL: http://www.ukrstat.gov.ua/operativ/menu/menu_u/prom.htm
10. L.V. Deineko, N.G. Gakhovich, O.M. Kushnirenko, Industrial policy as a key tool of development strategy (NAS of Ukraine, 2019)

Development and Use of Information and Digital Technologies at Ukrainian Enterprises

Marta Shkvaryliuk[1], *Liliana* Horal[2], *Inesa* Khvostina[2], *Alla* Maksymova[3], and *Vira* Shyiko[2]

[1]NBU, Editorial Department, 01001 Kyiv, Ukraine
[2]IFNTUNG, 76000 Ivano-Frankivsk, Ukraine
[3]Kryvyi Rih Vocational College of State University of Economics and Technology, 50000 Kryvyi Rih, Ukraine

Abstract. The paper considers the problems consider the problems of enterprises digitalization. Based on the research of the scientific literature, it is established that enterprises in the modern world need the active introduction of information and digital technologies to ensure the competitiveness of production and active development in the future. The analysis and assessment of the use and development of communication and information technologies by domestic enterprises is carried out. According to its results, it is established that the main areas of information and communication technologies implementation in domestic enterprises are cloud computing services, sources of "big data" for the analysis of "big data", 3D printing, external links to the Internet, own websites and electronic trade via the Internet, etc. It is determined that due to the rather intensive growth of the number of enterprises in the information and communication industry, the use of their developments in production is rather insignificant. Only 5% of enterprises during the study period used all the above information technologies in their activities. Based on the analysis, the problem areas of the process of implementation and development of communication and information technologies at domestic enterprises are identified and recommendations for improving the efficiency of information and communication technologies are provided. The MatLab Statistic Toolbox built into MatLab is used to determine the trends of the impact of digital innovations and the number of information and communication enterprises on GDP.

1 Introduction

The digital world evolves at a tremendous rate in the last decade. The development of the Internet, mobile communications and online services is a basic tool for shaping the digital economy. These processes affect all sectors of the economy and social activities, manufacturing, health care, education, finance, transport, and so on.

The increase in the efficiency of enterprises necessitates the optimization of the business process management system of the enterprise. This task cannot be solved today without the use of information systems and technologies, because the growth of information needs and the development of information services involves new priorities and requires the use of new management methods.

For Ukraine, it is absolutely necessary to move to an innovative model of enterprise development using the latest production technologies to realize the potential and effectively manage it. Ukrainian enterprises are undergoing significant transformations, in particular, they are taking place in the socio-economic system in recent years and cause significant changes in the definition of development priorities. This is especially felt in the IT sector, which is constantly evolving, using the latest information technologies, introducing new technology for enterprise management. Most companies understand the need for innovation in the strategy, but few manage to implement this area and bring it to a logical conclusion, paying considerable attention to the internal business processes of the enterprise and their optimization. The growth of information volumes, information uncertainty, the complexity of information management of business processes of the enterprise determines the use of the latest information technologies. In addition, many companies in connection with the COVID-19 pandemic have switched to remote work and most likely this model of domestic enterprises will remain in the near future as there is evidence that real resources are saved if the enterprises are skillfully and intelligently managed according to a well-established system and wise use of innovative information and digital technologies. Therefore, there is a problem of analysis and evaluation of the use and development of information and digital technologies in domestic enterprises, development of recommendations for improving the functioning of enterprises in modern conditions and forecasting GDP depending on the introduction of innovations in domestic enterprises and changes in the number of information and communication enterprises.

2 Literature review

The problem of the development and use of information and digital technologies in the enterprises

* Corresponding author: inesa.hvostina@gmail.com

taking place in a society under the influence of digitization was considered by many authors: Berisha-Shaqiri Aferdita, Berisha-Namani Mihane [1], Thanos Papadopoulos, Konstantinos N. Baltas, Maria Elisavet Balta [2], Urbinati Andrea, Chiaroni Davide, Chiesa Vittorio and Frattini Federico [3], Melissa Liborio Zapata, Lamia Berrah, Laurent Tabourot [4], Robert Eller, Philip Alford, Andreas Kallmünzer, Mike Peters, Antecedents [5], Aijaz A. Shaikh, Ravishankar Sharma, Heikki Karjaluoto [6], Douglas Carl Engelbart, Joseph Carl Robnett Licklider, Richard Lipsey, Joseph Stiglitz [6], V. Soloviev, S. Semerikov, V. Solovieva [7], Horal Liliana, Korol Svitlana, Khvostina, Inesa, Shyiko Vira [8], Anders Olofsson, G. Fransson, J.O. Lindberg [9], V. S. Litvinenko [10], Ning Hu, Zhihong Tian, Hui Lu,Xiaojiang Du Mohsen Guizani [11], G. Mamonova, N. Maidaniuk [12], B. He, KJ. Bai [13], etc.

Berisha-Shaqiri, Aferdita & Berisha-Namani, Mihane in the work [1] present the importance of information technology in economy, computerization trends and the opportunities offered by this technology where consumers have quick and easy access on the global market, as well as in information about goods and services [1]. Scholars [2] have highlighted the role of Digital Technologies (DT) in enhancing productivity and performance in Small and Medium Enterprises (SMEs). However, there is limited evidence on the use of DT for dealing with the consequences of extreme events, such as COVID-19. They discuss this gap by outlining potential research avenues and reflecting on the managerial implications of using DT within SMEs to deal with the repercussions of COVID-19 and securing business continuity. Urbinati, Andrea, Chiaroni, Davide, Chiesa, Vittorio & Frattini, Federico [3] by adopting a capability-based perspective to identify the patterns of actions that firms have to deploy to use and implement a set of digital technologies in their innovation process, the paper [3] provides a map of enabled and enabling capabilities ensuing from and required for their use and implementation. From a methodological point of view, we conducted an exploratory multiple case study analysis, which involved a sample of eight companies operating in different industries and characterized by different size, market share and organizational structure [3]. The aim of paper [4] is to evaluates the facility of digital matureness models in their part of assisting producers of smart production in their digital conversion journeys and offer a set of recommendations to develop the usability of the toolbox in this scenario. To get off this, an evaluation of a extraction of seven matureness models is performed, using some design element to the special case of producers of smart production. In accordance with the most suitable findings, the propositions suggested for the formers are related to the ought for a wider area for the toolbox, a prescriptive state, and a wide business promising in the designation of their greatness. Confirmation of the suitability of the propositions in the small and medium business scenario is also capable through an demonstration of the producer of the Arve Valley [4]. Drawing on the resource-based view, Robert Eller, Philip Alford, Andreas Kallmünzer, Mike Peters, Antecedents first search the effect of three key SME powers on digitalization: digital strategy, employee skills, information technology. Second, they assess the effect digitalization has on financial efficiency. We then search if digitalization intermediates the effect of powers on efficiency. The results of a review of 193 SMEs show how digitalization can effect SME efficiency, with the three powers positively relating to digitalization. And severally, digitalization importantly relates to performance, mediating the impact of information technology on realization. Nevertheless does't mediate the impact of digital strategy or employee skills on efficiency [5]. Scholars [6] have noted that digital innovation is transforming the technological landscape, entrepreneurial practices, and the behaviours, attitudes, and beliefs of consumers across the globe. This research note serves three purposes. First, it introduces digital innovation and enterprise as a fruitful area of research in the sharing economy. Second, it extends the agenda set by Yoo, Henfridsson, and Lyytinen (2010) on "new organising logics of Digital Innovation" to propose contemporary research questions for scholarly collaboration. Third, it attempts to move forward research in digital business from B2B, B2C, mobile-social contexts to emerging eco-systems that address current socio-economic trends. In proposing such a digital business research agenda, the authors reason why Action Design Research studies may be particularly suited for the iterative development, replication and sharing of findings in the form of artefacts such as use-cases [6].

3 Analysis and Methodology

Examining the dynamics of the number of economic entities from 2011 to 2019 (table 1, figure 1) we can conclude that during this period the total number of economic entities increases from 1600127 units in 2012 to 1941701 units in 2019, which indicates on the development of business activity in the economy. Calculating the number of enterprises by type of economic activity (table 1), we can determine that the largest share in the number of enterprises falls on those engaged in wholesale and retail trade; repair of motor vehicles and motorcycles and is 53% in 2012 and 43% in 2019. It should also be noted that in 2012 the next largest share of economic entities in the structure of the number of enterprises accounted for industrial enterprises and amounted to 7%, while in 2019 the second largest share of enterprises was occupied by economic entities engaged in information and telecommunication activity - 11%.

Figure 1 shows the dynamics of the number of enterprises by type of economic activity structure, excluding enterprises engaged in wholesale and retail trade; repair of motor vehicles and motorcycles.

Table 1. Number of business entities by type of economic activity in 2012-2019 years, units.

The type of economic activity	2012	2013	2014	2015	2016	2017	2018	2019
Agriculture, forestry and fishing	68497	71058	75660	79284	74620	76593	76328	75450
Industry	114028	121244	131491	135149	127069	123876	125859	130324
Construction	50830	52983	52189	55128	50208	50261	52531	56855
Wholesale and retail trade; repair of motor vehicles and motorcycles	843075	890658	988694	989064	910413	837797	818217	834159
Transportation and storage	87252	92366	111807	119037	109334	95815	90591	96235
Accommodation and food service activities	44085	52077	57553	58436	57696	57578	61761	69689
Information and communication	66568	86377	114355	116136	129704	146909	174622	206147
Financial and insurance activities	10342	11069	11888	12381	11865	10290	9739	10086
Real estate activities	81726	92013	97794	94077	90921	90553	93383	98361
Professional, scientific and technical activities	87431	97052	115123	131035	125133	121546	126100	134800
Administrative and support service activities	38769	40191	44144	47361	45988	46553	49206	53787
Education	7317	8467	10117	10873	11077	11656	13241	15429
Human health and social work activities	16479	18048	21114	21683	21583	22085	24961	30994
Arts, entertainment and recreation	9238	11620	13523	14887	13873	13045	13797	15168
Other service activities	74490	76847	86709	89908	86147	100587	109336	114217

Source: formed by the author on the basis of [15]

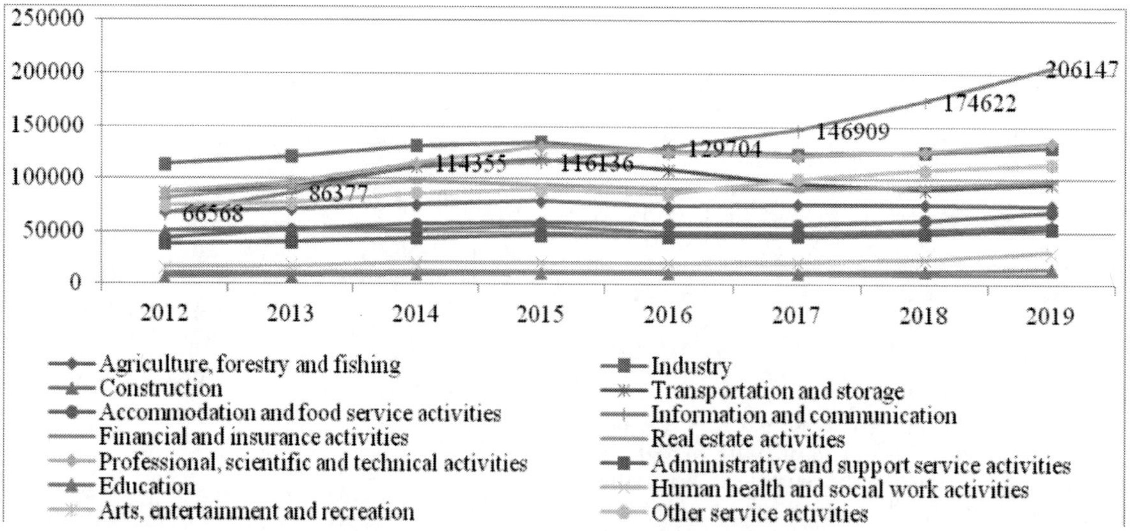

Fig. 1. Dynamics the number of business entities by type of economic activity in 2012-2019 years, units.

Source: built by the author

Thus, the growth of the share of enterprises engaged in information and telecommunications activities from 4% in 2012 to 11% in 2019 indicates the formation of active demand for their services and favorable conditions for their operation (figure 2).

The results of calculating the dynamics of the number of business entities during the analyzed period showed that at the end of the period the number of enterprises engaged in information and telecommunications activities grew most intensively. There was an increase in their number by 210% compared to the base year 2012 (figure 3). The number of enterprises engaged in educational activities increased by 111% in 2019 compared to the base year 2012 and the number of enterprises in the field of health care and social assistance increased by 88%. In general, we can note an increase in the number of enterprises in almost all activities, except wholesale and retail trade, repair of motor vehicles and motorcycles and financial and insurance activities, the number of which during the study period decreased by 1% and 2% respectively in 2019 compared to 2012.

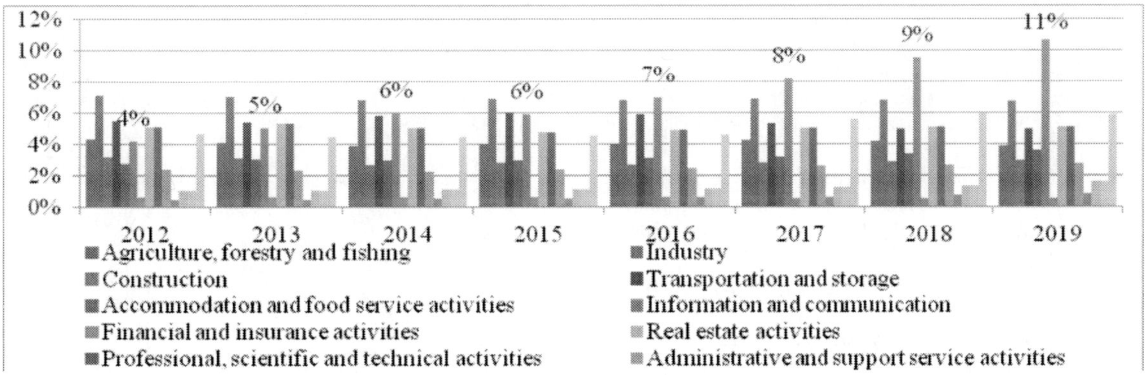

Fig. 2. Share of business entities by type of economic activity in total in 2012-2019 years, %.

Source: built by the author

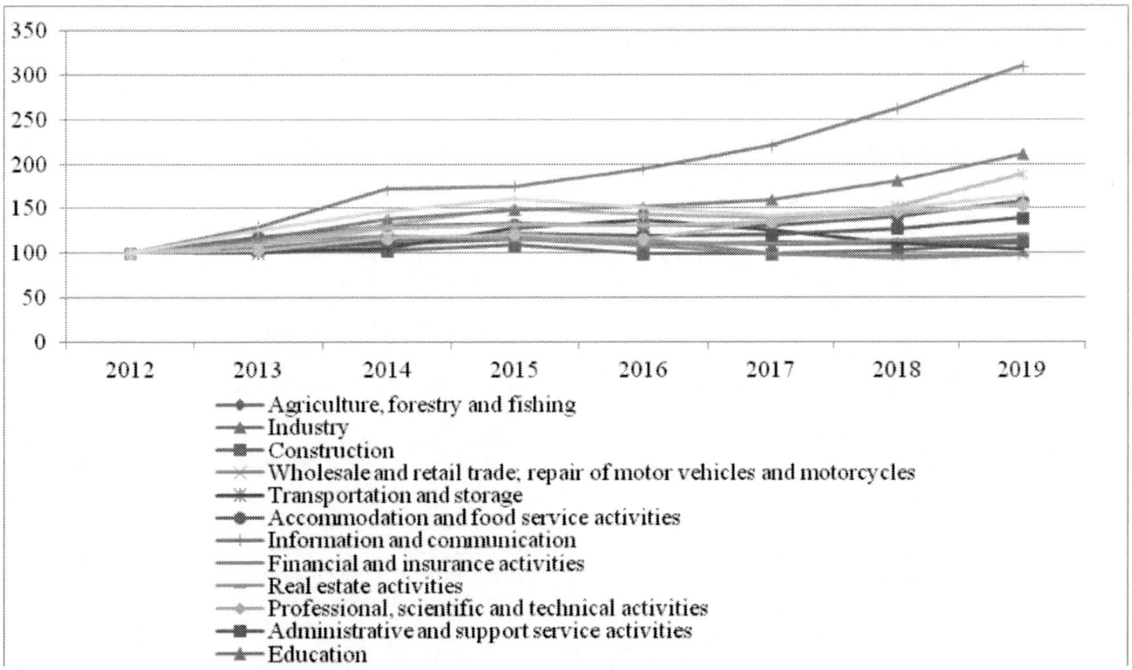

Fig. 3. Dynamics of number business entities by type of economic activity in total in 2012-2019 years, %.

Source: calculated and built by the author

According to the information provided on the website of Ukrstat for the period from 2017 to 2019, statistical observations are conducted on the intensity of the information and communication technologies usage at domestic enterprises. Thus, according to the observations, domestic enterprises in their activities use cloud computing services, sources of "big data", 3-D printing, information and communication technologies to provide invoices, carry out e-commerce via the Internet, actively establish external communication with the Internet and develop their own sites.

The use of these information and digital technologies and other types is a time requirement, which is declared in key documents: Industry 4.0 and domestic regulationsFor industrial enterprises, the transition to a digital economy has been defined as Industry 4.0 - the fourth industrial revolution. The use of the latest information technologies in business, economics, public administration and other fields is an extremely necessary condition for effective development. If this condition is not met at the appropriate level, there will be a slowdown in development and there will be an opportunity to get into the ranking of lagging behind and ineffective. The EU pays great attention to the development of the digital economy, believing that the level of the continent's economy will depend on how efficiently companies use digital technologies. Businesses that are not integrated into the digital space will not belong to the World Market.

Modern information technologies are computer processing of information according to pre-developed algorithms, storage of large amounts of information on various media, analysis and visualization of data and transmission of information at any distance in the shortest possible time. A new era has come - the era of "Big data" and "Data mining". 3D printers using viscous plastic instead of ink are becoming widespread. When working with a digital file, a 3D printer can "build" a three-dimensional object by printing out its "layers" in turn.

That is why in economics and business, information technology is used to process, sort and aggregate data, to organize the interaction of process participants and computers, to meet information needs, for operational communication at all stages and phases of the life cycle of the organization. In general, modern business to ensure a competitive position solves a threefold strategic task: first, it is necessary to establish closer relationships with suppliers and customers; secondly, to increase the level of own operational efficiency; third, to increase the competitiveness of products.

Table 2 shows the distribution of the number of enterprises by types of modern information and communication technologies used. Judging by the data given in Table 2, most domestic enterprises use external communication with the Internet in their activities. At the beginning of the study period, 39,528 companies were Internet users and their number increased to 43,758 units in 2019. The growth rate of enterprises using the Internet is about 10% per year. It should be noted that the largest number of enterprises that have used this technology is in the processing industry, wholesale and retail trade; repair of motor vehicles and motorcycles and their number increases annually during the study period. In their activities, most companies use fixed broadband Internet access.

An important trend is the development and use of companies' own websites. In 2017, 16,240 companies had their own websites and their number grew every year, and in 2019 it was already 17,856 companies, which indicates a 10% increase in the number of companies that have their own websites.

Table 2. The use of different types of information technologies in domestic enterprises

The type of information technologies used by enterprises	2017	2018	2019	Growth rates	
				2018/ 2017	2019/ 2017
Cloud computing services	4135	4831	5207	116,83	125,93
Sources of "big data" for the analysis of "big data"	10252	9188	9280	89,62	90,52
3-D printing		1219	1471	100,00	120,67
E-commerce via the Internet	10764	12059	12609	112,03	117,14
External connection to the Internet	39582	43303	43785	109,40	110,62
Own website	16240	17522	17856	107,89	109,95

Source: formed and calculated by the author on the basis of [15]

The next information and communication technology used by domestic enterprises is e-commerce via the Internet. And as can be seen from the table in 2017, this information and communication technology was used by 10,764 companies, there is an annual increase in the number of companies using this technology, namely in 2018 their number increased by 12.03% and amounted to 12059 units, and in In 2019, the growth rate was 117.14% of the base period, equal to 12,609 users. Over the past two years, there has been an increase in the volume of sold products (goods, services) received from trade through websites or applications from 228035634.7 thousand UAH in 2017 to 292731939.1 thousand UAH. in 2019, which is respectively 3.5% and 4.5% of the total sales of products (goods, services) of enterprises. It should also be noted that a larger number of domestic enterprises purchased goods or services via the Internet (except for orders sent by e-mail). Whereas the number of enterprises that received orders via the Internet for the sale of goods or services (excluding orders received by e-mail) was 2476 units in 2018 and 2440 units in 2019.

The trend of using "big data" sources for "big data" analysis is positive. Although at the end of the study period the number of enterprises that used this technology decreased compared to the base year 2017, but compared to the previous period their number increased from 9188 enterprises 2018 to 9280 enterprises in 2019. If we consider the distribution of the number of enterprises by sources of "big data" usage (Table 3), most companies analyze the "big data" using data obtained from smart devices or sensors. In 2017, the number of such enterprises was 3194 units, during the study period, their number decreased and in 2019 amounted to 2896 units, which is 10% less than the base period. During the study period encreases the number of enterprises that analyzed the "big data" based on geolocation data obtained from portable devices. At the beginning of the study period, their number was 1584 companies, and in 2019 their number increased by 18.31% and amounted to 1874 companies. Another source of "big data" is social media, which were used by 1,659 businesses at the beginning of the period, and at the end of the analyzed period their number decreased slightly and amounted to 1,600 companies in 2018 and 1,658 companies in 2019. To determine which companies are the most active "big data" users, a diagram was constructed with the distribution of the number of companies by industry. (Table 3).

As can be seen from Figure 4, the largest users of "big data" are enterprises in the processing industry (23.02%), wholesale and repair of motor vehicles and motorcycles (25.99%), construction (9.32%), transport, warehousing, postal and courier delivery (9.20). In general, domestic enterprises analyze "big data" on their own (71.5%).

Table 3. Number of enterprises that conducted "big data" analysis, according to "big data" sources

"big data" sources	2017	2018	2019	Growth rates	
				2018/ 2017	2019/ 2018
The data obtained from smart devices or sensors	3194	2917	2896	91,33	90,67
Geolocation data obtained from portable devices	1584	1697	1874	107,13	118,31
The data generated from social media	1659	1600	1658	96,44	99,94
Other sources	3815	2974	2852	77,96	74,76

Source: formed and calculated by the author on the basis of [15]

Cloud computing services were used by 4135 enterprises at the beginning of the analyzed period, it should be noted that in 2019 the number of such enterprises increased by a quarter and amounted to 5207 business entities.

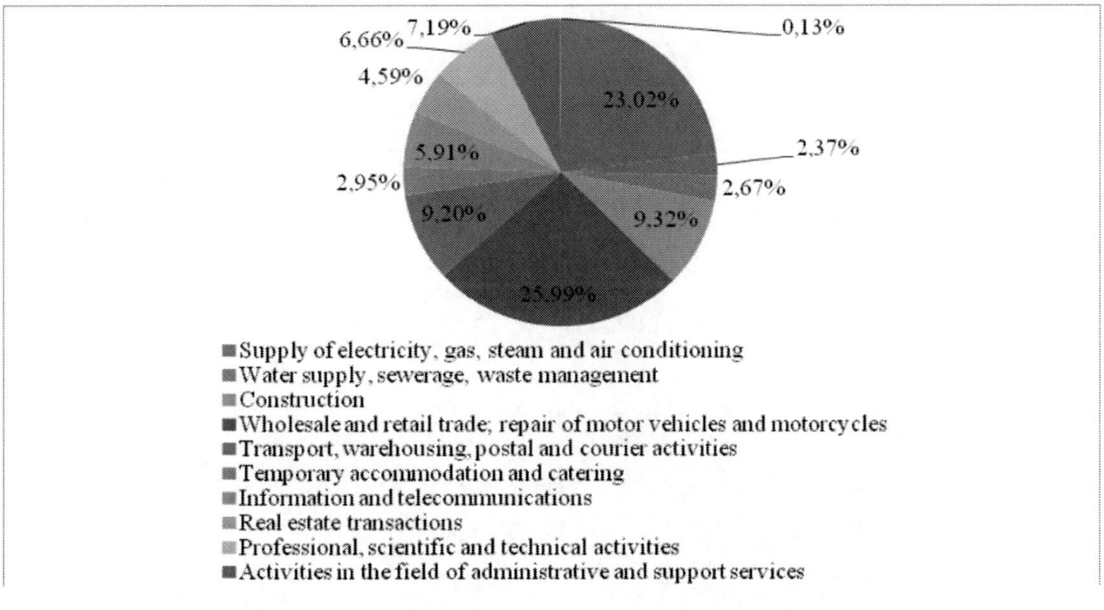

Fig. 4. Distribution of enterprises according to the level of "big data" usage in 2019

Source: built by the author

Examining the statistics of the types of cloud computing that enterprises use in their activities (Table 4), we can conclude that their number is growing every year. According to the information provided in Table 4, increases the number of enterprises purchasing file storage services from 1447 business entities in 2017 to 2139 in 2019 is growing the most, which is 47.82% more than in the base period. At the end of the period, the number of enterprises purchasing a database of enterprises increased by 40.92%, if in 2017 the number of such enterprises was 1449, in 2019 there were already 2042 business entities.

The number of enterprises that use customer relationship management programs in their activities is growing from 1,050 businesses in 2017 to 1,453 in 2019, which is 38.38% more than the base period. The number of enterprises purchasing e-mail services increased by 36% in 2019 and amounted to 2,972 business entities against 2,175 enterprises in 2017.

Table 4. Distribution of the number of enterprises that used cloud computing services in 2017-2019, units

Types of cloud computing services	2017	2018	2019	Growth rate	
				2018/ 2017	2019/ 2017
Email	2175	2548	2972	117,2	136,6
Office software	1932	2125	2445	110,0	126,6
Enterprise database hosting	1449	1664	2042	114,8	140,9
File storage service File storage service	1447	1788	2139	123,6	147,8
Financial or accounting applications	2413	2585	3010	107,1	124,7
Programs for customer relationship managemen t	1050	1223	1453	116,5	138,4
Computer power for the operation of enterprise software	1341	1501	1761	111,9 3	131,3 2

Source: formed and calculated by the author on the basis of [15]

On average, during the study period, 57% of enterprises in their activities purchased financial or accounting applications, 54% - e-mail services, 46% - office software, 38% - file storage services, 36% - file storage hosting, 32% - computer power for the operation of enterprise software and 26% - programs for relationship management.

Examining the statistics of the types of cloud computing that enterprises use in their activities (Table 4), we can conclude that their number is growing every year. According to the information provided in Table 4, increases the number of enterprises purchasing file storage services from 1447 business entities in 2017 to 2139 in 2019 is growing the most, which is 47.82% more than in the base period. At the end of the period, the number of enterprises purchasing a database of enterprises increased by 40.92%, if in 2017 the number of such enterprises was 1449, in 2019 there were already 2042 business entities. The number of enterprises that use customer relationship management programs in their activities is growing from 1,050 businesses in 2017 to 1,453 in 2019, which is 38.38% more than the base period. The number of enterprises purchasing e-mail services increased by 36% in 2019 and amounted to 2,972 business entities against 2,175 enterprises in 2017. Given the fact that domestic enterprises are actively using information and communication technologies, there is a growing need to increase computer capacity for the operation of enterprise software, so the statistics in Table 4 indicate that during the study period, the number of enterprises that increased computer capacity from 1341 enterprises in 2017 to 1761 in 2019, which is 31.32% more than the base period. At the end of the period, the number of enterprises purchasing office software increased by 26.55% from 1,932 enterprises in 2017 to 2,445 in 2019. There is an annual growth of enterprises that use financial or accounting applications in their activities, namely, if in 2017 their number was 2413, then in 2019 their number increased by 24.74% and amounted to 3010 businesses. Figure 5 shows the shares of enterprises in the total number of enterprises (4135 in 2017, 4831 - in 2018, 5207 - in 2019) by types of cloud technologies used in their activities.

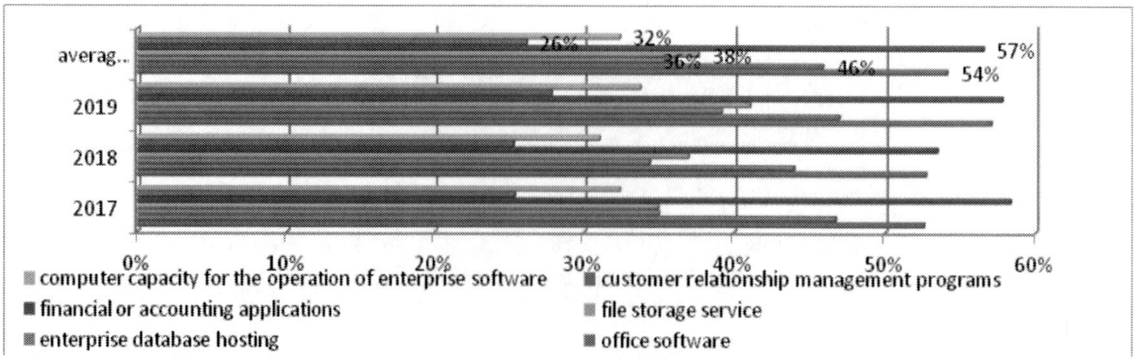

Fig. 5. Shares of enterprises in the total number of enterprises by types of cloud computing used in their activities.

Source: built by the author

The most active users of cloud computing services during the analyzed period were wholesale and retail enterprises; repair of motor vehicles and motorcycles and the manufacturing industry account for an average of 28.10% and 23.16%, respectively (Table 5). 9.88% of construction companies use cloud computing in their activities. From 7% to 6% of enterprises in the field of professional, scientific and technical activities, information and telecommunications, administrative and support services, transport, warehousing, postal and courier activities use cloud computing services. In other industries, less than 5% of enterprises use cloud computing in their activities (Table 5).

An important component of the of domestic enterprises development in the context of active digitalization is the development and use of their own websites to improve their activities. According to the analysis, it can be concluded that at the beginning of the study period, 16,240 companies had their own websites, and in 2019 they became 9.95% more, ie 17,856 (Table 2.). The largest share of the enterprise on average for the studied period, which have developed websites, falls on the processing industry (30.14%) wholesale and retail trade; repair of motor vehicles and motorcycles (26.26%). In all other activities, less than 10% of businesses have their own websites.

Thus, according to the data in Table 6, it can be concluded that the examined enterprises were mainly used for customer service (47% of the surveyed enterprises). 46% of elaborated companies use sites to quickly find them on social media. In 30% of companies, websites advertise vacancies or apply for vacancies online. 29% of the websites of enterprises are formed in such a way that visitors have the opportunity to form orders for goods and services online and in 26% of cases the sites have the ability to monitor the status of placed orders.

18% of the surveyed companies have the opportunity to supply products and services online on websites.

In less than 15% of companies, the developed websites have a personalized content of the website for regular or repeat customers and have the opportunity to train staff. Positive is the growing trend of all opportunities to use websites.

Table 5. Distribution of enterprises from their total number by types of economic activity that use cloud computing in their activities in 2017-2019

Type of economic activity	2017	2018	2019	average
Processing industry	22,47%	23,27%	23,74%	23,16%
Supply of electricity, gas, steam and air conditioning	1,86%	1,51%	1,71%	1,69%
Water supply; sewerage, waste management	2,08%	1,76%	2,07%	1,97%
Construction	9,72%	10,02%	9,89%	9,88%
Wholesale and retail trade; repair of motor vehicles and motorcycles	28,17%	28,48%	27,64%	28,10%
Transport, warehousing, postal and courier activities	6,63%	6,13%	6,28%	6,34%
Temporary accommodation and catering	2,85%	2,71%	2,98%	2,85%
Information and telecommunications	6,89%	6,91%	7,32%	7,04%
Real estate transactions	4,43%	4,33%	3,99%	4,25%
Professional, scientific and technical activities	7,47%	8,07%	8,07%	7,87%
Activities in the field of administrative and support services	7,23%	6,54%	6,16%	6,65%
Repair of computers and communication equipment	0,19%	0,27%	0,15%	0,21%

Source: formed and calculated by the author on the basis of [15]

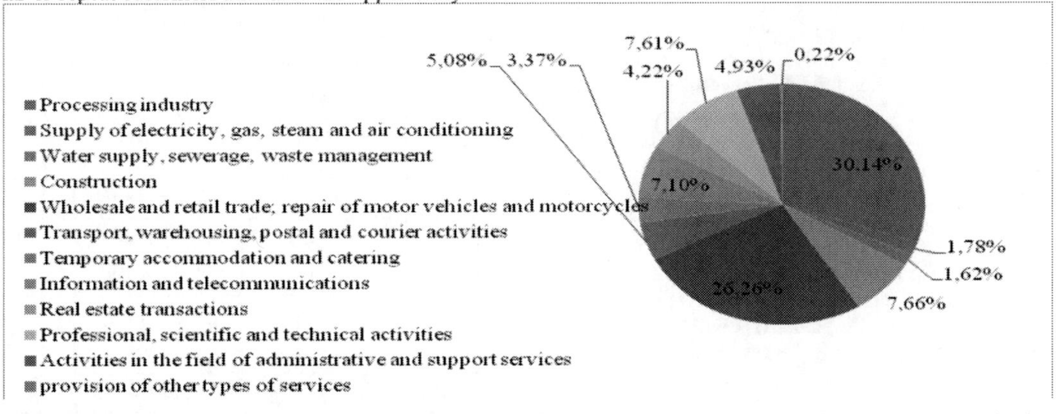

Fig. 6. Distribution of enterprises (on average for the study period) from their total number by type of economic activity, that have their own websites

Source: built by the author

Table 6. Distribution of enterprises according to the capabilities provided by the developed website.

Possibilities of using websites	2017	2018	2019
customer service	46%	47%	47%
delivery of products and services online	17%	18%	18%
the ability of visitors to place orders for goods and services online	27%	29%	29%
monitoring the status of placed orders	25%	26%	26%
personalized content of the website for regular or repeat customers	25%	12%	12%
links to the company's website on social media	42%	46%	46%
vacancy announcements or applications for vacancies online	28%	30%	30%
staff training	10%	11%	11%

Source: formed and calculated by the author on the basis of [15]

Domestic enterprises actively use 3D printing technology in their activities. Official statistics on the number of enterprises since 2018. And as shown in table 2 in 2018, 1219 companies used 3-D printing in their activities, and in 2019 there were 1471. And it should be noted that 52% of the examined companies carried out 3-D printing with their own 3-D printers, and the remaining 48% of enterprises used 3D printing services provided by other enterprises. The uses of 3-D printer products are shown in Figure 6.

According to the results of statistical surveys (Figure 7) in 2019, 48% of enterprises used printed prototypes and models for their own use, 37% of enterprises - printed on 3-D printer products for use in the production process, 28% of enterprises used printed prototypes or models for sale and 19% of businesses used 3-D printers to print goods for sale, except for prototypes or models. About 40% of manufacturing enterprises use 3D printers in their activities, less often they are used by those enterprises that repair computers and communication equipment.

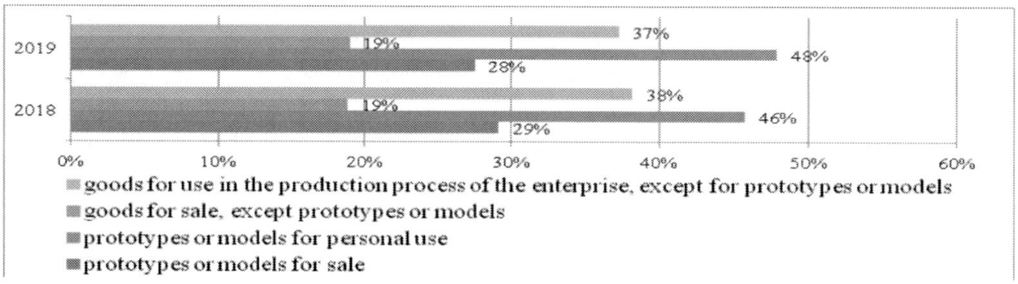

Fig. 7. Areas of products printed by 3-D printers usage

Source: built by the author

Having studied the processes of development and use of information and digital technologies at domestic enterprises, there is a need to determine the dependence of GDP on the number of enterprises that use information and digital technologies in their activities.

We will solve this problem in the MatLab environment. First, we form a matrix of input data and calculate the correlation coefficients using the function "corrcoef" (Table 7).

Next, we use the function regards (responses, DATA, 'model'), which is designed to calculate the parameters of the multiple regression model for the vector of the dependent variable responses values, the matrix of independent variables DATA, regression model 'model' . The function displays a graphical window with a set of statistics that are used to assess the quality of the multiple regression models. To select, it is necessary to note the relevant notes. Selected statistics with specified variable IDs will be calculated and exported to MATALAB (figure 8).

Table 7 Input date to determine the dependence of GDP on the number of enterprises that use information and digital technologies in their activities

Year	The share of the number of enterprises that implement information innovation, % (x2)	Information and communication (x1)	GDP, mln. UAH. (Y)
2012	13,6	66568	1 404 669
2013	13,6	86377	1 465 198
2014	12,1	114355	1 586 915
2015	15,2	116136	1988544
2016	16,6	129704	2385367
2017	14,3	146909	2 983 882
2018	15,6	174622	3 560 596
2019	13,8	206147	3 974 564

Source: formed and calculated by the author

```
>> x1=[13.6 13.6 12.1 15.2 16. 14.3 15.6 13.8]';
>> x2=[66568 86377 114355 116136 129704 146909 174622 206147]';
>> y=[1404669 1456198 1586915 1988544 2385367 2983882 3560596 3974564]';
>> X=[x1 x2];
>> R = corrcoef([y X])

R =

    1.0000    0.3916    0.9656
    0.3916    1.0000    0.2860
    0.9656    0.2860    1.0000
```

Fig. 8. Input date

Source: calculated by the author

matlab1.mat (MAT-file)

Workspace

Name	Value
adjrsquare	0.9258
beta	[-1.6125e+06;9.7403e...
beta_i	*3x8 double*
cookd	[0.6186;1.6484e-04;1....
covb	[1.3076e+12 -9.1710e...
covratio	[0.3194;2.6635;0.4269;...
dfbetas	*3x8 double*
dffit	[2.3472e+05;-3.1049e...
dffits	[1.7977;-0.0199;-2.330...
dwstat	*1x1 struct*
fstat	*1x1 struct*
hatmat	*8x8 double*
leverage	[0.4071;0.2673;0.5410;...
mse	7.2926e+10
r	[3.4186e+05;-8.5102e...
rsquare	0.9470
s2_i	[4.1879e+10;9.1133e+...
standres	[1.6441;-0.0368;-1.636...
studres	[2.1695;-0.0329;-2.146...
tstat	*1x1 struct*
X	*8x2 double*
x1	[13.6000;13.6000;12.1...
x2	[66568;86377;114355;...
y	[1404669;1456198;158...
yhat	[1.0628e+06;1.4647e+...

Fig. 9. Statistics with specified variable IDs

Source: formed and calculated by the author

According to the results of statistical processes, the following regression model was obtained:

$$Y=-1612467,74+20,29*x_1+97403,15*x_2 \quad (1)$$

Based on the characteristics given in Figures 8 and 9, it can be concluded that the formed model is adequate as the statistical characteristics are within acceptable limits. Thus, if the share of the number of enterprises that innovate increases by 1%, GDP will increase by 20.29 million. UAH under other constant conditions. If the number of enterprises in the information and communication industry increases by 1%, GDP will increase by UAH 97,403.15 million. However, examining the correlation between dependent and independent variables, it can be concluded that there is not a close relationship between the GDP function and the share of the number of enterprises that innovate, so it is necessary to conduct a stepwise regression using the stepwise function.

The results of the implementation of this function are shown in Figure 10. To do this, we should implement a step-by-step regression algorithm with the exception of variables (stepwise function), given that the equation should include only one of the closely related variables.

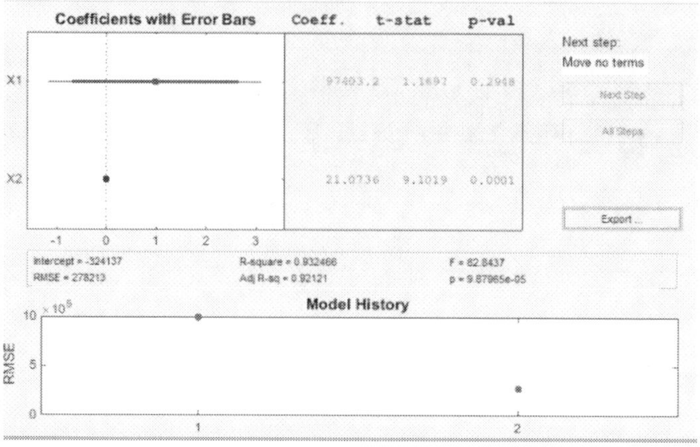

Fig. 10. Coefficients with Error Bar

Source: formed and calculated by the author

According to the results of the statistical processes, the following functional dependence was obtained:

$$Y = -324137 + 21.0736 * x_1 \quad (2)$$

After the implementation of the step-by-step regression process, it can be concluded that the increase in the number of enterprises in the information and communication sphere by 1 will lead to an increase in GDP by UAH 21.0736 million.

4 Conclusions and recommendations

After conducting a study of the development of industries, it was found that during the study period, the total number of businesses is growing, which indicates a revival of economic activity by the end of 2019. The largest growth rate of the business entities number is characterized by the information and communication industry, the growth rate of their number was 11% at the end of the period. Which indicates the intensification of the information and digital technologies development and, their introduction into the activities of economic entities? Therefore, there is need to assess the level of information and communication technologies usage in domestic enterprises. Research has shown that domestic companies use cloud computing services, big data sources to analyze big data, 3D printing, external links to the Internet, their own websites and e-commerce through the network, Internet, etc. The most intensive use is usually external communication with the Internet, but only 20% of Ukrainian enterprises had the maximum speed of a fixed broadband connection (from Mbps and more) to the Internet. This problem indicates the need to increase the speed of a fixed broadband connection (from 100 Mbps and more) to the internet. Regarding websites, only 35% of domestic enterprises had their own websites; the main areas provided by the company's own websites were customer service and the implementation of electronic links to company profiles on social media. This situation indicates rather undeveloped directions of this information communication technology, so it is necessary to expand the functionality of websites developed in enterprises. However, e-commerce is developing quite intensively, however, only 20% of the total number of enterprises purchased goods or services via the Internet and only 5% of domestic enterprises received orders via the Internet to sell goods or services. Therefore, e-commerce should be intensified. The positive point is the use of "big data" sources for the analysis of "big data" in the activities of domestic enterprises. However, in the realities of domestic enterprises, only 5-6% of enterprises analyze the "big data". Regarding cloud computing, only 10% of enterprises purchase cloud computing services, which accordingly indicates the need to develop this area of information and communication technologies.

Studies of the relationship between GDP and the number of enterprises in the information and communication industry have shown that there is a close relationship between the resulting and factorial feature, and the above equation (2) is adequate and can be used for forecasting.

References

1. B-S. Aferdita, B-N. Mihane. Information Technology and the Digital Economy. Med. J. of Soc. Scien. (2015). doi: 6. 10.5901/mjss.2015.v6n6p78
2. Th. Papadopoulos, K. N. Baltas, M. E. Balta The use of digital technologies by small and medium enterprises during COVID-19: Implications for theory and practice. Int. J. of Inf. Man. **55**, (2020) doi:10.1016/j.ijinfomgt.2020.102192.
3. U. Andrea, Ch. Davide, Ch. Vittorio, F. Federico, *The role of digital technologies in the innovation process,* in Proceedings of the 24th Innovation and

Product Development Management Conference, IPDMC, June 2017, Reykjavik, Iceland (2017)

4. M. L. Zapata, L. Berrah, L. Tabourot, Is a digital transformation framework enough for manufacturing smart products? The case of Small and Medium Enterprises, Proc. Manufact. **42**, 70-75, (2020), doi:10.1016/j.promfg.2020.02.024
5. R. Eller, Ph. Alford, A. Kallmünzer, M. Peters, Antecedents, consequences, and challenges of small and medium-sized enterprise digitalization. J. of Bus. Res. **112**, 119-127, (2020), doi:10.1016/j.jbusres.2020.03.004
6. A. A. Shaikh, R. Sharma, H. Karjaluoto, Digital innovation & enterprise in the sharing economy: An action research agenda, Dig. Bus. (2021). doi:10.1016/j.digbus.2021.100002
7. L. Horal, S. Korol, M. Havrylenko, I. Khvostina, V. Shyiko. (2020). *The Management of Business-Processes Strategic Sectors of Economy on Digital Transformation Conditions*, in Proceedings of the III International Scientific Congress Society of Ambient Intelligence ISC-SAI 2020, March 2020, Ukra.ine (2020)
8. V. Soloviev, S. Semerikov, V. Solovieva (2020). Lempel-Ziv *Complexity and Crises of Cryptocurrency Market*, in Proceedings of the III International Scientific Congress Society of Ambient Intelligence ISC-SAI 2020, March 2020, Ukraine (2020)
9. A. D. Olofsson, G. Fransson, J. O. Lindberg, A study of the use of digital technology and its conditions with a view to understanding what вЂ˜adequate digital competenceвЂ™ may mean in a national policy initiative. Educ. Stud. **46,** 727 – **743,** (2020). doi:10.1080/03055698.2019.1651694
10. V.S. Litvinenko, Digital Economy as a Factor in the Technological Development of the Mineral Sector. Nat Resour Res **29,** 1521–1541 (2020). doi: 10.1007/s11053-019-09568-4
11. Hu, N., Tian, Z., Lu, H. *et al.* A multiple-kernel clustering based intrusion detection scheme for 5G and IoT networks. Int. J. Mach. Learn. & Cyber. (2021). doi:10.1007/s13042-020-01253-w
12. G. Mamonova, N. Maidaniuk Mathematical Tools for the Internet of Things Analysis. Cybern Syst Anal **56,** 621–627 (2020). doi:org/10.1007/s10559-020-00279-w
13. B. He, KJ. Bai Digital twin-based sustainable intelligent manufacturing: a review. Adv. Manuf. **9,** 1–21 (2021). doi:10.1007/s40436-020-00302-5
14. M. Anshari, M.A. Arine, N. Nurhidayah. et al. Factors influencing individual in adopting eWallet. J Financ Serv Mark **26,** 10–23 (2021). doi:10.1057/s41264-020-00079-5
15. *http://www.ukrstat.gov.ua*

Transformation of the Higher Education System of Ukraine and Algeria in the Context of The Impact of the Fourth Industrial Revolution

Anna Levchenko[1*], *Dmytro* Plynokos[2], *Ilona* Tsarenko[3], *Abdelmalek* Boussadia[4], and *Seyf-Eddine* Bouaita[5]

[1]Economic Sciences, Economy, Management and Commercial Activity Department, 25006 Kropyvnytskyi, Ukraine
[2]Economic Sciences, Entrepreneurship, Trade and Exchange Activities Department, 69063 Zaporizhzhia, Ukraine
[3]Economic Sciences, Economy, Management and Commercial Activity Department, 25006 Kropyvnytskyi, Ukraine
[4]Economic Sciences, Entrepreneurship, Trade and Exchange Activities Department, 69063 Zaporizhzhia, Ukraine
[5]Economic Sciences, Economy, Management and Commercial Department, 25006 Kropyvnytskyi, Ukraine

Abstract. The paper focuses on the presence of a strong relationship between the level of education and unemployment in the EU and the OECD countries. A graphic confirmation of the change in the relative wages of employees depending on the level of education is given. The authors noted that in recent years there has been an expansion of the mass character of higher education, however, the latter has undergone transformations, in particular, the emergence of universities of a new format - University 21, School 35 and others. The author identifies the key features inherent in the field of higher education under the influence of the 4th Industrial Revolution. Of particular importance are the results on the existing imbalances in the field of higher education in the context of the transition to the new stage in the development of society - the 4th Industrial Revolution and the existing strategic gaps in the economies of the countries of the world depending on the level of development. The most optimal strategy for eliminating the existing imbalances in the field of higher education in the context of the 4th Industrial Revolution for the transition countries is proposed.

1 Actuality

The system of higher education in modern conditions of deepening the innovative orientation of the world's economies and internationalization processes is a unique area of state regulation, which can accelerate the economic development of the country, the pace of its development.

There is evidence of a direct link between education and employment and an inverse relationship between unemployment in the OECD and the EU. This phenomenon is connected with the fact that without education a person cannot be realized in the labour sphere, finding a job, which affects not only its way of life, but also causes economic and social losses to the state. Moreover, based on the analysis of OECD reports, we see that the advantage of higher education is inherent in both individuals who receive it and the state as a whole, which explains the trend of widespread higher education in the world in general and in Ukraine and Algeria in particular.

However, scientists state that support for the expansion of the mass of higher education has been justified by the governments of various countries as a necessity to solve the problem of youth unemployment. However, in the coming decades, the problem of "re-education", is a large number of young unemployed people with higher education, will appear for political power as the most important and most difficult aspect of its activities. And these difficulties are generally due to the fact that "the system of mass higher education, with the exception of a few elite universities, can no longer be considered as one that provides social mobility" [1, p.192-193].

The sharp increase in the number of higher education institutions (different levels of accreditation based on different forms of ownership) with a low level of teaching staff and underdeveloped material, laboratory, classroom facilities has negatively affected the quality of training - this quality is low and the state diploma of higher education now it is not always a criterion of professional competence and a guarantee of success in the labour market. The fact that there are many young people among the entrants, who at the time of admission have not yet decided on the choice of specialty they would like to receive in the process of study, also adds to the problems. The specialists are well acquainted with the current problems of enrolment of applicants, as most of them submit documents to several universities, to different faculties, to different specialties and to the last are not able to make a choice. In this case, of course, the final decision is often influenced by completely random factors. That is, for many it is not important to get a specific specialty, but the very fact of being in the university is significant. Moreover, the introduction of external evaluation for admission to the university has further deepened this problem, because, according to the analysis of data from the introductory info platform, future students do not care what specialty to master, the main thing is to

* Corresponding author: ilonka.tsarenko@gmail.com

enter to the university. and the development of professional skills and abilities.

In addition, in contrast to OECD countries, where the salaries of specialists with higher education are 2-4 times higher than those with basic education (Fig. 1), Ukraine is characterized by a maximum increase in wages by 30%, which indicates the inefficiency of the state regulation of higher education in Ukraine and actualization of changes in the conceptual foundations of such regulation at the present stage. The same situation is be seen in Algeria, in particular 40 % by the last 10 years [2].

The field of higher education is a powerful catalyst for the situation in the country, so its development is influenced by the state of political and economic situation, which in turn affects the efficiency of the entire higher education system, the image not only of higher education but also the level of competitiveness in the labour market, both nationally and internationally, given the openness of the world's economies.

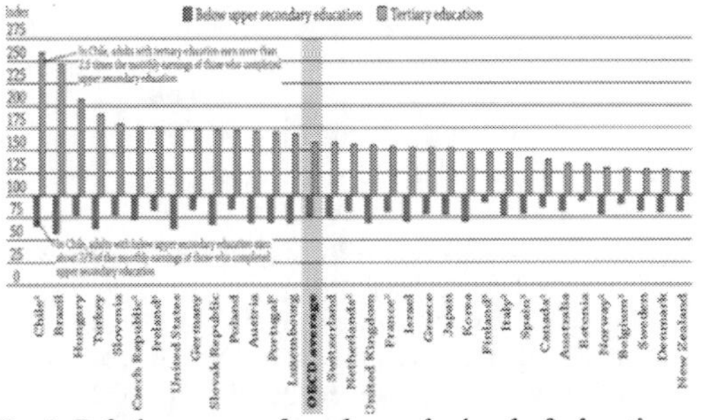

Fig. 1. Relative wages of employees by level of education (2012) 25-64 years old with income from employment; upper middle = 100 [3]

At the present stage of transformations, we consider it expedient to classify Ukraine and Algeria as a transit economy, which is characterized by structural changes aimed at the development of market institutions, economic liberalization, democratization of all processes in the economy, transition from industrial to post-industrial society (a society in whose economy the priority has shifted from the predominant production of goods to the production of services), research, organization of the education system and improving the quality of life, in which the class of technical specialists has become the main professional group and, most importantly, in which the introduction of innovations is increasingly dependent on the achievement of theoretical knowledge.

Post-industrial society presupposes the emergence of the intellectual class, whose representatives at the political level act as consultants, experts or technocrats "[4, pp. 102-118].

The Third industrial revolution, also called the digital revolution, began in the second half of the twentieth century on the creation of digital computers and the further evolution of information technology. Of course, today the most developed countries operate in the era of the Third Industrial Revolution.

According to Bell, "post-industrialism" is characterized by five feat ures: the transition from the production of goods to the production of services; predominance among employees of the "class" of professionals and technicians; the leading role of theoretical knowledge as the basis of innovation in economics, politics and social structure of society; focus in the future on methods of control and assessment of possible directions of technology development; decision-making on the basis of a new "intellectual technology" [4, P. 102-118].

All these features are inherent in the third stage of development of society, the era of informatization, the transformation of which is aimed at the reform processes of all sectors of the economy, including higher education in most countries with economies in transition. After all, transformations and reforms in the field of higher education are aimed at achieving a change of specialties, the ability to meta-qualify, the elimination of boundaries in the educational space, the need and access to lifelong learning, i.e to move from mass education to universal.

Thus, today, Ukraine is one of the countries with a strategic gap of the first level (GAP 1) in the development of higher education. Instead, when all the processes of reforming Ukraine are aimed at bridging the strategic gap of the first level (GAP 1), the developed countries move to a new stage of development - the Industrial Revolution 4.0, which features the mass introduction of cyber physical systems into production.

2 The basis of Industrial Revolution 4.0 and its impact

According to the WEF founder Klaus Schwab, the Industrial Revolution 4.0 is blurring the boundaries between the physical, digital and biological spheres. It is assumed that these cyber physical systems will be integrated into a single network, communicate with each other in real time, self-tune and learn new patterns of behaviour; will be able to build a production process with fewer errors, interact with manufactured goods and, if necessary, adapt to new consumer needs. The first steps in the world towards a new industrial revolution were cloud technologies, the development of Big Data collection and analysis methods, crowdsourcing, biotechnology, unmanned vehicles and medicine based on 3D printing, cryptocurrency Bitcoin and Blockchain technology.

The fourth Industrial Revolution was named industry 4.0. This term was first introduced at the German industrial fair in 2011 in Hanover. During this period, Germany approved two state programs: "Platform for Industry 4.0" and "Industry 4.0", both programs aimed at the strategic development of industry in the country (Germany: Industry 4.0).

The leading direction of these programs was the development of technologies and the Industrial Internet based on global communications. The key leitmotif of Industry 4.0 is total digitalization, the development of flexible production, the creation of "digital twins", the creation of cyber-physical systems (CPS), the unification of virtual and real worlds, augmented reality technologies, artificial intelligence (the New Industrial Revolution).

To date, such programs have already been developed and are actively implemented in many countries of the world, such as the USA, the Great Britain, France, Belgium, Japan, etc. (Horizon 2020; Factories of the Future). For example, in Germany, the Industry 4.0 program is one of 10 future projects with strategic competitiveness improvement, and programs aimed at developing research activities in the industrial sector (Fraunhofer Program) are being implemented. Australia is implementing a program, which is aimed at the developing the cooperation between academic organizations and industrial organizations to conduct the joint research projects with the aim of commercializing the innovative developments (Industry and Innovation Program).

In Japan, the Society 5.0 concept is being implemented, which is a continuation of the Industry 4.0 program, expanding it and adding a socio-cultural component to the industrial development, that is, it considers the solving of social problems through the integration of physical and cyberspace (From Industry 4.0 to Society 5.0).

According to the Club of Rome Report presented in 2017, despite the economic growth of the last 30 years in developed countries, society faces the following challenges: inequality, unemployment and climate change.

The gap between rich and poor is widening, millions of people are out of work, and real wages in many countries continue to fall, which can be closed by accelerating economic growth.

However, as Graham Maxton and Jorgen Randers point out, to continue the current path of development of the world economy is not only wrong, but also extremely unwise, because it will only worsen the problems of the world. According to them, the way out of the situation is to rethink the definition of paid work, fair taxation of business, restrict trade if necessary, introduce an unconditional basic income for the poorest third of the population and increase the duration of paid leave.

The impact of the Industrial Revolution 4.0 can be traced in the field of higher education with the emergence of a new format of universities - University 21, School 35 and others, which provide new approaches to the learning process, project-oriented methods, lack of teaching staff, universities in its usual form. All these changes to transform existing higher education systems into a new format are reflected in the strategic gap of the second level (GAP 2) that developed countries have at the present stage.

Moreover, the transformational transformations into New Format Universities within the Industrial Revolution 4.0 are based on the fact that such institutions should contribute to the development of society and the market, while based on the reduction of dependence on public funding. That is, they are gaining more and more social significance and financial autonomy.

Higher education within the Industrial Revolution 4.0 is based on the following components:

1. Change of organizational structure: change of business models; new format of faculties and departments; expansion of interdisciplinarity;
2. Change of accreditation procedure: new conditions of accreditation; "Floating" educational programs; training within the framework of accelerated innovative development;
3. Digital culture: the development of the diversity of the digital life cycle; new psychological distances (education without geographical borders, spaces); new forms of social communication; globalization of education; "Web education"; change of all types of information; problem-solving training; teacher - a guide to finding information.
4. Changing teaching methods: new approaches in teaching; new educational infrastructures; digital rights management.
5. Changes in learning: learning through distance platforms; new training infrastructures; personalized learning.
6. Individualism of learning through the model "Do it yourself" (DIY education).

Thus, while the reform process eliminates GAP 1 to achieve the level of development of developed countries, the latter, trying to move to the next stage, eliminate GAP 2, hence, as a result, Ukraine, like and Algeria after a certain period (T1) will in any case have strategic gap with developed countries in general and in higher education in particular.

Graphically we have the following in the form of fig. 2.

Therefore, in order to eliminate the disparity in the higher education system that may arise over a period of time (T1), it is necessary to develop a mechanism to ensure the competitiveness of higher education, which would not only eliminate GAP 1, but also to address the strategic gap of level 2, as a result, it will not only enable the implementation of effective reforms, but also help reduce the timeframe for achieving them, because in the transition to the Industrial Revolution 4.0, the speed of response to change, turbulence of the existing environment and accelerated innovation will be key.

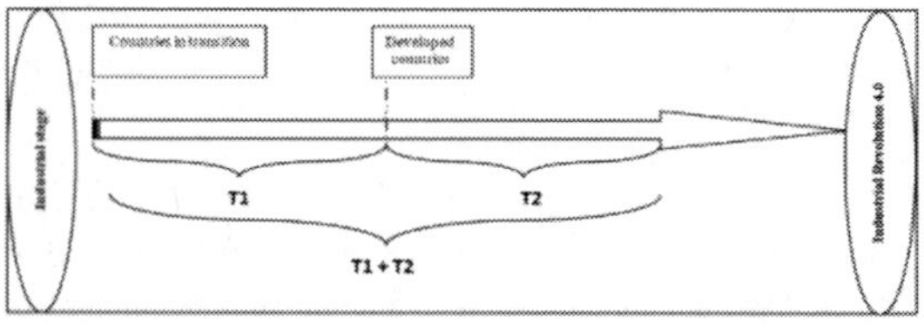

Fig. 2. Time chains of elimination of disparities in the field of higher education under the prism of the transition to the new stage of development of society – the 4th Industrial Revolution

Therefore, the higher education system of Ukraine and Algeria, having strategic reserves, must at the present stage develop a new strategy aimed at trends and trends in higher education under the influence of the Industrial Revolution 4.0, rather than the transformation of the post-industrial stage of development.

Moreover, the University of a new format, modifies such components as teaching, research, in particular:
1. (Teaching 4.0): textbooks, instruction and training provided by teachers; acquaintance with mass open Internet courses (MOOCs); cultivation of innovative methods; blended learning.
2. (Study 4.0): open innovation; evolutionary and revolutionary innovations; new research and development developments related to technical progress; reduction of innovation cycles.
3 (Services 4.0): university as a platform (UaAP); education as a service (EaaS); internationally related programs.

Previously, we have already analyzed the state of Ukraine's higher education system in the international aspect, where we studied the most popular assessment systems, among which there are ratings. Despite the fact that rating is one of the most popular methods of assessment, we consider it necessary to note the following:
1) the existing international rating systems provide for a balanced assessment of the results of various activities of universities with the dominance of the weight of achievements in research and educational activities. At the same time, improving, international methods are expanding in the direction of such components as: teaching and learning, research activity, internationalization, the availability of regional and socially-oriented programs, ie, as a result we note that methodological approaches are variable given the time requirements. their advantage over other approaches to evaluation;
2) the advantage is the focus of ratings on the assessment of consumers of educational services and employers, despite the likelihood of subjectivity;
3) among the shortcomings should be noted - a high probability of inaccuracy of data: on the websites of universities and organizations often lack information on higher education and science;
4) the lack of ratings is also a fairly high degree of subjectivity in determining the weight of the components of the assessment within the integrated assessment;
5) the bases of comparison in the rationing of the components of the integrated assessment are quite contradictory.

Thus, in determining the level of competitiveness of the assessment based solely on ranking, it is advisable to conduct a comprehensive assessment is not enough informative indicator.

3 SWOT-analysis

Therefore, we consider it necessary, with the help of SWOT-analysis, to analyze the main trends, prerequisites and consequences of higher education in Ukraine, to position the strengths and weaknesses of the national higher education system, as well as its threats and opportunities.

SWOT-analysis is one of the most popular methods of strategic management, formed as an abbreviation of four English words: Strengths, Weaknesses, Opportunities and Threats.

This method can be used to study the external and internal environment of any object of the economic system, including the higher education sector.

As a result, the following main points should be highlighted:
- Strengths: quite strong intellectual potential of both research and teaching staff and students, a wide network of universities in the country, which increases the availability of higher education among its consumers, facilitating the possibility of continuing education in European higher education institutions in connection with Ukraine's accession the 19th May in 2005 at the conference in Bergen to the "Bologna Process", the agreements of Ukraine on legal aid, abolishing the requirement of legalization of official documents on education, the possibility of studying for double degree programs with the European universities.
- Weaknesses: low level of funding for education, which reduces the motivation of research and teaching staff, as well as weak interaction with employers, which hinders the effective transfer of knowledge, problems with employment of graduates.
- Opportunities: participation in international grant programs and academic mobility, as strengthening activities within the framework of international cooperation and involving the academic community in updating the processes taking place in the national economy, providing educational services to foreigners (increasing the number of foreign students);
- Threats: demographic crisis, macroeconomic processes in the economy, which have a negative effect on the scientific and technological potential and living standards of the population - the ability to obtain higher education, increasing competition from the European universities.

It was stated that today the greatest threat to the further development of the domestic competitiveness of higher education in Ukraine, given the innovative transformations and strengthening of internationalization, is the growth of competition from European universities, a decrease in the potential contingent in connection with the external independent testing, the inconsistency of the material base with the transformational changes in the sphere of higher education, incompatibility of curricula, complexity their adaptation to the European and lower wages of the teaching staff poses a threat to the personnel security of national universities.

4 The current strategies of development the higher education system in the context of the Industrial Revolution 4.0

Thus, in the system of higher education in Ukraine, the strengths of the internal environment are dominated by the weakness, and the opportunities of the external environment are threats.

In this case, the recommended strategy is "Maxi-Maxi" - to use the strengths of the higher education system to realize the external opportunities. It consists in taking active steps to strengthen its position in the educational services market, including external, that is, increasing its share, diversifying products - opening the new specialties, providing paid services, creating an environment for science parks based on universities, commercializing scientific developments, support the generation of new ideas and the introduction of English-language teaching to increase the degree of internationalization of educational services. All this can be done with the help of an effective mechanism of state regulation of ensuring the competitiveness of higher education.

The current trends in economic processes, which are becoming increasingly turbulent, require, more and more often, the use of a wide range of mathematical techniques and tools to justify management decisions to determine the direction of further development of a particular activity. An important role is played by the methods of strategic analysis and forecasting, with the help of which, with a higher probability, it is possible to forecast indicators for the future.

To this end, we decided to perform a correlation-regression analysis of the final position of the U21 rating and its components to identify the most significant indicators.

To predict the factors influencing the final position of the country within the international rating, four main components of the rating were taken: the component "resources"; the "environment" component; the "communication" component; component "issue".

5 The study of empirical data

As a result of the study of empirical data, we have the opportunity to build a functional relationship between them and, as a result, to make a forecast for the future. The general position of the rating, which depends on a number of factors, can be expressed by nonlinear multiple regression. Estimates of the coefficients of which are calculated using the least squares method in matrix form.

As a result, we have built a mathematical model that looks like this:

$$y = 0.2387 + 0.2146x1 + 0.2115\,x2, + 0.2152x3 + 0.4315\,x4,$$ where

y is the overall position in the U21 rating;
x1 - component "resources";
x2 - component "environment";
x3 - component "connection";
x4 - component "issue".

Forming the structure of the regression model, on the one hand, should include all factors that have a statistical impact on the indicator, and on the other hand, take into account the need for linear independence between factors (in the presence of approximate linear dependence between at least two factors). between them multicollenarity).

The using of the Farrar-Glober method made it possible to establish the presence of multicollenarity in the model. Verification using the $\chi 2$ test showed the presence of general multicollinearity, with a reliability of $p = 0.96$.

The analysis of the correlation matrix allowed us to draw the following conclusion: there is a close relationship between the factors x1 and x3, as well as x1 and x4. In addition, it should be noted that the factor x4 has the highest level of influence on the value of the indicator y (ryx1 = 0.92, ryx3 = 0.92, ryx4 = 0.96)).

As a result, in order to eliminate multicollenarity, we consider it necessary to remove the factors x1 and x3 from the regression model.

Due to the test, the value of $\chi 2$ was obtained, within which we observe that the multicollinearity persisted, but its value decreased slightly. Next, we perform a t-test to determine the significance of the regression coefficients, which showed that all the parameters of the regression model are positive, ie significant (none of the factors can be removed from the regression model).

According to the results of verification by Fisher's criterion, a model was obtained that is adequate to the statistical data (F = 80,685, Fcrit = 1,045).

Based on the calculations of the mathematical model, we conclude that the most significant components that should be considered in the development of strategic priorities are:

- the share of female students in higher education;
- the share of women in the teaching staff of universities;
- data quality assessment. (For each quantitative series, the value is 2 if the data is available to accurately define the variable; 1 if there is some data related to the variable, but some informed adjustment is required; and 0 otherwise);
- the percentage of students studying in private institutions and the percentage of students who obtain the educational qualification level "Master";
- the political environment of the country and the level of regulatory support;
- the level of financial autonomy of state universities;
- the degree of satisfaction of the higher education system with the needs of a competitive economy;
- number of published articles of universities;
- the number of published articles of universities per capita;
- the number of universities in the country included in the Shanghai ranking;
- the degree of enrollment in universities as a percentage of the number of persons who received secondary education during the last five years;
- percentage of the population aged 25-64 with higher education;
- number of researchers (full-time equivalent) per 1 million population;
- the unemployment rate among people with higher education aged 25-64 compared to the unemployment rate for people with incomplete or secondary education without higher education.

The above complements the preliminary analysis and allows to determine the main strategic priorities for the development of a strategy to increase the competitiveness of higher education (Fig. 3)

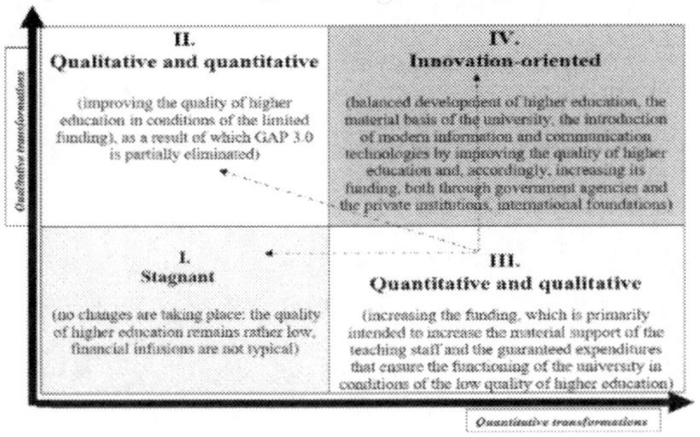

Fig. 3. Matrix of strategic priorities

Selection of strategic priorities of state regulation of competitiveness of higher education, which, in contrast to the existing ones, are based on possible scenarios of regulatory influence that determine the development strategy:

1) stagnant (no changes are taking place: the quality of higher education remains quite low, financial infusions are not typical);

2) qualitative and quantitative (improving the quality of higher education in conditions of limited funding), as a result of which GAP 3.0 is partially eliminated;

3) quantitative and qualitative (increase in funding, which, primarily, is designed to increase the material support of faculty and guaranteed expenditures that ensure the functioning of universities in conditions of low quality higher education);

4) innovation-oriented (balanced development of higher education, the material base of universities, the introduction of modern information and communication technologies by improving the quality of higher education and, accordingly, increasing its funding, both through government agencies and private institutions, international funds);

For example, Ukraine is characterized by the second strategic priority, according to which the amount of financial resources allocated to universities by both the state and private sources is relatively low, which significantly prevents quality improvement at the appropriate level, and funding is focused solely on providing wages, the amount of which is not motivating.

At the same time, the best option for Ukraine in the current period is a gradual improvement of quality through the introduction of public-private partnership, development of flexible regulatory support for financial injections into higher education and the ability to fulfill orders as paid services, which will further improve quality. higher education without significant financial infusions from the state and the basement for the fourth strategic priority - innovation-oriented, which provides for the balanced development of higher education by improving the quality of higher education and increasing its funding, both through government and private institutions, international funds , positioning of universities as full-fledged and integral participants of cluster development of territories, innovative shifts of economy.

However, when choosing this priority, significant sectoral and regional asymmetries and imbalances should be taken into account. After all, based on the preliminary analysis, there is an uneven distribution of universities in Ukraine, in particular the separation of flagship universities in the world rankings and regional universities that are not popular among applicants, so we consider it appropriate to distinguish marketing strategies of universities in such possible areas.

The penetration strategy is a strategy for most higher education institutions seeking to increase the number of students within the relevant (regional or national) market. Expanding market penetration is the most obvious strategy, its usual practical expression is the desire to increase the number of students. We believe that this strategy is typical, first of all, for regional universities.

The main disadvantage of this strategy of individual universities for higher education in general is that it is achieved mainly through the use of marketing tools during campaigning or by reducing the price of educational services, thus imitating this strategy in the development.

Market development strategy means the adaptation and introduction of existing educational services on the market of educational services for foreign applicants, ie the implementation of their export. For the successful implementation of the strategy it is necessary to confirm the presence of a contingent of foreign applicants seeking to receive such services. The main risk of such a strategy may be the lack of potential customers and the mismatch of services - needs. Regarding the aspect of the potential market of foreign applicants, then the universities of Ukraine, in particular technical and medical education, are considered as attractive for foreign citizens; the increase in the student body is mainly due to the independent influx of foreign students (reasonable price and availability of nostrification agreements).

However, it should be noted that in recent years, due to the unstable situation in the country (wax conflicts) and the refusal to nostrify the diplomas of Ukrainian universities, the strategy of "passive waiting" is less effective.

The implementation tool is the introduction of English-language teaching, which increases the influx of foreign applicants.

Product development strategy - the offer on the existing market of new specialties or specializations, the possibility of double diplomas and the possibility of foreign internships. This strategy is typical for the country's leading universities.

The diversification strategy is the most costly and risky strategy used to exhaust the opportunities for growth in existing markets, change market conditions, which is typical for Ukraine (departure of Ukrainian applicants abroad, in connection with the choice of more prestigious foreign universities), profitable opportunities and the high potential benefits of capturing a new market. Such a strategy within the framework of the development of the state mechanism can be applied in part, only to universities with a special status - universities that have entered the international rankings.

Thus, we believe that the isolation of separate marketing strategies, along with scenarios of strategic priorities will allow the most accurate direction of the existing opportunities of the university to increase their competitiveness, in particular, and higher education in general.

6 Conclusion

Thus, the Industrial Revolution 4.0 does not leave aside either the economy as a whole or the sphere of higher education, as the latter provokes the corresponding transformations, which are prerequisites for the further development of any sphere in the modern conditions. First of all, the appropriate transformations should be carried out in the field of higher education, given the key role of the tertiary sector in building an effective innovation economy and prospects for society as a whole.

References

1. Ph. Brown, R. Scase. *Higher Education and Employment in Post-Industrial Society*. The Student's Companion to Sociology. – Oxford: Blackwell Publ. (1997)
2. *Algeria Wages*. URL: https://tradingeconomics.com/algeria/wages
3. *Education at a Glance 2016*: OECD Indicators. URL: http://download.ei-ie.org/Docs/WebDepot/EaG2016_EN.pdf
4. D. Bell. *Notes on the Post-Industrial Society*. The Public Interest. **7**, (1967)
5. D. Bell. *The Coming of PostIndustrial Society: A Venture in Social Forecasting*. New York: Basic Book. Factories of the Future. URL: https://www.effra.eu/sites/default/files/factories_of_the_future_2020_roadmap.pdf (1973)
6. *From Industry 4.0 to Society 5.0: the big societal transformation plan of Japan*. URL: https://www.i-scoop.eu/industry-4-0- society-5-0/.
7. *Germany: Industrie 4.0*. URL: https://ec.europa.eu/growth/tools-databases/dem/monitor/sites/default/files/DTM_Industrie%204.0.pdf.
8. *Horizon 2020*. URL: https:// ec.europa.eu/programmes/horizon2020/en/.
9. J. Raven. *Competence in Modern Society*: Its Identification, Development and Rebase. Toronto: University of Toronto (1984)
10. K. Schwab. *The fourth industrial revolution. What it means and how to respond*. URL: https://www.foreignaffairs.com/articles/2015-12-12/fourthindustrial-revolution

Personal Security and Cyberspace in Time Of COVID-19 Modes of Working

Ewa Matuska[1*], and *Wojciech* Piestrzynski[1]

[1]Pomeranian Academy, Chair of Management, 76-200 Slupsk, Poland

Abstract. The paper presents the analysis of the quantitative research examining the level of the sense of security and the cognitive satisfaction connected with the intense use of digital technology as the result of the switch into remote or hybrid modes of working, especially in a group of high skilled workers. The study was conducted during second wave of the Covid-19 pandemics on the sample (N=246) of Polish employees representing Millenials' generation. The online questionary was disseminated among employees until 30 age working in different sizes of enterprises (micro, SMEs and big ones), located in northern regions of Poland. The results have showed the moderately negative impact of the intensive cyberspace presence in connection with new modes of working for the personal sense of security of the respondents. The sense of security of respondents didn't correlate with cognitive satisfaction flying from digital mode work. The found results can be used as recommendations for introducing more holistic well-being hr practices in organizations and also for implementing new kinds of nonfinancial benefits in talent management programs.

1 Introduction

"(...) Technology destroys jobs, but not the job itself"[1].

The digital economy, a concept introduced in the mid-1990s [2] is rapidly developing in 21st century together with so-called 4.0 economy [3] and both imply fundamental changes in the world of organizations and workplaces. New technologies, such as artificial intelligence, automation, virtual and augmented reality, big data processing etc., are connected with the reduction of traditional job places and have the basic impact on the way, shape and forms of work as well as its valuation [4]. Scenarios of this impact, designed in different parts of the world, such as Asia, America, or Europe differ significantly, but all of them assume inevitable changes for both organizations and employees [5]. The expected impact of the digital technology is connected directly with labour costs, and indirectly with job places and earnings of employees. According to research already by 2030 is expected that about a third of man-hours can be replaced by intelligent machines [5]. However, the digital transformation of enterprises occurred to be problematic component of the overall digital transformation of the economy. Research conducted in 2018 by McKinsey on a sample of 1,793 representatives of companies from all over world revealed that "a successful digital transformation is more difficult than any a known type of organizational change" [6]. The feeling of subjective personal threat in the upcoming era of artificial intelligence is accompanied by numerous doubts perceived at the organizational, micro and macroeconomic level. Personal security is usually associated with a sense of stability, durability, certainty and lack of threats, both in the physical and mental sphere[7]. Obviously, the state of personal security perceived by employees influences not only their well- being but also readiness to meet new technological challenges, to effectively cope with new organizational stressors and to take proactive attitude towards efforts of own upskilling or reskilling.

Questions such as: how will the development of modern digital technologies affect the global economy, national or regional labour markets? - bother entrepreneurs and political decision-makers. For the individual employees more vital seem to be considerations how the automation of work changes the work organization in enterprises and the job demands. In that conditions fundamental questions arise: whether the technical development, which requires systematic investment in training new professional competences of employees operating automated production systems, eliminate the economic rent from automation? The future sociological and psychological consequences of the automation and robotization of work, analysed both on a macroeconomic, microeconomic and individual scale are equally important. The common question arises: will there be enough jobs in the future for everyone interested? Probably not, and it means tensions in the labour market, especially among young people entering employment and not posing valued digital competences. These concerns express the state of a sense of personal and organizational (in)security caused by the development of cybertechnology. The sense of security in literature is assumed as a state of internal balance resulting from having a sufficient, from the point of view of individual, level of own goal-oriented activity [7]. Recently, the anthropocentric approach to the issues of safety and security systematically displaces the old approaches focused on the security of the state, regions, etc. The main focus of anthropocentric concept of security situates the personal security, team security or organizational security within the area of security culture [8]. The contemporary analysis of the threats that affect human safety requires the identification of diverse cause-and-effect relationships in the complex postmodern reality, both material and non-material one. Authors of the article used this concept to plan own research. Deloitte's experts in the report „The voice of the European workforce" from the mid of 2020 [9] warn that although digital technology is a key, it should not be considered more important than the needs and expectations of the same employees, especially in hard time of ongoing global crisis suddenly started in 2020 together with Sars-Cov2 pandemics. This unexpected turn in economy and social life was created first of all via biological threat (fear about the health and the life) but that threat only strengthened the threat connected with the expectation of job lost [10]. The issue of realistic economic threats, data of bankruptcy, minimized economic activity of the companies representing variable sectors, or growing up rates of unemployment, are the core part of the contemporary economic studies. The conclusions are that pandemic is going to disrupt labor market at least as hard as economic crisis [11]. At the moment the less attention is paid on the problem of the occupational stress caused by the risk of COVID-19 and its possible, short-term and long-term repercussions. However, the first observations and research data are collected also in that field, especially in the context of the new modes of working (flexible, remote, or hybrid work), threats of massive decruitment, challenges of using new digital tools, managing dispersed teams, etc.

From other side, pandemic hard time sped up the implementing of modern technologies in companies together with a remote (or hybrid) work mode commonly implemented by demand for social distancing [10]. All of it verified digital competences of employees, especially in a group of "knowledge workers" and caused their intensive professional development. The recent surveys done on the sample European workers show that people and their skills will play a critical role in overcoming the difficulties of adapting the workplace to the new, post-pandemic reality [10]. According to many sources the main forms of remote work are likely to persist long after COVID-19 is finished [12]. That will require from the side of employers' new costs investments, both in hard digital infrastructure as well as in human capital development, in deepening and broadening IT skills of employees in different job places. However, such structural switch also generates different risks accentuating inequalities and creating new psychological and emotional stresses among employees, including from social isolation. The "Future of work model" designed for post pandemic reality [13] recommends to include the strategy based on both people and technologies. The main assumptions underlying this approach is wide concept of employees' well-being where economics and sense of personal and social security are interconnected [14]. The pandemic crisis threats employees economic and social safety in a massive way because creates the grand danger for their health and life perspectives. While some workers realize only minor changes when switching work to their home office, for many it means they cannot do their jobs like they used to do before. In fact, blurred borders between work and life during remote, or hybrid work model bring a lot of pitfalls for employee's well-being. Being at home office they loosing time and space structure which in a traditional mode of work allows clearly divide between time for work and for private life. Home stops to be asylum and a place for relax. It is now a space of imperfect work conditions. Job place at home office usually doesn't meet ergonomic requirements, even the space is usually not enough wide as has to be shared with other family members, also working remotely. These limitations, compounded with deprivation from wide social contacts and favourite activities, easily create new kind of physical and psychological threats for employees' subjective well-being. They activate depression symptoms development, burn-out syndrome or at least feelings of loneliness and sadness [15]. In a such way, the physical and psychosocial organizational security context disrupted by current coronavirus crisis determines not only economic survival, but also basic conditions for development of individuals and organizations [14].

Employees representing particular work generations obviously differ in their reception of the current crisis situation, as younger and elder people differ in their health and life risks in confrontation to get the coronavirus disease [16]. Already it is well proved that the risk for death and severe illness with COVID-19 is best predicted by age. The likelihood of death increases with age (sharply after age 50) among those who contract the virus in all countries where this has been examined [17]. That means COVID-19 threats are evident for elder people, including elder employees. Consequently, from that point of the view, the pandemic situation could be interpreted as less stressful for the young cohort of employees (Millenials) as they belong to the less risky cohorts. Moreover, they are better prepared to fill remote or hybrid work because usually have higher IT skills than older workers. But the interpretation can be done also in an opposite way - the new modes of work imposed by pandemic situation are the source of the high psychological stress especially for the young employees as they are very involved in real interpersonal relationships (thus, more suffer with social isolation) and till now hadn't experience with the economic crisis in their professional life.

2 Cyberspace acceleration use in pandemic times

One of the contemporary definitions of cyberspace recognizes it as "the space for processing and exchanging information created by ICT systems (...)

together with the connections between them and relations with users."[18]. The activities of individuals and organizations in cyberspace consist reading, exchanging, modifying, and creating information. These operations are performed in the digital domain and require appropriate digital competences of the users. Also important are the mutual relations between systems and users, the two-sided connection of activities in cyberspace with activities in the "physical" reality (real world) and their mutual consequences (both positive, as well negative) [19]. The intensive use of cyberspace resources is a new dimension for human activities, enabling gathering large amount of data in one physical place, and processing them both for the implementation of task goals as well as for personal cognitive satisfaction. Proper acquisition, use and preservation of data resources become a priority not only in the effective management of organizations, but also in the effective fulfilment of daily employee's tasks at various workplaces. Data processing and interpreting creates the new information which becomes a value for the organization. The cyberspace and their opportunities fascinate especially young people who have usually sufficient digital skills for intensive use ICT channels in their social and professional networking. Working in pandemic reality substantially have increased their presence and operations in cyberspace.

The paper content has been conceptualized based on literature review and on the basis of the quantitative research via using non standardized questionary regarding the personal security and cognitive satisfaction [7].

1. *Does the experience of pandemic working modes influencing the personal sense of security of young employees?*
2. *Is the intensive operating in a cyberspace in conditions of remote or hybrid work modes perceived as cognitively attractive and personally secure?*

The above research questions inspired own authors' research dedicated to explore the issue how the intensive cyberspace activity caused by implementing remote, flexible or hybrid work modes is perceived by young employees in COVID-19 pandemic work circumstances. Exactly: in which way (if at all) new kinds of work connected with prolonged exploration of cyberspace and virtual reality influences self-evaluation of Millenials' well-being in terms their sense of security and cognitive satisfaction.

3 Research Design, Results and Discussion

The authors' first research assumption was that stressful work context implemented together with COVID-19 demanding for social isolation is perceived by young employees as a kind of personal oppression. Thus, it is reflected in the lower (than the average) level of their perceived sense of security regardless of the size of the organization they are working for (research hypothesis no 1). From the other side, as young people usually have sufficient IT skills and were born together with the Internet, they should value the possibility to spend more time in the cyberspace. The new modes of working should deliver them cognitive satisfaction and by thus way – positively moderate perceived level of their sense of security (research hypothesis no 2).

The study was conducted in a second half of Nov. 2020 when COVID-19 pandemics restrictions in Poland were again more strict and companies continued their mostly digitally based modes of working. The online questionary was disseminated among randomly selected sample (N=246) of employees until 30 age, representing different sizes of enterprises (micro, SMEs and large ones), located in northern regions of Poland (Pomeranian and West Pomeranian regions). The size of the company was determined on the basis of the number of people employed (up to 10 - micro-company, up to 50 - small company, up to 250 - medium company, over 250 - large company). The descriptive statistics of the sense of security for examined employees of the three groups of company sizes (A, B, C) are shown in Table1.

Table 1. Respondents self-evaluation regarding their level of sense of security*

Enterprise size	Number /N	Median /M	Standard Deviation /SD
Micro and small (A)	87	3,52	.96
Medium (B)	99	3.35	1,03
Large (C)	60	3.43	.83

Source: own elaboration

Considering the phenomenon of examined level of the sense of security among young employees, it should be emphasized that there were not discovered statistical differences in their intensity in the studied subgroups representing different size of workplaces. However, it also has to be remarked that all subgroups revealed at best moderate level of security, what is indicated by their median rates (M87= 3,52; M99=3,35, M60=3,43). The final picture of this phenomenon checked as analysis of variances of results delivered by the three investigated subgroups is justified as nonsignificant ($F = .76$ and $p= .46$). It is showed in a Table 2a.

Table 2a. ANOVA of the sense of security of employees in different size companies referred to intense use of digital mode of working

Source of variance	SS	df	MS	Test F
Between subgroups (A, B, C)	1.42	2	.71	.76
Inside of subgroups (A/B/C)	225.03	243	.92	

Sum	226.45	245		

p=.46
Source: own elaboration

When looking for answers to what pairs' treatments contributed to the above overall result, the post hoc Tukey test was used, but again no significant differences in the level of sense of security were established employees representing different size of companies (see Table 2b).

Table 2b. Post hoc Tukey's test for respondents' sense of security in different pairs of company size*

Pair	Tukey's statistics HSD Q	p values of Tukey's HSD Q	p
(A)versus (B)	1.75	.43	Nonvalid
(A)versus (C)	.83	.80	Nonvalid
(B)versus (C)	.71	.85	Nonvalid

Source: own elaboration

The above results basically confirm the assumptions of the first hypothesis (h1). The employees of the three investigated groups of company sizes did not significantly differ in the level of their perceived sense of security and they revealed the similar - moderate level of that subjective well-being measure. The expected negative (stressful) impact of the distant working modes occurred to be evident. However, the new working conditions have influenced the well - being of respondents as they didn't evaluate their sense of security on a high or a very high level, only on an average, or moderate level. The second research question was that maybe, the opportunity to spend more time in a virtual world compensate the social deprivation caused by pandemic situation? That question has referred to the possible correlations between personal sense of security and the cognitive attractiveness of the intense operating in cyberspace when filling work tasks. It was the topic of verification of research hypothesis number 2. The results obtained for that issue show Table 3a and Table 3b.

Table 3a. ANOVA of the sense of security of employees of different size companies referred to the cognitive satisfaction with digital mode of working

Source of variance	SS	df	MS	Test F
Between subgroups (A,B,C)	0.06	2	.03	.03
Inside of subgroups (A/B/C)	225.94	243	1.04	
Sum	253.00	245		

p=.97
Source: own elaboration

The general result of the analysis of variance for the variable: "the cognitive value of the intensive use of cyberspace in a new, pandemic work modes" - turned out to be irrelevant (F = .03 and p= .97).

When looking for answers to what pairs' treatments contributed to the above overall result, again the post hoc Tukey test was used but no significant differences in the level of cognitive satisfaction in different subgroups of company size were found (see Table 3b).

Table 3b. Post hoc Tukey's test for respondents' sense of security in different pairs of company size*

Pair	Tukey's statistics HSD Q	p values of Tukey's HSD Q	p
(A)versus (B)	.03	.89	Nonvalid
(A)versus (C)	.28	.89	Nonvalid
(B)versus (C)	.32	.89	Nonvalid

Source: own elaboration

The answer for the posted second research question is as follow: the new working modes imposed by pandemic conditions in a similar, moderately positive way, influence the perceived cognitive value flying from intensive contact with cyberspace among young employees regardless of the size of the company they are working for. However, the both studied variables - the sense of security and the sense of cognitive satisfaction in connection with new working conditions in a pandemic - turned out to be independent of each other. Summing - the obtained results doesn't implicate that expected cognitive value flying from intensive operating in a cyberspace in remote (or hybrid) mode of working compensate the young employees' slightly decreased sense of security.

Conclusion

The pandemic situation forcing employers to organize the work in a way to meet requirements of social isolation evolved the new forms of delivering work tasks. The most characteristic include the remote work and the hybrid work. Employees, from other side, had been confronted with the demand for a rapid adaptation to new work modes without the time for necessary preparation. Employers' usually didn't offer them special upskilling programs in digital technologies, nor dedicated change management programs. Thus, the employees had to manage themselves to new circumstances and quickly learn new behavioural patterns connected with jobs. Very fast the SarsCov2 disease occurred to be to the new kind of severe biological threat, and, in the same time, an important psychosocial organisational stressor. This agent is still aggressively acting on a global scale, with high speed of spread and, till now - without the definite deadline. Covid-19 crisis have revealed the key barrier in effective organizational activity. It occurred to be the lack of intense social interaction and face to face direct communication during implementing job tasks.

The Covid-19 situation created also new stressors for managers. They had to answer questions like for example: how to effectively manage dispersed teams? or how to support subordinates suffering with social isolation working in home offices? Current crisis is challenging both regular employees as managers. Confronting them with problem of lower motivation and engagement caused by pessimistic mood or by the feel being overloaded. Home office work stopped to be perceive as desirable benefit as it was forced by formally introduced restrictions. People who are pressed to stay at home long time easily fall into a state of discouragement, depression, losing their long- term goal orientation. Suffering with low mood, are less creative, less resistant for overcoming different obstacles and problem solving. Are not able effectively plan or implement own professional careers.

However, after close to one year of the pandemic crisis existence, it can be concluded that this abnormal situation creates also a kind of paradox opportunities, both for organizations as well for employees. Organizations have to redefine own business strategies being focused on searching for the possible new "lean" solutions and cutting costs. They have to meet new kinds of organizational and strategic management challenges, learn new patterns of leaderships. Employees have the chance for to checking and updating their digital competences, maybe even have to decide for reskilling. They collecting new kind of employee experience - the possibility to feel satisfaction with the intense using cyberspace during implementing their work tasks. That all create the new kind of the positive professional challenge which can make wider their careers options.

Summing up, the switch into the new, remote or hybrid modes of work, especially in case of high skilled job places, creates both threats and challenges, as well as new opportunities. For the same employees as well as for organizations. From the point of the view of individual employee it generates the new kind of psychosocial work-related stress which can activate the set of negative effects for his physical and psychical health. It is necessary that this new category of work connected risk should be included into the work safety and hygiene factors monitored and controlled by organization. It seems to be very important especially that as the new modes of work are predicted to stay as the new stable forms of work organization also in the post-covid times. Thus, companies have to consider how to enhance the well- being of their workforce, not only in a current pandemic times but also after crisis. In an investigated sample of the young employees, representing Millennial generation, the personal sense of security occurred to be on a moderate level and was not connected with the cognitive satisfaction expected because of the intensive work digital activity. This result confirms that the new modes of work are not neutral for the emotional and motivational psychical sphere of employees. They influence their subjective sense of security.

What is especially important from the side of management - the obtained results didn't differ in relation to the size of the workplace. It means, both big corporations as well as small and medium enterprises have to be aware how the new work conditions influence the well – being of their skilled employees. All sizes and types organizations have to rethink own talent management strategies and implement the new non-financial types of benefits. Today the flexible work modes including remote work or hybrid work are not benefits anymore. Necessary is to substitute them with the new proposals to gain the competitive position from attracting talents. Managers have to especially care about most effective, high skilled employees who positively verified their competences working according to new rules of distant work. And many of them probably recognized also new attractive career possibilities on fast changing labour market.

References

1. *Technology and the American economy: Report of the National Commission on Technology, Automation, and Economic Progress*, US Department of Health, Education, and Welfare, Feb. 1966.
2. Tapscott D., *The digital economy: promise and peril in the age of networked intelligence.* New York: McGraw-Hill, 1995
3. Hermann M., Pentek T., Otto B.,: *Design Principles for Industrie 4.0 Scenarios*, working Paper , 01/2015, http://www.snom.mb.tu dortmund.de/cms/de/forschung/Arbeitsberichte/Design-Principles-for-Industrie-4_0-Scenarios.pdf [accessed 15.12.2020]
4. Harari Y.N., *21 lekcji na XXI wiek (eng. 21 lessons for the 21st century)*, Publ. Wydawnictwo Literackie, Kraków 2018.
5. Manyika J, Lund S., Chui M., Bughin J., Woetzel J., Batra P., Ko R., Sanghvi S., *What the future of work will mean for jobs, skills, and wages*, Mc Kinsey Global Institute, Dec 2017, online: www.mckinsey.com [accessed 10 Dec. 2020]
6. McKinsey &Company, *Unlocking success in digital transformations*, McKinsey 2018, www.mckinsey.com [accessed 12 Dec. 2020]
7. Klamut, R. *Two – factor model of safety experience. Theoretical assumptions and empirical verification: Safety Experience Questionnaire*, Polish Psychological Forum, t. 24, no 3, 2019, p. 308–323.
8. Sokołowski, M. *The concept and the essence of personal security,* Kultura Bezpieczeństwa. Nauka – Praktyka – Refleksje, 2019, no 33, p. 117-130.
9. W. Gosling, M. Coppola, K. McCarthy, *May the workforce be with you: The voice of the European workforce 2020*, Deloitte Insights, online [accessed 12 Jan. 2021]

10. K. Pouliakas, J. Branka, *EU jobs at highest risk of COVID-19 social distancing: Will the pandemic exacerbate labour market divide?*", IZA DP No. 13281, May 2020.
11. https://www.eurofound.europa.eu/publications/blog/coronavirus-a-labour-market-earthquake(accessed: 8.01. 2021)
12. S. Lund, A Madgavkar, J. Manyika, S. Smit, *What's next for remote work: An analysis of 2,000 tasks, 800 jobs, and nine countries*, McKinsey Global Institute, Nov. 2020. www.mckinsey.com [accessed 18 Dec. 2020].
13. E. Volini,J. Schwartz, B. Denny, D.Mallon, Y. van Durme, M. Hauptmann, R. Yan, S. Poynton, *Returning to work in the future of work*, Deloitte Insights, May 2020 , online [acessed 7.Jan 2021]
14. Webb, D., Wills-Herrera, E. *Introduction* [in:] D. Webb and E. Wills-Herrera (eds.), *Subjective Well-Being and Security*, 1 Social Indicators Research Series 46, Springer Science + Business Media B.V. 2012.
15. Henning-Smith, C.(2020). COVID-19 poses an unequal risk of isolation and loneliness. Retrieved April 10, 2020. [accessed 23.11.2020 from https://thehill.com/opinion/healthcare/488215-covid-19-poses-an-unequal-risk-of-isolation-and-loneliness].
16. https://osha.europa.eu/en/highlights/coronavirus-disease-covid-19-outbreak-and-workplace-safety-and-health [accessed: 05. 02. 2021]
17. Crimmins, E. Age-Related Vulnerability to Coronavirus Disease 2019 (COVID-19): Biological, Contextual, and Policy-Related Factors, *Public Policy & Aging Report*, Volume 30, Issue 4, 2020, p.142–146
18. *Cyberspace Protection Policy of the Republic of Poland*, Ministry of Administration and Digitization, Internal Security Agency, Warsaw 2013.
19. Carnevale,J.B., Hatak, I., *Employee Adjustment and Well-Being in the Era of COVID-19: Implications for Human Resource Management, Journal of Business Research*, May 2020. DOI: 10.1016/j.jbusres.2020.05.037

Importance of the Innovative Business Models for the Future Success of the Company

Leyla Djuraeva[1*]

[1]Tashkent Branch of Moscow State University, named after M.V. Lomonosov, Tashkent 100000, Uzbekistan

Abstract. During the last decade, the market environment for the businesses has been influenced by digital developments and as a result, the focus on business model innovations has rapidly grown. Businesses started moving away from the classic method of creating competitive advantage like a new product development. Thus, this influenced business activities as well as companies' business models into innovations of new products and services, a new ways of company relationships with customers and employees. Customers started expecting to receive services at a level comparable to digital solutions. Simultaneously, it put pressure on organizations to reflect on their current strategy and discover new business opportunities at the early stages. Digitalization is a one of the significant reasons for innovation on the business model of companies. Currently, business model as a tool is becoming a very popular topic due to its benefits received by companies and this topic studied by various researchers. This paper is based on theoretical perspective and the study was done by analysing previous researches, articles and papers in the present context. The study emphasizes that business model innovation can be a basis of sustainable competition for companies and innovators may increase their returns 4 times than product and service developers.

1 Introduction

Current business world is influenced by tremendous factors, and it is very challenging to survive for companies in this rapidly changing environment. In 21 century, well-known Companies like Uber, Airbnb and Xiaomi became one of the leading companies in their respective industries because of their unique business models [1]. Nowadays, companies and organizations are keen for profitable growth, for being competitive and stay sustainable in the market. In major industries, innovations considered to be in the form of new products and services [2]. Of course, service innovations are vital for the business growth and considered as an important factor for competitive pressure, but innovations in a business model became on priority level. As of today, the effective business model for the companies are key to future success [3]. There is no future growth for a company, which cannot transform itself and develop a practical and effective business model, as it will lose its strength and die.

For every company, business model is considered as a logical key to perform, a system approach to create a value for its stakeholders. Thus, business models always existed and always will. Companies always had opportunities to create new business models that may change the standard rules of an industry, such as Kodak and Zara. However, taking into considerations of rapid speed of technological progress in current days, businesses also dictated by technological changes to do things in radically different ways [2]. Therefore, there is a tremendous increase in opportunities for innovation in business models, and same as the threats posed by innovations in competitors' business models.
A business model innovation is a new approach to create value for customers taking into consideration digital changes, and how this value is carried or how the business gains profits from the service/product offered to the customer.

Business models for a company also known as a plan or a strategy that describes value chain delivery to final customers [2]. In the practice, startups are keen to use business models as modeling tools to plan, des
ign and build their new projects. However, established and well-known companies also use them to develop and succeed to support their innovation process.
Professor Haim Mendelson from Standford University discusses the progress of business models and the impact of the developments in information technology. In his research, he illustrates, that by implementing it and digitalization the companies advances their business processes and becomes progressed in the market [3]. He also states that changes that are incorporated into the new business model of the company, will gain more benefits. As per professor Ricarts research paper, technology is main force to drive the change today, particularly information and communication technologies [3].

In the same research, author underlines, as it was mentioned above, business model concept emphases on the way the business creates value and adds revenues and profits. Based on this theory. He defines model into three core elements: a value-creation model, a profit model, and the logic of the business.

* Corresponding author: leyla-djuraeva@rambler.ru

Well known researchers, Raphael Amit and Christoph Zott stated that the focus area of business model has one of the significant positions for managers and academic researchers mainly for 3 reasons. Firstly, in the current world, Innovative business model indicates as an underutilized source of future value. Secondly, it gives competitor advantage, as other market players will face difficulties to imitate a complete innovative activity system than a single new product or process. Thirdly, management of the company should never challenge the effort of their competitor's in business model area as innovations considered a powerful competitive tool [4]. Actually, the business model in the digital economy grows into the success formula and it is gaining the power to establish a competitive advantage [5]. There are a huge number of examples in current business world to support these arguments, including such as examples of Dell Computers, IKEA and Southwest Airlines. Though there is an increasing number of research studies on this topic for last decade, researches usually consider Business models and tools from the technical aspects –elements of the business model, how can be innovated within the business model and how this innovation process can be implemented [5]. There are very few research studies about the consequences of innovations in business model in the business and its problems. Previous researchs on business model innovation has concentrated on how innovative changes in the company affect the business model [6]. Thus, there is a still distinct gap in research on interrelation of success and competitive advantage with innovative business model.

We have reviewed list of articles and empirical studies on this topic, analysed the opinions and statements of scholars which are presented in this paper below. This paper is constructed on the following way. Firstly, the theoretical background on this research topic will be stated, where the concept of the innovative business model and its importance, its influence on the business of the companies will be analyzed. After that, different views and results of the previous researchers will be compared and stated. Finally, the paper will provide summary on the results.

2 Research Purpose and Research Question

The issue of innovations in business model and its effectiveness were already identified almost decades ago in one of the oldest research papers of Zoe, Osterwelder [18 and 19]. Nowadays, reaction of management on changes happening from environment, technology or other factors and reflection of these alterations on the company's business model is vital [4]. Recently, a few scholars started to research the interrelation of changes in the market and business model innovation [13]. However, this topic on business model innovations still lacks of practical and clear analysis on how implementation of these changes in business model of the company occurs. Based on this research gap, proper research purpose is identified. The purpose of this thesis is to identify the effectiveness of innovative business models in the context of the company, benefits gaining from its implementation. In doing so, the researcher seeks to explore how successful can be companies which totally innovates their business model rather than invests on other projects. Consequently, the research question in this work will be: *Will the Companies succeed by implementing innovative business model ?*

3 Methodology

The purpose of this research is to examine how the innovations in the business models may influence the successfulness of the company in the market. To answer the research question, the researcher will examine the existing literature review and also qualitative research method is chosen to generate more accurate data. The Researcher will observe and create semi-structured interviewes with 2 largest companies in the developing country, which are successfully innovated existing business models.

4 Theoretical Background

In this section, we will review existing scholars and articles on this topic, empirical studies and summarize key ideas, statements which illustrate the importance of this topic nowadays.

4.1 Theoretical Background

The main purpose of the Theoretical Background section is to provide a reader with the literature overview in the respective areas of the business. Aligned with the research question and research goal, conceptual background offers a literature review in the field of innovation management and types of innovations, business model, , competitive advantage and at the end, the connection between competitive advantage and business model innovations. Also, different studies will be analyzed and compared on the effectiveness on the innovative business model of the current technologically progressed world.

Nowadays, change can be seen everywhere, the world is changing very fast, thus society has been under greater impact of them and trying to adopt to these changes. One of the global issues is significant changes in the technological sphere and transformation of major processes, activities and the business through digitalization. Digitalization is a new concept of last few decades and it plays major role in success of organization. Digitalization has spread been widely extended to every organization and has become an integral component of the modern world, as a result, it is hard to imagine an organization without implementation of innovative tools in current world [4, 7]

In the digital business world, information has been vital resource and newly established companies, startups in different industries, take advantage of this culture and economic change to perfection [18]. As a result, these new small companies are devastating the business of big organizations and increasing market from traditional brands.

In order to face successfully to market changes such as the case of the progressive rise of the digital business environment, existing organizations must be able to quickly shift and adopt changes. They have to maintain a variety of innovation efforts to operate more efficiently and deliver greater value to their customers [6].

As earlier described, digital technology can considerably contribute to a company's business model, which supports the companies' purpose; profit and growth [6].

One of the earlier researchers, Raphael Amit and Christoph Zott, provided in depth analysis on the topic of business model innovation and its importance for managers and academic researchers. Strategic discussions on the Business model and influence of digitalization and IT were enormous, and there is no still one defined compromise on this topic.

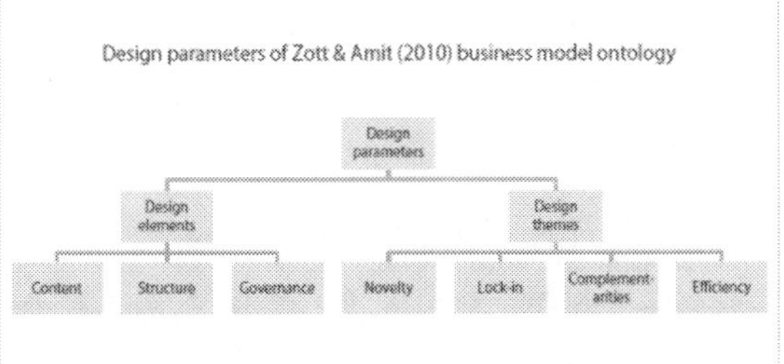

Fig. 1. Zott and Amit business model design parameters

These researchers conceptualize business model as a general system of activities correlated with each other and consequently, during the process of innovating business process 2 parameters: design elements and design themes for activity system should be done (for details refer to Fig 1.) [23]. The researchers explains that by interrelating each of these components shown in Fig 1, business model can be properly constructed in order to achieve greater efficiency [23]. They also presented 10 year-long research programs on business models and draw a parallel between digitalization and business model innovation [4]. Thus, business model innovation has been acknowledged as a new method of gaining competitive advantage for mature companies [3].

There are variety of business models that adopts innovations to stay digitalize. The study of digitalization and the effect of information technology on companies and their success have been critical to various researchers [13]. The Power of business models and their success has been analysed firstly through it related and mobile communication related industry [2]. In his research, the two cases Apple and HTC were investigated and results identified illustrates that implementation of new business model with innovations and digitalization assisted to derive revenue and boost income of the company. Innovations and business models with consideration of digitalization can be applied not only in the sphere of IT, but also can yield more benefits in other industries. As per research paper of Lee and Vonortas [9], a business model comprises of multiple components or elements. Though referred to previous studies such as Rayport and Jaworski [8] where business model is based on 4 choices, such as

- a "value proposition" or a value chain for niche customers;
- coverage of different type of products services or information (or combination of all of them) for covering market needs;
- An exclusive and secure resource system for benefit creation, such as , the "associated resource system";
- Appropriate model for financial system of the company, which covers, income and return models, shareholder value models, and future growth models.

In the same study, another researcher [3] proposed 4 components for digitalized business model framework:
product innovation or value proposition, customer relationship, infrastructure management, and financial aspects that can be found throughout the three former components, such as cost and revenue structures. To summarize their investigations, business model in new digital world may vary with different components, thus there is no clear definition of ideal business models for generation of income [4]. Another well known researchers Johnson and Christensen also supports idea of 4 components which interrelates and in sum delivers value to the customer [23]. The model is developed almost decade ago, but it is still considered as a basis for planning the business model innovations topic.

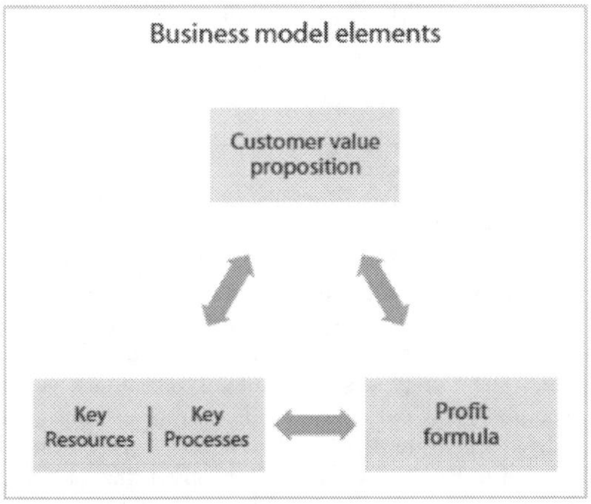

Fig. 2. Business model components from the Johnson&Christensen

Referring to the study of Johnson, Christensen and Kagermann [12], which has illustrated the four elements that in total create, deliver and capture value. These are considered as "customer value proposition" (CVP), formula for profit model, basic resources and vital processesThis model also illustrates that the companies oriented on value creation recognize that each business component has significant role for success in the model [23]. Change of any of them will lead to the loss or profit of the company as the value proposition may be affected.

The differences of Business model innovations related in digitalization has been studied and analyzed in depth already few decades ago. In the summary report of Lee and Vonortas, several examples were presented were value creation of companies differs in digital and physical economy. According to the study of Meredith and Schaffer [7], in the industrial economy the main purpose of the management is to develop the physical transformation process by implementing management techniques, while in the digital economy, digital data and it is vital and main input into a business transformation process.

The digitalization allowed growing to a number of new business models and these models can be used in birth traditional business and in more innovative businesses. Innovation in the digital world concept as a quick development of new business models, it can also rapidly cause existing businesses to become outdated [10,11].

According to the general data gathered through different studies and researches, business models of Amazon, Walmart and Target accounted for 76% of market capitalization in retail industries by 2007, earning more than $300 billion of value [14]. Furthermore, Comparable results have been proven in the airline industry, rental industry, digital communications and others, which illustrates the importance of proper developed Innovative Business model in business environment in current world.

Another interesting statement was found in the research of Eksell and Harenstam [6], were evaluation of new business models discussed. It has been stated that start-up methodology is an optimal way to approach market, based on specific requirements. 3 main success factors for it were prescribed in research as follows:

1. to do things differently from the status quo,
2. to bolster the competencies required for the new business model, and
3. to continuously and endlessly strive for the change to happen [6].

The authors also inputted that the results of research on value drivers, which comprise the successful model. They are classified as Innovations, lock-in, protection from competition, and efficiency [19,6]. The emphasis is on the key activities and key resources that outline a business area, estimating the output in market attractiveness as well as competitiveness. This evaluation gives the obligatory path to rearrange assets and gives ability to provide a unique product or service.

Another interesting approach, which is tested by Kim and Mauborgne [6] on how to innovate a business model called Creating Blue Oceans. The basis of the model is to take the industry standard and use it as a reference point when creating a customer-centric business model. The focus is to create an advantage against the market's standard value proposition. The Blue Oceans model is suitable and works well with the business model canvas as concept, which is applicable to the whole canvas model.

In order to be a successful business, organizations should be able to challenge themselves on a number of issues and also review the dynamics of the respective mechanisms which are:

- offered value to customers [strategic goals and value proposition];
-scope of offerings;
-pricing strategy;
-appropriate revenue model;
- calculated and estimated list of required resources and all expected costs to provide the value to final customer [6]. In case of Business Model Innovations, Organizations should be able to understand and capitalize on the distressing issues of the electronic world and e-commerce to renovate their business models for future success. Actually, the major researchers agreed on the statement, that Business model innovation nowadays is mainly based on proper use of new knowledge [technological and market based] that to design and implement an innovative way of product offerings and services for segmented customers [6]. One of the well-know cases of successful business model innovations that were able to capitalize on some of the troublemaking features of the Internet include channel innovation [e.g., Dell's build-to-order virtual integration model], process innovation such as Boeing virtual design and e-procurement, customer-relationship (e.g., Yahoo!), and affiliate network (e.g., Amazon) [6].

Rachinger and other authors in their study emphasised that, one of the methods and tools to renovate business model and make it more related to the digital world is implementation of E-commerce tool. It is useful to implement as a supporting innovation to boost existing business of the company. In that case, companies might be unsuccessful to recognize and profit from on many of the Internet's opportunities. Thus, in order to benefit from all advantages, the "disruptive attributes" or types of e-commerce innovation must be recognized and applied in a practical business model [6].

Furthermore, in theories, one of the successful and useful tools for development of business model and proper identification of key activities is known as the business model canvas structure [6]. The model can separate processes and key activities from total value proposition and representation of reasons for considering as key and non-key activities.

One of the main advantages of the business model canvas is the fact that it represents business plan in one page [14].

During the process of development of the business model, several ideas have be taken into consideration and one thing emphasized by Osterwalder and Pigneur [2] is the concept should be consumer oriented or as known customer centric. It means that all choices made regarding the value proposition, distribution channel, customer relationship and revenue stream all should tie back to your intended customers. These segments are directly linked to the customer segment and together they form the core of the business itself, remaining segments form a foundation for them to work but without a solid core structure, the model will not work [4].

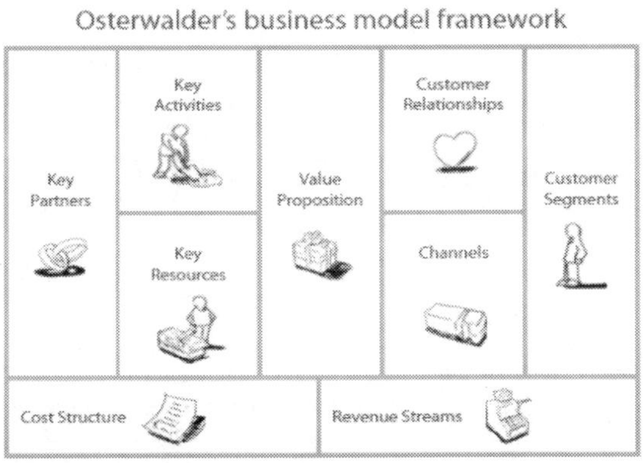

Fig 3. Osterwalder and Pigneur Business model innovations

However, in the studies of Osterwalder and Pigneur [19], it was also argued that Business model should be viewed as a holistic concept that embraces all organisations part:

1. Complex system
2. Holistic concept with many organisational parts
3. Functioning in dynamic business environment

The business model approach presented by Osterwalder and Pigneur is widely used by organisations and it also gives opportunity to be successful [23].

One of the famous and well used models is proposed by Boston Consulting group which consist of 2 elements value proposition and operating model [23]. In general 6 sub elements which are illustrated below in figure 4 helps companies to identify structural way to identify key areas to change and practical to develop relevant model.

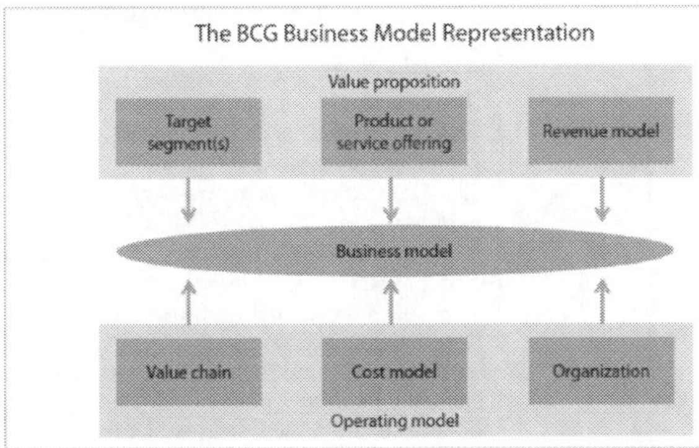

Fig 4 Boston Consulting group Business model representation

According to the research on digital transformation and its success factors [8], company should have a valuable and effective business model, with strong digital capabilities, and it is achieved by the implementation of these digital tools. However, they have respectively, severe impact on the operation processes, resources management, internal and external consumers of the organization [8].

Proper business model with consideration of innovative tools and digitalization will improve business processes of the organization, improve efficiency, quality and consistency, also it helps to reduce manual processes [7]. Digitalization helps in improving the business process efficiency, quality and consistency, which results in effective internal efficiency of organization; this can be achieve by eliminating the existing manual process and maintain better accuracy [8]. Taking into consideration the increasing importance of digital business models for modern enterprises and the fundamental changes they cause, the proposed theories will also affect future corporate management tools. In general, based on the pace of changes and innovations appearing in the market, total business management concepts, the requirement for in depth analysis for business model development and its implications should be reconsidered and updated.

The speed of these technological changes and use of these great challenges as an opportunity, made variety of business stronger and successful [9]. As it was stated in the article, the era of digitalization and being able to react on these changes, made start-up companies like Airbnb and Uber to multi-billion dollar companies, within a short period of time [6]. As it was mentioned above in Introduction section, the leading companies (Google, Uber, Amazon) have totally broke stereotypes and business rules in their respective industries by changing the basic building blocks of their business models [1]. After researching more closely on these cases, authors identified that the source of competitive advantage has change. Now, unique model is more applicable and valuable, rather that innovative products and services. Innovations on Business model facilitated these companies to gain higher sales, better profit margins and more cash flows than other players in the market. Actually, these companies are lack of clear value chain and without any fixed charges (in machinery, equipment's) [1].

As per findings of Economist Intelligence Unit (2012) which had survey with more than 4,000 executive managers around the world on the topic innovativeness in business [1]. According to findings, these senior managers highlighted that, they gave preferences to business models as a competitive advantage rather than developing new products and services. Additionally, one of the previous and first studies by IBM (2006) emphasized that due to rigorous global pressures, high level management shift the focus and interest towards to business model innovations. It was also mentioned there, that companies with best results in the market and who beaten the competitors and who grew on rapid speed, were mainly focused on successful business model creation.

In the same research done by IBM (2006), which is discussed in the paper of Bashir and Verma [1], it was stated that in 2006 major CEOs spent about 30% of their time in the endeavor to find best solution for business by innovating their business models.

The study also underlined that companies gained higher profits with the use of new innovative business. 40% of senior managers surveyed stated firms put efforts, energy and resources to be the first to drive the change and be a leading competitor in the market, rather than being the one who will follow others. Furthermore, list of benefits from integration of innovations in business model were also presented in this report, where more than 50%(surveyed) of the directors mentioned that it may reduce costs and allow a strategic flexibility. Also, more than 40% of executive respondents encouraged that innovative business model gives new market opportunities, helps to renew focus [1].

Another Researcher Eksell and Harenstam [6], in their report illustrated that in general, innovations in the business model leads to number of positive changes in the company. In their case study analysis, they identified that companies who innovated business model successfully, reduced costs, boost better performance. However, they also faced to major limitations and obstacles [6]. The Researcher mentioned that this process takes strong team and takes time. It is improper to follow market trends and try to quickly and temporary fix existing business model to be competitive in the market [6].

Furthermore, different academic researchers identifies the variety level of innovations in the business model [19]. While one definitions stating that its moving from old business model to new one, other scholars debating that it's the process of adding innovations into existing model. There is no right answer as the methods and approaches for each business model will vary during implementation [23]. Breiby and Wanberg, in their report on successfulness of the innovation of business models illustrated the challenges in the implementation of both methods. In order to make successful the process of innovation of business model, all proper stages (which is presented in their paper [23]) should be managed. Also, the importance of analysis of existing (current} model helps more to identify week point as it is challenging to forecast how far business model should be innovated and how will work [23].

The researchers in their scholarly written reports conclude that in current aggressive competitive world, business model innovation are more applicable and relevant to implement, as new products and others can easily copy products or services. Innovative Business model is an alternative solution for future growth, allows gaining sustainable revenues and helping to earn four times more returns than product innovators [1].

4.2 Results identified based on literature review

The scholars, which were stated on these papers, did a significant research on the issues related to innovations of business model and its influence to the core business. In the findings of the research conducted by Tesarova, the b2b large market companies benefited significantly by applying the innovations through digitalization to the core business model, they gained higher customer value and in major cases, the companies adapt incrementally changes by that got higher opportunities to grow [16].

The table below summarizes key points identified from the literature review. It illustrates the theoretical framework for the innovations on business model and its success factors.

Table 1. Summary of key points from Literature Review

Element	Key points
Business model	Business model is considered as a logical key to perform, a system approach to create a value for its stakeholders. Business models for a company also known as a plan or a strategy that describes value chain delivery to final customers A business model innovation is a new approach to create value for customers taking into consideration digital changes, and how this value is carried or how the business gains profits from the service/product offered to the customer
Classicification of business models according to researchers	1.Zott and Amit analyzed and presented that in a effective and successful business model is general system of activities correlated with each other and consequently, during the process of innovating business process 2 parameters: design elements and design themes for activity system should be done 2. Lee and Vonortas in their research, mentioned that a business model comprises of multiple components or elements which is about behavioural and structural correlation between them 3. Rayport and Jaworski researchers illustrated that business model is based on 4 choices, such as a "value proposition, coverage of different type of products services or information (or combination of all of them) for covering market needs; secure resource system and model for financial system 4. Johnson, Christensen and Kagermann illustrated the four elements for innovating business model that in total create, deliver and capture value. 5. Osterwalder and Pigneur, - argued that Business model should be viewed as a holistic concept that embraces all organisations part: 1. Complex system 2. Holistic concept with many organisational parts 3. Functioning in dynamic business environment The business model approach presented by Osterwalder and Pigneur is widely used by organisations and it also gives opportunity to be successful 6. One of the famous and well used models is proposed by Boston Consulting group which consist of 2 elements value proposition and operating model [23]. In general 6 sub elements which are illustrated below in figure 4 helps companies to identify structural way to identify key areas to change and practical to develop relevant model.
Features	Business model innovation consist of different types of innovation, based on it the scope in the model are changed New business models can fully integrate with

	the old business model or may partially replace it Researchers explain business model innovation as an cyclical process going over different stages of designing and implementing business models. For successful integration of innovations in business models, they should be competitive and hard to understand and copy for other market players.
Result from implementation of innovations in Business model	In case of successful implementation of innovation in the business model, the success of the company is its competitive advantage over market which help to gain higher customer satisfaction and respectively profit for the company

One of the newly published studies presented by famous consulting agency McKinsey, illustrates that major innovation in business comes after getting over the crisis [22]. In 2020, one of the largest crisis for the world is the COVID-19 pandemic, which affected almost everyone from all businesses to each person. Their current survey consisted of more than 200 organizations across industries, where more than 90 percent of managers said they expect the significant consequence from COVID-19, which will change business processes and the way the business is done. Furthermore, it will definitely influence the customer needs and preferences. The survey also identified that the key focus of the executives now is to sustain business continuity; they have to evaluate cutting costs, motivating productivity, and holding employees beside innovation to business growth As a result, innovations in business are least priority investment. However, the managers from conducted survey strongly believe that innovations in business are very important and as soon as the world is stabilized, companies will return to these investments. Moreover, McKinsey's survey and interviews conducted with business heads indicates that, innovations were not prioritized into their business as they plan to concentrate on following activities:

- Reinforce the core business of the company by following recognized opportunity spaces, maintaining cash and minimizing risk, until clarity of the future actions will be identified
- Adaptation of the core of the business to meet changed customer needs
- Clarify and rapidly address new opportunity zones
- Reevaluate portfolio of the innovation initiatives and allocate resources properly
- Build the basis for post crisis development in order to persist competitive in the recovery period

The businesses should take into consideration, that it will never operate as they used to do it in the past.

As the business model starts adapting to new market changes, the Competitive advantages changes dynamically, because the core competences that distinguish the business from others might become suddenly less attractive. While the rise of digital has been mounting similar pressures for more than a decade, the current crisis has significantly exacerbated and accelerated its disruptive force.

To summarize, the outcomes indicates that one of the reasons why business models are innovated is due to a fear of being replaced by a competitor. The change is happening on a daily basis. Consequently, companies who wisely created and developed their existing strategies and innovated existing business models with proper methods, are becoming successful. The advantage of innovative business model in the current work is very high for the successful companies.

4.3 Results identified from Interviews

The data collection based on the interview is mainly depend on the participants level of strategic insight about the company, his openness and knowledge. Researcher was able to conduct interview with deputy directors of the banks, as they are one of the representatives and knowledgeable person. Several management level respondents declined the request for the interview, thus only few was identified.

One of the Banks (Bank A) in Uzbekistan which is known in the market as a Mortgage Bank is opersting in the market for several decades. Currently. it is in the process of innovating its business model. As per discussion with the management, the reason for this decision was based on the governement level. The central Bank of Uzbekistan, requested the bank to be more technologically efficient and digitalize. The main problem of the Bank was, it was not customer oriented, the services were time consuming and in general the existing system was outdated. It tried to be competitive by inventing new products, new design of services, by hiring high skilled personnel. However, without updating and innovating the model itself, the results were not achieved. As a result, in 2017 the bank started working on changing the structure, innovating its business model.

The Deputy Director of the Bank stated that it is already 4 years and results are not achieved yet.

The obstacles they faced :

- Expensiveness of innovations on existing model
- Time consuming
- Due to digitalization and It excellence of new model, major other areas should be changed

On the Other hand, partial implementation of new innovative business model, already help bank to reduce costs and create better reputation in the market. According to the Banks predictions, by 2023 the bank will become in top 3 banks in the market as a technogically shift and highly profitable.

The second Bank (Bank B) is the largest government bank and is also in process of implementing innovations in Existing business model. The researcher had chance to have a interview with the head of Strategy department, which is in charge of this change project. The Bank is planning to innovate whole business model within 3 years.

The reason for implementing Innovative business model is to become highly competitive and gain higher profits. As per analysis of Business Intelligence Team of the Bank B, the forecasts done on the results of innovations in Business model shows very good results. The Bank B will become number one bank in the market for its stakeholders.

As per interview results, The Bank B decided to innovate its business model by creating customer centric. Currently, due to very high risks and difficulty level on innovations, bank decided to implement them in step-by-step mode.

5 Conclusion

The findings on the theoretical research by analysing the literature review is presented in this paragraph. After the review of all research and findings of studies and investigations, it was identified that in this changing market, the organizations have to be able to make a proper examination of the existing model and based on industry requirements, competition in the market. In general, Innovative Business model is a powerful tool for succeeding in business, for being updated in the market and to gain competitive advantage. It was also analysed that, old market strategies like Porters on new product development and optimization of value chain as an competitive has been effective advantage. In current social and digitalized era, the modern companies try to distinguish themselves in order to be profit oriented. Moreover, the startups, and companies which are influenced by technologies and who were able to implement digital mechanism into the business becoming more popular and showing to the market the win strategy [1]. As it was mentioned above, some of the big entities such as Google, Amazon and newly emerged in the last half decade Airbnb, Uber are changing the ground rules in the market by their innovative style of business. These technologically developed companies stress on mainly on the idea of innovating their business model, rather than concentrating on fixed value chains. They are result of successfully planned business with high margins and turnover. Accordingly, the studies show that in the era of technology and digital revolution, companies are more concentrated on the competent development of Innovative business model. The companies should not spent time and resources, and put high emphasis on the development of new products and services with an existing business model. Different studies are showing that it will not gain competitive advantage in the market. The study held by different researchers has reasoned that business model innovation can be functioned as a viable competitive advantage as imitation of entire the business model is not easy and affordable as imitation of new product in the competition [13].

Nevertheless, Innovative Business model tool is still not highly popular among companies. It is still under the process of full implementation and has to be vital for existing and new businesses to use for success and digitalize in the new economic era. All the opportunities and challenges organizations may face during the process of business model changes, for example, by using digital technologies was discussed. This summarized information signifies a valued, significantly relevant contribution to the sectors that absences on an empirical foundation. According to Amit et al. [19] strategies in the management level strategy has to be used business model frameworks in order to identify the correct business model innovation for the specific situation, and how to apply it, before proceeding with the implementation. The Success of implementation and success of the business in current digitalized economic conditions relies on both, getting the model right and making sure that the compulsory business does not offset the new model [6]. Furthermore, according to the results of the interviews conducted with the largest banks in the Uzbekistan, clarifies that Innovations in business models is very expensive and strategic tool and it should not be used only for competition, but for strategic growth of the company. In case of using wisely this method, the companies will benefit and succeed in the future, To finalize, a practical innovative business model will be able to cover the essential economic principles of the market (various costs and revenue/pricing models) and give company opportunity to minimize extra costs [13]. Also, they have to be able to detect and get the most out of on the disruptive attributes of Internet commerce. Moreover, it should be developed based on the general picture, vision and strategy of the company. Additionally, the inability to implement the innovative business models and not being able to catch the pace of new digitalized economic factors may lead to the failure of existing large businesses and organizations. Thus, nowadays, the digital world dictates the rules and creates great opportunities for the market, only if they are used properly. Therefore, the only mechanism that will have substantial value and effect in the near future is managing the technological shifts in the company's business, by innovating the business model upon the rivalry. Innovating the Business model is not easy process, it is more challenging, expensive and time consuming than the product and process innovation, but the paybacks derived out of it are also superior. It will not only sharply increase the profit of the company, but also will help to reduce costs and optimize the business itself [1] Moreover, it is very vital for CEOs to deeply analyse the existing model and identify the limitations, before taking opportunities and innovate the business model. Finally, the research on implications of business model innovation and its importance is still at a potential stage. There is a huge research gap in this area. Besides, there is a deficiency of a practical method for how companies should follow business model innovation in the digitalization framework. There is still lack of literature and researches, same as businesses, which are lack of the knowledge of how to perform business model innovation, which method to use and the prediction of results from these changes. Nevertheless, the existing studies are proving the innovations in business model will help to improve the future progress and sustainability of the business in this digitally changing world.

References

1. Bashir M., Verma R, Why Business Model Innovation Is the New Competitive Advantage. IUP Journal of Business Strategy. Mar2017, Vol. **14** Issue 1, p7-17. 11. (2017) Available at SSRN: https://ssrn.com/abstract=3175788
2. Ricart, J. E. "Business Models for the Companies of the Future." In Reinventing the Company in the Digital Age. BBVA (2014) https://www.bbvaopenmind.com/en/articles/business-models-for-the-companies-of-the-future/
3. Mendelson, H. "Business Models, Information Technology, and the Company of the Future." In Reinventing the Company in the Digital Age. Madrid: BBVA, (2014) https://www.bbvaopenmind.com/en/articles/business-models-information-technology-and-the-company-of-the-future/
4. C Zott, R. Amit, & L. Massa, The business model: Theoretical roots, recent developments, and future research (2010) DOI: ,http://doi.org/10.1177/014920631140626511 1.6285
5. D.J. Teece, Business Models, Business Strategy and Innovation. Long Range Planning (2010) DOI: https://doi.org/10.1016/j.lrp.2009.07.003
6. A. Eksell and A. Harenstam , Business Model Innovation for a digital future (2017) Available at: https://pdfs.semanticscholar.org/05eb/dc8412edd5f425cb27f84eace1baff5ec100.pdf
7. C. S. Lee, & N.S. Vonortas, Business Model Innovation in the Digital Economy. 1[3], 292-312. (2014) DOI: http://doi.org/10.4018/9781599049472.ch260
8. M. Rachinger, R. Rauter, CH. Muller, W. Vorrabor, E. Schirgi. Digitalization and its influence on business model innovation. EmeraldInsight Vol. **30** No. 8, pp. 1143-1160 (2018) DOI http://doi.org/10.1108/JMTM-01-2018-0020
9. C. S. Lee, & N.S. Vonortas. Toward an integrated model of strategy formulation for strategic technical alliances. (2004) DOI: https://doi.org 10.1504/IJTTC.2002.001790
10. OECD, "Exploring Data-Driven Innovation as a New Source of Growth: Mapping the Policy Issues Raised by "Big Data", OECD Digital Economy Papers, No. 222, OECD Publishing, Paris. (2013) DOI: http://doi.org/10.1787/5k47zw3fcp43-en.
11. OECD. "The digital economy, new business models and key features", in Addressing the Tax Challenges of the Digital Economy, OECD Publishing, Paris. {2014) DOI: https://doi.org/10.1787/9789264218789-7-en
12. C. S. Lee, Y.G. Chen and Y.H. Fan, Structure and Components of E-Commerce Business Model (2006) DOI : https://doi.org/10.4018/9781591407997.ch170
13. Tesarova M., Krmela A., Simberova I., Digitalization as an enabler of business model dynamics. (2020) DOI: https://doi.org/10.3846/bm.2020.562
14. U. Umihana, Business Model Innovation as a New Source of Competitive Advantage - A Case Study based on the Energy Sector., Research Gate (2014) DOI: http://doi.org/10.13140/RG.2.2.16812.26244
15. S.J. Berman, Digital transformation: opportunities to create new business models. Strategy Leadership. (2014) DOI: http://doi.org/10.1108/10878571211209314
16. R. Mustafa , Business model innovation. Journal of Strategy and Management, **8 (4)**, 342–367 (2015) DOI: http://doi.org/10.1108/JSMA-06-2014-0054
17. What-When-How, Business Model Innovation in the Digital Economy, https://what-when-how.com/information-science-and-technology/business-model-innovation-in-the-digital-economy/
18. R. Alt, & H. Zimmermann. Preface: Introduction to special section—business models. (2001) DOI: http://doi.org/10.1080/713765630
19. A. Osterwalder, Y. Pigneur. Business Model Generation: A Handbook for Visionaries, Game Changers, and Challengers. (2010) DOI: https://doi.org/10.1080/14719037.2020.1718187.
20. M. Dubosson-Torbay, A. Osterwalder, & Y. Pigneur, E-business model design, classification, and measurements. (2002) DOI: https://doi.org/10.1016/j.lrp.2009.07.003
21. C Zott, R. Amit, & L. Massa, The business model: Theoretical roots, recent developments, and future research (2010) DOI:http://doi.org/10.1177/014920631140626511 1.6285
22. BarAm J., Furstenthal L., Jorge F., and Roth E., Innovation in a crisis: Why it is more critical than ever (2020) Available at: https://www.mckinsey.com/business-functions/strategy-and-corporate-finance/our-insights/innovation-in-a-crisis-why-it-is-more-critical-than-ever
23. Breiby E. and Wanberg M., Successful business model innovation (2011) Available at: https://core.ac.uk/download/pdf/52108745.pdf

Modern Structural Level and Dynamics of Crimes with The Use of Computers, Automation Systems, Computer Networks and Electric Connection Systems

Sergiy Tkalichenko[1], *Valentyna* Khotskina[1*], *Zhanna* Tsymbal[1], *Victoria* Solovieva[1], and *Olena* Burunova[2]

[1]State University of Economics and Technology, Kryvyi Rih, Ukraine
[2]Science and Business Center ZAK, 96100, Poland

Abstract. Actuality of surveillance the cyber-criminality problem and its impact upon a society is proved out by the rapid increase in a quantity of such crimes and material losses accordingly. The statistical analysis of their number increases and the caused losses is made in the process of cybercrimes studying. It's revealed that besides the catastrophic number growth of such crimes, the relative size of losses is increased too. The analysis of the actual data for the nine years is made, on which basis the indicators' table of the cyber-attacks number, the general losses and the indexes of their dynamics is elaborated. The analysis of the struggle's state with cyber-criminality in our state is made. The recommendations are presented in the limits of the research for the increase of the information protection's reliability.

1 Introduction

The rapid development of the world information-communication technologies, being observed for the last two decades, is accompanied by the dynamic development of crimes in this field. Such development brings the negative phenomena of the new type – cyber-criminality – into our life. Besides the crimes, being specific for it, the cyber-crime presented the new possibilities of the traditional crimes' commitment and creates conditions for the realization of the principally new schemes and methods of the criminal activity. The criminals actually created the black market for the sale of drugs, weapon, the stolen goods, etc. with the help of the Darknet system.

The growth of the cybercrime's provision with the modern computing engineering, the means of the telephone communication with the access to the networks, the specific software form the threat not only for the crossing citizens in particular, but for the national security of the state in general.

2 Background

At present, in the times of information technologies, the identification of the cyber-criminality problem is gaining actuality. Correspondingly, it's necessary to construct the operating system of the cybernetics security guarantee at the state level.

The research materials of the cybersecurity problems are presented in the European Cybercrime Center [12], Norton Cybercrime Report, SecureWorks Cybercrime, FBI IC3Report, Globalstudy.bsa.org and the other sources.

To increase fighting efficiency related to such crimes, it is necessary to synchronize Ukraine's legislation with legislation of the countries, which have achieved considerable successes in combating cyber-criminality. The international rules introduced by ISO/IES 15408 standard, should be implemented into the state's legislation [1].

The different aspects of the problem are lighted up in the works of the leading professionals: the study of the international experience of information security [2, 3]; the information security's audit [4]; the hybrid aggressive threats [5]; the prevention of cyber-criminality [6, 7, 8]; the protection of the critical infrastructure objects [9]; widening of cyber-criminality in different branches (the protection of the data base, banking protection, the protection of the intellectual ownership, the protection from the pornography, electronic swindling, etc.) [10].

2.1 Problem Positing

The EU Commission presented the new Strategy of the EU Security Union on July, 24, 2020 with an emphasis on the protection of the critical infrastructure, the struggle with cyber-criminality, the opposition to the hybrid threats and the organized criminality. Such strategy has become the continuation of the complex measures of the previous years: the first European security strategy of 2003, the European agenda on security of 2015, the Global EU strategy of 2016, where the significant attention was paid to the problems of security.

The domestic realities of the cybersecurity sphere testify to a series of the important problems, preventing from the creation of the efficiently operating system of

*Corresponding author: khotskina_vb@ukr.net

opposition to the threats in the cyberspace. The following ones belong to such problems in the first turn: the terminological uncertainty, the absence of the proper coordination of activity of the corresponding government departments, the Ukraine's dependence on the program and the engineering products of foreign origin, the difficulties with the staff complement of the corresponding structural subdivisions [13]. The official statistics reflects not only the state of the criminality, but the state of its registration in the country. The high delitescence of such type of crimes is observed.

That's why the modern structural level and the dynamics of crimes with the use of computers, automation systems, computer networks and the systems of the electric connection are selected as the subject of the research.

2.2 Presentation of Materials and Results

The available classifications of the notion "cybercrime" from the position of the scientific understanding are various enough.

The growth dynamics of cybercrimes, according to the statistics of the Internet Crime Complaint Center [11, 14, 15, 16, 17], is presented in the Table 1, in fig.1.

According to the calculations of the professionals, the quantity's jump of all the cybercrimes took place in 2017. After that the quantity of the cybercrimes received the tendency to the sharp rise. Thus, according to the data of the cyber-police in Ukraine, it's fixed: 1795 cases in 2017, 1023 cases – in 2018, 2826 – in 2018, already 4263 cybercrimes – in 2019. Let's present some general-world tendencies. We consider it to be opportune, if we present the analytical indicators of the investigated process (Table 2, Fig. 2).

As the visual analysis shows, the growth dynamics of the crimes' number and their cost have the different character.

Let's use the methods of the index analysis for the more detailed analysis.

$$Ipg = \frac{\sum p_1 g_1}{\sum p_0 g_0}, \quad Ip = \frac{\sum p_1 g_1}{\sum p_0 g_1}, \quad Ig = \frac{\sum p_0 g_1}{\sum p_0 g_0}$$

where p1, p0 – the average cost of one crime (current and previous period), q1, q0 – the quantity of crimes (current and previous period), Ipq – the general index of losses, Ip – the general index of losses, due to the increase of the crime's average cost, Iq – the general index of losses, due to the number increase of crimes (Table. 3, Fig. 3).

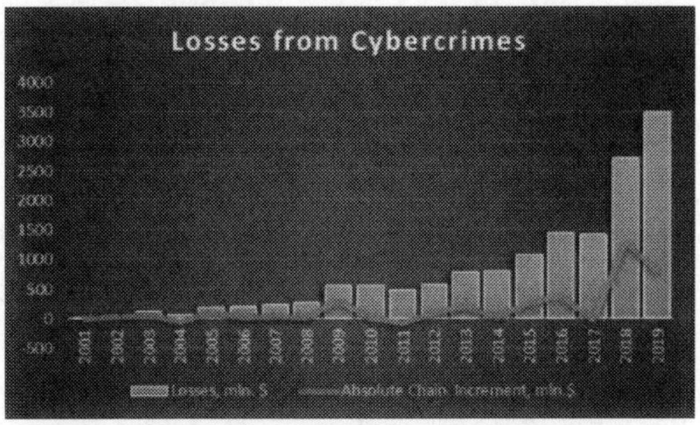

Fig. 1. Growth Dynamics of Losses from Cybercrimes

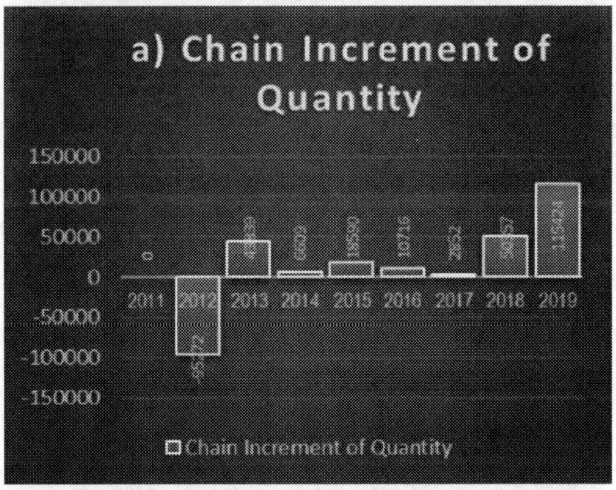

Fig. 2. Dynamics Indicators of Quantity Cybercrime (a)

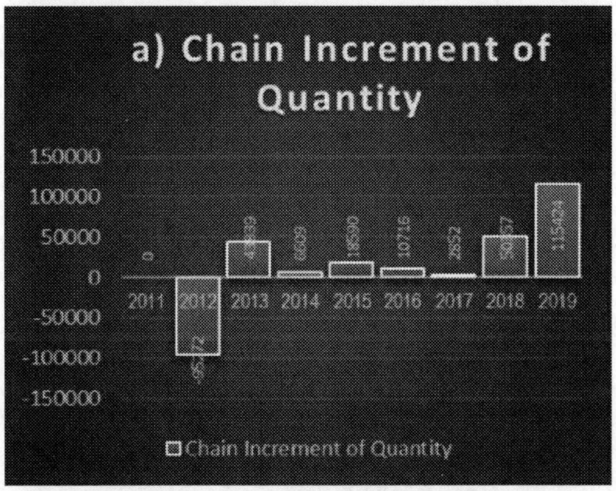

Fig 2. Dynamics Indicators, Average Cost of One Cybercrime (b)

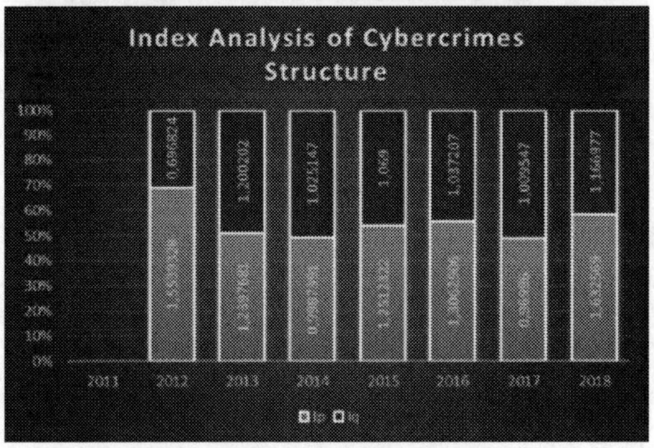

Fig. 3. Correlation of General Losses from Cybercrimes, due to Quantity and Average Cost of One Crime

Table 1. Growth Dynamics of Losses from Cybercrimes.

Year	Losses, mln. $	Absolute Chain Increment, mln. $	Year	Losses, mln. $	Absolute Chain Increment, mln. $
2001	17,8	0	2011	485,2	-78
2002	54	36,2	2012	581,4	96,2
2003	125,6	71,6	2013	781,8	200,4
2004	68,1	-57,5	2014	800,4	18,6
2005	183,1	115	2015	1070,7	270,3
2006	198,4	15,3	2016	1450,7	380
2007	239,1	40,7	2017	1418,7	-32
2008	264,6	25,5	2018	2710	1291,3
2009	559,7	295,1	2019	3500	790
2010	563,2	3,5			

Table 2. Dynamics Indicators of Cybercrimes' Quantity and the Average Cost of One Crime.

Year	Quantity	Losses, $	Average Losses, due to One Cybercrime	Chain Increment of Quantity	Chain Increment of Crime Price, $
2011	314246	485253871,00	1544,18	-	-
2012	218974	525441110,00	2399,56	-95272,00	855,37
2013	262813	781841611,00	2974,90	43839,00	575,34
2014	269422	800492073,00	2971,15	6609,00	-3,75
2015	288012	1070711522,00	3717,59	18590,00	746,45
2016	298728	1450700000,00	4856,26	10716,00	1138,66
2017	301580	1420555000,00	4710,38	2852,00	-145,88
2018	351937	2706400000,00	7690,01	50357,00	2979,64
2019	467361	3500000000,00	7488,86	115424,00	-201,16

Table 3. Index Analysis of Cybercrimes Structure.

Year	Quantity	Losses,$	Average Losses, due to One Cybercrime	Ip	Iq	Ipq
2011	314246	485253871	1544,18			
2012	218974	525441110	2399,56	1,5539328	0,696824	1,082817
2013	262813	781841611	2974,90	1,2397681	1,200202	1,487972
2014	269422	800492073	2971,15	0,9987391	1,025147	1,023855
2015	288012	1070711522	3717,59	1,2512322	1,069	1,337567
2016	298728	1450700000	4856,26	1,3062906	1,037207	1,354893
2017	301580	1420555000	4710,38	0,96996	1,009547	0,97922
2018	351937	2706400000	7690,01	1,632569	1,166977	1,905171
2019	467361	3500000000	7488,86	0,973842	1,327968	1,293231

Thus, the aggregate index of the general growth of losses is Ipq=1,308 (i.e., 30,8% in the average per one year). Such a growth is explained by the average increase (by 6,6%) of the crimes' quantity (Iq=1,066) and the sharp rise of the average cost of one crime – by 24% (Ip=1).

In order to fight efficiently with the cybercrimes, it's necessary to segment their demonstrations and to reveal the crimes, to which it's necessary to pay the maximal attention urgently, and to create the corresponding methods of struggle with them. The most dynamic types of the quantity of such violations, being revealed by the method of the index analysis, are represented in the Table 4.

*Corresponding author: khotskina_vb@ukr.net

Table 4. Indexes of the Most Dynamic Quantitative Types of Cybercrimes.

Type of Cybercrime	Average Index
Swindling with the Enquiry of Personal Data	2,230
Demanding of Illegal Profit by Intimidation	1,713
Forgery of Goods and Services	1,656
Lottery/Totalizator	1,390
Swindling in the Sphere of Medicine	1,293
Games of Chance	1,274
Breaking-Up of E-Mail, Accounts, etc.	1,257
Trust Abuse of Investors	1,228
Violation and Forgery of Copyrights	1,203
Confidentiality Violation of Personal Data	1,171
Computer Blocking by Attacks	1,159
Abuse on Confidence	1,106
Crimes on the Basis of Technical Support at Remote Access	1,090

The types of the cybercrimes, according to the "hardness" (which lead to the most losses) are also presented in the Table 5.

Table 5. Indexes of the Most Harmful Types of Cybercrimes.

Type of Cybercrime	Average Index
Swindling by Presenting Himself (Herself) as the State Official	2,700
Demanding of Illegal Profit by Intimidation	2,568
Swindling in the Field of Medicine	2,005
Crimes on the Basis of Technical Support at Remote Access	1,969
Terrorism	1,830
Swindling in Social Networks	1,821
Swindling with Real Estate	1,771
Computer Blocking by Attacks	1,744

All the key "classical" cybercrimes, committed with the help of the computer and the telecommunication technologies, which number grows every year, are present now in Ukraine in the whole scope.

As Oleksandr Grynchak, the first deputy head of the Ukraine cyber-police department, states, the most spread types of such actions in Ukraine are the following: the illegal access, the illegal catching, the interference into data, the abuse by devices, the swindling, connected with computers; the violations, connected with the children's pornography, etc. The swindlers create the sites and sell the not existent product more often. There are many crimes, concerning the defrauding of information from the cards and the online-crediting [18, 29].

According to [21], 4263 cybercrimes were registered in Ukraine in 2019, which caused losses for the sum of 28 mln.UHA, 17 mln.UHA of them were recompensed. The main part of such crimes is the following:

- crimes in the application sphere of computers, systems and computer networks, i.e. viruses, attacks and others – 1494;
- E-Commerce – 744;
- crimes, connected with the payment systems – 1641;
- lawless contents – 332.

The presented quantity of the revealed cybercrimes is scanty, compared with the European statistics. Let's state that the trustworthy statistics is almost absent, but the available one causes doubts.

The information of NCCC at CNSD of Ukraine [30], concerning the quantity of the cyber-incidents (Table 6), cannot but cause the amazement in the background of the insignificant number of the officially registered cybercrimes.

Table 6. Quantity of Fixed Cyber-Incidents (the 9-th of September – the 6-th of December, 2020).

Type of Cyber-Incident	Quantity
Scanning of Resources	15449264
BRUTEFORCE	4028226
Network Attacks	1184980
WEB-Attacks	1033221
Harmful Software	275981
Not-Sanctioned Access	83384
Spam	79261
HARVEST-ATTACK	18858
Exploits	2771
Fishing Attacks	813
DDOS Attacks	68

Thus, the very significant delitescence exists in the sphere of cybercrimes. The most part of cyber-incidents are not registered as crimes.

According to the data of the Ukraine's state statistics, the specific weight of the young people (aged to 30), convicted for the crimes in the application sphere of the electronic-computing machines (computers), the systems and the computer networks, was equal to the following: in 2010 – 55,1%, 2012 – 45,0%, 2013 – 36,7%, 2014 – 43,2%, 2015 – 41,9%, 2016 – 41,7%, 2017 – 40,5% [19]. To our mind, such statistics is not grounded enough. The National Agency on Struggle with Criminality (NCA) in Great Britain launched the #CyberChoices campaign for the cybercrime warning. The statistics pushed the NCA to such step: the average age of a cybercriminal reduced from 24 to 17 years old [22]. The age reduction of a cybercriminal was caused by the accessibility of the highly-technological means of communication for the youth and the availability of the mobile access to the networks.

The swindling with the services' use of the mobile communication operators, including the SMS-messages, has become traditional. It's reasonably to divide the offences with the use of the mobile telephone into the following groups: hooliganism (including the telephone terrorism; various types of swindling, aimed against the operators and the subscribers of the mobile communication and others).

The messages on minelaying acquired a special spreading for the last time. Such crimes lead to the pulling away of the police forces, the operation stoppage of the metro (underground), enterprises, the trade complexes, the educational establishments, the state institutions, resulting in the greatest losses. As they informed in the National police of Ukraine, 3730 anonymous messages on the minelaying of buildings and the infrastructure objects came in 2019. Only 750 similar minelayings were fixed last year. Thus, it may be said of the mining of up to five hundred objects in one message [25]. It's very difficult to follow such calls or messages, because, according to the data of the companies "Kyivstar" and "Vodafone Ukraina", 90–96% of Ukrainians use today the mobile communication anonymously. According to the data of the Inter-Bank association of payment cards, EMA, the ill-intentioned persons stole more than 275 mln UHA from the accounts of Ukrainians with the help of calls and SMS-messages. This figure reached 340 mln UHA in 2018, moreover, 80% of the assets were received by the thieves, who gained the card requisites by fraud over the telephone [24].

Besides the scanty number of the registered cybercrimes, it is worth paying the attention to the insufficient degree of their punishment. The data of the General Public Prosecutor's Office, concerning the investigation of a cybercrime, are presented in the Table 7 [26].

Table 7. Crimes in Application's Sphere of Electronic-Computing Machines (Computers), Systems and Computer Networks and Electric Connection Systems.

Year	**Quantity**		**Investigation is Closed**	
	Accounted Offences	Message on Suspicion is Delivered	Quantity	%
2013	595	256	331	55,63%
2014	443	207	237	53,50%
2015	598	263	411	68,73%
2016	865	472	420	48,55%
2017	2573	1272	605	23,51%
2018	2301	1608	169	7,34%
2019	2204	1481	182	8,26%
2020	2498	1675	51	2,04%

We stress on the fact, that besides the significant number of the closed investigations, the quantity statistics of the real punishments is absent.

It's worth emphasizing, that the cyber-criminality reasons are changeless for years. They may be characterized briefly in the following way: for the enterprises – the insufficient quantity of the qualified professionals in cybersecurity; the computer threats of the new type [28]; carelessness and incompetency of the personnel; for the crossing citizens – the excessive trustfulness, poverty (the use of the licensed software, the absence of the anti-virus protection), the Internet-incompetency, carelessness, etc.

Let's indicate that the inefficient struggle is realized with such crimes in Ukraine. For example, the National Bank launched the great program of struggle with the cyber-swindling in 2020. The Anti-crisis center of the business cybernetic protection at the Trade-Industrial Chamber of Ukraine spoke with the proposal to the Ministry of Education and Science that each academic year would begin with the lessons of cyber-hygiene [23]. The cyber-police launched the campaign of knowledge in cybersecurity [27]. The activity of the cyber-police grows every year and the promulgation of the operation results is realized at the professional conferences. As the counter-action to the telephone swindling and terrorism, the petition is located (on October, 19, 2017, on the Official Internet-Representation of Ukraine President) with the demand to oblige the operators of mobile communication to identify all the mobile numbers, according to the owner's documents. The struggle with such type of crimes is only starting.

A special attention is worth being paid to the cybercrimes, aimed at the undermining of the national security, and the global threats, connected with the hacker attacks, which have become the weapon in the hybrid Russian-Ukrainian war.

The demonstrations of such threats are the attacks at the objects of the state's strategic infrastructure, which may be examined as cyber-terrorism. The Ukrainian economy lost $466 mln (or 0,5% of the GDP) only, due to the Petya virus. As Ukraine is faced with the hacker attacks at the state resources every day, the Ukrainian Service of State Security (SSU) has an intention to strengthen its cybersecurity. "The penetration scales into the state information resources strike – the attacks take place almost every day. The conclusions are simple – we need to act immediately and systematically" [20].

The director of the Cisco Representation in Ukraine and the CIS countries for the work with partners and clients, Sergiy Martynchuk (Cyber Defence Congress 2K18) announced that the majority of the great cyber-attacks fulfilled not the economic, but the political and military tasks [20].

The Minister of the Internal Affairs of Ukraine, Arsen Avakov, stated at the ZOOM-conference "Digital Transformation of the State: Perspectives and Risks of Cybersecurity" (2020) that the number of cybercrimes in the state grew by twice and a half for the last five years. The cyber-police fixes the growth of the following types of crimes: the interference into the operation of information systems and their intentional damage; the illegal collection, storage, use and spreading of the personal data and information with the limited access; the creation of channels for spreading of weapon and drugs; the illegal financial operations, including the ones with the digital currencies; robbery and swindling in the Internet system; spam and the virus programs [20].

According to [30], the quantity of the cyber-incidents, connected with the critical infrastructure, reached 63505 cases, but with the bodies of the state government – 2938475 - for the three months of 2020.

The state tries to be opposed to such provocations. The experts of the National Coordination Center of Cybersecurity at the Council of National Security and Defense started to elaborate the Strategy of the Ukraine Cybersecurity. The cyber-police of Ukraine plans to increase the staff of the special agents in the sphere of opposition to cybercrimes of such type in 2021 [23].

Thus, the Situational Center of the cybernetic security provision on the basis of the Department of the Counter-Intelligence protection of the state interests in the sphere of the SSU information security, according to the NATO standards, was created in the SSU in 2018 with the support of the foreign partners. It was developed, according to the agreement on the realization of the Ukraine-NATO Trust Fund. More than $1 mln. were allocated for the project. Its key possibilities are revealing and reacting to the various online incidents, which allow prevent the cyber-attacks, determine their origin, analyze for the opposition improvement.

3 Conclusions

Cyber-criminality at present – is the real global threat, which may go out of any country of the world beyond the limits of the definite jurisdiction (in contrast to the other traditional types of economic crimes).

The COVID-19 pandemic promotes to the growth and spreading of cyber-criminality. At the same time, when the online swindling, the demanding and the sexual violence at children in the Internet are aimed at the separate groups of persons, the programs-demanders, in the first turn, undermine the operation of the organizations, including the hospitals.

The distance work increased the number of the potential victims of cyber-criminality. Working online from home, people are subjected to the bigger risk than at the usual mode of operation.

Thus, the introduction of digital technologies at enterprises, information technologies (IT) for information protection, causes the appearance of the new form of counteraction – cyber-insurance.

In the result of the research we have come to the conclusion, that Ukraine fights the above mentioned problems, but organization of such a fight compared to the highly developed countries is at the initial stage. As the necessary conditions for increasing fight efficiency with cyber-criminality can be named the following: the reasoned scientific analysis of such problems, the up-to-date legislative provision, the increasing financial support of appropriate organs for strengthening their personnel and technical potential.

References

1. International Standard, Access Mode ISO/IEChttp://isotc.iso.org/livelink/livelink/fetch/2000/2489/Ittf_Home/PubliclyAvailableStandards.htm
2. D. Gavlovsky, To the Problem of Counter-Action to the Use of Harmful Software Struggle with Organized Criminality and Corruption (Theory and Practice) **1**, 125–130 (2014)
3. O.V. Orlov, State Government of Professionals' Training in the Sphere of Cyber-Security, State Construction. Access Mode: http://kbuapa.kharkov.ua.
4. V.A. Romaka, A.E. Lagun, Yu.R. Garasym and oth.; Audit of Information Security: Texbook State Service of Ukraine on Extraordinary Situations: Lviv State University of Life Activity's Safety, NAS of Ukraine, Institute of Applied Problems of Mechanics and Mathematics, named after Ya.S. Pidstryhach, Lviv Spolom (2015)
5. D.V. Dubov, Cyber-Space as New Measure of Geo-Political Rivalry. Monograph NISR (2014)
6. M.O. Kravtsova, O. M. Lytvynov, Prevention of Cyber-Criminality in Ukraine: Monograph, Kharkiv Panov (2016)
7. S.V. Melnyk, V.I. Kaschuk, Actual Directions of Violations' Warning in Cyber-Space as Strategy's Component of State's Cybernetic Security. Information Security: Challenges and Threats of Modernity: col. of materials of sc.-pr. conf., Kyiv NPC NA SS of Ukraine, 5 April 2013
8. I.V Diorditsa, Notion and Contents of Cyber-Security's National System, Access Mode http://goal-int.org/ ponyattya-ta-zmist-nacionalnoi-sistemi-kiberbezpeki/
9. S.F. Honchar, G.P. Leonenko, O.Yu. Yudin, Methodological Bases of Elaboration and Introduction of Information Protection's Systems in Objects of Critical Infrastructure. Special Telecommunication Systems and Protection of Information **1** (25), 158–163 (2014)
10. Control of Struggle with Cyber-Criminality. Ministry of Internal Affairs of Ukraine. Access Mode http://mvs.gov.ua/mvs/control/main/uk/ publish/article/544754
11. Law of Ukraine "On Principal Bases of Ukraine's Cyber-Security Provision" (News of Verhovna Rada (NVR) **45**, (2017)
12. European Cybercrime Center – first year report, URL https://www.europol.europa.eu/content/european-cybercrime-center-ec3-first-year-report. Date of address: 29.05.2016).
13. Modern Trends of Cybersecurity Policy: Conclusions for Ukraine. Analytical Note,

National Institute of Ukraine Strategic Research, Access Mode http://old2.niss.gov.ua/articles/294/. Date of address: 31.01.2021)

14. Sergiy Tkalichenko, Valentyna Khotskina, Zhanna Tsymbal, Cyber-Criminality: Protection's Aspects of Modern Information Space. Advances in Economics, Business and Management Research, **129**. 137–143 (2020)
15. Internet Crime Report 2017, FBI's Internet Crime Complaint Center (IC3) URL: https://pdf.ic3.gov/2017_IC3Report.pdf /. Date of address: 31.01.2021)
16. Internet Crime Report 2013, FBI's Internet Crime Complaint Center (IC3) URL: https://pdf.ic3.gov/2013_IC3Report.pdf /. Date of address: 31.01.2021
17. Internet Crime Report 2019, FBI's Internet Crime Complaint Center (IC3), URL: https://pdf.ic3.gov/2019_IC3Report.pdf /. Date of address: 31.01.2021
18. What Do You Need to Know of Cyber-Criminals in Ukraine? URL: https://www.radiosvoboda.org/a/details/29031166.html. Date of address: 31.01.2021
19. Ukraine in Figures in 2017, State Service of Ukraine Statistics, URL:http://www.ukrstat.gov.ua/druk/publicat/kat_u/2018/zb/08/Ukr_cifra_2017_u.pdf. Date of address: 31.01.2021
20. Cyber Defence Congress 2K18, URL: https://itukraine.org.ua/cyber-defence-congress-2k18.html. Date of address: 31.01.2021
21. Report of Ukraine National Police Head on Operation Results of Department in 2019, URL:https://www.kmu.gov.ua/storage/app/sites/1/17-civik-2018/zvit_2019/zvit-npu-2019.pdf. Date of address: 31.01.2021
22. Cyber Choices: Helping You Choose the Right and Legal Path, NCA, URL: https://www.nationalcrimeagency.gov.uk/what-we-do/crime-threats/cyber-crime/cyberchoices. Date of address: 31.01.2021
23. Staff of Special Agents for Opposition to Cyber-Swindling Will Be Increased, URL: https://www.ukrinform.ua/rubric-economy/2805152-v-ukraini-zbilsit-stat-specagentiv-dla-protidii-kibersahrajstvu.html. Date of address: 31.01.2021
24. Telephone Swindling: How Do They Demand Money from Ukrainians and What to Do with IT? URL: http://chp.com.ua/ua/all-news/item/59212-telefonnoe-moshennichestvo-kak-u-ukraintsev-vymogayut-dengi-i-chto-s-etim-delat. Date of address: 31.01.2021
25. Who and Why "Mines" Massively Public and Administrative Buildings? Deutsche Welle, URL: https://p.dw.com/p/3Mjk3. Date of address: 31.01.2021
26. Report on Criminal Offences in the State, General Public Prosecutor's Office of Ukraine, URL: https://old.gp.gov.ua/ua/statinfo.html. Date of address: 31.01.2021
27. Cyber-Police Launched Campaign on Knowledge of Cybersecurity, United Portal of Ukraine MIA System Bodies, URL: https://mvs.gov.ua/ua/news/18914_Kiberpoliciya_zapustila_kampaniyu_z_obiznanosti_pro_kiberbezpeku.htm. Date of address: 31.01.2021
28. NHS Cyber Attack: Everything You Need to Know about 'Biggest Ransomware' Offensive in History (2017), Access Mode: https://www.telegraph.co.uk/news/2017/05/13/nhs-cyber-attack-everything-need-know-biggest-ransomware-offensive. Date of address: 03.01.2020
29. Global Research of Swindling Problems in Banking Sphere, KPMG International Cooperative, URL: https://assets.kpmg/content/dam/kpmg/ua/pdf/2019/11/Global_Banking_Fraud_Survey.pdf. Date of address: 31.01.2021
30. State of National Cybersecurity for the Last Three Months, NCCC at CNSD of Ukraine, URL: https://www.rnbo.gov.ua/ua/Diialnist/4761.html. Date of address: 31.01.2021

Foreign Direct Investment as a Driver of Economic Recovery and Business Innovation

José Manuel Macedo Botelho[1], *Irina* Maksymova[2], *Oleg* Padalka[3], *Mansur* Mamanazarov[4], *Volodymyr* Kulishov[2]

[1]Business Factory, Evora 7000, Portugal
[2]State University of Economics and Technology, Medychna street 16, Kryvyi Rih 50000, Ukraine
[3]National Pedagogical Dragomanov University, Pyrohova str., 9., Kyiv 01601, Ukraine
[4]National University of named after Mirzo Ulugbek, University street, 4. Tashkent city, 100174 Uzbekistan

Abstract. Global economy is expected to contract, consequence of a collapse of international trade with millions of failed businesses and lost jobs. Foreign Direct Investment (FDI) flows are fundamental to support the economic recovery. Developed economies and developing economies in particular must make a great effort to restore and increase capital inflows, especially in form of foreign direct investment. FDI inflows have long been the largest source of finance investment. FDI is an opportunity to support the crucial economic recovery. An internationalization model of FDI is built in order to be a driver to governments and firms to implement a success strategy to attract FDI to the country

1 Introduction

The COVID-19 pandemic increased the number of poor people in the world as no crisis had done before [1,2]. The economic crisis of 2020 had a big impact on businesses, families and put social protection and health systems at risk. The catastrophe also had serious impacts on the Fourth Industrial Revolution, on the digital economy, on human capital and underlined the gap between the economy and society [2].

The global economy is expected to drop by an impressive 4.3% in 2020. Millions of workers and people living on a livelihood are at risk and about 130 million people will fall into extreme poverty if the economic crisis perseveres [3].

Governments must define the short and long-term policies to support the health and social system and as well the businesses activities and families [4].

Governments must define short and long-term policies to support the health and social system as well as business and family activities [4]. The emerging market and developing economies need additional support due to their economic weakness in trade, in the social and health system and, especially, those that depend on tourism and exports. [4].

The recovery must be supported by countries and the international community after the health crisis has passed [4]. Policy choices should generate new investments in productive infrastructure, in digital economy and an increase of cash incentives for social protection for the families [1].

International trade collapsed in 2020 in developed and developing economies [1]. Major economies should support developing economies in order to bridge the gap between their economies caused by the COVID-19 crisis. International trade has a fundamental role in stimulating economic recovery [1]. However, the big question is the way how countries will balance the measures of recovery to achieve the 2030 Agenda for Sustainable Development's socio-economic growth goals [3].

FDI has fallen in the COVID-19 pandemic and is expected to decline further in 2021 and only begin to recover in 2022 [3]. The Organization for Economic Cooperation and Development (OECD) identifies a 50% decrease in global FDI in first semester of 2020 (comparatively to the second semester of 2019), the lowest level since 2013 [5] and United Nations Conference on Trade and Development (UNCTAD) predicts a drop of at least 40% [3].

2 Literature Review

The Second World War took place between 1939 and 1945 and since then there has been no crisis like the COVID-19 crisis with a major impact on emerging market and developing economies, the biggest recession in 60 years [6].

FDI is fundamental to the economic recovery of developing economies. Therefore, developing economies have to strain to increase FDI because FDI is the key to supporting the development of developing economies. It will be fundamental for a post COVID-19 recovery [6]. With the right policies FDI can provide financial stability and boost economic well-being for the benefit of businesses activities and families [26].

In particular, FDI has the advantage of being a catalyst for the transfer of technology and know-how from one economy to another and to the host economy

* Corresponding author: jmmbinternationalization@gmail.com

allows products to be boosted in international markets [13].

This article highlights a review of the empirical studies on FDI in order to explain the geographic distribution of FDI flows worldwide.

Botelho [7] made a systematization of the most relevant studies of the determinants of FDI since 2000. The study presents the author of the study divided in three dimensions: economic, institutional and business facilitators (table 1).

Table 1. Systematization of FDI determinants

Researchers (first author)	Year	Dimensions		
		1	2	3
Villaverde, J.	2015	a		
Acheampong, P.	2014	a	a	
Anuchitworawong, C.	2014	a		
Hoang, H.	2014	a		
Hui, E.	2014	a		
Jeong, H.-G.	2014	a	a	a
Jiao, P.	2014	a		
Kersan-Škabić, I.	2014	a		
Kowalewski, O.	2014	a		
Liu, K.	2014	a		
Penfold, M.	2014	a	a	
Sánchez-Martín	2014	a	a	
Tang, C.	2014	a		
Gamboa, O.	2013	a		
Hu, Y.	2013	a		
Ibrahim, O.	2013	a		a
Kokouma, D.	2013	a	a	
Larimo, J.	2013	a		
Rivera, C.	2013	a		
Tintin, C.	2013	a	a	
Bilgili. F.	2012	a		
Liargovas, P.	2012	a	a	
Merková, M.	2012	a		
Reschenhofer	2012	a		
Sharma, K.	2012	a		
Solomon, B.	2012	a		
Staats, J.	2012			a
Tembe, P.	2012	a		
Uwubanmwen A.	2012	a		a
Baek, K.	2011	a		
Liu, K.	2011	a		
Ali Khrawish, H.	2010	a		a
Ali, F.	2010	a	a	
Azam, M.	2010	a		
Bellak, C.	2010	a		
Casi, L.	2010	a		
Choong, C.- K.	2010	a		
Lin, F.-J.	2010	a		
Sharma, K.	2010	a	a	
Van Wyk, J.	2010	a		a
Dumludag, D.	2009	a	a	
Ledyaeva, S.	2009	a		
Ang, J. B.	2008	a		
Bitzenis, A.	2007	a		
Wijeweera, A.	2007	a		
Hadjit, A.	2006	a		
Ramirez, M. D.	2006	a		
Dar, H.	2004	a	a	
Janicki, H. P.	2004	a		
Cheng, L. K.	2000	a		
Legend: 1-Economic; 2-Institutional; 3-Business Facilitators.				
Source: Botelho, J. (2015). Doctoral dissertation.				

«The systematization of studies presented (table 1) divides the determinants into: economic, institutional and business facilitators. Villaverde & Maza (2015), Anuchitworawong & Thampanishvong (2014), Hoang & Goujon (2014), Hui & Chan (2014), Jiao & Jian (2014), Kersan-Škabić & Tijanić (2014), Kowalewski & Radło (2014) , Liu et al. (2014), Tang et al. (2014), Gamboa (2013), Hu (2013), Larimo & Arslan (2013), Rivera & Castro (2013), Bilgili et al. (2012), Merková et al. (2012), Reschenhofer et al. (2012), Sharma et al. (2012), Solomon & Ruiz (2012), Tembe & Xu (2012), Baek & Qian (2011), Liu & Daly (2011), Azam (2010), Bellak et al. (2010), Casi & Resmini (2010), Choong & Lam (2010), Lin (2010), Ledyaeva (2009), Ang (2008), Bitzenis (2007), Wijeweera et al. (2007), Hadjit & Moxon-Browne (2006), Ramirez (2006), Janicki & Wunnava (2004) and Cheng & Kwan (2000) developed studies on the determinants of FDI in countries or sectors of activity and concluded that the economic determinants that contributed to attracting FDI are the following: market size, market growth, market opening, agglomeration, taxes, qualified labor, economic stability (inflation), exports, natural resources and external indebtedness. Acheampong & Osei (2014), Jeong (2014), Penfold (2014), Sánchez-Martín et al. (2014), Kokouma & Xu (2013), Tintin (2013), Liargovas & Skandalis (2012), Ali et al. (2010), Sharma & Bandara (2010), Dumludag (2009) and Dar et al. (2004) identified in their investigations, in addition to the economic determinants already mentioned, institutional determinants (control of corruption, government effectiveness, political stability, normative quality, rule of law and voice and responsibility) that contributed to attracting FDI. Jeong (2014), Ibrahim & Hassan (2013), Staats & Biglaiser (2012), Uwubanmwen & Ajao (2012), Ali Khrawish & Zakaria (2010) and Van Wyk & Lal (2010) considered that facilitators of doing business are crucial to attracting FDI» (Botelho, 2015) [7].

The United Nations published an international investment framework/structure [26] that explains the differences in FDI inflows among countries and formulates policies to capture inbound investment, based in Transnational Corporations investment. The principal determinants of the location of FDI are divided in three dimensions: the policies made by governments and institutions to influence the attraction of FDI, the economic determinants and the business facilitators (table 3).

The Support Structure to FDI policy is made by governments and institutions and consists in a political, economic and social principles and measures that drive to a stability environment to attract FDI, regulations to ingress and operations, rules for functioning of markets, FDI international agreements, privatization policy, trade policy and tax policy [26]. Policies can prohibit the entry of FDI into a market and can make rules and regulations on the treatment of non-discrimination between domestic and foreign companies – and can also regulate the special treatment of foreign companies [26]. Policies can achieve a number of objectives to increase or decrease FDI inflows, such as investing in identified sectors of the economy, the country's origin of FDI and how FDI contributions are made [26]. FDI policies are generally associated with other policies to influence investors decisions to achieve the same results. However, the Support Structure to FDI policy is becoming less important because almost all countries are having the same strategy [26].

Instead, business facilitators are becoming relatively more important than the investment policy. They include investment incentives, post investment services and measures to minimize the «hassle costs» (related to corruption and efficiency of administrative processes) in the business environment. This set of measures aims to be more competitive in attracting FDI, since with the opening of markets, the country's policies are becoming similar. In addition, the country's policy measures are being more complex to reach private investors giving them business facilitators to attract investment [26].

After implementing the Support Structure to FDI policy, the economic factors are the most important dimension to attract FDI inflows. Economic factors have three motivations to attract FDI, the market-searching, the resource/asset searching and efficiency-searching (table 3) [26].

The pandemic crisis is expected to hardly affect developing countries, especially in Africa. Africa has 13 percent of the world's population and had more than 50 percent of world's extreme poverty in 2020. Least developed countries are expected to increase the number of poor by 33.4 million in 2020 and a change in the proportion of headcount ratio of 3.16 percent, based on the $1.90 per day poverty rate table 2) [3].

Table 2. Estimated changes in extreme poverty in 2020

	Change in headcount ratio (%)	**Change in number of poor (millions)**
East Asia and the Pacific	0.21	4.41
Europe and Central Asia	0.23	1.16
Latin America and the Caribbean	0.55	3.58
Miidle East and North Africa	1.22	4.91
Other high income	0.01	0.07
South Asia	1.25	23.28
Sub-Saharan Africa	2.74	31.17
Least developed countries	3.16	33.4
World Total	**0.88**	**68.57**

Source: UNCTAD, 2020 [3]

Table 3. Framework to FDI Attraction Determinants

Host country determinants	**FDI by factors**	**FDI Economic determinants in recipient countries**
I. Support Structure to FDI Policy	**A. Market-searching**	- size of the market
- political, economic, and social stability		- market growth
- regulations to ingress and operations		- enter in markets (regional and global)
- rules for functioning of markets		- country consumer preferences
- FDI international agreements		- structure of markets
- privatization policy	**B. Resource/as set-searching**	- raw materials
- trade policy (tariffs and non-tariff barriers)		- low-cost unskilled labour
- tax policy		- skilled labour
II. Economic determinants		- assets with advanced and innovative technologies
III. Business facilitators		- infrastructures (ports, roads, power, telecommunication)
- incentives and investment promotion	**C. Efficiency-searching**	- cost of assets listed in B
- after-investment services		- other costs (transport and communication costs)
- hassle costs (related to corruption, efficiency of administrative processes, etc.)		- associated of a regional corporate network

Source:[26]

3 Methodology

This research aims to identify the economic determinants that attract FDI and proposes an internationalization model that will be a future reference to implement internationalization strategies for governments, institutions, agencies and companies, that will contribute to the economic recovery using foreign investment as a motivator to do so.

The study is based in the review of empirical studies and has the purpose of find out what are the determinants of FDI, in order to discover the motivations that companies are looking for investment.

This study adopts the literature review research method. The research includes a systematization of the study of FDI determinants with 50 empirical studies on different countries around the world that differ in size, economic development, business environment, governance, infrastructure, trade openness, location and the identification of the main determinants that attract FDI. A qualitative approach will be adopted in the present study given the current context and the scope of the research. A qualitative research is more suitable for the study due to the distinct objective of providing a complete understanding and interpretation of a phenomenon [9].

Data used for the research comes from a variety of sources that includes various academic journals, books and publications.

3.1 Opportunities and Challenges to support Recovery

The COVID-19 crisis had a negative impact on the world project finance presented with consequences in the decline of FDI (table 4). In Developing Countries, between 2015-2019 projects of 417.7 billion of dollars were presented when in the period of 2010-2014 projects of 616.1 billion of dollars were presented, a decrease of -198.4 billion of dollars (-32% of investment).

The biggest drop was in the Power sector where the value of the investment was 73.4 billion dollars in the 2015-2019 period, a reduction of -89.4 billion dollars (-55%), compared to the period 2010-2014 where the investment amounted was 162.8 billion dollars (table 4).

Table 4. Project Finance by sectors (Developing Countries)

	Announced projects (billion dollars)		**Change (%)**
	2010 - 2014	2015 - 2019	
Total	**616.1**	**417.7**	**-32**
Number of projects	**478**	**676**	**42**
Power (excluding renewable energy)	162.8	73.4	-55
Number of projects	*144*	*117*	*-19*
Climate change mitigation	117.3	125.7	7
Number of projects	*80*	*334*	*318*
Transport services	310.2	191.6	-38
Number of projects	*209*	*178*	*-15*
Tele communications	8	4	-50
Number of projects	*10*	*8*	*-22*
Water and sewerage	17.8	23.0	29
Number of projects	*35*	*41*	*17*

Source: UNCTAD, 2020 [3]

In the Least Developed Countries case (table 5), between 2015-2019 projects of 33.5 billion of dollars were presented when in the period of 2010-2014 projects of 41.7 billion of dollars were presented, a decrease of -8.2 billion of dollars (-20% of investment).

The biggest drop was in the Transport services sector where the value of the investment was 9.9 billion dollars in the 2015-2019 period, a reduction of -13.1 billion dollars (-57%), compared to the period 2010-2014 where the investment amounted was 23 billion dollars.

Table 5. Project Finance by sectors (Least Developed Countries)

	Announced projects (billion dollars)		**Change (%)**
	2010 - 2014	2015 - 2019	
Total	**41.7**	**33.5**	**-20**
Number of projects	**40**	**79**	**99**
Power (excluding renewable energy)	6.7	9.5	41
Number of projects	*13*	*18*	*37*
Climate change mitigation	11.5	13	13
Number of projects	*15*	*39*	*170*
Transport services	23	9.9	-57
Number of projects	*10*	*18*	*88*
Tele communications	0.3	0.5	85
Number of projects	*1*	*1*	*-29*
Water and sewerage	0.2	0.6	142
Number of projects	*1*	*2*	*200*

Source: UNCTAD, 2020 [3]

These indicators highlight a decrease in the number of projects presented compared to previous trends before the 2015-2019 period, where the marked sectors grew more than 40 percent compared to the 2010-2014 period.

In the 2015-2019 period, the global value of projects decreased -32 percent in developing countries and -20 percent in least developed countries, especially due to the reduction in the size of projects in sectors such as energy.

Despite negative trends, investment in public health and the digital economy can be increased now and in the medium term. The European Investment Bank announced an agreement with the World Health Organization to strengthen financial support for the needs of the COVID-19 crisis.

These investments have the health system and the primary health-care services target in low-income and middle-income countries. The World Health Organization strategy is to develop diagnostic tests and vaccines for COVID-19 treatment [3].

Another challenge is to support the most dependent countries, those that depend on Official Development inflows, Remittances and FDI inflows (table 6).

Table 6. Most dependent countries on official assistance, remittances and FDI inflows (2019)

	Net Official Development Assistance inflows (2018)	Remittances (2019)	FDI inflows (2019)
	Millions of dollars		
Least developed countries			
Tuvalu	19	4	0.3
% of GDP	*44%*	*10%*	*0.7%*
Kiribati	74	20	0.5
% of GDP	*38%*	*10%*	*0.3%*
Central African Republic	656	..	26
% of GDP	*30%*	..	*1.1%*
Yemen	7 985	3 771	-371
% of GDP	*29%*	*14%*	*-1.2%*
Afghanistan	3 789	804	39
% of GDP	*19%*	*4%*	*0.2%*
Other developing countries			
Palau	84	2	22
% of GDP	*30%*	*1%*	*7.6%*
Nauru	32	7	..
% of GDP	*25%*	*5%*	..
Micronesia (Federated States of)	99	23	..
% of GDP	*25%*	*6%*	..
Marshall Islands	54	31	4
% of GDP	*24%*	*14%*	*1.9%*
Tonga	87	183	13
% of GDP	*19%*	*41%*	*2.4%*

Source: UNCTAD, 2020 [3]

The Official Development Assistance and Remittances have a huge importance in support the least developed countries and other developing countries as the table 6 shows. Some of these countries attract a few FDI inflows and they depend on the Official Development Assistance and Remittances, so they must be supported by the policies designed [3].

Governments must adapt their policies to respond quickly to the COVID-19 crisis. These actions must be effective in developing countries supporting short-term research and development and fiscal measures and also a long duration strategy. The moment has the opportunity to increase cooperation among countries, especially between developed and developing countries, to share technology and innovation through the creation of a network of centres of scientific excellence [3].

Government policies must pay special attention to the e-commerce sector to implement reforms and adequate investments to reach their potential. An UNCTAD research says that most of governments have e-commerce strategies and want to share them with their partners to increase digital compliance [3].

Cooperation should gather together various entities such as governments, international non-governmental organizations, research and development agencies and the private sector to increase digital commerce and outline a beneficial group strategy. Collaboration between small and large companies is essential to create an innovation environment, exploring new innovation models that strengthen companies' competitive advantages. Developing countries must build new innovation centres where they can concentrate all the knowledge involved in innovation and pay special attention to learn, assimilate and decentralize knowledge and technologies to other partners and to the market. Research and Development (R&D) plays a key role in responding to the health, economic and social consequences of the COVID-19 crisis [3].

3.2 Model

Economic factors are the most important category as locational FDI determinants [22]. This study takes as a conceptual anchor the most important FDI economic determinants namely natural resources, market size, growth market, market openness, skilled labour, education system, tax policy and innovation.

Natural resources. Looking back on history and more specifically in the 19th century, natural resources, mainly minerals and primary products for the industrialized countries of Europe and North America, have become the most important determinant of FDI. In the 20th and 21st centuries, natural resources remain important but the importance of the primary sector was overtaken by the secondary and tertiary sectors [26].

Dunning's eclectic theory holds that countries with more natural resources attract more FDI. Companies can increase their competitiveness by investing in markets that have higher quality natural resources and at a lower real cost than in the country of origin. This motivation is more important in the case of industrial companies that need to ensure low production costs and continuous sources of supply [12]. A set of empirical studies confirms natural resources as a determinant of FDI [10, 11, 16, 17].

Market size and growth. The size of a market is one of the main reasons that attract FDI, considering the size and population of that market. Large markets may have more companies which gives them greater benefits from economies of scale. [26]. Several studies [14, 21] concluded about the relevance of market size and market growth as a reason of attraction of FDI.

Innovation. Policies to increase investment in innovation systems are fundamental due to the creation of new assets [23].

Several studies [15, 23, 25] found evidence that the innovation is a determinant of attraction of FDI.

An internationalization model is built to support the strategy of economic recovery (figure 1).

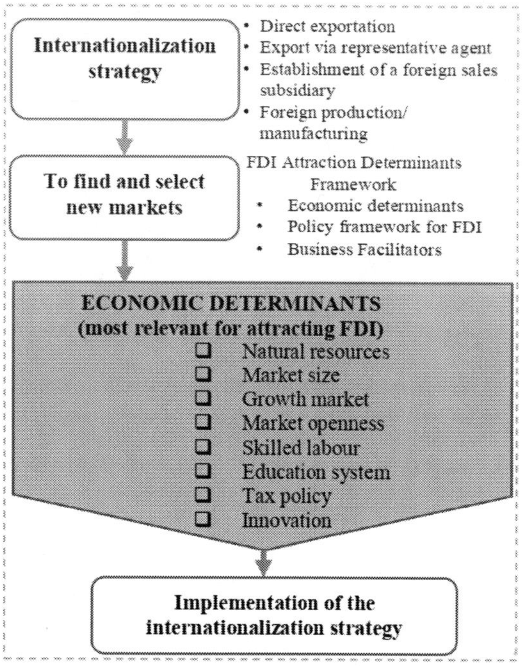

Fig. 1. Internationalization Model for Economic Recovery

Source: Made by the authors of the article

FDI flows are crucial to economic recovery in developed and developing economies. The Internationalization Model is an international reference framework to encourage FDI in economies affected by the crisis.

The first step in the Internationalization Model is to define the internationalization strategy which could be direct exportation, export via independent representative, establishment of a foreign sales subsidiary or investing in foreign production/manufacturing.

The second step in the Internationalization Model is to find and select new markets based in the Framework to FDI Attraction Determinants where the economic factors are the most important determinants [26] to attract FDI in host countries.

The Frame of Reference to FDI Attraction Determinants developed by UNCTAD has three dimensions (Economic factors, Support Structure to FDI Policy and Business facilitators). The Support Structure to FDI Policy and the Business facilitators are dimensions that are very similar in the context of the liberalization and globalization among countries and the Economic determinants become the most important determinants for attracting FDI.

The most important Economic determinants that must be analyzed by governments, institutions and companies, by each entity separately or by the various entities together, to find and select new markets are natural resources, market size, growth market, market openness, skilled labour, education system and tax policy. The countries that have the best indicators should be the selected.

The third step is to implement the internationalization strategy in the markets, previously selected, preferably by a strategic group formed by interests at national, sectorial and business level.

4 Conclusions

Due to a huge unpredictability, it is impossible to predict the consequences of the government policies on the economy. However, the economic recovery is going to happen in a short or medium term as the vaccination in the world against COVID-19 progresses.

The enormous task faced by national governments is to find the right solutions on national, industry and firm level. Economic activities seriously affected in some sectors, like tourism, should be replaced by secure activities with economic viability.

In future, the pandemic crisis will have multiple damage in several channels, in human capital due to closure of businesses, in the education system due to the loss of schooling, in the foreign investment due to a lower demand and a lower investment.

This study gives to governments, institutions and firms an internationalization model for attraction of FDI, that incorporates a Support Structure of FDI Policy, Business facilitators and the more important factors to support the attraction of FDI, the economic determinants, to help and accelerate the economic recovery.

References

1. Malpass, D. (2021). World Bank Group President. Countries Can Take Steps Now to Rebuild from COVID-19 Available at: https://www.worldbank.org/en/news/press-release/2020/06/02/countries-can-take-steps-now-to-speed-recovery-from-covid-19, accessed in 11.01.2021.
2. Schwab, K. (2020). World Economic Forum. The Global Competitiveness Report Special Edition 2020: How Countries are Performing on the Road to Recovery. Available at: https://www.weforum.org/reports/the-global-competitiveness-report-2020, accessed in 11.01.2021.
3. UNCTAD (2020). United Nations Conference on Trade and Development. Pandemic on trade and development. Transitioning to a new normal. Available at: https://unctad.org/system/files/official-document/osg2020d1_en.pdf, accessed in 12.01.2021.

4. WB (2020). World Bank. Countries Can Take Steps Now to Rebuild from COVID-19. Available at: https://www.worldbank.org/en/news/press-release/2020/06/02/countries-can-take-steps-now-to-speed-recovery-from-covid-19, accessed in 12.01.2021.
5. WB (2020a). World Bank. Reviving FDI flows is crucial to economic recovery in developing economies. Available at: https://blogs.worldbank.org/voices/reviving-fdi-flows-crucial-economic-recovery-developing-economies, accessed in 13/01/2021.
6. OECD (2020). Organization for Economic Cooperation and Development. FDI in Figures – October 2020. Available at: https://www.oecd.org/investment/investment-policy/FDI-in-Figures-October-2020.pdf, accessed in 12.01.2021.
7. Botelho, J. (2015). Internationalization of Enterprises: Contributions to Building a Model Support Analysis and Implementation Strategies for Internationalization. Doctoral dissertation. Evora University, Portugal.
8. Anuchitworawong, C. & Thampanishvong, K. (2014). Determinants of foreign direct investment in Thailand: Does natural disaster matter? International Journal of Disaster Risk Reduction. http://dx.doi.org/10.1016/j.ijdrr.2014.09.001.
9. Ritchie, J., Lewis, J., Nicholls, C. M., & Ormston, R. (Eds.). (2013). Qualitative Research Practice.
10. Ledyaeva, S. (2009). Spatial econometric econometric analysis of foreign direct investment determinants in Russian regions. World Economy, 32 (4), 643-666.
11. Cheung, Y.-W. & Qian, X. (2009). Empirics of China's outward direct investment. Pacific Economic Review, 14 (3), 312-341.
12. Dunning, J. H. & Lundan, S. (2008). Theories of foreign direct investment. in John H. Dunning and Sarianna M. Lundan (orgs.), Multinational Enterprises and the Global Economy, Cheltenham: Edward Elgar Publishing Limited, pp. 79-115.
13. OECD (2008). Organization for Economic Cooperation and Development. Benchmark Definition of Foreign Direct Investment, fourth edition. Available at: https://www.oecd.org/daf/inv/investmentstatisticsandanalysis/40193734.pdf, accessed in 12.01.2021.
14. Hansen, H. & Rand, J. (2006). On the causal links between FDI and growth in developing countries. The World Economy Vol. 29, Issue 1, pages 21-41, January.
15. Na, Lv and Lightfoot, W.S. (2006). Determinants of foreign direct investment at the regional level in China. Journal of Technology Management in China, Vol. 1, No 3, pp. 262-278.
16. Asiedu, E. (2005). Foreign direct investment in Africa: The role of natural resources, market size, government policy, institutions and political instability (April). University of Kansas - Department of Economics. Available at SSRN: https://onlinelibrary.wiley.com/doi/abs/10.1111/j.1467-9701.2006.00758.x, accessed in 16.01.2021.
17. Deichmann, J., Eshghi, A., Haughton, D., Sayek, S. & Teebagy, N. (2003). Foreign direct investment in the Eurasian transition states. Eastern European Economics, vol. 41, n.1, January-February, pp. 5-34.
18. Gorter, J. & Parikh, A. (2003). How sensitive is FDI to differences in corporate income taxation within the EU? De Economist, 151(2), pp. 193-204.
19. Noorbakhsh, F., Paloni, A. & Youssef, A. (2001). Human capital and FDI inflows to developing countries: new empirical evidence. World Development, 29 (9), pp. 1593-1610.
20. Zhang, K. H. (2001). What Attracts Foreign Multinational Corporations to China?. Contemporary economic Policy, Vol. 19, No. 3, pp. 336-346.
21. Shatz, H. & Venables, A. (2000). The geography of international investment. World Bank Policy Research Working Paper N°. 2338. Available at: https://openknowledge.worldbank.org/bitstream/handle/10986/18843/multi_page.pdf?sequence=1, accessed in 16.01.2021.
22. Lipsey, R. (2000). Inward FDI and economic growth in developing countries. Transnational Corporations, vol. 9, n° 1, Abril.
23. OECD (2000). Main determinants and impacts of foreign direct investment on China's economy. Working papers on international investment, Number 2000/4.
24. Nicoletti, G., Scarpetta, S. & Boylaud, O. (1999). Summary indicators of product market regulation with an extension to employment protection legislation. OECD, ECO Working Paper n 226. Available at: http://papers.ssrn.com/sol3/papers.cfm?abstract_id=201668, accessed in 16.01.2021.
25. Zhang, K. & Markusen, J. (1999). Vertical multinationals and host-country characteristics. Journal of Development Economics 59 (2): 233–252.
26. UNCTAD (1998). United Nations Conference on Trade and Development. World Investment Report 1998. Trends and Determinants. Available at: https://unctad.org/system/files/official-document/wir1998_en.pdf, accessed in 15.01.2021.

Digital Technological Solutions Are an Important Factor in The Effective Development of Higher Education in the Republic of Uzbekistan

Irina Zhukovskaya[1], *Sharafutdin* Xashimxodjayev[1*], and *Elena* Pilipenko[1]

[1]Tashkent State University of Economics, Department «Digital economy and Information Technology», Tashkent, Uzbekistan

Abstract. This article shows the main directions of improving higher education in the Republic of Uzbekistan in the context of the formation of the digital economy. The authors' research is based on real examples of the Tashkent State Economic University. The article proves that the digital transformation and change in the methods and means of teaching students and undergraduates in the context of the pandemic caused by the spread of COVID 19 infection require new technological approaches to teaching and managing the university from the teaching staff and university administration. In this article, the authors conclude that digital technologies individualize the educational process, diversify, develop students' independence, and develop creative thinking skills. Thus, the quality of education improves, and it is not only about the full and high-quality fulfillment of the requirements of the curriculum: classes begin to meet the personal interests and needs of students and undergraduates, which is undoubtedly an additional incentive to master the educational material. Thus, modern technological solutions based on the latest digital technologies contribute to an increase in the efficiency and quality of the educational process in a higher educational institution.

1 Introduction

The formation of the digital economy on a global scale has a great impact on the development of industries and spheres of national economic systems. Technologies such as cloud computing, blockchain, mobile Internet are becoming widespread, robotics, geonomy, new means of energy storage are actively developing, digital platforms, web services, big data processing and analytics technologies, etc. are becoming widespread.

In the Republic of Uzbekistan, a whole range of measures is being taken to introduce digital technological solutions in the industry and the sphere of the national economy. One of the important stages in the development of the digital economy was the Decree of the President of the Republic of Uzbekistan Sh.M. Mirziyoyev "On approval of the Strategy" Digital Uzbekistan 2030 "and measures for its effective implementation", signed on October 5, 2020.

Scientific research by scientists testifies that education is not just a way of transferring knowledge, it accumulates the cultural heritage of the nation, promotes the empowerment of a person, and forms his moral ideals. Education in general and higher education in particular, is one of the conditions for economic growth, increasing the material well-being of the country's population. The level of education has a significant impact on various aspects of human life: health, perception of culture and art, professional and political mobility, etc. It is no coincidence that in the Universal Declaration of Human Rights, the right to education is enshrined as one of the basic human rights.

The role of education in increasing labor productivity, increasing the production of goods and services, as well as increasing workers' income is assessed within the framework of the theory of the so-called human capital. Within the framework of this theory, human capital is assigned a central role, and the most important goal is the creation of such a socio-economic system that will ensure the development and use of the capabilities of each person. Human potential is the main part of the country's national wealth. In developed countries (Japan, Canada, USA, etc.), the share of human capital in the country's national wealth is 60 to 80%. However, a high level of education is not a guarantee of economic growth in the country. The reasons for the lag in the economic development of countries with a high level of education may be the following:

- irrational structure of education;
- ineffective use of the existing educational potential (for example, when a person is forced to perform work that does not correspond to his level of education);
- low level and inadequate structure of investment in education;
- low quality of education;
- insufficiently clearly chosen strategy for the development of education.

*Corresponding author: sharaf_x@mail.ru

In the modern period, the main directions of the development of higher education in the Republic of Uzbekistan are:

- comprehensive assistance to the professional education market;
- the priority of forming a "talent pool" for the development of promising markets and technologies, training of personnel for those industries that have not yet begun to present a wide effective demand for graduates (creating a scholarship fund for those students who will choose areas and specialties that are priority for the state);
- the formation of innovative and corporate cooperation as a form of integration of diverse educational institutions of higher and secondary vocational education, research institutes, basic enterprises and organizations, within which educational programs of different levels of vocational education will operate;
- the priority of the scientific function of the higher education system, which directly affects the change in the principles of organizing production;
- the increasing influence of higher education on the labor market through the creation of new technologies and the opening of new areas of social activity.

2 Literary sources and legal documents

In the Republic of Uzbekistan, the development of higher education in the context of the formation of the digital economy is receiving close attention. This is evidenced by the adoption of a number of government documents.

In particular, on October 5, 2020, the Decree of the President of the Republic of Uzbekistan No. UP-6079 "On approval of the "Digital Uzbekistan - 2030" strategy and measures for its effective implementation" was adopted [1], which emphasizes that without the development of digital infrastructure and raising the level of knowledge in the field of digital technologies, further development of the country is impossible. In addition, the digital development of industries and spheres of the national economy requires qualified personnel. In this regard, in the last 3 years alone, a number of decrees and resolutions of the President have been adopted, contributing to raising the level of the higher education system to a new, higher level. (Resolution of the President of the Republic of Uzbekistan "On measures for the further development of the higher education system" No. PP-2909 dated April 20, 2017 No. 18, art. 313, No. 19, art. 335, No. 24, art. 490, No. 37, art. 982; Resolution of the President of the Republic of Uzbekistan "On additional measures to improve the quality of education in higher educational institutions and ensure their active participation in the large-scale reforms being implemented in the country" No.PP-3775 dated 05.06.2018; Resolution of the President of the Republic of Uzbekistan "On measures improvement of the training system for qualified personnel in demand and the development of scientific potential at the National University of Uzbekistan named after Mirzo Ulugbek in 2019-2023"). In addition, the President of the Republic of Uzbekistan Sh.M. Mirziyoyev issued a decree "On approval of the concept of development of the higher education system of the Republic of Uzbekistan until 2030", which defines the key areas of comprehensive reform of higher education in the country. This document defines the key tasks of transforming higher education in Uzbekistan.

These include:

- increasing the level of knowledge, skills and abilities of future specialists who have the necessary competencies, high moral principles and are able to think independently;
- modernization of higher education;
-social and economic development based on the latest educational approaches.

The reform of vocational education also provides for the creation of "career centers" to assist students with internships and finding a job in the labor market, the formation of an employer base and the effective use of the potential of graduates based on the optimization of the structure of scientific and educational institutions.

The activities of the sphere of higher education are implemented in accordance with a set of government documents and are aimed at cardinal transformations of the teaching and educational process in universities in order to achieve high rating indicators at the world level.

As the present study shows, a wide range of works by domestic and foreign scientists is devoted to the application of modern information and communication technologies in education.

For example, in the works of such scientists as Akimova O.B. [2], Kvashin A.Yu., Kramarenko N.S. [12] it is shown that in the development of higher education in the digital era, technological security of the educational process and digital literacy of the teaching staff are of great importance.

The publications of foreign scientists Anderson T.D., Clark T., Detweiler R. [7], Dillon C., Hillman D. show that the use of methods and means of distance education has a significant place in the development of digital education. The same concept is expressed by such Russian scientists as Ardovskaya R.V., Bakalov V.P., Batygin B.S., Dikov N.R., Ivanchenko D.A. [9], Ignatova N.Yu. [10], Isaev Yu.V., Kanaev V.I., Kozlova N.Sh. [11], Komarova I.N, Mateev D.A., Muromtsev V.V., Prokhorova N.S. and etc.

In particular, we have studied the scientific works of such authors as Alimov R.Kh [3], Begalov B.A. [4], Bekmuratov T.F., Bobodzhanov A.B., Vertakova Yu.V., Tolstykh T.O., Shkarupeta E.V., Dmitrieva V.V. [5], Gayibnazarov B.K., Gulyamov S.S.[6], Dadabaeva R.A., Dzhanadilov Sh.U., Zhukovskaya I.E. [8], Kuznetsova S.A., Markova V.D.[12], Musaliev A.A., Odilov Shch.G., Swan M. [15], Xashimxodjayev Sh.I. [16] etc.

Analysis of literature sources also shows that scientists around the world are studying the phenomenon of the development of digital technologies.

In fig. 1. presents the forecast of analysts of the American analytical company McKinsey International Data Corporation (IDC), specializing in research of the information technology market. As can be seen from Fig. 1, the potential economic effect from the introduction of such technologies should be from 14 to 33 trillion USD.

Experience shows that digital technological solutions cover all sectors and spheres of the modern economy.

Higher education in the modern period is also unthinkable without the use of the latest digital technologies and mechanisms, as well as the capabilities of artificial intelligence.

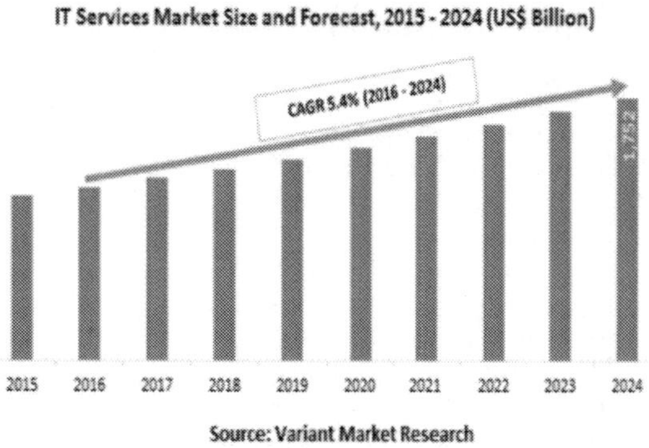

Fig. 1. Forecast of the growth of the economic effect obtained through the introduction of advanced technological solutions

Source: *data from the American analytical company McKinsey International Data Corporation (IDC).*

Analysis of literary sources shows that in the context of the digitalization of the world economy, the sphere of higher education needs new management and technological solutions that are designed to bring higher education institutions to the highest levels in international rankings.

The formation and development of the digital economy in Uzbekistan is inextricably linked with the development of information and communication technologies (ICT). Conducted research has shown that a distinctive characteristic feature is the all-round penetration of ICT into various spheres of public life. According to the results of the research, the specialists came to the conclusion that the successful development of the digital economy in a separate country is an important role for the state.

The development of the digital economy in the republic is caused by an increase in the field of those types of economic activity, which are associated with the production of intelligent and industrial products. The functioning of the digital economy creates the basis for innovation and the formation of human capital. In Uzbekistan, the basis of the digital economy is a set of industries in the sphere of information services. There is a growing dependence of the functioning of the economy on information. Moreover, we are talking about increasing the share of such areas as education, information and communication and intellectual services.

The modern stage of development of the digital economy in Uzbekistan is characterized by the period of state of affairs and high dynamics of development.

The modern stage of development of the digital economy can be assessed on the basis of the following characteristics:

1) Computerization of economic activity and the use of electronic internal computer interoperability on the basis of electronic document circulation;

2) Connections to the Internet are not only computers, but also the Internet-thins, directed at the creation of business models;

3) The use of ICT for the involvement of citizens in social activity, the development of new effective forms of interaction with the state authorities;

4) The use of effective feedback mechanisms in the framework of the electronic government to ensure the participation of citizens in public life;

5) Mobilization of citizens by means of the Internet for solving publicly important tasks, etc.

The basic institutions of the digital economy include electronic commerce, electronic employment, and electronic government, which form the institutional and educational institutions.

In the framework of the electronic commerce as the basic institute of the digital economy, IT companies enter into cooperation with the representatives of other companies, the state and the scientific community. The electronic commerce allows companies to more efficiently and flexibly carry out internal operations.

3 Modern approaches to the organization of education in a higher educational institution in the context of digital transformation

In the conditions of the formation of the digital economy, serious qualitative changes are taking place, which are reflected in the development of all sectors and spheres of the economy, including the modernization of the education system. Scientific research by scientists testifies that education is not just a way of transferring knowledge, it accumulates the cultural heritage of the nation, promotes the empowerment of a person, and forms his moral ideals. Education in general and higher education in particular, is one of the conditions for economic growth, increasing the material well-being of the country's population. The level of education has a significant impact on various aspects of human life: health, perception of culture and art, professional and political mobility, etc. It is no coincidence that in the Universal Declaration of Human Rights, the right to education is enshrined as one of the basic human rights.

Digital technologies are dramatically transforming the content of the taught disciplines and the way they are presented. This is not only electronic presentations or the use of video, but also direct connections to electronic databases, news, ongoing forums, video broadcasts. In modern higher education, virtual (VR) and augmented reality (AR) systems are increasingly used. VR and AR technologies are used in immersive education (IE) programs. Such programs include the use of modern information technologies in the learning process, which takes place inside various virtual worlds and simulations, often in a playful way [8]. This type of training helps to increase engagement, communication between students and interest in the subject, as well as the implementation of projects for real enterprises and organizations.

The digital transformation of the economy is based on the following technologies:

- data mining;
- business intelligence (eng. Business intelligence);
- machine learning (eng. Machine learning);
- artificial Intelligence;
- big data technologies;
- internet of Things;
- blockchain.

Let's consider the specifics of each of these technologies. Data mining - research and discovery by a "machine" (algorithms, artificial intelligence) in raw data of hidden knowledge that was not previously known, non-trivial, practically useful, available for human interpretation. Business intelligence - systems, tools and technologies for data analysis of organizations. Digital transformation is transforming business intelligence into intelligence. It is intelligent business intelligence that is the trend that determines the development of tools that effectively turn a "caterpillar into a butterfly". Machine learning is a class of artificial intelligence methods, a characteristic feature of which is not a direct solution to a problem, but learning in the process of applying solutions to many similar problems. The essence of the machine learning process is the use of algorithms that allow some technical system to draw conclusions from data without following certain rules. Artificial intelligence (AI; English artificial intelligence, AI): the science of creating intelligent technologies; the property of intelligent systems to perform creative functions that are traditionally considered the prerogative of a person. AI allows you to automate and intellectualize repetitive learning and retrieval processes through the use of data. AI functionality and technologies integrate into existing products and intellectualize them at different levels. AI combined with large amounts of data improves various technologies that are used in all areas of the economy. The Internet of Things (English internet of things, IoT) is the concept of a computing network of physical objects ("things") equipped with built-in technologies for interacting with each other or with the external environment, considering the organization of such networks as a phenomenon that can restructure economic and social processes, eliminating the need for human participation from part of actions and operations. Blockchain (English blockchain, originally a block chain - a chain of blocks) is a continuous sequential chain of blocks (linked list), built according to certain rules, containing information. In the digital economy, blockchain is a very promising digital technology.

The world of education and science is becoming global, the development of digital technologies leads to the removal of language barriers. Universities and individual teachers are actively entering the MOOC market (MOOC - Massive on-line open course) - an already established form of distance learning with open access on the Internet. The number of students who study remotely is increasing, since traditional educational programs often do not keep pace with the dynamics of technology development [17].

The education sector is also undergoing great changes in the Republic of Uzbekistan. In secondary educational institutions of Tashkent and in the regions, such systems as "Electronic Diary", "Assessment of Teachers", "Electronic Student" have begun to work. The Ministry of Public Education and the Ministry for the Development of Information Technologies and Communications were instructed to create 14 special schools for training programmers this year and 82 - next year [9]. In higher education institutions in the context of the pandemic caused by the spread of the virus infection COVID - 19, digital platforms were introduced, which made it possible to conduct online education in universities. Digital transformation implies improving the system of professional development of the teaching staff. In particular, on the basis of the Tashkent University of Information Technologies and IT-parks, a national system of training and certification of information technology teachers will be introduced. Obtaining a certificate upon completion of training will allow teachers to receive a 50 percent salary increase. School teachers will also be rewarded if their students successfully complete the One Million Programmers course [18].

At present, in Uzbekistan, within the framework of the implementation of the Presidential Decree "On measures for the further development of the higher education system", each higher educational institution establishes close partnerships with leading foreign universities and centers. The decree stipulates that well-established partnerships with foreign universities are capable of attracting at least 350 foreign highly qualified teachers and scientists to Uzbekistan to the educational process at universities annually. At the same time, work is underway to widely introduce advanced pedagogical technologies, curricula and teaching materials based on international educational standards into the educational process. Taking into account the prospects for the integrated development of regions and sectors of the economy, the needs of territorial and sectoral programs, target parameters of personnel training are formed in accordance with higher education, directions and specialties of training are optimized. Work is being consistently carried out to solve the problem of creating and introducing a new generation of textbooks into the higher education system, providing universities with modern educational and scientific literature. Work has been established to translate the latest foreign literature into the Uzbek language.

The digitalization of education makes it possible to gain access to information that was previously available only to experts and scientists. Most publishing houses specializing in educational literature have switched to electronic versions of textbooks and manuals. Thus, there is a transformation of teaching: students have the opportunity to study disciplines in any sequence, on any platform, using technical means and technologies convenient for them, and the task of the teacher is to help students navigate the vast amounts of information.

Thus, it can be noted that the transformation of higher education ensures that each student achieves the necessary educational results due to an individualized educational process, which in turn implies the use of a full range of digital technological solutions.

Currently, the Tashkent State University of economics (TSUE) has accumulated its own experience in using digital technologies in the educational process.

So, during the quarantine period caused by the COVID 2019 pandemic, educational activities at TSUE were carried out using a digital platform, on which teaching materials were placed by the teaching staff. Students studied educational resources displayed on the platform and submitted completed assignments for verification. Work on the basis of this platform allowed students to successfully pass tests in the studied disciplines at the end of the semester.

Practice has shown that distance learning is an interesting process based on the use of all kinds of gadgets. When organizing distance education, the most effective form of material placement is a digital platform. The educational material on the platform is presented in the form of modules, including methodological recommendations for studying the topic, visual and theoretical resources and explanations for practical tasks, links to the necessary literature.

In the course of analyzing the work of the platform for the semester, the university's leadership identified the positive sides and shortcomings in the functioning of the platform. The developed platform was easy enough for students to use. However, it was not possible to carry out laboratory tasks on this platform. In addition, the process of evaluating the practical assignments completed by the students, took a long time.

Assessment of the platform's functioning, analysis of international experience allowed the TSUE management to make a decision to introduce modern software Moodle (Modular Object-Oriented Dynamic Learning Environment) into the scientific process of a higher educational institution. Moodle is a free and open source platform.

The Moodle web application allows you to create a customized learning management system that allows teachers and students to interact effectively online.

Moodle was created at the Australian University of Technology, and from the outset was positioned as an open, easy-to-install and free educational site.

The advantage of Moodle is the ability to use it both on a stationary computer and on any mobile device with the ability to access the Internet.

The platform maximizes the preservation of the traditional values of full-time education. It is a modern, constantly evolving environment. The chat and forum system, which users actively use, makes it possible to quickly identify platform flaws and fix them.

This system is very mobile, it allows teachers to create all kinds of web courses and fill them with educational content. Elements of online courses are various interactive tasks, open-ended questions, text pages, dictionaries, links, files, cases and much more.

In addition, the following positive aspects of the Moodle program should be noted:

- multifunctionality and ease of use, both on the part of students and on the part of the teaching staff;
- the ability to customize and edit student accounts;
- each student can choose a convenient time for study and vary the presentation of the material, both in time and in speed;
- a huge set of components for effective exchange of information: a regular online lesson, chat, questionnaire, forum, terminological dictionary, etc .;
- all passed material, as well as intermediate and final control with teachers' comments, are saved in the system. You can return to them at a time convenient for the student;
- an objective assessment of the student's knowledge;
- the possibility of organizing an e-mail distribution at the request of the student;
- increasing the level of assimilation of educational material through the use of innovative teaching methods;
- the teacher for any course is constantly in touch with the student;
- the functionality of the platform allows you to conduct the learning process both with one student and with a whole group of students.

World practice shows that the improvement of the technical base of universities, the use of artificial intelligence methods, the development of a digital educational environment, the use of electronic scientific publications in various foreign languages of libraries around the world contribute to the achievement of high results in the preparation of competitive specialists.

Based on modern reality, the following priority areas of interaction between universities in Uzbekistan and foreign universities and research centers can be identified: - branches of foreign universities; - joint faculties; - joint educational programs of double degrees; - attraction of foreign specialists; - advanced training and internships. A very important decision in higher education today is the decision made in 2018, which grants the right to basic (leading) universities, based on the needs of customers, to independently develop and approve curricula and programs in agreement with the Ministry of Higher Education. In addition, in 2018, the universities of the Republic received the right to accept foreign citizens for bachelor's programs outside the admission quota through interviews, without passing tests. The credit-modular system is being implemented in a pilot mode for all forms of education.

In recent years, optimization of directions and specialties of education in universities, taking into account the prospective development of industries and regions, has become widespread in the Republic of Uzbekistan. For example, from the 2018-2019 academic year, the universities of Uzbekistan began training personnel in 66 new areas of bachelor's degree and 48 new specialties of magistracy.

However, it should be noted that with the many possibilities of digital technologies in the field of improving the educational process at a university, their integration into the educational process also causes certain problems.

First of all, these are the problems associated with technical equipment and the ability of each teacher and student to access the unified educational environment of the university from remote regions of the country.

The second problem is that it is necessary to constantly retrain the teaching staff of the university in accordance with the requirements of digitalization. In the context of the development of the digital economy, each teacher needs to improve their knowledge and skills in the field of digitization of educational and methodological material, the development of electronic educational publications, the creation of open educational courses, conduct online classes with constant improvement of communication skills.

Digital technologies contribute to the implementation of the "Digital University" concept, which includes several components at once:

- a complex of information systems for university management;
- implementation of the educational process online;
- effective management of the educational process based on the creation of conditions for taking into account the individual abilities of students;
- the formation of the key competencies of the digital economy among students, faculty, personnel serving the educational process of the university.

We would also like to note that the digital economy contributes to the development of interaction between the university and employers. This is clearly confirmed by the annual holding of Career Days at the Tashkent State Economic University. In particular, we can say that digital technologies help to identify promising professions, contribute to the formation of graduates' competencies based on the requirements of the economic market.

4 Conclusion

Today, digital technologies throughout the world are developing rapidly, being both an engine of economic growth and a sector that has already significantly changed and transformed economic processes in other industries and continues to influence the formation of a new type of economy based on knowledge, use of information and products of human intellectual labor. In the course of writing this article, the authors revealed that the telecommunications infrastructure in Uzbekistan, compared to neighboring countries, is developing relatively well and stably, but, at the same time, it is necessary to take measures to achieve a higher level comparable to the level of advanced foreign countries. Studies of modern scientists indicate that the successful cooperation of the Republic of Uzbekistan in the field of ICT with foreign partners indicates the great potential of Uzbekistan in this area. In addition, ICT analysts note that Uzbekistan, thanks to constant development, is potentially able to act as a regional "broadband backbone" and telecommunications hub for neighboring countries. Big data and blockchain technologies are becoming more and more widespread in industries and spheres of the economy, artificial intelligence systems and robotization are rapidly developing.

To achieve high results in the preparation of qualified specialists, it is necessary not only to increase the technical capacity of the university, but also to constantly improve the methods and forms of teaching in the context of the digitalization of the economy. And also to improve the research methodology on the effectiveness of methods, forms, teaching aids in an open educational space; their methodological substantiation, including the ratio of traditional and e-learning, contact and independent work of students; methods of monitoring progress and the formation of educational results in the context of the transfer of the educational process to the global network.

Thus, the transformation of the world space in the context of the digitalization of the economy has a great impact on higher education, contributes to the solution of problems in improving the qualifications of personnel, personalization of education, and also makes it possible to integrate into the global educational space using the methods and means of digital technologies and take high steps in the international ranking of universities.

Summing up the above, it should be noted that modern higher education performs a number of socio-economic functions. It helps a person to become a full-fledged, comprehensively educated, in-demand person, and provides conditions for successful social interaction. Thus, we can conclude that the quality of higher education determines the effectiveness of economic development. With an increase in the rate of economic growth, the need for highly qualified specialists who are able to develop and introduce new methods and technologies increases, which has a positive effect on the system of education development. In turn, the development of the education system, leading to an increase in the number of highly qualified specialists, contributes to economic growth.

References

1. Decree of the President of the Republic of Uzbekistan dated October 5, 2020 No. UP-6079 "On the approval of the" Digital Uzbekistan-2030 "strategy and measures for its effective implementation" // Narodnoye Slovo 2020, October 6. (2020)
2. O.B.Akimova. Theory and practice of professional pedagogical education: [collective monograph]. (2013)
3. R.Kh. Alimov U.T. Hayitmatov *Prospects for the development of the digital economy in Uzbekistan* in Proceedings of the International Congress "Ragamli iкthisodiets: itisodiy rivozhlanish tendency larini modelashtirish va zamonaviy ahborot-communication technologylarini llash istiкbollari." 2 December, 2019, Tashkent, Uzberistan (2019)
4. B.A. Begalov, I.E. Zhukovskaya Methodological foundations of the influence of information and communication technologies on the development of the national economy. Monograph. (2018)
5. Yu.V. Vertakova, T.O. Tolstykh, E.V. Shkarupeta, V.V. Dmitrieva Transformation of management systems under the influence of digitalization of the economy: monograph. (2017)
6. S.S.Gulyamov, A.T. Shermukhamedov *Innovative development of entrepreneurship with the framework of scientific and research project "Global economic challenges and national economy development."* September, 2018 Tashkent-Jakarta (2018).
7. R. Detweiler Educational technology can cransform lectures into tutorials and tutorials into adventures (the chronicle of higher education, 2004).
8. I.E. Zhukovskaya. (Open Education, M, 2016)
9. D.A.Ivanchenko. *The role of the Internet space in the formation of the educational information environment,* Electronic resource (https://textarchive.ru/c-1953970.html).
10. N.Yu Ignatova Education in the Digital Age: Monograph. (2017)
11. N.Sh. Kozlova N.Sh. Modern Technologies in Network Security (2018)
12. N.S. Kramarenko, A. Yu. Kvashnin. Psychological and organizational aspects of the introduction of digital education, or how the introduction of innovations cannot be turned into a "digital collective farm" (2017)
13. F. Kubareva (EDP Sciences, Les Ulis, 2019)
14. S.A.Kuznetsova V.D. Markova. Propulsion physics (Sciences, Les Ulis, 2018)
15. M. Swan M. Blockchain. (Publishing house "Olimp - Business", 2017)
16. Sh.I. Xashimxodjayev *The impact of digital transformation on economic processes in the Republic of Uzbekistan*. in Proceedings of the International Congress "Modern Management: Problems and Prospects". SPb (2019)
17. Bulletin of Science and Practice Vol. 5. No. 8. 2019 https://www.bulletennauki.com DOI: 10.33619 / 2414-2948 / 45.
18. Resolution of the President of the Republic of Uzbekistan "On measures for the further development of the higher education system" dated April 20, 2017 No. 1111-2909 No. 18, art. 313, no.19, art. 335, no.24, art. 490, no. 37, art. 982). // https://www.lex.uz

Business Innovation in The Hotel Industry

Tetiana But[1*], *Daria* Mamotenko[1], *Valentyna* Zaytseva[1], *Tetiana* Pulina[1] and *Tetiana* Bukoros [2]

[1]National University "Zaporozhye Polytechnic", Ukraine, 64, Zhukovsky Street, 69000, Zaporizhzhya, Ukraine
[2]European institute of further education, Za humnami str., 508/28 Podhâjska, 94148, Podhâjska, Slovak Republic

Abstract. Today, business innovation is a hot topic. Therefore, it was proposed to use a competence-based approach to assess staff as a business innovation in the hotel industry. The paper investigates the theoretical and methodological principles of hotel staff evaluation using a competency-based approach. Scientific approaches are critically estimated, and aspects applied to material incentives management for hotel staff are analyzed. The scientific and theoretical basis of managing the staff's material stimulation in the hotel business on the basis of the competency approach is investigated and practical tools of its implementation in the hotel business activities are estimated. Methodical approaches to evaluating staff competencies as a component of material incentives management have been developed. The structure of remuneration and the system of bonuses based on staff competencies evaluation have been formed, which will help increase the efficiency of personnel management in the hotel industry.

1 Introduction

The current situation in the hotel services market requires forming new innovatively effective approaches and management concepts, the implementation of which is aimed at consolidating the staff to achieve socio-economic goals of enterprises. This will harmonize the relations in the systems of "employee-employer", "employee-employee" and "employee-consumer", increase professional competence, productivity and quality of staff. For hotel business enterprises, these management aspects are relevant because the staff is directly involved in creating a hotel product as a priority factor in generating enterprise income, maintaining and strengthening its market position and ensuring competitiveness.

Modern aspects of staff evaluation using a competency approach are covered in the studies of M. Armstrong [6], McClelland D.C. [13]. Boyatzis R.E. [11] analyzed the nature and factors of competence and the existence of the competence structure. J. Raven [12] studied the system of features by which the types of competence can be used.

Barybina Y.O., Lysenko M.O. [14] claims that the Ukrainian hotel market is characterized by a combination of a mixed model of incentives – an effective combination of tangible and intangible incentives focused on staff development and training.

Melnychenko S., Bosovska M., Poltavska O. [4] offer to make decisions as for an employee by the job quality evaluation on the basis of the competence approach – to make individual plans for the competence and career development, as well as to rotate staff.

As a result of many studies, various approaches, procedures and methods for organizing and conducting staff evaluations have been developed. However, at the present stage of innovative development, staff evaluation requires using all scientific approaches, their generalization and combination in order to achieve greater efficiency, optimality and effectiveness in business innovations in the hotel industry.

The purpose of the study is to generalize and develop the theoretical foundations of the hotel staff evaluation using a competency-based approach and to develop practical recommendations for improving the management system of material incentives for hotel staff.

2 Result and discussion

The coronavirus pandemic has caused serious damage to the hotel business in Ukraine, and it may take more than a year for the sector to recover. According to the study, 93% of respondents confirmed a general decline in their hotel's revenue, in 21% of hotels gross revenue decreased by 25-40%, in a third of hotels – by 40-60%, in 30% of hotels such reductions reached more than 60%, only 4% of hotels showed an increase in annual revenue, and 3% remained at the same level. According to the data: 66% optimized their costs by reducing the staff, 2/3 reduced prices for their services, 63% of respondents improved their product and services (repairing, repositioning, updating standards and conditions with suppliers, changing equipment, etc.), 1/3 of respondents introduced digital and marketing tools, 27% introduced alternative services (coworking, renting rooms for offices, etc.), 7% decided to repurpose some of the premises, for example, for gambling establishments rent [1].

In these conditions, the issues of optimal use of labor, information and financial resources in the hotel business remain underexamined. The capabilities of the hotel

* Corresponding author: tatyanabut.zp.79@gmail.com

business as for cost savings, strengthening the role of security in customer service, automation of its activities, personalization of services are largely related to how well the resource potential is studied, and what will contribute to its development.

Innovative management methods in modern HR management used by the world and EU leading countries are primarily based on a competency-based scientific approach. In the process of personnel management based on the competency approach, HR specialists or auditors determine a system of parameters that combines all the necessary competencies that employees must have in accordance with job descriptions, and obtain results of assessing the current state of human resources management. This allows identifying the level of each component of the set of employees'competencies. Research based on this approach should take into account the fact that professional competencies should be understood as dynamic, because a person is constantly developing knowledge, skills, abilities, so we can assume that the competencies development is a certain innovative process of increasing the organization's efficiency as a whole by improving the level of employees' professional skills. This is also one of the main tasks of the personnel management system and human resources development.

Of particular note is the definition suggested by J. Raven who gives a detailed interpretation of competence and understands it as a phenomenon consisting of "a large number of components, many of which are relatively independent of each other" with some components belonging more to the cognitive sphere, and others – to the emotional,... ..these components can replace each other as components of effective behavior" [2].

Obviously the competency approach in personnel management is distinguished by its redirecting personnel management goals from solving operational personnel problems to tasks of a higher strategic level, which go beyond the usual responsibilities of the personnel management service. It is not enough just to increase knowledge, improve employees' skills, expertise and behavior. It should result in increasing productivity and organizational change that can boost the competitiveness and efficiency of the corporation as a whole. Therefore, the goals of personnel management in terms of the competency approach are formulated to show that these processes can improve the organization by achieving higher results, changing employees' behavior, increasing productivity and efficiency of the organization.

In fact, this is an innovative concept of personnel management which focuses not on the process or operational results, but on the mechanisms and management models based on the competency approach and their impact on the organization's long-term effectiveness.

A focus on organizational development priorities requires, on the one hand, shaping personnel management functions based on a competency-based approach that can help implement business strategy, and, on the other – highlighting the need to intensify mechanisms of employees' self-development and self-organization, since it is impossible to take an active part in improving the organization's activities without involving the developed knowledge and individual abilities. The task of the personnel management system on the basis of the competence approach is to create an innovative environment that supports and guides the employees' self-development.

The concept of competency approach is an integrated concept that forms the basic principles of personnel management of a modern organization. Such principles include:

1. The principle of systematization when the use of a competency-based approach to personnel management should involve interconnected elements: goals, objectives, personnel management processes and be focused on short-term and long-term organizational goals.

2. The principle of complexity when tactical and strategic decisions in the competency approach use should be developed in accordance with the relationships between different areas and aspects of personnel management.

3. The principle of relevance with activities in the field of using a competency-based approach to personnel management corresponding to the personnel situation, offering solutions to current personnel problems of the organization, based on best practices and modern scientific developments.

4. The principle of continuity with activities in the field of competence approach to personnel management focusing on the progressive training and employees' development in order to improve performance, build capacity for growth and advancement during one's work in the organization;

5. The principle of succession with expanding the dominant values, unique knowledge, skills and experience acquired by employees in the organization in order to improve performance, maintain and increase its competitive advantages.

6. The principle of advanced development with extending the professional horizons and improving employees' skills to create a stock of knowledge, skills and abilities that may be needed to solve complex problems or non-standard tasks of the enterprise in the future.

7. The principle of self-development with creating conditions for employees' self-learning and self-expression in order to activate development internal mechanisms which motivate to work more effectively, increase job satisfaction, unleash professional and personal potential more deeply.

8. The principle of efficiency when the results of activities in the field of competency approach to personnel management should provide the required level of economic, organizational and social effects, thus helping to increase the efficiency of the organization [3].

It is considered relevant to use the concept of competence in the personnel management of the hotel business, and apply the results of the competence analysis in order to improve the processes of selection and hiring, employee development and motivation.

During a long time of research, the scientific literature has formed a system of features that can be used to characterize the types of competence (Fig. 2.1)

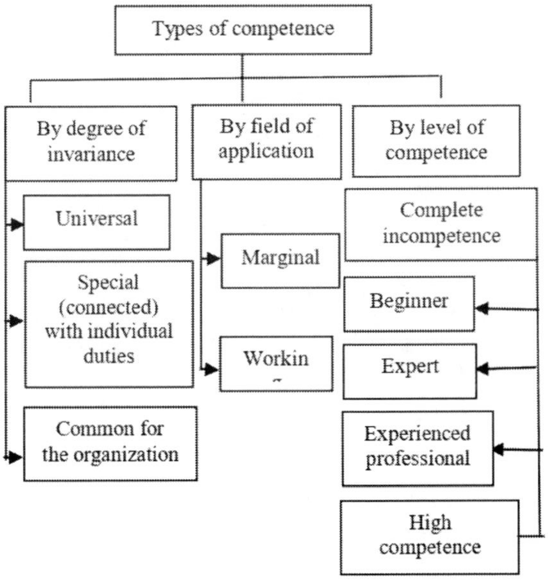

Fig. 1. Types of competence

In addition, competence can be common to the organization – it can be applied to all its employees. Or it can be applied to a group of related jobs in which work is similar but performed at different levels.

As hotel enterprises operate in the field of services, a significant part of which is provided to foreign consumers, we consider it appropriate to introduce an additional compensatory factor (criterion) – "level of foreign languages knowledge". It is known that the quality of hospitality services is the main prerequisite for forming demand for them, and therefore the key to the effective hotel business operation.

In this regard, the list of compensatory factors (criteria) for evaluating the positions of the hotel business includes the "consumer orientation". According to expert estimates, the list of compensatory factors (criteria) for the hotel industry can be systematized as follows Table 1.

The most important factors (criteria) for the hotel industry are the level of qualification, work experience, responsibility for decision-making, complexity and intensity of work, etc.

The conducted expert research of domestic practice of hotel enterprises determined the generalized list of compensatory factors (criteria) with validity of each of them according to strategic purposes and specifics of the enterprise's activity (matrix of job evaluation), formed the scale of factors (criteria) with their distinctive descriptions at various levels, and carried out a direct evaluation of each job Table 2.

Table 1. Compensatory factors (criteria) for evaluating the hotel business jobs and their validity.

No.	Compensatory factors (criteria)	Evaluation of factor (criterion) validity, points
1	Level of qualification (education)	100
2	Level of foreign languages knowledge	80
3	Work experience	140
4	Ability to search for and analyze information	80
5	Consumer orientation	80
6	Interpersonal communication skills	100
7	Work complexity and intensity	120
8	Responsibility for decision-making	130
9	Ability to manage staff	100
10	Impact on the result (error price)	50

When forming levels, it is necessary to have their calibrated sequence, which provides clear guidelines for assessing a particular factor.

Table 2. Matrix of job evaluation at the hotel industry enterprises.

No	Compensatory factor	Evaluation of factor validity, points	Distribution of factor evaluation in points by detection levels					
			A	B	C	D	E	F
1	Level of qualification (education)	100	20	40	60	80	90	100
2	Level of foreign languages knowledge	80	0	15	30	45	60	80
3	Work experience	140	0	20	50	80	110	140
4	Information search and analysis	80	0	16	32	48	64	80
5	Consumer orientation	80	0	10	20	30	50	80
6	Interpersonal communication skills	100	10	20	40	60	80	100
7	Work complexity and intensity	120	10	20	40	65	90	120

8	Responsibility for decision-making	130	15	30	50	75	100	130
9	Staff management	100	0	15	30	50	70	100
10	Impact on the result (error price)	50	0	10	20	30	40	50
		1000						

Consistent levels can be defined by describing specific skills, competencies or needs for a particular qualification, training or experience.

Job evaluation of the hotel industry was carried out by an expert committee on the example of PJSC "Salyut Hotel" in Kyiv in accordance with the matrices indicating the level of factor detection for each job. The expert committee included the enterprise's CEO, his deputy for organizational development, the head of accommodation service, the head of the personnel department, the authors as consultants. The results of job evaluation of PJSC "Salyut Hotel" are shown in Table 3.

Table 3. Job evaluation of PJSC "Salyut Hotel".

No.	Job at PJSC "Salyut Hotel"	Actual factor evaluation, points
1	CEO	950
2	Deputy CEO for organizational development	804
3	Methodist of the physical health room	240
4	System administrator	504
5	Secretary-assistant	322
6	Chambermaid	150
7	Cleaner	85
8	Linen-keeper	85
9	Head of the floors	372
10	Head of accommodation service	532
11	Doorman	165
12	Receptionist-cashier	286
13	Auxiliary worker	85
14	Driver	210

Then, it is necessary to analyze which jobs are included in the groups. For example, if, according to the results of the expert assessment, the value of jobs with different qualification characteristics is similar, they can be placed at the same level of the job structure.

Thus, the purpose of designing a hierarchical job structure based on their evaluation in the hotel industry is to create a flexible mechanism for managing staff motivation grounded on a clear and transparent system that will unite all existing jobs in the company into certain groups and divide them by value levels for the enterprise.

According to the proposed method of designing the hierarchical job structure (remuneration) of the hotel industry on the basis of a single competency model for all groups of staff, all jobs of the enterprise are divided into seven levels (G1… G7), each of which is relevant to certain pay ranges, and the latter, in their turn, correspond to five groups (A, B, C, D and E). An individual basic salary within the pay range determined for a given job should be based on evaluating a particular employee by quality criteria.

According to the developed job structure (remuneration) of the hotel industry, payment of an individual employee will be decided in accordance with the results of a comprehensive quality evaluation, which in turn is based on estimating the results of his/her work and assessing the level of competence that has a direct impact on these results.

In the scientific professional literature, such a payment model is characterized as a "contribution-oriented payment model" [4].

The "contribution-oriented" payment model can be effectively applied within the differential payment structure developed for hotel enterprises. In this case, personal reward will be formed on the basis of work results, competence and motivational incentives for developing competencies, efficiency, career trajectory [9].

Payment for contribution means payment for the obtained results and competence, as well as for the past and future work achievments (Fig. 2).

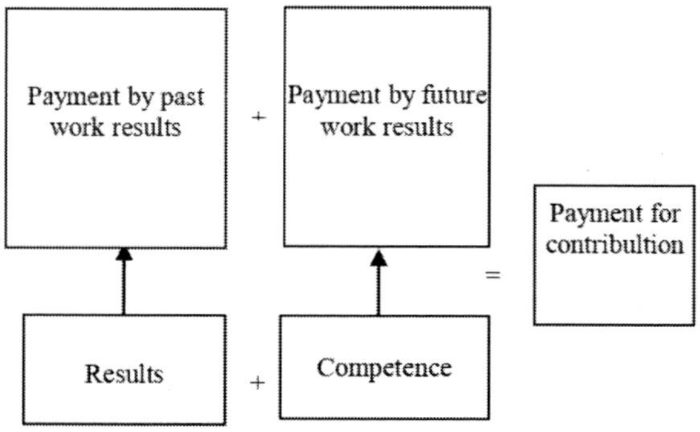

Fig. 2 "Contribution-oriented" payment model [4].

Thus, contribution-oriented payment is based on a mixed model of performance management – the assessment of input and output factors (competencies and results), and allows concluding that the level of payment for a particular employee in a particular job is determined by both past and future work results.

The results of job evaluation in PJSC "Salyut Hotel" and data on average job payments are given in Table 4.

The salary is assigned to the employee after analyzing the wage market within the range of a certain level of structure, and its specific amount depends on individual competence, work experience, level of efficiency and requires additional assessment of an individual employee's quality of work.

Therefore, each level of the structure (Gn) also needs to be subdivided into sublevels (Gjn,, where j is the number of sublevels of a certain pay range) which will correspond to certain salary estimates within a given range

As the results of the conducted analysis established the relationship between the hotel category and the level of payment, it is advisable to determine the average job salary among hotel enterprises in Kyiv with a 3-star category which corresponds to the category of PJSC "Salyut Hotel".

Table 4. Results of job evaluation of PJSC "Salyut Hotel" and average job payment.

Enterprise's department	**Job**	**Overall job evaluation, points (OHi)**	**Average job payment*, UAH (Sin)**	**Recommended basic part of average job payment**, UAH (ZPdin)**	**Basic salary of PJSC "Salyut Hotel", UAH (ZPactually)**
Management	CEO	2478	17350	12145	12300
Management	CEO deputy for organizational development	2150	15050	10535	10000
Accommodation service	Head of accommodation service	1457	10200	7140	7300
Accommodation service	Senior receptionist-cashier	1372	9607	6725	6800
Accommodation service	Receptionist-cashier	1036	7250	5075	5200

* The average estimate of an employee's average monthly salary in a similar job among the studied 3-star hotels in Kyiv according to 2019.

** 70% of the average estimate of an employee's average monthly salary in a similar job among the studied 3-star hotels in Kyiv according to 2019 data.

In addition, the research results of the structure of employees' average salary at 3-star hotels in Kyiv showed that its main share is less than 54.8%, while the share of allowances, surcharges, bonuses, other compensation and incentive payments averages 45.2 %. In modern market conditions and in the practice of economically developed countries' enterprises, the share of the fixed part of payment should be at least 70%.

In this regard, we propose to accept 70% of the average market payment among the studied 3-star hotels of Kyiv (Sin) as a basic part of the salary of the i-th profession in the labor market (ZPdin) in order to develop a new job payment system of PJSC "Salyut Hotel".

The other (30%) will be a variable part – the amount of bonuses, rewards, other compensation payments (ZPdin).

Specific salaries and bonuses should be assigned due to the staff competence and contribution by the following models in order to determine the amount of payment and appropriate strategies for the enterprise's activities in the regional labor market in the hotel business sector:

$ZP_i^{\text{вн}} < ZP_i^{\text{вн}}$ Defense strategy;

$ZP_i^{\text{вн}} = ZP_i^{\text{вн}}$ Follower's strategy;

$ZP_i^{\text{вн}} > ZP_i^{\text{вн}}$ Leader's strategy.

Thus, within one job, the range of salary increase is 50%, which is a strong motivating factor. The minimum salary of the lowest (first level) is not lower than state norms and guarantees, i. e. not lower than the state-established minimum salary.

The salary for a certain job is set within the range of a certain level of structure, and its specific amount depends on an employee's competence, his/her work experience, the level of work efficiency and requires additional assessment of an individual employee's quality of work.

Basic salaries are adjusted annually in accordance with changes in the labor market, inflation, changes in assessing the complexity of work performed, and so on. The salary grid, which is calculated in relation to basic salaries, is adjusted accordingly.

It is established that the structure of remuneration developed for hotel enterprises has seven levels and, accordingly, seven ranges of payment that intersect (overlap each other) by 30% in average.

The general payment structure at the hotel industry

after introducing a new system of basic salaries formation is as follows (Fig. 3).

At the hotel enterprises it is rational to accept the main part of the salary at the level of 70%, while the other 30% will be the variable part.

It is planned to accrue one amount of bonus for each employee, which will be considered as compensation for individual contribution and level of competence, the other – as a reward based on the enterprise's final results.

Thus, the application of the developed remuneration system on the basis of job evaluation and the staff competence will increase the efficiency of the hotel business and will help in the following areas:

1. To form a team of adaptive managers who are able to develop and implement a program of the hotel survival and development in changing conditions.

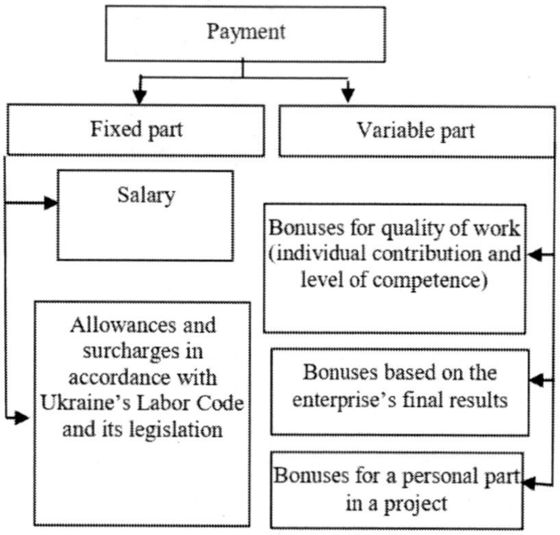

Fig.3. Improved overall payment structure in the hotel industry

2. To identify and preserve the core human resources of the hotel, i. e. managers, specialists and labour force of special value.

3. To restructure human resources in accordance with: organized transformations while restructuring; implementation of innovation processes; diversification; a complete reorganization.

4. To reduce socio-psychological tension in the team.

3 The paper's scientific significance

According to the analysis of statistics on the hotel services market in Ukraine, it was determined that it was the crisis that forced hotels to reconsider cost management and turn to a more flexible pricing policy, but what is more – to intensify work with hotel staff and consider them as the main resource that ensures the hotel's competitiveness. In this regard, the development of a job evaluation system on the basis of a competency-based approach is recommended as a way of increasing labor resources potential in the hotel industry.

Thus, introduction of business innovation based on jobs evaluation and staff competence creates prerequisites for improving efficiency, optimizing internal business processes, simplifying wage management and motivating employees on the principles of fairness, transparency, flexibility, social partnership.

4 Conclusions

A balanced (hierarchical) payment structure is formed to improve the organization of payment at the enterprise by introducing a transparent and flexible remuneration system based on individualization of employees' salary and determined not only by traditional parameters (an employee's experience, qualifications, experience), but also by his/her individual characteristics, his/her competence on the basis of abilities evaluation and their possible implementation.

In addition, a hierarchical job structure at the enterprise is formed to contribute to achieving business goals and strategies of the enterprise, investing effectively in increasing the cost of human capital, attracting and retaining the best professionals.

References

1. "Pandemiia koronavirusu zavdala serjoznykh zbytkiv hotel'nomu biznesu v Ukraini, na vidnovlennia sektoru mozhe pity ne odyn rik", [Online], available at: http://www.ukrstat.gov.ua/ (Accessed 1 February 2021), (In Ukrainian).
2. Raven D., Kompetentnost' v sovremennom obschestve: vyiavlenye, razvytye y realyzatsyia [Competence in modern society: identification, development and implementation], Kohyto-Tsentr, Moscow, Russia, (2002).
3. Tsurkan M. L. "Competence approach in the assessment of the personnel of the integrated holding" (2017), *Efektyvna ekonomika*, [Online], **7**, available at:
http://www.economy.nayka.com.ua/?op=1&z=5703, (Accessed 1 February 2021).
4. Tkachenko, T.I., Mel'nychenko, S.V., Bosovs'ka, M.V., Poltavs'ka O.V. Upravlinnia iakistiu posluh hoteliv: metodolohiia ta praktyka, [Hotel service quality management: methodology and practice], Kyiv. nats. torh.-ekon. un-t, Kiev, Ukraine, (2012).
5. Tkachenko T.I., Bojko M.H., Mel'nychenko S.V., Mykhajlichenko H.I., Vedmid' N.I., Poltavs'ka O.V. [ta in.]; in. Mazaraki A.A. (ed.), Stratehichnyj rozvytok turystychnoho biznesu [Strategic development of tourism business], Kyiv. nats. torh-ekon. un-t, Kiev, Ukraine, (2010).
6. Armstronh M. Praktika upravlenija chelovecheskimi resursami [Human resource management practices], in. Mordovyna S.K. (ed.), SPb: Piter, Russia, (2009).
7. Djevid K. MakKlelland "The concept of competence", [Online], available at: http://www.timetogo.ru/articles-

/what_is_competency.htm / (Accessed 1 February 2021), (In Ukrainian). (2009).

8. Bekh I. D. Kompetentnisnyj pidkhid uuchasnij osviti [Competence approach in modern education], Henezys, Kiev, Ukraine. pp. 21-24, (2009).
9. But, T. Pulina, T. Khrystenko, O. Zaytseva, V. Managing the Field of Reconstruction and Preservation of Historical and Cultural Complexes in Ukraine and Europe // *Lecture Notes in Civil Engineering 73*, pp. 709-720, Ukraine, (2020).
10. Shmygol, N., Schiavone, F., Trokhymets, O., .Zavgorodniy, R., Vorfolomeiev, A. Model for assessing and implementing resource-efficient strategy of industry CEUR Workshop Proceedings, 2713, pp. 277–294, (2020).
11. Boyatzis, R.E. The competent manager: a model for effective performance. New York: Wiley, p. 29, (1982).
12. Raven J. Dictionary of occupational titles. 4thed. U.S. Departament of Labor. Washington, D.S: U.S. Government Printing Office, (1989).
13. McClelland D.C. Testing for competence rathe then for intelligence. American Psychologist, 28. P.1 – 14. (1993).
14. Barybina Y.O., Lysenko M.O. “Modern methods of stimulating the staff of the hotel industry” Zbirnyk naukovykh prats' Cherkas'koho derzhavnoho tekhnolohichnoho universytetu. Seriia: Ekonomichni nauky 2 (40), pp. 64–68, (2015).

Transformation of Ukrainian Automotive Industry in the Context of Electromobility: Applying the Experience of Visegrad States

Yevhen Savelyev[1]*, *Vitalina* Kuryliak[1], *Konstantia* Darvidou[2], *Mariia* Lyzun[1] and *Ihor* Lishchynskyy[1],

[1]Department of International Economics, West Ukrainian National University, 46009 Ternopil, Ukraine
[2]University of Western Macedonia, 52100 Kastoria, Greece

Abstract. 2009 has marked the start of a decline in Ukrainian automotive industry. Its volumes neared insignificancy in 2020. The authors consider the issue of the industry's revival in the national economy. The basis of the concept is the creation of enterprises for the production of electric vehicles, including components, semi-finished products and accessories to them. It is suggested that investors can be attracted by developing cooperation of legislative and governmental institutions and local governments with world market leaders and taking measures to improve the national indicators in terms of the ease of doing business. It is proposed to focus the business philosophy of new enterprises on the use of vehicles rather than the field of production, as has been the case since the invention of the automobile. The choice of innovative areas should take into account the megatrends and use individualization, urbanization, artificial intelligence, mobility, digital culture, cooperation with other industries, offers of additional goods and services as its foundations.

1 Introduction

In 2021, Ukraine celebrates the 30th anniversary of its independence. In this time, the country's economy has undergone a transformation from a socialist system to a "trophy" in the hands of oligarchic economic actors. This happened when a small group of "new" entrepreneurs ("oligarchs" in the current terminology) captured large "trophies" – industries. This process was made possible because the privatization mechanism was adapted in such a way as to redistribute state property between the emerging oligarchs, which did not lead to the replacement of the less efficient owner (the state) with a more efficient one. As a result, Ukraine entered the 21st century as a country with predominantly raw material exports and weak innovative industries that could have otherwise brought significant value added. According to the latest results of the International Comparisons Program, the level of Ukrainian GDP in 2017 was equal to 504.4 billion US dollars (as per PPP). Ukraine has a GDP per capita of $11,871 PPP, which, in Europe and Central Asia, exceeds only Tajikistan and Kyrgyzstan.

Ukraine's long-term economic lag has various causes. However, the low level of development of the real economy, especially the processing industry is among the main, if not the main factor. The share of processing industry in the structure of gross value added has a steady downward trend and amounted to 11.6% in 2018 compared to 13.0% at the beginning of the decade (2010), and 19.4% at the start of the century.

This kind of deindustrialization determines the low level of competitiveness of the economy and requires the development of a radical strategy to increase the role of industry in GDP. The indicators of the world market leaders can serve as benchmarks. For instance, the share of industry in the structure of US GDP is 18.9%, for Germany it is 30.15%, for Great Britain - 19.0%, for Norway - 31.1%, for Sweden - 33.0%, for France - 19.1%, and for Japan - 29.7% [1]. Despite significant differences in these indicators (caused mainly by international specialization and the peculiarities of post-industrial development), they testify to the expediency of advanced industrial development as an important source of economic growth and job creation.

In order to determine the changes necessary to reverse the trend towards deindustrialization, the proverbial Gordian knot of Ukraine's economy must be found and cut before any significant influence on the country's GDP can be achieved. Today, the automotive industry is the "bottleneck" in the country's industrial sector. Its indicators over the last five years show an unfortunate trend: 2015 - 84.4%; 2016 - 99.0%; 2017 - 116.6%; 2018 - 115.3%; and 2019 - 96.8%. Within this industry, production of vehicles, trailers and semi-trailers is the weakest link, the indicators of which were respectively 94.4; 94.8; 111.6; 101.8; and 74.3.

However, the automotive industry is of great importance in the development of convergence of post-socialist countries and their European integration. This is evidenced by the experience of the Visegrad Four countries, which opened the market for multinational companies and created conditions for the development of knowledge-intensive industries. Their greatest success was the formation and development of mechanical engineering, especially in the automotive industry. In terms of vehicle production, the group is now a leader in the European and world economy. In the European

* Corresponding author: yev.savelyev@gmail.com

Union, the Czech Republic ranks 5th, Slovakia 7th, Poland 8th and Hungary 9th. In the world market they have 17, 21, 22 and 23rd positions, respectively.

The experience of the Visegrad Four countries is important for Ukraine because the development of the automotive industry introduces them to a market with high effective demand, one that is changing rapidly and growing at the same time. The development of this industry leads to the creation of a large number of enterprises specializing in the production of parts, semi-finished products and accessories, the provision of sales and logistics services. Most enterprises in the modern automotive industry can operate with up to 100 employees, which makes it possible to create jobs without creating large migration flows in the country. At the same time, tens, and over time, hundreds of thousands of jobs can be created in the industry. Finally, the technological capabilities of the automotive industry gradually form local, regional and international production networks, which become the basis for the country's inclusion in the system of European economic integration, as evidenced by the experience of the Czech Republic, Slovakia, Poland and Hungary.

2 Analysis of recent research and publications

The interest in the automotive industry has declined in the Ukrainian scientific literature in the last decade, which can be explained by the decline in domestic car production. However, despite the crisis in the automotive industry, there are publications that try to find the prospects for vehicle production in Ukraine. They are considered in terms of attracting foreign investment and increasing the competitiveness of the domestic model range, resource and infrastructure support necessary for the functioning of a competitive automotive industry and attracting the attention of investors, assortment policy transformations, state policy for the development of road transport market infrastructure [2; 3]. A monograph by I. Yu. Shevchenko offers a systemic view at the development of the automotive industry and a methodology for assessing the competitiveness designed by the author, as well as original spherical model and concept of the state program of industry development up to 2025 [4]. However, the author did not link their proposals to the use of market methods. The concept of competitiveness in a rapidly changing world economy was developed by Braja & Gemnik-Salwach, who proved that the growing role of innovative and destructive technologies is one of the most important factors driving these processes, and knowledge-intensive and innovative industries have become the main drivers of economic growth and economic competitiveness [5]. In recent years, the development of methodological problems in the context of rapid changes in the world economy has been accelerating. In December 2020, the European Automobile Manufacturers' Association published Roadmap for the deployment of automated driving in the European Union, which discusses the benefits and terminology of automated driving [6]. In summary, the analysis of existent publications shows a lack of scientific research that would systematically solve the problem of overcoming the crisis of the automotive industry and position it as a renewed modern production subsystem of Ukraine's economy, an important step for a country with a focus on accelerating European integration.

The aim of the study is to develop proposals for the revival of the automotive industry in Ukraine based on the introduction of electric vehicle production and creation of local and regional production networks with their gradual integration into global production networks and value added networks.

3 Research Results

3.1 Development of the automotive industry in the years of Ukraine's independence

During the 90s and the first decade of the 2000s, the automotive industry of Ukraine developed as a priority area that provided the solutions to social, economic, environmental, scientific, and technical problems while the market economy was still being established. About 90 thousand jobs were created in this industry and the same number of employees were engaged in related industries. After gaining independence, Ukraine had four plants for the production of vehicles, including cars (Zaporizhzhia and Lutsk), trucks (Kremenchuh) and buses (Lviv), and more than 60 companies that produced components.

Changes in the automotive industry that took place after Ukraine's independence, i.e. after 1991, were mainly precipitated by the relations with Russia. There were several assembly lines in Ukraine, including minibuses "Bogdan" in Cherkasy and cars "Volga" in Illichivsk (now Chornomorsk), low-tonnage cars "Gazelle" - in Simferopol and Lubny. These Russian car models were inferior in their competitiveness to Western brands and their production worsened the lag of the Ukrainian car industry in scientific, technical and innovative terms.

A relative success was achieved by the construction of the "Eurocar" plant for the production of VW Group cars near Uzhhorod, two kilometres from the border with Slovakia and Hungary, where there were global manufacturers of automotive components. In the context of the Eastern Partnership policy, equipment from the best European manufacturers "Transsystem" (Poland), "Chropynska Strojirna" (Czech Republic), "EISENMANN" (Germany) and others was provided to equip the enterprise. "Eurocar"'s innovativeness is also reflected in the uniqueness of software and technological innovations to ensure the flexibility of production processes. However, the plant managed to achieve only minimal profitability mainly due to limited orders of 10,000 cars, while the plant project was estimated for 50,000-100,000 cars. The mistake of the company's founders was to bet on the Russian market, which in Ukraine is problematic for political reasons.

Table 1. Production of cars and commercial vehicles in Ukraine in 2000-2019 [7].

Year	Cars	Commercial vehicles	Total	Change, %
2019	6254	1011	7265	9,7
2018	5660	963	6623	-22,9
2017	7296	2296	9542	81,2
2016	4340	924	5264	-36,1
2015	5654	2590	8244	-71,3
2010	75261	7872	83133	20,06
2009	65646	3649	69295	-83,6
2008	400799	22328	423127	5,1
2007	380061	22530	402591	39,7
2006	274860	20400	295265	36,8
2005	196722	19037	215759	15,4
2004	179098	7792	186890	73,2
2003	103000	4890	107890	100,6
2002	50393	3380	53773	69,0
2001	24995	6829	31824	1,8
2000	18124	13121	31255	63,0
1999	10136	9044	19180	

Ukraine's automotive industry has tended to grow since the early 2000s. In 2008, the production peaked at 423,127 cars. Since 2009, the trend has changed to an annual decline. Initially, the cause was the global crisis of 2008-2009. In 2019, the volume of production amounted to 7265 units, which were manufactured mainly at the plant "Eurocar" (table 1). Added to this is Ukraine's accession to the World Trade Organization and the adoption of its terms to reduce import duties on foreign cars from 25% to 10% and the waiver of a number of preferences for domestic producers. At the same time, illegal schemes of importing cars with foreign registration spread. Later, the decline in production was affected by the low competitiveness of Ukrainian cars in Western markets. Simultaneously, Russia waged a trade war with Ukraine in the eastern markets. Common methods were stopping car carriers, banning transit to Kazakhstan, blocking transit routes to the markets of Central Asia and the Caucasus.

3.2 Revival of the automotive industry of Ukraine: priority conceptual components

The crisis of the automotive industry of Ukraine has led it to a state that can be assessed as approaching zero. Therefore, there are grounds make reviving the industry a priority, as a country with an area of 607 thousand km2 and a population of more than 40 million, should not exist without its own automotive industry.

The problem of the revival of the automotive industry should not be understood as a matter of modernization or renewal of bankrupt enterprises. The task should be to establish modern production by corporate and private owners of the latest scientific and technical achievements in the automotive industry. It is clear that the latter will drive a hard bargain, to which the state must generally agree. It can be argued in advance that they will prioritize radical reforms of the judiciary and anti-corruption measures to ensure the security of their businesses. According to V. P. Prykhodko, "safe" investments create an opportunity to make capital investments and production savings at a level that guarantees sufficient rates of expanded reproduction, technological re-equipment and restructuring of the economy, which directly increases the potential protection of economic interests from threats of various origins. Given the scale of the project, creating an investment security mechanism is necessary to attract related producers as major investors.

Carrying out a large-scale transformation in the economy with the involvement of investors and world market leaders involves the implementation of measures to achieve good indicators in terms of the ease of doing business, since this indicator serves as a guide for making positive decisions about choosing a country for capital investment. Theoretically, it is considered a herald of socio-economic wealth [8]. Improving business conditions is particularly important for Ukraine, as it has never been ranked in the top 100 countries in the Ease of Doing Business Index compiled by the World Bank since 2006. However, in 2020 there was a significant improvement in this regard - Ukraine rose to 64th place in the ranking. Now there are reasons to establish a firmer place in the first hundred and continue to further increase the rating.

3.3 Global competition at the start of the revival of Ukrainian automotive industry

The world automotive industry is now on the verge of revolutionary changes. According to experts, by 2025, traditional automakers and suppliers should refocus on the challenges of global trends in electric mobility and significantly increase investment in the transformation of their business models. New competitors with new thinking and ideas will enter the market. This is in addition to the fact that, according to expectations, by 2030 more than 55% of new cars can be fully electrified, and in general, 95% of sales of new vehicles will have at least partial electrification. Moreover, up to 40% of mileage in Europe will be carried out by autonomous vehicles (without a driver) [9].

The drivers of change in the automotive industry are largely the technically well-educated young adults. It is the youth that should become the driving force behind the development of more sustainable and convenient scientific and technical solutions. They will have to rethink the whole system of economic relations in the industry, focusing not on the field of production, as has happened since the invention of the car and still continues to happen today, but on the use of vehicles. Given the megatrends, the choice of innovative areas will be based on individualization, urbanization, artificial intelligence, mobility, digital culture, cooperation with other industries, offers of additional goods and services. Given these trends, the European vehicle fleet is projected to decline from 280 million to 200 million units by 2030 [9].

Inclusion of Ukraine in competition with existing car manufacturers will happen in conditions of the difficult transition of existing enterprises to a new dimension. Many market leaders, including the Visegrad Four, end their production cycle in 2023, complicating the transformation process. Transnational corporations are choosing markets with cheaper labour for new investments. There are several economic studies showing that companies are considering moving production to countries with cheaper labour after the average wages in the region exceeds 1,200 euros [10].

In the context of transformation, the competition on the market is influenced by the fact that the cost of labour in the Eurozone is higher than in other countries. This encourages investors to look for alternatives to invest in. Among Ukraine's western neighbours, such a situation is emerging in the Visegrad Four. Slovakia, which is the only member of the Eurozone, has higher salaries than the rest of the Group, which is why in 2018 it lost the construction of a new BMW plant to its neighbour Hungary. At the same time, Budapest won the research contract for Jaguar Land Rover, despite the fact that in 2018 the company commissioned the only European plant outside the UK near the Slovak city of Nitra. Simultaneously in the neighbouring Czech Republic, BMW is building a research centre to test the management of autonomous vehicles. According to the Slovak Automobile Industry Association, the country has lost its competitiveness in the Visegrad Four region and is losing its advantage over the countries of Southern Europe [10].

Financial and economic readiness to start transformational changes is an important factor in competition in the automotive industry today. At the moment, it is not profitable for the owners of existing enterprises to restructure their production facilities to produce electric cars. They require a lag of time during which the most expensive assets can be depreciated, namely plants for the production of internal combustion engines and gearboxes. Under such conditions, outsider countries have obvious competitive advantages - they have nothing to lose. In view of this, it can be argued that Ukraine, given the entry of a strong investor in its market due to the time factor has a good chance of becoming an automotive state.

Thus, international competition in the field of car production is characterized by the accumulation of sufficient scientific and technical potential to carry out restructuring on market terms. Both current leaders in the automotive business and outsiders have a chance to compete for success in the vehicle market. Given the high cost of future transformations, their success will depend to a large extent on the interest and participation of governments. In this regard, it should be noted that the automotive industry is not a leader in investing in R&D. The top 10 companies in the world in terms of R&D costs include only Tesla Motors, which spends $ 0.7 billion (ranked 10th), while the spending of Alphabet (ranked 1st, specializing in software and internet) amounts to 12.3 billion dollars. At the company level, the prerequisite for success will be the choice of production location on the basis of a competent assessment of the factors of spatial, human, raw material and environmental potential. The final decision should be made taking into account the cooperation with central governments and local authorities.

3.4 Avenues for revival of the automotive industry of Ukraine

The revival of the automotive industry is a decade-long task of national importance, because it fundamentally changes the structure of the economy and convergence strategy, affects social institutions and the development of society. The scale of the task should be ensured from the very beginning by the adopted legislative and governmental actions. The experience of Poland is very significant in this context, as in 2003 it was among the top three countries offering the best conditions for the location of plants for the production of components for the automotive industry. This was followed by an increase in FDI inflows and the development of a sub-supplier reserve, and during 2007-2009 Poland became the second largest (after the Czech Republic) car manufacturer in Central and Eastern Europe [11].

The central government should initiate the invitation of the main investor. It is advisable to attract a multinational company that owns the brand with top sales in the Ukrainian market, namely Toyota, Renault, Skoda, Nissan, Hyundai, or Volkswagen. Regulatory documents should regulate the allocation of land for the creation of industrial zones to the parent company and manufacturers of parts and semi-finished products.

The global trend of transformation of the automotive industry indicates the impossibility of success without the use of public-private partnership. For example, the Automobile Industry Association of Slovakia appealed to the government to conclude a Memorandum of Understanding and Cooperation on creating conditions for sustainable development of the industry. This document should help maintain the competitiveness of the automotive industry in the long run. For Ukraine, which is developing the industry after the fall in production to almost zero, success without close cooperation with central government agencies is almost impossible.

Along with the choice of the main investors and the model range of the cars, it will be necessary to decide on the division of labour. The automotive industry in the narrow sense is a component of mechanical engineering. However, its functioning requires the creation of a kind of symbiosis of mechanical engineering, electrical and chemical industries. To ensure the production of components, the parent company must create a network of related companies and global suppliers. Given the global experience, generally for this purpose cooperation with small and medium-sized businesses, most of which have of 25 - 50 - 100 employees is established and developed. The developed system of automobile production has hundreds of subcontractors. In Slovakia, for example, there are about 700 companies operating in the automotive industry [12].

The creation of an enterprise for the production of electric vehicles will require the development of technology of numerous components for a new type of product. These include autonomous control systems, alternative drives, digital data transmission, braking systems, lane keeping assistance systems, medium capacity batteries that are charged through recuperation during braking, batteries for plug-in hybrids.

The division of labour among car manufacturers forms the geographical structure of the automotive industry. Local, regional and global production networks are created for the purposes of balanced and synchronous functioning of producers. However, this process of division of labour should not be final. Today, in the real economy, more and more attention is paid to global value-added networks. They make it possible to distribute individual production operations between different countries of the world, which is a means of ensuring a high level of competitiveness [13]. For Ukraine, this requires finding forms of cooperation with the Visegrad Four. They are, firstly, territorially the closest countries of the European Union, and, secondly, world leaders in the automotive markets. The formation of international production networks and global value-added networks in Central and Eastern Europe with the participation of Ukraine can be an effective direction for successful and accelerated development of the country's automotive industry

4 Discussion and conclusions: car production, a new business model - a new community

Revival of the automotive industry of Ukraine is a road from deindustrialization to industrialization of the country following the newest scientific and technical ideas focused on the future. That is why an electric car with different levels of automation and digitization should be the basis for this process. The mission of the automotive industry is changing rapidly, it can no longer be traditionally interpreted as meeting the growing needs of the population in mobility and approaching the level of economic prosperity of developed countries. At the same time, electrification in combination with digitalization encourages the establishment of new social relations and a new way of life. We do not deny the importance of traditional innovations: economical engine, reliable brake mechanism, high-speed car, attractive design, etc. However, in time there is a need for new innovations, such as autonomous driving and automated cars.

A comprehensive and rapid reorganization of the automotive sector will result in the emergence of new business models. Manufacturers and suppliers must offer consumer-oriented innovations to succeed in the markets. It should be assumed that a certain part of society, especially young people, does not have and does not need a personal car. Multinational companies are already developing models that will be effective in online car-sharing agencies to create driving communities on online platforms with intermediary functions for private travel. The greatest prospects are predicted for business models of car sharing in large cities, combined with public transport. Modern electric cars have great potential to create an inclusive society and expand opportunities for equal participation in public life, in particular with the use of a car without a driver.

Electric mobility experts are developing new concepts based on the fact that the future of the car is the future of mobility. Discussions on such a link are important for developing an automotive strategy. The development of the debate on the future of the automotive industry in the global dimension and in Ukraine in particular can be conducted based on new terminology: the coefficient of integration, infrastructure efficiency, inventory ratio and levels of autonomy.

Achieving the goal of reviving the automotive industry is a task of at least a decade for Ukraine. It will be impossible to do without reorienting the training of engineering and technical personnel for the production and maintenance of equipment of the future. Ukraine had a competitive engineering education in the Soviet system. Its comparative advantage lay in the harmonious combination of two main components - educational and scientific. However, it operated separately from manufacturing entrepreneurship and global innovation trends. In recent years, the government has increased government orders for engineering training. However, the increase in the number of specialists alone is not able to sufficiently ensure the country's international competitiveness in the context of the transformation of the world economy to the requirements of Industry 4.0.

The reform approach to engineering training should be based on the fact that Industry 4.0 will operate in a new economy that not only necessitates transformation, but may also pose a threat [14]. Engineers need to be able to implement a new approach to production, fully linking physical work to the digital world, automating processes and allowing machines to operate independently. Despite the view of the job automation's limited effect [15] (for Germany), most researchers predict a significant impact on employment and / or wages ([16] for the US; [17, 18] for six EU countries). It is proved that the pace of implementing the achievements of Industry 4.0 will vary in different industries and countries. Therefore, it can be argued that it will be determined by the readiness of the engineering to create and develop innovations.

If a new system of engineering training is to be created, it must be included in the processes of creating clusters, technology parks, business incubators, design bureaus, and small businesses. Accordingly, the diploma on engineering education should reflect the level of knowledge, practical skills, creative (innovative) potential of the university graduate. Such a model can be implemented in the framework of dual education, the implementation of which has begun in Ukraine, but does not yet have a sufficient scientific, methodological and regulatory background.

Acknowledgements

This article was prepared in the framework of cooperation on the research topic ME-05-2019 "B" "Creating the alliance of Poland, the Czech Republic, Slovakia, Hungary and Ukraine as the Visegrad Five: the context of regional economic integration", which is undertaken by West Ukrainian National University (WUNU) with the financial support from the State Budget of Ukraine. However, the support of WUNU is not an automatic support of the article's content, which reflects only the views of the authors. WUNU is not responsible for any use of the information provided in the article.

References

1. A. O. Zadoia, De stvoriuietsia VVP, abo industrializatsiia chy deindustrializatsiia? (Where is GDP created, or industrialization or deindustrialization?). European vector of economic development. 2(23), 27-41 (2017)
2. O. V. Boiko, V. V. Koval, in Proceedings of the International Scientific and Practical Conference "Development of Trade and Entrepreneurship in Ukraine: Trends and Prospects", Odessa, 2017. http://www.otei.odessa.ua/wp-content/uploads/2017/11/%D0%97%D0%B1i%D1%80%D0%BD%D0%B8%D0%BA_%D0%9C%D0%9D%D0%9F%D0%9A_25.05.2017.pdf#page=33
3. I. Yu. Shevchenko, Harmonizatsiia yak novitnii pidkhid do derzhavnoho rehuliuvannia rozvytku malosubiektnykh haluzei natsionalnoi ekonomiky (na prykladi avtomobilnoi promyslovosti Ukrainy). International scientific and practical conference, November 2020, 27-28. doi:10.30525/978-9934-26-005-6-4
4. I. Yu. Shevchenko, Rehuliuvannia rozvytku avtomobilebuduvannia Ukrainy: monohrafiia (Regulation of automobile development of Ukraine: monograph). (Kharkiv: KhNADU, 2019), p.556
5. M. Braja, A. Gemzik-Salwach, Competitiveness of high-tech exports in the EU countries. Journal of International Studies. 13(1), 359-372 (2020). doi:10.14254/2071-8330.2020/13-1/23
6. European Automobile Manufacturers Association, Automated driving. Roadmap for the deployment of automated driving in the European Union. (ACEA, 2020), https://www.acea.be/uploads/publications/ACEA_Automated_Driving_Roadmap.pdf. Accessed 14 Dec 2020
7. International Organization of Motor Vehicle Manufacturers. Production statistic. https://www.oica.net/production-statistics/. Accessed Jan 14 2021
8. A. L. Leal-Rodríguez, C. Sanchís-Pedregosa, Could the Ease of Doing Business Index be considered a predictor of countries' socio-economic wealth? An empirical analysis using PLS-SEM, 12(4), 229-243 (2019) doi:10.14254/2071-8330.2019/12-4/15
9. F. Kuhnert, Ch. Stürmer, A. Koster, Five trends transforming the Automotive Industry. (PricewaterhouseCoopers GmbH Wirtschaftsprüfungsgesellschaft, 2018) https://www.pwc.at/de/publikationen/branchen-und-wirtschaftsstudien/eascy-five-trends-transforming-the-automotive-industry_2018.pdf. Accessed 12 Dec 2020
10. B. Toma Ako Detroit? Domáci automobilový priemysel čakajú posledné dobré roky (2020). https://ekonomika.pravda.sk/firmy-a-trhy/clanok/538578-koniec-ako-detroit-domaci-automobilovy-priemysel-cakaju-posledne-dobre-roky/ Accessed Dec 12 2020
11. Wikipedia, Przemysł motoryzacyjny w Polce. 2021. https://pl.wikipedia.org/wiki/Przemys%C5%82_motoryzacyjny_w_Polsce. Accessed 15 Jan 2021
12. Udržateľnosť automobilového priemyslu na Slovensku. MôjElektromobil. 2019. https://www.mojelektromobil.sk/elektromobilita/infrastruktura/nabijacie-stanice/ Accessed 16 Dec 2020
13. L. Slušná, M. Balog et al., Automobilový priemysel na Slovensku a globálne hodnotové reťazce Slovenská inovačná a energetická agentúra (Bratislava, 2015). https://www.siea.sk/wp-content/uploads/files/inovacie/publikacie/studia_Automobilovy_priemysel_na_Slovensku_a_globalne_hodnotove_retazce_SIEA_web.pdf. Accessed 14 Dec 2020
14. S. Szabo, Transition to Industry 4.0 in the Visegrád Countries. (Luxemburg:European Union, 2020).
15. W. Dauth, S. Findeisen, J. Südekum, N. Wößner, German robots – the impact of industrial robots on workers. IAB Discussion Paper No. 30/2017, (Institute for Employment Research, 2017)
16. D. Acemoglu, P. Restrepo, Robots and jobs: evidence from US labor markets. NBER Working Paper No. 23285, (2017)
17. F. Chicago, G. Petropulous, D. Pichler, The impact of industrial robots on EU employment and wages: a local labour market approach. Bruegel Working Paper, 2, (2018)
18. Belhadi, A., Kamble, S., Jabbour, C. J. C., Gunasekaran, A., Ndubisi, N. O., & Venkatesh, M. (2021). Manufacturing and service supply chain resilience to the COVID-19 outbreak: Lessons learned from the automobile and airline industries. Technological Forecasting and Social Change, 163, 120447.

Criteria Evaluation of the Results of the Implementation of the Business Process Automation System

Kateryna Holikova[1*], *Svitlana* Matsyura[1], *Viktoriia* Radko[1], *Nina* Rudenko[2], and *Daryna* Zhukova[2]

[1]State University of Economics and Technology, Kryvyi Rih, Ukraine
[2]Kryvyi Rih Technical College of the National Metallurgicall Academy of Ukraine, Krivyi Rih, Ukraine

Abstract. Approaches and methods of qualitative and quantitative analysis of business processes are analyzed, their advantages and disadvantages at application in an estimation of business processes of the enterprises and the organizations of various spheres of activity are defined. A method for determining the economic efficiency of the implementation of an automated design system, as well as indicators and criteria for assessing the economic efficiency of the implementation of an information system in the reengineering of business processes. The main examples of quantitative indicators are considered, in particular: process indicators, product (service) indicators and consumer and stakeholder satisfaction indicators. In the course of the research, recommendations were developed for the creation of a generalized system of indicators for assessing the business processes of enterprises and organizations, regardless of their field of activity.

1 Introduction

The use of the latest information technologies in business, economy, state administration and the other branches is the extremely necessary condition for the efficient development. In case, when this condition is not observed at the proper level, the development slowdown will take place and the possibility to get into the rating of the lagging behind and the inefficient ones will appear.

The EU pays a great attention to the development of the digital economy, considering that the economy level of the continent will depend on the fact, how efficiently the enterprises will use the digital technologies. The enterprises, being not integrated into the digital space, will not belong to the World market.

Thus, it's clear, that the realization of control is not possible without using the intellectual information technologies for the efficient organization of business. It's necessary to create the corresponding information system of the organization, which would be unified in all the structural subdivisions and would help to take the strategic and tactical managerial decisions efficiently. Taking into account, that the concept of the strategic development is left to be the dominating concept of the modern company's development, it's necessary to use the intellectual information technologies at all the phases and stages of the strategic decisions-taking at the formation of the company's information environment [1].

The modern information technologies are the computer processing of information, according to the previously processed algorithms, the storage of big amounts of information on different bearers, the analysis and the visualization of data and the information transmission at any distance within the utmost minimal time. The new era has come – the era of "Big Data" ("Big Data era") and "Data Mining" (the intellectual analysis of data).

That's why the information technologies are used in the economy and business for processing, sorting and aggregating of data, for the organization of the interaction between the process participants and the computing engineering, for the satisfaction of the information needs, for the operative connection at all the phases and stages of the organization's life cycle.

In general, in order to provide the competitive positions, the modern business solves the following three strategic tasks:

- it's necessary to establish the closest relations with suppliers and customers;
- to increase the level of the private operational efficiency;
- to increase the competitiveness of the production, being manufactured.

The fulfillment of all these tasks is not possible without the integration of information systems and technologies into the sphere of business.

We propose to examine the achievements of these three tasks through the prism of the organization's business-processes with the use of the definite information technologies (digital tools) (Table 1).

We also demonstrated the tasks, concerning the rational cooperation between the business-structures and the executive bodies, banks and others in this Table and presented the information technologies that had been already introduced into life at present and reflect the digital transformation of the corresponding interrelations between the corresponding subjects too.

* Corresponding author: golikova_kp@kneu.dp.ua

Considering that, you may come to the conclusion that all the subjects, using the corresponding information systems, had an opportunity to feel the considerable positive effect at the definite interaction.

Table 1. Organizational business process groups and digital technologies (systems, tools, programs) for their transformation [1].

Business process groups organizations	Digital tools (systems, products, software solutions)
Task № 1. Establishing a closer relationship with suppliers and customers	
1. Interaction with suppliers	Systems SCM (Supply Chain Management)
2. Interaction with customers (consumers)	Systems CRM (Customer Relationships Management)
Task №2. Increase the level of own operational efficiency	
3. Resource management	ERP (Enterprise Resources Planning)
4. Business process management	BPM- systems (Business process management)
5. Analytical management component (data analysis)	Technologies Big Data, Data mining, OLAP-cube, cloud computing, Google Analytics
6. Providing modern employee technologies companies to effectively perform daily work	Office 365, Google doc, use of CRM systems, dashboards for key evaluation indicators
Task №3. Increase the competitiveness of products	
7. Product management (control, storage and providing the necessary information)	PLM (Product Lifecycle Management) systems and PDM (Product Data Management), digital sensors, GIS-technologies, etc.
8. Marketing activities	Digital internet marketing tools, Using Google Analytics tools Google Adwords; SEO and SMM
Tasks for rational cooperation between business structures and executive bodies, banks, etc.	
When registering / liquidating a business	Online service of the Ministry of Justice of Ukraine
State Tax Service, pension fund, social services, bank	Electronic reporting, electronic signature
Customs authorities	Single Window System

2 Background

The indicators of the business-processes are necessary to determine such targets, which are possible to be measured and controlled for the achievement of the defined purposes. The indicators of the business-processes are the indicator of the successful run of the processes and, on the contrary, indicate the presence of certain disadvantages. The indicators are the stimulus for the support of the existing level of standards, as minimum, and for the improvement in comparison with the competitors and the successful satisfaction of the constantly growing needs of the consumers, as maximum.

2.1 Problem Positing

As it's [2] mentioned in the work, the indicators' system of the process allows:

- estimate the state of the affairs at the initial stage (for the model "As is" ("As it is")), i.e. to define the starting point for the introduction of the improvement process;
- identify the processes, that require improvement;
- elaborate the forecast or the trend of the process development;
- compare the indicators' level of the researched enterprise with the analogous indicators of the competitors or with the average meanings in the corresponding branch of the economy;
- estimate the result of the already realized measures, concerning the improvement of processes;
- select the most corresponding tools for the further improvement.

The mostly-spread scientific approach to the classification and the differentiation of the estimation indicators of processes is the approach, offered by V.V. Repin and V.G. Yeliferov [3], based on the division of the processes analysis into the qualitative and the quantitative ones.

The generalization of these approaches allowed us determine the advantages and disadvantages in their use, presented in the Table 2.

Table 2. Advantages and disadvantages of existing approaches and methods of business process analysis [4,5,6]

Type of approaches and methods to the analysis of business processes	Pros and cons
Qualitative analysis of the business process	
SWOT analysis	**Pros**. Allows you to systematize information about the invoices of internal and external influence on business processes. SWOT-analysis is suitable for preliminary qualitative evaluation of the business process, its results can be useful for determining the characteristics of the process and the reasons for its low efficiency. **Cons**. SWOT analysis is characterized by a high level of subjectivity, its results are a weak basis for management decisions.
Analysis of process problems Ranking of processes	**Pros**. Managers and direct participants take part in the implementation of the methods in the survey process executors, which allows you to identify problem areas of the process. Allows you to determine the order of process improvement. **Cons**. The method allows to detect only the presence of problems, but does not allow to measure their depth. It is not possible to assess the problems from a financial point of view or from a point of view view of quality indicators.
Analysis of inputs and outputs	**Pros**. These methods allow you to determine the shortage or redundancy inherent in the inputs, outputs, resources

Function analysis Resource analysis Analysis of the process for compliance with typical and / or regulatory requirements	of the process. Their use allows to structure business processes, to make them more integral, to eliminate "bottlenecks", "to make healthy" the most poorly organized processes. **Cons**. All of these types of analysis are necessary for use, but none of them is not a tool for fundamental reengineering and implementation of deep changes in business processes, increasing the efficiency of business processes.
Quantitative analysis of the business process	
Simulation	**Pros**. The method allows to consider a large number of alternatives, to improve quality management decisions and more accurately predict their consequences. Allows you to find what you need solution in case of impossibility of conducting an experiment in real life. **Cons**. The complexity and high cost of using the method.
ABC analysis	**Pros**. Clarity of the analyzed indicators, accuracy of the calculated parameters, possibility of automation of the analysis. **Cons**. Requires restructuring of the cost accounting system in the enterprise, because on most companies do not have a system of process accounting of process costs.
Analysis of process, product and customer satisfaction indicators	**Pros**. Indicators characterize the main components of the business process from different points vision (financial, technical, etc.). Indicators carry objective information in numerical terms, outline the "problem areas" of the studied object. **Cons**. Process, product and customer satisfaction indicators are separated characteristics, on their basis it is difficult to determine the assessment of the effectiveness of the study object as a whole.

The conducted generalizations of the existing approaches to the estimation of business-processes testify to the fact that the qualitative analysis gives the possibility to get information of the run's peculiarities of processes, the presence of the problem sections, the "narrow places", the excessive or the missing functions of the processes. The qualitative analysis is the distinctly structurized characteristic feature of the business-process, which is the very important background for the further realization of the quantitative analysis.

The quantitative analysis, in its turn, allows get the numerical values that characterize the run of the business-process, define its state from the viewpoint of the technical requirements, the quality indicators. The quantitative analysis gives the possibility to measure the degree of the business-process efficiency, to calculate the profitability, to define the time characteristic features of the business-process. It's possible to set the normatives, to correct the strategic purposes, to foresee the further development of the organization, according to the results of the quantitative analysis [7].

As far as the indicators, by which the quantitative analysis of the process operates, are concerned, then, there are three groups of indicators here:

- the indicators of the business-process – the numerical values, characterizing the process run and the financial, the time, the resource and the human losses;
- the indicators of the product (service) – the numerical values, characterizing the product (service) as the result of the process realization. For example, the absolute scope of services, the scope of services in relation to the ordered or the needed amount, the number of mistakes or breakdowns at the presentation of service, the nomenclature of the presented services in relation to the necessary services;
- the satisfaction indicators of the process clients – the numerical values, characterizing the consumers' satisfaction degree with the process result. It's important to distinguish the satisfaction of the internal consumer or the external one with the process result and the satisfaction of the final consumer with the production or the service [3].

The indicators, characterizing the business-process, may be presented in the following way (Table 3).

Table 3. Types of business process indicators.

See indicators	Examples of indicators by type
Financial Indicators	The cost of the business process; Costs of raw materials and supplies; Labor costs of performers; Amortization; Heat and energy costs; Communication and acquisition costs information;
Technical Indicators	The number of business process features that performed in the workplace; Number of staff, including managers and specialists;
Temporary Indicators	The average process execution time in in general; Average downtime; The average execution time of individual business process functions;

O. Kochnyev and the iTeam [8, 9] company professionals offer the classification of the business-projects indicators, in the frames of the quantitative approach, with the separation of the four main groups of indicators: effectiveness, efficiency, productiveness, quality, – and the three auxiliary groups of indicators: activity of performers, losses of resources, profit (Table 4).

The evaluation's express-method of business-processes was offered by K.K. Chuprov in the work [10] in 2005. The evaluation system's following components of the business-projects' control efficiency are determined, according to this method: the efficiency indicators of the separate processes; the qualitative indicators of the enterprise's production; the indicators of the clients' satisfaction with the results of the enterprise's activity; the microeconomic and the financial indicators of the activity. The express-method of the business-processes' evaluation of K.K. Chuprov is founded on the principles and methods of the structural modelling, the object-oriented modelling, the analysis of cost, according to the processes, and the analysis of the key indicators.

Table 4. The main indicators of the business process (according to O. Kochnev).

The main groups of business process indicators and their essence	Examples of indicators
Effectiveness - the indicator is a measure of the result useful for the client process	Number of clients who applied; Sales volume; Number of concluded agreements; Production volume; Delivery time
Efficiency is an indicator characterizes the ratio of the achieved result and used resources	The cost of attracting a client; Sales conversion; Average check; Cost of production; Shipping cost
Productivity - an indicator that characterizes the ratio of achievements result and human resources used	Sales per person employee; Number of customers served by one employee; Production volume per employee; Revenue per employee; Profit per employee
Quality indicator - reflects the degree of customer satisfaction of the business process with the result	Product quality indicators (services): the degree of defect of the process products; number of defects, incomplete orders; marriage elimination costs; Service quality indicators: the number of returns and complaints on the products of the process; number of complaints and reclamations on the quality of service that received from customers; the number of emergency situations to be resolved the participation of the higher was required leadership

The work's author [10] suggests to use the methods of the functional modelling IDEF and DFD or the modelling methods of the integrated information systems ARIS, ORACLE, SAP/R3, BAAN for the construction of the processes' models. It's offered to calculate such indicators of the business-processes' efficiency on the basis of the constructed models of the processes as complexity, processability, controllability, resource-capacitance, regulatedness. The normative meanings of these indicators are calculated by the author [10].

To calculate these indicators, quantitative values of such basic parameters as:

- the number of levels of the business process system (Pl) should be identified;
- number of copies of business processes (Pc);
- the number of "gaps" in the process instances (Pg);
- number of business process classes (Pcl);
- number of business process owners (PO);
- the number of resources used to perform the process (R);
- the number of outputs in copies of business processes (Pout);
- number of regulatory documents (Preg).

The principle of calculating indicators according to the express method of assessing business processes K.K. Chuprov is presented in table. 5.

Table 5. Calculation of indicators in accordance with the express method of assessing business processes K.K. Chuprov.

Indicator	Calculation formula	Normative value
The complexity index is the ratio of the number of levels of decomposition process models to the total amount of selected processes, determines how complex is the hierarchical structure of business processes	Icomp = Pl / Pc	Icomp ≤ 0,66
Process index - the ratio of the number of "gaps" (lack of causal relationship between the subprocesses of the process that considered) in business processes to the sum of the classes of the business process, characterizes the integrity of the process, which can be defined as problematic, or as a process	Ipr = Pg / Pcl	Ipr < 1
Controllability indicator - the ratio of the number of process owners to the number of business process classes, characterizes the efficiency managing the owners of business processes that belong to them	Icont = PO / Pcl	Icont < 1
Resource intensity - the ratio of the number of resources used to "Outputs", the results of the business process, demonstrates efficiency use of resources in the implementation of a particular business process	Ires = R / Pout	Ires < 1
Regulability - the ratio of the amount of regulatory documentation to the number of classes of business processes, determines the level of regulation current processes	Ireg = Preg / Pcl	Ireg ≥ 1

After calculating the five performance indicators, an integrated indicator is calculated. If its regulatory value is within 1≤ΣK2, the business process is considered inefficient and in need of change. S.M. Kovalev and V.M. Kovalev [11] in 2005 proposed a method of analyzing business processes by critical success factors. The authors suggest such indicators for the selection of the prior business-processes as the importance of the business-process, the problemsness of the business-process and the possibility and the cost of the business-process changes' realization. It's suggested to evaluate the Critical Factors of Success (CFS), flowing out of the company's mission after its decomposition into the strategic purposes (targets), according to these criteria. Basing on the Pareto Rule, S.M. Kovalyov and V.M. Kovalyov offer them from the plural number of the strategic purposes.

The criterion of the business-process importance characterizes the contribution's weight of the definite business-process into the achievement of the company's strategic purposes. It's offered to define the importance of the business-process with the help of the matrix of the CFS comparison with the business-processes. The quantitative evaluation of the business-processes' importance is defined as the sum of the CFS, affecting the business-process and lying in the range from 1 to 8, or each CFS is given the weight coefficient, which characterizes its importance, and the evaluation scale of the business-process' correspondence, according to the criterion of importance, is elaborated. In order to estimate the problemness of the business-processes, it's worth to define the main problems of the business-processes previously and to evaluate the force of these problems, according to the scale from 1 to 5, to compare the results with the problemness scale of the business-processes, offered by the authors, by having estimated them from 1 to 5, or to elaborate the private scale of the problemness evaluation, considering the specific character of the organization's activity. The final coefficient of problemness is calculated as the sum of the problems' forces and the problemness degree of the business-process. The following step, offered by the authors, is to elaborate the matrix of the business-processes' ranging with the scales of importance and problemness, in order to define the priority of the business-processes. Those business-processes will be the most prior ones, which are characterized by the greatest level of importance and problemness. The authors of the methods propose to evaluate the possibility degree of changes realization in the frames of the three groups of the barriers, being able to prevent from changes introduction, they are – finances, the personnel and the legislation. It's necessary to estimate the value of the barriers, according to the scale from 1 to 5, after the definition of the main barriers in every business-process. The summary value of all the barriers is calculated after that, which will be the searching coefficient of the changes realization's probability. The priority of the business-processes is finally defined as the sum of importance, problemness and the possibility degree of changes realization. The most prior is the process, for which the sum of the indicators is the biggest one. The re-engineering of the processes, which operation is unsatisfactory at the present moment, is realized in the first turn, the attention is paid after them to the processes, having the average influence on the operation of the enterprise; the attention is paid to the processes, having the insufficient influence or working well enough, in the final turn.

It's reasonably to separate several methods of the results' efficiency evaluation of the information systems' introduction among the researched methodical instruments at the re-engineering of the business-processes, where the subjectivism element of the expert evaluation may be essentially reduced. Let's mark the following ones among them:

- the analysis of the efficiency key indicators and the evaluation of the effectiveness balanced indicators;
- the fuctional-cost analysis;
- the analysis of the investment's repayment;
- the evaluation of the possession's aggregate cost.

You may not find today many domestic companies, being able to introduce the integrated system of their activity's evaluation, based on the use of the presented models. It's worth recognizing, that the above-mentioned models of the financial result evaluation are not widely-known till the present time not only in Ukraine, but in the developed industrial countries, having essentially more experience in the use of the information systems at the realization of re-engineering [12].

Let's investigate the types of the economic efficiency that may be estimated at the information system's introduction of the re-engineering measures' monitoring at the industrial enterprise. It's necessary to notice that there are the absolute and the relative (comparative) economic efficiencies. As the scientist-economist L. Vodachek indicates, the analysis of the already selected (or even the realized) variant of the information system's introduction is conducted without any accounting of the possible alternatives in the first case. In the other case, the alternative strategies of the system's introduction are compared with each other from the position of the economic efficiency. You may also calculate the specific indicators of the efficiency (per one hrivnya of the contributed costs, per one employee and so on) [13].

2.2 Presentation of Materials and Results

According to the economic efficiency evaluation methods of business-processes re-engineering, analyzed by the author of the research, you should ascertain the absence of the united approach to the solvation of this problem.

Let's examine the calculation of the absolute efficiency of the information system's introduction at the realization of business-processes re-engineering. From the positions of the user (the final consumer) of the information system, the absolute efficiency of the introduction is defined as the relation of the difference between the obtained results (or the evaluation of these results in future) and the costs for the introduction to the corresponding costs for the introduction of the information system. If the enterprise has not been automated in general before the realization of the evaluation operations, it's enough to compare the activity results without any introduction before the re-engineering measures – R_1 at the corresponding zero costs ($S_1 = 0$) – with the results after the introduction at the realization of re-engineering – R_2 at the corresponding costs S_2. The simplified indicator of the efficiency (E) is calculated, according to the formula [14]:

$$E = \frac{(R_2 - R_1 - S_2)}{S_2} \cdot 100\% \qquad (1)$$

At this, the difference ($R_2 - R_1$) characterizes the obtained effect from the main activity of the enterprise after the project realization of the information system's introduction at business-processes re-engineering.

The value of the aggregate costs for business-processes automation at their further re-projection and the formation of the new business-process, as a rule, is composed of the following:

– the cost of the used computers;

– the cost of printers, the system equipment and the other devices;

– the cost of the program products or the licences for their use;

– the costs for the installation, introduction, adaptation, studying and the accompaniment of the program complexes;

– the evaluation of losses, due to the change of the control structures, the personnel re-organization and so on;

– all the current costs (for the electrical energy, premises, communication and so on).

In case, when the introduced information system of business-processes automation has been already at the enterprise, it's necessary to take into account the current costs for its maintenance and to compare the effects at the basic and the actual variants of the system's introduction. Due to that, we offer to define the economic efficiency from the introduction of the automated projection (designing) system at business-processes re-engineering in the following way:

$$E = \frac{(R_2 - R_1) - (S_2 - S_1)}{S_2 - S_1} \cdot 100\% \quad (2)$$

where ($R_2 - R_1$), ($S_2 - S_1$) – is the difference of the results ($R_2 - R_1$) and ($S_2 - S_1$), the difference of the costs at the new and the basic variants of the information system of the enterprise's business-processes automation, being subjected to re-projection. E characterizes the CFS efficiency coefficient, on which ground the criterial base of the economic efficiency evaluation of the information system's introduction is offered.

It's completely clear that if $E > 0$, then, the system's introduction was held effectively. How much effectively it was really done is shown by the relative indicator of the efficiency per one hrivnya of the aggregate costs, which is reflected by the quantity of the effect's hrivnyas, coming per one hrivnya of the realized costs (including the time factor). The use of the CFS relative indicator gives the opportunity to define the optimal variant of the automated projection (designing) system's introduction at re-engineering of business-processes at different budgets of the projects. In order to realize the expert evaluation of the measures, concerning the business-processes automation, the expert groups need the corresponding criterial base at the definition of the projects' economic efficiency. The indicators and the criteria of the information system introduction's economic efficiency at re-engineering of business-processes are offered in the Table 6.

Let's mention that the ranges of the economic efficiency indicators of the automated projection (designing) system's introduction at re-engineerng of business-processes are obtained, according to the results of the bench-marking research of the transformation/restructuring projects of the Ukrainian industrial enterprises of the machine-building, light, processing, coal and the oil-gas industries of Ukraine on the basis of calculation of the expected results from the introduction of these projects [14].

Basing on the presented Table 1, we separated the four meanings' areas of the specific efficiency coefficient of the automated projection system's introduction at re-engineering, forming the strong, the sufficient, the weak and the insufficient economic efficiency levels of business-processes automation at the realization of their further re-projection. It allows increase the quality of the realized evaluation and establish the more real economic efficiency's level of the measures, concerning the introduction of information systems [15, 16].

Let's indicate in total, that the presented criterial base may be also used at the definition of the economic effects from the synergetic influences on the economic potential of the industrial enterprise.

Table 6. Indicators and criteria for assessing the cost-effectiveness of information system implementation in business process reengineering.

Indicator	Value	Conclusion
Index efficiency	$Ief > 2$	We can talk about real success in business by increasing controllability of the enterprise on the basis of new information technologies, and, consequently, conducting effective reengineering of business processes
	$1 < Ief < 2$	This range of values of the specific coefficient of efficiency indicates sufficient efficiency of the project results during reengineering business processes and eliminate bottlenecks in business processes
	$0 \le Ief < 1$	This range of coefficient values indicates a weak economic efficiency and about possible negative tendencies in introduction of IS at conducting reengineering. It is advisable to talk about the impact internal and external factors that interfere with normal conducting reengineering of business processes, in particular about error set goals and objectives of business process reengineering, inefficient strategic planning, incomplete compliance with the implemented IP reengineering and technological processes at the enterprise, etc.
	$Ief < 0$	The reengineering project is unprofitable due to excess costs implementation of an information system over the expected results

3 Conclusions

The indicators of business-processes are the irreplaceable tool to control the achievement of purposes, to compare with the normatives, the standards and the achievements of the competitors, to control the satisfaction of the consumers. The use of the qualitatively elaborated system of the business-processes' indicators is the guarantee of the successful introduction of the process approach to the activity of any enterprise or organization. The system of indicators will allow define the efficiency or inefficiency of the processes' run, it will point out the low effectiveness and promote for the quick solvation of the problem questions in the very that part of the business-process, where it is necessary.

The modern researchers separate the qualitative and the quantitative analysis of business-processes. The use of the qualitative analysis methods of business-processes gives the possibility to structure the processes, to avoid the "narrow places", to define the improvement regularity of the processes. However, the existing methods of the qualitative analysis are intrinsic in the significant subjectivism and the definite "erosion" of the obtained information, not giving any opportunity to take the strategic managerial decisions on the basis of the results, received at the use of the qualitative analysis.

The methods of the quantitative analysis, in contrast to the qualitative one, operate by the facts, the numerical values, allow avoid the subjective evaluations and the mistaken interpretations, received, due to the insufficient formalization of the conclusions. At this, we grounded a series of both advantages and disadvantages in the run of the conducted investigations for the comprehensive use of these methods in the evaluation of business-processes of enterprises and organizations. Besides, the absence of the experience to calculate the costs in the processes stands as an obstacle on the use of such quantitative analysis method as the ABC method, but the introduction of the balanced indicators' system requires the essential enlargement of the costs for the marketing researches and the internal audit.

The express-method of business-processes evaluation, offered by K.K. Chuprov, is worth being referred to the advantages, due to the possibility to realize the evaluation of business-processes quickly enough, by avoiding the complex calculations and the subjective qualitative methods. However, the express-method of K.K. Chuprov may be used only on conditions of the previous description of the business-processes with the use of the modelling methods IDEF, DFD, ARIS. Besides, the use of the presented methods does not allow realize the deep evaluation, which may become an obstacle to the demonstration of the unobvious problems of the process.

The methods of the business-processes' analysis, according to the Critical Factors of Success, offered by S.M. Kovalyov and V.M. Kovalyov, are simple enough and understandable from the viewpoint of selection of the prior business-processes, but they require the closer definition of the notions: the force of the problem, the problemness of the business-process or it's necessary to make these notions more exact and to elaborate the criteria of their evaluation for each enterprise individually. Besides, it's necessary to define the coordination degree of the experts, while calculating the coefficient of concordation - for the methods, that foresee the application's possibility of the expert evaluations' method.

The key aspects of the economic efficiency evaluation of business-processes re-engineering at the industrial enterprise are examined. The economic efficiency increase ways of the enterprise's functioning at the realization of re-engineering are separated. The calculation of the absolute efficiency of the information system's introduction at the re-engineering realization of business-processes is examined. The method of the economic efficiency definition from the introduction of the automated projection system and also the indicators and the criteria of the economic efficiency evaluation of the information system's introduction at re-engineering of business-processes are offered.

References

1. I. Strutynska Information technologies of business organization - imperative of innovative development of business structures, Halytskyi ekonomichnyi visnyk, 2 (2018), pp. 40-49 available at: http://nbuv.gov.ua/UJRN/gev_2018_2_6
2. B. Andersen Byznes-protsessy. Ynstrumenty sovershenstvovanyia, Business Process Improvement Toolbox: Translated by Arynycheva S.V., RYA «Standarty y kachestvo», Moscow, Russia (2003)
3. V. Repyn, V. Yeliferov Protsessnyj podkhod k upravlenyiu. Modelyrovanye byznesprotsessov [The process approach. Business process modeling], RYA «Standarty y kachestvo», Moscow, Russia (2004)
4. B. Zuhaira, N. Ahmad, Business process modeling, implementation, analysis, and management: the case of business process management tools, Business Process Management Journal, Vol. 27 No. 1 (2020), pp. 145-183. https://doi.org/10.1108/BPMJ-06-2018-0168
5. A. Gunasekaran, B. Kobu Modelling and analysis of business process reengineering International Journal of Production Research Vol. 40 (2010), pp. 2521-2546. https://doi.org/10.1080/00207540210132733
6. R. Parthasarthy, S. Prakash Sethi The Impact of Flexible Automation on Business Strategy and Organizational Structure, Academy of Management ReviewVol. 17, No. 1 (1992) https://doi.org/10.5465/amr.1992.4279572
7. V. Tigareva, I. Stankevich Analysis of existing approaches and methods of evaluation of business processes of enterprises and organizations, Bulletin KrNU of Mykhailo Ostrohradskyi (98)

(2016), pp. 113-122 available at: http://visnikkrnu.kdu.edu.ua/statti/2016_3_113-3-2016-2.pdf.

8. V. Sytnyk Imitatsijne modeliuvannia, Imitation modeling, KNEU, Kyiv, Ukraine (1999)
9. S. Bratushka Imitation modeling as a tool for exploring complex economic systems, Transactions of Ukrainian Academy of Banking, no. 2 (2009), pp. 113–128.
10. K. Chuprov The Express-method of diagnostics of the business processes of the company, Konsul'tant dyrektora, no. 20 (2005), available at: cfin.ru/management/controlling/fsa/express.shtml
11. S. Kovalev, V. Kovalev Selection the business processes for optimization, Konsul'tant dyrektora, no. 5 (2005), available at: betec.ru/index.php?id=6&sid=39
12. L. Taranyuk Estimation of economic efficiency of carrying out reengineering of business processes at the enterprise Bulletin of Sumy State University. Economics series 2 (2009) pp. 168-178
13. L. Vodachek Strategy of innovation management at the enterprise: monograph, Ekonomika, 1989. pp. 205-210
14. K. Mironenko Estimation of economic efficiency of reengineering of business processes at the industrial enterprise, 5 (2) (2016), pp. 51-56 available at: http://nbuv.gov.ua/j-pdf/ecfipr_2016_5%282%29__14.pdf
15. M. Alles, G. Brennan, A._Kogan, M. Vasarhelyi Continuous Monitoring of Business Process Controls: A Pilot Implementation of a Continuous Auditing System at Siemens, Chan, D.Y., Chiu, V. and Vasarhelyi, M.A. (Ed.) Continuous Auditing (Rutgers Studies in Accounting Analytics), Emerald Publishing Limited, (2018), pp. 219-246. https://doi.org/10.1108/978-1-78743-413-420181010
16. A. Babkin, S. Kuzmina, A. Oplesnina, A. Kozlov Selection of Tools of Automation of Business Processes of a Manufacturing Enterprise, 2019 International Conference «Quality Management, Transport and Information Security, Information Technologies» (IT&QM&IS) (23-27 Sept. 2019) https://doi.org/10.1109/ITQMIS.2019.8928302

Access to Finance as a Driver Innovation

Iryna Vasylchuk[1*], *Natalya* Izmaylova[1], *Olga* Smyrna[1], *Tetiana* Petrishyna[1] and *Kateryna* Slusarenko[1]

[1]State University of Economics and Technology, Kryvyi Rih, 50000, Ukraine

Abstract. Innovation accelerates economic growth and becomes the key to success in the new post-industrial paradigm. The more innovation, the more a country's prosperity grows. Innovation is a risky and costly type of activity, so it requires significant financial resources attracted from different sources and by different methods. Traditionally, the entrepreneurial sector, which has sufficient amounts of various resources, is the most innovatively active. At the same time, the innovative activity of companies is under pressure from a number of factors. The purpose of the article is to analyze the factors influencing the innovative activity of industrial enterprises in Ukraine and to identify the relationship between the volume of innovation and economic growth of the country. For the analysis there were used the methods of descriptive, comparative and correlation and regression analysis. The sample is created at the basis of data from statistical reports on the research and development activities of economic entities in Ukraine for the period 2000-2019.It has been revealed that there is a steady decline in the number of workers in the innovation sphere in Ukraine, and compared to European countries, these indicators are much lower and tend to decrease further. This affects the place of Ukraine in the global ranking of innovation in a negative way. It is confirmed that there is a direct positive significant relationship between the costs of innovation and bank loans and foreign direct investment in the capital of Ukrainian companies, confirms the thesis of funding as a driver of innovation. It is confirmed for Ukrainian content that there is a positive relationship between innovation and economic growth.

1 Introduction

R&D and innovation is an important factor in a country's economic growth and prosperity, and contributes to its competitiveness in the modern world. Even small and resource-poor countries achieve a high level of economic development with the introduction of disruptive technologies and organizational and product innovations. In order to become successful in the new post-industrial paradigm, the state and business must significantly increase the volume of spending on innovation and science, create conditions for replication and multiplication of innovation, develop the entrepreneurship sector, inculcate the ideology of creativity, innovation and enterprise in society. An important condition for achieving a technological breakthrough is access to funding. And while public sector research institutions can rely on funding from the state budget, business can rely primarily on internal resources and funds raised from the financial market. In conditions of underdevelopment of stock market in Ukraine, and as a consequence, its inability to absorb financial capital and fulfill its mission of provider of capital, low volumes of business crediting by credit institutions, chronic shortage of own funds of enterprises as a result of permanent economic recession, business suffers from inability to activate and finance innovation activity.

The aim of the article is to analyze the factors influencing the innovative activity of industrial enterprises in Ukraine and to identify the relationship between innovation and economic growth of the country. From the above, the task of this study is to find answers to the following research questions:

RQ1. Do macroeconomic factors influence the innovative development of the country and the innovative activity of Ukrainian companies?

RQ2. Is the determinant of human capital significant for the level of innovativeness of the country's economy?

RQ3. Does company size influence the level of its innovative efficiency and the amount of expenditure on innovation?

RQ4. Does the company's access to financing from various sources and forms of attraction affect the amount of innovation expenditures?

RQ5. Is there a positive relationship between the R&D expenditure on innovation of industrial companies and the economic growth of the country?

To find answers to these questions, there were used descriptive, comparative and correlation-regression analysis methods. The sample is created at the basis of data from statistical reports on the research and development activities of economic entities in Ukraine and on the Statistical Yearbook of Ukraine for the period 2000-2019.

1.1 Literature review

At the beginning of the study it is necessary to define the main terms. In scientific publications, the innovative activity of companies is understood as an activity, action

* Corresponding author: vasylchuk_ip@duet.edu.ua

or set of activities aimed at creating a new product or its significant improvement, in order to obtain commercial benefits and competitive advantages of innovation-active enterprise [1-5]. It is important to emphasize that it is an activity aimed at the use and commercialization of R&D results and predetermines the release of new competitive goods and services to the market [report]. So, innovative activity leads to development. In turn, the innovative development of the enterprise is defined as positive qualitative changes in the state of the enterprise as a result of the implementation of innovative activity and effective use of innovative potential [2]. It can be assessed by the level of implementation of innovation, namely by the achieved level of practical use of inventions, research results, technologies and other innovations [3]. There are innovatively active and innovatively capable enterprises. Innovation-active enterprise is an enterprise that was engaged in innovative activities (developed or implemented new or improved products and innovative processes) during a certain period of observation [6].

The innovative ability of the enterprise should be understood as the ability to carry out innovations at the expense of the models of the corresponding resources - financial, human, physical, organizational, and so on. The absence of a certain type of resource in sufficient volumes reduces or eliminates the company's ability to carry out innovative activities.

At the same time, implementation of scientific research, introduction of innovations requires appropriate expenses. The innovation expenditure is understood as the cost of the enterprise to implement innovations both new to the enterprise, and new to the market, including: internal R&D, acquisition of R&D results, machinery, equipment and software, other external knowledge and other costs [6]. The innovative activity of the enterprise depends on the influence of various factors. Ukrainian scientists noted that among the factors hindering the activation of innovative activity of domestic enterprises the main ones are such as imperfect legislative base, lack of financial resources and weak incentives for implementation of inventions, research and technology [6]. In the study [3] based on the results of the survey, the author identifies such factors as the lack of own funds of the enterprise (15.6%); high costs of innovation (13.3%); lack of funding from external sources (8.4%); dominant market role of individual enterprises (7.7%); difficulties in finding partners for innovation (5.5%); low demand for innovative products or services (4.6%). In another study [7] we find that the main barriers to innovation activity of Ukrainian enterprises are as follows: lack of own funds (83.0% of surveyed enterprises), high costs of innovation (55.9%), insufficient financial support from the state (56.6%), high economic risk (38.9%), long payback period of innovation (34.6%). In addition, the development of innovation activity was restrained by the imperfect legal framework (37.7%), lack of demand for the products (16%) and qualified personnel (17.2%). As wee can see, the overwhelming majority of scientists identify the factor of lack of financial resources from various sources as one of the most significant factors of influence on the innovation activities of enterprises.

In their studies domestic scientists use mainly descriptive and comparative methods of analysis, while the vast majority of scientists base their research on the use of other analytical methods, in particular methods of econometric analysis. A significant number of publications [8-11] are devoted to assessing the relationship between innovation and economic growth. Scientists have found that innovation activity is equally influenced by economic growth and other macroeconomic factors. This means that in practice both innovation activity and economic growth can condition each other and, therefore, there is a possibility of feedback between them. Also for different jurisdictions the factors of influence on innovation are investigated [11], [12-13]. Thus, K. Širec and D.Močnik [14] evaluated the relationship between firms' innovation activities and the education of their owner-managers, firms' international orientation and their growth aspirations. M. Bhattacharya and H. Bloch [13] investigated how firm size, market structure, profitability and growth affect the innovation activities of small and medium-sized businesses. A. Pecea et al. [9] used such factors as the number of patents, number of trademarks, R&D expenditures as determinants of economic growth and innovation. In [15-16], considerable attention is paid to the influence of such factors as external direct investment, access to credit resources, the state of development of the stock market and the like on the innovative activity of companies. The authors [15] discovered that the companies with participation of the foreign capital or having good access to domestic bank loans are more innovatively active, than others.

However, existing studies have not carried out a quantitative assessment of the impact of determinants on the innovative activity of Ukrainian companies and did not study the relationship between innovation and economic growth and well-being of the country.

2 Hypotheses Development and descriptive data analysis

2.1 Innovation and macroeconomic indicators of economic development

Innovation activity, as well as investment activity, in countries depends on the trends of global and national economies and fits into certain cycles. According to the research [17] at the global level, historically it has been noticed that business R&D expenditure, investments in intellectual property and venture capital moved in line with GDP, slowing down noticeably during the economic downturn of the early 1990s, early 2000s and 2009 [18]. Ukraine isn't an exception and trends of innovation activity development in the country as a whole and business R&D activity also move cyclically (Fig.1). Thus, between 2000-2006 the economy of Ukraine was on the rise, GDP was gradually growing, and against this background the indicators of innovative activity of Ukrainian companies were gradually

 https://doi.org/10.1051/shsconf/202110001020

increasing. In addition, after the victory of the "orange" revolution in Ukraine (2004), international investors became more interested in Ukrainian companies, which contributed to a surge in their business activity, improvement of the investment climate and inflow of funds. In the period before the global financial crisis (2006-2008) the volume of foreign direct investment in Ukraine increased significantly, the interest rates on loans decreased significantly, which led to an increase in lending to the economy and the total amount of available financial resources (Fig. 2).

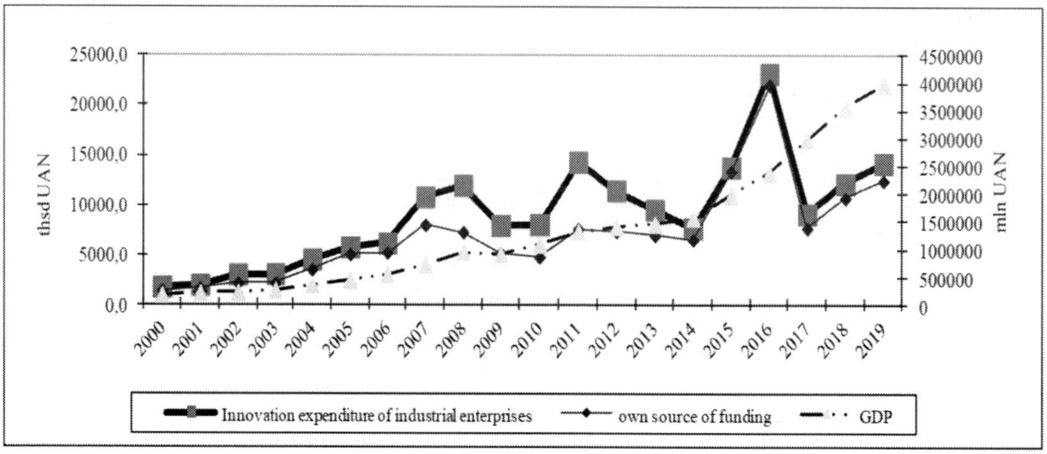

Fig. 1. Ukrainian GDP, business R&D expenditure and innovation expenditure of industrial enterprises by source of own funding in 2000-2019
Source: constructed by the author according to [6], [19-21], [23]

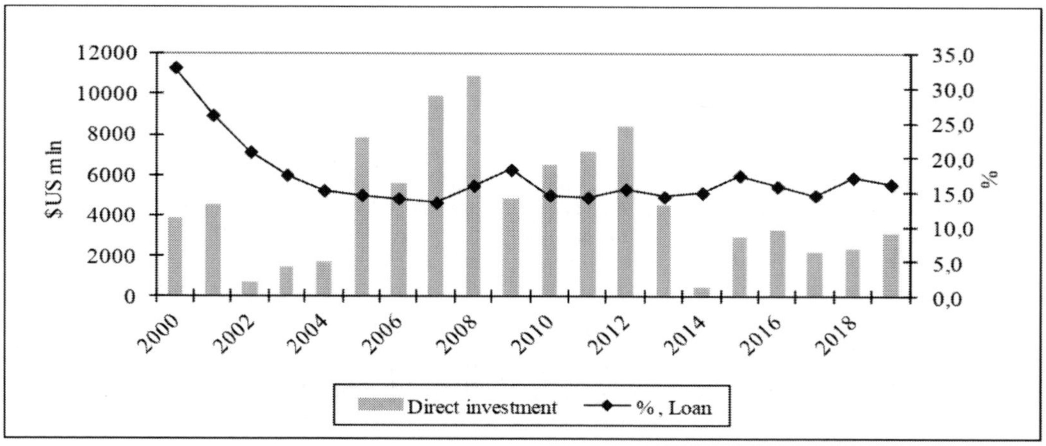

Fig. 2. Foreign direct investment and % on loans in Ukraine in 2000-2019
* 2000-2004 with interest accrued; including loans to the government sector; 2005-2019 loans to nonfinancial corporations
Source: constructed by the author according to [6], [19-23]

In the period after the financial crisis (2009-2010) the volume of business R&D expenditure dropped significantly, and already in 2011 there was a significant "rebound", although in next years (2012-2014) research expenditure decreased again, which was due to domestic political and economic factors (annexation of Crimea, military actions in the East of Ukraine). In 2016, Ukraine's economy finally stabilized, the demand for the main export products of domestic producers intensified, which caused a "surge" of spending on innovation activities. However, in next years, the volume of innovation expenditures already had a moderate upward trend. Thus, the analysis allows us to confirm the *research question RQ1* and claim that innovative expenses of companies are influenced by trends in the economy, including the main macroeconomic indicators. We can see (Fig. 2) an inverse relationship between the volume of foreign direct investment and interest rates on loans, which confirms the stimulating role of the interest rate and the weighted monetary policy.

To assess the general state of innovativeness of the country's economy the share of total intramural R&D expenditure in GDP is calculated, the level of which in Ukraine is too low compared to European countries and has a decreasing tendency (Fig. 3). Thus, if in 2010 the share of R&D expenditure was 0,75%, for the last 10 years it has decreased to 0,43%, whereas for the totality of European countries during the same period the indicator has increased from 1,92% to 2,12%.

The level of innovativeness of the country's economy can be evidenced by its rating in global indices. In particular, in 2019, according to the "Global Innovation Index – 2020" calculated by the international business school INSEAD, the World Intellectual Property Organization and Cornell University, Ukraine took 45th place out of 130 in the ranking of the most innovative countries in the world, remaining at a relatively low position, and maintained 2nd place among "countries with income below average" in the sub-index [24].

If we analyze the total expenditure in R & D in Ukraine according to their distribution by sectors of activity (Fig. 4), we can see that traditionally the main driver of innovation in Ukraine is business, the share of which is more than 58% of the total costs. So, we can consider that the success of innovative activity of the business sector, in particular industry, to a great extent determines the success of innovative development of the country as a whole. Therefore, the results of innovative activity of industrial enterprises in Ukraine require more detailed research.

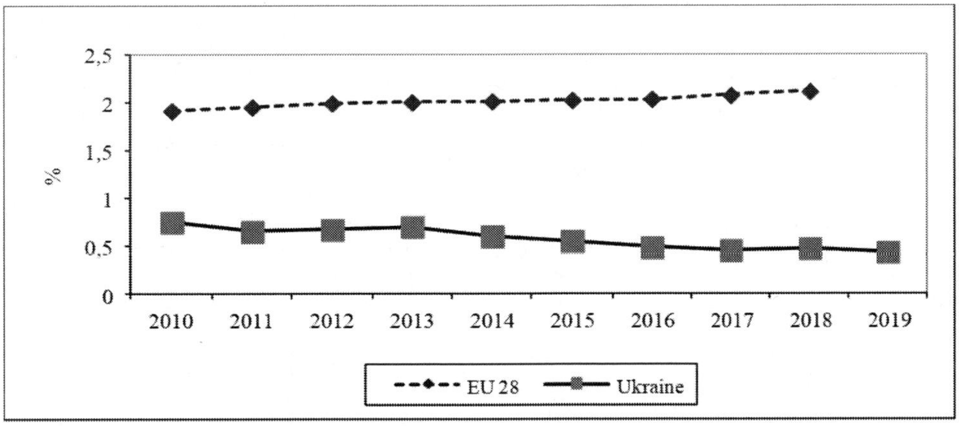

Fig. 3. Share of intramural R&D expenditure in GDP in 2010-2019
Source: constructed by the author according to [19]

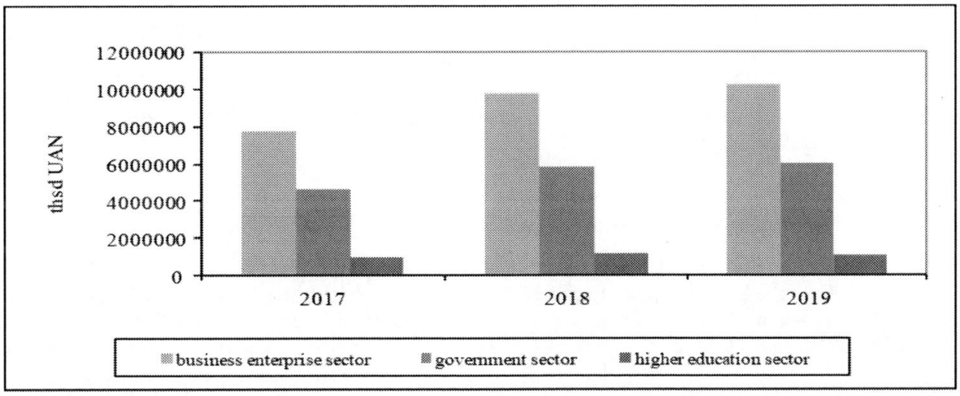

Fig. 4. Intramural R&D expenditure by sectors of performance in Ukraine in 2017-2019
Source: constructed by the author according to [19]

2.2 Innovations and human capital

One of the important factors influencing the innovative activity of business and the country as a whole researchers determine a sufficient number of employees with a high level of education. It has been confirmed [9, 14] that the innovation activity and potential of enterprises is directly influenced by the education of managers and researchers, in particular above. Higher education provides the opportunity to possess more modern knowledge and research competencies, to be erudite on a wide range of issues, to be able to constantly learn and seek necessary information. In addition, the more high-level workers are engaged in R&D research, the more new products and technologies they can implement and produce more innovative products. In Ukraine there has been steady decline in the number of employees and researchers engaged in R&D over the past 10 years (Fig.5).

Thus, the number of highly qualified researchers decreased by more than 2.6 times from 133.7 thousand people in 2010 to 51.1 thousand people in 2019, and their share in the total employed population at the end of 2019 is only 0.31%.

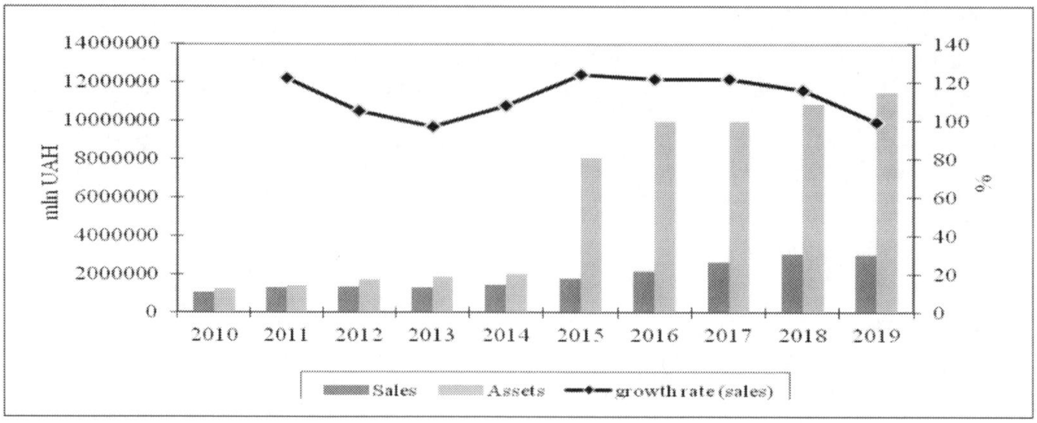

Fig. 5. Dynamics of the number of employees and researchers engaged in R&D in Ukraine in 2010-2019
Source: constructed by the author according to [6], [19-21]

So, we can see a significant reduction in human capacity for research and innovation, which does not strengthen the position of Ukraine in the ranking of innovative countries and worsens its ability to innovate.

A comparative analysis of the number of workers and researchers involved in research in Ukraine and Europe per 1000 people employed found a low level of this indicator. For example, in 2010 in EU 28 the rate of workers was almost 18 workers per 1,000 population and researchers more than 11.5, in 2019 the rate rose to 21.6 and 14.0 for workers and researchers, respectively.

While Europe is increasing its human capacity for innovation, in Ukraine the indicator for workers is almost half as low in 2010 (9.5 people), and in 2019 the gap increases to 4 times and is 5.8 workers per 1000 people of population [6]. We observe the same trends for researchers, that is, there is a deterioration of scientific and human capital potential due to many reasons, including the outflow of "brains" to other countries.

The connection between the number of innovation workers and the share of R&D expenditure in GDP can be shown by the results of comparative indicators calculations (Table 1).

Table 1. Comparative indicators Share of intramural R&D expenditure in GDP and Number of R&D personnel per 1000 persons (EU-28 vs Ukraine)

№	Indicators	Years			
		2010	2015	2017	2017/2010
1	Share of intramural R&D expenditure in GDP (EU-28), %	1,92	2,03	2,08	1,08
2	Share of intramural R&D expenditure in GDP (Ukraine)	0,75	0,55	0,45	**0,60**
3	*Excess (multiplicity)(line 1 / line 2)*	2,6	3,7	4,6	1,78
4	Number of R&D personnel per 1000 persons (EU-28)	17,9	20,4	21,6	1,21
5	Number of R&D personnel per 1000 persons (Ukraine)	9,5	9,1	5,8	**0,61**
6	*Excess (multiplicity) (line 4 / line 5)*	1,9	2,2	3,7	1,95

Source: calculated by the author according to [6]

The calculations show that in 2010 R&D expenditure in GDP in EU-28 exceeded the same indicator for Ukraine by 2.6 times, and already in 2017 the excess was 4.6 times. The growth rate of the indicator for Ukraine was 60% compared to 2017 before 2010. The excess of the number of innovative workers per 1000 people in the EU over the same indicator for Ukraine in 2010 was 1.9 times, and already in 2017 - 3.7 times. The growth rate of this indicator for Ukraine was 61% compared to 2017 before 2010.

Thus, we can see that the fall in the number of employees in the sphere of R & D directly leads to a decrease in the share of spending on innovation in GDP which provides an answer to the *research question RQ2* and confirms the importance of the determinant of human capital for the level of innovativeness of the economy.

2.3 Innovation, Firm Size and Growth

Studies [9, 11-13] show that there is a direct link between firm size, firm growth and innovation activity. It is proved that large companies have much more opportunities and resources (financial, human, technical, technological, etc.) to conduct research in the development of new technologies, products, organizational innovations, than small companies. Large companies have sufficient amounts of liquid funds and long-term capital, more often have balanced portfolios of clients and suppliers, and a higher "appetite for risk".

Therefore, large companies often win in the competition, although sometimes small firms can be more innovative, for example in the field of IT - technology. At the same time, large international companies have the opportunity to use and implement the developments of «parent» companies, which also provides a competitive advantage. Analysis of large companies shows that for many of them investment in R&D remains a priority even in times of unfavorable economic conditions. This allows them to achieve strategic benefits, enter new markets and strengthen their positions in the industry after the crisis [25]. Researchers have proved a positive relationship between the growth of the firm and an increase in its innovative activity for companies from different jurisdictions. Extrapolation to the macro level confirms the following dependence: the more developed economy, the more innovations in it and vice versa [17]. So, it is possible to consider the size of the company as a factor of influence on innovative activity and innovative ability of the company.

The size of the company can be assessed by indicators of the size of its assets or the volume of revenues.

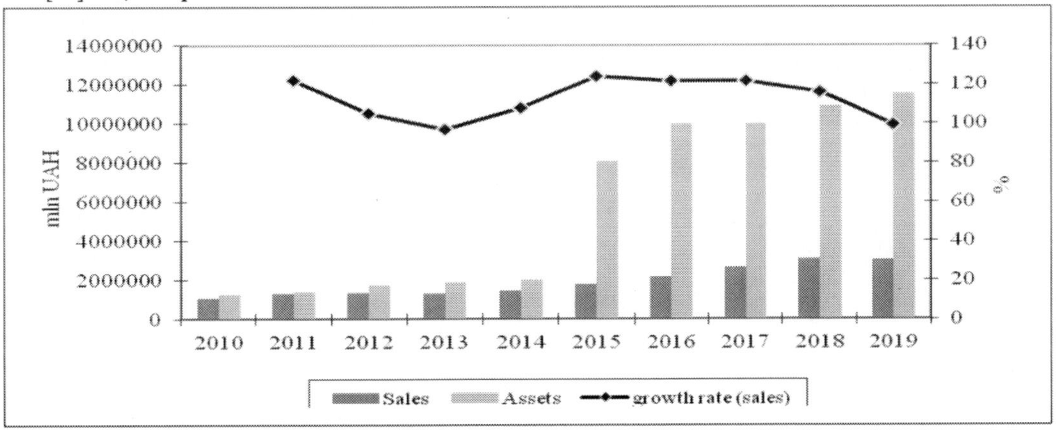

Fig. 6. Dynamics of company size indicators in Ukraine for 2010-2019
Source: constructed by the author according to [19-20]

Analysis years (Fig. 6) shows a gradual increase in revenues, beginning in 2013, although in 2019 we can see a drop in revenues. At the same time, the size of companies' assets grew very significantly after 2015, almost 4-fold compared to 2014, and almost 6-fold in 2019. Since quite often the size of assets of industrial enterprises grows at the expense of increasing the volume of accounts receivable, which is essentially deferred income, so to estimate the size of Ukrainian enterprises should choose the indicator of the volume of proceeds from the sale of products. Based on the above and to obtain an answer to the *research question RQ3* we put forward a hypothesis:

H1: there is a positive relationship between companies' size and total innovation expenditure of industrial enterprises (determinant – companies' size (sales)).

2. 4 Innovation and access to financing

The availability of funding for research and development, implementation of innovations and development of innovative products is undoubtedly an important factor in enhancing the innovative activity of companies. In researches [15-16], [26-29] the direct positive connection between development of the financial market, foreign direct investment, access to finance, and innovation activity on the example of Chinese companies is proved. Scientists note that the cost of innovation requires significant amounts of financial resources received from various sources. For long-term and high-cost developments associated with significant risks, it is quite difficult to be limited to domestic sources of financing. Therefore, the more developed is the financial market, the more opportunities for companies to finance scientific development, the more innovative products can be obtained on the output. The study [17] explicitly states that the lack of funding sources due to imperfect capital market and other reasons can lead to an insufficient level of investment in innovation. In addition, the technological risk associated with innovation may be too high for entrepreneurs and investors, so it is important to have a portfolio of financial resources diversified by sources, methods and instruments.

Analysis of the sources of funding for innovation in Ukraine has shown that the main source of funding for innovation of industrial enterprises is domestic resources, the share of which ranges from 52.9% in 2011 to a record 97.2% in 2015 and 87.7% in 2019 (Fig. 7)

A detailed analysis of the structure of sources of funding for innovation expenditures of industrial enterprises (table 2) revealed that the share of budget financing (from all levels) is too low, but tends to increase - from 1.2% in 2010 to 3.7% in 2019. The share of bank lending is also insignificant, only 6% in 2019, although in 2011 the share of loans in financing enterprise innovation was more than 38%.

We can see that the banking sector in Ukraine in recent years does not fulfill its role in lending to the economy, but acts as an active player in the market of government bonds.

Foreign investors are reluctant to invest into domestic enterprises, same way other forms and methods of financing innovative activity of companies are not being used. In fact, neither credit nor stock market is functioning in Ukraine, which makes entrepreneurs count only on their own resources. In such a situation the ability to implement "disruptive" technologies and develop innovative products in domestic enterprises is limited, which "conserves" the country's economy in the old industrial paradigm and dooms it to lagging behind.

https://doi.org/10.1051/shsconf/202110001020

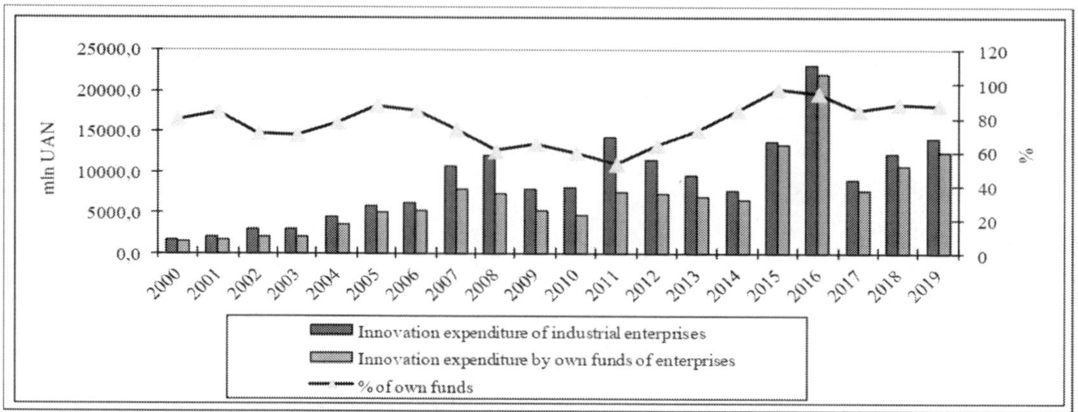

Fig. 7. Dynamics and share of financing of investment expenditures at the expense of own funds of enterprises in Ukraine in 2010-2019
Source: constructed by the author according to [6], [19-21]

Table 2. Structure of financing sources for innovation of industrial enterprises in Ukraine in 2010-2019

Source of funding	Years									
	2010	2011	2012	2013	2014	2015	2016	2017	2018	2019
Total innovation expenditure, %	100	100	100	100	100	100	100	100	100	100
of them due to										
- own funds of enterprises	59,3	52,9	63,8	72,9	85,0	97,2	94,9	84,5	88,2	87,7
- state budget funds	1,1	1,0	2,0	0,3	4,5	0,4	0,8	2,5	5,2	3,9
- funds from local budgets	0,1	0,1	0,1	1,6	0,1	0,3	0,4	1,0	0,1	0,8
- funds of resident investors	0,4	0,3	1,3	1,3	0,1	0,6	0,6	3,0	0,9	0,5
- funds of non-resident investors	30,0	0,4	8,7	13,1	1,8	0,4	0,1	1,2	0,9	0,3
- loans	7,8	38,3	21,0	6,6	7,3	0,8	2,7	6,5	3,9	6,0
- funds of other sources	1,3	7,0	3,0	4,2	1,2	0,3	0,5	1,3	0,8	0,8

Source: grouped by the author according to [6]

The importance of sufficient funding for innovation and research and easy access to various forms and sources of financing innovation for business is recognized at the global level. The "Global Innovation Index 2020" [17] asks as a headline question: Who Will Finance Innovation? As countries and innovative companies use a wide range a result, the study notes that in today's environment, of mechanisms to finance innovation, in particular corporate venturing, intellectual property (IP) marketplaces, microfinance, crowdfunding, and Fintech solutions [17].

In the conditions of Ukraine it is possible to rely so far on the traditional mechanisms - the funds of shareholders (investments of residents and non-residents in the share capital) and creditors, which form a pool of available resources for the company. On this basis, we formulate hypotheses to obtain an answer to the *research question RQ4:*

H2: there is a positive relationship between amount of loans granted to non-financial corporations and total innovation expenditure of industrial enterprises (determinant – loans).

H3: there is a positive relationship between volumes of foreign direct investment and total innovation expenditure of industrial enterprises (determinant – foreign direct investment).

2.5 Economic Growth and Innovation

It is almost axiomatic that innovation has a positive effect on economic growth and welfare both at the level of individual economies and individual business entities. Since J. Schumpeter, numerous scientific publications [8-11], [17], [30-32] have expressed the opinion and received empirical evidence that innovation acts as a driver of economic development for both developed and emerging economies. Innovation, R&D expenditure and investment in technology are prerequisites for competitiveness and progress, and through them sustainable economic growth [9]. Thus, innovation activity has a great influence on the development of the country's economy and is the basis for economic, technological, political, ecological and social development of society [31].

At the same time, it has been confirmed that it is

economic growth that can increase the level of innovativeness of the economy, so scholars [8], [10] postulate a two-way causal relationship between innovation and economic growth. Accepting the view that there is a two-way causal relationship between economic growth and innovation and it is relevant to countries with different levels of economic development we put forward the following hypothesis:

H4: there is a positive relationship between total innovation expenditure of industrial enterprises and Gross Domestic Products (determinant - total innovation expenditure of industrial enterprises).

3 Research Method

According to the hypotheses raised and to the results of descriptive and comparative analyzes for Ukraine, the study has analyzed the relevant factors of innovation for the Ukrainian industrial companies. The independent changes that were selected for the regression analysis have already been tested for other jurisdictions and are based on the results of previously published prior empirical studies [9-11], [13], [15-16], [26-30]. This study has employed six variables: total innovation expenditures of industrial enterprises (TIE), sales of industrial enterprises (approximates the size of the company), bank loans to non-financial corporations (BL), foreign direct investment (FDI), share of innovatively active enterprises in the total number of industrial enterprises (SIE) and Gross Domestic Product (GDP). The dependent variable for testing *hypotheses 1-3* is the indicator of total innovation expenditure of industrial enterprises, and for hypothesis 4 is the indicator of GDP. The sample is created at the basis of data from statistical reports on the research and development activities of economic entities in Ukraine and on the Statistical Yearbook of Ukraine for the period 2000-2019 [6], [19-23].

Since there is a time lag between the innovation expenditures and the results of their implementation that has to be reflected by GDP growth, the total innovation expenditures of industrial enterprises with a lag of TIE (t - 1) has be included into the model to test *hypothesis H4.*

To create the data sample for the regression analysis, information on the total innovation expenditure of industrial enterprises was obtained for the year 1999 till 2020, and for GDP numbers up to 2020 (preliminary estimates).

The regression analysis was chosen as main research method to test hypotheses about the relationship between 1) total innovation expenditure of industrial enterprises and their determinants and 2) total innovation expenditures of industrial enterprises and GDP. Data were analyzed using pooled regression with multiple explanation variables. Following model specifications support our analysis:

Model 1

$$TIE_t = \alpha_i + \beta_{1t} * Size_t + \beta_{2t} * BL_t + + \beta_{3t} * FDI_t + \varepsilon_i \quad (1)$$

Model 2

$$GDP_t = \alpha_i + \beta_{1i} * TIE_{t-1} + \beta_{2t} * SIE_t + \varepsilon_i \quad (2)$$

Where TIE represents: total innovation expenditure of industrial enterprises; Size show a volume of revenue from companies sales as the size of the companies; BL are the amount of bank loans granted to non-financial corporations; FDI is foreign direct investments.

4 Results and Discussion

Hypotheses testing is based on regression analysis. To answer the questions about the determinants defining the amount of the total innovation expenditure of industrial enterprises, a regression analysis was performed according to the identified *Hypotheses 1-4.* Descriptive statistics are given in Table 3.

The results of the calculations show that for the indicator of total innovation expenditure of industrial enterprises the average value for the entire period of the study is only 9035.5 million UAH, while the variability of this indicator by the coefficient of variation is 58%, which is one of the least variable indicators of the group of indicators.

Table 3. Descriptive Statistics

Variable	Min	Max	Mean	SD
Total innovation expenditure of industrial enterprises	1760,1	23229,5	9035,5	5257,9
Sales of industrial enterprises	182700	3045202	1194533,1	916769,0
Bank Loan	18594	859740	443706,3	316740,4
Foreign direct investment	410,0	10913,0	4608,9	3020,5
The share of innovatively active enterprises in the total number of industrial enterprises	8,2	16,6	12,8	2,2
GDP	211200	3974564	1508060	1228486

The lowest variability is only share of innovatively active enterprises in the total number of industrial enterprises - 17% on the variation coefficient.

This confirms the conclusions of the preliminary descriptive analysis of the low and relatively constant level of expenditures on innovation among the industrial enterprises of Ukraine. The most variable indicator for the whole period is GDP with the variation coefficient of 0.81, and in second place is the indicator Sales of industrial enterprises with the variation coefficient of 76%. If we compare the amount of innovation costs and the amount of revenues of enterprises, we can see that they do not account for more than 0.8% of revenue. The share of innovative enterprises in the total value of all industrial enterprises is at most 16.6%, which means that only every 6th enterprise is engaged in innovation and its future development.

The next step is to present the results of a multiple regression analysis for the most common factors in the literature that influence the volume of innovation expenditures of Ukrainian industrial enterprises (*model 1*). In describing the results of the regression analysis with three independent variables for total innovation expenditure (Table 4), it should be noted that bank loans (coefficient = 0.0764 * standard error = 0.0052) and foreign direct investment (coefficient = 0,0448 **, standard error = 0.2498) were significant factors.

Table 4. Results of multiple regression (Dependent = Total innovation expenditure of industrial enterprises)

Variables	Coef	SD	t-stat	p-value	Result
Constant	515,7303	1790,606	0,28802	0,7771	No hypothesis
Sales of industrial enterprises	0,0014	0,0018	0,7337	0,4737	Not confirmed
Bank Loans	0,0099	0,0052	1,8948	0,0764*	Confirmed
Foreign direct investment	0,5439	0,2498	2,1775	0,0448**	Confirmed
Adj. R2	0,6343				
F-stat	11,9839				
F-stat (p-value)	0,00023				

Note: *Significant at 10%, **Significant at 5%, ***Significant at 1%

It is obtained that the regression coefficient for both variables is positive, although not significant. This means that the increase in the volume of bank loans and foreign direct investment has a positive effect on the volume of innovation expenditures of Ukrainian industrial enterprises and leads to economic growth. In addition, the adjusted coefficient of determination (Adj. R2) for the model at the level of 63% indicates sufficient explanatory power of the estimated model.

This confirms the previously obtained results of empirical analysis [15-16] for companies from other jurisdictions and confirms the conclusions of the report "Global Innovation Index – 2020" about the importance of the factor of access to finance. The descriptive analysis conducted to find the answer to the *research question RQ4* also confirmed the importance of attracting funds in sufficient amounts, obtained from different sources and by different methods, to finance innovation. This result confirms the conclusions of Ukrainian scientists [3], [33], who note that the main factors constraining the implementation of innovation policy in enterprises are primarily financially based. The analysis showed (see table 2) that financing of the overwhelming majority of innovation costs by Ukrainian enterprises occurs at the expense of internal sources, which is due to both the risk of innovation activity and the lack of opportunities to attract funds from investors and creditors. As noted in the study [3] low level of financing of innovative activities by domestic investors due to the lack of available funds, the high degree of uncertainty and risk inherent in innovative activities and the lack of legislative incentives for innovative activities.

Thus, we can conclude that *hypotheses H2-H3* are confirmed for Ukrainian companies. But the *hypothesis H1* about the relationship between the amount of innovation costs and the size of the company was not confirmed for the Ukrainian context, which contradicts the previous research. The reasons may be both the insufficiency of the sample for the analysis and the fact that small technology companies in Ukraine are more active in implementing innovation.

Next, we tested the hypothesis of a positive impact of innovative development on the economic development of the country, which was evaluated by the GDP indicator. Two determinants that characterize the degree of innovativeness of the economy, namely the share of innovatively active enterprises in the total number of industrial enterprises and total innovation expenditure of industrial enterprises, were selected for evaluation (Table 5).

Table 5. Results of multiple regression (Dependent = GDP)

Variables	Coef	SD	t-stat	p-value	Result
Constant	-1855662	1168431	-1,58817	0,130673	No hypothesis
Total innovation expenditure of industrial enterprises	134,0784	40,94259	3,27479	0,004467**	Confirmed
The share of innovatively active enterprises in the total number of industrial enterprises	168482,7	97395,56	1,729881	0,101764	Not confirmed
Adj. R2	0,504				
F-stat	10,661				
F-stat (p-value)	0,000999				

Note: *Significant at 10%, **Significant at 5%, ***Significant at 1%

According to the results of the test it was found that only one independent variable turned out to be significant, namely, the absolute amount of innovation expenditures of the previous period shows a positive connection with the level of gross domestic product (coefficient = 134.0784**, standard error = 40.9426).

Despite the lack of significance for other coefficients, the value of the adjusted coefficient of determination (Adj. R2) for the model at the 50% level is significant. Results of testing of *hypothesis H4* confirm that growth of investments into innovations, and in this case expenses on innovations, allows increasing opportunities to buy or develop for introduction of new technologies, to incur expenses on development of innovative products that in aggregate causes realization of more demanded new products or technologies. The increase in sales of innovative products also directly affects the size of the country's GDP.

Based on the calculations, we can derive a multiple regression model for innovation costs with weighting coefficients:

$$TIEt = 515,7303 + 0,0014 * Size\ t + 0,0099 * BL\ t + 0,5439 * FDI\ t, \quad (3)$$

and a model for factors influencing economic development:

$$GDPt = -1855662 + 134,0784 * TIE(t\text{-}1) + 168482,7 * SIE\ t \quad (4)$$

Least squares regression is based on several assumptions, in particular: linear dependence between variables, independence of regression residuals, normal distribution of variables and regression residuals, homoscedasticity and absence of multicollinearity. These assumptions were tested using appropriate tests for the two model specifications presented above. The White's test was used to check for heteroscedasticity, and the variance inflation factor was used to check for multicollinearity, for which there is a rule that its value should not exceed 10. The value of the White's test and the variance factor were within the permissible limits for both cases. Thus, the permissibility of using the regression method for the analysis of the hypotheses was confirmed.

It should be noted that the conducted analysis has limitations due to the limited types and volumes of collected data, as statistical reports in Ukraine do not allow obtaining more detailed and diverse information about the innovative activity of enterprises to erect it into an equal panel of data (both for individual indicators and in the context of periods). This was repeatedly noted in previous studies for Ukrainian content [2]. Also Ukrainian companies do not have the obligation to publish different information on innovation activity indicators in the annual reports, does not allow forming a representative data panel. There is a discrepancy in the methodology and time frame for the data from the published reports of the NBU and the stock market, which also does not contribute to the formation of a comprehensive and balanced sample.

The choice of certain indicators as determinants of influence on innovation expenditure and GDP can be debatable for this research. Thus, innovativeness of economy can be estimated by indicators of introduction of innovations, volumes of realization of innovative products or their share in the total realization of products, according to some authors [2], [5] are more exact indicators. In addition, in some previous studies [9], the authors propose to use the GDP per capita indicator to assess the economic development of the country. There are also publications in which other methods of analysis are used to assess the influence of factors, in particular, in our opinion; the analysis carried out in this study by different methods (descriptive, comparative and regression methods) allows us to draw reasonable conclusions.

6 Conclusions

There is a two-side relationship between economic growth and innovation - the higher the amount of funds allocated to innovative activities, the higher is the effectiveness of such activities, and therefore, the higher is the volume of the country's GDP, and vice versa. It is believed that the more developed and richer the country, the more innovative it is and more is directed to innovation. The overwhelming majority of scientists, practitioners and politicians define availability of financial resources for scientific research and implementation of innovations as the key factor influencing the innovative activity of companies. It is the limited access to finance, the lack of free own funds does not allow companies and countries to develop innovations and condemns to poverty.

The analysis carried out in the work for the Ukrainian content provided answers to the research questions and confirmed that the state of innovation activity in the country is influenced by a complex of factors, including: macroeconomic factors, including monetary factors, the state and level of human capital development, access to funding from different sources. Since in Ukraine the greatest contribution to innovative activity is caused by the entrepreneurial sector, the research of the relationship between innovation and its determinants is carried out for industrial enterprises.

It was proved that the reduction in the number of employees engaged in innovation is directly related to the reduction of innovative activity of enterprises. It has been established that in Ukraine there is a steady decline in the number of workers involved in innovation, and in comparison with European countries, these indicators are much lower and tend to decrease further. This negatively affects the place of Ukraine in the global ranking of innovation. The relationship between the size of the company and the cost of innovation has not been tested.

It was found that there is a direct positive significant relationship between the volume of costs of innovation and the volume of bank loans and foreign direct investment in the equity of Ukrainian companies, confirms the thesis of access to finance as a driver of innovation. In addition, it was found that the reduction of interest rates on loans has a positive effect on the volume of lending to the economy and provides an increase in spending on innovation. Therefore, the practical value of this study is to substantiate the directions of development of the state strategy for innovation development, including the development of the financial market. To diversify the portfolio of financial resources available to finance innovative activities of companies, it is important to more widely introduce various methods and forms of financing (venture capital, microfinance, crowdfunding, financial technology) and attract funds from various participants, such as investment companies, non-profit organizations, various state and international funds of the relevant profile.

The direction for further research can be defined by studying the state of innovation activity in certain types of activities and depending on the size of the company.

References

1. M. Berdar, O. Yevtushevska, Baltic Journal of Economic Studies, **6** (2), 25-33 (2020) DOI: https://doi.org/10.30525/2256-0742/2020-6-2-25-31
2. L. Voloshchuk, Problems of the economy, **4,** 269-274 (2014)
3. 3. I. Gerasimova, Economic space, **137** 55-67 (2018). DOI 10.30838/P.ES.2224.250918.55.227 http://prostir.pdaba.dp.ua/index.php/journal/article/view/301
4. I. Dovba, O. Djugovanetz Collection of scientific works of Uzhhorod National University, **1**(24) 150-161 (2020) DOI: 10.24144/2078 – 1431.2020.1(24).150–161
5. B. Seniv, Ja. Paranjuk, Economic space, **120** 69-79 (2017)
6. 6. Scientific and innovative activities Ukraine 2019. Statistical publication (State Statistics Service of Ukraine, Kyiv, 2020)
7. V. Orlova, O. Kuzmenko, O. Sergeeva, European vector of economic development, 1(20) 112-120 (2016)
8. R. Maradana et al., Journal of Innovation and Entrepreneurship **6**:1 (2017) DOI 10.1186/s13731-016-0061-9
9. A. M.Pece et al., Procedia Economics and Finance **26** 461 – 467 (2015)
10. R.P. Pradhan, M. B. Arvin, J.H. Hall, M. Nair, Applied Economics Letters, **23**(16), 1141–1144. (2016)
11. A. Barrichello, E. Gomes dos Santos, R. S. Morano, Innovation & Management Review, **17**(3) 307-320 (2020)DOI 10.1108/INMR-04-2019-0040
12. R. Stanisławski, Sustainability, **12** 3290 (2020) doi:10.3390/su12083290
13. M. Bhattacharya, H. Bloch, Small Business Economics, **22** 155–162 (2004). https://doi.org/10.1023/B:SBEJ.0000014453.94445.de
14. K. Širec, D. Močnik DOI: 10.5772/intechopen.71843 https://www.intechopen.com/books/entrepreneurship-development-tendencies-and-empirical-approach/determinants-of-established-entrepreneurs-innovative-activity-in-northern-and-western-europe
15. S. Girma, Yu. Gong, H. Gürg Article in *The World Bank Economic Review* (Kiel Institute for the World Economy, 2008)
16. K. Xu, Ch. Geng, X. Wei, H. Jiang, Journal of Business Economics and Management, **21** (4) 1010–1034 (2020) https://doi.org/10.3846/jbem.2020.12727
17. S. Dutta, B. Lanvin, S. Wunsch-Vincent (World Intellectual Property Organization, 2020) https://doi.org/978-2-9522210-8-5
18. Global Innovation Index http://www.globalinnovationindex.org/analysis-indicator
19. Statistical Yearbook of Ukraine 2019. Statistical publication (State Statistics Service of Ukraine, Kyiv, 2020)
20. Statistical Yearbook of Ukraine 2011 Statistical publication (State Statistics Service of Ukraine, Kyiv, 2012)
21. Statistical Yearbook of Ukraine 200**9** Statistical publication (State Statistics Service of Ukraine, Kyiv, 2010)
22. Monetary statistics NBU http://www.bank.gov.ua › files › 3.3-Loans.xls
23. GDP of Ukraine 2020 https://index.minfin.com.ua/ua/economy/gdp/
24. Ukrayina pokrashchila svoyu poziciyu v rejtingu globalnij innovacijnij indeks-2020, https://www.kmu.gov.ua/news/ukrayina-pokrashchila-svoyu-poziciyu-v-rejtingu-globalnij-innovacijnij-indeks-2020
25. K. Kovtunenko et al., *Innovative economics: theoretical and practical aspects*: monograph (Kherson: Green DS, 2017)
26. J. R. Brown, G. Martinsson, B.C. Petersen, European Economic Review, **56** (8) 1512–1529 (2011) https://doi.org/10.2139/ssrn.1684731
27. A. Alam, M. Uddin, H. Yazdifar, R&D Management, **49** (1), 21–32 (2019) https://doi.org/10.1111/radm.12303
28. X. F. Cao, Soft Science, **28** (12), 73–78 (2014) https://doi.org/10.3969/j.issn.1001-8409.2014.12.017
29. P. Chen, M. Yang, Communication of Finance and Accounting, **26** 57–61 (2018) https://doi.org/10.16144/j.cnki.issn1002-8072.2018.26.014
30. W. Hadhri, R. Arvanitis, H. M'Henni, Journal of Innovation Economics & Management, **3** (21) 77-107 (2016)
31. L. Mamatova, Reporter of the Priazovskyi State Technical University, Economic sciences, **36** 279 – 291 (2018)
32. M. Cetin, International Journal of Economic and Administrative Studies, **6** (11), 1–16 (2013)
33. A. Kasych, K. Kanunikova, Investments: practice and experience, **22** 21-24 (2016)

Research of Net Users by Means of Big Data Technology

Irina Vdovychenko [1]*, *Svitlana* Khotskina [1], *Mykola* Khodukin [2], and *Ilyos* Mamanazarov [3]

[1] National University, Kryvyi Rih, Ukraine
[2] State University of Economics and Technology, Kryvyi Rih, Ukraine
[3] Ministry of Economic Development and Poverty Reduction of the Republic of Uzbekistan, 100174, Uzbekistan

Abstract. The presented article offers the review of means, methods and possibilities of one of the most welcomed modern information technologies – Big Data. The popularity of the technology is grounded in the work. The sources of the big data, the speed of their appearance, amounts of information in the world are examined. The main specializations of the employees for the operation in the field of the structuralized and non-structuralized data of big amounts and the significant variety are marked. The scientific and practical directions and the structure of Big Data are analyzed. The examples of the Big Data use's technology in the world are examined. The research results of the big data, received from the Ukrainian sites, are presented in the article. The possibilities of the programming language R, as one of the leading instrumental means of the Big Data technology, were used in the work. The research aim was to define the attendance frequency's dependence of different sites by the young people, aged from 17 to 35 years old, upon their sex, provision of job, incomes and age. The authors present the obtained results and offer directions of the activity, where they may be used.

1 Introduction

The amount of information grows very quickly today. The scope of information, used by the modern individual, are not only books, magazines, films and web-pages. Besides it, the vast number of data, appearing in the operation process of enterprises, banks, institutions and so on, is hidden from the user. The following ones refer to them:

- ✓ data of all-possible accounting;
- ✓ information of production output and sales;
- ✓ histories of transactions in banks;
- ✓ data of the sick persons and medicaments at the hospitals;
- ✓ data of the location and routes of all the cars in the auto-parks;
- ✓ vast quantity of research results;
- ✓ data of observation of planets and stars;
- ✓ data from the cameras, fixed at the enterprises, plants, in the roads and streets of the cities to observe the traffic of people and cars;
- ✓ data are contained in telephones, TV sets;
- ✓ and so on and so forth.

"The humankind produces the gigantic quantity of data through the nets, the public transport and the Internet-purchases without any stop. Their volumes impress the mind. We load 95 million of images and video, 340 million tweets and 1 milliard of documents every day. We produce 2,5 quintillions of bites per a day in general" [5].

The analytical company IDC presented the report in 2012, where it was predicted, that the information volumes would be doubled every 2 years during the next 8 years. The data amount in the world will reach 40 ZB (1 ZB = 10^21bite) for the nearest 7 years and it means that every resident of the Earth will have 5200 GB of data at the disposal.

The analogical data are presented by the company IBS, which tells that the amount of the digitalized information grows along the exponential curve and, according to the forecasts, the mankind will form 40-44 Zettabites of information by 2020. The data represent the vast amount of information of different type: pictures, video, multimedia, text, geo-data, web-magazines, machine code. All the information is located in various storehouses and is hardly subjected to the analysis with the help of the traditional methods. The specialized technologies are used for that.

The traditional methods of information analysis cannot overtake the vast volumes of the constantly growing and restored data, - the fact, which finally opens the road for the Big Data technologies. The term Big Data has the relation to the selection of data, which size exceeds the possibilities of the standard data bases in processing, storage, control and analysis of the information. It's the instrument and the method of data analysis.

2 Background

The very term Big Data began to be used several years ago. The term "big data" was introduced by Clifford Lynch, the editor of the magazine "Nature", in 2008. The problem of the rapid growth of the world volumes of information was examined in the magazine. It's clear,

* Corresponding author: vivin2015@knu.edu.ua

that big data existed before as well. According to the words of the professionals, the most part of the data flows of more than 100 GB per a day belongs to the Big Data category.

The first investigations on the data separation from data were conducted already long ago. The mathematical theory, concerning the reveal of the definite data on the indistinct field of data was elaborated still in the 70-80-s. The practical use of the theory was limited by the possibilities of the computer engineering.

The Big Data notion today has already entered the everyday life reliably throughout the world. The most part of these data is collected through the Internet.

Big Data — are the totality of approaches, methods, tools that are used for the operation with the structuralized and non-structuralized data of the significant amounts and great variety for the solvation of the definite tasks and objectives. They are welcomed on conditions of the continuous growth of information. "Big data serve as an alternative to the traditional systems of the data base control and to the decisions in the frames of Business Intelligence. These methods, that allow fulfill the distributed processing of information, may be applied both to the vast selections of data (for example, the contents of all the pages in the Internet) and to the small ones. Big data have an important significance, because the big quantity of data leads to the more exact analysis, which, in its turn, guarantees the more efficient decisions-taking" [5].

The main difference of the big data lies in: their amount, speed and variety.

The data were selected from one source and were kept in the single format for quite a long while. At present, the data may be of any forms and sizes: text, sound, video, graphics. Thus, Big Data grant the possibilities for the use of the most varied data and their combinations, for the creation of the new methods of data collection and processing in future. The unification of the heterogeneous information storehouses for the construction of the complex inquiries will be efficient.

The important characteristic feature of the big data is the speed of the new information generation, it's also important, how rapidly the data are accumulated from different systems for the next processing.

The Big Data sphere may be conditionally divided into two categories – the elaboration, i.e. the creation of systems for calculations and processing of the data scope (Big Data Engineering) and the analytics, i.e. consolidation and the analysis of data (Big Data Scientist / Data Analyst). They are different in the character of tasks.

Working with Big Data, you should be able to process the vast information scopes and to interpret the results, to able to fulfill the previous processing of data, to create clusters, to analyze information, to comprehend the mathematical statistics, the analytics, the theory of algorithms, the numerical methods, the theory of probabilities and to have good knowledge of the higher mathematics.

You may separate the three main specializations of the employees for the operation in the field of data processing:

- ✓ Data Engineer
- ✓ Data Scientist
- ✓ Data Manager

Data Engineer – is a professional in projecting of such data processing systems that operate with the petabites of data. He masters all the modern technologies and approaches in the sphere of data processing: MapReduce, Hadoop, Spark, Aerospike, Redis, Storm and so on.

Data Scientist - "data researcher" – is able to find the objective regularities in the big data scopes, knows the sphere of machine training well, masters confidently of such instruments as R, Weka, Python + Scikit-Learn + Pandas. The very Data Scientist is able to get the maximal benefit from data and to project the algorithms that will give answers to the necessary questions. But the main force of the data researcher - is in the knowledge of the subject area: he knows where to find so much required "gold grains" [6].

The thorough intellectual analysis of big data allows see the hidden regularities, being unseen for the limited human perception. It gives the unprecedented development possibilities for all the spheres of our life: the state control, education, medicine, telecommunications, economy, transport, production and so on.

You may separate the following directions in the sphere of Data Science:

✓Data Mining – allows solve the tasks of the prognostication, the probability of events and results.

✓Text Mining – allows find the regularities in the text, define its topics automatically and determine its psychological implication.

✓Processing of pictures – allows find images on photos, recognize the text on the picture, reveal diseases, according to photos and so on.

✓Processing of audio-signal – gives an opportunity to give the commands and to receive responses by voice.

✓Recommendation systems – provide the service, corresponding maximally to the interests of the user;

Data Manager – the professional, controlling the project competently. He (she) is obliged to know the possibilities of modern technologies, to understand the subject sphere, to master the special terminology and also to have good skills in the control technologies of information projects.

The Corporation EMC conducted the research, in the process of which it was revealed that the use of Big Data leads to the essential improvement of the decisions-taking processes, increases the competitiveness of the companies and simplifies the control of risks.

Big Data — are the system, the qualitative transition to the composition of the values' chains, based on knowledge. According to the effect, it may be compared with the appearance of the accessible computer engineering at the end of the last century.

The Big Data technologies are often examined for the solvation of the following tasks [1]:

- ✓ prognostication of the market situation;
- ✓ marketing and optimization of sales;
- ✓ efficient segmentation of clients;

- ✓ improvement of goods and services;
- ✓ taking of more grounded decisions on the basis of Big Data analysis;
- ✓ optimization of the investments portfolio;
- ✓ increase of labor productivity;
- ✓ efficient logistics;
- ✓ monitoring of the fixed capital's state.

The similar use of Big Data is evident and clear. But the presented technologies give more opportunity for the realization of the serious psychological tests. You may obtain the most interesting conclusions on the basis of the data synthesis from the social nets, due to the revealing of the hidden regularities.

The principally new effect from the massive use of this approach in the data processing starts to become distinct for the last years. The scientists are searching for the hidden correlations between the researched phenomenon (object, process) and the thousands of the other factors, where the vast statistics, having been accumulated for the long years, was used as the initial data. The use of these empirically closed regularities promises the progress in the development of many scientific directions.

The complex modern Big Data models reveal still more often certain irrational-fantastic, at the first sight, dependencies that allow have a look at far away beyond the borders of the famous scientific picture of the world [1].

Due to that, Big Data are sometimes called as "the new astrology of XXI century". And this is the result of the smooth transition from the quantity of information to its quality, when the machines become capable to reveal the principally new dependencies, being inaccessible beforehand for the limited human consciousness.

Big Data today are demanded most of all in both: trade and business. Big Data allow earn not only the additional money, but are actively used for the more efficient control of society.

For example, the so-called Predictive Policing has been already used in the USA for the third year. This is the specialized computer system of the USA police, being elaborated on the principles of the "big data" analysis, which helps to predict the splashing time of crimes and also the definite areas of their escalation. Such an approach has quite definite results – the noticeable decline of criminality has been already fixed in the cities, where the presented system was introduced. For example, the number of the arrests in Santa-Crus was increased by 55 %, but the quantity of robberies and the cars stealing was reduced by 10-15% [1].

No doubt, the predictive abilities of Big Data have the impressive possibilities. The company Farsite fulfilled the forecast by the Big Data technology before the presentation with the cinema reward "Oscar". The analysis was made on the basis of the data scope, prepared on the historical cut of last 40 years of the American cinema industry, the statistics of presentation with the different cinema rewards, the frequency of the actors/films quoting in press. As a result – the exact hitting of the target for all the nominations. And there are many similar examples.

2.1 Problem Positing

At present, Big Data is a special approach – the ideology of information processing, being used for the processing of big scopes of the "raw" data. The evident transition occurs from the initial function into the new functional form. If they were simply the calculation methods that allowed process the fantastic volumes of entering data beforehand, then, at present, they are the complex self-studying algorithms, which ideally allow not only efficiently "press" the information, but are capable of becoming complicated and improving independently. After such a qualitative jump, the instruments, created on the basis of these technologies, become the general and the universal ones, penetrating far beyond the borders of the Internet. Big Data technologies open the new approaches, technologies, the possibilities of use [4].

The analytical companies point out the interesting distortion in the use of Big Data: many world energetic companies already collect and accumulate the statistical information, spending hundreds of millions dollars for that. Only 1% of the allocated sums is spent in general for the very processing and data analysis at that. Naturally, the real effect from the Big Data is practically absent at such position of the affairs [2]. This paradoxical fact reveals the main problem of the Big Data current state, its narrowest place. According to the calculations of the IBS company, only 1,5% of the accumulated data amounts had the information value in 2013. It's clear that to collect all the data one after the other is useless, the priorities are needed and they are defined by the task-setting.

It's interesting that education and medicine, where the relatively small investments into the "big data" begin to give the sensible return already now, are mostly separated by the analysts among all the possible directions of growth. In the opinion of some experts, the massive introduction of Big Data into these branches may increase the quality level of human life in the nearest terms [3].

As it has been already said, the amounts of information grow rapidly (by 50% yearly). The most part of the information, being kept in the world, is non-structuralized, i.e. not suitable for research and use. Big Data – is the tool, which allows adapt to this scope of information.

The classical SCDB, such as Postgres, MySQL, Oracle have no such flexibility in scaling (scoping) at the processing of big data amounts and the further processing becomes impossible at the increase of the amounts.

The spreading of the Big Data technology changes greatly the format of interaction between the consumers in all spheres of life: retail, medicine, transport, insurance, banking sphere. The analysis of big data helps to know more of the preferences of clients, gives an opportunity to make the personalized offer and to recommend the product or the service, being necessary here and now [5].

2.2 Aim of Research

The big scope of data became to accumulate during the boom of the social nets.

According to the statistics, nearly 2 milliard inquiries are processed by the Facebook every day. Google receives 5,5 milliard inquiries every day. And this is not the limit as the amounts of information grow constantly. The obtained data are analyzed and used for the increase of the contents quality. Due to the thin adjustment of the news ribbon of Facebook, (Facebook has never revealed the operation principle of the news ribbon algorithm, at this, the company restores it regularly), only those posts get into it, which the user will probably like. "Almost everything is taken into account: everything you reacted with your like or the sad like, the fact that you commented, what message you have opened, whether you watched the video to the end, whether you read the post text till the end. The contents, capable to interest the user with the most probability, gets into the ribbon finally. The algorithmic ribbon was included into Instagram, VC and the other social nets after Facebook. There are some working actions in YouTube too: likes, subscriptions to the channel, comments and reviews to the end" [7].

"You may compare any action in the Internet with the traces on the clean snow. Online-purchases, photos, activity in the groups in the social nets, the lists of friends and even the likes "reveal" the user. The history of the searching inquiries also characterizes the user; you may define the sex, the age, the circle of interests. Smartphones send data on the movements. The interested companies scrupulously accumulate the received information and buy it [8]. It gives the powerful push for the development of the Big Data analysis.

According to the above-mentioned, the decision to use the Big Data technologies for the analysis of the users of the Internet sites becomes urgent.

The tasks, which are needed to be solved at putting the Big Data projects into practice, are not only technological. Data collection is not the most complex problem. Every user leaves many "finger-prints" in the Internet that may be used for the analysis. But you may not always receive any benefit from the data, due to the lack of ideas.

You may represent each social net in the view of the mathematical graph, as the totality of the units and ribs between them. It's accepted that the unit is the channel, but the rib, between the two units, is information, by which the channels exchange. The size of the unit depends on the number of subscribers, but the width of the rib – on the scope of information, passing between them.

In order not to "drown" in the sea of information, we made the selection, according to the following criteria:

1. channels in the Ukrainian language;
2. channels with the number of subscribers of more than 500.

We reduced the number of channels by that.

In order to facilitate the further work, it's advisably to divide the channels, according to the categories, but this is not enough too. As Victor Delisov advises, in order to solve this problem, it's necessary to use the cluster analysis which may be conditionally presented in the following formula:

$$Y=F\,(N,A,C), \tag{1}$$

where Y – type of behavior (cluster);

N – number of recalls in the period t_1 and t_2;

A – speed of reviews' selection for $(t_2 - t_1)$;

C – coefficient of the category reposting;

F – function [9].

We would receive the definite number of clusters, which allow us estimate the channels and information in result.

2.3 Practical Realization and Research Results

The research, conducted by us, is the analysis realization of the big data, received from the Ukrainian sites, which use the information agencies (sport.ua, 112ua.tv, pravda.com.ua, rozetka.com.ua, bigmir) net, work.ua, OLX.ua and etc.). In order to fulfill the research, it's supposed to use the possibilities of the programming R language as one of the leading instrumental means of the Big Data technology [4].

It's planned to analyze the attendance frequency dependence of different sites by the young people, aged from 17 to 35 years old, upon their sex, job provision, incomes and age.

We planned to define the other dependencies between those indicators too. Big Data, using the vast information scopes of various data about the net users, find the completely non-evident correlations between the sex, income and the inclination to the Internet use.

The part of the big data, needed for us to make the analysis, are located in the special storehouses DB and KB and are ready for the use. But the access to them is not always granted freely. The special information companies – the suppliers of information – are used. They suggest the definite types of services to the consumers.

The collected data are formed by us as a table of the text information. The table consists of four columns (indications):

- sex of a young man (POL: female = 0 and male = 1);
- occupation: (ZAN: student = 1, working person = 2, not working person = 3);
- incomes in UHA. (DOH)
- age (VOZR):

Collected data on 700,000 people.

The initial data for the analysis have the following view: see Fig.1.

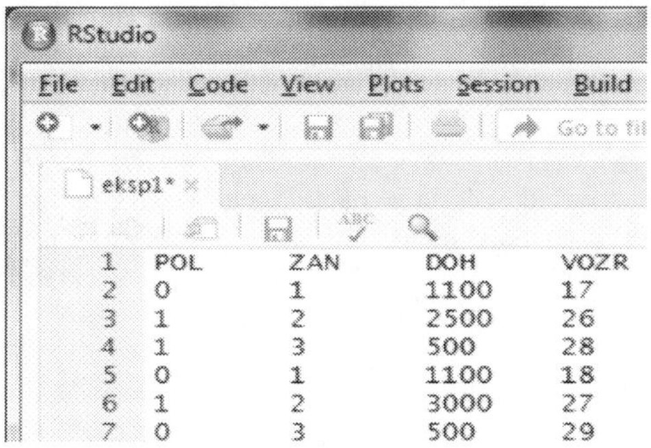

Fig.1. Part of Data, Collected for Analysis.

```
> str(data)
'data.frame':   51 obs. of  4 variables:
 $ POL : int  0 1 1 0 1 0 1 0 1 0 ...
 $ ZAN : int  1 2 3 1 2 3 1 2 3 1 ...
 $ DOH : int  1100 2500 500 1100 3000 500 1100 4000 500 1100 ...
 $ VOZR: int  17 26 28 18 27 29 19 28 30 20 ...
>
```

Fig.2. Structure of Data File.

Let's select data only for the female sex, i.e. (SEX = 0): data.f <- data[data$SEX ==0,]

```
> data.f <- data[data$POL == 0,]
> head(data.f)
   POL ZAN  DOH VOZR
1    0   1 1100   17
4    0   1 1100   18
6    0   3  500   29
8    0   2 4000   28
10   0   1 1100   20
12   0   3  500   31
>
```

Fig.3 Data Are Only for Female Sex.

As one of the analysis variants, you may imagine the following lines. Let's create the separate object with the data for men, being older than 25 years old: data.m.big <- data [data$SEX == 1 & data$AGE > 25,]

```
> data.m.big <- data[data$POL == 1 & data$VOZR > 25,]
```

```
> str(data.m.big)
'data.frame':   16 obs. of  4 variables:
 $ POL : int  1 1 1 1 1 1 1 1 1 1 ...
 $ ZAN : int  2 3 2 3 2 2 2 3 3 2 ...
 $ DOH : int  2500 500 3000 500 5000 6000 7000 500 500 5000 ...
 $ VOZR: int  26 28 27 30 30 31 33 35 30 35 ...
```

Fig.4. Results of Selection: Data Only for Men, Being Older than 25 Years Old.

Let's define the specific weight of a person (the relation of an individual's age to his/her incomes) — WEIG.R: data$WEIG.R <- data$AGE/data$INCstr(data$WEIG.R)

```
> data$VES.R <- data$VOZR/data$DOH
> str(data$VES.R)
 num [1:51] 0.0155 0.0104 0.056 0.0164 0.009 ...
>
```

Fig.5. Specific Weight of a Person.

Our data, with the additional criterion, have the view, represented in Fig.6:

```
D:\BigData\eksp_new.txt

POL ZAN DOH VOZR VES.R
1 0 1 1100 17 0.0154545454545455
2 1 2 2500 26 0.0104
3 1 3 500 28 0.056
4 0 1 1100 18 0.0163636363636364
5 1 2 3000 27 0.009
6 0 3 500 29 0.058
7 1 1 1100 19 0.0172727272727273
8 0 2 4000 28 0.007
9 1 3 500 30 0.06
10 0 1 1100 20 0.0181818181818182
11 1 2 5000 30 0.006
12 0 3 500 31 0.062
13 1 1 1100 21 0.0190909090909091
14 1 2 6000 31 0.00516666666666667
15 0 3 500 32 0.064
16 0 1 1100 22 0.02
17 1 2 7000 33 0.00471428571428571
18 0 3 500 34 0.068
```

Fig.6. Part of the Widened Data.

The programming R language gives the possibility for the fast calculation of the main parameters of the descriptive statistics: Min, Max, Median, Mean, 1st Qu., 3rd Qu., NA's and others. [9].

```
>
> summary(data$VES.R)
    Min.  1st Qu.   Median     Mean  3rd Qu.     Max.
0.003625 0.008375 0.018182 0.029150 0.060000 0.070000
```

Fig.7. Data Statistical Indicators of Net Users.

Let's analyze the criterion of incomes (INC). summary (data$WEIG.R)

```
> summary(data$DOH)
   Min. 1st Qu.  Median    Mean 3rd Qu.    Max.
    500     500    1100    2327    4000    8000
>
```

Fig.8. Statistical Characteristics of INC. Criterion.

Let's calculate the average meanings of incomes separately for all the combinations of the age and sex.

> tapply(data$INC, list(data$SEX, data$AGE), mean). The result is in the Fig. 9

```
> tapply(data$DOH, list(data$POL, data$VOZR), mean)
    17   18   19   20   21   22   23   24   26   27   28        29   30   31
0 1100 1100 1100 1100 1100 1100 1100 1100 5000   NA 5000  500.000  500  500
1 1100   NA 1100   NA 1100 1100   NA   NA 2500 3000  500 4833.333 2000 3250
    32   33   34   35
0 4250 2250 3800 4000
1  500 7000   NA 2000
>
```

Fig.9. Average Meanings of Incomes Separately for All Combinations of Age and Sex.

Let's define the average arithmetical number for the occupation, separately for men and women. tapply(data$OCC, data$SEX, mean)

```
> tapply(data$ZAN, data$POL, mean)
       0        1
1.870968 2.200000
>
```

Fig.10. Results of Mean Function for Occupation, Being Differentiated, according to the Sex.

Let's define the average arithmetical number for the occupation separately, according to the age categories. tapply(data$OCC, data$AGE, mean)

```
> tapply(data$ZAN, data$VOZR, mean)
      17       18       19       20       21       22       23       24
1.000000 1.000000 1.000000 1.000000 1.000000 1.000000 1.000000 1.000000
      26       27       28       29       30       31       32       33
2.000000 2.000000 2.333333 2.500000 2.833333 2.666667 2.666667 2.333333
      34       35
2.400000 2.500000
>
```

Fig.11. Average Arithmetical Number for Occupation Separately, according to Age Categories.

Let's calculate the standard deviation: (standard deviation) > sd(data$WEIG)

```
> sd(data$VES)
[1] 0.02465665
>
```

Fig.12. Indicator of Standard Deviation.

Let's calculate the dimensionless coefficient of the variation (CV):

> 100*sd(data$WEIG)/mean(data$WEIG)

```
> 100*sd(data$VES)/mean(data$VES)
[1] 84.58492
>
```

Fig.13. Coefficient of Variation.

We calculated the correlation coefficients of Pearson, Spearman, Kendall, the coefficient of variation, the standard deviation and the other statistical characteristics, defined the correlation degree between the variables [10]. We used 13 methods of the analysis in the process of the research. Due to the use of the programming R language as one of the leading instrumental means of the Big Data technology, we received the curious results, serving as the sufficient basis for the conclusions. Therefore, basing on the final results of more than 85 calculations, you may come to the following conclusions.

3 Conclusions

Our expectations of the fact that incomes depend significantly on the age, were not confirmed: there is some dependence, but the weak one. There is no link between the incomes and the sex of the young people. It's seen, according to the results of the calculations, the men visit sites more seldom.

We received the following results in the process of the scrupulous analysis of the data, taking part in the research:

1. Sites are visited by the men, being older than women;
2. There are more working ones and not working ones among men, there are more students among women;
3. Women have the higher incomes after they are 30 years old;
4. It is seen, according to the results of the calculations, that men visit sites more seldom than women;
5. You may come to the conclusion that the young people after 28 years old visit sites more often;
6. 39,2% of men and 60,8% of women take part in the experiment;
7. The older the individual is (at the age from 17 to 35), the higher are his/her incomes;
8. There is a weak link between the sex and occupation. Many other conclusions were received at the use of the Big Data technology that may be used for the following:

✓planning of the advertising production, designed for the definite auditorium of the net users;
✓at the study and the decision of social problems;
✓at the elaboration of the thematic (subjects') sites;
✓for the career-guidance work in the net;
✓for the employers;
✓at the decision of the gender problems;
✓at the organization of leisure-time of the young people and others.

The analysis of the big data allows get the exact information of the net users. The application of the new technological possibilities, being revealed by the Big Data, helps the corporations to become still more client-oriented. The analysis of big data helps to know more of the preferences of clients, gives an opportunity to make the personalized offer and to recommend the product or the service, being necessary here and now.

References

1. A. Dobrychenko. *How to Study Big Data — Experts Answer.* https://tproger.ru/experts/learn-big-data/ Access - 10.01.2021.
2. O. Petrov. *Specialist on Big Dat.* https://rb.ru/opinion/resheniya-data-science/ Access - 08.12.2020.
3. T. Susan. *Big Data: All You Need to Know.* PC Week/RE. № **25**. (2012).
4. I.N. Vdovychenko. *Big Data Technologies and Their Use for Analysis of Net Users.* III International Scientific-Practical Conference "Information Security and Computer Technologies". Kropivnytsky. P.189-191. (2018).
5. P.D. Ivanov, V. Vampilov. *Big Data Technologies and Their Use at Modern Industrial Enterprise.* Engineering Journal: Science and Innovations. Iss. **8**. (2014).
6. I. Savchuk. *Big Data – Technology, Giving Birth to New Type of Business.* Magazine "Bit". **3**. (2014).
7. I. Popov. *How Media Corporations Use Big Data.* https://rb.ru/opinion/media-big-data/ Access - 11.10.20.
8. A. Rychkov. *How FinTech Uses Big Data.* https://rb.ru/opinion/big-data-v-fintech/ Access - 20.12.20.
9. V. Delisov. *How we learned to track emerging trends with mathematical precision.* https://rb.ru/opinion/martech-advlab/
10. R. I. Kabakov. *R in Action. Analysis and Visualization of Data in R Language*. Moscow. DMK Press. (2014).
11. G. Groulmund, K. Uikem. *R Language in Tasks of Science on Data: Import, Preparation, Processing, Visualization and Modelling of Data.* Dialectics - Williams. Moscow. (2019).

Improvement of the Salary Organization on the Basis of the Personnel Professional-Qualification Level Evaluation Method

Oksana Yelisyeyeva[1], *Nadiia* Shmygol[2*], *Nataliia* Kutova[3], *Iryna* Stolietova[4] and *Jan* Polcyn[5]

[1] Department of Statistics, Accounting, and Economic Informatics, Oles Honchar Dnipro National University, 49000, Dnipro, Gagarin Avenue, 72, Ukraine
[2] Department of Entrepreneurship, Management of Organisations and Logistics, National University «Zaporizhzhia Polytechnic», 64 Zhukovsky street, Zaporizhzhia, 69600, Ukraine
[3] Department of Economics and Business Administration, Kryvyi Rih National University, 50027, Kryvyi Rih, Vitaly Matusevich str, 11, Ukraine
[4] Department Of Digital Economy And System Analysis, Kyiv National University Of Trade And Economics, 02156, Kyiv-156, Kyoto str., 19, Ukraine
[5] Stanisław Staszic University of Applied Sciences in Piła, 64-920, ul. Podchorążych 10, Piła, Poland

Abstract. It is justified the use of a generic integral indicator and its graphical interpretation for the enterprise staff incentives system establishing, which makes it possible to determine the influence of group indicators as well as to establish the functional links between indicators. The integral index of personnel incentives efficiency and influence factors are determined. The most important quantitative factors that affect the high efficiency of staff incentives are defined. The relevance between the results of enterprises and indicators of the effectiveness of staff incentives is determined. The educational and professional quality of the staff is determined by means of a score. The employee motivational profile diagram is constructed. The staff assessment system has been developed to improve staff skills. The based on a competent and attributive approach review of grades and categories for staff is proposed.

1 Introduction

In today's economy, the main condition for the competitiveness of enterprises is the productive activity of the staff. Staff is a priority strategic resource of an enterprise and a key factor in its development and competitiveness.

The assessment of the staff performance at any enterprise or organization is a process of examining the quality of the workforce, identifying the strengths and weaknesses of the organization's staff, and it also becomes the basis for the improvement of the individual working capacity of the worker, the improvement of his qualification, and sometimes even the requalification.

Competitive growth in the modern economic conditions of Ukraine requires the efficient assessment of personnel at enterprises. High-quality professional activity of employees affects the successful activities of enterprises and the results of their activities. Therefore, there is a need to use the latest technologies for personnel management and assessment of its activities.

It becomes necessary to establish a qualitative method of staff assessment to determine and improve the effectiveness of their activities, the aim of which is not only the study of the employees skills' formation, using, developing as well as the defining a concrete result according to certain criteria, but also the contributing to the development of certain measures to improve staff reproduction.

Thus, the trends of the economic situation development in Ukraine and the domestic enterprises activity current realities impel the search for the requirements of the present approaches to the assessment of the personnel quality. This is the reason for the relevance of the research direction.

2 Actual scientific researches and issues analysis and the research objective

Personnel management is a «set of mechanisms, principles, forms and methods of interaction in the formation, development and operation of the enterprise personnel, which is implemented as a number of interrelated areas and activities» [1]. Ukrainian scientists V. Danyliuk, V. Petyuk and S. Tsynbalyuk believe that personnel management is "the part of the personnel management functional sphere as the main mechanism of the organization» [2, p. 43]. T. Alexandrov in his work determines the personnel management as the set of influences on the organizational behavior of people aimed to the activating of the untapped professional and spiritual opportunities to solve the tasks of an enterprise [3].

The personnel management is a component of the enterprise management system, management of strategic, priority, unique resources. It has specific

* Corresponding author: nadezdash@ua.fm

features as a management system. Firstly, it is implemented through functions and subfunctions, and this set is constantly changing depending on social relations. So, recently, more and more scientists are paying attention to the study of knowledge management, strategic staff development on account of the the information society development,.

Secondly, the management system should cover all areas of HR activities (Human Resource) on the basis of science-based personnel management strategy. Several studies can be distinguished among the studies in this area of scientific knowledge. So, O. Sardak in the work [4, p. 135] offers the following elements of the personnel management system structure:

- subsystem of HR-landmarks and planning;
- subsystem for ensuring and organizing the work of staff;
- personnel marketing subsystem;
- subsystem of personnel formation and use;
- subsystem of development and activation of HR-potential.

L. Poshelyuzhna considers the following subsystems of personnel management that take part in the activities of domestic enterprises: analysis and planning of personnel; selection and hiring of staff; staff evaluation; organization of staff training and advanced training; certification and staff rotation; staff motivation; accounting of enterprise employees; organization of labor relations at the enterprise; creation of working conditions; social development and social partnership; legal and information support of the personnel management process [5, p. 127].

T. Korsakova [6] identifies the subsystems of the enterprise personnel management system: analysis and planning of personnel; hiring and accounting of personnel; motivation; assessments; personnel development management; creation of working conditions; information support of the personnel management system; development of organizational management structure; legal support.

G. Dudukalo examines the functioning of the personnel management system, which provides for the implementation of the following functions [7]: analysis and planning of personnel; staff placement; certification and evaluation; organization of labor relations; staff motivation; creation of working conditions; information support; staff development and training.

Of course, there are scientists who study certain functions of personnel management. For example, when constructing a scheme of organizational and economic mechanism for managing the development of enterprise personnel, N. Markova includes the following components in the managed subsystem: career guidance and adaptation, personnel planning, personnel evaluation, career development, staff motivation, professional training, social development [8, p. 201].

All components of the management system, which are considered by scientists, depend on the choice of the scientific approach: functional, component, complex, attributive, processing, which determines the composition of the personnel management system.

The works of many domestic and foreign scientists, in particular Balabanova L.V., Grishnova O.A., Pakulin S., Shlyaga O.V. and others, are devoted to a study of the forms and methods of personnel assessment. As a result of these studies, various approaches, procedures and methods are developed for the organization and conduct of staff assessment. Foreign scientists such as Douglas Macgregor, I. Ansoff, M. Mescon, etc., are also studying the problems of staff assessment., The results of the studies are the theoretical and methodological bases of staff assessment. However, there is a need to define the modern features of staff quality assessment methods when applying them in the conditions of the enterprises.

The goal of the research is to identify current features of staff assessment. In order to achieve the goal, the following objectives were solved: Staff incentives were evaluated by means of an integral indicator, taking into account the material, social, professional and organizational groups of indicators; Staff qualifications were assessed; The characteristics of its introduction in domestic enterprises were analyzed; Grades and categories for the staff were reviewed on the basis of competent and attribute-based approaches.

3 Formation of a system for evaluating staff incentives

The modern personnel management system is based on the fact that people are the most important economic resource of the enterprise, and its source of income, competitiveness and prosperity [9]. The achievement of objectives in the process of enterprise personnel activities determines the success of the enterprise and the achievement of high results. After all, ensuring the desire of employees to perform their duties effectively is one of the important tasks of management, and it is necessary to form and develop a system of incentives that can maximize the effectiveness of enterprise management.

In order to improve competitiveness and productivity, it is necessary to develop an incentive mechanism that precisely combines and complements the means of material and non-material incentives, promote the formation of certain stereotypes in the employee about the usefulness of the chosen activity, its prospects and the possibility of obtaining advantages for himself and for society as a whole [10]. Wherein it should be noted that job incentives are provided at two levels: the first is the stimulation of each individual, and the second is the stimulation of the entire workforce as a means of spreading synergistic effect., The manager should identify incentives to increase productivity at both levels. The staff manager should establish an effective incentive system as a part of the staff management system [11, 20-25]. Thus, the stimulation of work involves the creation of conditions in which active work becomes a necessary condition for satisfying the important and socially determined needs of individuals and for forming their motivation for work. The incentive is an integral function of the management of the enterprise's personnel, it is realized effectively only when the condition of development of all other functions is fulfilled. The establishing a staff incentive assessment system that is balanced in terms of accuracy, objectivity, relevance, simplicity, convenience and intelligibility is a complex multidimensional process [12]. Evaluation is an integral part of the staff management system, the process of determining the staff members performance effectiveness and the achievement of organizational objectives, or the process of assessing the quality of position staff (abilities, skills, motives) requirements or job [13]. Staff assessment has many objectives. The most common is the classification developed by Douglas Mac Gregor, a well-known human resources development expert, which takes into account: administrative, which consists in making personnel decisions on an objective and regular basis - Staff placement, movement and remuneration; Informative - providing managers with the necessary data on the number and quality of staff; Motivational - orientation of employees towards improvement of work in the direction needed for the organization [14].

Judging from the opinion of the analyst, the establishment of economic indicators system for calculating an integral measure of staff incentives is an important stage for the accuracy of the results obtained and is a condition for the operation of the enterprise. The based on a competent approach formation of an expert group taking into account the rank of the posts on makes it possible to form a group, in accordance with the job responsibilities as well as to build a system of personnel assessment in accordance with the objectives of the enterprises. The establishment of a staff incentive system begins with the implementation of the allocation of indicators to groups [15]: subsistence needs; social needs; recognition needs; self-realization needs. Subsistence needs are represented by material stimulation, social needs by social stimulation, recognition needs by professional stimulation, self-realization needs by organizational. Thus, according to certain expert estimates, four groups were formed according to the indicators of motivation of the employees: material, professional, social, organizational. In order to form a system of incentives for enterprise personnel, we propose the use of a composite integral indicator, which is the set of integrated group indicators of the corresponding incentive means [16]. It was calculated according to the formula:

$$I_{stimul} = \sum_{i=1}^{4} \beta_i \times I_{stimuli} \qquad (1)$$

where β_i – coefficient of influence on the value of group integral measures of staff incentives; $I_{stimuli}$ – group integrated staff stimulation measures; i – the number of means of staff incentives.

The integrated indicator of staff incentive assessment allows to determine the impact of group indicators of incentives on the integrated indicator, establishes functional links between indicators. The level of functional links between staff incentive indicators can be divided into ranges of values: high inverse connection ∈ [-1;-0,5], middle inverse connection ∈ (-0,5;0), no connection – 0, middle direct connection ∈ (0;0,5], high direct connection ∈ (0,5;1]. Graphically, the level of functional connection between the indicators of staff stimulation in three-dimensional space can be traced from the height (at direct high connection) to the decline (at inverse high connection).

The industrial enterprises of Kryvyi Rih were selected for the research: PJSC «Kryvyi Rih Central Ore Repair Plant» (PJSC «KCRZ»), PJSC «Kryvyi Rih Iron Ore Plant» (PJSC «KZRK»), PJSC «Kryvyi Rih Mining Equipment Plant» (PJSC «KZGO»).

A graphical interpretation of the integral measure of staff incentives on the example of the studied enterprises (fig. 1, fig. 2, fig. 3, fig. 4).

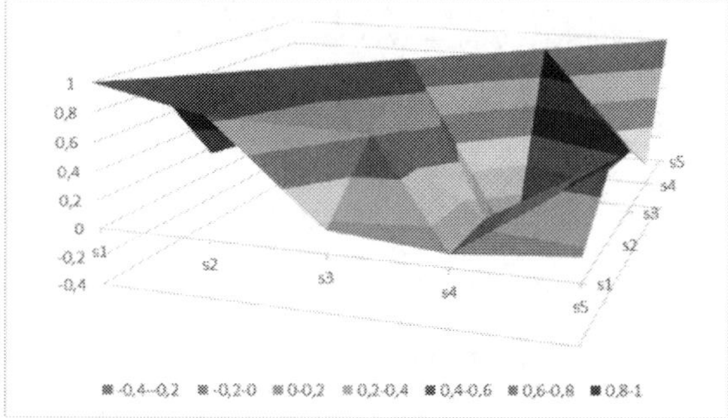

Fig. 1. A Graphical interpretation of the group integrated indicator of personnel incentive assessment «Material Incentive» (I_{mater}) on the example of PJSC «KZRK»

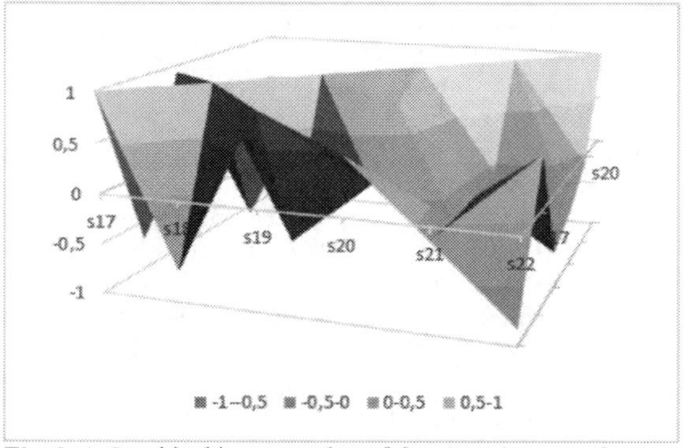

Fig. 2. A Graphical interpretation of the group integrated indicator of personnel incentive assessment «Organizational Incentive» (I_{org}) on the example of PJSC «KZRK»

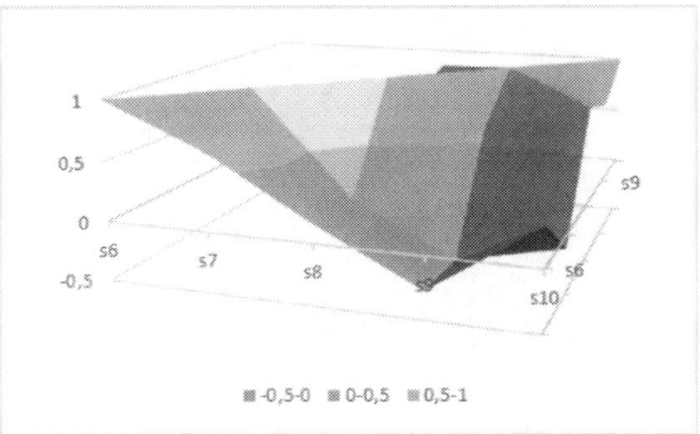

Fig. 3. A Graphical interpretation of the group integrated indicator of personnel incentive assessment «Professional Incentive» (I_{prof}) on the example of PJSC «KZRK»

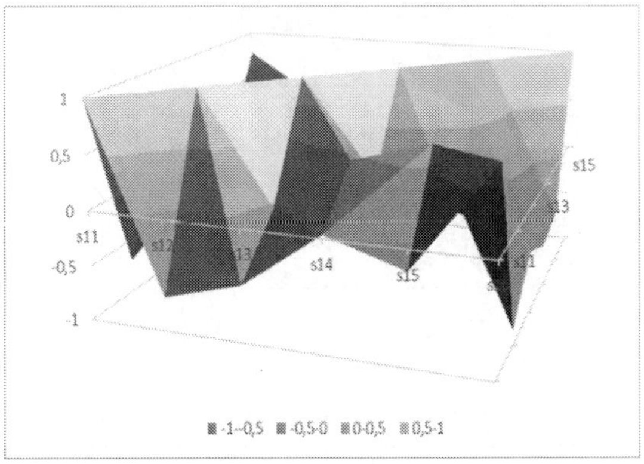

Fig. 4. A Graphical interpretation of the group integrated indicator of personnel incentive assessment «Social Incentive» (I_{soc}) on the example of PJSC «KZGO»

In the table 1 presents the value of the integrated indicator of staff incentives evaluation for three enterprises for 2016-2020.

Table 1 The value of the integrated indicator of staff incentives evaluation for enterprises for 2016-2020.

Enterprise	Years				
	2016	2017	2018	2019	2020
PJSC «KCRZ»	0,37	0,35	0,33	0,32	0,3
PJSC «KZRK»	0,61	0,58	0,56	0,51	0,52
PJSC «KZGO»	0,43	0,46	0,44	0,46	0,45

Source: formed by the author on the basis of his own research

Taking into account the above results, we note that the most important factors influencing the efficiency high level providing of staff incentives are the indicators of material incentives [17].

Based on the values of the integrated indicator, it is determined that the indicator of evaluation of staff incentives of PJSC «KZRK» is the highest (average value Istimul = 0.556), which can not be accompanied by a significant share of material incentives to increase staff productivity. The lowest level is observed at PJSC «KCRZ» (average value Istimul = 0.334). This value of the indicator is due to the low level of productivity, poor system of information and motivational support for the effectiveness of staff incentives. None of the three researched enterprises was included in the group with a high level of personnel incentives, that is, the incentives used at the enterprises are not effective, therefore, it is necessary to adjust the personnel incentive mechanism.

The integrated indicator use makes it possible to increase the objectivity of the incentives assessment, that directly affects the efficiency of staff performance.

It is proposed to use the method of assessing the professional qualification level of staff on the basis of a competency-based approach in order to improve the organization of remuneration as a means of material incentives. Herewith, it is necessary to select such indicators as: qualification, level of education, work experience in the profession, mastering additional professions and determine their weight on the basis of expert analysis. Forming an expert group on the basis of a competency-based approach, taking into account the ranks of positions, allows you to form a group that, in accordance with job responsibilities, has the opportunity to build a system of personnel assessment in accordance with the objectives of enterprises.

Specific requirements for members of the expert group leave their mark on the principles of their selection. To determine the competence of experts, it is necessary to assess the following qualities: competence, creativity, attitude to expertise, conformism, analytical, collectivism, self-criticism, reliability [18].

4 Assessment of educational and professional quality of staff

The improved quantitative assessment of the professional and qualification level of the personnel for review of categories differs in simplicity, calculation – differs in universality and it is available for necessary information base at the enterprises of branch.

We use a method based on scoring assessment, which allows you to assess the qualification level of staff to assess the educational and professional quality of staff.

Construction of the diagram of the employee motivational profile allows to build a system of the personnel qualification assessment for the purpose of qualification increase and professional level for needs of the enterprise in modern conditions of external environment turbulent changes.

In our view, it is also necessary to propose a review of the grades and categories for staff based on a competent and attributive approach. The evaluation of the personnel according to this method allows to compare the existing competences of the employees with the competences that the enterprise needs [19]. The essence of this approach is to stimulate the desired economic behavior of each employee to achieve the goals of the enterprise.

But this approach can only be applied if the capacity of the enterprise's personnel is developed through an

 https://doi.org/10.1051/shsconf/202110001022

incentive mechanism. All this requires constant monitoring of the impact of the relevant organizational stimulation measures on the social and psychological climate of the collective and the productivity The assessment of a worker social and professional quality makes it possible to quantify, by means of scores, the essential characteristics of both the worker and the work he performs.

We separate the four components of the socio-professional quality of the company's staff: qualification, education, length of service in the profession, age. These components create the necessary prerequisites for the performance of the respective functional duties. Qualifications, experience, education, age are selected to assess the quality of staff – and these components have an undeniable advantage over other personal characteristics (attributes) of the person because they can be quantified and documented in the operational and statistical reporting of the enterprise. Age is a demographic attribute as a natural quality of the person; the level of qualification, work experience at the enterprise, level of education are acquired qualities of the employees. The employer can influence through an effective system of incentives.

Let's consider each component of the socio-professional quality of staff. The information base is the documents of staff management services.

The main indicator that directly affects the quality of work performed is qualification.

To assess the qualification, we propose to use an integrated indicator that includes the following components:

existing tariff class, category – $T_{tar.cl.}$;

mastering additional professions (for each - 1.4 points) – T_{add};

mastering related professions (for every 1.2 points) – $T_{rel.}$;

creative potential, measured by the amount of data for the last five years of innovation proposals (1.1 points) – $T_{cr.pot.}$.

Assessment of staff qualifications (A_{qua}) is carried out according to the formula

$$A_{qua} = T_{tar.cl.} + 1{,}4 \times T_{add} + 1{,}2 \times T_{rel} + 1{,}1 \times T_{cr.pot} \quad (2)$$

Other components of socio-professional quality of staff (work experience at the enterprise, education and age) are assessed by experts on a scale from 0 to 10 - low - 0 ÷ 4, medium - 4 ÷ 7, high - 7 ÷ 10.

According to the length of service at the enterprise, all employees are divided into four groups, each of which corresponds to a certain score (Table 2).

Table 2. Staff groups by length of service at the enterprise (A_{length}) and its grade points

Staff groups	**Grade points**
With less than 3 years of work experience	3,0
With more than 3 years of work experience, but less than 8 years	6,0
With more than 8 years of work experience, but less than 15 years	8,0
With more than 15 years of work experience	10,0

Source: calculated by the authors

According to the level of education, employees are divided into four groups, then the assessment of the potential of the company's staff is carried out in accordance with table 3.

Table 3 Staff groups by level of education (A_{edu}) and its grade points

Staff groups	**Grade points**
With general secondary education	4,0
With the level of education "Bachelor, Junior Bachelor"	6,0
With the level of education "Master (Specialist)"	9,0
With a scientific degree	10,0

Source: calculated by the authors

By age, all employees are divided into six groups, each of which corresponds to a certain score (Table 4).

Table 4. Staff groups by age (A_{age}) and its grade points

Staff groups	Grade points
Over 18 years old, but less than 24 years old	4,0
Over 24 years old, but less than 34 years old	7,0
Over 35 years old, but less than 49 years old	10,0
Over 50 years old, but less than 55 years old	9,0
Over 55 years old, but less than 59 years old	8,0
Over 60 years old	5,0

Source: calculated by the authors

At the same time, the proportion of the components of the indicators of the socio-professional quality of staff, taking into account expert analysis, is equal: qualification ($Q_{qua.}$) – 0,4; work experience at the enterprise ($Q_{w.exp.}$) – 0,28; education (Q_{edu}) – 0,22; age (Q_{age}) – 0,1.

After calculations of indicators of components of socio-professional quality of the personnel it is possible to calculate the level of socio-professional quality on the enterprise for the average worker. ($\overline{\overline{Q}}_{soc-prof}$):

$$\overline{\overline{Q}}_{soc-prof} = \frac{\left(\sum_{i=1}^{n} Q_{qua\,i} \times A_{gr\,i}\right) \times K_j}{n} + \frac{\left(\sum_{i=1}^{n} Q_{exp\,i} \times A_{gr\,i}\right) \times K_j}{n} + \frac{\left(\sum_{i=1}^{n} Q_{ed\,i} \times A_{gr\,i}\right) \times K_j}{n} + \frac{\left(\sum_{i=1}^{n} Q_{ag\,i} \times A_{gr\,i}\right) \times K_j}{n} \quad (3)$$

where $\overline{\overline{Q}}_{soc-prof}$ – the level of socio-professional quality for the average employee of the enterprise; $Q_{qua.}$ – indicators of the level of qualification of staff with a total number; Q_{exp} – indicators of staff work experience with a total number; Q_{edu} – indicators of staff education with a total number; Q_{age} – indicators of the age of the staff with a total number; A_{gr} – assessment of gradations of indicators of socio-professional quality of staff; K_j – the coefficient of weight of component of socio-professional quality of staff; i – gradation of indicators; j – component of socio-professional quality of staff; n – number of employees in this category.

Assessment of the employee socio-professional quality factors by experts. The weight coefficient of the staff socio-professional quality components (K_j), which is determined by experts, and the calculation are presented in table 4.

Table 4. Assessment of the weight coefficient of the employee socio-professional quality components

Experts (m)	Components of socio-professional quality of the employee (n)				
	Q_{qua}	Q_{exp}	Q_{edu}	Q_{age}	$\sum$
1	0,4	0,3	0,2	0,1	1
2	0,3	0,3	0,3	0,1	1
3	0,4	0,4	0,1	0,1	1
4	0,3	0,3	0,2	0,2	1
5	0,3	0,3	0,2	0,2	1
6	0,4	0,3	0,2	0,1	1
7	0,4	0,2	0,2	0,2	1
8	0,5	0,2	0,2	0,1	1
9	0,5	0,2	0,1	0,2	1
10	0,4	0,3	0,2	0,1	1
The sum of the ranks obtained for each factor, $\sum x_i$	3,9	2,8	1,9	1,4	8,6
Deviation from the average sum of ranks, $x-\bar{x}$	3,04	1,94	1,04	1,4	7,42
Quadratic deviation of the sum of ranks, $(x-\bar{x})^2$	29,37	3,76	1,08	34,21	68,45
Average assessment gradation by factor, A_{gr}	0,39	0,28	0,19	0,14	1

Source: calculated by the authors

Notes: Q_{qua} – indicators of the level of qualification of staff with a total number;

Q_{exp} – indicators of staff work experience with a total number;

Q_{edu} – indicators of staff education with a total number;

Q_{age} – indicators of the age of the staff with a total number;

A_{gr} – assessment of gradations of indicators of socio-professional quality of staff;

i – gradation of indicators.

The Kendall concordance coefficient was calculated using MS Excell, W = 0,681 – the consistency of experts is strong.

Thus, the range of changes in the level of socio-professional quality of the indicator for the considered enterprises ∈ [0 ÷ 10,0]. Thus, the indicator value of the staff socio-professional quality is within the range 0 ÷ 4 – low (zone C), at 4 ÷ 7 – middle (zone B), at 7 ÷ 10,0 – high (zone A).

5 Assessment of socio-professional quality of employees

We will make calculations according to the proposed improved method of assessment of the socio-professional quality of the staff of PJSC «Kryvoriz'kyy zalizorudnyy kombinat» (PJSC «KZRK»). The assessmet is accomplished for more than 200 workers according to formula 1, then the results are summarized (table 5).

Table 5 Assessment of the qualifications of the personnel of the division at PJSC «KZRK»

Employees	Components of qualifications and their grade points				Overall assessment qualifications
	Tariff grade	mastering additional professions	Mastering related professions	Creative potential	
1	4	0	0	0	4
2	6	1,4	1,2	0	8,6
3	3	0	0	0	3
4	5	0	1,2	1,1	7,3
5	6	2,8	1,2	0	10
6	4	0	0	0	4
7	3	0	0	0	3
8	6	1,4	0	1,1	8,5
9	4	0	0	0	4
19	3	0	0	0	3
11	5	0	0	1,1	6,1
12	5	0	0	0	5

Source: calculated by the authors

Employees received the highest grade points of qualification and mastery of additional professions. Thus, the highest assessment of qualification of the employee number two, who has the highest tariff category, has mastered an additional specialty and has a related profession and the employee number eight, who has the highest tariff category, has mastered an additional profession and has developed creative potential. Assessment of socio-professional quality of employees of the unit was carried out according to formula 2 (Table 6).

Table 6. Assessment of socio-professional quality of employees of the unit at PJSC «KZRK»

Employees	Components of socio-professional quality of the employee								Overall grade
	Qualific a-tion		Work experienc e at the enterpris e		Educatio n		Age		
	Grade points	the weighting coefficient of the component	grade points	the weighting coefficient of the component	grade points	the weighting coefficient of the componenti	grade points	The weighting coefficient of the component	
1	4		6		9		7		5,96
2	8,6		10		6		9		8,46
3	3		3		6		7		4,06
4	7,3		8, 0		6		9		7,38
5	10		10		6		5		8,62
6	4		6		9		5		5,76
7	3	0,4	6	0,28	9	0,22	10	0,1	5,86
8	8,5		10		9		8		8,98
9	4		8		6		7		5,86
10	3		3		6		10		4,36
11	6,1		10		6		5		7,06
12	5		8		9		8		7,02

Source: author's development

The following conclusions can be drawn from the results of assessing the socio-professional quality of employees.

The employees numbered 2, 4, 5, 8, 11, 12 are included in the zone of high assessment of social-professional quality [7÷10,0]

All other employees are included in the zone of average value of the indicator.

There are no employees with a low value of socio-professional quality in the research unit. Based on the results of the assessment, a Pareto diagram of the socio-professional qualities of the employees of the structural unit at PJSC «KZRK» was constructed on fig.1. The Pareto diagram clearly shows the distribution of values of indicators of socio-professional qualities of the employee of the structural unit at PJSC «KZRK» in descending order of their frequency.

The line of total values of the calculated indicator on the additional axis reflects the percentage of the accumulated amount of the result. «Zone A» entered 6 employees, for whom it is desirable to apply incentive methods for further professional development. Thus, employees numbered 5, 4, 11, 12 are recommended to increase the components of qualifications, in particular, to master additional and related professions and develop their creative potential. This is especially true for employees numbered 5 and 11, who have more than 15 years of work experience at the enterprise and can use the accumulated production experience to provide, for example, innovation proposals. 6 employees entered Zone «B». Staff managers need to increase the impact of incentives on the components of the socio-professional quality of these employees. The third and tenth employees have critical value of indicator of socio-professional quality, accordingly: 4.06 and 4.36, due to short work experience at the enterprise (less than 3 years), low level of qualification and level of education «Bachelor, Junior Bachelor». If the influence of the appropriate means of stimulation is not applied, this indicator may become critical, and these workers will enter «Zone C».

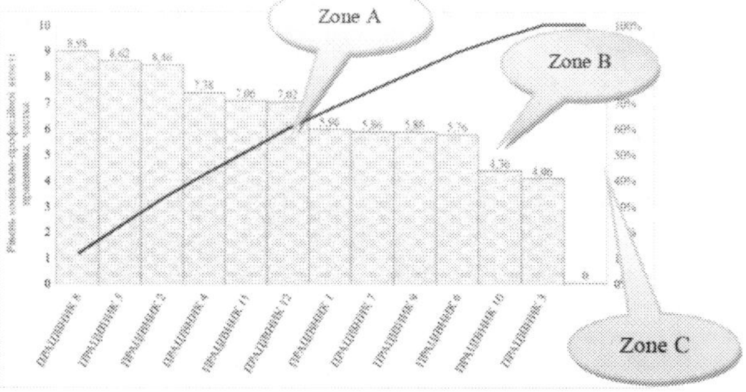

Fig. 5. Pareto diagram of socio-professional qualities of the employee of the structural unit at PJSC «KZRK»

Source: author's development

6 Discussion of research results

It should be noted that the composite integral indicator in the formation of the incentives system for the employees of enterprises, is a set of group integrated indicators of the appropriate means of incentives: material, professional, social, organizational.

It is possible to determine the effectiveness of the incentive mechanism based on the integral indicator calculation of staff incentive assessment.

The use of an integral indicator makes it possible to increase the objectivity of the of incentive measures assessment, what directly affects the efficiency of staff.

It is established that one of the measures of the improved economic mechanism of stimulation is the revision of the personnel qualification assessment on the base of competence and attributive approaches.

Four elements are identified in assessing graduation and the weight of socio-professional quality indicators, namely: qualifications, education work experience by profession and age. A four-pronged system is used for the assessment of qualifications: tariff grade, mastering additional professions by the worker, mastering related professions by the worker and creative potential. The proposed quantitative assessment of staff development is characterized by the simplicity, calculating - universality and availability of the necessary existing statistics of the personnel department in all communications enterprises without the involvement of a specially trained worker.

Conclusions

It is the formation of an effective staff incentive system is based on the proposed tools, namely the calculation of the integral indicator of staff incentives, construction of motivators and demotivators maps, motivation profile, assessment of socio-professional quality of the employee on the basis of integral indicators, the construction of regression models. That allows to identify problems related to the development of personnel of enterprises and allows to make appropriate management decisions for the introduction of motivational means to increase the quality and productivity of work.

The introduction of an improved economic mechanism of personnel incentives in enterprises has led to the formation of an updated incentive system and corresponding changes in Collective Agreements, the building a system of private pension provision, staff qualification assessment system based on the socio-professional quality of employees, and the building a model of incentives for productivity of work. It is the construction of a diagram of the motivational profile of the employee that allows to build a system of assessment of staff skills as well as to improve the qualification and professional level for the needs of the enterprise in today's turbulent environment. According to the results of employees socio-professional quality assessment of the unit at PJSC «KZRK», half of the employees entered the zone of high assessment of the socio-professional quality. All other employees entered the middle quality zone. There are no employees in the study unit who have a low value of the socio-professional quality indicator. Based on the results of the assessment, a Pareto diagram of socio-professional qualities of employees of the structural unit at PJSC «KZRK» was constructed.

References

1. Menshikova, M. (2008) On the issue of personnel management as a system and integrated subsystem of organization. Kursk: Kursk state. University. URL: http://scientific-notes.ru/pdf.
2. Danilyuk, V., Petyuk, V., Tsinbalyuk, S. Personnel management: teaching method. manual. K.: KNEU, 2006. 398 c.
3. Alexandrova, T. (2005) Personnel management: guidelines for practical training. O.: GOU OSU. 63 p.
4. Sardak, O. (2011) The position of personnel marketing in the management of employees of enterprises. *Economic analysis*. Vip. **8**, Part 2. P. 303-307.
5. Poshelyuzhna, L. (2010) Peculiarities of modern personnel management in domestic enterprises. URL: http://www.nbuv.gov.ua/portal/Soc_Gum/inek/2010_1/163.pdf/
6. Korsakov, D. (2011) Organizational diagnostics of the personnel management system of the enterprise. *Library Bulletin*. **4**. - URL : http://www.nbuv.gov.ua/portal/Soc_Gum/Ecoroz/2011_3/e113kors.pdf.
7. Dudukalo, G. (2012) Organizational and economic preconditions for the development of the machine-building industry in terms of structural transformation. Collection of scientific works of Cherkasy State Technological University. Series: economic sciences. CHI. - P. 45-51.
8. Markova, N. (2012) Formation of organizational and economic mechanism of personnel development management. *Business Inform*. **6**. C. 200–202.
9. Korytko, S. (2012) Formation of the system of labor motivation at the enterprise. *Student Scientific Bulletin "Manager INFO»*. URL: http://kerivnyk.info/2012/05/korytko.html
10. Menshikova, M. (2008) On the issue of staff management as a system and integrated subsystem of organization/ Kursk: Kursk state University. URL: http://scientific-notes.ru/pdf.
11. Danilyuk, V., Petyuk, V. (2006) Tsinbalyuk S.O Staff management: teaching method manual/ K.: KNEU. 398 p.
12. Mazur, N. (2014) Stimulation of the productivity of the enterprise's staff on the basis of expert assess. *Problems of rational use of socio-economic and natural resource potential of the region: financial policy and investment.* Collection of scientific works: Issue XX, **2**. Jubilee. Kyiv, SPP/Rivne, NUWMES. P. 173-181.
13. Balabanova, L., Sardak, O. (2011) Staff management: textbook. K.: CEL. 468 p.
14. Grishnova, O. (2010) Personnel assessment: modern approaches to ensuring efficiency. *Human resource management: problems, theories and practices*. **7**. p. 42-50.
15. Pakulin, S., Kushnir, T., Chorna, M. (2006) New methodical approach to determining the wages of workers of the enterprise. East. **2(74)**. p. 139-140.
16. Yeliseyeva, O., Kutova, N. (2017) Personnel management performance assessment system based on the integral indicator. *Bulletin of the National Academy of Statistics, accounting and auditing*. Quarterly collection of scientific works. **1-2** (52-53). Kyiv. P. 46-54.
17. Yeliseyeva, O., Kutova, N. (2017) Assessment of the mechanism of stimulation of the enterprises. Scientific journal *Business Inform*. **5**. Kharkiv. 2017. P.76-81. URL: http://www.business-inform.net/export_pdf/business-inform-2017-5_0-pages-76_81.pdf
18. Introspective analysis. Methods and means of expert assessment: monograph. below red. E..Petrova. Kherson: Grin D.S. 2011.1168 p.

19. Shlyaga, O., Bilous, A. (2013) Approaches to staff assessment at enterprises. URL: http://www.zgia.zp.ua/gazeta/evzdia_5_126.pdf.
20. Gudz, P., Gudz, M., Sydoruk, I., Kunaiev, A. (2019) Financial component of estimation of efficiency of management of enterprise on basis of BSC. *Collection of scientific works Financial and credit activity: problems of theory and practice.* **4**, № 31 http://fkd.org.ua/article/view/190804
21. DOI: https://doi.org/10.18371/fcaptp.v4i31.190804
22. Zavidna, L. and other. (2019) Strategy of Innovative Development as an Element to Activate Innovative Activities of Companies. In: Academy of Strategic Management Journal. https://www.abacademies.org/articles/strategy-of-innovative-development-as-an-element-toactivate-innovative-activities-of-companies-8385.htm
23. Shmygol, N., Galtsova, O., & Varlamova, I. (2018). Developing a methodology to assess the environmental and economic performance index based on international research to resolve the economic and environmental problems of Ukraine. Baltic Journal of Economic Studies, **4(4)**, 366-374
24. Zos-Kior M., Kuksa I., Samoilyk I., Storoška M. Methodology for assessing the countries' globalization development. Economic Annals-XXI. **11-12**. 2017. P. 4-8.
25. Kwilinski, A. (2018). Mechanism of modernization of industrial sphere of industrial enterprise in accordance with requirements of the information economy. Marketing and Management of Innovations, **4**, 116-128. http://doi.org/10.21272/mmi.2018.4-11
26. Savchenko, T., Basiurkina, N., Rodina, O., & Kwilinski, A. (2019). Improvement of the assessment methods of product competitiveness of the specialized poultry enterprises. Management Theory and Studies for Rural Business and Infrastructure Development, **41(1)**, 43-61. https://doi.org/10.15544/mts.2019.05
27. Savchenko, T., Basiurkina, N., Rodina, O., & Kwilinski, A. (2019). Improvement of the assessment methods of product competitiveness of the specialized poultry enterprises. Management Theory and Studies for Rural Business and Infrastructure Development, **41(1)**, 43-61. https://doi.org/10.15544/mts.2019.05

Determination of the Clustering Potential of the Machine-Building Industry in Zaporizhzhia Region

Tetiana Pulina[1], *Nadiia* Shmygol[1*], *Nikolai* Somych[2], *Oleksii* Narivs'kyi[1] and *Peter* Gudz[1]

[1] National University «Zaporizhzhia Polytechnic», 64 Zhukovsky street, Zaporizhzhia, 69600, Ukraine
[2] Poltava State Agrarian Academy 1/3 Skovorody str., Poltava, 36003, Ukraine

Abstract. The possibilities of creating cluster associations of enterprises in the sub-branches of the Machine Building Industry in the Zaporizhzhia region by determining the clustering potential are considered. The potential of clustering of the engineering industry enterprises of the Zaporizhzhia region was assessed by using the localization coefficients, production per person and specialization of the industries. According to the results of the assessment of the clustering potential of the machine-building industry in Zaporizhzhia there are such five sub-sectors were identified in which the creation of clusters is possible: the production of computers, electronic and optical products, the production of electrical equipment, the production of machinery and equipment that are not assigned to other groupings, the production of motor vehicles, trailers and semi-trailers and other vehicles, the production of furniture and other products; repair and installation of machinery and equipment. It is proved the need to create a electric equipment production cluster in Zaporizhzhia region.

1 Introduction

The strategic goal of developing machine-building industry enterprises in Zaporizhzhia region is to form a structure that would meet the existing natural resource and production potential, as well as ensure the competitiveness of final products.

The engineering industry in 2018 included 208 enterprises. The volume of their production amounted to 2,8435.7 million. or 13.0% of the total industrial production in the region and 14.0% of the volume of the machine-biulding industry in Ukraine. As of January 1, 2019, foreign investors ' funds in the machine-building industry amounted to UAH 117.23 million., or 28.1% of the total amount of foreign investment in Zaporizhzhia region.

Mechanical engineering is a knowledge-based sector of the economy, the financial condition of which largely depends on the state level of industrial development. The machine-building industry plays a significant role in the formation of the most important macroeconomic indicators of the region.

But industry enterprises are characterized by a imbalance between the availability of production capacities and real production volumes, which negatively affects the results of production activities. At the same time, the aging process of the equipment is constantly progressing. In addition, the deterioration of the financial and economic condition at most enterprises of the industry occurred due to the lack of a production structure optimized for real production volumes and product nomenclature, the deterioration of production capacities, and the discrepancy between the number of workers and real production volumes. High level of depreciation deductions, competition from the side of imported products. Therefore, the machine-building industry requires a radical restructuring. An achieving the strategic goal of the development of machine-building enterprises in Zaporizhzhia region is possible only if the leading sectors of the machine-building industry will be clustered.

2 Actual scientific researches and issues analysis and the research objective

A number of scientific works of domestic and foreign scientists and researchers are devoted to the study of determining the potential of clustering industries in individual countries [1-3, 11-25]. . To determine the clustering potential, Vinokurova M.V., Honcharova, K.V., Єrmishina, A.V., Turganbaev, E.M. etc. propose to apply coefficients of localization, production per person and specialization of industries that are widely used in the regional economy. Yermisina A.V. and Kozlova M.V. recommend analyzing statistical indicators to determine the competitiveness of the industry. Turganbaev E.M recommends to use national tables «costs-production» for indentification industrial clasters. [1-4]. But the analysis of research of the problem of determining the clustering potentials of industries in different countries revealed the lack of a methodically based approach to determining the clustering potential of an individual industry, which determines the need and relevance of stusying the problem. So the task statement is determination of the

* Corresponding author: nadezdash@ua.fm

clustering potential of the Machine-Building Industry in Zaporizhzhia region and its individual sub-sectors.

3 The possibility of creating cluster associations in the sub-sectors of the Machine-Building Industry in Zaporizhzhia region

Let's consider the possibilities of creating cluster associations in the sub-sectors of the Machine-Building Industry of Zaporizhzhia region by determining the potential for clustering the machine-building industry of Zaporizhzhia region and its individual sub-sectors.

The potential of clustering existence is the availability of competitive advantages of industries, enterprises, infrastructure organizations located in the region and the possibility of combining and using these advantages to increase the competitiveness of the region. Currently, a unified methodology for determining the evaluation of clustering potential has not been developed. To determine the clustering potential of the machine-building industry of Zaporizhzhia region, we will use the method of N.V. Vinokurova [1], which consists in calculating the coefficients of localization, production per person and specialization of industries. If the calculated coefficients are close or more than one and tend to grow, then clusters in these industries can be created.

Table 1 shows statistics to determine the clustering potential of the machine-building industry of Zaporizhzhia region [5-8].

We will determine the localization coefficient (C_L) of the machine-building industry of Zaporizhzhia region. The localization coefficient determines the localization (concentration) of production in the region, the level of the industry development in the region, its significance for the region's economy.

The localization coefficient of this production in the region is the ratio of the specific gravity of this sector of the economy in the structure of production in the region to the specific gravity of the same sector in the structure of production in the country.

The localization coefficient of the machine-building industry of Zaporizhzhia region is calculated by using the formula (1):

$$C_L = \frac{VMEP_z}{VIP_Z} * 100 : \frac{VMEP_U}{VIP_Z} * 100; \quad (1)$$

Table 1 Statistical data to determine the clustering potential of Zaporizhzhia region engineering industry 2016-2018, in million UAN

Indicators	2016 year	2017 year	2018 year
UKRAINE			
Total volume of industrial products sold, million UAH	2158030	2625862,7	3045202
Engineering	131352	168281,9	207204
production of computers, electronic and optical products	12366,1	13783,3	16975,2
production of electrical equipment	26594,6	32986,9	43098,9
production of machinery and equipment not classified in other groupings	50105,3	59767,7	70237,3
production of motor vehicles, trailers and semi-trailers and other vehicles	42285,8	61744,0	78365,0
production of furniture and other products; repair and installation of machines and equipment	43650,3	56383,9	67706,2
The number of permanent population (at the end of the year) of Ukraine, million people.	42,5	42,4	42,2
Gross domestic product (in actual prices), UAH million	2385367	2983882	3560596
Zaporizhzhia region			
Total volume of industrial products sold, million UAH	152863	197567,8	220108
Engineering	18214,8	26390	28435,7
production of computers, electronic and optical products	1580,2	1013,8	1523,8
production of electrical equipment	6003,7	7624,5	10537
production of machinery and equipment not classified in other groupings	3048,7	4183,4	4597,6
production of motor vehicles, trailers and semi-trailers and other vehicles	7582	13568,3	11777,1
production of furniture and other products; repair and installation of machines and equipment	5721,9	5203,5	4432,7
Resident population (annual average), million	1,75	1,73	1,72
Gross regional product, UAH million	104323	130377	147076

where: $VMER_Z$ is the volume of mechanical engineering products sold in Zaporizhzhia region, million UAH;

VIP_Z - the volume of sold industrial products in Zaporizhzhia region, million UAH;

$VMER_U$ - volume of engineering products sold in Ukraine, million UAH;

VIP_U - volume of sold industrial products in Ukraine, million UAH

Table 2 shows calculations of localization coefficient of the machine-building industry of Zaporizhzhia region [5-8].

Table 2 Calculations of localization coefficient of Zaporizhzhia region engineering industry

Indicators	Years		
	2016	2017	2018
$\frac{VMEP_z}{VIP_z}$	0,119	0,134	0,129
$\frac{VMEP_U}{VIP_Z}$	0,061	0,064	0,060
Localization coefficient	1,95	2, 09	2,15

For three years the average indicator of the localization coefficient of enterprises of the machine-building industry in Zaporizhzhia region was 1.99. This is good indicator to create a cluster.

The per capita production coefficient (C_{pp}) is calculated as the ratio of the specific gravity of the economic activity of the region in the corresponding structure of the economic activity of the country to the specific gravity of the region population of the population in the country according to the formula (2):

$$C_{pp} = \frac{VMEP_z}{VMEP_U} * 100 : \frac{P_r}{P_c} * 100; \qquad (2)$$

where: $VMEP_Z$ is the volume of mechanical engineering products sold in Zaporizhzhia region, million UAH;

$VMEP_U$ - volume of engineering products sold in Ukraine, million UAH;

P_R - the population of the region;

P_c - the population of the country.

Table 3 shows calculations of per capita production coefficient of the machine-building industry of Zaporizhzhia region [5-8].

The average value of the per capita production coefficient was 3.66 for three years which indicates the possibility of creating clusters.

Table 3 Calculation of per capita production coefficient of the machine-building industry in Zaporizhzhia region for 2016-2018

Indicators	Years		
	2016	2017	2018
$\frac{VMEP_z}{VMEP_U}$	0,139	0,159	0,141
$\frac{P_r}{P_c}$	0,041	0,041	0,041
Per capita production coefficient	3,39	3,87	3, 43

The coefficient of specialization (C_s) is defined as the ratio of the specific gravity of the region in the country for this type of economic activity to the specific gravity of the region in the country's GDP (3):

$$C_s = \frac{VMEP_z}{VMEP_U} * 100 : \frac{GRP}{GDP} * 100; \qquad (3)$$

where $VMEP_Z$ is the volume of mechanical engineering products sold in Zaporizhzhia region, million UAH;

$VMEP_U$ - volume of engineering products sold in Ukraine, million UAH;

GRP – gross regional product UAH;

GDP is gross domestic product UAH.

Table 4 shows calculations of the specialization coefficient of Zaporizhzhia region engineering industry [5-8].

Table 4 Calculation of specialization coefficient of enterprises of the machine-building industry in Zaporizhzhia region.

Indicators	Years		
	2016	2017	2018
$\frac{VMEP_z}{VMEP_U}$	0,139	0,159	0,141
GRP/GDP	0,044	0,044	0,044
specialization coefficient	3, 15	3,61	3,20

The average value of the specialization coefficient has been 2.98 for three years which indicates the possibility of creating clusters.

Integral coefficient of clustering potential is determined by formula (4):

$$C_{kl} = \sqrt[3]{C_l * C_{pp} * C_s}, \qquad (4)$$

The integral clustering potential coefficient allows us to evaluate the industry's prospects for creating cluster associations of the enterprises (table 5)

Table 5 Integral coefficient of clustering potential

Indicators	Years		
	2016	2017	2018
Localization coefficient	1,83	2, 16	2

Specialization coefficient	3, 5	4	3,5
Per capita production coefficient	3,5	4	3, 5
Integral coefficient of clustering potential	2.94	3,39	3,0

The average value of the integral clustering potential coefficient was 3.11 for three years, which indicates the possibility of creating clusters.

Thus, we can conclude that the machine-building industry of Zaporizhzhia region has a clustering potential sufficient to create clusters. To calculate the coefficient of localization for the sub-branches of Zaporizhzhia region engineering industry, we will consider the volume of sold products in the production of engineering for 2016-2018 in Ukraine and Zaporizhzhia region.

Table 6 shows calculated production localization factors for sub-sectors of the machine-building industry of Zaporizhzhia region.

Table 6 Calculation of localization coefficient for enterprises of the machine-building industry of Zaporizhzhia region

Indicators	**Years**		
	2016	**2017**	**2018**
Engineering	1,83	2, 16	2
production of computers, electronic and optical products	1,56	1,97	1,56
production of electrical equipment	1,31	3,07	1, 78
production of machinery and equipment not classified in other groupings	1,16	1,68	1, 41
production of motor vehicles, trailers and semi-trailers and other vehicles	1,39	2, 92	1,94
production of furniture and other products; repair and installation of machines and equipment	1,85	1,22	1,41

Table 7 shows calculated per capita production coefficient for the sub-sectors of the machine-building industry of Zaporizhzhia region.

Table 7 Per capita production coefficient of sub-branches of the machine-building industry in Zaporizhzhia region

Indicators	Years		
	2016	2017	2018
Engineering	3,5	4	3, 5
production of computers, electronic and optical products	1,31	1, 8	1, 17
production of electrical equipment	5, 48	5,6	3, 63
production of machinery and equipment not classified in other groupings	1, 47	1,64	1, 42
production of motor vehicles, trailers and semi-trailers and other vehicles	4, 35	5,32	3, 64
production of furniture and other products; repair and installation of machines and equipment	3, 18	2, 23	1, 58

Table 8 shows the calculated coefficients of specialization of production for sub-sectors of the metallurgical industry of Zaporizhzhia region.

Table 8 Specialization coefficient for sub-sectors machine-building industry of Zaporizhzhia region

Indicators	Years		
	2016	2017	2018
Engineering	3, 5	4, 0	3, 5
production of computers, electronic and optical products	2, 92	1, 68	2, 17
production of electrical equipment	5, 16	5, 25	5,91
production of machinery and equipment not classified in other groupings	1,39	1,6	1, 58
production of motor vehicles, trailers and semi-trailers and other vehicles	4, 09	5, 02	3, 63
production of furniture and other products; repair and installation of machines and equipment	2, 99	2, 11	1, 58

Table 9 shows integral coefficient of clustering potential for Zaporizhzhia region engineering industry sub-sectors.

Table 9 Specialization coefficient for sub-sectors engineering industry of Zaporizhzhia region

Indicators	Years		
	2016	2017	2018
Engineering	2.94	3,39	3,0
production of computers, electronic and optical products	1, 93	1, 82	1, 63
production of electrical equipment	3, 98	4, 64	3,77
production of machinery and equipment not classified in other groupings	1,34	1,64	1, 47
production of motor vehicles, trailers and semi-trailers and other vehicles	3, 28	4, 42	3, 07

production of furniture and other products; repair and installation of machines and equipment	2, 99	2, 47	1, 52

After analyzing the data of tables 5,6,7,9 it can be seen that the localization coefficients, per capita production coefficient, specialization coefficients, integral coefficient of clustering potential of the five sub-sectors of the Zaporizhzhia region engineering industry are more than one. And in the sub-sector of electrical equipment production, these determined coefficients are the highest.

It is proposed to create a cluster association of enterprises in the sub-sector of the electrical equipment production. We consider it expedient to include such regional enterprises of energy engineering to this cluster, nameles: «Zaporizhtransformator» PJSC, «MGT» PJSC, «Zaporizhzhia superpower transformers plant» PJSC, SE "SPC “Iskra”, «ZZVA» PJSC, «Zaporizhzhia plant «Preobrazovatel» PJSC, "Motor Sich" PJSC, SE «Ivchenko-Progress», SIA "RENKO".

It should be noted that «Zaporozhtransformator» PJSC is the largest specialized enterprise of world transformer building and carries out development, supply and service maintenance of the entire range of oil, dry power and general purpose distribution transformers, special transformers, components for them (cooling plants, voltage regulators, etc.), higt amperes current conductor of closed type.

«MGT» PJSC produces distribution transformers with a capacity of up to 1000 kVA and power transformers with a capacity of 10-1000 kVA.

«Zaporizhzhia superpower transformers plant» PJSC specializes in the manufacture of power dry and oil transformers, as well as in the repair of transformers of all types.

Thus, the involvement of these enterprises in the energy engineering cluster will make it possible to carry out a comprehensive replacement of the outdated transformer fleet of the energy sector within the Zaporizhzhia region.

The need to participate in the cluster project of «ZZVA» PJSC is due to the fact that the enterprise is a producer of products that ensure the transmission of electricity, especially over long distances, and its distribution to end users.

The usefulness of in involving "Zaporizhzhia plant Preobrazovatel" PJSC to the cluster is due to the fact that the company is a manufacturer of power electronics that are able to regulate and convert energy from one type to another, optimal. The need to use power electronics is due to the fact that the use of electricity in the form in which it is produced, that is, in the form of alternating current with a frequency of 50 Hz or impossible under the conditions of some technologies.

Thus, the use of modern products of "Zaporizhzhia plant Preobrazovatel" PJSC will make it possible to turn the generated electricity into the state that specific consumers need, meet various technological needs while reducing its losses.

State Enterprise "Scientific and Production Complex “Iskra” is a leading developer and manufacturer of ground-based radar technology for Ukraine’s defense industry. The Complex includes a Design Bureau and a Engineering Plant.

In addition, we consider it necessary to also include in the cluster of "Motor Sich" PJSC which is capable of producing cogeneration stations. The use of cogeneration plants on the basis of heating boilers can increase the efficiency of fuel use to 90-92%, because the additional equipment of boiler houses with electric power generation modules allows without increasing gas costs for heat production to receive "as a gift" electricity for an amount that corresponds to the cost of about 20% of the spent fuel.

Thus, the involvement of «Motor Sich» PJSC in the energy engineering cluster will allow to introduce energy-saving technologies in the sector of urban heating plants with the prospect of obtaining additional income from an increase in electricity production in the region.

Note that «Motor Sich» PJSC has the ability not only to produce cogeneration stations, but also with the help of SE "Ivchenko-Progress" to develop various modifications of cogeneration plants to fully meet the needs of various consumers. In addition, the production capacity of «Motor Sich» PJSC is currently loaded by only 25-30%, which creates significant opportunities for the perspective growth of the enterprise and increase its share in the domestic market, which today accounts for no more than 5% of the total supply of «Motor Sich» PJSC.

SIA "RENKO" Scientific and Industrial Association was the first to present an energy efficiency program in Zaporizhzhia region [9]. The main focus of the program is the consistent introduction of energy-saving technologies in the residential sector and the municipal sector of the region; estimated amount of capital attraction 100-130 million UAH per year. One of the areas of implementation of this program is the construction of a network of cogeneration stations with a total capacity of up to 5 thousand. MW (capacity of one - 3.4 MW).

The implementation of this direction will contribute to improving the efficiency of the power system due to the creation of a competitive conditions in the field of thermal and electric energy generation.

Note that we consider it necessary to include in the proposed regional cluster enterprises that are transfed and distributed the electric energy, which will act as end users of the cluster's products, primarily «Zaporozheoblenergo» PJSC.

It should be noted to increase innovation, regional research institutes can play an active role in the cluster. It is these institutions, which have significant potential, but are not enough in demand under the conditions of the domestic present, that can act as a catalyst for the innovative development of cluster association [10].

Involvement in cooperation within the cluster of specialized scientific institutions will make it possible to intensify research and design work in the direction of creating innovations (development of new technologies, improvement of technological processes, etc.), as well as will contribute to accelerating the pace of implementation of research results at enterprises, which will have a positive impact on the processes of design and production of new types of products by cluster participants.

This interaction within the cluster will make it possible to establish a close connection in the chain "science - industry (production complex) - market." We note that the ability to develop and use innovative technologies is one of the most important factors in strengthening the position of the cluster. In addition, the nearby enterprises and institutions within the project will contribute to the accelerated exchange of existing knowledge, experience, ideas, developments and strengthening of production cooperation.

Specialized scientific institutions in the field of mechanical engineering such as «Zaporizhzhia plant «Preobrazovatel» PJSC, SE «Ivchenko-Progress», «All-Ukrainian» Institute of Transformer Engineering» PJSC should be involved in the energy engineering cluster.

The involvement of scientific institutions will make it possible to intensify activities in such areas as the development of new technologies, research, innovation, the generation of ideas, the development of new products ordered by cluster members, as well as to ensure the optimal combination of existing production capacities and the modern base of scientific research and development.

To increase the efficiency of the cluster, we also propose to involve the educational sector in cooperation, primarily Zaporizhzhia Polytechnic National University, Zaporizhzhia National University (ZNU), which have practical experience in training specialists in the field of mechanical engineering (including electrical engineering), energy and energy saving. Higher education institutions have the opportunity to provide training on orders of project participants; develop and implement special training, retraining and development programmes, etc. This will create additional opportunities to receive profit from the orders of cluster members. It is also possible for students to carry out scientific, thesis and term papers in areas that are most interested for enterprises, as well as carry out scientific research ordered by entrepreneurs. In turn, cluster enterprises will become a good practice base for students with the provision of the necessary materials for student research and perspective employment opportunities.

The involvement of regional banks in the financial sector cluster project will provide additional financial resources for the cluster to perform its tasks. In addition, regional banks can get several positive effects from participating in the project:

- Strengthening its own position in the conditigons of increased competition in attracting and servicing accounts of new customers (especially large industrial enterprises);
- the possibility of increasing financial resources, obtaining significant profits by increasing operations and the client base;
- the possibility of obtaining income from direct participation of cluster activities.

As part of a single cluster program, the participating bank will be able to manage most of the financial resources of the cluster and profit not only from lending to members of the association, but also from managing its financial flows, thus acting not only as a source of financing, but also as a financial operator of the cluster.

An important role in the implementation of the cluster project in the region should also be played by state and non-state public organizations, which can provide legal support for the relations of project participants, help in attracting additional sources of financing, deal with the distribution of risks, provide information and educational services, contribute to a significant dissemination of information about cluster projects, etc.

Therefore, in the field of production, we can expect the development of fundamentally new types of products, especially taking into account the needs of domestic consumers, increasing the specialization of production, reducing the production cycle of production due to the interaction of participants, and increasing the requirements for the quality of cluster products.

The main areas of development of the association in the production and technical and technological spheres include:

- increasing of the range of produced products;
- improving the quality, reliability and convenience of maintenance;
- creation of new constructions, equipment for the needs of domestic energy and industry;
- improvement of existing constructions with the aim of technical improvement of their parameters, namely reduction of material consumption, labour intensity and losses based on introduction of new technological solutions;
- meeting the requirements of international and national standards for products, meeting the specific needs of customers.

In the non-production sphere, we can expect the development of partnerships, the development of new interaction schemes, deepening cooperation between enterprises, increasing the availability of information, which will increase the cost of joint research and development, the introduction of new technologies for managing production and sales processes, financial processes, improving the quality of business processes, and developing cluster infrastructure.

At the same time, we believe that the specialization of the cluster is an important factor that will contribute to the most effective manifestation of cooperative communication within the association. The formation of a unified cluster association management system will facilitate the coordination and integration of management activities of individual participants into a single complex. This will ensure a more flexible and rapid response of the cluster to changes in the internal

https://doi.org/10.1051/shsconf/202110001023

environment, which means that it will improve its competitive position.

In addition, the intensification of information flows within the cluster will contribute to the accelerated dissemination of existing knowledge and technical innovations, which can be considered as an important competitive advantage of the cluster. We emphasize that these facts will help improve the position of both the joint cluster brand and the perspective self-development of individual member brands.

It should be noted that cooperation within the cluster of scientific and business sectors will contribute in the future to the reorganization of scientific institutions in the direction of increasing scientific research for specific purposes, first of all, to meeting the current and promising needs of various groups of potential clients of the cluster.

4 The structure of the power engineering cluster of Zaporizhzhia region

A more specific structure of the energy engineering cluster of in the Zaporizhzhia region is pointed out in Figure 1.

In general, within the existing cluster, it is possible to expect an improving of the competitive position of its participants due to an increase in their share in the domestic market (expansion of the consumer base of cluster products), as well as the opportunity to offer consumers an updated nomenclature and a wider range of products, compared to the assortment of individual participants to cluster.

We emphasize that one of the consequences of the effective work of the proposed cluster may be an increase in the cost of cluster enterprises and an increase in capitalization, which, in turn, will contribute to increasing the investment attractiveness of both enterprises alone and the association as a whole.

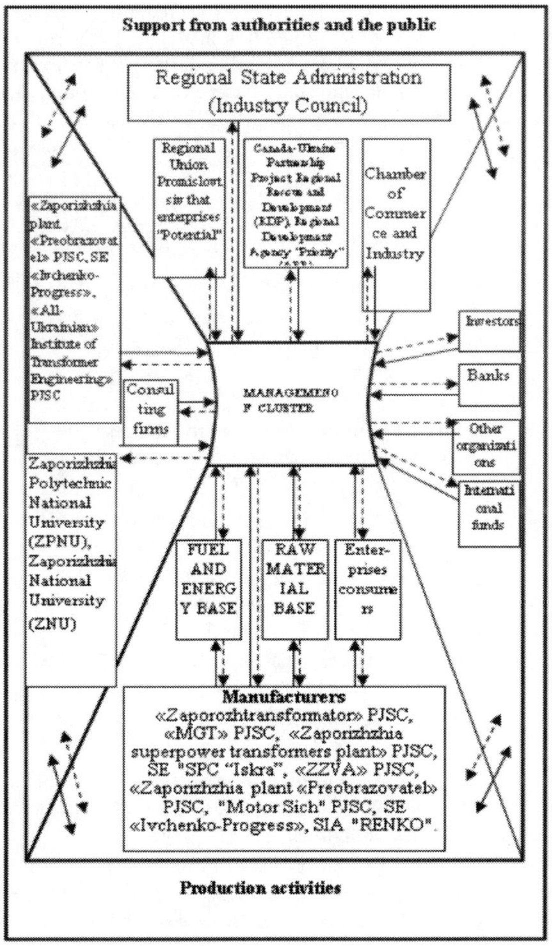

Fig. 1. The cluster structure based on electric equipment production in Zaporizhzhia region.

Due to the scale of the association, it is also possible to enter directly the international financial level (international banks, exchanges, insurance companies, funds, etc.) in order to place their own shares and obtain the necessary investment resources.

Conclusions

Based on the results of the study, a significant potential was identified for the creation of cluster associations of enterprises of the machine-building industry of Zaporozhye region, the assessment of which was carried out using coefficients of localization, production per person and specialization of industries. According to the results of the assessment of the clustering potential of the machine-building industry of Zaporozhye region, five sub-sectors in which the creation of clusters is possible have been identified. namely, the production of computers, electronic and optical products, the production of electrical equipment, the production of machinery and equipment not classified elsewhere, the production of motor vehicles, trailers and semi-trailers and other vehicles, the production of furniture and other products; repair and installation of machinery and equipment Proved the need to create a cluster of electrical equipment production in Zaporizhzhia region.

It was found necessary to intensify innovation and transition to an innovative development model. The cluster development strategy is substantiated by the innovation strategy model of Zaporizhzhia region's industrial complex.

References

1. Vinokurova M.V. (2006), "Competitiveness and clustering potential of industries in the Irkutsk region", *Jeko.,* vol. 12, pp. 73–93. Russia.
2. Honcharova, K.V. (2009), "Improving the methodology for determining the clustering potential of industries in a particular region", *Ekonomichnyj prostir: Zbirnyk naukovykh prats', PDABA,* vol. 29, pp. 80–88.
3. Єrmishina, A.V. (2002), "Competitiveness of the region", *Strategii,* vol. 1, pp. 116–118.
4. Turganbaev, E.M., Kozlova, E.M. (2009), "Identification of regional industrial clusters and assessment of their structural impact on the region's economy (on the example of EKO)", *Upravlenie bol'shimi sistemami,* vol. 25, pp. 139–178.
5. Statistical Yearbook of Ukraine (2018), in. Vernera, I. Ye (ed.), Konsul'tant, Kiev, Ukraine.
6. Official site of Derzhkomstat, [Online], available at: http://www.ukrstat.gov.ua. / (Accessed 9 January 2021), (In Ukrainian).
7. Statistical Yearbook of Zaporizhia region (2018), *Main Department of Statistics in Zaporizhia region,* in. Shejko M. (ed.), Zaporozhye, Ukraine.
8. Official site of the main department of statistics in the Zaporozhye region [Online], available at: http://zp.ukrstat.gov.ua/ (Accessed 9 January 2021).
9. Nadiia Shmygol, Olena Cherniavska, Tetiana Pulina, Ruslan Zavgorodniy (2020), "Economic assessment of the implementation of the resource-efficient strategy in the oil and gas sector of the economy on the basis of distribution of trade margins between extracting and processing enterprises", *Polityka Energetyczna, Energy Policy Journal,* vol. 23(3), pp. 135–146.
10. Narivs'kyi, O.E, Yar-Mukhamedova, G., Temirgaliyeva, E., Mukhtarova, M., Yar-Mukhamedov, Y. (2016), "Corrosion losses of alloy 06XN28MDT in chloride-containing commercial waters International Multidisciplinary Scientific GeoConference Surveying Geology and Mining Ecology Management", *SGEM,* vol. **1,** pp. 63–70.
11. Zavidna L. and other. Strategy of Innovative Development as an Element to Activate Innovative Activities of Companies. In: Academy of Strategic Management Journal, 2019. https://www.abacademies.org/articles/strategy-of-innovative-development-as-an-element-toactivate-innovative-activities-of-companies-8385.htm
12. Shmygol, N., Galtsova, O., & Varlamova, I. (2018). Developing a methodology to assess the environmental and economic performance index based on international research to resolve the economic and environmental problems of Ukraine. Baltic Journal of Economic Studies, 4(4), 366-374
13. Kwilinski, A., Dalevska, N., Kravchenko, S., Hroznyi, I., Kovalenko, I. (2019). Formation of the entrepreneurship model of e-business in the context of the introduction of information and communication technologies. Journal of Entrepreneurship Education, 22(1S), 1528-2651-22-S1-337: 1-7. Retrieved from https://www.abacademies.org/articles/Formation-of-the-entrepreneurship-model-of-e-business-1528-2651-22-S1-337.pdf
14. Mykoliuk, O., Bobrovnyk, V., Fostolovych, V., Kucherova, H., Nataliia, P. Modelling the Level of Energy Security at Enterprises in the Context of Environmentalization of Their Innovative Development. 2020 10th International Conference on Advanced Computer Information Technologies, ACIT 2020 - Proceedings, 2020, pp. 621–625, 9208996
15. Kucherova, H., Pokataieva, O., Bilska, O. Modelling tax consciousness evaluation in the context of economic development uncertainty. Periodicals of Engineering and Natural Sciences, 2019, 7(2), pp. 567–579
16. Gryshko V., Zos-Kior M., Zerniuk O. Integrating the BSC and KPI systems for improving the efficiency of logistic strategy implementation in construction companies. International Journal of Engineering & Technology. Vol. 7, № 3.2. 2018. Special Issue 2. P. 131-134.
17. Zos-Kior M., Kuksa I., Samoilyk I., Storoška M. Methodology for assessing the countries' globalization development. Economic Annals-XXI. №11-12. 2017. P. 4-8.
18. Kuksa I. Adaptive accounting of production activities of transnational companies in the conditions of globalization / I. Kuksa, L. Vasyurenko, A. Litvinov, O. Lytvynova, I. Shtuler, M. Zos-Kior, O. Naholiuk. Accounting. №6.2020. P. 913–922.

Smart Specialization of Innovative Development of Functional Territories: Strategic Experience and Potential of Ukraine

Hanna Kucherova[1*], *Inessa* Khvostina[2], *Dmytro* Ocheretin[3], and *Olha* Bilska[1]

[1]Classic Private University, Chair of Economics, 69002 Zaporizhzhia, Ukraine
[2]Ivano-Frankivsk National Technical University of Oil and Gas, Department of Applied Economics, 76019 Ivano-Frankivsk, Ukraine
[3]Zaporizhzhia National University, Faculty of Economics, 69600 Zaporizhzhia, Ukraine

Abstract. The article examines the strategic experience of introducing smart specialization in Ukraine using the methods of strategic analysis, financial analysis, generalization, statistics, critical thinking. It has been proved that the state priorities of Ukraine's strategic development have changed. The concept of providing state aid to industries that cannot withstand competition has been changed to the concept of identifying specialized functional areas based on the principles of reasonableness, sustainability and inclusiveness of development, the ability to integrate into value chains of world markets. The above more locally meets the requirements of the competitive environment, minimizes the risks of scattered financing, increases the level of effectiveness of strategic planning, innovative and technological development of the state as a whole. The inconsistency of the state and regional strategic support for the introduction of reasonable specialization was established against the background of determining the effectiveness of the use of such forms of cooperation as the cluster model and cross-functional interaction. The need for standardization and greater regulation of the smart specialization procedure at both the conceptual, methodological and practical levels is determined. Ukrainian smart specialization is constrained by the insufficient potential for the implementation of regional development strategies, the urgent need to adapt the method of assessing smart specialization in the format of entrepreneurial discovery to the Ukrainian realities of types of economic activity, insufficient funding of potential functional territories, low initiative, diversification and self-organization. The delay in the implementation of reasonable specialization of activities in Ukraine is explained by the need for additional time to reconfigure the socio-economic system at all levels and to understand the territorial needs, the prospects of choosing from existing opportunities and its consequences.

1 Introduction

In the context of active information technology development, European initiatives have identified a range of new prerogatives and challenges to modern regions and industries of developed and developing countries. The development concept was transformed from innovative to smart by declaring the concept of smart specialization for the functioning and development of regions, thereby maximizing the angle of view on modern socio-economic processes and the competitive potential for their improvement. In practice, this approach allows not to scatter funds, but to focus funding on truly unique objects or entities, whose potential is characterized not only by innovation, self-development, but also by the ability to self-integration into chains of added value formation on world markets.

Ukraine has received further European initiatives and is gradually introducing the concept of smart specialization. The European Union has not established clear rules and requirements for the introduction and irealizatation of smart specialization. The restriction concerned only the strategic vector, platform and general assessment methodology. And this is understandable, since smart specialization is a weakly structured category, and the specificity of its formalization and determination methodology depends on the characteristics of the subject or object of assessment, so the approach must be adaptive.

On the one hand, the permissive nature of the introduction of smart specialization does not limit, but on the other requires the immediate development of new skills, knowledge, and most importantly - the formation of a new vision of those very territories and industries, the principles of their development. The non-regulation of the procedure for introducing smart specialization will eventually reveal as a test separately those subjects that, in fact, are already functioning in a new, innovative, unique way, but are limited in development by certain parameters.

Smart specialization of economic activity introduces the identification and diversification of assets such as knowledge, experience, resources, business models, mechanisms of functioning, innovation, and form functional territories of interregional and transnational levels by means of cross-functional interaction.

* Corresponding author: kucherovahanna@gmail.com

Thus, the scientific approach substantiates the principles of entrepreneurial discovery, the government forms a vision for assessing the opportunities, sentiments, trends, needs and prospects of smart specialization, business and society determine the entrepreneurial choice of market opportunities. The effect of qualitative and structural transformations as a result of smart specialization of activities is multidimensional, therefore, it is not only a quantitative assessment of trends and structural indicators.

Thus, the priority steps in the countries have been taken, the factors that inhibit and promote the implementation of smart specialization have already been identified, which requires careful study and research in each country separately.

1.1 Ralated works

The formation and development of an innovative economy is the only priority way of developing socio-economic systems to ensure their life in the long term. The newest European direction is the introduction and development of smart specialization as a policy of unity of three basic principles: rationality, resilience and uniqueness of growth [1, 2]. This concept is still only spreading, but it was first mentioned in the materials of the Knowledge for Growth - K4G concept, where its content was interpreted as a promising strategy for encouraging investment in areas that should complement the region's production assets to create its internal potential and interregional comparative competitive advantages in the future [1, 2]. Since the approach is based on the principle of searching for an entrepreneurial discovery, its implementation will ensure a qualitative change in the production function of the industry, will reveal the potential of the functional territory by means of using modern knowledge and technologies. The foundation of smart specialization lies in the concept of global growth and is presented in the European Union strategy document "Europe 2020. A European strategy for smart, sustainable and inclusive growth". The importance of the regional potential for the implementation of smart specialization is outlined in the provisions of the European Union document "Regional Policy for Smart Growth in Europe 2020". The main goal of smart specialization is the formation of a list of activities with high added value, which would strengthen the competitive position of the territory, pulling up related industries with less potential, and financing would be concentrated on the most promising activities.

Given the vagueness of the concept and the absence of a rigid and principled methodology, the EU's arsenal contains only a set of tools and a recommended methodological approach, the features of which are arbitrary choice and depend on the capabilities and goals of individual regions. There is also a lack of approved quality and efficiency criteria, statistical data in general. The effectiveness of the implementation of smart specialization today is assessed only by expert means.

Modern scientists study smart specialization in the context of the fourth innovation spiral, the interdependence of the knowledge economy, science and the competitiveness of industries, regions [3,4,5,6]. Separate studies are devoted to the advantages and disadvantages in the practical implementation of the smart specialization policy [7,8,9], the factors influencing it [10,11], in particular the actions of universities [12] and the institutional structure [13], study the experience of implementing relevant measures in different countries [14,15], the features of the strategic promotion of the concept [16,17], the assessment of various implementation policies [18], the typology of regional diversification models [19], and so on.

The problems of the formation of competitive advantages of territories [20], the search for a synthesis of regional, cluster, structural and innovation policies of territories and the state as a whole [21], the implementation of innovative and structural diversification of socio-economic processes [22] and the importance of introducing individual strategies of smart specialization at the regional level [23]. The results of generalization by the European Union of monitoring, expert assessments, surveys of stakeholders of smart specialization are presented in [24], where a positive attitude is proved and the prospects of this direction of increasing the innovativeness of activities in Ukraine are assessed. Today, the experience of various countries is of scientific and practical value, in particular, the mechanism for introducing smart specialization launched in Ukraine. State priorities have already been identified and there is already a certain strategic experience in the implementation of certain measures, which requires a thorough study.

1.2 Strategic experience of Ukraine

During a long period of the country's development, the following is noted: the necessary macroeconomic strategic documents and measures are approved at all levels, but the expected result is not achieved. The reasons in each case are different, including insufficient funding, its fragmentation, lack of action plans or their partial or complete failure to implement. This is confirmed by the results of the analysis of individual programs for the creation of innovative instructions, reforming state policy in this area. Also, park structures have been created, a significant number of which, with some exceptions, have already existed for a long period only on paper or operate on a limited basis without a real return on investment and effort.

Repeated attempts were made to approve a single strategic document that would generalize the ways of solving issues regarding the innovative development of the country's economy; drafts of innovative development strategies were developed, which were never approved in the proposed version. In the future, some of their provisions were partially used as the basis for the development of the next project. For example, the draft Strategy of innovative development of Ukraine for 2010—2020 in the conditions of globalization challenges included certain provisions of previous documents. However, the document was never approved, although it was adopted by the Cabinet of Ministers of Ukraine and submitted for consideration by the Verkhovna Rada of Ukraine.

The foundations of reasonable specialization in all spheres of life are laid down in such a strategic document of Ukraine as the Concept of development of the digital economy and society of Ukraine for 2018 - 2020 and approval of the action plan for its implementation. In 2019, the Strategy of the development of the innovative activity sphere for the period up to 2030 [25] was approved, which ensures the organization of activities in building a modern innovation ecosystem for the long term and is the basis for transforming new fresh ideas into innovative products, increasing the level of innovativeness of the economy. In accordance with the provisions of Strategy 2030, it is proposed to apply new approaches in determining territorial priorities, including with respect to ensuring the quality of connections between participants in the innovation process by introducing digital technologies, applying the concept of smart specialization to implement the exclusive aspects of each territory.

Despite the recommendations of the European Union, Ukrainian specialists have artificially and expertly identified the following list of regional specializations: resource materials; bioeconomics and biotechnology; energy and power engineering; aerospace technology; information and communication technologies; healthy society. The Ukrainian understanding of smart specialization focuses on the search for a reasonable specialization of regions, competitive advantages, uniqueness and ways of decentralization within the boundaries of the industrial complex (the draft of the Strategy for the development of the industrial complex of Ukraine for the period until 2025), although the vision of the European Union is broader than one industry.

Overcoming sectoral limitations on the basis of Strategy 2030, it is planned to form strategies for regional innovative development, taking into account the needs and potential of the regions, and the Ministry of Education and Science of Ukraine, together with the central executive authorities, is determined to develop and submit a draft action plan for 2019 - 2021 to the Cabinet of Ministers of Ukraine within four months. years for the implementation of the Strategy [25], starting from 2022, to develop and submit for consideration by the Cabinet of Ministers of Ukraine draft three-year action plans for the implementation of the Strategy approved by this decree.

To assess the effectiveness of the implementation of the Strategy, it is envisaged to carry out both control over the implementation of action plans and monitoring their impact on various sectors of the economy. The monitoring procedure is developed by the Ministry of Education and Science of Ukraine together with the Ministry of Economic Development of Ukraine and other interested structures. The draft action plan for the implementation of the Strategy for the development of the sphere of innovation for 2019 - 2021 was approved by the order of the Cabinet of Ministers of Ukraine and published on the official website of the Ministry of Education and Science of Ukraine on November 29, 2019, but at the moment the monitoring procedure is not officially available.

Separate regional strategies for innovative development have not been approved today, and the provisions of the Strategy 2030 are included in regional strategies for the period 2021-2027, approved in accordance with the Draft State Strategy for Regional Development for 2021-2027 "Human-Centered Development and Unity", the need to prepare arose due to the expiration of the previous document [27] and the need to adjust regional policy in the face of new challenges.

The Ministry of Regional Development, Construction, Housing and Communal Services of Ukraine (from 29.08.2019 renamed to the Minister for Communities and Territories Development of Ukraine) on the official website on 13.03.2019 announced the start of work on the draft State Strategy for Regional Development of Ukraine until 2027.

At the moment, the regional development strategy is being developed in accordance with paragraph 1 of the fourth part of Article 8 of the Law of Ukraine "On the Foundations of State Regional Policy" and paragraph 5 of the Resolution of the Cabinet of Ministers of Ukraine dated 11.11.2015 No. 931 "On approval of the Procedure of development of the State strategy for regional development of Ukraine and the plan of measures for its implementation, as well as monitoring and evaluation of the effectiveness of the implementation of the said Strategy and plan of measures" [28] and taking into account the interests of the subjects of regional development in the State Strategy for Regional Development of Ukraine. It should be noted that the Draft of the 2021-2027 Regional Development State Strategy, formed in accordance with the requirements of the updated procedure for the development of regional strategies [28], included the need to use smart specialization in the implementation of the strategy, which is a new tool for realizing regional innovation potential based on the synthesis of the knowledge economy and local advantages, and makes it possible to determine the exclusivity of subjects, industries, regions. Technically, the approach provides for the definition of a set of competitive advantages, regional assets, scientific and innovative potentials, the use of which is effective only in the conditions of cross-functional and intersectoral interaction of all stakeholders (government, business, society , academic society). At the same time, clause 7 of Order 932 [29] determines that what is meant is that "regional strategies are developed for the duration of the State Strategy for Regional Development of Ukraine and must comply with its provisions" and in fact, regional strategies for the period 2021-2037 were developed mainly in accordance with the requirements provisions of the Draft of the 2021-2027 Regional Development State Strategy, taking into account the needs and potential of the regions and the Strategy 2030. So, basically, regional strategies for 2021-2027 were approved in the period from December 2019 to

March 2020, that is, even before the approval of the 2021-2027 Regional Development State Strategy. It is worth noting the Strategy 2021-2027 for the Dnipropetrovsk region, the draft of which was approved on May 08, 2020 based on the provisions of the already approved State Strategy, and not its draft. Since the overwhelming majority of regional strategies were based only on the draft state strategy, then with the approval of its provisions, regional strategies should be completely revised in accordance with state strategic updates.

As a result of the analysis of the content of 27 existing regional strategies, it was found that the best probability of ensuring the achievement of strategic goals was determined based on the results of the SWOT analysis of regional development. In addition, they adhere to the legal requirements regarding the inclusion of at least one strategic goal of the regional strategy, which is determined on the basis of smart specialization and provides for the innovative development of priority types of economic activities in the region, the disclosure of the potential of functional territories, and an increase in the level of competitiveness of the regions.

However, although all regions approved plans for the implementation of the strategy, the implementation plan and the procedure for monitoring the state Strategy 2030 have not been approved. That plans requires approval, since, according to the requirements, regional strategies are based on certain state priorities for strategic development and without monitoring the effectiveness of the action plan, the provisions of state strategic documents will be declarative.

Also, the improvement of already approved regional strategies continues. For example, in August 2020, proposals were made to the Development Strategy of the Poltava Oblast till year 2027 to amend the socially oriented project - ensuring the availability of social services regardless of where you live. This confirms a flexible and adaptive approach to the strategic development of regions.

1.3 Experience in the operation of functional territories

Ukraine is introducing a smart implementation approach with the support of the European Commission's Joint Research Center directorate, which has developed analytical reports on the economic and innovation potential of three pilot regions: Kharkiv, Odesa and Zapirizhzhia oblasts, where, with the support of The Joint Research Center (JRC), an innovative approach is being introduced to implement a strategic planning – smart specialization (S3) and a comprehensive assessment and analysis of data is carried out according to the methodology of the European Union for the research of economic sectors, their priority is determined according to certain criteria, the best of which will be involved in the process of entrepreneurial discovery – the second stage of introducing smart specialization.

So, according to the Strategy of Regional Development of the Zaporizhzhia Region for the period up to 2027, Strategic Goal 2 is defined - Competitive economy of the region (which develops on the basis of smart specialization) in the national and global space. It should be noted that the selection was made thoroughly and thoroughly, taking into account the available potential of types of economic activity and further opportunities for the development of related industries. and, if according to the results of a study by a group of experts from the Joint Research Center (JRC) of the European Commission, 19 types of economic activities were identified that can be recognized as having innovative potential in the Zaporizhzhia region, then according to the results of an in-depth study, only 8 types of activities were identified that can form the basis for the implementation of smart specialization in the Zaporizhzhia region. However, as a result, only 5 types of economic activities are included, namely: the production of endoprostheses, cables, equipment, motors, and so on.

In the Odesa oblast, also take palace the very carefully approached the choice of types of economic activity using smart specialization. While the initial selection included 198 industries, in total only 5 industries in the Odesa oblast showed both economic and innovation potential. The name of the strategic goals of the region was proposed: "Formation of an export-oriented food industry on the basis of innovations and the development of proposals for operational goals and objectives aimed at achieving a specific strategic goal." The priority direction of development was identified as "organic cultivation of vegetables, fruits and berries, production of processed products from organically grown products (canned vegetables, canned vegetables, pastes, sauces)". One of the effective forms of cooperation of all participants in the agrarian market, in order to minimize costs at all stages of production and marketing of agricultural products, a cluster model has been defined, which is budgetary in implementation and is based on the organizational mobility of the structure and coherence of interactions.

The Development Strategy of the Kharkiv Region for the period 2021-2027 and the Action Plan for 2021-2023 for its implementation define a separate goal: "to build a competitive and smart specialized spatial economy with high added value." Based on the assessment results, the following types of economic activities with high economic and innovative potential were identified: "production of basic pharmaceutical products and drugs, production of machinery and equipment, energy equipment, production of vehicles, trailers and semi-trailers and other vehicles".

The list of Regional Development Strategies until 2027 is given on the official website of The Ministry of Communities and Territories Development – 19 strategies, despite the fact that their overwhelming majority has been approved.

It should be noted that an updated approach to regional and sectoral development opens up new opportunities specifically for initiative regions, subjects of priority sectors of the economy, which positively affects the further formation and development of

https://doi.org/10.1051/shsconf/202110001024

functional territories that create added value and continue the chain of its formation, even at the global level.

Thus, the state's attention is focused on the economic locomotives of the region, support for the development of which will stretch the associated related objects and sectors of the national economy, which will ensure the formation of a stable in time and specialized structure of the system of high-quality innovative economy of the state.

Thus, the approach of the Cherkasy region to achieve innovative changes in the economy of its region is interesting. Despite the fact that the region was not included in the three "pilot Ukrainian regions", on its own initiative, Cherkasy region registered on the smart specialization platform and established a connection with an interested party in the European Union in order to prove competitive advantages, hidden opportunities and prospects for unique development. This once again proves the promise of self-organization and initiative in matters of strategic innovative development of regions in the direction of smart specialization.

At the same time, the Strategy for Balanced Regional Development of Vinnytsia Oblast for the period up to 2027 defines the directions of smart specialization, namely: "the development of agricultural enterprises, the field of information and communication technologies and the development of health and medical tourism in the following areas: treatment heart diseases, musculoskeletal system and neuralgia, health rehabilitation of ATO / OSS participants ". And it is indicated that only after the adoption of the strategy implementation plan at the state level, the measures for the development of innovations in the region will be concretized. By this time, "dialogue platforms" will be formed for the implementation of intersectoral and intersubjective interaction of individual functional territories.

In 2019, within the framework of the project "Smart Specialization of Industry as a Trigger of Innovative Changes", the vision of 50 subjects of the industrial sector of Ukraine (sectors of the light, pharmaceutical, food industries) and government authorities was investigated by means of sociological research, namely, surveys [24]. The types of activities were selected according to the principle of determining growth prospects when applying the concept of smart specialization, the present high innovative development and a significant contribution to the formation of added value [24]. The study revealed a hierarchy of sensitivity to intellectual specialization from the R&D stage, production and its greening, logistics, human resources, trade and so on [24]. Engineering, robotics, automation, data analysis, cloud technologies, artificial intelligence, blockchain [24], which are used in various activities to solve various, even opposite problems, are defined as the most promising technologies in terms of intelligence.

Thus, at the level of awareness in the Ukraine, smart specialization is rated at a high level and it is seen as a large-scale implementation process. The sequence of implementation of the concept of smart specialization by Ukrainian specialists and practitioners is seen in the first place at the level of functioning, and only then – at the level of development, which is logical for the promotion of the existing launch of new business processes.

However, the absence of an approved Action Plan for the implementation of the Strategy 2021-2027 disorients the development of industries and hinders the formation of functional territories. The lack of regulation of the smart specialization procedure, the limited range of certain criteria and monitoring, the lack of statistical reporting makes it impossible to substantiate reliable estimates of the implementation of smart specialization activities. In addition, the issues of financial support and self-financing remain very relevant for Ukraine.

1.4 Financial and economic potential

One of the main conditions for smart specialization is financial opportunity, since the realization of the potential is impossible without financial and economic support of the functional territory at the proper level. Consider the key financial indicators by type of economic activity in Ukraine.

The mining industry of Ukraine (Table 1) is not able to provide short-term liabilities (except 2018), since the Current Ratio<100%. Also, Cash Ratio<20%, therefore, cash and their equivalents are not enough to ensure the current obligations of the subjects of the complex. Value of index Equity-to-Assets<50%, therefore, the complex is characterized as insolvent. However, the net profit margin rises to 4.4% in 2019.

Table 1. Mining and quarrying in Ukraine [30]

Index	2015	2016	2017	2018	2019
Current Ratio, %	89,6	93,3	98,7	103	84,7
Cash Ratio,%	1	1	1,9	1,3	1
Equity-to-Assets, %	18,1	16,9	18,4	33,8	30,3
Return on assets, ROA %	-0,1	0,1	0,8	3,4	1,6
Net margins, NPM %	0	0,3	1,8	4,5	4,4

The processing industry (Table 2) is able to provide short-term liabilities, Current Ratio>100%, but not with its own funds and their equivalents, Cash Ratio<20%. The industry's copywriting capacity increases sharply, but does not reach the standard value. Profit margin has almost halved.

The liquidity of agriculture, forestry and fisheries (Table 3) is decreasing, but as of 2019 it is more than twice the standard value. However, own funds are not enough to cover current liabilities. In general, the industry is solvent, the stability margin is strong. In the context of identifying unique subjects or technologies, it can become an object of smart specialization.

Table 2. Processing industry in Ukraine [30]

Index	2015	2016	2017	2018	2019
Current Ratio, %	140,5	137,6	133,4	130,9	127
Cash Ratio,%	2,2	2,2	2,2	3	2,7
Equity-to-Assets, %	2,8	2,8	2,8	38,2	40,7
Return on assets, ROA %	4,4	4,4	4,4	2	2,1
Net margins, NPM %	3,9	3,9	3,9	1,5	1,7

Table 3. Agriculture, forestry and fisheries in Ukraine [30]

Index	2015	2016	2017	2018	2019
Current Ratio, %	357,9	356,5	305,2	258,7	233
Cash Ratio,%	7,9	6,5	4,3	3,1	2,3
Equity-to-Assets, %	79	77,9	76,6	72	69,6
Return on assets, ROA %	20,9	14	9,4	7,1	3,4
Net margins, NPM %	25,5	19,4	12,2	8,9	3,9

The trade and repair industry (Table 4) provides short-term liabilities, but cash and cash equivalents are not sufficient to meet current liabilities. The complex is characterized as insolvent, although the net margin is increasing.

Table 4. Wholesale and retail trade, repair of motor vehicles and motorcycles in Ukraine [30]

Показник	2015	2016	2017	2018	2019
Current Ratio, %	116,6	117,6	116,5	114,8	115,5
Cash Ratio,%	4,5	4,6	4,3	3,6	3,8
Equity-to-Assets, %	16,3	17,6	18,1	17,4	19,8
Рентабельність активів, ROA %	1,5	2,1	2	2,2	2,5
Чиста маржа, NPM %	0,7	0,8	0,9	0,9	1,1

In the short term, the transport and storage sector (Table 5) is liquid only in 2017-2018. Current obligations are not secured. Solvency has not reached the standard value, and profitability has tripled.

The housing and catering industry (Table 6) short-term liabilities cover almost a whole period. However, it does not cover current liabilities with its own funds. Insufficient level of solvency. The net profit margin has tripled.

Table 5. Transport, storage, postal and courier activities in Ukraine [30]

Index	2015	2016	2017	2018	2019
Current Ratio, %	96	99,4	100,7	100,8	93,6
Cash Ratio,%	4,3	3,5	4,4	4,1	3,9
Equity-to-Assets, %	36	37,6	36,6	39,1	38,4
Return on assets, ROA %	0,3	0,8	0,6	0,9	1,3
Net margins, NPM %	0,4	0,8	0,5	0,8	1,2

Table 6. The housing and catering industry in Ukraine [30]

Index	2015	2016	2017	2018	2019
Current Ratio, %	99	102,3	99	101,6	100,5
Cash Ratio,%	7,3	8,4	9,3	7,6	7,4
Equity-to-Assets, %	46,8	40,7	40	30,2	40,3
Return on assets, ROA %	0,3	0,9	0,8	1,5	1,3
Net margins, NPM %	0,2	0,5	0,4	0,6	0,6

The sector for the supply of electricity, gas, steam and conditioned air (Table 7) in terms of liquidity and solvency did not reach the standard values. There is an unprofitable activity.

Table 7. Supply of electricity, gas, steam and conditioned air in Ukraine [30]

Index	2015	2016	2017	2018	2019
Current Ratio, %	88,5	88,5	84,2	85,4	75,5
Cash Ratio,%	4,1	3,5	3,2	3,1	2,8
Equity-to-Assets, %	21,3	13	9,6	4,9	5,3
Return on assets, ROA %	-2,4	-2,1	-1,2	-0,7	0
Net margins, NPM %	-2,6	-1,5	-1,2	-0,8	0,1

Against the background of the failure to reach the standard liquidity values, the industry of water supply, sewerage, waste behavior (Table 8) is solvent. However, unprofitable activity is observed during the period under review.

The construction industry (Table 9) provides short-term liabilities, but current liabilities are not covered by its own funds. Also, the industry is characterized by insolvency. Net margin is insignificant, increased by 50% over the period and is 1.2% in 2019.

Table 8. Water supply, sewerage, waste management in Ukraine [30]

Index	2015	2016	2017	2018	2019
Current Ratio, %	95,2	85,9	94,6	88,9	93,6
Cash Ratio,%	2,8	3,8	4,9	3,3	4
Equity-to-Assets, %	71,3	65,8	66,1	64,8	67,1
Return on assets, ROA %	-0,9	-1	-0,2	-0,5	0
Net margins, NPM %	-1,9	-2,5	-0,4	-0,9	0

Table 9. The construction industry in Ukraine [30]

Index	2015	2016	2017	2018	2019
Current Ratio, %	108,6	112,3	107,8	107,5	108,4
Cash Ratio,%	2,9	4,6	3	2,3	2,3
Equity-to-Assets, %	19,6	21,4	20,5	19	18,2
Return on assets, ROA %	0,5	1,1	0,9	0,9	1
Net margins, NPM %	0,8	1,1	1	1,1	1,2

The field of information and telecommunications (Table 10) provides short-term liabilities, but covered current liabilities with its own money only in 2019. Solvency decreased and in 2019 did not reach the standard value. The industry's profitability rose to 3.3% in 2019.

Table 10. The field of information and telecommunications in Ukraine [30]

Index	2015	2016	2017	2018	2019
Current Ratio, %	152,8	156,2	139,2	133	116,3
Cash Ratio,%	12,2	11,4	16,9	17,6	20,2
Equity-to-Assets, %	60,6	56,5	56,3	53,4	48,6
Return on assets, ROA %	0,7	1,4	3,9	3,6	6,9
Net margins, NPM %	1,1	2,1	2,6	2,9	3,3

For the sphere of financial and insurance activities (Table 11), the data exceed the standard indicator of the Current Ratio by several times, but the current liabilities are not secured with own funds in 2019. On the whole, the industry is fairly solvent. The profit margin fluctuates at 1.9% in 2019.

The real estate sector (Table 12) is liquid and solvent, but it does not cover current liabilities with its own funds. The share of net profit in sales increased to 3.3% in 2019.

Table 11. Financial and insurance activities in Ukraine [30]

Index	2015	2016	2017	2018	2019
Current Ratio, %	634,5	702,2	593,9	556,5	508,3
Cash Ratio,%	26,3	30,3	32,3	23,9	19,2
Equity-to-Assets, %	86,2	87,3	84,7	83,8	82,6
Return on assets, ROA %	0	0,1	0,1	0,1	0,1
Net margins, NPM %	1,9	2,3	1,6	1,6	1,9

Table 12. Real estate operations in Ukraine [30]

Index	2015	2016	2017	2018	2019
Current Ratio, %	105	103,1	101,7	101,7	103,6
Cash Ratio,%	1,4	1,5	3,4	3,4	2,9
Equity-to-Assets, %	68	65,6	64,8	62,2	64,5
Return on assets, ROA %	0	0,1	0,2	0,5	1,2
Net margins, NPM %	0	0,5	0,8	1,9	3,3

During the period under review, the sphere of professional, scientific and technical activities (Table 13) is liquid. However, it covered current liabilities with its own funds only in 2016, and reached solvency in 2018. Profitability increased to 1.4% in 2019.

Table 13. Professional, scientific and technical activities in Ukraine [30]

Index	2015	2016	2017	2018	2019
Current Ratio, %	137,5	135,8	136	135,6	125,2
Cash Ratio,%	17,3	22,4	19	18,8	12,5
Equity-to-Assets, %	48,7	47,3	48,6	50,8	46,4
Return on assets, ROA %	0,3	0,8	0,9	0,9	0,8
Net margins, NPM %	0,9	1,5	1,6	1,7	1,4

Despite the liquidity and ability to pay in the education sector (Table 14), the net profit margin decreased from 1.9% to 0.3% over the period.

In general, one can argue about the liquidity of the prevailing majority of Ukrainian industries, but there is a problem with covering current liabilities with own money and their equivalents. Solvent are agriculture, forestry, fisheries, water supply, sewerage, waste management, information and telecommunications, financial and insurance activities, real estate transactions, education.

Table 14. Education in Ukraine [30]

Index	2015	2016	2017	2018	2019
Current Ratio, %	162,1	123,3	131,6	119,2	115,1
Cash Ratio,%	30	47,7	44,7	24,9	34,9
Equity-to-Assets, %	59,9	62,3	63,1	69,7	63,6
Return on assets, ROA %	2,6	4,4	4	1,4	0,7
Net margins, NPM %	1,9	4,1	2,4	0,9	0,3

The share of net profit in sales is insignificant, but it is increasing; it is decreasing only in education, agriculture, forestry, fisheries, and the processing industry. The unprofitable sectors are the supply of electricity, gas, steam and conditioned air, the supply of water, sewerage, and waste management. Thus, in order to implement the concept of smart specialization of functional territories of Ukraine, it is imperative to ensure positive trends in the values of key financial and economic indicators by means of synthesis of self-organization of business entities and state support. This will have a positive effect on improving the business climate in the country [31-32].

Conclutions

In the context of European initiatives in building an innovative economy, the main role is played by the formation of smart specialization of cross-functional economic activity of the territory. In Ukraine, in this context, a new Strategic long-term document of innovative development has been developed, new requirements and priorities for innovative development of Ukraine have been developed. The advantages of the new state strategy include a departure from the territorial structure and the formation of ten functional territories of different types, the development of which is focused on the renewed goals and the existing potential, the disclosure of which is provided for by appropriate measures. In the future, it is proposed to overcome the ineffectiveness of the previous strategic documents by linking the implementation of state strategic standards, requirements, plans with the volume of state funding. Therefore, in the future, only actual smart entities and their cluster-functional associations will be provided with funding. The delay in the adoption of the State Strategy disorientated regional strategic development, as a result of which regional strategies were formed only on the basis of the provisions of the draft state strategy. However, regional documents are currently being revised taking into account the new challenges of our time, new European foundations in terms of introducing smart specialization of the activities of the regions based on the use of the proposed methodology of the European Union, which is based on the principles of SWOT analysis and building a flexible structure of intersectoral interactions. One of the effective forms of cooperation is the cluster model, which is budgetary when implemented and is based on the organizational mobility of the structure and sustainable interaction. Thus, smart specialization boils down to identifying those functional territories with different types of economic activity, which are as close as possible to European standards in terms of technological, socio-ecological and economic development, are unique, can function synchronously, compete with dignity and integrate themselves into the global value chain, shape the city's budget and labor market, thereby implementing a smart new strategic development policy. Such entities do not require significant funding to bring them to the European market, which is relevant given the budget deficit in Ukraine. For the maximum possible disclosure of their potential, only a well-organized organizational and informational model of identification and presentation in the world market, initiative and interaction with European stakeholders are lacking, which can be overcome at the state level with minimal costs. Problematic issues remain the quality of the formation and implementation of the regional development strategy, the adaptation of the method for assessing smart specialization to the Ukrainian realities, the level of funding only for those functional territories whose activities are practically implemented, comply with the standards and really have innovative and unique significance.

References

1. P. McCann, R. Ortega-Argilés, Reg Stud, ***49*(8)**, 1291-1302 (2015).
2. P. Balland, R. Boschma, J. Crespo, D. Rigby, Reg Stud, ***53*(9)**, 1252-1268 (2019).
3. E. Carayannis, E. Grigoroudis, Foresight, ***10*(1 (eng))** (2016).
4. A. Varga, T. Sebestyén, N. Szabó, L. Szerb, Reg Stud, **54(1)**, 1-12 (2018).
5. L. Höglund, G. Linton, R&D Management, ***48*(1)**, 60-72 (2018).
6. S. Krammer, Technol. Forecast. Soc. Change, ***123***, 95-107 (2017).
7. R. Capello, H. Kroll, Eur Plan Stud, ***24*(8)**, 1393-1406 (2016).
8. H. Kroll, Eur Plan Stud, ***23*(10)**, 2079-2098 (2015).
9. P. McCann, R. Ortega-Argilés, Eur Plan Stud, ***24*(8)**, 1407-1427 (2016).
10. R. Hassink, H. Gong, Eur Plan Stud, ***27*(10)**, 2049-2065 (2019).
11. M. Trippl, E. Zukauskaite, A. Healy, Reg Stud (2019). DOI: 10.1080/00343404.2019.1582763.
12. M. Markkula, H. Kune, Technology Innovation Management Review, **5(10)**, 7-15 (2015).
13. M. Benner, Eur Plan Stud, ***27*(9)**, 1791-1810 (2019).
14. B. Asheim, M. Grillitsch, M. Trippl, *Smart specialization as an innovation-driven strategy for*

economic diversification: Examples from Scandinavian regions. In Advances in the theory and practice of smart specialization, pp. 73-97, Academic Press (2017).

15. Y. Petrenko, E. Vechkinzova, V. Antonov, Insights into Regional Development, **1(2)**, 118-128 (2019).
16. D. Foray, *The economic fundamentals of smart specialization strategies*. In Advances in the theory and practice of smart specialization, pp. 37-50, Academic Press, 2017.
17. E. O. Olaniyi, M. Reidolf, Journal of Security & Sustainability Issues, **5(2)**, 213–227 (2015).
18. S. Radosevic, *Assessing EU smart specialization policy in a comparative perspective*. In Advances in the theory and practice of smart specialization, pp. 1-36, Academic Press, (2017).
19. K. A. Piirainen, A. N. Tanner, L. Alkærsig, *Regional foresight and dynamics of smart specialization: A typology of regional diversification patterns*. TECHNOL FORECAST SOC, ***115***, 289-300, (2017).
20. D. Foray Structuring a policy response to a "Grand Challenge", URL: http://ec.europa.eu/invest_in_research/pdf/download_en/selected_papers_en.pdf.
21. D. Foray, P. David, B. Hall, *Smart specialization. From academic idea to political instrument, the surprising career of a concept and the difficulties involved in its implementation*. MTEI Working Paper No. **2011.001.** Lausanne, Ecole Polytechnique Federale de Lausanne (2011).
22. D. Foray, X. Goenaga, *The Goals of Smart Specialisation. Luxembourg*, Publications Office of the European Union (2013).
23. D. Foray, European Journal of Innovation Management, **17**(4), 492–507, (2014).
24. M. Sychevskiy, L. Deineko, O. Kushnirenko, O. Tsyplitska, O. Deineko, *Implementation of Smart Specialisation for Regional Industrial Development*. In III International Scientific Congress Society of Ambient Intelligence 2020 (ISC-SAI 2020), pp. 323-331, Atlantis Press, (2020).
25. Stratehiia rozvytku sfery innovatsiinoi diialnosti na period do 2030 roku. Rozporiadzhenniam Kabinetu Ministriv Ukrainy vid 10.07.2019 № 526-r [Development strategy of the sphere of innovation activity for the period up to 2030. By the order of the Cabinet of Ministers of Ukraine dated July 10, 2019 No. 526-r], URL: https://zakon.rada.gov.ua/laws/show/526-2019-%D1%80#Text [in Ukrainian].
26. Derzhavna stratehiia rehionalnoho rozvytku na period do 2020 roku. Postanovoiu KMU vid 06.08.2014 № 385 [State strategy of regional development for the period up to 2020. Resolution of the Cabinet of Ministers of Ukraine dated 06.08.2014 No. 385], URL: https://zakon.rada.gov.ua/laws/show/385-2014-%D0%BF#Text [in Ukrainian].
27. Ministry of Development of Communities and Territories of Ukraine (2019). State Strategy for Regional Development for 2021 – 2027 "Human-centered Development and Unity". Ministry of Development of Communities and Territories of Ukraine website, URL: http://www.strategy.kharkiv.ua/docs/SSRD.pdf. [in Ukrainian].
28. Pro zatverdzhennia Poriadku rozroblennia Derzhavnoi stratehii rehionalnoho rozvytku Ukrainy i planu zakhodiv z yii realizatsii, a takozh provedennia monitorynhu ta otsinky rezultatyvnosti realizatsii zaznachenykh Stratehii i planu zakhodiv: Postanova Kabinetu Ministriv Ukrainy vid 11 lystop. 2015 r. № 931 [On approval of the Procedure of development of the State strategy for regional development of Ukraine and the plan of measures for its implementation, as well as monitoring and evaluation of the effectiveness of the implementation of the said Strategy and plan of measures: Resolution of the Cabinet of Ministers of November 11, 2015 No. 931]. URL: https://zakon.rada.gov.ua/laws/show/931-2015-p/ed20181117#n76 [in Ukrainian].
29. Poriadok rozroblennia rehionalnykh stratehii rozvytku i planiv zakhodiv z yikh realizatsii, a takozh provedennia monitorynhu ta otsinky rezultatyvnosti realizatsii zaznachenykh rehionalnykh stratehii i planiv zakhodiv: Postanova Kabinetu Ministriv Ukrainy vid 11 lystopada 2015 r. № 932. Ofitsiinyi visnyk Ukrainy, 2015 r., № 92, st. 3131 [The procedure for developing regional development strategies and action plans for their implementation, as well as monitoring and evaluating the effectiveness of the implementation of these regional strategies and action plans: Resolution of the Cabinet of Ministers of Ukraine dated November 11, 2015 No. 932. Official Gazette of Ukraine, 2015, No. 92 Art. 3131] [in Ukrainian].
30. Ofitsiinyi sait analitychnoi systemy dlia komplaensa, analizu rynkiv, dilovoi rozvidky i rozsliduvan Youcontrol [Official site of the analytical system for compliance, market analysis, business intelligence and investigations Youcontrol]. URL: https://youcontrol. com.ua/(accessed: 05.11.2020). (in Ukrainian)
31. Ocheretin D., Los V., Kucherova H., Bilska O. *An alternative approach to modeling the country's business climate in conditions of limited information*, in E3S Web of Conferences, **166,** 13024, (2020).
32. A. Uchitel, H. Kucherova, A. Didenko, O. Kravets, Y. Honcharenko, *Scenario forecasting information transparency of subjects' under uncertainty and development of the knowledge economy*, M3E2-MLPEED 2020, Odessa, Ukraine, July 13-18, **2713**, CEUR-WS.org, online (2020),

URL: http://ceur-ws.org/Vol-2713/paper05.pdf

Cultural and Mental Peculiarities of Innovation Processes' Development in Ukraine in the Context of Industry 4.0

Andrii Karpenko[1*], *Tetiana* Naumchuk[1] and *Natalia* Karpenko[1]

[1]"Zaporizhzhia Polytechnic" National University, 69063, Zaporizhzhia, Ukraine

Abstract. The article deals with cultural and mental peculiarities of innovation processes' development in Ukraine in the context of Industry 4.0. There has been presented a comparative rating of some world countries according to NRI, ICT, GOI, IDI, GII indices and the position of Ukraine over the past decade has been stated. The main results of innovation activity in Ukraine that lie in low innovative activity have been characterized. The factors that hinder innovation activity are insufficient level of educational training, low activity in offering innovative ideas, no incentive to create something new etc. In addition, a significant negative impact on innovation activity is exerted by cultural and mental factors of the population, a considerable part of which does not have entrepreneurial skills. There have been substantiated some socio-cultural and mental characteristics of Ukrainians that determine their innovative activity, namely reduce the competitiveness of the country's human capital both at the personal and national levels. The solution of the problem has been proposed which implies cooperation between three main components of the innovative economy – "science – education – production" as a common system.

1 Introduction

Changes in the modern world are taking place at a very rapid pace, are mostly radical in nature and are accompanied by the emergence of self-driving cars, 3D printers, nanotechnologies, etc. Leading positions in the modern economy are occupied by industries that are based on the use of information technologies. For a long time, the scientific community has been talking about the onset of the Fourth Industrial Revolution, which will change the labor market, economy and education in general: "the Fourth Industrial Revolution can "robot humanity" and for many people it will irreparably change the way their work, environment, family life and identity itself look" [1].

The concept of Industry 4.0 provides increasing automation of absolutely all processes and stages of production: from product design to delivery to the end user, taking into account the requirements of the latter. Industry 4.0 also implies a well-established production without human intervention, ensuring maximum efficiency [2], which is rather contravercial though.

Industry 4.0 is an established global innovation program aimed at making manufacturing facilities more intelligent, efficient and flexible [3]. Technologies 4.0 are among the priority factors of competitiveness and the leaders of industrial enterprises have a justified need for more investment in the development of new technologies, as all / most of industrial enterprises depend on digital transformation [2].

We cannot but agree that "given the complexity and scope of industrial digitalisation, much of the academic research focuses on technological and organisational problems. Addressing "human factors" and in particular conceptualising and measuring human readiness for digitalised work receives less attention, and remains an early-stage project. Nevertheless, this project is crucial to the success of Industry 4.0 because aside from a few 'dark factory' scenarios, humans are considered more adaptive than machine entities and vital to future production". The technology has made possible new products and services that have generated significant transformations in both personal and professional life, emphasizing the interaction between machines and people [4]. Accordingly, a person should change and at the same time perceive the environment, resources and attitude to education and production in a new way.

It should be noted that the Industrial 4.0 Revolution inevitably causes some disadvantages. The reduction of creativity may occur due to the automation of machines and the disappearance of human capital from the production technology. Human capital can generate controversy with its creativity, while devices are optimized, set up, and programmed to perform, despite the errors that may occur [5].

To be competent in the modern world, to adjust to the changing environment, people have to acquire new skills which contribute to the human capital. Among them the researchers identify *technical knowledge* and abilities, which imply computer programming / coding abilities, data and information processing and analytics, specialized knowledge of manufacturing activities and processes, organizational and process understanding, interdisciplinary / generic knowledge about technologies, statistical knowledge; *managerial competencies*, which involve creativity, entrepreneurial thinking, problem / conflict solving, decision making, research and analytical skills, efficiency orientation;

* Corresponding author: karpenko.a.v2@gmail.com

social competencies, which include intercultural and language skills, communication skills, leadership skills, ability to be compromising and cooperative, to work in a team, to transfer knowledge and accept change [6].

Besides, the notion of entrepreneurial activity intermingles with the national culture, which is the set of shared values, beliefs, and expected behaviors. Researchers distinguish three broad research streams that address national culture and entrepreneurship. The first research stream focuses on the impact of national culture on the aggregate measures of entrepreneurship such as national innovative output or new businesses created. The second stream addresses the association between national culture and the characteristics of individual entrepreneurs. Within this stream of literature, researchers have examined the values, beliefs, motivations, and cognitions of entrepreneurs across cultures. The third stream explores the impact of national culture on corporate entrepreneurship [7].

In the study the dialectical and empirical methods were applied to consider the evolution of cultural and mental traits of the population, the consequences and possible reserves of innovation activity's growth. Methods of comparative analysis and dynamics series were used to compare actual data of the reporting and previous years, identify changes in indicators and parameters of innovative enterprises' functioning. Materials from international organizations for evaluating indices and ratings of Ukraine and individual European countries, as well as the state statistics service of Ukraine, were used for the analysis.

2 Main results of the research

2.1 The level of innovativeness of Ukraine among European countries

An important indicator of a country's potential and development opportunities in the information society is the network readiness index (NRI), which characterizes close relationship between ICT development and economic well-being. ICTs today permeate all areas of activity and improve their quality and effectiveness. The leading role of ICTs in developing innovation, increasing productivity and competitiveness, diversifying the economy and stimulating business activity is increasingly recognized today, which contributes to development and is possible with the effective use of intellectual assets of human potential.

According to the Global Information Technology Report published by the World Economic Forum, Ukraine ranked 71st among 143 countries in 2015 in terms of ICT development based on the Network Readiness Index (NRI) (fig. 1). Singapore and Sweden were the leaders of the rating on the network readiness index in 2015-2016, while Ukraine was located between Trinidad and Tobago and Kuwait.

In 2020 Ukraine took the best position in the rating, which proves positive changes in the national economy. In addition, since 2019, the NRI has been revised to reflect how technology and people need to be integrated into effective governance structure in order to make the right impact on our economy, society, and environment. NRI has become focused on digital transformation and is calculated based on 60 variables.

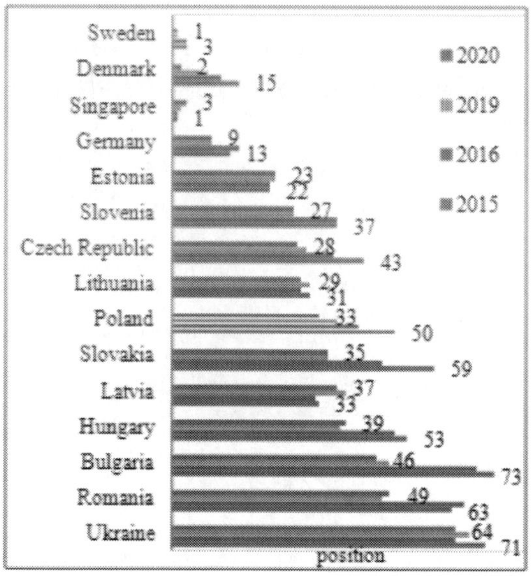

Fig. 1. Ukraine's position among different world countries in terms of Network Readiness Index in 2015-2020.
Source: built on the basis of [8, 9, 10]

The presented comparative rating of some world countries proves significant reserves in Ukraine for improvement, since European countries have significantly better conditions for the use of ICT and, accordingly, the use of intellectual assets of human potential [11].

The reason for Ukraine's low position in the rating is still the lag in terms of components that characterize the political and regulatory environment and the low level of ICT use by the government, and the advantage is the availability of ICT. The components of the network readiness index of Ukraine represent a clear upward trend, but the growth rate in the world for the development of network readiness is slightly higher than in Ukraine, which confirms a low level of current changes that do not contribute to the highest level of use of intellectual assets of human potential.

The ICT Development Index (IDI) of Ukraine for 2010-2017 increased by 1.21, but the country's position in the corresponding rating worsened from 69 to 79 (176 countries of the world in 2017 and 167 countries of the world in 2010) (fig. 2). That is, despite positive changes in the resource supply and level of ICT use, as well as the development of ICT skills, their growth rates are lower than required by modern world changes. This is what proves the insufficient level of use of intellectual assets of human potential in comparison with the developed countries.

Ukraine shows positive dynamics in certain indicators of the ICT Development Index (ICT Development Index – IDI), based on the assessment of access to ICT, the use of ICT, as well as ICT skills, i.e. practical knowledge of these technologies among the population. An increase in the share of families who have a computer; an increase in the share of residential premises connected to the Internet, the share of the population who use the Internet are the factors, which give grounds to assume more intensive use of ICT, including in the educational sphere. According to the ICT Development Index for 2015-2017, the first places were occupied by South Korea, Denmark, Sweden.

In 2010-2020, Ukraine's position in the Global Opportunity Index (GOI) rating, which takes into account a set of factors that contribute to attracting foreign direct investment (main economic indicators; ease of doing business; quality of regulation; rule of law), worsened by 5 positions.

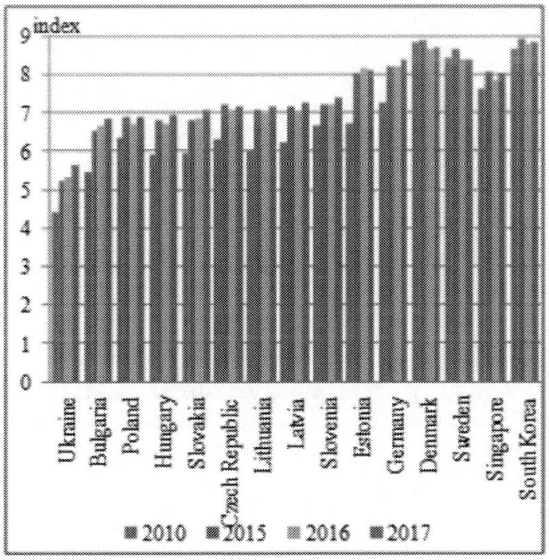

Fig. 2. Dynamics of the ICT development index (IDI) in Ukraine, selected CIS countries and Eastern Europe.
Source: based on data from the International Telecommunication Union [12]

However, according to the ICT Development Index in Ukraine for 2008-2017, the position worsened by 20 positions (fig. 3).

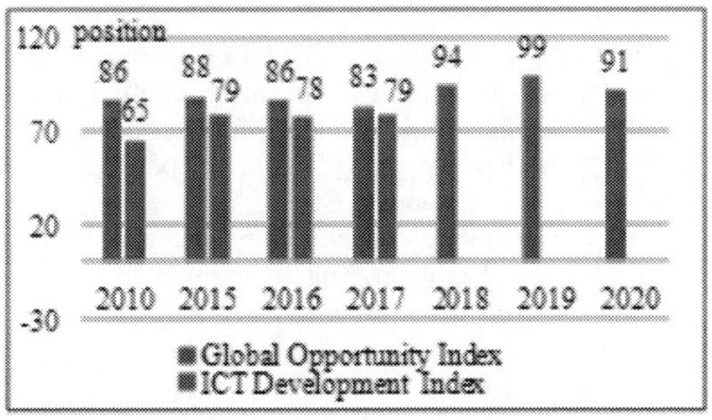

Fig. 3. Global Opportunity Index (GOI): changing Ukraine's position in the rating.
Source: based on data from the International Telecommunication Union [12]

According to the Global Innovation Index (GII) in 2020, covering 131 economies of the world, Ukraine took one of the highest positions in the last 10 years – 45th place (fig. 4).

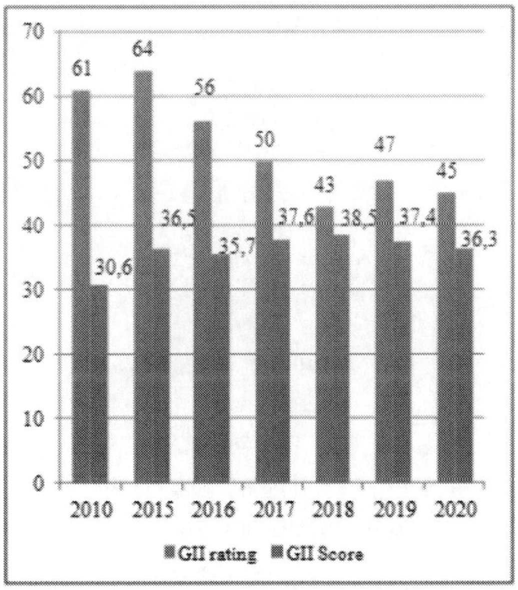

Fig. 4. Dynamics of the Global Innovation Index (GII) and rating of Ukraine.
Source: built on the basis of data [13]

The study of the problem shows that in recent years the improvement of positions is due to a high coefficient of innovation efficiency, that is, the ratio of the result obtained to innovative resources. Ukraine has demonstrated the highest indicators of innovation in education, science and business, but institutions and infrastructure remain the least innovative.

In 2018, Ukraine ranked first in the below – average income group, and in 2017 – the second one after Vietnam. Switzerland, Sweden, the Netherlands, the United States, Great Britain, Denmark, Singapore, Finland, Germany and Ireland remain the leaders for several years in a row, while Ukraine lags behind the countries of Eastern Europe (fig. 5).

Human capital is the basis of Ukrainian innovation competitiveness, which in 2010-2020, although it decreased in the rating from 36 to 39 positions, but consistently remains decisive in the formation of the GII. Its indicator reduced due to the decrease in public spending on education and science. The fundamental factor hindering innovative development is the low level of R&D costs, which leads to the search for other sources of funding and the migration of scientists outside of Ukraine. Insufficient development of infrastructure and innovation chains, unfavorable political situation and environmental sustainability also have a negative impact. A significant role also belongs to creativity, which in 2020 was on the 44th position (intangible assets, creative goods and services, online creativity) (fig. 6).

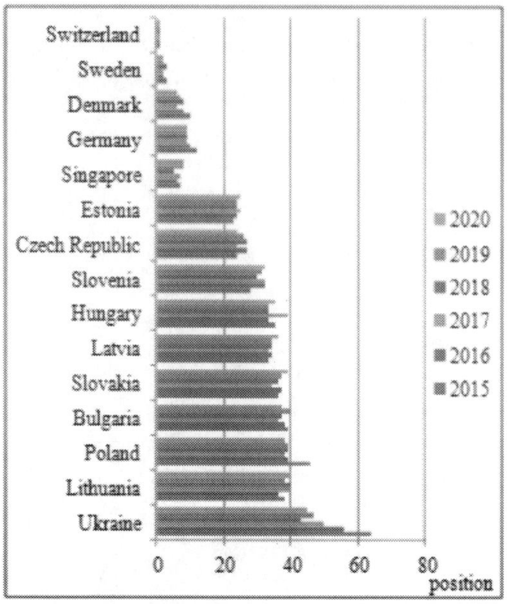

Fig. 5. Rating of the Global Innovation Index (GII) of Ukraine and some world countries.
Source: built on the basis of data [13]

The greatest positive changes for the entire period were achieved in terms of the effectiveness of scientific research (moving Ukraine from 75 to 25 positions in the rating, including in terms of creation, influence and dissemination of knowledge). According to the business experience sub-index, Ukraine has risen by 20 positions and ranked 54th in 2020 (in terms of the number of knowledge workers, innovative connections, and knowledge perception).

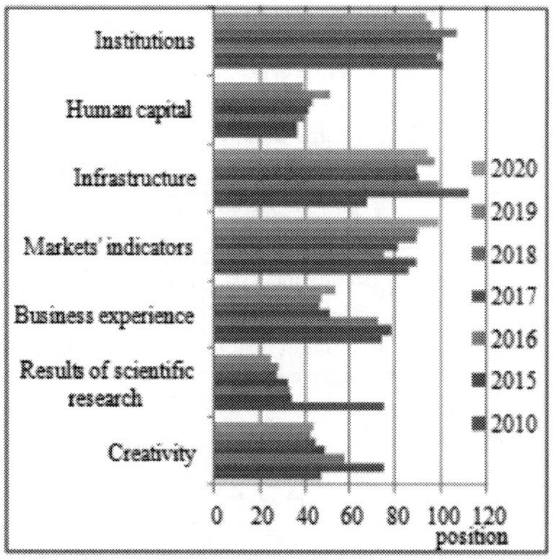

Fig. 6. Dynamics of sub-indices of the Global Innovation Index (GII) for Ukraine for 2010-2020.
Source: built on the basis of data [13]

Infrastructure development and market indicators (in terms of loans, trade and competition, investment), as well as the "institutions" sub-index (political environment, regulatory environment, business environment) still have the worst results of the reform in Ukraine.

2.2 Main results of innovation activity in Ukraine

In 2000-2019, the number of enterprises implementing innovations decreased by 530 units, or more than 2 times. The share of industrial enterprises that introduced innovations increased by 2.3%. Fluctuations in these indicators are partly due to changes in the research methodology, although they have never approached the threshold value (25%) and are much lower than the same indicator in developed countries (70-80%) (fig. 7).

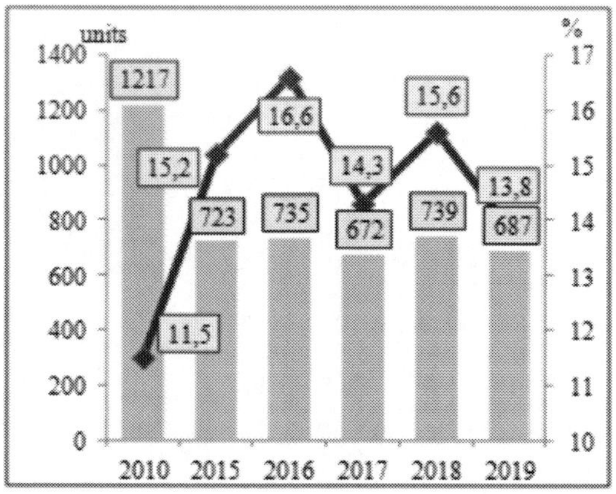

Fig. 7. Number of enterprises that implemented innovations (units) and its share in the total number of industrial enterprises (%).
Source: built on the basis of data [14]

In 2010-2019, the number of enterprises selling innovative products in Ukraine decreased by 522 (from 964 to 442), the share of which in the total volume of industrial products is very low (decreased from 3.8 to 1.3%).

A significant obstacle to increasing the volume of commercialization of innovations in Ukraine is the low innovative activity of entrepreneurship (fig. 8).

In 2005-2017, the share of enterprises that were engaged in innovative activities and introduced innovations increased by 2 and 2.3 percentage points, respectively, while the share of enterprises that sold innovative products new to the market decreased by 2.2 percentage points. That is, in the domestic industry, mainly innovative products are introduced, which are new only for the enterprise.

Consequently, the indicators of innovation activity in Ukraine are low compared to current global trends, and low level of potential demand for innovative products and return on investment explains small volume of innovation implementation.

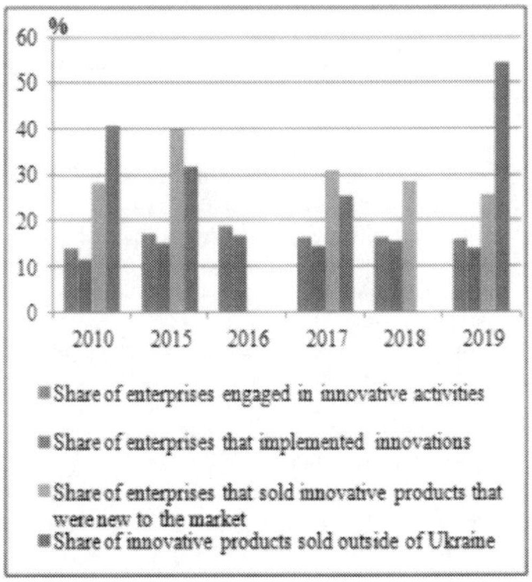

Fig. 8. Dynamics of innovative activity of industrial enterprises in Ukraine, %.
Source: built on the basis of data [14]

In 2010-2019, there is an ambiguous trend in Ukraine, when peak volumes occurred in 2008. indicators of innovation activity in Ukraine are low compared to modern global trends, respectively, therefore, the indicators of innovation implementation in Ukraine are too low (fig. 9).

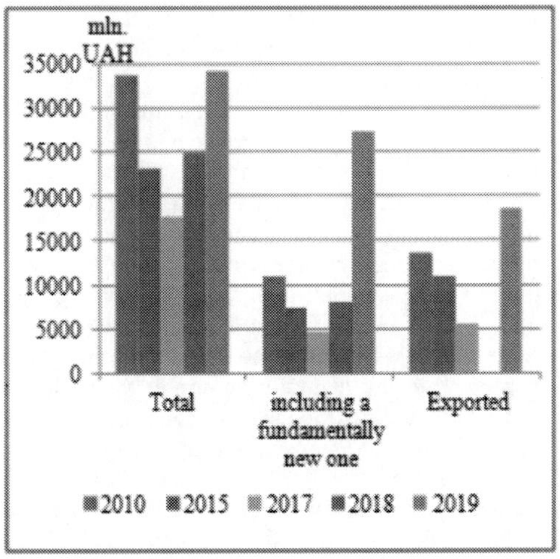

Fig. 9. Volume of innovative products sold in Ukraine, mln. UAH.
Source: built on the basis of data [14]

In recent years, the share of enterprises that implemented innovations in Ukraine has decreased. Among the areas of innovation implementation, the development of innovative types of products prevails, the number of which decreased by 260 units in 2010-2019. The volume of new technological processes implemented increased by 275 and generally did not have any drastic changes.

After some recovery in 2015, the number of technologies implemented, including low-waste and resource-saving ones, fell below the level of 2010. Perhaps, not the least role here was played by the fact that enterprises began to purchase fewer new technological processes (fig. 10).

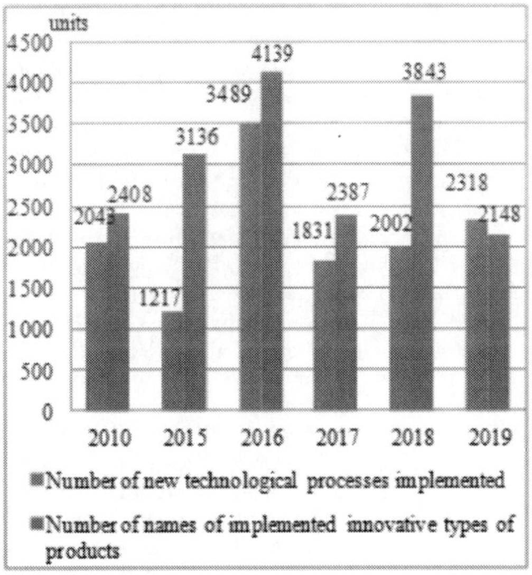

Fig. 10. Volume of innovation implementation at industrial enterprises of Ukraine, units.
Source: based on data from the State Statistics Service of Ukraine [14]

It is worth noting that many enterprises do not use innovative developments due to a number of factors that hinder innovation activity: insufficient level of educational training; low activity in offering innovative ideas; there is no incentive to create something new.

These indicators of innovation activity of industrial enterprises in Ukraine indicate the lack of NIS development, a stable development strategy of the country, which causes instability and situativeness of the development of innovative factors of economic growth. In general, the indicators of innovative and technological development of the real sector of the economy have remained almost stable in recent years. This highlights the existing pattern regarding the importance of the stabilizing role of innovation factors in economic development.

2.3 Cultural and mental characteristics of Ukrainians

In the modern conditions of world economy globalization, sharp increase in competition, spread and development of innovations, attitude to the main productive force of society – man, to his value and motivational activities is changing. Nowadays competitiveness and development are mainly determined by the availability of innovative proposals and their rapid implementation. The only factor of production that has a creative component and acts as a generator of innovative ideas and their performer is a person (human resources). In this regard, there is a need to form and implement motivating measures at the macro, meso and micro levels of economic activity to activate the innovative activity of human resources. The issue of creating a favorable environment that would encourage the population to innovate and perceive innovative changes becomes urgent.

On the one hand, Florida (2002) in his work "The Rise of the Creative Class" [15] puts a great emphasis on creativity laying in the conducted activities or processes beyond currently accepted standard and being the decisive source of competitive advantage. The so-called "creative class" members have certain lifestyles distinct from other classes, such as: individuality, diversity, authenticity, identity, casual instead of intimate social interaction, and self-defining.

The three key factors that influence regional growth are 3T's (tolerance, which is the openness to creative people worldwide; talent is represented by the relative concentration of the creative class and Talent Index (the percentage of bachelor holders); and technology the proxies for which are Innovation Index (patents granted per capita) and High-tech Index) [16].

On the other hand, motivations for becoming an entrepreneur have generally been categorized as either push/pull situational factors or personal characteristics. A number of studies have come to the conclusion that new venture initiation often occurs as a result of situational pushes or pulls that include frustration with present life-style, childhood, family environment, education, age, work history, role models, and support networks. Some individuals are pushed into entrepreneurship by negative factors such as dissatisfaction with existing employment, loss of employment, and career setbacks [17].

Innovation activity in Ukraine is officially one of the priorities, however, its support and incentives by the state are often only declared. Due attention is not paid to the formation and increase of innovation activity. In addition, a significant negative impact on innovation activity is exerted by cultural and mental factors of the population, a significant part of which does not have entrepreneurial skills and or is not able to use them.

In such conditions, increasing the innovation activity of the population and business entities is one of the most pressing issues for the development of the Ukrainian economy, since innovations create the basis for the competitiveness of the country and individual business entities. Significant priorities of innovative activity of enterprises are innovative labor and intellectual assets of human potential, determined by the innovative activity of the individual.

Economic behavior of a person and his innovative activity is determined by many factors, among which the leading role is played by public consciousness, which is formed under the influence of generally accepted values, traditions, socio-cultural and ethnic characteristics. Among the various components of behavioral influence on the life of a person and an individual nation, an important role, in our opinion, is played by mentality.

We could associate mentality with corporate culture. These two notions correlate at macro and micro levels and imply the following aspects:
- behavioural aspects in the communication of people: language, customs, and traditions, rituals;
- collective norms: unspoken standards and values adopted in the organization/society;
- official rules: publicly stated principles, which the organization/society tries to follow;
- formal philosophy: a variety of ideological and political principles governing the activities of the organization/society;
- unspoken rules of the organization/society;
- climate: the feeling transmitted to the organization through the external environment and the forms of interaction between employees;
- established skills: demonstration of certain skills in the performance of the task;
- the way of thinking: the way of thinking shared by employees of the company/members of the society, which determines the perception, thoughts, and language of the organization/people;
- a common understanding: arises when members of an organization/society communicate with each other;
- metaphors and integrated symbols: ideas, feelings, and images, through which the organization/society itself characterizes [18].

We consider it reasonable to use the term of "economic mentality", as the concept reveals the activity of a person (collective or community) in order to receive a certain reward. Since the country's economy is based on economic activity (any socially useful activity, the results of which are sold as a commodity), the use of the concept of "economic mentality" will be important in this direction.

We can say that the economic mentality is a certain consciousness of a person or the population as a whole, which develops historically, manifests itself in the unity of conscious and unconscious values, norms and attitudes reflected in the labor behavior of the population and expresses a system of value orientations and self – determination of the individual in the social, legal, political, historical and professional space.

In the post-Soviet space in general, and in Ukraine in particular, the mentality is one of the main problem that greatly affects the current economic situation. That is, significant transformations in the society and high rates of world knowledge are constantly integrated into the modern mentality and become one of the dominant factors of a developed economy [11]. Since the transformation of mentality is a long-term process, that is why the Ukrainian society is characterized by a certain conservatism of individual and social consciousness with its positive and negative characteristics, which to a certain extent hinders the socio-economic development of the country.

Let's consider through the prism of mentality the conditions for the development of intellectual assets of the human potential of Ukraine, namely, the education of society and the availability of highly qualified personnel.

According to the World Bank ratings, Ukraine has continued to occupy the first positions in terms of general education of citizens in recent years.

The high level of the education index (average and expected duration of study) has a positive impact on the image of the country, since the current development of the economy depends on the ability to use new knowledge and innovations. The high education index of Ukraine indicates a high education of the population, however, the lag in coverage and quality of education is noticeable.

Ukraine has received such a high level of education of the population precisely because of its mental characteristics, since traditionally it has always been appropriate for the population to think about the determining role of education (especially higher education) in ensuring a successful career, even if this was not always confirmed by real knowledge and professional skills. Such priorities were significantly influenced by the system of power formation during the planned economy. Accordingly, the significant popularity of higher education, which was formed at the mental level of the population in the absence of effective state monitoring of the quality of education, regulation and planning in the educational sphere, caused the discrepancy between the quantitative and qualitative characteristics of the domestic education system and the needs of the economy. The current problem of Ukrainian society is precisely that a high level of education in domestic practice is not always identical to professionalism and cannot be the key to improving the economic situation.

As a result, the problems of disparity between supply and demand in the labor market and the discrepancy between the education system and the needs of the labor market in terms of qualification level and professional structure are constantly becoming more acute in Ukraine. This leads to a shortage of employees of certain professional groups, in particular workers. The structure of modern domestic education is characterized by a shift in vocational education in favor of higher education, which has led to an asymmetry of the labor market. The actual lack of a link between the level of education received and the level of income has led to a formal attitude to receiving education and a significant decrease in its quality. Accordingly, the level of intellectual assets of human potential remains insufficiently variable or slowly increases. Consequently, significant opportunities for further innovative development are lost, which can lead to irreversible destructive phenomena in the domestic economy.

Innovation activity in Ukraine remains very low. Peculiar are a number of negatives: paternalism of the majority of the Ukrainian population; social and political conformism; disregard for health and health care system; presence of "equalization" in wages and its low level, as a result – low quality of work and general negligence; lack of a clear national idea [19]. These traits significantly reduce the competitiveness of the country's human capital both at the personal and national levels, and do not contribute to the development of intellectual assets of human potential.

An important component of the new type of national mentality in the digital economy, in our opinion, is the susceptibility to change, awareness of the importance and necessity of innovative development for further progress. Scientists and practitioners are optimistic that there are still many people among the domestic population who are capable of innovative behavior and innovative activities. Taking this into account, an important task is to prioritize the development and implementation of comprehensive measures to improve the level and quality of higher education and stimulate innovative activity of the population and the corresponding development of intellectual assets of human potential. This, in our opinion, will contribute to the development of an innovative society, increase the volume of production and commercialization of innovations.

2.4 Development of higher education and its impact on the Ukrainian economy

Education, which is a process of mastering knowledge and skills in the context of socialization of the individual, aimed at reproducing the intelligence of the nation, determines its key place among other economic sectors. The level of education is a reflection of economic progress and ensures the country's competitiveness in the global market. Therefore, in the context of the digital economy in Ukraine, one of the determining conditions for development is education. In interaction with the labor market it can create prerequisites for the formation of human capital and commercialization of innovations as the basis for effective socio-economic innovation development of the country, since knowledge and qualifications are important factors for ensuring and implementing the socio-economic progress of society.

Education is not only a connecting, but also a constructive link in the system of three main components of the innovative economy – "science – education – production". The educational potential acts simultaneously as a source of replenishment of science with personnel and as the main factor in mastering them, modern knowledge necessary to ensure the economic, social and cultural development of society based on the use of advanced achievements of science, technology, innovation [20].

The processes of modernization of the domestic economy and society as a whole, focused on the development of market relations in the digital economy, significantly affect the development of educational institutions. Accordingly, the processes of commercialization in the educational sphere are spreading, which gradually requires the existence of new rules of interaction, status-role relationships, value and normative formations [21]. However, modern transformations are accompanied by negative trends in the development of the education system and the use of accumulated intellectual assets. This is due to the devaluation of knowledge and diplomas, social insecurity of intellectual labor and a decrease in public requests for specialists, weakening and loss of ties between educational and industrial structures, etc. [22].

According to the national strategy for the development of education in Ukraine for 2012-2021 [23], education belongs to the most important directions of the state policy of Ukraine, because it is a strategic resource for socio-economic, cultural and spiritual development of the society, improving the well-being of people, ensuring national interests, strengthening international authority and forming a positive image of our state, creating conditions for self-realization of each individual. However, it is very slow to implement certain priorities in real life. As a result, the unsatisfactory state of domestic education and its inconsistency not only with the European and global, but also with the domestic labor market.

A significant innovative component of the domestic educational space is technical training, which can be roughly represented by four groups: vocational, special technical, higher technical and postgraduate education. Today, it is not customary to classify technical specialties of higher education in Ukraine as elite, since the humanities are in the lead. However, if we assess the real picture in the labor market of Ukraine and its regions, about 40-50% of the unemployed are young people, most of whom have mastered humanitarian specialties. In addition, the basis of the economic system – its industry – suffers significant losses, so we can confidently say that the overwhelming number of humanitarians in comparison with engineering workers does not benefit the state as a whole and separate region.

In the Ukrainian society, there is no demand for high-quality education in the field of exact and engineering sciences, as evidenced by entrance campaigns in recent years. However, it is known that in the modern world, the competitiveness of the state is determined by the number of physicists, mathematicians, engineers, designers and technologists.

In Ukraine, the largest share of people with full higher education is concentrated in the financial sector (59.7%), public administration (59.5%) and education (47.3%) [24]. Accordingly, these areas are characterized by high social status and economic prestige, especially for young people with higher education. As a result, the training of specialists in these areas exceeds the demand for them from the national economy, and in technical specialties, on the contrary, it does not meet the demand.

Taking into account the current situation, the Ukrainian government in recent years has increased state order for technical specialties (electronics, energy, mechanical engineering, radio engineering, radio-electronic devices and communications, metallurgy and materials science, chemical technologies and engineering, specialties related to aviation, transport infrastructure) and reduces management, economics and law. However, even these measures do not make a proper shift.

Negative for Ukraine is not just the process of oversaturation of the labor market itself, but also the wasted part of the state budget (training by state order), which is used to train workers and specialists who are not competitive in the labor market. However, until now, the country has not created an effective monitoring system that would contribute to forecasting and marketing of educational services. The system of vocational education is negatively affected by insufficient involvement of employers and other social partners in the development and approval of educational standards, curricula and programs, and the lack of incentive lending to educational institutions for obtaining equipment and other equipment. The main reasons for problematic situation with the graduates' employment are the following: slow pace of structural changes in the economy, which leads to significant imbalances in the structure of demand in the labor market; increased competition in the labor market of professions and specialties for which training is conducted in educational institutions; non-compliance of the received specialty with the needs of the labor market; lack of practical experience or work experience in the specialty.

An important direction for the development of higher education and improving its quality, as well as the formation of an innovative society in Ukraine, is the integration of science, higher education and business as a process of accumulating resources for the development of the knowledge economy [25]. The formation and effective use of the intellectual basis of human capital takes place in these processes. Integration is able to solve modern socio-economic problems of the domestic economy while uniting around the solution of common urgent tasks, since it is accompanied by the establishment, complication and strengthening of important ties. It allows you to get a number of positive effects (stimulating innovation activity, producing and commercializing innovations, optimizing the use of resources and achieving the highest level of well-being), although it reduces the level of intersectoral competition.

It is proved that integration is mutually beneficial for each of the interacting parties, since it is determined by the needs of society: for scientific organizations (development of the material and technical base and acceleration of commercialization of scientific developments; exchange of innovative developments; expansion of access to information resources); for higher education institutions (attracting teachers, students to research work on the most pressing development problems; improving the level of education by providing access to the latest equipment; strengthening the human potential of higher education institutions and increasing the level of their material support; development of the material and technical base of scientific research; optimization of the educational process; attraction of additional financial resources; rapid implementation of research results in the educational process; further employment of graduates); for production (training on the basis of higher education institutions' qualified personnel who are able to be involved in the production process with a minimum adaptation time; direct access to developments necessary to create competitive products, and the ability to adjust them in accordance with needs); for the state (the ability to more efficiently allocate limited resources and activate the economic activity of the population and subjects of the national economy).

Thus, close cooperation between higher education and business is a realistic way of innovative development of the country and an effective means of ensuring competitive production with highly qualified personnel. The formation and development of integration links of science, education and production into a common system can organically combine and ensure more effective functioning of each of these spheres of human activity and will contribute to the development of intellectual assets of human potential.

Conclusions

To conclude, the presented comparative rating of some world countries proves significant reserves in Ukraine for improvement, remaiming still at lower positions of using intellectual assets of human potential than required by modern world changes.

The indicators of innovation activity in Ukraine point out the lack of a stable development strategy of the country, which causes instability and situativeness of the development of innovative factors of economic growth. In general, the indicators of innovative and technological development of the economy's real sector have remained almost stable in recent years. This highlights the existing pattern regarding the importance of the stabilizing role of innovation factors in economic development.

It has been determined that the cultural and mental characteristics of Ukrainians have a significant impact on the development of intellectual assets of human potential, innovation activity and the effectiveness of innovation activities. In particular, the negative impact of the historical heritage of Ukrainians remains noticeable, when private property was officially absent and a negative attitude to it in the society and business activities was formed. The real disproportionate distribution of benefits, which remains today, has led to a significant loss of trust in the modern Ukrainian society and hinders the development of cooperation, collaboration and integration of various institutions: within the public, education and science, business and government. And today, despite the high level of education coverage, scientific research (being fundamental ones, the findings are sometimes difficult to perceive by the society, do not take into account market demands and global trends, or lack the possibility of practical usage), the level of implementation of intellectual property objects in practice and commercialization of innovations remain low.

In the context of radical technological, economic, and social changes that are taking place in the society, a traditional role of universities as research centers that create new knowledge and training centers for highly professional personnel should be supplemented by an important function - innovative one, which allows to fully integrate the higher education system into society, "open" it to production and the labor market. In other words, there will be more favorable conditions for the formation and effective use of intellectual assets of human potential.

Modern high education institutions that have the resources to fully implement innovation process should become active participants of innovation activity. The latter should be aimed at activating the innovative thinking of students, postgraduates, scientists, and teachers. Higher education institutions should become centers for the emergence and spread of technology transfer processes. That will provide them with the role of one of the key elements in the innovation infrastructure, which acts as a link between science and business.

References

1. K. Schwab, *The Fourth Industrial Revolution* (World Economic Forum, Geneva, 2016). ISBN 978-1-944835-01-9
2. V. Vanzha, A. Kukharuk, Staging the process of enhancing of competitiveness of machine-building enterprises in the conditions of Industry 4.0. Cur. Probl. of econ. and manag. **14** (2020)
3. F. Orellana, R. Torres, From Legacy-Based Factories to Smart Factories Level 2 According to the Industry 4.0. Intern. J. of Comp. Integr. Manuf. **32** (4-5), 441-451 (2019). doi: 10.1080/0951192x.2019.1609702
4. T. J. B. Blayone, R. van Oostveen, Prepared for work in Industry 4.0? Modelling the target activity system and five dimensions of worker readiness. Intern. J. of Comp. Integr. Manuf. **32** (4-5), 441-451 (2019). doi: 10.1080/0951192X.2020.1836677
5. V. Sima, I. G. Gheorghe, J. Subi´c, D. Nancu, Influences of the Industry 4.0 Revolution on the Human Capital Development and Consumer Behavior: A Systematic Review. SUSTDE. **12** (2020). doi: 10.3390/su12104035
6. A. Łupicka, K. Grzybowska, Key Managerial Competencies for Industry 4.0 - Practitioners', Researchers' and Students' Opinions. Logistic and Transport. **3**(39), 39-46 (2018). doi: 10.26411/83-1734-2015-3-39-4-18
7. J. C. Hayton, G. George, S. A. Zahra, National Culture and Entrepreneurship : A Review of Behavioral Research. ETP. 26(4), 33-52 (2002). doi: 10.1177/104225870202600403
8. S. Dutta, B. Lanvin. The network readiness index 2020. Accelerating Digital Transformation in a post-COVID Global Economy. Portulans Institute. (2020). ISBN 978-1-63649-055-7

9. S. Dutta, B. Lanvin. The Network Readiness Index 2019: Towards a Future-Ready Society. Portulans Institute. (2019). ISBN: 978-0-578-62273-6
10. S. Baller, S. Dutta, B. Lanvin. The Global Information Technology Report 2016 Innovating in the Digital Economy. World Economic Forum, Geneva (2016). ISBN: 978-1-944835-03-3
11. A. V. Karpenko, *Rozvytok intelektual'nykh aktyviv lyuds'koho potentsialu: teoriya ta praktyka* (The Development of Intellectual Assets of Human Potential: theory and practice). (FOP V. V. Mokshanov, Zaporizhzhya, 2018); ISBN 978-617-7520-52-7
12. Measuring the information society : report. Geneva : International Telecommunication Union, https://www.itu.int/en/ITU-D/Statistics/Pages/default.aspx
13. The Global Innovation Index, https://www.globalinnovationindex.org/Home
14. Official site of (2020) the State Statistics Service of Ukraine. http://www.ukrstat.gov.ua. Accessed 20 Sept 2020
15. R. L. Florida, *The rise of the creative class: and how it's transforming work, leisure, community, and everyday life* (Basic Books, New York, 2002). ISBN 978-0-465024-76-6
16. E. Junarsin, Creativity and Competitiveness. Glob. **8**, 1–13 (2009)
17. S. L. Mueller, A. S. Thomas, Culture and entrepreneurial potential: a nine country study of locus of control and innovativeness. J. of Bus. Ven. 16, 51–75 (2001). doi: 10.1016/S0883-9026(99)00039-7
18. A. Karpenko, K. Basenko, Highly Effective Corporate Culture as an Instrument of Talents' Attracting and Retaining. Balt. J. of Econ. Stud. 3(4), 101-106 (2017). doi: 10.30525/2256-0742/2017-3-4-101-106
19. O. Bilenko, Mentality as an institutional factor in the development of human capital. Instit. Vect. of Econ. Devel. **5**(2), 26–34 (2020)
20. V. I. Kutsenko, V. P. Udovychenko, Ya. V. Ostafiichuk, *Suspilnyi vektor staloho rozvytku: teoriia ta praktyka* (Social vector of sustainable development). (Lozovyi V. M., Chernihiv, 2013)
21. M. V. Klov, Commercialization of higher education: characteristics and development trends. Grani. **5**, 85-91 (2014)
22. A. A. Pakulina, Innovative development of the higher education system, which promotes the commercialization and transfer of technology. Econ. Scope. **70**, 266-274 (2013)
23. *Natsionalna stratehiia rozvytku osvity v Ukraini na 2012–2021 roky* (National strategy for the development of education in Ukraine for 2012-2021), http://guonkh.gov.ua/content/documents/16/1517/Attaches/4455.pdf
24. D. P. Melnychuk, Educational structure of the population: ways of optimization in the context of preservation and development of human capital of Ukraine. Ukraine: asp. of work. **8**, 21-26 (2009)
25. V. S. Ponomarenko, O. V. Raievnieva (eds.), *Intehrovana systema «Vyshcha shkola – biznes struktura» : metodolohiia ta kontseptualni zasady pobudovy* (Integrated system "Higher school - business structure": methodology and conceptual principles of construction). (INZHEK, Kharkiv, 2011)

Micro-Sized Enterprise Management During the Epidemic. Qualitative Analysis

Marcin Kęsy[1*], *Dariusz* Pawliszczy[2], *Liudmyla* Morozovska[3], *Nataliya* Navolska[3], *Oksana* Apostolyuk[3]

[1]University of Economy in Bydgoszcz, 85-229 Garbary, Poland
[2]Mayor of the Gromadka commune, 59-706 11-go Lisstopada, Poland
[3]West Ukrainian National University of Ternopil, 46009 Lvivska, Ukraine

Abstract: The epidemic has affected the business situation in business entities, especially micro-sized enterprises, which mainly work in the service sector. In the event of a crisis, it is the service sector that is at high risk for business. To improve the financial condition of companies throughout Poland, a so-called "anti-crisis shield" was prepared, in addition, the governing bodies of the Lubuskie Voivodeship prepared an assistance plan for entrepreneurs. Its counterpart at the local level was assistance directed at companies in Zielona Góra. The cognitive value of the article is enriched by a thematic study done at micro-sized enterprises operating in the service sector in the territory of the Lubuskie Voivodeship with offices in Zielona Góra.

1 Introduction

The epidemic situation, which gradually, although unexpectedly, arose throughout Europe, affected all aspects of society's daily life. Changes have taken place at all levels of social functioning. In the education system, the direct form of studies was urgently and immediately replaced by indirect one, using distance learning. Due to the organizational load and legal restrictions associated with the Covid-19 virus, the health service was forced to introduce teleconsultation in the event of the first contact between a doctor and a patient and to limit the volume and intensity of the healthcare provision due to the system's orientation towards the process of diagnosing and treating patients with coronavirus.

All these changes in the functioning of a large number of business entities led to the fact that social life has moved from offices, workplaces or streets to homes. The result was a decrease in domestic demand, which in April was lower by 24.7% than in the same period last year [18]. A similar situation was recorded the next month. The exacerbation of the crisis is critical for the labor market, leading to the rise in unemployment levels in May to 6.0% against 5.4% this March, which is about 100 thousand unemployed in absolute values.

Forecasts for the market are not favorable, and macroeconomic data show that by the end of 2020 there will be about 2 million unemployed, or more than 10%, and the unemployment rate will not fall below 8% by 2023, in addition, we should not count on an increase in the minimum national living wage [9]. This will directly affect the decrease in domestic demand, which, with a possible reduction in inward investments and a decline in the export of goods and services, according to the French Minister of Economy and Finance Bruno le Maire, will result in a situation that the world economy is threatened with a recession and a financial crisis similar to the crisis of 2008 or even 1928 [12].

Among enterprises, the most vulnerable and the least resilient are micro-sized enterprises, which do not have sufficient savings, the so-called financial maintaining. Thus, they may not survive the period when the demand for their services will decrease, while constant costs will not decrease.

2 Features of the micro-sized enterprise

Enterprises, according to the Law on Entrepreneurs, are divided into four types: micro-sized, small, medium-sized and large [13]. Taking into account the criterion of business size, micro-sized enterprises employ an average of less than 10 employees each year and there are achieved an annual net turnover from sale of goods, products and services, as well as financial transactions, which does not exceed the amount equivalent to 2 million euros in PLN, or the amount of its assets balance sheet drawn up at the end of one of these 2 years, which does not exceed the amount equivalent to 2 million in PLN [6, p.13].

The author, characterizing micro-sized enterprises at the management level, took into account 7 factors: management functions, information system, management models, the role of intuition, the relationship between information and decision-making, formalization of problems and delegation of powers

* Corresponding author: marcin.kesy@byd.pl

[11, p. 363], as well as a low degree of diversification of goods and services produced, poor recognition of external environment – consumer market, competition, suppliers [6, p. 19-20].

By means of these features, it is possible to distinguish a micro-sized enterprise from a small, medium-sized and, above all, a large enterprise, in which the organizational structure is more complex, often three-level, which affects the distribution of roles between supervisory personnel and employees in vertical and horizontal projections. On the other hand, at micro-sized enterprises there is personal management, without taking into account the functions of the lower management level. The manager has direct and individual contact with the employee. Due to low level of employment and often homogeneous activities, information systems at micro-sized enterprises do not play a prominent part, their functions are mainly limited to e-commerce or accounting. The centralized management model characterizes micro-sized enterprises in which the head, manager manages authoritarianly and according to the principle of "top-down", and the decisions are not delegated below the level of the organization, therefore it should be noted that the activities associated with the delegation of powers are limited. Through direct contact between employees and the manager, the formalization of problems at the enterprise is at a low level, most tasks are transferred orally, without a written record. Hence, we conclude that at micro-sized enterprises, internal procedures are minimized, there is a low degree of formalization of problems. Often, business with fewer than 10 employees is a family business, with the founder or descendants of the enterprise in charge and family members acting as employees. Family business managers apply a business management method that is largely based on intuitivity and experience, which in turn develops into operational knowledge [4, p. 44]. The activities of micro-sized enterprises are characterized by a low degree of diversification, which means that they produce homogeneous products or services, and due to institutional constraints, expenditures and competencies, they do not expand the company strategy by expanding the range of goods or services. Doing business is fraught with considerable uncertainty and operates on the basis of "business intuition" rather than economic data and information. Entrepreneurs know little about the business environment in which they operate. Most often this is a perfectly competitive market, which by definition is very changeable.

A significant group of entities belonging to micro-sized enterprises are the so-called family enterprises [3, p. 9-10; 5, p. 20], which also affects their nature.

The development of an enterprise, including a micro-sized, small or medium-sized enterprise, is one of the most complex processes occurring in the course of its operation. As a rule, being the most eagerly awaited effect of purposeful and organized economic activity for interested parties, it belongs to the highest level of organizational goals [8, p. 12].

3 Evolution from industrial economy to knowledge-based economy

Globally, regionally and locally, we are experiencing changes that are shaping a new type of society based on knowledge and information [7].

Describing the changes taking place in the world, he identified three turning points, which he called waves [10, p.33]. These were turning periods in the nature and structure of the economy and society. The 1st wave is called the agrarian era, in which the dominant sector was agriculture. The next, 2nd wave was the era that marked the beginning of the industrial revolution and lasted about 200 years. It was characterized by the predominance of industry in the manufacturing industry and the participation of workers in employment. A group of workers appeared, who came from the impoverished peasantry, the landlord or the bourgeoisie. The third wave, called the age of the information-oriented society, has become the age of knowledge. At this stage of change, the most important sector is services, and the end product is information and knowledge. Services, which have become the dominant sector of the economy, are becoming increasingly important.

In the last few years, there have been many transformations in all spheres of socio-economic life. The transformation of the production era into a knowledge-based economy is forcing a change in perceptions of the world. Traditional factors of production, such as land, labor and capital, are no longer the only key differentiators of business development; currently knowledge and information determine the company's position at the market [1, 11].

The data shown in Table 1 confirm that the largest market share is made up of micro-sized enterprises, there are about 2 million business entities operating throughout the country. The small and medium-sized business sector comprises 99.9% of all companies operating at the Polish market. According to the distribution of companies by the type of activity, attention is drawn to the fact that in the case of small and medium-sized businesses, the dominant sector is services, and then trade. However, for large enterprises the industrial sector predominates, and the next area of activity is services. The conclusion is that micro-sized enterprises are usually service companies that operate homogeneously with a low degree of diversification, an extra feature of which is the low value of assets and an overexposure of the market situation with limited financial liquidity.

Table 1. The share of enterprises at the market by the number of employees

	Poland		**Lubuskie Voivodeship**		**Services**	**Trade**	**Construction**	**Industry**
Enterprises	Absolute values	%	Absolute values	%				
Micro-sized	about 2 million	93.6	about. 103 thousand	94.9	53.1	23.9		9.4
Small	about 53.8 thousand	5.5	about 4.600	4.2				
Medium-sized	about 15.3 thousand	0.8	about 840	0.8				
Large	about 3.6 thousand	0.1	about 130	0.1	30.7	13.7	3.7	51.9
Total	2.07 million	100.0	about 109 thousand	100.0				

Source: Report of the State of the Small and Medium-Sized Business Sector 2019 Polish Agency for Enterprise Development 2019 p. 15-19.

4 Epidemiological situation in Poland and immediate forecasts

The measures taken so far to manage the epidemiological situation can generally be divided into two stages. The first was the introduction of complete social isolation through the presence of citizens in places of residence, restricting the movement of people in public places or at work. The second stage was the removal of restrictions on movement, the introduction of so-called personal hygiene measures in the form of the requirement to wear masks in these places and the so-called public distance: 1.5 meters in a public building, 2 meters outside buildings.

The above actions have led to a situation where the increase in infected people is linear rather than exponential. Thus, the infrastructure and medical staff are sufficient to combat the effects of the epidemic in the country. However, the negative effect of smoothing the incidence rate is the continued steady increase in new cases of coronavirus carriers. 1.5 months ago, the author predicted that the daily increase in the incidence was from 250 to 350 people, which was confirmed in the consecutive days and weeks. Currently, every day increase about 7-8 thousand cases.

Figure 1 shows the epidemiological situation in Poland as of July 16 this year. In Lubuskie Voivodeship, there are 261 people infected with the coronavirus, of which 146 recovered and 1 person died [2].

Fig.1. The COVID-19 pandemic in Poland.

Source: https://wroclife.pl/, 23.01.2021r.

5 Measures of support of micro-sized enterprises

The reaction of Poland to the deterioration of the economic situation of enterprises was 4 laws prepared and adopted by the Sejm of the Republic of Poland in the form of the so-called "anti-crisis shield" marked 1.0, 2.0, 3.0 and 4.0 [14].

Two of these documents are intended for micro-sized enterprises, ie anti-crisis shields 2.0 and 3.0. Within the framework of state aid, assistance measures are carried out, presented below:

1. exemption from contributions of the Social Insurance Office for a period of 3 months,
2. assistance for persons engaged in entrepreneurial activity (PLN 2,080 or 1,300),
3. additional financing of part of the cost of salaries for employees and social security contributions due on this salary,
4. additional financing of the salary of employees,
5. installments or payment of arrears in the Social Insurance Office later,
6. advances can be paid by PIT / CIT,
7. low-interest loan from the labor fund (PLN 5,000).

Further measures of support to entrepreneurs have been taken by the local government of Lubuskie Voivodeship for micro-sized enterprises:

1. Lubuskie entrepreneurs support vouchers (the total amount of assistance is 30 million zlotys), of which the assistance for 1 micro-sized enterprise is from 30 to 120 thousand zlotys. The assistance has two goals: investment and revolving, or one goal: investment. The maximum funding level for the voucher is 95% of eligible costs. Receipt of applications in an extraordinary mode.

Another unit at the regional level that supports micro-sized entrepreneurs is the Labor Office of the Voivodeship, which has offered the following assistance:

1. Support to entrepreneurs in the field of job protection from the Fund for Guaranteed Employee Benefits.

Within the framework of the anti-crisis shield, 1,040 enterprises were supported in the territory of the Lubuskie Voivodeship, which made it possible to secure 49,078 workplaces. The total cost of the assistance was 174,356,120 zlotys [19].

The district labor offices operate one level down and they have prepared assistance for micro-sized enterprises, consisting of a low interest loan from the Labor Fund or additional financing of part of the cost of wages and social security contributions.

On the other hand, the authorities of the city of Zielona Góra have prepared a proposal for enterprises located in Zielona Góra, which includes:

1. 6-month deferral of real estate tax, after which the members of the commission make a discretionary decision,
2. exemption from exhibition and lane fire charges for catering establishments,
3. reducing the amount of rent for the lease of city premises into a symbolic one – 1 zloty

The total cost of assistance is estimated at 30 million zlotys.

6 Research methodology

The theoretical analysis realized in the earlier part of the article has been complete with empirical analysis. The researchers created the research goal: Problems which are related to functioning in a pandemic environmnet. To released the goal we used informations from owners of firms.

All of criterions of the analysis was presented later in the text. The hypothesis was built on five criterions. It's mean: A new pandemic situation is negatively affect the financial and economic condition of service firms.

Media reports, own analysis of the macroeconomic situation and observation of isolated situations in enterprises suggested that the epidemiological situation negatively affects the economic and financial situation of enterprises, especially those working in the service sector. Based on this, the author of the research suggested that the subject of the research will be micro-sized enterprises that operate in the service sector. In turn, the topic of the research is the assessment by entrepreneurs of their own situation in the context of the market situation in the environment in which their enterprises operate.

Through organizational, time and financial capacity, the research was limited to non-randomly selected micro-sized enterprises operating in the city of Zielona Góra in service-related areas. The best research method was the interview method, which used a measurement tool – an interview questionnaire, which consisted of open-ended questions. Therefore, the author's interpretation served as the basis for cognitive analysis.

The companies included in the research were divided into 4 groups in accordance with the Polish Classification of Activities.

1. Activities involving accommodation (Polish Classification of Activities I-55): hotel business, short-term accommodation.
2. Activities of tour operators (Polish Classification of Activities N-79): travel agencies, means of transport.
3. Individual service activities (Polish Classification of activities s-96): esthetic, hairdressing, funeral.
4. Construction (Polish Classification of Activities F-41): buildings, finishing works.

The researches were conducted at 9 enterprises, 1 from each type of activity. The main task set by the author of the research was not to describe the situation in the business, but to show the position of enterprises in diverse fields in order to be able to build a general idea of the position of markets and build a future picture of the situation at enterprises in six months.

In this research, the author focused on 5 planes. The assessment covered: demand for services, unit cost (price), income generation (especially variable costs), and business development prospects in the short term and financial liquidity (financial maintenance of activities for the next 3 months).

The research used the interview method, and as a measuring tool, the interview questionnaire, which consisted of open-ended questions concerning all 5 planes of analysis. The survey was conducted within 5 days (May 5-9 this year), which allows us to conclude that all business entities had the same business situation.

7 Presentation of the results of own and others' research

The author conducted the primary research at 9 micro-sized enterprises. Table 2 presents the results of the research taking into account the field in which the company does business.

The results of the research were confirmed by the secondary research, which the author presented in graph form, and by the description of the National Association of Purchasing Managers' Index [21], which is an indicator of financial activity.

The results represented in Table 2 show that in 7 out of 9 surveyed enterprises there is a decrease in *"demand for services provided"*, in the other 2 enterprises there is stagnation in this regard. The author believes that the decrease in demand for services is the result of social exclusion and search cheaper substitutes by consumers because of doubtful future and for reasons of costs saving. Only in the case of beauty and hairdressing services there was no decrease in the services provided, probably due to the nature of the activity.

Table 2. Conclusions of the primary research.

Subjects	Hotel Business	Short-term accommo-dation	Travel Agencies	Means of transport	Esthetic	Hairdressing	Funeral	Building	Finishing works
demand for services (1-3)	decline	decline	decline	decline	stagnation	stagnation	decline	decline	decline
price of services (1-3)	decline	decline	stagnation	decline	stagnation	stagnation	stagnation	increase	stagnation
expenses for income generation (1-5)	growth of fixed expenses	without changes	without changes	reduction of variable expenses	growth of variable expenses	growth of variable expenses	without changes	without changes	without changes
time perspective (1-4)	negative	negative	indefinite	negative	indefinite	indefinite	indefinite	indefinite	indefinite
financial liquidity (1-3)	at risk	at risk	at risk	at risk	in question	in question	at risk	in question	in question

Source: own elaboration on the basis of primary researches.

Another criterion of analysis, that is, *"the price of services"*, differentiated in the responses of entrepreneurs. Thus, for the primary real estate market there was an increase in the unit price per 1 square meter. This is the result of the growing public interest in alternative forms of allocation of accumulated funds. In three cases, there was a decrease in the price per unit of services, resulting from a high level of supply with a simultaneous decrease in demand for services.

Research in the area of *"expenses for income generation"* has shown that entrepreneurs have often shown no change in this regard. Hairdressing and beauty services bear large variable expenses. This change is due to the obligation of service providers to provide a workplace in accordance with sanitary requirements.

Taking into account the *"time perspective"*, entrepreneurs take the market situation in "black colors" and, therefore, their business situation in the next 12 months. Entrepreneurs notice that they are not out of touch with the environment and depend on it, an imbalance in the transport industry can lead to an increase in prices expressed in logistical terms, and problems with suppliers will lead to an imbalance in the activities carried out and the search for new sources of components or necessary goods for services.

The result of the analysis is the *"assessment of financial liquidity"* of business entities. So, at 5 enterprises there is a threat to the operation of business, entrepreneurs believe that this is due to a decrease in demand and margin per unit of service, with little or no provision of financial capital. In other subjects, threats in the environment are noticeable that may affect the enterprise performance, however, during the exploration period they have not yet formulated their opinion, they feel doubts about the future, but this is still "in question".

The result of someone else's research, presented in Figure 2, illustrates how Internet advertising companies relate to the market situation in the context of the Covid-19 epidemic. Almost half of the respondents answered "definitely yes", and 38% - "most likely yes".

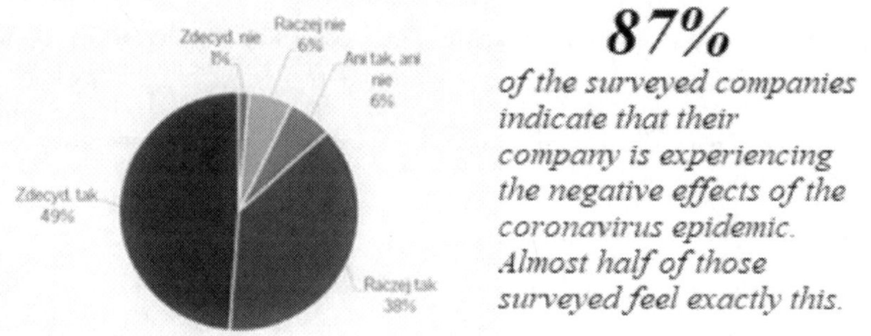

Fig. 2. Someone else's research on the impact of the epidemic on business situation in online advertising

Source: wyborcza.biz/Gieldy/.../iab-polska-87-firm-z-branzy-reklamy-online-odczuwa-negatywne.html, ISB News : 04/28/2020, 05/08/2020.

The National Association of Purchasing Managers' Index for industry shows a picture of the financial activity of managers. It is valued in the range of up to 100 pt, and 50 pt are assumed to be neutral. The results above 50 points are positive, the closer to 0 points, the worse the score. The National Association of Purchasing Managers' Index in Poland in March 2020 fell to 42.4 points, in April it reached 31.9 points (this is the lowest indicator since the beginning of the research in 1998), in May - 40.6 points, in June – 47.2 p. The change +6.6 in June shows that managers perceive the business situation more positively. However, the result has not yet reached the so-called neutral result, that is, 50 pt. The last time a neutral indicator of the National Association of Purchasing Managers' Index received by the Polish economy was in October 2018.

Table 3. Comparison of the National Association of Purchasing Managers' Index in Poland, Germany and France

State/month	03/2020	04/2020	05/2020	06/2020
Poland	42.4	31.9	40.6	47.2
Germany	45.4	34.5	36.8	44.6
France	43.2	31.5	40.6	52.3

Source: own elaboration based on https://pl.tradingeconomics.com/, 07/15/2020.

To compare National Association of Purchasing Managers' Index in Poland, the author cited the results of the National Association of Purchasing Managers' Index obtained in France and Germany. Managers in France are increasing their financial activity; however, the National Association of Purchasing Managers' Index in Germany is at a lower level than the Polish National Association of Purchasing Managers' Index. This is a very adverse outcome for the Polish economy, as Polish trade exchange with its western neighbor is on an upward trend and in 2019 amounted to a record 123 billion euros. That in the situation of a decrease in the financial activity of German companies may have a significant impact on a decrease in goods turnover, this, in turn, will lead to a reduction in financial resources in the turnover of the Polish economy.

8 Conclusions of the research

The presented research results indicate that entrepreneurs are experiencing uncertainty about the near future due to decrease in price and demand, which, in turn, leads to the oscillation in the financial economy at enterprises.

Service enterprises, especially micro-sized enterprises, do not have the financial capital needed to survive in an economic downturn at the market. State aid from the financial side is, perhaps, insufficient, but it has a positive response and interest from entrepreneurs who notice problems in formal solutions, not adapted to the current legal situation.

What will happen to the economy in the coming months? Some economists predict deterioration in macroeconomic indicators that may occur in late September or early October. The time required for the economy to surmount the crisis is estimated at 2-3 years.

References

1. K. Beyer, *From the agrarian era to the knowledge-based economy*, Research and works of the Faculty of Economics and Management, No. 30 (2012)
2. R. Czystowski, *Coronavirus in Lubuskie Voivodeship: the number of new cases as of July 16,* Gospodarka300gospodarka.pl/live/koronawir
3. us-lubuskie-liczba-nowych-zachorowan (2020)
4. ed. D. Jegorow, *Person-Economy. Cooperation-Development. Local and Global Perspective,* Civis&Sun Solution Publ. (2013)
5. T. Kupczyk, M. Stor, *Competence Management. Theory, Research and Practice of Business*, University of Economics in Wrocław (2017)
6. K. Safin, *Family businesses: the essence and strategic behavior*, Oskar Lange Academy of Economics (2007)

7. M. Sidor-Rządkowska, *Small Business HR, Management* Wolters Kluwer, Warszawa (2010)
8. E. Stolarczyk-Ambroziak, *Continual learning in the process of creating a learning society and a knowledge-based economy*, www.mentor.edu.pl/
9. artykul/index/numer/2/id/12 (2020)
10. A. Szplit, J. Fudaliński, P. Markiewicz, Ch. Smutek, *Organization development strategies*, Antiqua Publ. (2002)
11. J. Szymczak, *2 million of unemployed by the end of 2020? According to the NFZ, we will face economic stagnation,* OKO.press (2020)
12. A. Toffler, *The third wave*, PIW Publ. (1985)
13. O. Torres, P. A. Julien, *Specificity and Denaturing of Small Business*, International Small Business Journal, Vol. 23, No 4 (2005)
14. *France heading for worst post-war downturn, finance minister warns*, France 24 with AFP and Reuters, www.france24.com (2020)
15. *Law on Entrepreneurs*, 03/06/2018, Art. 7 (2018)
16. *Law on amendments to the Law on particular solutions related to the prevention, counteraction and control of COVID-19, other infectious diseases and emergencies caused by them, as well as some other laws,* 03/31/2020, Legislative Bulletin, prov. 568 (2020)
17. *Law on specific support tools in connection with the transmission of the virus SARS-CoV – 2*, 04/16/2020, Legislative Bulletin, prov. 695 (2020)
18. *Law on amendments being made to some laws on protective activities in connection with the transmission of the virus SARS-CoV – 2*, 05/14/2020, Legislative Bulletin, prov. 875 (2020)
19. *Law on allowances for the assessment of bank credits provided to entrepreneurs affected by the consequences of COVID-19, and on a simplified procedure for approval of the contract in connection with the onset of COVID-19*, 06/19/2020, Legislative Bulletin, prov. 1086 (2020)
20. Main Statistical Office (2020)
21. Labor Office of the Voivodeship in Zielona Góra, zielonagora.praca.gov.pl (2020)
22. wyborcza.biz/Gieldy/…/iab-polska-87-firm-z-branzy-reklamy-online-odczuwa-negatywne.html, ISB News from (2020)
23. PMI By IHS Markit, https://www.markiteconomics.com/public?language=pl (2020) https://pl.tradingeconomics.com/ (2020)

The Ethnomatematics and Ethnolinguistics of the Cultural Aimara – Peru

Elena Yucra Yucra[1], *Oscar* Mamani Aguilar[2], and *Luis* Morales-Aranibar[3*]

[1,3]Professor at the National Intercultural University of Quillabamba, Cusco, Peru.
[2] Professor at the Jorge Basadre Grohmann National University, Tacna, Peru.

Abstract Current research on cultures worldwide has been recovering the knowledge of our ancestors, which has been lost over time and education in the Spanish language does not seek to recover them. The objective of the research is to describe the ethnomathematics and ethnolinguistics of the Aymara culture in the district of Pomata, Puno. The type of research is qualitative-ethnographic. The methodology applied was the interview, survey, participatory observation. Aymara culture currently practices ethnomathematical knowledge to designate the "*Jilawi*" ages of people, of the time by "*Pacha chimpu*"; the stars are recognized, the morning star "*Qhantati ururi*", the Andean new year *"Machaqa mara"* is celebrated in June, the "*Yapuchañanaka*" crops; the harvest "*Yapu apthapi*", ethnogeometry uses it for designs of their houses, marking animals, such as the triangle "*Kimsa k'uchu*", square "*Pusi k'uchu*", rectangle "*Pusi wiskhalla*"; they measure the cereals "phuqthu" by hand, the land by "*Yunta*". Aymara ethnolinguistics began with the "*k'inchu*", which were colored threads with knots, and each color had a meaning: white means living in harmony; black mourning and protection; lead property record; coffee registration and marking of animals; They learned to read and write clandestinely, some serving priests, others through military service.

1 introduction

The knowledge of ethnomathematics and ethnolinguistics are developed by indigenous peoples according to their needs, experiences and relationships with the world of which they are part [1]. Ethnomathematics as a science was formulated for the first time by D'Ambrosio in 1984 [1] [2], currently this science has different ways of naming its concept, but none has achieved the acceptance of researchers. Ethnolinguistics was first described by [3] [4] and in his article he associated ethnolinguistics in the Anuario de Letras XIX. These two sciences are related and each culture has its own way of expressing it, and many of them are still valid and others are being lost; If this knowledge is alive, it is because the elderly people still retain it, and share the information with their children, grandchildren, for example the use of some names of animals in the area such as the alpaca, the llama of which They use wool and its different colors; the forms of stones as an accounting tool for their animals [5], among many events that make us think about the use of natural resources that they used in their daily lives and the Peruvian state should not let this knowledge be lost, as many countries like Mexico, Chile, Argentina among others, are recovering their ancestral knowledge as they name it [6] [7] [8] [9].

The Aymara converse with nature, for them all have life and live in harmony with their environment (trees, animals, hills, among others) [10], it is also observed that the inhabitants of the communities define their lands, belongings, changes social, adaptations; information that are essential to understand and characterize different cultures [11] and in this way they can be rescued.

Ethnolinguistics is the relationship of language with the speaking community that links with its own culture [12], where language, culture and society converge, therefore, the object of this science was to articulate the knowledge about their daily lives manifested by their language [13], the Aymara did not know the alphabets of Spanish, in order to save that knowledge and record it in time, they had their own ways of calculating, managing time and space [14].

Currently, this type of work serves to describe and document information in this way to be able to delve into the languages of cultures in academic terms [15]. These types of research have grown in recent years especially in the sociolinguistic field in terms of numbers and attitudes of its speakers [16]. The effects of the inclusion of ethnomathematics and ethnolinguistics are one of the central axes of education, we believe that thinking about the training of indigenous teachers, who teach should be a fundamental proposal in multicultural countries, which will allow the inclusion of policies that protect this knowledge and do not get lost in time [17], this would imply creating new teaching techniques under this

*Correspondencia: luis.morales@uniq.edu.pe

perspective as [18] which presents a teaching method where characteristic approaches of ethnomathematics are identified, used in the academic world to analyze the observations of the Participants remaining as evidence of their conceptions, similarly [19] describes a pedagogical experience on ethnomathematics that focuses within the pedagogical process, also in studies they describe the experiences and the use of these experiences in their daily lives as they did [20] In the south of Chile.

One of the current problems of the Peruvian educational system in the face of Aymara culture, lacks basic studies that allow revaluing it, so there is an immediate need to carry out studies with the aim of contributing to the recovery and development of intercultural education. , we question that the ancestral knowledge of Aymara ethnomathematics and ethnolinguistics is currently being gradually lost.

The article rescues the practice of the cultural and linguistic identity of the Aymara, as they do in their studies [21] [22] [23] [24], which raise awareness among academics to revalue and teach children, youth, and adults to to continue practicing, valuing and strengthening their cultural roots, since there is currently a problem that is the massive migration of the youth population out of the areas where they were born, leaving only the elderly population. Therefore, this article opens up researchers to develop more research related to culture at the local, regional, national and international levels.

With the interest of seeking answers from the different studies of the ethnomathematics and ethnolinguistics of Aymara culture, which will contribute to intercultural education at all levels of the educational system, we propose to describe the ethnomathematics and ethnolinguistics of Aymara culture in the district of Pomata, Puno.

2 Materials and methods

The research scope was carried out in the province of Chucuito, in the district of Pomata, geographically located in the department of Puno, according to the 2017 census, the district of Pomata has a population of 8544 inhabitants and is located at an altitude of approximately 3824 masl In the present work, 24 people between men and women were taken as a sample, with whom they lived together in order to obtain the information. Bilingual people who understand Spanish and Aymara were selected. Semi-structured surveys were applied referring to cultural, agricultural, livestock and commercial activities and who have a good knowledge of Aymara culture [10].

In the research, the ethnographic method was applied with the technique of participatory interview and observation, in order to carry out a qualitative and comprehensive investigation. In the dimension of analysis, the setting or context, description of each activity and systematization of the data sequentially were prioritized [15].

Results and discussion

Ethnomatematics aimara

Aymara can determine the age of people from the time they are in the womb to old age, for example the pregnant woman is called "*Usuri warmi*", they can also determine if the woman is between 4 or 5 months of pregnancy, cataloging it as "*Chika usuri*", 6 months pregnant as "*Tirsu usuri*", they can also determine the baby of approximately 8 to 10 months as "*P'uru wawa*", he drinks from one year to two years as "*T'aqa wawa*", It has also been shown that they can determine if a boy or girl suffers a biological delay by cataloging it as "*Ipi wawa*", and they are able to determine when a boy or girl is skilled in their biological development by calling it "*Ch'iki wawa*".

Also in school age they are called "*Yuqalla*: boy" and "*Imilla*: girl", more specifically 6-year-old boys as "*Jisk'a yuqalla*" and the girl of the same age as "*Jisk'a imilla*", between the age of 7 to 12 "*Jach'a yuqalla*: boy" and "*Jach'a imilla*: girl". The ways of categorizing young people are peculiar due to their physical capacities, abilities, being a leader among other aspects, so the young man is called "*Wayna*" and the young woman as "*Tawaqu*", more specifically at 15 to 17 years old. They call "*Q'axu wayna*: immature male" and "*Q'axu tawaqu*: immature woman". Those who take religious marriage are called "*Chacha / warmi*" these people are recognized by the community, in this way they may have some responsibility and they are told "from today onwards you are in the large group" which according to their language is "*jichha uruta uksaruxa, jach'a tamanxtawa*", we can delve further into table 1.

Table 1. Knowledge of ages from pregnancy to Aymara old age

Spanish	Aimara
One month pregnant	***Phaxsi usuri***
Nine months pregnant	***Usuri phuqhata***
Newborn baby	***Asu wawa***
The boy or girl from 3 to 5 years	***Irqi wawa***
The child without parents	***Iñu wawa***
7 to 12 Years old (Male)	***Jach'a yuqalla***
7 to 11 years old (female)	***Jach'a imilla***
Young from 18 to 25 years old	***Achachi wayna***
From the marriage	***Chacha / warmi***
Married adult up to 50 years (Male)	***Chuymani chacha***
Married adult up to 50 years (Woman)	***Chuymani warmi***

The Aymara have their own ways of controlling time, being knowledge that they have inherited from their parents, grandparents over time, animals are a way of guiding themselves, such as the crowing of the rooster or "*Wallpa aru*" they announce the dawn, and its songs determine the time, in this case there are three songs, the first song indicates that it is 2 am "*Nayra wallpa aru / maya wallpa art'a*", the second that it is 3 am "*Taypi wallpa aru / Paya wallpa art'a*" and the third which is 4 am "*Qhipa wallpa aru / Kimsa wallpa art'a*". The dawn between 4 to 5 am when it tends to clarify, the Aymara call it "*Q'axatatti*", they also determine it by the song of the birds who leave their nests to look for food and call it "*Willjta qhanatatti*", more knowledge in Table 2.

Table 2: Aymara day and night time measurement

Spanish	Aimara
6:00 pm. next to dark	***Jayp'uthapi***
9:00 pm. Bedtime	***Ikintaña***
12:00 midnight	***Chika aruma***
02:00 a.m. First crow of the rooster	***nayra wallpa aru / maya wallpa art'a***
Dawn or new day	***Urjti / urutatti***
6:00 am. Sunrise	***Inti jalsu***
07:00 a.m. Breakfast time	***Alwa manq'asiña***
07:00 a.m. Start of farm work	***Yapu luraña qallta***
12:00 pm. Break time and lunch	***Chika uru***
1:30 p.m. Start of agricultural activity	***Jap'utuqiru yapu lurawi qalltaña***
4:30 pm. Pick from pasture to corral	***Uywa anakthapi***
5:00 pm. Break after work	***Jayp'u samaraña***

The Aymara settlers demonstrate a wisdom to control the time, as well as the braying of the donkey and the crowing of the rooster, they call it "*Chika uru*" estimating that it is noon, that is, it is 12:00 pm to 12:30 pm This indicates that they should go to eat, it is evidenced that the villager is very strict with their meal times, similarly at 10 in the morning they take a break from their farm work, which they call "*Yapu lurañana samarawi*".

The Aymara know and observe the stars at night and at dawn table 3, it is in this way that they control the time generally between 0:00 a.m. to 6:00 a.m.

Table 3. Aymara astronomical knowledge

Spanish	Aimara
Constellation Flame Centaurs	***Qawra nayra***
Oval rectangular shaped stars	***Qãna***
Cross-shaped stars	***Kurusa wara wara***
Evening star	***Jayp'u ururi***
Morning star	***Qhantati ururi***

In the past, the Aymara had tracts of land, they called it "grocery stores" because they had no owner. The Aymara settlers use part of their body such as fingers, feet, arms, or things to be able to measure lengths, since they lacked measuring and writing instruments, an example when they measure 6 cm this is equivalent to 3 fingers, when they measure from 6 to 8 cm they join four fingers that they call "*Maya t'axlli*" or if they measure a fourth, which is the distance between the thumb and the little finger, calling it "*Maya chhiya*", the arms or strokes use it to measure rivers, roads and terrain, etc. , the measure of a half meter they measure it through the elbow to the tip of the middle finger naming it "*Maya mujlli*".

Table 4. Aymara calendar

Spanish	Aimara
Sunday fair day, considered as the first day of the week (main fair)	***Althapi uru***
Rest day (Saturday)	***Samaraña uru***
The first week of the month	***Phaxsi qallta***
End of the month	***Phaxsi tukuya***
Andean New Year (June)	***Machaqa mara***
Product storage (June and July)	***Manq'a apthapi***
Bean sowing begins (August 1)	***Sata qallta***
Sowing (September, October and November)	***Yapuchañanaka***
Sowing goose	***apilla sata***
Planting quinoa	***jiwra phawa***
Sowing barley	***siwara phawa***
Potato planting	***ch'uqi sata***
Dry season (November to December)	***Awti phaxsi***
Rainy season (December to March)	***Jallu pacha***
Harvest beans, quinoa, barley, potato, goose (April to May)	***Yapu apthapi***
Cold season making chuño (June and July)	***Juyphi pacha***

In the harvest of tubers, the Aymara keep them in holes of approximately 1.2 m calling it "*Qaxa*", they are generally built on the ground, another way to measure the potatoes is in "*Phina*" that contains 4 arrobas of potatoes, the stem of beans or barley that they gather with their arms they call it "*Marqa*", and when the rope is used with their arms they know it as "*Pichu*". The wool of animals such as alpaca, sheep, they process it on a spinning wheel to make the yarn they call it "*Qaputa*", where the unit is "*lluchhu*", where a pound is removed 5 conicoidal,

Usually bartering is done at the fairs they attend, the exchange of fruits such as oranges, bananas, breads, matches among other products for their agricultural products that they bring, which have their own ways of measuring for example five wielded with the two Open palms of chuño, quinoa, among others, exchange for a bread to this process they call it "*Phuxthu*", another way of measuring is wielded with a single palm of the hand, for example, five handfuls of their products exchange it for a banana to this process denominate "*Jach'i*", other forms of measurement are observed in table 5.

Table 5. Aymara Measurement Forms

Spanish	Aimara
The land distributed by number of furrows	***suka***
Land of the crop of products is measured by	***Yunta***
Land calculated by seeds	***Jarphi***
10 drums	***Jach'a wakulla***
5 drums	***Taypi wakulla***
3 drums	***Jisk'a wakulla***
4 gallons	***Yuru***
Wool equivalent to a pound or half a pound	***Jawi***

The Aymara man uses geometric figures for the construction of his house, markings on animals or in clothing and they are shown in detail in Table 6.

Table 6. Aymara geometry

Spanish	Aimara
Triangle	***Kimsa iskina***
Square	***Pusi iskina***
Rectangle	***Pusi iskina wiskhalla***
Circular shaped ear piercing in cattle	***K'illpha***
Triangular shaped ear pieces in cattle	***Llawi***

Ethnolinguistic aimara

Learning Spanish was forbidden in the Aymara, it was forbidden because the mistis and landowners did not allow it, to add to that that in ancient times there were no basic tools to learn to read and write (paper, pen, ink) that is why people did not know read or write, our study showed that the Aymara could record information through the "***K'inchu***", being the most important discovery of our work and that this research is only the beginning of more research that we suggest to be given in more depth to revalue and save these ancestral knowledge.

The "*K'inchu*" consisted of moorings of threads or pitas that contained five or six threads, similar to the "*Kipu*", they were made of alpaca and llama wool that had different shades table 7, and that could be evidenced thanks to the replicas of the elderly settlers, generally the "*K'inchu*" went behind the door of the houses.

Table 7. Meaning of the Aymara "*K'inchu*" colors

Colour	Meaning
White	Live in tranquility, harmony
Black	Grief, mourning, sadness, protection from rain, frost
Lead	Property registration
Coffee	Register and mark animals

The color black in Aymara culture has many meanings apart from those shown in table 6, for example when someone assumes the position they have to wear black their entire period to protect their pacha, in this case it has been evidenced that his partner also has to dress in black with the intention of giving prosperity to his population, as well as representing that his family has had a loss of a relative and the time of being in black clothes is from 6 months to a year , this way of life has passed through time and is currently observed. The brown color is used in the marking of animals, it is represented by putting a pita around the neck of your cattle.

The Aymara peasants were mistreated by the mistis who lived in the urban area and knew Spanish. The humiliation and exploitation was mainly due to the fact that the majority were illiterate, being the `` beginning for the learning of the illiterate Aymara.

To learn Spanish, the Aymara had to suffer a lot and had to build strategies such as being a servant of the priests where they learned to read and write, another serving the barracks, the priests also recommended children to be educated, in ancient times only there was a school and the only ones who could access were the children of the gamonales or upper class people.

The Aymara settlers in their inscriptions were inspired by their natural resources such as trees, vicuñas, their habitat such as their pampas, hills, their lines express their feelings to protect them as part of their heritage, which expresses their identity.

Conclusion

The Aymara culture has an important ethnomathematical knowledge, which allowed them to estimate times, seasons and spaces; Using the animals and the stars as indicators, to measure they used their fingers, strokes, they could determine the time of the fetus in pregnancy, the age of the child at 6 years *"Jisk'a yuqalla"*, of the Young from 18 to 25 years "*Achachi wayna*"; Animals like the rooster are very important because their song announces the dawn "*Wallpa aru*", when hearing the braying of the donkey and the crowing of the rooster noon "*Chika wallpa aru*" estimates that they are between 12:00 to 12:30 pm, indicating that they should go to eat, they had an important knowledge of the stars, their movement during the night and early morning, they recognize, the evening star "*Jayp'u ururi*" and the morning star (Venus) "*Qhantati ururi*" , its fairs, seasons and transcendent activities play an important function, through the different months they determine its agricultural calendar. The Andean New Year is called "*Machaqa mara"* being the month of festivity in June, the sowing of potatoes, beans, oca, quinoa are between September, October and November calling it "*Yapuchañanaka*", the harvest time of these products between April to May calling it "*Yapu apthapi"*, the lands were distributed by number of furrows "*suka*", the Chacareo Terrain is measured by "*Yunta*", the knowledge of geometry was important in their daily life, the triangle they call it "*Kimsa k'uchu*", squared

"*Pusi k'uchu*", rectangle "*Pusi wiskhalla*", being observed in the construction of their houses, marking of animals or in clothing.

The development of Aymara ethnolinguistics was prohibited by mistis and landowners, there was no paper, pen or ink, however we discovered that if there was a way to record information and it was the "*K'inchu*", which were made of ties of threads or pitas of colors where each one had its own meaning, the white color means living in tranquility, harmony; the black mourning, sadness, protection from rain, frost; the lead registration of goods and coffee which was to register and mark animals; The mistreatment and humiliation were important situations for learning Spanish, and they suffered a lot to learn it. They had to be servants of priests, serve the barracks, and some were recommended by the same priests so that they could study in schools where only class children could enter High or gamonales, the Aymara were inspired by their natural resources and their environment where they express conservation and their identity.

Acknowledgement

As a result of the work, we thank the Aymara native inhabitants, who provided us with very valuable first-hand information, recovering the wisdom of the people, as well as the Andean culture researchers, who correctly guided us to the research team. We also thank the native resident Juan Quispe Escobar, for his contributions to the thesis work of the master's degree in ethnomathematics and ethnolinguistics of the inhabitants of the Pomata district and thanks to that work we were motivated to carry out this research article.

References

1. Peña, P., Tamayo, C., Parra, A. (2015). Una visión latinoamericana de la etnomatemática: tensiones y desafios. *Revista Latinoamericana de Investigación en Matemática Educativa*, vol. 18, núm. 2, pp. 137-150.
2. D'Ambrosio. (1985). Socio-cultural Bases for Mathematics Education. Campinas: UNICAMP.
3. Coseriu, Eugenio. (1978). La socio y la etnolingüística. Anuario de Letras XIX, 5-30.
4. Aroca, Armando. (2015). Diseños Prehispánicos, Movimientos y Transformaciones en el Círculo y Formación Inicial de Profesores. Bolema: Boletim de Educação Matemática, 29(52), 528-548.
5. Bermejo Paredes, Saúl, Maquera Maquera, Yanet Amanda, & Bermejo Gonzáles, Luz Yohana. (2020). Procesos de educación intercultural y autosegregaciónindígena en los aimaras de Puno, Perú. Revista Historia de la Educación Latinoamericana, 22(34), 19-43.
6. Ochoa, R., Peláez, J. (1995). La matemática como elemento de reflexión comunitaria Pueblo Tule. 1. ed. Antioquia: Asociación de Cabildos Indígenas de Antioquia, Editorial Lealon, p.147.
7. Aroca Araújo, Armando. (2014). La lógica de elaboración en los diseños de los platos de las culturas prehispánicas de Nariño. Revista U.D.C.A Actualidad & Divulgación Científica, 17(2), 587-596.
8. Gentil Guegia, Húber Castro Cáliz, Laura Calambás, Arnulfo Guegia, Carlos Alberto Pacho, Edilma Díaz & Carlos Gueguia. (2012). Matemáticas en el mundo Nasa. Ministerio de Cultura de Colombia, Bogotá.
9. Jaramillo, D., Berrío, L. (2011). Prácticas sociales y prácticas escolares en la escuela indígena: ¿una dialogía posible?. *Horizontes*, Brasil, v. 29, n. 1, p. 89-99.
10. Apaza Ticona, Jorge, Alanoca Arocutipa, Vicente, Ticona Alacona, Cesario, Calderon Torres, Alfredo, & Maquera Maquera, Yuselino. (2019). Educación y alimentación en las comunidades aymaras de Puno. Comuni@cción, 10(1), 36-46.
11. Aroca, Armando, & Cauty, André. (2017). Dificultades Metodológicas en la Investigación sobre Pensamiento Matemático Indígena y su Paradójica Educación Matemática. Bolema: Boletim de Educação Matemática, 31(58), 841-860.
12. Bartmiński, Jerzy. (2012). Aspects of Cognitive Ethnolinguistic. London: *Equinox publishing*.
13. Teillier, Fernando. (2015). Configuración de realidad y lengua: bases de una epistemología del mapunzugun. Tesis de Doctorado en Lingüística. Concepción, Chile, Universidad de Concepción.
14. Alanoca Arocutipa, V., Mamani Luque, O., & Condori Castillo, W. (2019). El significado de la educación para la nación Aymara. Revista Historia De La Educación Latinoamericana, 21(32), 227-246.
15. Llanquinao Llanquinao, G., Salamanca Gutiérrez, G., & Teillier Coronado, F. (2019). Aprendizaje del mapunzugun desde metodologías propias: perspectivas y avances para la revitalización de lenguas originarias. Revista Historia De La Educación Latinoamericana, 21(33). https://doi.org/10.19053/01227238.9918
16. Castro-Inostroza, A., Rodríguez-Nieto, C. A., Aravena-Pacheco, L., Loncomilla-Gallardo, A., & Pizarro-Cisternas, D. . (2020). Nociones matemáticas evidenciadas en la práctica cotidiana de un carpintero del sur de Chile. Revista Científica, 39(3), 278–295. https://doi.org/10.14483/23448350.16270
17. Osorio, Carolina Tamayo. (2018). Licenciatura en Pedagogía de la Madre Tierra, etnomatemática y formación de profesores. Ciência & Educação (Bauru), 24(3), 759-777. https://doi.org/10.1590/1516-731320180030014

18. Albanese, Veronica, & Perales, Francisco Javier. (2020). Mathematics Conceptions by Teachers from an Ethnomathematical Perspective. Bolema: Boletim de Educação Matemática, 34(66), 1-21. Epub April 17, 2020.https://doi.org/10.1590/1980-4415v34n66a01
19. Santos, Jorge Alejandro, Bernardi, Lucí dos Santos, & Nascimento, Márcia. (2020). Algoritmos y sistemas de parentesco: aproximaciones etnomatemáticas en la formación de profesores indígenas. Bolema: Boletim de Educação Matemática, 34(67), 628-650. Epub August 14, 2020.https://doi.org/10.1590/1980-4415v34n67a14
20. Becerril Tello, Merit Nefernefer. (2011). Alcances y limitaciones del método etnográfico aplicado al estudio de la cultura de los grandes simios en cautiverio. Cuicuilco, 18(50), 55-77.
21. Gondim, Diego de Matos. (2020). O trabalho de campo na/para/com Etnomatemática como possibilidade de uma pesquisa afecção: potências do devir. Bolema: Boletim de Educação Matemática, 34(68), 1077-1104. Epub November 30, 2020.https://doi.org/10.1590/1980-4415v34n68a12
22. Rosa, Milton, & Orey, Daniel Clark. (2019). Ethnomathematics and the responsible subversion of its pedagogical action: an investigation based on three anthropological approaches. Revista Brasileira de Estudos Pedagógicos, 100(254), 191-210. Epub May 16, 2019.https://doi.org/10.24109/2176-6681.rbep.100i254.3939
23. Martín Camacho, J. (2020). La Etnolingüística como disciplina científica. Propuesta de definición y ámbitos de estudio. Actas do XIII Congreso Internacional de Lingüística Xeral, Vigo 2018, 584-591
24. Sunzuma, Gladys, & Maharaj, Aneshkumar. (2020). In-service Secondary Teachers' Teaching Approaches and Views Towards Integrating Ethnomathematics Approaches into Geometry Teaching. Bolema: Boletim de Educação Matemática, 34(66), 22-39. Epub April 17, 2020.https://doi.org/10.1590/1980-4415v34n66a02
25. Fernandes, Filipe Santos. (2019). Formação de Professores de Matemática em Licenciaturas em Educação do Campo: entre cartas, epistemologias e currículos. Bolema: Boletim de Educação Matemática, 33(63), 27-44. https://doi.org/10.1590/1980-4415v33n63a02

Theoretical Design of the Formation of Professional Communicative Competence of Future Non-Language Experts in the Process of Learning a Foreign Language

Oleh Lazariev[1], *Yuliia* Fernos[1*], *Nataliia* Komisarenko[1], *Liudmyla* Movchan[1], and *Olena* Oliinyk[1]

[1]Uman National University of Horticulture, 1 Instytutska Str., 20305 Uman, Ukraine

Abstract. Competence-based technologies used for teaching a foreign language in non-linguistic higher education institutions, put at the center of all education system student competence, providing comfortable, conflict-free conditions of its development, realization of natural and acquired potentials. The construction of a competence-based educational process presupposes the orientation of the training space on the competence sphere of students. The studied experience makes it possible to identify competence-based "platforms", as well as the main "landmarks" for teaching a foreign language to future experts of non-linguistic specialties. Not only the general English language course, which is quite universal for all specialties, but also the professionally-oriented course, which depends on the chosen specialty in the higher education institution, acquires great importance in the training of future non-language experts. The competence orientation of teaching foreign languages to students of non-linguistic specialties is to ensure proficiency in a foreign language within the framework of both everyday communication and in the specialty at a creative and professional level. Theoretical design of the formation of professional communicative competence of future non-language professionals in the process of learning a foreign language confirms the feasibility of the proposed method, allows us to theoretically justify its effectiveness.

1 Introduction

The changes taking place in our country and in the world require other approaches to learning a foreign language for students of non-linguistic higher education institutions. Prospects for the development of education are due to both the use of innovative methods and technologies in the meaningfulness of the educational process, and the growth of the teacher's professional competence [1].

The working conditions of foreign language teachers have changed significantly over the last decade. Compulsory for all curricula and textbooks, strict requirements for the organization of the pedagogical process are now a thing of the past; the way is open for choice, initiative and creativity of a teacher. The freedom gained by the teachers imposes many additional obligations: to correctly choose a general learning strategy, specific techniques that correspond to personal characteristics, requests and capabilities of students, the conditions under which training takes place; to select textbooks that are recommended to students as basic and additional, programs and information resources that can be used as a basis, etc.

The development of related sciences, the emergence of psycholinguistics, social psychology, theory of activity gave rise to the development of activity-based competence and communicative technology of teaching a foreign language. The focus is on the formation of communicative competence, taking into account the competencies of students and communicative motivation of the educational process.

This means that communication of ready-made knowledge, cramming, reproduction of other people's thoughts cannot be the leading forms of education nowadays. Consciousness in teaching communication is the ability to navigate the characteristics of the audience, to correlate the tasks, forms and means of your message in a proper way; to predict its influence, possible reactions of the audience or partner.

However, all this is only a general starting point that characterizes the modern approach to teaching a foreign language, its specific implementation depends on the teacher, his psychology, temperament, language skills and similar characteristics of his students. The teacher himself has the right to decide whether his students need an oral introductory course, what type of motivation he can rely on in the first place – communicative, cognitive, aesthetic or game[2].

Achieving the main goal of teaching foreign languages to students is possible, in our opinion, only in terms of communication. Thus, communication is a means to an end, a means of learning. The educational task is the formation of students' competencies, which means the possession of a number of skills and abilities that allow them to communicate in everyday and professional spheres. Education of the future expert is considered by us as education of the man of culture and highly qualified expert (in this case by means of a foreign language), capable to solve work problems and

* Corresponding author: fernosyulia@gmail.com

to create proper cultural environment around himself/herself.

If the teacher communicates easily and attracts others to communication, then it is advisable to focus on the development of oral speech by means of dramatization, role-plays, etc. The implementation of this task expands the horizons of students in the field of literature, history, general scientific knowledge, increases their general culture, as well as the culture of thinking, communication, language.

In accordance with these subsystems, it is possible to determine interrelated landmarks that can form the basis of teaching a foreign language in non-linguistic higher education institution in the context of competence-based education.

2 Literature review

Studying the issues of teaching a foreign language, one should focus more on the student's independent creative activity, on the search for the optimal strategy for mastering the language, on the use of his own hidden reserves and potentials. New methods of teaching a foreign language, developed in the 90s, are aimed at this training, which reveals the possibilities of the future expert. There are a number of psychological theories and concepts that are directly related to the process of mastering a foreign language, in particular the theory of gradual assimilation of knowledge and formation of scientific concepts (P. Halperin); psychological theory of collective-distributive educational activity (V. Rubtsov, H. Zuckerman, etc.); psychological theory of formation of educational activity (D. Elkonin, V. Davydov); psycholinguistics in domestic modification (A. Leontiev, A. Zalevskaya, etc.). All these theories and concepts, which eventually developed into the approach first described by the genius psychologist L. Vygotsky, not only do not contradict each other, but can and should lead to the development of new aspects of methods of teaching foreign languages [3]. For example, problem-based learning (PBL) is becoming more and more relevant. This is done by setting problematic tasks (find, learn, understand, express an opinion, argue, oppose, etc.), search and orientate in ways to solve them so that the student can come to an independent conclusion and, on this basis, to activity. Management of educational activities is implemented in the direction from receptive to productive skills. However, some scholars argue that "The sense-making process that PBL is supposed to be then becomes non-sense when the group is not able to recognise and correct misinterpretations. In situations like this we need critical questions, but students seldom ask critical questions" [4].

In the context of a student-centered approach to teaching foreign languages in foreign theory and practice of teaching, the concept of learner autonomy and autonomous self-directed learning was proposed by H. Holec [5]. Its main provisions are as follows. The autonomy of a student in the process of mastering a subject is understood as "the ability to take responsibility for one's educational activity" is defined by H. Holec through the concept of "educational competence". This concept is a personal characteristic and can be considered as a broad student's ability to make decisions, reflection, analysis of the learning situation and his experience, awareness of himself as a subject of learning activity, responsible for its result.

The most adequately productive skills are realized, as a rule, only in interpersonal communication. "By the nature of solving educational problems management should be implemented according to the principle – from receptive to productive, in reception – from visual to auditory modality, from emotionally neutral to emotionally charged, from personality-oriented to socially-oriented, from socially symmetrical to socially asymmetric" [6].

Researcher M. Kabardov believes that according to their individual characteristics, those who study foreign languages are of two types: "communicative type" with developed auditory memory, once in a foreign language environment, they quickly hear the basic stereotypes of language communication. Students of the so-called "non-communicative type" must understand the language system, they have a predominant visual memory, they need to see a foreign text, so it is easier for them to understand its content than to perceive it by ear [7].

The experience of teaching foreign languages in higher education establishments allows us to draw conclusions about the great potential and feasibility of using role-based communication in teaching. Role theory has been used in research related to the use of role-playing games in foreign language teaching [8-10].

3 Results of the study

The use of competence-based technologies will contribute to solving the problem of intensifying the process of teaching a foreign language to students of non-linguistic higher education establishments. Technology can become competence-oriented under two conditions: the availability of competence content emanating from students and teachers, as well as step-by-step implementation. The concept of competence-based education is an attempt to combine humanitarianism and technocraticism. Competence-based technologies used in teaching a foreign language in non-linguistic education establishments, put in the center of the entire education system the competence of the student, ensuring comfortable, conflict-free conditions for its development, and realizing natural potentials. Student's competence in this technology is a priority of the object, the goal of the education system, and not a means to achieve any abstract goal. Competence-based technologies are aimed at diverse, free and creative development of the student. They are more flexible and varied, related to specific problems and situations of individual, personal development of the subjects of the educational process: students and teachers. A teacher needs to organically combine two systems: teaching the subject and working with students.

Nowadays, it is a completely natural process of mastering new technologies, in which traditional methods, and developmental learning, as well as competence-based elements of teaching can be used. In this combination, the teacher of a foreign language of non-linguistic education establishment gets an opportunity for personal orientation of various stages of educational activity in the use of proven methods (including personality-oriented, consciously-practical, problem-solving, communicative, intensive learning and game modeling).

The construction of a competence-based educational process involves the orientation of the learning space on the competence sphere of students. Training is based on dialogues of various levels: formal dialogue – dialogue as a form of communication between participants in the educational process; meaningful dialogue – presentation in a dialogical form and content of educational activities; semantic dialogue – dialogue as a way to establish value-oriented unity.

In the conditions of competence-based teaching of a foreign language, each student of the group is actively involved in the communication process, gets a real opportunity to meet socially important needs in achieving prestige, status, attention and respect from others. The teacher creates a tone of communication saturated with joy, benevolence, satisfaction with the success achieved and confidence in the final result of training.

Having studied the previous experience, we will consider the features of competence-based "platforms" for teaching a foreign language to future experts of non-linguistic specialties.

Modular platform. A separate module of the studied educational material is necessarily accompanied by rating control of students' knowledge and skills. The basis for the formation of modules is the curriculum of the discipline, consistent with the government educational program. Control of modules includes: modular control, tests or exams.

Expansion platform. Teaching with increased didactic units provides a general vision of the topic, a holistic perception of the material. Several lessons on one common theme are united by a cross-cutting plot.

Communication platform. The main participants in the process are the teacher and the student. The relationship between them is based on cooperation and equal language partnership.

Semantic platform. The main thing is to create emotional and psychological attitudes, the ability to influence the feelings of students, building the drama of the lesson.

Interaction platform. The work is organized in pairs of variable composition.

Problem platform. The teacher creates a problem situation and students solve it independently, as a result of which there is a creative mastery of professional knowledge, skills, abilities and the development of thinking abilities.

Business game platform. Students imitate reality. This helps fit the educational process in the context of their real life as future experts. Students act out situations aimed at recreating and assimilating social and professional experience, in which self-management of behavior develops and improves.

Intensive learning platform. Techniques are used that activate the conscious and subconscious processes of the psyche to create a large and strong language base. Tasks that motivate communication are developed. The use of this technology involves an increase in the number of hours.

Advance platform. Advanced knowledge and tasks, such as those that precede the perception of a foreign language or reading a text in a foreign language, contribute to the effective preparation of students to learn new material.

Information and technical platform. Use of computer, Internet, video, audio, multimedia, interactive means, etc. One of the organizational foundations is the digitalization of education, the use of information technology and software for learning a foreign language.

Remote platform. In response to the challenges of global circumstances that dictate the organization of distance and blended learning, effective forms of learning foreign languages are being developed in conditions of forced and planned extracurricular learning.

The principles of the competence-based approach and the orientation of learning to the development of a student's secondary linguistic personality are correlated with the subsystem "learning objective". The following principles are correlated with the subsystem "learning content": situationality, modeling, dual function of educational communication. The principles of homogeneous formation of subgroups, the activity basis of learning, and group influence determine the learning process. The principle of orientation towards a profession/specialty corresponds to the subsystem "learning outcome" [11].

According to the proposed principles, we will consider the main landmarks for teaching a foreign language to future non-language experts.

Personal landmark. The teacher and the student are like-minded people. Human communication begins with establishing contact. Initial communicative adaptation and readjustment of those who communicate are needed. In such conditions, complexes, internal tension are removed, confidence and mutual understanding come.

Direction landmark. Learning a foreign language is aimed at developing the features of the secondary language personality, that is, the teacher not only teaches the student to behave with the help of a foreign language as a native speaker, but also develops his willingness to carry out textual activities in the language studied, the ability to adequately interact with other cultures.

Situational landmark. When learning a foreign language, it is important to select and organize material based on situations and communication problems that interest each student. Motivation to speak arises only in the reproduced situation, as a result of which they speak.

Modeling landmark. The amount of knowledge of the country is very large, threfore, it is necessary to select a sufficient amount of material to present the culture of the country in a concentrated form. In this case, the content of the language consists of problems, not topics. In order to minimize and streamline linguistic knowledge, the language system is given in a model form: schemes, tables, comparisons, etc.

Double landmark. The dual function of educational communication is that this communication is planned for the teacher. The teacher guides the communication in order to form, practice and consolidate the skills and abilities of all types of speech activity. For students, communication is a goal, since with the help of language they can go beyond educational situations.

Differentiation landmark. Students take placement test, which determines the level of their knowledge of a foreign language. According to this level, educational subgroups are formed.

Activity landmark. The activity basis of learning is expressed in the external and internal (mental) activity of the student, passive and active stock of knowledge. The volume of independent, group and collective forms of learning is increasing, while the traditional frontal work, in which language activity is shown by the teacher, and students are active only if they are asked, is being reduced.

Group landmark. The individuality of each student is revealed through communication. Interpersonal interactions can be considered in three directions: communication between teacher and student, communication between teacher and the whole group, communication between students. The educational subgroup should have a psychological microclimate that allows you to effectively identify and reveal the capabilities of each student.

Professional landmark. This orientation is based on profession and specialty. The curriculum of teaching a foreign language to students of non-linguistic specialties focuses on learning the basics of business communication and the language of the specialty. In higher education institutions, students are gradually immersed in the language and essence of the profession or specialty. They learn to communicate in a foreign language and read literature on their future specialty, deal with business documents, annotate scientific texts, etc.

In the methodology of teaching foreign languages, the importance of a role is considered in the structure of educational and speech situations, where it is determined by the fact that the role can contain information in the estimated volume of a language product, the sphere of oral communication, language material necessary to formulate an utterance, the subject content of a message and other characteristics of the speech act. The role is considered as an important component of the learning-speaking situation, which focuses on its basic parameters and the corresponding concept of conversational practice. Game conditions are always correlated with real conditions, this makes it easier to find patterns of behavior and means of verbalizing statements. In these conditions, the information that the student receives from communication partners is also involved in the individual cognitive process. B. Lomov believes that "this, first of all, determines the higher efficiency and originality of the dynamics of cognitive processes in communication" [12]. As a result, the student's attitude to the subject, which appears to him not only as an object of knowledge, but also as a means of satisfying needs in certain relationships with group members, changes. In other words, the effect of solidarity, which in fact characterizes any collective activity, becomes a personal motivation for everyone's activity.

Role-playing at a later stage of learning can be combined with a business game. Business games are a type of learning, in which the acquisition of knowledge, the formation of skills and abilities are superimposed on the canvas of professional activity, in training it is presented in a certain model form. Makoviei O. M. states that "the advantage of business games lies in combining both theory and practice, contributing to the formation of professional knowledge and practical skills. Games increase the interest in the subject being studied, as they are accompanied by positive emotions." [13].

However, transferring business games to the field of teaching foreign languages, one should realize that business games must be a logical and natural continuation of role-playing games in conditions when students are linguistically and professionally ready to implement foreign language professional activity. Thus, business games should be introduced in the last years of non-linguistic higher education institutions, if the previous stage of training was conducted in the form of role-playing games, which are necessary in the business game, when students' attention is focused on the professional aspect. In this case, it will be justified by the fact that in the last stages of study, students already have an idea of their future professional activity and must prepare for it. It is role-playing games that teach the norms of verbal communication, provide mastery of those communicative units that are necessary for participation in a business game, when students' attention is focused on the professional side of verbal behavior and language activities.

Describing the competence-based organization of foreign language teaching of future non-language experts, we must not forget that it characterizes only one side of a complex dialectical unity, consisting of the interaction of organized in a certain way educational material and the process of teaching itself. Thus, the adequacy of the organization of educational material for the learning process becomes of paramount importance. The organic unity of the subject and the process in teaching a foreign language was first proposed by G. Lozanov, who claims that the main educational text is a polylogue, and the participants in this polylogue are the students themselves [14]. According to the story line of the whole course, each student gets a "new role" wearing a "new mask" with its responsibilities, based on which a proper demeanor is chosen. A student must rebuild relationships with all unfamiliar members of the group within the norms and rules that are set in the new organization, i.e. in the group of tourists, or in the group of experts who came to the congress, or in a group of students from any country who came to study in this higher education establishment, etc. In the practice of such training there

 https://doi.org/10.1051/shsconf/202110002002

is no conflict between "student-mask" and "student-student", because the educational texts contain, at least symbolically, positions, values and personal attitudes of actors, which do not contradict the worldview of students and contributes to disclosure of positive qualities, in particular, creative imagination, etc.

The significance of the introduction of the legend-mask according to A. Leontiev [2], is that it allows the student to "express himself/herself", which generates a psychotherapeutic, psycho-hygienic effect, as it frees students from "clamps", from fear of mistakes, feelings of shame and inconvenience. Psychological barriers are removed, students' creative opportunities are opened and activated. Changing the names and distribution of legends in the group is to some extent a methodological technique, since the phonemic and lexical-grammatical organization of the "autobiography" of the group members allows to put phonemes in semantic opposition, and the development of "legends" in vital situations partially compensates for the lack of an introductory phonetic course. Situations of educational dialogues are selected in such a way as to simulate the speech behavior of each, unite the group with a single activity of verbal communication. This joint language activity, "embedded" in the textbook, forms a team and creates a favorable psychological climate for joint activities [15].

For example, one of the variants of modern technology of teaching students of humanitarian universities, proposed by L. Shestoperova: "Learning is a system of generalized structured knowledge, based on clearly constructed functional modules, as well as differentiated rules of formation and operation of grammatical phenomena for reading and translating special popular science and special literature" [16]. In the training of future non-language experts, not only the general English course, which is quite universal for all specialties, but also a professionally oriented course, which depends on the chosen specialty of the higher education institution, is of great importance. Business English is predominantly taught in the Master's program.

In addition, due to the lack of hours of teaching a foreign language, it is being intensified – the types of work with text are expanding, the number of creative exercises is increasing in comparison with lexical and grammatical ones. Creative opportunities are manifested not only in the course of ordinary text-based training sessions, but also in the preparation of messages, in the expansion of topics, more complex vocabulary is mastered. The nature of the dialogues is also changing – professional topics are discussed: the economic and professional components of labor, the conclusion of contracts, the criteria and principles for hiring, the types of activities of the organization, management, and the like. Such a topic not only significantly expands the language capabilities of students in a professional sense, but also contributes to the growth of future experts' confidence in their capabilities, the formation of professional communicative competence, an increase in interest and motivation in learning a foreign language, for its more in-depth assimilation [17].

In parallel, work should continue on reading periodicals or books in a foreign language in the specialty, watching films, news, social networks, Internet resources. At this stage, the student's task is not only to reveal the content, but also to analyze what has been read. The learning process includes group discussions, comparison with their own knowledge, identification of ambiguities. Practical classes are held in the form of round tables, debates, seminars where certain professional problems are discussed, and the like. The topic and nature of such conferences is determined both by the teacher and offered by the students of the educational subgroup. For example, "Farming in Ukraine and GB", "Agroecology", "Students' scientific researche", "Future career", "Job hunting" and others. If the subgroup is already accustomed to this type of activity, then it not only enthusiastically takes part, but also prepares for this type of activity, especially willingly selecting and deeply studying a large amount of material on the selected topic. It is precisely such a diverse multilateral educational activity that not only helps to study and master a foreign language, but also stimulates research work, independence, develops creative thinking, professional communicative competence [18].

However, if the main task was to provide the student with language training for a short trip abroad in the shortest possible time, it would be advisable to use G. Lozanov's method [14] or methods that go back to it, developed, for example, by G. Kitaygorodskaya and I. Shekhter [19, 20]. Practice shows that in a couple of weeks of intensive classes using this technique, English can be mastered at a level sufficient for effective communication in various life situations – at the airport, hotel, shop, on the street, etc. But if a student does not go to an English-speaking country in the near future and does not have intensive language practice, after a maximum of six months, everything learned will become passive vocabulary. A separate approach is required for effective preparation for the single entrance exam students take to get their master degree in Ukraine, where the tasks textually do not meet either the curriculum requirements or areas focused on the specialty. And what ultimately turns such an exam into a "dropout tool" is the unjustified time limit of up to one hour for all exam tasks. If English is required for professional purposes, and the student intends to travel, for example, to the United States for a long time, in this case it requires a different teaching method that allows you to gain a basic knowledge of the language, as well as in-depth and expanded knowledge of it. Thus, not all foreign language courses are professionally oriented in terms of methodology.

The distribution of educational material over time should take into account not only the methodological conditions and requirements, but also the attitude of students towards it. In the context of the competence-based orientation of teaching foreign languages, it is necessary to take into account the interest of students in one or another topic, the desire to develop it or move on to another. It should be noted that the focus on students' interest in the distribution of educational material contributes to the strengthening of trust between the teacher and students and increases the responsibility of students for the results of work.

Under this approach to learning, students develop their skills related to intellectual and mental processes: to receive, analyze, compare, develop, systematize information focused on a specific educational goal; to express one's own opinion from what has been read or listened to, to generalize information; make a plan, formulate abstracts, compile a summary; prepare reports, presentations, essays or short messages; search for and extract the most important information, according to a certain educational task.

Theoretically, after completing such a competence-based course, students should be able to: conduct a conversation in the conditions of everyday, professional and business communication – at meetings with foreign colleagues during negotiations, when concluding agreements and signing contracts, at seminars, webinars, workshops, round tables, conferences, symposia, congresses, etc.; have the skills to speak on different topics and in different situations – a story, a short information, clarifications, explanations, a report, a consultation; process business correspondence – draw up business letters, send and receive information through modern means of communication; read, translate and understand the main content of original professional texts, extract necessary, important and useful information.

In this study, the positive role of languages is that they can be used outside the national framework of their functioning as: one of the means of expanding access to information, its dissemination regardless of national borders; a means of understanding and studying the diversity of cultures and civilizations, ways and results of their interactions in the world flow of culture; means of interethnic and intercultural cooperation and cooperation in science and social activities of people, contribute to the search for common ways to solve common problems, ways out of conflicts arising in human relations, society and nature; a means of overcoming national and cultural limitations in various spheres of society; a means of preserving, disseminating and developing world culture; one of the means of removing the linguistic barrier in ensuring the human right to free movement in the world, choosing a place to study and work, leisure, participation in various international organizations. But the most important function of language is to serve as a tool for the development of global thinking and a means of teaching intercultural communication.

In turn, the professional portrait of an expert consists of a set of professional, competence, socio-psychological and personal qualities. In an integrated form, these qualities can be identified, written off and interpreted as professional communicative competencies: the ability to interact, to integrate with experts in related and other specialties; ability to manage the workforce; ability to organize people; follow the ethical and legal norms of society; presence of axiological formations – ideals, values, priorities, motivations, principles, ambitions; knowledge of ways to resolve conflicts, the ability to conduct business negotiations; knowledge of cultural, national, socio-psychological characteristics of the behavior of people of a region, country; mastery of communication techniques, language skills, forms of language communication. The effectiveness of the process of formation of professional communicative competencies depends on a number of conditions: the organization of foreign language learning in the context of future professional activity – vocational training; application of a system of measures to increase motivation, development of students' need for professional communicative competencies; development of the mechanism of pedagogical support; development of educational and methodical support. All of the above creates the basis for the formation of professional communicative competence of future non-language professionals.

4 Conclusion

Thus, the theoretical design of the formation of professional communicative competence of future non-language experts in the process of learning a foreign language confirms the feasibility of the proposed method, theoretically justifies its effectiveness, including measures to increase students' motivation to master professional communicative competence, a set of techniques and tools for the formation of skills and abilities of verbal communication, organization of the learning environment.

In order to teach a foreign language using a professional method, it must, first of all, meet several interdependent points: adequacy of learning goals and conditions; ensuring the acquisition of all aspects of language (and not only learning vocabulary and consideration of grammatical phenomena); the possibility and availability of technology to monitor the effectiveness of training.

Drawing on the pedagogical experience of teaching a foreign language to students of non-linguistic specialties on the basis of a competence approach, it is necessary to begin the educational process in the first year to determine the personal interests of students and their prior language training. Based on these data, you can form subgroups and select a curriculum. Experience of teaching in non-linguistic higher education institutions, interviews with students and questionnaires showed that students working under CBL realize the importance of a foreign language in becoming a professional. Basically, such a program is aimed at modeling learning, which stimulates the development of competencies of students who act as subjects of the educational process in a foreign language.

Thus, the competence-based approach of teaching foreign languages to students of non-linguistic specialties is to ensure knowledge of a foreign language in the framework of both everyday communication and specialization at the creative and professional level. The integrity of the learning process depends on the interconnection of all its components. The target setting for the development of a competent expert strengthens the emphasis on educational tasks with a focus on dialogues about world and domestic cultural values

Mastering a foreign language without getting acquainted with the culture of the country, the mentality of people who speak this language can not be complete. In other words, it is necessary to master not only the language itself, but also the worldview of those who speak it. This does not mean that one should think and act like the people of the country whose language is being studied. But you should put yourself in the place of an English-speaking interlocutor to know and understand at least a minimum of what surrounds him in everyday, social and professional life.

So, the process of mastering a foreign language becomes economical and effective if the student understands why and how he learns, and realizes the communicative value of each language unit that he masters. It is also important to establish a link between practical and developmental goals. Here, the developmental goal is considered not as an expansion of horizons (educational aspect), but as the development of intelligence, respectively, the development of intellectual operations, which underlie the functioning of language skills and abilities.

Thus, the above allows us to assert that the competence-based organization of educational material provides a clear relationship between the academic discipline and the educational process, contributes to the effective implementation of the goals of teaching a foreign language and the formation of professional communicative competence of future non-language experts. One of the main provisions of the competence-based approach to teaching a foreign language is the focus on developing the competence of the student as an active subject of educational activity and comprehensive preparation for the continuous process of education, self-development and self-improvement throughout life.

References

1. A. Dubskikh, N. Zerkina, in *Current trends in the development of education system*, ed. by Zh. Murzina, G. Nikolaieva, N. Tolstov (Cheboksary, 2018), p. 108. doi: 10.31483/a-28
2. A. Leontiev, FL at school. **2,** 27-32 (1986)
3. L. Vygotsky, in *The collected works of L.S. Vygotsky,* ed. by R.W. Rieber & A.S. Carton. Thinking and speech, vol. 1 (New York, Plenum Press., 1987), pp. 39–285
4. Dirk Humalda, Research in HM. **Volume 6 (2),** 207-212 (2016). doi: 10.1080/22243534.2016.1253292
5. H. Holec, *Autonomy and Foreign Language Learning*, 2nd edn. (Pergamon, Oxford, 1979).
6. I. Zimniaia, *Psychology of teaching a non-native language.* (Moscow, 1989), p.97
7. M. Kabardov, Ye. Artsishevskaia, Psych. issues. **1,** 34-49 (1996)
8. A. Kotkovets, Journal of the National Technical University of Ukraine "KPI": Philol. and Educ. Studies. **7,** 43-49 (2016)
9. Rafsanjani Ali Akbar, Andhika Ratna, Budiana Haris, VEL, **Vol 1,** No 1, (2020). Available at http://jurnal.unswagati.ac.id/index.php/ELTERA/article/view/2340
10. O. Honcharova, A. Krasulia, in *Innovations and traditions in language training of students.* Proceedings of the international scientific-practical seminar, Kharkiv, 6 December 2019, pp.91-93
11. I. Stasiukevich. Dissertation, Stavropol State University, 2005. p.109
12. B. Lomov, Psych. journal. **Vol.1, 1,** 23-45 (1980)
13. O. Makoviei, Automobile Transport. **Vol. 43,** p.119 (2018). doi: 10.30977/AT.2219-8342.2018.43.0.118
14. G. Lozanov, Dissertation, RUDN University, 1970
15. G. Kitaigorodskaia, *Activating learning activities.* (Moscow, 1982)
16. L. Shestoperova, *A practical course on the main sections of English grammar.* (Moscow, 1995)
17. O. Lazariev et al, in *Vision 2025: Education Excellence and Management Innovations through Sustainable Economic Competitive Advantage,* ed. by Khalid S. Soliman. Proceedings of the 34-th IBIMA Conf. 13–14 Nov. 2019. (Madrid, Spain, 2019), pp. 11099–11107
18. O. Lazariev, Yu. Fernos, Pedagogy of creative personality formation in higher and general education schools. **Issue 73, vol 2**, 77-81 (2020). doi: 10.32840/1992-5786.2020.73-2.14
19. G. Kitaygorodskaya, *Methodology for intensive teaching of foreign languages* (Moscow, 1986)
20. I. Shehter, in *Methods and psychology of intensive learning* (Moscow, 1981)

Development of Future Managers' Resilience as a Condition for Efficiency and Reliability of Management Activities

Volodymyr Kravets[1], *Halyna* Meshko[1*], *Oleksandr* Meshko[1], *Alina* Leskiw[2], and *Nataliia* Habrusieva[3]

[1]Ternopil Volodymyr Hnatiuk National Pedagogical University, Ukraine
[2]Bronisław Markiewicz State Higher School of Technology and Economics in Jarosław , Poland
[3]Ternopil Ivan Puluj National Technical University, Ukraine

Abstract. The article is devoted to the study of the future managers' resilience development problem at the stage of Master's training. The state of development of the specified problem in psychological scientific literature and practice was analyzed. The essential approaches to interpreting resilience were clarified and the factors of resilience, considering the analysis of the scientific literature, were analyzed. The method of determining neuropsychological resilience, the risk of maladaptation in stress "Prognosis", adapted method of L. Wasserman "Diagnosis of neuroticism" and SACS-questionnaire and models of coping behavior, developed by S. Hobfoll, were used to measure the ability to maintain psychosocial adaptation during stress and to diagnose the level of neuroticism, neuropsychological resilience to stress. The results of the study served as a basis for finding alternative ways of developing future managers' resilience. The program of future managers' resilience development, focused on the formation of productive interaction and constructive overcoming the professional difficulties skills; mastering constructive coping strategies; learning methods and technics of self-regulation; formation of assertive behavior and positive thinking, self-efficacy of the individual, was substantiated and developed. The main ways of the offered resilience development program's realization were outlined. The role of the course "Occupational Health" in the development of future managers' resilience was developed and characterized.

1 Introduction

New requirements to the manager's personality constantly appear under the current conditions of development, reform, and modernization of society, the rapid leap into the new reality of quarantine restrictions because of coronavirus infection. A modern manager needs the ability to adapt to rapid changes in living conditions. He needs the skills and qualities necessary to work in new realities, distant work technologies, virtual communications, distant team management, etc. A manager needs something that gives psychological strength to hold on and withstand the challenges of modern society, constructively overcome life and professional difficulties, recover from stress. Therefore, the basis of quality life and successful management is high-stress resistance and resilience. They help to adapt more easily to changes, increase the level of viability of the manager, and contribute to the preservation of team spirit and the formation of a favorable socio-psychological climate of the organization.

Stress resistance is one of the personality traits that ensures the success of overcoming professional stress. Stress resistance determines how long a specialist can be in an uncomfortable environment without harm to physical and mental health. Stress resistance helps the specialist to protect his personality from disintegration; various disorders create a basis for inner harmony, high efficiency, and determines the state of his professional health [1].

If stress resistance means how easily we can cope with stress, resilience depends on how quickly we can get back in shape after the stress and at what time we can maintain a positive psychological attitude. Resilience is a person's ability to cope with difficult life events and recover from difficulties or stress [2]. Therefore, it is extremely important to develop such a quality of a specialist as resilience, which allows to overcome the negative consequences of encountering adverse circumstances and serious professional difficulties, to prevent destructive behavior, to ensure the quality of further life and professional activity.

It is important to develop both stress resistance and resilience simultaneously and in unity, because a specialist will be able to quickly deal with stressors, but will "go away" for a long time, or recover quickly from stress, but when you meet a new stressor again experience a destructive surge of emotions.

All this actualizes the professional stress resistance and resilience formation problem of the future managers at the stage of education in a higher education institution during the Master's training in particular. Our previous research was devoted to the problem of future heads' of educational institutions' professional stress resistance formation. [3].

2 Literature review

* Corresponding author: hal-meshko@ukr.net

Modern science developed a significant arsenal of tools for managing stressful situations, ways to optimize functional states, methods of preventing negative emotional states [4, 5, 6]. But research is mainly aimed at preventing the occurrence of stress in professionals, increasing the level of their stress resistance. Insufficient attention is paid to the formation of stress resistance and resilience of future professionals. The possibilities and ways of developing resilience in future managers are unexplored.

In the scientific literature, the terms "psychological resilience", "stress resistance", "stress resistance", "stress tolerance", "frustrating tolerance", "viability", and "viability" are used as synonyms for resilience.

Stress resistance is understood as a structural-functional, dynamic, integrative property of personality and the result of the transactional process of collision of an individual with a stress factor, including self-regulation, cognitive representation, objective characteristics of the situation, requirements to personality [4, p.74] Stress resistance provides the ability to carry out professional activities, withstand significant intellectual, volitional and emotional stress in stressful conditions while maintaining a state of psychophysical balance and avoiding the consequences of harmful effects on activities and their own health. Indicators of stress resistance are emotional stability, self-control, self-regulation, psychological stability, frustration tolerance [5, p.27].

The concept of resilience was empirically discovered more than 40 years ago. The term "resilience" in psychological, scientific terminology was introduced in the works of E. Werner and R. Smith [7]. In the scientific literature, personality quality and resilience are interpreted as the ability to maintain resilience to external and internal threats without losing the pace of development. [8, p.160]. Resilience is a dynamic process that involves positive adaptation in the context of significant problems [9].

O. Khominich [8] analyzes the approaches to defining the concept of "resilience" in domestic and foreign psychology, understanding this phenomenon as a property of the individual and as overcoming stress. The scientist proposes to use the concept of "resilience" in domestic psychology to denote a dynamic process in the contexts of adaptation and self-regulation.

The term "resilience" is generally associated with the ability of the psyche to recover from adverse conditions [10]. Resilience plays an important role in the ability and formation of post-traumatic personality growth [11]. Highly resilient individuals have strong coping skills, but they cannot cope with the psychological consequences of trauma. They also do not have the inner motivation to feel new experiences, which brings many new opportunities for personal change, which offers the most traumatic situation with its emotional tension. [12].

In a study by G. Lazos, [13] the data of modern scientific researches concerning the resilient qualities of the person were analyzed. The scientist substantiated and developed a theoretical and methodological model of psychotechnics of resilience development and proposed the structure of a theoretical model of resilience through its main constructs risk factors, protective factors, and vulnerability factors.

Scientists focus on the diagnosis of both the general level of resilience and protective (resilient) factors, resources, competencies, the developed psychotechnics of influence. In his article, [13] G. Lazos gives examples of effective psychotechnologies of development and formation of resilient personality traits. Most of the psychotechnics are based on a combination of different psychotherapeutic methods and technics. Modern resilience psychotechnologies include spiritual practices, physical procedures, recommendations for maintaining and promoting health, and so on.

The purpose of the research is to identify and analyze opportunities and ways to develop resilience in future managers at the stage of the Master's degree. The purpose of the study was specified in the following tasks: 1) to determine the state of development of the problem of development of resilience of future professionals; 2) to diagnose the formation of resilient personality traits of the future manager; 3) taking into account the results of diagnostics to develop a program for the development of resilience of future managers at the stage of master's degree.

3 Research methodology

The model of resilience proposed by G. Lazos is important for our study [13], comprising four separate phases: a confrontation with risk factors (with traumatic events); activation of protective factors and vulnerabilities; the interaction between protective factors and vulnerabilities; creating resilience or going into a maladaptation.

In research, works devoted to the study of factors influencing the development of resilience are of interest. In the scientific literature, there are factors that affect resilience: a friendly attitude to themselves and their abilities; ability to make realistic plans and implement them; development of communication skills, in particular, assertiveness (this is the ability to defend oneself in a polite and friendly way, to express one's feelings and thoughts without offending or violating the dignity of other people); nurturing values and planning life according to them; development of own "support system": meaningful and deep relations with close people, with nature, with God, and also with itself; nurturing hope; developing the ability to solve problems [2].

Foreign scientist A. Masten [14] created a list of resilient factors and corresponding adaptive systems: effective holding in childhood and the quality of education; close relationships with other adults; the presence of close friends and romantic partners; intellectual development and problem-solving skills; self-control, emotion regulation, planning; motivation for success, personal effectiveness; faith, hope, meaning of life; effective training; efficiency in relations with neighbors, colleagues.

Positive thinking plays an important role in the development of resilience. Positive thinking helps you to manage difficult, conflict situations and your efforts to resolve them. M. Moltz calls positive thinking a universal intellectual tool, the essence of which is to strengthen the positive image of one's "I" through successful deeds and actions and not does not allow it to deteriorate or collapse in case of failures or mistakes in behaviour, activities, life [15]. The scientist names two main reasons why the image of one's own "I" is the key to a successful life: 1) all actions, feelings, deeds, even abilities are consistent with the image of one's own "I" through unconscious and conscious mechanisms of self-regulation; 2) if the holistic image of one's own "I" changes, then life and professional problems, including complex, stressful, traumatic ones, which are consistent with this new image, are solved much easier, without undue effort. Therefore, M. Moltz recommends the formation of positive thinking, starting with the search for oneself, with their own self-acceptance and adequate self-esteem. It is also important to study the principles of positive thinking and constantly adhere to them, change your worldview, a worldview toward optimism for positive thinking to become a way of life and activity.

The growth of such personal formations as intelligence, reflection, self-attitude, motivation to succeed, self-efficacy, will contribute to achieving a high level of formation of positive thinking [16, 17]. Therefore, the development of positive thinking of future managers involves the formation of a creative attitude to management, the ability to find joy in achievements, insignificant successes and achievements, conscious control of their thoughts and emotions, strengthening faith in their own strengths, capabilities, success, the ability to be optimistic in life and professional situations, even complex and extreme.

The technology of changing the self-concept and forming positive thinking, developed by V. Kaloshin, is important for our research. It involves the following stages: awareness of the need and possibility of a positive self-concept; relaxation; subconscious programming (change of self-concept); repeat the last two stages over a period of time (at least 21 days) [18]. A similar approach is presented in the work of B. Tracy, which proposes a five-element (five-stage) process of gaining strength to become highly effective, positive thinking: the idea of the ideal → translation into a verbal form → visual representation → emotional perception → implementation [17].

Creating a program for the development of resilience S. Osborn [19] suggested focusing on the targets of psychological / psychotherapeutic interventions to enhance human resources to promote resilience to life's problems, rather than focusing on and exploring their conditions (post-traumatic stress disorder, exhaustion, burnout, etc.). This provision was the basis for the development of a program for the development of resilience of future managers and served as a guide for the content of training practice.

To identify opportunities and ways to develop the resilience of future managers, to develop a program for the development of resilience of future managers at the stage of Master's degree, it is important to diagnose the formation of resilient personality traits of the future manager.

4 Research methods

To diagnose the level of the neuroticism of the personality of future managers, neuropsychological stability in stress, the ability to maintain psychosocial adaptation during stress we used: adapted method of L. Wasserman "Diagnosis of the level of neuroticism" [20, p.193 -194]; method of determining neuropsychological stability, the risk of maladaptation in stress "Prognosis" [20, p.195-198]; questionnaire of SACS-strategy and model of coping behavior, developed by S. Hobfoll [20, p.230-238]. 84 applicants for higher education were involved in the study of the second (master's) level of specialty: 073 Management of Volodymyr Hnatiuk Ternopil National Pedagogical University and Ivan Franko Drohobych State Pedagogical University (Ukraine).

The results of the study formed the basis for the development of a program for the development of resilience in future managers.

Adapted method of L. Wasserman "Diagnosis of the level of neuroticism" [20, p.193-194] provides answers "yes" or "no" to each of the 30 judgments. The more positive responses received, the higher the level of the neuroticism of the individual.

The use of the method "Determination of neuropsychological stability, the risk of maladaptation in stress" Prognosis " [20, p.195-198] requires a "yes" or "no" answer to each of the 84 statements. The number of correct answers, according to the key, was counted. The higher the number of points scored, the higher the level of neuropsychiatric instability. The obtained points correlate with the conditional scale of neuropsychological stability. The higher the points on the conditional scale, the higher the level of neuropsychological stability and the lower risk of maladaptation under stress. The methodology contains a scale for checking the sincerity of the answers, which allows determining the reliability of the results.

Questionnaire SACS-strategy and model of coping behavior, developed by S. Hobfoll [20, p.230-238] contains 9 subscales, each of which reflects a particular model of behavior in a complex (stressful) situation. Respondents were asked to choose the answer ("no, not at all", "yes, quite right", "difficult to answer", "rather no than yes", "rather yes than no") to 54 statements that characterize their behavior in a difficult (stressful) situation. When calculating the results of the survey, we use a key for the representatives of communicative professions, which makes it possible to identify the predominance of one or another model of behavior in a tense (stressful) situation.

5 Research results

The results of our study using the method of determining neuropsychological stability, the risk of maladaptation in stress "Prognosis" indicate a significant percentage of

 https://doi.org/10.1051/shsconf/202110002003

future professionals (38.1%) with a low level of neuropsychological stability and a high risk of maladaptation in stress (figure 1).

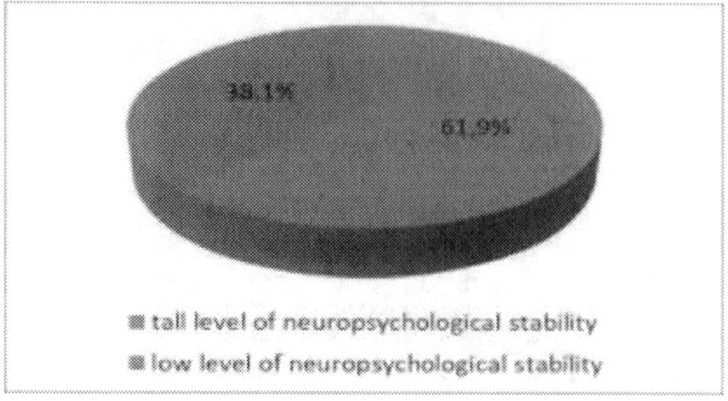

Fig. 1. The results of our study using the method of determining neuropsychological stability, the risk of maladaptation in stress "Prognosis".

Using the adapted method of L. Wasserman "Diagnosis of the level of neuroticism" [20, p.193-194] made it possible to identify the predominance of future managers with an average level of neuroticism (52.4% of respondents). The low level of neuroticism, which is characteristic of 25.1% of respondents, shows emotional stability, a positive background of feelings (calm, optimism), initiative, self-esteem and self-sufficiency, independence and social courage, ease of communication. While the high level of neuroticism, found in 22.5% of respondents, is evidence of pronounced emotional excitability, lack of initiative and passivity, egocentric personal orientation, communication difficulties, increased susceptibility to mental tension and stress (figure 2).

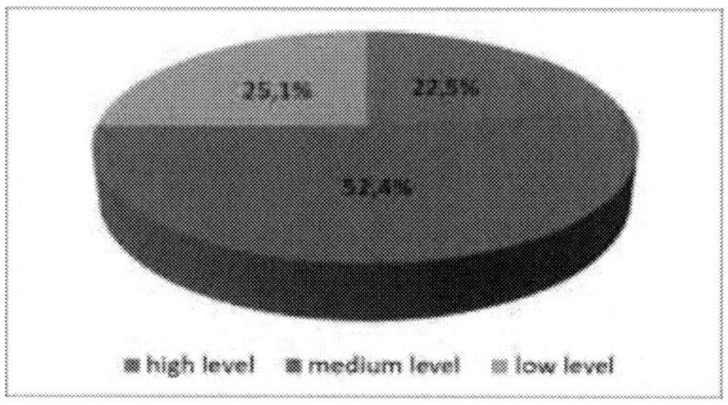

Fig. 2. Distribution of future managers by level of neuroticism.

Research using the SACS-strategy questionnaire and coping behavior model [20, p.232-234] made it possible to diagnose the predominance of unconstructive models of behavior of future managers in a stressful situation.

The results of the study show that undergraduates have a small number of constructive models of overcoming stressful situations, which is expressed in low rates of assertive behavior (dominated by 2.4% of respondents), entering into social contact (11.9%), seeking social support (14.3 %), in higher rates of caution (predominant in 16.7% of respondents), aggressive (19.04%) and manipulative actions (26.2%). Impulsive actions are typical actions for 2.4% of respondents, avoidance - 4.8%. antisocial -2.4% (figure 3).

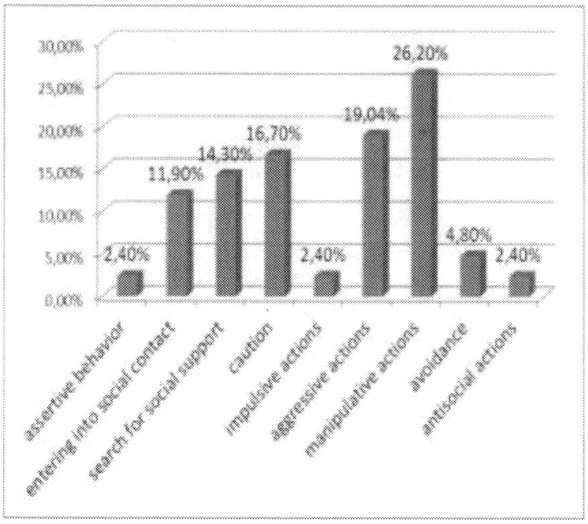

Fig. 3. The predominance of behavior patterns of future managers in a stressful situation.

Coping helps maintain psychosocial adaptation during stress or in difficult situations. Studies show that "successful" managers differ from "unsuccessful" by more different models of overcoming difficult (stressful) situations. High rates of assertive behavior, social contact, and social support, and low rates of aggressive and antisocial behavior [21]. Models of their behavior are characterized by activity, prosociality, and flexibility. For "unsuccessful" managers, these figures are just the opposite [21].

6 Discussion

The results of diagnosing the level of personality neuroticism, neuropsychological resilience in stress, the ability to maintain psychosocial adaptation during stress give grounds for concluding the low level of formation of resilient qualities of future managers and the need and assumptions about the possibility of targeted influence on students' resilience development on the stage of their Master education in a higher education institution.

Taking into account everything that was mentioned above, the program of resilience of future managers was developed by us, which involves (figure 4, 5):

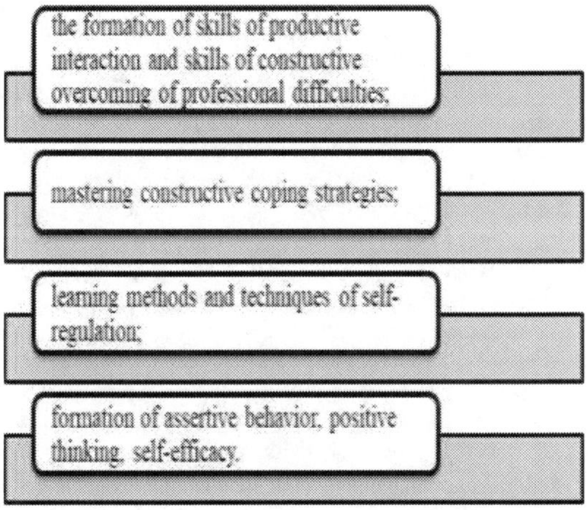

Fig. 4. Program of development of resilience of future managers.

The development of resilience of future professionals involves the elimination of potential sources of stress and neurotic components in their internal and strengthening resources that will contribute to resilience to life's difficulties and occupational difficulties.

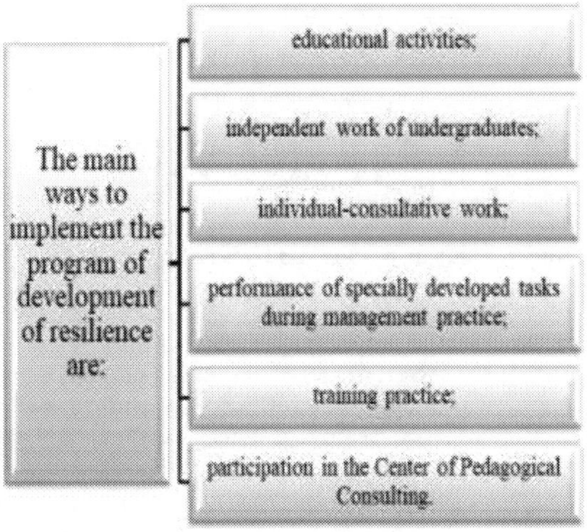

Fig. 5. The main ways to implement a resilience development program.

The course "Occupational Health" developed by us is directed on the formation of competence of health care, professional stress resistance, and development of resilience of undergraduates [22]. The objectives of the course are: mastering by undergraduates a system of knowledge about the essence of occupational health, its criteria, determinants; mastering the basics of stress management in management; mastering modern health technologies, psychotechnologies of self-healing, technologies of self-rehabilitation; motivation to find their own systems of recovery and harmonization of their own inner world [22].

The course "Occupational Health" (3 ECTS credits, 10 hours of lectures, 20 hours of practical classes) contains two content modules: "Occupational health as a condition for effective management" and "Ways and means to preserve and enhance occupational health". Fundamentals of professional self-rehabilitation ". The program of the course includes the study of the following topics:" Determinants of occupational health"; "The impact of the profession on the health of the manager. Professional deformations, destructions, ways of their prevention and elimination"; "Occupational stress. Emotional burnout"; "Stress management in management. Methods of formation of professional stress resistance"; "Orthobiotics in professional activities"; "Harmony of personality and professional health"; "Technologies for maintaining and strengthening occupational health. Occupational self-rehabilitation. Psychotechnologies of self-healing» [22].

In the process of studying this course, future managers perform several tasks aimed at the formation of health competence and the development of their resilient qualities:

1. Create a "Manager's Occupational Health Code".
2. Make two "Memo for the manager" (optional): Psychotherapist yourself! How exactly ?; Recipes for mental balance; Occupational stress medication; How to protect yourself from professional stress; How to deal with "erosion of the soul"; How not to burn in the flames of the profession; How to become a self-effective person; The path to assertive behaviour; How to get rid of thinking "red pencil". How to overcome professional difficulties with the least loss of health, Professional longevity: the path to it.
3. Prepare an essay on one of the proposed topics: a) Laws of the reasonable lifestyle of the manager; b) Constructive overcoming of professional difficulties; c) Constructive and destructive reactions to failure and occupational health; d) Positive thinking, sanogenic thinking: two sides of the same problem; e) Sanogenic potential of the individual, positive mental position, their activation in difficult situations of professional activity. Professional achievement stress and workaholism; Coping strategies in management.
4. To identify their own problems, external and internal resources to maintain and strengthen their health based on the results of self-observation, self-diagnosis, testing the level of stress resistance, the tendency to emotional burnout, level of neuroticism, level of anxiety, etc.. To develop an individual program for maintaining and strengthening occupational health. To analyze internal reserves and external conditions that will facilitate its implementation.
5. To create a portfolio on the problems of maintaining and strengthening the manager's professional health.
6. To develop and present projects "Own self-healing system" and "Manager's psychological first aid kit" and other tasks.

We have also developed and implemented an anti-stress training program [23]. The major tasks of anti-stressing are aimed at launching self-concept, constructive communication, and self-regulation of stress, positive thinking. The activities of the training group are aimed at the formation of reflection, the ability to recognize their emotional state, motives, consequences of actions; positive perception of oneself and acceptance of oneself, one's strengths and weaknesses, assertive behavior; activation of psychological mechanisms of self-regulation, development of readiness for self-changes and self-development, the formation of skills to find own resources in difficult situations.

The activity of the training group involves the actualization of personal and neuro-psychological resources of its members, including active strategies for overcoming behavior, awareness and reduction of the destructive influence of accentuations and temperament, resolution of interpersonal conflicts. Future managers learn to model behavior in stressful situations, understand their own emotions, and use different psychotechnics to focus on the positive.

The participation of future managers in the work of the training group contributes to the knowledge of their strengths and weaknesses, the gradual elimination of neurotic components of their own inner world, improving interaction with other people. The training creates conditions for the development of such qualities as communicativeness, reflexivity, assertiveness, tolerance, and mastering of subjective self-influence by psychotechnics. Introspection during the training allows to separate negative emotional reactions from negative habitual patterns of self-destructive behavior that destroys health, reduces the degree of emotional negative reaction to situational influences.

During the management practice, undergraduates perform specially designed tasks aimed at developing the ability to adequately assess their own strategy of behavior in a difficult situation, the ability to diagnose the level of self-regulation, the level of neuropsychological stability, and susceptibility to stress, ability to overcome professional difficulties. Students test the effectiveness of developed programs for the preservation of professional health, in particular, the ways of regulating the emotional state, relaxing emotional stress, removing fatigue, achieving mental equilibrium.

It should be noted that in the program of development of resilience, an important place is given to individual psychodiagnostic and psychocorrective conversations-consultations with undergraduates. They are held at the Center of Pedagogical Consulting [24]. Such conversations-consultations help to identify personal resources of stress resistance and resilience, the causes of destructive behavior in conflict and difficult situations, the causes of pathogenic thinking, insecure behavior.

The Center's of pedagogical consulting activities are focused on the diagnosis of individual-typological and personal characteristics of undergraduates, their interpretation, assistance in finding resources of resilience and selection of constructive coping strategies, assistance in mastering methods of professional stress management, mastering stress management. Master's student's training, webinars, masterclasses, round tables are held within the functioning of the Center of pedagogical consulting. [24,25].

7 Conclusions

Thus, resilience is an important factor in ensuring the efficiency and reliability of management. Diagnosis of the level of the neuroticism of future managers, neuropsychological resilience in stress, the ability to maintain psychosocial adaptation during stress, confirmed the need for the formation of their resilience at the stage of a Master's degree in higher education.

For the development of resilience of future managers: the formation of skills of productive interaction and constructive overcoming of professional difficulties, the formation of strategies to protect against stress, learning methods, and technics of self-regulation, the formation of positive sanogenic thinking are important.

The main ways of realization of the offered program of development of resilience are outlined: educational activity, independent work of undergraduates, individual consultative work, training practice, performing specially developed tasks during administrative practice, participation in the activity of the Center of pedagogical consulting.

The proposed program of resilience development is focused on strengthening the resources of the personality of the future specialist, which will contribute to their resilience to life's troubles and professional difficulties. Our course "Professional Health" and anti-stress training played an important role in future managers' resilience development.

Availability of acmeological educational environment in the institution of higher education and maturity of lecturers' acme-synergetic position are prerequisites for the implementation of the programme for the development of resilience of future managers. The lecturers developing the resilient qualities of the students must be the embodiment of the standards themselves; otherwise, the goals are not just failed, but the opposite result may be achieved.

We see the prospects for further scientific research in testing the effectiveness of the developed program for the development of resilience, justification, and development of technologies for the formation of the resilience of future managers in the digital space of formal and non-formal education.

References

1. H. Meshko, O. Meshko, N. Drobyk, O. Mikheienko, Psycho-pedagogical training as a mean of forming the occupational stress resistance of future teachers. Proc. of the ICSF 2020, Kryvyi Rih, Ukraine, May 20-22, 2020. E3S Web of Conferences **166** (2020), https://www.e3s-conferences.org/articles/e3sconf/abs/2020/26/e3sconf_icsf2020_10023/e3sconf_icsf2020_10023.htm. Accessed 26 January 2021
2. Ed. S. Bozhyk, Yak plekaty u sobi rezylientnist ta chomu tse vazhlyvo? (How to cultivate resilience and why is it important?) (Fizis: Psychological literacy, Lifestyle, 2020), https://fizis.net/psyhologichna-gramotnist/iak-

plekaty-u-sobi-rezylientnist-ta-chomu-tse-vazhlyvo. Accessed 26 January 2021

3. G.M. Meshko, O.I. Meshko, Formuvannia profesiinoi stresostiikosti maibutnikh kerivnykiv zakladiv osvity na etapi mahisterskoi pidhotovky (Formation of professional stress resistance of future heads of educational institutions at the stage of Master's training). Humanitarium **43** Issue 1: Psychology (2019), pp. 103-112. doi: 10.31470/2308-5126-2019-43-1-103-112
4. V. Korolchuk, Dissertation, *Psykholohiia stresostiikosti osobystosti* (Psychology of stress resistance of the person) Institute of Psychology G.S. Kostyuk Academy of Pedagogical Sciences of Ukraine (Kyiv, 2009), p.513
5. Kudinova M.S. Porivnialnyi analiz poniat «stresostiikist» ta «emotsiina stiikist» (Comparative analysis of the concepts of "stress resistance" and "emotional stability"). Theory and practice of modern psychology **1** (2016), pp. 22–28
6. W. Pilecka, *Odporność psychiczna jako wyznacznik zdrowia,* eds. by H. Wrona-Polanska and J. Mastalski. Promocja zdrowia (Krakow, Wydawnictwo Uniwersytetu Jagiellonskiego, 2009), pp. 109-122
7. E. Werner, R. Smith, *Overcoming the odds: high risk children from birth to adulthood* (Ithaca, New York, Cornell University Press, 1992)
8. O.M. Khaminich, Rezilientnist: zhyttiestiikist, zhyttiezdatnist abo rezilientnist? (Resilience: viability, viability or resilience?) Scientific Bulletin of Kherson State University. Series: Psychological sciences **6** T. 2 (2016), pp.160-165
9. D. Fletcher, M. Sarkar, Review of Psychological Resilience. European Psychologist **18** (2013), pp. 12-23 https://www.researchgate.net/publication/263651506. Accessed 26 January 2021
10. A. Borucka, *Koncepcja resilience. Podstawowe założenia i nurty badań*, ed. W. Junik. Resilience (Warszawa, Parpamedia, 2011), pp. 11-28
11. G.P. Lazos, Posttravmatychne zrostannia: teoretychni modeli, perspektyvy dlia praktyky (Post-traumatic growth: theoretical models, new prospects for practice) Current problems of psychology **1** Issue: 45 (2016), pp. 120-127
12. R.G. Tedeschi, L.G. Calhoun, Posttraumatic growth: Conceptual foundations and empirical evidence Psychological Inquiry **15** (2004), pp. 1–18
13. Lazos G.P., Teoretyko-metodolohichna model rezylientnosti yak osnova pobudovy psykhotekhnolohii yii rozvytku (Theoretical and methodological model of resilience as a basis for constructing psychotechnology of its development) Organizational psychology. Economic psychology **2-3 (17)** (2019), pp.77-89 doi: https://doi.org/10.31108/2.2019.3.17.9
14. A. Masten, *Ordinary magic: resilience in development* (New York, The Gulford press, 2014)
15. Molts M. Y., *- eto Ya, ili Kak stat schastlyvym (- it is Me, or How to become happy)* (Moskva, Mysl, 1994), pp. 192
16. Pil N.V. *Udivitielnyie riezultaty pozitivnoho myshleniia (Amazing Results of Positive Thinking)* (Mynsk, OOO «Popurry», 2003), pp. 320
17. Treisi B., *Izmieni myshlieniie — i ty izmienish svoiu zhizn (Change your thinking - and you change your life)*, 2nd edn. (Minsk, «Popurri», 2006), pp.384
18. Kaloshyn V.F., Yak sformuvaty pozytyvne myslennia (How to form positive thinking) Praktychna psykholohiia ta sotsialna robota **3**. (1998), p. 27
19. J. Osborn, Seven salutary suggestions for counsellor stamina Journal of Counselling & Development **82** (2004), pp. 354–364
20. B.A. Smirnov, E.V. Dolgopolova, *Psykholohiia deiatelnosti v ekstremalnykh sytuatsiakh* (Psychology of activity in extreme situations), 2nd ed. (Kharkiv, Humanitarian Center, 2007), p. 292
21. G.S. Nikiphorov, M. A. Dmitrieva and V.M.Snetkov (eds.), *Praktikum po psikholohii menedzhmenta i profesionalnoi deiatelnosti* (Workshop on the psychology of management and professional activity) (SPb., Rech, 2001), p. 448
22. Profesiine zdorovia: elektronnyi navchalno-metodychnyi kompleks dystsypliny dlia pidhotovky mahistriv spetsialnosti 073 Menedzhment (Occupational health: electronic educational and methodical complex of discipline for preparation of masters of specialty 073 Management) (Website of the Ternopil Volodymyr Hnatiuk National Pedagogical University), https://elr.tnpu.edu.ua/enrol/index.php?id=1796. Accessed 26 January 2021
23. Antystres-treninh (Antistress training) (Website of the AIPo), http://www.airo.com.ua/antistres-trening. Accessed 26 January 2021
24. Polozhennia pro Tsentr pedahohichnoho konsaltynhu (Regulations on the Center of Pedagogical Consulting) (Website of the Ternopil Volodymyr Hnatiuk National Pedagogical University), http://tnpu.edu.ua/about/public_inform/upload/2020/Polozhennia_pro_Tsentr_konsaltynhu.pdf. Accessed 26 January 2021
25. Tsentr pedahohichnoho konsaltynhu. Fakultet pedahohiky i psykholohi Ternopilskoho natsionalnoho pedahohichnoho universytetu im. V. Hnatiuka (Center of Pedagogical Consulting. Faculty of Pedagogy and Psychology, Ternopil Volodymyr Hnatiuk National Pedagogical University), http://fpp.tnpu.edu.ua/pidrozdili/centr-pedagogichnogo-konsaltingu . Accessed 26 January 2021

Cognitive Metaphors of Covid-19 Pandemic in Business News

Tetyana Kozlova[1*]

[1]Zaporizhzhia National University, 69069 Zaporizhzhia, Ukraine

Abstract. The research considers the COVID-19 pandemic cognitive metaphors conveyed by means of the English language in business news. The interpretation of metaphor goes beyond its traditional understanding as a rhetorical device. The approach is consistent with a cognitive theory claiming that metaphor is a mental instrument to reflect the way we reason and imagine the world. The paper provides a brief theoretical framework of the research, discusses the concept, role and types of cognitive metaphor. It deals with particular cases of metaphoric representations of the pandemic selected from *The Financial Times*, an international daily with focus on business and economic affairs. The results of the study reveal a variety of lexical means to express the dynamic image of the pandemic that exhibits a gradual shift from the military metaphor to variant interpretations. The findings prove the pervasiveness of metaphor in business and mass media communication, its significance to understand difficult situations, efficiently communicate ideas and influence the audience.

1 Introduction

The outbreak and consequences of the COVID-19 pandemic will leave a deep mark in the consciousness of people all over the world. Because of its unexpectedness, rapid pace and global scale, the pandemic has forced significant changes into our lives. Under such circumstances, immediate response, and efficient prevention have become the greatest challenges faced by the world community. The enormity of the challenge issued by the dreaded pandemic is being discussed, explained, rationalized and interpreted in numerous publications relating medicine [1, 2] and other fields of science, politics, economics [3], education, culture [4, 5] to name but few. In addition to that, there exists a sizable academic literature discussing linguistic and communicative aspects of the pandemic crisis [6–8], addressing the ongoing events, tackling cognitive and emotional response to the unpredictable circumstances, and explaining how coronavirus outbreak communication is being handled by political, mass media and scientific communities.

1.1 Background

The studies with a focus on applied linguistics perspective were carried out to discover and explain new coinage related to Covid-19 situation, its influence on other languages and problems arising in the translation and coordination of terminology [9–11], generate taxonomies of terms with the help of corpus analysis and estimate word frequencies [12–14], collect and systematize massive Covid-19 related text data [15]. The findings shed light on the specificity of scientific and medical language which is significant in specialist and everyday discourse.

The findings revealed an unprecedented and rapid (within 3 months) growth in the frequency of pandemic-related words (*coronavirus*, *corona*, *COVID-19*) compared to lexical items connected with recent political and social events (*Brexit*, *impeachment*) [13]. It was discovered that linguistic changes accelerated with the increase in names for social response and consequences (*social distancing*, *self-isolation*, etc.), economic impact (*lockdown*), and distant communication (*zoom*) [14].

The noticeable result of such linguistic change is the qualitative and quantitative expansion of the English vocabulary. The corona-related vocabulary amounts to approximately 500 items [11] which find equivalents in other languages. The quantitative change can be traced through linguistic creativity of speakers. According to Haddad & Monterero-Martinez [9], a tremendous vocabulary expansion should be attributed to solving communication needs in specialist and everyday communication by filling lexical gaps. The data from various studies [9–15] indicated that most productive types of vocabulary development included metaphoric and metonymic transfers. Affixation, compounding, abbreviation, clipping and conversion prevailed over other word-building processes.

However, it is not only naming gaps that require to be filled. A more challenging task precedes creating naming labels. It is crucial for interlocutors is to cognize new aspects of reality, systematize new and old experience, evaluate things and events. The researches found that such concepts as disease, Covid-19, pandemic became prevalent in various types of discourse today. Having studied framing of Covid-19 in Twitter communication, Wicke & Bolognesi [16] showed that the discourse around the pandemic made use of the "war", "monster", "storm", and "family" metaphors. The conclusion was made about "a metaphor menu"

* Corresponding author: ethstlab@yahoo.com

facilitating the communication of various aspects connected with Covid-19.

Semino's [17] findings about the appropriateness of "fire" metaphors in communication about contagion and public health measures resulted from the analysis of news articles in English.

Other researchers analysed "war" metaphorisation of Covid-19 to discover the diverse arsenal of means of its manifestation in language and explain the diversity by socio-political individual variables speakers', such as political orientation [18], or by universal character of the war rhetoric [19–22].

Although much work has been done to date, more studies need to be conducted to ascertain whether different discourses and their various genres determine the type of metaphors they employ how they may vary.

1.2 The purpose and methodology of research

The purpose of this study is to investigate metaphoric representations of the Covid-19 pandemic in business news articles in English.

It is hypothesized that the universal metaphor "disaster is war" can manifest as a set of alternatives to communicate senses and ideas particularly significant for mass media interpretation of business affairs.

The material used in this research consisted of 125 metaphoric manifestations of the Covid-19 pandemic. The extracts were selected from *The Financial Times* (FT) articles placed online between February 2020 and January 2021.

The analysis of cognitive metaphors of the Covid-19 pandemic was based exclusively on the data collected from FT. This source was considered valid and credible for the following reasons. Firstly, FT is a respected international daily. Secondly, with its focus on business and economic affairs, it is generally regarded to be an authority in these subjects. Thirdly, in spite of the fact that it is primarily targeted at the readers interested in finance, FT is a newspaper with a wide coverage of topics attracting audience from various fields of life.

As to methods of material selection, it should be mentioned that modern "conceptual metaphor theory has no explicit methods for identifying conceptual metaphors" [23]. That is why a set of criteria was employed at the stage of empirical data selection. It was devised to ensure the presence of the lexical units *Covid*, *Covid-19*, *coronavirus*, *corona*, *pandemic* in the contexts, and the presence of some word/s the written utterance that could be taken figuratively (metaphorically), not literally. The choice also had to be made as to the appropriateness of the selected samples and the presence of cross-domain ties found in the expressions. To decrease the uncertainty and find systematicity in selected expressions, the following features were considered: similarity or polarity in meaning (*struggle*, *fight*; *win*, *lose*); shared semantic component/s and relatedness to the particular semantic field (*fighter*, *hunter* "someone who attacks); shared collocates (*animal hunter*, *vaccine hunter*, *Covid hunter*).

The collected samples covered a wide range of aspects of the pandemic including its outbreak and unfolding, the scale of spread, preventive measures, impact on business. The whole set was subjected to descriptive, contextual, semantic, and structural analysis employed in combination with conceptual analysis to make judgements about the mappings and the types of metaphors. Finally, the data were systematised to discover the source domains of metaphorisation and linguistic means of metaphoric manifestations.

1.3 Paper structure

The paper is structured as follows. Section 2 gives a brief account of the theoretical basis for the research and systematizes the key principles of the theory of cognitive metaphor, its types and significance in communication.

Section 3 presents the analysis of metaphoric interpretations of the Covid-19 pandemic disaster and means of their manifestation in business news. It also discusses variant metaphoric conceptualisations of Covid-19 pandemic elaborated by speakers from different but related to the military domain spheres of knowledge. Furthermore, this section deals with the business news rhetoric and expressivity of means employed to communicate about the pandemic.

2 Theoretical basis for the research

2.1. Traditional and innovative definitions of metaphor

There is an agreement among traditional and cognitive linguists on the importance and pervasiveness of metaphor. The history of the theory of metaphor has its roots in the distant past.

The earliest accounts about metaphor are found in the works of Aristotle and other ancient philosophers who treated metaphor as a rhetorical figure, an element of speech decoration and the demonstration of eloquence skill. Such approach was determined by the social environment in ancient Greece and established democracy when all political innovations, rise to power, endurance of laws and political achievements could be, though not secured, but assisted by argumentation, pervasiveness, impressiveness of communication of necessary ideas. Metaphors played a significant role in adding suggestive power to rhetoricians' and critics' elegant speeches, "the grand, also called "lofty," … florid speech that impresses with sound" [24]. Since then, the influential potential of metaphors has never been questioned.

In later studies, metaphor was defined in terms of traditional linguistics as a path of meaning development that occurs due to the associative transfer on the basis of similarity between two entities in question. Thus, linguistic metaphor is associating between two referents which resemble each other. From that classical point of view, metaphors are based on various types of similarity (shape, position, colour, etc.) and discussed as similies,

 https://doi.org/10.1051/shsconf/202110002004

personifications, spatial images, transfers of sensation and the like.

Considered as a product of logical ties, metaphor is treated as a naming technique with characterising function.

Lakoff and Johnson [25] paid attention to the mental/conceptual spaces and how they operate with knowledge domains. Fauconnier and Turner [26, 27] discovered How they can preserve structures and make permanent correspondences, or how they can blend them. In contemporary linguistics, metaphor is tackled from a cognitive point of view. It is acknowledged that metaphor is an instrument of thinking, reasoning about the world.

The interplay occurs between the source domain of and the target domain of metaphor. The former contains some knowledge or experience with which new knowledge is identified. The latter is the sphere to which metaphor is applied. The structural correspondences between the source and target domains are defined as mappings (or metaphor maps).

The theory of metaphor was enriched by Kövecses who expressed doubts on whether literal language existed at all, argued that contextual metaphors were also conceptual ones [28], developed the theory of the scope of metaphor [29]. He discovered that metaphors can be derived from several domains in order to interpret an abstract concept.

These and other contributions into the theory of cognitive/conceptual metaphor added to the understanding of the variety of metaphors.

2.2 Types and significance of cognitive metaphors

Cognitive metaphors break into several types depending on what sources and mechanisms they employ to represent a more complex idea: conduit metaphors that deal with viewing communication, ontological metaphors and personification that the most basic as they allow to interpret abstract ideas, events, etc. as entities and objects, structural metaphors that structure one concept in terms of another, orientational metaphors that allow to ground our experience of the wold with the spatial experience.

The role of cognitive metaphors is crucial in our life. They enable appropriate orientation in the environment, systematisation of our experience, provide contiguity of knowledge and cultural values. Despite their representative nature, they are not deprived of descriptive potential. They help us to depict the world, build up images, compare and recognise various properties of things around us. They are also of importance for the creative and cognitive development of humans.

As to communication, cognitive metaphors carry out a set of functions. They link and harmonise individual cognitive processes with those of the social group. They become persuasive performatives responsible for implicit inclination, involvement, invitation, etc. Metaphors are invaluable suggestive tools shaping social behaviour and thought.

Their psychological role is no less significant. With the help of analogy carried by metaphors, speakers are less vulnerable to cognitive disorders, better cope with stressful circumstances, avoid communicative failure. In sum, metaphors are mediators between the world and the human, between the society and an individual. For mass media communication, metaphoric effects are vital in the sense that metaphorical patterns become models of thought and behaviour of the recipients of information.

3 Results and discussion

The epidemic disaster has influenced the way we think and speak about the world, society and various spheres of life, particularly business and economy, environment, health issues. Moreover, our perception and reasoning of the crisis shapes the way we communicate about it. Most appropriate and common way of interpreting the pandemics is in terms of the war. The discourse of business news makes use of both the common and military parlance around the current disaster.

3.1 Military metaphor: the epidemic as a common foe

The world is seen as a war zone where people are fighting with the disease: "*Europe battles to contain surge in Covid-19 cases*" (FT, 29 July 2020), "*...countries fight Covid-19 resurgence*" (FT, 25 Dec 2020), "*We need to aggressively stop the spread now*" (FT, 5 Dec 2020), "*hospitals and intensive care units are struggling to cope.*" (FT, 21 March 21 2020). In business news, the current events relating to the coronavirus outbreak are represented in terms of an armed conflict with the help of the words denoting a fight, an armed conflict, pressure or the use of force (*battle, beat, buckle, combat, conquer, defeat, fight, struggle*). For example, "*War on superbugs must follow defeat of Covid-19*" (FT, 20 Dec 2020), "*...the reality of Britain's "wartime economy" in the era of coronavirus*" (FT, 27 March 2020).

Experienced difficulties and great efforts made to win the battle and overcome the pandemic are interpreted as shooting from a gun – " *'... bullet' to beat Covid-19*" (FT, 29 July 2020), while a record of the virus spread is seen as following the target – "*coronavirus tracked*" (FT, 25 Dec 2020). In metaphoric representations, warnings of danger are portrayed as alarms given by a loud noise or flashing light: "*US states sound alarm on Covid-19 hospitaliazations*" (FT, 5 Dec 2020).

Reporting on coronavirus-related deaths has also added to the image of a military disaster with enormous loss of life: "*Death toll surges*" (FT, 22 March 2020), "*... worst day for mortalities in escalating European outbreak*" (ibid.).

Preventive measures are described as actions demanded by law: *order mandatory testing, impose measures, urge mask use* and other expressions are frequent in the texts of daily business news. Both in cases of armed conflicts and epidemic disasters, restrictive measures are taken against regional instability as well as protection of people and property. Among other things, pandemic restrictive defence includes Covid-19 border control, lockdown, isolation, distancing: "*World on lockdown: West closes borders and orders isolation*" (FT, 18 March 2000), "*… western economies took drastic measures to limit public movement on Monday, closing borders, shutting down retailers and ordering citizens to stay in their homes in an urgent effort to arrest the spreading coronavirus pandemic*" (ibid.)

The key conceptual feature "common", that is "the same enemy in a lot of places or for a lot of people" is explicitly expressed by adjectives meaning "relating to the whole world" (*global, international, world-wide,* etc.), nouns denoting "the act of working together" (*cooperation, collaboration,* etc.). For example, "*Global cooperation is needed to beat the virus*" (FT, 29 Jan 2021). Grammatical means such as the marker of the plural of nouns (*-s*), the plural pronouns (*we, us, they, them*) and determiners (*our, their*) are also employed to communicate the idea of "common foe": "*Governments must restore these freedoms when the virus is eventually beaten*" (FT, 29 Jan 2021), "*… we'll be shooting to get to a year's immunity*" (FT, 23 Sept 2020), "*To our relief and surprise, the number of cases have started to come down…*" (FT, 29 Jan 2021). While the use of noun plurals and adjectives with cooperative meanings is unambiguous, the plural pronoun *we/us* can vary according to the context. It can be either semantically inclusive (i. e. including the addresser(s)/speaker(s) and the addressee) or exclusive (i. e. excluding the addresser(s)/speaker(s) but including other people as the addressees): "*We won't remember much of what we did in the pandemic*" (FT, 14 Aug 2020). However, news reporters seem to favour explicit methods and make a clear reference by adding other words (*we all*) or employing generic uses of personal pronouns to refer to people in general. For instance, "*…what we all feel about the high number of… the global pandemic and the rapidly evolving economic crisis*" (FT, 4 Jan 2021), "*The next pandemic: where is it coming from and how do we stop it? As deforestation and climate change increase spread of new diseases, meet the virus-hunters trying to prevent the next Covid-19*" (FT, 29 Oct 2020). In contrast, the concepts of loneliness and isolation during the pandemic are successfully verbalized by the nominative and accusative personal pronouns (*I, me*): "*I live alone, I work alone, I'm hundreds …*" (FT, 20 Nov 2020), "*Loneliness and me. Millions of us were living with this curse before the pandemic. How can we break it?*" (ibid.).

Overall, the military metaphor has appeared to be a remarkably efficient tool in reasoning and reporting about the pandemic. The prior knowledge of war as a conflict between particular actors which is characterized by violence, social disruption, and economic destruction is linked to new experience. Inasmuch as cognizers bridge what they already know (an armed conflict) to a new entity (the Covid-19 pandemic), they create a new cognitive space that results from framing the target domain (the pandemic) in terms of the source domain (the war). Hence, if the Covid-19 pandemic is interpreted with the help of the military metaphor, then it will produce an image of "the war against the common aggressor". However, there are diverse applications of the "common foe" description in business news reports.

3.2 Variant conceptualizations of the pandemic as a foe

Variant conceptualizations of the Covid-19 pandemic appeal to different but related cognitive domains. They differ in the sphere of knowledge that is prioritized by business news reporters who target to create eloquent images of the pandemic: knowledge and experience in military affairs, hunting, games, killing a monster, and dealing with liquid are not alike. Despite that fact, these domains enable speakers to establish associative links with the image of "common foe" and provide corresponding mappings between the source domains and the target domain of the metaphor. In other words, variant conceptualizations of the Covid-19 pandemic have resulted from one-to many ties between the target domain and the source domains.

The relationship between the actors of the situation (the pandemic versus humanity, society, business, economy, etc.) is alternatively considered as competition in which one party becomes victimized ("*victim of the pandemic*" (FT, 29 Jan 2021)). Let us look into specific cases:

In the hunting metaphor, the actors of the pandemic situation come up as the hunter (scientists, investors) and the game/wild beast (the virus): "*Hunting for new viruses has become more difficult during an actual pandemic, but it has never been more important*" (FT, 29 Oct 2020), "*on the hunt for a cure*" (FT, 14 May 2020), "*Investors' hunt for coronavirus rebound stocks*" (FT, 11 Feb 2020), "*Investors hunt for alternative data to track coronavirus shock…*" (FT, 18 Feb 2020).

Being metaphorically considered as game players, the actors of the pandemic situation are conceptualized as the competing participants: "*Vaccine makers prepare for game of Covid cat and mouse. Manufacturers and regulators need to be ready if shots prove less effective…*" (FT, 10 Jan 21), "*Asia plays a long game on Covid vaccine rollout*" (FT, 15 Dec 2020), "*… attempts to play down Covid-19*" (FT, 29 Oct 2020).

The virus disaster has reshaped the way we talk about society, politics and economy dividing businesses and agencies into winners and losers: "*… winners and losers in the Covid economy*" (FT, 10 Oct 2020), "*Three ways the banks will be winners from Covid recovery*" (FT, 16 Nov 2020), "*This year, Covid-19 has brought some of the most powerful countries in the world to their knees*" (FT, 29 Oct 2020), "*Covid-19 unmasks weakness of English public health agency*" (FT, July 22, 2020).

Another way to communicate the struggle against the pandemic is the metaphor "a fight with a monster": "*Coronavirus turns the City [London business center] into a ghost town*" (FT, 27 July 2020). "*'Silver bullet' to beat Covid-19 unlikely*" (FT, 29 July 2020) seems to be a transparent allusion to the stories about werewolves and vampires fired to death with silver bullets by vampire-hunters. This conceptualization is a very interesting instance of the cognitive frame development. It is based on the mapping between the vampire lore and the contemporary knowledge about the evil of pandemic. Furthermore, it provides a convincing evidence to the existence of conceptual ties among different domains of knowledge and experience. By this the consistency of mapping is achieved within the frame representation of the active agent as a hunter / fighter (cf. *animal hunter*, *vaccine hunter* and *Covid hunter* as semantically close expressions implying "someone who is trying to find and get the desired/stated thing"). The explanation that can be suggested for such a metaphor of the pandemic is that cognitive metaphors are not single, independent tools of our cognition but complex mental images embracing sets of associated featured.

3.3 The expressive language of metaphoric manifestations

The findings about the linguistic means of metaphoric manifestations showed that reporters employed typical for the news discourse expressivity. As is clear, expressive vocabulary and structures enable addressers to encode their emotions and evaluations of the current situation. Expressivity provides efficiency in expressing the addressers' intentions as well as their individuality. In addition to that, expressivity of speech allows to achieve loftiness in communication. What follows is the systematization and examples of the expressive means found in the selected contexts:

– words with inherent expressivity and mostly negative meanings (*dreadful* "causing shock and suffering", *monstrous* "very cruel", *vicious* "showing an intention to hurt badly" as in "*We can defeat this invisible and vicious adversary [coronavirus] – but only with global leadership*" (FT, 25 March 2020)), which may be intensified in the context (*extraordinary* "very strange, unusual" as in *extraordinary crisis*; *curse* "speak angrily" as in *live with this curse* (FT, 21 Nov 2020));

– words with inherent expressivity and intensive meanings (*to surge* "to increase suddenly and greatly", *to hit* "to move one's hand with force; to produce a negative, unpleasant effect", *to thrive* "to grow and become successful"; *boom* "a sudden increase in something" as in "*Europe fears as coronavirus surge threatens to overwhelm hospitals*" (FT, 23 Oct 2020), "*Nations look into why coronavirus hits ethnic minorities so hard*" (FT, 29 Apr 2020));

– colourful phraseological units (*get your act together* "organize and deal with something effectively", *bleak future* "without anything to make one feel happy or hopeful, *be on one's knees* "be weak" as in "*Covid brings China's high-growth rental industry to its knees*" (FT, 18 Nov 2020)) and their occasional modifications (for example, the idiom *to have the stomach for* "to be brave or determined to do something dangerous or unpleasant" was transformed into *to lose the stomach* "to become exhausted because of doing something difficult" as in "*America is losing the stomach to fight Covid-19*" (FT, 11 Jun 2020));

– names of imaginary entities (*ghost* "the spirit of dead, transparent image", *to haunt* "(of a ghost) to appear in a place repeatedly, cause anxiety or suffering" as in "*If Covid-19 is not beaten in Africa, it will return to haunt us all*" (FT, 25 March 2020);

– phrasal verbs (*to lose out* "to not have an advantage as other people have" as in "*The kids aren't alright. How generation Covid is losing out*" (FT, 17 Nov 2020); *to bounce back* "to return to a usual state after having a problem" as in "*... if the virus is not defeated in Africa, it will only bounce back to the rest of the world*" (FT, 25 March 2020)).

The following quote is an example of how neutral words get involved into utterances with expressive vocabulary and acquire intensification to communicate about the pandemic disaster: "*Gap between financial markets and global economy yawns wider ... We have a monster mash-up of the Great Depression in size, the crash of ... effects of measures*" (FT, 24 Apr 2020). Owing to the proximity of distance between the expressive (*yawn wider*, *monster*, *mash-up*, *crash*) and neutral vocabulary, the latter become emotionally charge. Thus, "*effects of measures*" should be interpreted as "ineffectual and unable to produce good results under the circumstances".

In the following passage, the involvement of the intensive verb *surge* provides expressivity for the whole utterance, suggesting complete destruction of most powerful (*major*) economic systems: "*Pandemic triggers surge in business start-ups across major economies*" (FT, 29 Dec 2020).

4 Conclusions

The results of the research proved the importance of cognitive metaphors in different spheres and genres of communication. The study of cognitive metaphors of the Covid-19 epidemic in FT daily provided evidence to the fact that the target domain of metaphors in question is actual for the coverage of various topics, directly or indirectly related to the themes of financing and business. It can be explained by the drastic and global scale influence of the epidemic on the economic sphere.

The prevalence of military metaphor can be explained by the universal nature of the concept "disaster is war". However, in internationally oriented business communication, reporters suggest the idea of the pandemic as a "common foe", implying the necessity for cooperation in coping with the crises stimulated by the disaster. Although this scenario demonstrates alternative metaphoric representations (Covid is an animal to be hunted, a competitor/player to be defeated, a monster to be destroyed), they are attracted by the central idea of the common enemy.

The linguistic means involved to manifest metaphors demonstrate expressivity typical of the news discourse.

Business news articles are an interesting and influential genre of mass media communication. It is an awesome challenge to extend the research of cognitive metaphors in business news reported by means of various languages in a variety of cultural settings. Further research into cognitive metaphors of the Covid-19 pandemic calls for historical and comparative study broadening the perspective on the similarities and differences in the experience, behaviour and practices of coping with a disaster and communicating about it.

References

1. A. M. Baig, M. Chronic COVID Syndrome: Need for an appropriate medical terminology for Long-COVID and COVID Long-Haulers. Jrn. of Med. Virol. (2020). doi:10.1002/jmv.26624
2. D. Ricks, *100 Questions & Answers About Coronaviruses* (Jones & Bartlett Learning, Burlington, 2021)
3. D. Seedhouse, *The Case for Democracy in the COVID-19 Pandemic* (SAGE Publications, London, 2020)
4. A. Bradshaw, J. Hietanen, *The Dictionary of Coronavirus Culture* (Repeater Books, London, 2020)
5. J. S. Spence, *Covid-19, Exposition and Poetry on Coronavirus: An Education in Literary Notes Uplifting* (Clever Fox Publishing, Chennai, 2020)
6. D. Lilleker, I.A. Coman, M. Gregor, E. Novelli (eds.), *Political Communication and COVID-19: Governance and Rhetoric in Times of Crisis* (Routledge, New York, 2021)
7. H. D. O'Hair, M. J. O'Hair, *Communicating Science in Times of Crisis: Coronavirus* (Wiley-Blackwell, New York, 2021)
8. S. Molthagen-Schnöring, Communicating Science in Covid Times: Why communication with "the public" is not the goal of science communication. Eleph. In the Lab (2020). doi:10.5281/zenodo.3956336
9. H. Haddad, S. Monterero-Martinez. COVID-19: a metaphor-based neologism and its translation into Arabic. JCOM **19**(05), (2020). doi:10.22323/2.19050201
10. L.-Ch. Chen, K.-H. Chang, H.-Y. Chung A Novel Statistic-Based Corpus Machine Processing Approach to Refine a Big Textual Data: An ESP Case of COVID-19 News Reports. Appl. Sci. **10 (**5505), (2020). doi:10.3390/app10165505
11. H. Ma, L. Shen, H. Sun, Z. Xu, L. Xou, S. Wu, A. Fang, J. Li, Q. Qian. COVID Term: A Bilingual Terminology for COVID-19. Res. Sq. (2020). doi: 10.21203/rs.3.rs-30923/v1 (in press)
12. T. Stewart, July 2020 update: scientific terminology of Covid-19. (OED, 2020), https://public.oed.com/blog/july-2020-update-scientific-terminology-of-covid-19/. Accessed 14 Dec 2020
13. Corpus analysis of the language of Covid-19 (OED, 2020), https://public.oed.com/blog/corpus-analysis-of-the-language-of-covid-19/. Accessed 10 Dec 2020
14. Using corpora to track the language of Covid-19: update 2 (OED, 2020), https://public.oed.com/blog/using-corpora-to-track-the-language-of-covid-19-update-2/. Accessed 10 Dec 2020
15. COVID-19 Open Research Dataset (CORD-19). Version 2020-05-02 (Allen Institute for AI, 2020), https://pages.semanticscholar.org/coronavirus-research. Accessed 10 May 2020
16. Ph. Wicke, M. M. Bolognesi, Framing COVID-19: How we conceptualize and discuss the pandemic on Twitter. PLoS ONE **15**(9). doi:10.1371/journal.pone.024prevalent0010
17. E. Semino, "Not Soldiers but Fire-fighters" – Metaphors and Covid-19. Health Commun. **36**(I), (2021). doi:10.1080/10410236.2020.1844989
18. M. Silvera-Roig, Lecture: Covid-19 and Conceptual Metaphor Theory: Disease is War (Complutense University of Madrid, 2020), https://youtu.be/jyiWRpd_3qU. Accessed 28 Sept 2020
19. D. Fernández-Pedemonte, F. Casillo, A. I. Jorge-Artigau, Communicating COVID-19: Metaphors We "Survive" By. Tripod. **47**(2), 145-159 (2020)
20. B. Nerlich. Metaphors in the time of coronavirus, https://blogs.nottingham.ac.uk/makingsciencepublic/2020/03/17/metaphors-in-the-time-of-coronavirus/. Accessed 14 Dec 2020
21. A. Sirait, K. Seva, Th. R. Lingga, Conceptualizing Corona Virus Metaphor in Media Headlines: a Cognitive Semantic Approach. Eng. Jrnl. Lit. Utama **5**(1) (2020). doi:https://doi.org/10.33197/ejlutama.vol5.iss1.2020.477
22. C. Oswick, D. Grant, R.Oswick, Categories, Crossroads, Control, Connectedness, Continuity, and Change: A Metaphorical Exploration of COVID-19. JABS, **56**(3), 284
23. R. W. Gibbs, Jr. and M. Perlam, in *Cognitive Linguistics: Current Applications and Future Perspectives*, ed. by G. Kristiansen, M. Achard, R. Dirven, F. J. R. de Mendoza Ibáñez (Mouton de Gruyter, Berlin, 2006), p. 215
24. K. Cmiel, *Democratic Eloquence: The Fight Over Popular Speech in Nineteenth century America*, University of California Press, Berkley, 1991), p. 21
25. G. Lakoff, M. Johnson, *Metaphors We Live By* (The University of Chicago Press, Chicago & London, 1980), p. 3-147
26. G. Fauconnier, M. Turner, *The Way We Think: Conceptual Blending and the Mind's Hidden*

Complexities (Perseus Group, New York, 2002), pp. 17-59, 279-298

27. G. Fauconnier, *Ten Lectures on Cognitive Construction of Meaning* (BRILL, Leiden, 2018), pp. 94-117
28. Z. Kövecses, in *Extended Conceptual Metaphor Theory* (Cambridge University Press, Cambridge, 2020), pp. 93-116
29. Z. Kövecses, *Metaphor and Metonymy at the Crossroads Metaphor* (de Gruyter Mouton, Berlin, 2003), pp.79-92

Communicating the Foreign Policy Strategy: on Instruments and Means of Ministry of Foreign Affairs of Ukraine

Natalia Sheludiakova[1*], *Bahodir* Mamurov[2], *Iryna* Maksymova[1], *Kateryna* Slyusarenko,[1] and *Iryna* Yegorova[1]

[1]State University of Economics and Technology, Kryvyi Rih 50000, Ukraine
[2]Bukhara State University, Bukhara200114, Uzbekistan

Abstract. The worldwide practice demonstrates that foreign policy of the states is often directed by a some set of strategic priorities referred to as foreign policy strategy. By communicating this strategy, states declare their intentions and prime concerns, which both aid their internal policy integrity and international perception as a consistent actor. Modern technology and media enable states to conduct systematic outreach or at least sporadically cover of their foreign policy priorities. The purpose of the article is to characterize the problems of application of communication tools in the foreign policy process and determine their role in forming a positive image of Ukraine in the modern world. As the mission of providing of foreign policy is vested in Ministry of Foreign Affairs of Ukraine, its communication practices and tools are examined. The raised issue involves the consideration of modern aspects of diplomacy and communication relations, including the concepts of public diplomacy, as well as media and digital diplomacy. Their tools and practices have been embraced by the Ministry of Foreign Affairs of Ukraine in order to convey some strategic foreign policy directions and state-branding messages, as the well-rounded foreign policy strategy is absent in Ukraine.

1 The significance of communication of foreign policy priorities

1.1 The foreign policy strategy problem in Ukraine

The modern world is at the stage of fundamental transformations, where the leading states are rethinking their priorities and conceptual principles of interaction with others. The process of transition from a bipolar system of international relations to a multipolar one, in which the "poles" are also embedded in the system of global interdependence, is unfolding. The world is in constant search of a new security system architecture capable of overcoming the problems of terrorism, extremism, and the proliferation of weapons of mass destruction.

Dynamic changes and civilizational shifts across the globe and the profound transformations of international relations have made the world less stable and less predictable. These changes affected Ukraine as well as its role and position among other nations.

For a long time, Ukraine has been shaping its foreign policy priorities, based on the need to respond to external factors and threats, rather than pursuing its own long-term goals and opportunities. The country went through a period of "multivectorism" as a strategy for maneuvering in the conflicts between the West and Russia, and a period of declarative course towards European and Euro-Atlantic integration, which was not marked by any real steps towards implementation. The Ukrainian establishment could not come to a systematic and comprehensive definition of Ukraine's priorities in foreign policy. Given the complexity of relations between the world's leading actors, foreign policy strategy is one of the tools to adapt to the new setup and position the state in the modern international environment, as it includes the state's vision of the future, defining methods to achieve this vision and dealing with global trends and challenges of the international community. In addition, it is important to give a signal to the world community about the direction of its movement by communicating its foreign policy strategy.

Given the political instability in Ukraine, involvement in one of the largest current armed conflicts in Europe, the question of defining and communicating Ukraine's foreign policy strategy is more acute than ever. Conceptualization of the state's foreign orientation involves determining ways to ensure the goals set for the state, the mechanism of control and decision-making, ways of cooperation with other states, good governance aimed at the population, position on various issues of world politics. An equally important aspect is the coordination of activities at the international level with state policy in other areas, in which Ukraine often fails.

The foreign policy strategy in theoretical framework is consistent with the concept of grand strategy as a set of practices aimed at achieving national goals. The concept is often used in three distinct meanings: 1) in

* Corresponding author: sheludiakova_na@duet.edu.ua

reference to deliberate, detailed plan devised by individuals; 2) to define an organizing principle that is consciously held and used by individuals to guide their decisions; 3) to refer to a pattern in state behavior [1]. But it should be mentioned, that such differentiation is an analytical tool rather than an actual divide of the meaning as each provides a valuable contribution to the understanding of the quite broad concept of grand strategy.

The modern understanding of grand has been largely formulated in the works of Paul M. Kennedy, John Lewis Gaddis, Charles Hill, Michael Howard and others. In particular, according to Paul M. Kennedy, "the crux of grand strategy lies therefore in policy, that is, in the capacity of the nation's leaders to bring together all of the elements, both military and non-military, for the preservation and enhancement of the nation's long-term (that is, in wartime and peacetime) best interests" [2]. John Lewis Gaddis determines the grand strategy as "calculated relationship of means to large ends". According to the scholar, it's about how one uses whatever one has to get to wherever it is one wants to go [3].

The grand strategy in fact includes the state's vision of itself in the future, awareness of its many goals and limited resources to achieve them, taking into account the global balance of power and finding the state's place in it. Therefore, the grand strategy is an important context and prerequisite for the formation of the state's foreign policy strategy.

In Ukraine, the main principles and directions of foreign policy at the legislative level are articulated quite abstractly in several laws (On the principles of domestic and foreign policy, On national security etc.), and strategic documents (Information Security Doctrine, National Security Strategy of Ukraine). There is no single and comprehensive document yet, however, some laws prescribe a number of doctrines, the development of which is currently underway, including the Strategy of Foreign Policy of Ukraine [4].

Although the state's strategy in international relations may be perceived by monitoring the total amount of government action and not being officially announced, the world practice shows that communicating a strategical views and goals is an important tool for establishing internal dialogue with society and interaction with foreign states. The direction of foreign policy should be clear to both domestic society and foreign partners. A clear definition of the vision of state's position in the network of world relations contributes to the establishment of mutual understanding, the growth of mutual trust with other states, as well as the strengthening of national identity. When state is communicating about its priorities in foreign policy, it proves the consistency of its policies and effectiveness in achieving national interests in the face of new opportunities and threats of the modern world.

US foreign policy, although not defined in a separate document, is still reflected in the Joint Strategic Plan [5]. In the Russian Federation, a strategic document, the Concept of Foreign Policy, which is approved by presidential decree, is prepared and constantly updated. In Latvia, strategic plans are an important part of the Ministry of Foreign Affairs' annual reports on implemented and planned activities. Poland adopts a Foreign Policy Strategy, which is tied to the period of the government's work and is constantly reviewed [6]. The Strategy of Actions for five priority areas of development of the Republic of Uzbekistan in 2017-2021 is in force in Uzbekistan today [7].

Official documentation of strategic foreign policy priorities is optional, and many states do not have such documents. What matters is not so much the document as clearly formulated foreign policy goals that affect the consistency, initiative of the state in the international arena, its readiness for crises and the emergence of new opportunities.

In view of this, the issue of coverage and explanation of the principles and main directions of Ukraine's foreign policy acquires special significance as an important part of its strategic communications and diplomatic activities.

1.2 Diplomacy and its ties to communication

For a long time, diplomacy remained a matter of a narrow circle of people and was a rather secret affair. However, the revolutionary changes that have taken place in the spheres of politics, international relations and mass communication have led to the creation of new conditions for the functioning of diplomacy. The development of new communication technologies enabled the media to cover almost all important events around the world, and the spread of the Internet has made the transmission and dissemination of information to large audiences instant. As the main source of information for the public, the media has greatly contributed to changing the nature and sources of power and influence, both domestically and internationally [8]. Contrary to realists' beliefs, "soft power" has entered the international arena in full force, based not on military and economic power, but on the ability to achieve desired results through effective communication and the functioning of institutional instruments [9].

Certainly, such paradigm shifts could not occur only due to the development of the communication sphere. They have been reinforced by relevant transformations in politics and international relations, in particular the process of democratization in many societies, so we can agree with Eytan Gilboa's statement about the interdependence and convergence of changes in politics and communication. The increasingly obvious relationship between these areas became the subject for research, including those in the realm of international relations. Thus, attempts to reflect the influence of the media on the sphere of diplomacy, resulted in coining the new terms: the world saw the concepts of "CNN-effect", "Teledemocracy", "media diplomacy", "teleplomacy", "real-time diplomacy" and others. The development of digital technologies, the transition of mass communications to online mode and the digitalization-related changes in the communication practices of diplomatic missions have brought the concept of "digital diplomacy" into scientific use [10].

A significant part of literature on political communication is devoted to comprehension of the media impact on the practice of foreign policy making and diplomacy. One of the most elaborate analyses is presented by Eytan Gilboa, who distinguishes three models of uses of the media as a major instrument of foreign policy and international negotiations. In Gilboa's study they are: public diplomacy, media, and media-broker diplomacy.

Within the model of public diplomacy, the researcher considers the media as one of the channels of communication among other means (cultural and scientific exchanges; participation in festivals and exhibitions; building and maintaining cultural centers; teaching a language etc.) [11] Today, public diplomacy is an integral and effective mechanism of foreign policy and national security of the world's leading states, as well as international organizations and institutions.

Public diplomacy is defined as "the process by which direct relations with people in a country are pursued to advance the interests and extend the values of those being represented" [12]. In other words, it refers to the process of communicating of the government with a foreign audience aimed at conveying the government's understanding of its ideas and ideals, institutions and culture, as well as national goals and current policies. Traditional public diplomacy covers a wide range of activities of policymakers in order to gain public support for their policies, promoting national interests and cultivating their own values abroad. The media and their information capabilities are, of course, part of its arsenal, but they are also accompanied by specially organized meetings, speeches, visits, support for certain groups or organizations abroad, i.e. all public aspects of foreign policy.

Today, the scientific literature insists on the formation of a "new public diplomacy", the emergence of which is explained by the updating of diplomatic practices in response to social change. Modern public diplomacy is conceptually based on two main characteristics: 1) multi-actor approach, i.e. the recognition of various actors in international relations, above and below national governments both within the country and abroad; 2) the formation of relations between these actors at different levels through dialogue and networking [13].

The second model of media-diplomacy interaction by Eytan Gilboa, media diplomacy, embraces "officials' uses the media to communicate with state and nonstate actors, to build confidence and advance negotiations, and to mobilize public support for agreements" [11]. Media diplomacy activities include current and specially organized media events, press conferences, interviews, information leaks, as well as the use of media resources for creation of a favorable atmosphere for negotiations or setting the necessary tone for interstate cooperation.

According to Yoel Cohen "media diplomacy includes all those aspects of public diplomacy where the media are involved as well as others not associated with public diplomacy including the sending of signals by governments through the media, and the use of the media as a source of information" [14].

Eytan Gilboa called the third model "media-broker diplomacy", where journalists serve as temporary mediators in international negotiations. In contrast to media diplomacy, in this model, journalists are tasked with initiating and conducting significant diplomatic moves, negotiating terms and conditions [11].

2 MFA mastering the new diplomacy instruments

According to both the world practice and domestic legislation, the task of formation and implementation of the state policy in the area of external affairs as well as information support is vested (not exclusively) in the government authority responsible for the nation's foreign policy – in Ukraine, Ministry of Foreign Affairs (MFA).

When reviewing the communication policy of the Ministry of Foreign Affairs of Ukraine, we should note that public diplomacy has long been out of the attention of the ministry. According to experts, it is only in 2015 when Ukraine began to master this new direction of foreign policy. Over the last five years, there has been an institutionalization of this trend and it acquired a systemic character [15]. The Department of Public Diplomacy was established under the MFA, which was later transformed into the Department of Communications and Public Diplomacy. At this time, the active work of the Ukrainian Institute affiliated with the MFA was also initiated and launched. The mission of the institution is "to strengthen Ukraine's international standing through the means of cultural diplomacy" [16].

Probably, it is the initial stage of formation of public diplomacy that determines rather vague understanding of the tasks assigned to it, as the strategies of public diplomacy (as well as the strategies of individual projects and institutions) have not yet been adopted. Therefore, the main tasks of this direction of foreign policy are the promotion of interests in the world information space and the creation of the brand "Ukraine".

One of the means that can be attributed to public diplomacy is the communication of the "Ukraine NOW" brand, which is carried out by the joint efforts of Ukrainian ministries and departments. In addition, the formation of a positive image of Ukraine among foreign audiences is the main task of the launched site "ukraine.ua".

These and other projects, as well as most information campaigns initiated by the MFA on social networks, show that Ukraine is taking steps in what is sometimes called "selfie diplomacy" as part of nation branding practiced in social media. According to researchers, an attempt to proactively manage and promote the country's image through digital diplomacy channels can be effective in tackling stereotypes [17]. By publishing attractive content, the MFA and other government bodies try to emphasize Ukraine's cultural heritage, its traditions and modern and vibrant present, instead of a picture of an underdeveloped country where armed conflict is taking place.

In recent years, the Ministry of Foreign Affairs of Ukraine has demonstrated its readiness and efforts to master new instruments of diplomacy, including digital diplomacy. Diplomats and officials began to use social media to interact with the public, to create new projects based on collaboration and co-creation with different audiences.

Digital diplomacy is often regarded as a component of public diplomacy, given that the former provides an opportunity to communicate and engage a wide variety of audiences. The emergence of the concept of digital diplomacy was provoked by the need to reflect the role of information technology, including the Internet and social media, in the implementation of foreign policy. This term still has no conventional definition, and is often used interchangeably with terms such as "e-diplomacy", "cyber diplomacy", "virtual diplomacy", "social diplomacy" etc.

According to one of the most common definitions, digital diplomacy is "the use of the Web, ICTs, and social media tools to engage in diplomatic activities and carry out foreign policy objectives" [18]. By definition, digital diplomacy is carried out through digital platforms and tools - websites, blogs, social networks and more. Facebook and Twitter are especially popular with politicians and officials, and attachment to these platforms has given rise to the new terms Facebook diplomacy and Twiplomacy.

There are two main stages in the development of digital diplomacy: Diplomacy 1.0, which means the early stages of a very limited use of information technology (mostly for the presentation of information on websites and information exchange via e-mail); Diplomacy 2.0 means the stage of use of interactive platforms, user-generated content, social networks, etc. by diplomatic missions.

Yet, taking up the digital communication tools entails the need to accept the individual and collective loss of control of public messaging and audiences, given the instant dissemination of knowledge that the Internet has brought with it [19].

The Ministry of Foreign Affairs of Ukraine has official accounts on Facebook, Twitter, Instagram, YouTube, LinkedIn with appropriate content variations. Official messages and relevant information are usually published in the former two, which are recognized in most countries as important channels for the expression of official information.

Moreover, the Twitter account of the MFA is in English, which is due to the low popularity of the social network among Ukrainians and the focus on foreign audiences, including foreign media. In addition to expressing an official position on certain issues, which according to the Twitter message format should be quite short and distilled, the MFA account serves as a platform for retweets of personal posts of Ukrainian and foreign officials on Ukraine's foreign policy and its place in international relations. This reveals to some extent the personification of Twitter diplomacy, and the interaction between the countries here takes quite specific forms of personal contacts, negotiations and their subjective assessments by the leaders of the countries, foreign ministers, officials, etc. Along with the fact that diplomacy has always been a matter in which the personal qualities of negotiators weigh a lot, we can also mention another modern reason for this move. Social media, such as Facebook or Twitter, make it possible to involve large audiences in a dialogue in which one of the parties must be a specific individual, rather than the collective image of the country or agency on whose behalf the dialogue is conducted.

Another aspect of digital diplomacy is worth mentioning, namely its far-reaching impact when speaking about audiences. In conditions of globalization and widespread use of information technology, the line between domestic and foreign audiences is increasingly blurred. On the one hand, it facilitates the communication process - instead of two "fronts" of communication (internal and external) there is a single platform with a unified position, and on the other hand, it puts forward the requirements of tailoring universal messages understandable to different audiences. Therefore, an important component of the MFA's communication policy, which can be described as "internal digital diplomacy", cannot be ignored. As a result, the Ukrainian public is an equally important consumer of information and a stakeholder for the Foreign Ministry. Digital diplomacy in this case can be used to perform several functions: the study of public opinion on the foreign policy course and international relations of the country; explaining the problems and challenges facing the country, as well as alternative ways to solve them; cultivating public support for the chosen foreign policy course, etc [10].

The digital channels are used to implement main direction of public diplomacy of the Ukrainian Foreign Ministry. The prominent direction is countering anti-Ukrainian propaganda and combat stereotypes about Ukraine and its citizens among foreign audiences. Increasing Ukraine's information presence in foreign media and social networks, which can be considered as part of national branding in digital space. This direction includes information campaigns initiated or supported by the Ministry of Foreign Affairs: #CorrectUA, #WearVyshyvanka, #MyUkraineIs, #MadeInUA, #Fight4Truth, etc. Campaigns on the Second World War (#NeverAgain, #WeRemeberTheFact), ones honoring the memory of the victims of the Holodomor of 1932-1933 in Ukraine are aimed at promotion of better understanding of the history of Ukraine by the international community and articulation of Ukrainian interpretation of important historical events. The platform for such campaigns was the information resources of the Ministry of Foreign Affairs, diplomatic missions of Ukraine, they also involved parliaments and governments of partner countries, representatives of governmental international organizations etc.

An important tool of communication of the Ministry of Foreign Affairs in public space, and at the same time, one of the means of interaction with the media is its website, in particular the Press Center department. Speeches, statements, comments of the highest diplomatic officials are published here, which express the official estimations of the ministry on this or that issue. As a rule, such messages are dedicated to the position of the ministry, current events or responses to them within the foreign policy implementation. Therefore, analyzing the content of the site, you can both track the strategic priorities of Ukraine by their implementation in the real steps taken to follow foreign policy course, and perceiving the conceptual bases and strategic priorities articulated in program statements and speeches.

Despite the fact that the site is available to the general public, its materials often enter the public information space through the Ukrainian and foreign media, which use them as a source of information in news reports or as an informational occasion. This gives grounds to talk about a two-stage process of dissemination of information, where the media act as an intermediate link between the Ministry of Foreign Affairs and the public.

An important element of the information content of the Foreign Ministry's website, as well as publications in online media that link to this content, is the publication of statements certifying the common positions of Ukraine and foreign partners on the process of peaceful settlement of the conflict in Donbass, condemnation people in the temporarily occupied territories of Ukraine, etc.

These materials were supported by information campaigns aimed at drawing the international community's attention to the problems of occupation of certain territories of Ukraine: #StopRussianAggression, #CrimeaIsUkraine, #FreeUkrainianPoliticalPrisoners. A new large-scale information campaign "Isolyatsia: must speak" was launched to reveal the facts of torture and inhumane activities in occupied Donetsk.

Despite the rather active practices in the field of public diplomacy, the communication activity of the Ministry of Foreign Affairs of Ukraine also has a number of shortcomings.

First of all, despite attempts to practice new approaches to communication, the core diplomatic practices remain unchanged. Analysis of the content of the Foreign Ministry's channels on social networks proves that the diplomatic craft has not become more open and transparent to general public. The information space includes traditional distilled materials, news releases, travel updates, latest statements, advisories. Often such items are not designed for active interaction with the audience, and work in the mode of "informing", not "communicating". That is to say, packaging has changed, but not messaging. Finally, given a rather conservative nature of the diplomatic profession, the established protocol and the significant consequences of mistakes, this point cannot be unequivocally attributed to "drawbacks".

Events in the international arena cannot always be planned in advance, and the ability to respond promptly and in a timely manner to challenges and crises in the information space is a prerequisite for the effective operation of modern diplomacy. The entry of external agencies into the online space for communications, along with the benefits, also set certain requirements. Official accounts are constantly receiving the attention of the international community, which in the event of a crisis awaits instant clarification, reactions, responses to challenges, and delays in the high-speed world are seen as a failure. The same applies to the recently introduced project approach to addressing particular problems and tasks within the services of the Ministry of Foreign Affairs and affiliated institutions. However, in reality, a significant number of bureaucratic obstacles and the rigidity of work plans and procedures hinder the requirement of efficiency and adaptability in the work of these institutions as they do not allow some variability. For example, in 2019, during the internal political clashes in the United States, in which Ukraine was involved, the diplomatic services failed to react in time and direct unplanned efforts to restore Ukraine's image. Thus, the United States accounted for only 2% of the implemented projects of the Ukrainian Institute, which, in fact, is tasked with forming a favorable image of the state [20].

In the absence of a coherent and complete foreign policy strategy, as a "master plan" and main reference in the activities of the Ministry of Foreign Affairs of Ukraine, the communication strategy of the ministry also seems somewhat inconsistent. In addition, when shaping the strategy, the traditional formation of the perception of Ukraine by foreign audiences through domestic political narratives should be taken into account.

3 Conclusions

Communication on foreign policy priorities is an important component of building Ukraine's image as a stable, predictable, trustworthy partner. Ukraine has not yet formulated a full-fledged foreign policy strategy as part of the "grand strategy" of the state, and only the first steps have been taken to define it - certain components of such a strategy have been identified and enshrined in law, such as principles of foreign policy, national security, etc., government plans with short-term goals and priorities are adopted, which, however, give some idea of Ukraine's further movement in the international arena. However, in the absence of an accomplished and well-rounded foreign policy strategy, communication policy is somewhat inconsistent and chaotic.

The main mission of implementing the foreign policy course, as well as articulating its priorities and forming a positive image of the state, is entrusted to the Ministry of Foreign Affairs and the diplomatic corps of Ukraine, which nowadays are working under the conditions of paradigm shifts in diplomacy.

Advances in communication and technology have significantly changed the working conditions of diplomats and foreign ministries. Technological innovations have reduced the distance between the countries of the world and accelerated the speed of communication, which requires diplomats and leaders of countries to be increasingly efficient in foreign policy decisions. Media and new information technologies open opportunities for diplomats to communicate with a global audience. Diplomats have the opportunity to convey political messages to huge audiences from around the world, instead of a rather narrow diplomatic corps. Diplomacy is becoming more open, and the form of political dialogue is being changed, with the mass media and social networks often acting as mediators.

The Ministry of Foreign Affairs' arsenal in communicating Ukraine's foreign policy strategy and shaping its image includes a wide range of information and communication tools that are commonly used in public and media diplomacy. One of the means of expanding the information and communication presence of Ukraine in world public opinion is the use of tools of modern digital diplomacy: the use of social media, networking, selfie diplomacy, information campaigns and more.

Significant use of social networks, which allow the use of user-generated content, interactive interaction with the audience, reaching the widest audiences, indicates that the Ministry of Foreign Affairs intends to implement a horizontal (interactive) model of communication. By using these tools, Ukrainian diplomats have the opportunity to spread more information to larger audiences with greater speed and efficiency. On the other hand, the immediacy and continuity of information flow on social networks require officials to react immediately to certain events, which can often draw reckless and ill-considered decisions or positions.

By resorting to a fairly wide range of means of public and media diplomacy, Ukraine is trying to combine two main approaches to this phenomenon. The first is based on the traditional perception of diplomacy as a hierarchical and mostly intergovernmental process, which is characterized by downward information flows and adherence to the theoretical foundations of strategic communications. The second approach is based on the consideration of foreign policy as one that is carried out through complex networks, and in which the public is not the object of information and communication influence, but a full partner in the diplomatic process. Without diminishing the importance of the first perspective, we can say that Ukraine is making efforts to adapt the second.

References

1. N. Silove, Beyond the Buzzword: The Three Meanings of "Grand Strategy", Security Studies, **27** **(1)**, 27-57, (2018) DOI: 10.1080/09636412.2017.1360073
2. *Grand Strategies in War and Peace*, ed. by P. Kennedy, New Haven, Yale University Press (1991).
3. J.L. Gaddis, What Is Grand Strategy? Prepared as the Karl Von Der Heyden Distinguished Lecture, Duke University, February 26, 2009, the keynote address for a conference on "American Grand Strategy after War", https://tiss-nc.org/wp-content/uploads/2015/01/KEYNOTE.Gaddis50thAniv2009.pdf. Accessed 12 Jan 2021
4. Ukaz Prezydenta Pro rishennya Rady natsional'noyi bezpeky i oborony Ukrayiny vid 14 veresnya 2020 roku "Pro Stratehiyu natsional'noyi bezpeky Ukrayiny". Zaklyuchni polozhennya (Decree of the President on the Decision of the National Security and Defense Council of Ukraine "On the National Security Strategy of Ukraine". Final provisions) https://zakon.rada.gov.ua/laws/show/392/2020#n12 Accessed 14 January 2021
5. Joint Strategic Plan FY 2018 – 2022, U.S. Department of State U.S. Agency for International Development, February, 2018 https://www.usaid.gov/sites/default/files/documents/1870/JSP_FY_2018_-_2022_FINAL.pdf Accessed 12 January 2021
6. Polish Foreign Policy Strategy 2017-2021 https://www.google.com/url?sa=t&rct=j&q=&esrc=s&source=web&cd=&cad=rja&uact=8&ved=2ahUKEwj9x8iHqb7uAhUXHHcKHeb4BfMQFjAAegQIAhAC&url=https%3A%2F%2Fwww.gov.pl%2Fattachment%2F869184c0-bd6f-4a20-b8af-a5c8190350a1&usg=AOvVaw0Ozc2-vsL5HWIfm-_F54eB Accessed 12 January 2021
7. The Strategy of Actions for five priority areas of development of the Republic of Uzbekistan in 2017-2021 http://strategy.gov.uz/en Accessed 12 January 2021
8. H. Mowlana, Toward a NWICO for the Twenty First Century, Journal of International Affairs, **47**, 59-72 (1993)
9. J. Nye, Soft Power, Foreign Policy, **80**, 153-171 (1990). doi:10.2307/1148580 P.164
10. E. Gilboa, Digital Diplomacy, in *The SAGE Handbook of Diplomacy*, ed. by C.M. Constantinou, P. Kerr, P. Sharp, pp. 540-551 (London, Sage, 2016)
11. E. Gilboa Diplomacy in the media age: Three models of uses and effects, Diplomacy and Statecraft, **12(2)**, 1-28, (2001) DOI: 10.1080/09592290108406201.
12. Ch. Paul, *Strategic Communication: Origins, Concepts and Current Debates* (Praeger, 2011), p.220
13. E. Huijgh. in *The SAGE Handbook of Diplomacy*, ed. by C.M. Constantinou, P. Kerr, P. Sharp, pp. 437-450 (London, Sage, 2016)
14. Y. Cohen, *Media Diplomacy* (London: Frank Cass, 1986), p.7.
15. Ukraiins'ka pryzma: Zovnishnya polityka 2019. Analitychne doslidzhennya, (Ukrainian prism: Foreign policy 2019. An analytical research) (HO «Rada zovnishn'oï polityky «Ukraïns'ka pryzma», Fond im. F. Eberta, Kyiv, 2020)
16. Ukrainian Institute: mission https://ui.org.ua/mission/en_Accessed Jan 15 2021
17. Manor, E. Segev, America's selfie: How the US portrays itself on its social media accounts, in *Digital diplomacy: theory and practice*, ed. by C.Bjola, M.Holmes. (Routledge, 2015), pp. 89-107
18. Sandre, *Twitter for Diplomats* (Geneva: DiploFoundation, 2013) http://isdi.esteri.it:

4300/ISDI%20ALLEGATI/Twitter%20for%20 diplomats.pdf Accessed 21 Jan 2021
19. Sabrina Sotiriu. Digital diplomacy. Between promises and reality, in *Digital diplomacy: theory and practice*, ed. by C.Bjola, M.Holmes. (Routledge, 2015), pp. 33-51.
20. Kontseptsiya zovnishn'oyi polityky Ukrayiny. Ekspertni rekomendatsiyi, ed.by Ye. Haber, S. Korsunskyi, H. Shelest. (2020).

The Role of the System of Exercises for Speaking Development

Zulfizar Karimova [1*]

[1]Tashkent branch of MSU named after M.V. Lomonosov, Amir Timur str., 22, Uzbekistan

Abstract. This article highlights the need to use a system of exercises for the development of oral speech. Also, in this paper, various points of view of scientists are considered, information is analyzed and own arguments are given regarding the use of a system of exercises in English classes to improve oral speech. The relevance of this article lies in the fact that it gives the concept of some types and kinds of exercises, considers several types of exercises, as well as techniques with which teachers can effectively improve the oral speech of students. In the conclusion, there are practical examples of exercises that correspond to different levels of students' training.

1 Introduction

In the last decade, in connection with the orientation towards the practical mastery of a foreign language, the idea has been increasingly persistent that we teach in educational establishments not language, but speech (oral and written), consequently far-reaching conclusions are drawn from this.

Before deciding whether such a change of concepts is fair, it is necessary to try to find out what caused it. It is known that in the distant past, when the methodology of foreign languages began to crystallize as an independent pedagogical branch of science, teaching a language was understood primarily and mainly as teaching its grammar. Speech training was carried out as needed outside the secondary school. Then, at the end of the last century, when there was a significantly greater need for persons who practically speak a foreign language, methods appeared that equated teaching a foreign language with teaching a native one. Moreover, since the child learns the native language without specially studying its grammar, they tried to teach a foreign language, without telling him any rules (neither lexical nor grammatical). Representatives of these methods, known in the history of the method as natural, without being very sophisticated in the science of language, announced in broadcast advertisements about a new method of teaching a language without learning its grammar.

Since grammar is very abstract, and its study usually does not attract people who do not have a penchant for philology, this way of teaching a language (without studying its grammar) turned out to be very attractive and therefore won a large number of supporters. But since grammar is one of the integral parts of the language and therefore it is impossible to learn a language without learning its grammar, therefore, it was only a question of whether to learn it consciously or not, that is, the way children learn their native language. One of the most significant gaps in the methodology of teaching foreign languages is the lack of an ordered system of exercises, although very interesting attempts have been made to create it.

2 Methods

First of all, we should mention the system of exercises proposed by the representative of the direct method, B. Eggert and his work "Regularities underlying exercises for teaching foreign languages", which systematizes the views of this direction in the methodology. B. Eggert established a number of regularities, of which the following are still, of the greatest interest today:

a) the nature of the exercises should match the acquired skills, i.e. to teach the student the comprehension skills you need to offer him appropriate special exercises, for teaching speaking — exercises for speaking and so on. As each exercise being focused on the development of any speech activity, at the same time contains elements common to other activities, for the development of this language activity can be used and not special exercises.

b) the meaning and use of language forms is best learned in the course of activities and in appropriate situations that are close to students and arouse their interest.

c) Global assimilation of language forms should be accompanied by an analysis of their components.

d) any new language phenomenon is better assimilated as a result of comparison with what is already known, including from the native language.

e) any new linguistic phenomenon is better understood if it is placed in different contexts.

f) when learning a language, the main role belongs to oral exercises, which are necessary for the development of any type of speech activity.

All exercises B. Eggert divides into exercises in speech activity (speaking, listening, reading, and writing) and aspect exercises for mastering vocabulary, grammar, and phonetics.

Also of great interest is the system of exercises developed by G. Palmer, mainly for reproductive language acquisition. G. Palmer distinguishes between receptive and reproductive exercises. The former are divided into unconscious and conscious exercises. The task of unconscious exercises is to draw students ' attention to the General content of the phrases spoken by the teacher, regardless of their form and meaning of individual words. This type of exercise also includes exercises in the execution of orders (imperative drill). Conscious exercises should focus students ' attention on specific language phenomena (phonetic, grammatical, and lexical). The second type of exercises (receptive and reproductive) involves not only perception, but also reproduction of speech by students. There are three types of exercises: repetition following the teacher, conditional conversation, in which students mainly answer questions posed by the teacher, and natural conversation on various topics of everyday life. To the system of exercises described above, G. Palmer added wildcard tables, which then played an important role in subsequent indirect methods.

The first methodologist who paid special attention to the development of the theory of the system of exercises was Professor I. A. Gruzinskaya. She built her system of exercises based on the goal of teaching: to teach students a comprehensive mastery of the language (oral speech, reading and writing). To do this, she considered it necessary to create exercises that would cover all aspects of the language (vocabulary, grammar, and phonetics) and correspond to the mental processes that occur during its assimilation — recognition, memorization, and reproduction. In accordance with this, I. A. Gruzinskaya put forward the following requirements for the system of exercises:

1. Sequence of exercises. Every new linguistic phenomenon must pass three stages of assimilation: a) receptive, this offers exercises in identifying, selecting, and grouping;

b) semi-reproductive — semi-free playback of this linguistic material, which are recommended a variety of exercises by analogy with the partial modification of the sample;

c) reproductive, which is the final stage of mastering of language skills and requires the ability to arbitrarily (without prompting) to reproduce a particular phenomenon of language to express his thoughts.

The last two stages-semi-free reproduction and random reproduction-in addition to exercises aimed at consolidating certain types of language knowledge should also include all the variety of exercises aimed at creating active skills.

2. Consistency in the application of various types of exercises. For each year of study, reference types of exercises should be selected, which should be repeated from lesson to lesson. Along with them (but in a smaller number), other exercises can be offered that would help to master this language phenomenon from all sides, in its various aspects.

3. Continuity of exercise types. Along with the obligatory continuity and repeatability of some of the main types of exercises, it is necessary to steadily expand their range and modify their content. I. A. Gruzinskaya distinguishes the following four types of exercises:

1. phonetic-spelling;
2. grammatical;
3. vocabulary;
4. exercises in creating active skills.

Within these types, the sequence is strictly observed, which is formulated in the above requirements: first, receptive exercises are given, where the student's task is to learn or choose a phenomenon from the language material given to him. They are followed by semi-productive exercises in which the sample is supplemented or partially modified. Last of all, the student is offered exercises in which he should independently reproduce the completed language material.

A more detailed acquaintance with the system of exercises proposed by I. A. Gruzinskaya leads to the conclusion that the main attention is paid to language exercises, after completing which students will be able to master oral and written speech within the requirements of the program.

In recent years, a number of studies have been devoted to the exercise system. Some develop the ideas of I. A. Gruzinskaya, such as Z. M. Lyubimova, N. B. Sokolova, P. B. Gurvich and others. They pay considerable attention to aspects of the language. Other scientists, such as, M. S. Ilyin, B. A. Lapidus, etc, build their system of exercises on a functional basis, giving preference to speech exercises.

P. B. Gurvich divides exercises into formal, formal-meaningful, meaningful-formal and meaningful. He considers most exercises with isolated words to be formal (word analysis, memorizing words, declension of nouns, etc., conjugation of verbs, grouping words, etc.).

Formal-content or formal-meaningful exercises are exercises that focus on the form associated with a particular content context (selecting words to describe something, paraphrasing, adding missing words, creating situations with these words, etc.).

Content-formal or meaningful-formal exercises are exercises in which the main attention is paid to the content, but at the same time the educational nature of speech actions is realized, the so-called conditional-communicative exercises. Content exercises are exercises in which all attention is focused on the content, and students do not receive any special lexical tasks ("natural" dialogue, retelling, narration, etc.).

Let us now briefly discuss the systems of exercises based on the functional principle. So, M. S. Ilyin, for example, distinguishes between two main types of exercises: pre-speech and speech exercises. According to M. S. Ilyin, the first type of exercises includes all exercises related to the development of grammatical and phonetic skills, as well as non -automated components of the act of speech-the choice of a sign in accordance with the situation. The second type includes exercises that develop speech.

B. A. Lapidus distinguishes two types of exercises in the system of exercises: the first is exercises for purposeful activation of language material and the second is exercises for non-stimulated activation of language material in the conditions of natural speech practice or approaching it. The first type includes: purely training exercises, elementary combined exercises (statements in accordance with a request or order, speech on a given structural skeleton, etc.) and combined exercises (without food with a given language material).

The second type of exercises includes exercises for unregulated activation of language material in the context of speech practice at a higher level (discussion, etc.).

B. A.Lapidus rejects exercises that require formal operations, and those where students ' attention is focused only on learning the form. In his opinion, it is also impractical to widely use exercises in which logical action (for example, opening parentheses) is not combined with speech.

In conclusion, the author lists the criteria for the effectiveness of exercises that he has five: the degree of mental activity of students; the degree of compliance with psychological conditions exercise conditions, typical of real communication; the degree of independent working on the production of the phenomenon; the efficiency of the exercises; coverage of students in the group exercise.

Consider the opinion of another scientist, E. I. Passov, who contrasts the speech system with the language system and believes that only communicative exercises are needed to master the language. They are based on three principles: the principle of imitation of the speaker's communicative task, the principle of analogy in the assimilation of grammatical forms, and the principle of the connection of form with its function in speech.

Communication exercises differ in their composition, method of implementation, and attitudes. According to the composition, there are five types of exercises: binomial (consisting of two replicas), three-term, expanded, complex and thematic. According to the method of execution there are four types of exercises: imitative, wildcard, paraphrasing, and reproductive. Exercises are divided into four types: interrogative, stating, negative, and motivational.

3 Results

However, despite a number of advantages that the works have listed above, they also have some common disadvantages:

1. The system itself is not defined and, as a rule, the requirements for it are not formulated in general: in particular, it does not show how its individual elements are related to each other.

2. The described systems do not take into account all the mental processes that are necessary for language acquisition and speech development.

3. It is not defined what types of exercises make up the system, and it is not shown what the specifics of each type of exercise are (their similarities and differences). Types and kinds of exercises are mixed.

4. The completeness of the system is violated, because, firstly, it is often reduced to two types of exercises: language and speech, or preparatory and speech, and, secondly, some types of exercises (mechanical, bilingual) are rejected or underestimated without sufficient grounds and the role of others (monolingual, communicative) is exaggerated.

Obviously, a system of exercises for teaching a foreign language should be understood as a set of types and kinds of exercises that are related to each other by purpose, material and method of their implementation and arranged according to the principle of composition and subordination. Each element of the system is irreplaceable and represents a binomial opposition.

Types of exercises are divided into types, which should be arranged in order of increasing difficulties. The following requirements should be applied to the exercise system. It should:

1. develop cognitive processes of the individual (thinking, memory, imagination);
2. consider the purpose, material, and method of performing exercises;
3. develop and ensure the assimilation of all aspects of the language and all types of language activities;
4. include all types and kinds of exercises, each of which should have its own specifics (irreplaceability of system elements);
5. communicate between elements of the system according to the principle of composition or subordination; each element of the system must represent a binomial (binary) opposition;
6. arrange kinds and types of exercises according to the degree of increasing difficulties. When classifying exercises, as a rule, terms with a negative characteristic should not be used.

The system of exercises should be aimed at mastering all aspects of the language, because one-sided preferential attention to one of them — grammar or vocabulary-leads to the fact that speech is impoverished or full of errors that are later difficult to correct. The completeness of the system is achieved only if all analyzers participate in speech development from the very beginning of training, and oral speech is supplemented by reading and writing.

Within the system, you should distinguish between types and kinds of exercises that are not hierarchically related. Kinds of exercises are divided into types, which in turn can be represented by various options.

They are practically used by teachers when teaching the language. Not being able to show how the system described above functions in different types of academic institutions, we analyzed the functioning of the exercise system in non-linguistic universities, where the material studied is more stable; we will try to show the effectiveness of using the proposed exercise system for the development of oral speech among students of non-linguistic Universities.

4 Discussions

However, it seems appropriate to cover some general issues first. These include the following: 1. Motivation for learning. 2. Use of technical means. 3. The uniformity of the exercises in each class. 4. The ratio of the new and the already traversed material and the amount of exercise. 5. The nature of the exercises depends on the language being studied.

1. Motivation for learning

The success of practical mastery of a foreign language depends on how motivated it is in the eyes of students. Motivation can be natural or artificial. Natural motivation can be created only in the senior years, when students have already mastered the basics of the language and will be able to use them in their practical activities — read literature in a foreign language and listen to podcasts and videocasts, radio and TV.

In junior years of university, motivation should be created artificially, namely by interest in the learning process itself. This interest can include three components: encouragement in the form of a good grade, interest in the subject, in the process of mastering new knowledge, skills and abilities, and the associated satisfaction with the results obtained.

Encouragement in a foreign language classes is no different from encouragement in other subjects and therefore does not require any additional explanations. Interest in the subject "Foreign language" has its own characteristics. In other Humanities subjects, the content of the science being studied, even if it is prepared for school education in an academic subject, is of interest. The science of language (General or particular linguistics) is not studied at school, and the theoretical information on phonetics, grammar, and lexicology that is communicated to students is extremely brief and sketchy, and therefore, obviously, does not arouse much interest in them. However, when teaching a language, we introduce students to certain activities, which can arouse interest if the teacher uses appropriate methodological techniques, and students feel the results of their activities.

Therefore, certain techniques and a specially designed system of exercises can arouse interest in this activity. Exercises are interesting if they involve mental work, are feasible and accessible. However, not all exercises require students to be mentally active. When learning a language, some phenomena have to be memorized mechanically, without the participation of mental operations.

This means memorizing the forms of words that cannot be explained without involving the history of the language, which, as is well known, is not studied in non-linguistic Universities. Special mention should be made of the methodical techniques used by the teacher. Some teachers have a tendency to turn foreign language classes into a game, as the well-known German methodologist M. Walter, but this is hardly advisable, firstly, there are not always conditions for this, and secondly, the student should be aware that a foreign language is as important a subject as any other is, and it should be treated seriously as well.

To arouse interest to classes, you can sometimes involve a game, but this should be an episode, and not the norm. The norm should be such methodological techniques that would appeal to the active thinking activity of students. Interest in classes can be enhanced by skillful use of technical means.

2. Use of technical means

Technical means are divided into visual (projector), audio (tape recorder, radio, speaker) and audio-visual (TV, tablet). The use of technical means, firstly, significantly improves visual and auditory perception and memorization of the studied language material, secondly, facilitates speech development and, thirdly, contributes to the involvement of all students in the work. They can be used both in the classroom and during extracurricular activities when students perform individual tasks. The latter make it possible to take into account the individual characteristics of students (their abilities, home conditions, etc.).

In the course of classes, technical means can be used, first, to concretize the studied language phenomena, second, to create appropriate situations that facilitate the process of speaking a foreign language, and third, and perhaps most importantly, to master the ability to understand speech by ear. We have already mentioned above that understanding speech by ear is a huge incentive for practical mastery of oral speech, and here technical means are absolutely indispensable.

Mastering the understanding of speech by hearing is associated with understanding different sound colors, different tempo, timbre of voice (male or female), intonation, i.e. those conditions that the teacher cannot create without resorting to technical means. Not to mention that not all teachers are proficient in idiomatic speech in the taught foreign language. In addition, the perception of the teacher's speech occurs without those hindrances that are characteristic of mechanical speech, namely, they will have to deal with most of all in practical life. By listening to a dubbed movie or podcast and understanding its content, students begin to feel the practical benefits of learning a foreign language — an incentive that can hardly be overemphasized. It increases interest in classes, which, in turn, affects the quality of learning the material and contributes to its involuntary memorization.

3. The uniformity of the exercises

One of the significant disadvantages of the exercises offered by many text books is that they are either too monotonous and are reduced from lesson to lesson to a repetitive formula: "Read, translate and learn words" or, on the contrary, are too diverse — new exercises appear in each part, and old ones are forgotten. The only exceptions are mechanical, aspect, and choral exercises, which are no longer necessary in senior years. As for the types and kinds of exercises, they should be distributed according to their difficulty. The difficulty of the exercise depends, depending on the kind and type of exercise, on the material to be learned, and on the thought and mnemic processes involved in the exercise.

For each course, you need to determine the leading type of exercise, which can be represented by different types and supplemented with other types, but not the main ones for this course.

https://doi.org/10.1051/shsconf/202110002006

4. The ratio of new and already completed material and the number of exercises

The ratio of new and old material is extremely important for its assimilation and is closely related to the problem of repetition and the strength of mastery, skills and abilities. In each lesson, exercises that contain material that you have already completed must be several times larger than those that contain the new material. New material should always be based on the old, and each time it should be offered in a new form, so as not to lose the charm of assimilation. One of the biggest problems is that the material being studied is not well understood, and therefore it is necessary to conduct special classes dedicated to the repetition of what has been passed.

K. D.Ushinsky was right when he wrote: "Repetition in order to remember what has been forgotten already shows a lack of learning and generally poor teaching in school. Obviously, the issue of revising the distribution of educational material is long overdue. It would seem that, for example, the lexical material is very small. If this dictionary is divided by the total number of hours devoted to learning a foreign language in non-linguistic Universities, then less than three new words are added to each lesson. If we further compare this number with the number that intensive students can remember in one lesson — 200 (although they are in completely different conditions), it will still become clear that the point is not in the amount of material, but in the inexpediency of the methodological techniques used, in an insufficiently thought-out system of exercises. After all, almost all universities complain that even this more than modest minimum of words (1500) students learns only 50%. Often, we have to deal with the fact that students spend a lot of time doing exercises, not because they are difficult and contain a lot of new material, but because the material they have learned is poorly understood, forgotten, and the old material is again perceived as new. The teacher completed the program at one time, but the student did not acquire any knowledge, skills and abilities. Memorizing the material and acquiring the necessary skills and abilities to use it is also associated with the number of exercises offered by the teacher. This question has not yet been completely tested experimentally and depends more on how well the student has mastered the previously learned material.

The experience of composing textbooks has shown that the material is fully absorbed,even if it has 100% repeatability. This principle is not easy to observe in all textbooks, but if you agree with the recommendations on the subject set out above, then compliance with this principle is quite possible with the concentric development of this topic. If now it is difficult to name the exact number of exercises that should be given to students for each lesson, then we can still say that it can be increased compared to what is offered in current textbooks.

There are two other factors that should be taken into account when drawing up exercises. First, that all reproductively acquired material is then assimilated receptively, i.e. that reproductive exercises are followed by receptive ones. Secondly, to take into account both voluntary and involuntary attention of students. The first requirement is due to the fact that all material intended for reproductive assimilation must also be assimilated receptively. Earlier, when describing reproductive and receptive exercises, we have already found out that each of these types of exercises has its own specifics, its own function, and one type of exercise cannot be replaced by another if this material must be learned both reproductively and receptively. As for voluntary and involuntary attention and the associated voluntary and involuntary memorization, there is some underestimation of students ' perception of the material that is not consciously memorized. As the latest psychological studies and University experience show, a student remembers some of the language material involuntarily, without special memorization, due to the fact that this material is of interest to him or is repeatedly repeated in various situations. When composing exercises, this feature of memory should be taken into account. To ensure that this process does not occur only spontaneously and chaotically, it is necessary to provide in advance what material should be memorized, and what the student can remember without spending special effort on it. It seems that for mastering grammatical phenomena, it is preferable to use arbitrary attention. Some parts of the vocabulary may be remembered involuntarily. Pronouncing norms must first be acquired consciously and only during subsequent training can serve as an object of involuntary attention.

5. The nature of exercises depending on the language being studied

5 Practical exercises and conclusion

The proposed system of exercises can be used when learning any language, but the ratio of individual types of exercises and their specific weight depend on the characteristics of the language being studied. For example, when learning German with its fairly well-developed system of declension and conjugation, much attention should be paid to morphology, and therefore the proportion of aspect and mechanical exercises will be greater than when learning English. French occupies an intermediate position, although it tends more towards English. When teaching French and English, syntactic language exercises will prevail over those given for mastering the syntax of the German language.

When teaching French and English, writing exercises play a greater role than when teaching German, since the spelling of these languages is based on a historical principle, and there are no rules in modern language (especially in English) that would help to spell a word correctly. In French, the correct spelling of words is also associated with the distinction of verb forms, plural nouns, etc. Significant differences are also observed in the vocabulary. In German, the composition of vocabulary for at least the first two years of study is largely determined by the grammar being studied.

The choice of many verbs depends on the knowledge of the case system of nouns, as well as their belonging to the weak or strong conjugation. In English, the use of certain adverbs is impossible without mastering the corresponding verb forms. In the French language sequence for the learning of verbs is dictated by their belonging to a particular type of conjugation. The choice of words also depends on their usage in the language being studied. For example, verbs with different prefixes are very common in German, which is not observed in either English or French. Verbs like faire, prendre, and take are widely used in French and English, prendre, take, while their counterparts in German are not so widely used.

In order for the teacher to have a clearer idea of how to use the proposed system of exercises in practice, the following examples of exercises are given.

These exercises do not represent a system of exercises for a particular course. It can only be shown in a textbook that provides instruction not only in oral speech, but also in reading and writing. This list of exercises has the following objectives:

a) to show what kinds and types of exercises should prevail in a particular class; the number of language exercises, in particular, is gradually decreasing;

b) to illustrate the continuity of exercises between classes;

c) to demonstrate that the difficulty of exercises does not always depend on the type, but rather on the chosen type or variant, so that the order of exercises is often determined by them, and not by the type of exercise.

In our list, the order of types of exercises is determined by the prevailing type of exercises for this class. Almost every one of the presented examples of exercises (except for specially specified ones) can be performed both orally and in writing. It is up to the teacher to decide how to implement it. Since the article describes a system of exercises for teaching oral speech, the number of oral exercises will prevail, as well as the number of monolingual exercises, although bilingual (translated) exercises should also be given in a certain proportion in all courses. For the most part, each type of exercise is represented by only one option, but this does not mean that there should not be several of them. It is not possible to show all the exercise options due to their large number. The number of options will depend on how well you have studied the completed material.

These exercises are not designed for studying the entire material of the foreign language program, but are only samples, by analogy with which other versions of exercises can be compiled for mastering other language phenomena and developing other skills.

I. Types of exercises

Comparison

1. The repetition of questions and answers.

2. Reading in chorus (behind the teacher).

3.Making suggestions based on wildcard tables (mechanical and creative).

4. Learning by heart.

5.Making suggestions based on a sample (without replacing it).

6. Drawing up suggestions based on a sample (with the replacement of individual parts).

Transformation

1.Replacing individual parts of the offer.

2.Converting the form of words.

3.Translation from Russian to English.

Extension

4.Exercises with missing words.

1. Please, repeat:

I. What is your name? — My name is Mike.

2. Do you play tennis? — Yes, I do.

3. Does he play tennis well? — Yes, he does.

2. Read in unison: We see Ann. We see her desk. We see her books. We take her books. We have her books.

3. Make suggestions:

Give me his pencil! Do not give Kate her notebook! We want to play games in the morning he likes to read books after study she wants to go out on Sunday.

4. Learn it by heart:

I take my book, he takes his book, she takes her book, you take your book.

5. Make suggestions using the following structure:

I like to...

1.I like to speak English. 2. I like to play tennis. 3. I like to read about animals

6. Make sentences that are related in meaning to the data: It is cold today. I cannot go to the forest. I read a book at home.

7. Replace individual parts of the following sentences:

Mike often comes to my house in the evening. He often comes to see me in the evening. He often comes to see me in the afternoon.

8. Replace the highlighted words with personal pronouns in the correct form: 1. Does Kate play games with her little sister? 2. Do you go to school with Mike? 3. Do you give your little brother your pictures?...

9. Translate into English:

We go to the University in the morning. We are going to class. In English lessons, we read and write. We play football after study.

10. Complete the missing preposition in or into:

1. I like to sit under a tree ... the forest in summer. 2. After dinner the boys run ... the yard and begin to play football. 3. They like to play football the yard.

II. Types of exercises

Likening 1. Reading in chorus (behind the teacher). 2. Memorizing by heart. 3.-Making suggestions based on wildcard tables or a sample. 4. Question-and-answer exercises.

Conversion 5. Form conversion. 6. Replacing individual parts of the offer. 7. Change of offers (due to changes in the situation) 8. Translation from Russian to English.

Expansion 9. Exercises with omissions. 10. Adding suggestions.

1. Read it in unison:

He came to London late on Saturday night and went to a hotel. The next morning he got up early. After breakfast, he remembered that he had to send his sister a telegram. Therefore, he sent the telegram with the address of the hotel and then made his plans for the day.

2. Learn it by heart:

be — was, eat — ate, bring — brought, give — gave, buy — bought, go — went, come — came, meet — met, do — did, read — read

3. Make suggestions a) using the wildcard table: I didn't go in the garden yesterday he didn't play on the excursion last week they didn't work in the yard last Sunday

б) by sample: The boys couldn't play football, because it began to rain.

4. Answer the following questions: 1. Did Mr. Black remember his address in New York? 2. Did Mr. Black's sister know his address in London? 3. Did she send him a telegram with his London address?

5. Put the selected nouns in the plural: The families of many workers live near their factory. The boy in my Russian circle knows a story about animals. The woman made the dress quickly.

6. Replace parts of the following sentences. 1. My brother came home late from work yesterday, so I could not play chess with him in the evening. 2. There were no trams or buses yesterday evening, so we had to walk all the way home. ...

7. Change the following sentences to reflect the changed situation:

The wind began to blow. The man who walked near the lake was cold and put on his coat. Suddenly the sun began to shine...

8. Translate it into English: In the summer, I was in a holiday home. There was a small river near it and on the other side of the river, there was a forest. We went there every day. ...

9. Complete the missing verbs: Yesterday ... Saturday. I ... home early from university. I ... two letters and ... them on a bench for my sister. After supper I ... her dolls and ... them on my table. I ... to ... their faces. ...

10. Complete the following sentences: Last Saturday was the first of January, and we had a party at... We also had a party at... After breakfast I helped mother... I took the chairs out of all the rooms and put them...

III. Types of exercises

Conversion 1. Translation from Russian to English. 2. Replacing parts of the offer. 3. The transformation of the shape of words. 4. Replacing some questions with others.

Extension 5. Exercise with missing words. 6. The story in connection with this situation. 7. Add suggestions. 8. Question-and-answer exercises.

Reduction 9. The implementation of the proposals. 10. Retelling with abbreviations and paraphrases.

1. Translate it into English:

1. Misha runs very fast. - Misha runs very fast.

2. Nina usually plays in the garden. — Nina is currently playing in the garden.

3. Mom always takes a break after work. — Hush! Mom is resting after work.

4. Anna draws best of all. — What are you drawing, Anna?

2. Replace the highlighted nouns with others: 1. I have opened the window. 2. He has cleaned his clothes. 3. We have washed our hands. ...

3. Put verbs in the past tense (Past Indefinite Tense): All the cities in Greece (to send) their best athletes to the city of Olympus, to compete in the games. Thousands of people (to come) from all parts of Greece to see the games. At that time cities in Greece (to be) not friends with each other. However, at the Olympic Games people (to be) friends. All wars between the cities (to stop)...

4. Replace the questions posed to the pictures with other ones.

5. Insert the missing verbs:

1. Now they ... their hands. — They ... already their hands. 2. Now they ... their dinner. — They ... already their dinner. 3. Now they ... chess. — They ... already chess.

(Students perform this exercise by first looking at a book and then listening.)

6. Tell us what you will be talking about with the guests from England who will visit your office next week (come up with 8-10 suggestions).

7. Finish the following sentences:

1. I cannot go with you because... 2. He ran all the way home because... 3. We have not seen him because...

8. Answer the questions:

1. Why are you laughing? 2. Why are you smiling? 3. Why are you angry with me? 4. Why are not you working?

9. Shorten the following sentences, keeping their main meaning:

1. We have many friends everywhere in the world. 2. Nobody saw him yesterday at school. 3. Mother refused to go to the country with us last Sunday. ...

10. Tell the following text in your own words, make possible abbreviations and substitutions:

The city of Sparta was in Laconia, so people sometimes gave the Spartans the name Lacons. The Lacons never spoke much and they taught their children not to use more words than they needed. "If you listen more and speak less," they said, "you will learn many things. People that talk too much are usually not very clever!" So, it became a tradition in Laconia to try to use less words. And even now we say that an answer in not many words is a laconic answer.

IV. Types of exercises

Extension 1. Adding suggestions. 2. The distribution of the text. 3. Answers to questions.

Conversion 4. The conversion of one form to another. 5. Translation from Russian to English.

Abbreviation 6. Shortening the text.

1. Complete the following suggestions:

1. ... because I have no money. 2. ... because I have not enough time. 3. ... because you were wrong. 4. ... because she refused to go with me. 5. ... because she was angry with her. ...

2. Please distribute the following suggestions: When I came to London, I went to see my uncle. I found him with his friend. My uncle invited me to have lunch. After lunch, his wife called him to the telephone. "I'm sorry," he said when he came back, "I must leave you now. I will try to be back soon. Don't go away!"

3. Answer the questions:

1. Where do you usually spend your holidays? 2. Where did you go last year? 3. What do you do after school? 4. Why do you learn English? 5. Whom do you sometimes write letters to? ...

4. Convert Present Continuous to Past Continuous:

When he came... (I am listening to the radio, I am writing my letter, He is reading a book...)

5. Translate into English:

6. Shorten the following text:

Jimmy lived alone in a little house on a hill. It was in the middle of an endless forest, and Jimmy did not have many visitors. He was a railway worker, and he had to take care of the line that came up the hill to the house and went down the other side. Trains roared up the hill and then down: they never stopped, and Jimmy paid no attention to them...

V. Types of exercises

Extension 1. Supplement offers. 2. Answers to questions. 3. Creating a dialog.

Abbreviation 4. Shortening the text. 5.Abbreviating text with possible conversions.

Conversion 6. translation from Russian to English.

1. Complete the following suggestions with as many options as possible:

1. The work was finished in time because... 2. The party was enjoyed by many girls though... 3. The letter was read twice but... 4. I was very tired, so...

2. Answer the following questions in two or three sentences:

1. What kind of schools are there in the your country? 2. What are you going to do when you finish school? 3. What English books have you read? 4. Who is your favourite English writer?...

3. Create a dialog (a foreigner stopped you and asked for directions...).

4. Shorten the following text: On the tenth day of my adventure, I received an invitation from the American ambassador and went to his house. He had read about me in the newspapers, of course, and was very glad to meet me. He invited me to a dinner party that he was giving that same evening. ...

5. Shorten the following text, by making possible conversions: In 1917, the United States joined the other imperialist countries in the struggle for world power. John Reed criticized it. He insisted that it was a Wall Street war, not a war of the American people. He gave lectures against the war, his voice was heard everywhere, and the capitalists were frightened. ...

6. Translate it into English:

VI. Types of exercises

Assimilation 1. Making up a story similar to the one you read.

Transformation 2. Expressing the same thoughts by other linguistic means.

Abbreviation 3. Drawing up a plan for the text.

Expansion 4. Story according to plan. 5. Creating a dialog for this situation.

1. Read the text and create a similar story using the data provided in the illustration.

2. Express the same thoughts differently (in other words and constructions):

According to some historians, there was a young man in Rome who won the admiration of the whole city by his ability as a swimmer. When some people said "he swims like a fish", others used to answer: "No, the strongest fish swims like him!" *3. Make a plan for the text you read.*

4. Tell us the text for this plan

5. Create a dialog using the specified situation.

Therefore, in order to successfully develop oral speech, it is necessary to create certain conditions: knowledge of the topic, knowledge of the relevant language material, understanding of the situation and the presence of an incentive for utterance.

However, we must not forget that all speech preparation, i.e. most of the exercises aimed at developing oral speech skills, is performed during extracurricular activities, in the form of external or internal speech, in the form of language or speech exercises. Without them, it is impossible to speak in class, it is impossible to have an educational or natural dialogue.

When determining the purpose of teaching oral speech, you should finally have a clear idea of the persons who will conduct the conversation and where it will take place — in the country of the language being studied or in their own country.

References

1. B. Eggert, *O Bungsgesetze im frcmdsprachlichen Unterricht*,Leipzig, 186. (2006)
2. I. Gruzinskaya, *Exercises in 5-8 grades of secondary school*, Foreign languages at school, (1998)
3. P. B. Gurvich, *Fundamentals of teaching oral speech at language faculties,* Vladimir State Pedagogical Institute, **II.** (2004)
4. E.I Passov, *Communicative exercises*. – M (2001)
5. K.D. Ushinskiy, *Man as a subject of education. Experience in pedagogical anthropology*, Pedagogical course programs for women's educational institutions, 544. (2005)
6. M. Fleming, F. Goullier, E. Thürmann, H. Vollmer, *The language dimension in all subjects A Handbook for Curriculum development and teacher training*, Council of Europe. (2015)
7. Sh. Torky, *The Effectiveness of a Task- Based Instruction program in Developing the English Language Speaking Skills of Secondary Stage Students*, Cambridge, 4th ed, 23. (2006)
8. G. Escudero; D. J. Cutiopala; Janneth A. Caisaguano; L. Gallegos, *A comprehensible overview of EFL students' drawbacks to produce oral communication*,**41**,3. (2020)
9. A. Aripova, Kh. Khodjayeva, N. Yuldasheva, *Methods, aspects and components of teaching the*

Uzbek (Russian) language as a foreign language on the experience of foreign students, Advance Scientific Research, 2019 DOI: http://dx.doi.org/10.31838/jcr.07.04.75

10. A. Al-Roud, *Problems of English Speaking Skill that University Students Encounter from Their Perspectives*, British Journal of Education, Society & Behavioural Science, ***18* (3)**, 1–9. (2016)

DOI: https://doi.org/10.9734/bjesbs/2016/28404

11. M.S. Ilyin, *Fundamentals of Foreign Language Exercise Theory*, Moscow, 152. (2018)
12. B. A. Lapidus, *Intensification of the process of teaching a foreign language oral speech*, Moscow, 55. (1970)
13. U. Yuldashova, M. Bozorova, N. Mansurova, Z. Ahmedova, A. Shoazizova, *Language teaching methods: theory and practice*, Journal of Critical Reviews, Vol 7, Issue 5. (2020) DOI: http://dx.doi.org/10.31838/jcr.07.05.157
14. M. Kh. Akhmedova, *Problems of typology of exercises for the development of foreign language speech of students, **2,** 480-483.*(2015)
15. L.R. Duncan, C.R.Hall, P.M. Wilson, *Exercise motivation: a cross-sectional analysis examining its relationships with frequency, intensity, and duration of exercise. Int J Behav Nutr Phys Act* **7,** 7. (2010) DOI: https://doi.org/10.1186/1479-5868-7-7
16. E. Dorgu, *Different Teaching Methods: A Panacea for Effective Curriculum Implementation in the Classroom*, International Journal of Secondary Education, ***3*(6)**, 77. (2016) DOI: https://doi.org/10.11648/j.ijsedu.s.2015030601.13
17. H. A Alharbi, *Improving Students' English Speaking Proficiency in Saudi Public Schools.* International Journal of Instruction, ***8*(1),** 105–116. (2016)
18. A.M. Gaifutdinov, *Guiding didactic principles in the knowledge of the domestic pedagogy of the second half xx -the beginning of the xxi century,*Historical and pedagogical journal, **1**, 112. (2014)
19. S. Al. Hosni, *Problems and Difficulties of Speaking That Encounter English Language Students at Al Quds Open University*. International Journal of Humanities and Social Science Invention, ***5*(12)**, 96–101. (2016)
20. L. A. Hendra & C. Jones, *Motivating learners with immersive speaking tasks.* Part of the Cambridge Papers in ELT series, 16. (2018)
21. A. Lin, *Bilingual Education in Different Contexts: Principles and Practice*, Bilingual Education, 27. (2009)DOI:10.5790/hongkong/9789622099586.003.0002
22. S.Bentin, T.Allison, A. Puce, E. Perez, & G. McCarthy, *Electrophysiological studies of face perception in humans.* Journal of Cognitive Neuroscience, 8, 551-565. (2010) DOI: 10.1162/jocn.1996.8.6.551
23. H. Fry, S. Ketteridge, S. Marshall, *A Handbook for Teaching and Learning in Higher Education Enhancing Academic Practice*, 202. (2008)
24. Merryanna L. Swartz, Masoud Yazdani, *Intelligent Tutoring Systems for Foreign Language* Learning, Springer, Berlin, Heidelberg, (2012) DOI: https://doi.org/10.1007/978-3-642-77202-3
25. M. I. Hossain, *Teaching Productive Skills to the Students: A Secondary Level Scenario*, BRAC University, Dhaka, Bangladesh, 34. (2015)

The Discursive Opposition and the "Power of Discourse" in the Ukrainian Mass Media

Iryna Synytsia[1*]

[1]Dalian University of Foreing Lanquages, Dalian, China

Abstract. The subject of this publication is related to such directions of modern linguistics as critical discourse analysis, media linguistics, and text semantics. The article examines the features of the media discourse of Ukraine on the example of antonymic semantic relations that are the result of modern discursive practice. The media discourse of modern Ukraine demonstrates the facts of the emergence and functioning of the pseudo-antonymic opposition. The research is based on economic topics articles published in Ukrainian online publications during 2019-2020. The texts of the publications are written in Ukrainian or Russian. All articles are devoted to the problem of "betrayal or victory". The author of the article asserts that the active functioning of the Ukrainian lexemes "betrayal" and "victory" in the media discourse influenced not only the emergence of unusual oppositional relations between these lexemes. The active use of pseudo-antonyms in speech contributes to the destruction of traditional axiological values of Ukrainians and involves them in pseudo-discussions. Thus, first of all, the "power of discourse" is manifested. The author speaks about the formation of an independent discourse "betrayal or victory" in the Ukrainian media discourse. Pseudo-antonyms explode the pragmatic intentions of the sender of the message and are a means of manipulative influence on the linguistic consciousness of the recipient.

1 Introduction

In recent decades, humanity has faced challenges, some of which were created by itself. Both in naive and special scientific mastering the surroundings, a person records the results of his knowledge in primitive forms (such as elementary tools, sayings, proverbs, fairy tales, etc.) or in high-tech, modern examples of human skill. However, whatever it is, consciously or unconsciously striving for progress, a person finds himself at a crossroads. On one side of the road are predictable advantages, on the other – not expected, not foreseen by anyone in advance disadvantages. Such a dichotomous situation can be found in various spheres of human life: political, business, economic, social, domestic, cultural, etc. At the same time, we consider the opposite, which we observe both in the actions themselves and in their results, to be one of the regularity of human consciousness.

The traditions of the binary worldview are inherent in man as Homo sapiens. Binary perception of the world provokes our mental actions to divide everything that surrounds us into bad and good, good and evil, light and darkness, progressive and regressive, etc. Each of us has inherited this vision of the environment from our ancestors and is implementing it daily in relation to the facts of the new reality. Such a division is so ingrained for a person that he sometimes does not think about the essence of the comparing units, concepts, and facts. Such a person's attitude to the world around him provokes him to change his views, to revise already established principles, attitudes, etc. We consider this ambivalent attitude as the root cause of the changes in the questions that a person asks himself in the world cognition process. Modern philosophers record a change in important questions for humanity: previously, humanity was interested in what are the laws of nature, now – why are the laws of nature such, and not others, and whether there is any "intelligence" in them, etc. Undoubtedly, at the beginning of the 21st century – the age of the information society – humanity is properly ripe for a new understanding of the problem [1].

The explicit or implicit use of this ability of human consciousness in manipulative strategies in various discursive practices, especially in mass media discourse, is quite natural. Modern discursive studies of mass media combine the communicative, pragmatic, stylistic, etc. analyses of both the communicative process itself and its participants – the addressee (sender) and the addressee (recipient).

«The person who speaks» (the concept of Claude Hagege «L'homme de paroles») [2] as a social being has a rich linguistic resource of his native language, which he can use in various communicative situations. The language decisions made by «the person who speaks», that is, the choice of certain language means, depends both on the communicative situation as a whole, and on a variety of intralinguistic and extralinguistic factors. The choosing of certain language means by «the person who speaks» helps researchers to understand their communicative and pragmatic purposes and intentions.

* Corresponding author: isynytsa@ukr.net

2 Our contribution

It is well known that the mass media are the most dynamic, open, widely available resources in any country in the world. Their focus on the maximum reach of the addressee is the reason that determines many of their features, including pragmatic and linguistic ones. The most substantial distinctive media discourse features for this study are the following:

- group correlation (the author shares the views of his group);
- publicity (focusing on mass addressee);
- disens orientation (creating conflict with its following discussion);
- staging and mass orientation (impact on several groups simultaneously).

That is why scientists around the world have chosen mass media discourse as the object of the researches. Taking into account the communicative and pragmatic features of the media discourse, we turned to the analysis of the discursive and communicative features of the texts of financial and economic topics of Ukrainian periodicals of recent years.

The purposes of this research are to identify and describe examples of the linguistic implementation of the discursive opposition "betrayal – victory" (Ukr. "зрада – перемога" ['zrada – peremoga']), built on pseudo-paradigm relations of antonymy, to argue for the discursive-pragmatic conditionality of the mentioned opposition, to determine its functional purpose in the analyzed type of discourse.

Text is not only the elementary source of information. Text and word are means of manipulating the process of perception, evaluating informative flows, and forming a worldview. That is why the discourse power is primarily a social power.

3 Background

3.1 The degree of knowledge of the problem

The subject of this publication is related to such directions of modern linguistics as critical discourse analysis, media linguistics, and text semantics. At the same time, it is easy to see that despite the diversity of the principles of language learning today, they are based on system-structural methods of describing language phenomena. The importance of using this approach was noted by A. Reformatsky: "Since we believe that language is not an ideology, but a tool, and moreover a tool of a special kind, which does not have a structure like any material tool (axe, plow, combine), but a structure and a system organization, for all speakers, the first task is to practically master this tool in its given state" [3]. And although, according to A. Reformatsky, language is not an ideology, nevertheless, it is language that is used as tool of power.

3.2 Power and discourse from the linguistic point of view

The relationship between power and discourse has become a special subject of analysis and has formed an independent direction in discursology – the critical analysis of discourse (henceforth CDA). The history of the formation of this linguistic direction is considered in a number of works [4]. Without analyzing the stages and different schools of CDA, we will mention the names of those scientists whose studies have influenced the formation of this linguistic direction.

First of all, let us remember that any scientific direction is based on the main methods and principles of analysis. N. Fairclough and R. Wodak have summarized the basic principles of CDA most completely and precisely. In one of their articles, they named the seven tenets of CDA:

1. CDA addresses social problems
2. Power relations are discursive
3. Discourse constitutes society and culture
4. Discourse does ideological work
5. Discourse is historical
6. The link between text and society is mediated
7. Discourse analysis is interpretative and explanatory
8. Discourse is a form of social action [5].

None of the works on CDA today is complete without mentioning the researches of T. van Dijk. According to his own and Ruth Wodak definitions, his studies have shaped the social-cognitive approach in the analysis of discourse [6; 9]. Based on the concepts of power, history, and ideology, T. van Dijk justifies the need to study the abuse of power and the reproduction of inequality through ideologies. One of the main points of his theory says that "power is related to control, and control of discourse means preferential access to its production and hence to its contents and style, and finaly to the public mind" [7].

In modern discuorse studies, the concept of "the discoursive power" is primarily associated with the ability to express power and influencein the process of communication. At the same time, scientists interpret the produced influence in two ways. Some of them talk about the power locations: where can power be "behind the discuorse", or the power is "in it". According to Fairclough, the idea of "power behind discourse" is that the entire social order of discourse is folded and held together as a "hidden effect of power" [8]. So what is hidden is not the power itself, but how it manifests itself in the discourse.

One way or another, it is language in its influencing power that is defined as the power center of the social communicative space. From this perspective, researchers at the macro and micro levels of discourse investigate how "opaque as well as transparent structural relationshis of dominance, discrimination, power and control" manifest themselves in language. According to Ruth Wodak, the purpose of critical discourse analysis is to critically examine social inequality and how it is expressed or signaled through linguistic means or in discourse [9]. If there is no direct explicit statement of the arguments or if the texts are not transparent, the researchers reconstruct them. The description of the hidden meaning is the result of this reconstruction. In any case, language is precisely the means that not only provides communication, but also allows you to achieve ideological understanding/misunderstanding.

The potential multidimensional nature of language study creates the conditions for a comprehensive analysis of a particular discourse. Ruth Wodak emphasizes that critical discourse analysis has never been presented or attempted to be a single or specific theory. The researcher insists that the research methodology in the CDA is very diverse. She states the fact that scientists who work in this direction of discursive analysis use a wide variety of methods. These methods are based on different theories and on various data. Some researchers in the field of CDA also use grammatical approaches [9].

In the first version of the article «Critical Linguistics and Critical Discourse Analysis», Ruth Wodak considers the social construction of meanings to be one of the constitutive principles of critical discursive analysis. We accept the scientist's opinion that meanings arise as a result of the interaction between the reader/hearer of the text and those who speak/write this text. Meanings "are always subject to more or less closely enforced normative rules (for instance, generic rules), and to the relations of power obtaining in this interaction. Many different conscious and subconscious motives and planning procedures are relevant in text production and text comprehension, which result in manifest and latent meanings, cognitive and emotional aspects of discourse" [10].

The comprehensive analysis, which covers all levels of creation and organization of discourses (lexico-semantic, phonetic, stylistic, rhetorical, communicative-pragmatic, compositional-structural, formal, etc.) and uses different linguistic directions methods, is characteristic of the critical discursive studies of T. van Dijk, R. Wodak, V. Cherniavskaya, F. Bacevich, O. Selivanova, O. Semenec, L. Shevchenko, etc.

The integrative approach is also characteristic of the representatives of the French school of discourse analysis. The works of M. Foucault, L. Althuesser, J. Lacan, R. Barthes, A.-J. Greimas, M. Pesce, and others, as well as the French-Swiss linguist P. Serio, implement this approach in the study of history, philosophy, psychology (psychoanalysis), and linguistics. The researchers focus on the social meaning of communication. In this aspect, K. Levi-Strauss, R. Barth, M. Foucault, J.-F. Lyotard, J. Baudrillard, and others consider language and symbolic space as the main dimension of social life. Studying different discourses, they are united in understanding communication through the prism of symbolic creativity of social actors. Representatives of this discoursology direction attempt to identify and to describe the patterns of influence of the total symbolic language system on a person. Thus, we observe the shift of research attention from formal linguistic features to their symbolic meanings associated with socio-cultural and psychological phenomena [11].

One of the important statement of M. Foucault's theory is the statement of the discursive nature of human consciousness. The scientist speaks about the epochs of the development of knowledge, which are characterized by historically changing forms of scientific discourse – epistemes that reflect certain ways of knowledge and determine the conditions for the possibility of thoughts, theories and sciences in each period. The norms of the discourse of a certain episteme determine both the speech behavior and the thinking of a person. Therefore, within each science, there is an interaction of the "will to know" and the "will to power".

Foucault describes three stages of formation of modern "European mentality" with corresponding major episteme: Renaissance (XVII), classical (rationalism of the XVI –XVIII centuries) and modern (the end of XVIII– the beginning of XIX). Each of them characterized by its correlation of "words" and "things". The scientist traces the stages of these relationships: from their identification and interchangeability (word-symbol), mediated relationship through thinking, in the space of representation (word-image), to the mediation of words and things in the modern episteme "language", "life", "work" (word – sign in the system of signs). In the era of postmodernism, the word is focused and closed on itself, exists by itself, which causes a crisis in the relationship between "words" and "things" [12].

In M. Foucault's vision, the "system of discoursivity" manifests itself in the fact that it presupposes the possibility or impossibility of the appearance of a certain type of utterances and actions [13].

3.3 Features of studying the media discourse

The foundations developed by the classics of the discourse theory are developing further today. The forward movement is carried out in different directions. This is evident from a number of dissertation studies and scientific publications. The most productive, we think, are the studies of various types of discourse, in particular media discourse. It is the discourse of the mass media that we find most interesting for the analysis of the "power of discourse".

Because the source of the material in this publication was mass media discourse, we want to clarify its interpretation by the French linguist Elizabeth Le. The researcher talks about three aspects of studying it: 1) discourse as the use of language; 2) discourse as the "implantation" of certain ideas into the public consciousness; 3) discourse as the interaction of social groups and individuals. Therefore, researchers, according to the author, have the following requirements: first, since discourse is the use of language, its study involves the analysis of the text, covering different areas: syntax, vocabulary, "voices of polyphony", functions, various ways of modalizing individual components in order to clarify or, conversely, obscure some details; secondly, the discourse forms, "implants" representations and makes it necessary to analyze the production of texts and their interpretation; thirdly, discourse as the interaction of social groups and individuals should be investigated in connection with the social structures of the culture of a given society [14].

The discourse of the mass media of Ukraine is studied by Ukrainian and foreign scientists in various aspects: formal-structural, sociolinguistic, cognitive, typological, etc. The Ukrainian media discourse of different periods was studied by S. Yermolenko, S. Sokolova, S. Bybyk, L. Shevchenko, M. Zhovtobryukh, A. Zagnitko, A. Koval, G. Solganik, G. Chernenko, A. Grygorash, I. Filatenko, N. Rudnichenko, K. Serazhym, O. Serbenska, O. Styshov, etc. [15].

3.4 Discursive opposition as a ground to generate pseudo-beliefs

According to the observations of many linguists, speech provides us with a unique material for analysis, which allows us not only to observe various modifications of word usage, but also to draw conclusions about those processes that go beyond the traditional ideas about the language system. One of these phenomena, which we found in the modern socio-political and media speech practice in Ukraine, attracted our attention.

The subject of the analysis in this publication is the opposition "betrayal – victory", which is quite actively used in publications on various topics.

The material for the study was formed from articles published in the online versions of periodicals of Ukraine during 2019-2020 and written in Ukrainian, Russian or both languages. From the total amount of material for analysis in this work, publications on financial and economic topics were selected.

3.5 History and usage of opposition "betrayal or victory"

This pair of lexemes appeared in the socio-political and media discourse of Ukraine around the middle of 2015. According to our observations, it is August–October 2015 that its first fixations are dated. At the same time, some bloggers talk about the active use of this opposition already during the Revolution of Honor. The data collected by us registers its appearance in the Ukrainian periodicals in the indicated period.

It is essential to note that texts fix the contraposition of two completely different phenomena, named by lexemes, which form their own antonymic pairs in the lexical and semantic system of the Ukrainian language "betrayal – victory".

In Ukrainian and Russian, victory (перемога / победа [peremoha / pobjeda]) means "Success in battle, war, in the struggle for something, achievement as a result of the struggle, overcoming something" [16]. The lexical and semantic antonym of this word, registered in the dictionaries of antonyms of Russian and Ukrainian languages, is the word defeat (поразка / поражение [porazka / porazenije]): "1. Defeat of the army in battle, bringing it out of the state of combat capability… 2. Failure in the struggle for something, in some business, etc.". Such an opposite of the results of actions called by these lexemes is also reflected in their semantic opposition, which is realized by a pair of antonyms victory – defeat (Ukr./Rus. перемога – поразка / поражение – победа [peremoga – porazka / pobjeda – porazenije]).

The opposition, which is offered to us by the texts of the media publications of Ukraine, is represented by the lexeme betrayal (Ukr. зрада [zrada]). Moreover, this Ukrainian word, which in the Russian language corresponds to the two lexemes treason (Rus. измена [izmena]) and betrayal (Rus. предательство [predatelstvo]), functions as the second component of the opposition in publications in both Ukrainian and Russian languages. The dictionary describes the following meanings of the word зрада: "1. going over to the enemy's side; treachery, betrayal... 2. violation of loyalty in love, friendship. ... 3. giving up your beliefs, views, etc.". This lexeme is connected by antonymic relations in the Ukrainian and Russian language systems with the lexemes faithfulness (Ukr./Rus. вірність/верность [virnist/vernost]) and devotion (Ukr./Rus. відданість/преданность [viddanist/predanost]) [17]. The noun faithfulness (Ukr. вірність[virnist]) is formed from the adjective faithful (Ukr. вірний [virnyj]) in the meaning of "who deserves trust; constant in his views and feelings; devoted".

As we can see, the semantic structures of the lexemes victory (Ukr./Rus. перемога / победа [peremoha / pobjeda]) and betrayal (зрада [zrada]) do not contain any grounds for their semantic opposition. But acknowledgment that the authors see the opposition relationship between these words, may be a combination of these lexemes with the conjunctions or (Ukr./Rus. чи/или [chy/ili]), which is more often used to express the separation relations, emphasizing the relationship of mutual exclusion, incompatibility [18]. Moreover, according to our observations, the separative-oppositional conjunction or (Ukr./Rus. чи/или [chy/ili]), in media discourse is used more often in the headlines, less often in the texts of publications, sometimes both in the title and in the text, for example:

a) in the headline: ["Zmicnenya hryvni -2019: zrada chy peremoha"] (Ukrainian transliteration), [Eng.: Strengthening of the hryvnia-2019: betrayal or victory] (24.01.2020. Ekonomichna Pravda) [25];

b) in the text: ["Arbitrazhnyj trybunal OON z morskoho prava vynis pershe rishenya u spravi "Ukrayina proty Rosiyi". Ale rishennya polovynchaste: vin vyznav yurysdykciyu lyshe za chastynoyu zvynuvachen. To ce zrada chy peremoha? <...>"] (Ukrainian transliteration), [Eng.: The UN Arbitration Tribunal for the Law of the Sea issued the first decision in the case " Ukraine versus Russia". But the decision is half-hearted: he admitted jurisdiction only on part of the charges. Is this a betrayal or a victory? <...>] (25.02.2020. Yevropejska pravda) [26];

c) in the headline and in the text: ["Peremoha chy zrada? <…> Vyznacheni tranzytni potuzhnosti na pyat rokiv, za jaki Hazprom harantovano maye zaplatyty. Ale obsiahy tranzytu mozhyt buty i bil'shymy, prosto potribno bude dodatkovo jih bronyuvaty (i platyty za znachno vyshchym taryfom) na rik/kvartal/misyac/dobu napered. Tryvalist kontraktu tezh mozhe buty prodovzhena. Peremoha chy zrada? Hazprom hotiv odyn rik, my – desiat, ziyshlysya poseredyni <…>"]

(Ukrainian transliteration), [Eng. Victory or betrayal? Transit capacities have been identified for five years, for which Gazprom is guaranteed to pay. But transit volumes may also be large, you just need to book them additionally (and pay at a much higher rate) for a year/quarter/Month/Day in advance. The duration of the contract can also be extended. Victory or betrayal? Gazprom wanted one year, we wanted ten, and we agreed in the middle. <...>] (23.12.2019. Novoe Vremya) [27].

Characterizing antonymy as a phenomenon of the language system K.V. Taranenko notes that antonymy is based on some common reasons that lie in the very nature of human thinking [19]. L.M. Polyuga defines antonymy as the most essential and characteristic for all native speakers of the language. The scientist notes that antonyms best express the contrast of meanings in speech [20]. The analyzed example of opposition, which is actively used in the mass media discourse of Ukraine, does not fit into the traditional notions of antonymy with its different types described in the works of Yu. Apresyan [21], L. Vvedenskaya [17], V. Turchyn [22], etc. At the same time, we have the opposition of absolutely different phenomena of objective reality, which is actualized in the media discourse.

Researchers of antonymic relations have repeatedly said that antonyms reflect the complex perception of objects of reality by the human brain, processing information about them based on comparisons with other objects or with the same ones, but in different conditions. Choosing an antonym reflects the influence of extralinguistic factors: the situation in which a person encounters an object, his prior knowledge of the object, his own, and more often imposed on him stereotypes in relation to the object, his own perception characteristics, etc. That is, the subjective factor, or rather the pragmatic component of communication, plays an important role in creating an antonymic pair.

We have noticed that the lexeme victory (Ukr. перемога [peremoga]) in the Ukrainian mass media discourse of this period loses its positive connotation, which is actualized in the antonymic pair "victory – defeat" (Ukr. перемога – поразка [peremoga - porazka]). By choosing the modality of doubt, the authors of publications violate the readers' confidence in the assessment of the facts they are talking about, for example: ["Yakoyu bude serednya zarplata v 2020 roci <...> avtory ne zhaduyut kurs 25 hrn/dol, yakuy tumchasovo zmicnyvsya pislya pryhodu nerezedentiv na rynok oblihaciy vnutrishnioyi derzhavnoyi pozyky (OVDP) i zrostayuchych nadhodzhen vid zarobitchan. Same vin dopomih v dolarah otrymaty taku ocinku. Chy ce mozhna vvazhaty peremohoyu? <...>"] (Ukrainian transliteration), [Eng. "What will be the average salary in 2020 ...the authors do not mention the rate of 25 UAH/USD, which temporarily strengthened after the arrival of non-residents on the market of domestic government loan bonds (government bonds) and growing revenues from employees. It was he who helped to get such an estimate in dollars. Can this be considered a victory? <..>"] (1.08.2019. Novoe Vremya) [28]. We also find the loosening of axiological values in the opposition of peremoga to the negative phenomena of the economic sphere. In the following article, the doubt set in the title is also supported in the text of the article: ["OVDP: "peremoha" chy vidstrochka vyroku <...> Dehto iz fahivciv z investyciy reklamuye OVDP, yak "bezpechnyi instrument dlya pochatkivciv investoriv", yihni kolehy po cehu svyatkuyut velyki "peremohy" pislya kozhnoho aukcionu, <...> ne varto vyprobovuvaty patriotychnyi orhazm kozhnoho razu, koly uryadu vdayetsya vzyaty v borh pid velychezni vidsotky. Taka "peremoha" bilshe shozha na vidsrochku vyroku <...>"] (Ukrainian transliteration), [Eng. "Government bonds: "victory" or postponement of the sentence .... Some investment professionals advertise government bonds as a "safe tool for novice investors", their colleagues in the shop celebrate big "victories" after each auction, ... you should not experience a patriotic orgasm every time the government manages to borrow at huge interest rates. Such a "victory" is more like a postponement of the sentence"] (26.11.2019. Novoe Vremya) [29]. The results of state measures in the economic sphere are very rarely unambiguously evaluated. Therefore, the use of pseudo-orientations in the form of the opposition "betrayal or victory" (Ukr. "зрада чи перемога" [zrada chy peremoga]), in our opinion, corresponds to the ambiguity and complexity of the phenomena that are evaluated by the authors. The authors of articles thus force their readers to doubt, to change their points of view.

The same thing is observed with the lexeme betrayal (Ukr. зрада [zrada]), when using which the unambiguous negative assessment is erased. The modality of the doubt "is this true or not?" shakes the established ideas about these concepts as in these examples:

a) ["<...> Ta ne varto dorikaty derzhavi za ce, a prohrash spravy ne varto pospishno nazyvaty "zradoyu". Sprobuyemo rozibratysya, zhcho oznachaye ce rishennya SOT i zhcho daye Ukrayini shansy na uspishnu apelyaciyu <...>"] (Ukrainian transliteration), [Eng. "...but you should not reproach the state for this, and the loss of the case should not be hurriedly called "betrayal". Let's try to understand what this WTO decision means and what gives Ukraine a chance of a successful appeal"] (8.04.2019. Yevropejska pravda) [30];

b) ["Chy ye "zrada" v zmini prohramy kredytuvannya MVF dlya Ukrayini"] (Ukrainian transliteration), [Eng. "Is there a "betrayal" in changing the IMF lending program for Ukraine" (8.05.2020. Ekonomichna Pravda)] [31].

Statement a) contains the construction "do not be hasty to name..." (Ukr. не варто поспішно називати...), which in a soft, unobtrusive form, without any imperative-ultimatum connotations, recommends thinking carefully before qualifying a particular phenomenon with the help of a popular pseudo-opposition. Statement b) has the form of a question that makes the reader think about its meaning. With the help of these simple speech constructions, the discourse influences the reader, exerting its pressure on him in assessing certain realities of modern life, in this case from the financial and economic sphere.

Lexico-semantic and other features of the media discourse in Ukraine are largely due to the social and political events of this period. We interpret the opposition that appeared in Ukrainian periodicals at a certain moment in the development of society as a manifestation of the "power of discourse". In connection with the above, the opinion expressed by A. Shekhovtsov in a post on the Facebook page is quite reasonable. The author characterizes the discourse "betrayal vs victory" (Ukr. зрада vs перемога [zrada vs peremoga]) not only as "intellectually flawed", but also as "harmful from the point of view of the medium-term strategy of the country's democratic development". The reasons for this lie in the potential for making mistakes. These errors are explicated by the semantics of the false opposition inherent in this discourse and conditioned by the Orange Revolution. According to the author of the publication: "Both "zrada" and "peremoga" are precisely detached observations of the behavior of political elites, so ... both narratives lead to catastrophe. Ukrainian civil society should go beyond the discourse of "zrada vs peremoga" and not just control the political elites, but also directly participate in the process of democratization of the country" [23].

Today, we can no longer determine the exact date and reason for the appearance of this pseudo-oposition. But appeals to the binary perception of reality, built on false guidelines, sometimes have negative consequences. For example, a Japanese researcher of mental personality states claims that mental personality disorders such as borderline personality and narcissism are associated with binary thinking: "The results indicated that thinking dichotomously may lead to wide-ranging personality disorders" [24].

4 Conclusion

The neutralization of positive / negative connotations is the result of a certain type of word usage practice that implements specific pragmatic goals. The use of pseudo-antonyms, firstly, destroys the logical opposition in the system of value concepts" good-bad", and secondly, disorientates Ukrainian and Russian languages speakers and affects their perception of events. We consider the destruction of traditional paradigmatic connections between lexemes as one of the means of manipulating the language consciousness of a person.

That is, the repeated reproductions of pseudo-antonyms in the speech which describe problematic situations provoked several negative processes for the development of the public consciousness of Ukrainians. Trying to give answers to the questions raised by the authors of publications about the presence of "betrayal or victory", the addressee unwittingly joins the imaginary process of fighting for justice: for victory or against betrayal, although the opposition itself is not logically justified. Not every victory can have positive results, just as treason does not always lead to negative consequences. This can be proved by numerous examples of both ancient and modern domestic and foreign history. We can say about the formed discourse "zrada vs peremoga", which we characterize as a discourse of detachment, falsity, erroneousness. A pair of pragmatic pseudo-antonyms, the contraposition of which is the result of consistent use in mass media resources, creates a false effect of active participation in social processes.

References

1. M.S. Kravets, Philosophy of Perception of the World in the Information Age. Scient. J. of UNFU 28(1), 163-167 (2018). https://doi.org/10.15421/40280131
2. Claude Hagege, *L'homme de paroles. Contribution linguistique aux sciences humaines* (Fayard, Paris, 1996)
3. A.A. Reformatsky, Synchronous description principles of language, in Reformatsky A.A. Linguistics and Poetics (Moscow, 1987), p. 20, http://www.philology.ru/linguistics1/reformatskiy-87a.htm
4. V.E. Chernyavskaya, *The discourse of power and the power of discourse: problems of speech influence* (Flint: Nauka, Moscow, 2006, in Rus.); Zh.O. Krasnobaeva-Chorna, *Discursive practice.* (DonNU, Donetsk, 2008, in Ukr.); Brian Paltridge, *Discourse Analysis* (Bloomsbury Academic, London, New Delhi, New York, Sydney, 2012); The Handbook of discourse analysis, ed. by Deborah Schiffrin, Deborah Tannen, and Heidi Hamilton. (Blackwell Publishers Ltd, Malden, Oxford, 2001). doi:10.1002/9780470753460
5. N.L. Fairclough and R. Wodak, Critical discourse analysis, in *Discourse Studies. A Multidisciplinary Introduction*, vol. 2, ed. by T.A. van Dijk (Sage, London, 1997), pp. 271–280
6. T. van Dijk, in *Handbook of Discourse Analysis*, ed. by John Richardson & John Flowerdew (Routledge, London, 2016), pp. 26–43
7. T. van Dijk, *Discourse and Power* (Palgrave Macmillan, New York, 2008), p. viii.; T. van Dijk, *Ideology and discourse. A multidisciplinary Introduction*, (Pompeu Fabra University, Barcelona, 1998), p. 8
8. N. Fairclough, *Language and power* (Longman, London, 1989), p. 43–76
9. R. Wodak, in *Handbook of Pragmatics*. Amsterdam; Philadelphia: Benjamins, vol. 10, 2006, pp. 1–24
https://doi.org/10.1075/hop.10.cri1
10. R. Wodak, in *Handbook of Pragmatics*, (Benjamins, Amsterdam; Philadelphia, 1994). https://www.benjamins.com/online/hop/articles/hop.m.cri1#s3
11. N.K. Kravchenko, *Practical Discourse Studies: Schools, Methods, Techniques, Contemporary Discourse Analysis* (Volynpolygraph, Lutsk, 2012); Serio P. (ed.), *Quadrature of Meaning: The French School of Discourse Analysis* (Progress, Moscow, 1999, in Rus.); O. Semenets, Discursive

practices of Social Communications: Creativity of the Subject of Speech and Mechanisms of Self-organization, in Words in the Context of Time, vol.1, ed. by V.I.Iychankava, (BSU publ.center, Minsk, 2014), http://elib.bsu.by/handle/123456789/106190

12. M. Foucault, *Words and Things. Archeology of the Humanities* (A-cad, St. Petersburg, 1994, in Rus)
13. M. Foucault, *Archeology of Knowledge* (Solomiya Pavlychko Publishing House "Osnovy", Kyiv, 2003, in Ukr.), p. 207
14. E. Le, Linguistic Analysis of Political Discourse: the Language of Articles about the Chechen War in the American Press, J. Polis. Political studies, 2, 95 (2001)
15. O.P. Polishchuk, O.I. Svintsitska, Infosphere of Ukraine: peculiar properties mass-media discourse in the context of aesthetic information, in J. History. Philosophy. Religious studies, 2, 56-59 (2008). http://eprints.zu.edu.ua/9927/1/ИФР%202008.pdf; L. Shevchenko, D. Dergach, D. Syzonov, Medialinguistics: Dictionary of terms and concepts (VPC "Kiev University", Kiev, 2014) http://medialing.spbu.ru/upload/files/file_1452013437_9323.pdf; S. Sokolova, S. Bybyk, Ukrainian Language in Modern Mass Media Communication: Problems of Speech Culture, Stylistics and Sociolinguistics, in: Ukrainian language: Nauk.-theoret. J., 2, 133-145 (2014); O. Styshov. *Dynamics of the lexical composition of the modern Ukrainian language* ("Authority", Belaya Tserkva, 2019) https://shron1.chtyvo.org.ua/Styshov_Oleksandr/Dynamika_leksychnoho_skladu_suchasnoi_ukrainskoi_movy_Leksykolohiia_Leksykohrafiia.pdf, etc., etc.
16. Dictionary of the Ukrainian language: in 11 volumes (Naukova Dumka, Kyiv, 1970-1980). The meanings of words of the Ukrainian language are given in the article according to this dictionary. http://sum.in.ua
17. L.A. Vvedenskaya, *Dictionary of Antonyms of the Russian language* ("Astrel Publishing House": "ACT Publishing House" LTD., Moscow, 2003); L.M. Polyuga, *Dictionary of antonyms of the Ukrainian language*, ed. by L.S. Palamarchuk, 2nd edn. (Dovira publ., Kiev, 2004)
18. A.P. Zahnitko, *Theoretical grammar of the modern Ukrainian language. Morphology. Syntax* (LLC "PKF "BAO"", Donetsk, 2011), p.665
19. K.V. Taranenko, *Pragmatic potential of antonymy of the Ukrainian language* (UMSF, Dnipro, 2017)
20. L.M. Polyuga, *Dictionary of antonyms of the Ukrainian language*, ed. by L.S. Palamarchuk. 2nd edn. (Dovira publ., Kiev, 2004), p. 6–9
21. Yu.D. Apresyan, *Lexical Semantics* (School "Languages of Russian Culture", Publishing Company "Eastern Literature" of the Russian Academy of Sciences, Moscow, 1995)
22. V.V. Turchyn, V M. Turchyn, Linguistic Directions in the Study of Antonymia, in J. Scientific Notes. Series "Philology", 19, 337–341, (Publishing House of the National University "Ostrog Academy", Ostrog Issue, 2011)
23. Anton Shekhovtsov, Facebook page, https://www.facebook.com/anton.shekhovtsov/posts/10206455150513353?pnref=story
24. Oshio A. An all-or-nothing thinking turns into darkness: Relations between dichotomous thinking and personality disorders, in J. Jap. Psych. Res., 54(4), 424–429 (2012). doi:10.1111/j.1468-5884.2012.00515.x
25. Zmicnenja hryvni -2019: zrada chy peremoha (Strengthening of the hryvnia-2019: betrayal or victory), in Ekonomichna Pravda, 2020, 24.01 https://www.epravda.com.ua/columns/2020/01/24/656229/
26. Serhiy Sudorenko, Olena Zerkal: Mayemo nareshti zrozumity, shcho pochalasya insha era vidnosyn z Rosiyeyu (We must finally understand that a different era of relations with Russia has begun), in Yevropejska pravda, 2020, 25.02 https://www.eurointegration.com.ua/interview/2020/02/25/7106774/
27. Yuri Vitrenco, Peremoha chy zrada? (Victory or betrayal?), in Novoe Vremya, 2019, 23.12 https://nv.ua/ukr/opinion/gazovi-peregovori-pro-shcho-domovilisya-gazproma-ta-naftogaz-novini-ukrajini-50060819.html
28. Anatoliy Amelin, Yakoyu bude serednya zarplata v 2020 roci (What will be the average salary in 2020), in Novoe Vremya, 2019, 1.08 https://nv.ua/ukr/opinion/yakoyu-bude-serednya-zarplata-v-2020-roci-novini-ukrajini-50035058.html
29. Ivan Kompan, OVDP: "peremoha" chy vidstrochka vyroku (Government bonds: "victory" or postponement of the sentence), in Novoe Vremya, 2019, 26.11 https://nv.ua/opinion/ovgz-kupit-kak-na-etom-zarabotat-novosti-ukrainy-50055685.html
30. Olena Omelchenko, Porazka z shansom na revansh (Defeat with a chance for a rematch), in Yevropejska pravda, 2019, 8.04 https://www.eurointegration.com.ua/experts/2019/04/8/7094887/
31. Vladyslav Rashkovan, Chy ye "zrada" v zmini prohramy kredytuvannya MVF dlya Ukrayini (Is there a "betrayal" in changing the IMF lending program for Ukraine), in Ekonomichna Pravda, 2020, 8.05 https://www.epravda.com.ua/columns/2020/05/8/660265/

Cross-Cultural Analysis: Representation of Some Aspects of a Parent-Child Relationship (on the Examples of English and Uzbek Proverbs)

Feruza Mamatova[1] *

[1]JMCU, Foreign Languages Department, 89 Qiyot, Uzbekistan

Abstract. The research paper deals with the problems of modern linguistics such as linguistic picture of the world which is realized principally by researching culturally marked linguistic phenomenon. A parent-child relationship is mostly studied by sociology, however, the fact that variety of speech in a parent-child relationship which is reflected in a linguistic image of the world makes it an object for linguistics. In addition to this, the study of the reflection of family relationship in the language enables to carry out a cross-cultural analysis by tools of linguoculturology. As data of the research English and Uzbek proverbs were selected from different sources. Analysis of phraseological units of English and Uzbek proverbs related to a parent-child relationship enabled to reveal similarities, differences, unique and specific features of this type of tradition. The periphery of this phenomenon comprises such notions as "parents are irreplaceable people", "parents' love" and "child's behaviour at different ages", "child associations" and others. Proverbs create a clear imagination of a parent- child relationship that has enough connotations expressed in the language. The analysis of the research may be implemented in cross-cultural studies, translation lessons and can be useful for a further research in this area.

1 Introduction

Anthropocentrism enables to get acquainted with the culture of a certain ethnos through the language. It is claimed that interconnection of language and culture is a bridge between person's word outlooks, mentality, nationality, national peculiarities and a language, since all of them are reflected in the language a person speaks. As anthropocentric direction in linguistics has attracted the scientists' attention there have been done a great deal of investigations where culture and language are the principal subjects. Traditions being considered one of the essential part of the culture of any nation as a rule, are reflected in the language too. Family traditions are a considerable part of traditions and they are inherent to the cultures of English and Uzbek nations. The study results show that a parent-child relationship reflected in proverbs vividly express culture-specific features of English and Uzbek family traditions. In order to reveal similarities, differences and unique features of some family traditions reflected in the English and the Uzbek linguoculturology, we have addressed to the proverbs of these languages. This, it is hoped, would assist the reader in comprehending the application of proverbial peculiarities in English and Uzbek.

Under the concept of family traditions are not only celebrations, holidays and ceremonies, customs but they also include a traditional relationship passing from one generation into another. A person's relatives no matter whether close or distant they have a certain place in his/her life. Kinship according to its degree of closeness can be divided into blood relatives and in-laws. In this research we aimed to make a comparative analysis of relations among these types of relatives in the English and Uzbek languages. "Family, family bonds and kinship relations form a nation's culture and show its peculiarities and assist to realize widely the national, linguistic image of the world" [29]. As the issue of family relations touch not only sociological science but also, cultural and linguistic ones, it is natural to approach to this problem from the perspectives of linguoculturology. Linguistic units possessing cultural-national features attract researcher's attention due to their study of linguistic image of the world of a native speaker. Phraseological and paremiological units which embody national-cultural worldview obtain the role of cultural stereotypes [25]. The research demands to clarify who exactly relate to the group of blood relatives. A Russian linguist, historian gives following classification of blood relatives:

1) Me and my siblings;
2) My father, my mother and their generation;
3) My grandparents and their generation;
4) My children; 5) my all grandchildren [24].

For children a family is one of the main habitats, development and psychological formation. It is family which creates relationship between family members. Parents are the principal sources of developing interpersonal relationship for their children (interrelation of parents and children, relationship of parents and ancestors). This issue has attracted the attention of a great number of scientists. So, what is a parent-children relationship? Relationship of parents and children is explained by Dunn& Brown [7]. As they claim, this form of relationship enables a child to understand emotions, since in the conversation of parent and a child emotional speech is inevitable when a small child tells the parents various situations, cases with full of joy, surprise and other feelings. Another scientist claims that sincere conversation in the a parent- child relationship guarantees respect, trust and security in psychology of a child [3]. A number of studies are devoted especially to the importance of a parent-child communications in the family. Despite the parents' love they must promote their capabilities to communicate with children and to listen to them. Communication depends on two interlocutors and it must not be stopped in any case Hisham Altalib, AbdulHamid AbuSulayman, & Omar Altalib [10]. The quality of relationship highly depends on a parent-child communication, parental feelings of aggravation Moore, Kristin & Busby, Andrea & Bandy, Tawana [23]. Another group of scholars support the idea that the earlier a parent-child relation is formed the more successful relationship is built up, since communication is associated with emotional survival, comprehension of emotions [8]. Parents vary, however, in the extent to which they discuss emotions with their children. Relative to insecure parents, secure parents may feel more comfortable engaging in discussions with their children about positive and negative emotions, and as a result, children of secure parents may have greater opportunities to discuss and think about emotions. Although most work on predictors of children's emotion understanding has focused on either parent characteristics, such as depression [25]. One longitudinal study provides evidence for a link between parents' attachment and children's emotion understanding [27]. English and Uzbek languages relate to different language families that makes the research not ordinary. Different cultures, mentality, life experience of these two nationalities are reflected in the language, specifically in proverbs. Contrastive and linguoculturological analysis of English and Uzbek proverbs about a parent-child relationship is new for linguistic research, since a close examinations are carried out to find out culturally specific aspects of the under study issue. There exist a variety of stereotypes about a parent-child relationship in the English and Uzbek nations. These stereotypes are reflected in the language, specifically in the proverbs, sayings and idioms. At this point it is logical to discuss the capability of proverbs as a linguistic phenomenon in revealing cultural features of a certain ethnos. To illustrate, Louis [17] argued that "proverbs are a kind of linguistic instrument, a rhetoric device by which people attempt to get other members of their culture and society to see the world and behave in a common way". According to Burger proverbs reveals the unchanging features of the mentality, as they live for many centuries and have a various frequency of use in the language [Burger]. Proverbs are representatives of such significant categories for national cultures as the style of life, traditions, customs, word outlook, character, habits and appearance [Jahr, Scwarz-Frisel].

Family relations are characterized by specific indication as love, help, and support in matters related to the social and physical needs of the family members. These features of family relations are rooted in the public consciousness and are reflected in such notions as "parents' love", "value of children". It is also necessary to point that family relations differ from other interpersonal relations, as they are subject to regulation by society, which gives the family relations an element of responsibility and duty. The experience of many generations and universal values, which are the basis of family relations, are reflected in numerous proverbs [11, 12, 18]. Russian linguist Vorkachov states that "a unit of knowledge and conscience shared by a large group of people who speak the same language, with a fixed linguistic form (expression), which is marked by distinct ethno cultural specifics" [32].

2 Literature review

In linguistics, the terms comparative and contrastive have come to mean two different approaches: the first one focuses on the similarities between two (or more) sets of the same class of items, while the contrastive is mainly concerned with explicating, studying, describing and explaining the differences between linguistic items on the synchronic plane. In paremiology however, these terms are often used as synonyms [22]. It should be noted that by comparative, i.e. diachronic research proper, both in linguistics and in proverb study, it is suggested to understand diachronic or historical study which takes into account the evolution of a text, or the chronological change of some aspects of a proverb text or group / class of proverbs.

Approaches to studying a parent-child relationship have been made from different perspectives such as psychiatry [6] and psychology [2]. This issue has attracted the attention of sociologists, psychologists, sociolinguists and linguists for many years. On the basis of literary texts, mass communication publications, poems Lavitskiy describes the concept "Children" of Vitebsk region in Belarus in his scientific article. He investigates children's life from different prospective of life in which the role and place of parents are inevitable. Go Pintin [28] approaches the issue as a family tradition and compares Chinese and Russian a parent-child relationship basing on the proverbs and sayings of the two languages. Biktagirova [4] carries out comparative analysis of the issue in the concept "Family" on the materials of systematically different languages as English, Turkish and Tatar languages. Adamova [1] investigates the issue in her research dedicated to interpersonal relationships of English and Lak (Northeast Caucasian language) languages. Another research studies the concepts "Mother, father and children" as the representation of the world of "one's" in the Kuban phraseology [5]. Studying a parent-child relationship in linguoculturology enables to get acquainted with a linguistic image of the world of a certain nation.

Comparative-linguoculturological analysis of a parent-child was carried out by Gasanova & Mayorova [9] who compared Lezgin languages – Agul, Rutul and Tsakhur using proverbs of this language as a tool of the research. Another lingucultural analysis by Gaychenya [14] was made on the basis of the concept "Family". Comparative method revealed culture specific features of Russian and Spanish a parent-child relationship.

3 Methodology

The research analysis started from description of a parent-child relationship in the English and Uzbek languages. Scientific articles, psychological, social science resources that exist in these languages, served to find the most reliable facts about different aspects of a parent-child relationship such as love, behavior, associations in the family, peculiarities of father-children and mother-children relationships and others. The core of the study requires carrying out comparative analysis of English and Uzbek proverbs relevant for a parent-child relationship; therefore, the next step was to select them from proverb dictionaries. After selection, we tried to reveal similarities, dissimilarities, specific features and uniqueness of a parent-child relationship specific to the English and Uzbek family traditions. It is known that proverbs reflected in the language speak for the mentality, life experience, world outlook, culture of a certain nation. For that reason, comparative method of proverbs was carried out to find out linguoculturological peculiarities of a parent-child relationship reflected in the proverbs. Analysis of statistical method in linguistics provided with exact quantity of proverbs in the two languages that reveal this or that aspect of the research issue.

3.1 Data collection.

In order to collect as many proverbs as possible we addressed to a number of dictionaries in both languages [19, 20, 29, and 30]. More than 50 proverbs in English and 56 proverbs in Uzbek have been selected for further contrastive analysis.

4 Results and Discussions

The principal focus of the research was to analyse parent-child relationship proverbs within English and Uzbek society. To achieve this objective we picked English and Uzbek proverbs in which a variety of a parent-child relationship is revealed. Comparative analysis illustrate that there exist potential and important similarities and dissimilarities and specific features between English and Uzbek a parent-child relationship. The results of the research are provided in details in the tables. According to the study results, there were revealed proverbs about having a child in the family. Both Uzbek and English proverbs reflect the happiness of a child being in the family, besides we found an English proverb with the expression of disagreement and discomfort of parents in having a child as well (see table1).

Table 1. Proverbs that reflect parents' attitude to having children.

(The quantity of proverbs is engaged in figures)

Parent's attitude towards having children	Pleased with having children	Desirable amount of children	Desired time of having childre n	Attitude to the absence of children
English	5	1	1	2
Uzbek	6	1	1	12

It can be seen from table 1 that both English and Uzbek are pleased to have children, almost the same number of proverbs about the joy, happiness of having children indicate that it is specific for both societies to show positive attitude towards children in the family. Here are examples: A babe in the house is a well – spring of pleasure [19] or Uzbek proverb Odamning mevasi bola (A human's fruit is a child). The most distinctive aspect between English and Uzbek proverbs is the presence of some English proverbs expressing a merely positive attitude of having children and the absence of proverbs with such connotations, for instance: He that has a wife and children has given hostages to fortune or Children certain cares, but uncertain comforts. Another evident difference is reflected in the amount of proverbs expressing the matter of absence of children in the family, 1 proverb in English (The house without children is cemetery) and 12 proverbs in Uzbek, the reason for this can be explained that Uzbek society is considered to be a child-loving nation (Болали уй жаннат-боласиз уй миннат; Болали уй бозор, боласиз уй мозор [29] and others). In the section of table 1 the amount of proverbs concerning the issue of desirable amount of children is the same in both languages. However, they contradict in the meaning, whereas Englishmen believe that a lot of children require a lot of effort whereas the Uzbek people consider that every child in the family no matter how many they has his/her own place in the heart of the parents: Eng. Many children, many cares, no children, no felicity and Uzb: Ўнта бўлса ўрни бошқа, қиркта бўлса қилиғи бошқа. The same case was revealed in the section which illustrates the desired time of having a child. While some English proverbs advise to start a family as soon as possible, since a big difference between the age of parents and children may negatively impact on child's destiny: Late children, early orphans, it is not troublesome issue for Uzbek when to have children, any time suits them: Фарзанд билан давлатнинг эрта-кечи йўқ (There is no time limited to have a child and to be rich).

Table 2 mentions the presence of the love in a parent-child relationship. Comparison of parent's love in the proverbs showed not a considerable difference in two languages, Uzbek proverbs about father's love is slightly more than in English, the similar amount about mother's love proverbs may be seen.

Table 2. Love and affection in a parent-child relationship (The quantity of proverbs is engaged in figures)

	Comparing father's and mother's love	Father's love	Mother's love	Children's love
Eng.	1	1	1	1
Uzb.	3	4	1	NA

In Table 2 we found one English proverb about children's love (The best love is that of children); however we did not come across any Uzbek proverbs expressing a child's love. In the process of investigation of parent- child proverbs, we revealed a number of proverbs which denote metaphorical meaning of richness, welfare in different images of parents if they possess children. For English, two proverbs out of three that we found express the state of richness of having children is close to the image of poor parents and one for all parents, whereas in Uzbek proverbs children are richness for every parent despite their status of being rich or poor. For example, Children are the parents' riches, Children are poor men's riches, Uzbek proverbs: Давлатнинг боши фарзанд (The beginningof richness is children), Давлатинг – ўғил- қизинг (Daughters and sons are your richness). See table 3:

Table 3. Children are parents' richness (the quantity of proverbs is engaged in figures)

Social classes the parents belong	Poor parents	Rich parents	Both
English	1	1	1
Uzbek	NA	NA	3

Analysis of table 4, which is about what parents associate children with, showed that children are assimilated only with positive things in both linguoculturology. Examination of this aspect of a parent-child relationship revealed several similarities as well as differences and uniqueness. Similarities were found in associations of children with sweets, sweet food in the English and Uzbek languages. For example, Children are the sweetest things of all to own, in the Uzbek proverbs children are associated with honey, sweets, e.g: Фарзандим - асал-қандим (My children are my honey and sweets). The statistics in figures is illustrated in the following table:

Table 4. The things children are associated with (The quantity of proverbs is engaged in figures)

Associations	Eng.	Uzb.
Sweets	1	2
Paradise	1	NA
Parts of body	NA	2
Nature	NA	1
Heaven	1	NA
Pleasure	1	NA
Power	NA	1

As can be seen from the table 4, associations in the languages differ considerably, 90 % of associations are dissimilar. This is explained by the life style, mentality, values, culture of each nation. To illustrate, for Uzbek children are future assistants in the household chores when they live together a strength that motivates parents to live, work and generally exist in this live, besides this, no Uzbek family can imagine his/her life wothout children, therefore, children for Uzbeks are associated with a human's eyes: Фарзанд – дилбанд, фарзанд- канот (Child is a soul mate, child is a wing); Ўғил билан қизинг – мисоли икки кўзинг (Your son and daughter are your two eyes); Куч билан фарзанд – белнинг қуввати (Power and child is strength for body). As for English, children are pure, naïve, clean beings, so the proverb where children are not heaven is not reflects peace, clearness that a child has.

Considering table 5, that illustrates English and Uzbek proverbs about parents' not objective attitude to their children, it was revealed that every that illustrates English and Uzbek proverbs about parents' not objective attitude to their children, it was revealed that every parent stand up for their children, protect them being sure that their child is the most and the best of all, this case cannot be a culture specific, it is mostly a social phenomenon. In this state, Uzbek and English proverbs have exactly the same connotations in proverbs, as well as equalize in quantities. The followings can be proof for above mentioned facts: He, whose father is judge, goes safe to trial; the crow thinks her own bird fairest and Uzbek proverbs Қўнғиз боласини оппоғим дер, типратикан боласини юмшоғим дер (The beat says that his own insect's body is white and a hedgehog says that his child is soft). Concerning how children's treat their parents depends at the age they are the moment of the relationship. Here, we revealed that English proverbs determine that attitude of children change as they grow up, the changes happen from minor to major, from physical to psychological that may cause serious, social, psychological damage to parents whereas children at early ages, according to English, can bring only mere physical and emotional difficulties which are considered only joy and happiness for parents, for instance Little children, little sorrows; Children suck their mother when they are young, and the father when they are old; Children when they are little make parents fools, when they are great they make them mad. For Uzbek, proverbs demonstrate that changes in children's behaviour has a physical aspect, the older the child gets the more parents get tired of taking care their children, for instance: Бола югурса кексанинг оёғи зирқирар (When child starts going, it makes an old parent's leg ache). The detailed data is given in the table below:

Table 5. Parents' subjective attitude towards their own children and age based a parent-child relationship (the quantity of proverbs is engaged in figures)

Parent-child interrelation	English	Uzbek
Parental subjective attitude	7	6

 https://doi.org/10.1051/shsconf/202110002008

Changing features of children's behavior based on age			
	Physical	**Psychological**	**Financial**
English	4	3	1
Uzbek	3	NA	NA

5 Conclusion

According to Ghazala proverbs are "special, fixed, unchanged phrases which have special, fixed, unchanged meanings" [12]. Meider claims that and their traditional function is didactic as they contain "wisdom, truth, morals and traditional views"[16]. The principal focus of the research was to carry out analysis the English and Uzbek a parent-child relationship in the frame of paremiological fund of the two languages. In order to reveal any potential considerable similarities and dissimilarities between English and Uzbek a parent-child relationship, proverbs selection and comparative-linguculturological method was used. Data analysis found out that there exist significant differences in the consideration of:

(1) Attitude towards having children
(2) Love and affection in the relationship
(3) What parents associate children with
(4) Belief that children are parents' wealth changes in the child's behaviour based on his/her age.

The research of our data has illustrated that a parent-child relationship is culture specific and is reflected in the linguistic picture of the world mostly by means of proverbs. Dealing with the different types of a parent- child relationship within English and Uzbek cultures showed that cultural factors play an important role in this issue. The research of our data has illustrated that a parent-child relationship is culture specific and is reflected in the linguistic picture of the world mostly by means of proverbs. Dealing with the different types of a parent- child relationship within English and Uzbek cultures showed that cultural factors play an important role in this issue.

References

1. Adamova S.M. *Poslovitsi I pogorki otrajayushiye mejlichnostniye otnosheniya v yazykah razlichnyh kultue: Lakskom i angliyskom* [Proverbs and sayings reflecting interpersonal relationship iin the languages of different culture: Lak and English] Extended Abstract of Candidate Dissertation, Makhachkala, "Firma KIT" IP Dagermanov I.D. (2015).
2. Bagga I., Saini S. *An analysis of a parent-child relationship as perceived by adolescents*. Adv. Res. J.Soc. Sci., **8** (2): 188-193, (2017). https://doi.10.15740/HAS/ARJSS/8.2/188-193.
3. Bretherton, I. *Communication patterns, internal working models, and the intergenerational transmission of attachment relationships*. IMHJ,11(3), 237–252, (1990).
4. Biktagirova Z.A. *Kontsept "Semya" v paremiologii angliyskogo, turetskogo i tatarskogo yazykov* [Concept "Family" in the paremiology of English, Turkish and Tatar kanguages] Extended Abstract of Candidate Dissertation, Kazan, Mnojitelniys tsentr Instituta istorii AN RT Publ., (2007).
5. Burdun S.V. *Kontsepti mat',otets, deti kak reprezentanty mira "Svoih" (na materiale kubanskoy frazeologii)* [Consepts mother, father, children as representatives of the world "One's" (on the materials of Kuban phraseology) PhN. 13(4), (2020). https://doi.org/10.30853/filnauki.2020.4.2
6. Chen Q., Du, W., Gao Y., Ma Ch. *Analysis of Family Functioning and A parent-child Relationship between Adolescents with Depression and their Parents*. Shanghai Arch Psychiatry, **29**(6). 365-372, (2017). 10.11919/j.issn.1002-0829.217067
7. Dunn J., Brown J. *Affect expression in the family, children's understanding of emotions, and their interactions with others*. Merrill Palmer Q. 1994; 40:120–137. [Google Scholar]
8. Dunn, J., Brown, J., Slomkowski, C., Tesla, C., & Youngblade, L. *Young children understand of other people's feelings and beliefs: Individual differences and their antecedents*. Child Dev, **62**, 1352–1366, (1991).
9. Gasanova M.A., Mayorova G.V. *Lingvokulturologicheskiy analiz kontsepta "Rebyonok" v poslovisah I pogovorkah agulskogo, tsahurskogo i rutulskogo yazykov* [Linguoculturological analysis of the concept "Child" in proverbs and sayings of Agul, Rutul and Tsakhur languages, USN, vol. **8**, (2016).
10. Hisam A., AbdulHamid AbuSulayman, & Omar Altalib, *A parent-child Relations: A Guide to Raising Children*, 528-529, (2012).
11. Herbrand, F. *Intercultural competence. Competitive advantage in a globalizing economy*. Bern, Stuttgart, Vienna: Verlag Haupt (2010).
12. Hofstede, G. *Local thinking, global action*. Berlin: German paperback publishing house, (2010).
13. Jahr, S. *Feelings and Emotions structures in expository texts*. Berlin: Walter de Gruyter, (2000).
14. Gaychenya A.A. *Kontsept "Semya" v russkoy I ispanskoy yazykovih kartinah mira. Lingvokulturologicheskiy analiz* [Family concept in Russian and Spanish language pictures of the world. Linguo-cultural analysis] Language and Text langpsy.ru 2018, vol. **5**, no 4, 68–77, (2018).
15. Ghazala, H. *Translation As Problems and Solutions:* A Coursebook for University Students and Trainee Translators. Beirut: Dar waMaktabat Al-Hilal, p.138, (1995).

16. Laible, D. J., Thompson, R. *A. Mother– child discourse, attachment security, shared positive affect, and early conscience development*. Child Dev, 71, 1424–1440, (2000).
17. Louis, C. *Proverbs and the Politics of Language. Proverbium 17:* p. 172-194. Yearbook of International Proverbs Scholarship. The University of Vermont (2000).
18. Maletzke, G. *Intercultural Communication. For interaction between people of different cultures*. Opladen: West German publishing house, (2013)
19. Manser M. *The Facts on File Dictionary of Proverbs*, New York, InfoBase Publishing, p 185, (2007).
20. Margulis M., Kholodnaya A. *Russko- Angliyskiy Slovar Poslovits i Pogovorok* [Russian- English dictionary of Proverbs and Sayings], McFarland & Company, Inc., Publishers Jefferson, North Carolina, and London, p.26-39, (1974).
21. Mieder, W. *Proverbs: A Handbook*. USA: Greenwood Press, p 3, (2004).
22. Mieder, W. & Dundes, A. Preface. In W. Mieder & A. Dundes (Eds.), *The Wisdom of Many. Essays on the Proverb*. 2nd ed. (pp. vii-xiii). New York: Garland Publishers, (1994).
23. Moore K., Busby A., Tawana B. *Parental relationship quality and child outcomes across subgroups*, Child Dev, **13**, (2011).
24. Olderogge V.A. *Malayskaya sistema rodstva//Rodovoe obshestvo*. Materialı İssledovaniya. M., (1951).
25. Raikes, H. & Thompson, Ross. *Family emotional climate, attachment security and young children's emotion knowledge in a high risk sample*. Br J Dev Psychol, **24** (89), (2006)https://doi.org/10.1348/026151005X7042.
26. Schwarz-Friesel, M. *Language, Cognition and Emotion: New Directions in Cognitive Science*. In M. Schwarz-Friesel. Language-cognition-culture, pp. 277–301.Stuttgart: UTB, (2007).
27. Steele, M., Steele, H., Johansson, M. *Maternal predictors of children's social cognition: An attachment perspective*. J Child Psychol Psychiatry, **43**(7), 861–872, (2002).
28. Pintin G. *Reprezentatsiya semeynih traditsiy v russkoy i kitayskiy yazikov kartinah mira* [Representation of family tradition of Chinese and Russian in the linguistic pictures of the world]. Extended Abstract of Candidate Dissertation, Ufa, Redaktsionno-izdatelskiy sentr Publ., (2018.
29. Salimova L.M. K voprosu ob izuchenii russkoy yazikovoy kartiny mira (po "Besedam o russkoy kulture" Yu.M.Lotmana. Filologicheskiye nauki. Voprosi teorii i praktiki, 1(31), (2014).
30. Shomaqsudov Sh., Shorahmedov Sh. *Ma'nolar mahzani* [Treasure of meanings] Tashkent, Ozbekiston Milliy Ensiklopediyasi, pp 167-428, (2009).
31. Speake J. *The Oxford Dictionary of Proverbs*: New York, Oxford University Press, p 536-539, (2008).
32. Telia V.N. *Russkaya frazeologiya: Semanticheskiy, pragmaticheskiy i lingvokulturologicheskiy aspekti*. M.: Shkola "Yaziki russkoy kultury", 288-289, (1996).
33. Vorkachov, S. G. *Lingvoconceptology and cross-cultural communication:sources and aims* Philologicheskiye nauki, pp.76-83,(2005).

Linguistic and Communicative Aspect of Business Education in A Modern Economic University

Larysa Bondarchuk[1], *Nataliia* Boiko[1], *Tamara* Podgurska[1], *Viktoria* Tovstenko[1], and *Nina* Kryvych[1]

[1]Kyiv National Economic University named after Vadym Hetman, Kyiv, Ukraine

Abstract. The paper considers specific language and communication projects conducted at the Kyiv National Economic University named after Vadym Hetman, which promote the humanization of business education, provide practical implementation of knowledge and skills of students of an economic university, increase professional, linguistic and cultural competencies of future professionals. Moreover, humanitarian projects are based on the interdisciplinary and cross-cultural foundations, that enriches the process of studying business disciplines, promotes intercultural ties, forms patriotic, civilizational and universal values.

1 Introduction

The task of a modern university is to educate a competent person who not only has knowledge and high moral qualities, but is able to act independently, in a non-standard way and creatively in various life situations, applying their knowledge and experience, taking responsibility for their own activities.

The Kyiv National Economic University named after Vadym Hetman has been among the leaders in training highly qualified specialists in the fields of economics, public and corporate governance, and economic law for many years. The University is actively working to provide the public demand for professional training of graduates, constantly monitoring the labor market, studying the requirements of employers for the specialists' qualifications and starting new educational programs, including implementation of joint educational programs with foreign higher education institutions and introduction of educational franchising. The University actively implements current programs of professional training, retraining, advanced training; it improves the methodological, organizational, legal and financial bases to provide the academic mobility of students, graduate students and teachers, as well as to develop double and joint graduation programs. The modernization of the educational process by updating the content of educational programs, their educational and methodological support, implementation of creative educational programs and innovative authors' courses, development and use of distance learning courses, video lecture courses for leading university teachers and teachers from partner foreign higher educational institutions is constantly carried out. The expansion of the range of selective courses, the improvement of the mechanism and procedure for their selection, the introduction into the educational process of active, interactive, game, problematic and other modern innovative technologies, forms and methods for preparing students for fuller development of their personal potentials are implemented. Leading scientists, specialists of various organizations and institutions, practitioners are actively involved in the educational process.

Business education has to consistently form students' economic thinking and consciousness, to educate reasonable needs, to develop abilities in a certain type of activity, to cultivate a careful attitude to the nature, to equip specialists with theoretical knowledge and practical skills in the culture of work, to form the needs of economic analysis, the vision of perspective, to learn to master the basics of the scientific organization of labor in market relations, to form in students the desire and need to improve the results of their work in educational and socially useful activities on a daily basis.

The great opportunities in ensuring a high level of business education are provided by the application of the method of educational projects. Finally, the ability to create, implement or participate in the projects is a significant vital competence of a person, the basics of which can be optimally formed and developed under the conditions of personal training at an educational institution.

1.1 Project method

The project work is actively used at the Kyiv National Economic University named after Vadym Hetman for studying various disciplines. In the course of project activities, students learn to: plan their work, predicting its results in advance; analyze and compare facts; argue their own judgments; make a decision; establish social contacts, use various sources of information; distribute responsibilities, interact; create a real "end product";

* Corresponding author: larysa.bondarczuk@ukr.net

present the results to the audience; evaluate their activities and the activities of partners. The project activities are characterized by joint work of a teacher and a student with equal status of both parties.

The project activities are aimed at solving the following psychological-pedagogical and organizational-pedagogical tasks: the development of cognitive interests, intellectual and creative abilities of students; socialization of students, development of key competencies communicative, cooperative, informational, etc.; the formation of active life position; creating an atmosphere of cooperation.

A. Tsymbalaru emphasizes on the three directions of interpretation of the term "project" in the pedagogical science. "First, as a preliminary one, approximate text of the full document (draft program, etc.) is… Second, the project means a certain action, a set of activities that have a common program, provide targeted actions, organizational form, etc. (publishing project, television project). Third, the project is defined as a completed cycle of productive activities (collective, group, individual, etc.), as a form of building a joint purposeful activity of people" [7]. This understanding of the project, as the researcher notes, led to the separation of certain stages of its development: the design phase, the technological phase (implementation), and the reflective phase (summarizing – evaluating the results, drawing conclusions). Therefore, the author concludes that the essence of the concept of "project" is interpreted in two important aspects: firstly, as a result of the activity – the product (productive aspect), and secondly, as an innovative form of joint activities aimed at achieving a certain result (activity aspect) – and defines the category "project" as "a system of implementation of planned successive actions aimed at achieving a certain intended result" [7].

The project method aims at a comprehensive and systematic study of the problem and the development of a specific end product; involves obtaining a practical result, not just a search for certain information. The project is primarily the result of collective efforts of the performers, so at the final stage of activity it involves the reflection on joint work, analysis of its completeness, depth, information support, and creative contribution of each participant.

The method of projects as a method of planning appropriate (purposeful) activities in connection with the solution of educational problems in real life circumstances began to take shape in the late nineteenth century in the US as a "problem method". It was developed in the pedagogical viewpoint and experimental work of John Dewey within the humanity trend in philosophy and education. The term "project method" first appeared in 1908–1910 in the reports of Massachusetts school authorities and was adopted from the agronomic literature. The project method contained the ideas for building learning on an active basis, through appropriate activities of students in relation to personal interests. The project method is based on the concepts of learning through "action" and "free education". J. Dewey's motto was an attempt to help students develop their abilities by collaborating on a particular problem in a collective search and creating a collective project. It was extremely important to show a child their personal interest in gaining knowledge, where and how they may need it in life. The students gained experience and knowledge in the process of planning and implementation of practical tasks which gradually became more difficult during training.

The project method is a genuine product of the American education movement. It was described by William Heard Kilpatrick in his essay "The project method". Professor W.H. Kilpatrick, managed to improve the system of the work on projects, theoretically substantiate the method based on the principle "education is life, not preparation for life"; he offered the first classification of projects (projects created, consumable projects, problem-solving projects, exercise projects). At that time, the project was meant the target act of activity, which was based on a child's interest [11]. Michael Knoll has traced the project method to architectural education in sixteenth-century Italy and to engineering education in eighteenth-century France. In his articles, Knoll wrote: "Unlike traditional methods, projects focus on the application rather than the transfer of specific knowledge or skills, and more carefully than lectures, demonstrations or recitations, they aim to increase intrinsic motivation, independent thinking, self-esteem and social responsibility".

In Ukrainian education, the relevance of this problem was studied by O. Makarenko as a result of his innovative pedagogical activity, he came to the conclusion about the design of personality as a subject of pedagogical work. This opinion has also been repeatedly claimed by V. Sukhomlynsky whose multifaceted pedagogical heritage is permeated by the idea of human design. The definition of the essence of design as a pedagogical phenomenon is quite complex, because the design system and the pedagogical process itself are extremely complex and valuable. The very word "project" in translation from Latin means "thrown forward, intention, plan" and so on.

In the modern pedagogy, the projects are classified according to a certain dominant typological feature. Following K.S. Polat, researchers distinguish the following projects according to the dominant method or type of activity:

a) applied projects with clearly defined results of the participants' activities (for example, a draft document, a dictionary, a reasoned explanation of any phenomenon, etc.). These projects involve a thorough understanding of the structure, the distribution of functions between the participants, the design of results, their further presentation and external review;

b) research projects aimed at organizing students' activities aimed at solving creative problems with previously unknown results, which involve certain stages of work (reasoning of the relevance of the research topic, subject and object, defining goals and objectives, identifying search methods and sources of information, hypotheses, identifying ways to solve the problem, data collection, analysis and synthesis, discussion and design of the results, presentation of communication or report, identification of new problems for further analysis);

c) information projects that are aimed at studying the characteristics of any processes, phenomena and objects, and provide their analysis and generalization of the identified facts. The structure of such a project is similar to the structure of a research one, which is often the basis for their integration;

d) role-playing projects, when the participants perform certain roles (literary or fictional characters) specified by the nature and content of the project; the participants mimic social or business relationships complicated by hypothetical game situations. The structure of these projects is just outlined and remains open until completion [4].

In real practice, the projects of mixed types are mostly used, which combine the features of several types of projects and are practice-oriented. These projects not only allow to gain new knowledge, but also contribute to gaining certain social experience.

An important role at all stages of the activity is given to the teacher who, on the one hand, is the organizer of the project and provides development and training of the students, and on the other, is an equal member of the working group and sets his own goals, analyzes the situation, offers interesting ideas for discussion. The teacher can share their own life experience, help students find diverse sources of information about the research problem, help to contact with specialists, government officials, etc.

The analysis of practical experience of the project activity organization allows to allocate the following main stages: preparatory (semantic and technological aspect), planning of project activity (problem statement, development and planning of certain action), research of the problem and the choice of a way of its solution; activity (problem solving, implementation of the action); presentation of the results; reflective stage (evaluation and analysis of the results).

At the initial (preparatory) stage it is important to motivate students for research work. The stage of preparation for work on the project involves the formation of an initiative group, search or selection of a problem among a number of topical issues, specification of the topic in terms of the reality of implementation, relevance, and social significance. This stage aims at theoretical training of the project participants. It is during this stage the students learn to analyze sources, communicate effectively, and discuss problems with the help of the teacher.

The stage of problem selection and its research involves the analysis of a wide range of problems that need to be solved, and the choice of one of them, which will be the object of study. This stage aims not only at gathering information on the problem of research, studying of sources and accumulation of materials, but also at direct learning and gaining experience of direct activity of the students.

The main task of the stage of choosing the way to solve the problem and develop a plan for its solution is planning made by students on the basis of systematization and analysis of the material given for problem solution. The purpose of this stage is to determine the most optimal solutions of the problem, which are supported by the majority of the group.

The implementation of the problem solution stage involves an attempt to fully or partially realize the proposed action plan of the team, organizing various actions, such as publications in the media; appearance on radio and television; appeals to public organizations; involvement of commercial structures for coworking, etc.

The result of project activities can be any event or specific product: presentation, tour, plan, concert, public speech, book, article, master class, etc. The success of the project largely depends on the design of its results. The presentation of research materials to a wide audience is a necessary prerequisite for working on the project. Successful presentation of research results can give impetus to further problem solution.

Work on a project cannot be considered effective without the students' analysis of the course and results of their activities at different stages of research, i.e. without reflection. At the final stage of project, it is advisable to organize a discussion in which students will be able to assess the project activities as a whole and the personal contribution of each to the common work, to identify shortcomings and discuss ways to overcome them.

2 Language and communication projects at an economic university

Language project "Week of Ukrainian Literature and Language." The Department of Business Linguistics of the Kyiv National Economic University named after Vadym Hetman annually implements in November a language project dedicated to the Day of Ukrainian literature and language. During a week, students of various faculties, teachers and university staff actively participate in the events dedicated to the Day of Ukrainian literature and language. There are competitions on essay writing, university dictation, an auction of ceremonial speeches; students present their own advertising video projects "Choose Ukrainian!" There are poetic translation studios. Everyone joins the flash mob "Do not miss the nice words!" The events "Poetic travels #ЛІРИЧНІЕКОНОМІСТИ" (#LYRICALECONOMISTS), "#ЗАБЕТКОВАНИЙКНЕУ" (#KNEUALPHABET), "#ПІДВІШЕНАКАВА" (#SUSPENDEDCOFFEE), "#КНЕУВСЬКИЙДИКТАНТ" (#KNEUDICTATION) arouse great interest among the students. The first stage of the all-Ukrainian language and literature competition of Taras Shevchenko and Petro Jacyk is also starting at this time. These events cultivate aesthetic taste, love for the native language, country and people, enrich with new knowledge and approaches in the efforts to improve professionally. In November 2020, the traditional project was implemented online on the Zoom platform.

Language projects of this type contribute to the development of students' sustainable skills of language culture of modern business communication, increase the level of communicative competence of future specialists, and form the national language personality of the future economists. The students gain practical experience of language and professional communication due to the language projects of this type and cooperation with teachers of profile departments of international management, international economics, international trade, and economics of enterprises.

Polish language and communication project "Skąd jesteśmy?" ("Where are we from?"). The study of Polish language at the Kyiv National Economic University named after Vadym Hetman started in 2014. Polish projects implemented during the study of this discipline promote the acquisition of communication skills in various areas of professional activity and life. It is advisable to use language etiquette in the situations of real language communication, and to introduce the basic economic terms in Polish.

Ukrainian public organization "Poruch" ("Near") in partnership with the Polish Association "Time Space Identity" and the Department of Business Linguistics of the Kyiv National Economic University named after Vadym Hetman implemented the project "Skąd jesteśmy?" ("Where are we from?"). As part of the project "Where are we from?" through a demonstration of the history of 20 Polish families from Szczecin (Poland), an attempt was made to find the answer to the question: "Where are we from?" To support this idea, Polish producer Adam Kuzycz-Berezowsky and his team have created a film of the same name where the problems of identity, time and space are explored. This film shows people's lives, their thoughts, feelings, experiences, the search for answers to the question: "Skąd jesteśmy?" The video conference before the screening was attended by the co-author of the film, Mr. Antoni Sobolewski, who told about the history of the film "Skąd jesteśmy?", about its shooting and the producer's plans. All this took place in Polish, which allowed future economists to better feel the peculiarities of communication in Polish, enrich their vocabulary and acquaint with the culture of Poland. The students also tried to express their thoughts and impressions from the film "Skąd jesteśmy?" In Polish. This film asks each of us the question. And we already know the answer: "We are from where we are, where our parents are, where our family is, where our country is…" The projects of this type contribute to the improvement of the Polish language culture, the study of the common history of the Polish and Ukrainian peoples, the establishment of international contacts, and form a sense of beauty of future economists.

Language and communication project "Days of Poland at the Kyiv National Economic University named after Vadym Hetman". The Department of Ukrainian language and literature of the Kyiv National Economic University named after Vadym Hetman within the framework of this annual project always plans a lot of interesting and informative events. Competitions for the best essay in Ukrainian or Polish are carried out the students represent their own views on modern relations between Poland and Ukraine, share their vision of the future of both countries. The best essays are published in the Polish newspaper "Dziennik Kijowski". At the practical classes on the Polish language, students make presentations and video presentations in Ukrainian or Polish about famous Polish writers, economists, public figures, historical figures; cities and villages of Poland, special places or tourist objects, places in Kyiv related to famous Poles or Polish realities, Polish-Ukrainian relations in different historical periods, famous Ukrainians in Poland, etc. Polish films (by Andrzej Wajda, Jerzy Hoffmann, Krzysztof Zanussi, etc.) are shown, followed by writing reviews about them. As part of the celebration of the Days of Poland at the Kyiv National Economic University named after Vadym Hetman, students fulfill a flash mob related to Poland and the Polish language. To acquaint the community of the Kyiv National Economic University named after Vadym Hetman with Polish culture and its prominent representatives, students prepare quotes about famous Polish writers, politicians, economists and public figures. The students give these "Polish flashcards" to their colleagues from different courses, representing the Polish language course in our university. As part of the project, the students take part in the cultural events of the European day of languages in Kyiv, where they attend trial lessons in the Polish language, screenings of European cinema, literary discussions, workshops, competitions, games. The students learn more about Polish cultural centers in Kyiv, get acquainted with Polish book novelties, and actively promote the Kyiv National Economic University named after Vadym Hetman where they study the Polish language. The students are happy to participate in various projects of the Kyiv National Economic University named after Vadym Hetman, related to the Polish language and culture, since they see real prospects for the implementation of the acquired language knowledge in the practice of future professional activity.

Language project "Certification course on modern Polish language business communication". In 2016, the Kyiv National Economic University named after Vadym Hetman offered a new project "Certification course on modern Polish language business communication" for the students studying "International Economics" and "International Business".

The purpose of the certification course "Modern Polish language business communication" is to update the knowledge acquired during the study of the basic course of the Polish language in a specific area of business communication. The program of the course includes: lectures and presentations "Polish language in business communication", a series of seminars "Polish language of modern mass media" (on the example of the experience of Polish language media in Ukraine "Dziennik Kijowski", "Krynica"), language training "Practice of working with Polish-speaking audience" (experience of the Community of Poles in Ukraine), a series of lectures "Polish in the diplomatic sphere" (lectures from the Consular Section of the Embassy of the Republic of Poland in Ukraine), "Polish in tourism" master classes from the managers of the Polish Tourist Organization in Kiev), online lecture "Web-session of questions and answers about the Polish language in business" (Mr. Albert Nowacki, Doctor of Philosophy from the John Paul II Catholic University of Lublin, Poland), online lecture "Polish language in science" (representative of the NAS of Ukraine Lesya Lazarenko, Berlin), coaching seminar "Polish language in business" (from representatives of the International Union of Polish Entrepreneurs in Ukraine), etc. Master classes "Fundamentals of translation from Polish into Ukrainian of economic texts", "Fundamentals of translation from Ukrainian into Polish of economic texts" are also planned. At the end of the course, the students prepare presentations of group language cases. The course is completely free, and all students participating in the project receive certificates [14].

After mastering the course, the students gain not only knowledge of modern Polish economic terminology, but also real practical skills of using the Polish language in business, which is an indisputable advantage for further career growth of future economists of European level.

Project "Paths of Trypillian civilization". The Department of Regional Studies and Tourism of the Kyiv National Economic University named after Vadym Hetman implements the project "Paths of Trypillia civilization", among a number of professional disciplines.

Each region in Ukraine at the present stage turns the pages of its history and culture in order to bring them into the category of tourist destinations [8].

The Trypillian civilization is one of the oldest ones in the world. It developed in Ukraine 7000 years ago. The Trypillian culture (it got its name from a settlement near the village of Trypillya in Kyiv region, discovered and studied by archaeologist Vikentiy Khvoyka in the XIX century), its level of development and quality of life impress modern archaeologists [12].

Upon the training of future tourism managers, a student project was created to promote the Trypillian culture. The project has the following goals: resuscitation of the improperly forgotten route through the Trypillian region of Kyiv region, which will promote the development of the tourist business of the region; formation of self-awareness among the youth on the basis of national dignity and respect for their own origins due to the involvement of historical and cultural potential of the Trypillian heritage; acquisition of professional skills in the development of the excursion route by students of the profile specialty "Tourism" upon the implementation of the project stages; conducting field theoretical and practical lectures for the students and teachers on the Trypillian culture.

Participating in the project, the students try to develop a holistic tourism product for different groups of tourists. It consists of two parts, separate and interdependent in terms of organizational activities and venues.

The first tourist offer are excursions for school and student groups "Paths of the Trypillian civilization" on the route "Kyiv-Trypillya-Khalepya-Vytachiv-Novi Bezradychi-Kyiv". The one-day educational and entertainment tour lasts 9 hours. The target audience are tourists of all ages, students, modern youth, who are interested in the historical and cultural heritage of Ukraine, Kyiv region and the Trypillian civilization.

The second tourist offer are field lectures for schoolchildren with master classes on the topic: "The secrets of Trypillia." The plan of this tourist event includes a short presentation of the phenomenon of the Trypillian civilization with video materials prepared by the Kyiv National Economic University named after Vadym Hetman students; acquaintance with the applied culture of Trypillia (the students of the Kyiv National Economic University prepare and conduct master classes in creative groups, where they fabricate pottery with Trypillia symbols, paint linen, wood, paper blanks or clothes with the Trypillian ornaments).

The main activities provided by the project are the following: educative, educational, creativity of youth and schoolchildren in the cultural sphere, educational and productive activities of the students (organizational, excursion, marketing activities, advertising, etc.).

In parallel with the implementation of the project, the teachers conducted classes in the disciplines "Local history", "Geography of tourism", "Tourist country studies", "Fundamentals of excursion activities", "Geography of culture". The students studied the scientific works of famous scientists who investigated the Trypillian civilization: V. Khvoyka, O. Olzhych, M. Videiko, V. Chabanyuk, other historical materials, artifacts of the historical museums of Trypillia, Kyiv, and Vienna. Also during the study of these disciplines, there were master classes and meetings with the folk artist and master of reconstruction of Trypillia ceramics L.I. Smolyakova in her private museum "Dyvosvit Trypillya" [13].

As a result, a scientific conference was held on the topic: "Trypillia the civilization of creators". Among various areas of the event at this conference, our students demonstrated a master class on sculpting the Trypillian souvenirs.

The students specialized in "Tourism" also participated in various thematic events: festivals of the Trypillian culture, cycling and festival "The Trypillyan miracle" organized by the local community of Novoukrainka, master classes on sculpting the Trypillian clay products during thematic events and exhibitions. During the preparation of the excursion route, trainings and visits to the settlements on the route were carried out: Novoukrainka, Trypillya, Khalepya, Vytachiv, the exhibitions of the V. Khvoyka's museum, the regional archeological museum of the Trypillian culture in the village of Trypillya, the settlements of the village of Shcherbanivka and Divych-gora, etc.

The highlight of the proposed route will soon be the center "Renaissance" in the village of Novi Bezradychi, where, among other unsurpassed exhibitions, a pottery workshop and built a Trypillian furnace are created. A separate segment of tourists was also invited to get acquainted with the amazing hotel and restaurant complex "Trypillya Sun" where one can get an interesting mini-tour, see a master class of famous chefs, and try and taste the Trypillian dishes.

After studying the transport infrastructure of the route, its possibilities and shortcomings were analyzed, and certain recommendations for drivers were developed. The above mentioned objects are included into the developed excursion route of a one-day tour, which was first tested for students of the Department of Marketing of the Kyiv National Economic University named after Vadym Hetman. During the project implementation, the teachers of the department and students took part in the seminar of history teachers "Innovative approaches in teaching the Trypillian topics in school courses of disciplines of secondary schools." There were also field lectures with a demonstration of master classes on the Trypillian topics for the schoolchildren of Kyiv.

The project "Paths of the Trypillian civilization" developed by the students and teachers of the Kyiv National Economic University acts as a means of organizing axiological (value-oriented) space and significantly affects the development of personality by expanding the accessibility of historical and cultural heritage for different target audiences. The creative activity provided by the project gets new qualitative characteristics through the immersion into the life and cultural traditions of Trypillia, and plays an important role in the personal and professional growth of future economists.

The project "Paths of Trypillian civilization" will help and promote the following aspects:

- preserving the integral state of the natural heritage of a region as a tourist resource;
- introducing the innovative methods and practices of language, communication, historical, cultural and educational activities with a focus on a certain age group;
- increasing the efficiency of operation of existing tourist infrastructure facilities, as well as the creation of new jobs directly or indirectly related to the project, in particular: guides, workers of catering establishments, specialists of creative workshops, employees of transport and travel agencies, museums;
- improving the quality of services in order to increase Ukrainian tourist flows to the destination sites and to attract foreign groups to acquaint young people of foreign countries with the historical and cultural heritage of Ukraine;
- supporting the domestic producers of goods and services, due to the involvement in the complex of the project program of processes for the manufacture of related to the theme of the Trypillia souvenirs, confectionery, clothing, toys, accessories, etc.;
- creating the conditions for tourist travel with appropriate excursion content and field lectures with master classes for children and youth with disabilities and for socially vulnerable groups;
- intensification of educational work provided by the relevant programs of secondary schools through excursions, training sessions (on the history of Ukraine, drawing, work, etc.), extracurricular educational activities.

Hlukhiv project. A long-term student project is also the so-called "Hlukhiv project" with a working title "Conceptual approaches to the formation of a holistic tourism product of the Hlukhiv city: a study of the current state, problems and prospects for its development." The purpose of the project is to develop conceptual approaches to creation of an integral tourist product of the Hlukhiv city (Ukraine) on the basis of empirical data.

A number of tasks were set and solved to develop the project:

- the tourist potential of the city for creation of a competitive tourist product directly under the expeditions of teachers and students in Hlukhiv was investigated;
- the necessity of the popularization of the objects of historical and cultural heritage of Hlukhiv is highlighted;
- proposals on the development of tourist infrastructure of the city with cultural, sports and other events that attract the attention of young people;
- the need to improve existing and introduce new thematic tours and tourist routes has been indicated;
- the proposals for the formation of a positive image of the city at the national and international levels have been put forward.

Both students' projects were presented at the All-Ukrainian competition of the student works in the specialty "Tourism" in Kharkiv.

When teaching courses "Local History", "Tourist Country", "Urbanism and Tourism", the students are also offered the following medium-term projects with demonstration of educational presentations: "Draft business card of the state", "Draft tourist booklet of the country", "Tourist attractiveness of the European, Asian, and American cities", "Prague, Warsaw, Kyiv (or other cities) the similarities and differences of the modern urban development", etc.

The project "Infographics of the European tourist cities." One of the ways to systemize a large amount of tourist and geographical information and to reduce its volume and simplify perception is the infographics. At present, this branch of design has not become widespread in Ukraine. The infographics is a graphic design of information through visual demonstration, aimed at clear presentation of a significant amount of information and enhancing its perception.

It is worth noting that the use of infographics is just starting to gain popularity in the tourism business. That is why the project "Infographics of the European tourist cities" offered to future tourism managers is relevant and timely and creates new opportunities to achieve the goals: the development and promotion of tourism products.

The purpose of creating and using the infographics in the field of tourism is the rapid spreading of information in an innovative way. The generalized information is presented in the form of symbols and icons. A part of the visual image is also a short text that accompanies it. When an image interacts with a text, two components are perceived simultaneously, visually and verbally.

As known, a person receives about 90% of information through sight, that is, the visual perception of information is more effective, has a greater impact on a person and allows to more effectively present the needed information. Also, the visual form of information is much faster because it is perceived 60,000 times faster than text. People remember only 10% of what they heard, 20% of what they read, and about 80% of what they saw or accomplished.

This presentation of information about the city in the course of studying, for example, the discipline "Tourism of urban agglomerations", has the following advantages: an effective focus on the specific elements of the presentation; processing of a large amount of information without significant information overload; obtaining optimal information about the city in the shortest possible time; highlighting comparisons, relationships, dynamics of upgrade or decline of urban infrastructure; minimization of the material and its aesthetics.

Using the infographics in tourism, it is advisable to effectively and concisely provide the specific information about the city: the popularity of a tourist route, possible options for getting from Kyiv to the tourist centers (transports, distances, ticket prices); number of tourists (inbound or outbound); the state of the city's infrastructure; the availability of the subway and other transports, fares; CityCard; the most popular destinations, attractions, museums; tickets, benefits; pricing policy of hotels of different categories; average check in catering establishments of different types; excursion services (buses, walking excursions); advantages and problems of the city; entertainment, leisure, holidays, festivals, shopping, souvenirs, etc.

The information in the form of infographics about the most popular European tourist centers was posted on the student tourist portal and tested on the personal experience of traveling students; it aroused interest not only among the students of the Kyiv National Economic University and individual tourists, but also aroused professional interest of travel companies, as evidenced by their positive feedback.

3 Conclusions

Thus, as M.O. Trygubenko points out, on the one hand the business education is a dynamic system of knowledge, practical actions, aspects, methods and psychological and pedagogical technologies, which is constantly evolving and improving; on the other hand, it is an open socio-economic system. that takes into account individual, collective, local, regional, social and national aspects, forms a personality for living in a business environment and an active personality that develops. In a modern economic university, the language and communication projects in business education combine all these aspects, contributing to the practical study of professional disciplines.

The use of language and communication projects in order to humanize the educational process is a promising and multi-vector direction of the modern business education. These projects give the modern Ukrainian economics education certain national and distinctive features, and allow to adopt the experience of world and European educational practice. Specific language and communication projects implemented at the Kyiv National Economic University named after Vadym Hetman increase the professional, linguistic and cultural competencies of future specialists.

References

1. S. Grabovenska, Competitiveness of territorial markets of tourist services in Ukraine, Lviv, 222 (2017)
2. L. L. Zalizniak, At the origins of the ethnic division of mankind, NDIU, Kyiv, **7-25** (2018)
3. J. Dewey, Experience & Education, New York, (1938)
4. K.S. Polat, M. Y. Bukharkina, M.V. Moiseeva, A.K. Petrov, New pedagogical and information technologies in the education system, M., 272 (2009)
5. A. Cymbal, Pedagogical design, Kyiv, School World, 128 (2009)
6. M. Videiko, N. Burdo, E. Slesarev, M. Videiko, Excavations near the village. Trypillia, Archaeological research in Ukraine 2018: yearbook, Institute of Archeology of the National Academy of Sciences of Ukraine, Kyiv, 62-65 (2020)
7. M.O. Trygubenko Business education as a social phenomenon of modernity and a factor of entrepreneurial success, Culture of the peoples of the Black Sea region, **205**, 125–128 (2011)
8. Yu.Ye. Kirilov, V.G. Granovska, L.O. Aleshchenko, Formation of competitive advantages of subjects of tourist branch, Economics of agro-industrial complex, **5**, 45 (2020)
9. A. Cairo, Functional art: introduction to infographics and visualization, Ukrainian Catholic University, Lviv, **17**, 349 (2017)
10. G. Gutek, Philosophical, ideological, and theoretical perspectives on education, New York, (2014)
11. W. H. Kilpatrick, The project method, New York, (1929)
12. A. Zborovsky, The discoverer of Trypillia culture, Kyiv, **12**, 10 (2013)
13. Z. Egorova, Vikentiy Khvoyka is an outstanding self-taught archaeologist, Voice of Ukraine, Kyiv, **9**, (2019)
14. N. K. Misyats, Modeling of interdisciplinary connections in Polish classes languages in higher education, Educational innovative technologies in the process of teaching disciplines, Zhytomyr, 153-157 (2004)
15. M. Knoll, The project method: Its vocational education origin and international development. Journal of Industrial Teacher Education, **34**, 59-80 (1997)
16. M. H. Bleeke, The project: From a device for teaching to a principle of curriculum. Diss., University of Wisconsin-Madison (1968)

17. J. K. Sood, New Directions in Science Teaching, Kohli Publishers, Chandigarh, 146-149 (1989)

Pedagogical Technologies for the Development of the Speech Competence of Future Specialists in Non-Linguistic Universities

Sayyora Saidjajalova[1]

[1] Tashkent Branch of Moscow State University named after M. V. Lomonosov, Amir Timur, 22, 100000, Tashkent, Uzbekistan

Abstract. The main task of modern society, the change in educational policy in general and in the field of teaching speech disciplines are discussed in this article. An analysis of pedagogical practice and the results of special studies suggest that the problem of using these technologies to develop the speech competence of future specialists of non-linguistic universities in the process of teaching the Russian language has not yet been adequately reflected in the research of scientists. It is also said that many problems in higher education, in particular, the use of pedagogical technologies in teaching the Russian language, are poorly solved in both theoretical and practical aspects.

1 Introduction

Language is the most important means of communication, without which the existence and development of human society is impossible. The changes taking place today in social relations, means of communication require an increase in the communicative competence of students, improvement of their speech training so that they can exchange thoughts in various situations in the process of interacting with other communication participants, while correctly using the system of language and speech norms.

The challenges facing by higher education today require the improvement of the educational process aimed at increasing the creative potential of students' educational activities. The basis for ensuring high quality education is the use of effective educational technologies, the integration of the educational process and scientific activities and the involvement of students in it, the individualization of teaching and upbringing, the creation in a higher educational institution of a trusting atmosphere of creative cooperation, continuous self-development and self-realization of students in the educational process.

One of the main tasks of modern society is to change educational policy in general and in the field of teaching speech disciplines as well. The top priority in education today is not so much the acquisition of a large amount of knowledge by students as the development of their intellectual and creative potential, which allows them to produce new knowledge in the future. In this regard, the principles of traditional education are being reformed, largely due to the social order.

Teaching speech skills is implemented within the framework of a student-centered approach, which involves the individualization of training, taking into account the level of language proficiency, the level of intellectual development and individual psychological characteristics of students. It means that the center of learning is the student's personality, his motives, goals, and a special psychological makeup, which together determines the nature of learning. This approach forms the student's activity, his readiness for learning activities, for solving problem problems through partnership with the teacher.

In modern learning conditions, such qualities of students should be developed as: the ability to independently acquire knowledge and apply it in practice; the ability to think critically and creatively, to produce new ideas;

the ability to work with information, analyze it and highlight the main thing; the ability to work in a team, coordinate with others and avoid conflict situations.

Pedagogical technologies, which have recently been widely used in modern education, are aimed at the formation and development of the above qualities of students.

2 Methods

Pedagogical technology is the design of the educational process, based on the use of a set of methods, techniques and forms of organizing training and educational activities that increase the effectiveness of training, the use of which has a clearly defined result. With the mastery of any new technology begins a new pedagogical thinking of the teacher: clarity, structure, clarity of methodological language, the emergence of a reasonable norm in methodology.

The use of pedagogical technologies focused on productive learning activities implies an increase in cognitive activity, interest in knowledge, the

Corresponding author: sayka_1971@mail.ru

development of initiative, creative activity, the creation of an interactive educational environment, a combination of independence and responsibility of the student as a subject of educational and cognitive activity for the process and result of the study.

Thus, the relevance of the study is determined by the social need for specialists with a high quality of professional language competence and the need to subject a detailed study of the process of its development in the system of multilevel education in a modern university. The analysis of pedagogical practice and the results of special studies suggest that the problem of using these technologies for the development of future specialists in non-linguistic universities speech competence is in the process of teaching the Russian language has not yet been sufficiently reflected in the research of scientists. Despite the available research in the aspect of increasing the effectiveness of teaching with the use of pedagogical technologies in higher education, many problems, in particular, the use of pedagogical technologies in teaching the Russian language are poorly resolved both in theoretical and practical aspects. The incompleteness of the theory of an integral and continuous system of multi-level language training of future specialists, the numerous needs of modern university practice, the need to generalize best practices, systematize the language training of future specialists in non-linguistic universities, the lack of a specific methodology for the introduction of pedagogical technologies and their improvement in the process of teaching Russian language and determined the choice of the research topic «Pedagogical technologies for the development of the speech competence of future specialists in non-linguistic universities». Compliance of the topic with the priority areas of scientific research in the Republic.

Political, economic and social problems occurring in the Republic of Uzbekistan lead to radically new approaches to various spheres of society, including the field of education. Our state, which has embarked on the path of democratic development, needs highly educated, comprehensively developed citizens, who are, first of all, an individual. Personality upbringing is a didactic priority task assigned to the educational system of the Republic by the Law on Education and the National Program for Personnel Training. That is why the use of modern technologies in pedagogical activity is one of the priority areas of education in our country. This is due to the need to train highly qualified specialists capable of actively participating in the economic, social and political spheres of our state, solving national economic, legal, educational and other tasks that are essential for our state.

The Republic of Uzbekistan is currently entering the stage that can become an important stage in the realization of its economic potential. In the country, the demand for skilled labor is growing at a faster pace than the supply, and the discrepancy between the professional knowledge and skills of university graduates and the needs of the market is obvious. In addition, while the development of innovations and technologies is an important factor of economic growth in any modern economy, in the Republic of Uzbekistan it is important to improve the correspondence of the knowledge and skills of university graduates to the needs of the market, as well as to create and strengthen an innovative infrastructure that contributes to improving the quality and market significance of activities. universities in general and stimulating the use of creative and innovative approaches to learning. The Republic of Uzbekistan has chosen and is implementing a course towards building a democratic rule of law and civil society. The main goal and driving force of the reforms carried out in the republic is a person, his all-round development and well-being. The most important condition for the development of the country is the functioning of a perfect training system based on the development of the modern economy, science, culture, technology, technology. The implementation of the National Program for Personnel Training provides for a radical reform of the structure and content of continuing education. A radical change in the existing systems of education and personnel training in the course of the implementation of the National Program is carried out based on social experience and the achievements of modern scientific thought, advancing scientific and methodological support of the educational process at all stages, in all forms and types of educational institutions of the continuous education system. Among the urgent tasks of pedagogical science at the present stage is the scientific provision of goals, content, methods, means and organizational forms of education, training and personal development based on the use of the achievements of science, technology, and advanced technologies. The state policy in the field of personnel training provides for the formation of a diversified personality of a citizen through the system of continuous education. In this regard, the individual acts in the system of continuous education and training, both as a consumer and as a producer of educational services. In particular, a person as a producer of educational services takes part in the transfer of knowledge and experience in the process of education, activities in the field of material production, science, culture and services. Individuals in the republic have the right to choose a professional educational program to realize their creative potential. Higher education occupies a special place in the system of continuous education. Higher education based on general secondary, specialized secondary, vocational education is an independent type of continuous education system and is carried out in accordance with the laws of the Republic of Uzbekistan "On Education" and "On the National Program for Personnel Training". In accordance with the National Program for Personnel Training, the main goals and objectives of higher education have been identified, one of which is to provide highly productive education and training of qualified personnel on the basis of modern educational and professional programs.

Therefore, the main goal of the National Program is as follows: a radical reform of the education sector, a complete rejection of its ideologized blinkering, and the creation of a National system for training highly qualified personnel at the level of developed democratic states that meets the requirements of high spirituality and morality. The content of the research includes issues of the development of society and education, the regulatory framework of public education and legislative norms, advanced educational technologies and pedagogical skills, the use of information and communication technologies in the educational process, modern approaches and innovations in teaching practical Russian language, modern methods of organization educational process and is aimed at the formation of knowledge, abilities, skills, qualifications and competencies. At the present time, there is a need for a new teaching model, built on the basis of modern pedagogical technologies, which implements the principles of student-centered education. Pedagogical technologies, considered as one of the components of a holistic learning system, not only facilitate access to information, open up opportunities for variability of educational activity, its individualization and differentiation, but also allow to organize the interaction of all subjects of learning in a new way, to build an educational system in which the student was would be an active and equal participant in educational activities.

The introduction of new pedagogical technologies in the educational process makes it possible to activate the learning process, implement the ideas of developing education, increase the pace of the lesson, and increase the volume of students' independent work. Education based on pedagogical technologies meets the following principles:

the principle of modularity ensures differentiation and individualization in teaching, independence of students, grading, variability, structuring, blocking of educational material; the principle of systematic dosing of educational material, consistency of work with it, economy of study time, generalization of training; the principle of problematicity includes a conglomeration of the following components: reflection, correction of actions when applying rules in new writing conditions, motivation, conscientiousness, strength, consistency of conclusions when faced with conflicting and inconsistent facts from the field of spelling; the principle of cognitive visualization contains the requirements of the principles of accessibility, visibility, cognitive activity, aesthetic and emotional orientation of training. The introduction of pedagogical technologies into teaching practice is carried out through the use of software and pedagogical tools. Existing modern teaching aids should be included in the educational process not by chance, not in isolation from each other, but in a certain system. The tasks of modernizing education cannot be solved without the optimal introduction of pedagogical technologies in all its areas. The use of pedagogical technologies gives impetus to the development of new forms and contents of traditional types of activity, which leads to their implementation at a higher level.

The modern teacher is faced with the problem of finding a new pedagogical tool. In this regard, there is an increasing need for language teachers who are proficient in pedagogical technologies, who are able to design, model new ideas and directions in the teaching practice of higher education.

Considering pedagogical technologies for the development of the speech competence of future specialists in non-linguistic universities, we note that the issue of the possibilities of technologizing the educational process for many years has been the subject of scientific disputes due to the probabilistic nature of its course. However, scientists agreed that the educational process can be recognized as taking place at the technological level if reproducibility and measurability of its results are guaranteed.

In contrast to standardized content, teaching techniques can vary. Depending on the set learning goals and their various combinations, a certain set of learning models is formed. Pedagogical technology is a project and implementation of a system of sequential deployment of pedagogical activities aimed at achieving the goals of education and personal development of students, guaranteeing the reproduction of learning outcomes regardless of the teacher's personality. The quality of training students of non-linguistic universities for professional activities largely depends on personality-oriented, developmental, teaching and educational technologies. This takes into account the psychological and pedagogical characteristics of students, their motivation to obtain a profession.

The theories of pedagogical systems and pedagogical technologies are devoted to the works of V.P. Bespalko, V.V. Guzeeva, V.I. Zagvyazinsky, M.V. Klarina, V.V. Kraevsky, H.H. Mikhailova, G.K. Selevko, S.A. Smirnov and others. The technologies of students' project activities are reflected in the works of E.S. Polat, M.B. Pavlova, M.B. Romanovskaya, I.D. Chechel; in the concept of problem learning by Yu.K. Makhmutova, in the classification of teaching methods Babansky, M.I. and others.

The works of Professor M.Kh. are devoted to the issues of pedagogical technologies and pedagogical skills in national education. Tokhtakhodjaeva, Doctor of Pedagogical Sciences, Professor V.I. Andriyanova, M.G. Voinova, as well as Russian scientists such as B.L. Zankova, V.V. Davydova, P. Ya. Galperin, D.B. Elkonin and others. The work of V.G. Aseeva, L.I. Bozovic, A.K. Markova, A.A. Reana, P.M. Jacobson and others.

The works of foreign and domestic authors Yu.E. are devoted to the issues of self-actualization, personal orientations. Alyoshina, L. Ya. Gozman, M.V. Zagika, M.V. Crozat, A. Maslow, R. May, K. Rogers, F. Pearl, E. Shostrom and others.

Ideas, methods and approaches in the field of continuous, multilevel professional education were developed in the scientific works of A.T. Glazunov, E.F. Zeer, I.P. Smirnova, E.V. Tkachenko, F.T. Khamatnurov and others.

3 Discussion

The main purpose of the study is to develop and improve effective methods of using pedagogical technologies for the development of speech competence of future specialists from non-core universities. Modern pedagogical technologies serve as a tool for the implementation in practice of new learning systems that have higher educational characteristics in comparison with traditional didactic systems. On the one hand, they focus on achieving the required level of efficiency and quality of education, on the other, while maintaining frontal forms of education as the main ones, they provide each student with the opportunity to study according to individual programs, taking into account in full his cognitive abilities, motives, inclinations and other personal quality. The accompanying task of the research is to characterize the applied pedagogical technologies of teaching speech skills within the framework of the course «Culture of speech» at the faculties of "Mining" and "Energy mechanics" of the Navoi State Mining Institute. The methodological basis of the study was the state documents on the education of the Republic of Uzbekistan, the works of domestic and foreign authors on this issue.

Research methods. The following methods can be used at various stages of the study:

- theoretical - understanding the problem and the subject of research, modeling the pedagogical process, theoretical generalization of research results, taking into account specific conditions and new facts;

- empirical - comparative-comparative analysis of the results of ascertaining and formative experiments, observation, conversations, questionnaires, pedagogical experiment;

mathematical - registration, mathematical and statistical processing of the results.

The scientific novelty of this paper is that:

- the work provides a substantiation of the role and place of pedagogical technologies in the classroom on the subject «Culture of speech» for students of non-linguistic universities;

- the principles of using pedagogical technologies as effective methods contributing to the development of speech skills of students of non-linguistic universities are identified and substantiated;

- the set of pedagogical technologies has been identified and substantiated, in which the foundations of the speech communicative competence of students are formed;

The theoretical significance of the study is that it:

- the organizational and substantive parameters of the implementation of the tasks of pedagogical technologies - training at a university were theoretically comprehended;

- the possibility of using pedagogical technologies in teaching language skills is substantiated at the scientific level.

The practical value of the study lies in the fact that a scientifically grounded and experimentally proven method of teaching speech skills using pedagogical technologies helps to increase the general and subject level of learning, can be used in the development of methodological techniques for mastering speech skills; research materials can serve as the basis for the creation of an educational and methodological complex and a textbook on the development of speech skills for specialties in non-core universities.

As in other countries, higher education in Uzbekistan involves the training of qualified specialists for various spheres of public life and sectors of the economy - scientific, economic, technical and others. The educational process involves the systematization of knowledge and acquired skills, orienting students towards solving theoretical and practical problems in the vector of the chosen specialization with the creative use of the achievements of modern thought and technology.

The state educational standard of higher education defines:

general requirements for the quality of training, the content of education;

necessary and sufficient level of preparedness of students and general qualification requirements for graduates of educational institutions.

The main feature of the program is the continuity of education. Each person has the opportunity to acquire knowledge, professional skills and specialties throughout his life. This program creates an effective mechanism of the educational system, in which the main components are present:

- the formation of a free-thinking person, his moral, spiritual and physical development;

- respect for the individual, disclosure of creativity and abilities;

- progressive training, obtaining professional skills and full self-realization of the individual in life.

In this regard, I would like to draw your attention to such a concept as rhetorical competence. It is no secret that the effectiveness of the assimilation of material by students is influenced by the availability, clarity, and accuracy of teachers' use of turns of speech, rhetorical techniques, which may include not only speech figures, but also elements of artistry. But do all teachers possess this knowledge and skills: the skills of public speaking, building effective speech. Unfortunately, we cannot give a positive answer to this question with complete confidence. Students often face cases of poor teaching of courses. And this is not about teaching methods, but about the speech skills of teachers. And this happens not because the teacher does not know his subject, he does not possess the skills of building coherent, logical, effective speech, which is the main obstacle for a fruitful dialogue "teacher-student". The solution to this problem could be the inclusion of such a subject as rhetoric in the obligatory part of the curricula of universities in all areas without exception.

Today rhetoric is no longer even an academic discipline, but a professional one. In the modern world, people who have the skills of oral speech, both public and dialogical, prosper. And to a greater extent, a teacher must have such skills - the ability to logically correctly, reasonably and clearly build oral speech. These skills are developed during the study of the course "Rhetoric".

Today on the Internet you can find many links to various schools of public speaking. People are interested, trainings take place. Therefore, there is demand. People want to communicate competently, build dialogues, speak in public, convey their thoughts to the audience, convince and convince. And this fact confirms the necessity of introducing the Rhetoric course into the curricula of universities, particularly in pedagogical universities.

Rhetoric is one of the most ancient philological sciences, in European culture it originates from the ancient Greek rhetorical schools, where it was formed as an academic subject, later as a science. Initially, the subject of rhetoric was live speech, verbal skill, eloquence. The first works on rhetoric were written in the 5th-3rd centuries BC (Gorgias, Lysias, Plato, Aristotle).

Oral speeches did not reach us, but the doctrine of these speeches was expounded in one of his dialogues by Plato - a disciple of Socrates and the greatest thinker of the ancient world. In rhetorical teaching, the most important attention was paid to the content of speech, its construction, verbal expression, memorization and mastery of public speaking, speech was closely related to poetics, the doctrine of styles, logic, philosophy.

In Rome, poetry, science, and rhetorical skill were highly valued. The rhetorical schools in Rome flourished, Mark Tullius Cicero, Julius Caesar, Mark Junius Brutus are still recognized as unsurpassed orators.

The social status of the speaker rises, recognition of his role in social life reaches extraordinary heights. It was at this time that entire treatises on the speaker were written. The supreme example of this is M.T. Cicero (106-43 BC) combined the skill of an orator, a poet, a statesman, and the scholarship of a rhetorician. His works - "On the orator", "Orator", "Brutus". He is both a philosopher and the author of the treatises On Fate and On the Nature of the Gods. Cicero asserts: the properties of an orator are activity, courage, sociability, versatility of interests, an orator is a thinker, and in anxious situations - a warrior.

The role of rules, training in the field of eloquence is increasing. Knowledge of rhetorical techniques and lengthy exercises become necessary. This line in Roman rhetoric was especially elaborated by Marcus Fabius Quintilian (AD 36-96). The famous "Twelve Books of Rhetorical Admonitions" by Quintilian have no analogue in the history of rhetoric, they are still used by all orators.

The famous German historian, author of the multivolume “History of Rome” Theodor Momsen wrote about this work of Quintilian as follows: “The textbook of rhetoric compiled by him ... is one of the best works that have come down to us from Roman antiquity; he is distinguished by delicate taste and fidelity, simplicity of feeling and image; lively and relaxed, instructive and exciting; in general, all this work is, according to the author's own idea, the complete opposite of the abundant phrases, but poor in thought, the literature of that time ”[3]. The essence of rhetoric, according to Quintilian, is the union of thought and word. The aesthetic factor plays an important role in the skill of eloquence. Cicero put speech at the head of all things, the laws of nature, the duties of people, for knowledge of all these things enters the life of people through speech. Cicero writes: “Before whom do people tremble? Who are they looking at, shocked when he speaks? Who is admired? Who is considered almost a god among people? The one who speaks harmoniously, shining with bright words and vivid images, introducing even into prose itself a certain poetic dimension, in a word, beautiful” [6].

Speaking about higher education today, we mean the formation of personal competencies, which are formed in the process of mastering the humanitarian, social, psychological and pedagogical disciplines and include: communication, the ability to conduct training, counseling, research, forecasting, application of knowledge in practice, etc., therefore , communicative competence is an integral part of the competence approach in determining the content of education.

At the present stage of education, special attention is paid to the formation of communicative competence. The structure of communicative competence includes the following elements: linguistic, linguistic, speech, cultural and rhetorical competence.

In this article, we will consider the issues of forming the rhetorical competence of specialists from non-core universities.

The very concept of "competence" is associated with the modernization of the higher education system in line with the competence approach.

However, most researchers consider only one or several aspects of pedagogical activity, and, therefore, pay attention to the formation of one or another quality that is part of professional competence.

Initially, the term competence was used in pedagogy only in relation to the study of a foreign language and the ability to express one's thoughts by means of this language.

The term "competence" was first used by N. Chomsky, a professor at the University of Massachusetts (United States of America) in 1965. By competence, he understood "the speaker's knowledge of the language", distinguishing it from the "use of language." However, there is still no unambiguous understanding of the term “competence”: “... on the one hand, this is the goal of learning, on the other, its content, on the third, the result, and on the fourth, a means of achieving the learning outcome”.

The competence of a specialist in a particular area forms the specialist's readiness to perform a professional role and makes it possible to solve production problems. And since the solution of professional tasks in many areas is impossible without building a system of communication links and effective communication, we can confidently assert that without the formation of rhetorical competence, full-fledged training of a specialist is impossible.

And then, rhetorical competence can be defined as a body of knowledge about communication, which is realized in the sphere of increased speech activity of a specific professional or social group; rhetorical skills that can be applied in standard and non-standard speech situations.

Rhetorical competence is the highest level of communicative competence that ensures a person's rhetorical activity [7].

Rhetorical competence - the ability to create, pronounce the original text of a certain speech genre, to show empathy in accordance with the addressee's need, the purpose of the speech and the situation of public speech.

For the formation, development and improvement of rhetorical competence, it is necessary to master such rhetorical skills as:

- invention of thought;
- the formation of goals, objectives, the choice of a genre and the corresponding type of speech for effective speech communication;
- definition of the type of addressee, especially communication with him;
- the disposition of thought, the logic of presentation, the choice of the most acceptable way of expressing consent, expressing refusal, refutation;
- determining the appropriateness of the use of a particular language unit in a given speech situation;
- skillful use of rhetorical techniques for effective communication with the aim of influencing, convincing the addressee.
- skillful use of intonational highlighting of keywords, etc.

The formation of speech culture as an integral part of professional culture includes communicative competence, which allows you to work successfully in the chosen field of activity.

The effective formation of the rhetorical component of the communicative competence of students in non-core areas of training is possible with the implementation of the system-activity approach in teaching.

Successful verbal behavior is not only correctness, accuracy, consistency, it is also non-verbal influence as a means of transmitting information; general requirements for the appearance of the speaker, his clothes, hairstyle, demeanor in front of an audience; this is the use of facial expressions and gestures, etc.

The concepts of rhetoric and culture of speech are close in meaning in relation to rhetorical competence.Today, the curricula of non-core universities of the Republic does not include the discipline "Rhetoric", the content and goals which are aimed at mastering not only philological knowledge, but also mastering speech skills, forming rhetorical competence. It is necessary to develop programs for the gradual formation of the rhetorical competence of future specialists from non-core universities, a set of means and methods, techniques for the formation of culture, etc.

The success of a modern person in any activity directly depends on his communicative competencies.

By developing the skills of independent expression, we teach our young people to think critically, create, remember, speak, because it is these forms of activity that ensure the successful implementation of all other skills and abilities of an educated person.

Modern society and the state impose increased requirements on the level of training of university graduates who possess the necessary professional competencies at a high level, including communicative competence. New tasks for the formation of personality have been set for education today. "Speech education" of a university graduate affects his competitiveness in the labor market, successful socialization in the future. However, in recent years, teachers, psychologists, linguists, specialists in other fields of knowledge have noted that the level of students' proficiency in speech competence is not high enough and does not fully meet the requirements for a specialist in any field today, although this competence is the most important component of professional training.

This means that a university graduate must be able to choose a strategy and tactics of communication, establish verbal contact and adjust his behavior in accordance with the situation and communicative intention; be able to reasonably state your point of view and conduct a discussion in accordance with the principles and rules of a constructive dispute; master the tactics of countering manipulative speech influences.

It should be noted that at present, the problem of developing the speech competence of students of non-linguistic universities in the classroom in language disciplines has not been sufficiently developed, because these disciplines are aimed at studying the structure of the language, comparative analysis, translation, etc., but not teaching speech skills. Here it is appropriate to refer to the classifier of directions and specialties of higher education of the Republic of Uzbekistan. The area of the specialty "Humanities" is represented by the following areas:

- Philology and language teaching (by language);
- Literary criticism (by languages and areas of research);
- Linguistics (by language);
- Classical philology (by type);
- Textology and literary source studies;
- Theory and practice of translation (by language);
- Comparative linguistics, linguistic translation studies (by language).

It also establishes the following requirements for the content of educational disciplines in the humanitarian block:

- develop and supplement the knowledge gained on the basis of general secondary and secondary specialized, vocational education; to form a scientific and humanitarian worldview, high spirituality and democratic culture, economic, legal and creative thinking, conviction and social and political activity based on the idea of national independence and democracy, national and universal values;
- ensure the focus of education on its organic unity with history, philosophy, folk traditions and customs; preserve and enrich the culture of the peoples of Uzbekistan, respect for the history and culture of other peoples;
- develop humanism, patriotism and internationalism;
- to ensure the orientation of training and education on the formation of an independently thinking, capable of making decisions, comprehensively developed, harmonious personality.

4 Results

Proceeding from these requirements, it would be advisable to add the direction of "speech education" to the existing directions of the humanitarian block. it is clear that in order to teach the speech skills of specialists in non-linguistic universities, specialists with speech education are needed who are able to teach the skills of speech interaction, methods of argumentation and refutation, as well as knowledge of the basics of speech culture, on which the skills of conducting a cultural dialogue, orientation in an ever-increasing flow of information are based; able to teach the construction of public speaking and debate, etc. And to solve this problem, we must solve a number of tasks:

- to identify the components of speech competence, the assimilation of which involves the development of students' ability to build a speech work in accordance with a specific situation;
- to create a structural model for the development of students' speech competence, including a competence-based and personality-oriented approach, forms of education, teaching aids and assessment tools;
- to develop a set of tasks / study guide for the development of speech competencies of specialists in non-linguistic universities.

In our country, improving the system of continuous education, improving the quality of educational services is a priority task outlined in the Decree of the President of the Republic of Uzbekistan "On the Strategy of Actions for Further Development". Consequently, ample opportunities have been created for conducting practical research to improve the level of methodological training of future specialists, ensuring the integration of the personnel training system with the priority areas of development of science and technology.

The "Strategy of actions for the further development of the Republic of Uzbekistan" identified priority tasks for improving work in the field of state youth policy, bringing up a harmonious, intellectually developed young generation, bringing the quality of education in higher educational institutions to the level of compliance with the current and future needs of the individual, society and the state.

The solution to this problem could be the introduction of a practical course of rhetoric into the curricula of non-core universities, and as part of the course, the creation of textbooks and teaching aids in accordance with the qualification requirements for a university graduate in the Republic of Uzbekistan.

And here we must understand that rhetoric is not a linguistic course in the Russian language. Rhetoric is distinguished by its goals, objectives, content and structure. In this regard, we are faced with an urgent and socially significant task of creating modern pedagogical technologies for the development of speech skills of specialists. And these technologies should contribute to the development of the personal orientation of the educational process. And here the question already arises about not "what to teach?" But "how to teach?" (A.B. Khutorskoy).

Speech skills are formed only as a result of purposeful work based on special methods and techniques. This work can be realized only within the framework of a special academic subject with its own content, logic of presentation and technology.

Today, as never before, the issue of "speech education" has become relevant. We must understand that along with professional skills (hard skills); we must improve speech skills (soft skills).

As you know, the activities of educational institutions are aimed at mastering hard skills, and there are certain levels of difficulty in mastering professional skills. So why are there no mastery levels for mastering soft skills?

There are certificates and diplomas to confirm hard skills. Soft skills are not certified and can only be proven in practice. Soft skills are learned more slowly than hard skills. To master Soft skills, systematic practical exercises are required.

For this, it is necessary to improve the methods of teaching speech disciplines. As you know, there are two areas of research in linguistics: in one area, language systems are studied, in the other, speech. Unlike linguistics, rhetoric deals only with speech, not language.

In our work, we set the task of improving speech skills, i.e. the object of study is speech activity, speech communication and behavior. Modern rhetoric is not the study of the structure of the language, and not even the study of the culture of speech as such, rhetoric combines knowledge about a person from the theory of communication, psychology of communication, linguistics, ethics, sociology, logic, philosophy, etc.

Today there is a lot of literature on rhetoric, scientific works. We, given the mentality of our society, must adapt it to the new needs of our time.

A specialist meeting modern requirements should know:

- features of speech use in writing;
- basic principles and rules of effective speech communication, stages of creating a speech work;
- basic principles and rules of effective speech communication;

Be able to:

- choose a strategy and tactics of communication, establish verbal contact and adjust their behavior in accordance with the situation and communicative intention;
- build a monologue speech, guided by the rules of the rhetorical canon;
- to reasonably state their point of view and conduct a discussion in accordance with the principles and rules of a constructive dispute;
- analyze speech actions from the point of view of their success / failure and draw practical conclusions from this analysis.

Own:

- skills of effective use of expressive means of the Russian language in different communication situations;
- tactics of countering manipulative speech influence;

- terminological apparatus of modern rhetoric (neorhetoric)
- must demonstrate the ability and willingness to apply the acquired knowledge in practice.
- to achieve the set goals, it is necessary to solve certain problems, which includes:
- improving the technology of developing the speech competence of future specialists;
- creation of a model for the development of speech competence of students of non-linguistic universities, as well as the development of a substantiated and special set of practical tasks / tutorial for the development of speech competence.
- development of a structure for improving technologies for the development of speech competence of future specialists.

5 Conclusion

Today there are a number of technologies that contribute to the development of skills and abilities. These are the so-called interactive teaching methods. Relevant in modern teaching methods is the use of interactive teaching methods. But this does not mean that teachers of language disciplines should give up everything that has been accumulated over many years. The specificity of language learning is such that it is a means of studying all other disciplines.

The goals of language learning, as you know, are personal, meta-subject and subject, which in turn have such sub-goals as the ability and readiness for self-development, interdisciplinary knowledge, cognitive, communicative, etc.

The most effective in teaching the speech aspect of the language is the competence-based approach. When developing competencies, we must pay special attention to the formation and mastery of various types of speech activity; the basics of the culture of oral and written speech; implementation of the speech-thinking process; creation and reproduction of speech in accordance with the target communicative setting. Also important is the cultural approach, i.e. mastering the language by mastering the norms of speech etiquette, the culture of interethnic communication, reflecting the historical experience of the people.

The teacher's innovative activity is designed to search for new interactive teaching methods. The main techniques of the interactive methodology are language learning in the community, learning in collaboration.

Various teaching methods can be used here. Yu.K. Babansky offers methods of organizing educational activities; methods of stimulating and motivating cognitive activity; method of control and self-control. A.V. Dudnikov offers inductive and deductive methods and their combination.

L.P. Fedorenko offers such a method as sources of knowledge, which uses such techniques as questions (direct analytical, synthetic, rhetorical, suggestive, etc.); reception of drawing up a plan on the go; recording the main provisions of the message; compilation of reference notes, tables, diagrams, etc.

A number of methodologists offer analysis of texts, presentation, composition as a type of work on the development of speech.

M.I. Makhmutov provides a rationale for problem learning as a didactic system and type of learning. The problem learning method is presented in the form of a structure: problem task + problem situation.

It is a flexible combination of different methods of teaching a language based on a culturological aspect, applied in the classroom together with a variety of teaching technologies, that will help motivate students and become effective in mastering languages in the future. At the same time, common to all interactive methods is the use of individual and group activities, and, of course, the collective work of students is a compulsory activity.

The use of interactive methods develops a culture of participation in dialogue, a culture of constructive debate, competent argumentation, helps to establish emotional contacts between students, develop creativity, the ability to think outside the box and be able to defend their interests, form teamwork skills, providing high motivation in self-development and self-study, active life position, disclosure of creative potential, which is an integral prerequisite for the formation of speech skills.

To date, an analysis of rhetorical knowledge shows how relevant this problem is and is not yet sufficiently understood and studied in the educational process.

References

1. A.S. Belkin, V.V. Nesterov Pedagogical competence. Ekaterinburg, (2003)
2. N.L. Goncharova, *Categories "competence" and "competence" in the modern educational paradigm*. Collection of scientific papers of the North Caucasus State Technical University. Series "Humanities.", **5**, (2007)
3. A.V. Khutorskoy, *Technology for designing key and subject competencies*, Eidos Online Journal, (2005)
4. L.N. Gorobets, "Pedagogical rhetoric": theory and practice of teaching non-philological students of rhetorical competence: monograph. - Armavir: Publishing and Publishing Center of ASPU (2007)
5. L.P. Aksenova, Fundamentals of the rhetorical culture of the teacher. Ekaterinburg (2002)
6. Appendix No. 2 to the Resolution of the Cabinet of Ministers of the Republic of Uzbekistan dated August 16, 2001 No. 343 "On the Approval of State Educational Standards of Higher Education".
7. V.I. Annushkin, Rhetoric. Introductory course: textbook. M.: Flint; Science, 224 (2018)
8. Y. V.Vorsina, T. A. Snigireva, *To the Issue of the Competence For- mation Criteria*, The Education and Science Journal DOI: https://doi.org/10.17853/1994-5639-2012-10-67-78

9. O. Ukpokodu, *Developing Teachers' Cultural Competence: One Teacher Educator's Practice of Unpacking Student Culturelessness*, Action in Teacher Education, 432-454, **33** (2012) DOI: https://doi.org/10.1080/01626620.2011.627033

10. K. Cushner, *Teacher Education*, The SAGE Encyclopedia of Intercultural Competence, (2017) DOI: http://dx.doi.org/10.4135/9781483346267.n264

11. Fritz K. Oser, F. Achtenhagen and U. Renold, *Competence Oriented Teacher Training: Old Research Demands and New Pathways*, Competence Oriented Teacher Training, Brill Sense (2006) DOI: https://doi.org/10.1163/9789087903374_002

12. B. Cherednichenko, N. Hooley, T. Kruger, R. Mulraney, *The practice of beginning teachers: Identifying competence through case writing in teacher education*, Australian Journal of Teacher Education, **22** (1997) DOI: http://doi.org/10.14221/ajte.1997v22n2.3

13. R. Gillies, *Teachers' and Students' Verbal Behaviours During Cooperative Learning,* The Teacher's Role in Implementing Cooperative Learning in the Classroom, 238-257 (2007) DOI: 10.1007/978-0-387-70892-8_12

14. John H. Sloan, *Understanding McLuhan: Some implications for the speech teacher and critic*, The Speech Teacher, **17**, 2 (2009) DOI: https://doi.org/10.1080/03634526809377665

15. Elizabeth Norwood A rationale for effective teacher education in speech communication, (2009) DOI:https://doi.org/10.1080/03634527509378151

16. M. Hochmuth, *Great teachers of speech*, The Speech Teacher, **4(3)** (2009) DOI:https://doi.org/10.1080/03634525509376741

17. III. Wayland Maxfield Parrish, *Great teachers of speech*, The Speech Teacher, **4(3)** (2009) DOI:https://doi.org/10.1080/03634525509376741

18. R. K. Tucker, The speech teacher in American industry (2009) DOI:https://doi.org/10.1080/03634526009377092

19. M. Kinnane, Promoting speech education, (2009) DOI: https://doi.org/10.1080/03634526609377547

20. M. Ogilvie, Student teaching in speech (2009) DOI:https: //doi.org/10.1080/03634525309376586

21. A.A.Volkov, *The course of Russian rhetoric: a manual for theological educational institutions*, Russian linguistic journal, **2(6),** 35 (2016)

22. A.A. Volkov, Theory of rhetorical argumentation, KDU Publishing House (2013)

23. I.M. Osmolovskaya, *Innovation and teaching practice*, Public Education, **6**, 182-188 (2010)

24. N.N. Simonenko, *Management of educational services using innovative teaching methods*, Bulletin of the Pacific State University, **2,** 201-206 (2012)

25. L.V. Cherepanova, Reflective approach in teaching Russian language, Science (2006) DOI: https://doi.org/10.23683/1995-0640-2017-3-202-215.

26. Appendix No. 2 to the Resolution of the Cabinet of Ministers of the RUZ dated August 16, 2001 № 343 "On the approval of state educational standards of higher education."

Economic Culture of Students' Communication in Market Conditions

Abdukhakim Mamanazarov[1*]

[1]Tashkent Branch of M. V. Lomonosov Moscow State University, Tashkent 100060, Uzbekistan

Abstract. The article substantiates the pedagogical need to develop economic culture among students as Uzbekistan is entering the developed market economy; develops classification and functions of economic culture in student's personality; determines parameters of economic culture development among youth; reveals components and criteria of economic culture among students of higher education institutions; presents methods of economic knowledge expansion; determines culture indicators of entry into economic relations; specifies the content, means, forms, methods and technologies of economic culture development among students; facilitates mechanisms of pedagogical correction based on assessment and diagnostics of the economic culture level formed among students; presents statistically derived methodology of organizing experimental work and its efficiency aimed at developing economic culture among students; designs the research and methodological guidance for improving the training process aimed at developing economic culture among students.

1 Introduction

In the context of globalization, the need to modernize economic education at higher school in terms of personal economic culture development has increased. Enrichment of the economic education content creates conditions for economic culture development and economic knowledge acquisition. Reforms aimed at economic education modernization, its integration with world educational standards, social and economic development are being implemented in such developed countries as the United States, the Russian Federation, China, Germany, France, Japan, South Korea and Singapore. Raising the level of economic culture of society, adapting to economic activity, changing attitudes towards labour, entrepreneurship and business require the introduction of modern pedagogical technologies in the training process.

The results of economic education analysis in the world and our country show that updating of the economic education content in accordance with modern requirements, and pedagogical process enrichment based on the Concept of National Culture are important for developing students' economic culture essential in their preparation for professional activity. In the system of higher education, teaching various topics related to economic issues creates the basis for economic culture formation among participants of the training process. The development of future specialists' economic culture will become the basis for formation of professional competencies in their future activities. As a result, this will have a positive impact on socioeconomic processes in society.

Socioeconomic changes in Uzbekistan, development of market economic relations require formation of high moral economic culture based on the principle of acceptability through the economic knowledge gaining by young people. There is a problem of cultural and spiritual relationship coordination of a younger generation under various effects produced by the developed market economy. The Strategy of Actions for the Further Development of the Republic of Uzbekistan is focused on upbringing fit, mentally and intellectually developed young people, thinking independently, loyal to the motherland, facilitating democratic reforms and increasing their social activity for developing civic society, which, in turn, also shows future specialists' contribution to economic culture development.

The research is aimed at developing technologies for economic culture formation among non-economic students at higher education institutions under conditions of established market relations and improving the mechanisms of their implementation.

1.1 Review of foreign scientific researches

Researches into economic culture development among youth related to formation of their economic knowledge, skills and competencies are carried out in leading research centers and institutions of higher education, for example, Lomonosov Moscow State University, St. Petersburg State University, Moscow State Pedagogical University (Russian Federation), University of Chicago (the USA), Belarus State Economic University (Belarus), Polonia University in Czestochowa (Poland), Kiev National Economic University named after V. Getman (Ukraine) and Rezekne Academy of Technologies (Latvia).

Some results are achieved thanks to global research into the economic culture formation among student

* Corresponding author: abduhakim_bazarovich@mail.ru

youth. In particular, methods of economic culture formation among young people are developed (Lomonosov Moscow State University); directions of students' economic knowledge enrichment responding to social requirements are defined (St. Petersburg State University); the role of new pedagogical technologies in teaching economic subjects is investigated (Moscow State Pedagogical University); market relations impact on making relevant economic decisions and their social advantages and disadvantages are studied (University of Chicago); psychological and pedagogical conditions that ensure efficiency of economic culture formation among students are revealed, practical guidance is presented (Belarus State Economic University); scientific approaches to the socioeconomic and political structure of society and its impact on economic culture are developed (Polonia University in Czestochowa); the concept of consumption culture is integrated into the economic theory and social sciences, and a model of developing consumption culture among future consumers (youth) is created (Kiev National Economic University named after V. Getman); the mechanism combining such students' skills as tourism, services, entrepreneurship and business administration with practice is developed (Rezekne Academy of Technologies).

Development of students' economic culture in leading world universities is carried out in the following priority areas: improving pedagogical mechanisms of organizing economic education activities in the higher education system based on innovative approaches; developing economic culture of future specialists based on the requirements of developed market relationship; improving the content of economic education by means of expanding opportunities for university students to adapt to the economic life of society. Research is conducted to develop students' understanding of the human capital nature and to form competencies for expanding the scope of social relations on this basis.

1.2 Literature review

The issues of improving the economic education content through expanding adaptability to the economic life of society are studied by economists, educators, psychologists and philosophers, namely by E. J. Dolan [1], A. B. Mamanazarov [2], James D. Gwartney, Richard L. Stroup, Russell S. Sobel, David A. Macpherson [3], Paul L. Heyne, Peter J. Boettke, David L. Prychitko [4], Ulrich Ph., Lehr U. [5], A. H. Maslow [6], R. B. Reich [7], D. R. Shaffer [8], Stuart H. Brody [9], D. Hirsch [10] and F. Cornali, S. Tirocchi [11].

In the works mentioned above, some research is carried out in the field of creating a stable attitude towards human capital, preparing for work in the developed market environment, enriching economic knowledge and thinking of students. However, the topic Development of Economic Culture among Students of Higher Education Institutions (Taking Non-Economic Students as an Example) as an independent object of research has not been studied yet.

1.3 Research methods.

In the course of the research, methods of theoretical and empirical analysis, observation, diagnostics, interviewing, questioning, modeling, pedagogical experiments, mathematical statistics, and generalization of results are used.

2 Theoretical and methodological principles of students' economic culture formation

Graduates of higher education institutions must/should have a certain level of economic culture because they enter into economic relations in the course of work in various spheres of public life (Fig. 1).

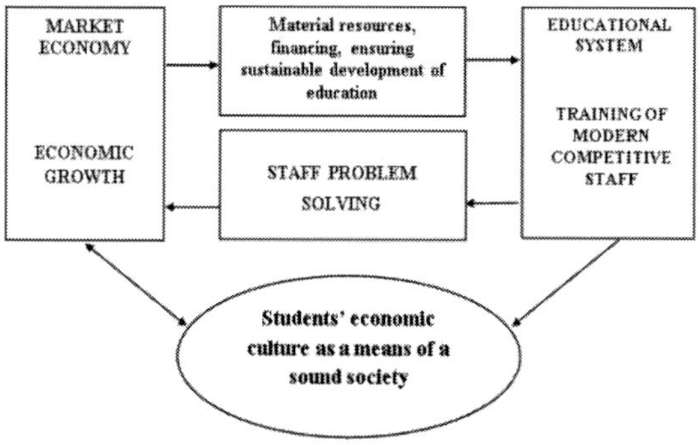

Fig. 1. Interdependence between economy and education during economic culture formation.

The conditions of modern communication substantiate the scientific need for training specialists possessing up-to-date economic, political, legal and professional knowledge which is successfully applied for the benefit of society including: development of curricula for consistent provision of economic knowledge of students in all higher education institutions; providing the educational process with educational materials and didactic projects that combine modern economic knowledge; providing the teaching staff of higher education institutions with the latest economic knowledge during their continuing professional development.

In modern scientific literature, economic culture of communication is studied from a philosophical, social, economic, psychological and cultural point of view. In fact, it reflects the economic behaviour of subjects, which is determined by the laws and principles of the economic system.

Economic culture is a subjective expression of objective reality based on creation of a certain type of economic behaviour that leads to practical goals and actions in the economic sphere being a pursuit of personal interests.

Due to the fact that development of students' economic culture is a complex and multifaceted process, it is impossible to successfully complete this task during academic hours allocated for the course Economic Theory. For non-economic education, it is possible to organize an optional course called Basics of Personal Consumption and Savings. Its programme, forms and methods of teaching should be based on the requirements of modern economic culture.

In order to improve the process of economic culture development, it is required: firstly, to present economic culture specifying its interdisciplinary relations and emphasizing its economic content; secondly, to enrich Economic Theory course with materials that characterize economic culture; thirdly, to introduce an elective course called Basics of Personal Consumption and Savings in the training process; fourthly, it is recommended to introduce modern pedagogical technologies and interactive methods in all universities contributing to development of economic culture among students.

Today, the following standards and principles of economic culture are important: freedom of property, freedom of entrepreneurship, active high-quality labour, the use of safe technologies, labour efficiency, humanism of economic activity, innovation, moral and legal basics of economic activity, forecasting, etc. They reflect important features and processes of cultural expression in the relationship of business entities.

In this regard, the leading indicators of society's economic culture are: the worldview and professional competence of the subjects involved in the economy; key elements of the economy, qualitative and quantitative classification of growth rates; optimality and stability of the state economic infrastructure; diversified, developed and reliable financial system in accordance with subjects of society; variety of forms of ownership and associated economic mobility; variety of government forms that satisfy mutual interests of a person, society and a state; criteria for moral and legal regulation of the economic relationship system; compatibility of operating culture, material goods and citizens' consumption culture; research and development, humane and ecofriendly economic activity of society.

The components of a person's economic culture include:

1) knowledge (a set of economic ideas about production, exchange, distribution and consumption of material goods) and practical skills;

2) economic thinking (allows for understanding the essence of economic events and processes, putting into practice mastered economic concepts, analyzing specific economic conditions);

3) economic orientation (needs, interests, and motives arising from human activity in the economic field);

4) methods of administration;

5) rules regulating human relations and behaviour (thriftiness, discipline, wastefulness, irresponsibility, extortion, fraud, etc.).

The need to develop students' economic culture is explained by the following:

- social and individual development of students, i.e. moral skills required for solving many problems in daily living, formation of spiritual and economic qualities, development of personal resources, motives, as well as self-knowledge, the ability to adequately assess the readiness for various socioeconomic activities;

- formation of a future specialist's economic attitude, i.e. economic issues of macro- and microeconomic realia understanding, upbringing of an owner – a leader capable of making correct and informed decisions when planning a family household;

- promoting awareness of civic duties and responsibilities, in particular, those of a citizen of a democratic state able to analyze cause-and-effect relationship of economic processes and form the ability of impacting social problem solving with personal views.

In our opinion, goals of economic education at higher education institutions include formation of economic thinking basics among students allowing them to avoid emotional, inconsistent ideas about the issues of the state economic policy; creation of favourable conditions for students' entry into relationship of the developed market economy in order to understand the changes taking place and expected in society; training of an economically literate citizen who is able to analyze economic processes and take an active part in economic activities of the republic; formation of an economic point of view that justifies their actions; acquisition of the ability to make economic decisions independently (as a citizen, an employee, a consumer) in practice, development of the ability to master and apply economic knowledge to analyze and explain economic events and situations.

2.1 Pedagogical tools for developing students' economic culture

Development of criteria and their indicators creates certain difficulties due to the fact that the process of forming economic prospects for students is complex and multifaceted. The existing criteria include acquisition of theoretical knowledge; possession of the economic worldview; development of economic thinking; readiness for effective economic activity.

Criteria for formation of economic attitudes among students are availability of economic knowledge; understanding of the essence of economic processes; interaction with subjects of the economic environment and influence; free orientation to the current economic environment.

The study results in identifying the following criteria for determining students' economic culture include availability of cognitive information; the system of values; involvement in intellectual and creative activities; practical work.

Economic knowledge acquired by students should cover two stages of activity. Nevertheless, it is required to selectively include knowledge necessary for students' professional and daily life activities in the curriculum. In most higher education institutions, basic economic knowledge is gained at the Economic Theory course. The level of economic knowledge development among students shows that essential economic skills are formed via case study seminars, heuristic and synergistic exercises, and moderation methods. These are determination of cause-and-effect relationship in economic processes; trend analysis of economic reality development in society; critical thinking; highlighting prioritized economic problems; the ability to analyze economic events.

The conducted practical and theoretical analysis suggests specific techniques, methods and technologies for developing economic culture among students. Using these methods, techniques and technologies we tried to develop a certain level of skills and competencies of our students:

Level 1 encompasses students able to reason about certain economic issues, make decisions, and present them adequately. These students can take an inclusive approach to problem solving.

Level 2 identifies students knowing how to reason and make decisions. Students will be able to solve a problem that requires an analysis of the suggested economic situation and make a decision by choosing specific options. Doing that they use different assumptions; summarize information from different sources; draw conclusions based on two or more sources of information.

Level 3 (Support) identifies students who can solve economic problems. Such students work with clearly defined information and cannot use other sources.

A level below level 3 identifies students who lack sufficient problem-solving skills. These students do not fully understand the problem and do not put enough effort into solving it.

The following methods can be used for development of economic culture among students:

1. A problem situation simulated to activate personal knowledge needed for its solution and facilitate students' thinking by setting corresponding educational tasks. The main function of the problem situation is to ensure educational material mastering when difficulties arise. The problem situation includes questions, figures, and graphs related to a phenomenon under study and ways to solve it in the form of incompletely stated ideas.

2. Generalization, coding, optimization of knowledge by using conventional signs, symbols, diagrams, graphs, tables as the main method of teaching and their subsequent deepening in student's mind, complete rethinking, thus providing interaction between a teacher and a student.

3. Independent search for knowledge as a teaching method that allows students to take into account existing knowledge and individual characteristics, set tasks that stimulate creative research, give advice on economic activity, and evaluate the results of the educational process.

2.2 Model of the pedagogical process aimed at developing students' economic culture

Modern methods of teaching economics are aimed at achieving the following goals: rapid economic knowledge acquisition; effective organization of economic processes; group economic knowledge acquisition by students; students' adaptation to the microcosm in the developed market economy; provision of insight into the economy of the republic, its current state.

The main features of interactive methods used for development of students' economic culture are:

1) free discussion;
2) free presentation of material;
3) the small number of lectures, the large number of seminars;
4) students' initiative;
5) group assignments;
6) continuous monitoring during a semester;
7) written assignments.

The above-mentioned methods, techniques and technologies promote formation of economic knowledge to a certain extent, which is the basis for students' economic culture.

In the course of experiments aimed at developing students' economic culture, the following shortcomings are identified: inconsistent systematization of modern economic knowledge in the programme/syllabuses, textbooks and study guides; insufficient implementation of pedagogical activities while involving students in economic relations; poor knowledge about operating economic concepts, categories, laws; a lack of creative approaches to solving economic problems and situations; difficulties in analyzing and applying new economic information; a low level of formation of emotional and value-based attitudes in the economy; a lack of self-confidence in making individual and collective decisions; insufficient formation of skills for assessing economic reality while applying knowledge to practice.

We rely on the following principles of pedagogical activity improvement aimed at economic culture development among students: consideration of specifics and dynamics of the developed market economy; the inclusive approach to providing students with economic knowledge; gradual improvement of curriculum requirements taking into account constant development of economic knowledge; strengthening practical aspects of economic culture development among students; the integrated innovative approach to organizing the process of developing students' economic culture.

Observations have shown that if economic activities are set, imitated and analyzed together with students, they can be effective for development of economic culture. For this purpose, economic education uses various trainings, business games, disputes, discussions, presentations, cases, work in small groups, etc. Individual work with students helps to form economic culture during the educational process through carrying out other types of extracurricular activities.

While working on components of cognitive and informational and practical activities, the implementation of corrective measures aimed at their economic culture development, the following aspects should be considered:

I. Academic work: including tasks aimed at developing economic, logical and analytical thinking of a problematic nature, which requires determining not only economic, but also social efficiency of decisions made in the content of academic disciplines.

II. Educational process: analyzing less complex cases aimed at developing the creative potential of students; introducing training activities that allow working on the project called Effective Use of Uzbekistan's Resource Potential within the framework of the course Economic Theory.

III. Extracurricular activities: expanding students' understanding of professions in economic and management fields; demonstrating remarkable examples of life achievements; highlighting valuable, important issues; conducting small researches within the student scientific community (preparing reports for conferences on economics, writing essays on interdependent economic and legal, economic and environmental, economic and political problems) and their involvement into activities organized by group tutors.

Extracurricular activities suggest the tasks that require study of additional literature, as well as research tasks the solutions to which have not been published.

Teaching students of higher education institutions includes various forms and methods, with more attention being paid to extracurricular and independent learning. Dissemination of technical teaching aids, new information technologies, and various free data sources can significantly expand the scope of the traditional academic process.

Using traditional forms of academic activity, students can acquire new knowledge, but their cognitive abilities need stimulating.

In addition to traditional and non-traditional forms of education, out-of-class (extracurricular) activities are especially important for economic culture formation, which is proven by practice and our surveys. Expanding the content of economic courses would allow organizing clubs, scientific circles, and round tables taking into account students' interests and enriching the content of classroom studies.

In the above classes, special attention was paid to determining the requirements for the level of students' training, namely:

- a wider range of knowledge allowing students to demonstrate their abilities at large. The tasks can be either written or oral, individual or group, test, formal, journalistic, requiring preliminary preparation or those that can be completed in class;

– a variety of assignments requiring a wide range of topics covered by the economics course. This enables assessing the level of economic thinking, its completeness and depth, proper use of economic terms, the ability to use knowledge gained elsewhere;

– complexity of tasks corresponding to the level of students' training. They should not be too difficult, but simplified tasks cannot be allowed either, as this negatively affects students' interests in the subject and their cognitive activity;

– excitement of interest provided by tasks with heuristic elements requiring creative approaches to their solution. These tasks should be prepared beforehand. Students' involvement can be encouraged by topicality of the tasks associated with certain current events both in the country and abroad.

2.3 Efficiency of economic culture development among students of higher education institutions

The process of students' economic culture formation was organized on the basis of the following experimental plan: determination and analysis of the initial level of students' economic culture in the experimental and control groups; determination of the experimental content; examination of forms, methods and technologies of economic culture formation; determination of the pedagogical conditions of the studied process; generalization and presentation of experimental results accompanied by analysis of positive and negative aspects; development of the economic culture formation model; making adjustments to achieve the desired result and eliminate deficiencies; introduction of programmes aimed at forming economic culture in experimental groups; comparison of the results obtained in the experimental and control groups with the predicted results; creation of the final version of the model of economic culture formation.

In order to develop economic culture among students, we organized elective courses and extracurricular courses during the experimental process.

We relied on students' choice when organizing elective courses aimed at developing economic culture. This course programme is aimed at developing students' skills to make decisions in compliance with the business protocol.

The experiment involved students of Andijan State University, Bukhara State University and the affiliated branch of Lomonosov Moscow State University in the city of Tashkent. A total of 407 students took part in the experiment: 220 of them studied economics following the current programme, and 187 students studied economics on the basis of the modernized programme in terms of the economic culture development.

The results of the formative experiment are a set of tested pedagogical forms, methods and technologies selected in accordance with the stages of economic culture development and the results of corrective work. This allows managing economic culture development in accordance with theoretical, psychological and pedagogical mechanisms, and confirms efficiency of economic education.

As a result of the experiment, the economic culture development system was formed (Fig. 2).

The research results allow for concluding that the level of economic culture development and its correspondence to the existing economic relations in society has increased in the experimental groups. The developed programme of economic education and teaching methods ensure high-quality acquisition of knowledge, their application in practice significantly impacts development of economic culture among students (Table 1).

 https://doi.org/10.1051/shsconf/202110002011

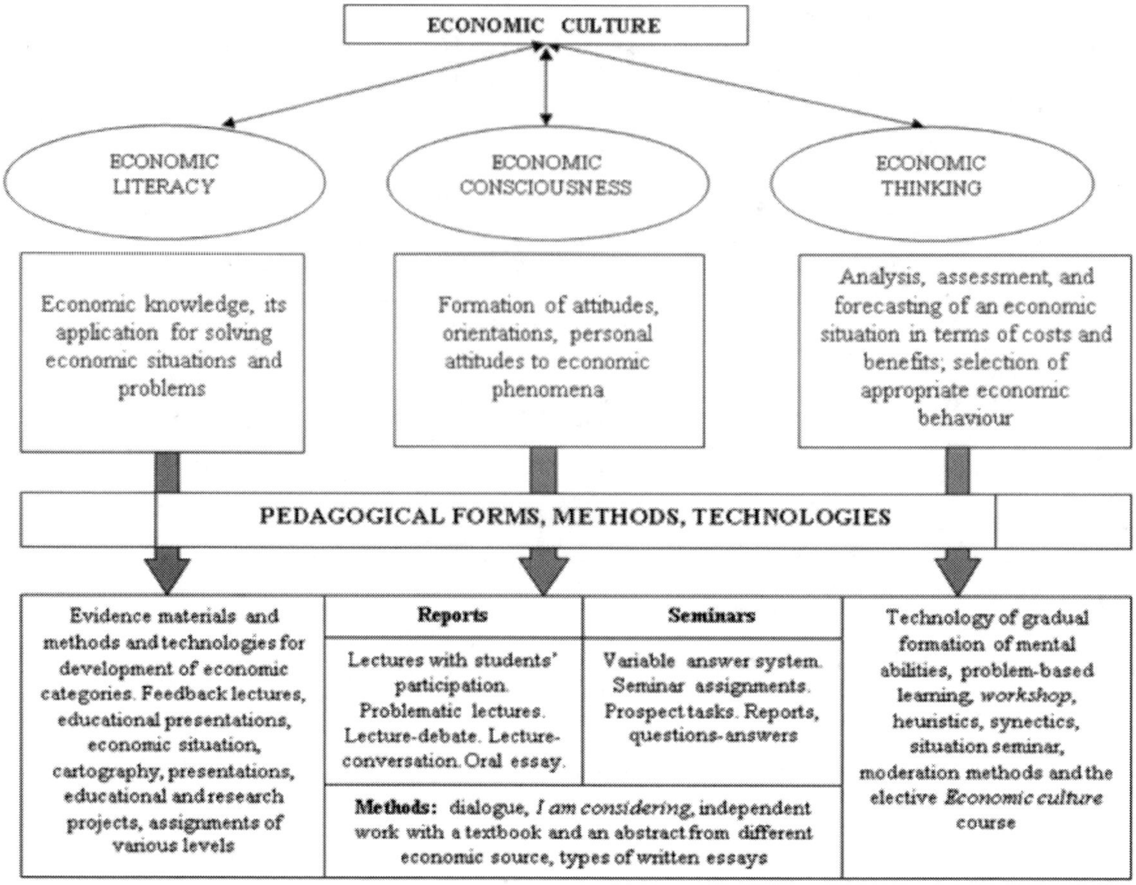

Fig. 2. Classification and functions of students' economic culture development.

Table 1. The efficiency level of the economic culture development among students at the beginning and at the end of the experiment.

Terms	Groups	Number of students	Formation levels in numbers and percentage			
			High	Above average	Average	Low
At the beginning of the experiment	Experimental group	216	19	41	95	61
			8.9%	18.9%	44.1%	28.1%
	Control group	191	24	48	73	46
			12.3%	25%	38.4%	24.3%
At the end of the experiment	Experimental group	216	71	84	43	18
			32.8%	38.8%	19.9%	8.3%
	Control group	191	25	70	50	46
			13.1%	36.6%	26.2%	24%

The results of Table 1 allow determining the average level of knowledge acquisition that reveals the level of efficiency of economic culture development by applying the Student and Pearson mathematical and statistical method $\chi 2$. The brief essence of the issue is as follows: there are two general data sets.

One of them is the average score of the students in the experimental group, and the other is the average score of the students in the control group. The grades are considered to have normal distribution. This assumption is valid, since the conditions for approximation to the normal distribution are simple and they are satisfied.

Based on the above tables, the hypothesis H1 is selected, it shows efficiency of teaching students in the experimental and control groups and the hypothesis H0 contradicts it.

In accordance with the indicators of the above table, the diagrams corresponding to these samples are as follows (Fig. 3, 4).

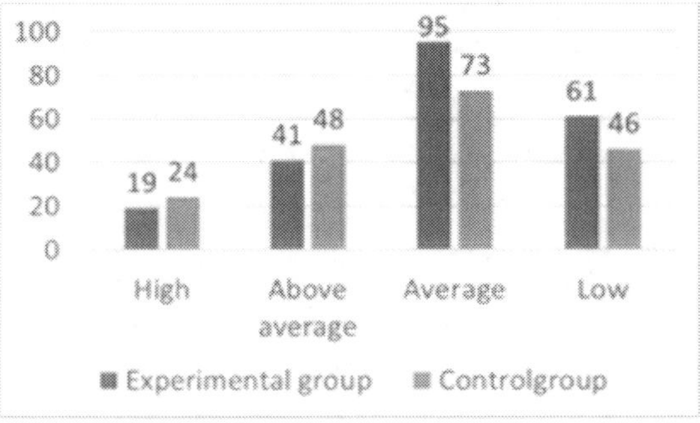

Fig. 3. At the beginning of the experiment, %.

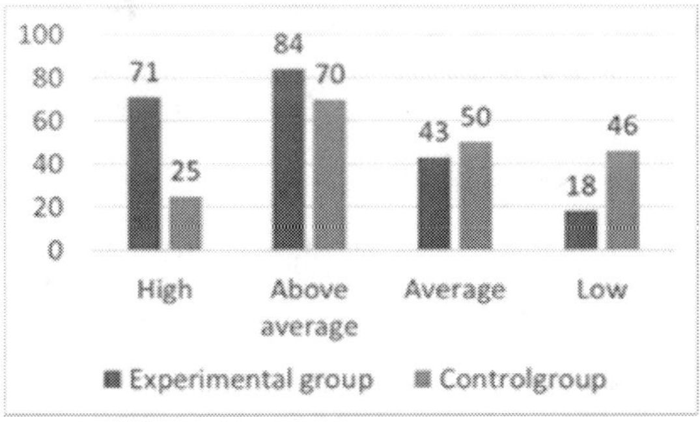

Fig. 4. At the end of the experiment, %.

Based on the results obtained, mathematical statistical analysis was performed to determine the standard deviation, sample variance, variance indices, the Student's selection criterion, degrees of freedom based on the Student's criterion, the Pearson's acceptability criterion, and reliable deviations from the results for the case at the end of the experiment. They are shown in Table 2:

Table 2. The parameters of the economic model and mathematical statistical analysis of the experiment results.

$\overline{X}$	2.09	2.97
$\overline{Y}$	2.25	2.39
S_x^2	Efficiency is not achieved at the beginning of the experiment	0.2491
S_y^2		0.9779
C_x		1
C_y		3
$T_{x,y}$		7.34
K		285.7
$X_{n,m}^2$		36.09
Δ_x		0.07
Δ_y		0.14

From the results obtained, it is known that average efficiency in the experimental group has increased by 14.5% compared to the control group.

On the basis of the above results, when calculating the qualitative indicators of the experimental work, the following is achieved.

It is known that $\overline{X} = 2{,}97; \overline{Y} = 2{,}39; \Delta_x = 0{,}07; \Delta_y = 0{,}14$.

Here quality indicators are:

$$K_u = \frac{(\overline{X}-\Delta_x)}{(\overline{Y}+\Delta_y)} = \frac{2{,}97-0{,}07}{2{,}39+0{,}14} = \frac{2{,}9}{2{,}53} = 1{,}15 > 1;$$

$$K_b = (\overline{X}-\Delta_x) - (\overline{Y}-\Delta_y); \quad (1)$$

$$K_b = (2{,}97-0{,}07) - (2{,}39-0{,}14) = 0{,}65;$$

$$K_b > 0.$$

The obtained results show that the criterion for assessing the training efficiency suddenly increases, and the criterion for assessing the level of knowledge exceeds zero. It is known that the efficiency in the experimental group grew higher than that in the control group.

It should be noted that economic culture is higher among those who studied Economics after updating the programme regarding economic culture development than among those who studied Economics following the standard programme.

When solving financial problems, students' economic knowledge is formed, while their socially significant relations are moderately polished.

3 Conclusion

As a result of the research, the following conclusions are drawn:

1. In developed market conditions of Uzbekistan, economic culture development of a person with a strong position is of particular importance. This requires regular acquisition of economic knowledge; development of students' economic knowledge, consciousness and thinking; formation of competencies for interpersonal economic relations. To achieve this goal, it is necessary to provide not only students of economic faculties, but also future non-economic specialists with new economic knowledge in order to give them a clear idea of socioeconomic development.

2. To carry out such pedagogical activity, first of all, it is necessary to enrich the content of economic education and develop mechanisms for organizing this process considering international experience. Students are taught that economic culture is a subjective expression of objective reality based on the creation of a certain type of economic behaviour that leads to practical goals and actions in the economic sphere being a pursuit of personal interests.

3. Economic culture development is an integral part of national culture. Economic culture performs certain functions in the community, such as translation,

selection, innovation, regulation, worldview, humanism, axiology, adaptation, stabilization.

4. In order to train a competitive specialist possessing up-to-date economic, political, legal and professional knowledge and apply it successfully satisfying the community's needs, the following problems are to be solved: development of curricula for consistent provision of economic knowledge at all higher education institutions; provision of the educational process with training materials and didactic projects that comprise modern economic knowledge; provision of the university staff with the latest economic knowledge in the process of their continuing professional development.

5. The psychological and pedagogical approach to development of students' economic culture made it possible to single out a number of components in its structure. The description of economic culture components along with psychological and pedagogical features of economic culture development created the basis for solving the problem of identifying the assessment criteria of the students' economic culture level. The research has defined the following criteria: availability of cognitive information; the system of values; intellectual and creative activity; practical activities.

6. Along with solving the economic problems assigned to the students, they have the opportunity to acquire knowledge that forms the basis of economic culture: the process of acquiring new knowledge is the main point for economic culture development; problem assignments ensure successful mastering of economic culture elements; problem solving tasks also allow students to control the level of their economic culture.

7. Teaching economic concepts and patterns prepares university graduates for social life, in which they form psychological stability in relation to possible difficulties associated with unemployment, competition, their work and profession, change of residence.

8. The conducted practical and theoretical analysis has become the basis for specifying methods, techniques and technologies to develop students' economic culture.

9. Correction of pedagogical activity aimed at developing students' economic culture was based on certain principles: taking into account specifics of the developed market economy and development dynamics; the inclusive approach to the presentation of the economic content; gradual improvement of the requirements for the curriculum in terms of continuous development of economics; facilitating practical aspects of economic culture development among students; the integrated innovative approach to organizing the process of economic culture development.

10. The results of the experiment have shown that use of innovative methods in the pedagogical process aimed at developing economic culture among students helps to elaborate their economic thinking, interpretation of current economic reforms, and navigate in modern economic life making optimal decisions in any economic situation.

11. It is possible to expand the sphere of students' economic culture by choosing the content of extracurricular activities of certain complexity. For this purpose, special extracurricular courses are included in the training programme. The content of special courses is inextricably linked with that of classroom studies.

12. The experiment has been carried out relying on methodological approaches to its organization. Theoretical and empirical data indicate that quality and efficiency of economic education and training of specialists for future social and practical activities depend on the development level of economic thinking and consciousness that characterize students' economic culture.

The results of the experimental work have confirmed the correctness of the idea expressed in the research.

Based on the above conclusions, we provide the following recommendations:

- introduction of the elective course Theory and Practice of Economics at non-economic departments of all higher education institutions on the territory of the republic;
- introduction of regular seminars at higher education institutions on the topic The Role of Individual Culture in Development of the Economic Life of Society taking into account sustainable development of the economic life of society;
- introduction of the latest economic concepts into the content of academic disciplines at higher education institutions on the basis of the integrative approach;
- organization of trainings, round tables, press conferences on economic issues for the teaching staff and students of higher education institutions.

References

1. Edwin G. Dolan, The Austrian Paradigm in Environmental Economics: Theory and Practice, Quarterly Journal of Austrian Economics, 17 (2) (2014)
2. A. B. Mamanazarov (ed.),, An interdisciplinary approach to the process of forming an economic culture (Fan va texnologiya, Tashkent, 2017).
3. James D. Gwartney, Richard L. Stroup, Russell S. Sobel, David A. Macpherson, Microeconomics: Private and Public Choice (Cengage Learning, 2014)
4. Paul L. Heyne, Peter J. Boettke, David L. Prychitko, The Economic Way of Thinking: Pearson New International Edition (Pearson Education, 2013)
5. Ulrich Ph., Lehr U., Economic effects of an E-mobility scenario – input structure and energy consumption, Economic Systems Research, 32:1, 84-7, (2020) doi:10.1080/09535314.2019.1619522
6. Maslow A. H. A theory of human motivation. Psychological Review, 50(4), 370–396. (1943). doi:10.1037/h0054346
7. Reich R. B. (ed.), The System: Who Rigged It, How We Fix It (Pan Macmillan UK, 2020)

8. Shaffer David R., Social and Personality Development (United States, Cengage Learning, 2008)
9. Brody Stuart H., The Law of Small Things: Creating a Habit of Integrity in a Culture of Mistrust (Berrett-Koehler Publishers, 2019)
10. Hirsch D., Cultural context in marketing communication on international market, Marketing of Scientific and Research Organisations, 12, (2014) doi: 10.14611/minib.12.02.2014.08
11. Cornali F., Tirocchi S. Globalization, Education, Information and Communication Technologies: What Relationships and Reciprocal Influences?, Procedia - Social and Behavioral Sciences, 47 (2012) doi: 10.1016/j.sbspro.2012.06.949

Translanguaging as an Aspect of ESP Acquisition in Non-Linguistic Universities

Kira Vyshnevska [1], *Olga* Bratanych[1], *Sergii* Skydan[1], *Olena* Hushko[1*] and *Zulfizar* Karimova[2]

[1]State University of Economics and Technology, Department of Foreign Languages, 50005 Kryvyi Rih, Ukraine
[2]Tashkent branch of MSU named after M.V. Lomonosov, Amir Timur str., 22, Tashkent, Uzbekistan

Abstract. This article seeks to develop Translanguaging as a theory and an efficient ESP classroom practice. We consider Translanguaging at linguistic and linguodidactic (spontaneous, or preplanned) levels. Our findings may be of certain practical value for ESP teachers, hesitant about acceptance of Translanguaging for building rapport with students, creating a positive learning environment and aiding their acquisition of target language. We contextualize Translanguaging in the ESP classroom realities of the State University of Economics and Technology (Ukraine), National University of Tashkent, Bukhara State University (Uzbekistan) and Samara State Technical University (the Russian Federation). The study was based upon the online survey of teachers' attitude towards Translanguaging and their practice of translanguaging in ESP classrooms. In order to obtain reliable information and to explore the topic rigorously we applied a combined-method design, integrating quantitative and qualitative approaches. The paper provides empirical results on the role of translanguaging in ESP classrooms, the expediency of using the translanguaging strategy in ESP classes and its "legitimization" in the methodology of teaching, the extent of using Translanguaging by university teachers and students, evaluations of Translanguaging shortcomings and benefits. Studies have shown that translanguaging is both an efficient learning tool and a useful teaching method if pedagogical activities are designed thoroughly.

1 Introduction

Globalization, intensification of migration processes, including in the world network space, created both possibility and necessity to study two or more foreign languages, aroused interest in different ethno cultures, their specifics, points of contact, features of interaction. In this regard, modern linguodidactics actualizes the problems of studying the contacts of languages and cultures, the patterns of formation of bi- and polylingual personality, the specifics of simultaneous and consistent development of different linguistic cultures. A new aspect of the study of this issue is fixed in the concept of "translanguaging", which has relatively recently entered scientific circulation in various fields of humanities.

Effective study of the essence of translanguaging in foreign science began in the 1990s, and at the turn of the century this phenomenon attracted the attention of linguists, culturologists, teachers, methodologists.

The translanguaging (TL) related concepts such as "translingualism", "translingual skills", "translingual personality" are studied in the most effective way in linguistics, sociolinguistics, literary studies, advertising theory, translation studies, culturology, theory and practice of foreign language teaching. In spite of growing number of research regarding necessity of understanding these concepts in order to face epoch of globalization, only a few studies explain ways to apply them in the foreign language education [1].

Therefore, the study of translanguaging in the methodology of teaching foreign languages is of paramount theoretical and practical significance nowadays. This research direction is today eagerly supported in foreign linguodidactics [2-8] but has hardly been studied thoroughly in domestic scientific works. In order to fill the gap our study is aimed at examining the implementation of translanguaging in ESP context with all the challenges, criticism and creativity.

2 Background to the study

While translanguaging has grown in global popularity recently, turning out to be more diverse, dynamic and democratic than "monolingual" competence, numerous educators consider it to be a threat, capable of destroying the instructional "English only" space and discouraging the use of target language (L2) by students. Others see translanguaging as a primitive crutch which is used to facilitate meaning making among students and teachers.

Being neither blessing no curse, translanguaging is a method of teaching and communicative practice, which has got both following advantages (facilitating understanding and learning, effective exploiting similarities and differences between home and target language structures, capitalizing on the students' existing knowledge and language awareness) and disadvantages such as preventing students from acquiring communicative competences, which are considered to be

* Corresponding author: pygmalion.utu@gmail.com

"treasured icon" of ESP, hindering thinking in English, encouraging off-task behavior.

There are numerous studies examining FL teachers' attitudes towards using translanguaging. The corresponding surveys cover different geographic areas and nationalities: García & Leiva, (Spain) [9], Jeff MacSwan (America) [10], Hornberger & Link (Australia) [7], Noella Charbonneau and Michiko Weinmann (France) [11], Rasman (Indonesia) [12], Jiajia Eve Liu, Yuen Yi Lo, Angel M.Y. Lin, (China) [13]; Flores & Rosa, (Latin America) [14] amongst others.

However, studies are very few investigating EFL teachers' from former socialist republics attitudes towards translanguaging practice. It might be due to the fact that monolingual immersion approach had been proclaimed as incontestable and having irrefutable value in ESP teaching for most post-Soviet universities. Therefore, present study is to be of certain practical value for ESP teachers, hesitant about acceptance of translanguaging for building rapport with students, creating a positive learning environment and aiding their acquisition of target language.

3 Literature Review

Since the end of the 19th century a monolingual approach to language teaching, such as the direct method, communicative language teaching, and immersion method have been adopted as the most effective English teaching methods. The monolingual approach either banned or minimized the use of L1 in the classroom, encouraging use of the target language. Therefore, using students 'native language (L1) was eagerly criticized in ESP classes.

Until recently, the monolingual approach has been rarely challenged. However, there is a wide range of classes currently in which monolingual exclusiveness lost its appeal. Moreover, some researchers began to find fault with target-language-only theories and highlighted that there could be positive reasons for translanguaging in the classroom [4, 7, 8].

The term "translanguaging", or TL for short, was introduced in the 1980s by Cen Williams to denote the planned and systematic use of two languages during the class. However, both the term itself and the related concept gained popularity decades later, in part, due to published research by Canagarajah and Ofelia García among others. Translanguaging is understood by Canagarajah as a "smooth synergistic transition from one linguistic culture to another, resulting in some merging, without complete assimilation and linguistic-cultural identity of language users" [1].

Since then the term "translanguaging" has been used in various ways by different scholars. A researcher in Applied Linguistics Li (2018) offers educators and researchers a conceptual definition of translanguaging practice and its relevance to language teaching [6]. Li points out that Williams made it clear from the beginning that unlike code-switching, translanguaging is not simply a set of linguistic structures. It is a dynamic practice that involves different named languages and language varieties but more importantly, a process of knowledge construction that makes use of, but goes beyond, individual languages. It concerns effective communication, function rather than form, cognitive activity, and language production.

R. Tsokalidou and E. Skourtou propose to use the term "translanguaging" to denote a process where one language is used to strengthen knowledge of another language and increase students 'proficiency in two languages. The researchers also note that through the use of translanguaging there is a balanced development of students 'speech skills in two languages [15]. The conceptual basis of the translanguaging technology is the student-centered organization of the educational process and its arranging on a multilingual basis. The main principles of the organization of ESP educational process using the technology of translanguaging are the principle of reflexive orientation of the educational process, the principle of connection of learning with life, the principle of consciousness and freedom.

Garcia, O., Ibarra Johnson et al [3] contributed to the field of bilingual and second language education by highlighting the key understandings and approaches to translanguaging practice in educational context and emphasizing the role of translanguaging as a tool for teaching complex content. They argue that "When we make space for students to use all the linguistic resources they have developed to maneuver and navigate their way through complex content, a myriad of learning opportunities open up. Rather than watering down our instruction, which risks oversimplification and robs students of opportunities to engage in productive grappling with texts and content, translanguaging better enables us to teach complex content, which in turn helps students learn more successfully"

Sembiante (2016) [4] discusses how bi / multilingual speakers draw upon their complex linguistic repertoire for communicative purposes in different contexts arguing that "translanguaging recognizes speakers" hybrid use of language in alignment with a variety of social purposes and communicative settings and acknowledges their ability to adapt to these diverse sociolinguistic situations.

Thus, the review of the literature on translanguaging reveals numerous benefits of using translanguaging as pedagogy.

4 Translanguaging in the ESP classroom

The issue of translanguaging in the foreign language classroom is becoming trendy. In terms of foreign language learning TL functions at two "levels":

- Linguistic - involves combining elements from different languages both in oral and written speech.
- Linguodidactic - involves alternating between a native language/s (L1) and a foreign language (L2) in different parts of a cognitive task.

In the cultural context of Ukrainian non-linguistic Universities these languages are Ukrainian/ Russian (Russian only on the part of Russian speaking students) and a foreign language. Both levels of TL coexist in the ESP classroom. Let's look at typical examples of using

TL at the linguistic level in "Teacher→Student/s" and "Student↔Student/s" modes.

Linguistic level: English- Ukrainian (transliteration in bold) translanguaging.

- **A teacher:** while explaining grammar:

*Present Simple **(a ne** Present Continuous) **zazvychay vykorystovuyetsya z diyeslovamy stanu**: agree, believe, enjoy, hate, know, like, love, mean, prefer, remember, seem, think, understand, and want. **Napryklad**: I do not agree with you. (**A ne** I'm not agreeing with you.) Dogs do not like cats. (**A ne** Dogs are not liking cats. '?') They want to stay at home. (**A ne** They're wanting to stay at home.)*

- **Students:** when discussing and assigning roles for a simulation based on a job/professional problem.

***Ty budesh** a crane driver, **a ya budu** Head of the rolling mill.*

Such communication should not be perceived as something shameful, wrong, forbidden, because it helps to solve the learning task in the best possible way. Moreover, we believe that the linguistic level of TL can be used as a strategy for teaching a foreign language not only in speaking, but also in writing.

It is a well-known fact that the classical methodology, in which translation exercises played an important role, is a thing of the past. Instead it has been replaced by the communicative method, which focuses on the possibility to communicate and which emphasizes speaking and listening. But nevertheless, classical L1-to-L2 translation exercises have a significant impact on the development of students' foreign-language communicative competence, and they should not be abandoned altogether. There are a number of benefits of translation exercises:

- complex use of grammar and vocabulary;
- forming the ability to build sentences in a correct way;
- creating speaking and listening comprehension base.

However, students with a low level of L2 proficiency cannot cope with tasks for translation from L1 to L2 without using an electronic translator and receive poor marks for completing / not completing such work, which reduces the motivation to learn a foreign language and causes a feeling of their own inferiority.

Therefore, why not give such students an opportunity to do a translation exercise at the TL linguistic level and grade such work, even if it is at the lowest level. This would be translation at the level of foreign words in the context of a L1 sentence. Hopefully, as students' foreign-language competence increases, there will be more and more foreign words, their grammatical features will appear, for example, the plural, the possessive case of nouns, the degree of comparison of adjectives, the particle to before the infinitive, etc. Here are some examples to illustrate this strategy:

***Sentence in L1:** Metaly – shchil'ni materialy tomu, shcho mizh atomamy v metalakh mala vidstan.* (Ukrainian transliteration).

***Sentence in L2:** Metals are dense materials because there is a small distance between atoms in metals. (English).*

Linguistic level of written translanguaging.

*Metal – **shchil'ni** materials **tomu, shcho mizh** atom **v** metal **mala vidstan**.* [Ukrainian (transliteration in bold) - English].

*Metals – **shchil'ni** materials **tomu, shcho mizh** atoms **v** metals small **vidstan**.* [Ukrainian (transliteration in bold) - English]

*Metals are **shchil'ni** materiasls **tomu, shcho mizh** atoms **v** metals small **vidstan**.* [Ukrainian (transliteration in bold) - English].

*Metals are **shchil'ni** materiasls because between atoms in metals small **vidstan**.* [Ukrainian (transliteration in bold) - English].

Translanguaging in ESP writing is not proposed here just as a self-regulating mechanism in which bilingual students can engage while planning, drafting or taking notes but as a strategy to be used in teaching L2 writing itself. The strategy helps students leverage their multiple linguistic resources to convey more information, express more ideas, while achieving a wider use of general and professional words.

There are two different types of linguodidactic translanguaging which can be defined in the corresponding studies [16, 17]. The first is spontaneous, or unplanned, translanguaging, occurring in classrooms where linguistic diversity is valued and seen as a resource for learning. It can take the form of students translating/explaining to each other and the teacher allowing L1 and L2 mix in communication, giving students with different English proficiency level opportunity to participate freely and fully in the class learning and socialisation.

The second type is planned translanguaging, where the teacher uses the approach purposefully and strategically, makes considered decisions about the learning objectives, taking into account many students having a good mastery of academic content and literacies in their L1, but having low English proficiency level. Planned translanguaging can be exemplified in three following activities:

1. Creating a rich experiential context by discussing and experiencing how to annotate and summarise articles on finance and banking/ international economics/ law or any other narrowly focused study, necessary for writing a degree project. Translanguaging, used for translating and elaborating key terms raises students' interest and encourages their involvement.
2. Reading, note-making analyzing and evaluating research articles to learn about their structure and moves, to be aware of the language features of each section of research articles. Translanguaging, used for translating, elaborating and exemplifying important/difficult terms or content for developing multilingual vocabulary and syntax inquiry to promote metalinguistic awareness.
3. Engaging students in textualizing the experience by writing a short essay to analyze and evaluate a research article. Translanguaging facilitates students' academic reading and writing.

5 Benefits and challenges of incorporating translanguaging in

everyday teaching practice (empirical evidence)

The study was based upon survey conducted at the State University of Economics and Technology (Ukraine), National University of Tashkent, Bukhara State University (Uzbekistan) and Samara State Technical University (the Russian Federation).

The language policy of universities aims to improve the students' proficiency in both the state language and foreign languages, in particular English, as the international language of higher education and science. The universities have majors in which students learn two foreign languages at an advanced level. Students in most undergraduate majors study the discipline Foreign Language of Speciality (ESP) and Master's degree students study Professional Foreign Language (ESP), for which only 2-3 academic hours per week are allotted. This is a typical situation for non-linguistic universities in Ukraine, Uzbekistan and the Russian Federation. The groups are heterogeneous; some students speak a foreign language at a level below average. University teachers of foreign languages are in active search of methods and approaches to increase the level of foreign language proficiency of their students.

In the case of "Only English" ESP class there is a paradoxical situation where you have to learn something unknown with the help of the unknown. ESP at a non-linguistic institution is very different from the Every Day English students learned in high school. ESP includes the study of professional content by means of a foreign language. Academic texts are written in a scientific style and contain complex grammatical phenomena and professional vocabulary whose semantization in a foreign language can take a long time and a foreign language teacher who is not a specialist in economics, law or metallurgy is not always able and obliged to do so.

Thus, there are contradictions in a non-language university, in particular:

Table 1. Questionnaire on the use of the home language (HL) within ESP foreign language class of non-linguistic universities

Questions	Always	Sometimes	Never
Do you use HL within a foreign language class? If so, when / for what:			
To manage the learning process;			
To explain grammar rules;			
To semantize vocabulary;			
To check reading comprehension;			
To consolidate the grasp of covered material when doing some vocabulary and grammar exercises (finding matches, filling gaps);			
To explain the purpose of educational activities;			
To resolve conflict situations;			
Other:			
Do you allow your students to communicate in their native language:			
To discuss creative assignments;			
To prepare for pair/team engagement activities (dialogues, debates, role playing, brainstorming);			
To make up a plan of monologue speech;			
Other:			
Name 3 – 4 advantages of using the native language within a foreign language class			
Name 3 – 4 disadvantages of using the native language within a foreign language class			

number of academic hours allocated for ESP study as well as the insufficient level of foreign language proficiency of a significant proportion of students versus the requirement of the Ministry of Education and Science of Ukraine, the Ministry of Higher and Secondary Specialised Education of the Republic of Uzbekistan, The Ministry of Science and Higher Education of the Russian Federation to conduct classes in a foreign language exclusively.

- The complexity of the ESP discipline content versus insufficient level of foreign language proficiency.
- The large volume of learning material versus the limited time available for its study.
- The requirement to democratize learning versus the oppression of the natural human desire to communicate, express oneself and assimilate certain content by all the known verbal means.

We believe that one of the strategies that can eliminate these contradictions is translanguaging strategy. We should emphasize, that in spite of this strategy having long been used in the practice of both higher and secondary school teachers (we prove it in the further presentation of the study results), most teachers are embarrassed and even ashamed of admitting this fact. They definitely had an interest regarding translanguaging in the classroom and responded to our survey on an entirely voluntary basis, but the most preferred to stay anonymous, since reference to the home language in a foreign language class is perceived as evidence of the teacher's professional incompetence and inability to communicate in a foreign language.

That is why our study and above given questionnaire aim to draw attention to the problem of expediency of using the translanguaging strategy in ESP classes and its "legitimization" in the methodology of teaching, provide empirical results about the role of translanguaging in ESP classrooms, add to our understanding of the extent of using translanguaging by university teachers/ students and the attitudes of practicing teachers towards translanguaging strategy.

The participants in this study comprised 50 teachers, each answered a questionnaire. As shown in Figure 1, the ESP teachers participating in this study represented three different nationalities, with the largest group (about 60 %) comprised of Ukrainians. The second highest percentage (25%) were Uzbekistan teachers from Tashkent and Buhara. About 15% of teachers represented the Russian Federation. Everyone, participating in the study possessed a foreign language-related postgraduate diploma, master's or PhD degree. The students' level in English varied substantially.

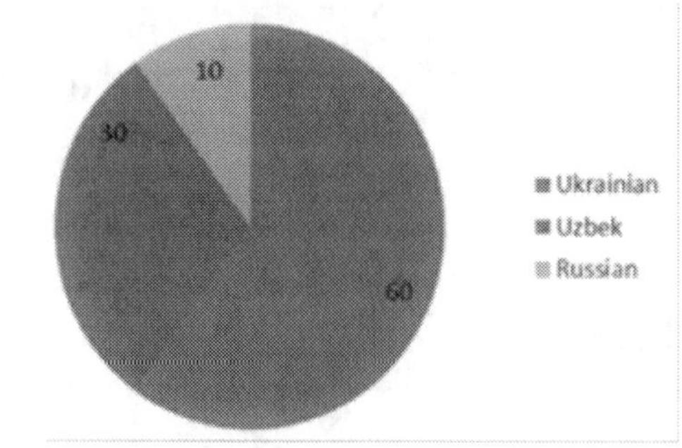

Fig. 1. Teachers' translanguaging (%).

The study applied a combined-method design, integrating quantitative and qualitative approaches in order to obtain reliable information and to explore the topic rigorously. 50 participants completed a survey online during the fall of 2020. To ensure the questionnaire items content reliability and validity, it was piloted to five English teachers to check the wording and the clarity of its items. After piloting the participants' feedback was taken into account and some questionnaire items were modified to provide better understanding.

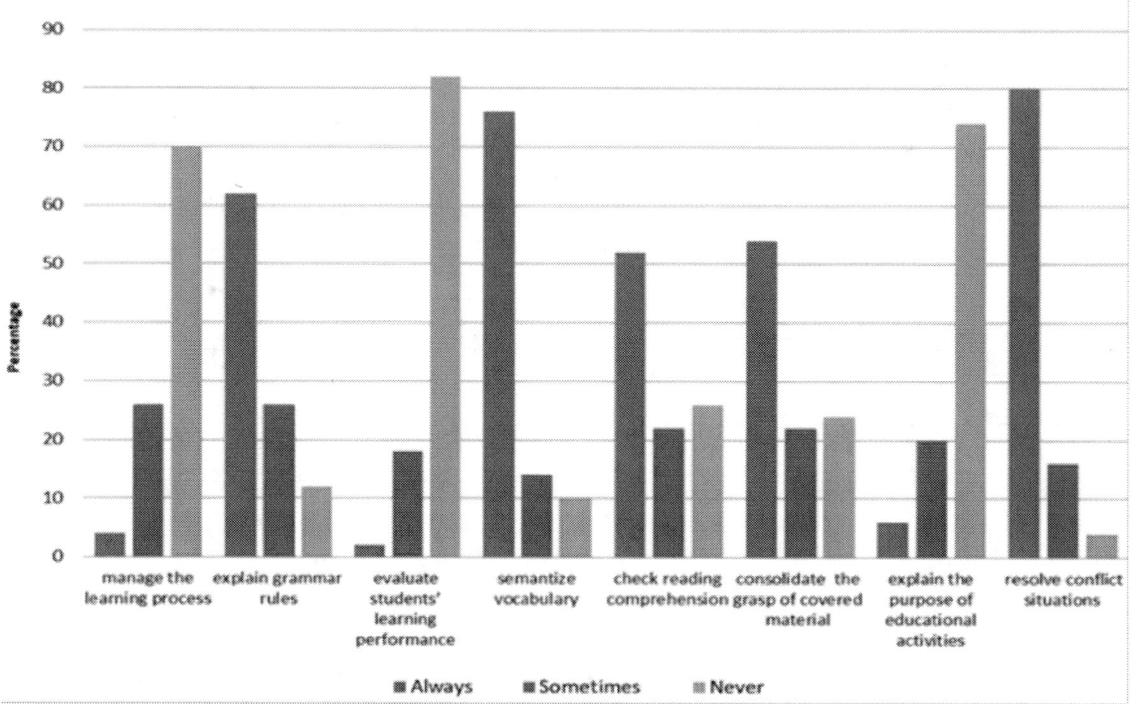

Fig. 2. Frequency of translanguaging by EFL teachers.

The final version of the questionnaire comprised ten multi-item Likert scale questions, including an open question, providing the participants the opportunity to add further comments, as well as ten short background questions. Finally, the participants were asked for their contact details, providing they were willing to participate in the follow-up interviews. The average time taken to complete the questionnaire was about 20 minutes. The questionnaire explored teachers' attitudes towards applying translanguaging approach in the professional context; therefore, it should be pointed out that the data represents reported practices, which can be far from actual ones. Concerning the follow-up interviews, the main target was to provide rigorous comprehension of participants' practices and attitudes towards translanguaging in the ESP classroom. Thus, Fig. 2 shows how often ESP teachers use TL regarding their educational aims and objectives.

https://doi.org/10.1051/shsconf/202110002012

In conformity with the questionnaire results presented in Figure 2, more than 82% of teachers reported that they never used translanguaging while evaluating their students, 74 % avoid translanguaging for explaining the purpose of educational activities and 70% under no circumstances apply it for managing the learning process correcting. When examining the results of "always" option, it turns out that the teachers tend to apply much translanguaging in their ESP classroom for resolving conflict situations (80%) and semantizing vocabulary (76%). However, for the "sometimes" option, as few as about 26% of teachers reported that they do not mind using TL to explain grammar and manage the learning process as well as to check reading comprehension and consolidate the grasp of covered material.

The next stage of research was to see how often students use TL in the ESP classroom and with what purposes.

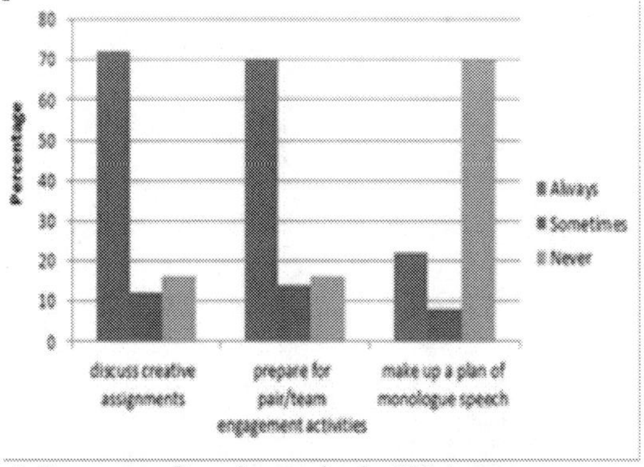

Fig. 3. Frequency of translanguaging by EFL learners.

Definitely, ESP of university students is a conscious process, involving analytical thinking, previous language learning experience and awareness. Therefore, teachers reported that the majority of their students used translanguaging to some extent in the ESP classroom, with the main functions of discussing creative assignments and preparing for pair/team engagement activities such as dialogues, debates, role playing, brainstorming (72 % and 70 % respectively). It was also apparent that a large percentage of students (70%) are not inclined to applying TL for making up a plan of monologue speech.

Of paramount interest were the interviewees' evaluations of translanguaging shortcomings and benefits. The data presented in Figure 4 show that most teachers are in agreement with three following points: (1) TL helps get rid of anxiety and frustration to create a stress-free learning environment which is pure emotional response (the highest percentage at 84 (%); (2) translanguaging allows parallelizing, comparing target language knowledge on grammar/vocabulary structure to their existing home language knowledge and introducing the main differences in grammar and pronunciation, (levels 78 (%); (3) translanguaging encourages complete comprehension of grammar and vocabulary (76%), which relates to pure cognitive skill. The highest level of disagreement was found out for the point, which applies to using translanguaging for helping students work together (68%). The teachers were likely to come to a larger agreement about psychological and cognitive reasons for using translanguaging than about practical reasons for its use, such as saving time (62 % only). Five of the interviewed teachers highlighted a need to use translanguaging at certain points, such as "explaining the meaning of abstract words, introducing the main differences in grammar and pronunciation", "using richer and more authentic texts, which mean more comprehensible input and faster acquisition" during class. One teacher referred to TL use as a "friendly asset people bring to the task of FL learning". Teacher A stated that TL contributes to "all-newly-acquired FL items taking roots in our minds which are eventually deep enough to function independently of the HL". Interviewee B focused on "TL techniques allowing teachers to use richer and more authentic texts, which means more comprehensible input and faster acquisition".

It is obvious from Figure 5 that the majority of teachers (more than 76%) have strong belief that using translanguaging reduces the opportunities for proficient learners to speak and practice English, prevents creating effective foreign language environment, which may have a negative impact on whether they think in English. The highest disagreement level shown in the survey was for the last point, which relates to causing faulty parallelism (52% did not accept it). During the interviews, one teacher stated that their opposing the use of TL is based on her firm belief that "teacher and students fail to observe the distinctions between equivalence of form, semantic equivalence, and pragmatic features, and thus oversimplify to the point of using crude and inaccurate translation" (Teacher C).

The survey results initiated hot debates and gave opportunity to claim that the use of TL in ESP teaching and learning English is inevitable. Being skillfully applied, it fosters linguistic, cultural and intellectual vigor of students with different level of command of language.

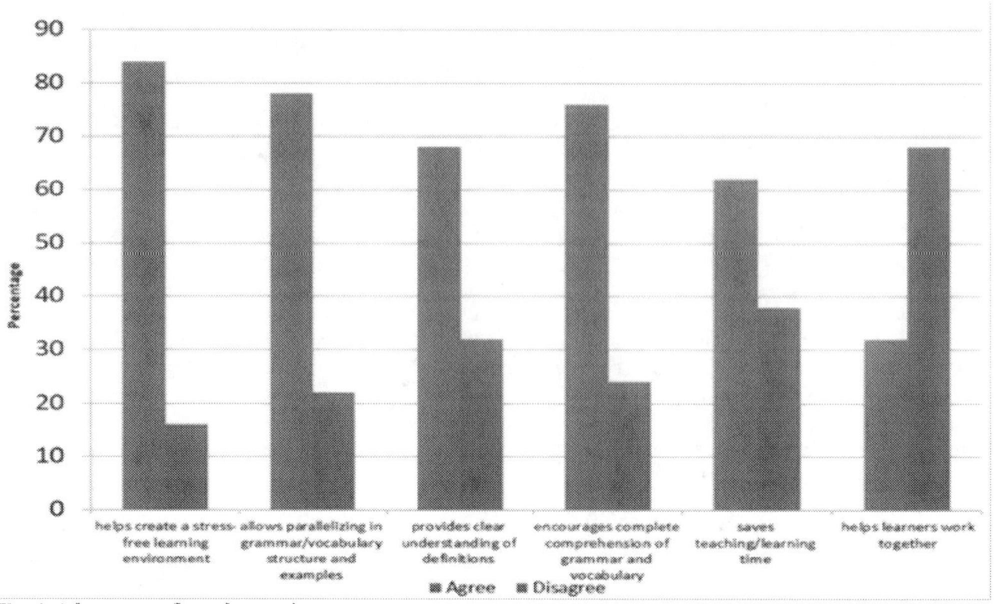

Fig. 4. Advantages of translanguaging.

Figure 5 shows teachers' opinion about disadvantages of using TL techniques in the learning process.

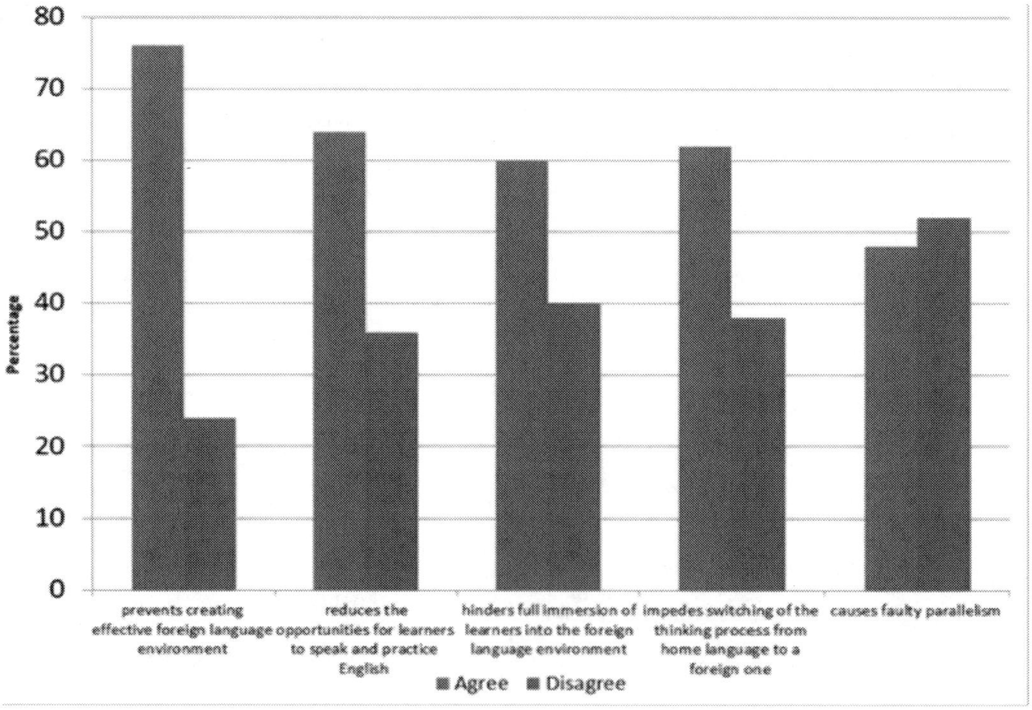

Fig. 5. Disadvantages of translanguaging.

The HL linguistic experience often becomes the foundation of their future learning. Thus, the teachers involve translanguaging judiciously and methodically in bilingual and multilingual classes with care and concern. The well-directed and appropriate use of lexical and syntactic balance between the HL and English promotes retention, understanding of linguistic affinity and transition of one language knowledge and skills to another.

6 Conclusion

Implications from the study indicate multiple benefits from creating a learning environment where translanguaging is encouraged. The potentials of translanguaging appeared to be limitless. The use of translanguaging in ESP classrooms is justified as flexible strategy and essential tool for students' making meaning of text, getting guidelines for their activity through the assignment exploring and expanding content, facilitating acquisition of complex linguistic knowledge, making and maintaining dialogue, effective accomplishment of jigsaw tasks. It is worth emphasizing, that translanguaging should only be used to help establish knowledge in the target language, facilitate interpersonal interactions, and increase efficiency. In no way should translanguaging be granted the same status as a target language in the ESP classroom. None of the supporters endorse its unlimited use. On the contrary, it should be borne in mind that excessive translanguaging deprives students of appropriate significant amount of target language input.

Research studies revealed that translanguaging is both an efficient learning tool and a useful teaching method if pedagogical activities are designed thoroughly. Students use translanguaging to facilitate their process of comprehension and to reduce any stress and inhibitedness which may arise due to their limited language proficiency. Teachers use translanguaging to promote a deeper and fuller understanding of the subject matter, help the integration of fluent speakers with low proficiency learners, consolidate foreign language knowledge which students have acquired, such as vocabulary, sentence structures, and grammar aspects.

It should be pointed out that it is next to impossible to define appropriate universal volume of translanguaging used by teachers because it depends on students' proficiency levels and teaching purposes. Taking into consideration all pros and contras, identified in a survey, we can assert that translanguaging is an overwhelmingly effective instrument which should not be either abandoned or denied in ESP classrooms. Ignoring supportiveness of translanguaging is nothing but a crucial mistake, which no skilled foreign language teacher should make.

References

1. A. Suresh Canagarajah, *Codemeshing in Academic Writing: Identifying Teachable Strategies of Translanguaging*, Modern Language Journal, 95, 3 (2011). doi: https://doi.org/10.1111/j.1540-4781.2011.01207.x
2. A. Creese & A. Blackledge, *Translanguaging in the bilingual classroom: A pedagogy for learning and teaching*, Modern Language Journal, 94, 103–115 (2010). doi: https://doi.org/10.1111/j.1540-4781.2009.00986.x
3. O. García, S. Ibarra Johnson, and K. Seltzer, The Translanguaging Classroom: Leveraging Student Bilingualism for Learning, Philadelphia, PA, Caslon, 196, (2017). doi: https://doi.org/10.1002/tesq.401
4. S. Sembiante, *Translanguaging and the multilingual turn: Epistemological reconceptualization in the fields of language and implications for reframing language in curriculum studies*, Curriculum Inquiry, 46, 45-61 (2016). doi: https://doi.org/10.17763/0017-8055.85.2.149
5. O. García, & A.M. Lin, *Translanguaging in bilingual education*, Encyclopedia of language and education, 5, (2017). doi: 10.1007/978-3-319-02258-1_9
6. W. Li, *Translanguaging as a practical theory of language*, Applied Linguistics, 39 (2), 9–30 (2018). doi:10.1093/applin/amx039
7. N. Hornberger, & H. Link, *Translanguaging and transnational literacies in multilingual classrooms: A biliteracy lens.*, International Journal of Bilingual Education and Bilingualism, 15(3), 261–278 (2012). doi: 10.1080/ 13670050.2012.658016
8. E. Caruso, *Translanguaging in higher education: Using several languages for the analysis of academic content in the teaching and learning process*, Language Learning in Higher Education, 8 (1) (2018). doi: https://doi.org/10.1515/cercles-2018-0004
9. O. García, & Camila Leiva, Theorizing and Enacting Translanguaging for Social Justice. Heteroglossia as Practice and Pedagogy, 199-216 (2014). doi: 10.1007/978-94-007-7856-6_11
10. Jeff MacSwan, *A Multilingual Perspective on Translanguaging*, American Educational Research Journal. 54, 167-201 (2017). DOI:10.3102/0002831216683935.
11. M. Weinmann, N. Charbonneau, *Enabling Translanguaging in the French Language Classroom*, Translanguaging: Opportunities and Challenges in a Global World, 10 (2020). doi: https://doi.org/10.18192/olbiwp.v10i0.3816
12. R. Rasman, *To translanguage or not to translanguage? The multilingual practice in an indonesian EFL classroom*, Indonesian Journal of Applied Linguistics, 7(3), 687 (2018). doi: 10.17509/ijal.v7i3.9819
13. J. Eve, L. Y. Yi Lo, Angel Lin, Translanguaging pedagogy in teaching English for Academic Purposes: Researcher-teacher collaboration as a professional development model, System 92:102276, The University of Hong Kong (2020). doi: 10.1016/j.system.2020.102276
14. N. Flores, R. Jonathan, *Undoing Appropriateness: Raciolinguistic Ideologies and Language Diversity in Education*, Harvard Educational Review, 85 (2): 149–171 (2015). doi: 10.18192/olbiwp.v10i0.3816
15. Tsokalidou, Roula & Skourtou, Eleni, *Translanguaging as a Culturally Sustaining Pedagogical Approach:* Bi/Multilingual Educators' Perspectives. (2020). doi: 10.1007/978-3-658-28128-1_13.
16. J. Y. Iversen, *"Translanguaging" and the implications for the future teaching of English in Norway*, Nordic Journal of Modern Language Methodology, 7(1):50-66 (2019). doi: 10.46364/njmlm.v7i1.520

https://doi.org/10.1051/shsconf/202110002012

17. A. M. Y. Lin, Y. Y. Lo, *Trans/Languaging and the Triadic Dialogue in Content and Language Integrated Learning (CLIL) Classrooms*, Language and Education, 31(1), 26-45 (2017).

Business Communication in the Context of Improving the Personnel Policy of a Modern Organization

Oleksandr Temchenko[1], *Nataliia* Shevchuk[2*], *Volodymyr* Musienko[1], *Nataliia* Borodina[3] and *Oleksandr* Drachuk[1]

[1]State University of economics and technology, Kryvyi Rih, Ukraine
[2]National Technical University of Ukraine "Igor Sikorsky Kyiv Polytechnic Institute", Kyiv, Ukraine
[3]Bilotserkivsky Institute of Continuous Professional Education, Bila Tserkva, Ukraine

Abstract. The article considers a comprehensive study of priority areas of staff motivation. Based on the systematic approach the practical ways of implementation a motivational mechanism as a key component of improving personnel policy in a banking institution were developed. The role and organization of the motivational mechanism and stimulation of labor activity in the personnel management system in the modern organization were covered. The existing methods, techniques and indicators for assessing the personnel motivation systems taking into account the financial stability of the banking institution were analyzed. Improved systems of work motivation in a modern domestic bank were proposed taking into account the monitoring and forecast of indicators of its financial and economic condition. Motivational approaches were evaluated based on the development of an effective system of material incentives and compensation (intangible) staff package in order to increase the interest of employees in obtaining the maximum result of activities in the studied bank. A comprehensive program of organizational, economic and managerial measures to improve the system of socio-economic motivation of staff in the bank was formed in accordance with the strategic guidelines of the institution.

1 Introduction

In a socially oriented market management system the problem of work motivation systems development becomes important. Implementation of the domestic economy development innovative model requires the creation at the state level of appropriate systemic conditions for progressive motivation and stimulation of staff work. Business communication helps very good in establishing a motivational system. Communication is a global motivating system that helps to strengthen such motives as: recognition of merits, striving for achievement, self-improvement.

The purpose of the work is a comprehensive review of business communication with coverage of personnel policy and identification of staff motivation priority areas in a modern organization, review of basic methodological approaches and practical ways of motivational mechanism development based on the study in a banking institution.

To this goal achieving the following tasks were solved:

- to investigate the role and organization of the motivational mechanism and stimulation of the staff labor activity in the context of personnel policy management in a modern banking institution;
- to analyze the existing methods, techniques and indicators of assessing the use of monetary and intangible staff motivation in banking institutions;
- consider the assessment of motivational approaches based on the development of an effective system of material incentives and staff compensation package with the aim of increasing the interest of employees in obtaining the maximum activity result in the studied bank;
- to offer improved systems of labor activity motivation in a modern domestic bank, taking into account the monitoring and implementation of effective organizational and managerial measures;
- to form a comprehensive program of organizational and economic measures to improve the system of staff socio-economic motivation in the bank in accordance with the strategic guidelines for the development of the banking institution.

The following scientific research methods were used in the article: system analysis, abstraction and analogy - to deepen the research methodology of the business communication formation process in a modern organization; economic and statistical (groupings, graph-analytical, correlation-regressive, main components), factor and functional analysis, classification, typology, comparative analysis - to assess the effectiveness of personnel policy and the dynamics of monetary and intangible motivation of bank staff, determining its impact on the institution final results, the effectiveness of economic incentives for staff, determining the optimal structure of the motivational mechanism; graphic - to visualize the issues considered.

The scientific novelty of the results is to develop a sound methodology for comprehensive assessment of the business communication mechanisms effectiveness and intangible incentives for bank employees that allows to

* Corresponding author: nata520522@gmail.com

diagnose personnel policy and socio-economic staff motivation based on the results of the banking institution in today's difficult business conditions.
This paper will consider the motivation of modern banking institutions employees.

2 Analysis of recent research and publications

Business communications - intergroup and interpersonal communications, the specifics, structure and functions of which are determined by the scope of business relations they are used.

Business communication allows an organization to share goals with shareholders both inside and outside the organization. Also, thanks to communication, the organization can gain the loyalty of employees and customers. Thanks to an effective communication system, the organization can calm angry customers and impress new ones [16].

Many scientific publications have been devoted to the study of problems and prospects of personnel policy formation. In particular, we consider it appropriate to single out the works of Parkhimchyk E.P. [15], Adizes I. [1], Bayrachna O.K. [2], Oyvental A.V. [14], Mamotenko D. Y. [11], Hrytsai A.M. [8], Udovenko T.S. [17], Lepeyko T.I. [10], Bortnik S.M. [5], Mornell P. [12], Berkova K. [3], Delamare le Deist F [6].

Thus according to Parkhimchyk E.P., personnel policy of the organization - is formulated (orally or in writing) principles, priorities, norms, rules of work with personnel obligatory for all participants of the personnel management process aimed at achieving organization tasks and strategic goals, and are used taking into account constant changes in internal organizational conditions and requirements of the external environment [15].

To implement the management functions of planning, organization, motivation, leadership and control, it is very important to maintain communication between stakeholders of the organization. Communication helps to deal establish between different departments of the organization, so information in modern organizations should be transmitted faster than before [13].

Analyzing scientific sources, we can conclude that scientists study mainly general trends in the development of theoretical principles and methodological approaches to the efficiency of staff activity and the formation of motivational mechanisms. Moreover the problems of formation of communication and motivation management systems, including intangible, insufficient attention is paid.

Communication processes, especially business communication, are carried out with certain goals, intentions, so the parties seek to provide not just data, but, in their opinion, ready-made information, freely or involuntarily processed, composed, formulated in such a way to obtain the expected result, behavior, reaction.

The main tasks of business communications are:

- creating and maintaining an image;
- study of external influence on the activities of the entity;
- establishing relations between the entity and target audiences;
- creating a favorable socio-economic climate and a favorable environment for doing business;
- assistance to the entity of economic activity in obtaining accurate and timely information;
- search, study and discovery of new business opportunities;
- establishing communications within the organization.

Business communications of the entity of economic activity play an important role in the implementation of strategic goals of the organization. Communicating properly formulated messages of the entity to its target groups forms a favorable basis for activities in the system of economic relations, helps to gain public support to protect the interests and reputation of the entity of economic activity. In the theory and practice of energy management, monitoring and planning systems are used in Western Europe, the USA, Japan (Pooley John, 2005; Jones Phil, 2004), the purpose of the creation and operation of Monitoring and Targeting Systems, in addition to operational management of the efficiency of the use of fuel and energy resources, is also monitoring of the real results achieved in implementing energy saving projects at a certain management object [2].

3 Theoretical and methodological bases of the banking institutions employee motivation program development

In the economic literature, the concept of "motivation" is interpreted from the influence standpoint not only internal but also external factors in relation to man and their interaction that motivates man work to achieve personal goals and goals of the organization. According to this understanding, the purpose of motivational management is to form a set of conditions that motivate a person to take actions aimed at achieving the goal with maximum effect [4].

Methods of staff incentives can be very diverse and depend on the operation of the incentive system, the overall management system of personnel policy and the characteristics of the organization. Consider ways to improve work motivation. They are combined in several independent directions. Wages play a great role in motivating, but the constant increase in wages does not help to maintain labor activity at the appropriate level and increase productivity. To stimulate employees you can also use the system of internal benefits to the enterprise workers, intangible (non-economic) benefits and staff privileges. Equally important is the creation of a favorable social atmosphere and a variety of activities increasing the content of work, independence and responsibility of the employee stimulating his skills growth. Involving employees in the enterprise management also increases their motivation because in this case the problem of alienation from the organization and its leaders is solved.

JSC CB Privatbank is the largest Ukrainian universal commercial bank operating under the license of the National Bank of Ukraine № 22 dated October 5, 2011, focused on servicing individuals and corporate clients of all forms of ownership.

Table 1. Results of PrivatBank's activity for 2017-2019.

Indicators	**2017**	**2018**	**2019**
Annual financial result, UAH million	-23914	12789	32609
Assets, UAH million	254805	278048	309723
Total equity, UAH million	24793	31464	54529
Total liabilities, UAH million	230012	246584	255194
Total liabilities and equity, UAH mln.	254805	278048	309723
Remuneration fund, UAH million	4559	5729	7300
Average number of employees, persons	22670	22600	25797

Looking at the table you can see at once that starting from 2018 economic indicators are improving, especially the large increase in profits in 2019.

For a more thorough analysis maps of competitive advantages were constructed according to the Ministry of Finance [9].

The x-axis indicates the level of credits, the y-axis indicates the size of deposits. 1.

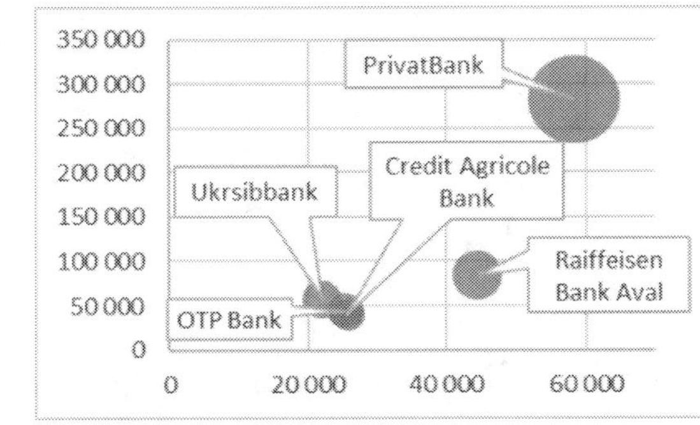

Fig 1. Map of competitive advantages by the amount of loans and deposits.

In terms of loans and deposits the closest competitor of PrivatBank is Raiffeisen Bank Aval. Less competitive there are Ukrsibbank, CrediAgricol Bank and OTP Bank that very close to each other.

In the following map of competitive advantages the x-axis indicates the level of stress resistance; the y-axis indicates the size of depositors loyalty.

On this map the leading banks have changed. PrivatBank has low indicators of stress resistance and depositors loyalty. The banks with the highest stress resilience and depositors loyalty are Ukrsibbank, OTP Bank and Raiffeisen Bank Aval. There are unusual indicators in CrediAgrikol Bank: stress resistance is high while depositors' loyalty is low.

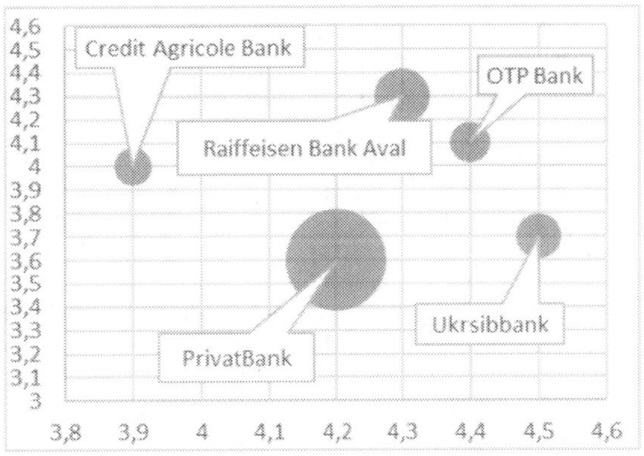

Fig. 2. Map of competitive advantages in terms of stress resistance and depositors loyalty

For a bank function well it needs a balanced personnel management system. To assess the system it is first necessary to examine the financial condition of PrivatBank, the factors influencing the bank's activities to determine competitors, strengths and weaknesses of the bank.

Table 2. Indicators of financial and economic condition of PrivatBank for 2017-2019.

Indicators	2017	2018	2019
Return on assets, %	-9,39	4,60	10,53
Return on capital, %	-96,45	40,65	59,80
Autonomy ratio	0,097	0,113	0,176
Liquidity, %	17,26	17,90	16,21
Financial leverage ratio	9,28	7,84	4,68
Reliability ratio	0,11	0,13	0,21
Equity participation in asset formation ratio	0,10	0,11	0,18

In 2017 the bank's indicators are bad but since 2018 the indicators are improving. And in 2019 the increase in return on assets and return on capital is particularly noticeable.

Currently JSC CB "PrivatBank" uses eleven ways of intangible motivation of employees as one of the important components of personnel policy of the institution:

- motivational meetings;
- meetings with significant dates;
- training: refresher courses, internships, seminars and conferences;
- the right to choose;
- possibility of feedback. It is necessary to give employees the opportunity to express themselves - wishes, comments on the work organization;
- equipped places for rest. It is important for almost all employees to be able to have lunch not at the desk but in a separate room, so as not to disturb others and relax;
- informing about the achievements and evaluations of employees;

- competitions and contests;
- personal working area. It can be a sign on the door, a card on the table, a badge - these are the tools for intangible motivation of employees;
- discounts on services or products of the company;
- encouraging corporate events: a festive banquet or a trip to nature, a corporate outing to the skating rink or to the cinema (in the absence of quarantine restrictions).

Intangible motivation of PrivatBank involves the creation of an appropriate motivational climate in the team, working conditions and intangible incentives. The efficiency of intangible motivation in the team depends primarily on the leader. The working climate in the team is created by the bank's managers by:

- providing opportunities for professional and career growth; staff rotation;
- informing employees about the goals, objectives of the bank;
- the evaluation of the work of subordinates (both positive and negative);
- giving the employee the opportunity to make decisions independently within their powers;
- creating conditions for "immersion" in work;
- ensuring a favorable psychological climate in the team; open encouragement of the employee for the achieved results;
- giving the employee the opportunity to participate in the decision-making process;
- support of the initiative employees; organization of mentoring; delegation of responsibilities and powers.

Special departments (sectors) for work with personnel are engaged in development of motivational strategy of the bank. Most modern banks, namely PrivatBank, use the following methods of motivation [7]:

- remuneration system: salaries of JSC CB "PrivatBank" employees consist of rates and bonuses. The piece-rate bonus system of payment is used;
- rating - is the final indicator characterized the efficiency of a particular employee in comparison with his colleagues within one profession;
- the social package is a set of benefits, privileges and incentives provided by the bank to its employees;
- motivational climate: in PrivatBank intangible motivation involves the creation of a motivational climate in the team, working conditions and intangible incentives;
- staff training and development: today PrivatBank has one of the most effective staff training and development systems.

The issues of staff training and development in PrivatBank are coordinated by PrivatUniversity and training centers [18]. Each training center has coaches - employees of the bank who teach and transfer their experience to younger employees.

4 Research results

To implement the strategy of future development, five areas have been selected, in particular: to update the competitive offer, create a corporate segment, implement a cost optimization program, strengthen IT, and improve risk management.

All these innovations will make the bank more efficient and will have the following effects:

- accelerating the bank's work with customers and improving the service model;
- competitive offer will increase due to customer segmentation, organization update and remuneration;
- cost optimization by diagnosing activities and identifying potentials;
- strengthening the IT sphere, hiring professional experienced specialists and their stimulation;
- renewal of the risk management committee.

To implement all these measures an improved system of staff motivation and research into the impact of the new system on the bank's activities is needed. The system of such incentives is offered:

- raising wages is a simple and effective way;
- development of a system of career growth combined with an increase in wages, it will attract the attention of ambitious people;
- indirect financial assistance (payment for medical services, assistance with rent, etc.) that will help make work in the bank more attractive;
- holding corporate holidays will help to strength friendly relations in the team and increase employee loyalty;
- conducting training courses for employees will meet the needs of ambitious employees for self-training and improvement - a positive effect for the bank is the presence of a more qualified employee.

For the successful functioning of the motivational system it is necessary to develop communication channels so that the employee has an idea of the main aspects of the enterprise and structural units. Thus at the stage of employment a person should have information about the level of wages, career opportunities that is indirect financial assistance and opportunities for self-improvement available in the organization. Such information will significantly increase the attractiveness of work for future and existing workers. We must also not forget that you need to constantly inform employees about the future development strategy of the enterprise. That is why it is very important to develop both internal and external sources of information to maintain the image of the organization.

One of the most important factors in the success of the organization is the involvement of employees in the strategy development and management of the organization. Each employee should have a role in the development of the organization that will greatly increase the employee's interest in the success of the company. That is why reliable communication channels are needed to disseminate all the necessary information.

Given the use of all proposals a forecast of financial and economic indicators of Privatbank for 2021 was made.

Table 3 shows that the new system of staff motivating using business communication will help PrivatBank to increase its profits in 2021 by 100.10% that is 32642 thousand hryvnias. It increase also the reliability and autonomy ratio, liquidity and generally the economic condition of the bank.

Table 3. Forecast of Privatbank's performance indicators taking into account the improvement of personnel policy for 2021.

Indicators	2019	2021	Absolute	Relative, %
			2019-2021	2019-2021
Annual financial result, UAH million.	32609	65251	32642	100,1
Assets, UAH million.	309723	456009	146286	47,23
Total equity, UAH million	54529	88631	34102	62,54
Total liabilities, UAH million	255194	367377	112183	43,96
Total liabilities and equity, UAH mln.	309723	456009	146286	47,23
Remuneration fund, UAH million.	7300	10162	2862	39,20
Average number of employees, persons	25797	39601	13804	53,51
Return on assets,%	10,528	14,309	3,781	35,91
Return on capital,%	59,801	73,620	13,819	23,11
Autonomy ratio	0,176	0,194	0,018	10,40
Liquidity,%	0,162	0,192	0,030	18,63
Leverage ratio	4,680	4,145	-0,535	-11,43
Reliability ratio	0,214	0,241	0,028	12,91
Equity participation in asset formation ratio	0,176	0,194	0,018	10,40

In addition, the innovations will have a positive impact not only due to improved economic performance, but also due to the fact that the new personnel management system provides for the improvement of employees' skills that means that the overall professional level of the bank's employees will increase. It is also effective to increase wages through a career growth system that encourages the employee to work better and provide creative ideas to improve the bank's performance. Due to effective business communication people's interest in work will increase, due to that a significant increase in the number of employees is planned. In general we can say that due to the developed system of communications the motivation of employees increases significantly. Due to greater participation in the development of the personnel policy the attractiveness of the banking institution for new potential highly qualified professionals is growing.

5 Conclusion

Thus based on the conducted research of practical ways of the motivational mechanism development the necessity of reasonable stimulation of labor activity of the personnel in the system of management of modern banking institutions is proved. The existing methods and indicators of assessment of financial and economic condition and using of motivation systems by personnel in banks are analyzed. The assessment of motivational approaches based on the development of an effective system of material incentives and compensation (intangible) staff package in order to increase the interest of employees in obtaining the maximum result of activities in the studied bank is considered also.

In the context of continuing scientific research in the chosen area it is advisable to offer further study of opportunities for improvement the system of work motivation in a modern domestic bank taking into account monitoring and development of management measures based on the results of economic and mathematical modeling. This will form a comprehensive program of increasing financial stability and implementation of organizational and economic measures for business communication to improve personnel policy in difficult conditions of quarantine restrictions, in particular the system of socio-economic motivation of staff in the bank in accordance with strategic guidelines.

References

1. I. Adizes. *How to overcome management crises.* Diagnosis and solution of management problems. M.: BestBusinessBox, 286 p. (2007)
2. O.K. Bayrachna. *Personnel policy of the organization in the system of functional training of management staff.* Ukrainian Journal of Applied Economics. Volume **5**. № 1. P. 342–348. (2020)
3. K.Berkova, P. Krpalek, K. Krpalkova Krelova (Czech Republic). *Future specialists in economics: development of practical skills and competencies in higher education from the point of view of international employers.* Economic Journal XXI **176** (3-4), 91-98. (2019)
4. N.M. Bogatska. *The value of staff motivation to achieve the goals of the enterprise.* Regional Economy. № **6**. P. 11 - 17. (2009)
5. S.M. Bortnik. *Strategic management of enterprise personnel development and modern approaches to assessing its level.* Economic Journal of the Lesia Ukrainka East European National University: a journal. Lutsk: Vezha-Druk. № **2**. P. 33-39. (2015)

6. F. Delamare le Deist, J. Winterton. *What is competence?* International Human Resources Development, vol. **8**, No. 1, 27 - 46 March. (2005)
7. O.P. Dyakiv. *Internal environment of corporate social responsibility.* Materials of the international scientific-practical "Imperatives and innovative mechanisms of ensuring decent work in the conditions of formation of the new economy" Kyiv: KNEU, P. 346–348. (2017)
8. A.M. Hrytsay. *Theoretical and methodological principles of the enterprise personnel policy formation.* Scientific Bulletin of Poltava University of Economics and Trade. № **1** (63). P. 148–155. (2014)
9. https://minfin.com.ua/banks/rating/?date=2020-10-01
10. T.I. Lepeyko, O.M. Mironova. Personnel management of the enterprise in conditions of uncertainty (behavioral approach): monograph. Kharkiv: Ed. KhNEU, 236 p. (2015)
11. D. Y. Mamotenko. *Development of personnel strategy of the organization in modern conditions.* Scientific Bulletin of Kherson State University. №**14**. P. 49–52. (2015).
12. P. Mornell. *Hiring efficiency technologies.* New system of personnel evaluation and selection. Moscow: National Space Bank; Good book, 264 p. (2005)
13. S.Tulchynska, N. Shevchuk, V. Chornii, B. Chornii. *Using a methodical approach to the evaluation of attractiveness investment resources for electricity distribution companies.* Scientific Bulletin of National Mining University . Issue **2**, 130-135. (2018). http://nvngu.in.ua/jdownloads/pdf/2018/02/NVNGU 02_2018_Tulchynska.pdf.
14. A.V. Oyvental. *Implementation of personnel strategy of the organization.* Problems of management theory and practice. №**4**. P. 61- 63. (2017)
15. E.P. Parkhimchik. Personnel policy of the organization: [textbook. Aid.]. Minsk: GIUST BSU, 128 p. (2011)
16. M. Radovic, A. Salamzadeh The Importance of Communication in Business Management Researchgate.net
17. T.S. Udovenko, C.M. Nevmerzhytska Formation of effective personnel policy at the enterprise [Electronic resource]. Access mode: http://knutd.com.ua/publications/pdf/Ukrainian_editions/ Nevmerzicka2015022807.pdf
18. V.D. Yakubenko. *Participatory management in the socialization of corporate relations.* Scientific Bulletin of the Chernivtsi Trade and Economic Institute KNTEU: Coll. Science. wash. Vip. IV. Chernivtsi. P. 19-24. (2003).

Plurilingual Approach to Improving Lexical Competence of Non-Linguistic Majoring Students

Olha Chyzhykova[1], *Oksana* Romanenko[1], *Iryna* Bazarenko[1*], *Larysa* Dzevytska [2], and *Ainur Jgiyatker* Baitelessova [3]

[1] State University of Economics and Technology, Kryviy Rih 50000, Ukraine
[2] Kryviy Rih Donetsk Law Institute of Ministry of Internal Affairs of Ukraine, Kryviy Rih, 50000, Ukraine
[3] Association of Silk Road Universities and Consulting companies, Almaty, Kazakhstan

Abstract. This paper presents innovative approaches to improving lexical competence of non-linguistic students. The authors analyze the concept of terminological vocabulary in professional discourse, paying special attention to peculiarities of teaching vocational-oriented vocabulary to non-linguistic majoring students. Pedagogical technologies based on plurilingual approach in teaching professionally oriented vocabulary are offered in this article. The authors disclose the main features of acquiring professionally oriented vocabulary in distance learning and its effect on the learning process. Furthermore, the best practices for the English-German-French combination are explored in teaching non-linguistic students professionnally oriented vocabulary from a plurilingual and communicative perspective. The findings also show that these activities not only motivate learners to be engaged in meaningful language learning experience, but also encourage them to improve their plurilingual communicative competence, therefore enrich their professionnally oriented vocabulary. The focus lies on exploring the use of plurilingual approach as an effective means for improving professional linguistic competence. Recommendations for teachers and non-linguistic students on mastering their professionnally oriented vocabulary are presented in this paper.

1 Introduction

Modern processes of globalization of the world economy require effective professional interaction in the study of foreign languages, including second foreign ones. It is particurlarly important for non-linguistic majoring students who will become in future the specialists in different fields of our national economy. Good knowledge of a foreign language in modern society is recognized as an imperative attribute of any specialist namely of those who will work in the sphere of international economic relations.

It is undisputable the fact that any graduate of the university should not only receive special language knowledge in all directions of the chosen specialty, but also be able to use best its vocabulary and grammar constructions, work with all types of material on the chosen major and, additionally, know and use the techniques of referencing and annotating. It also acquires increasing professional value of communication with foreign colleagues.

Thus, knowledge of foreign languages affects career moves and the development of professional qualities, thereby ensuring to a large extent the competitiveness of specialists on the labour market.

In order to reveal their professional talents and opportunities in the process of international interaction, graduates of universities, in accordance with current educational standards, need to be able to continue their studies and conduct professional activities in a foreign language environment.

The graduates of non-linguistic higher educational establishments should be ready to participate in both all-Ukrainian and international innovative projects as well as to use the advanced scientific achievements of foreign scientists, information about which is available on the Internet for those who know and use special terminology in their professional activity. Bearing in mind that the knowledge of special terms of any modern industry is aimed not only at satisfying internal, national needs of developing the national economy, but also at expanding external, international contacts, we understand that the international function of professional terminology is increasingly growing.

Accordingly, it is advisable to teach terminology in accordance with the scientific principles of humanization, which are manifested in the fact that students learn about the cultural characteristics of the country being studied, acquiring the norms of communication and behavior characteristic of another national culture and understanding that the development of any new term depends on solving the language issues in intercultural people interaction.

1.1 The concept of terminological vocabulary in professional discourse

A new terminological vocabulary is included and enshrined in professional discourse as a result of the cognitive activity of a specialist in a particular industry, which consists in conceptualizing and categorizing new and existing different types of knowledge. The complex nature of cognition is due to the multiplicity of objects of the surrounding world, their difference as well as the degree of cognition and choice in consciousness, which

* Corresponding author: irinabazarenko@gmail.com

implies the relationship and integration of various cognitive levels.

The main ones include three: empirical perception (perceptual-substantive activity), conceptual understanding and interpretative-evaluation understanding. At the same time, the analogue capabilities of human thinking make it possible to effectively integrate new knowledge into already existing system, which is reflected at the verbal level, namely, when using terms in professional discourse. Of particular importance is updating the language unit of its synthagmatic properties based on the value of the lexical concretizer.

The development of new syntactic meanings is determined by the concepts system of certain sciences and is an objective phenomenon due to the fact that the cognitive activity of consciousness contributes to the isolating new features of known concepts through theoretical thinking. Such expansion of the component composition in terminology leads to new paradigmatic and synthagmatic meanings of lexemes, known primarily to those skilled in different fields of knowledge. Accordingly, these words can be implemented in different terminology systems.

The problem of communicative actualizing terminological vocabulary in discourse today cannot be considered unambiguously, but it should be noted that the dynamism of modern life associated with the rapid development of science and high technology contributes to the disappearance of clear boundaries between the pre-scientific and scientific (conceptual) world pictures. The formation of new scientific concepts objectively causes the further development of the terminological meaning and its enrichment in connection with the reflection of new reality signs.

Resulting from the relationships of the form and content, the modified termological meaning has a certain new form of expression. Thus, the principle of correlation with current understanding the scientific picture of the world is one of the fundamental ones in the study of terminology.

Acquiring terminology systems is an important step in forming lexical competence of non-linguistic students. *Lexical competence,* in its turn, means the knowledge of the vocabulary and also the usage of this knowledge in different contexts [5, 14]. Thus, M. S. Thirumalai (2002) underlined that it also means "having the knowledge of how the vocabulary is used in sociolinguistic and sociocultural context" [19]. According to Muhammad Din & Mamuna Ghani (2018) "*Lexis* or vocabulary in its turn refers to an item of meaning. A foreign or the second foreign language learning involves the processes of comprehension, acquisition and recalling of lexical units as it constitutes the base of the learning process" [17].

K. Caro and N. R. Mendinueta emphasized that "lexis is one essential component of language and language development. Limited lexical knowledge can lead the learners to frustration and demotivation. To reach a higher level of development in the four basic communication skills, learners should have a basis of lexis that allows them to do so. Lexis is broader and engulfs vocabulary, lexemes and lexical items. Lexical competence encompasses lexical knowledge and its use in diverse contexts" [14].

Some scientific principles proposed for the study of professional terminology in universities should contribute to the effective assimilation of terminology for successful professional communication in a foreign language, which reflects modern trends and requirements for higher education.

It is worth noting that at present, the teacher's task is not only to equip students with modern knowledge, but also teach them to obtain this knowledge on their own, be able to absorb it based on what has already been studied. To achieve these goals, different training systems were introduced in universities, for instance, the Moodle station training system.

1.2 Innovative approaches to teaching non-linguistic students professionnally oriented vocabulary

The acquired knowledge helps students to master the terminology norms not only of their native language, but also of any foreign language, which provides further successful professional communication. The basic principle of studying professional terms was the principle of communicative actualizing terminological vocabulary in discourse. Today, one of the most promising areas of studying sectoral terminological vocabulary is cognitive-discursive, which is based on "the definition of language as a cognitive process carried out in communicative activities and is provided by special cognitive structures and mechanisms in the human brain " [7].

Scientists [1, 12, 17] analyze numerous approaches on improving professional lexical competence of the learners. Among them:

1) text-based strategies with the usage of computer-assisted language learning (students are working with a computer-based package, including annotated texts; the introduction to lexicology; a personal data base of vocabulary, a number of different texts chosen for reading by students themselves including literary works and websites) [12];

2) foreign languages learning outside the classroom environment which is programmed for vocabulary learning, when students do extracurricular activity (watching television, working with the Internet, etc.) [12];

3) development of students' lexical competence (students are given tasks to improve their own understanding of the foreign language lexicon, and first of all, to be aware of what they are faced with while reading and listening: instead of guessing wildly about the meaning, or disregarding unfamiliar lexical units, they now get the tools to analyse difficulties they have faced and to learn from the experience for future educational activities) [12];

4) active usage of a given list of devices and a dictionary may be of great assistance in order to facilitate students' effort to find the correct lexical item [1];
5) using phraseological units, embedded clauses, if-clauses may be helpful [1];
6) task-based strategies and activities should be implemented in the process of foreign language learning [17].

1.3 Peculiarities of teaching vocational-oriented vocabulary to non-linguistic students

Considering the current state of studying vocational-oriented vocabulary at non-linguistic universities, some points should be noted. Students who already have a sufficient base for proficiency in English after two first years should learn to use it in professional environment, because the program requires further professionalization of training. Unfortunately, second year students do not yet have clear understanding of their future profession and what their range of duties or responsibilities will be. This complicates the process of mastering business foreign language.

However, among the competencies, the formation of which is provided for by the process of learning a foreign language, lexical competence occupies a leading place. Since lack of grammar skills does not significantly affect the understanding of foreign language information, but misunderstanding of vocabulary can greatly worsen the process of perceiving new material by students of non-linguistic specialties. A student from the beginning of his studies at the university should be prepared to acquire professionally-oriented terms, use them in learning and in further professional activity.

Regarding the second foreign language, it should be noted that the acquisition of English professional vocabulary and its practical skills improve the second foreign language learning, since English knowledge acquiring ensures mastering of necessary luggage as lexical knowledge, which in the future will act as a solid basis for second foreign language competence formation.

According to the program the students begin to study the second foreign language when they already understand the peculiarities of their future profession. The fact that students master it in parallel with a second foreign language learning facilitates to a large extent their preparation for the perception of professional terminology.

Besides, there are some other ways of motivating students to second foreign languages learning. Modern computer technologies have greatly influenced all aspects of our lives, and most importantly, provided students with free access to information and communication.

Practice shows that due to the availability of mass media, expansion of boundaries in getting new information, the influence of the Internet, which has long crossed all possible borders, modern students are easily attracted to foreign language communication, they are steadily looking for interpersonal contacts, and if they exist, without difficulty enter into verbal relations with peers, ready to discuss with them both household problems and work issues.

Modern life prompts that young people would be involved in the future profession from the early beginning of their student life. Programs, courses, internships, and even just communication on professional topics - all this creates stable prerequisites for the successful formation of lexical competence. However, in order to direct the acquisition of foreign language vocabulary into the professional course, it should be useful to develop a strategy for the effective mastering of professional terminology from the early beginning of a foreign language learning.

The main teacher's task is the creation of motivational settings, the preparation of educational and methodological support, the implementation of direct leadership and the management of each student's independent work on educational material. To solve the problem of ensuring the necessary motivational mood of students, the teacher must first of all disclose (identify, justify, classify) the goals of studying this discipline and show the need, utility, significance of mastering this composition of special knowledge and skills.

For instance, the rating system for assessing students' knowledge including regular monitoring of each student's knowledge and skills, with the assignment given in accordance with the the aim of training has proved its efficiency. It should be the main part of the complex of continuous multilevel education and is designed to organize the process of learning based on educational technologies, which strongly activates the independent students' work.

Due to this fact, individualization of training with the development of self-reliance skills should be the main educational principle of distance technology. Noteworthy that distance learning foresees autonomy in a broad sense, including self-government, self-control, self-esteem, self-organization, self-motivation, self-confidence, self-realization, etc.

In the context of distance learning, students have access to various sources of information, and it is important to help each of them compile an educational set that best suits their individual characteristics. Therefore, early differentiation of training related to traditional educational structures and access to individual training using methods of remote access to educational programs are important.

Training tools used in distance learning technology include the following tools:

- a printer textbook designed for independent training and containing, in addition to methodological materials, a glossary, a scientific examination of the educational material, as well as a number of exercises for their acquiring;
- a video lecture providing educational material in dynamics and using auditory and visual channels of information perception, as well as an audio lecture that enhances the assimilation of the material with the help of emotional influence on the listener (overview training);
- wall-mounted educational materials of the corridor and office type, which are changed as students move

through the studied discipline (immersion in the educational environment);

- active seminars in the form of discussions, round tables, role-playing games, business games and other game forms, imitating various game and professional situations and involving the active presentation and use of knowledge and skills by each student;
- problematic lectures aimed at the development of cognitive activity of students;
- tools for monitoring work, written and oral;
- standard tests for each specialty in the form of questions covering all educational material for multiple-choice comprehension;
- computer-based curricula and tests in a subject or part thereof, which determine the level of knowledge in each didactic unit and give the student and teachers a holistic picture of knowledge acquired in a particular field.

With the help of educational computer programs, various skills are trained and thereby individual assimilation of algorithms of various actions takes place, to name just a few:

- Case technology, which is a close analogue of distance learning technology (when a student receives a special set of teaching materials (case) for self-studies and periodically consults with teachers in regional training centers created for this purpose;
- TV-technology, the use of which foresees television lectures and consultations with teachers on line;
- Network technology based on the application of educational materials sent to students for consultation by means of Internet technologies.

Thus, a vocational-oriented approach creates a sustainable motivation for learning a foreign language as a means of improving professional knowledge. When teaching a certain specialty, a business foreign language, in other words professional language, should be combined with future specialists' professional activities. Professional language needs to make it necessary to acquire a certain amount of terminology in the process of studying at universities, which is most reimbursable for specific situations of professional communication. Therefore, solving the problem of developing professional lexical skills requires a systematic and integrated approach to the choice of educational methods, technologies and means.

1.4 Integrated Plurilingual Approach to Improving Lexical Competence of Non-Linguistic Students

M. González-Davies (2017) defines "an *Integrated Plurilingual Approach* as a plurilingual learning strategy to improve foreign language learning and plurilingual communicative competence" [18]. This approach is aimed to foster advanced language learning on the basis of plurilingual competence in order to promote communicative competence among students [2, 5].

According to C. Ernest [2], plurilingual students are able to use more frequently "metalinguistic awareness and strategies such as risk taking, learning from mistakes, translating, adapting words or understanding general meaning" in comparison with monolingual ones. "*The plurilingual speaker* is communicatively competent in several languages, at whatever level of proficiency shaped by her different languages and her experiences of languages, because these languages naturally intertwine within the plurilingual mind" [2].

According to C. Ernest [2], "a plurilingual approach towards languages based on IPA should imply the development of communicative competences and the need to abandon a monolingual approach to languages and favour the language repertoire of students, intercultural competence, a positive attitude towards language(s) and language learning" [2].

She also supports the idea of the text as the basic learning instrument and thinks concept-based learning to be very useful for learning foreign languages.

With regard to all mentioned above, we consider plurilingual approach as an effective tool in improving professional lexical competence of future economists that should be used in the process of learning foreign languages according to the needs of modern globalized world.

2 Our Contribution

The given research shows how important the study of professional terminology by non-linguistic students is especially in the conditions of modern globalized world. As already noted, formation of lexical competence is considered to be one of the essentaial parts of teaching non-linguistic majoring students the first and the second foreign languages in higher educational establishments. Our contribution is the analysis of current state of foreign languages learning and the research of modern approaches to acquiring professional terminology.

As different approaches to professional terminology training are shown in the article, namely a computer-assisted, text-based approach providing students with a computer-based package, plurilingual approach is defined as a reliable tool in teaching students to work with professional terms. Its objectives are defined as the next: to advance language learning and plurilingual communicative competence.

Taking all this into account we are trying to define the best methods of improving and fostering the formation lexical competence of non-linguistic majoring students with the aim of their best preparation for further professional acitivity.

Therefore, the objectives of the given research are to find the best ways of providing the most effective training of non-linguistic students for the use of professional foreign language, to give comparative analysis of professional terminology peculiarities of three European languages for better assimilation and use of the terms in professional communication, consider pedagogical technologies to be used by university teachers and work out the exercices for mastering better understanding of professional technology, providing its right use in professionally oriented texts with regard to the similarity of language system of different European languages. This is the first attempt to define the best approaches to professional terminology learning based on plurilingual approach to acquiring professionally oriented vocabulary.

2.1 Pedagogical Technologies based on Plurilingual Approach in Teaching Professionnally Oriented Vocabulary

It should also be emphasized that pedagogical education today requires pedagogical technologies that ensure the effective training of specialists who are competitive in the market.

We suggest such activities aimed at improving and fostering the development of economic lexical competence of non-linguistic students as:

1) working with the synonyms

Table 1. Working with German synonyms

der Verbraucher	machen
wirtschaftlich	wertvoll
teuer	der Nachfrager
der Topmanager	ökonomisch
tun	der Geschäftsmann
der Kaufmann	der Generaldirektor

Table 2. Working with English synonyms

shareholder	executives
inexpensive	charge cards
entrepreneur	co-owner
corporation	company
managers	stockholder
co-proprietor	cheap
credit cards	employer

2) finding the proper ending

Table 3. Finding the proper ending in German

1. Der Unternehmer und der Verbraucher müssen die Grundfragen der Wirtschaft	a. die Wirtschaft
2. Der Mensch auch beeinflusst	b. lösen
3. Das Wessen der Wirtschaft bildet	c. eine Selbstverständlichkeit
4. Der Umgang mit Geld ist für jeden heute	d. Schulden bezahlen, Verbindlichkeiten begleichen
5. Man kann mit Geld	e. eine bequeme Einrichtung, um miteinander in Austausch zu treten
6. Überall ist Geld	f. das Rationalprinzip

Table 4. Finding the proper ending in English

1. Advertising is	a. to live within one's income.
2. Financial advisers and counselors are	b. a poor product or service into a good one.
3. Business people pool	c. in business to help people manage their money wisely.
4. Production Department	d. their money, efforts and talents to organize a partnership.
5. It's a good idea	e. any paid form of nonpersonal presentation and promotion of products, services, or ideas
6. Advertising cannot turn	f. is responsible for manufacturing the goods that the company sell

3) working with word derivatives

1. (Finance) reward isn't the only reason for people to go to work.
2. You can select 1,2,3, or 6 months interest (pay) intervals.
3. An (employ) shall not refuse to conduct labour negotiations.
4. PR manager's (responsible) are to create and maintain the best possible image of the company.
5. Let's sign this (agree).
6. He was a (fail) as a businessman.
7. The country exports a large number of different (produce).
8. If you want to get this position you must be very (skill).
9. Soon he was (promotion).
10. From the very beginning Sue was a (success) businesswoman.

4) explaining the meaning of the words and word expressions (ask your partner to help you understand the meaning of one of the words unknown to you):
franchising, franchiser, franchisee, employee, sole proprietorship, corporate tax, stock exchange, shares, interests.

5) making up a story, using the following tips:
to start a business, investment capital, a business plan, pool managerial talents together, to get a loan from a bank, advertising campaign, competition, special offers, to gain a profit, to pay the interest.
Here are the examples of exercises for mastering the knowledge of professional terminology according to the language learned on the basis of plurilingual approach.

Table 5. Exercises for improving plurilingual professional terminology

English professional terminology	German professional terminology	French professional terminology
1. Explaining the meaning of the words and word expressions (ask your partner to help you understand the meaning of one of the words unknown to you). 2. Making up a story with the given bank of words and word combinations.	1. Drafting stories (initiate your own story creation in groups and then translate into English in order to facilitate understanding of the languages). 2. Working with synonyms.	1. Guessing the meaning of the word without using the dictionary. 2. Defining the meaning of the given word from the context. Checking the guess. 3. By means of the analysis of the grammar characteristics defining what part of speech

3. Strategies for translating professionally oriented texts (translate with the use of online translation tools and answer thought-provoking questions: is it possible to literally translate word by word? Is it the same to translate from German into English than from English into Ukrainian; suggest degrees of fidelity in translation by referring to communicative versus literal translation). 4. Create a real-life communicative situation from professional field for other students to solve an issue and then explain the thinking process aloud (in any language(s) that they choose - German or French). 5. Read the following text in three languages (English, German and French) and compare them (look for similarities and differences and find "false friends").	3. Finding the proper ending. 4. Working with word derivatives. 5. Explaining the meaning of the words and word expressions (ask your partner to help you understand the meaning of one of the words unknown to you). 6. The plurilingual guessing game (choose a term in German from the whiteboard and say it in native language and ask your partner to say that word in English paying attention to its professional use). 7. Reading a professionally oriented text in German and giving a summary in English. 8. Replacing a term in a professionally oriented text (give an English equivalent for a term in German).	the given word is. Giving grammatical features. 4. Comparing the use of the given word in English and French, paying attention to its professional use. 5. Giving the examples of the use of the given lexical unit. Making "word chains" with the given term. 6. Making sentences using the term in the context of its professional use. 7. Analyzing the origin of the given term. 8. Looking for the examples of the use of given terms in professionally oriented texts. 9. Concentrating efforts not on theoretical explication but on the practical use of the given term in professionally oriented texts.

2.2 Acquiring professionally oriented vocabulary in distance learning

Distance learning requires more careful approaches to the development of new lexical units. On the one hand, it complicates the process of acquiring skills in working with foreign language vocabulary, on the other hand, the fact that the educational process is forced to move from classrooms to virtual space contributes to the development of new methods of training using the Internet, social networks, virtual space and encourages teachers to seek for new effective methods of working on professionally-oriented vocabulary.

Given this, here are the examples of exercices aimed at learning professional terminology:

1) One of the distance education disadvantages is the lack of interactive communication, which is limited in the virtual classroom. Thus, the creation of such conditions is possible with the proper organization of the educational process. If desired, teachers can diversify the process of acquiring new lexical units while doing the given exercises: guess the meaning of the word that will appear on the screen, try to explain the meaning of the word, make sentences with given words, provide its synonyms, choose the examples explaning the use of the lexem, evaluate the best use of the word in the context, try to guess the word by its characteristics, explain the difference in the use of other meanings. It is quite effective to compile a lexical dictionary of a specialty if the purpose of the lesson is to refresh in memory and recall professional terms studied at the beginning of the previous semester. Students themlelves make up the alphabet of professionally oriented vocabulary. Another effective way to learn a word is to use the phrase matching exercise to handle a given term. For example, the list of phrases that can be compiled with the word "money" is endless.

2) The experience of using computer capabilities in the course of classes is interesting. Writing training can be achieved by offering to write phrases in a chat in a limited period of time, inviting students to write in turn. Thus, their skills in writing without mistakes, grammar skills and the knowledge of the correct meaning of the word are checked. As it turned out, the exercise in which the teacher invites students to act as a teacher is popular among students. Therefore, students begin to better understand the difficulties faced by the teacher in the process of online learning. It is well known that every foreign language activity offered for students may be supported by different Internet and social networks abilities. Using sites, blogs, chat rooms students can not only publish the results of their written tasks but also discuss publicly the problems connected with their further professional activity.

3) While working on the topic "Advertising" it would be useful to give students the task of considering advertising in accordance with its goals and the main objective of attracting customers. The essay on the topic "My Favorite Ad," which explains how the selected version of the ad meets the requirements of advertising goods, and the discussion of given examples would be very useful for best acquiring the economic terminology. It should be noted that the feedback is rather important in such exercices. It usually provides the better understanding of learning professional terminology. It also should be noted that working with advertising material, it would be advisable to use advertising in printed publications, giving students the task of describing and analyzing samples of printed advertising. Moreover, here you can analyze the language tools that are used by the authors. Thus, advertising and commercials are illustrative examples of language

development by manufacturers for better sale of goods. It will be useful to show students that language serves as a means of attracting customers and the engine for selling goods with which the market is now oversaturated.

Conclusions

The research conducted has shown that plurilingualism proved to be an effective means in improving professional linguistic competence.

The aim of providing non-linguistic majoring students with better acquisition of professional terminology can be achieved while subjecting to the given conditions:

1) lexical competence should be an integral part of the language competence of non-linguistic students; thus, its formation must be organized within the process of language knowledge acquisition in higher educational establishments;

2) the knowledge of business terminology even with the lack of grammar skills will help non-linguistic majoring students to understand information and literature on professional topics better as well as provide students with reliable instruments for mastering their specialty in future;

3) while learning the second foreign language it is advisable to analyze professional terms with regard to the plurilingual approach considering the common background of European languages and understanding their roots;

4) in learning process, it would be useful to take into consideration practical usage or character of professionally oriented vocabulary more than grammar prerequisites of their use;

5) assimilation of professional vocabulary of the second foreign language which helps students understand the strong ties not only between languages but also between economic systems in globalized modern world;

6) increasing motivation of students by using innovative techniques (role playing, case study, project-based learning, presentations, brainstorming, group work);

7) special attention must be paid to students' creativity development, particularly in the conditions of distant learning;

8) comparative method may be of great importance and usefulness while studying the first and the second foreign languages and working with professional terms in the fields of economy, finance, management and law with reference to English, an active use of international words in practice;

9) already acquired knowledge and practical skills of the first foreign language should be transferred to the second one;

10) students should be given necessary academic feedback.

Drawing on the above, we therefore believe that the research area chosen is very actual one and has its own prospects.

References

1. C. Demir, Competence in lexical boosters and nativeness in academic writing of English: The possible relation, Journal of Language and Linguistic Studies, **13**, 593, (2017). https://www.researchgate.net/publication/329310788_Competence_in_lexical_boosters_and_nativeness_in_academic_writing_of_English_the_possible_relation
2. C. Ernest. A plurilingual approach to language teaching and learning in Catalonia: Using heritage languages in the additional language classroom. http://hdl.handle.net/10803/403850
3. C.Ahmimed , S.Quesada-Montano, Intercultural dialogue - a Tool for Young People to Address Exclusion in Southern Africa, Journal of Intercultural Communication, **50**, (2019). https://www.immi.se/intercultural/).
4. C.Hadfield, J.Hadfield,Teaching grammar creatively. Creativity in the English language classroom, London: British Council,**51**,(2015). https://englishagenda.britishcouncil.org/sites/default/files/attachments/pub_f004_elt_creativity_final_v2_we b-1.pdf
5. Council of Europe, A Common European Framework of Research for Languages, Language, Teaching, Assessment, Cambridge, (2001). https://rm.coe.int/16802fc1bf
6. Council of Europe, From linguistic diveristy to plurilingual education: Guide for the development of language education policies in Europe, (2007). https://rm.coe.int/16802fc1c4 doi: 10.1558/lhs.v11.1.00000
7. E. S. Kubryakova, Language and knowledge.On the way to obtaining knowledge of language: Parts of speech from a cognitive point of view. The role of language in knowing the world, Languages of Slavic culture, 560, (2004).
8. E. S. Yemelyanova, Terminology and professional communicative culture of agricultural specialists, Kharkiv, (2013).
9. E.S. Melyanova, Ukrainian-Russian-English-French Explanatory Dictionary of Basic Terms on Energy and Electrical Engineering, Kharkiv, (2012).
10. E.S. Polat, A.E. Petrov, Distance learning: what should he be, Pedagogy, **7**, 29(2010).
11. H.Cherednichenko, L.Shapran and L. Kunytsia, Methods of Teaching Foreign Professionally-Oriented Reading for Students of Non-Linguistic Higher Educational Establishments, Teaching Languages in Higher Educational Institutions, **17**, 150 (2010).
12. J. Lachlan Mackenzie, Mike Hannay, Developing lexical competence: a computer-assisted, text-based approach, Cauce: Revista de filología y su didáctica,

ISSN 0212-0410, **24**, 541, (2001). https://www.researchgate.net/publication/28169507_Developing_lexical_competence_a_computer-assisted_text-based_approach (Emerging Sources Citation Index, MLA - Modern Language Association Database)

13. J. Richards, & T. Rodgers, Approaches and methods in language teaching. (3rd ed.). Cambridge: Cambridge University Press, (2014). doi:10.1017/cbo9780511667305.
14. K. Caro, N. R. Mendinueta. Lexis, Lexical Competence and Lexical Knowledge: A Review, Journal of Language Teaching and Research, **8**, 205, (2017). DOI: http://dx.doi.org/10.17507/jltr.0802.01 http://www.academypublication.com/ojs/index.php/jltr/article/view/jltr0802205213/0
15. K. Y. Averbukh, Differentiation of means of special nomination and differentiated approach to their description in dictionaries, Col. of scientific works on lexicography,**12**,150, (2011).
16. L. Achkasova, E M. Pozdnyakova , Revisiting Teaching of the Second Foreign Language in the Higher Educational Institution, International Research Journal, **10**, (2017). https://doi.org/10.23670/IRJ.2017.64.003
17. M. Din & M. Ghani, Developing Lexical Competence Through Literature: A Study of Intermediate Students of Pakistan, International Journal of English Linguistics, **8**,164, (2018).
18. M. González-Davies, The use of translation in an Integrated Plurilingual Approach to language learning: teacher strategies and best practices, Journal of Spanish Language Teaching, **4**,124, (2017). DOI:10.1080/23247797.2017.1407168
19. M. S. Thirumalai, An Introduction to TESOL, LANGUAGE IN INDIA, 2, (2002). http://www.languageinindia.com/april2002/tesolbook.html
20. O. Esteve, F. Fernández, E. Martín-Peris and E. Atienza, The Integrated Plurilingual Approach: A didactic model providing guidance to Spanish schools for reconceptualizing the teaching of additional languages, Linguistics and the Human Sciences, **11.1**, **(**2015).
21. O. García, R. Otheguy, Plurilingualism and translanguaging: commonalities and divergences, International Journal of Bilingual Education and Bilingualism, (2019). https://doi.org/10.1080/13670050.2019.1598932
22. P.I. Pidkasisty, O.B. Tychtchenko, Computer technologies in the system of distance learning, Pedagogy, **5**, 7, (2010).

Legal Assessment and Speculation on the Topic of Collaborationism in Ukraine. Digital Dimension

Andrii Shaikan[1], *Valentina* Shaikan[1*]

[1]State University of Economics and Technology, 50005, Kryvyi Rih, Ukraine

Abstract. The majority of modern researchers of the history of WWII and Great Patriotic War (1939-1945), analyzing the events that took place seventy years ago speak about the extreme danger of Nazism, since its aggression, in the case of the Nazis' victory, would cause a civilizational decline for considerable part of mankind, and would lead to humanitarian catastrophe. A significant number of innocent people got into the post-war repression and their fates were crippled by war and totalitarianism. The author of this paper does not in any way desire to justify activities of collaborators or enemy aiders and deserters, as certain historian critics sometimes mistakenly imagine. Our research aims to show – but not to justify – motives of deeds of a person in critical social settings and incentive factors, which may under certain conditions align one's behavior with a specific course of action. We believe that any statements concerning the inadmissibility of the Nuremberg International Trial rulings review are incompetent and illegitimate.

1 Introduction

Nowadays, it is not a secret anymore, that the existing digital information space is an area of unprecedented struggle of states for the right to interpret history and consequently the impact on national legal systems. The struggle is taking place at all levels of information supply and dissemination. Often without direct access to the influence of national legal systems, opponents resort to powerful multi-pronged information attacks, distortion of facts, interpretations in order to manipulate the consciousness of society regarding the legal assessment of certain events in history. The issue of legal evaluation and research into the content of the phenomenon of collaborationism in Ukraine during the Hitler occupation of 1941-1944 is no exception. Speculation on this issue is for decades a permanent feature of the Russian Federation's propaganda system. The Russian government's attraction to everything related to the history of the Second World War made the issue of collaborationism in Ukraine in 1941-1944 almost the main factor in building a modern coordinate system in relations between Russia and Ukraine. In order to justify their actions in the Crimea and in eastern Ukraine, the Russian authorities are deliberately manipulating historical facts concerning the manifestations of collaborationism in Ukraine by trying to resolve several issues at once. First of all, to strengthen the propaganda effect on the consciousness of their own society by creating a corresponding cliché to the perception of Ukrainian as traitors and fascists. Secondly, to exert a similar influence on the consciousness of Ukrainian civil society, destroying its national idea. Thirdly, to prevent the international community from recognizing the struggle of the Ukrainian people for independence during the Second World War as a party to the conflict. the propaganda machine of the Russian authorities is trying to tarnish the entire Thus, the propaganda machine of the Russian authorities is trying to tarnish the entire national movement of Ukraine equating it to the banal treason of individual Soviet citizens of Ukrainian origin and trying to give it a proper legal assessment. Success in this case would allow the Russian authorities (in their unfounded opinion) to raise the question of the randomness of Ukraine's statehood as an independent country. Such propaganda rhetoric has intensified especially since the beginning of the armed conflict between Russia and Ukraine in 2014. Since that time the digital information space has become the point of a massive offensive by Russia's propagandists. Examples of mass propaganda shows and provocative speeches can now be easily found on the Internet [1].

How to resist such an information invasion, defending their state positions both domestically and internationally? One of the strongest tools in this struggle is to conduct open, unbiased research on the issues of collaborationism in Ukraine during the years of Hitler's occupation, and to make such results widely public. At the same time, providing a legal assessment of such a phenomenon as collaborationism will make it possible to resolve one of the most difficult issues of the future reintegration of the temporarily occupied territories of Donbas and Crimea. But in order to make such an assessment fairly, it is necessary to refute the myths surrounding this problem. There are already enough tools and materials for such work.

Today the world community has access to the materials of the Nuremberg Trials (Tribunal) of war criminals of World War II. However, the root causes of

* Corresponding author: kaf_sgd@kneu.dp.ua

the military conflict and war are impossible to find in these documents. This is not surprising, at least because international legal acts after the world conflicts, and the Second World War was no exception, are written by the victors. According to A. Lysenko "international law does not always adequately reflect the realities of the past, leaving ample room for speculation, distortion and outright falsifications". And so it is, otherwise, how to explain that the International Tribunal at Nuremberg shifted all the blame for the war only to Germany, Italy and Japan. We now know that before the war the Soviet Union actively collaborated with Nazi Germany in the military, political and economic spheres. Today, historians regard the secret protocols and agreements of 1939 as a legal prologue to World War II. And more than that, both foreign and Ukrainian historians believe that collaborationism in Ukraine and the countries of Western Europe are phenomena of a different kind [2, p.7-8].

Unfortunately today the "Left" fail to notice the fact that during the war the leaders of the Reich declared that Ukrainian nationalism was more dangerous for Reich than communism, and even offered to hand over the power in Ukraine to the Communists, who because of their hatred for the Ukrainian independence supporters would give them away to the German authorities [3, P. 122]. Both Soviet historiography and today/s pro-Soviet one, which is especially characteristic of modern historiography of the Russian Federation, produced and are still producing myths of total betrayal and collaborationism of all the Ukrainian population, all areas of Ukrainian movement for independence and beyond. Everything that is not in the interests of the functioning totalitarian system: UNO, UIA, RLA, RONA, "Cossack" organizations and movements, voluntary police units (with great number of various representatives of the peoples of the USSR), the volunteers of the Wehrmacht, the SS, and others are dumped and mixed improperly in one pile. Moreover, incontrovertible evidence of the criminal nature of diplomacy of the Third Reich and the Soviet Union on the eve of World War II is ignored.

These facts were promulgated by famous French scholar of the twentieth century, Raymond Aron in his fundamental work "Peace and war among nations": "Diplomacy and strategy of the Third Reich, as well as diplomacy and strategy of the Soviet Union, were between two wars, ideological and imperial and have virtually no nationality.

In the period from 1939 to 1945 the nation lost the unity that used to be characteristic of them from 1914 to 1918. A great number of ideological traitors such as the Germans who gave the advantage to their country defeat over Hitler's victory, the Russians, who fought against the regime, considered to be oppressive and even the French, who dreamt about German victory ... - shows that the nations were no longer believed to be of the highest value and a single principle of political organization by all people [4, p.283]".

As you can see, a significant number of innocent people got into the post-war repression and their fates were crippled by war and totalitarianism. We believe that any statements concerning the inadmissibility of the Nuremberg International Trial rulings review are incompetent and illegitimate, since these decrees were taken by winners, who "take it all". Most Ukrainian scientists share the opinion that research of World War II and Great Patriotic War is already complete; but this assumption is a premature one. We fully agree with the fact that there is enough for historians to research and write about: Ukrainian historiography has just started researching historiosophy of the last world war.

Among the issues of public concern are such complicated social occurrences as collaborationism, aiding the enemy, desertion, and other negative social occurrences, which possess a particular psychological facet. Ukrainian historians are also interested in the origins of these occurrences-phenomena, and in the determining factors that may, under certain conditions, lead to their emergence. The majority of modern researchers of the history of WWII and Great Patriotic War (1939-1945), analyzing the events that took place seventy years ago speak about the extreme danger of Nazism, since its aggression, in the case of the Nazis' victory, would cause a civilizational decline for considerable part of mankind, and would lead to humanitarian catastrophe.

Within a short time (the period of occupation) Eastern Europe, as well as Western, encountered incredible violation of ethical norms, witnessed and experienced inhuman attitude of Nazi occupiers towards public morals, science, education, and culture in general. Ukrainian society during the Nazi occupation and after the liberation from the Nazis in 1943-1944 experienced various manifestations of deviant behavior: from socially tolerable violations of moral and ethical standards (everyday crudeness, displays of selfishness, friend betrayal), and misdemeanors to felonies: murders, thefts, robbery, participation in punitive actions, etc. Reaction of Ukrainian population to Hitler's occupation regime varied from the very beginning: a certain part (the lesser one, by the way) gladly greeted the occupants as their liberators; the majority – and today it is not a top secret – resorted to timeserving since it was important to preserve the lives of children, and there was the problem of finding food, clothing, medicine, work, housing, etc. Certain part of population (paradoxically, often supported by these opportunists) started underground and guerilla struggle against the Nazi invaders. Partisans and underground fighters also needed food, as well as clothing, medicine, safe houses, and insiders among the civilian population.

It is well known that no occupier can do without the locals' help; no occupation system is viable without engaging local professionals of various levels. So did the Nazis in the occupied Ukrainian territory: in parallel with their own bodies of vertical administration they created bodies of local government, engaging the Ukrainian population for staffing them.

In a flash, a wartime person was put in a situation, demanding a choice: who to support. The occupants left no time for making this choice.

Researchers of complex occurrences-phenomena as, for example, collaborationism and aiding the enemy, cannot but consider the existence of multivariate, alternative possibilities and justification of subjective choices of goals, ways and manners of actions. It will be subjective to divide people into "good" and "bad" without a serious analysis of their actions; and interpretation of historical phenomena without analytics will be a step backwards.

In the 21st century the historiosophy of occurrences-phenomena naturally became the primary object of historical studies. It is well known that all occurrences of social process share an extremely complex structure, and under the influence of a specific historical context the leading role may belong to social-political, cultural-ideological, religious and other factors and contradictions that induce the emergence of a variety of options and pluralism. We believe that this approach helps to understand the causes of emergence of social-psychological occurrences in historical process. Research of any human activity, motives of behavior in general and during Nazi occupation in particular is impossible without analysis of specific features of human thought and moral-psychological image of a person. The authors to justify activities of collaborators or enemy aiders and deserters, as certain historian critics sometimes mistakenly imagine.

Our research aims to show – but not to justify – motives of deeds of a person in critical social settings and incentive factors, which may under certain conditions align one's behavior with a specific course of action.

2 Our contribution

Life of any person comprises critical, active and creative periods. If a person's critical period of life falls within the period of regression and stagnation of society, it cannot but affect his or her fate. It is known that between wars as well as during the war the most part of Ukrainian population was experiencing those unstable periods of its development – repression and political trials against the "public enemies", constant struggle with the "extremes" and manifestations of "bourgeois nationalism" among Ukrainian creative intelligentsia, "hvylevism", "volobuyevschina", "skrypnykovschyna", "shumkism", etc. Ukrainian population suffered numerous "cleansings", famines, collectivization, dispossession of the kulaks, relocation, industrialization, Sovietization of Western territories.

Thus, the socialization of most Soviet citizens took place in an unstable period of the development of society, when the individual value systems and social behavior were also changing by force of circumstances. It is well known that before the war some people in Ukraine, in order to avoid arrest, consciously reported their neighbors and coworkers. During the war one could witness the same situation: to survive, people were choosing to collaborate with the occupants.

According to psychologists, regardless of age, religion, ideology, the perennial problem in human relations are contradictions between the ideal of social-psychological tolerance and actual strategies of individual behavior, which can be addressed by tolerant actions [5, p. 16]. The majority of Ukrainian believers found themselves in this situation when they were faced with the dilemma: who to support? People could not behave indifferently. One had to decide. On the one hand, according to Christian ethics, war and violence are wrong, and on the other hand – the priests called their congregations for patience, loyalty, and support for the "new" regime. We do not believe one can condemn such a position of the clergy and the faithful, who saved their own lives and many lives of their compatriots by being tolerant. But it is impossible to justify the actions of those priests, who worked closely with the occupiers, collecting a variety of information and passing it to the Gestapo.

The authors of this paper believes that archival documents and materials from Ukrainian archives allow to assert that under the Nazi occupation a certain part of Ukrainian population (different nationalities: Ukrainians, Russian, Poles, Crimean Tatars, Armenians, and others) with the dominant traits of negative identity unambiguously chose the abnormal course of personal development in society – that is, embarked on a criminal path. Thus, individuals with uncertain personal and civil identity who were prone to conformism became the primary "victims" of criminal collaboration. It is well known that the majority of young people born in Ukraine between the years 1923-1924 were raised in the spirit of the Soviet system rules and ethics of unconditional worship of authorities, dogmatic acceptance of existing practices, rules and regulations, as well as passive, opportunistic adoption of behavior standards. It was such individuals who became an easy "prey" for Nazi propaganda both at the front and in the occupied territory. People who have long been marginalized in society, or reticent individuals reduced to hopelessness also succumbed to the propaganda.

A person who has lost the meaning of life, who has irrevocably wasted years sometimes tries to make up for his losses and even to take revenge for his failure. These individuals are known to have treated prisoners of war, women and children particularly badly, and to have participated together with the German police in punitive actions against the partisans and local population.

According to psychology of motivation, there are two kinds of individuals: people with high and low (with dominating desire to avoid difficulties) achievement motivation. Under the conditions of war and occupation, when transformation of positive needs and values into negative becomes a commonplace occurrence for certain part of society, a person rapidly becomes asocial. This fact was taken into account by the occupiers; that's why various instructions and rules of conduct were prepared for Wehrmacht soldiers and police in the East.

The occupational authorities did not give the population time for reflection, and demanded immediate decisions regarding people's loyalties – whether they accept the "new" order, or oppose it, or try to compromise with it. The answer depended on each individual's value priorities and spiritual culture. Prevalence of base moral values led to treachery, murder, and inner spiritual meltdown. We assume that the investigation of negative occurrences-phenomena (treachery, desertion, voluntary surrender, punitive actions, etc.) that occurred during the war is supported by psychosocial theories of personality – psychoanalysis, theory of personal self-esteem coefficient by American philosopher and psychologist William James, and Skinner's "operant" reactions. The occupiers' propaganda actively used the possibility of amplifying or decreasing public reactions by punishment or encouragement, aiming to destroy universal human precepts of coexistence and to promote asocial behavior. It divided society and sparked animosity between nations and peoples.

Following such secret instructions as closing hospitals for "aborigines", discontinuing disease control of typhoid and tuberculosis, supporting immorality, if it did not harm the Germans, the occupiers pursued a policy of enslaving people, "educated" the population. The affect of inadequacy is also a source of anti-civil behavior. In extreme cases, negative behavior of delinquent individuals leads to criminal activities. Such individuals are prone to social instability and immoral actions, aggressive and extremely cruel. On the eve of war thousands of criminals were released from Soviet prisons. Those who remained in the occupied territory, adapted, too; they offered their services as prefects and policemen, hiding their past; participated in punitive actions against partisans and population.

Understanding empirical material on theoretical level is required today. Historians are expected not only to provide confirmation, but also to do research based on thorough knowledge of specialized literature and sources as well as theoretical and methodological tools that allow disclosing historical occurrences. Time requires to leave behind traditional layouts of historical materials and to consider the multidimensional manifestations of historical occurrences. Use of the results of philosophers', sociologists', economists and psychologists' work guarantees the possibility of objective and stereoscopic study of every historical occurrence.

History testifies that collaborationism (cooperation with the enemy) exists as long as humanity. A historian will probably not be supported by his compatriots if he proves that in the past their ancestors also cooperated with the invaders. This is only natural, since from ancient times love for Motherland, patriotic feelings and defense of native land used to be each society's foundation of national identity.

Contemporary Ukrainian historians define collaborationism as cooperation of certain representatives of Ukrainian nation with occupational authorities; this cooperation on individual level was characterized by participation in local government and the auxiliary police, controlled by the occupiers; unequivocally assert that there was no mass collaboration among Ukrainian population. Distinguished Western European historian W. Kosik and Canadian historian O. Subtelny also argue that cooperation with Nazi occupiers in the Ukrainian territory took place at individual level (allowing some Ukrainians to survive during the Nazi occupation), but collective political cooperation on the orders of the Ukrainian government or any political party did not take place. According to Prof. S. Kulchytsky, well-known in Ukraine and beyond, collaborationism is defined by a relationship of two unequal parties and is only possible if both parties agree to establish such relationship. Struggle of Ukrainian nationalists for an independent and united Ukraine excluded collaborationist nature of relationship with the German occupiers; only cooperation at the institutional level is collaborationism. S. Kulchytsky argues that during WWII War collaborationism did not exist in Ukraine, and cooperation on a personal level can not be qualified as treachery. I. e., as one can see, today there is no unanimity on this issue yet.

While researching the topic of collaborationism in the territory of the Reichskommissariat Ukraine and in war zone during WWII the authors of this paper made the following definite conclusions [6; 7; 8; 9; 10;11]: firstly, the phenomenon of collaborationism in Ukraine became possible as a result of people's adapting to extreme conditions of war and occupation. Complex dialectic of political and socio-economic processes, which occurred in ethnic Ukrainian territory while it was part of several states before and during the war, catalyzed collaboration among Ukrainian population; secondly, concept comprises following statements:

a) collaborationism is defined as a complex multidimensional social-psychological and ethical phenomenon, resulting from interactions of subjects of communication process (separate individuals, a certain part of population) with the occupiers; that is, cooperation with enemy under the conditions of occupational (Nazi) regime; collaborationism is a voluntary or forced cooperation with the enemy in political, military, economical, every day, cultural and other areas, which possesses criminal traits; aiding the occupiers is voluntary or forced assisting the enemy (enemy's military or civil authorities) in carrying out military, political, economical, cultural activities, aimed at reinforcing the regime; labor of workers, peasants, employees of administrative and other establishments, whose activity during the Nazi occupation of Ukrainian territory was aimed at survival and lacked the traits of crime should be considered as such that is not prosecuted;

as for the b) typology of collaborationism, in Ukraine it manifested in a wide range – from household and administrative levels to military, economic, cultural, and individually political forms; political collaborationism is cooperation with the enemy on ideological foundations; administrative collaborationism is a form of cooperation with the enemy of loyal individuals from local authorities, whose actions were monitored and guided by the occupiers; military collaborationism is a voluntary or forced service in the military units of Nazis and their satellites; economic collaborationism means cooperation in any branches of economy benefiting the enemy; cultural collaborationism involves cooperation with occupiers in the spiritual realm, which facilitated spreading sentiment of allegiance among the population, promoting supremacy of the Aryan race, was consistent with interests of the occupying authorities and contributed to raising morale of the occupiers; everyday collaborationism is connected to establishing friendly relations between occupiers and the locals and comprises no crime;

c) collaboration as a phenomenon reproduces socio-psychological and ethical reality in society during martial law, defines and regulates behavior of both individuals and population in general, which has already been in a state of conformity, and promotes organizing various forms of cooperation and relations with the occupiers.

Erroneous tactical and strategic policy of leaders of the Organization of Ukrainian Nationalists on the eve and on initial stage of Germany's war against the Soviet Union affected all further activity of Ukrainian movement for independence as well as overall assessment and significance of the struggle for an independent state.

Situational, temporary collaboration with the Nazis for tactical reasons discredited long struggle of the Ukrainian Insurgent Army (UIA), the Organization of Ukrainian Nationalists (OUN) for Ukrainian Independent United State (UIUS).

Conciliatory stance of one of the branches of the OUN-M (Melnykists), the Ukrainian Central Committee (UCC), Hetmanists and collaborators, whose activity, in varying degrees, has been associated with the occupation regime, has been extrapolated to the OUN and UPA in general – and that is, in our opinion, unfair and unreasonable.

Violent actions of the Security Service (SS) of the OUN-B (Banderivtsy) in the fight against the enemies of Ukrainian statehood, collaborators and aiders of occupiers, UPA deserters, Soviet activists and other categories of population had nothing to do with collaborationism.

Multidirectional activity of the collaborating local authorities satisfied the occupiers. Due to their help, Nazis controlled situation in the occupied territory. However, activities of collaborators can not be assessed unambiguously – solely positively or negatively – just because all the local authorities were established as subsidiary bodies to impose a "new" Hitler's regime. On the other hand, due to socio-cultural activities of collaborators, a certain part of the population was able to survive in the face of brutal occupation.

Totalitarian system morally crippled generations of people; their ideological confusion, political uncertainty, socially amorphous nature were the main (primary) causes for collaboration of part of Ukrainian population.

One of the distinguishing features of collaboration in Ukraine as compared to Western Europe was the formation of the Eastern European military units (Ukrainian security police battalions, Voluntary Cossack Corps, Ukrainian National Army, etc.) not on the basis of Nazi ideology, but based on anti-Bolshevik and independence ideologies.

Political (at the individual level), military, administrative, economic, cultural, everyday collaboration and aiding, involving a certain part of Ukrainian society (working class, peasantry, intelligentsia, military, clergy) were caused by political landscape of the time and largely resulted from past cooperation between the USSR and Germany.

Fully legitimate and generally accepted in international practice prosecution of war criminals in the Soviet Union was accompanied by violation of civil rights, politicization of justice, application of collective responsibility, ignoring presumption of innocence, etc.

Conclusion

Ukrainian collaborationism is a unique phenomenon just because, on the one hand, it was collaboration with the occupiers for various reasons in all areas of life in wartime, which was typical for a wide range of collaborators. On the other hand, it was a struggle for independence of Ukrainian people (primarily concerning participants of the Ukrainian national liberation movement). Despite the fact that each of the branches of Ukrainian liberation movement used its own ways of struggle, all of them, at different times, took into their ranks outright collaborators – those who by that time had already tarnished themselves by participating in punitive actions against civilians and prisoners of war.

Clearly, such collaborators concealed such activities and disguised themselves as fighters for Ukrainian independence. Famous Ukrainian researcher of activity of Ukrainian Division "SS-Galicia" A. Bolyanovsky writes about it, too, claiming that Ukrainian police in the years 1943-1944 used to be one of the recruiting bases of Ukrainian Insurgent Army. That is why, in our opinion, it is important to approach the issue of Ukrainian military collaboration individually – to avoid mistakes and to properly determine whose activity was criminal, and whose was not. Not every collaborator was a war criminal. It is known that under international law war criminals are considered those who took part in crimes against humanity, outright traitors or those who accepted Nazi ideas and actively helped the occupiers to impose their expansionist programs, or pursued their own mercantile interests, trying to preserve their lives at any cost, ignoring the interests of their nation – though sometimes being far from Nazi ideas. There were also individuals who combined both cases.

Today it is no secret that Ukrainian combat units of the Wehrmacht and SS troops were manned with both natives of Eastern and Western Ukraine. Thus, the first anti-tank squad "Vilna Ukraina" under command of Petro Dyachenko as well as the squad of Taras Borovets (Bulba) comprised 80% of Eastern Ukrainians and 20% of Halycians. Yet the opposite thing did happen, too: the squad under command of

Pituleya comprised 70% of Western Ukrainians and 30% of soldiers from Eastern divisions and guard battalions as well as Ukrainian police. And it is not a coincidence. It is known that the combat formations of Ukrainians in the Wehrmacht were manned to make the majority either "Easterners" or "Westerners". In the Red Army most manpower comprised "Easterners" who fought against the UPA, where the majority consisted of "Westerners". Nazis threw "Westerners" into battle against "Easterners" using their ideological differences and further escalating their confrontation.

Specific feature of Ukrainian military collaboration, according to the authors, are also the extreme unscrupulousness, ideological confusion, amorphous nature and uncertainty of the part of Ukrainian population, which voluntarily agreed to cooperate with the Wehrmacht and occupational authorities (police, for example). Therefore it is not a coincidence that a significant part of Ukrainian policemen switched from Nazis to Ukrainian nationalists, and then again to the Germans; and when the Red Army arrived they joined Soviet troops or Soviet partisan detachments. And all this is confirmed by a large number of archival documents of various origins, among which there are also diaries of the members of Ukrainian national liberation movement.

It is this phenomenon that a young banderivets from OUN Alexander Povshuk wrote about in his diary (we deem it important to convey an excerpt from his diary in the original – Ukrainian – language): "When they return to Bolsheviks – they harm Germans. Police, whom Germans kept as their hounds, and who shot their own parents, brothers and sisters, were disbanded; so now they turned into patriots of their folk (partisans, who are many in the woods). Brother Ilko is there, too. I love the truth.").

This dates to April, 30, 1943. As already mentioned, there are quite a lot of such examples – or rather, just the right number to lead to the following conclusion: the totalitarian system morally crippled an entire generation of people who were not able to understand who was the greater enemy – either Nazism or Bolshevism; it was a generation (this only applies to the collaborators!) of unprincipled, cruel, selfish individuals who were capable of quick adapting and serving whoever was in power. Unfortunately, such individuals occurred within the ranks of OUN, UIA, the Red Army, Soviet partisans, the NKVD, and even more – among voluntary Ukrainian police and security units. Similar unprincipled and indifferent to their Motherland-Ukraine individuals occurred among the Ostarbeiters. This is evidenced by a mere fact from the memories of a former Ukrainian Ostarbaiter, who worked in a car workshop near Berlin and met there at the end of the war young guys from Ukraine who did not care who to serve, since the main thing for them was the availability of "good food".

Certain part of Soviet population who had already lived under the Soviet rule for 20 years before WWII was demoralized and depoliticized. Only this fact can explain the mass desertion of soldiers from the Red Army, Soviet partisans from squads and units, from 'voluntary' police subdivisions, mass surrender of Soviet servicemen, especially early in the war. This is confirmed by both Soviet and the Nazi archival documents; for example, by the report of Field Police Chief of Army Group "South" from September, 10, 1943 on the conduct of "volunteers" – "Hiwi" and soldiers from local population – on their licentiousness and casual attacking local population, which had a negative, as evidenced by the German soldiers themselves, impact on relationship between population and the Wehrmacht.

The main reasons for Ukrainian collaboration were ideological rejection of Bolshevism, confusion, fear of the power and cruelty of occupiers, desire to survive whatever the price, mercantile interests. At the expense of "volunteers" and "voluntary" units the Nazis not only replenished the Wehrmacht with "cannon fodder", but also significantly reduced the influx of Ukrainian population to the ranks of the Ukrainian Insurgent Army.

With the assistance of local collaborators occupiers managed to quite rapidly start the production on Ukrainian enterprises of light and food industries for the Wehrmacht and the Reich; mechanical engineering, chemical, metal processing factories and workshops were accommodated for the repair of motor vehicles, tanks, aircraft, aircraft engines, armored vehicles and production of projectiles for the Wehrmacht. Reconditioned power plants also worked exclusively to benefit the occupiers. All in all, the operation of labor force aimed only at serving the Wehrmacht and the Reich.

However, the authors believe that labor activity of the majority of ordinary workers, peasants and officials, who for various reasons remained in the occupied territory, can not be at all attributed to neither collaborationism nor aiding the enemy, since this activity was not a criminal one, but a way of survival and the result of the people's adaptation to the extreme and existential conditions of war.

However, a significant number of innocent people with destinies crippled both by war and totalitarianism were repressed, too. We believe that any statements concerning the inadmissibility of the Nuremberg International Trial rulings review are incompetent and illegitimate.

References

1. Maslova V., Poletaeva P., Onichyk E. Ukrainian information: why Kiev intends to rewrite "Wikipedia" (2020).
2. Lysenko A. Ukrainian Insurgent Army - the Phenomenon of the Nation's History, Materials of the All-Ukrainian scientific conference, Ivano-Frankivsk (2003)
3. Croitor B. Ukrainian Insurgent Army, the phenomenon of the nation's history, Materials of the All-Ukrainian scientific conference, Ivano-Frankivsk (2003)
4. Raymond Aron. Peace and war among nations, Kyiv, «Univers» (2000)

5. Shvachko O. V. Social'na psyhologija, Kyiv, Vyshha shkola (2002)
6. Shaikan V. Socio-political and economic causes of collaborationism in the territory of the Reichskommissariat "Ukraine" and the military zone during World War II: Monograph, Kryvyj Rih, «Mineral» (2004)
7. Shaikan V. Collaborationism in the Reichskommissariat "Ukraine" and the war zone during World War II: Monograph, Kryvyj Rih, «Mineral» (2005)
8. Shaikan V. The ideological basis of collaborationism in Ukraine during the Nazi occupation, Kyiv, Vyd. dim «Kyievo-Mogyljans'ka akademija» (2006)
9. Shaikan V. The ideological struggle in Ukraine during World War II, Kyiv, «Naukova dumka» NAN Ukraine (2010)
10. Andrii Shaikan, Valentyna Shaikan. Kollaborationism and complicity in the history of mankind reflected in schlars and writers works, Scientific jornal of Polonia University (2018)
11. Valentina Shaikan, Kateryna Datsko, Nina Ivaniuk, Alim Batiuk. Collaborationism and War Crimes as Phenomena in the Information Society, Proceedings of the III International Scientific Congress Society of Ambient Intelligence 2020 (2020)

Digital Transformations in Agri-Food Systems: Innovation Drivers and New Threats to Sustainable Rural Development

Olena Borodina[1], *Oksana* Rykovska[1*], *Oksana* Mykhailenko[1], and *Oleksii* Fraier[1]

[1]Department of Economy and Policy of Agrarian Transformations, Institute for Economics and Forecasting of National Academy of Sciences of Ukraine, Kyiv, 01011, Ukraine

Abstract. The paper proves that sweeping digital transformations are a global trend in agri-food development. Significantly improved economic efficiency, rational use of natural resources, operational exchange of relevant information, new markets, and economic opportunities under modern climate change are possible owing to digital transformations. It is grounded that the elimination of world hunger based on the provided internal food security and secured rural communities should be the primary goal of the digital innovations in agri-food systems within Sustainable Development Goals (SDGs). The paper states that along with the significant benefits of digital technologies, the great destructive impact on the overall societal development is possible due to the corporate monopolization of digital processes. Thus, there is the risk of the development of food systems, which are characterized by decreasing in the food supply, loss of biodiversity and rural livelihoods. Proposals to mitigate the threats of digitalization, strengthen internal food security and enhance the development of rural communities through information and communication technologies (ICT) are substantiated. The specific of the agri-food digitalization in Ukraine is outlined, which is determined by the duality of the organizational structure of agriculture (corporate and individual sector) with a tendency to the concentration of land, power, and financial resources in favor of corporations expanding opportunities for digitalization. Appropriate safety measures to mitigate the negative impact on the development of small producers, rural areas, and Ukrainian society are identified.

1 Introduction

The introduction of digital technologies is a global trend of modern world development that has no alternatives. Digitalization is cross-border in nature and free of any restrictions on the area of application. Its rapid growth is explained by a range of solutions for efficient management, rational use of limited natural resources, improved transparency, information exchange, and so on. Meanwhile, the implementation of digital technologies towards better profit margins and economic benefits destroys existing food chains, restricts freedom of choice, and increases inequalities in social development. An adverse effect is also shown in that the involvement in the appropriate processes does not require personal consent. Even opponents of the digital transformations being registered regular users become participants by default. Digital technologies bring people to new chains of responses developing massive databases. For instance, participating in different platforms means subject/user's (companies, governments, citizens) respond to numerous requests with digital tools (message in Twitter, like a Facebook post, send an e-mail, use cloud storage service, hail a ride with Uber, create new apps, etc.). Such data "reserves" go through algorithmic computation to further commercial use [1].

At the 43-rd session of the Human Rights Council, the United Nation Secretary-General António Guterres noted, "the digital age has opened up new frontiers of human welfare, knowledge and exploration" [2]. However, the use of new technologies leads to violations of human rights and privacy through surveillance, repression, online harassment, and the spread of hatred. Advances in technologies, such as facial recognition, robotics, digital identification, and biotechnology must not lead to weakening human rights, rise in inequality, or discrimination [2].

The digital reconstruction of food systems amid growing food insecurity under the COVID-19 pandemic is being activated with the focus on the concentration of power and development of globalized value chains [3].

Agri-food digital transformations show that consumers will have to continue to participate in the data exchange, delegating control over their food preferences to corporate structures. Developments "behind-the-scenes" in Amazon's corporate headquarters or Samsung, Nestlé, Pepsi research laboratories are oriented towards using information technologies to secure future profits of food producers.

* Corresponding author: Oksana.Rykovska@gmail.com

Results of their experiments will be reflected in alterations of the internal super-markets layouts, restrictions in flavors or styles of food consumption, impacts on customers' experience of buying food, and strengthening the food industry expanded power to shape farms' activity worldwide [1].

The abovementioned warns of the possible emergence of food systems where supply is quite limited and significantly dependent on consumer characteristics [4], which, in turn, will contribute to the loss of biodiversity in food and agriculture that currently degrades at an alarming rate because of ecosystems destruction [5].

The purpose of this paper is to reveal the relationship between digital transformations and digital inequity in agri-food systems centered on the harmonization of rural social and economic relations as well as on the equal access to food produced by small family farms that can contribute to the mitigation of the effects of uncontrolled agri-food digital transformations.

The **information resources** for the study are comprised of publications of UN [6, 7], FAO, IFAD, UNICEF, WFP, and WHO [8, 9], OECD [10], data of LAND MATRIX [11], State Statistics Service of Ukraine (the economic activity of large-scale agricultural enterprises, private and farming households) [12], studies of national and foreign researchers.

2 Background

2.1 Global situation.

Climate change, conflicts, recessions, the spread of transboundary animal diseases and pests have exacerbated the food crisis, reduced the well-being of rural communities, and increased the number of hungry people around the globe. The COVID-19 pandemic has aggravated existing problems of malnutrition and food insecurity. One of the key factors inducing severe food crises and particularly hunger is conflicts [13].

According to FAO (2020), there are about 690 million people, or 8.9% of the world's population who suffered from hunger last year. Over the previous year, the number has increased by 10 million people and almost by 60 million people in the last five years respectively. The growing tendency is also confirmed with the indicator of food insecurity: about 750 million people or almost every tenth planet inhabitant suffered from hunger in 2019. Approximately two billion people do not have permanent access to sufficient, safe, and nutritious food. If the trend continues, the global number of starving people may exceed 840 million people [9].

The results of the one-third term spent to implement the Agenda for Sustainable Development show that the world is out of schedule to achieve the global Goals by 2030 [7]. Before the COVID-19 outbreak, the progress had been uneven, and more proactive measures were needed to settle the situation. The pandemic abruptly disrupted implementation towards many of the SDGs and, in some cases, turned back decades of progress [6]. Preliminary assessments show that the last year of the COVID-19 pandemic may increase the famished population from 83 to 132 million people [9].

2.2 An FAO initiative to reduce poverty based on digital transformations.

Early in 2020, the Food and Agriculture Organization of the United Nations (FAO) presented a new Hand in Hand Initiative [8] with a focus on regions with high agricultural potential. The Initiative should unite the capacities of developed countries in the transfer of new digital knowledge and technologies to those with extreme levels of poverty and hunger. Partner selection may refer to attracting new donors or private sector investments, as well as identifying investment gaps to work with existing donors, multi- and bilateral agencies.

The Initiative aims to help poor and the least developed landlocked countries, the least developed small island states, developing countries, countries experiencing food crises, as well as large countries with significant outbreaks of destructive phenomena. It is based on a multilateral approach focused on accelerating the growth of agriculture and sustainable development of the participating countries through the use of digital technologies. The focus lies in strengthening the capacity of food systems and their ability to provide quality food and a healthy diet for everyone, as well as to increase households' incomes to reduce extreme poverty.

The introduction of state-of-the-art instruments, including advanced methods of geospatial modeling and analytical tools, such as the Geospatial Data Platform, the Data Laboratories to support innovations in statistics have been envisaged in the framework of Initiative implementation. They provide high-quality, accessible, time-bound, and reliable data that help countries make sound strategic decisions based on objective information. To create the platform, FAO has gained the support of leading tech companies and open data providers, including Google, IBM, and the World Bank. FAO provides coach training activities for participating countries to ensure the tools of the Hand in Hand Initiative are available and widely used. The FAO's new form of cooperation is being introduced based on the Initiative and under the direct leadership of participating countries interested in accelerating agricultural transformation, promoting sustainable rural development, ending poverty (SDG 1), eliminating hunger and all forms of malnutrition (SDG 2).

Dietary optimization remains a global challenge that dictates the urgent need to create sustainable food systems adequately addressing current challenges. Called to raise public welfare, food digital transformations may improve the situation significantly.

2.3 Overview of current trends in digitalization of agri-food systems.

Over the last years, digitalization extended in all parts of the agri-food system, such as agricultural production, processing, storage, packaging, distribution, food exchange, etc. Digital transformations are characterized by the use of both relatively simple apps such as map data and direct online marketing, and complex digital innovations used in agribusiness "to change the business model suggesting the development of new profits and opportunities to create a novel type of value chains" [14].

It is worth mentioning that currently introduced digital innovations change the modern design of agri-food systems at a rapid pace. Among them digital land registers; sensors in farm robotics; harvesting robots; genetic modifications; information control in global value chains; artificial intelligence in plant growing and animal care, and round-the-clock health monitoring; satellite determination of fish concentration and distribution of fishing rights; automated trade and food distribution; e-commerce; development of personalized diets; etc.

Digital transformations in agri-food systems integrate advanced technologies (artificial intelligence, sensors, robotics, drones, etc.), devices and communication networks into one system towards use in production, management, processing, and marketing. As such, agricultural production, energy and resource management, and economic operations are highly efficient with these technologies promoting the creation of new markets and economic opportunities under climate change (considering climate-adapted methods of management). Digital transformations seem too attractive to the private corporate sector in terms of improving productivity, increasing production and exports, excluding and replacing human labor with robotics. Thus, such technologies and infrastructure of digital transformations in agriculture are practically usurped by corporations, focused mainly on profit maximization rather than food security.

Shaping the global food system, corporate investments involve practices of non-transparent automated data capture, which complicate democratic control over corporate activities. The ethical responsibility for the security and reliability of the food system they form is entirely entrusted to software developers, lawyers, and managers.

The public concern about such seizure by a limited number of global digital industry players has activated debates about the role of "Big Data" in the food systems reformation. On the one hand, "Big Data" is seen as a kind of basis for the development of new views, prospects, and opportunities; on the other, it is being criticized for specifics of data collection and management, the commodification of many aspects of everyday life and lack of attention to digital technologies in society in terms of the labor process and respect for fundamental civic rights and freedoms.

The asymmetry of the data collection/capture process is a means of capitalist "accumulation by dispossession", that colonizes everyday life in previously impossible ways [4]. As such, a new discriminatory relationship system evolves between data that shape managerial decisions on human nutrition and a man who needs a safe and diversified diet as a precondition of physical existence.

2.4 How digital transformations contribute to food security, rural development, and hunger elimination?

The need to feed the world stems from the lack of access to food resulting from the social, economic, and political power imbalance. Meanwhile, food security is closely linked to social marginalization, poverty, and gender inequality. The power within the agri-food system is exercised primarily through the concentration of land resources, market power and the capitalization, which allow large producers to exert a dominant influence on price, affordability, food quality, and access to productive resources. Such processes narrow agricultural employment for members of rural communities, restrict their livelihoods, deepen poverty, provoke hunger, and so on.

The balanced non-discriminatory arrangements in food systems are a crucial element in fighting hunger. Serious inequalities in access to land, water, capital, health services, and education significantly inhibit rural development and malnutrition elimination. Digital and tech solutions do not have appropriate responses to the raised question. Handling a power imbalance has a much greater impact on the living standards and nutrition of rural communities than digital transformations of agricultural corporations and increased productivity based on information systems.

Digital transformations with no other purpose than to increase agricultural productivity, lead to the concentration of corporate political and economic power, come with depriving rural communities of resources and knowledge, become a threat to sustainable development at all, and, consequently, unable to solve the global problem of hunger. Alternatively, the support of innovative processes in small producers and local communities to be activated towards the development of sustainable, healthy, and equal food systems. This is the only way to secure food sovereignty and achieve the economic well-being of small farms.

Small producers and marginalized social groups can benefit significantly from digital innovations since the expanded window of opportunities becomes available: digital land management provides land tenure security; the satellite distribution of fishing rights ensures the transparency and safety of fishers; blockchains connect producers with consumers simultaneously excluding intermediaries; digital agriculture helps reduce input costs under increasing irrigation and production efficiency; e-commerce promotes the development of new markets and farm sales channels [15].

Such agri-food digitalization, to be noted, should become a conscious choice of society and be enshrined in the relevant state policy. Digital apps and related solutions should be focused primarily on meeting the needs of small producers, rather than protecting corporate interests focused on increasing agricultural and food sales (including exports) and making consumers contingent upon this basis.

As part of FAO's new global Hand in Hand Initiative the equality, digital inclusion, and the protection of the rights of farmers, family farms, indigenous peoples, and other people working in rural areas should be promoted. During the implementation of programs and projects related to digital transformations, it is necessary to recognize and protect individual and collective economic rights to digital innovations for indigenous peoples, peasants, fishers, pastoralists, food workers, consumers, etc. Appropriate regulatory mechanisms for the digital food and agricultural economy should be introduced at the international level to avoid the destructive impact of monopolistic information ecosystems and the concentration of digital power; encourage efforts combination by international organizations and donors to protect family farms and authentic knowledge from data capture; to promote the food sovereignty of small food producers and indigenous people by supporting the implementation of digital agriculture policies and practices.

The United Nations should provide the necessary support through its organizations and programs to bridge "the digital gap between and within developing countries. Developing countries - in particular, the least developed countries - and vulnerable groups such as the rural population and women benefit significantly less from the development of new information technologies than people in industrialized countries do" [16].

2.5 Digital transformations in agri-food systems in Ukraine.

The spread and application of digital technologies in the agri-food of Ukraine have special characteristics associated with a distinctive duality of the agricultural organizational structure and unequal development of two sectors: corporate, much of which is represented by powerful horizontally and vertically integrated structures (agricultural holdings); individual, represented by two types of individual subjects: farms and farming households.

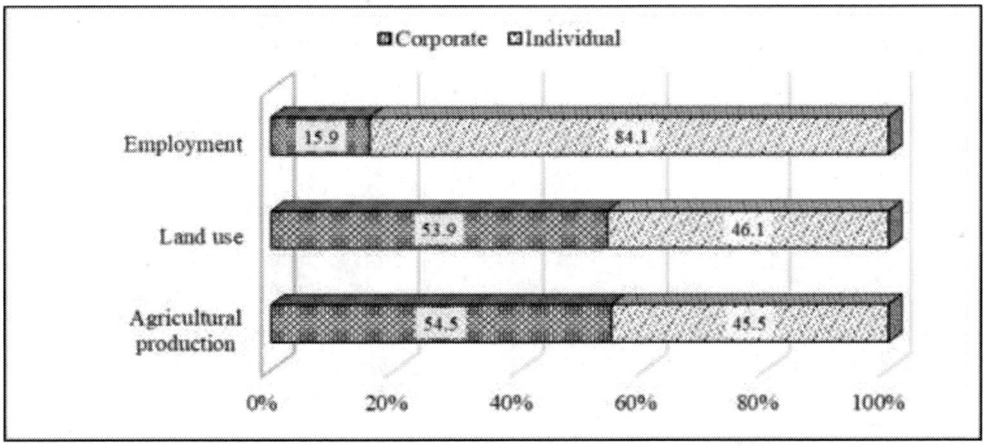

Fig. 1. Share of corporate and individual sectors by indicators of agricultural production in Ukraine, 2019
Source: Agriculture of Ukraine. Statistical Yearbook. State Statistics Service of Ukraine, Kyiv, 2020 [12].

The corporate and individual sectors produce approximately the same number of agricultural outputs (54.5% and 45.5%, respectively) (see Fig. 1). Agricultural cultivated area here is of nearly-equal size (53.9% and 46.1% respectively) [12]. However, the number of employed economically active rural population in the individual sector is dramatically higher than those employed in the corporate sector. The labor employment imbalance is caused by many factors, including the active corporate implementation of industrial technologies and, as a consequence, the release of a significant number of workers. Moving to a self-employed sector for the majority of the economically active rural population is a forced step since they have no other alternative for employment in the home area. Self-employment in farming households is less productive, largely based on manual labor, and provides a minimum level of subsistence needs. Hence, digitalization can significantly expand the scale of negative impact on the social dimensions of rural development. It requires additional state regulations.

According to the independent global land monitoring initiative Land Matrix, over recent years there has been a significant increase in foreign investors in the agriculture of Ukraine, including UK companies (Rohini Finch, 30 th. ha; Tommy International Trading Limited, 23 th. ha, etc.), Saudi Arabia (Public Investment Fund (PIF) of Saudi Arabia, 189 th. ha; Saudi Grain and Fodder Holding Co. LC, 34 th. ha, etc.), the United States of America (NCH Capital Inc., 360 th. hectares, TNA Corporate Solutions LLC, 132 th. ha, etc.), and many other countries [11]. The People's Republic of China is particularly interested in Ukrainian lands (particularly the Xinjiang Production and Construction Corps (XPCC), the China National Corporation for Overseas Economic Cooperation) [11].

Well-known in Ukraine agricultural holdings being registered in "tax havens" and with the support of transnational financial institutions currently consolidate significant amounts of land, including Ukrlandfarming - 500 th. ha, Kernel - 530 th. ha, MHP S.E. – 370 th. ha, Astarta Holding M.V. - 235 th. ha, and others [11]. Expert evaluations show that the total area of land under cultivation away back in 2011 was about 5.1 mln ha or 12% of the total farm area of Ukraine [10].

Table 1. The largest agro holdings by land size in Ukraine, 2020

Agro holding	Registration country	Landbank, th. ha
Kernel Holding S.A.	Luxembourg	530
Ukrlandfarming PLC	Cyprus	500
MHP S.E.	Cyprus	370
AP Group	USA	300
Astarta Holding N.V.	Netherlands	235
Continental Farmers Group, beneficiary SALIC (UK) Ltd.	United Kingdom	195
TNA Corporate Solutions LLC	USA	132
HarvEast, under the control of System Capital Management (SCM) and Smart Holding	Cyprus	127

Source: based on sample survey [11].

Agricultural corporations being backed by significant resources and funds, access to state support, bank loans, including funds of international financial institutions, such as the World Bank, the European Bank for Reconstruction and Development, the European Investment Bank. rapidly implement digital innovations. They include updating the park of tillage machinery, equipped with computerized control, maps for differentiated seed sowing, GPS monitoring for crops, optimized fertilization, and machinery movement, automated control of technological operations, agricultural modeling. The use of complex analytical systems allows companies to summarize land characteristics, create field maps, track machinery and daily productivity, monitor photosynthesis intensity, form a real-time forecast of crop yields, etc [17].

The availability of such technological solutions for small producers in Ukraine is extremely limited, which leads to discrimination, unequal competitive positions, and affects minimizing farm producers in the individual sector, raising rural unemployment, increasing migration, and so on. Due to the high level of costs of digital innovations, neither the domestic banking sector nor international financial institutions cooperate with small producers [18]. In order to prevent deepening inequality in Ukraine's agricultural development through digital transformations, state measures should be developed. Priority is to be given to the support of small producers in implementing digital advanced technologies to maintain competitiveness and prevent monopolization in the agri-food market, the introduction of economic control instruments, including taxation, over intensive corporate agricultural production. Preservation of the rural way of life and biodiversity, support of the population in the traditional industries, an increase of peasants' incomes, and improvement of their living standards should become a state priority of digital technology dissemination in agriculture (Fig.2).

Fig. 2. The conceptual framework for the digital development of Ukraine's agri-food system
Source: developed by the authors based on the Concept of Digital Economic and Social Development of Ukraine for 2018 – 2020 [19].

3 Conclusion

The majority of politicians and corporate actors are currently inclined to recognize digitalization achievements as an unalterable perspective of social and economic progress. The rapid development and application of information technologies, herewith, lead to changes in living standards, work, production, social interaction, trade, environment, public policy, and governance.

Strategy design of digital transformations calls for awareness of the process and its consequences for people and society. Moreover, it should be noted that such technologies are used under national and global inequality in access to basic goods and services, as well as information and digital technologies. Keeping mentioned inequalities will lead to the reproduction and deepening of existing discrimination models within the framework of the new technology introduction.

The impact of digital transformations in agri-food systems on food, social and economic, and ecological security has not yet been properly assessed. Digital technologies in food and agriculture may potentially be useful or dangerous depending on the preconditions (context) and purposes of their application. On the one hand, digital technologies open up innovative opportunities in the agri-food sector, and, on the other, new challenges for society are being actualized due to the uncontrolled application of these technologies.

In an environment where agricultural corporations tend to seize control over technology, markets, consumer preferences, and able to implement the latest technologies and innovative products themselves, the main task of the world community is to support small farms, indigenous peoples, and other populations in their efforts to use digital opportunities. to provide themselves with means of subsistence in rural areas. The country's population to be guaranteed with traditional safe food, preserved biodiversity, and sustainable development of ecosystems, etc.

Targeted government policies on the protection of interests balance and the implementation of regulatory mechanisms for digital changes in the food and agriculture to avoid monopolization of digital food systems and the concentration of digital power in Ukraine should be introduced as strong measures against the undesirable impact of digitalization on rural development, human communities, and society as a whole. Dissemination of information about the benefits and consequences of digital transformations, public awareness about the impact of these processes on daily nutrition will contribute to informed decision-making at the micro, macro, and global levels.

References

1. A. Fraser, The Digital Revolution, Data Curation, and the New Dynamics of Food Sovereignty

Construction. The Journal of Peasant Studies, Vol.**47**, pp. 1-19 (2020). DOI: https://doi.org/10.1080/03066150.2019.1602522

2. United Nations, The Highest Aspiration. A Call to Action for Human Rights (2020). Available: https://www.un.org/sg/sites/www.un.org.sg/files/atoms/files/The_Highest_Asperation_A_Call_To_Action_For_Human_Right_English.pdf
3. Open Call for Engagement to Respond to the UN Food Systems Summit (October 2020). Available: http://www.csm4cfs.org/wp-content/uploads/2020/10/EN-Open-Call-on-UN-Food-Systems-Summit-12-October-2020-1.pdf
4. J. Thatcher, D. O'Sullivan, and D. Mahmoudi., Data Colonialism Through Accumulation by Dispossession: New Metaphors for Daily Data. Environment and Planning D: Society and Space, Vol. **34** (6), pp. 990–1006 (2016). DOI: https://doi.org/10.1177%2F0263775816633195
5. J. Bélanger and D. Pilling, The State of the World's Biodiversity for Food and Agriculture. FAO Commission on Genetic Resources for Food and Agriculture Assessments, Rome (2019). Available: http://www.fao.org/3/CA3129EN/CA3129EN.pdf
6. United Nations, The Sustainable Development Goals Report. UN Department of Economic and Social Affairs, NY (2020). Available: https://unstats.un.org/sdgs/report/2020/The-Sustainable-Development-Goals-Report-2020.pdf
7. United Nations, Transforming Our World: the 2030 Agenda for Sustainable Development (A/RES/71/1). Available: https://sdgs.un.org/sites/default/files/publications/21252030%20Agenda%20for%20Sustainable%20Development%20web.pdf
8. Food and Agriculture Organization of the United Nations. Hand-in-Hand Initiative (2021). Available: http://www.fao.org/hand-in-hand
9. FAO, IFAD, UNICEF, WFP and WHO, In Brief to The State of Food Security and Nutrition in the World 2020. Transforming food systems for affordable healthy diets. FAO, Rome (2020). Available: http://www.fao.org/3/ca9699en/CA9699EN.pdf.
10. Review of Agricultural Investment Policies of Ukraine. OECD (2015). Available: https://www.oecd.org/eurasia/competitiveness-programme/eastern-partners/Agricultural_Investment_Policies_Ukraine_ENG.pdf
11. Land Matrix. Available: https://landmatrix.org/data/by-investor-name/?country=804&order_by=-deal_size
12. Agriculture of Ukraine. Statistical Yearbook. State Statistics Service of Ukraine, Kyiv (2020). Available: http://www.ukrstat.gov.ua/druk/publicat/kat_u/2020/zb/09/zb_sg_Ukr_2019.pdf
13. Monitoring Food Security in Countries with Conflict Situations. A Joint FAO/WFP Update for the United Nations Security Council, Issue 5, 40 p. (2019). Available: http://www.fao.org/3/ca3113en/CA3113EN.pdf
14. M. Rudenko, The Impact of Digital Technologies on Agricultural Production: a Methodical Aspect. Scientific Notes of Taurida National V.I. Vernadsky University. Series: Economy and Management, Vol. **30**(69), pp. 30-37 (2019). DOI: https://doi.org/10.32838/2523-4803/69-6-28
15. N. Trendov, S. Varas, and M. Zeng, Digital Technologies in Agriculture and Rural Areas - Status Report, Food and Agriculture Organization of the United Nations, Rome (2019). Available: http://www.fao.org/3/ca4985en/CA4985EN.pdf
16. L. Basserman, Digitalisation in Agriculture (2019). Available: https://www.welthungerhilfe.org/news/latest-articles/2019/digitalisation-in-agriculture/
17. O. Borodina, Industry 4.0: Are We Moving Forward? Bulletin of the Gold Meir Institute, Vol. **7**, pp. 35-37 (2017).
18. O. Rykovska and O. Fraier, Digital Technologies in Rural Economic Activity. In Inclusive Rural Development in Ukraine. Kyiv, Institute for Economics and Forecasting, NAS Ukraine, pp. 89-98 (2020).
19. The Concept Digital of Economic and Social Development of Ukraine for 2018-2020, Cabinet of Ministers of Ukraine (2018). Available: https://www.kmu.gov.ua/storage/app/uploads/public/5a7/c21/c86/5a7c21c867bde621479479.doc

Legal Support of Non-Cash Settlements in Economic Activity

Serhii Braichenko[1], *Sergii* Dzholos[2], *Artem* Kolisnyk[3], *Oksana* Koshulko[4*], and *Larysa* Varunts[5]

[1]Legal Sciences, Cherkasy Educational-Scientific Institute of Banking University, 18028 Cherkasy, Ukraine
[2]Legal Sciences, the Bohdan Khmelnytsky National University of Cherkasy, 18031 Cherkasy, Ukraine
[3]Legal Sciences, Cherkasy Educational-Scientific Institute of Banking University, 18028 Cherkasy, Ukraine
[4]Economic Sciences, Alfred Nobel University, 49000 Dnipro city, Ukraine
[5]Legal Sciences, Kharkiv National University of Internal Affairs, 61080 Kharkiv, Ukraine

Abstract. The article presents the exploring of legal support of non-cash settlements in economic activity. In the article, the complex analysis of the tendencies of the legal regulation of the cashless transfers in commercial activity has been done. Also, the article has been prepared some recommendations on improvement of non-cash settlements, taking into the consideration fact that there are different types of payments under modern economic conditions, which change the approaches to the specific features, protection, and the peculiarities of their usage. Besides, in the article, some problems of the legal regulation of cashless settlements have been investigated according to the papers of international academics from different countries. Also, some legal documents of the state Ukraine in this direction have been analysed in the paper. Among the documents was the Instruction on Cashless Payments in Ukraine in the National Currency; the Civil Code of Ukraine, etc. Therefore, the legal support of non-cash settlements in economic activity is critical and valuable because all the support, as well as all the conditions of the implementation of the modern informational and telecommunication technologies in the activities of the financial institutions, may increasing of the clients' demand on the services of the cashless transfers.

1 Introduction

The legal support of cashless transfers in economic activity is a very important problem because modern business activity needs a stable and common mechanism of payment.

Therefore, the study and investigation of the new type of transfers and their methods is an actual thing.

The institute of the cashless settlement regulates commercial and civil relations among the client and the bank, as well as administrative and financial and informational relations, connected with the circulation of documents in the paper and digital form due to that fact that during the executing of the noncash transfer appear the documents, prescribed by the legal regulations of the National bank of Ukraine, and usually there is the exchange of the information among banks and payment systems about the transaction of money, which are on the operating account of the payer and the payee.

The legal regulation of the cashless settlements in the commercial activity plays a very important role in the modern world with its modern digital world economy, because the noncash transfers are not only the commercial or legal problem but also the political one, which is connected with the economic safety and demands the strategic approaches to its solution.

So, the correlation of the legal regulation of transfers and the modern legal support of transfers in the commercial activity is a very important problem, and its solution will allow discovering the peculiarities and specific features of the legal regulation of the payments and to define the main directions of the improvement of the commercial and legal regulation of the payments under the modern conditions.

The article aims to make a complex analysis of the tendencies of the legal regulation of the cashless transfers in the commercial activity and the creation of the recommendations on their improvement, taking into the consideration the fact that there are different types of payments under the modern economic conditions, which change the approaches to the specific features, protection and the peculiarities of their usage.

The problems of the cashless settlements and their legal regulation were investigated in the papers by J. Carles Maixé-Altés, E. Mourelle [1], D. Garcia-Swartz, R. Hahn, A. Layne-Farrar [2], M. Krueger, F. Seitz [3], O.V. Dzera [5], I.A. Bezklubyi [6], I.M. Domanetska, M.S. Petrushenko, O.V. Fedusenko, A.O. Fedusenko [8], N.M. Pantielieieva, N.V. Rogova, S.M. Braichenko, S.V. Dzholos, A.S. Kolisnyk [9], N. Pantielieieva, S. Zaporozhets, N. Nagaichuk, O. Bartosh [10], N.M. Panteleeva [11], V.F. Kuzmin [12], Bhaskar Chakravorti, Benjamin D. Mazzotta [15], H.Sheffield [16], J. Henley [17], F. Mokhtar [19], D. Frisby [20], M. Smith [21], C. Quinn [22], A.S. Bohdanovych [23], L.A. Novoselova [24], A.M. Borysova [25], and others.

With all respect to the scientists, who work in this field, it should be noted that the commercial and legal regulation of the payments in the commercial activity needs the next study to make our knowledge deeper and to improve it because the level of the regulation of

* Corresponding author: oksanakoshulko2015@gmail.com

the payments in the commercial activity is not sufficient and some theoretical problems of the legal regulation of the payments among the subjects of the commercial activity are awaiting on the solve.

2 The results of research

First, to understand the essence of the cashless settlements it is necessary to formulate their definition. According to the "Instruction on the cashless transfers in Ukraine in the national currency", the cashless settlement is a transfer of funds from the account of the payer to the account of the payee, as well as the transfer of money on the accounts of the recipient of funds, made by the bank on the demand of the natural bodies and legal entities, who pay money in cash to the cash department of the bank [4]. These transfers are made by the bank on the grounds of the payment document in a paper or digital form.

To understand the legal nature of the legal relations in the field of cashless money transfers, it is necessary to characterize such their elements, as their subjects, objects, matter, and legal facts, as the grounds of the appearance, changes, and termination of these legal relations [5].

Firstly, the order of the client of the bank to transfer money from one account to another one is a ground of the appearance of the legal relations on the cashless settlements. At the same time, the order addressed to the bank, is the unilateral transaction [6, p. 250], which should be stated in the written or digital form.

Therefore, the cashless transfers need the existence of the payer's account, created according to his contract with the bank. At the same time, some people may partially disagree with this statement, because the bank account is not always necessary to make the transfer of money. Article number 1089 of the Civil Code of Ukraine, 2003, confirms this and says:

"Under the payment order, the bank shall take responsibility to transfer by a payer's order a certain sum of money at the expense of the funds placed on his account in this bank to the account of a person (recipient) determined by the payer in this or another bank within the term determined by the law, or the banking rules, unless another term is stipulated by the agreement or customs of business practice. Provisions of this Paragraph shall be also applicable to the relations connected with the fund's transfer via the bank by a person having no account in this bank unless otherwise is stipulated by the law, banking regulations or results from the essence of the relations" [7].

Speaking about the German experience, Malte Krueger and Franz Seitz say that in the macroeconomic context the internal costs of businesses, connected with the cash and non-cash payments, are equal for at least 2-2,5% of GDP [3, p. 34-35].

Secondly, the bank, the payer, and the payee are the subjects of the legal relations on the cashless transfer. It should be noticed that in the different types of payments they might have their special names (nominated bank, payer, recipient of the payment, drawer of check, holder of cheque, the declarant of the letter of credit, beneficiary, etc.).

At the same time, according to the current legislation of Ukraine, payment systems and other financial institutions, which provide service of the money transfer, are also the subjects of the cashless settlements. After the appearance of the cryptocurrency, the peering payment systems, which do not have the central administrator and the issuer of the means of payment, are also the subjects of the cashless settlements.

These payment systems are based on the computer network of the "peer-to-peer" type (or P2P). They are based on the equality of their participants, and their elements may be connected among themselves, in the contrast to the traditional architecture, when only the separate category of participants, the servers, may provide other participants with certain services. There are no "clients" and "servers" in the pure "peer-to-peer" network, but there are only the equal nodes, which are at the same time the clients and the servers in the relation to the other nodes of the network. This model of the network interaction differs from the traditional architecture "client-server", where the connection occurs only between the clients and the central server. Such an organization allows keeping working capacity under the conditions of any configuration of its available participants. However, sometimes there are some P2P-networks, which have servers, but their role is not to provide clients with services, but to keep the information about the services, provided by the clients to the network [8, p. 124].

Thirdly, the object of the cashless settlements is the money funds. The function of payment, peculiar to the cash, maybe realized during the payment, made by the non-cash money, which is accounted for on the bank account. The same may be said about the digital money and cryptocurrency.

It is necessary to say few words on such new phenomenon as electronic money and cryptocurrencies.

There are different ideas on the nature of the "electronic money". They may be considered as the units of value, which are kept on the electronic devices. Also, they may be considered as a new type of money. Some authors say that the electronic money is the mean of payment. Other scientists say that the electronic money is the monetary obligation of the person, who issued them, and so on.

The characteristic features of cryptocurrencies are the following:

this money has the digital form;

cryptocurrencies, including the bitcoin, are the units of account, but, at the same time, their legal status is not regulated by the laws, legislation and regulatory acts of many different countries and states of the world;

its legal status is undefined;

its emission is made by the decentralized program;

its issuer is a private company or person;

its mechanism of exchange is a decentralized direct electronic exchange;

they may be used only inside some specific group of people or inside a certain virtual commonwealth, and the members of this virtual group or commonwealth should have a special and specific network, and this network is not the usual part of the existing systems of payment of the traditional type;

the issuer of the cryptocurrency defines the offer of the electronic money, and his will plays the main role in this context;

the possibility to redeem is not guaranteed;

the stage of the cryptocurrency formation is characterized by the absence of the supervision and by the lack of the regulation;

the operations with the cryptocurrencies are connected with the high level and a great number of the risks, and these risks may have different types, including the operational risks, credit risks, legal risks and even the liquid risks, etc.;

the cryptography, as well as the mutual and common trust among the participants of this virtual commonwealth, support the necessary level of the stability in the questions of the payment; but, at the same time, the participants of this virtual commonwealth are not responsible for the volatility of the cryptocurrency as a rate of exchange.

The scientists say that there are many positive qualities and characteristics of the cryptocurrencies. The first one is connected with the unique existence of each electronic coin, which may not be used the second time. The second positive quality of the cryptocurrency is the high velocity of the financial settlement. The third one is the high level of security, provided by the transactions, made with the usage of the cryptocurrency. The fourth one is the absence of the supervision in this field. The fifth positive quality of the cryptocurrency is their suitability to be used in the electronic and real payments for the goods, services and works without the assistance of the intermediaries. These five positive qualities of the cryptocurrency together make the sixth one – its efficiency. At the same time, it is necessary to understand, that cryptocurrencies also have the negative characteristics and qualities. One of them is the anonymity, which takes place in the process of the settlement (for example, it will be very hard to police to find and to prove the guilt of the hirer of a contract killer, who pays by cryptocurrency for the murder). The next one is the high volatility of the cryptocurrency price. And, of course, the next one Achilles' heel of the cryptocurrency is its turnover not on the territory of the certain state or worldwide, but only inside the certain virtual commonwealth of the people, who risk to invest their real money into the risk project of the cryptocurrency [9, pp. 413-414].

Also, it is necessary to say that creation of the cryptocurrencies led to the existence of the specific situation, when the cryptocurrencies and the traditional national currencies exist together. This situation changes the essence of the modern economic relations, as well as the movement of funds, in accordance with the tendency of the liberalization in all spheres of the social and economic life. Theoretically, it may reduce the harmful tendencies of the postmodern economy. But, at the same time, the great changes in the circulation of money and in the monetary relations, connected with the spread of the cruptocurrencies, make a serious danger to the traditional financial system of the state [10].

It is necessary to remember that today the cryptocurrencies may be used as the payment instrument, when one of the parties wants to by, and another one wants to sell the real estate or the car, or some other goods. It should be added, that a virtual space permits to make the sale contract, which involve the participants (natural persons and legal entities) from the different countries of the world, and the object of this purchase and sale contract may be different goods, services and works, including the traditional and non-traditional one. We must say that the stock markets, the virtual space, different commonwealths of the cryptocurrencies, the cooperation of the private persons and the financial markets, etc. are developing in our time very fast, and it leads to the growth of the Initial Coin Offering (ICO) market. It helps to find investments for the different business and commercial projects and startups with the help of the digital electronic platforms of the crowdfunding and by other means and methods [11].

The new methods of the investment generation appear and develop in our time because the cryptocurrencies-investors make the fresh ground in this sphere. Also, we must not forget about the existence of the Initial Exchange Offering and about the existence of the Initial Fork Offerings. The first one provides the primary stock offering, and the second one – the primary futures offering. However, it should be noted that the weakest element in the new system of the economic relations is the relations and the cooperation between the private persons and the state on the international, as well as on the national level. But, at the same time the Venezuelan experience shows that the issuance of the national cryptocurrency (Petro) may help to overcome the foreign and international sanctions and the economic crisis. The central banks as well as the governments of many countries, for example such as Norway, Sweden, Canada, the USA, China and others, are thinking about the pluses and minuses of the issuance of the national electronic and digital currencies. The National Bank of Ukraine also tests the project, called the "E-hryvnia" [9, p. 414].

In our time, we may foresee the creation of the free economical system as well as the free markets in future, and this process, probably, will be completed with the help of the new approaches to the economy, based on the possibilities and features of the cyberspace. At the same time, nobody may imagine all aspects and all features of these changes, and that is why today we have different attitudes to and evaluations of the cryptocurrencies and other novelties of modern digital economy. Some people distrust them, and others invest all their store of money into these projects. At the same time, the state needs the stability, and that is why different state authorities make attempts to reduce the risks and threats connected with the undermining of the state privilege to issue money, loss of the benefits of the seigniorage, decrease and depreciation of the traditional national currency, reduction of the money velocity, troubles to the efficiency of the state money policy, and its increasing complexity, inability of the state to control the policy and relations in the field of money circulation, possibility to use the cryptocurrency for the illegal purposes (for example arms and drug trade, etc.) [10].

In many different countries, some modern and new technologies may have the lack of the state regulation. For example, the cryptocurrency not always conform to the essence, nature and definition of the "electronic money", given by the legislation, due to the facts that the issuer of the cryptocurrency is not obliged to repay it; the crytpocurrencies may be issued by different bodies, but not by the one center of their issuance; the cryptocurrencies are not connected and bound with the defined non-cash or cash monetary funds. Also, the traditional relations, rights and duties, prescribed by the civil legislation to the cashless resources (for example, the right of claim), may not be applied to the cryptocurrency, because its legal status is not regulated enough by the laws or bylaws. Besides that, the transactions, made by the means of cryptocurrencies, unlike with the cashless transactions, are controlled and supported only by the means of cryptography and the owners' registers, but they are not monitored or controlled by any state organs or commercial banks.

So, in this context, it is very important to answer the question, how and why the participants of the cryptocurrency commonwealth may trust it? The honour is the concept of the feudal epoch, and unfortunately it is forgotten in our time, so the cryptography, the accounting of all transactions as well as the impossibility to change the electronic records without the traces of interference provide the cryptocurrency with the necessary security and produce the trust among the people, who participate in the cryptocurrency commonwealth. The transactions are confirmed by the digital electronic signatures of the former and future owners. The transaction procedure is secured by the blockchain, and the transaction in the cryptocurrency is authorized by the digital electronic signature. The circulation of the cryptocurrency is provided by the addition of the new and the next databases, which are added to the chain during the performance of the transaction, made in a certain cryptocurrency.

At the same time, the payment in cryptocurrency is connected with the specification of the new address of the public character, made by the owner with the usage of his personal or private key. Such transactions are anonymous, because their subjects are not personalized and their personal data is not available to the third person, unlike with the information about the transaction, which is available to persons, who are not the participants of the certain transaction. The virtual money units are kept on the virtual electronic purse on the owner's computer or, sometimes, on the remote server. Cryptocurrencies have the virtual nature and essence, so, they may not be the taken or given into the loan as the money of the traditional type, also the cryptocurrencies may not be the object of the obligation, because they are not the material object and the things, which may be measured by the generic parameters and characteristics in the physical world. Under these conditions, the Ukrainian Civil Code (as well as the civil codes of many other countries) currently does not regulate the civil rights and duties, connected with the cryptocurrencies or virtual money [9, p. 415].

Fourthly, the rights and duties of the subjects are the matter of the legal relations in the field of cashless transfer [12, p. 8]. Their sources are:

- Agreements, because article 1071 of the Civil Code of Ukraine, 2003, "Grounds to Write-off the Funds from the Account" says that "The bank may write off the funds from the client's account based on his instruction. The funds may be written off from the client's account without his instruction based on the court's decision and also in cases determined by the agreement between the bank and the client" [7].
- Laws and bylaws, different legal regulations, banking regulations, and customs of business practice.

The essence of these rights and duties defined in the article no. 1066 of the Civil Code of Ukraine, 2003, which says:

"Under a bank agreement, the bank shall be obliged to accept and enter into an account opened for a client (the account owner) monetary fund's receivable by him, to fulfill the client's instructions on recalculation and issuance of the respective sums from the account, and to complete other account transactions.

The bank shall be entitled to use the funds on the client's account ensuring his right to freely dispose of these funds.

The bank shall have no right to establish and monitor the directions of the client's funds use and establish other restrictions on his right to dispose of the funds at his discretion that is not stipulated by the agreement or by the law.

Provisions of this Chapter shall apply to the other financial institutions while concluding a bank account agreement by them following the issued license as well as to the correspondent accounts, and other bank accounts unless otherwise is established by the law" [7].

Also, it should be added that the bank shall have no right to establish and monitor the directions of the client's funds use and establish other restrictions on his right to dispose of the funds at his discretion, except the cases, when some restrictions on the property rights on the monetary means on the bank account are stipulated, for example, in the cases, prescribed by the Law "On the Prevention and Counteraction to the Legalization (or Laundering) of Proceeds from Crimes, Sponsorship of Terrorism and of the Spread of Weapons of Mass Destruction" [13].

In addition, one of the reasons to reject to perform the transaction may be the suspicion, that the operation is connected with one of the crimes, defined in the Ukrainian Criminal Code of 2001. In this context it is necessary to say, what acts in our time may be considered as a crime, according to the current criminal legislation.

The article No. 209 of the Ukrainian Criminal Code of 2001 punishes the legalization (or laundering) of the money and other property, which were got (or obtained) by the criminal means [14].

The article No. 209-1 of the Ukrainian Criminal Code of 2001 prescribes the punishment for the intentional violation of the requirements of the legislation on prevention and counteraction to legalization (laundering) of proceeds from crime, terrorist financing and financing of proliferation of weapons of mass destruction" [14].

The article No. 222 of the Criminal Code of Ukraine, 2001, punishes such act, as the fraud with financial resources [14].

In the context of the modern realities and conditions in the world, very popular is the article No. 258-5 of the Criminal Code of Ukraine, 2001, which punishes the financing of the terrorism [14]. At the same time, any lawyer knows that according to the article No. 27 of the Ukrainian Criminal Code, the accessory is one of the forms of the complicity in crime. So, probably, the addition to the Criminal Code of the new article, the article No. 258-5, which punishes the financing of terrorism, contradicts with the logical principle, known from the XIV century as the "Occam's razor", which says that any entity or concept must not be multiplied without the great and reasonable necessity.

Also, it is necessary to remember that many other crimes may be made with the usage of the cryptocurrencies.

Fifthly, the object of the cashless transfers is some act, committed by the subjects of these legal relations, and its result is a payment, made by the transfer of funds from the account of the payer to the account of the payee on the grounds of fee and equivalent commodity relations.

Therefore, the specific informational relations, connected with the circulation of the financial and bookkeeping documents, which contain the information about the payer and the beneficiary of money, appear during the debiting of monetary resources from an account and the crediting funds to an account. At the same time, money, as an object of these relations, does not move from the payer to the payee.

It is necessary to add that modern society is in the economic state where financial transactions are not carried out with tangible money in the form of banknotes or coins. More and more payments are made by transferring digital information between two or more participants of the transaction [15]. At the same time, there is a discussion about the term "cashless society", which is moving towards economic legal relations when money will exist only in digital equivalent. This means that the legal tender is registered and exchanged only in electronic digital form. This method makes it possible to reduce the cost of using the national currency, and also makes it possible for the state to make the control over transactions.

This idea is being discussed primarily because there is a tendency in the world towards a wider use of digital methods of registering and exchanging money in the sale of goods and investments, as well as transactions that have historically taken place. Financial transactions with cash are now often carried out in electronic form [16]. Some governments continue to impose limits on the amount of transactions if there is reason to doubt their legality [17].

As the observations of US scientists show, cashless systems can create problems for people who trust more to the money in cash. These categories include certain groups of the population, such as the poor, the elderly, illegal immigrants, youth, criminals. Electronic transactions require a bank account and skills in using of the payment system. Many people in poor areas do not have or have the limited access to the banking services. In the United States, nearly a third of the population does not have a full range of basic financial services. The research, made by the Federal Deposit Insurance Corporation, shows that approximately 25 percent of households with an annual income of less than $ 15,000 do not have an open bank account. In the United States, more than 7% of the population does not have a bank account, and in the same time this level exceeds 20% in some towns and in some locations in the countryside and even 40% in some areas [18].

During the implementation of the Smart Nation initiative, it was noted that Singapore is steadily moving towards a cashless economy. In this country, more than 14% of the population is over 65 and they still use cash as the only one method to pay. This is due to the lack of skills in the field of the digital payment methods, unwillingness to solve and eliminate problems associated with the recovery of the lost cards and passwords. They are also a more vulnerable category of the population, which is attacked by fraudsters [19].

It is necessary to say that under the conditions of the digital economy, the payments will be tracked by government agencies. Through tracked transactions, organizations will have access to personal information. Digital transactions are vulnerable because they leave the digital traces. These transactions allow businesses to build customer databases based on their spending and preferences. The problem of database mining depends on the style of the country movement towards a cashless society. Nowadays, we can see how trading companies accumulate personal data and consumer preferences of the population [20]. The possibility to collect this data may have positive and negative consequences. Firstly, the presence of a customer profile can positively affect on the quality of a personal approach to each client, on the guess and predict of his or her desires. The negative consequences of the collecting of the consumers' personal data include the risks that this information may be used for the illegal purposes.

This increases the risks of unauthorized hacking in the cases, if the payment transactions are stored on the servers. Financial cyber attacks and digital crimes also create great risks in the cases of the refusal of the payments in cash. Many companies have losses from information leakage, including the leakage from the payment systems. Electronic accounts are still quite vulnerable to unauthorized access and transfer of funds to another account or unauthorized purchases [21].

Also, it should be noted that attacks on telecommunications infrastructure or accidental shutdowns significantly impede the operation of the electronic payment systems. It shows the significant difference of the electronic payment systems from the cash transactions, which may be made with a minimal infrastructure [22].

D.D. Garcia-Swartz, R.W. Hahn, A. Layne-Farrar say that the appearance of the cashless society was foreceed by the scientists in the 1950s, after the appearance of the first general-purpose banking cards, and, of course, now, in the beggінning of the XXI century we are much more closer to the cashless society, than it was before [2].

Carles J. Maixe-Altes and Estefanía Mourelle wrote a significant scientific work. The authors propose the indicator, which helps to understand that a certain society is cashless. On their opinion, the indicator which shows that a certain society is cashless is based on the reflection of the proportion (or the percent) of transactions made with the usage of the cashless instruments from the total amount and value of the completed transactions in a certain society [1, p. 13]. The authors emphasize on that fact that the situation in the Spanish society has been changed very much during the last 20 years, and that now Spain is much closer to the cashless society, that 2 decades before [1, p. 12].

D.D. Garcia-Swartz, R.W. Hahn, A. Layne-Farrar say in their research on the cashless society that the minor transactions are often made in cash, but the large payments in our time are usually made in the non-cash form; the authors say that cashless payments encrease the welfare of society due to that fact that they are cheaper than the payments in cash (in particular to the customer, but not always to the merchants), but at the same time the authorities of the USA, Australia and the European countries recognize the danger of the great policy interventions, made by the different international digital payment systems in the markets [2, p. 195-196].

So, it should be added that it is necessary to understand the legal nature of cashless transfers in commercial activity to define the peculiarities of its legal regulation.

3 Legal relations in the field of cashless settlements

The majority of the legal relations in the field of cashless settlements may not exist separately, in a pure manner, because they exist in the connection with the performance of the civil or commercial obligations. The contractual ground of the relations among the banks and other financial institutions and their clients shows the existence of the civil-legal element in the cashless transfers. At the same time, the complicacy of the legal mechanism of the cashless transfer does not allow us to say that they are only the system of the civil-legal constructions. It is necessary to say that the free expression of the will, which is peculiar to the branches of the private law, has limited character in the field of the institute of the payments in the cashless form. The imperative method of the legal regulation dominates there, and it is peculiar to the financial law. It is expressed in such aspects, as:

- The procedure of the cashless settlement is strictly regulated by the imperative rules of the financial law, defined in the special legislation.
- Cashless transfers are the elements of the banking operations and the expression of the functioning of the banking system.

Therefore, they are one of the elements of the object of the legal regulation of the financial law. At the same time, the National Bank of Ukraine plays a great role in the field of the regulation of cashless transfers.

The institute of the cashless transfers regulates not only the material (civil-legal) relations but also the organizational and administrational relations [23, p. 392], as well as the commercial and informational relations.

The legal regulation and provision of the cashless transfers in commercial activity based on such obligatory criteria:

1. Cashless transfers are made by the parties of the civil-legal obligation through the operating, current, or other bank accounts, which permit to make payments according to the client's order. This statement is a partly right in our time due to the existence of digital money and cryptocurrencies.
2. The participants of the payment may choose and stipulate in the agreement any form of the payment, provided by the law, the banking regulations, and customs of business practice. Banks may not refuse clients to make these operations unless otherwise established by the law or by the bank account agreement when the bank has a right to refuse to carry out a financial operation.
3. In the payments in the field of civil-legal relations, the debiting of monetary resources from the account may be made by the order of the owner of the account, except to the cases, prescribed by the current legislation or by the agreement between the bank and the client. The will of the owner of the bank account may be expressed in the form of the direct order to the bank to transfer funds or by the cash payment order, as well as in the form of the written consent to make a payment on the demand of the third person.
4. The bank shall have no right to determine and to control the directions of the client's funds use and establish other restrictions on his right to dispose of the funds at his discretion that is not stipulated by the agreement or by the law.
5. The bank, which participates in the payment, based on the civil-legal obligation of the counter agents, is not a part of this obligation. The bank is a part of the bank account agreement, and it is responsible before its

client only for the performance of this agreement.

6. It is not allowed to limit the client's rights to dispose of funds on his account, except the cases of the seizure of accounts or the suspension of operations on accounts in the cases, prescribed by the law, including the cases, when the transaction is the object of the financial monitoring.
7. The payments from accounts may be made in the presence of the funds on the payer's accounts or at the expense of a bank loan provided to the payer.
8. The non-cash payments should be based on the documents of the prescribed form [24, pp. 47-49].

The operations of any banking institution and the entire banking sphere are full of informational relations, and their main aspect is the information [25, p. 77] because during the performance of the cashless transfers appear the documents, prescribed by the legislation of the central bank, and in the majority of the cases takes place the exchange of the information among the banks and payment systems about the transfer of money, which are kept on the operation accounts.

4 Conclusions

It should be noted that under the conditions of the implementation of the modern informational and telecommunication technologies in the activities of the financial institutions, the appearance of the electronic trading, the appearance of the new financial instruments and electronic banking systems, the development of the market of payment services is characterized by the increase of the clients' demand on the services of the cashless transfers.

At the same time, the recommendations on the improvement of the legal regulation of cashless transfers in the commercial activity should be based on the complex analysis of the interests of the participants of the market and the state.

Also, it is necessary to add, that cashless settlements are based on the exchange of the documents, established by the state and prescribed by the legislation, which contains the information about the movement of funds.

The non-cash payments are characterized by the contractual form, which is based on the rules of commercial and civil law.

At the same time, the rules of the financial and the administrative law prescribe the demands to the information, which serves to the circulation of the non-cash funds.

Also, it is necessary to add, that the non-cash settlements are the new institute in the economic activity, so it is necessary to understand its economic and legal essence, to avoid the potential risks, which may make damages to the interests of the state, of the society, and of the person.

The existence of the different methods of the non-cash settlements and digital payment systems, as well as the existence of the different types of the cryptocurrencies, which are not controlled by the state in the appropriate forms and degrees of inspection, make some risks. So, these modern technologies may be used as the instruments to achieve the illegal purposes, such as the development of the shadow economy, terrorism, human trafficking, drug trade, trade in arms, tax evasion, bribery, corruption, etc.

The existence of the non-cash settlements, of the digital payment systems, and of the different types of the cryptocurrencies undermines the state monopoly of issuing money, and even the status of the national currency as the legal tender.

In the pre-modern society the non-cash settlements, the digital payment systems, and the different types of the cryptocurrencies should be considered as the mixture of theft, fraud and coinage offence. Of course, such evaluation of these modern technologies is not always acceptable under the modern conditions, but the state should do all the best at least to prevent their usage in the criminal and other socially harmful purposes.

So, the state should adopt a system of regulations on the non-cash settlements, the digital payment systems, and the different types of the cryptocurrencies, to define the punishments for their violations, and to create the competent state agencies, which will realize the state policy in this field.

In other case, under the conditions of the development of the cashless society the state will not be able to realize its regulatory and protective functions, to protect the human rights, freedoms and interests in the economic field, and it will be transformed into a virtual state, which has no any real influence on the social life, which will lead to the economic crisis, to the poverty, to the collapse of the society, to anarchy in the state and to the fall of the statehood and the violation of the law and order.

That is why, the legal regulation of the non-cash settlements is a very important economic, legal, social and even political problem, which should be solved by the state.

References

1. Maixe-Altes, J. Carles and Mourelle, Estefanía. Cash and non-cash payments in a long-run perspective, Spain 1989-2014. Department of Applied Economics, University of A Coruna, A Coruna, Spain. Retrieved from https://mpra.ub.uni-muenchen.de/72590/ MPRA Paper No. 72590 (2016).
2. Garcia-Swartz, D.D., Hahn, R.W, Layne-Farrar, A. The Move toward a Cashless Society: A Closer Look at Payment Instrument Economics. Review of Network Economics No. 5(2), 175-198. (2006).
3. Malte Krueger and Franz Seitz. The Importance of Cash and Cashless Payments in Germany: Overview and first estimates. The usage, costs, and

benefits of cash - revisited: International Cash Conference 2014. Frankfurt am Main: Dt. Bundesbank, p. 15-73 (2014).

4. The Order of the National Bank of Ukraine "On the Adoption of the Instruction on Cashless Payments in Ukraine in the National Currency" No. 22, adopted on the 21st of January, 2004. Retrieved from https://zakon.rada.gov.ua/laws/show/z0377-04 [in Ukrainian].
5. Dzera, O.V. (Eds.). Contract law of Ukraine: special part. Kyiv: Yurinkom Inter [in Ukrainian] (2009).
6. Bezklubyi, I.A. Banking transactions: civil law problems. The Taras Shevchenko National University of Kyiv [in Ukrainian] (2005).
7. The Civil Code of Ukraine No 435-IV, adopted on the 6th of January, 2003. Retrieved from https://zakon.rada.gov.ua/laws/show/435-15#Text [in Ukrainian] (2003).
8. Domanetska, I.M., Petrushenko, M.S., Fedusenko, O.V., & Fedusenko, A.O. Simulation model of peering network node functioning // Managing the development of complex systems, No. 12, p. 124-132. Retrieved from http://nbuv.gov.ua/UJRN/Urss_2012_12_27 [in Ukrainian] (2012).
9. Pantielieieva N.M., Rogova N.V., Braychenko S.M., Dzholos S.V., Kolisnyk A.S. Current aspects of transformation of economic relations: cryptocurrencies and their legal regulation // Financial and credit activity: problems of theory and practice. Vol. 4, No. 31. pp. 410-418 (2019).
10. Pantielieieva, N., Zaporozhets, S., Nagaichuk, N., & Bartosh, O. Transformation of financial intermediation in the context of spread of digital trends // Bulletin of the National Academy of Sciences of the Republic of Kazakhstan, No. 3, pp. 144-152 (2019).
11. Panteleeva, N.M. Innovative potential and characteristics of the mechanism of attraction of financing in new technological projects through the issue and sale of new cryptocurrencies to market participants // Proceedings of the 3rd International Conference: New forms of money and financial assets: formation, prospects, risks. (pp. 153-158). Kyiv: KNEU [in Ukrainian] (2017).
12. Kuzmin, V.F. (1971). Raschyotnye pravootnosheniya v novyh usloviyah hozyajstvovaniya [Settlement legal relationships in the new economic conditions]. Extended abstract of candidate's thesis. Moscow [in Russian].
13. Law of Ukraine Law "On the Prevention and Counteraction to the Legalization (or Laundering) of Proceeds from Crimes, Sponsorship of Terrorism and of the Spread of Weapons of Mass Destruction" № 361-IX, adopted on the 6th of December, 2019. Retrieved from https://zakon.rada.gov.ua/laws/show/361-20#Text [in Ukrainian].
14. The Criminal Code of Ukraine, adopted on the 5th of April 2001 No 2341-III. Retrieved from https://zakon.rada.gov.ua/laws/show/2341-14#Text [in Ukrainian].
15. Bhaskar Chakravorti, Benjamin D. Mazzotta. The cost of cash in the United States (2013) / The Fletcher School Tufts University. Retrieved from https://web.archive.org/web/20161201110118/http:/fletcher.tufts.edu/CostofCash/~/media/Fletcher/Microsites/Cost%20of%20Cash/CostofCashStudyFinal.pdf
16. Hazel Sheffield (2015). The UK is getting closer to becoming a completely cashless society. The Independent. May 21, 2015.
17. Henley, Jon (June 4, 2016). Sweden leads the race to become cashless society. The Guardian (June 4, 2016). Retrieved from https://www.theguardian.com/business/2016/jun/04/sweden-cashless-society-cards-phone-apps-leading-europe
18. Corporation for Enterprise Development (2016).The Most Unbanked Places in America. Retrieved from https://web.archive.org/web/20160820103339/http://cfed.org/assets/pdfs/Most_Unbanked_Places_in_America.pdf
19. Mokhtar, Faris (28 August 2017). Govt will seek to ensure no one is left behind in cashless drive. Today (August 27, 2017). Retrieved from https://www.todayonline.com/singapore/govt-will-seek-ensure-no-one-left-behind-cashless-drive
20. Frisby, Dominic (21 March 2016). Why we should fear a cashless world. The Guardian. Retrieved 8 April 2018.
21. Smith M. (1 January 2018). Forever 21: Hackers breached payment system for 7 months. CSO. Retrieved from https://www.csoonline.com/article/3245069/forever-21-hackers-breached-payment-system-for-7-months-no-encryption-on-pos-devices.html
22. Cristina Quinn (2018). As Society Becomes Increasingly Cashless, Is Massachusetts Ready? GBH News. February 13, 2018. Retrieved from https://www.wgbh.org/news/2018/02/13/news/society-becomes-increasingly-cashless-massachusetts-ready
23. Bohdanovych, A.S. Cashless settlements using electronic means of payment as a complex legal sub-institution. Journal of Kyiv University of Law, 4, 391-394 [in Ukrainian] (2013).
24. Novoselova, L.A. Cash settlements in business. Moscow [in Russian] (1996).
25. Borysova, A.M. Directions and prospects for improving information banking relations]. Young scientist, 1, 75-78 [in Ukrainian] (2014).

Electronic Cases as An Element of the Administrative Procedure in The Provision of Administrative Services

Oleg Dubinskiy[1], *Olena* Lomakina[1], *Oleksandr* Sikorskyi[1*], and *Azamat* Kudaibergenov[2]

[1]*Admiral Makarov National University of Shipbuilding,* 54025, Ukraine
[2]*Karakalpak State University named after Berdakh,* 230112, Republic of Karakalpakstan, Uzbekistan

Abstract. The article is devoted to the study of the current state of legal regulation of the institute of electronic case as an element of administrative procedure in Ukraine and the practice of its implementation on the example of the procedure for providing administrative services through the Centers of administrative services. The main problems of the institute and possible ways to solve them are considered, as well as prospects for the development of administrative procedure, based on international experience and regulatory framework of states, including - members of the European Union in the study area.

1 Introduction

After gaining independence, Ukraine took the direction of updating the legislative basis in all branches of law, streamlining all spheres of society. The globalization processes, which caused rapid changes in the protection of the legal rights, freedoms and interests of the citizen, influenced the formation of the mechanism of the system of administrative services. With the beginning of the administrative reform in our state, the issues of improving the quality of the provision of administrative services, which are provided by public authorities and local governments, are becoming increasingly important.

Despite many fundamental studies on the subject, questions about the definition of public services and questions about procedure of administrative services still remain quite controversial and therefore require a deeper analysis, the study of the specific features and improvement of not only theoretical basis, but also the consolidation of research data on legislative level.

The problems of the study of administrative services, their legal regulation and ways of improvement were analyzed and studied in their works by such scientists as: Andrushko P.P., I.V. Boyko, A.T. Zima, Glushkov V., Evtikheev I.I., Lazarev B.M., Kruglikov L.L., Korzhansky M.I., Melnik M.I., V.P. Timoshchuk, Khavronyuk M.I., Galakhova A.V. and others.

However, given the realities of the ever-growing role of information telecommunication technologies in state and municipal administration, informatization of society at all levels and the peculiarities of the functioning of our state in conditions of quarantine, the issue of implementing electronic government in the provision of administrative services seems to be relevant. Indeed, it is the in-depth and systematic implementation of the latter that will fully ensure the implementation of the concept of state policy for improving the quality and availability of administrative services, transparency, efficiency and openness of the activities of the subjects of their provision.

So, according to the information posted on the official website of the Cabinet of Ministers of Ukraine, the most important area in the field of electronic governance is electronic services. Currently, the Government Portal, which serves as a "one-stop shop" for all online services, already has 118 e-services. Among them are such socially important ones as the registration of a childbirth allowance, business registration services, services in the land and construction sectors. Also, important sets of electronic services for drivers and carriers were launched, available respectively in the carrier's and driver's electronic cabinets. It is summarized, that only in 2018, the level of use of electronic services in Ukraine has tripled [1] and continues to grow in the future, given the increase in the categories of people using information and communication technologies in everyday life and business.

However, some problematic issues of the introduction of e-government in the provision of administrative services were covered in the works of the mentioned scientists fragmentarily, within the framework of a more general issue. In particular, in this direction special attention should be paid to the formation of an electronic case at the time of provision of administrative services, because the case is a key element of the administrative procedure at all stages, which, given the above, determines the relevance of this study.

2 Main part

2.1 Legal regulation of e-government and provision of e-services in Ukraine

It is worth noting that the legal prerequisites for the introduction of electronic government and the provision of electronic services by public authorities in Ukraine are due to the adoption of the Laws of Ukraine

*Corresponding author: alex.nic.gm@gmail.com

"On electronic documents and electronic document management" dated May 22, 2003 No. 851-IV and "On electronic digital signature" dated 22 May 2003 No. 852-IV.

Developing the provisions of these laws, taking into account domestic experience in this area, in 2016 formed the Concept for the development of electronic services in Ukraine, approved by the Order of the Cabinet of Ministers of Ukraine dated November 16, 2016 № 918-r. In the provisions of this concept, the electronic service is defined as an administrative and other public service provided to the subject of circulation in electronic form by means of information, telecommunications, information and telecommunications systems. The text of the document also stipulates that reducing the number of documents required from the subject of application through the introduction of interdepartmental electronic interaction, opening access to state information resources and introduction of electronic forms of interaction between subjects and subjects of administrative services are the main ways and means of solving problems, related to the development of the electronic services system in Ukraine [2].

The next stage of improving Ukrainian legislation in 2017 was the adoption of the Law of Ukraine "On electronic trust services" from 05.10.2017 № 2155-VIII, settled relations in the provision of electronic trust services of electronic identification [3] and approval by the Order of the Cabinet of Ministers of Ukraine from 20 September 2017 № 649-r of the Concept of e-government development in Ukraine (hereinafter - the Concept) [4].

According to the Concept, e-government is a form of organization of public administration, contributes to increasing the efficiency, openness and transparency of the activities of state authorities and local governments using information and telecommunication technologies to form a new type of state focused on meeting the needs of citizens [4].

Taking into account the advantages of e-services technologies, the main measures to ensure the development of e-government in Ukraine in the direction of modernizing public services and developing interaction between government, citizens and business using information and communication technologies are:

- the introduction of electronic services, including administrative ones, in all spheres of public life, as well as the provision of integrated electronic services for life and business situations;
- implementation of the one-stop-shop principle by ensuring the development and functioning of the Unified State Portal of Administrative Services as a single point of access for individuals and legal entities to electronic services;
- development of electronic public procurement, electronic contracts and invoices, electronic auctions;
- encouraging the use of electronic services by individuals and legal entities [4].

The implementation of the provisions of the Law of Ukraine "On electronic trust services" took place by introducing the Resolution of the Cabinet of Ministers of June 19, 2019 № 546 in Ukraine functioning of the electronic identification system [5], which fully implements the idea of remote provision of administrative services in electronic form.

Further development in the field of "digitalization" of administrative services in Ukraine is the introduction of the Resolution of the Cabinet of Ministers of Ukraine "Issues of the Unified State Web Portal of Electronic Services of the Unified State Portal of Administrative Services" from December 04, 2019 № 1137 Portal "Action" - as a service by means of which the idea of "the country in the smartphone" is introduced. The provision also stipulated that from January 1, 2020, the Unified State Web Portal of Electronic Services will also perform the functions of the Unified State Portal of Administrative Services [6].

It should also be added that in accordance with the Law of Ukraine "On Amendments to Certain Legislative Acts of Ukraine Concerning Optimization of the Network and Functioning of Administrative Service Centers and Improvement of Access to Administrative Services Provided in Electronic Form" of Novrmber 03, 2020 №943-IX Unified the state portal of administrative services has been liquidated as such, which does not meet today's requirements. Information to it is mainly entered manually. The portal has an outdated interface and software, which makes it impossible to implement modern services (such as using the portal via a smartphone). In addition, it provides only administrative services, which are only part of other electronic public services, and therefore has become an extra link between the applicant and the Unified State Web Portal of electronic services [7].

Also, among the important innovations of the law were: the elimination of Administrative Services Center formed by district state administrations and the provision of services only in the Administrative Services Center of local governments; provision of services not through the Administrative Services Center, but in the Administrative Services Center (that is, administrative services will be provided by representatives of the subjects of administrative services located in the Administrative Services Center, administrators or through administrators through their interaction with the subjects of administrative services).

A certain step towards the introduction of the obligation to conduct electronic cases during the provision of administrative services was taken with the adoption of the Resolution of the Cabinet of Ministers of October 28, 2020 No. 1035. Thus, the Resolution provides that from March 1, 2021 applications for administrative services on certain social issues are accepted by officials of the executive body or the center for the provision of administrative services exclusively with the formation of the electronic case [8].

2.2 Analysis of the current state of implementation of electronic cases as an element of administrative procedure

Taking into account the fact that the formation of a case during the provision of administrative services, including electronic, is inextricably linked with the procedure for their provision and that there is no separate legislation that would define the procedure for such provision in a systematic form, it is advisable to consider the issue under study through the prism current legislation on administrative services.

In national legislation, the issue of regulating the administrative procedure is also assigned a significant role.

This issue becomes especially relevant in the context of a comprehensive reform of the system of public administration and local self-government.

One of the examples of the regulation of the basic principles of administrative procedure at the level of the law is now two drafts of Law of Ukraine "On Administrative Procedure" dated 09.06.2015 [9], however, it never came into force. In 2020, the draft of law "On Administrative Procedure" was finalized, developed by the Ministry of Justice of Ukraine and approved by the Cabinet of Ministers of Ukraine. From the moment of its registration by the Verkhovna Rada of Ukraine dated May 14, 2020 under number 3475, the process of passing the legislative procedure of this bill began [10].

In our opinion, the basic category of "administrative procedure" is a problematic or insufficiently resolved issue in these draft laws. Although the very concept of "administrative procedure" of the 2020 draft law duplicates the definition of the 2015 draft law, that is, it discloses the procedure through the concept of administrative proceedings (Paragraph 4,5, Part 1, Article 2 of the draft Law). There were also attempts to codify administrative and procedural activities in Ukraine, while forming unified principles for the activities of officials of public authorities.

Analyzing the provisions of the Law of Ukraine "On Administrative Services", we determine that it contains an article, dedicated to the order of the provision of administrative services, but does not contain a clear set of actions of subjects of legal relations in the field of administrative services.

The draft of Administrative Procedural Code of Ukraine on December 3, 2012 also contains the definition of "administrative procedure", which is the procedure for administrative proceedings determined by legislation [11]. Such a definition is not advisable, since it does not reflect the main features of the concept at all and leads to a misunderstanding and interpretation of this term.

An interesting idea is represent by V.P. Timoshchuk, who notes that the administrative procedure is the procedure established by the current legislation for the consideration and resolution of individual administrative cases by administrative bodies [12, p.24].

A.S. Lagoda details the concept and defines the administrative procedure as the procedure established by law for the consideration and resolution of individual cases by executive authorities and local self-government bodies, the ultimate goal is the adoption of an administrative act or the conclusion of an administrative contract [13, p.4].

Electronic services are generally a novelty for Ukrainian legislation and the definition of the administrative procedure for the provision of electronic services is not disclosed at all. So, most scientists consider the administrative procedure for the provision of electronic services as a set of sequential actions associated with the implementation of assigned tasks and powers to provide electronic services, which are performed using information and communication technologies by the relevant official in accordance with the current legislation [14, p.124].

The current legislation, namely the Law of Ukraine "On Administrative Services" contains the term "procedure" in Paragraph 8 of Part 1 of Art. 4 and Part 5 of Art. 8 of this Law. Thus, Paragraph 8 of Part 1 of Article 4 contains the principle of rational minimization of the number of documents and procedural actions, on which the state policy in the provision of administrative services is based [15, p.8, p.1 of Art.4]. This principle implies a reduction in the number of documents and actions that are required to provide a public service and aims to simplify the procedure for obtaining administrative services.

In turn, in Part 5 of Art. 8 indicates that the technological map of the public service contains information on the procedure for the provision of public services, namely:

1) the stages of processing an application for the provision of a public service;

2) the responsible official;

3) structural units responsible for stages (action solutions);

4) the timing of the stages (actions, decisions) [15, p.5 of Art.8] . This norm implies elements of streamlining the procedure for the provision of public services.

It follows from this that the technological map of an administrative service is a list of elements of the improvement of the procedure for the provision of public services.

The Resolution of the Cabinet of Ministers of Ukraine dated January 30, 2013 No. 44 "On approval of the requirements for the preparation of a technological card of public services" contains the stages of providing administrative services, which can be considered stages of an administrative procedure, because each stage includes the procedure for performing actions when solving an administrative case [16]. It follows from this that the structure of a procedure is made up of stages, and stages are an element of stages.

Moreover, it is worth noting that Paragraph 6 of this Resolution states that the stages of processing an application for the provision of a public service include:

1) registration (execution) of the appeal of the subject of circulation;

2) development of an appeal and registration (approval) of the result of the provision of public services by structural divisions and officials of the

subject of the provision of administrative services, in particular, through interaction of structural divisions of the subject of providing administrative services, the subject of providing administrative services with other bodies to obtain documents, information, decisions, answers required for the provision of public services, as well as with administrators - when providing public services through the center for the provision of administrative services, indicating the deadlines for such actions;

3) the issuance of the result of the provision of public services and its registration [16].

In turn, A. N. Bukhanevich notes that such a mechanism is not acceptable, and even goes beyond the procedure for providing administrative services [17, p.129].

V.P. Tymoshchuk, noting that such information should not be contained in the technological map, since the latter is not intended for subjects of circulation. Filing a complaint represents the beginning of new stages in the course of the case; therefore, it should only be briefly said about the mechanism for appealing the result of the provision of the public service [18, p.139].

Thus, we can say that the procedure for the provision of public services conditionally includes three stages, which are divided into the appropriate stages: registration of the appeal, consideration of the appeal and issuing a response to the appeal. Moreover, the formation of the case is carried out already at the initial stage of the administrative procedure.

It should also be added that in accordance with Part 1 of Article 9 of the Law of Ukraine "On Administrative Services", administrative services are provided by the subjects of the provision of administrative services directly through the centers for the provision of administrative services and / or through the Unified State Portal of Administrative Services [15, p.1 of Art.9].

Thus, when applying for an administrative service to the Center for administrative services, the first stage of the procedure for providing such a service will contain the stages that are determined in the Model Regulations of the Center for administrative services. Thus, registering the application, the administrator: checks the compliance of the incoming package of documents with the information card of the public service, if necessary, assists the subject of the application in filling out the application form; compiles a description of the incoming package of documents; registers the incoming package of documents by entering data into the registration log (in paper and / or electronic form); draws up a letter on the progress of the case in paper and / or electronic form; finds out the subject of providing administrative services, which is competent to decide on the case; sends the incoming package of documents to the subject of administrative service [19]. So, we can conclude, that the form of the case for the provided administrative services is determined individually.

At the same time, the organizational and legal basis of electronic document management is established by the Law of Ukraine "On electronic documents and electronic document management" dated May 22, 2003 No. 851-IV.

According to Article 10 , the sending and transmission of electronic documents are performed by the author or intermediary in electronic form by means of information, telecommunications, information and telecommunications systems or by sending electronic media on which this document is recorded [20].

The law also provides for the procedure for obtaining electronic documents. Thus, an electronic document is considered to be received by the addressee from the time the author receives a message in electronic form from the addressee on receipt of this electronic document by the author, unless otherwise provided by law or prior agreement between the subjects of electronic document management [20].

If a preliminary agreement between the subjects of electronic document circulation does not determine the procedure for confirming the fact of receipt of an electronic document, such confirmation can be carried out in any order by automated or otherwise in electronic form or in the form of a document on paper. The specified confirmation must contain data on the fact and time of receipt of the electronic document and on the sender of this confirmation [20].

If the author does not receive confirmation of the fact of receipt of this electronic document, it is considered that the electronic document has not been received by the addressee [20].

Unless the author and the addressee have previously agreed otherwise in writing, the electronic document is considered to be sent by the author and received by the addressee at their location (for individuals - place of residence), including if the information, telecommunications, information and telecommunications system with which the the document is located elsewhere. The location (residence) of the parties is determined in accordance with the law [20].

The second stage of the procedure for providing public services covers not one but several stages with different performers and is quite extensive and contains a large number of procedural points. Thus, in paragraph 6 of the Resolution of the Cabinet of Ministers of Ukraine of January 30, 2013 No. 44 "On approval of the requirements for the preparation of technological card of public service", it is noted that the stages of processing the application for public service are determined in accordance with regulations establishing the procedure public service and unless otherwise specified by law [16]. This indicates that there are cases when certain types of administrative services provide for a different number of stages at this stage, so it is necessary to consolidate this procedure in the relevant legislation.

The third stage of the procedure for the provision of public services includes the following stages:

1) the subject of providing administrative services: forms the output package of documents; transfers it to the Center for administrative services; notes this in the case letter;

2) the administrator of the center: informs about the result of providing the state service to the subject of the address in the way specified in the description of the incoming package of documents; registers the outgoing package of documents by entering the relevant information in the letter of the case, as well as in the relevant register in paper and / or electronic form; transfers the original package of documents to the subject of application in person against a receipt (including his authorized representative) upon presentation of an identity document and / or certifying his authority, or in cases provided by law, is transferred by other acceptable to the subject of application [19].

Since the electronic provision of services is introduced, it would be appropriate at this stage to provide the possibility for a person to pay the funds needed to receive the service in electronic form.

In addition, the process of providing electronic services should include a stage of appeal against the actions or inaction of the authority that provided the service. The implementation of a person's appeal of actions or inaction of a public administration body, for example, administratively, should also take place electronically with the possibility of filing an electronic complaint and monitoring the results of such a complaint [14, p.127].

There is a certain problem of providing administrative services to determine the subject composition of the relevant legal relations, as there is a question of determining the range of subjects of administrative services, the competence of these subjects and the range of powers, in turn reflected in the order of cases.

The Law of Ukraine "On Administrative Services" defines the subjects of administrative services - the executive body, another state body, the authority of the Autonomous Republic of Crimea, local government, their officials, state registrar, the subject of state registration, authorized in accordance with the law to provide administrative services [15]. At the same time, their powers in a systematized form are absent and prescribed for each of them separately in their own normative legal acts, which in their number constitute a very large accumulation of regulatory and legal information, which leads to misinterpretation and application and creates conflicts. Moreover, today, the number of entities providing administrative services is quite large, and this leads to difficulties for the relevant entities applying for a certain public service.

Some administrative scientists believe that this composition of the subjects of administrative services is narrow and does not contain a complete list of subjects of administrative services. Moreover, the Concept for the Development of the System for the Provision of Administrative Services by the Executive Authorities, approved by the order of the Cabinet of Ministers of Ukraine dated February 15, 2006 No. 90-r, provided for the presence of state enterprises, institutions and organizations among the subjects of administrative services, who are not subjects of power [21].

I.B. Koliushko also expresses his opinion on this issue, noting that the circle of subjects of providing administrative services is limited only by executive authorities and local governments and instead of including enterprises and institutions in the circle of subjects of providing administrative services, they were simply removed from the list of subjects of providing administrative services, which leads to the spread of bureaucracy [18, p.9].

Even despite the presence of certain information on the Internet, it is not always easy to find information about whether the subject provides a certain administrative service and indeed it is this subject for the provision of administrative services that is endowed with appropriate powers and does not duplicate the activities of other subjects of public administration and the like.

2.3 International experience in implementing electronic cases as an element of administrative procedure

Scientist Yu.M. Zhuk notes that today there is an urgent need to introduce foreign experience in the provision of administrative services, which should be used to improve the existing Ukrainian model and bring it in line with European standards [22, p.59]. In particular, most administrative scientists are inclined to believe that there is a need to put into practice a real dialogue between Ukraine and a number of European countries to implement and study in practice the experience of providing services, rather than formally copying one or another element of the system.

From the point of view of the functioning of the system for the provision of administrative services in electronic form, Poland has an interesting experience. In this country, a centralized system of Resident Service Departments (WOM) has been established, which is a guarantee of proper quality of services and speeding up administrative procedures [22, p.63].

A special feature is the eWOM Internet resource, which contains document forms for accessing any of the 18 local departments in the districts. Moreover, the site contains extended information about: precinct departments, customer service standards, the page of the citywide Citizen Service Center, regulatory framework, means of communication with other institutions and the like. At the same time, there is a special page "Alphabetical list of cases", which provides information on 268 administrative cases and services.

For each service there is detailed information: card number (application form to be filled in); documents required for submission; the place of filing documents and obtaining a decision on the case, the procedure for payment for services, the deadline for providing consumers with a response, etc. [23, p.80].

The structure of the WOM also includes: points of consideration of appeals of consumers and provision to them of administrative services on administrative questions, cash desks for payment of administrative services, information points, points of insurance, provision of photocopying services and photo services [23, p.80].

In the process of solving the case and receiving public services in the Department, the person carries out the appropriate list of actions. In particular:

- receives a form of the document on the case in the Information point of the department or on electronic resources;
- fills in the form of the document on the case in the Information point of the department or on electronic resources;
- attaches to the form the documents on the case in the information point of the Department or on electronic resources the necessary documents that are required;
- receives a receipt for the service, pays at the box office in cash, by card, or by transferring funds to the specified account [23, p.81].

Canada's experience in providing administrative services is also quite useful. First of all, it should be noted that Canada has three clearly separated levels of government: federal, provincial, and municipal. Thus, the largest number of services is provided at the provincial level.

Also in Canada, the term "administrative services" is not used, but the term "public services" ("government services") is used. They are divided into "internal" (government for government) and "external" (for citizens, business). These services include regulatory measures and information services [24].

The most common service integration in Canada is database integration. A novelty is the consideration of comfort for the consumer of the service, because if it is possible to submit an application or receive services via the Internet or by telephone, then this method of receiving or providing the service is available.

In Canada, special attention is now paid to five channels for applying for services: telephone, personal visit, autonomous interactive service centers, Internet, mobile communications. Moreover, mobile (outbound) offices are being introduced in Canada [24].

The creation of a single digital market for public services is a modern direction of reforming the provision of electronic services in the EU countries. It should be noted that the market of electronic public services in the countries of the European Union is acquiring a rapid pace of development, and the necessary task today is to ensure compatibility between the systems for the provision of electronic public services of various EU member states [25, p.101].

The provision of cross-border administrative services in the EU countries is stipulated by the Regulation of the European Parliament and of the Council "On electronic identification and trust services for electronic transactions in the internal market and repealing Directive 1999/93 / EU" dated 23 July 2014 No. 910/2014, Decision of the European Parliament and Council "On the creation of a program of compatibility of solutions and a common framework for European public administrations, enterprises and citizens (ISA2 program) as a tool for modernizing the public sector" dated November 25, 2015 No. (EU) 2015/2240 and the like.

To meet the requirements of these regulations in the EU introduced the program "STORK", which "based on ID technology is available in each country and provides systematization of data on common technical and legal schemes, provides understanding between users with each other" [26, p.15].

In addition, in order to provide cross-border electronic services in the EU, there is an electronic service "SPOCS (Simple Procedures Online for Cross-Border Services)", which provides simple online procedures for cross-border services, electronic portal e-CODEX (e-Justice Communication via Online Data Exchange)", created to provide electronic interaction in the field of justice through online data exchange, as well as the program "epSOS", which provides access to cross-border health services, and the electronic service "PEPPOL (Pan-European Public Procurement) Online)", introduced for the purpose of online access to public procurement. The current state of cross-border service provision by administrative bodies in the EU is characterized by the integration of separate electronic services to provide these services into a single web portal, the functions of which will provide a wide range of public services in electronic form throughout the EU. In particular, the portal "e-SENS (Electronic Simple European Networked Services)" is currently being developed, combining the functions of the above-mentioned electronic services and formed to provide cross-border public services in electronic form using common and reusable technical components" [27, p.1].

3 Conclusion

After analyzing the above material, we can identify the following problems associated with the formation of electronic cases:

- the lack of systematic legislation on the procedure for providing administrative services, and therefore the order of formation of cases, including electronic;
- the question of determining the range of subjects of administrative services, the competence of these subjects and the availability of a range of powers, in turn, is reflected in the order of formation of cases;
- insufficient implementation of information systems in the implementation of electronic document management and their departmental branching, because for most subjects of power e-government is understood only as the informatization of certain documents.

Most of these problems can be solved by adopting the Administrative Procedure Code of Ukraine, which should include current scientific works and take into account aspects of modern knowledge about man, society and the state, the foundation and patterns of their development, as well as the interaction of the subject of administrative services with a person, because all the sphere of rendering of administrative services is directed on realization of the rights and freedoms of the last.

It would be expedient to provide in such an act the obligation to keep an electronic form of record keeping during the provision of administrative services and to introduce sufficient technical support for this process. In addition, the conceptual apparatus of the draft Law of Ukraine "On Administrative Procedure" dated May 14, 2020 under number 3475, namely, "administrative procedure", requires improvement, because in our opinion the understanding of the legislator of this category has not changed in all similar bills since the first bills. National scientists have developed a number of alternative interpretations of the concept of "administrative procedure", which are more in line with the modern paradigm.

At the same time, we should not forget about the "human" factor, to overcome which it is necessary to carry out proper selection of personnel and constantly improve their professional level.

References

1. E-government is the key to reforms in Ukraine. Government portal website. URL: https://www.kmu.gov.ua (access date 28.01.2021)
2. On approval of the Concept for the development of the electronic services system in Ukraine: Order of the Cabinet of Ministers of Ukraine dated November 16, 2016 No. 918-r. URL: https://zakon.rada.gov.ua (date of access 28.01.2021)
3. On electronic trust services: Law of Ukraine dated 05.10.2017 No. 2155-VIII. URL: https://zakon.rada.gov.ua (date of access 28.01.2021)
4. On approval of the Concept of Development of Electronic Governance in Ukraine: Order of the Cabinet of Ministers of 20.09.2017 No. 649-r. URL: https://zakon.rada.gov.ua (access date 28.01.2021)
5. On approval of the Regulation on the Integrated Electronic Identification System: Resolution of the Cabinet of Ministers dated June 19, 2019 No. 546. URL: https://zakon.rada.gov.ua (date of treatment 01.28.2021)
6. The issue of the Unified State Web Portal of Electronic Services of the Unified State Portal of Administrative Services: Resolution of the Cabinet of Ministers of Ukraine dated 04.12.2019 No. 1137. URL: https://zakon.rada.gov.ua (date of access 28.01.2021)
7. On amendments to some legislative acts of Ukraine regarding optimization of the network and functioning of centers for the provision of administrative services and improvement of access to administrative services that are provided in electronic form: Law of Ukraine dated 03.11.2020 No. 943-IX. URL: https://zakon.rada.gov.ua (date of access 28.01.2021)
8. On amendments to some resolutions of the Cabinet of Ministers of Ukraine: Resolution of the Cabinet of Ministers of 28.10.2020 No. 1035. URL: https://zakon.rada.gov.ua (access date 28.01.2021)
9. On the administrative procedure: the draft Law of Ukraine of 09.06.2015. Ips.ligazakon: website. URL: https://ips.ligazakon.net (access date 28.01.2021)
10. On the administrative procedure: the draft Law of Ukraine of 14.05.2020 No. 3475. VRU: website. URL: https://zakon.rada.gov.ua (access date 28.01.2021)
11. Draft of Administrative Procedure Code of Ukraine. VRU: website. URL: https://zakon.rada.gov.ua (access date 28.01.2021)
12. V. Timoshuk Administrative procedure and administrative services. Foreign experience and proposals for Ukraine, P. 24 (2003)
13. A. Lagoda Administrative procedure: theory and practice of application, P. 4 (2007)
14. I. Tishchenkova Forum of Law, Administrative procedures for the provision of electronic services by public administration in Ukraine, **No. 2**, pp. 124-129 (2017) URL: http://nbuv.gov.ua (accessed 28.01.2021)
15. On administrative services: Law of Ukraine of 06.09.2012 No. 5203-VI. URL: https://zakon.rada.gov.ua (access date 28.01.2021)
16. On approval of the requirements for the preparation of the technological card of public services: Resolution of the Cabinet of Ministers of Ukraine of 30.12.2013 No. 44. URL: https://zakon.rada.gov.ua (access date 28.01.2021)
17. A. Bukhanevich Law and Society, The concept and essence of administrative procedure, **No. 5**, pp. 126-131 (2015)
18. V. Timoshchuk Scientific and practical commentary to the Law of Ukraine "On Administrative Services", P. 9, 139 (2013)
19. On approval of the Model Regulations of the Center for Provision of Administrative Services: Resolution of the Cabinet of Ministers of Ukraine of 1.08.2013 No. 588. URL: https://zakon.rada.gov.ua (access date 28.01.2021)
20. On electronic documents and electronic document management: Law of Ukraine of 22.05.2003 No. 851-IV. URL: https://zakon.rada.gov.ua (access date 28.01.2021)
21. On approval of the Concept of development of the system of providing administrative services by executive authorities: Resolution of the Cabinet of Ministers of Ukraine of 15.02.2006 No. 90-p. URL: https://zakon.rada.gov.ua (access date 28.01.2021)
22. Yu. Zhuk Efficiency of public administration, Provision of administrative services to the population: foreign aspect, Vol. 1/2 (46/47), pp. 59-69 (2016)

23. V. Timoshchuk Administrative services; Swiss-Ukrainian project "Support to decentralization in Ukraine - DESPRO", P. 63, 80-81 (2012)
24. V. Timoshchuk Canadian experience in organizing the provision of administrative services. Local Democracy Foundation. Foundation of Local Democracy: website. URL: http://fmd.kh.ua (access date 28.01.2021)
25. Ya. Mikhailyuk, A. Pukhtetska Administrative services in the countries of the European Union and Ukraine: comparative legal aspect, P. 101 (2016)
26. STORK: making life easier, public services more secure // Research * Eu Focus Magazine, No. 12, P. 14-16 (2012)
27. E-SENS: easy digital public services within Europe. e-SENS: website. URL: https: // www.esens.eu (accessed 28.01.2021)

Digital Anti-Corruption Tools and Their Implementation in Various Legal Systems Around the World

Andrii Halai[1], *Viktoriia* Halai[2], *Roksolana* Hrechaniuk[3], and *Kateryna* Datsko [2*]

[1] Member of the Board of the Association of Legal Clinics of Ukraine, Ukraine
[2] State University of Economics and Technology, Ukraine
[3] West Ukrainian National University, Ukraine

Abstract. Just as there is no ideal society, there is no society without corruption. Fighting corruption is one way to achieve the UN-approved goals of sustainable development. The article summarizes the process of formation and implementation of international anti-corruption standards and their features in different national legal systems. It is determined that the adoption and recognition of standards does not in itself solve the problem, which is why the world is constantly looking for ways to combat corruption, testing new tools and mechanisms to influence corruption risks. This is confirmed by numerous projects funded by international donor organizations. Information technologies create endless opportunities not only for economics development (robotics, blockchain technology, neural networks, artificial intelligence, quantum virtual and augmented reality), but also for the fight against corruption: from e-government and courts, digital public services and tools that report about the facts of corruption, to the means of influencing society through social networks. The article analyzes scientific and professional research in the field of digitalization of anti-corruption activities and substantiates the relationship of these concepts from the standpoint of the impact of IT technologies on combating and preventing corruption. The article summarizes modern technological advances and their implementation in various national legal systems, defines the role of information technology in the fight against corruption and forms an understanding of information technology as a tool to control and combat corruption.

1 Introduction

Corruption is already recognized as one of the main problems of the world economy. According to the UN, the annual amount of bribes worldwide is $ 1.6 trillion. At the same time, the world economy is losing another $ 2.6 trillion due to corrupt activities, which account for 2.7 percent of world GDP [1].

EU Commissioner for Home Affairs Cecilia Malström said in a 2014 report that corruption cost the European economy around € 120 billion a year. These calculations are based on a study of data on corruption in 28 EU member states [2].

Corruption is a multifaceted phenomenon that should be considered from a legal, economic, political, sociological and even psychological point of view.

A number of studies conducted by the World Bank have examined the relationship between the effectiveness of combating corruption and the development of digital technologies. These studies point to the positive impact of increasing government transparency and the quality of public services and the negative impact of political processes in the country, for example, during the election period, when voters are influenced through social networks and the uncontrolled process of collecting personal data.

Modern scientific and specialized research on ways and methods of combating corruption is based on fundamental research on the concept of corruption and the factors that affect it.

The work of scientists who laid the foundation for the study of corruption at the professional level is devoted to the essence of corruption, the historical background of its emergence, its philosophy and impact on society. Among them we can specify the works of Klitgaard, R., Friederich, C., Huntington, S.

Back in 1988, the American economist Robert Klitgoard summarized the factors that stimulate corruption in the already known "formula of corruption":

$$\text{Corruption} = \text{Monopoly} + \text{Freedom of Action} - \text{Accountability} \quad (1)$$

In his study, he investigated and proved the destructive effects of corruption and demonstrated how bribery, extortion, fraud, kickbacks and conspiracies have led to backward economies, uncontrolled oligarchic influence and political instability [5].

The above formula of corruption illustrates and makes it possible to analyze the degree of influence of various factors on the process of preventing and combating corruption, in particular when looking for effective tools in anti-corruption work

* Corresponding author: datsko_kp@duet.edu.ua

https://doi.org/10.1051/shsconf/202110003005

2 Recognition of the destructive impact of corruption in the world's legal systems

In legal systems, corruption is a crime. The start of international recognition of combating corruption took place in 1975, when the world economy entered a new stage of development and rapid growth of transnational corporations. It was then that the Resolution against Corruption was adopted, which is practiced by transnational and other corporations. This document called on the governments of all countries to take the necessary measures at the national level to prevent and combat corruption.

Later, the world will determine the need to criminalize corruption and establish strict accountability for its manifestations.

A significant international achievement in the fight against corruption has been the adoption of the UN Convention against Corruption by the United Nations General Assembly. This document laid the foundation for most national laws around the world. As of 2019, the Convention has been ratified by 186 countries.

The document is not binding, i.e. a country will not be sanctioned for non-compliance with the Convention and to be obliged to comply with it. But the Convention sets common standards of conduct and work standards of government that will help combat corruption in the country.

The UN Convention does not define corruption, but defines standards of conduct, signs of modern governance and contrasts them to corruption, promotes common principles: honesty, transparency, impartiality of government.

Corruption is a worldwide problem, and the world did not immediately realize its devastating impact. For example, the corruption scandal with the private company Enron, which deceived key partners whose shares on the stock exchange were overvalued because they published distorted financial statements. For its operations, Enron created, according to various estimates, about 600-700 companies located in offshore areas located in the United States and on the islands. For many years, they had an auditor who turned out to be corrupt. This audit firm, which no longer exists after this major corruption scandal, was closed (USA, 2002) [3].

The scandal helped the United States completely overhaul anti-corruption legislation, and in 2002 a new law was passed, the Sarbanes-Oxley Act, which tightened liability for corruption and financial fraud.

2014 marked the beginning of the struggle for development, economic growth and overcoming corruption in Ukraine. National anti-corruption legislation was initiated in the wake of the resonance of the events of the Revolution of Dignity in 2014, which revealed numerous corruption risks in the country. The Law of Ukraine "On Prevention of Corruption" was adopted with the involvement of the public opinion, based on international standards and the peculiarities of the national legal system. Today, the country has a well-developed anti-corruption structure, and at the same time, numerous problems in its work due to lack of transparency of the government and evolutionary unwillingness to recognize some activities as corruption, uncontrolled funding of political parties and, consequently, their impact on anti-corruption bodies.

The world did not immediately understand the importance of fighting corruption, and so International Anti-Corruption Standards have come a long way, based on numerous corruption scandals and a gradual understanding of the need to criminalize corruption.

It is advisable to conduct a critical analysis of global corruption research conducted by relevant anti-corruption organizations and the world scientific community.

Profile NGOs investigate corruption at the level of global sociological and analytical research. Experts from the world's most authoritative anti-corruption organization, Transparency International, say: "To sweep away corruption, we must first understand it. That is why we investigate what exactly causes corruption and what acts against it" [4].

Transparency International assesses corruption by calculating the annual Corruption Perceptions Index (CPI) and classifies countries and territories by perceptions of the level of corruption in the public sector. CPI is the most well-known indicator of corruption worldwide. The index assesses corruption only in the public sector, identifying one of the main reasons for the corruption development - lack of control over the financing of political parties, reduced transparency of public administration and the process of providing public services. The calculation of the Index is based on 13 surveys of relevant organizations, including a survey conducted by the World Bank and the International Monetary Fund.

Such studies are a tool for both the government and the professional community, as their results provide an opportunity to assess the effectiveness of anti-corruption work and the tools involved at the national level.

3 The impact of digitalization on anti-corruption processes

Digitalization is the introduction of digital technologies in all spheres of public life: from the usual communication between people to the development of high-tech enterprises, it is the transition from the real to the virtual world, the so-called online world.

The influence of digital technologies is felt in all areas, as confirmed by modern researchers.

The world of digital technologies, which we enter, is not only a new logical stage in the development of the technological sphere of mankind, but also the entire existing legal and socio-political reality. There are no generally accepted and harmonized definitions and legal definitions yet, but digital technologies are already rapidly capturing bridgeheads for the attack. Digitalization is becoming the most important factor in the economic growth of any country's economy and in general is a modern trend [6].

The development of IT technologies often occurs during global crises. In particular, due to the global pandemic of COVID-19 a variety of electronic services and tools for communication, education and even the provision of medical services has rapidly developed. In response to the health emergency, governments have introduced new tools, such as COVID-19 special information portals, hackathons, electronic medical supplies, virtual medical consultations, self-diagnosis programs, and more.

Many countries have rapidly deployed distance learning programs as well as home schooling.

The rapid development of digital technologies is creating not only positive changes, but also increasing risks for the legal systems and security of states.

The World Bank has assessed the challenges governments are facing in fighting corruption. The report of this study emphasizes that as a result of the global COVID-19 pandemic, billions of dollars are borrowed and spent by countries, bypassing standard accountability procedures to manage the COVID-19 situation, leading to an increased risk of corruption. To promote greater accountability, governments need to articulate their actions, apply rules, address violations, and cope with the issues of transparency as soon as possible. The main tools of fiscal transparency, citizen involvement and social responsibility are becoming increasingly important [7].

Expert on social programs O. Pyshchulin from the Ukrainian Center for Economic and Political Studies named by O. Razumkov identifies the following threats to digitalization:

- the possibility of concentrating power in the market and strengthening monopolies;
- capture of new markets by transnational corporations;
- destabilization of the monetary system;
- increasing dependence on leading companies in the field of information and communication technologies [8].

These threats are factors that increase the corruption risks of the state, because the creation of monopolies increases market control by individual companies, reduces competition, promotes the development of the oligarchic class and, accordingly, influence on political parties and government – everything abovementioned is considered corruption risks. That is why these risks should be taken into account for developing or transition countries, despite the fact that digitalisation contributes to the development of the economy, and thus the living standards of citizens.

The impact of digitalization on the fight against corruption is unclear, the threats and factors that may contribute to the growth of corruption risks of the state require separate studies of the potential impact of digitalization on the perception of corruption and the effectiveness of IT technologies in combating it.

In 2018, the Latin American Conference discussed mechanisms to reduce corruption and restore public confidence in public institutions, including the automation of service delivery processes, governance and accountability.

"The digital revolution is not only a revolution of the economy, but also of the state. Latin American society is becoming younger, more urbanized and more united, and this transition offers countless opportunities to use the digital world and adapt institutions to the demands of civil society," said Carlos Santiso, Director of Digital Innovation at CAF (Latin American National Development Bank).

The African authorities already recognize that corruption is a constraint on the achievement of the Sustainable Development Goals, and at the same time makes it impossible to overcome poverty, high child mortality and hunger. In 2020, the IMF's African Department conducted a global empirical study in 26 African countries, surveying 23,000 people and examining the impact of digitalization on perceptions of corruption in the country and trust in Africa's tax authorities [9].

Today, the African government has high hopes for IT technologies that reduce the human factor in the delivery of public services, as it promotes transparency of government and the civil service as a whole.

The results of the study indicate that the use of digital technologies helps to create direct and more transparent channels of communication between a civil servant and a citizen of the country. The results of the study also touched upon the impact of the Internet on the perception of corruption. This part of the study indicates that the population trusts the government more when it has free access to the network (during 2020, African countries systematically blocked the Internet for the population).

The use of information technology helps to increase perceptions of corruption, trust in government and can suggest new tools to increase government transparency and prevent corruption.

4 World experience in the use of IT technologies to combat and prevent corruption

The main goal is to automate standard, paper, outdated systems and operations used to interact with citizens. This should make public services open, accessible and convenient for citizens.

International experience already has a number of digital solutions, the use of which helps to ensure transparency, openness of government and equal access to public services for the population

4.1 E-government and convenient government services

The level of development of e-government in the world is assessed by the United Nations Department of Economic and Social Affairs (UNDESA) since 2003 according to the EGDI (EGovernment Development Index), which is conducted every two years. The EGDI index is a comprehensive indicator that combines three sub-indices - online services, telecommunications infrastructure, human capital and is used to measure the readiness of governments to use information and communication technologies to provide quality information and public services to the people, business and their use in the work of the authorities themselves [10].

In addition to the e-government development index, the EGDI uses the EPI (EParticipation Index), which is based on three components: e-information, e-consultation and e-decision-making, to assess the provision of interactive information services to citizens. On the top of the ranking of 193 UN member states on the level of digital government by 2020, taking into account the volume and quality of online services, the state of telecommunications infrastructure and existing human potential, are Denmark, the Republic of Korea and Estonia, followed by Finland, Australia, Sweden, UK, New Zealand, the United States, the Netherlands, Singapore, Iceland, Norway and Japan.

Among the least developed countries, Bhutan, Bangladesh and Cambodia have become leaders in the development of digital government, moving from the middle to the highest group of the e-government development index (EGDI) in 2020. Mauritius, Seychelles and South Africa are on the top of the government ranking in Africa. Overall, 65 percent of Member States are at a high or very high level of EGDI [11].

Latin American countries are characterized by an overly bureaucratic process of public service delivery and a high level of corruption. According to Transparency International, in 2017, 30% of Latinos paid a bribe for access to public services, which is almost 30 million people in the region. Such indicators have helped to find new solutions that already have a positive impact.

For example, Mexico, Peru, Brazil and Argentina are already united in the process of improving the system of public access to quality public services. These countries have introduced integrated portals to a number of public services. That is, the population already has the opportunity to register a number of services online. However, online service delivery rates are low. This activity was investigated by the Wait no more project [12]. The 2018 draft report indicates that, despite significant efforts, the process is being hampered by the fact that not all civil servants and government officials are ready for such changes. Some of them are afraid of losing their jobs, others – of losing systemic income from illegal benefits in the process of providing public services.

On January 01, 2019, the Order of the President of Argentina came into force, which introduces the so-called e-government or "paperless government". The order came into force gradually, and government agencies were given a year to train and translate a number of procedures online. To implement it, the government has trained 227,900 users of the new system.

Portugal, which has introduced e-government through the Simplex program, has a successful experience [13]. The country uses this tool to modernize the process of providing public services, increase openness and facilitate the interaction of legal entities and individuals in obtaining public services. Experts acknowledge that the success of this program is based on the fact that it is constantly being improved through discussions with the public and relevant organizations. The program's website has a window for public appeals with suggestions, recommendations and comments, which gives the opportunity for everyone to be heard. This process puts the citizen at the center of public services, forms in them an understanding that their voice is heard and gradually changes the psychology of civil servants.

A similar approach to de-bureaucratization of public services has been introduced in Brazil since 2017, where the program "sSimplifique!" is functioning [14].

Countries that are on the path of civil service and judicial reforms are also hoping to digitize processes and eliminate the influence of the human factor in the quality, speed and accessibility of public services. Ukraine has been actively starting the reform process since 2014. Such work has been introduced at the system level since 2016, when Ukraine approved the Digital Agenda of Ukraine until 2020. In 2019, the Ministry of Digital Transformation was launched in the country. The strategic goals until 2024 are to ensure the availability of 100% of public services for citizens and businesses online.

Today, the country's public services portal Diya is already operating in the country. The portal makes it possible to register a number of services online for citizens (benefits, certificates, registration of a child in kindergarten, etc.) and for business (business registration, business closure, reporting, ordering certificates). At the same time, the problem remains that not everyone is ready to use such services as citizens and civil servants. Therefore, the queues in the departments of social protection, unfortunately, are still maintained.

Successful practices for such improvements in public service delivery already exist in the following countries: the United Kingdom, France, Canada and Israel. The results of the work of governments in this direction were presented at the summit of state technologies [15].

We believe that the spread of such practices can contribute to the development of interesting startups and create competition for the state in providing relevant services. And competition is a factor in economic development, and hence the path to quality public services for citizens.

The list of successful cases of countries that have recently begun their path of automation of public services can be continued. All decisions differ in terms of implementation, number and characteristics of the country's population, features of the political and economic situation in the countries and the level of technology development. However, the implementation of such decisions is a step that changes the consciousness of citizens and civil servants themselves.

4.2 Use of open data

Open data can increase the transparency of government by disclosing data on budget procurement, land ownership, education and medicine, crime statistics, financial reports of government agencies and the state, government contracts, international procurement, environmental statistics and more.

At the international level, the process of introducing open data by governments is regulated by the International Open Data Charter. The Charter is an international initiative that promotes cooperation, the adoption and implementation of common principles, standards and best practices for the disclosure and use of open data around the world.

The public is also working to develop and promote open government data. The World Wide Web has created the Open Data Barometer, which summarizes the Global Indicator of How Governments Publish and Use Open Data for Accountability, Innovation, and Social Impact.

Among the tools to combat corruption of the anti-corruption NGO Transparency International is the maintenance of a register of unscrupulous participants in public procurement, February 2013 [17].

The greatest positive effect from the introduction of information technology was observed in the UK, Canada, France and the USA. Sweden, Denmark, Korea, and Japan were in the top ten. These are countries that also have a high level of technology development and the Corruption Perceptions Index, according to Transparency International research.

At the same time, countries with significant problems in terms of integrity and achieving the Sustainable Development Goals have a positive impact on the introduction of IT in the public sector.

For example, Mexico has expanded the breadth and depth of open budget data through its state-of-the-art portal on financial transparency, which includes public procurement and infrastructure investment. Mexico City became the first city to present its purchases in an open format. In turn, relevant NGOs use this information to study government and raise red flags to uncover once-hidden practices and patterns.

According to O.V. Altsyvanovych and Ya.Yu. Tsymbalenko, deserves attention in the US and EU e-procurement and electronic reporting on public procurement, where the e-procurement system is considered a tool that increases accountability and transparency in public administration [18].

The use of open data has made it possible to establish a transparent public procurement process in a number of countries. A successful case is currently working in Ukraine. Starting the Prozzoro system. The efficiency of the software product in 2020 indicates that the country has saved UAH 43 billion in Prozorro tenders and earned UAH 11 billion in Prozorro auctions [19].

Similar software tools have been introduced in Romania, Spain, and Brazil during public procurement of medicines.

Open data ensures transparency of government and, accordingly, helps reduce the level of corruption risk of the state and civil servants; expanding opportunities for business development and participation in government procurement, and consequently increasing competition in the market of goods and services

4.3 Use of blockchain technologies in the fight against corruption

Blockchain technologies alone are difficult to imagine as a tool in the fight against corruption, but their current use raises hopes that they can provide an asset tracking process and control government contracts. However, there are legal problems with the use of blockchain technology and its impact on public relations.

Research on the impact of blockchain on social change is also the subject of attention of relevant organizations. As a tool for resolving information asymmetries, Blockchain may be able to empower people who do not have sufficient services and help to solve societal problems. This study was conducted as part of the BLOCKCHANGE project [20]. The project explored the impact of blockchain technologies on the social sphere in three areas: human resources, logistics, contracting and public services.

In 2018, the National Research Council of Canada (NRC) began piloting a new blockchain based on Ethereum for more open public procurement administration. For example, Canadians can obtain government funding data through this smart contract.

With blockchain technology, India has improved the process of tracking land ownership by combining technology with land registries. Switzerland is improving access to public services. And in the United States, electronic voting in elections is already underway.

All of these processes are closely linked to the collection and transfer of personal data, so in addition to the benefits, there can be challenges for the entire system that involve identity theft, cybercrime and fraud.

4.4 Use of IT technologies in the work of anti-corruption bodies

Anti-corruption infrastructure exists in every country in the world. However, in different countries its structure, influence of power, methods of work and responsibility for corruption offenses are different. It depends on the level and perception of corruption in the country, the will of the top leadership of the state, the degree of public involvement in the work of anti-corruption bodies and, in particular, the level of digitalization of a number of processes in the state.

For example, in Denmark, which has a relatively low level of corruption and a high Corruption Perceptions Index, no anti-corruption body has been created, but criminal liability has been established for corruption offenses. A similar example is Finland, which also holds some of the highest positions in anti-corruption rankings. This country also does not have a developed anti-corruption infrastructure. Germany, France, Italy have more developed anti-corruption structure, these countries due to their historical features (development of mafia groups, business that conducts extensive international activities) at due time, were forced to take drastic measures and increase the level of anti-corruption bodies. Today, the countries are an example of a fairly successful experience in the fight against corruption.

The basic principles on which governments base their fight against corruption are:

- transparency of the public sphere;
- control over the financing of political activity;
- personal civic activity.

All these principles will be able to work effectively only with the digitalization of individual processes. Of course, countries that are leaders in the development of IT technologies have introduced electronic registers, e-government, e-court long ago. Their effectiveness has already been proven.

Let us consider the successful practices of involving IT technologies in the work of anti-corruption bodies, in the countries that are at the stage of formation of the fight against corruption.

Ukraine's experience shows that the adoption of anti-corruption legislation and the launch of anti-corruption infrastructure is not enough for its effective operation. Together with the factors of influence of power and level of competence of employees of anti-corruption structures, such structures work more effectively when they involve IT technologies that process large amounts of data, provide services for public information on successful cases of the body, allow to receive reports of corruption offenses, etc. The country has already launched the following technological solutions to improve anti-corruption work: public Unified State Register of Declarations of Persons Authorized to Perform State or Local Self-Government Functions (for submission, verification and analysis of information on persons whose activities are subject to declaration); the electronic register of financial statements of political parties is already working in test mode; the Unified State Register of Persons Who Have Committed Corrupt or Corruption-Related Offenses has been launched. Strategic plans also include the launch of an anti-corruption portal and the further development of open data on anti-corruption structures.

In Slovenia, the electronic register of declarations was launched in 2011, and from this time declarations can be made online. For example, in Macedonia there is also an electronic register, but declarations are still submitted on paper. Through the relevant websites of anti-corruption structures in the country, online training of target groups of the public and civil servants is conducted.

Through official websites and social networks, anti-corruption bodies around the world carry out educational work with the population and organize online training for civil servants.

Information technologies have a significant impact on ensuring effective work with whistle-blowers, because in this part it is necessary to ensure reliable protection of communication channels, anonymity of the user and at the same time effective communication to verify the received information.

5 Conclusion

Technology is becoming a key ally in the fight against corruption. New technologies and open data now allow anti-corruption people to detect, prevent and even predict corrupt practices that in the past could have been hidden behind a curtain of paper document opacity.

Technology is becoming a reliable ally of governments seeking to combat corruption, because combined with their political will, technology can change the policy-making process of public service delivery. Digitalization allows states to establish effective anti-corruption bodies and adapt to the requirements of the digital society and achieve the goals of sustainable development.

The introduction of anti-corruption tools in the digital world is impossible without the openness of governments, free access to networks and, consequently, information from state institutions, which, in our opinion, will build public confidence in government and improve the quality of public services.

At the same time, their implementation should take into account a number of features of each country and factors that directly or indirectly affect the effectiveness of the use of IT technologies. The study of anti-corruption experience of countries in different legal systems of the world makes it possible to form a list of factors that should be taken into account when implementing the relevant systems. After all, sometimes such tools can become a tool in the hands of corrupt government officials or government officials. That is why it is important to assess national anti-corruption strategies, features and degree of digitalization of other related processes, the level of skills of performers and the population in the use of information technology, public support.

The study also pointed to the need for constant dialogue with relevant NGOs to improve such systems and find new technological solutions to prevent, combat corruption and develop quality public services.

Thus, among the digital means of combating corruption in the studied legal systems of the world the following ones can be distinguished:

1) creation of an e-government system;

2) provision of digital public services with the use of information and communication technologies;

3) creation of tools to detect manifestations of corruption, including: e-declaration, electronic bidding for public procurement, electronic reporting;

4) use of social networks to detect cases of corruption in the public sector;

5) use of the Internet for free access to public information and the opportunity to obtain the necessary information on the activities of public bodies, through open access, e-mail, etc.

References

1. UN: Hundreds of Billions of Dollars Lost to Governments Due to Financial Crime that could be spent on social development (2020) URL: https://news.un.org/ru/story/2020/09/1386622
2. Štefan Šumah Corruption, Causes and Consequences (2018) URL: https://www.intechopen.com/books/trade-and-global-market/corruption-causes-and-consequences
3. Demonstrative collapse: how the history of the American Enron reshaped the corporate legal field of the world (2008) URL: https://mind.ua/publications/20183712-pokazovij-krah-yak-istoriya-amerikanskoyi-enron-perekroyila-korporativne-pravove-pole-svitu
4. Research. To end corruption we must first understand it. That's why we look at what causes corruption and what works against it. URL: https://www.transparency.org/en/research
5. Klitgaard, R. Controlling Corruption. University of California Press (1988)
6. Sosnin O. Digitalization as a new reality of Ukraine. Legal Bulletin of Ukraine (2020) URL: https://lexinform.com.ua/dumka-eksperta/tsyfrovizatsiya-yak-nova-realnist-ukrayiny/
7. Sasana Kijang, Jalan Dato Onn, Kuala Lumpur Enhancing Government Effectiveness and Transparency The Fight Against Corruption, Malaysia (2020)
8. O. Pushchulina Two sides of digital technologies: "digital dictatorship" or maintaining stability. (2019)
9. Rasmane Ouedraogo and Amadou N.R. Sy IMF Working Paper African Department Can Digitalization Help Deter Corruption in Africa? (2020) URL: https://www.imf.org/en/Publications/WP/Issues/2020/05/29/Can-Digitalization-Help-Deter-Corruption-in-Africa-49385
10. E-Governance for Accountability and Participation (EGAP) (EGAP). URL: https://egap.in.ua/en/
11. United Nations E-Government Survey (2020) URL: https://publicadministration.un.org/egovkb
12. Benjamin Roseth, Angela Reyes, Carlos Santiso Cataloging-in-Publication data provided by the Inter-American Development Bank Felipe Herrera Library, USA (2018)
13. Simplex. Official site URL: https://www.simplex.gov.pt/livro
14. Simplifique. Official site URL: http://www.simplifique.gov.br/
15. E-govtech summit 2020. Official site URL: https://www.govtechsummit.eu/agenda
16. The Open Data Barometer Official site URL: https://opendatabarometer.org/?_year=2017&indicator=ODB
17. Anastasia Kalinina Corruption and public procurement: international legal instruments. Adviser in the field of public procurement (2015) URL: https://radnuk.com.ua/zhurnaly/koruptsiia-ta-derzhavni-zakupivli-mizhnarodni-pravovi-instrumenty-2/
18. Altsivanovich, Ya.Yu. Tsymbalenko Ya. Yu. International and European experience of avoiding corruption risks during public procurement. Aspects of public governance. № 10. V. 6. P.62-73. (2018)
19. Ukraine saved UAH 43 billion in Prozorro tenders and earned UAH 11 billion in Prozorro auctions. Sales in 2020. (2020) URL: https://prozorro.gov.ua/news/-ukrayina-zaoshchadyla-43-mlrd-grn-na-tenderah-u-prozorro-i-zarobyla-11-mlrd-grn-na-aukcionah-prozorroprodazhi-u-2020-roci
20. Blockchain for social change. Official site of the project URL: https://blockchan.ge/

Legal Aspects of Remote Trial

Vladyslav Yamkovyi[1*], *Oksana* Telychko[1], *Iryna* Mima[1], and *Nina* Ivaniuk[1]

[1]State University of Economics and Technology, 50005, Kryvyi Rih, Ukraine

Abstract. The introduction of new technologies, as well as digital is a new logical stage in the development of society, affecting all areas of humanity, including legal. The digitalization of society is becoming the most important factor in economic growth and is a modern trend in the development of any country in the world. Ukraine is no exception to this list, as it has chosen to bring its national legal system closer to the European Union's legal system and international legal standards, which is a complex and lengthy process. Unfortunately, so far there are no generally accepted and harmonized definitions and legal definitions for the legal regulation of introducing digital technologies in the Ukrainian legal system. This necessitates a more thorough study of implementing digital technologies in the legal field. This article is devoted to the problems concerning legal regulation of remote court proceedings by videoconference, which is designed to simplify access to justice and lead to savings in material and procedural means in court proceedings. The purpose of this article is to identify problematic issues regarding legal regulation of distance litigation in Ukraine and to propose mechanisms for its improvement as well as to harmonize it with international legal standards.

1 Introduction

Electronic technologies have become an integral part of the entire society life, the basis for building information flows at the local, national and international levels around the world, including work of the judiciary in Ukraine.

The use of electronic technologies in the administration of justice in Ukraine was enshrined in the Law of Ukraine "On the Judiciary and the Status of Judicial Bodies" № 2453-VI of 07.07.2010 [1], so amendments were made to all procedural codes for the introduction of automated document management in courts. Informative and impartial distribution of cases between courts, providing individuals and legal entities with information on the status of cases in which they participate.

Subsequently, in 2011 the Law of Ukraine "On Amendments to Certain Legislative Acts of Ukraine in Connection with the Ratification of the Second Additional Protocol to the European Convention on Mutual Assistance in Criminal Matters" №3529-VI of 16.06.2011 [2] in force at that time the procedural code of Ukraine [3] was supplemented by Article 85-3 "Application of telephone conference and videoconference during the investigative action". This provision provided that in case it is impossible for participants of investigative or other procedural action to appear at the body of inquiry, investigator, prosecutor or in the court at the place of proceedings as well as in case of the need to ensure the safety of people involved in criminal proceedings or other reasonable grounds, investigative or procedural action with their participation may be carried out by telephone or video conference. A telephone or video conference was to be conducted on behalf of an inquiry body, investigator, prosecutor or court and could be used during the interrogation of a witness, expert, suspect, accused and defendant, face-to-face, identification, reproduction of the situation and circumstances of the event.

With the repeal of the old CPC of Ukraine and on 13.04.2012 a new version adoption of the Criminal Procedure Code of Ukraine [4] was introduced by Art. 336 entitled "Conducting procedural actions by videoconference during court proceedings", which stipulated that court proceedings may be carried out by videoconference during broadcasts from another room including those located outside the court premises (remote court proceedings). The decision to conduct remote court proceedings is made by the court on its own initiative or at the request of a party or other participants in criminal proceedings. If the party to the criminal proceedings or the victim objects to implementing remote court proceedings, the court may decide to carry it out only by a reasoned decision justifying the decision taken in it. The court does not have the right to make a decision on remote court proceedings in which the accused is outside the court premises, if he or she objects. The technical means and technologies used in remote court proceedings must ensure the proper quality of image and sound, compliance with the principle of publicity and openness of court proceedings as well as information security. The course and results of procedural actions carried out in the mode of videoconference are recorded by means of video recording technical means. A protected person may be questioned by videoconference with changes in appearance and voice that are unrecognizable.

*Corresponding author: advokatvladyslav@gmail.com

Subsequently, the Law of Ukraine "On Amendments to Certain Legislative Acts of Ukraine Concerning Participation in a Court Conference by Videoconference" № 5041-VI of 04.07.2012 [5] relevant amendments to the possibility of holding a court hearing by videoconference were introduced into the Commercial Procedure Code of Ukraine (hereinafter - CPC), the Civil Procedure Code of Ukraine (hereinafter - CPC) and the Administrative Judicial Procedure Code of Ukraine (hereinafter - AJPC). These codes were supplemented by articles entitled "Participation in a court hearing by videoconference" of similar content, which provided that the court on its own initiative or at the request of a party or other party to the proceedings may decide to participate in a court hearing by videoconference. However, such participation in a court hearing by videoconference of a party or other party to the proceedings could necessarily take place only in the premises of another court, where it was necessary to ensure a videoconference.

In 2013, the project "Electronic Court" was launched, which continued to spread the use of new technologies and informatization of the Ukrainian judiciary.

According to the Strategy for the Reform of the Court Organization , Judiciary and Related Legal Institutions for 2015-2020, adopted by the Presidential Decree of Ukraine dated 20.05.2015 № 276/2015, one of the main judicial reform areas is to ensure the widespread use of information systems (IS)for providing greater number of e-justice services; to create electronic management in information systems courts, in particular to introduce full-fledged electronic systems, including electronic document management systems, and case tracking (to higher instances), electronic notifications, electronic calls, electronic cases (in some cases), electronic payments, audio and video recording of meetings [6].

In addition, on December 15, 2017, the new Commercial Procedure Code of Ukraine, the Civil Procedure Code of Ukraine and the Administrative Judicial Procedure Code of Ukraine came into force, which provides for the introduction of the Unified Judicial Information and Telecommunication System (hereinafter - UJITS), the operation of which should begin 90 days after publication by the State Judicial Administration of Ukraine in the newspaper " The Voice Of Ukraine "and on the web portal of the judiciary the announcement on the establishment and maintenance of the UJITS, which has been postponed for more than three years. Unfortunately, until now, participation in the court hearing by videoconference within the court premises is carried out only with the help of technical means operating in the courts.

This shows that at the state level, considerable attention is paid to ensuring the conduct of court proceedings remotely in order to facilitate access to justice, but in the absence of the functioning of the UJITS, these opportunities are used to a limited extent.

2 Analysis of foreign experience and international practice in the field of remote litigation

The use of videoconferencing during court hearings has become a common practice throughout the world. Therefore, the study of foreign experience in this regard is an integral part of the study on the improvement of national legislation. The analysis of foreign and international legal norms on this issue becomes especially relevant.

The Convention for the Protection of Human Rights and Fundamental Freedoms of 1950, better known as the European Convention on Human Rights, stipulates in Article 6, paragraph 1, that everyone in determining his or her civil rights and obligations or in bringing any criminal charges against him or her, has the right to a fair and open hearing within a reasonable time by an independent and impartial tribunal established by law [7].

The general right to judicial protection is also enshrined in the International Convention on the Elimination of All Forms of Racial Discrimination 1965 (Articles 5, 6) [8]; Convention on the Rights of the Child 1989 (Articles 12, 37, 40) [9]; UN General Assembly resolution No.40/33 [10].

A number of provisions on ensuring simplified access of citizens to effective justice and speeding up the consideration of cases are enshrined in the regulations of the European Council , including:

- Recommendation No.R (81) 7. Of The Ministers Committee To Member States. On Measures Facilitating Access To Justice (Adopted by the Committee of Ministers on 14 May 1981 at its 68th Session), which sets out principles and provisions for simplifying and expediting judicial proceedings [11];
- Recommendation No. R (84) 5. Of The Committee Of Ministers To Member States On The Principles Of Civil Procedure Designed To Improve The Functioning Of Justicel (Adopted by the Committee of Ministers on 28 February 1984 at the 367th meeting of the Ministers' Deputies), according to which judicial authorities must have the most up-to-date technical means to enable them to administer justice in the most efficient way, in particular by facilitating access to various sources of law and by speeding up the administration of justice [12];
- Recommendation No. R (95) 13. Of The Committee Of Ministers To Member States Concerning Problems Of Criminal Procedural Law Connected With Information Technology (Adopted by the Committee of Ministers on 11 September 1995 at the 543rd meeting of the Ministers' Deputies), which established that information on case law in all areas of law and in all regions should be disseminated through one or more automated systems, which, in turn, would facilitate the provision of online court hearings [13].

These recommendations are general in justifying the need to introduce provisions for conducting court hearings online, the feasibility of their holding as well.

According to Opinion № (2011) 14 of the Consultative Council of European Judges "Justice and information technologies (IT)", IT should be a tool or means to improve the administration of justice, facilitate user access to courts and strengthen the guarantees established by Art. 6 of the European Convention on Human Rights, access to justice, impartiality, independence of the judge, fairness and reasonable time for consideration of the case. Video conferencing can facilitate security hearings and the taking of testimony from witnesses and experts. Particular attention should be paid to ensuring that conference calls and the taking of evidence in this way do not infringe on the guaranteed right of defense, encourage the use of IT to strengthen the role of the judiciary in upholding the rule of law in democracies; IT should not interfere with the authority of a judge and undermine the guidelines set out in Art.6 of European Convention on Human Rights [14].

According to Zub O. the introduction of video conferencing in the American judicial system, called "telejustice", has allowed not only to reduce the cost of maintaining the judiciary, but also to speed up the trial as a whole. According to the Pew Charitable Foundation, which conducted a study on the effectiveness of introducing telejustice in the United States, consideration of cases, which were usually heard for about 120 days, now took no more than 10 days, which was caused by the lack of need for personal (in-person) appearance of the participants in the courtroom. Money savings from the use of teleconferencing in US criminal justice, such as the Philadelphia Pennsylvania Court Court, amounted to $ 550,000 per month in 2011 because prisoners were not taken from custody or imprisonment to court, but used videoconferencing connection. At the same time, providing courtrooms with equipment needed for video conferencing in Pennsylvania cost $ 4.2 million [15].

Videoconferencing has also become an effective tool that has facilitated and accelerated cross-border litigation in European Union countries. In the context of European e-Justice, video conferencing has been introduced in a new concept - interstate video conferencing.

In Finland, for example, there are several types of video conferencing equipment: (a) for court hearings, a complete set, with a high resolution that provides high-quality recording and screen images; (b) for preliminary hearings, a separate set for conference rooms; (c) for hearing witnesses, a basic set of terminal, camera and microphone.

Under the rules of civil procedure in England, the court may order hearings of any dispute or part thereof to be heard by videoconference. If one of the parties objects to the application of the remote trial or the right to be heard or the right to be notified of the time and place of the trial is violated as a result of the videoconference, the videoconference shall not apply [16].

The German Code of Civil Procedure provides for open-air hearings using video conferencing systems. There are two types of such listeners: remote participation of the parties and remote reception of the shown witnesses. The hearing in the case of remote participation of a party is possible only with the consent of both parties by videoconference. This hearing should be accompanied by the simultaneous mutual transmission of image and sound from one place to another, and vice versa. The second case provides that in agreement with both parties the court may find a court decision on how to allow a witness, an expert to testify from other places. But this practice of using new technologies is a quicker exception to the rules of the court hearing, and therefore preference is given to the traditional hearing for the participation of all people in the courtroom [17, 22].

A slightly different approach to the use of video conferencing exists in Japan, where the installation of video conferencing systems is witness-oriented. According to Art.204 of the Civil Procedure Code of Japan, the court may remotely interrogate a witness using a method that provides mutual exchange of video and audio information, due to the age or psychological state of the witness, or if there are reasons to believe that direct participation in the hearing will cause psychological suffering to the interrogated or put pressure on the witness [17, p. 22-23]

This suggests that despite the differences between different legal systems, video conferencing and remote justice have played an important role in ensuring simplified access to justice for individuals.

3 Analysis of the state of scientific research

The issue of using videoconferencing in the judiciary in the Ukrainian scientific community has been considered only partially, in the context of research on the use of information technology in the judiciary. The vast majority of authors in their works pay considerable attention to the definition of "videoconferencing" and the legal regulation of its conduct. The vast majority of works concerns criminal proceedings.

Sizonenko A. defines videoconferencing as a two-way video, audio communication between two points, the distance between which does not matter, in real time (excluding time zones), in one of which there are persons conducting and participating in the investigation, and in another - those with whom the procedural action is carried out [18, p. 96]. However, such a definition artificially narrowed the scope of videoconferencing only to cases of investigative actions.

On the other hand, Smirnov M. defined videoconferencing as a special procedure during which communication in the form of audio and video information exchange between remote participants of the investigative action takes place at a distance (remotely), but in real time [19, p. 97].

An analysis of international legal acts and foreign scientific publications shows that in the law enforcement practice of most UN member states the terms "videolink" (or "video link"), "videoconference" (or "video conference") are used to refer to remote litigation; "Videoconferencing" (or "video-conference link") [22].

The Ukrainian legislation actually borrows this terminology, but currently there is no single approach to its definition and disclosure. Thus, in criminal proceedings, the term "remote court proceedings" is used, which is absent in other court proceedings. At the legislative level, the concept of "videoconferencing" is not disclosed, there are no terms "videoconferencing system", "technical means of video recording", "videophonogram", neither procedure for making and storing copies of videophonograms, nor the procedure for making archival copies to create working copies in case of their insufficiency, damage or destruction is regulated. This greatly complicates the use of video conferencing.

However, according to Bogunov V. the introduction of the protection possibility and their rights exercise in court by videoconference is justified by such factors as saving time and money of participants in the process, which are spent on travel (or delivery) to court [23].

Garievskaya M. considers that the advantages of holding court hearings by videoconference also include the ability to speed up the trial and save money for both the judiciary (for example, to send summonses to the hearing) and for participants in the process (including business trips) [23, p. 60]

Among the so-called "disadvantages" or "difficulties" of conducting court hearings by video conference Ozernyuk G. highlights the following: the need for technical equipment of courts with special facilities for organizing a court hearing by videoconference, the availability of high-speed Internet; the need for awareness of the participants of the process with the latest technologies and the ability to use them; legislative unresolved issues on participation in the trial by videoconference, in particular, outside the courtroom; loss of the verbal side of communication with the participants of the process with the participation of an interpreter, etc. [25, p.27].

Among the so-called "disadvantages" or "difficulties" of conducting court hearings by video conference Ozernyuk G. highlights the following: the need for technical equipment of courts with special facilities for organizing a court hearing by videoconference, the availability of high-speed Internet; the need for awareness of the participants in the process with the latest technologies and the ability to use them; legislative unresolved issues on participation in the trial by videoconference, in particular, outside the courtroom; loss of the verbal side of communication with the participants of the process with the participation of an interpreter, etc. [25, p.27].

According to Kravtsova T., Kornienko M. the advantages of conducting videoconferences outside the court include ensuring social isolation, security of participants, judges and court staff, and among the disadvantages - new opportunities for abuse of procedural rights by unscrupulous participants [26].

After analyzing these views, it should be noted that the videoconferencing system is undoubtedly an important innovative breakthrough in the domestic justice system, but has not found proper application in practice due to a number of difficulties and legislative gaps, one of which is delays in the start of UJITS. However, a more complete legislative settlement can remedy this situation.

4 The current state of regulation and ways to improve it

Normative legal acts, which at the national level enshrine the possibility of participation in court proceedings by videoconference and the exercise of the right to justice, are primary regarding the rights of participants in court proceedings to participate in remote proceedings, so their improvement is an important aspect.

Today, all procedural codes, without exception, stipulate that, in accordance with the law, the UJITS ensures the recording of the trial and the participation of the participants in the trial in a court session by videoconference. Persons who have registered official e-mail addresses in the UJITS may perform procedural actions in electronic form only through the Unified Judicial Information and Telecommunication System, using their own electronic digital signature, equated to a handwritten signature in accordance with the Law of Ukraine "On Electronic Digital Signature" (hereinafter - EDS Law) [27], unless otherwise provided by the relevant procedural code. Peculiarities of the use of electronic digital signature in UJITS are determined by the Regulations on UJITS. However, the relevant regulations have not been adopted for more than three years.

And therefore it is necessary to be guided by provisions of codes. Thus, the possibility of participating in a court hearing by videoconference is provided by Art.197 of the Civil Procedure Code [28], Art. 212 of the Commercial Procedure Code of Ukraine [29], Art.195 of the Administrative Judicial Procedure Code of Ukraine [30], which stipulates that the parties to the case participate in the hearing by videoconference outside the courtroom using their own technical means and electronic digital signature in accordance with the requirements of the Regulations on UJITS.

In addition, in connection with the adoption of Regulation (Eu) No 910/2014 of the European Parliament and of the Council of 23 July 2014 on electronic identification and trust services for electronic transactions in the internal market and repealing Directive 1999/93 / EC [31] The EDS Law has expired and has been replaced by the Law of Ukraine "On Electronic Trust Services" [32], and therefore the provisions of the procedural codes regarding the identification of persons in the UJITS should be brought in line with current legislation on electronic trust services.

Therefore, prior to the introduction of the UJITS and legislative changes, court hearings may be held by videoconference only if the party to the proceedings is in the courtroom closest to his or her place of residence and equipped with special technical means.

The situation has changed in the context of a global pandemic, as a result of which the Code of Civil Procedure, the Commercial Procedure Code and the Administrative Judicial Procedure Code of Ukraine were amended, according to which during the quarantine established by the Ministers' Cabinet of Ukraine to prevent the spread of coronavirus (COVID-19), the parties to the case may participate in the court session by videoconference outside the court premises using their own technical means. The confirmation of the person involved in the case is carried out using an electronic signature, and if the person does not have such a signature, in the manner prescribed by the Law of Ukraine "On the Unified State Demographic Register and documents confirming citizenship of Ukraine, identity or special status" [33] or State Judicial Administration of Ukraine.

Based on this, the State Judicial Administration of Ukraine by order of 23.04.2020 № 196 approved a new version of the Procedure for working with technical means of videoconferencing during the hearing in administrative, civil and commercial proceedings with the participation of the parties outside the court [34]. This has greatly simplified the participation of individuals in remote litigation. This Order stipulates that in order to participate in a court hearing by videoconference, the party to the case must:

- pre-register using your own electronic signature in the Video Conferencing System (hereinafter - the System) at the link on the official web portal of the judiciary of Ukraine vkz.court.gov.ua or with the help of other available to the court and litigants means to ensure court hearings by videoconference and meet the requirements of the law;

- check his / her own technical means for compliance with the technical requirements defined for work with the System selected for videoconferencing;

- if the participant does not have an electronic signature (or the System does not allow registration using an electronic signature), he/she must pre-register in the System using a login and password or using other means of registration provided by the selected System. In this case, the decision on the possibility to confirm the identity of the party to the case, which does not have an electronic signature as well as the decision on choosing the System to be used for court hearings by videoconference,

- submit an application for participation in the court hearing by videoconference outside the court premises not later than five days before the court hearing, obligatorily stating: the name of the court; court number; the date and time of the court hearing in which the participant wishes to participate in the videoconference; your last name, first name and patronymic; your status in a court case; the e-mail address used by them for registration in the System; the name of the system proposed for video conferencing; court telephone number; a note on the presence or absence of an electronic signature.

The application is submitted to the court and sent to other participants in the case in the manner prescribed by procedural law, by permitted means of communication.

These provisions have temporarily but not definitively resolved the issue of unresolved participation in court hearings by videoconference outside the court premises prior to the implementation of the UJITS Regulation, but the need for its adoption at the legislative level has not disappeared. Therefore, the legal regulation of remote litigation needs further improvement.

5 Conclusion

As a result of the research, the following conclusions can be drawn:

1. The advantages of remote court proceedings are the acceleration of the trial, saving time and money for both participants in the process and the state in the face of the judiciary, ensuring social isolation in a global pandemic, ensuring the safety of litigants.
2. In order to overcome the difficulties associated with conducting videoconference hearings, it is necessary to provide courts with adequate technical equipment of the premises in sufficient numbers to organize videoconferencing trials and with uninterrupted high-speed Internet; to introduce clear and effective mechanisms for the legal regulation of all court proceedings; it is essential to ensure the functioning of the Unified Judicial Information and Telecommunication System as soon as possible.
3. Bring the provisions of the procedural codes regarding the identification of persons in the UJITS and the videoconferencing system by means of an electronic digital signature in line with the current legislation on electronic trust services.
4. In order to unify the terminology in the legal regulation field of remote court proceedings and the use of videoconferencing to establish at the legislative level the following terms and concepts:

Remote litigation is litigation by videoconference during a broadcast from another room, including one located outside the courtroom.

Video conferencing is a telecommunication technology of interactive interaction of two or more remote participants in court proceedings with the possibility of exchanging audio and video information in real time.

Video conferencing system is a set of hardware and software (according to the link on the official web portal of the judiciary of Ukraine vkz.court.gov.ua) or other means available to the court and litigants, providing court hearings by video conferencing and meet the requirements of the law.

Video recording technical means of the course and results of procedural actions are a set of software and hardware and devices that provide proper audio and video recording, storage, copying (duplication) and use of information that reflects the process of video conferencing.

Video phonogram is video and audio recording, which is formed directly during the video conference and recorded by technical means of video recording and is the source material for making working and archival copies.

 https://doi.org/10.1051/shsconf/202110003006

Archival copy of a video phonogram is a recording of a video phonogram copy from technical means on a video recording medium, which has the status of the original and is intended for long-term storage. An archival copy can be used to create working copies in case of their insufficiency, damage, destruction, etc.

Working copy of the video phonogram is a recording a copy of the video phonogram from the technical means of video recording on the video recording medium, which has the status of the original and is attached to the case file.

A video medium of videoconferencing is a disk for laser reading systems on which a video phonogram is recorded, which is an appendix to the minutes of the court hearing.

5. The procedure for working with technical means of videoconferencing during a court hearing, the procedure for holding a court hearing by videoconference, the procedure for making and storing a technical record and its copies shall be subject to determination at the legislative level; the procedure for submitting an application for participation in a court hearing by videoconference, the procedure for the distribution of responsibilities for the implementation of organizational measures related to the functioning of the videoconferencing system; ensuring the possibility of cross-border proceedings in different countries as well as participation in the court hearing by videoconference of the participants in the process, who are abroad.

References

1. Law of Ukraine № 2453-VI of July, 07, 2010, «On the Judiciary and the Status of Judges», https://zakon.rada.gov.ua/laws/show/1402-19#Text,
2. Law of Ukraine №3529-VI of June, 06, 2011, «On Amendments to Certain Legislative Acts of Ukraine in Connection with the Ratification of the Second Additional Protocol to the European Convention on Mutual Assistance in Criminal Matters», https://zakon.rada.gov.ua/laws/show/3529-17#Text
3. Law of Ukraine № 1001-05 of December, 31, 1960, «Criminal Procedure Code of Ukraine, https://zakon.rada.gov.ua/laws/show/1001-05#Text
4. Law of Ukraine № 4651-VI of April, 13, 2012, «Criminal Procedure Code of Ukraine», https://zakon.rada.gov.ua/laws/show/4651-17#Text
5. Law of Ukraine № 5041-VI of July, 04, 2012, «On Amendments to Certain Legislative Acts of Ukraine Concerning Participation in a Court Session by Videoconference», https://zakon.rada.gov.ua/laws/show/5041-17#Text
6. Decree of the President of Ukraine № 276/2015 of May, 20, 2015, «Strategy for reforming the judiciary, the judiciary and related legal institutions for 2015-2020», http://zakon2.rada.gov.ua/laws/show/276/2015/print1509894126567793
7. The Convention for the Protection of Human Rights and Fundamental Freedoms https://www.echr.coe.int/Documents/Convention_ENG.pdf
8. International Convention on the Elimination of All Forms of Racial Discrimination, Adopted and opened for signature and ratification by General Assembly resolution 2106 (XX) of 21 December 1965 entry into force 4 January 1969, in accordance with Article 19 https://www.ohchr.org/en/professionalinterest/pages/cerd.aspx
9. Convention on the Rights of the Child, Adopted and opened for signature, ratification and accession by General Assembly resolution 44/25 of 20 November 1989, Entry into force 2 September 1990, in accordance with article 49 https://web.archive.org/web/20100611182141/http://www2.ohchr.org/english/law/crc.htm
10. United Nations Standard Minimum Rules for the Administration of Juvenile Justice ("The Beijing Rules") Adopted by General Assembly resolution 40/33 of 29 November 1985, https://www.ohchr.org/documents/professionalinterest/beijingrules.pdf
11. Recommendation No. R (81) 7. Of The Committee Of Ministers To Member States. On Measures Facilitating Access To Justice. (Adopted by the Committee of Ministers on 14 May 1981 at its 68th Session), https://rm.coe.int/168050e7e4
12. Recommendation No. R (84) 5 Of The Committee Of Ministers To Member States On The Principles Of Civil Procedure Designed To Improve The Functioning Of Justice1 (Adopted by the Committee of Ministers on 28 February 1984 at the 367th meeting of the Ministers' Deputies) https://rm.coe.int/CoERMPublicCommonSearchServices/DisplayDCTMContent?documentId=09000016804e19b1
13. Recommendation No. R (95) 13 Of The Committee Of Ministers To Member States Concerning Problems Of Criminal Procedural Law Connected With Information Technology (Adopted by the Committee of Ministers on 11 September 1995 at the 543rd meeting of the Ministers' Deputies), https://rm.coe.int/16804f6e76
14. Opinion No.(2011)14 of the Consultative Council Of European Judges «Justice and information technologies (IT)» Adopted by the CCJE at its 12th plenary meeting (Strasbourg, 7-9 November 2011), https://rm.coe.int/168074816b
15. O. Yu. Zub, Videoconference as a means of simplification of civil proceedings, ***Bulletin of VN*** *Karazin Kharkiv National University*. Series: Right, 17, (2014), http://nbuv.gov
16. Sir Brahmaiah Maroju. Video Conferencing in Judicial Proceedings – An Innovative Practice // Centre for Innovations in Public Systems, of May, 2020, http://www.cips.org.in/public-sector-26systemsgovernmentinnovations/documents/vide

o-conferencing-injudicialproceedings-an-innovative-practice.pdf.

17. Miklos Kengyel; Zoltan Nemessanyi. Electronic technology and civil procedure : new paths to justice from around the world, 329, (2012). http://books.google.ru
18. Syzonenko A. Video communication as a means of investigative actions outside Ukraine, *Law of Ukraine. 3, (*2003).
19. Smirnov MI Features of mutual legal assistance in criminal cases using the method of videoconferencing: dis…PhD, 268, (2005).
20. Pashkovsky, MI Videoconference in criminal proceedings, *Legal life of modern Ukraine*: International scientific conference of teaching and graduate students (Odessa, May, 21-22, (2010)
21. Bortun M. Current issues of interrogation by video conference, *Bulletin of the National Academy of Prosecutors of Ukraine*, 1, 34, (2014).
22. Chernychenko I. The concept and procedural-legal essence of videoconference in the criminal proceedings of Ukraine, *Visnyk of Lviv National University*. The series is legal, 59, (2014)
23. Bogunov V. Videoconferencing will allow participants to save time and money, *Law and Business, 35, (*2012).
24. Garievskaya MB Videoconference in civil proceedings, *Private law and entrepreneurship,* 14, (2015*)*..
25. Ozernyuk GV Legal regulation of court hearings by videoconference under the laws of the European Union, 1, (2020), http://nbuv.gov.ua/UJRN/srjusc_2020_1_6
26. Kravtsova T., Kornienko M. Court hearings by videoconference, *Ukrainian lawyer*, 4-5, (2020). https://www.asterslaw.com/ru/press_center/publications/court_hearings_via_vidvideoconferen_in_ukraine/
27. Law of Ukraine № 852-IV of May, 22, 2003, «About electronic digital signature», https://zakon.rada.gov.ua/laws/show/852-15#Text
28. Law of Ukraine № 1798-XII of November 6, 1991, «Commercial Procedural Code of Ukraine», https://zakon.rada.gov.ua/laws/show/1798-12#Text
29. Law of Ukraine № 1618-IV of March, 18, 2004, « Civil Procedure Code of Ukraine», http://zakon5.rada.gov.ua/laws/show/1618-15
30. Law of Ukraine № 2747-IV of July, 6, 2005, «Code of Administrative Procedure of Ukraine», https://zakon.rada.gov.ua/laws/show/2747-15#Text
31. Regulation (Eu) No 910/2014 Of The European Parliament And Of The Council of 23 July 2014 on electronic identification and trust services for electronic transactions in the internal market and repealing Directive 1999/93/EC, https://ec.europa.eu/futurium/en/system/files/ged/eidas_regulation.pdf
32. Law of Ukraine № 2155-VIII of October, 05, 2017, «About electronic trust services», https://zakon.rada.gov.ua/laws/show/2155-19#n534
33. Law of Ukraine № 5492-VI of November, 20, 2012, «About the Unified State Demographic Register and documents confirming the citizenship of Ukraine, identity or special status», https://zakon.rada.gov.ua/laws/show/5492-17#Text
34. Order of the State Judicial Administration of Ukraine № 196 of April, 23, 2020, «About modification of the Order of work with technical means of videoconferencing during court session in administrative, civil and economic processes with participation of the parties out of court premises», https://zakon.rada.gov.ua/rada/show/v0196750-20#Text

Electronic (Online) Voting in Ukraine: Realities and Prospects

Yuliia Lomzhets[1*], *Iryna* Dmytruk[1], and *Ihor* Dubinskyi[1]

[1]Admiral Makarov National University of Shipbuilding, 54025, Ukraine

Abstract. The article considers the process of introduction of electronic voting in the electoral systems of modern countries and the condition of implementation of electronic democracy instruments in Ukraine. The main normative legal acts that have been adopted in recent years and the international experience of using the online voting system are analyzed. Prospects for further introduction of electronic voting in Ukraine, the main problems and possible ways to solve them are highlighted.

1 Introduction

The formation of Ukraine as a modern democratic, social and legal state encourages it in all ways to ensure the realization of human and civil rights, defined by the Constitution of Ukraine and international law. A significant role among them is played by a group of political rights, because according to Part 3 of Art. 21 of the Universal Declaration of Man, the will of the people must be the basis of the power of government; that the will must be manifested in periodic non-falsified elections [1].

Despite the fact, that Ukraine gained independence almost thirty years ago, our state is still in the process of finding an optimal electoral system. That is eloquently evidenced by the fairly frequently changing electoral legislation. Thus, the modern bill was submitted to the parliament on October 2, 2015, but it was developed on the basis of the draft of Electoral Code of Ukraine, which found a positive opinion of the Venice Commission and was registered in March 2010. And only on November 7, 2017 Verkhovna Rada adopted the Electoral Code in the first reading. In the future, the project was called "the largest in history", because from the deputies received about 4.5 thousand of amendments and the volume of the act was very large (460 pages).

Only on December 19, 2019 Verkhovna Rada adopted the Electoral Code as a whole [2].

However, society does not stand still and with the development of scientific and technological progress it changes, and its information components are constantly evolving and inextricably linked with the daily functioning of both man and the state as a whole.

As Yu. Danko and N. Belotserkovskaya aptly point out, in the era of the information society, when the amount of information and knowledge in the life of society increases, the role of information technologies in public and political-legal relations increases accordingly. The application of new information and communication technologies is changing political life, public discourse and tools for mobilizing citizens of many countries around the world [3, p. 61].

It is worth saying, that the issue of introducing of the system of electronic elections and approaches to their organization has been studied by such foreign and native scientists as A. Barikova, N. Belotserkovskaya, A. Gotun, Yu. Danko, N. Gritsak, H. Kokhalik, A. Konstantinovskaya, A. Nasibulin, R. Nachos, V. Peter, I. Polovko, I. Sidenko, N. Titovskaya, G. Sheludko, M. Chevallier, M. McGaley, PJ Gibson, M. Remmert and others.

Unfortunately, the COVID-19 epidemic and as a consequence of the introduction of quarantine in our state have shown the need for further legislative introduction of e-democracy, in particular, in the electoral process. In turn, this necessitates a comprehensive, in-depth study of the current state of implementation of electronic voting in the electoral system of Ukraine and the prospects for its development, taking into account international experience.

2 Main part

Turning to the main material of the study, it is worth saying, that today in Ukraine the development of e-democracy and e-elections is still under discussion, but the issue of introducing e-voting is not new. The legal preconditions for the functioning of the electronic election system were created by the introduction of the laws of Ukraine "On Information" of October 02, 1992 № 2657-XII, "On Electronic Documents and Electronic Document Management" of May 22, 2003 № 851-IV and "On Electronic Digital Signature" of 22 May 2003 № 852-IV, that now is invalid.

On June 10, 2011, the draft law of Ukraine "On the Concept of Implementation of the Electronic Voting System in Ukraine" № 8656 was submitted to Verkhovna Rada of Ukraine.

* Corresponding author: ylomzhets@gmail.com

According to Section 3 of this bill, the introduction of electronic voting technology along with the traditional procedure has the following advantages:

- significant reduction of risks of pressure on the will of citizens;
- ensuring the possibility of voting within a few days;
- avoidance of negative phenomena, in particular forgery of ballots, photographing of counted ballots, their change in order to make them fictitious etc;
- providing the possibility of conducting an electronic referendum and bilateral communication between the state, local governments and citizens;
- increasing the number of citizens and, above the all, young people participating in the voting;
- significant savings of material resources [4].

At the same time, the main emphasis in the text of the draft was made on the last point, and the introduction of the electronic voting system was proposed to be carried out in five stages.

Thus, in particular, the author of the bill singled out the following sequence of necessary actions:

1) development, processing and adoption of the state program;

2) implementation of economic calculations and estimation of one-time expenses for formation of budgets of two levels, and also attraction of other material resources;

3) ensuring the implementation of the program taking into account international experience and with maximum use of native intellectual potential;

4) bringing the current legislation of Ukraine in line with established norms and requirements, carrying out experimental testing of electronic systems on models and analogues;

5) development of a full-scale working project and its implementation [4].

The document also outlined the practice of using electronic voting in countries such as Switzerland, Estonia, Kazakhstan, the United States, India, Brazil, Australia and South Korea, and the technical requirements for the voting system. However, the bill was withdrawn, although the argumentation of the benefits of implementing such a system remains relevant today.

The next step in implementing the e-voting system in Ukraine was the adoption in November 2017 the Order of the Cabinet of Ministers of Ukraine "On approval of the Concept for the development of e-democracy in Ukraine and the action plan for its implementation" № 797-r. The concept provided the formation of the foundations for the establishment of various forms of e-democracy in two stages.

Among the main provisions that were related to the formation of the electronic voting system at the first stage (2017-2018) were the following:

- formation of regulatory support;
- ensuring the availability of e-democracy tools for all individuals;
- improvement of the tool for electronic petitions and consultations, the mechanism of openness of information on the activities of public authorities;
- formation of the regulatory framework for the introduction of electronic voting, as well as the electronic electoral process, electronic referendums and electronic plebiscites;
- development of electronic identification of individuals and legal entities in state information and telecommunication systems [5].

Later, at the second stage (2019-2020), it was assumed:

- the introduction of network services in the field of e-democracy and the creation of conditions for their proper provision;
- direct implementation of electronic voting, as well as the electronic electoral process, electronic referendums and electronic plebiscites;
- the spread of e-democracy, the formation of knowledge and skills in using its tools among the population;
- ensuring further implementation and accessibility of e-democracy tools;
- carrying out a study of the state development of e-democracy in Ukraine;
- ensuring coordination and control over the development of e-democracy in Ukraine [5].

In turn, to implement the provisions in June 2019, the government approved the "Action Plan for the implementation of the Concept of e-democracy in Ukraine for 2019-2020", which details the sequence of actions, the timing of their implementation, responsible persons and partners. At the same time, the creation of a regulatory framework to ensure the introduction of electronic voting was expected in the III quarter of 2020 [6].

In part, one of the positive results of the action plan was the introduction of the Resolution of the Cabinet of Ministers of Ukraine "Issues of the Unified State Web Portal of Electronic Services of the Unified State Portal of Administrative Services" dated December 04 2019 № 1137 Portal "Action" as a service through which the idea of "state in a smartphone" is realized.

This provision defines the goals, objectives, functional capabilities and subjects of the Unified State Web Portal of Electronic Services Portal "Action" contained in it and the procedure for their introduction, as well as other issues of its operation.

The provision also assumed that from January 1, 2020, the Unified State Web Portal of Electronic Services will also perform the functions of the Unified State Portal of Administrative Services. In turn, the web portal has an official address on the Internet diia.gov.ua, and you can use the Portal "Action" for free without any time limit.

Among the main tasks of the Action Portal, the position defines the following:

- provision of electronic services (including administrative) with the receiving and using, if necessary, of information from national electronic information resources, which is necessary for the provision of such services;
- creation and functioning of the electronic cabinet of the user on the web portal, and also access through the specified cabinet of users to the information from

national electronic information resources, including information about the person;

- submission of electronic appeals through a web portal, etc;
- providing official electronic correspondence during the provision of services, consideration of appeals and administrative cases (cases considered by executive authorities, local governments, their officials, other entities that are legally authorized to perform public authority functions, to make decisions of individual action aimed at establishing, changing, terminating or exercising the rights and obligations of a person, as well as the protection of his rights), as well as when considering using the web portal other issues;
- payment of the administrative fee for the provision of administrative services, fines for administrative offenses, state fees, other payments;
- settlements of other public (in particular housing and communal) services;
- providing the subjects of the appeal with information on the progress and results of the provision of services, consideration of the appeals and the admin in real time;
- receiving by the subjects of the appeal the results of the provision of e-services, consideration of appeals and administrative cases;
- filing complaints on the results of services, consideration of appeals and administrative cases, providing official electronic correspondence during appeal procedures, providing the appellant with information on the progress and results of the complaint in real time, as well as receiving the specified results;
- formation by means of a web portal, including with use of the mobile application of the Portal "Action", digital images of documents which the person can present on the smartphone through the specified mobile application;
- monitoring and evaluation of the quality of services provided using the web portal, in the Centers for the provision of administrative services or by the subjects of consideration of appeals directly;
- protection of data (including personal) of the specified web portal from unauthorized access, destruction, modification [7].

Equally important was the normative introduction in Ukraine the functioning of the electronic identification system. Thus, the Resolution of the Cabinet of Ministers of June 19, 2019 № 546 approved the Regulations on the integrated electronic identification system, which established the procedure for determining, defining the structure, functioning of the integrated electronic identification system, its creation and use.

The system is an integral part of the information and telecommunication infrastructure, providing electronic interaction of the subjects of interaction with the users of the system, and the owner of the system is the state [8].

A key element of electronic identification was the creation of the Resolution of the Board of the National Bank of Ukraine of March 17, 2020 № 32 BankID System of the National Bank of Ukraine (the BankID System of the NBU).

According to the provisions of the above resolution, the BankID System of the NBU is a national electronic remote identification system, performs accounting functions and provides electronic remote identification of individuals by transmitting personal data of users by the subscriber to the subscriber - service provider, and also records the number and volume of services remote identification [9].

According to the information, posted on the official website of the NBU, the BankID System currently has 33 subscribers, including:

- 14 banks (CB "Privatbank", JSC "Sberbank", JSC "Alfa Bank", JSC "FUIB", Monobank (JSC "Universal Bank"), A-Bank (JSC "Accent-Bank"), JSC "Alliance", JSC "FORWARD Bank", JSB "Pivdenny", JSC " Kredobank", JSC JSB "Radabank", JSC "Motor-Bank", JSC "Bank Vostok", JSC "Idea Bank");
- 19 other institutions (service portals) [10].

Thus, after the implementation of all these measures, the government expected to provide the basic elements of e-democracy in all spheres of public life. But, given the timing and huge range of necessary actions, insufficient development of information and telecommunications resources and security tools, storage of information on servers, which must have strong protection against any interference, it is impossible to fully implement the above plan.

The second important step in the process of formation of electronic voting was also the adoption of the above-mentioned Electoral Code of Ukraine dated December 19, 2019 № 396-IX.

In accordance with Article 18, the Central Election Commission may decide to introduce innovative technologies, hardware and software in the organization and conduct of elections in the form of an experiment or pilot project on:

1) voting of voters at the polling station with the help of technical means and software (machine voting);
2) conducting the counting of votes with the help of technical means for electronic counting of votes;
3) drawing up protocols on the counting of votes, results and results of voting using the information-analytical system [11]. Analyzing this norm, we can conclude that, despite the presence in international experience of several forms of electronic voting, the legislator ignores them, including online voting, and considers only one its form.

Today, there are three main forms of electronic voting in the world:

- voting via the Internet (online) - voters use a personal computer with an Internet connection, which receives the voter's voice and transmits it to another computer, remote access computer;
- voting in the booth with the help of "electronic ballot boxes" - the interface of such a system can be a touch screen or a scanner that reads the mark, placed by the voter in the ballot. Then the voice is registered and stored in the device for e-voting. Thus, citizens

vote in their traditional polling station only with the use of information technology;

- voting using a mobile phone, which can combine pre-specified forms [12, p. 77].

At the same time, the use of "electronic ballot boxes" remains the most widespread form of electronic voting. This type of voting is popular because "electronic ballot boxes" can operate autonomously, without being connected to power grids and communications, which is especially important in countries where polling stations are located in sparsely populated areas or in areas with limited access to telephone lines and the absence of the Internet. [13, p. 17].

The electronic voting system was first introduced in Brazil. In 1996, the system was successfully tested in municipal elections in 57 cities, by 2000 the system was applied in municipal elections throughout the country, and in 2002 this experience was applied in the world's first electronic national elections of the head of state. The elections were observed both by representatives of different countries of the world (USA, Japan, Mexico, Venezuela) and by representatives of the international organization "Transparency International", who confirmed that the voting procedure excluded the possibility of fraud and abuse during the expression of will [14, p. 52].

The first experiment with online voting was carried out in 2000 in the state of Oregon, USA. Over time, voting technologies have improved. Complexes for electronic voting have been widely used in local and federal elections in the United States. In 2002, the Federal Law "Help America Vote Act" was passed. The purpose of the Law was to reject paper ballots, lever devices (voters press the lever next to the candidate's name) and punch-card machines (holes are punched on special cards next to the names of candidates) [15, p. 59].

The experience of some Asian countries is in terms of the use of electronic means in the conduct of elections. So, in 2005, for all political elections in Korea, an electronic voting system was introduced, the tests of which began in 2004. Korea became the first Asian country to vote in all political elections electronically. In trial mode, the new voting system began operating in 2004. In turn, already in 2005, the entire state document circulation was digitized, and in 2006 a database of basic civil documents was created to improve the efficiency of the government, which made it possible to reduce costs. In 2007, the government established a national information network to monitor the import / export of goods by rail and road through customs. [4]

Asian countries, including Japan, Taiwan and Hong Kong, have also launched pilot projects for e-voting systems. In June 2002, residents of the city of Niimi, about 500 km from Tokyo, became the first in Japan to use the electronic electoral system. According to the Associated Press, more than 15,000 people made their choice by placing their fingers on the touch screens of voting machines with using the biometric parameters of voters [4].

The most preferable and close to Ukraine example in the field of legislative implementation of electronic voting is Estonia. Since the introduction of Internet voting in 2005 in local elections, this voting method has been used here 7 times: in October 2005, 2009 and 2013 in local elections, in 2007 and 2011, in parliamentary elections in 2009 and 2014, in elections in European Parliament. The number of voters who voted online also grew over the years from 2% in 2005 to 31% in 2014, despite massive hacker attacks, which, during the 2007 elections, did not undermine the confidence of the country's citizens in e-voting [3, p. 66].

India's experience is also interesting and quite unusual for the modern understanding of electronic voting. For example, the Indian Voting Machine, or EVM (Electronic Voting Machine), was developed in 1989-90. The device consists of two modules - a control unit and a console for voting. The console has a list of candidates, next to each of which is a vote registration button. After pressing it, the input device is locked until the next voter, and the fixed voice is entered into the memory cell of the corresponding candidate. The control unit is entrusted with the functions of the general support of the process of issuing the total number of citizens who voted and the final announcement of the election results. For security reasons, electronic machines are not connected to any networks and central databases, and voting results are recorded by members of the regional election commission. Thanks to such technical solutions, the electronics of the voting machine do not require any operating system, all command codes are sewn directly into the chips, and the system is therefore considered extremely resistant to unauthorized interference, in particular hacker attacks [4].

But such an electoral system is categorically not accepted by the opposition, as EVM is manufactured by state-owned enterprises and those controlled by the "ruling party". It has been shown that, contrary to the assurances of the manufacturers and the election commission about the "resistance of the machine to hacking", the motherboard is easily and without loss of functionality removed from the EVM case, chip code is read, and chips can be changed to reprogrammed. As a result, such a modified machine does everything as it should, but eventually removes a certain share of votes from all memory cells and adds them to the votes that are "needed", guaranteeing victory to the right candidate and maintaining the total number of voters who voted [4].

The experience of Switzerland is interesting, where in 2003 an experimental electronic voting through the "E-voting" system took place for the first time. According to the plans of the Swiss government, the absolute majority of citizens, even those who were outside their country, could already vote in the next parliamentary elections in 2015 on the Internet. Switzerland is constantly expanding the use of e-voting. Today in Geneva, about 30% of local voters carry out online voting. In order to improve the security of the electoral process, an electronic voting procedure with authorization under an individual code was introduced [16, p. 171].

Also, no less important is the experience of the Swiss city of Zug, in which in 2018 an experiment with the use of technology in the field of cryptographic protection - Blockchain, which allows creating open and protected data registries in the electoral process [17], was successfully completed.

Another successful experiment in 2018 was the online elections in West Virginia, USA, which were supported by the private project Voatz. This project is a platform for blockchain-based Internet voting, which can then be used at the state level. The project is currently being tested. Also a popular project in the United States in this area is the Follow My Vote project. The developers seek to create a platform for voting, which will achieve transparency in the election, without compromising the confidentiality of voters, as well as to provide mathematical algorithms to obtain an accurate election result [18, p. 83].

It should be noted that the Central Election Commission of Ukraine also tested the NEM blockchain protocol for storing polling station protocol data in 2014 during the presidential elections in Ukraine [18, p. 83].

In turn, the government of Kazakhstan was also interested in the formation of an electronic voting system and in September 2000 electronic elections were held. By Presidential Decree, a State Government Working Group was established to develop changes and additions to the country's legislation. In 2004, electronic equipment was purchased and operators were trained. In April 2004, amendments to the Law on Elections were adopted, which provide for an electronic voting procedure. Appropriate additions have also been made to administrative and criminal legislation [4].

Taking into account the fact that Ukraine has long taken the course of European integration, an analysis of the normative acts of the Council of Europe is important for the legislative support of the functioning of the electronic voting system in Ukraine.

According to the Council of Europe Recommendations on Legal, Operational and Technical Standards for Electronic Voting No. 11 dated September, 30, 2004, a number of principles must be observed for the use of electronic voting systems. The elections must be accurate, with the guarantee of transmission of the electronic vote to the central network without any changes. It is necessary to adhere to the democratic principles of traditional elections - direct, free, general, with the provision of procedures for secrecy and transparency. When conducting electronic voting, it is important to adhere to the privacy policy and ensure that it is impossible to identify the voter with his vote. The last important principle is the possibility of verification and audit of votes, when each voter will be able to check the correctness of accounting for his vote [9].

To the requirements described above, J. Gibson, M. McGaley and M. Chevallier added a number of criteria for the proper functioning of the electronic election system. In particular, an electoral vote cannot be detected and changed during its transfer to the election commission. Only registered voters have the right to vote, and only once, a second vote will be considered election fraud. The voting site must be protected from hacker attacks and protect the personal data of voters. The state must ensure the impossibility of pressure on citizens through convictions; introduce a ban on the purchase / sale of votes with criminal responsibility [18, p. 5; 19, p. 4].

The Cabinet of Ministers of the Council of Europe has outlined general standards for regulating the e-voting procedure, which reflect the basic principles of democratic elections and strengthen confidence in the e-voting procedure. They are the guiding lines for the widespread implementation of the electronic voting system in the member states of the European space.

The standards can be divided into three types: legal, defining international and national regulations governing the election procedure; operating standards that govern the organization of elections and the voting procedure itself; technical standards to ensure the compatibility of software and electronic devices. The selected requirements must be applied at all stages of the elections: announcement of elections, registration of voters, registration of candidates, campaigning, voting procedure, calculation of results and their protection [20, p. 13].

These provisions were updated in 2017 with the adoption of CM / Rec (2017) 5, which clarified the definition of electronic voting to include electronic ballot counting. However, other electronic voting tools that are used during the electoral process, in particular electronic registers, methods of processing information about voters, counting votes, transfer of results, were not included with the provisions of the above recommendations [23].

However, the regulation of the electoral process, first of all, has a national prerogative, and therefore its international legal principles are contained in the national constitution and / or national electoral legislation, and in some countries local elections are governed by local legislation. However, in general, all normative legal acts (supranational, national and local) contain general principles that are enshrined in international documents.

With this in mind, the Council of Europe's electoral administration and civil society in its publication "Digital Technologies in Elections. Question, conclusions and prospects" summarized the characteristics of the national legislation of European countries on the issue of electronic voting. Thus, the authors found that the introduction of technologies in the electoral process was carried out in two waves. Legislators first regulated the use of simple technologies (paper and mechanical), particularly in the 1960s and 1970s in Germany, the Netherlands and France. Later, the legal framework was "updated", taking into account technological progress, and mainly concerned the use of devices for electronic voting or electronic counting in the 1990s. In turn, the countries that introduced Internet voting in their electoral process developed special regulatory acts, although they were formed by analogy with the existing "paper" systems, namely, voting by mail (Switzerland or Estonia at the beginning of 2000) [24, p.12].

However, the authorized bodies of Austria recognized that the legal regulation of Internet voting does not comply with the constitution for its insufficient detail and as such that cannot allow members of election commissions to fulfill their duties without technical assistance [24, p.13].

The jurisprudence of the highest national courts has also played an important role in clarifying the practical significance of electoral principles in the context of their application in the e-voting procedure. The decisions of the constitutional courts of Germany (2009) and Austria (2011) attracted particular attention. The jurisprudence has demonstrated the importance of interpreting the principles in terms of their transformation into detailed requirements for technologies and helped to form a general agreement on the need for detailed regulation of digital technologies. It showed that the same principles can be interpreted in very different ways, which ultimately leads to different results depending on the factual, historical, cultural characteristics. The preparation of such an interpretation must be put on the legislator/regulator, and not on technical workers or technology suppliers [24, p.13].

Thus, the Netherlands returned to paper voting in 2008, the main reasons were the danger of secrecy of the voice and a large dependence on sellers and certified agencies. As already noted, Germany in 2009, declared the e-voting procedure unconstitutional due to a lack of publicity. Unlike Norway, which in 2014, temporarily suspended the process of introducing e-voting in the context of the security problem [16, p. 171].

Also, despite the existing regulations and the positive experience of countries, a common problem for all is the threat to the security of election results, in particular, possible hacker attacks on election servers. Ukraine has already had precedents with insufficient IT-protection of state structures from such threats. A striking example is the actual paralysis of the activities of public authorities due to infection with the "Petya" virus in June 2017 and the information about the leakage of personal data of citizens from the digital state service of the "Action" application in April 2020. And although about the latter, the Cyber Police Department of the National Police of Ukraine did not reveal the facts of cyber attacks, but in fact there was a seizure of personal data of citizens, although outdated. In addition, the practice of much more technically developed states also testifies to the danger of such situations.

Such an example is the scandal surrounding the possible interference of Russian hackers in the 2016 US presidential elections, when, before the start of the procedure, as a result of a cyber attack on the servers of the US Democratic National Committee (DNC), came almost 20,000 emails written or received by committee officials. The hackers also attacked the Democratic Congress Committee (DCCC). American Internet security experts blamed the attacks on Russian hacker groups called Cozy Bear and Fancy Bear. [21]

According to Microsoft vice president of user security Tom Burt, in 2019 the latter named group of hackers tried to gain access to the personal data of political advisers working for both the Democratic and Republican parties, as well as representatives of the expert community - in general, they felt attacks on themselves. more than 200 organizations [21].

Also, the problem of such a system using the global Internet for Ukraine is the insufficient level of access of the electorate to the network. This manifests itself in two aspects:

- first, high-quality Internet in remote settlements in our state, unfortunately, is absent today;
- secondly, the complexity of the e-voting system for some categories of Ukrainian, in particular the venerable age, due to the lack of knowledge and skills necessary for electronic expression of will, which creates a threat of dividing society into those who have a sufficient amount of information and those who do not owns and therefore is limited in its rights.

A. Weyer aptly notes that a weak point in the electoral system and elections through the Internet in Ukraine can be considered the difficulty of identifying a particular voter, which may result in two options:

1. In the first situation, a citizen-voter at an convenient time, before election day, receives an individual electronic key or password that will allow him to register on the server to vote. If such a key is issued, like a passport, for a long period of time and, accordingly, will be stored for a long time with one owner, the reliability of passwords is reduced. They can be stolen from the voter himself or from a database for storage. Thus, the amount of time taken to store the password is inversely proportional to its reliability;
2. In the second scenario, the password is given to the voter immediately before the election and is valid once. With such a system, the reliability of the password does not cause complaints, but the fact that the voter before each local or national election has to go to the polling station, where individuals will give him a password, at least halves the usefulness and convenience of electronic voting as such [26 , p. 266].

Thus, we can identify the following main problems of the introduction of electronic voting in Ukraine:

- imperfection of the legal regulation of the electronic voting procedure, calls into question the observance of the general principles of the electoral process (anonymity of voting, publicity, etc.);
- threat to the security of the election results, due to insufficient technical support of the procedure;
- insufficient level of informatization of the Ukrainian electorate.

Nevertheless, despite a number of these shortcomings, the electronic online voting system in the electoral process also has a number of advantages, including more accurate results and faster counting of votes; preventing abuses in the counting of results at polling stations by reducing human interference; the ability to vote from abroad; the possibility of voting for sick voters and people with disabilities; reducing the number of spoiled ballots, as electronic electoral systems can warn voters about invalid votes; saving money by saving time of employees, reducing the cost of producing and distributing ballots, in order to avoid costs of transporting ballots; reduction of incidents of sale of votes, excluding voting by one voter in several constituencies [3, p. 69].

3 Conclusion

It can be concluded, that modern technologies and their introduction into the political system of many countries of the world, including Ukraine, change the vector of development of society as a whole, and become an effective tool for attracting citizens to decision-making through the introduction of e-democracy mechanisms. The use of such technologies in the electoral process is acquiring a wide range of development, especially through electronic voting, which, even with certain disadvantages and economic difficulties, is becoming more widespread in world democracies.

It is also positive for Ukraine that e-voting is a significant saving of financial resources aimed at:

- ensuring the electoral process;

- reducing the risk of infection, since the contact of citizens is minimal;

- the care for the environment is manifested due to the decrease in the use of paper information carriers.

Today Ukraine has adopted basic regulatory legal acts and partially implemented the first technical capabilities that can be used to further ensure the application of electronic online voting and e-democracy, but the ultimate prospect of introducing an electronic voting system in Ukraine is questionable in the coming years. Since the main problem is not only the lack of a full-fledged legal framework, but also the vulnerability of native electronic systems from negative information influences, viruses, hacker attacks and the possibility of manipulation and outside interference in the results of the expression of the will of citizens.

That is why, before seriously approaching the implementation of electronic voting, it is necessary to clearly clarify the issue of fundamental legal regulation of the introduction of an electoral mechanism for electronic online voting in Ukraine and means of protection, storage of information on servers that would have powerful protection against any hacker interference and implement incentive measures informatization of all categories of society.

References

1. Universal Declaration of Human Rights of December 10, 1948. URL: https://zakon.rada.gov.ua (access date 28.01.2021)
2. Electoral Code of Ukraine. Wikipedia: website. URL: https://en.wikipedia.org (access date 28.01.2021)
3. Yu. Danko and N. Belotserkovskaya Modern Society, Electronic Voting: Possibilities of Ukrainian Reception, **№ 1 (17)**, pp. 61, 66, 69 (2019)
4. Draft Law of Ukraine "On the Concept of Implementation of the Electronic Voting System in Ukraine" dated 10.06.2011 № 8656. Ips. ligazakon: website. URL: https://ips.ligazakon.net (access date 28.01.2021)
5. On approval of the Concept for the development of e-democracy in Ukraine and the action plan for its implementation: Order of the Cabinet of Ministers of 08.11.2017 № 797-r. URL: https://zakon.rada.gov.ua (access date 28.01.2021)
6. On approval of the action plan for the implementation of the Concept for the development of e-democracy in Ukraine for 2019-2020: Order of the Cabinet of Ministers of June 12, 2019 № 405-r. URL: https://zakon.rada.gov.ua (access date 28.01.2021)
7. Question of the Unified State Web Portal of Electronic Services of the Unified State Portal of Administrative Services: Resolution of the Cabinet of Ministers of Ukraine of December 4, 2019 № 1137. URL: https://zakon.rada.gov.ua (access date January 28, 2021)
8. On approval of the Regulations on the integrated electronic identification system: Resolution of the Cabinet of Ministers of June 19, 2019 № 546. URL: https://zakon.rada.gov.ua (access date 28.01.2021)
9. On approval of the Regulations on the BankID System of the National Bank of Ukraine: Resolution of the Board of the NBU of March 17, 2020 № 32. URL: https://bank.gov.ua (access date 28.01.2021)
10. Created prerequisites for the implementation of a commercial model of using the BankID System of the NBU. National Bank of Ukraine: website. URL: https://bank.gov.ua (access date 28.01.2021)
11. Electoral Code of Ukraine: Law of Ukraine of 19.12.2019 № 396-IX. URL: https://zakon.rada.gov.ua (access date 28.01.2021)
12. G. Sheludko Recent Studies of Social Sciences. Section: Politics, Electronic voting as a kind of selective information and communication technologies: foreign and domestic experience, p. 77 (2015)
13. A. Barikova Electronic state: a new management efficiency, P. 17 (2016)
14. A. Gotun Journal of Taras Shevchenko National University. Series: Philosophy. Political Science, The use of new information technologies in the electoral process: world experience and practice in Ukraine, **№ 89-90**, P. 52 (2008)

15. P. Nachos Comparative and Analytical Law, Electronic Voting in Ukraine: Prospects for Implementation and World Experience, **№ 1**, P. 59 (2018) URL: http://pap-journal.in.ua (access date 28.01.2021)
16. H. Kokhalik Efficiency of public administration, World experience of introduction of electronic democracy: problems and achievements, **№ 42**, p. 171 (2015)
17. Switzerland's first municipal blockchain vote hailed a success. *SWI swissinfo.ch* : website. URL: https://www.swissinfo.ch/eng/cryptovalley-_-switzerland-s-first-municipal-blockchain-votehailed-a-success/44230928 (date of access 02.21.2021)
18. A. Chernyshev Actual problems of public administration, Internet voting as a tool of public administration in Ukraine, **No. 3 (79)**, pp. 79-84 (2019)
19. Counsil of Europe. Recommendation on legal, operational and technical standards for e -voting - explanatory memorandum. Rec. (2004) 11. URL: https: // www. coe. int / t / dgap / goodgovernance / Activities / Key - Texts / Recommendations / LinkEvotingRec _ en. asp (access date 28.01.2021)
20. Chevallier M. Internet voting: Status, perspectives and Issues. In TU-e-Government Workshop, P. 5 (2003)
21. McGaley M. & GibsonP.J. E-voting: a safety critical system, P. 4 (2003)
22. Remmert M. Towards European Standards on Electronic Voting. In Electronic Voting in Europe. P. 13 (2004)
23. Recommendation CM / Rec (2017) 5 of the Committee of Ministers to member States on standards for e-voting. URL: https://rm.coe.int/0900001680726f6f (date of access 02/21/2021)
24. A. D. Maurer Digital technologies in elections. Question, conclusions and perspectives, pp. 10-12 (2020)
25. Microsoft: Russian hackers are again interfering in the US elections. *Deutsche Welle* : website. URL: https://www.dw.com/uk/microsoft-rosiiski-khakery-znovu-vtruchaiutsia-u-vybory-v-ssha/a-54890149 (date of access 02/21/2021)
26. A. Weyer Current policy issues, Advantages and disadvantages of voting on the Internet, Vol. 60, P. 266 (2017)

Digital Society: Regulatory and Institutional Support of Electronic Governance in Modern Realities

Dmytro Arabadzhyiev[1], *Yana* Popovych[1*], *Iryna* Lytvynchuk[2], *Kanat* Bakbergen[3] and *Yurii* Kyrychenko[1]

[1] National University «Zaporizhzhya Polytechnic», 69063 Zaporizhzhya, Ukraine
[2] Polissia National University, 10008 Zhytomyr, Ukraine
[3] Kazakh-German polytechnic college, 160400 Kentau, Republic of Kazakhstan

Abstract. The article highlights the opportunities for the introduction of e-government in the institutional system of society on the basis of digital, information and communication technologies (on the example of Ukraine). It develops the concept of digitalization from the point of that information, knowledge and communication technologies constitute the system of social and legal infrastructure of society. It demonstrates that digital society forms the basis of the safety of modern globalized world, expressed in quality indicators of "e-government". It is proved that E-government in the legal system of the country involves: first, creating conditions for interaction between government and citizens through the use of modern information and communication technologies; secondly, the availability and openness of state registers; third, the introduction of e-democracy tools. It is determined that there is a steady tendency in the Ukrainian society to spread e-government tools in the legal sphere through the provision of administrative information services as well as the use of e-democracy technologies. It is also justified that potential of e-democracy technologies is strong enough to create conditions for e-government in legal sphere through technological and technical support of legal services, improving the level of computer iteracy of the population, increasing accessibility and openness of public information.

1 Introduction

Modern information era in global social development of mankind requires rapid response to challenges in the social, economic, political and legal space. In a broader sense, a new type of society is being formed, which is based on the concept of the information society. "A new society is a society of mental labor based on the application of human knowledge to everything that is produced and to ways how it is produced. New ideas will be the main source of wealth" [5, p.264-265]. Information society is a society where information has become the most important resource, the production and distribution of which serve as socially and personally significant economic, political and cultural activities. In this society information is perceived 1) as a commodity, 2) as a means of managing the individual and society, 3) as an instrument of power, 4) as a weapon in economic and political struggle.

The creation of the population social protection effective system in Ukraine is one of the priority tasks of the state social policy, provides for the social services system modernization for the population in accordance with new socio-economic challenges, the population dominant needs, the development of the information society, requires the introduction of innovative technologies for the social services provision, based on dissemination and application of e-government and administration methodology.

The analysis shows that the issue of the social services system effectiveness for the population has been and is the subject of scientific research and heated debate in sociological discourse. One of the main approaches to research is the structural-functional approach, the central problem of which is the institutionalization of the providing social services system (R.Merton, N.Smelzer, T.Parsons). No less important is the role of subjective orientation theories (M.Weber, N.Luhmann, J.Habermas, E.Giddens, P.Bourdieu).

It can be stated that there is an important scientific problem, which lies in the contradiction between the spread of social services' modern practices in the system of electronic services, spread in the most developed societies of the world, on the one hand, and the lack of available knowledge about the plurality of forms, ways of using and mechanisms for implementing social services in the system of electronic services in the social protection field in Ukraine at the present stage of social and technological transformations, on the other hand.

At the stage of transition to an information society, it is no longer the problems of further technicalization of society, as it was believed a few years ago, that come to the fore, but the problems of its intellectualization, the creation and implementation of new social technologies based on the effective use of the main society - knowledge strategic resource.

"Digital government" by virtue of global trends and practices of public administration is defined as a system that consumes or as a state organization of electronic government, a separate management model

* Corresponding author: popovych.jana@gmail.com

at the level of the state, created using information and communication technologies. Considering the above, digital government is viewed as a form of governance or as an organization of governance and thanks to which communicative, information technologies are used to increase efficiency in accessibility, transparency in public administration and a complex of administrative and social services for people and the population as a whole, for the state and for business. There are some problems have been identified that impede the implementation and development of e-government in the social services system for the population as the provision of e-social services: firstly, the lack of a scientifically grounded methodology for the e-social services provision, their list, standardization, mechanisms and social subjects of their provision, which is fixed and regulated at the legislative level; secondly, the lack of material, technical and technological equipment of social protection institutions and social services for the population and uneven access to the Internet and modern computer facilities of certain territories and categories of the population; thirdly, the absence of an advanced training system and social workers education in the social services provision in a remote form.

While in previous types of society the main share of capital was concentrated in the production of material resources necessary for the life of human society, in the information society this share is concentrated in the production and distribution of information. Another definition was proposed by the Commission of the European Union: "information society is a society in which human activities are carried out on the basis of the use of services provided through information technology and communication technologies" (EUC, 1993).

The interpretation of the concept was developed by scientists who formulated the definition of the information society as an ideal market society, in which education and information shape every person who aware of collective constraints; moreover, as a society of ideal planning, where the centre receives true communication from the members concerning their goals and desires and forms its own structure and position accordingly [9]. Information and participation in management are developed as a single process. In the information society group plans reflect social and cultural aspirations to a greater extent than before. At the same time, the external pressure will increase. In these circumstances, "only the government, which has decent information, will be able to promote development and guarantee the independence of the country" [9].

On the basis of the information society background M. McLuhan introduces as a leading concept "electronic society", the defining elements of which are information and communication technologies and information networks. From that moment we can introduce creating a digital analogue of society - "e-society".

Thus, the concept of information society where knowledge and information and communication technologies constitute the system of social and legal infrastructure for the interaction of state and people (in particular through the use of electronic communication systems) is the basis of modern realities. In modern management theory and practice information technologies are embodied in the concepts of "e-government" and "e-government", which are implemented in the legal system of modern society.

Therefore, the purpose of the article is to identify opportunities for the implementation of e-government into the legal system of Ukrainian society on the basis of digital and communication technologies.

The methodological basis of the study are general scientific methods of cognition of social phenomena and processes (analysis, synthesis, generalization, classification) and sociological methods of obtaining empirical data (mass and expert sociological survey). The empirical basis of the study are the results of:

1) a mass sociological survey among the population of Ukraine aged 18 years (n = 1000, November 2020). The sample is quota, randomized, distributed by sex, age, region and place of residence (city and countryside). The error is not more than 2.2%;

2) expert survey among legal scholars and civil servants (n = 200, November 2020).

2 Materials and Methods

Theoretical and methodological approaches consider the concept of "e-government" as a service, as a state electronic organization and as a model of public administration based on the use of information and communication technologies. For instance, some experts consider e-government as the application of information and communication technologies in public administration combined with organizational change and new skills, aimed to improve the quality of public services and democratic processes as well as to strengthen support for public policy (EUC, 1993). Others defines e-government as a way of organizing state governance with the help of local information networks and segments of the global information network, which ensures the functioning of certain services in real time and makes daily communication with official institutions as simple as possible [10, p.131].

The legal framework for e-government is predominantly declarative in nature, as the relevant regulations are strategic and, in some cases, directly related to the implementation of the Association Agreement with the EU. However, the introduction of e-government technologies at an appropriate level and, accordingly, their legal support is not instantaneous and requires both time and significant efforts related to the adaptation of the existing Ukrainian legislation.

On the website of the Supreme Council of Ukraine there are normative legal acts on informatization (since 1993). In February 1998, as is known, the Law of Ukraine “On the National Informatization Program” and the Law of Ukraine “On the Concept of the National Informatization Program” were adopted.

Among the normative documents adopted before this date, one should mention the Decree of the

https://doi.org/10.1051/shsconf/202110003008

President of Ukraine "On the state policy of informatization of Ukraine" dated May 31, 1993 No. 186/93. Decree of the President of Ukraine "On the establishment of the National Agency for Informatization" dated March 13, 1995 No. 206 / 95 Decree of the President of Ukraine "On measures to ensure the activities of the National Agency for Informatization" dated May 27, 1995 No. 390/95 Decree of the President of Ukraine "On the Regulation on the State Communications Committee of Ukraine" dated December 11, 1997 No. 1352/97, as well as Resolution of the Cabinet of Ministers of Ukraine "Informatization Issues" dated August 31, 1994 No. 605, Resolution of the Cabinet of Ministers of Ukraine "On the National Agency for Informatization Issues under the President of Ukraine" dated July 10, 1995 No. 505.

So, in matters of state policy of informatization, there is a leading, organizing role of the President of Ukraine. The government, pursuant to the above-mentioned decrees, has been implementing the policy of informatization.

Actually, the first mention of the term "electronic government" in domestic legislation is contained in government documents, namely in the order of the Cabinet of Ministers of Ukraine "On Approval of the Concept of Formation of a System of National Electronic Information Resources" dated May 5, 2003 No. 259-r and in the Report of the Cabinet of Ministers Ukraine Supreme Council of Ukraine on the state and development of informatization in Ukraine in 2003 (one of the directions of development of the regulatory framework of the sphere of informatization in 2003 was the introduction of electronic management technologies).

At the beginning of 2007, the Law of Ukraine "On the Basic Principles of Development of the Information Society in Ukraine for 2007-2015" was adopted. Subsequently, by the order of the Cabinet of Ministers of Ukraine dated August 15, 2007 No. 653-r, an action plan was approved to fulfill the tasks provided for by this law.

In the period after the approval of the Concept for the Development of Electronic Governance in Ukraine, a number of important regulatory and legal acts were adopted. So, in Article 3 of the Law of Ukraine "On Information" (as amended in 2011), the electronic government development is defined as one of the main directions of the state information policy implementation.

With the aim of introducing into the public administration bodies activities a system of electronic interaction between executive authorities in 2011-2012 The Cabinet of Ministers of Ukraine adopted the Order "Issues of implementing the system of electronic interaction of executive authorities" dated December 28, 2011 No. 1363-r and the Resolution "Some issues of electronic interaction between executive authorities" dated July 18, 2012 No. 670.

In 2012, the Government of Ukraine approved the Concept for the Creation and Operation of an Information System for Electronic Interaction of State Electronic Information Resources, which was to become the main component of electronic governance. Later, on July 11, 2013, by order of the Cabinet of Ministers of Ukraine No. 517-r, an action plan for its implementation was approved, and already in 2016 - a regulation on the specified system.

On May 15, 2013, by order of the Cabinet of Ministers of Ukraine No. 386-r, the Strategy for the Development of the Information Society in Ukraine, calculated until 2020, was approved. Electronic governance is defined as one of the main directions of its implementation.

In the context of Ukraine's aspirations to integrate into the European Union, the introduction of European norms and standards for the state information policy implementation is of particular importance. Thus, within the framework of the implementation of the Association Agreement, Ukraine must ensure the comprehensive development of e-government in accordance with European requirements. In turn, by the Decree of the President of Ukraine "On the Strategy for Sustainable Development" Ukraine - 2020 "No. 5/2015 dated 12.01.2015. The e-government development is defined as one of the top-priorities of reforming the public administration system.

The adoption of the Concept for the e-government development in Ukraine (hereinafter - the Concept) at the end of 2010 in no way contributed to the understanding of both state bodies and ordinary citizens of what exactly the term "e-government" means and which components need to be developed. This, in turn, led to a significant lag of Ukraine from the global rates of electronic government development.

According to the order of the Cabinet of Ministers of Ukraine No. 649-r dated September 20, 2017, the new text of the Concept for the electronic government development in Ukraine was approved. The purpose of the Concept is to determine the directions, mechanisms and timing of the effective e-government system formation in Ukraine to ensure the interests and needs of individuals and legal entities, improve the public administration system, increase the competitiveness of the ability and stimulate the socio-economic development of the country. Then, when the previous concept was mostly declarative.

Since in the current legislation e-government is considered as a component of the informatization field, the e-governance technologies introduction is possible through the widespread use of digital technologies, thereby ensuring the public services modernization and the development of interaction between government, citizens and business as the main stakeholders of relations. That is, there is a digitalization of public relations.

The concept provides for the introduction of electronic governance using digital technologies in three key areas: management services modernization and the development of interaction between government, citizens and business; public administration modernization; management of the electronic governance development, thereby improving the quality and effectiveness of the task's implementation in this area.

E-government is part of a more global phenomenon - e-governance. The strategic goal of e-government is to support and simplify the management process for all stakeholders: government, citizens and businesses through information and communication technologies. In Ukraine the Concept of e-Government Development defines the following: "Digital government" - can be called the government of public administration, helps the development, accessibility and transparency of local and national authorities that used information technology, communication technologies, contributing to the creation of a type of state that did not exist until now, the purpose of which is to fulfill the population needs [4]. It means that e-governance is a combination of e-government with e-democracy, in particular the use of information and communication technologies to accelerate democratic processes [2, p.121].

Scientists define the following stages of building e-governance in most countries: 1) *"e-governance"* (*«government on-line» (GOL)*), which is carried out on the basis of the use of information and communication technologies; 2) *"electronic government" ("e-government")* as information interaction of public authorities and society with the use of ICT; 3) "e-governing". At this stage, the maximum combination of the two components "e-governance" and "e-democracy" is possible, which leads to improved interaction in the whole society in the processes of state decision-making, state regulation and provision of universal government services [6].

The rapid development of communication technologies, information technologies of our time, the development of the Internet makes it possible in the legal aspect to use various methods of obtaining or providing services in the field of electronic form, namely those that are most popular when creating a state, that is focused on the transition to a digital state and electronic services on the part of the state: efficiency, openness and transparency of public administration, to provide remote administrative and legal services for people, society, state and business.

At present, in world practice, the following channels or their combinations are mainly used for the provision of legal services: personal reception of citizens; telephone (voice) service and call-centres; web portal; Email; SMS and other messaging services; mobile portal (mobile site); mobile application; social networks; online legal services; intermediaries based on public-private partnership. A special place among these channels is occupied by the tools of e-democracy - submission of e-petitions, electronic proposals for the adoption of legislation (public discussions), e-voting. Thus, we can observe the slow digitalization of the legal system, which concerns both - individual branches of law and the interaction between the state and society as a whole.

Leading world experience is important for the success of e-government in developing countries (especially in the field of law). According to many experts, the Estonian experience of implementing e-government is the most acceptable.

The rapid development of the digital society in Estonia was provoked by the use of secure data exchange systems and the implementation of electronic identification - eID, which led to the promotion of the following electronic services: *«e-Banking»* (1996), *«e-Tax Board»* (2000), *«e-Cabinet»* (2000), *«m-Parking»* (2000),*«e-Geoportal»* (2001), *«e-School»* (2003), *«e-Ticket»* (2004), *«e-Police»* (2005), *«e-Notary»* (2006), *«e-Business»* (2007), *«e-Health»* (2008), *«e-Prescription»* (2010), *«e-Residency»* (2014). Currently, Estonian citizens have access to such electronic services as e-taxes, e-health, e-banking, e-school, e-elections, ID-ticket [8]. It demonstrates the active phase of e-government implementation in this country. Most of the population was so positive about the concept of e-government that the traditional forms of public affairs in Estonia become "a thing of the past" [1], which indicates the priority of e-government in the social, economic, political and legal spaces.

Currently, a number of electronic services for the provision of legal services operate in Ukraine as well. Thus, the Cabinet of Electronic Registers was created on the website of the Ministry of Justice of Ukraine, among which the following can be distinguished: State Register of Real Property Rights, Unified State Register of Legal Entities and Individual Entrepreneurs, Register of Public Associations, Unified State Register of Court Decisions, Unified register of notaries, etc.

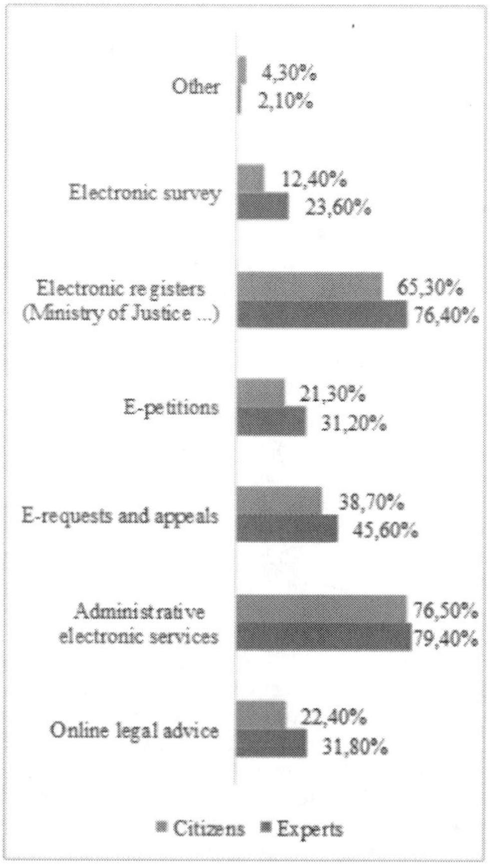

Fig. 1. Evaluation of the most accessible and widespread e-government tools in the legal system (multiplied options were allowed)

To identify opportunities for the use of e-government tools in the legal system of Ukraine, there were conducted sociological and expert surveys (November, 2020).

The survey revealed a higher-than-average level of awareness about the possibility of providing (receiving) electronic services in the legal field among experts and citizens (more than 50%). Fig. 1 presents an assessment of the most accessible and widespread e-government tools in the legal system of Ukraine.

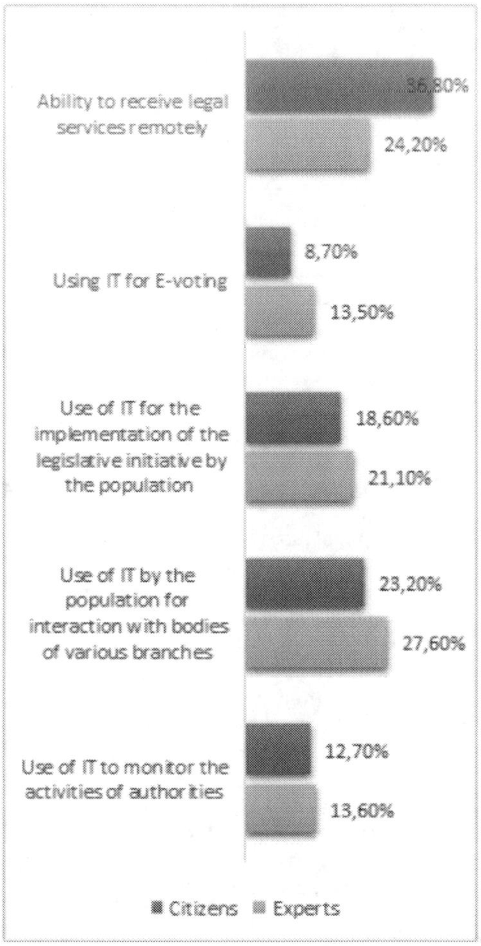

Fig. 2. Definition of e-democracy as a field of application of e-government in the legal system of the state (multiplied options were allowed)

In general, the assessments of experts and citizens coincide in a hierarchical structure. The most common and available are administrative electronic services, electronic registers, inquiries and appeals. The rapid development of modern information and communication technologies and the Internet already allows the use of various forms of providing and receiving electronic services in the legal field, which are becoming widespread due to the state policy of digitalization of public services ("state in smartphone"). Among them the most popular services are electronic forms, obtaining information on court decisions, property rights register etc. Therefore, the survey indicators show that more than half of the surveyed citizens already had such experience (54%). In particular, it was proposed to define the essence and substantive features of e-democracy as a field of application of e-government in the legal system of the state (Fig. 2).

The results of the survey among citizens indicate the importance of e-democracy to receive open and accessible electronic legal services. Interesting figures we get according the use of ICT in voting various levels: for experts they are the possible tools of e-democracy in Ukrainian society, but citizens often are less familiar with international experience of electronic forms of voting. It should also be noted that only 20% of respondents believe that the essence of e-democracy is the use of ICT in the implementation of legislative initiatives by citizens, which may threaten to discredit this form of democracy (widespread disregard for public opinion, petitions, appeals, etc.).

Fig. 3 presents the factors for optimizing the implementation of e-government in the legal system of the state.

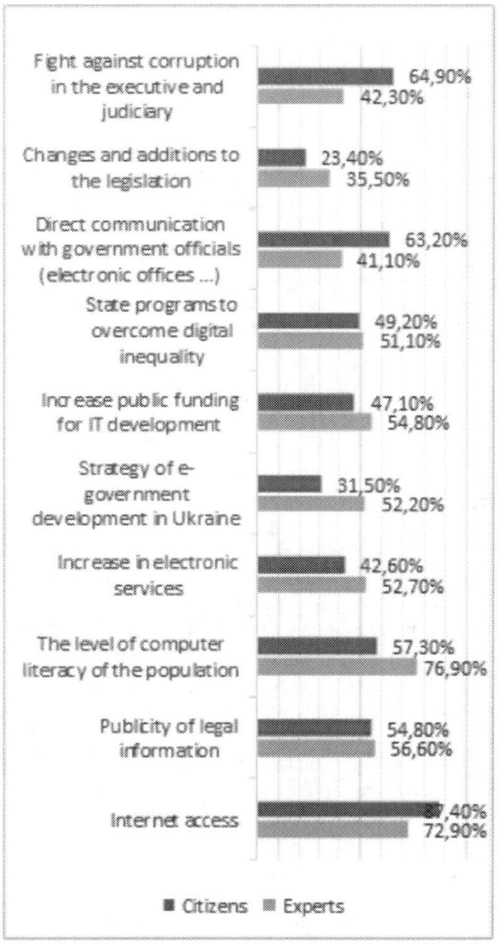

Fig. 3. Factors for optimizing the introduction of e-government in the legal system of the state (multiplied options were allowed)

For the interviewed citizens, the factors of optimizing the process of implementation of digital technologies into the legal system of the country are: accessibility and coverage of the Internet, the fight against corruption and secrecy of government, the possibility of free interaction with the authorities, improving computer literacy; for experts - improving the computer literacy of the population, the availability and coverage of the Internet, expanding the scope of electronic public services, the formation of a state strategy for the development of e-government in Ukraine.

Conclusions

Based on the results of theoretical and empirical analysis, we have made the following conclusions.

1. The concept of information society, in which knowledge, information and communication technologies constitute the system of the state social infrastructure (including the interaction of state and society through the use of electronic communication systems) is the basis of modern development models and finds practical application in e-government algorithms.

2. E-government in the legal system of the country involves: first, creating conditions for interaction between government and citizens through the use of modern information and communication technologies; secondly, the availability and openness of state registers; third, the implementation of e-democracy tools.

3. There is a steady tendency in the Ukrainian society to spread e-government tools in the legal sphere through the provision of administrative information-services, as well as the use of e-democracy technologies. In those circumstances factors for optimizing the introduction of digital technologies in the legal system of the country are: 1) technological and technical support for the provision of legal services (Internet coverage of territories, overcoming digital inequality, etc.); 2) raising the level of computer literacy of the population and forming a positive perception of the e-government and e-democracy concept; 3) increasing the level of accessibility and transparency to public information.

The study also shows that the legal component of digital activity in the context of social assistance in Ukraine is weak, which makes it recommended to legalize "electronic social services" in the legal field as soon as possible.

For the implementation of electronic services and their provision in the legal field, it is necessary to develop a conceptual system-activity model of the system for the electronic social services provision, the main dimensions of which are: 1) scientific and methodological dimension (scientific substantiation of the social services provision in electronic and online form, development of standards for the electronic services provision, etc.); 2) regulatory dimension (legislative support and regulation of the providing electronic social services process); 3) institutional and organizational dimension (the creation of the social services organizational structure in the e-government system at all levels of the social protection sphere functioning - national, regional, local) 4) measurement of material and technological support for the electronic social services provision; 5) measurement of social and public management of the system for the electronic social services provision; 6) measuring the formation of a criteria system for the social efficiency of social services as the electronic social services provision. Based on research, it is possible to improve that will increase the electronic social services system efficiency, the e-government development in the field of social services in general areas, through the implementation of a set of social mechanisms, the owner of which provides appropriate optimization solutions for remote social services. the following groups: 1) social and managerial mechanisms (organizational and managerial, institutional, normative); 2) mechanisms of social partnership (social and information-communication interaction of subjects of social services provision); 3) mechanisms of social marketing (social advertising, PR-technologies in the social sphere); 4) educational and personnel mechanisms (educational and training programs for the population, training or refresher courses for social workers).

Based on the study, it can be argued that the effectiveness of the functioning of the system for the electronic social services provision, the electronic management development in the field of social services for the population in general, depends on the implementation of a set of social mechanisms, which in fact represent the directions for optimizing the implementation of social services in a remote form, among which are highlighted the following groups: 1) social and managerial mechanisms (organizational and managerial, institutional, regulatory) 2) mechanisms of social partnership (social and information and communication interaction of subjects of the social services provision); 3) mechanisms of social marketing (social advertising, PR technologies in the social sphere) 4) educational and personnel mechanisms (educational and training programs for the population, training or refresher courses for social workers).

References

1. E. Arkhipova, N. Dmitrenko. Experience of e-government implementation in Estonia and its introduction in Ukraine. A young scientist, 11 (2015)
2. N. Dragomiretska. Theoretical analysis of communicative activity of a civil servant: monograph. Odessa (2005)
3. T. Lower. Theoretical principles of e-government. Investments: practice and experience, 20 (2017)
4. On approval of the Concept of e-government development in Ukraine: Order of the Cabinet of Ministers of Ukraine № 649-r dated 20.09.2017 URL: http: // zakon.rada.gov.ua.

5. D. Tapscott D. Electronic-digital society society: pros and cons of network intelligence. K.: ITN Press; M.: Refl-buk (1999)
6. M. Shvets, R. Kalyuzhny, V. Tsymbalyuk, V. Gavlovsky. Ukraine on the way to the information society. Legal informatics, 1 (2011)
7. E-government Council of Europe URL: http://www.coe.int/T/E/Com/Files/Themes/e-voting/definition.asp
8. e-Estonia – e-Governance in Practice. URL: http://ega.ee/wp-content/uploads/2016/06/e-Estonia-e-Governance-in-Practice.pdf (2016)
9. S. Nora, A. Minc. The Computerisation of Society. A Report to the President of France. Cambridge, L. (1980)
10. Pogrebnyak. E-government and e-governance: concepts and principles of functioning. Law and innovation, **№3**, 7 (2014)
11. Law of Ukraine "On Social Services" (Supreme Council of Ukraine (VVR), **№ 45**, st.358. http://zakon.rada.gov.ua/cgi-bin/laws/main.cgi ? nreg = 966-15 (2003)

Communication of the Public Authorities and Self-Governmental Organizations: Challenges of the Digital Society

Valentyn Orlov[1]*, *Victoriia* Poliakova[2], *Valentyna* Shylova[1], *Alina* Kuzmych[1], and *Olena* Leshchenko[3]

[1]State University of Economic and Technology, Kryvyi Rih 50086, Ukraine
[2]Executive Committee of Kryvyi Rih City Council, Kryvyi Rih 50050, Ukraine
[3]Kryvyi Rih foundation of the future, non-governmental organization, Kryvyi Rih 50000, Ukraine

Abstract. The article actualizes the problem of communicative interaction between public authorities and self-governing organizations as stakeholders of civil society. The understanding of the essence of self-government, typology of self-government organizations is generalized; the experience of separate legal systems on maintenance of proper communications of the power and self-government organizations in the conditions of development of digital society is studied. The research is aimed at solving the scientific problem of substantiating the conceptual foundations of building public communication systems of government and self-government organizations through the creation of technological platforms for constant constructive dialogue of self-government organizations with local authorities and local government. It is concluded that there are systemic obstacles to communication between the government and civil society. Overcoming such obstacles will ensure a qualitatively new, partnership nature of communication between government and society. The article for the first time raises the issue of creating platforms for communication between the government and self-governing organizations to ensure the participation of citizens in public administration decisions. The results of the study can be the basis for the formation of a new paradigm of the relationship between public authorities and self-governing organizations.

1 Introduction

The problem of communication between government and society is original and was solved in accordance with specific historical conditions of society based on two aspects - political and philosophical vision of the need for such communication and communicative capabilities of society (tools of communicative interaction), which technologically corresponded to the level of social relations, tangible and intangible culture, the nature of the political system, etc.

With the transition to a digital society, the technological conditions of the communication process have greatly improved; digital technologies have virtually removed the communication barriers that previously existed given the level of technological development of society. Examples of such progress are new definitions of the science of public administration such as e-government, smart state, state in the smartphone, which define the technological essence of the new stage of development of public administration.

However, as practice shows, technology alone does not solve the problem of communication barriers in the "power-society" chain. They significantly change the nature and methods of such communication, encourage the government to determine the challenges of the digital society, but do not change the main contradiction between the authoritarian nature of government and state monopoly in this area, and trends in self-government. These contradictions consist not only in the replacement of centralized structures with decentralized ones, but also in changes within the subject of public administration itself. We mean the involvement in the decision-making processes and governance mechanisms of other stakeholders of civil and, consequently, digital society, which should become not just consumers of management services, but direct participants in management processes.

1.1 Related Work

1.1.1 Self-government as the logic of social development

The logic of social transformations in the age of digital society is determined by the gradual movement towards civilized forms of democracy, which, in contrast to the era of the crisis of representative democracy, involve broad involvement of citizens in the decision-making process. Thus, representative democracy in the digital age is increasingly saturated with elements of direct democracy.

Although these trends are more relevant to developed representative democracies, there are successful examples of the expression of political will for such reforms in the post-Soviet countries. In

* **Corresponding author:** valentin.orlov.000@gmail.com

particular, in Ukraine, where in January 2021 the parliament adopted the Law of Ukraine "On Democracy". This law actually fills the legal vacuum that has arisen in the country as a result of the recognition by the Constitutional Court of Ukraine that the law "On the All-Ukrainian Referendum" adopted in 2012 contradicts the Constitution. This law not only restores the mechanism of democracy through a referendum, but also allows for the initiation of referendums on the repeal of laws or their individual provisions. From the point of view of our research, it is interesting that political parties and public organizations can initiate a referendum.

Another innovation in the legislation is that electronic procedures, including electronic voting, are also being introduced for the referendum. To this end, the bill, based on the study of foreign legislation and the experience of its application to electronic voting, primarily in European countries, provides for the creation of an automated information and analytical system.

It is also interesting that this is the first bill in a series of laws on democracy. Bills have already been prepared and submitted to the parliament on the mechanism of local referendums, popular legislative initiative, the possibility of recalling any official or deputy by early termination of their powers, the possibility of vetoing a law or any normative act of local self-government, implementation of the mechanism of consideration and decision-making on the basis of electronic petitions and a number of others.

The deepening combination of the norms of representative democracy with the norms of direct democracy implies an increase in the volume of communication contacts and the quality of public communication. Increasingly, citizens are taking the initiative in the communication process, and their intermediaries in this process are increasingly organized by civil society - self-governing organizations. Thus, we see the acquisition of broader subjectivity by self-governing organizations that are interested in the implementation of public initiatives and in the broad advocacy of the interests of citizens in the process of exercising power.

This process is long, and the agents of the influence of citizens on public power, peoples, as the only source of power in democratic forms of political regime, are various institutions of civil society, which are more organized and better prepared for such a mission.

All these stakeholders are united by one common feature that distinguishes them from others, including citizens - they are organized (primarily motivated, that is united by a common purpose and have the appropriate resources), and in addition, also self-governing (that is have an internal structure that differs significantly from the mechanism of the state and in essence tends more to local self-government).

Against the background of such trends, the research interest in such organizations is quite understandable, as their phenomenon and role in political and public administration processes, and even more so in the processes of communication with the authorities, is insufficiently studied.

Moreover, the process of state interaction with these stakeholders of change is important, because in modern conditions communication problems are a serious obstacle to the implementation of self-governing organizations in the legal field of public administration and their institutionalization as actors representing citizens in state-building processes.

The study of scientific works on the spread of direct democracy in the world as a whole and the communication of self-governing organizations with the government, in particular, allowed us to draw a number of interesting conclusions that formed the basis of this study.

First of all, we should turn to the basic things that serve as a goal, a strategy for social development, and which govern the communication between government and civil society as a tool to achieve the goal. All state power comes from the people. This is the basis of the Swedish management system. These are the constitutional definitions of the basic principles of direct democracy, which determine the nature of a democratic political regime. Thus, the Swedish Constitution stipulates that everyone has the same rights and is free to study how politicians and public institutions exercise their power. In the Constitution of Ukraine, this provision is implemented through the thesis that the only source of power in Ukraine is the Ukrainian people. He has full power, but delegate's part of his powers to other bodies or officials who, on his behalf and in his interests, exercise these powers in the manner and within the limits prescribed by constitutional norms.

Some constitutional and legal systems also enshrine the principle of decentralization of power, which, in our opinion, has a common vector with the introduction of direct democracy. Decentralization is determined by direct democracy and develops forms of government in which self-government is a key principle of local government.

It is worth noting the opinion of Tony Porter and Karsten Ronit, who consider self-government an important alternative to government regulation. At the same time, they rightly note that self-governing subjects are rarely completely autonomous from state power [1, p. 41-72].

Indeed, the current practice of public administration in the context of self-government and decentralization of power does not distinguish between self-government and the state definitively. They are interacting elements of a single mechanism of public administration, which participate in the processes of exercising power and a priori communicate with each other. The essence of these two elements of public power is that they received it as a result of legitimizing the political choice of the people (the population of a certain administrative-territorial unit).

At the same time, we should agree with the authors, who conclude that, the function of the state to activate self-governing mechanisms at different stages of the political and public administration process [1].

The state is interested in the development of self-governing structures, as they naturally ensure the functioning of public administration structures at the local level of public authority.

Against this background, in the digital society there is a process of updating the communications of government and society through its agents - self-governing organizations, which is manifested in the significant digitalization of the communicative process - from direct expression of will.

In her study, Kieron O'Hara characterizes digital reality as capable of influencing the political subsystem of society. According to her, digital technologies transform the material state of mankind and the benefits of such transformations may be political. Exploring the concept of digital modernity, in particular the possibilities of digital network technologies, she draws attention to the fact that the narratives of digital modernity shape reality through actors who make political, authoritative decisions [2].

From this point of view, public communications act as a mechanism for citizen participation in the management of public affairs. According to Tina Tomazic and Katja Udir Mišič, social media is one of the tools for activating the position of citizens in modern society. They draw attention to their role in communication between Parliament and citizens in terms of ensuring the self-government of local authorities. In their study, the authors present interesting results of a survey on the use of social media on the websites of the parliaments of the Member States of the European Union. 57% of such entities in 2016 already used Facebook, Twitter, YouTube, video streaming, calendar and RSS (Rich Site Summary) messages [3].

The technologization of the process of exercising political power and the participation of civil society and citizens directly in this process is also manifested in such a technological tool as electronic elections. In this regard, the position of Dr. Silvano Moeckli, who considers digital technology as a tool of political communication through the use of the Internet in election campaigns to embody the political will of citizens [4], deserves attention.

Thus, self-government is a key characteristic of local government structures in the digital society. The self-governing nature of these structures and their focus on interaction with civil society determine the process of their communication with citizens and the role of mediator in relations with the state. In these communications, a key role is played by the technologies of the digital society, which determine the methods of communicative interaction between government and society.

1.1.2 Our Contribution

This article presents the results of the analysis of the problem of communication between public authorities and self-governing organizations against the background of the development of the digital society. Based on the generalization of observations and based on their own experience of communication with local authorities, the authors critically considered the practice of building public communications in the subsystem "public authorities - self-governing organizations". Our contribution is to clarify the essential characteristics and classification of self-government organizations, identifies contradictions in communication between government and civil society, and clarifies the functions of self-government organizations in ensuring communication between government and citizens, government and society.

The article proposes a new solution to the problem of communication between the government and civil society agents in order to strengthen the opportunities for involving citizens in decision-making processes in the public sphere.

1.1.3 Paper Structure

The presented text of the article is organized in such a way that first the problem is posed in general, reveals the essence and characteristics of communication between government and self-governing organizations, the conditions in which processes take place, and then reveals the specifics of self-government in the public sphere as a phenomenon, and self-governing organizations and ways of forming models of public communication between the government and self-governing subjects of civil society.

2 Background

2.1 Specifics of self-government in the public sphere

In scientific works V. Averianov, G. Atamanchuk, M. Baitin, L. Grygorian, R. Dal', L. Kask, A. Kim, V. Lisnychyi, R. Maksakova, N. Nyzhnyk, M. Orzikh, V. Sirenko, M. Studienikina, A. Selivanov, A. Trachuk, M. Tepliuk, P. Usmanova, A. Yugov, the term "Public Power" is an element of scientific research and to this day has not received an unambiguous interpretation.
So R. Maksakova [5] defines the term "public authority" as the activity of local governments at the regional and municipal levels, in cooperation with the government and government at the national level, and emphasizes that the term is related to the activities of individuals and legal entities in the performance of their public duties and functions. V. Lisnychyi [6] reflects the term "public power" as a state organization of political life, carried out with the help of the state apparatus. Also, in his works, V. Lisnychyi notes that the state in the exercise of public power is based on material and moral impulses that affect the effectiveness of the exercise of power.

In Volume 6 of the Legal Encyclopedia [7] the term "public power" is identified with the term "democracy" and defines certain types of public power, they include: democracy, state power, local government. A similar opinion is shared by O. Skakun [8], who notes that public authority is the so-called "voice" of society and has open principles of its activities.

In his scientific works, A. Kim [9] clearly distinguished between the concepts of "public power", "state power" and "public power" and noted that public power is born of political relations. In turn, L. Grygorian [10] also distinguishes between the concepts of "public authority" and "public authority".

A. Yugov [11] considered the term "public authority" as a system of public and public participation in resolving matters concerning their territories and in resolving issues of public interest. The scholar emphasized that public power is a mass social power, which is described and supported by law.

Investigating the work of A. Selivanov [12], we conclude that he considered public power as a dual phenomenon: as state power and as the power of local government (municipal power).

According to M. Tepliuk [13] is directly a power exercised in the interests of subordinate bodies or organizations that represent public power and are public power.

Thus, the term "public authority" can be interpreted in terms of the mechanism of implementation, the nature of existence, the nature of functioning. The fact that there is currently no unequivocal opinion on the interpretation of the term "public authority" is also confirmed. Summarizing the views of scholars, we can determine that public power is a modern form of public power and is the main organizer of public, political and state interaction, which arose to address socially significant issues and governance processes.

Turning to the terminology of the concept of "self-governing organization", we note that the theme of the essence, formation, functioning, development of self-governing organizations in Ukraine in their scientific works studied by such Ukrainian scientists as O. Bielokurova, G. Zelen'ko, M. Lirchuk, A. Matviichuk, M. Obushnyi, M. Stavnychuk, O. Sungurova, V. Tsvyh and Yu. Shemshuchenko.

Defining the term "self-governing organizations" consider the opinion of the scientist M. Lirchuk. In his scientific works, he notes that self-governing organization is the highest form of civic non-political associations, which are characterized by the establishment of standards, norms and rules of its members - various social actors, and establish mechanisms of interaction with the state.

According to A. Matviichuk [14] self-governing organization is an element of civil society, which must have constructive cooperation with the state on the regulation of relevant activities and has a non-governmental nature.

In general, agreeing with the authors of the analyzed works, we pay attention to the specifics of their communication with the authorities, which follows from the self-governing nature of such entities.

Thus, it can be determined that self-governing organizations are institutions that are created to meet public interests and needs, to organize the public, to promote the development of one or more activities of society.

If we consider the content of communication processes, we should dwell on the approach of O. Sungurov [15], who described seven types of communication process between public authorities and self-governing organizations:

1. The model of the gardener provides acceptance of organizational and legal bases at the state level;
2. Partnership model provides for parity of public authorities and self-governing organizations;
3. The architect's model provides for the initiation and involvement of self-governing organizations in the creation of new structures of public authorities.
4. The paternalistic model provides for the autonomy of self-governing organizations.
5. The model of drive belts, characterizes the dominance of the interests of public authority over the interests of self-governing organizations.
6. Model of ignoring, which comes into force when partial or complete ignoring of self-governing organizations by public authorities.
7. The model of struggle against the enemy presupposes public disobedience and confrontational relations between self-governing organizations and public authorities.

An important theoretical and methodological value is also the concept of O. Belokurov [16], who pointed to three models of interaction of self-governing organizations with public authorities, to which he refers:

- Normative model (liberal model, full-fledged interaction);
- Legitimation model (participation of self-governing organizations in the development of territories);
- Instrumental model (active involvement of self-governing organizations in solving socially significant problems).

It should be noted that self-governing organizations do not have the power and are not a body of public authority, although they operate in the public sphere, because they are part of civil society. Therefore, the term "public authority" distinguishes elements of the public sphere that are institutionalized in government (eg., local government) from self-governing organizations that are not endowed with public authority: associations, associations of apartment building co-owners, public organizations and associations, political parties, trade unions and other independent institutions.

The development of civil society in Ukraine and the strengthening of the democratic system led to an increase in the level of conscious participation of citizens in state-building processes, which was preceded by a number of reasons: the state creates such social relations, where objects and subjects, seeks to increase its role and importance in the public sector. Thus, communication processes are an important aspect of a modern democratic society and enable:

- The presence of a certain stabilizing tool in society;
- Openness of the government to the public through self-governing organizations;
- Participation of self-governing organizations in the decision-making process.
- Participation of self-governing organizations in the development and discussion of draft regulations, as well as the initiation of such work;

- Control over the observance of norms and laws by the subjects of the relevant relations;
- Creating a development strategy for any area;
- Participation of self-governing organizations in the development and discussion of draft regulations, as well as the initiation of such work;
- Involvement in state-building processes.

Based on the above, we can distinguish four stages of communication between self-governing organizations and public authorities, namely:

1) Information process (access to information, mutual information);

2) Counseling (discussion);

3) Dialogue (two-way communication, cooperation);

4) Partnership (division of responsibilities between the two parties to the process).

Thus, an active dialogue between self-governing organizations and public authorities creates a positive environment for the improvement of existing processes and actions, as well as creates a positive environment for society, which leads to:

- Decision-making with a focus on socially significant needs;
- Increasing public confidence in public authorities;
- Reduction of the corruption component in public authorities;
- Improving the quality of public services;
- Successful development of any territory.

Thus, we are convinced that a significant part of research interest in this issue is focused on studying the mechanisms of functioning of self-governing organizations in the context of their involvement in the process of public authority. In this process, in a digital society, an important place is occupied by communicative interaction between participants in the process.

2.2 Communication of public authorities and self-governing organizations

Communication in the context of digital social transformations is one of the most important tools of public administration given the technologization of society and social connections within it. It is especially important for us to consider the essence of this process in the system of public power - self-governing organizations.

Analyzing the legal and managerial aspects of state policy in the field of cooperation between civil society and the executive, O. Turii determines that civil society as a self-governing institution plays a transformational and protective function in preventing the threats of authoritarianism. According to the author, volunteer, charitable and voluntary movements are the tools for implementing such functions, and the example of the Ukrainian state in recent years is another proof of this. The author emphasizes the need to create effective mechanisms for integrating the communicative efforts of government and society in solving common problems [17].

Indeed, the lack of communication entails an imbalance in the unity of the holistic formation of power as an organic phenomenon. It is organic, because the current system of power is a rather subtle matter, absolutely sensitive to challenges and threats, and therefore one that no longer acts as a mechanism, but as an organism. Such organicity makes the processes of information exchange extremely important. Because they connect the body and integrate its capabilities.

M. Tokar analyzes the potential of interaction between authorities and public organizations in the development of civil society and concludes on the need to integrate the experience, initiatives and capabilities of these actors, draws attention to the state-building importance of close cooperation between them [18].

Of course, such integration must be technological. That is to have a technological basis for its implementation. Such technologies are offered by the digital society. The emphasis is on electronic communications, which are becoming comprehensive, quite accessible and, most importantly, effective.

O. Goncharenko and L. Neskorodzhena give an example of electronic communication between the government and self-governing organizations in the field of culture; conclude that this is a consequence of a creative approach to public administration processes. According to them, public associations in the field of culture perform representative and protective functions, as they have the opportunity to create charitable organizations to finance cultural programs, participate in the development of regulations on cultural activities, promote a stable dialogue between the government and such a specific area. , which is the culture [19].

Agreeing with this statement, we can assume that the creative nature of public communication affects the communicative processes in other spheres of public life. This is satisfied by the experience of European countries, where integrated models of communicative interaction between government and society have been formed. This experience is extremely valuable for young democracies that have just embarked on the path of digital transformation of social organization.

S. Lizakowska studies the specifics and conditions of effective interaction between civil society and public authorities in EU member states. It establishes that cooperation based on common interests and values of citizens and social groups with public authorities is an effective model, tested by the experience of European democracy [20].

It should be emphasized that the thesis of common interests is a key, system-forming factor of such interaction. One can see from the recent events in Russia, related to mass protests against the current government, that one of the reasons for the protest mood is the lack of a common vision of the country's development prospects between the current government and civil society. In Russia, civil society, despite its low institutionalization of the government and the practical absence of the opposition, nevertheless actively influences the government. The latter does not want to communicate with the opposition other than through repression.

What does the experience of the totalitarian system of governing Russia show? First of all, that in the political sphere this state continues to remain in the past, without passing into the digital age with its social transformations. And it's not about technology, but about the fact that, unfortunately, they do not determine social and political change. It is this contradiction that unbalances the political system and slows down the development of a new type of social relations in Russia. And it is necessary to constantly improve communication in the system of power relations. This is what many researchers say.

For example, the scientific work of P. Shpyha and M. Khmelnytska is about improving and ensuring the effectiveness of the process of interaction between government and the community through the development of a modern information and communication system. The authors argue that the purpose of public administration is to organize the joint activities of people, their individual groups and social organizations, ensuring coordination and interaction between them [21].

Thus, for the purposes of our study, it is important to identify the contradictions that arise in the communication of self-governing organizations and government:

- Contradictions between the public nature of the goals of the state and the private interests of representatives of individual self-governing organizations, based on differences of opinion on the prospects for society;
- Contradictions between the powers of public administration bodies and the public nature and volunteer nature of the activities of self-governing entities;
- Contradiction between the interests of specific representatives of public political power and the views of the general population;
- Contradictions between the broad possibilities of the digital society in creating communication channels for interaction with civil society and the limited methods of government in the dissemination of information messages;
- Contradictions between the ubiquity of public power and the limited nature of authoritarian communication with society;
- The contradiction between the widespread development of information technology and the lack of willingness of the authorities to use their capabilities in communication with the public structure;
- Contradictions between the technological nature of the process of communication between the authorities and political ways of interaction with civil society.

These contradictions are in varying degrees inherent in the process of communication between government and self-governing organizations in any social system. The degree of contradiction depends largely on the political perfection of power and the development of democratic institutions.

2.3 Formation of communicative models of interaction: the experience of Ukraine

Taking into account the existing systemic contradictions in the communication between the government and civil society allows not only to see the causes of the problems, but also to correct the mistakes of the government in its relations with stakeholders. This is especially important at the level of individual local communities, where the most relevant decisions are made, where communication seems to have the least barriers, but no fewer problems.

Two processes are important backgrounds for Ukraine in the transition to digital society:

- Decentralization;
- Decommunization.

How do these processes affect the formation of a national culture of communicative interaction between government and society?

Decentralization provides local authorities with more material resources to address local issues. It is clear that some of these resources can be used to establish effective communication with stakeholders. Indeed, the communication influences of local authorities have recently become more organized due to the concentration of financial resources. However, has the new government effectively disposed of these resources? Unfortunately, there are in most cases not. The main direction of the funds allocated for communication is spent on the so-called coverage of the activities of local authorities, local governments. Moreover, the nature of such coverage is mostly personalized. In other words, the authorities do not communicate in order to get feedback from the population, but in order to maintain their positive image. From the point of view of strategic communications, this is ineffective, because the image formed as a result of one-way communication is unstable, and on the other hand, this process will increasingly require increasing investment.

In addition, the methods of involving citizens and their representatives - self-governing organizations - in dialogue with the authorities are also one-sided, in the form of public councils with the authorities, which perform an advisory function and are very limited in the discussion process, especially decision-making, system of local government.

And here is the time to mention decommunization. Since a significant part of regional and local political elites tend to the vestiges of the communist past, they are the bearer of the worldview formed in the past, it is quite logical that the methods of communication of the authorities tend to this past. And this is against the background of new information technologies and the development of modern information space.

Thus, the problem of the current state of power communications in Ukraine is the lack of a holistic strategy, future-oriented and mentally formed on the European models of communicative interaction.

Such a one-sided assessment of the media as a tool of political and managerial interaction hinders the development of elements of civil society at the local level. Such communication is sporadic, mediated by

 https://doi.org/10.1051/shsconf/202110003009

the interests of the authorities, completely devoid of elements of feedback.

P. Krister B. Anderssona, K. Chang, A. Molina-Garzónb in their study Voluntary leadership and the emergence of institutions for self-governance draw attention to such an important aspect as the selflessness of leadership, which is important in sparse communication. Their conclusions boil down to the fact that the effect of selfless leadership in self-governing organizations is much greater than in bureaucratically organized systems of power [22].

In our opinion, the communication activities of public administration are not identical to political communication, which is one-sided and manipulative. We see public communication, especially of local authorities, as a tool for building relationships with citizens that have long-term consequences, rather than situational political commitment based on a high level of trust in government. This model of public communication is focused on long-term transparent dialogue, publicity of discussions, and debatable nature of discussion of contradictions, which comprehensively implements the mechanism of citizen participation in local affairs management, independent, adapted to realities, public interest-oriented system of communicative interaction.

There are several options for solving the problem - from an information hub, an Internet portal that would become a communicative platform for dialogue between government and public self-government organizations to standardize the results of local referendums (polls) and search for direct democracy technologies in a digital society.

The search for forms of such institutions is characteristic of young democracies, where the paradigm of social communication is gradually shifting from one-sided socialization to more interactive. As a result, freedom of speech develops; the public can actively communicate with the government and vice versa. Since the government is almost incapable of communicating with all social groups, it makes sense to use the experience of implementing the concept of social institutions as a communication center between the government and its public, which in Surabaya is called Kelompok Informasi Masyarakat (KIM) [23].

3 Conclusion

The system of communication interaction in the public sphere is a complex, dynamic phenomenon that is transformed in accordance with the content and nature of social relations. In the digital society, such transformations change not only the forms of communicative interaction, but also significantly adjust the nature of the relationship between government and civil society.

The tendencies of self-government that accompany the development of social systems in the digital society actualize the problem of interaction of political power with self-governing organizations that act as stakeholders of power and mediate its relations with citizens.

Self-governing organizations are a fairly structured phenomenon, the diversity of manifestations of which gives grounds for the conclusion that they can be classified. Despite such diversity, self-governing organizations, as autonomous entities, are able to be the subjects of political interaction, to be involved in the processes of public administration and to oppose the authorities. Such involvement in political and managerial processes allows us to conclude that their communication with the authorities is conditional.

Research interest in the problems of communicative interaction between the state and self-governing organizations is justified by the prospects of developing decentralized forms of democratic political regime with the tools of direct democracy, which balance and contribute to the renewal of public governance.

The practice of relations between the government and self-governing organizations requires understanding and scientific substantiation. The article makes such an attempt and proposes a new partnership type of relationship between these entities, as well as possible forms of interaction between them in the practice of the new public government.

The practical orientation of this study is revealed on the one hand by the interaction of civil society actors with the government, and on the other hand - requires built-in interactive relationships between governments and public authorities on a qualitatively new basis and in new social conditions.

Prospects for further research in this area may be the search for and scientific justification of forms and methods of institutional interaction between self-governing organizations and government agencies, the study and generalization of the experience of their participation in public administration decision-making processes.

The world experience of building models of communicative interaction between government and the public sector can be adapted to the conditions of young democracies with their post-Soviet rudiments and specific legal systems. The choice of such models is a good practice of good governance as a philosophy of modern public administration.

References

1. T. Porter, K. Ronit, Self-Regulation as Policy Process: The Multiple and Criss-Crossing Stages of Private Rule-Making Policy Sciences 39(1), 32 (2006), doi:10.1007/s11077-006-9008-5
2. K. O'Hara, The contradictions of digital modernity. AI & SOCIETY 35, 197-208 (2020), doi:10.1007/s00146-018-0843-7
3. T. Tomazic, K. Udir Mišič, Parliament-citizen Communication in Terms of Local Self-government and Their use of Social Media in the European Union (2019), doi:17.4.1057-1079
4. S. Moeckli, So functioning directly Democracy Stuttgart, 2018

5. R. Maksakova, Constitutional and legal problems of organization and implementation of the constituent power in Ukraine, Zaporozhye, 2012
6. V. Lisnychyi, Conceptual foundations of political science: theory and history, method and didactic materials, Kyiv, 2001
7. Yu. S. Shemshuchenko (ed.), Legal encyclopedia: in 6 volumes (Kyiv, 2013)
8. O. Skakun Theory of state and law (encyclopedic course) (Kharkiv, 2005)
9. A. Kim, Gosudarstvennaya vlast i narodnoe predstavitelstvo v SSSR, Tomsk, 2015
10. L. Gregorian, Democracy in the USSR, Moscow, 2012
11. A. Yugov, Constitutional principles of public power: the Russian dimension, Moscow, 2015
12. A. Selivanov, Public power and the citizen in the conditions of application of judicial administrative jurisdiction (problems of theory and practice), Law of Ukraine, 8, 28 (2016)
13. M. Tepliuk, Public power in Ukraine: definition and constitutional and legal nature, Law of Ukraine, 5, 6-22 (2019)
14. A. Matviichuk, Self-governing organizations as elements of civil society and principles of their creation, Gileya: scientific bulletin, 83, 397-401 (2014)
15. A. Sungurov Models of interaction of state authorities and structures of civil society: Russian experience, 9 (2018)
16. N. Yargomskaia, E. Belokurova, M. Nozhenko, Why NGOs and authorities need each other. Models of interaction in the regions of the North, Public policy: issues of soft security in the Baltic region, St. Petersburg, 2014, ed. MB Gorny
17. O. Turia, State policy in the field of civil society cooperation, executive and local government: legal and managerial aspects. Public Administration Aspects, 6, 38-44 (2018). doi:10.15421/151837
18. M. Tokar, Potential of interoperability of the state, local self-government and public organizations in Ukraine: integration of experience, initiatives, opportunities, Efficiency of public administration, 64 (2020)
19. O. Goncharenko, L. Neskorodzhena, Self-regulation of culture: the role of public associations and electronic communication, Bulletin of the National Academy of Management of Culture and Arts, 4, 121-126 (2018)
20. S. Lizakowska, European standards of interaction between public authorities and civil society organizations, Public Administration and Regional Development, 8, 422-443 (2020)
21. P. Shpyha, M. Khmelnytska, Digital communications between government and community: problems, risks, barriers. Public Administration: Theory and Practice. 2, 13-20 (2019). doi: 10.36030-2311-6722-2019-2-13-20
22. P. Krister, B. Anderssona, K. Chang, A. Molina-Garzónb Voluntary leadership and the emergence of institutions for self-governance PNAS, Proceedings of the National Academy of Sciences of the United States of America, 44, 117 (2020)
23. Dinas Komunikasi dan Informatika Kota Surabaya, Buku Panduan KIM Kota Surabaya, 2017

Social Aspects of State Regulation of Agricultural Production

Lesia Barabash[1]*, *Mykhailo* Malovanyi[1], and *Alla* Osipova[1]

[1]*Uman National University of Horticulture, 20305 Uman, Ukraine*

Abstract. State regulation of agricultural production in Ukraine is aimed at increasing production and ensuring food security of the country and its population. However, the low level of social development, as well as social problems, which in rural areas among the population have now become particularly acute, reduce the potential of this area of the national economy. That is why the question arises of finding active incentives, tools and mechanisms that would encourage the rural population to act and create a promising basis for improving its social status. One of the main existing problems of state regulation of agricultural production in Ukraine is the provision of financial incentives to large agricultural producers. And although such a policy has allowed to increase export potential, but has led to the destruction of rural areas and the impoverishment of the rural population. The intensive activities of these producers on the basis of intensive technologies have led to significant imbalances in the development of the industry with the prevalence of crop production and the decline of livestock, reduced soil quality, as well as rising unemployment of the rural population. This has led to an increase in the shadow sector through illegal activities and hidden income.

1 Introduction

State regulation is one of the mechanisms of influence of the administrative apparatus on the development of the economic environment of the country. And depending on which industry or sphere of activity will become a priority of state regulation, such will be the vectors of development of the entire national economy.

Ukraine is a country with a strong agricultural potential. Therefore, it is quite obvious that agricultural production should be a priority for the interests of the state in general and state regulation in particular.

Currently, among the priorities of state regulation of Ukraine, implemented through reforms in the country, are sustainable long-term economic, social and environmental development, as well as innovation, institutional and comprehensive growth [1]. These goals, indeed, are general, because they do not contain relevant specifics and cover all possible aspects of state activity. However, we consider it necessary to focus exclusively on the social components of state regulation in a particular area – agricultural production.

During the long 30-year period of functioning of state regulation of agricultural production in Ukraine, social aspects were leveled, which provoked the decline of rural areas and the outflow of productive citizens. The lack of adequate social infrastructure, adequate social protection and development prospects only exacerbates the crisis.

1.1 Problem Statement

The issue of strengthening the social direction of state regulation in the field of agricultural production in order to strengthen its development is currently one of the most relevant, especially in the post-pandemic period, when the entire economy has suffered significant losses. That is why the object of the study is the peculiarities of state regulation of the development of agricultural production in Ukraine in the social context.

The subject of the study was the financial and economic relations that arise at the intersection of the interaction of the state and agricultural producers in order to improve the social situation of the rural population in particular and rural areas in general.

1.2 Related Work

The problem of state regulation of the development of agricultural production in order to improve the social development of rural areas is not unique to Ukraine. It faces all nations of the world and its main goal should be to eradicate poverty by achieving an adequate level of food supply in case of population growth.

«Achieving food security cannot be accomplished without ending poverty in developing nations», said Frances Bekele and Isaac Bekele, [2] while noting the innovative change in agricultural production technologies and approaches to state social policy to reduce rural poverty.

Randy Stringer & Prabhu Pingali also emphasizes the need to develop agricultural production in line with innovation [3]. The authors note that a balanced state policy to regulate agricultural activities can help attract additional investment in this sector of the national economy, which will encourage increased production, improve food security, reduce poverty and

*Corresponding author: Lesly-z@ukr.net

 https://doi.org/10.1051/shsconf/202110004001

unemployment, as well as optimize the social and environmental situation as in rural areas in particular and in the state as a whole.

Y. Kustepeli, Y. Gulcan, M. Yercan, et al. focus on the possibilities of activating the social component in rural areas and the cooperative component of agricultural production [4]. As a result of the surveys, the authors have found that agricultural cooperation can significantly affect the growth of the social component and income of the rural population.

I. Zastozhnikova emphasizes the need to intensify the social priorities of state regulation in the agricultural sector: «… the task of regulating agriculture… should be derived… from the objective provisions of development and focus on the criteria of economic and social efficiency» [5].

O. Khodakivska and O. Mohylnyi emphasize the significant negative consequences of the current state regulation in Ukraine with the support of agricultural holdings [6].

In general, it should be noted that modern financial and economic systems, in this period of pandemic crisis and expected post-pandemic economic downturns, are actively striving to strengthen the social component. And as soon as possible this trend should be implemented in agricultural production as a basic component of food security.

2 Research results

2.1 Current state of state regulation of agricultural production in Ukraine

Since 1990 and until now, state regulation of agricultural production in Ukraine has pursued the goal of creating a strong market for agricultural products. Among the areas of implementation of these goals – direct budget subsidies to reimburse agricultural producers for parts of the costs and to support the development of industries, as well as state target development programs for individual programs. The instruments used to achieve these goals include budget financing, price support, mechanisms for preferential lending and taxation, risk insurance and financial support for leasing procedures. However, the volume of financial support for agricultural producers was quite low, as in most post- Soviet countries (Fig. 1).

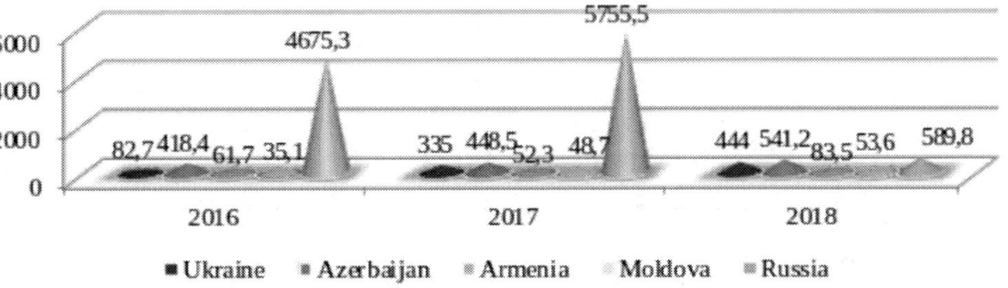

Fig. 1. Volumes of budget financing of agricultural production in some countries of the post-Soviet space [7]

At the same time, the methodological basis of state incentive processes is the legislative protection of property rights and information and legal services for agricultural producers, as well as the implementation of the coercion system under antitrust law, standardization and certification of products and more.

However, there is a significant drawback to all of the above: access to the proposed programs and subsidies is enjoyed mainly by large, in some cases medium-sized farmers, while small ones remain outside the incentive field. The actual consequences of these actions were the impoverishment of rural areas, reduced welfare of the rural population, as well as soil degradation and imbalances in the development of the agricultural sector.

In general, the goal of increasing sales of agricultural products abroad and increasing export potential was achieved (Fig. 2).

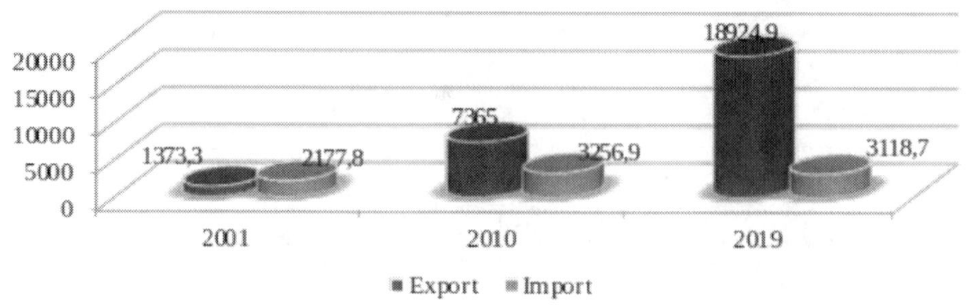

Fig. 2. Dynamics of volumes of export and import of agricultural products of Ukraine, million dollars USA [8]

Despite the general positive, the overall growth in agricultural exports was achieved through growth in crop production. After all, since 1991, the increase in sown areas of cereals and legumes amounted to 647 thousand hectares, sunflower - 4327 thousand hectares. At the same time, the sown areas of sugar beets (-1336 thousand ha) and potatoes (-224 thousand ha), vegetables (-25 thousand ha), fruit and berry crops (-617 thousand ha) decreased significantly.

Livestock sales remained low: during the same period the number of livestock and poultry also decreased: cattle – by 21.9 million heads, pigs – by 13.9 million heads, sheep and goats – by 7.7 million heads, poultry – by 43.4 thousand heads.

In addition to the fact that the export potential of the state is not realized in full force, negative changes are observed in the food security of the state, which provokes an imbalance in the diet of the population (Table 1).

Table 1. Consumption of food by households in Ukraine, kg on average per month per person [9]

Type of food	2001	2010	2018	Minimum consumption rate	Rational consumption rate
Meat and meat products	2,8	5,1	4,9	4,3	6,7
Milk and dairy products	17,3	19,2	19,1	28,4	31,7
Eggs, pcs	16,0	20,0	19,0	19,0	24,0
Fish and fish products	1,4	1,8	1,4	1,0	1,7
Sugar	3,3	3,0	2,7	2,7	3,2
Oil and other vegetable fats	2,0	1,8	1,5	0,7	1,1
Potato	11,1	7,7	6,3	8,0	10,3
Vegetables and melons	9,0	9,5	8,9	8,8	13,4
Fruits, berries, nuts and grapes	2,2	3,7	3,8	5,7	7,5
Bread and bakery products	10,7	9,3	8,3	7,8	8,4

Based on the previous data, we see that during the study period, the population of Ukraine feels the need for almost all of these foods. The volumes of consumption of production and processing of plant products are approaching rational norms: bread and bakery products and sugar, exceed the rational norms of consumption of oil and other vegetable fats. At the same time, there is a significant shortage of products of animal origin - meat and meat products, dairy products, eggs.

That is, we have a clear idea of the need to increase production of livestock products in particular and quality agricultural products in general.

One of the striking results of the state's disincentive influence on the social aspects of Ukraine's agricultural production is a reduction in the number of jobs and livestock, and thus an increase in unemployment in rural areas (Fig. 3).

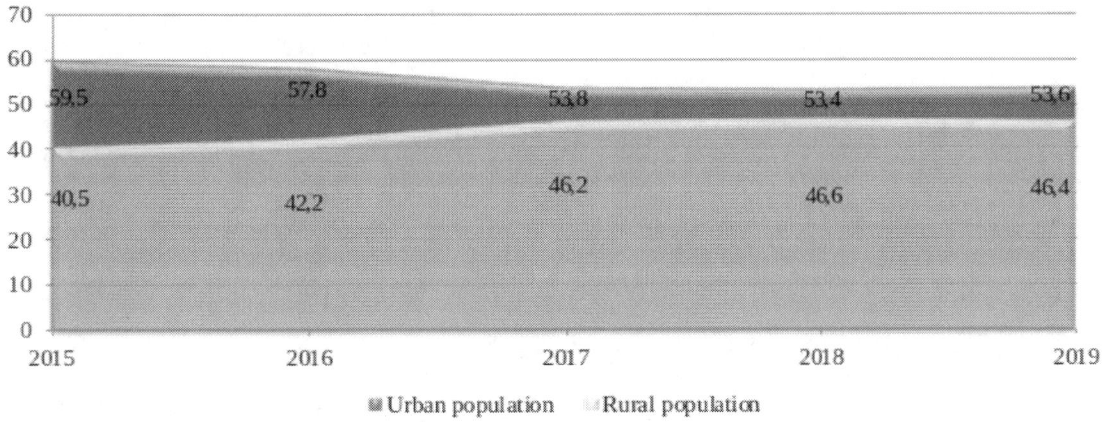

Fig. 3. The share of the unemployed population at the place of residence in the total number of unemployed in 2014–2018, % [10]

Also, in addition to all the above, during the entire period of measures to stimulate the development of agricultural production, funds were never provided to improve the social component of this sector of the economy. As a result there was an increase in hidden employment (about 39%), a high level of consumer price index for agricultural products, a decrease in investment activity in the agricultural sector and more.

2.2 Directions for strengthening social trends in agricultural production in Ukraine

Based on the current situation and existing foreign experience, we believe that the basis for the development of agricultural production in Ukraine in the near future, as well as the main vector of social policy in rural areas should be to stimulate small forms of agricultural production, which are now personal farms.

The reason for this statement is that rural households independently sell their products, and the demand for such products, which are organic, is constantly growing. As a result, the hidden employment and illegal income is appearing which is not covered in the statistics.

In general, over the past five years, there has been an increase in the number of small and micro enterprises in the agricultural sector (Fig. 4).

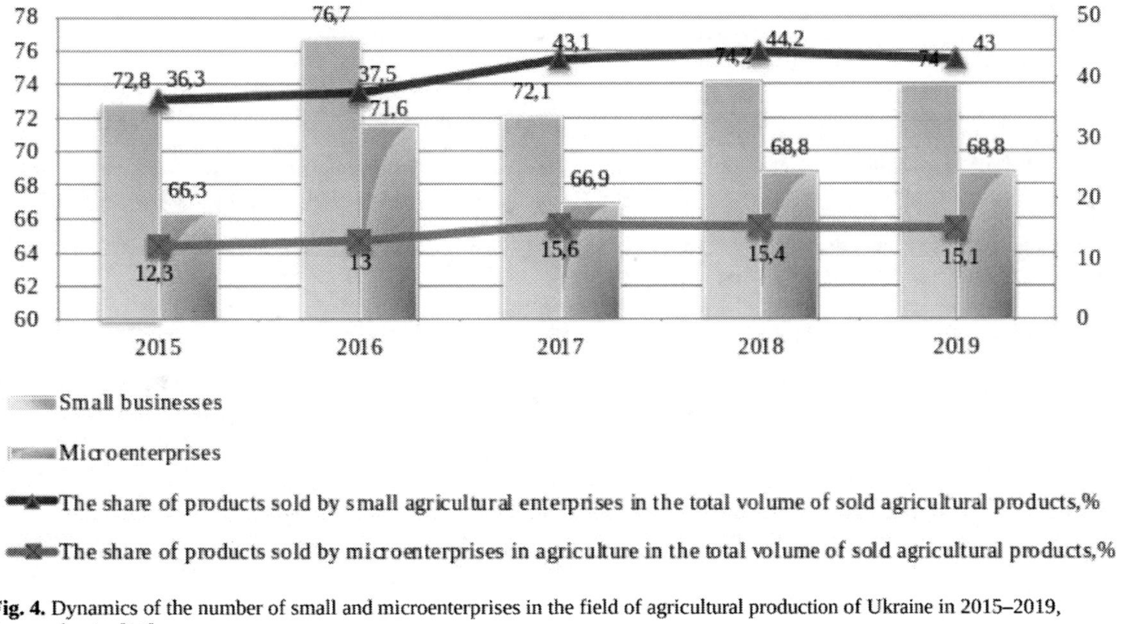

Fig. 4. Dynamics of the number of small and microenterprises in the field of agricultural production of Ukraine in 2015–2019, thousand units [11]

In accordance with the data of fig. 4, we can state a fairly high volume of agricultural products sold by small and microenterprises, in the total scale of agricultural products sold in the industry – 40.8% and 14.3% on average for the study period.

It should also be noted that the role of personal farms in agricultural production is very important (Fig. 5).

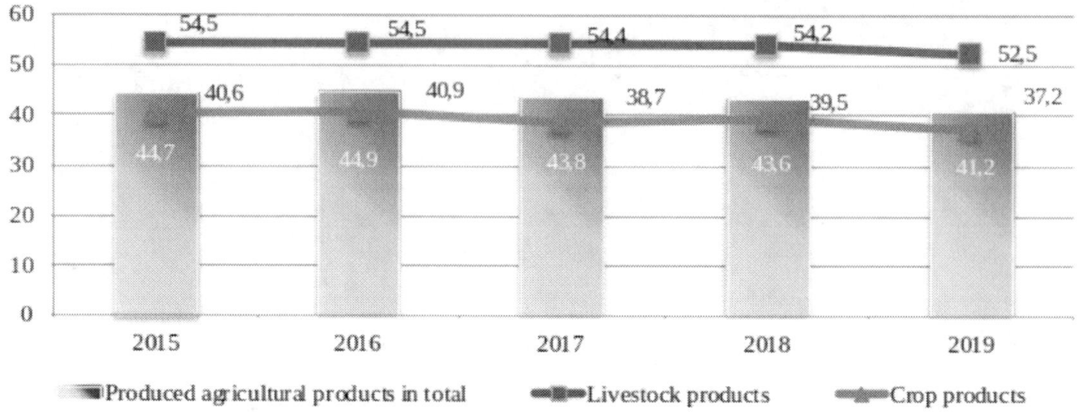

Fig. 5. Production of agricultural products by households (personal farms) (share in total agricultural production of Ukraine) in 2015–2019, % [8]

Shown in Fig. 5 data prove that personal farms are potentially promising participants in the process of effective stimulation of agricultural production in Ukraine at the present stage. In addition, their inclusion in the range of objects of state regulation will be the first step towards the formation of a social policy for the development of rural areas of a new type.

In our opinion, the necessary reform of social policy in rural areas in the framework of state regulation of agricultural production should be carried out by encouraging small farmers to unite in territorial production clusters in order to implement productive agricultural activities. At the same time, such tendencies will correspond to the conditions of the decentralization reform, which envisages granting greater independence, primarily financial, to territories and local self-government bodies.

Thus, personal peasant farms, within the administrative-territorial unit, should unite according to the profile of the activity and carry out production activities in the profile associations - territorial production clusters. In 2019, there were 735 operating agricultural service cooperatives in Ukraine; the number of dairy cooperatives was fixed at 230 units. There are separate cooperatives for tillage and harvesting, fruits and vegetables, grains, meat, and other services. However, their significant disadvantage is that they bring together farms from different regions, while increasing the cost of production at least at the expense of transport costs, as well as the lack of a closed production cycle. [12].

Given the peculiarities of agricultural activities, we believe that the territorial production cluster should unite the following participants: personal farms by activity (crop, livestock, harvesting, fish farming, etc.), processing plants, storage facilities, higher education institutions and research institutions.

Carrying out agricultural activities with the help of territorial production clusters will increase the production of environmentally friendly organic products, create stable markets, reduce logistics costs (and hence the price of finished products), introduce virtually waste-free production, and most importantly ensure basic social goals and increasing the level of employment in rural areas and the level of welfare of the rural population. Therefore, the model of the territorial production cluster, in our opinion, will look like this (Fig. 6).

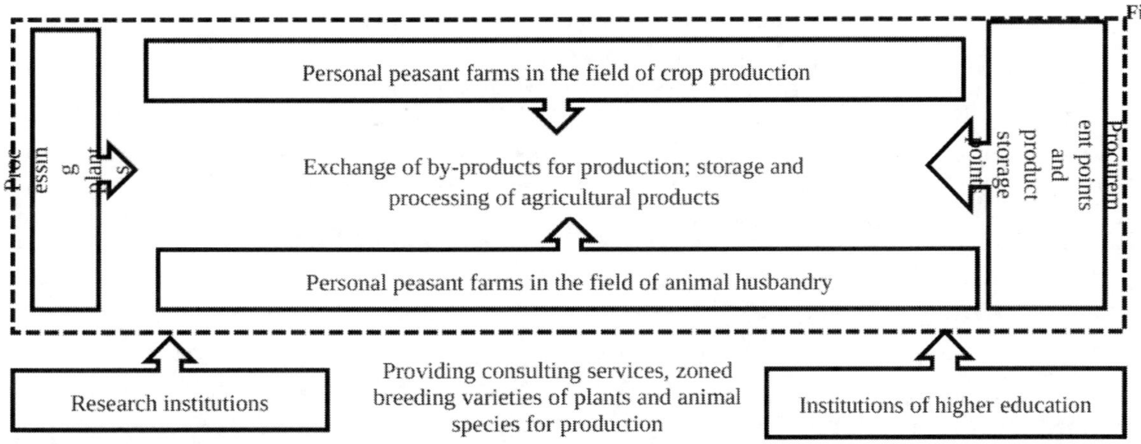

g. 6. Model of formation and functioning of territorial production cluster

In order to encourage peasant farms to join such an association, the state, for its part, will provide incentives in the form of cheaper loans, special insurance products and preferential taxation of certain businesses, especially those working in agricultural production, which are not popular in potential manufacturers through a long production cycle. Local authorities, in turn, must create a favorable social and logistical infrastructure, a positive tax microclimate (in terms of local tax benefits), to search for markets.

The peculiarity of the proposed model is that the production cluster includes research institutions and institutions of higher education, which provide small agricultural producers with raw materials for activities: biological assets, zoned breeding varieties of crops, animal species and more.

Microinsurance should become a promising element of stimulating small agricultural producers to production activities within territorial production clusters. Practiced in foreign countries exclusively for the category of low-income people, in this case it should involve small farmers in the financial market of the state. Therefore, to stimulate agricultural production, microinsurance should be carried out under the following conditions (Fig. 7).

Incentives for small farmers through microinsurance will help reduce the level of risks, both industrial and personal. To create a proper basis and form a field of social and financial protection, the state will act as a guarantor of insurance payments in the event of insurance events and will cover the costs of producers on insurance premiums for property and production insurance (natural risks, fires, theft, etc.). Manufacturers will conclude personal insurance contracts (life, health, accident) at their own expense. Insurance rates for this type of insurance for this group of policyholders will be regulated by the state in agreement with insurance companies.

3 Conclusions

The analysis of indicators of state regulation of agricultural production in Ukraine showed an underestimation of the potential of small and micro production in this area. In particular, at the end of the study period there was an increase in the number of micro-enterprises in rural areas - by 2.5 thousand units, as well as the share of agricultural products sold by them of their own production in total agricultural production - by 2.8%. At the same time, a negative indicator is the 5.9% increase in the number of unemployed in rural areas, as well as the low level of consumption by the population of Ukraine of necessary products produced in agriculture.

Based on the existing prerogatives and problems, as well as in light of the growing urgency of social issues in state regulation of agricultural production, we consider it appropriate to intensify financial and organizational components as necessary components for the formation of a proper social microclimate in rural areas. In particular, the creation of regional clusters will encourage small and micro-producers to be active, and micro-insurance will provide an opportunity to minimize the potential risks of production activities.

Fi

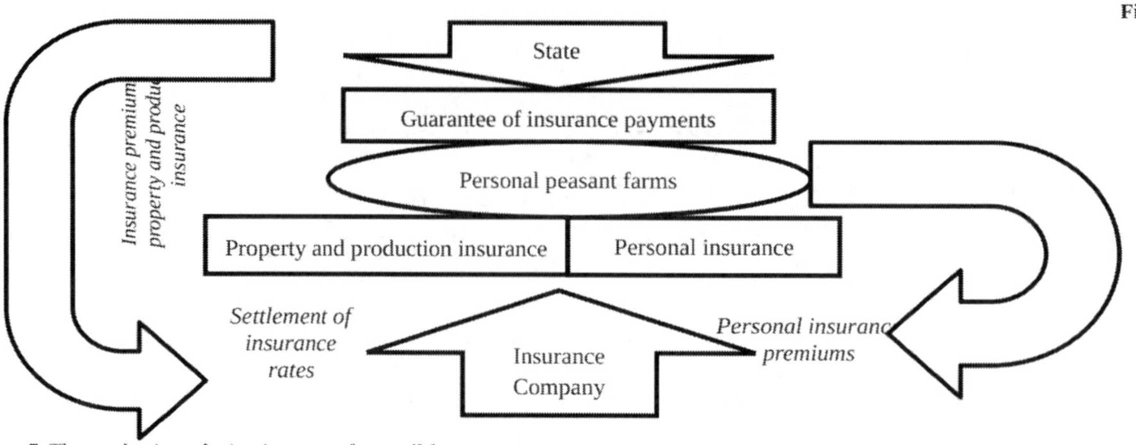

g. 7. The mechanism of microinsurance for small farmers

Implementation of state regulation in order to stimulate the development of agricultural production, taking into account social aspects through these specified mechanisms will contribute to the formation of an active social basis for further transformation:

- reducing the unemployment rate of the rural population;
- formation of a network of self-sufficient agricultural formations – territorial production clusters capable of functioning in the full cycle of agricultural production;
- improving the level of product quality;
- stimulating the financial, ecological and energy culture of agricultural production;
- balanced development of agricultural production;
- increasing the level of welfare of the rural population;
- formation of social infrastructure of rural areas.

The proposed directions of development of small agricultural producers take into account the peculiarities of the modern economy of Ukraine and are aimed at reaching a compromise between the expectations of such producers and the expectations inherent in the state. Due to the realization of the existing potential of the small and microsector of agricultural production, the real vectors of social state regulation will be realized in practice and the basis of priority of the social principle of state action will be formed.

References

1. V.V. Tomareva-Patlakhova Investments: practice and experience. **6**, 38–43 (2016).
2. F. Bekele, I. Bekele (2017) https://www.researchgate.net/publication/309417376_Social_and_Environmental_Impacts_on_Agricultural_Development Accessed 5 Jan 2021
3. R Stringer & Prabhu Pingali **1(1),** 1-5 (2004)
4. Y. Kustepeli, Y.Gulcan , M. Yercan, et al. Ann Reg Sci (2020) https://doi.org/10.1007/s00168-019-00965-4 Accessed 15 Jan 2021
5. I.V. Zastozhnikova Scientific intelligence in state and municipal administration **1**, 121–135 (2013)
6. O. Mohylnyi, O. Khodakivska Viešoji politika ir administravimas public policy and administration **17** (4), 526–538 (2018)
7. Review of agri-food trade policy in the post-Soviet countries 2017-2018. FAO. Rome. 2020. http://www.fao.org/documents/card/en/c/ca7674ru Accessed 22 Dec 2020
8. Statistics Service of Ukraine. http://www.ukrstat.gov.ua/operativ/operativ2006/sg/sg_rik/sg_u/rosl_1991-2019_u.html Accessed 10 Jan 2021
9. State Statistics Service of Ukraine. Consumption of food in households. http://www.ukrstat.gov.ua/ Accessed 12 Jan 2021
10. Statistics Service of Ukraine. http://www.ukrstat.gov.ua/operativ/menu/menu_u/rp.ht Accessed 11 Jan 2021
11. State Statistics Service of Ukraine. The volume of output (goods, services) of enterprises by type of economic activity. http://www.ukrstat.gov.ua/ Accessed 14 Jan 2021
12. Agravery May 16 (2019). https://agravery.com/uk/posts/show/kilkist-fermerskih-gospodarstv-zrosla-do-46-tisac Accessed 4 Jan 2021

Implementation of Electronic Services in The System of Population Social Protection

Anna Zastelo[1], *Tetiana* Serha[1], *Vladislav* Koval[1], *Yana* Vikhliaieva[1], and *Nataliya* Shuberty[1]

[1]National University «Zaporizhzhya Polytechnic», 69063 Zaporizhzhya, Ukraine

Abstract. The study assesses the status and opportunities for the introduction of electronic social services in Ukraine on the example of Zaporizhzhya region. The peculiarities of providing and receiving e-social services based on the analysis of the main models of e-government and individual countries within the selected models: Anglo-American, Continental European and East Asian. The ambivalence in the perception of the population's readiness to receive electronic social services among experts and clients has been determined. The results of the comparative analysis revealed some statistically significant differences in determining the reasons for the slowdown in the process of implementing the provision of electronic social services. The main factors that will contribute to the development of electronic social services are highlighted.

1 Introduction

Today, Ukraine follows the practice of foreign countries, not only in political processes but also in governing the country. The development of Ukraine is also envisaged in the creation of e-government, in the development of the provision of electronic social services.

The purpose of the study is to assess the state and possibilities of introducing electronic social services in Ukraine (on the example of the Zaporizhzhia region).

The research objectives:

1) to determine the features of the experience of providing (receiving) electronic social services;

2) to determine the factors for optimizing the development of social services in the system of electronic services.

To study the features of the state and opportunities for the introduction of electronic social services in the Zaporizhzhia region, the following was carried out: 1) an expert survey among managers and employees of social services and social service institutions (n = 100, August 2020) 2) a mass sociological survey among clients of social services and social service institutions service (n = 300, Aug 2020). Characteristics of the sample of a mass sociological survey among social services clients and social service institutions (n = 300): 1) by gender: men - 30.3%; women - 69.7%; 2) by age: 15-30 years old - 24.6%, 31-50 years old - 61.5%, 51 years and older - 13.9%; 3) by type of settlement. Zaporizhzhia - 40.2%, regional center, countryside - 59.8%. Characteristics of an expert survey sample among managers and employees of social services and social service institutions (n = 100): 1) professional work experience: up to 3 years - 23.4%; from 3 to 10 years old - 46.8%; from 10 years old - 29.8%; 2) by type of settlement, Zaporizhzhia - 57.4%, regional center, rural area - 42.6%.

2 Results

The development of the information society in Ukraine and the introduction of ICT in all spheres of public life create fundamentally new opportunities for social work. In the information society, work in the social sphere requires knowledge and skills of effective search, accumulation, processing, storage, presentation, transmission of data using computers and Internet networks. In the context of the introduction of e-government in Ukraine and the process of public services` digitalization, the problem of technologies incorporation for the provision of electronic social services into the social work sphere is actualized.

During the restructuring of e-government in Ukraine, it is very important to study and borrow the best experience of foreign countries, which will help identify promising ways for the further development of e-government in Ukraine, and in particular the introduction of e-services provision to the population. Let us analyze the features of the main models of e-government and individual countries within the framework of the selected models: Anglo-American, Continental European and East Asian.

Anglo-American e-government model.

The Anglo-American model is typical for countries such as the United States, Canada, Britain and Australia. Typical features of the Anglo-American model arise from the peculiarities of the territorial-political structure and traditions of statehood: a high level of independence of territorial administrations (states) in conjunction with a strong central government. The model is designed to create electronic

* Corresponding author: zastelo@ukr.net

government for countries with a large territory and fairly uniform Internet access.

The Anglo-American model, on the one hand, allows for a significant difference in the processes of local e-government, and on the other hand, it clearly standardizes the rules for the exchange of information "citizen-government" at the national level. The system of local government in the states is based on the administrative-territorial division, therefore, the schemes of self-government in different states are different [1].

An important component of the success of the Anglo-American e-government model implementation has become significant reforms of the entire structure of the state's work with information flows and the introduction of responsibility system for the government agencies informatization.

Electronic government in the United States began with the creation of the President of the United States website in 1993. The main purpose of which was to inform the country population about the activities of the Presidential Administration via the Internet. However, already in early 1994, a working group on e-government was established, which included more than 80 representatives from 46 government agencies and bureaus. This group has developed an e-government strategy document aimed at modernizing government performance. This group concluded that the federal government can significantly improve the efficiency of serving its users in the coming period of up to 24 months, focusing on 23 high-priority, government-wide initiatives. [2] These initiatives ensured the delivery of services to citizens in terms of minutes or hours, instead of days or weeks.

The main US government portal [3] was launched in 2000. It brings together many sites at the level of integration links, contains complete information about government resources, services and forms for citizens, business and government. Online services featured on the US government web portal are provided by redirecting to the appropriate sites. To solve the problems of integrating interdepartmental projects, the US government relies on the concept of federal corporate building. The main goal of the federal information technology architecture is to provide conditions for the interdepartmental development of processes, standards and information exchange [4]. Later the e-government law was adopted in the United States, (*the E-Government Act of 2002*), which was aimed at expanding access to information about the government, as well as providing information and services to citizens, employees, other agencies and legal entities [3].

Public discussions can serve as a striking example of wide public involvement in the e-government system. In the United States, the obligation to conduct discussions of draft regulatory acts is enshrined at the legislative level in the Law on Administrative Procedure [5], which, however, leaves it to the discretion of government institutions to determine the forms of communication with the public (usually a certain time is given to send comments on draft decisions).

A second potentially useful means of enhancing citizen participation in influencing policy-making is through electronic petitions (e-petitions). As a rule, the Internet is better suited for using this tool, thanks to which the public and interest groups can initiate a collective appeal, gather a group of its supporters who agree to support it with their signatures and send it to the appropriate government agency or to the government.

In a direct democracy, citizens' participation in the decision-making process is usually carried out by voting. E-voting is one of the techniques for improving and strengthening (quantitatively and qualitatively) public participation in the process of policy formation and decision-making at all levels of government: local, regional, national. This e-democracy tool aims to enhance the democratic and legitimate governance through the use of ICTs to directly determine the will of the voters.

Thus, the e-government model is determined by the structure and traditions of public administration in the United States. The path of the information society formation is determined by the general model of socio-economic development, in which the functions of the state are reduced to a minimum, and the activities of individuals - to a maximum, therefore, the main priorities in the provision of electronic services are; firstly, the introduction of e-democracy tools (discussion, petitions, e-voting), and secondly, the interaction of the state with business and public associations, the sphere of social security and social services is covered to a lesser extent.

The UK e-government initiative is a fundamental element of the government's modernization program. E-government sets the strategic direction for public sector transformation by introducing business models that take advantage of the latest technology. Four principles of e-government in Great Britain: 1) compliance of public services with the needs of citizens; 2) ensuring greater accessibility of the state and its services; 3) the principle of social equality in access to public services; effective use of information [6].

The Continental European e-government model.

The Continental European e-government model operates in the countries of the European Union (EU). The Continental European e-government model was developed taking into account the fact that the European Union includes 28 countries with different languages, cultural and political traditions. At the level of an individual EU member state, the elements of e-government can be very different both in terms of the technical platform use and in terms of the interaction organization between the state and the citizen (e-voting, services to the government, payment of taxes and fines, etc.). At the same time, between the members of the European Union, for the development of the e-government system, the minimum necessary legal framework was agreed. It is expressed in the form of directives of the European Commission, all member states have implemented into national legislation.

At the level of the supranational structures of the European Union (the European Parliament, the

European Commission, the European Court of Justice), the continental European model of e-government is due to the clear unification of interaction between supranational structures and intrastate structures of the member states of the Union. Also, according to this model, a citizen of any European Union state can, through the electronic government system, apply directly to the pan-European authorities in real time.

All EU countries have adopted laws on access to information. The requirements of the European Union for legislation governing e-government systems cover both technical aspects (digital signature, e-commerce) aimed at simplifying the creation of e-government systems, and ethical and political issues (distribution of government orders, access to information), define new responsibilities "Responsibilities state institutions for the protection of citizens' rights and organizations in the new context of the provision of public services by electronic means.

The continental European model of e-government is characterized by: the presence of supranational institutions, the recommendations of which are binding on all EU countries; a high degree of European countries integration is manifested in a single pan-European information space; strict regulatory legislation in the information sphere. The management and activities of national governments and supranational structures in this model are due to the use of high technologies with a focus on the needs of citizens - users of information networks and systems. When it comes to building an e-state, the EU focuses on achieving social cohesion between member nations and the European community as a whole. The Information Society for Europeans is a means of preserving the diversity of Europe. For Europe, investments in information society technologies are, first of all, investments in people.

The path chosen by the countries of Europe differs from the American one. In 2000. the leaders of the European Union adopted the Lisbon Strategy [7] in order to make the EU a dynamic and competitive knowledge-based economy in the world. Information and communication technology has been a key component in achieving these goals. A special program *"e-Europe"* [8] was developed to realize the goals of the Lisbon ICT Strategy.

In the program "e-Europe" 2000-2002. there were three main goals: a cheaper, faster and safer Internet, investing in people's skills, and increasing Internet usage. Later, these goals were changed in the following EU programs for the implementation of e-government [9, 8] to provide public services online, a dynamic e-commerce environment, ubiquitous broadband access and secure information infrastructure [10]. Essential documents in this area are also *"The Role of E-Government for the Future of Europe"* [11] and "E-Government in Europe: the state of affairs" (*"E-Government in Europe: The State of Affairs "*[12]), presented at the European conference in the city of Cernobbio (Italy).

The European Union has introduced a system of continuous monitoring of the characteristics of the following electronic services to citizens and businesses: 1) services to citizens: provision of public information and responses to inquiries; assistance in declaring taxes; job search through the employment service; provision of social assistance, payment of unemployment benefits, per child, reimbursement of medical expenses, tuition fees; issuance of personal documents (passport, driving license, car registration, filing applications for construction, informing the police (theft, etc.), providing access to public libraries (availability of catalogs, search tools), issuing birth and marriage certificates, accepting applications to higher education institutions, informing about a change in residence, services related to medicine (interactive consultations, requests for treatment in a specific place, the availability of services in a hospital), etc. 2) services to business are carried out in such areas as social contributions from wages workers' wages; corporate taxes and declarations; value added tax; registration of a new legal entity; reporting statistics; Customs declaration; obtaining permits related to environmental protection; government procurement and the like.

As part of the *Europe2020* strategy, one of its seven *flagship initiatives* was adopted - *the Digital Agenda for Europe* [13], which is an action plan aimed at achieving a dual goals: ensuring the global competitiveness of the European Union through the development and implementation of ICT, as well as the spread of digital technologies in all spheres and layers of the European society.

Today, the "Digital Agenda for Europe" is the main current project document for the development of the information sphere of the European Union, which identifies 101 events within the *"seven pillars"* around which it is planned to reboot the economy of the European Union in order to get the most out of digital technologies for citizens and businesses. These key areas are [13]: creation of a single digital market; achieving interoperability (interoperability) of software and content, optimization of relevant standards; increasing user confidence and safety; development of high-speed internet; development of scientific research and innovation; spreading digital literacy and IT skills; IT for solving social problems.

Consider the introduction and development of e-government using the example of France.

The formation of the information society in France is characterized by a number of profound transformations. Among the main stages of France e-government introduction, one can single out the stage of strategic planning and the stage of direct implementation. During the first stage, which began in January 1998, a general strategy for the formation of an "information society" in the country and the construction of "electronic government" structures was agreed and approved by all government bodies. The goal of this program was to accelerate the formation of the information society and make it accessible to everyone. The program took into account: the real possibilities that France has for building a nationwide and publicly accessible system of information services; positive and negative experience of the world, accumulated in the process of creating such systems;

basic requirements and parameters of the plan "Electronic Europe 2002".

The stages of the *PAGSI* plan (government action plan for the creation of an electronic society) included: opening by the end of 2000 public access to government services and documents; organization of high-speed Internet access; support for training, research; accelerating the development of information and communication technologies (ICT); each ministry was given the task of developing projects of websites containing information about their activities and the services they provide.

The priority tasks of the program included the development of the following key directions of the country informatization: 1) in the field of education: creation of a specialized educational network in the national Internet zone; 2) in the field of promoting culture and art in France: electronic copying of works of art and other elements of cultural heritage, the creation of virtual museums and libraries; 3) in the activities of public administration bodies: information technology (IT) modernization of public services in order to provide enterprises and citizens with access to their information resources and provide an opportunity to interact with them using the Internet; 4) in conducting scientific research and introducing innovations: supporting the development of new technologies and industrial innovations in order to develop national science, industry and the economy in general; - in matters of jurisprudence and law: solving the problems of legislative regulation of the activities of the state, business and citizens on the Internet; 6) in the e-commerce sector: promoting the development of legitimate commercial transactions on the Internet [14].

The second stage of e-government development in France was aimed at the timely and almost complete implementation by 2002 of all the main provisions of the *PAGSI* program, which allowed the new French government, headed by Prime Minister Jean-Pierre Raffarin, to announce the end of the first building stage an information society in the country and the transition to the next, second stage of e-government, which included: 1) creation, on the basis of the existing program, a system of "complete electronic" state administration, advanced by world standards, due to the introduction of a highly efficient mechanism of the State-Consumer (*administration-to-consumer*), which will ensure the possibility of providing citizens and enterprises with all public services via the Internet; 2) the transition of the entire state administrative apparatus to the use of absolutely paperless technologies and a fully computerized mode of operation, due to which budget savings, according to the calculations of the Ministry of Finance specialists, amount to at least 7 billion euros per year [14].

In 2003, the French Ministry of e-Governance and Administrative Reforms developed and issued a strategic plan ADELE (*ADministration ELEctronique*). It includes strategic and tactical action plans that should be reviewed annually and updated accordingly. The *ADELE* strategic plan has a detailed structure for the development of the e-government of France. This concept is based on the following principles: government should focus on citizens, not bureaucracy; effectiveness; innovativeness; transparency and control by citizens; confidentiality; compatibility of services and information systems [15]. The *ADELE* main task is to implement e-administration through achieving a high level of accessibility and ensuring interactivity, allowing users to perform all administrative procedures and services online.

Let consider the e-services provided to businesses and citizens in France as a result of France e-government introduction. The assessment methodology and the level of online service provision are determined by the EU Council, according to which the following information is provided. The main public services for citizens are: income taxes (makes it possible to submit a declaration, report, pay taxes) job search by the Employment Agency (makes it possible to register at the labor exchange); social insurance (makes it possible to receive social assistance) personal documents (makes it possible to issue a passport or driver's license); car registration (makes it possible to register your car); building permit; declaration to the police (makes it possible to file a statement with the police) public libraries (makes it possible to find a book in the online library) certificates of birth, marriage (makes it possible to order a marriage certificate or the birth of a child); admission to the University; change of residence address; health-related services (makes it possible to make an appointment with a doctor), etc. [14].

France's e-government strategic program is built on the following fundamental principles: 1) continuous assessment of user needs. Creation of advisory centers throughout the country with the aim of increasing popularity, establishing a dialogue, presenting a new program (counseling is carried out by an experienced group of people); 2) ensuring the availability of e-services for every citizen. Use of service terminals especially in rural areas, simplification of administrative procedures; 3) the establishment of trust between the state and citizens. Confidentiality and control of personal data by users 4) provision of an effective and full range of administrative and social services, keeping government spending under control, increasing the provision of new services without increasing costs.

The East Asian e-government model.

The East Asian e-government model is typical for countries such as: Singapore, Japan, South Korea, Taiwan. The Eastern model is based on a specific management style, an Asian type of corporate culture and a multilayered system of public administration, organized according to the principle of a hierarchical pyramid. Its representatives seek to introduce an alternative Western approach, which is based primarily on the assertion of their own value orientations in industrialization, informatization and social development. It is based on cooperation between the state and the market, an attempt to establish a connection between the cultural values inherent in the supporters of Confucianism, accompanied by social changes. The model is good for monocultural states, even large in terms of territory, with a high population

 https://doi.org/10.1051/shsconf/202110004002

density and generally accessible high-speed Internet access.

Within the Eastern model, Japan, "Asian tigers" and China are distinguished. The concept of "Japanese economic miracle" is textbook. The country's successes in the development of the information society as a whole coincide with the successes of the United States in this area. One of the most important factors in their achievement is significant expenditures on research and development, the high priority of ICT in solving the problems of the country's socio-economic development. Overall, Japan's e-government program pursued two goals: to provide user-centered management services and to implement public administration. Real progress has been made in empowering individuals and businesses to apply and register for virtually all national administrative procedures. So, since 2005, 96% of national administrative procedures are available online, for example, such as tax registration and request to the social insurance system [16].

The information development of the "Asian tigers" (South Korea, Taiwan, Singapore and Hong Kong) is based on the so-called model of economic cooperation between the state and the market - public-private partnership. The success of these countries is based, in particular, on state intervention in decision-making in the field of large private investments and on the active participation of the state in the national information infrastructure creation. The problems of information development, which the governments of these countries are focusing on, are the constantly growing competition in the production and implementation of the latest ICTs, the associated potential loss of a certain market segment or jobs, as well as the problem of ensuring equal access to information resources.

In the concept of electronic government in South Korea, the Government for Citizens (G4C) program is being implemented, which is aimed at simplifying the procedures for interaction between government agencies and citizens of the country and legal entities. The program is focused on minimizing direct contacts between officials and citizens of the country and business entities. Currently, the government offers a significant number of online integration services as a one-stop-shop.

So, for more than two decades since the practical implementation of e-government technologies in various countries, significant experience has been accumulated, as a result of which resources and time have been saved and a high-quality assessment has been provided for the results of a comprehensive assessment of the e-government development. "United Nations E-Government Survey 2018 In Support Of Sustainable Development" from the United Nations, - in 2018 Ukraine took 82nd place among 193 countries that participated in the assessment [17]. Compared to the 2016 rating "United Nations E-Government Survey 2016 E-Government For The Future We Want" (held every two years) - Ukraine has decreased in the 2018 rating by 20 positions, which indicates certain systemic problems in many areas of the implementation of electronic management in Ukraine (Fig. 1).

The use of successful practices of foreign countries is becoming a necessary factor in achieving the goals of e-government, preventing possible problems of informatization of public authorities and the development of the information society.

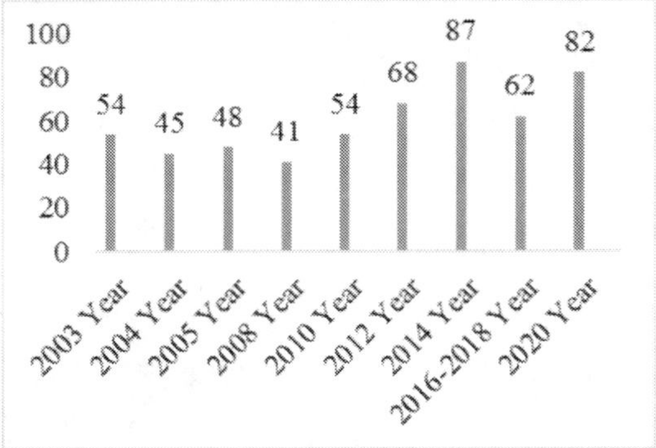

Fig. 1. Dynamics of Ukraine's readiness for e-government according to various UN ratings *"E-Government Survey"* for 2003-2020

The development of e-governance in Ukraine is primarily associated with the creation of a regulatory framework for the development of the information society in the context of European integration processes. Ukrainian researchers identify several stages in the implementation of e-government in Ukraine.

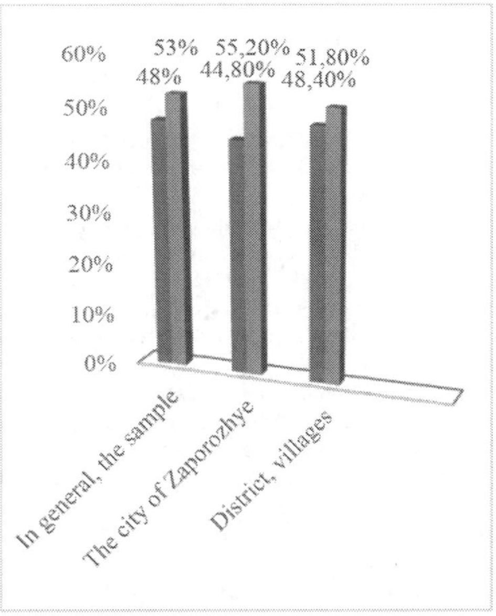

Fig. 2. The experience of receiving electronic social services among clients

Experience in providing (receiving) electronic social services. The rapid development of the modern information use and communication technologies and the Internet already allows the use of various forms of providing and receiving electronic social services, is spreading due to the state policy of public services digitalization ("the state in a smartphone"), among which the electronic registration of social services is in special demand. benefits, allowances, pensions, etc., therefore, the indicators of the study indicate that almost half of the respondents have already had such an experience (Fig. 2-3).

59% of the surveyed social workers noted that they already have experience in providing electronic social services, and there were more of them in the regional centers of the Zaporizhzhia region than in the regional center (67% and 43%, respectively), but these indicators are determined by those forms of electronic the tools that they use in their professional activities (Fig. 3).

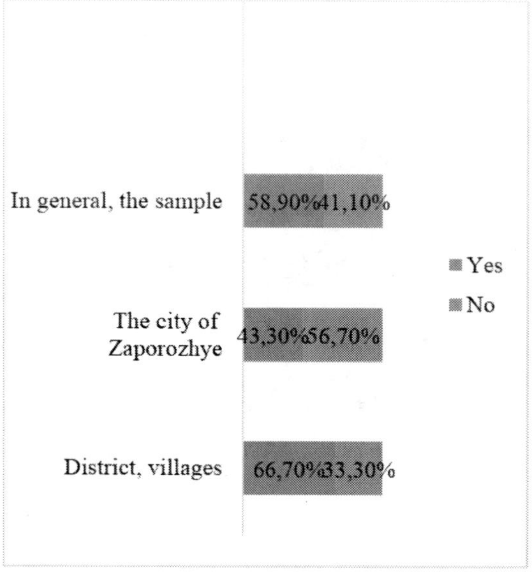

Fig. 3. The experience in providing electronic social services among experts

Among the electronic tools that social workers use in their professional activities, the most common are: Google services (60%), electronic inquiries and inquiries (45%), the website of a social service institution (42%) and social networks (40%). Tools that directly enable open communication in the process of providing social services (communication online services) are used by only 30% of social lawyers (more of them in Zaporizhzhia city).

So, it can be noted that at this stage of development of the provision of electronic social services, the tools that are used by ordinary Internet users (Google services and social networks) and the opportunities provided by the format of the site of the social service institution, among which electronic requests and requests dominate to receive social services (which is more common in district centers of social services and social services).

If the identification of the experience level in receiving and providing electronic services determines the real state of modern information application and communication technologies in social work and social services (in this case, at the regional level), then the indicator that determines the possibility of their use is the assessment of the readiness of both suppliers and recipients of development social service systems using Internet technologies and remote communication. In this aspect, an expressive indicator is the assessment of the population readiness to receive electronic social services from the point of view of both the recipients themselves and from the position of social services providers.

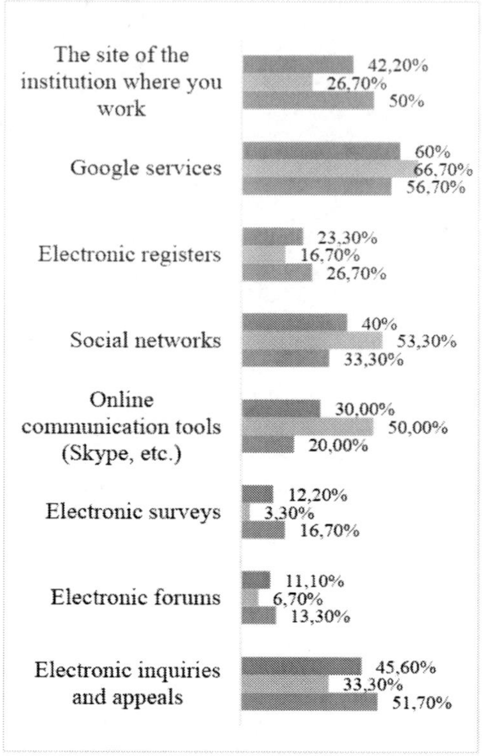

Fig. 4. The experience in using electronic tools in professional activities among experts (several options could be chosen)

According to the results of the study, a positive assessment of the population readiness to receive electronic social services is 32% of the interviewed experts and clients, that is, the points of view of recipients and social services providers in this matter coincide; negative assessment of readiness is also almost the same (40% among interviewed clients and 36% among experts). 30% of the respondents have defined their position and possibly represent a group of social service entities with a sufficient level of awareness and awareness of the possibilities of using e-government tools in the social service system.

According to the place of residence and work and experienced research results have some statistical discrepancies. So, among the experts representing social services and social service institutions in the regional center, 20% of respondents gave a positive assessment of the population readiness, in district institutions - 38%. Distribution by negative assessment of readiness: Zaporizhzhia - 37% and regional centers - 37%.

On the other hand, the population itself (as clients of social services and social service institutions) evaluates their own readiness to receive electronic social services differently (Fig. 3.10). Thus, residents of Zaporizhzhia are generally ready for the introduction of electronic social services (41% of respondents gave a positive assessment and 34% - a negative one), and clients living in regional centers and rural areas believe that the population is not ready for digital transformation in the social service (42% gave a negative rating and 29% gave a positive rating).

So, we can observe an ambivalence in the perception of the population's readiness to receive electronic social services among experts and clients: social workers of regional centers and clients of social institutions from the regional center are optimistic, and social workers of Zaporizhzhia and clients of social institutions from regional centers are negative. This dissonance of ideas, in our opinion, reflects the following trends: first, the optimistic attitude of social workers in the periphery, associated with the ability to simplify the provision of social services in remote settlements and villages, thereby reducing the burden; and secondly, the pessimistic attitudes of the population of the districts of the Zaporizhzhya region, which is represented by the older age category, do not have both subjective motivation for remote communication and objective conditions for receiving electronic services (lack of the Internet, technical support, ICT skills, etc.).

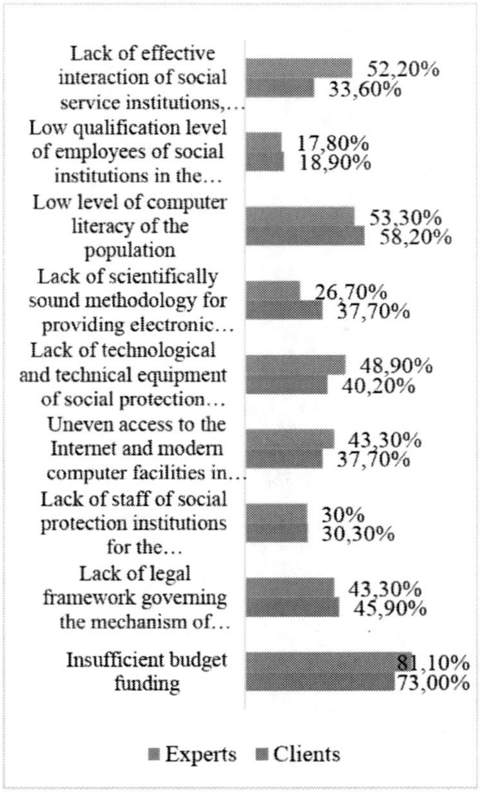

Fig. 5. The main reasons for the slowdown in the implementation of the electronic social services provision in Ukraine

Factors for optimizing the social services development in the system of electronic services (Fig. 5).

A comparative analysis of experts and clients survey results found some revealed some statistically significant differences in determining the reasons for the slowdown in the implementation of the electronic social services provision. So, for experts, such factors are: insufficient budget funding, low level of computer literacy, lack of well-established government-private-public interaction in the development of the electronic social services provision in the region, low level of technical support and uneven access to the Internet, and there are more of these among social workers of social service institutions in the Zaporizhzhya city.

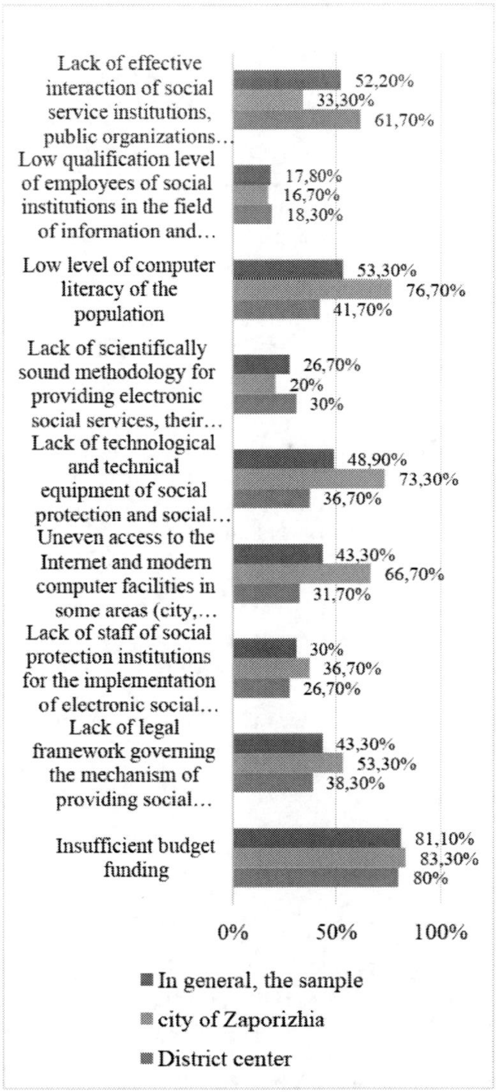

Fig. 6. The main reasons for the slowdown in the implementation of the electronic social services provision in Ukraine (experts)

For clients of social services, the main factors that will contribute to the development of the electronic social services provision are: an increase in budgetary financing of the social service system, computer, lack of a legislative framework and scientific justification for the implementation of a social services system for the population in the context of e-government development.

According to the social services` interviewed clients` age categories, can single out the actual directions of optimizing the process of introducing electronic social services. For the age category from 15 to 30 years old, such factors are: technological and technical support and computer; from 31 to 50 years - equal access to the Internet, a high level of computer literacy, substantiation of the methodology for the provision of electronic social services; for the population of 50 years and older - sufficient funding and material and technical provision of social service institutions, advanced training of social workers.

3 Conclusions

So, based on the results of expert and mass sociological surveys, the following conclusions can be drawn:

- firstly, half of the respondents (both clients of social services and social workers) already have experience in social services in electronic form, to a greater extent associated with the general trend of electronic government development in Ukraine through government Internet portals;

- secondly, at this stage of development of the electronic social services provision, most of all are used tools that are available to all Internet users, and those opportunities provided by the format of the social service institution site, among which electronic requests and requests for social services dominate;

- thirdly, a positive assessment of the population readiness to receive electronic social services is 32% of the surveyed experts and clients, that is, the points of view of recipients and providers of social services in this matter coincide;

- fourth, the main factors of slowing down the provision of electronic social services development in the region are: insufficient budget funding, low level of computer literacy, lack of well-established public-private-public interaction in the development of the electronic social services provision in the region, low level of technical support and unevenness access to the Internet.

References

1. M.S. Mikhrovska, *Legal scientific electronic journal*, 7, 272-275 (2020)
2. O. Onufriienko, *Public administration and local self-government,* **1(40)**, 21-33 (2019)
3. The U.S.A. Government's Official Web Portal. http://www.usa.gov
4. V.V. Yvanov, A.N. Korobova, *Gosudarstvennoe i municipal'noe upravlenie s ispol'zovaniem informacionnyh tehnologij* (State and municipal management with the use of information technology). (INFRA-M, Moscow, 2018)
5. R. Lindner, G. Aichholzer, *European E-Democracy in Practice* (2019) https://link.springer.com/chapter/10.1007/978-3-030-27184-8_2
6. V.S. Politanskyi, *Scientific Bulletin of Public and Private Law,* **2**, 29-35 (2018)
7. Lisbon Strategy http://cordis.europa.eu/programme/rcn/843_en.html
8. A.O. Serenok, *Pressing problems of public administration,* **2(50)** (2016)
9. Yu.O. Kovalenko, *Reporter of the priazovskyi state technical university*, **35**, 271-280 (2018)
10. B Warf , *IJEPR,* **7(4)**, 61-77 (2018)
11. J.M. Sánchez-Torres, I. Miles, *European Journal of Futures Research*, **5** (2017)
12. eGovernment in the European Union 2016 https://joinup.ec.europa.eu/sites/default/files/inline-files/eGovernment_Factsheet_European_Union_June_2016_v_7_04.pdf
13. I. Bertschek, J. Ohnemus, *ZEW policy brief,* **2** (2016)
14. E-Gouvernement: le digital au service du développement économique et du bien-être de vos citoyens http://sofrecom-production.s3.amazonaws.com/2017/03/15/11/18/06/751/Livre_blanc.pdf
15. Features of e-government, http://www.egovinterop.net/Res/9/eGovInterop%2006%20%20Programme%20final.pdf
16. B. Warf, *e-Government in Asia:Origins, Politics, Impacts, Geographies* (Chandos Publishing, 2016)
17. UN study: e-government. The use of e-government to build a sustainable and flexible society (2018), https://publicadministration.un.org/egovkb/Portals/egovkb/Documents/un/2018-Survey/E-Government%20Survey%202018_Russian.pdf

E-government Potential in Social Service Delivery: Regional Context

Vasyl Popovych[1*], *Serhii* Shcherbyna[1], *Halyna* Barshatska[1], and *Olena* Baluchtina[1]

[1]National University «Zaporizhzhya Polytechnic», 69063 Zaporizhzhya, Ukraine

Abstract. The article identifies the potential for introducing a system of electronic social services in the context of electronic government development in Ukraine. The definition of an electronic social service as a social service is given, fully or partially provided with the help of social protection institutions online services and the population social services, to individuals, certain social groups who are in difficult life circumstances. It has been determined that the development of modern information use and communication technologies and the Internet already allows the use of various forms of providing and receiving electronic social services, is spreading due to the state policy of public services digitalization ("the state in a smartphone"), among which a special demand is registration in electronic the form of social benefits, benefits, pensions and the like. According to the results obtained, the greatest potential for introduction into the social service system is possessed by social services, psychological online consultations, distance learning and socio-economic services. Among the factors for optimizing the implementation of electronic social services, the most significant identified are the following increases in budget funding, computer population, information support for the introduction of the electronic social services system through social advertising, increasing the level of technical support and access to the Internet.

1 Introduction

The modern information age, the challenges of globalization, and now pandemic threats (like the COVID-19 pandemic) require the introduction of digital (electronic, remote) forms of providing public services, including social ones. The concept that justifies the feasibility of using information and communication technologies in public administration is the concept of electronic government.

The British "White Paper: Modernising Government" examines this issue from the point of view of form: the structure and composition of services required for ordinary users and non-governmental organizations, expanding the range of services provided, while ensuring full coverage of citizens with public services, improving the use of information [11], and the EU programs for the introduction of e-government [8, p.8] provide for the priority of providing public services (both administrative and social) online, the widespread distribution of broadband access and secure information infrastructure.

The Concept for the Development of Electronic Governance in Ukraine provides the following definition: "E-government is a form of public administration organization that helps to increase the efficiency, openness and transparency of public authorities' activities and local governments using information and telecommunication technologies to form a new type of state focused on satisfaction needs of citizens" [6]. According to Yu. Solomko, "e-government provides for a way of organizing state power using systems of local information networks and segments of the global information network, which ensures real-time operation and makes the daily communication of a person with the authorities as simple and accessible as possible" [7, p.136]. T. Nizhniy, having analyzed publications on e-government, defined e-governance as a form of public administration organization that contributes to increasing the efficiency, openness and transparency of the activities of public authorities and local self-government bodies using information and telecommunication technologies to form a new type of state focused on meeting needs citizens [4, p.113], that is, a service state, the basis of which is the provision of services, including on the basis of information and communication technologies, is expressed in such a concept as "e-state (which, by the way, is already being implemented in many countries, for example in Estonia [1, 9]).

Hence, we see that e-government appears in a general sense as a form of public administration / governance organization, which, thanks to the widespread use of information and communication technologies, contributes to an increase in the efficiency, openness and transparency of public administration activities, the provision of a complex of administrative and social services for a person remotely.

Thus, the purpose of the article is to identify the potential for the introduction of electronic social services in Ukraine (on the example of the Zaporizhzhia region). The objectives of the study were: 1) to identify the awareness level of the electronic

* Corresponding author: popovic.vasil@gmail.com

social services provision; 2) determination of the possibilities for the provision of social services in remote (electronic) form at the regional level; 3) determination of factors for optimizing the development of social services in the system of electronic services.

In the main electronic document of Ukraine, which was approved on September 20, 2017, "Concept for the development of electronic government until 2020", the need to introduce online on average 100 of the most important services for citizens and business was noted. The key tasks are the implementation of a single portal for the provision of all electronic services from one resource, as well as the implementation of electronic contracts and the popularization of all services.

However, the lack of reliable and accessible electronic identification remains a barrier to the development of electronic services and other forms of electronic interaction. Therefore, it is necessary to introduce MobileID, which will make electronic identification more widespread and popular among citizens. The key task remains to fill the unified demographic register and issue ID-cards with digital signatures, as well as the introduction of electronic trust services.

All public services (state, municipal, administrative, social, management) that are provided in electronic form using information and communication technologies should be considered electronic.

Social services are one of the subtypes of public services that differ from management and administrative services in that: aimed at a limited target audience, namely families, children, youth, certain social groups or individuals who are in difficult life circumstances and need an outsider help; don`t provide for the registration of an administrative act as a result of the public service provision; can be provided both individually and in group; by-laws expand the possibilities of receiving social services, providing for the possibility of receiving them without a written application and even anonymously (for example, paragraphs 3.7, 3.9); in the practice of public administration are regulated by a special law.

We offer the following classification of electronic services: by type of electronic submission, by field of activity, by subject of publication, by consumers, by place of receipt from the point of view of the client and from the point of view of attraction to the electronic service.

1. By the content of the electronic services provision. According to the Procedure for the provision of information and other services using the electronic information system "Electronic Government", four types of electronic information services provision are distinguished: informing (providing directly information about state (administrative, social) services); one-way interaction (the user is provided with the opportunity to receive an electronic form of the document); two-way interaction (the possibility of processing the electronic form of the document, including identification is provided); carrying out transactions (electronic implementation of decision-making capabilities and their delivery). From a technological point of view, the provision of services of the first and second types has now been completely solved, and the problems of providing services of the third and fourth types (requiring identification of the parties to electronic interaction and associated with the use of electronic digital signatures) require their solution.

2. By the field of activity, electronic services are subdivided into the following information services; consulting services; services for the preparation of political decisions or laws; services for interaction between institutions and organizations; assistance and assistance services; services maintenance of the state order; services for the implementation of the supervision functions and control by government agencies.

3. by the subject of the services provision are divided into those provided by the central executive authorities, local government bodies, local government bodies and subjects of delegated powers.

4. By consumers, electronic services are divided into services for public authorities (use of information registers, provision of electronic signature status, etc.), for individuals, for legal entities.

5. According to the place of receipt are divided into: public authority; "Single window"; government portal.

6. By the form of receipt: fully automated and partially automated: fully automated - services in which the relevant processes (acceptance of applications, provision of services, payment, delivery, etc.) are performed in electronic form; personal contact and participation of a person (civil servants) in the provision of such services are minimized or completely absent; partially automated - services in which some of the processes are performed electronically, and some - manually.

7. Based on the results of the receipt, they may either provide for the registration of an administrative act, or not provide for its registration.

On the development of criteria for the electronic social services quality, then here can be the criteria that are defined in the Concept for the development of the providing administrative services system by the authorities, "which should serve as a model for administrative services in electronic form." These criteria, according to the Concept, are: 1) efficiency - meeting the needs of an individual or legal entity in administrative (social) services; 2) timeliness - the provision of administrative (social) services within the time period established by law; 3) accessibility - the actual ability of individuals and legal entities to apply for administrative (social) services; 4) convenience - taking into account the interests and needs of service holders in the process of organizing the administrative (social) services provision; 5) openness - unhindered receipt of information necessary for obtaining a state (social) service, which is posted on information boards in administrative bodies, institutions of social protection and social services for the population on their websites and printed in official publications and booklets; 6) professionalism - the appropriate qualifications level of employees from the administrative governing body for managing social protection and social services for the population.

In our opinion, the process of providing electronic social services should proceed as follows: 1) there is a preparation and placement of information about a social service by the relevant management body for social protection of the population on the official website and / or in the Unified State Portal of Administrative Services and Social Services; 2) individuals and legal entities are informed about the procedure for the social services provision; 3) receiving and registering an electronic application from a person for receiving a social service using information and communication technologies is ensured: using information and communication technologies, a person is identified; remotely, electronic payment for the service provision is carried out; 4) the received electronic application of the person is transferred to the departmental information resources using information and communication technologies: the fact of the electronic application transmission is electronically recorded; control over the consideration of an application for the electronic service provision using information and communication technologies; provides the opportunity to provide electronic documents required to receive the service; 5) the electronic service provision is completed by the person receiving an electronic document or receiving it in paper form at the time specified in the electronic queue.

2 Materials and Methods

The methodological basis of the study is made up of general scientific methods` of social phenomena and processes` cognition (analysis, synthesis, generalization, classification) and sociological methods for obtaining empirical data (mass and expert sociological survey). The empirical base of the study is the results of: 1) a mass sociological survey among clients of social services and social service institutions in the Zaporizhzhia region (n = 400, November 2020). The sample is quota, randomized, distributed by sex, age, place of residence (urban and rural areas). The error is no more than 2.2%; 2) an expert survey among managers and workers of social services and social service institutions (n = 100, November 2020).

In the field of social services, e-government, in our opinion, may be the most effective model of e-government called "Service Agency" (*Agency on Demand*), the essence of which is to provide electronic social services.

Let's try to define the concept of "electronic social service", relying on O. Bernazyuk's interpretation of an electronic public service [2, p.198]: an electronic social service is a social service that is fully or partially provided through the online services of social protection and social services for the population, individuals, individual social groups who are in difficult life circumstances, cannot overcome on their own and need outside help, is carried out through the use of information and telecommunication technologies, ensures its efficiency, accessibility, convenience, mobility of receiving social services.

In the context of the e-government development in Ukraine, the most common services today are administrative services related to the execution of documents and obtaining information from public authorities and local governments. Regarding the system of population social protection, the most widespread are such social services as registration of housing subsidies, benefits, social payments, that is, mainly socio-economic services, and social information services (information on the activities of social protection institutions, employment, etc.) [2].

The current state of the provision of legally defined social services in electronic form and the possibility of their implementation and development is characterized as follows: 1) socio-economic services are provided through the electronic services of government sites [5], and the provision of such social services is associated with remote processing of documents for receiving social assistance , pensions, social benefits and the like; 2) legal services through online services in the field of social services relate to the areas of persons in difficult life circumstances legal protection, which can be provided in electronic form (online consultations, inquiries, correspondence); 3) social and medical services provide an information component - displaying on the websites of institutions and social service institutions a list of medical organizations, specialists and services that can be received by persons in difficult life circumstances; 4) the provision of social and domestic services presupposes the availability of information on the list of available social and domestic services, their tariffication and the possibility of receiving free of charge, information on the qualifications of specialists providing social services; 5) a feature of the electronic technologies use in social and educational services provision is the introduction of SMART-learning technologies [10]; 6) the psychological services provision in electronic and online forms is possible: firstly, as an appeal through the line of trust via Skype in emergency and crisis situations and circumstances; secondly, as the provision of the first psychological counseling; third, as an additional form of psychological support for the client; fourthly, as a form of psychodiagnostics examination of the client; 7) today employment services can be obtained in electronic form on the websites of the State Employment Service, other commercial and non-commercial organizations.

In the context of the introduction and development of e-government in Ukraine, the sphere of social protection and social services for the population, in particular, is at the initial level of using online services. Now the following electronic services are available on the website of the Ministry of Social Policy of Ukraine: 1) assistance with the birth of a child; 2) the appointment of subsidies; 3) the system for calculating pension provision [3]. The Information and Analytical Management System for Social Support of the Population of Ukraine (E-SOCIAL), which is being created within the framework of the project

"Modernization of the Social Support System for the Population of Ukraine" for funds from the International Bank for Reconstruction and Development loan No. 8404-UA, is under development.

3 Results and Discussion

To highlight the possibilities and potential of using e-government tools in the system of providing social services, a massive sociological and expert survey was conducted (November, 2020).The survey revealed an above average level of awareness of the possibility of providing (receiving) electronic social services among experts and clients (more than 50%), with the majority of the surveyed clients of social services in Zaporizhzhia aged 15 to 30 years.

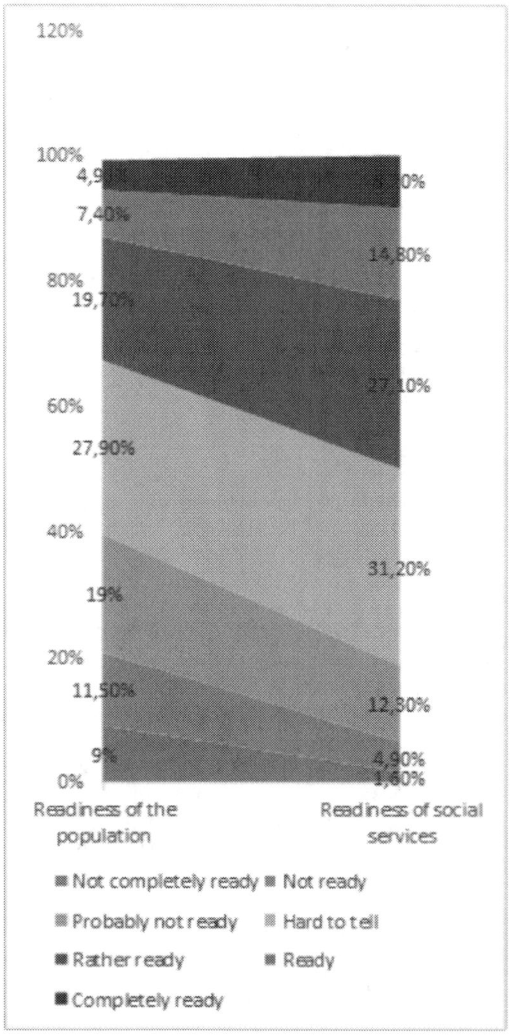

Fig. 1. Assessment of the social workers readiness to provide electronic social services and the population to receive them (clients)

The fig. 1 shows the results in a comparative measurement of the population and social services readiness to provide social services in electronic (remote) form from the point of view of the interviewed clients: 50% of clients positively assess the readiness of social services to provide electronic social services.

So, clients of social services believe that social services are able to provide electronic social services (50% of them were identified), while 40% of the interviewed clients note that the population is not ready to receive social services remotely, which may be due to the pessimistic attitudes of the population, does not have both subjective motivation for remote communication and objective conditions for receiving electronic services (lack of Internet, technical support , ICT skills, etc.).

With regard to the population, we can say that there is a lack of trust in digital services. Several factors influence this attitude of the population: first, there is no motivational component for the population to want to communicate remotely; and secondly, technical problems persist to this day. The point is that there are still places in Ukraine where the Internet is absent or the signal level is extremely weak, most consumers of services in the social field simply do not know how to use personal computer programs and so on. In parallel, it is possible to note an ambivalent attitude towards the adoption of electronic services in the social sphere on the part of the experts who were interviewed. Having made some similarities, we can state the following: consumers of social services in regional centers, where there are fewer technical problems, have a positive attitude to the electronic services provision, but the professionals in social work and the clients of the city perceive negatively and their attitude to this type of service is negative.

Such an ambiguous view indicates the following: a positive attitude of social workers in the region, associated with the ability to simplify the social services provision in remote settlements and villages, thereby reducing the burden; negative perception of the population from the Zaporizhzhia region districts, which is represented by the older age group, do not have both subjective motivation for remote communication and objective conditions for receiving electronic services (lack of the Internet, technical support, ICT skills, and so on.)

Today, a list of certain electronic services associated with the system of social protection and social security of the population is already functioning. The respondents were asked to identify the 4 most accessible (Fig. 2).

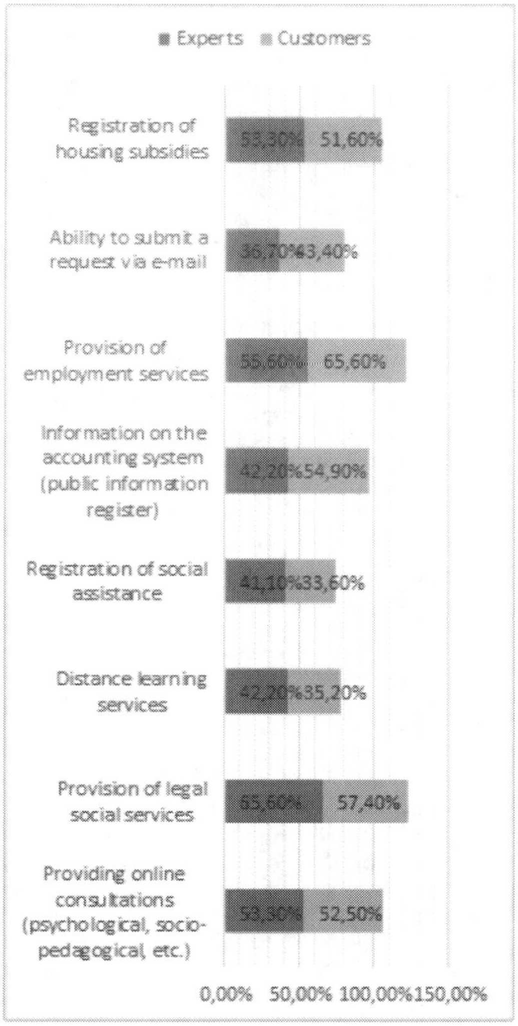

Fig. 2. Availability of electronic services for the population

Thus, experts define the following electronic services for the population, which, in their opinion, have a high level of accessibility: 1) the ability to submit an electronic request; 2) distance learning services; 3) providing online advice and registration of housing subsidies; 4) registration of social assistance and employment services; for clients of social service institutions, these are: 1) distance learning services; 2) submission of an electronic request; 3) registration of social assistance; 4) registration of housing subsidies and online consultation.

Deserves special attention to determine the assessment of the social services implementation defined at the legislative level in electronic form. The respondents were asked to rate the possibility of providing the following social services in electronic form on a 5-point scale (Fig. 3).

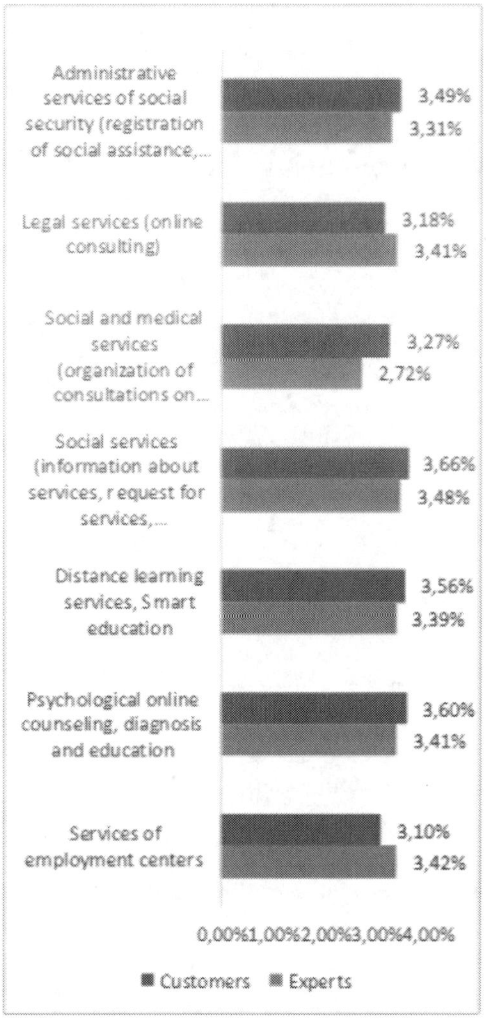

Fig. 3. Assessment of the providing social services possibility in electronic form (average values)

Experts, based on their own practical experience in the social services provision, assessed the possibilities of introducing e-services as follows: the highest indicator, according to social workers, is social services (such as filing an application and a request for their receipt), online legal advice, psychological and socio-pedagogical services, and so thinks the majority of social service institutions employees in Zaporizhzhia city than the regional centers of the Zaporizhzhia region. For clients, social services, online psychological consultations, distance learning and socio-economic services (as an opportunity for remote registration of social assistance, payments, benefits, etc.) have a high rate of social services implementation assessment in electronic form.

If we consider consumers of social services by virtue of age categories, we can dwell on the following: the presence of an actual direction for optimizing the process of introducing electronic social services. For the population aged 15 to 30, we are talking about the following factors: technological and technical support and a computer; the population aged 31 to 50 has a high level of literacy in the field of technology, have access to Internet resources, as regards the methodology for the provision of digital services in the field of the social sphere; for people 50 years and older - sufficient funding and material and technical support and technical social service institutions provision, social workers advanced training. The results of the study show that more accessible (and possibly received in electronic form) for young people are legal services, processing of social assistance and online counseling for the middle age category - distance learning, electronic requests and social assistance processing, for the older age category - electronic requests and processing of housing subsidy. In this context, V. Popovych said that the bilateral interaction between the state and society can be formed through effective modern tools [12, p.320].

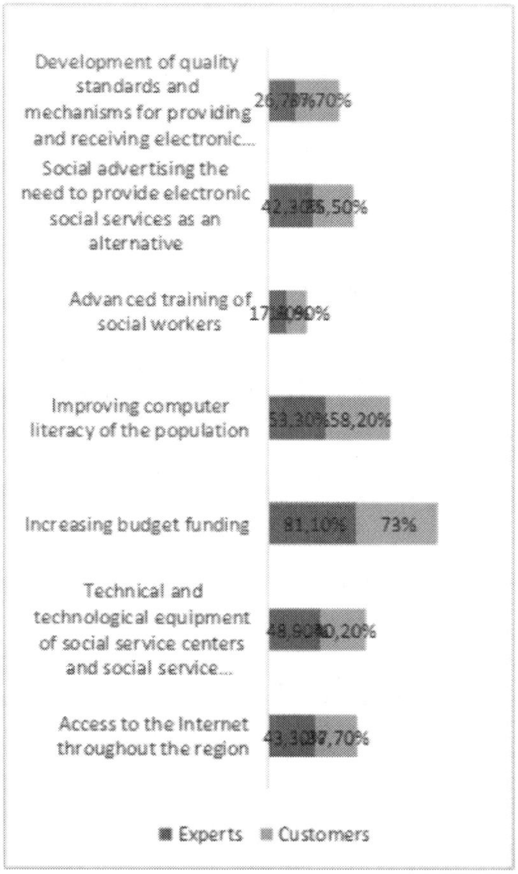

Fig. 4. Factors for optimizing the process of introducing the electronic social services provision (it was possible to choose several options)

Deserves special attention to determine the assessment of the social services implementation defined at the legislative level in electronic form. The respondents were asked to evaluate the possibility of providing the following social services in electronic form.

The fig. 4 presents the factors for optimizing the implementation of social services in a remote (electronic) form. A comparative analysis of the survey results of experts and clients revealed some statistically significant differences in determining the factors for optimizing the process of introducing the electronic social services provision. So, for experts, such factors are: an increase in budgetary funding, computer population and social workers, information support for the introduction of an electronic social services system, an increase in the level of technical support and technical support and unhindered access to Internet resources, and this is most closely monitored among specialists in the social sphere of Zaporizhzhia city. The main points that were determined by consumers of social services are those that can influence the increase in the level and improvement of the provided electronic social services. We are talking about improving the situation with state support in the field of finance, computerization, the lack of a regulatory framework that would regulate the electronic social services provision at the legislative level and, of course, scientific justification for the implementation of the electronic social services system introduction of a system of electronic social services through the development of quality standards and mechanisms for their provision.

For clients, social services, online psychological consultations, distance learning and socio-economic services (as an opportunity for remote registration of social assistance, payments, benefits, etc.) have a high rate of assessment of the social services implementation in electronic form. For clients of social services living in regional centers and rural areas, a significant high indicator of the social possibility and medical services in electronic form is primarily related to the organization of consultations on the prevention and development of possible organic personality disorders, support of their health, the implementation of preventive, treatment and recreational activities, registration of electronic requests for their receipt at the place of residence.

Conclusions

Based on the results of theoretical and empirical analysis, we made the following conclusions.

1. The definition of electronic social service as a social service is given, fully or partially provided with the help of online services of state authorities, social protection institutions and social services for the population, and the most affordable services for today are distance learning services, the provision of online consultations, registration of housing subsidies, registration of social assistance, employment services.

2. According to the results obtained, social services, online psychological consultations, distance learning and socio-economic services (as the possibility of remote registration of social assistance, payments, benefits, etc.) have the greatest potential for implementation into the social service system.

3. Among the factors for optimizing the implementation of electronic social services, the most significant ones were found - an increase in budget

funding, computer population and social workers, information support for the introduction of electronic social services system, an increase in the level of technical support and access to the Internet.

4. The level of accessibility in terms of methods` awareness and possibilities of obtaining (providing) services in the social sphere by electronic resources was found among experts and clients above average, and among social workers there is a sufficiently high level of support for the information implementation and communication technologies for remote social services. A separate part of the respondents who took part in the study (these are clients of the social sphere and specialists in the field of social work) provided information on their experience in the social services provision in electronic form, which is largely associated with the general trend of e-government development in Ukraine through government Internet portals. In the modern world, including in Ukraine, the electronic form of the social services provision most uses those tools that are available through the Internet - resources for a wide range of users. The same opportunities for clients are provided by the official websites of social service institutions, among which a social survey was conducted. Of all the services in the electronic context, the readiness to receive electronic social services received a positive assessment among experts and the population - 32% of the interviewed experts and clients.

References

1. Modernizing Government. Presented to Parliament by the Prime Minister and the Minister for the Cabinet Office by Command of Her Majesty. March 1999. The Stationery Office. URL: http://www.e-gorod.ru/eproject/ whitepaper/4310.htm (1999).
2. E-Europe 2005: An information society for all / An Action Plan to be presented in view of the Sevilla European Council. Brussels, 22 p., (2002).
3. On approval of the Concept of e-government development in Ukraine: Order of the Cabinet of Ministers of Ukraine № 649-r from 20.09.2017 p. URL : http:// zakon.rada.gov.ua. (2017).
4. Solomko Y. E-government: the concept, essence, principles and directions of development. Efficiency of public administration. **Vol. 2(55)**. 135-143. (2018).
5. Lower T.V. Theoretical principles of e-government. Investments: practice and experience. **№ 20**, 112-116, (2017).
6. Arkhipova EA, Dmitrenko NO Experience of e-government implementation in Estonia and its implementation in Ukraine. *A young scientist.* **№11**, 148-152, (2015).
7. E-Estonia – e-Governance in Practice. URL: http://ega.ee/wp-content/uploads/2016/06/e-Estonia-e-Governance-in-Practice.pdf (2016).
8. Bernazyuk O. Administrative electronic services: concepts and conditions of implementation in Ukraine. *Entrepreneurship, management and law.* **№5**, 196-199, (2019).
9. Public services portal iGov. URL: https://igov.gov.ua/service/97/general (2020).
10. Lynch M. 20 Top Virtual Reality Apps that are Changing Education: March 11, 2017. URL: https://www.thetechedvocate.org/20-top-virtual-reality-apps-that-are-changing-education/ (2017).
11. Ministry of Social Policy of Ukraine. URL: https://www.msp.gov.ua/main/Eservices.html (2020).
12. V.Popovych, F.Ragimovb, V.Kornienko, I.Ivanova and B.Zoriana. Development of social and communicative paradigm of public administration in the field of social networks. International J. of Data and Net. Sc., **4**, 319–328, (2020). doi: 10.5267/j.ijdns.2020.6.001

Problems of Improving the New Paradigm of Social Processes Management in Modern Ukraine

Volodymyr Taran[1*], *Vitaliy* Dudaryov[1], *Hanna* Koval[2], *Vasyl* Popovych [1], and *Jan* Holonic [3]

[1]National University «Zaporizhzhya Polytechnic», 69063 Zaporizhzhya, Ukraine
[2]Lviv state university of life safety DSNS Ukraine, 79000 Lviv, Ukraine
[3]Univerzita Komenského v Bratislave, 81499 Bratislava, Slovak Republic

Abstract. Theoretical analysis and practical application of the management system of social processes is impossible without a clear methodological basis, which is part of the paradigm. The paradigm is usually based on a system of principles that determine the prospects for the development of science. An important achievement of modern methodology of science is the theoretical substantiation of the position on the historical variability of the paradigm. This is due to the steady process of development of social practice, the expansion of the problem field of science and the improvement of research methods. It follows that any paradigm, including management, exists only within a certain time. Then it inevitably undergoes rethinking and gives way to a new paradigm. Based on this, the purpose of this article is to try to identify the most promising semantic landmarks around which to improve the new paradigm of management of social processes, and the task - to record such paradigmatic changes in the system of principles.

1 Introduction

Ukraine today is a young state, despite the millennial tradition of its own statehood. The fact of gaining its independence was recorded by the Declaration of the State Sovereignty of Ukraine on July 16, 1990 and acquired its final legal form after the Supreme Council of Ukraine approved the Act of Independence of Ukraine on August 24, 1991. The whole recent period of the Ukrainian state development is characterized by undoubted achievements. However, there are many miscalculations due to both objective and subjective circumstances. One of the effective levers of the state successful development is its personnel policy, especially in the field of civil service, which allows to implement strategic and tactical tasks of the state.

State personnel policy, as a component of social processes management - is a strategic activity for goal setting and software formation and development of human, labor, human resources as the main prerequisite for building a democratic, legal, social state. It determines the place and role of personnel in society, goals, objectives, the most important areas, priorities and principles of work with personnel, the main criteria for their evaluation, ways of rational use of human resources of the state. That is, the state personnel policy is the main strategy and tactics of the state in the field of formation, development and rational use of the country labor resources.

State personnel policy is a complex and dynamic institution. It depends on the type of state and the form of political system. That is, we can talk about differences in personnel policy in different socio-economic formations (caste, class, religious, party, ethno-national, etc. approaches), as well as in authoritarian, totalitarian, aristocratic, oligarchic or democratic forms of political government. Modern Ukraine as a transitional state clearly demonstrates the complex process of transforming the personnel policy paradigm, which in reality (rather than formally) has not yet fully complied with the principles of a democratic society and retains elements of clannishness, kinship, personal devotion, regionalism, etc. previous management culture and the influence of the oligarchic system.

State personnel policy is a complex and dynamic institution. It depends on the type of state and the form of political system. That is, we can talk about differences in personnel policy in different socio-economic formations (caste, class, religious, party, ethno-national, etc. approaches), as well as in authoritarian, totalitarian, aristocratic, oligarchic or democratic forms of political government. Modern Ukraine as a transitional state clearly demonstrates the complex process of transforming the paradigm of personnel policy, which in reality (rather than formally) has not yet fully complied with the principles of a democratic society and retains elements of clannishness, kinship, personal devotion, regionalism, etc. previous management culture and the influence of the oligarchic system.

Theoretical analysis and practical application of the management system of social processes is impossible without a clear methodological basis, which is part of the paradigm. In the philosophical context, the latter is understood as a basic conceptual scheme, a model of problem statement and solution, research methods used by scientists during a certain historical period. The

* Corresponding author: zastelo@ukr.net

paradigm is usually based on a system of principles that determine the prospects for the development of science, scientific industry or discipline, ways of solving problematic issues, parameters of scientific searches, criteria for the truth of research results, and the like. An important achievement of modern methodology of science is the theoretical substantiation of the essential features, structure of the paradigm, types of paradigms, as well as patterns and mechanisms of their change, which is primarily due to the steady process of development of social practice, the expansion of the problem field of science and the improvement of research methods. The analysis of these issues is reflected in the works of such leading domestic and foreign scientists as T. Kuhn, S. Tulmin, M. Alexandrov, E. Babets, P. Drucker, V. Ivanov, V. Kryzhko, M. Moiseev, G. Osovska., O. Osovsky, A. Furman, F. Khmil, Y. Yakovets and others

Thus, the position on the historical variability of the paradigm is one of the key conclusions of scientists. This means that any paradigm, including the managerial one, exists only within a certain time frame and is inevitably subject to rethinking, losing its methodological role and giving way to a new paradigm. Based on this, the purpose of this article is to try to identify the most promising semantic landmarks around which the new paradigm of social process management has to improve, and the task is to fix such paradigmatic changes in the system of principles.

2 Results

Speaking of the need for a paradigmatic rethinking of the principles of social governance, we can usually proceed from two levels of paradigmatic models.

The first is narrowly professional, which is based on the development and application of universal (including the most modern and innovative) principles, concepts and methods of influencing the object of social management (in this case society) in order to use its potential to achieve the goal. It is obvious that the models of this level have their own logic of development and application, focused mainly on solving the "technical" side of management. However, in the context of intensification of crisis phenomena, which today cover almost all spheres of public life, such a methodological approach is limited.

The second: worldview, which is based on the most fundamental principles methods and laws that the system of social management uses to implement the strategy of society, taking into account the current most progressive trends of civilization and national levels. Such paradigmatic models can maximally harmonize the relationship "subject - object" in the system of social management and contribute to the effective implementation of management decisions that are ultimately able to meet the interests of society as a whole.

In fact, leading Ukrainian scientists V. Andrushchenko, M. Mykhalchenko, F. Kanakh, V. Lobas, V. Kremen, S. Pazynych, O. Ponomarev, S. Farenik, S. Gudkov K. Shumovska and others are actively engaged in the search for worldview bases of the new paradigm of social process management. Their theoretical achievements indicate the existence of serious obstacles that stand in the way of forming a new concept of management paradigm in Ukraine. This is due to a number of reasons of general theoretical and specific practical nature:

- significant differences in the interpretation of basic concepts and features of management science itself: it studies management as a phenomenon that occurs in systems of different types, which are also in constant dynamics;
- the absence in the social world of laws similar to the laws of nature; here the subject of the study is constantly changing, which means that the ideas that were true yesterday may literally at one point become incapable, and even more so - erroneous [1];
- conditions of functioning of paradigms that "operate" in a certain system of space-time coordinates, where space determines a certain socio-cultural field (traditions, way of life, education system, mentality, etc.), and time - features of a specific historical moment;
- socio-cultural circumstances of the functioning of Ukrainian society, which for the whole post-Soviet period has remained a transitional society, where the horizons of the destination of its historical progress retain final uncertainty, which negatively affects the process of choosing worldview principles of social management paradigm. The latter to some extent explains why effective management paradigms that have already been tested in other countries and do not always justify the expected results in Ukrainian realities.

At the same time, the process of searching for worldview paradigmatic bases of social processes management taking into account the complex of modern problematic phenomena is marked by significant achievements, which in general express the theoretical principles of the humanistic scientific approach. This approach is focused on a radically new system of worldviews and values in comparison with the paradigmatic approach of Soviet times.

The system of government used in the Soviet Union was based on the principles of Eastern culture and was characterized by rigid centralism in economic life on the basis of state property, domination of the party-state nomenclature, lack of real economic freedom and democracy. The transition to market methods that correspond to another model of society: the existence of various forms of ownership dominated by private, market, competition, limited role of the state, expanding the rights and freedoms of economic and social actors, etc., has led to radical changes in society as a whole. This, in turn, led to a new management paradigm. According to K. Shumovska, the new management paradigm should be understood as the replacement of the Soviet (authoritarian style of leadership) with a market (democratic style of leadership) [2], which, nevertheless, is in the process of further improvement.

Such a paradigm shift in the management system orients us to shift the emphasis in understanding the essence of man. Thus, if the command-administrative model of government was based on the concept of "economic man", moreover, fully controlled by government and administrative structures, the market-democratic model emphasizes the man "social", "spiritual and creative".

Sociality in this case involves not only the ability to live and incorporate into society, but also the value category, which means the mechanism of harmonization of the individual's relationship with the group, its ability to share the values of this group and to represent the interests of this group. In a broader context, it is about the awareness of society as a complex entity, united on the basis of harmony of individual, group and social interests, the realization of which takes place in a single legal field. Hence, the effectiveness of management of social processes depends on the ability, guided by the principles of solidarity and subsidiarity, to take into account the presence of different views in society and the ability to unite them around common program objectives. The lack of such mutual understanding leads to distrust of the authorities, unwillingness to accept managerial decisions, inability to concentrate on achieving social goals, weakening statehood. Thus, in the sense of paradigmatic importance for governance, sociality is meaningfully consistent with the principles of the welfare state, which, in turn, is organically linked to the socio-market organization of the economy, developed institutions of civil society and the rule of law. This implies a focus on a certain system of principles - justice, tolerance, competitiveness, freedom, democracy, and initiative, responsibility, which are explained and enshrined at the legislative level in accordance with the course of development in the direction of the chosen state civilization model. In the sense of paradigmatic importance for management, sociality corresponds in content to the principles of the welfare state, which, in turn, is organically linked to the social-market organization of the economy, developed institutions of civil society and the rule of law. This implies a focus on a certain system of principles - justice, tolerance, competitiveness, freedom, democracy, and initiative, responsibility, which are explained and enshrined at the legislative level in accordance with the course of development in the direction of the chosen state civilization model. And this course should be clearly fixed at the level of state ideology, which should represent not only a theoretical substantiation of the basic principles of a certain state self-development strategic model, but also determine the value orientations of the main state institutions management activities in full accordance with the provisions of the basic legal documents developed and adopted by the Verkhovna Rada of Ukraine.

At the same time, it is necessary to take into account the historically determined influence of different civilizational traditions on the specifics of the formation of administrative relations in the Ukrainian state. We are talking about European (persona centric paradigm / democratic, when a person is seen as the highest value and goal (goal) for the system, when the system serves man, creates maximum opportunities for its existence and development) and Asian (system-centric / authoritarian, when the system prevails over man, when the system has only rights, and a person only responsibilities, when a person for the system - a means, not an end) traditions with appropriate value orientations, which differently determine the priorities of the individual or team in the system of social relations. And even clearly stated in the basic legislative documents, Ukraine's orientation towards the European system of values does not deny the presence of the Asian tradition elements, manifests itself both at the level of public consciousness in the form of paternalistic sentiments, and in power structures, for which the principles of authoritarianism and monopoly sometimes look very tempting. At the same time, the choice of one of the above paradigmatic models cannot be an end in itself. Undoubtedly, the new management model in Ukraine should be based primarily on a system of values that embody the features of the historically formed Ukrainian national culture and Ukrainian national and political identity. Based on this, the relationship of participants in management activities with the cultural environment as a center of national-specific, in which, moreover, in purely managerial terms, the authority of the informal organization and the informal world is no less important than the requirements of regulations, rules, instructions from the administration. Thus, it is a question of harmonization of mutual relations between all components of administrative system within the limits of uniform national sociocultural whole.

The attitude to man as an expression primarily of the spiritual and creative principle allows us to consider the inclusion of her (man) in the management system from other positions. The essential features of such positions are due to the content of this category as a whole and its components. An important point in this context is what spirituality determines moral measurement human life, it is a living source virtues man, her morally capacity and higher value. She (spirituality) is ontological them environment man, or understood as the ability and need to assimilate and transform the external and internal world according to the laws of Truth, Goodness, Beauty in their unity .In this regard, we can say that spirituality is an important means of self-improvement and the individual and society as a whole, including as components of the management system, focusing them on the ongoing process of improving knowledge, expanding management experience, compliance with moral norms.

Creativity is understood as the more important meaning of human activity, which aims to create something fundamentally new, not similar to the previously created, activities that involve the discovery of new for this individual, and also as property of the person which is shown in all spheres of its life. In addition, it is openness to the perception and implementation of innovative approaches, without which success cannot be guaranteed in any area of social activity, including management. Spirituality and creativity are inextricably linked, their formation takes place in parallel, and they ultimately complement each other, which allows you to use them in a single conceptual formation, which is aimed at a large number of external and internal influences. At the same time, one more point should be taken into account: the focus on the priority of the spiritual and creative potential of a person is extremely relevant and justified in view of the fact that the main subject of the future - information - society should be the "class of intellectuals" who have the main capital of the information age - scientific knowledge, combined with creativity and innovation, and the most important properties of human capital - moral attitudes. In principle, we are talking about a tendency towards ever closer interaction and interdependence of human capital, information capital, social capital and cultural capital as a necessary condition for the development of a modern information society in the broadest sense of the word [3]. As for the problem we are investigating, taking into account the characteristics of the spiritual and creative factor in conjunction with sociality, we can talk about the possibility as a whole to more effectively influence management decisions in the context of the development of an independent, democratic, social, legal state.

Identifying the features of social processes in the current environment, we cannot ignore the problem of hybridity, which has recently significantly strengthened its paradigmatic and methodological role, including directly or indirectly in the search for more effective approaches to managing these processes.

The theoretical substantiation of the objective nature of the hybridization process, which has a long history, has been continued in modern research, carried out mainly at the intersection of sciences (R. Knox, M. Brun, J. Martin-Barbero, W. Mignolo, Iha-ba Hassan, R. Bart, J.-F. Lyotard, R. Penrose, P. Gilroy, H. Bhabha, P. Verbner, D. Perkins, F. van Kappen, R. Marutyan, L. Voloshin, A. Doroshkevich, V. Gorbulin, etc.). They significantly expand our understanding of the mechanisms of such a process and show that the universal relationships, interactions, interpenetration and interdependence of different systems are carried out not only within their own structural organization or within a particular form or level of organization of the material world levels of organization of the universe as a whole. And this gives a better understanding of the role of hybridity in social processes and trends in the modern local and global world.

Despite the significant number of scientific developments that are of interest in this particular context, it is enough to refer to a few. First of all, this is a new theory of worldview - synergetics, which has taken a special place in the light of the formation of a post-non-classical picture of the world with its pluralism and dialogism ideas. Created by the German scientist G. Haken in the second half of the twentieth century, it is also known under other names - the theory of dynamic chaos in the United States (M. Feignbaum), the theory of dissipative systems in French-speaking countries (I. Prigogine), the "theory of self-organization" in Russian literature, which, however, not only do not contradict the main direction of the new science - the study of the processes of self-organization of phenomena through the use of methods of nonlinear thermodynamics, but also substantially supplement some of its provisions. It should be noted that although synergetics originated in the depths of quantum mechanics, thermodynamics, physics of nonequilibrium states and the theory of dynamical systems, it is considered an interdisciplinary science, and the subject of his interests reaches macroscopic processes and multilevel coordination, affecting various branches of natural and social cognition, including including the scope of the network information space. Moreover, synergetics blurs the insurmountable boundaries between physical and chemical processes, on the one hand, and biological and social, on the other. Synergetics focuses on the functioning of objects as open systems. Since an open system is a system that exchanges matter, energy and information with the environment, it is changeable and dynamic. Such instability of the system presupposes that it loses its previous state and acquires a new qualitative state under the influence of the environment or interaction with another / other system / systems. The latter determines the endless process of "struggle" between its two states - equilibrium and nonequilibrium - dissipative, associated with an increase in entropy (degree of complexity, chaos, uncertainty). Non-equilibrium, in turn, is able to go back to equilibrium as a new stationary state of the system [4]. So, each new stationary state, as a manifestation of the next stage of self-organization of the system, represents its new quality, acquired as a result of interchange with the environment, and can be considered as a certain form of hybridity. Moreover, in such a hybridity, through gradual internal structuring, which ensures its identity as an object with itself, there is a transformation of its previous dynamic "chaotic state", or dynamic chaos (M. Feignbaum) into "order", which, in turn, is a prerequisite for the next chain of transformations from a state of chaos to order, etc. A hybrid can also be considered as the next stage or bifurcation point in the system development. Finding a parallel between a hybrid and a bifurcation point makes it possible to talk about the multivector nature of the further possible development of any hybrid, often unpredictable, or even about the possibility of the final self-destruction of the existing system. Here, such qualitative changes in systems are also possible, which make it possible to carry out the transition from one level of organization of matter to another.

So, M. Eigen uses the principle of self-organization of matter as the main principle for biological evolution and substantiates the hypothesis about the origin of life. At the same time, he also connects it with the mechanism of movement from dynamic chaos to order within the process of transition from inanimate to living [5], which implies the emergence of certain transitional hybrid systemic formations.

Actually, the evolutionary process of the living things development as a whole can be viewed from this angle. A striking confirmation of this fact can be the symbiogenesis of L. Margulis, who added a third -

 https://doi.org/10.1051/shsconf/202110004004

development through symbiosis in two evolutionary directions of the global formation of life - random gene mutations and large-scale gene exchange through the global exchange network. The essence of this idea lies in the fact that symbiosis is considered as the ability of different organisms to live in close interaction with each other, which leads to the formation of new forms of life [6]. And such life forms can also be considered as hybrid in nature, which, at the same time, are certain steps in the process of its endless hybridization.

The theory of living systems and the systemic vision of life by F. Capra and P. Louise advanced scientific thought further and made it possible to reveal the universal nature of hybridity and the mechanisms of hybridization inherent in all not only biological, but also social systems, are a prerequisite for their formation, functioning and self-development, influencing both on the process of preservation of its own identity by each system, and on the nature of intersystem interaction, which does not exclude elements of the struggle for survival and domination.

The mechanism of hybridization at the social level is quite complex, because here, along with objectivity and spontaneity, the subjective factor begins to play an increasing role. It is the active purposeful activity of the social subject that ultimately determines the forms, direction of development and dynamics of those hybrid processes that can be influenced by man. It is clear that this should take into account the natural trends in the development of a particular system, including causation, as well as the presence of a factor of chance, unpredictability. Thus, in social processes, chaos and order are also subjected, in connection with which we can speak of artificial, planned, controlled chaos or organized, established order. As well as the conscious choice of attractor or "supertractor" (V. Bryansky) at the bifurcation point, which can be identified with the moment of making a management decision. Interestingly, at the level of social cognition, chaos and order are not perceived as mutually exclusive, but rather as conceptual spaces where intuitive ideas, concrete scientific interpretations and their philosophical understanding intersect, which proves a synergetic approach. Yes, and the very terminological apparatus of social synergetics, as one of the methodological foundations of management theory, is multilayered and the degree of abstraction of concepts, and their origin. This is the concept of natural sciences (bifurcation, fluctuation, homeostasis, entropy, metabolism, balance, imbalance, etc.); the concept of general scientific areas that study the laws of organization, management, technical modeling - information theory, cybernetics, systems theory (system, structure, organization, information, feedback, open and closed systems, self-regulation, etc.); philosophical categories to describe more general laws and mechanisms of the development process (randomness and necessity, variability and stability); the concept of social disciplines that emphasize the specifics of the social system (order, turmoil, revolution) [7].

Interesting from the point of view of the characteristics of social objects, in our opinion, are the reasons of self-organization of socio-historical systems mentioned by G. Haken: change of global influence on the system from the environment; increasing the number of system components; mixing of components; sudden change of parameters that control [8], which, as you can see, contains signs of hybridity. Therefore, such large-scale social phenomena as unrest, revolution, perestroika, modernization, reform, etc. (as manifestations of the current state of Ukrainian society), understood through the prism of a synergetic approach, are essentially transitional - hybrid, as they contain a set of different elements that define them. Intermediate unstable state of the organization on the way to further systemic certainty, and therefore require appropriate adjustments in the management system.

Another important ideological paradigmatic principle of managing social processes can be considered eco-socialism, which follows from the concept of co-evolution (Vernadsky, P. Yerlikh, P. Raven, N. Timofeev-Resovsky, J. Tinbergen). Its actualization is due to the fact that in the conditions of globalization the problem of mutual relations and mutual influence becomes more acute human life in its bio-psycho-techno-social integrity with the natural environment, and this leads to the formation of a new quality of the universe. Violation of the balance between the components of such integrity entails deformation of the society social structure and the way of its life [9], negatively affecting in the end the political and the economic spheres of society as a whole. Such a contradiction of these processes highlights the need to create and implement the concept of a sustainable (global) society with an appropriate strategy and governance model capable of harmonizing such contradictions.

This model, which clearly correlates with the paradigm of survival (not in biological, but in its social sense) as a component of the philosophy of management (S. Gudkov) [10], focused on achieving the optimal balance between the three components of development-economic, social and environmental. Moreover, it is precisely hybridization that actively covers both each of the specified components, and intercomponent communications with all specific features of action of the hybridization mechanism carefully described within the limits of the synergetic approach. Implementation of the "global goals" of such development involves the development of management decisions in the direction of not only optimization and interconnection of economic growth, social integration and environmental protection, but also the coherence of international and national programs in economics, politics, law, technology, science, education, morality, etc., which means the adoption of a new model of world civilization, requiring the maximum level of management culture. The strategy for such development was enshrined in the UN General Assembly Resolution of September 25, 2015 [11], which is reflected in the legislative framework of Ukraine in 2019 [12] and confirmed the country's readiness to move to integrated economic, social and environmental development.

The strategy of sustainable development is based on a number of models, which, while remaining in its value coordinate system, offer specific management ways of practical implementation of the proposed ideas. Among them, a special place is occupied by models of "green" economy, which based on the theory of R. Goldsmith on the limitation of the assimilation capacity of the environment and the need to reconsider the attitude to the economy as a whole. The difference between the "green" economy and the dominant economic paradigm today is to focus on innovative industrial production, convergence of economy and morality, overcoming the unilateral dependence of society and the environment on the economy and its synchronization with both social and natural processes [13] - use of environmentally friendly, resource- and energy-saving technologies, renewable energy sources, intangible nature management, which requires significant corrections in the management of social processes at different levels.

One of the ways to implement the strategy of a sustainable global society is the model of the "blue economy", the ideologue of which is Gunther Pauli. It is a modern business philosophy based on the logic of globalization, which represents an updated paradigm of harmonious coexistence of man and the environment, ecological and social systems. G. Pauli promotes waste-free production based on cyclicity, where the waste products of one production become the raw material basis of another. Thus, the process of coexistence and productive interaction of economic, ecological and social spheres is harmonized [14].

A variant of such an approach is the "cradle to cradle" model of M. Braungart and W. McDonagh, which involves the reuse of materials or elements of such materials to create other materials. This model works in parallel with their concept of "waste is food", by which they understand that waste from one system or process can become food or raw material for another system or process. This model is based on the principle of metabolism (metabolism between the body and the environment, which is the basis of the mechanism of hybridization) - biological and technical. A substance circulating in biological metabolism and subject to biological decomposition becomes food for other living organisms. Materials that are not biodegradable, are considered as technical raw materials and create technical metabolism [15]. Obviously, the implementation of such advanced technologies requires the formation of new thinking and appropriate organizational and managerial decisions.

3 Conclusions

Based on the above, we can conclude:

The need to form a new paradigm of management of social processes in Ukraine is due to a set of problems associated with the long transition state of Ukrainian society and the need to find the most effective management solutions to address these problems in conjunction with leading global trends. The basic worldview principles of the new paradigm of social process management can be the following: the priority of sociality and spiritual and creative essence of man, which contributes to the harmonization of relations within Ukrainian society and activation of creative potential of all members; awareness of hybridity as a universal means of interconnection and interaction of various intra system and intersystem elements, as well as global and national to improve their own social model; bio-psycho-techno-social integrity. These principles in general, the strategies of development of Ukrainian society, taking into account its national specifics and general tendencies of development of world civilization, correspond to the greatest extent.

References

1. L.L. Lazebnyk, V.O. Voitenko, *Economics Bulletin of Dnipro University of Technology*, **1**, 22-29 (2020)
2. S.O. Perminova, *Public administration: improvement and development,* **1** (2020) http://www.dy.nayka.com.ua/?op=1&z=1550
3. V.Iu. Lysak, V.M. Semendiak, *Global and national economic problems*, **21,** 548-552 (2018)
4. R. Condorelli, *Open Journal of Applied Sciences,* **06(07),** 422-448 (2016)
5. S.A. Seheda, *Economy and state,* **10,** 14-22 (2018)
6. L. Margulis, Endosymbiosis https://evolution.berkeley.edu/evolibrary/article/history_24
7. V. Taran, *Regional Revue*, **1**, 186-198 (2020)
8. H. Khaken, P. Plat, V. Ebelynh, Yu. Romanovskyi, *Ob ystoryy synerhetyky. Obshchye pryntsypы samoorhanyzatsyy v pryrode y v obshchestve* (About the history of synergetics. General principles of self-organization in nature and in society*)* (Institute of Computer Research, Yzhevsk, 2017)
9. R. Gloverberg, *Texila International journal of management* https://www.texilajournal.com/management/article/1389-the-theory-of
10. S.O. Hudkov, *Bulletin of KNTEU*, **1**, 45-55 (2018)
11. H.T. Tereshkevych, *Democratic governance*, **22** (2018) http://dv.lvivacademy.com/issue/view/9942
12. Decree of the President of Ukraine of September 30, 2019, №722 "On the Sustainable Development Goals of Ukraine until 2030" https://www.president.gov.ua/documents/7222019-29825
13. M. Bublyk, M. Bay, The concept of green economy and its difference from the traditional economic model (2016) http://www.economy-confer.com.ua/full-article/2055/
14. G. Pauli, *The Blue Economy 3.0: The marriage of science, innovation and entrepreneurship creates a new business model that transforms society* (XLIBRIS, Bloomington, 2017)
15. F. Trampenaars, P.H. Cooberg, Trompenaars Fons. 100 key management models and concepts (Mann, Ivanov and Ferber, Moskow, 2019)

Forecasting Career-And Competence Indicators of a Social Worker in the Context of Digital Transformations of the Society

Mariia Kuzmina[1], *Viktor* Kuzmin[1], *Yurii* Mosaiev[2], *Natalia* Karpenko[1*] and *Kyryl* Tarasenko[1]

[1]"Zaporizhzhia Polytechnic" National University, 69063 Zaporizhzhia, Ukraine
[2]Municipal Institution of Higher Education "Khortytsia National Educational and Rehabilitation Academy" of Zaporizhzhia Regional Council, 69000 Zaporizhzhia, Ukraine

Abstract. The article is devoted to the problem of forecasting the career and competence indicators of a social worker in the context of digital transformations of the society. The paper analyzes the social and historical context of the formation of social workers' career competence indicators. The article compares the analogue and digital system of career competence indicators of a social worker.

1 Introduction

Career indicators of a social worker, like any specialist in various spheres of social existence, directly depend on the professional and life competencies they have acquired. As the competences have already been acquired, the person has the opportunity to implement them in the sphere of his/her own professional activities. Taking into consideration that nowadays there exists a practice of uniting professional and other aspects of life, so the process of living combines the mentioned above aspects in one discourse. Taking into account the specific character of social work which includes both social and practical activity as well as its historical peculiarities we can outline the necessity of investigating career and competence indicators and predicting its modifications. The specific character of Ukrainian society and other societies of the modern world is that the process of digital transformation is becoming more and more active.

The aim of the article is to investigate the process and peculiarities of forecasting career and competence indicators of a social worker in the context of digital transformations of the society.

2 Research methods

The basis of research includes general scientific methods such as induction, deduction, extrapolation and others. Method of sociological modeling has been used according to algorithms of structural functionalism, integral sociology of P. Sorokin, elements of the theory of social conflicts of R. Darendorf in order to analyze the forecasting of career and competence indicators of a social worker in the context of digital transformation of society. Method of social and mental assessments has been used to study the national peculiarities of the formation of career and competence indicators of a social worker.

3 Theoretical aspects of forecasting career and competence indicators of a social worker

Social work is a relatively new area of social activity. In spite of the fact that it traces its history since ancient times, its formation started in the beginning of the XXth century, exactly when Mary Richmond presented namely a professional approach to social work. It was she who managed to formulate and implement the first technologies and protocols of practical social work as well as work out its theoretical basis [1]. Every textbook and monograph on the History and Theory of Social work contains at least one chapter devoted to this outstanding woman, but it should be taken into consideration that the methodology of social work was based then mainly on Psychology, Social Pedagogics, Pedagogics, Defectology, Sociology etc. This situation is caused by the fact that social work was formed as a multiparadigmal sphere of practical activity aimed at providing social assistance and social services for the people who appeared to be in difficult life circumstances

In addition, one should keep in mind that appropriate methodology of social work as a scientific discipline was formed much later on [2]. Only after that social work was formed as an appropriate scientific and academic discipline and the field of practical activity. It should be emphasized that social work was based on charity, and in many countries of the world, social work as a professional activity was based namely on the basis of individual charity of patrons [3]. In countries with dominant protestant traditions, the philosophy of social responsibility of business has become the basis for the formation of an organized social assistance system.

The bright example is the activity of Andrew Carnegie aimed at developing this philosophy [4]. All

* Corresponding author: natalyakarpenko14@gmail.com

these facts indicate that the formation of social work as a scientific discipline and sphere of practical activity did not happen systematically but even sometimes chaotically. That is why there were no specialists who would be systematically engaged in providing social assistance as people started to organize and implement social work and household social assistance at the bidding of the heart. All of this shows, however, that the main competencies of the first social workers were empathy and personal interest in spreading humanism and principles of mutual assistance. All these factors caused that the formation of social assistance and social service system all over the world was held according to the principles which did not suit the formation of professional competencies of social workers, especially since the issue of forecasting these professional competencies was not actual yet.

Under the notion “career” we mean social and labor relations of an individual throughout his/her life. In the context of social work, a special feature of a career is the high level of social responsibility and empathy of participants of the social process. A special feature of a career in the context of social work is that the career growth process is not obvious due to updating changes of social problems, and a social worker is highly likely to need retraining, which can lead to horizontal social mobility and even to a decrease of social status.

Carried out analysis demonstrates there is no universal and generally accepted interpretation of the notion "career". At the same time there is no universal conceptual content of this notion. The change in the general approach to this notion in its conceptual sense has happened within the period of time: we mean a transition from the concept "career which leads to the promotion in the professional sphere" to the concept "life is a career". This of course influences the researchers' approach. Thus, Ukrainian and Russian sociologists define "career" and "career strategy" as moving up the ladder, first in the professional sphere, and then in life in general, while western scientists define career movement as moving in a vertical straight line.Taken into account this feature, E. Shane has developed a classification system for career strategies, which is used by most sociologists in the world and is generally accepted. Taking into account this system of classifications and certain features, we conclude that the conceptual essence of a career strategy is mainly the choice of an individual aimed at success in professional activities and in life in general.

The conceptual content of the notion "career", which is relatively new in Sociology, is not clearly defined, although the problem of career has already been reflected as an object for consideration. Taking into account the definitions of the concepts "career" and "career strategies" proposed by sociologists, we can determine the conceptual essence of these concepts. The analysis of scientific publications and scientific works on this topic makes possible to characterize the term "career" in a conceptual sense, clearing out two vectors of its reception: the first one is formed on the basis of definitions proposed by the classics of Sociology, and characterizes it as a process of professional growth, which determines the general need for self-development and self-improvement; the second one is defined by the majority of sociologists of the XXth century, life is a career, it is an uninterrupted process in a certain direction of professional and personal growth [5].

If we take into account the social portrait of the first social worker, we can observe a gender and age disproportion in this aspect. Gender disproportion and dysfunction is connected with the fact that in the late XIXth and early XXth centuries, a social vector to a high level of empathy among women and a social demand for their participation in charitable activities was the mainstream. The mentioned above social vectors and social demands in the middle of the XXth century turned into social stereotypes that have not yet been overcome in the most developed countries of the world yet, not to mention the countries with a lower level of development. In addition to the gender disproportion one could easily observe an age disbalance among the first social workers. Social work included mainly younger and older employees. It can be easily explained that that the two mentioned above categories of the population showed the highest level of empathy. Young people showed it because of the so called “youth maximalism” and humanitarian education, and older people because of the desire to compensate the low level of empathy in the period an active professional career.

Overcoming the conscientious lack of professionalism of social workers and their involvement in the social service system began after the monopolization of the sphere of social assistance and social services by the state. In the 20s of the XXth century this tendency occurred in many countries of the world, especially in countries that have chosen the socialist and paternalistic ways of developing the social sphere. In countries mentioned above, instead of social work, they increased the development of active social policy, providing at the legislative level revolutionary changes in the field of providing social standards by ensuring the right to maternity leave, an eight-hour working day and leave with financial support [6]. In this situation, social work became more and more bureaucratic, which led in the countries of the socialist vector to the total bureaucratization of all spheres of social assistance activities in future, meanwhile, a prototype of the system for predicting career and competence indicators of a social worker were implemented. Due to the lack of digital infrastructure, we will consider this stage as analogue one.

Under analogue format of forecasting career-competence indicators of a social worker, we consider the formation of career-competence indicators of a social worker by analogues of the development of (socio)political and (socio)economic processes of social being and the main trends in the professional engagement and interaction between social institutions within a particular society.

The important element in the formation of professional competencies of a social worker is the model of social policy which dominates in this or that country. For the countries with a dominant social democratic model of social policy (Germany, Austria, Denmark), it is accepted to form the competencies of social workers based on solidarity, solidarism, social responsibility and a high level of personal empathy. In the context of the Anglo-Saxon model of social work (Great Britain, USA, Australia), it is accepted to form the competencies of a social worker based on the principles of updating the problems of a specific social problem and specific professionalism within a particular sphere of social services. Meanwhile for the Latin model of social work (Italy, Spain and Portugal) it is normal to base on Catholic religious ethics, religious solidarity and the principles of the social state.

The main features of the analogue format of career and competence indicators of a social worker are as follows:

- Implementation of instructions of the state leadership and local governments as a counterbalance to the needs of people who find themselves in situations of extreme difficulties. Namely this factor has led to a low level of trust to a particular social worker and social work in general in many countries of the world. Within the framework of authoritative and totalitarian regimes, social workers have become full-rate agents of the anti-popular system. That is why in countries where the above mentioned principle dominated, people were mostly unwilling to use social services and to deal with specific social workers in general;
- The high level of formalism and lack of flexibility in the work of a social worker with a specific case led to the fact that the act of helping a particular person or family has turned into a small bureaucratic operation, which appeared to be an imitation of the providing of social services or social assistance. As a result, the domination of this analogue feature of career and competence indicators of a social worker leads to low efficiency of both his specific activity and the social service system as a whole;
- Differentiation of contact with a specific person regarding the age, gender, and social status of the client. This analogue factor was that the client's achievements for the country, state system, or gender and even age could lead to social inequality of social service clients to the social worker, who could provide high - quality social assistance for certain clients, while others were deprived of it. For example, in the USSR, better social assistance was provided for veterans of World War II and socialist labor. And it happened both in the sphere of social assistance and social protection. It became particularly visible in the field of Health Improvement and the possibility of getting rare goods and services firstly.

The examples of individual analogue career and competence indicators of a social worker indicate the inefficiency and non-objective character of social workers' activities stimulated in some countries. Of course, this factor in the activities of social workers in developed countries of Europe and North America differed much from the social sphere in countries with authoritarian and totalitarian regimes, but it also had a lot of drawbacks. Therefore, positive changes in the system of the analogue system of career and competence indicators of a social worker started with changes within the framework of the "person-state" system. These changes became possible due to theoretical developments of the French School of postmodernism and, in particular, the works of M. Foucault, who proclaimed the deconstruction of traditional knowledge and the transition from their rational forms to emotional ones [7].

It was the verystage which was marked in the Humanities and social sphere as a stage when interpretation dominated formal content. So if taken in the paradygm of an analogue organization of social work, a social worker was a specialist who formally provided a particular form of social assistance, that means the professional competencies of a social worker were reduced to a step-by step application of certain regulations in this area. This form of analogue competence was characteristic not only for social work, but also for other spheres of social being and professional activity.

That is why in the paradigm of French postmodernism [7], the possibility to interpret not only situations from their everyday life and to make decisions based on their own competencies was offered to each individual person. But in this case they got the same opportunity. This situation was familiar to social workers either. It was the time when the idea of forecasting and forming professional competencies of both social workers and representatives of other specialtiesstarted its formation. The main reason for this state of affairs was the situation of the beginning of a rational attitude to stuff politics both at state-owned enterprises in the countries of the socialist vector and at private enterprises in the countries of the capitalist world. Of course, this trend was current for the social sphere. This led to a new wave of volunteers. People began to trust social institutions and institutions more, and they began to show a high level of empathy for people who found themselves in difficult situation, making this project either independently or together with representatives of social institutions, with social workers directly.

Later on the possibility of social workers in their activities became flexible and began to depend in some aspects on the level of emotional intelligence. This category means the ability to express emotions and perceive the emotions of others [8]. Namely in the early 80s of the XXth emotional intelligence became one of the leading competencies of a social worker. Since passive empathy has already existed and required increased emotionality according to the trend of social development in Western civilization, and therefore today social workers should have a high level of emotional intelligence. Since there is a high level of emotionality and the ability to develop ones own emotional intelligence and the colleagues'emotional intelligence and people who are in trouble. Modern psychological and sociological concepts consider emotional intelligence as an important universal life-and-professional competence that makes possible to establish communication in your professional activities and personal life.

4 Practical aspects of forecasting career and competence indicators of a social worker in the context of digital transformations of the society

Forecasting the career and competence indicators of a social worker gained its significance in the context of digital transformation. Over the past 30years total digital changes have become a strong tendency and they are, unfortunately, can't be observed in Ukrainian social work. At the same time, it is impossible to tell about the high level of implementation of new technologies into the social sphere of other countries, and is not only Ukrainian burning issue, but also has a regional context at least.

The 90s of the XXth century contributed much to the professional competencies of social workers: the so called office-competences were added to the list. Due to the fact that social work in any country has never been a highly paid profession, and therefore this sector mainly includes people over middle and (before) retirement age. Therefore, in the era of information technologies, social workers have not been able to master these skills due to age-related difficulties and lack of information culture. If we consider Ukrainian realities, we can draw attention to the fact that Ukrainian society missed the global information transformation which took place in the early 90s. Our country was engaged in the restructuring of social processes and the majority of the actively young generation did not integrate into the Information Society, which was founded in North America and Europe.

So in the 90s, there was only a tendency to develop the information society as well as system social programming both at the interregional, national and regional levels, and already in the middle of the first decade of the XXI century, both Ukrainian society and the social sphere found themselves in a situation where all social groups except young people were not integrated to the global level of it. This allowed young people to take advantage of a number of social ladders that allowed them to increase their social status by means of professional implementation. At the same time, this trend has not developed in the social sphere yet. Since young people did not evaluate the prospects for the development of the social sphere as a sphere of professional implementation, and the lack of basic equipment for the implementation of these skills until the middle of the first decade of the XXth century led to the fact that the older generation of social workers did not want to implement computer competencies in practice, and the younger generation simply did not realize the perspectives and appropriate tools for it.

A particular solution to this problem was to provide computer equipment, to connect existing social workers to the network at the end of the first decade of the XXI century and to organize trainings for the stuff of social institutions and social services. At the same time, professional requirements turned out to be minimal for social workers and were reduced to the necessity to use office programs, print out text documents and simple tables. So, most Ukrainian social workers in the second decade of the XXI century aqiured only computer skills at the level of an office worker of a commercial company of the late 90s of the XXth century. Though ideally, their skills should have been extended to the level of primitive programming of local network products for the social sphere and developing or updating mobile content for various social service clients. Surely, social workers would not have to become full-rate programmers, but their professional skills could contribute to the formation of software and mobile applications if provided a developed information culture and appropriate information ecosystem within the social sphere. Social workers from different social institutions could have formed a request for appropriate software products. If such a system had been functioning for the past two decades and this would have created a number of relevant IT professional competencies for social workers. It is also important to ensure the formation of a highly effective culture among social workers. It is proposed to use motive the game in adapting new social personnel and in building a workflow for employees [9].

If all the mentioned above had been implemented, so we would have the following results now:

- The professional skills of an ordinary social worker would include computer skills which could be higher than that of an ordinary office worker due to the existence of global, local and regional computer specialized networks that social workers would use to create standard documents in the sphere of assessing the quality of social work technologies with different groups of social service clients, depending on circumstances;
- Appropriate resources and algorithms would be created to test, aprobate and modernize existing methods of social work. Moreover, in all the years of Ukrainian independence, many topic related conferences, seminars, round tables and other analogue scientific events are held, where various topical issues for the social sphere are still being constantly discussed. But practically, the problems we have mentioned have hardly been updated in this sense. This shows, however, that social specialists had a large number of current organizational problems and the topic of creating network resources to make a full-rate information "community" has never been actual for them. The situation has not changed later on, at the beginning of the second decade of the XXI century, whatever social sphere could be chosen;
- Total information system development at the beginning of the XXI century was expected to initiate the formation of a request for programming social processes in order to solve the problems of specific people who appeared to be in difficult life circumstances. Although social workers could be actively involved in the process, this problem is actual

 https://doi.org/10.1051/shsconf/202110004005

in the research by Bulgarian scientists Venelin Terziev and Magip Georgiev, they raise and update this problem both at the national and European levels [10]. Without updating this problem today, social work as a whole and specialists who ensure its functioning may find themselves out of actual social processes taking place in modern Ukrainian society and cease to be in demand in modern and postmodern innovative social practices. This disfunction will finally lead to the marginalization of social work of both social practices and institutions, as well as individual specialists involved in its functioning;

- The era of mobile applications continued the total information-developing of social processes, and social workers, due to the importance of their own specialty to ensure the harmonious development of society, should have been the first to master it. Due to the fact that this social aspect has not been implemented, as a result, today there are dozens of mobile applications in the smartphone of every Ukrainian that can help a person learn a foreign language, order a taxi, food, communicate with friends and relatives, but at the same time there is no way to get social services or at least to consult about it. This state of affairs indicates brightly the lack of a system for forming professional skills of social workers in the field of Information Technology.

The analysis is of advanced pedagogical experience has shown a wide range of research in the system of development and formation of students' competence. It has demonstrated that the formation of students ' competence happens under the influence of a number of factors and reasons among which practical training is quite important. It is shown that implementation of practical training tools by teachers in the educational process will improve the professional training of students in accordance with the requirements of the labour market and potential employers. Analysis and assessment of practical results of the formation of students' competence by means of practical training have demonstrated a number of trends. Thus, the analysis of the results of the PISA-program which has been worked out by the Organization for Economic Cooperation and Development (OECD) showed the highest indicators of the average score in Mathematics, Reading and Science in Austria, Belgium and Germany, compared to the average score in the OECD countries. It was found that the average score of students in Mathematics in Ukraine in 2018 is 21 points lower than the average score in the OECD countries, Reading is 36 points lower, and Science is 20 points lower. Analysing the dynamics of students, who get higher education in the EU member states in general and in German-speaking countries in particular, the tendency for increasing of this indicator has occured. Thus, in 2018, compared to 2015, the number of students in Belgium, Luxembourg and Germany increased significantly (2.09%, 2.09% and 4.80% accordingly), while in the EU member states the growth was only 1.65%. Analysis of the dynamics of changes in the number of students as a percentage of the population aged 20-24 years in German-speaking countries showed the highest level of students getting higher education among the population in Belgium aged 20-24 [11].

Nowadays one of the options for improving the situation in this area would be the way of implementing optional courses for the students (social work), who will be engaged in creating mobile applications, or at least establishing links between students and young specialists in the field of social work and technical specialties, especially in the areas of programming, system analysis etc. Of course, this will not lead to their becoming programming specialists in future, but it will create a common interdisciplinary field that will expand the horizons of social workers in the information sphere and programming. In the longperspective, this would give a significant increase in the technical competencies of social workers. This system would be especially effective if employers set social workers the task not only to acquire such knowledge and competencies, but also to encourage social workers with financial support. Due to the lack of it in this area and the lack of appropriate knowledge, such competencies are hardly to be found among social workers.

One of the options for spreading technical and informational skills among social workers is to create social worker positions in social services with the skills of a technical administrator and a social network worker. It may seem that the first position is not relevant due to the existence of such a position in social services, which is replaced by professional programmers or technical specialists. Our idea is to replace these specialists with social workers with technical or programming skills. This replacement should take place due to the fact that technical workers fulfill this work in a standard way without taking into account the specifics of the social sphere.

On the one hand, it may seem that the technical specifics should not differ within the social sphere and, for example, administrative management, but a technical worker with the skills of a social worker has the ability to adjust all technical settings not only according to technical protocols, but also according to the logic of the development of social processes that occur in society and trends in the development of the social sphere. The problem under consideration is relevant because today we need to involve social workers in creating new mobile applications. It would be aimed at consulting and providing social assistance in a smartphone. And today this is a time requirement that may become not up-to date, as mobile applications are developing rapidly, and neither in Ukraine nor in the vast majority of countries of the world has not developed appropriate mobile applications yet, but even a concept on this issue hasn't been formulated.

In our opinion the main concept of creating mobile applications for social services can be expressed to the following items:

1. These mobile apps must be created in cooperation between mobile application programming specialists and social workers. This cooperation should combine the experience of programmers and practical

social workers in order to better develop and continue their functioning. Realising namely this aim, training courses and educational programs in the field of social work should provide acquiring competencies in system analysis, programming theory and, surely, visualization of information and educational products.

2. In addition to the main aim, mobile applications must have a pedagogical component. According to Kh. Tangirov, mobile applications and other information technologies play one of the most important roles in the modern educational process and their potential is far from being totally used [12]. Mobile applications related to the functioning of social services with different groups of clients should include both informational and educational materials. That is why social workers should develop competencies in training important skills among social service clients that will help them overcome difficult life circumstances in their lives. In general, online-consultations and teaching skills in the same form enable practical social workers to make their work interactive and up-to-date.

3. Taking into account the fact that many social workers during their studies in higher education institutions mostly received theoretical knowledge and did not have the opportunity to test their knowledge in practice, so in order to avoid this case, social workers should be given a task that contains the necessity for practical skills [11]. Despite certain educational drawbacks, future and current social workers take full responsibility for the lack of sufficient practice. Since it often happens that social workers, learning mostly theory, do not show initiative in the field of obtaining additional practical training. This process can be divided into two phases. The first one is when future social workers, while taking introductory, training, industrial, professional and other types of practice, are not interested in gaining practical experience themselves, as they want to get functional responsibilities that are as simple as possible to perform and do not require enough time to complete. This cognitive position affects the overall competence negatively.

So, the lack of a social worker's constant desire to improve not only by learning theoretical knowledge, but also by mastering practical knowledge is a negative sign and not a desire to work efficiently. Another way to avoid the opportunity to learn new knowledge is simply absence of desire to learn more, including also practice.

4. The problem mentioned above is typical both for social work students and for practical social workers with appropriate education and experience in social services and other institutions providing social services. In this case, there is a problem of lack of necessary professional competencies among young social workers who are starting their professional career and social workers who, due to the peculiarities of their own professional activities, could not get professional competencies that are vitally important for their work. In our opinion, this problem can only be solved by simulating specific cases of providing social assistance and social services. To fulfill it, one need to create appropriate social motivation, such as stimulants for airplane pilots and medical stimulants, which are based on communication with virtual patients. For most young social workers, such a form of mastering professional competencies would be very useful as it is common in form and resembles a game, and may even be a full-rate mobile game with its own character (Fig. 1).

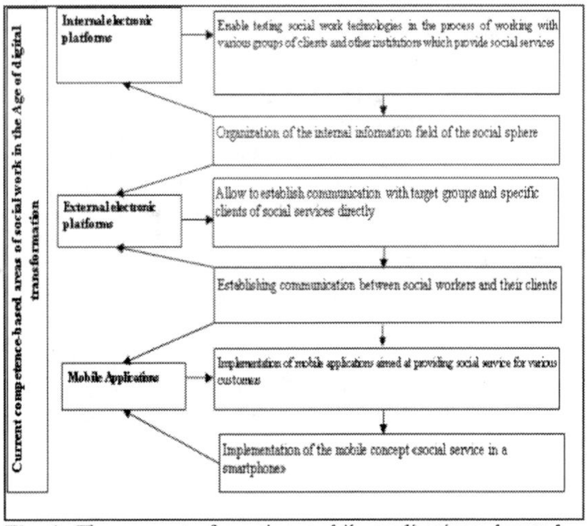

Fig. 1. The concept of creating mobile applications devoted to social services activities in the Age of digital transformation

From a practical point of view, it promotes the creation of game content on mobile devices devoted to social services and the formation of an innovative set of visualizations of the most common effective social practices that are important for a person to overcome certain social difficulties. Nowadays, we can assure that there are some existing examples of such social content in the world, but we also realize the high probability of its updating. This state of affairs gives us the opportunity to predict that for the implementation of such projects, such competencies of a social worker as the need to create positive ways out from difficult life circumstances by using of certain technologies of social work. In addition, a social worker must independently or by cooperating with his colleagues predict the probability of a positive or negative result of using a particular technique based on the peculiarities of the difficult life situation in the life of a particular person.

Nowadays in order to form the students' competence in the academic process it is necessary for the teachers to fully apply not only the well-known means of practical training (implementation of module educational program, forming the necessary level of theoretical basis, implementation of educational, training, pre-graduate and other kinds of practical component in the process of educating students, the usage of information, innovative and interactive technologies, training students in accordance with the requirements of the labor market and employers, providing cooperation between higher educational institutions within the framework of student exchange programs), but also implementing the so-called "social-services" game content on mobile devices and the formation of an innovative complex of visualizations of the most well-known, effective social practices which are extremely important for a person to overcome life difficulties.

Conclusions

Forecasting the career and competence indicators of a social worker is an important component in the development of social work. This has become particularly important in the era of digital transformation. New information technologies are rapidly changing the social reality and setting new competence guidelines for representatives of various specialties, including social workers is very important. Modern social workers, despite the absence of clear rules and competencies, need to master them, including Information Technology and Programming, as social sphere today requires high-quality technological solutions to provide high-quality social assistance and modern social services in working with various groups of people.

References

1. M. J. Austin, Social Service Review, vol. **92**, 4, (2018). doi: 10.1086/701278
2. J. Steve Hothersal, Eur. J. of Social Work, vol. **22**, (2019). doi: 10.1080/13691457.2018.1499613
3. S. Mapp, J. McPherson, D. Androff, S. Gatenio Gabel, Social Work, vol. **64**, (2019), https://doi.org/10.1093/sw/swz023
4. A. Zhulina, Performing Philanthropy from Andrew Carnegie to Bill Gates, vol. **23** (2018). doi: 10.1080/13528165.2018.1533762
5. V. V. Kuzmin, The Bull. of Y. Mudryi National Law University. Series:Philosophy, philosophies of law, political science, sociologyVol. 1 No. 48 (2021). doi: 10.21564/2075-7190.48.224813
6. A.A. Gumenyuk, Klio. **3** (87), (2014), p. 65-72
7. J, Guilhaumou, Presses Universitaires de Provence (2016), p. 267-289
8. J. Wena, Intern. J. of Hospit. Manag., vol. **81**, (2019), p. 120-130
9. A. Karpenko, K. Basenko, Balt. J. of Econ. Stud. **3**(4), 101-106 (2017). doi: 10.30525/2256-0742
10. K. Tangirov, Mental Enlightenment Scientific-Methodological J., (2020), p. 76-84
11. M.O. Kuzmina, O.L. Protas, T.V. Fartushok, Y.M. Raievska, I.B. Ivanova, Intern. J. of Higher Educ., **9**(7), (2020), p. 279-288
12. Kh. Tangirov, Didactical possibilities of mobile applications in individualization and informatization of education. Ment. Enlight. Sc.-Met. J. **1**, 76-84 (2020)

Use of Innovative Technologies in Training of Future Social Workers

Valentyna Sanzharovets[1], *Tetyana* Petrenko[1], *Iveta* Mietule[2], *Larysa* Platash[3] and *Liubov* Kostyk[3]

[1]National Aviation University, 1 Liubomyra Huzara ave., Kyiv, Ukraine
[2]Rezekne Academy of Technologies, 115 Atbrivosanas aleja, Rezekne, LV-4601, Latvia
[3]Chernivtsi National University named after Yuriy Fedkovich, Kotsyubynskoho St., 2, Chernivtsi, 58012, Ukraine

Abstract. The article considers the actual problem of using the innovative technologies during the educational process in higher education institutions for training of future social workers. Professionalism in the activities of a social worker is in constant motion. Actions that take place in a society require constant professional changes in methods and technologies of social work. Social work is inseparable from society as a whole and is aimed at the population. Designing in social work is a constructive, complex, creative activity, the essence of which is in advanced reflection of reality in order to transform the personality of oneself, own life situation and the surrounding world with the help of a social worker. Social work refers to the category of innovative, creative technologies, because it involves a change of reality. It is based on the existing life situation, which can be unified, mastered and refined. A social worker is the initiator of innovation, the purpose of which is to create, modernize or maintain in a changed environment material or spiritual value of people, recognized as positive in its social value.

1 Introduction

Modern Ukraine, during the period of its development, requires higher education sphere to prepare a free, socially active, communicative, responsible, professionally trained person, capable of self-development and a specialist who knows his own path, chooses it consciously and responsibly. The training of such specialists requires the use of the latest technologies and teaching methods in the educational and pedagogical process of higher education institutions.

The formation of professional social work in our country is associated with the search for new approaches, development, implementation of innovations in the life of society and activation of innovative processes in the social sphere. This is especially important in today's socio-economic conditions of the cardinal renewal of all life's foundations, of the need to adapt a person to ever-changing circumstances and for his effective social protection.

Today, the world is experiencing crisis phenomena that have affected all spheres of human life without exception. Thus, it should be noted that economics, politics and culture are three interdependent levers: the development of one of them promotes the development of the other two and vice versa - the inhibition of at least one has the effect of total destructive processes.

Moreover, constructive or destructive processes in these areas affect the life of society as a whole and the individual in particular.

Social work, as defined by the European Association of Schools of Social Work, is an academic discipline and professional activity that promotes social change and development, social cohesion, and the empowerment and independence of individuals, involving individuals, groups, and societies to the deciding the problems of persons and families who are in difficult life circumstances and improving the quality of their lives and well-being, on the principles of social justice, respect for human rights, collective responsibility and respect, counteracts the factors of social exclusion and promotes respect for human rights.

The most important goal of social policy is to ensure more complete satisfaction of the members of society with their vital needs, because it is on this occasion that social relations are formed in the first place, and it is on this basis that the social reproduction of social groups and individuals takes place. At the same time, when it comes to meeting the vital needs of people in the implementation of social policy, we mean their satisfaction is not through the own efforts of members of society. Social policy embodies society's concern for people in need of support and assistance. In other words, social policy is aimed at providing assistance and support to people from the social system in which they belong, by using the social potential available in the relevant social system.

* Corresponding author: valyasan@ukr.net

However, it should be noted that the main purpose of social work is to promote the normalization of human life, meet and harmonize its social needs, harmonize the needs of the individual and society.

Today, society has many theories of art and creativity, which help to see beyond the standard look at existing problems and offer their non-traditional solutions. The basic preconditions for creative work, the development and implementation of innovations in the field of social work are social needs, as well as social problems. These arise in the life of society, its individual layers and individuals, and require for its solution new forms, methods, means, models, programs and projects of social work and social assistance.

2 Theoretical Aspects of a Research

Students of higher education institutions are graduates of the 11th form, which can be attributed to the youth. Juvenile age is the most sensible in terms of upbringing respect, self-esteem and its development, etc. Young people of this age are trying to grasp the basic principles and laws of morality, to deduce the basic rules of living together in society. At the same time, young people tend to overestimate the level of their own knowledge and abilities, while their life orientation is not yet well-formed. Here is "Me" blurred, while continuing to seek their own identity.

The peculiarity of adolescence is vulnerability and inconsistency in judgments, vital observations and conclusions. Children of this age combine high demands on others (parents and teachers) and uncritical attitude towards oneself, their own thoughts, actions.

For primary school students, there are typical crises that are explained by changes in social status, adaptation to new conditions of life, new rights, demands and responsibilities, namely:

✓ The crisis of professional choices, based on the dissonance between the unconsciously chosen profession and the need for education;

✓ crisis of dependence on the parent's family;

✓ crisis of intimate-sexual relations;

✓ identity crisis.

In this situation, young people need appropriate support and assistance from parents and teachers. An important factor in the education of self-esteem among young people is the formation of a demanding to oneself, which determines not only self-knowledge, self-esteem, but also the desire to work on oneself, the self-realisation, including in the professional sphere. In order to feel confident in a situation of constant change and to respond adequately to them, the social worker must increase his creative potential, be able to help the client to activate his creative skill and to direct them to solve the problem. All this determines the task of professional training of a social worker - the development of a future specialist, ready for the creative exercise of professional activity.

Social workers and social specialists provide advising and counseling to individuals, families, groups, communities and organizations in the event of social and personal difficulties. They help clients develop the necessary skills and provide access to appropriate resources and support services needed to respond to problems related to unemployment, poverty, disability, addiction, deviant behavior, including criminal, family problems, etc.

Their main tasks are:

• conducting an assessment of the needs of social services clients;
• interviewing clients individually, as a family or in groups, to assess their condition and problems, as well as to determine the necessary services;
• analysis of the client's situation and identification of alternative approaches to solving problems;
• provision of social services to their recipients in accordance with the concluded agreement on the provision of social services and individual plan;
• providing counseling, mediation services, group classes to assist clients in developing the necessary skills and knowledge needed to solve their own social and personal problems;
• planning and implementation of client assistance programs, including crisis intervention and referral to institutions that provide financial and legal assistance, housing, medical treatment and other services;
• participation in the investigation of cases of negligence or abuse and taking measures to protect children and others at risk;
• representing the interests of different groups of clients in the community, as well as lobbying to address the problems they face;
• development of prevention and intervention programs to meet the needs of the community;
• liaising with other social, educational and health care providers who work with clients to provide information and provide feedback on the general condition of clients and their achievements, etc .;
• formation of publicly available information resources that contain information about the activities of social service providers and provide access to these resources by placing them on information stands in the premises of social service providers, in the media, on the Internet.

Professional training of specialists in social work is based on qualification requirements that combine professional competence, intellectual, personal and moral qualities, which are an indicator of the success and effectiveness of their activities. Representatives of this profession must have the established moral qualities and properties, because it is morality that regulates the human relations between people in different situations on a democratic basis.

Morality, as spiritual quality of a person, the need to bring good to people, is the basis of the personality of a social worker, and it is trained during the educational process in higher educational institutions. These are the moral qualities of a specialist that largely determine the behaviour of the client, his social well-being and psychological state. And this, students can learn while studying the subjects “Social work with Families”, “Social work with different groups of customers”, operating with the Code of Ethics of a social worker.

By teaching disciplines, we offer students to fill the wheel of moral qualities, and to specify the main eight qualities of a specialist. The student first assesses how many points, according to this personal point of view, this quality is important. Then the person assesses the number of points, which specify the presence of this feature in itself. Often between the desired and available there is a difference in a certain number of points. Everyone then determines what needs to be done to become kinder, more tolerant, etc. That is how we start the path towards improving or gaining the moral qualities of a future specialist.

The main ethical principle of social worker’s activity is respect for the dignity of every person. The principle of respect for the dignity of each person should be understood as a special moral attitude towards the person, recognition of his rights and opportunities.

Students can be offered the following form of work. Each participant determines the main moral quality that he considers necessary for a decent living, work, communication with the environment, and writes a statement on paper using a certain feature. Then I change the sheets, and the students get notes of another. The task before us – is to reveal the content of the sentence. For this we can suggest to come up with a story or to give an example. Next we compare the written content with the idea, which was enclosed in the statement, first source.

Disturbance against human dignity are physical violence, disrespectful attitude, injustice, inequality, rudeness, psychological pressure and trauma, etc. Every person is extraordinary and unique, which must be taken into account by social workers without any manifestation of disregard for the individual. Everyone has the right to self-realization, which does not lead to a violation of similar rights of other people. Social work is incompatible with direct or indirect coercion of clients to any action, even in favour of the client or his close social environment.

Communication with the client is the integral part of social worker’s activity. Unfortunately, we often see a decline in the culture of communication. We often see professionals who are impatient with clients with special needs, the elderly, and whose communication does not seem to look like civilized [8].

Studying with students "Social Gerontology", we offer them to prepare proverbs and sayings about old age and elderly age, to pick up excerpts from the works of art we discuss. Students are offered to make a conclusion about the attitude towards the elderly people, to provide forms of work with the gerontological group of the population.

The effectiveness of social work depends not only on professional attitude to the customer, but also on the attitude to oneself. Only those who respect themselves are fair and respect others. Word is a way of communicating with others, the way of transmitting information, own thoughts, desires and requests. The word can get you interested, cheer up, heal. Each spoken word has a different level of vibration and has a certain energy and emotional appeal. Words of a social work specialist are a powerful tool that affects the level of personal energy of the client, emotional balance and health. Words with low vibrations, negative references that restrict beliefs negatively affect the state of health and human behaviour. The first step is to fill life with words with high vibrations is awareness. Through the customer's observation of his phrases, which he describes himself or his life, the social worker determines the methods and technologies that will be used in his work in the conduct of the case. A person who is accustomed to inventing others in his own failings sees only the injustice of destiny and does not pay attention to his own disadvantages needs special attention. Therefore, in order to change the internal world of such a client, it is necessary to start with value orientations, to pick up special words, signs, focusing on the perfection of the individual. The words of the social worker are the paint he applies and teaches with the help of creativity and meditations to expand the client's outlook. Intelligently and consciously choosing methods of creativity in social work are needed for client to realize his life and desire to change it. [6].

A social worker has to deal with different categories of clients, and regardless of age, social status, financial position, the social worker should place the client's personality and his interests in the centre of his professional activity. Adherence to the principles of a person-oriented approach, the creation of a supportive atmosphere in the learning process will allow not only to develop the students' creative abilities, but also to transfer to future specialist practical principles of working with people aimed at creating favourable conditions for a person. Customer orientation is one of the important principles of social work. Such interaction with students should be learned not only in theory, but also in practical examples, based on their experience.

Creativity, i.e. the ability to creativity, the culture of social work is so clearly and distinctly characterized. "Creativity" in its very first meaning is the creation of a new person.

The social worker, who addresses to the problem of creativity, thinks in the categories of "eternity" and "beginning". Creativity in terms of social work is the life force, the ability of a specialist – as the bearer of culture to respond to the challenges of time.

The tasks of social work are:

1. Obtaining necessary knowledge in the field of creative and innovative activities in the social work system.

2. Mastering the experience of innovation activities, development and implementation of social projects.

3. Development of the autonomy of thinking in the light of obtaining new knowledge in the field of the theory of creativity and spirituality.

From the doctrine of Volodymyr Vernadsky for social work, it is clear that the spirituality of man carries a distinct imprint of social existence, as all practical actions of human are the individual expression of historically composed social practices of mankind. In an explicit and implicit way, the noosphere idea is presented in various aspects and continues in the studies of the 20th century. It is submitted that the distinction between traditional societies and modern changes is connected by changes in the field of methods and technologies, science and scientific innovations, which inevitably entails the change of all other spheres of human life.

3 Results and discussion

Every new rethinking of the radical concepts of social organization and development of social relations always piques interest and is the subject to lively discussions. And if society was studied as a phenomenon without an external mental environment - immersion in the meta-spiritual sphere, then with the discovery of the intelligent universe - the noosphere, obviously there happens a coup in the human thought. According to Vernadsky, society is the only social organism whose internal organization is a set of definite, characteristic various relationships, in which the human mind is involved. Particularly characteristic is Vernadsky's metaphorical description of the social life of a person, which is as impossible outside the society, as the life of a plant, pulled from the ground and thrown into dry sand. Public life penetrates through the psychic inside the biology of the individual, which in such a transformed form acts as the basis of his mental, conscious life.

Famous Stanislav Losev is valeologist of the master class, specialist of the international level, his methods are approved by the International scientific practical conferences and tested by many years of practice. He, as a member of the National Union of Writers of Ukraine constants: "At the university, we established a clear pattern between the psychological state of man and his fate: if the son is offended by his father - he will not have normal professional activity. If by his mother - he will never create a normal family life. And here we learned how to work with these programs and to correct them with key phrases. In life there is no reason to be unhappy. The most complicated conflicts and failures of life are only the possibility of nature to awaken man to action. Moreover - the stronger the opportunities of the individual - the more difficult the circumstances that make him or her act in the right direction" [2].

The whole system of relations with the world in spiritual culture is distributed in the base matrix, which allows the social experience to be broadcasted, closing it and introducing into the matrix of a social unit, such as personality. Modern practical science has proved that our thoughts are "alive" and they can be heard. Every time a person thinks about something, he emits a specific vibrational energy. Thoughts have different vibrations and strength. The more energy a certain thought has, the greater its strength and the ability to attract different events in our reality. In accordance with the law of attraction, energy attracts energy. Therefore, a person attracts to oneself the circumstances, events and people who are at the level of the dominant vibrational frequency. Understanding that people form their own thoughts around reality will help them make decisions and start living on high vibrations. The more goodness and light a person brings into the world - the more attracts. With this method, a social worker will help the client change its life circumstances, look at them from another direction and understand the best life scenario for him.

One of the first commandments of a social worker is "a glass of cold water, timely submitted to the traveller is more valuable and noble than a dozen lunches donated to a person who can afford to pay for them". Here is bearing in mind the practical philanthropy, spirituality and creativity, which is the starting point in the search for ways and leverage to achieve the goals. It is through self-esteem that the system of human values or desires in life is determined. Self-esteem is an interesting system phenomenon which attributes the value ascribed to the individual based on various parameters. The time that a person, social worker gives another person is invaluable. It's impossible to return it, so it's so important to take care of it. Following the commandments of philosophy, a thinking and enlightened social worker will be able to understand the truth and help people, groups and communities. Generating the enthusiasm of a specialist in social work, and "enthusiasm", says Edward Bulwer-Lytton, "is a spirit of sincerity and truth ...". Actions in human life mean more than words and they even make humans worthy. There are laws of the universe that work when a person is disappointed. It may be due to the fact that person thinks the world is guilty of something. Life, by its nature, is not just or unfair. Life is just life. The quality of our live is determined not by external factors, but by the way we perceive them. Release those things that have already served, or are beyond our control. It is from this position that a social worker is asked to create the reality of the client. No one can take more than he can have, whether it is love, devotion, time, work or money. Thus, the social worker must have the knowledge, skills and abilities that can give the client a solution to life issues.

Students are invited to learn and develop using metaphorical associative cards and board games. The cards are created as a social worker tool. Each card provides answers to the most important questions of social work clients. With the help of the proposed drawings, the specialist has the opportunity to awaken the client's imagination, the subconsciousness and to contribute to solving his psychological problems. They can be used during diagnostic, educational, training, consulting work, etc. A set of proposed cards and techniques for working with them are presented in a way that it can be used in the work as by professional social workers, helping clients to solve their life difficulties, promote their personal growth, also by the people without special education who are willing to use this kit for self-development and self-improvement. Working with the log of metaphorical associative cards "Know Yourself and the World", you can use different techniques, namely: "Who am I?", "Communication", "Technique of fulfilling the desired", "The Wheel of Life", etc.

By mastering the technique "Who am I?", the client randomly selects from the log five cards that correspond to his status in the past, in the present and in the future, also how people perceive the person and what person wants to be. Then it is suggested to compare the cards, find common and different. The conclusion is that a person wants to change oneself. If work is conducted with children, then they can be offered to make a fairy tale or story about the heroes in cards.

In the "Communication" technique, the client is offered to select two cards, which he associates with his friend and with the person who has problems in communicating. Then cards are selected that are associated with the emotions that arise in the process of communication. You can choose two more cards that will show how the client wants to be perceived, etc. [7].

Working with "difficult" teenagers one can use the technique of "Work with the problem" where you offer to form a problem or question, then to choose three random cards. Considering each card one can interpret what blocks solving the problem, what can help and what you do to solve the problem.

Thus, one of the main tasks of professional education at the present stage is not only the transfer of knowledge, skills and abilities in the process of professional training, but also the development of the personality traits of the future specialist. Formation of a world-view based on multi-vector solutions, desire to move away from stereotypes in thinking, tolerance to thoughts dissimilar to their own, largely determines the creativity of the future specialist in social work.

Leo Vygotsky wrote on the problem of creativity: "Creative activity we call such activity of a person who creates something new. It is the creative activity of man that makes it a human being drawn to the future, creates it and modifies it".

4 Conclusion

Almost every person in modern society falls into difficult living conditions, where her psyche faces the impossibility of rethinking and adequately responding to a problematic stressful situation. That is, falling into an unusual, extreme, extraordinary situation depends on the personal organization, how the person will react to what has happened. Client of social work (man caught in difficult life circumstances) - as a collective individual acquires various social roles, that's why social work is the best possible indicator of creative energy and a social worker – is the specialist with the methods and technologies and the necessary knowledge to help people.

So, the future specialist in the social sphere must combine professional and moral values, have the knowledge, skills and abilities, be a true professional. Therefore, the task of teachers of higher educational institutions is in the creative approach to the process of education and training, in the use of innovative technologies in the pedagogical process.

Thus, in the process of revival and constructive changes required by the very objective reality and actuality, social work, as a science, as a type of professional activity, as an academic discipline, plays an extremely important role. A special role is given to social workers who receive professional education in higher education, differ in their views on a particular social problem or situation, which allows them to find the most effective ways to overcome them.

The progressive changes taking place in society today, on the one hand, promote self-disclosure, self-determination and self-affirmation of man, on the other - reveal a lack of goodness and mercy.

References

1. Alston M. Social work, climate change and global cooperation // International Social Work. 2015. Vol. 58(3). P. 355-363.
2. Findley P. A., Pottick K. J., Giordano S. Educating Graduate Social Work Students in Disaster Response: A Real-Time Case Study // Clinical Social Work Journal. 2017. Vol. 45:2. P. 159-167.
3. Harris C., & Boddy J. The Natural Environment in Social Work Education: A Content Analysis of Australian Social Work Courses // Australian Social Work. 2017. Vol. 70 (3). P. 337-349.
4. Petrenko. T. Sanzharovets V. Modern technologies in the training of social

specialists. Bulletin of the National Aviation University. Series: Pedagogy. Psychology: Collected Science Ex-K: National Aviation University, 2018.-Issue 2 (13) .- 38 p.

5. Lombard A. Global agenda for social work and social development: A path toward sustainable social work // Social Dialogue. 2016. Vol. 14. P. 6-15.
6. Ganna Koval, Valentina Sanzharovets and Yulia Shatylo 2020 Social Works in Certain Areas of Donetsk and Lugansk Regions 2020 *III International Scientific Congress Society of Ambient Intelligence 2020* (ISC-SAI 2020). Atlantis Press. Advances in Economics, Business and Management Research 129, 361-369 doi:10.2991/aebmr.k.200318.045
7. Sanzharovets V. and Shatylo Y. 2020 Psychological features of the development of communicative negotiation skills in future social workers. Scientific journal of the National Pedagogical University names M P Drahomanov Series 12 Psychological sciences: Collection of Scientific Works. (Kyiv: Publishing house of NPU names M.P. Drahomanov) Vol 10 (55) 151 p doi: https://doi.org/10.31392/NPU-nc.series12.2020.10(55).10
8. Sanzharovets V. Features of pedagogical methods of social work with military personnel. Bulletin of the National Aviation University. Series: Pedagogy. Psychology -K., 2017.-Issue 2 (11). - 192s.
9. Semigina T. International social. robot: global priorities of social development and innovations of social robot // Bulletin of APSV. 2015. No. 1. S. 6-11.
10. Nesmith A, Smyth N. Environmental Justice and Social Work Education: Social Workers' Professional Perspective // Social Work Education. 2016. Vol. 34:5. P. 484-501. 66.
11. Nikku B. R. Living through and responding to disasters: Multiply roles for Social Work // Social Work Education. 2015. Vol. 34:6. P. 601-606.
12. Teixeira S., Krings A. Sustainable Social Work: An Environmental Justice Framework for Social Work Education // Social Work Education. 2015. Vol. 34(5). P. 513-527.
13. Dominelly, L., ed. (2018). The Routledge Handbook of Green Social Work. Abingdon: Routledge. 2018. 628 p.
14. Platash LB, Prodanchuk T. Professional training of social workers as a scientific problem / Science of the third millennium: searches, problems, prospects of development: materials of the II International scientific-practical Internet conference (April 25-26, 2018): collection of abstracts. - Berdyansk: BSPU, 2018. - Part 1.– 333 p. - P. 251-252.

Inter-Community Cooperation at a Regional Level: Methods of Assessing Migration and Institutional Ties

Sierhiei Sakhanienko[1], *Natalia* Kolisnichenko[2], *Hanna* Koval[3*], *Oksana* Dulina[4], and *Temirbay* Zhiemuratov[5]

[1]Odessa Regional Institute for Public Administration of the National Academy for Public Administration under the President of Ukraine, Odessa, 65009, Ukraine
[2]Odessa Regional Institute for Public Administration of the National Academy for Public Administration under the President of Ukraine, Odessa, 65009, Ukraine
[3] Lviv State Univerrsity of Life Safety, Lviv, Ukraine
[4]Odessa Regional Institute for Public Administration of the National Academy for Public Administration under the President of Ukraine, 65009, Ukraine
[5] Berdakh Karakalpak state university, 230100 Karakalpak Republic, Uzbekistan

Abstract. The article reveals the assessing methods of inter-community cooperation at the regional level regarding its migration and institutional ties. The recommendations on the assessment of inter-municipal relations are focused on the development of appropriate strategies for cooperation between local communities. The article defines the algorithm of analysis and evaluation of inter-municipal relations. The essence and peculiarities of the relevant stages of cooperation are examined and revealed. The components of the assessment of inter-municipal relations and methods of their definition are offered. The matrix and the corresponding coefficients of inter-municipal ties are examined.

1 Introduction

At the current stage of public administration reform in Ukraine the local communities are required to increase the efficiency and effectiveness of local self-government authorities in addressing all issues of local importance, especially of administrative and social nature, of territory development etc. This is mainly relevant for rural areas, because quality and standards of living (level of social and economic development) in urban and rural areas differ significantly. Today, most rural local communities are not able to ensure proper solution of issues related to the livelihoods of their residents. This is mainly because of the lack of local self-government own resources, especially at the level of those local communities that have not yet practice municipal consolidation within the voluntary amalgamation of local communities. However, the analysis of amalgamated local communities clearly shows that in many cases they have not become self-sufficient communities. That does not provide the expected qualitative improvement of the living conditions.

Meanwhile, one of the ways to increase the level of social and economic development of rural areas is a combination of various types of resources based on cooperation of local communities. Its legal framework was acknowledged in 2014 after the adoption of the Law of Ukraine "On Cooperation Between Local Communities" [1]. Despite the fact that local communities differ significantly in number of residents, area, structure of local economy, it is often more profitable to solve the issues of local importance together with other local communities than to implement them independently. The organization of inter-municipal cooperation in order to address the issues of local importance ensures self-sufficiency of local communities (necessary for sustainable development), creates conditions for economic and social well-being of the residents.

At the present stage, the practice of local communities cooperation in Ukraine is mostly fragmentary one. The strategic and program documents of local communities do not comprise the issues of inter-municipal cooperation, which inhibits the development of ties between them. In Ukraine 121 agreements were concluded (the data of the Ministry of Regional Development, Construction, Housing and Communal Services) according to the Register of Agreements on Cooperation between Local Communities (January 11, 2018) [2]. Most of them are aimed: i) to implement the joint projects on coordination and implementation of activities of cooperating entities; ii) to accumulate communities resources for some definite period of time. The other forms of cooperation are less popular (e.g. delegating one or more tasks to one of the subjects of cooperation with the transfer of relevant resources or the formation of joint utilities, institutions and organizations by the subjects of cooperation). Such form of cooperation as the formation of the subjects of cooperation of a joint governing body for the mutual implementation of the statutory powers is not used at all. The prevalence of local communities cooperation is also noteworthy: almost half of the agreements on cooperation are

* Corresponding author: koval.gv@ukr.

concluded in Poltava region, and in some regions of Ukraine they are not concluded at all.

The potential effectiveness of local communities' cooperation actualizes research and methodological relevance of inter-municipal interaction, search of new forms and models of cooperation between the municipalities in the sphere of social and economic development and in the field of management.

2 Theoretical framework

A rather small amount of literature is devoted to the institute of cooperation between the local communities because of its short-term introduction into the practice of municipal governance. The relevant works are aimed to provide a scope-wide theoretical analysis of cooperation between local communities [See, e.g. 3; 4; 5], the legal basis of inter-municipal cooperation [See, e.g. 6; 7], foreign experience [See, e.g. 8; 9], forms of cooperation [See, e.g. 10; 11; 12], some aspects of social and economic [See, e.g. 13; 14; 15] and managerial effectiveness of inter-municipal interaction [See, e.g. 16; 17]. In 2017, the first dissertation was defended in Ukraine [18], which comprehensively analyzed the current state and prospects for further development of cooperation between local communities as one of the priority areas of national local government reform.

But despite the emergence of a wide complex of systematic studies on cooperation between local communities, a number of disputable issues regarding this form of inter-municipal governance remains unsolved. In particular, the analysis of the needs of cooperation between specific local communities and regions is not examined. That is why there is a need in effective and tested methodology for assessing the potential and development of such cooperation.

The objective of the article is to develop the guidelines for the assessment of inter-municipal relations at the regional level aiming to work out an appropriate strategy for the cooperation of local communities under specific conditions.

3 Discussion and results

Inter-municipal cooperation is an organizational and legal manifestation of cooperative relations between local communities and their bodies. The communities develop their cooperation in many ways, including by [19]: advocating in policy dialogue with governments; improving mechanisms to fund their operations and increase their independence and responsiveness to community needs. The forms of inter-municipal cooperation, depending on their organizational and legal content, can be both contractual (making agreements on cooperation, joint activities), provided by the Law of Ukraine "On Cooperation between Local Communities" and associative (creation of inter-municipal associations), provided by the law of Ukraine "On Associations of Local Self-government" [20]. However, there is a variety of forms of inter-municipal cooperation between local communities and their bodies: joint consultations; joint coordination, advisory and/or consultative bodies, working groups (without legal entity status), etc.

In fact, in practice of any self-government body there are various forms of cooperation at least with the neighboring local communities. This circumstance, as well as the processes of voluntary amalgamation of local communities, involve the analysis of the factors that determine the cooperation of local communities and the monitoring of this cooperation.

A required prerequisite for the development of cooperation between local communities is the analysis of the degree of inter-municipal relations and the identification of attraction of local communities to each other.

However, there are many difficulties in conducting a qualitative analysis due to the lack of research on this issue, the weak development of municipal statistics, the lack of tested methods for assessing inter-municipal relations. As a result, it is proposed to develop the guidelines for the analysis of inter-municipal interaction and its assessment in relation to the region as a whole, sub-region (district) or a particular set of local communities. While developing the topic on inter-municipal relations of Zhmerynka raion of Vinnytsia region, an attempt was made to develop an appropriate methodology. This technique involves a certain algorithm for analysis and evaluation of inter-municipal relations:

Step 1. Provide comprehensive analysis of social and economic development of the selected area.

Step 2. Study the basic program documents of local communities.

Step 3. Identify and evaluate inter-municipal ties.

Step 4. Reveal the directions and forms of cooperation development between local communities.

At the first stage, a comprehensive analysis of the social and economic development of the selected area (as a whole area), which includes, in particular, determining the budget supply, investment component, infrastructure development, social status in local communities, as well as appropriate rankings.

The second stage involves the study of the main legal and program documents of local communities, which are examined to identify the existing interactions with other local communities and the degree of their development. This stage allows to determine whether local communities have: i) strategic guidelines for the development of inter-municipal relations; ii) the goals of these interactions, which are set out in the program documents.

At the third stage, it is proposed to identify and assess inter-municipal relations within the following aspects:

- between local communities within several neighboring raions;
- between local communities within the raion and the local community of the city of oblast importance (in the case when the city of oblast importance is the raion center);
- between local communities within one raion.

At this stage, it is necessary to identify the types of inter-municipal relations, including resource, migration, economic, infrastructural, social, institutional relations. At the same time, the quantification of resource, economic, infrastructural and social links is complicated by the lack of required data in State and municipal statistics. That is why the analysis of these types of ties is of expert nature. Migration and institutional ties can be quantified while using the following instruments.

Migration ties. It is recommended to use the following indicators to quantify inter-municipal migration ties:

- absolute indicators that characterize the scale and power of migration - the number of those who arrived, left or both of them (migration turnover);
- coefficients of migration ties intensity, which allow to establish the degree of ties' intensity between local communities according to the proposed scale of values;
- the coefficient of effectiveness of migration ties, which shows the amount of migration exchange between each pair of local communities.

The application of the whole set of indicators gives the most accurate assessment of migration ties, and in order to prevent possible randomness in the assessment, the calculations should be made over several years.

The number of population which arrive and leave a local community in statistical collections is given in one figure, regardless of the place of departure and arrival. Thus, the data on inter-municipal migration can be obtained only on the basis of the analysis of primary material of the territorial agency of the State Statistics Service of Ukraine and the State Migration Service of Ukraine.

The indicators can be presented in tabular form for the analysis of migration relations between local communities. Such a table is a square matrix of order *m* (*m* is equal to the number of local communities of "departure" and, accordingly, the local communities of "arrival"). The peculiarity of the obtained matrix is that its elements, which are located on the main diagonal (which is passing from the upper left corner to the lower right corner) are meaningless, because the local community of "departure" coincides with the local community of "arrival". Thus, in any pair of local communities there is one direct response and one feedback.

To calculate the intensity factor, it is advisable to use the coefficient of intensity of migration ties (CIMT), proposed in the 70's by L.Rybakovsky [21]. In contrast to the general coefficients of migration intensity of the population, CIMT is used to analyze migration between local communities and characterizes the scope, results and intensity of migration exchange between administrative-territorial units. In addition, CIMT allows to compare the levels of population's mobility of different (regarding their rank and size) local communities, to identify the dynamics of migration between the territorial units which are examined. However, the complexity of calculations and the amount of necessary information did not make CIMT a popular indicator in the migration analysis (though it is absolutely adequate).

Current statistics on the migration of the population are used to calculate the intensity coefficients of the migration ties between local communities. In the proposed method one finds the proportion of all "departure" local communities in the migration of the population with each "arrival" local community separately. At the same time one calculates the proportion of each "arrival" local community in the total population of all local communities. Then the first series of values (share of local communities in migration) is divided by the second series of values (share of local communities in the population). The corresponding formula for the calculations is as follows:

$$K_{ij}=\frac{M_{ij}}{\sum_{i=1}^{m} M_{ij}} \div \frac{S_i}{\sum_{i=1}^{m} S_i} = \frac{M_{ij}\sum_{i=1}^{m} S_i}{S_i \sum_{i=1}^{m} M_{ij}}, \quad (1)$$

where

$\mathbf{K_{ij}}$ - intensity coefficient of migration ties between local communities;

$\mathbf{M_{ij}}$ - the number of migrants who arrived from the "departure" local community ***i*** to the "arrival" local community **j**;

$\mathbf{S_i}$ - the population of the "departure" local community ***i***;

m - the number of all "departure" local communities.

It is proposed to divide the obtained values of coefficients into 5 groups. The first group includes coefficients with values up to 1; the second group - from 1 to 2; the third - from 2 to 4; the fourth - from 4 to 8 and the fifth from 8 and above. According to the proposed method, the ties of each group are defined as insignificant, noticeable, medium, high, higher.

After that, for the most complete assessment, the coefficient of effectiveness of migration ties (CEMT) is calculated. It shows the value of migration exchange between each pair of local communities:

CEMT = (Number of persons who left/Number of persons who arrived) (2).

The matrix of average values of intensity coefficients of migration relations between local communities of Zhmerynka raion and neighbouring raions and cities of oblast importance in 2018 (table 1) can be calculated using the data from the State Statistics Service of Ukraine. Similarly, migration ties between individual local communities within a raion and/or a sub-region are assessed, although the territorial agencies of the State Statistics Service of Ukraine do not yet conduct statistical analysis of individual communities within raions. This necessitates the introduction of municipal statistics, which, in particular, is the objective of the Register of Local Communities, the need for which is determined by the legislation of Ukraine.

Table 1. The matrix of average values of intensity coefficients of migration relations between local communities of Zhmerynka raion and neighbouring raions and cities of oblast importance

Departure	Arrival								
	Zhmerynka city	Communities of Zhmerynka raion	Vinnytsya city	Communities of Vinnytsya raion	Communities of Bar raion	Communities of Tyvriv raion	Communities of Lityn raion	Communities of Shargorod raion	Communities of Derazhnyans'k raion
2016									
Zhmerynka city		1,4	4,6	1,3	0,3	0,1	0,2	0	0
Communities of Zhmerynka raion	6,6		13,2	1,7	0,3	0,5	0,4	0,2	0,1
Vinnytsya city	0,2	-		6,9	0,3	0,6	0	0	0
Communities of Vinnytsya raion	0,6	0,9	28,5	0	0,5	0,2	0	0,2	0
Communities of Bar raion	0,7	0,3	3,7	1,9	0	0	0,2	0	0
Communities of Tyvriv raion	0,1	0,4	6,3	2,1	0	0	0	0,2	0
Communities of Lityn raion	0	0	7,5	3,4	0,2	0	0	0	0,4
Communities of Shargorod raion	1,6	1,1	4,2	1,3	0,3	0	0	0	0
Communities of Derazhnyans'k raion	0	0	0,9	0,4	0	0	0,2	0	0

Assessment of inter-municipal institutional relations. Institutional ties are determined by the territorial agencies of the ministries and other central executive bodies (territorial agencies of central executive bodies - TACEB), state and municipal institutions, enterprises, organizations that serve several or all local communities of the territory presented at the level of the local community.

In order to determine the intensity of interaction for this type of ties, the formula, which determines the intensity of any phenomenon was used: in the numerator there is the number of the cases of a special kind, in the denominator is the main set of cases. Thus, the intensity coefficient of inter-municipal institutional relations will be equal to the ratio of the number of TACEB, state and municipal institutions, enterprises, organizations that serve the local community *i*, but located in the local community *j*, to the total number of TACEB, state and municipal institutions, enterprises, organizations under consideration:

$$K = \frac{V_{ij}}{S}, \quad (3)$$

where

K - intensity coefficient of institutional ties;

V_{ij} - the number of TACEB, state and municipal institutions, enterprises, organizations that serve the local community **i,** but are located in the territory of the local community **j**;

S - the total number of TACEB, state and municipal institutions, enterprises, organizations under examination.

This ratio is calculated for each pair of local communities under examination and shows the degree of institutional dependence of local communities on each other.

In order to calculate the total intensity coefficient of institutional ties, we use the formula (4):

$$K_{\cdot} = \frac{\sum V_{ij}}{S} \quad (4)$$

The general intensity coefficient of inter-municipal institutional ties characterizes the intensity degree of ties between all considered local communities together. Being based on the results of the analysis of all types of ties, a matrix of inter-municipal ties is built. It reflects the number of existing ties for each pair of local communities. Being based on the data of the matrix, the level of local communities' interaction development is estimated: periodic ties (the coefficient is less than 0.2), regular ties (the coefficient is 0.2-0.5), intensive ties (the coefficient is from 0.5 to 1).

In particular, the intensity coefficient of institutional ties between local communities of Zhmerynka raion and neighboring raions and cities of oblast importance is characterized by the following (Table 2), which is facilitated by the inter-raion nature of some TACEB, state and municipal institutions, enterprises and organizations.

Table 2. The intensity coefficient of institutional ties between local communities of Zhmerynka raion and neighboring raions and cities of oblast importance

	Zhmerynka city	Communities of Zhmerynka raion	Vinnytsya city	Communities of Vinnytsya raion	Communities of Bar raion	Communities of Tyvriv raion	Communities of Lityn raion	Communities of Shargorod raion	Communities of Derazhnyans'k raion
Zhmerynka city		0,9	0,7	0	0	0	0	0	0
Communities of Zhmerynka raion	0,9		0,7	0	0	0	0	0	0
Vinnytsya city	0,7	0		0	0	0	0	0	0
Communities of Vinnytsya raion	0	0	0,9		0	0	0	0	0
Communities of Bar raion	0,5	0	0,4	0		0	0	0	0
Communities of Tyvriv raion	0,3	0,1	0,3	0,1	0		0	0,1	0
Communities of Lityn raion	0,4	0	0,3	0	0	0	0	0	0
Communities of Shargorod raion	0,4	0	0,4	0	0	0	0	0	0
Communities of Derazhnyans'k raion	0	0	0,2	0	0	0	0	0	0

The examination of cooperation agreements of local communities allows to identify the most difficult issues of local importance.

The analysis of the activity of existing inter-municipal organizations allows to determine the most developed types of ties of local communities at this level.

The degree of the development of local communities' interaction can be assessed by: i) interviewing village, town, city mayors, heads of raion councils, ii) an expert survey of heads of executive bodies of local councils. This information allows to determine the preconditions for inter-municipal interaction, to reveal the existing ties between local communities and to reveal the opinion of the representatives of local communities about the priority areas of cooperation. The use of this method allows to diagnose the understanding of the nature of cooperation of local communities as a function of local self-government, to identify the spatial attraction of settlements for common interests and problems.

The questionnaire (offered within this method) consists of two parts. The first part contains a list of issues of local importance of local communities. Regarding each issue, the respondent is asked to indicate how actively his/her local community participates in inter-municipal interaction.

The second part of the questionnaire includes four open-ended questions on the problematic and promising areas of cooperation between local communities:

• Which specific projects on cooperation of local communities do you participate in?

• Which local communities do you have to cooperate most often with?

• What local communities would it be desirable to cooperate with and in what areas?

• Are there any problems in your local community that you cannot solve independently?

The data processing is carried out for each local community separately, for raion or sub-region as a whole.

At the final, fourth, stage of the offered algorithm the generalization and interconnection of results in an estimation of local communities' ties of all kinds is carried out; the explanation of preconditions for development of strategic directions and the mechanism of inter-municipal cooperation development in a region is carried out.

4 Conclusions

Therefore, the cooperation of local communities at the regional level should be based on the analysis of existing migration, institutional, etc. ties between the communities. Ignoring the latter can lead to the artificiality of any form of inter-municipal cooperation. However, relevant information that is not included into the relevant government classifiers is lacking to analyze the existing ties. This implies the need to develop an appropriate methodology. The proposed methodology i) can be used in the process of projects on cooperation between local communities' development and ii) disseminated through the existing associations of local governments for further use and improvement.

References

1. On cooperation of local communities: Law of Ukraine №1508, June 17, 2014 (2014).
2. Register of agreements on cooperation of local communities. URL: http://www.minregion.gov.ua/napryamki-diyalnosti/regional-dev/rozvytokmistsevoho-samovryaduvannya/reyestr/
3. O.I. Grindey, Intermunicipal cooperation is an innovative form of local development. Scientific works [Petro Mohyla Black Sea State University of the Kyiv-Mohyla Academy]. Series: Public Administration. V. 263, I. 252. 54-58 (2015)
4. P.M. Lyubchenko, Cooperation of local communities: theory and practice. Problems of legality. I. 129. 50-59 (2015).
5. Yu.P. Sharov, Potential and tools of cooperation of capable communities under decentralization. Aspects of public administration. 4. 102-110 (2015).
6. H.V. Prikhodko, Intermunicipal cooperation: procedural and legal aspect. Scientific Bulletin of the Academy of Municipal Administration. S.: Law. I. 1. 149-158 (2011).
7. Yu.S. Rogozyan, Agreement on interregional cooperation: basic principles and project development. Investments: practice and experience. 24. 64-68 (2015).
8. K.O. Zakomorna, Intermunicipal cooperation as a trend of constitutional reform in post-socialist states-members of the European Union. State Building and Local Self-Government. I. 33. 90-101 (2017).
9. V.D. Poltavets, Inter-municipal cooperation in France. Experience and lessons. Public administration: theory and practice. 2. 214-218 (2013).
10. A. Gelych, Formation of joint communal enterprises, institutions and organizations as a form of cooperation of local communities. Legal Ukraine. 10. 67-71 (2014).
11. V.V. Mamonova, Organizational pluralism of cooperation in local self-government as a reflection of integration processes in the territory. Public administration: theory and practice. I. 3. 60-65 (2013).
12. V.D. Poltavets, Forms of voluntary amalgamation of local self-government bodies as a principle of inter-municipal cooperation. Efficiency of public administration. I. 37. 295-299 (2013).
13. S. Bugil, Strengthening the resource base of municipalities through partnership and cooperation. Bulletin of Odessa National University. Series: Economics. V. 21. I. 9. 116-119 (2016).
14. Yu.V. Kozachenko, Cooperation of Local communities within the framework of infrastructural transformation. Investments: practice and experience. 20. 67-71 (2017).
15. N.D. Chala, Inter-municipal cooperation in managing economic space: social and financial sections. Scientific Bulletin of the Academy of Municipal Administration. Series: Economics. I. 10. 184–194 (2011).
16. D.G. Bezugliy, Project approach in development management and cooperation of enlarged local communities. Aspects of public administration. 9. 64-70 (2015).
17. V. Demchyshen, On some aspects of the development of inter-municipal cooperation as an innovative tool in the implementation of State regional policy in Ukraine. Viche. 24. 6-9 (2010).
18. N.A. Kostina, Development of cooperation between local communities under reforming local self-government in Ukraine: dis. Ph.D. Classical private university. Zaporizhzhya. 227 (2017).
19. Making development co-operation more effective: How development partners are promoting effective, country-led partnerships. Part II of the Global Partnership 2019 Progress Report (2019). URL: www.oecd.org/dac/effectiveness/Part-II-of-the-Global-Partnership-Progress-Report.pdf
20. On associations of local self-government bodies: Law of Ukraine №1275-VI. 16.06.09. (2009).
21. L.L. Rybakovs'kyi, Regional migration analysis. M.: Statistics. 278 (1973).

Modern Demographic Processes in Ukraine, Factors of Influence

Victoria Yavorska[1*], *Vitaliy* Sych[1], *Ihor* Hevko[2], *Inna* Shorobura[3], and *Olexandra* Dolynska[3]

[1]Odesa I.I. Mechnykov National University, Champaign lane, 2, Odesa, 65058, Ukraine
[2]Ternopil Volodymyr Hnatiuk National Pedagogical University, M. Kryvoniis st., 2, Ternopil, 46027, Ukraine
[3]Khmelnytsky Humanities and Pedagogic Academy, Proskurivs'koho Pidpillya St, 139, Khmelnytskyi, 29013, Ukraine

Abstract. The presented article gives the analysis of the modern demographic indicators such as life expectancy, age structure and fertility rate of the population in Ukraine and their geographical differences. The aim of the research is a spatial analysis of the demographic situation to determine the development trends of the population of the regions of Ukraine. In the conditions of depopulation and gradual aging of labor force in Ukraine requirements to its qualitative characteristics, first of all a state of health and average life expectancy grow. Ukraine belongs to the countries with high intensity of depopulation processes that is caused by a number of various factors of economic, socio-cultural, institutional character. Large-scale and protracted depopulation is now a component and one of the most pronounced manifestations of the general crisis demographic situation in Ukraine. The tendencies of influence of average life expectancy on indicators of population aging are investigated in the work. The regional analysis is carried out and territorial differences in indicators of average life expectancy and indicators of population aging are defined. The reasons for such a tense situation have been identified. The paper emphasizes that the development of really effective measures aimed at bringing the oblasts out of the deep demographic crisis and slowing down the progressive depopulation should focus not only on reducing mortality, but mainly on improving the health of the population, which can correct high mortality. These demographic trends increase global demographic asymmetry, give a special sharpness and new "sound" to the demographic problem in the global context, and, in addition, have long-term socio-economic consequences for countries with depopulation and deepening aging. In view of this, there is now a need for a comprehensive scientific understanding of the phenomenon of depopulation and assessment of its possible prospects, the development of a strategy to influence depopulation processes by means of socio-demographic policy. In our country, the study of depopulation factors, its nature and origins, analysis of the peculiarities of the deployment of depopulation and assessment of its consequences have become particularly relevant.

1 Introduction

At the turn of the XX-XXI centuries in the socio-geographical development of mankind there were profound transformations that affected all aspects of his life. An important factor in new approaches to population problems has been the powerful demographic crisis in Ukraine, which has started a general depopulation since 1991.

The aim of the research is a spatial analysis of the demographic situation to determine the development trends of the population of the regions of Ukraine. The strengthening of regional differentiation and the need for scientific substantiation of the policy of "response" to the negative consequences of the gender - age structure of the population require the separation of this problem as an independent and as a specific object of study of social geography. Geographical understanding of this process involves the identification of spatio-temporal features of the gender - age structure of the population, awareness of this process for different socio-territorial systems; elucidation of factors that cause territorial differences; substantiation of measures aimed at improving the living standards of people and reducing the negative consequences of gender and age structure for various socio-territorial entities, etc.

2 Analysis of previous researches and publications

During last decades it has been discussed numerous issues concerning demographic situation in Ukraine which is characterized as an open demographic crisis. A fundamental contribution to the study of the given issues were done by Ukrainian scientists such as V.O. Dzhaman, O.G. Topchiev, V.A. Popovkin, V. K. Simonenko, M.S. Dnistriynskiy, K.V. Mezentcev, Z.S. Varnaliy, M.I. Dolishniy, V. Malinovskiy, M.P. Demchenko and many other domestic scientists

* Corresponding author: yavorskaya@onu.edu.ua

who analyzed in detail the features of modern geodemographic development of Ukraine as a whole and its individual oblasts, substantiate possible options for further transformation of demo-reproducing processes.

3 Research methodology

The age structure of the population is one of the important geodemographic indicators, which characterizes the geodemographic situation and its territorial differentiation [1]. In the practice of demographic research, the analysis of age structures of the population by enlarged age groups is widespread. In this case, depending on the ratio of these three groups in the total population, there are three types of age structure. *Progressive* - characterized by a large proportion of children in the total population. This type of age structure is usually characterized by high rates of natural population growth. *Stationary* - is characterized by an almost balanced proportion of children and the elderly. This type of age structure is characterized by a small natural increase in population, or it is at a constant (stationary) level. *Regressive* - characterized by a fairly large proportion of the elderly. This type of age structure of the population corresponds to a narrow reproduction of the population, and sometimes it is characterized by depopulation, which takes place in Ukraine and its oblasts today [2, 3].

Results and discussion

According to this methodology, Volynska, Zakarpatska, Ivano-Frankivska, Lvivska, Odeska, Rivnenska, Khersonska and Chernivetska oblasts belong to the progressive type of age structure where the share of younger age groups exceeds the share of the elderly (2019) (Table 1). The Ternopilska oblast belongs to the stationary type, where the share of persons of retirement and working age is the same and the rest of the oblasts of Ukraine are referred to the regressive type of age structure of the population.

The difference between the oblasts in terms of the share of the population in real working age (15-65 years) is insignificant. Its average level in the country is 70.5%, and indicators in the oblasts range from 72.6% (Kharkivska oblast) to 67.8% (Volynska oblast). The part of the working population in Kharkivska, Zaporizka, Mykolaivska, Odeska, Sumska, Khersonska, Dnipropetrovska oblasti are above the average.

The oblasts differ more in the share of the population in the older age groups (over 65 years), which has an amplitude from 19.0% (Chernihivska oblast) to 11% (Zakarpatska oblast). The average rate of this group for Ukraine is 15.3%. Below the "norm" this indicator is in Zakarpatska, Rivnenska, Volynska, Chernivetska, Ivano-Frankivska, Odeska, Lvivska, Khersonska, Mykolaivska, Crimea, Kharkivska oblasts. This group of oblasts has a correspondingly lower demographic burden.

The under working age group (0-14 years) has an average level in the country of 14.2% and ranges from 11.9% (Luhanska oblast) to 19.3% (Rivnenska oblast). The highest shares (19.3% -16.6%) of the under working age group are in Rivnenska, Zakarpatska, Volynska, Ivano-Frankivska and Chernivetska oblasts (Table 1).

Table 1. Indicators of the ratio of age groups of the oblasts of Ukraine (2019) *

Oblast**	Population by age (‰)		
	0-14 years	15-65 years	65 years and older
Ukraine	14.2	70.5	15.3
Avtonomna Respublika (AR) Krym	-	-	-
Vinnytska	14.9	67.8	17.3
Volynska	18.7	68.2	13.1
Dnipropetrovska	13.6	70.7	15.7
Donetska	-	-	-
Zhytomyrska	15.6	68.1	16.3
Zakarpatska	18.8	70.2	11.0
Zaporizka	13.1	71.1	15.8
Ivano-Frankivska	16.7	69.3	14.0
Kyivska	14.5	70.0	15.5
Kirovohradska	14.1	68.9	17.0
Luhanska	-	-	-
Lvivska	15.6	70.1	14.3
Mykolaivska	14.4	71.1	14.5
Odeska	14.9	70.9	14.2
Poltavska	13.1	69.9	17.0
Rivnenska	19.3	68.3	12.4
Sumska	12.6	70.8	16.6
Ternopilska	15.7	68.6	15.7
Kharkivska	12.3	72.6	15.1
Khersonska	14.8	70.8	14.4
Khmelnytska	14.9	68.5	16.6
Cherkaska	13.4	69.1	17.5
Chernivetska	16.6	69.5	13.9
Chernihivska	12.7	68.3	19.0

* Compiled according to the State Statistics Service of Ukraine [4]

** The names of the units of administrative division of Ukraine are represented in writing according to the [11]

On the one hand, it is a sign of increased demographic load, and on the other - better provision of their own labor resources. Further distribution of oblasts by population to under working age is as follows. The average level of this indicator (15.7% -14.1%) is in Ternopilska, Zhytomyrska, Lvivska, Odeska, Vinnytska, Khmelnytska, Khersonska, Kyivska, Mykolaivska, AR Krym, Kirovohradska oblasts. The smallest share (13.6% -11.9%) of the under working age population is observed in Dnipropetrovska, Cherkaska, Zaporizska, Poltavska, Sumska, Kharkivska, Donetska and Luhanska oblasts. In the current demographic crisis, these oblasts will experience a certain shortage of their own labor resources.

For every 1,000 people aged 15-64 in 2019, there were 419 people, including 202 people aged 0-14 and 217 people over 65 (Table 2).

Table 2. Demographic load on the population of Ukraine aged 15-64, 2019 (per 1,000 people) *

Oblasts**	Urban settlements			Rural settlements		
	0-14 years, 65 years and older	0-14 years	65 years and older	0-14 years, 65 years and older	0-14 years	65 years and older
Ukraine	376	185	191	522	243	279
AR Krym	-	-	-	-	-	-
Vinnytska	367	202	165	596	238	358
Volynska	381	236	145	568	319	249
Dnipro-petrovska	397	186	211	514	229	285
Donetska	-	-	-	-	-	-
Zhytomyrska	389	211	178	594	258	336
Zakarpatska	396	253	143	441	276	165
Zaporizka	382	174	208	492	221	271
Ivano-Frankivska	372	216	156	503	263	240
Kyivska	358	199	159	556	221	335
Kirovohrad-ska	397	188	209	548	234	314
Luhanska	-	-	-	-	-	-
Lvivska	379	204	175	508	254	254
Mykolaivska	375	183	192	477	246	231
Odeska	378	183	195	478	268	210
Poltavska	376	178	198	525	203	322
Rivnenska	353	224	129	580	344	236
Sumska	354	171	183	545	193	352
Ternopilska	356	209	147	546	246	300
Kharkivska	351	161	190	487	201	286
Khersonska	394	192	202	444	237	207
Khmelnytska	350	203	147	618	239	379
Cherkaska	367	180	187	565	214	351
Chernivetska	351	198	153	509	272	237
Chernihivska	358	177	181	681	205	476

* Compiled by the author according to the State Statistics Service of Ukraine [4]

** The names of the units of administrative division of Ukraine are represented in writing according to the [11]

In urban settlements, the load on the working population was 376 people, and in rural areas - 522 people. And in the villages there are more dependant people than working age people (demographic load is 1.5 times higher than in cities).

This situation in our country leads to such a phenomenon as "maintenance", when unemployed people are dependents of workforce. Along with this, the aging process of people should not be considered as a purely negative phenomenon. It is necessary to effectively use the experience of the elderly, to create conditions for their full life and activities.

Demographic load is calculated as the ratio of the under-working age population (up to 15 years) and post-working age (after 64 years) to the real working age population (15-64 years) per 1,000 people. In general, for Ukraine this indicator is 419 people and has amplitude of fluctuations in the oblasts from 475 people (Vinnytska oblast) to 376 people (Kharkivska oblast). It is significant that for rural settlements the demographic load is almost 1.4 times higher than in urban areas (Table 2).

According to the level of demographic load, the oblasts are distributed as follows (2019). Kharkivska, Zaporizska, Mykolaivska, and Odeska oblasts have the minimum demographic load (410 people). The second group with a moderate demographic load (410-420 people) is formed by Sumska, Khersonska, Dnipropetrovska, Odeska oblasts. A close to average level of demographic load (420-440 people) is observed in Zakarpatska, Lvivska, Kyivska, Poltavska and Chernivetska oblasts. Ivano-Frankivska, Cherkaska, Kirovohradska and Ternopilska oblasts belong to the group of oblast with a high demographic load (440-460 people). Rivnenska, Chernihivska, Volynska, Zhytomyrska, Khmelnytsks and Vinnytska oblasts have the highest demographic load (over 460 people).

The issue of "loading" the working population with the dependent population is important for many aspects of life [5]. The pension fund is formed by social taxes. In addition to children and the elderly, this social group of "dependents" also includes people with special needs. The development of effective measures to bring the oblasts out of the deep demographic crisis and slow the progressive depopulation should focus not only on reducing mortality, but mainly on improving the health of the population, which can correct the high mortality rate [6].

Retirement age is one of the key issues in the discussion of any pension system. It serves as a kind of indicator of the degree of financial stability of the pension system of social justice in society. Therefore, it is not surprising that in the course of pension reform in Ukraine, many discussions arise around the issue of rising the retirement age. Experts have very different views on this issue. If alone express the opinion of a gradual increase in the retirement age, while others categorically deny the expediency of such transformations [7].

The higher level of population aging in rural areas causes a greater demographic load than in urban settlements. At the same time, if in urban settlements the general demographic load increases annually, then in rural areas there is a decrease in the load. As the population in urban settlements is more than twice the population in rural areas, the tendency to increase the demographic load in the country as a whole is determined by the growth of this indicator in the urban population.

The next indicator of the characteristics of the population in Ukraine is the indicator of life expectancy at birth, which means the number of years that a given generation should live on average, assuming that throughout the life of the generation, when it passes from one age to another, mortality will be equal to the current mortality rate in certain age groups [8].

Average life expectancy is called expected because it is a prognostic indicator that shows at what age people, who born in a given period, will die on average, if the mortality rate in all age groups remains throughout their next life - from birth to complete extinction of their generation will remain the same as in this period. In this way it is possible to characterize all mortality rates for a given year or period, one generalized figure. The latter indicator is free from the influence of age differences and is suitable for comparing different oblasts for different periods.

Characteristics of life expectancy and average age of the population are an integral indicator of all structural characteristics of mortality, a statistical quintessence of the level of viability of the population. The negative process of natural reduction occurs against the background of the general loss of population and deformation of its age structure, the number of which is reduced in reproductive age - that is, the aging of the nation [9].

Ukrainians live 10 years less than residents of the countries of European Union (EU) and many countries of the Commonwealth of Independent States (CIS). In particular, Ukraine ranks are 150th among 223 countries in the world in terms of life expectancy. The average life expectancy in Ukraine is 69 years, in the EU-74 years, and in the CIS countries: in Georgia - 76.7, in Moldova - 70.8, in Belarus -70.6, in Uzbekistan -71.9. Moreover, the average life expectancy of healthy living in Ukraine is 59.2 years, and in the European Union - 67 years. According to the State Statistics Committee, every tenth Ukrainian does not live to 35 years, and every fourth - to 60 years [4].

Authors consider the dynamics of changes in life expectancy in Ukraine, in the late 40's - first half of the 60's in Ukraine there was significant progress in reducing mortality and increasing life expectancy in Ukraine. Thus, during the period 1950-1964 years, the average life expectancy at birth increased in men by 11.7, and in women by 7.7 years (in men from 56.2 to 67.9, and in women from 66.8 to 74.5 years). According to the average life expectancy of the population of Ukraine in the first half of the 60's entered the top ten countries. Thus, in 1964 in Japan the average life expectancy was 67.2 years for men and 72.6 - for women, in the United States was 66.8 years for men and 73.6 - for women, in France - 68.0 years and 75.1 years, respectively [10].

More than two-thirds of the increase in life expectancy between 1950 and 1964 years was due to a reduction in infant and child mortality between the ages of 1 and 14. The highest growth rates of life expectancy occurred in 1950 - 1959 years, when their annual growth averaged 0.8 years for men and 0.5 years for women.

Positive changes in reducing mortality in Ukraine have been made possible by mass prevention and immunization of the population, the introduction of sulfamide drugs and antibiotics in the practice of health care, in particular their availability to the general public. A significant role also belongs to the purely demographic determinant - the huge losses of the adult population in the 30's and 40's from famine, repression and war [10].

Unfortunately, the reserves for reducing mortality, which were in effect until the mid-1960s, were largely exhausted, and over the next 20 years the mortality rate increased and life expectancy decreased. The main role in the deterioration of the mortality situation was played by the increase in the age probability of death in men from the age of 20 and in women from the age of 40. As a result of these trends, life expectancy has decreased significantly. The minimum values of its indicators fall on 1980-1981 years (for men - 64.6 years, for women - 74.0 years). During the period 1964-1981years, the average life expectancy of men decreased by 3.3 years, and that of women by 0.5 years (Table 3).

Table 3. Dynamics of life expectancy at birth of Ukraine (years) [4, 10]

Years	Total	Including	
		men	women
1958-1959	69.8	66.1	72.6
1961-1962	70.9	67.3	73.6
1965-1966	71.6	67.7	74.5
1969-1970	70.8	66.5	74.3
1979-1980	69.7	64.6	74.0
1985-1986	70.5	65.9	74.5
1989-1990	70.7	65.9	75.0
1991-1992	69.3	64.2	74.2
1995-1996	66.9	61.4	72.7
1999-2000	67.9	62.4	73.6
2000-2001	68.3	62.8	74.1
2005-2006	68.1	62.4	74.1
2009-2010	70.4	65.3	75.5
2018-2019	69.3	63.8	74.9

Since the mid-1980's there has been a certain break in the unfavorable dynamics of mortality in Ukraine: the probability of mortality began to decline compared to the first half of the 1980's in almost all age groups, life expectancy increased in 1985-1987 by 1.7 for men and 0.9 years in women. After a twenty-year break, the average life expectancy of women in 1989 yr. exceeded the maximum values of this indicator by 0.5 years in the 1960s. However, it was not possible to consolidate the positive direction of the tendencies to reduce mortality. In the late 1980s and early 1990s, the negative trends in mortality that dominated the demographic sphere until 1985 are gradually recovering: there is an increase in the age probability of death in men and women from the age of 15. Life expectancy in 1992 compared to 1987 decreased by 3.0 years for men and 1.4 - for women.

For the entire period up to the mid-1990s, the average life expectancy of the urban population was on average 1.5-2.0 years higher than that of the rural population, and there is a tendency for them to equalize.

From 1990 to 2010 years there is a decrease in the overall life expectancy of the population of Ukraine by 0.3 years (from 70.7 to 70.4 years), for men - by 0.6 years (from 65.9 to 65.3 years)), and women, on the contrary, increase - by 0.5 years (from 75.0 to 75.5 years). In 1989-1990 years., the life expectancy of women exceeded the life expectancy of men by 9.1 years, and in 2009-2010 years - by 10.2 years. In the period from 2010 to 2019 years, these changes amounted to 1.1 years.

Of great concern is the significant reduction in life expectancy, which is the main criterion for the health of the population and the level of human development, as it is an integral embodiment of the processes of reproduction of both health and the population itself. The average value of this indicator today in Ukraine reaches 69.3 years and is the lowest among the countries of the European space. The worst performance on this parameter recorded in Zhytomyrska, Odeska, Mykolaivska, Dnipropetrovska, Chernihivska, Kirovohradska, Chernihivska oblasts (Table 4).

The best life expectancy indicators are observed in Ternopilska, Lvivska, Chernivteska, Ivano-Frankivska, Khmelnytska and Rivnenska oblasts. In other words, life expectancy is firmly held in the Western of the country, despite the fact that world practice shows longer life expectancy in economically developed oblasts (Fig. 1).

Fig.1. Territorial differentiation of life expectancy in 2019

The average life expectancy in Ukraine in 2018–2019 was 69.3 years, including 70.0 years in urban settlements and 67.8 years in rural ones.

In all oblasts of Ukraine, the average life expectancy of women significantly exceeds the same figure for men, and this difference is much greater than in economically developed countries (minimum - 6.8 years in Chernivetska oblast and maximum - 13.8 years in Chernihivska oblast) (Table. 4). Characteristically, 89% of the oblast variation of this difference is explained by the male mortality regime, and the relationship is inverse - the higher the male mortality in the oblast, the smaller the gender difference in life expectancy at birth (Fig. 2).

Table 4. Average life expectancy in the oblasts of Ukraine in 2019*

Oblast**	Average life expectancy (years)		Difference of years
	men	women	
AR Krym	63.7	74.8	11.1
Vinnytska	64.9	75.8	10.9
Volynska	63.7	75.8	12.1
Dnipropetrovska	62.2	73.5	11.3
Donetska	61.8	73.7	11.9
Zhytomyrska	61.8	74.2	12.4
Zakarpatska	64.8	73.7	8.9
Zaporizka	64.2	75.0	10.8
Ivano-Frankivska	66.4	77.0	10.6
Kyivska	62.3	74.4	12.1
Kirovohradska	61.9	73.8	11.9
Luhanska	62.7	74.1	11.4
Lvivska	66.3	77.0	10.7
Mykolaivska	62.3	73.3	11.0
Odeska	63.3	72.9	9.6
Poltavska	63.8	74.9	11.1
Rivnenska	64.0	75.6	11.6
Sumska	63.5	75.0	11.5
Ternopilska	66.6	77.4	10.8
Kharkivska	64.9	75.2	10.3
Khersonska	62.3	73.8	11.5
Khmelnytska	64.4	75.9	11.5
Cherkaska	64.4	75.4	11.0
Chernivetska	66.9	73.7	6.8
Chernihivska	61.3	75.1	13.8

* Compiled by the authors according to the State Statistics Service of Ukraine [4, 10]

** The names of the units of administrative division of Ukraine are represented in writing according to the [11]

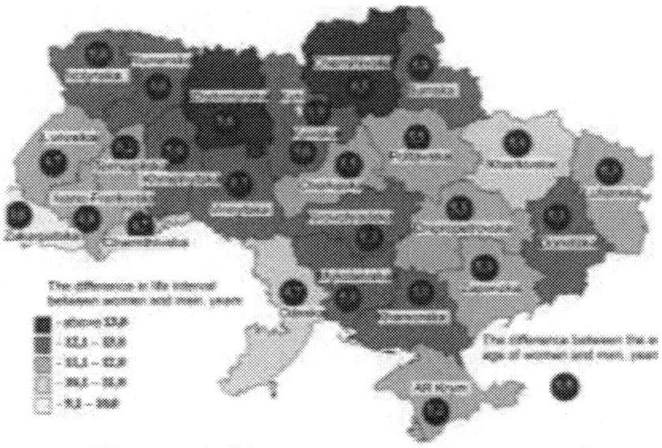

Fig. 2. Differences in life expectancy and average age of women and men in Ukraine*

* Compiled by the authors according to the State Statistics Service of Ukraine [4]

Differentiation of life expectancy at birth varies for women from 72.9 years in Odeska oblast to 77.4 years in Ternopilska oblast and for men from 61.3 year in Chernihivska oblast to 66.9 years in Chernivetska oblast.

In general, the life expectancy of the population of Ukraine is quite stable, for the male population it fluctuates from 62 to 64 years, for women the limits of variation vary from 73 to 76 years (Fig. 1). The life expectancy of the entire population of the country is approaching 70 years.

Life expectancy in urban settlements is 2.2 years higher than in rural ones (69.0 vs. 66.8). This is typical for all oblasts of Ukraine, except for the Zakarpatska oblast, where life expectancy in rural settlements is 0.9 years higher than in cities. The largest excess of life expectancy in urban settlements over rural ones was recorded in Chernihivska (5.5 years) and Zhytomyrska (4.4 years) oblasts.

The average life expectancy of men is 11.1 years less than that of women - 63.8 and 74.9, respectively (Fig. 2). This gap is larger in rural settlements (11.9 years) than in urban ones (10.6 years). The largest excess (13.8 years) is observed in the Chernihivsky oblast, and among the rural population it is 15.9 years. The smallest differences in the life expectancy of men and women are typical for Chernivetsky oblast - 6.8 years (for the general population) and 9.6 years - for the rural population [4].

The next indicator that significantly affects the gender and age structure of the population is the total fertility rate in the oblasts of Ukraine, as shown in Table 5 and Fig. 3 in 2019 by type of settlement had the following territorial differences.

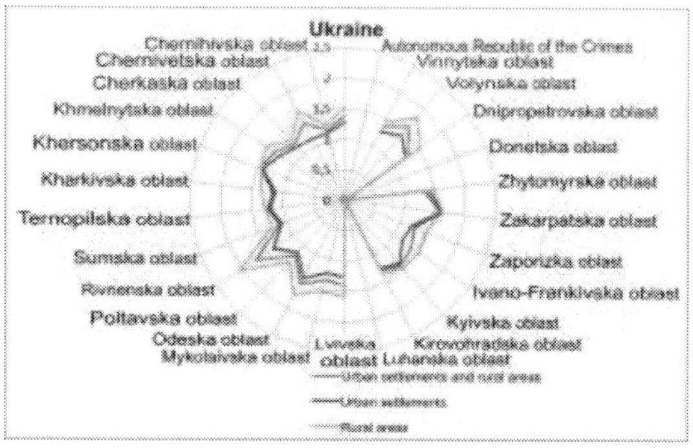

Fig. 3. Total fertility rate indicator in oblasts of Ukraine (2019)

With regard to rural areas, the group of oblasts with the highest total fertility rate in such oblasts as Odeska, Rivnenska and Volynska is distinguished in these oblasts, there is an almost simple type of reproduction. The second group of obalsts in which the total birth rate is close to a simple type of reproduction - Mykolaivska, Chernivetska, Dnipropetrovskf, Khmelnytskf, Zhytomyrska and Zakarpatska oblasts. Moderate indicators of the total birth rate in rural areas were observed in such oblasts as Zaporizhska, Ivano-Frankivska, Chernihivska, Vinnytska, Kirovohradska, Lvivska, Khersonska, and Kyivska.

Table 5. Fertility rate indicator in Ukraine and oblasts (2019)*

Oblasts**	**Total reproduction rate (per 1 woman)**		
	urban and rural settle-ments	**urban settle-ments**	**rural settle ment s**
Ukraine	1.374	1.283	1.522
AR Krym	-	-	-
Vinnytska	1.355	1.268	1.438
Volynska	1.617	1.421	1.82
Dnipropetrovska	1.287	1.236	1.527
Donetska	-	-	-
Zhytomyrska	1.415	1.345	1.503
Zakarpatska	1.641	1.617	1.649
Zaporizka	1.25	1.185	1.448
Ivano-Frankivska	1.358	1.164	1.507
Kyivska	1.412	1.38	1.457
Kirovohradska	1.301	1.247	1.366
Luhanska	-	-	-
Lvivska	1.392	1.263	1.576
Mykolaivska	1.337	1.228	1.533
Odeska	1.543	1.429	1.739
Poltavska	1.213	1.132	1.342
Rivnenska	1.735	1.4	2.049
Sumska	1.13	1.079	1.239
Ternopilska	1.253	1.205	1.279
Kharkivska	1.168	1.126	1.348
Khersonska	1.432	1.407	1.435
Khmelnytska	1.385	1.273	1.53
Cherkaska	1.231	1.171	1.297
Chernivetska	1.426	1.136	1.653
Chernihivska	1.181	1.138	1.262

* Compiled by the authors according to the State Statistics Service of Ukraine [4, 10]

** The names of the units of administrative division of Ukraine are represented in writing according to the [11]

The lowest indicators of the total birth rate in rural areas were observed in Poltavskaa, Sumska, Chernihivska, Cherkaska, Ternopilska oblasts.

The situation is somewhat different with regard to the distribution of oblasts according to the total fertility rate for urban areas:

- the highest indicators are typical for the Volynska and Zakarpatska oblasts;
- moderate indicators of the total birth rate in urban areas were observed in such oblasts as Odeskaa, Khmelnytska, Vinnytska, Kirovohradska, Lvivska, Khersonska, Ternopilska;
- the lowest indicators of the total birth rate in urban areas were observed in Ivano-Frankivska, Chernihivska, Chernivetska, Cherkaska oblasts.

In 2019, the lowest indicators of this rate were 1.1-1.2, that is, they were already approaching that limit beyond which there is no simple reproduction of the population.

It should be noted that the birth rate remains low, as it compensates only 59% of the reproduction of the population of Ukraine [4].

Based on the study of indicators influencing the formation of gender and age structure of the population of Ukraine, analysis of demographic factors, taking into account the current demographic situation, we came to a number of conclusions.

Conclusions

Gender - age structure of the population is one of the important demographic indicators, which characterizes the demographic situation. Characteristics of the age structure of the population and the ratio of men and women (in general and by individual age groups) are needed for the calculation and evaluation of labor resources, the contingent of preschool and school age children, retirees, as well as to plan the range of consumer goods. The gender structure of the population of Ukraine of 2019 is dominated by women; the share of women is 53.9%, and men - 46.1% [4]. The relationship between men and women in different age groups is different. The modern social norm of balancing is considered to be the age of 50 years. The age of balancing between the male and female population of Ukraine according to 2019 data is 30 years. The huge shortage of older men was affected by the losses in the war of 1941-1945. The change in the ratio of gender groups and its imbalance was also influenced by the different intensity of migration of men and women. High mortality among the male population, especially those of working age, has too great an impact on the gender imbalance. Therefore, reducing male mortality is one of the main tasks of socio-demographic development, which should be addressed at the national level.

The age structure of the population of Ukraine has a clear tendency to aging. The identified negative trends in the formation of gender and age structure of the population of Ukraine (increase in the share of elderly people, decrease in demographic potential) put forward urgent tasks to overcome the demographic crisis, which is possible only with a significant increase in living standards, targeted influence on social transformation, infrastructure, ensuring uniform living standards in all oblasts of the country, in urban and rural areas. This will provide basic social guarantees for the vast majority of the population and reduce the quantitative and qualitative losses of society from the demographic crisis.

References

1. V.O. Dzhaman, V.P. Krul, H.Ya. Cherniukh *Demoheohrafiia: Navchalnyi posibnyk* (Demogeography: Textbook). (Ruta, Chernivtsi, 2002)
2. Ptoukha Institute for Demography and Social Studies of the National Academy of Sciences of Ukraine (2021), https://www.idss.org.ua. Accessed 22 Jan 2021
3. O.H. Topchiiev, V.V. Yavorska, N.V. Dimova *Heodemohrafiia: rehionalnyi demohrafichnyi rozvytok Ukrainy* (Geodemography: regional demographic development of Ukraine). (BMB, Odesa, 2014)
4. Official site of the State Statistics Committee of Ukraine (2021), http://www.ukrstat.gov.ua. Accessed 20 Jan 2021
5. E.M. Libanova, O.V. Makarova, O.V. Pozniak, P.Ye. Shevchuk, V.S. Shyshkin *Zainiatist ta rynok pratsi* (Employment and labor market). (NASU, Kyiv, 1999)
6. T.S. Morshchenok (2009), *Demohrafichni peredumovy formuvannia trudovoho potentsialu* (Demographic prerequisites for the formation of labor potential) http://dspace.nbuv.gov.ua/bitstream/handle/123456789/2856/st_44_29.pdf. Accessed 27 Jan 2021
7. I.V. Ponomarenko, Study of the impact of population aging on the number and structure of economically active population in Ukraine and European countries. Efficient economy, **1** (2013)
8. V.S. Steshenko, V.P. Piskunov, Demographic knowledge: current status and development prospects. Demography and social economy, **1** (2012)
9. E.M. Libanova, S.I. Pyrozhkov, N.S.Vlasenko *Demohrafichna kryza v Ukraini: yii prychyny ta naslidky: Monohrafiia* (Demographic crisis in Ukraine: its causes and consequences: Monograph). (Kyiv, 2003)
10. V.V. Yavorska *Rehionalni heodemohrafichni protsesy v Ukraini* (Regional geodemographic processes in Ukraine). (Aksioma, Kamianets-Podilskyi, 2013)
11. Toponymic guidelines for map and other editors, for international use: Ukraine (State Service of Geodesy, Cartography and Cadastre, Kartographia, Kyiv, 2

Problems of Formation of Ecological Networks for Environmental Safety of Odesa Region

Victoria Yavorska[1*], *Kateryna* Kolomiyets[1], *Valentina* Trigub[1], *Ihor* Hevko[2], and *Olexandra* Chubrei[3]

[1]Odesa I .I. Mechnykov National University, Champaign lane, 2, Odesa, 65058, Ukraine
[2]Ternopil Volodymyr Hnatiuk National Pedagogical University, M. Kryvoniis st., 2, Ternopil, 46027, Ukraine
[3]Yuriy Fedkovych Chernivtsi National University, M.Kotsiubynskyi st., 2, Chernivtsi, 58012, Ukraine

Abstract. Currently, the concept of sustainable development of nature and society is gaining relevance, a key aspect of which is the development of the ecological network. In Ukraine, there is a regulatory framework for the formation of an ecological network of three levels - Pan-European, National and regional. One of the important problems is that in the developed schemes of regional eco-networks should be interconnected to the eco-networks of neighboring regions and countries. The main features of the ecological network of the Odesa region are due to its coastal position and location mainly in the Steppe, partly Forest-Steppe landscape zones. The region includes vast areas of coastal territories and coastal waters - coastal zones, which concentrate unique protected areas. Mandatory basis for the formation of ecological networks is land use. The proposed concept of geoplanning is based on planning developments of the main components of the territory: the natural environment; population; economic activity. The basic characteristics of the natural environment are the landscape map and physical and geographical zoning of the territory. Maps of resilience of the natural environment to man-caused load, natural and ecological potential of the territory, levels of ecological and economic balance have already been drawn up. This series of maps for the needs of planning the territory of Ukraine should be continued by project maps of national and regional ecological networks, as well as maps of ecological capacity of the territory for population settlement, various economic activities and the general level of economic development in general. For the needs of spatial planning it is necessary to emphasize the levels of anthropogenic and urban pressures on the natural environment in the settlement of the population. Allocation of water fund lands in kind and strict regulation of their use is the main prerequisite for the formation of ecological networks of Ukraine and its regions.

1 Introduction

Ukrainian geographers took the experience from European countries regarding territory planning. Formational problems of ecological networks for environmental safety, ecological networks in country and regions are mentioned as the most ferocious and demanded in Ukraine. The current Ukrainian legislation provides general legal foundations for the formation of ecological network of environmental safety in the country and its regions. Primarily the Law of Ukraine: “On the national Program for Creating the National Ecological Network of Ukraine for the years 2000-2015” (1989-III, from 21.09.2000) also “Ecological Network of Ukraine” (1864-IV, from 24.06.2004). The Cabinet of Ministers of Ukraine resolution in year 2000 obliged Council of Ministers of ARC and regional state administrations to develop concepts and programs of regional ecological networks as the components of national environmental networks. Further development of schemes for territorial planning began regionally. Our study overlooks the problems of national and regional territorial planning and environmental activity in Ukraine.

The development and realization of national environmental strategy occurs in the context of transition of our country to the model of sustainable development according to decisions of UN Conference on environment and development (3-14 June 1992 Rio de Janeiro), Johannesburg Summit (2002) and political guidance of Pan-European process “The environment for Europe” require the greening of public consciousness, the introduction of a system of professional environmental training, raising the level of environmental education and culture, in general, the implementation of a balanced state ecological policy.

The law provides for the creation of a framework of ecological security of the country as a single system of territories subject to special protection, which are composed of lands of nature reserves, health and recreational purposes, historical and cultural purposes, as well as lands of forest and water fund, areas of natural vegetation agricultural, residential and public buildings and other purposes. The law defines the formation of national and regional ecological networks as one of the

* Corresponding author: yavorskaya@onu.edu.ua

most important prerequisites for sustainable socio-economic development of the country.

Ecological network (eco-network) is a single territorial structure of protected and undisturbed areas, which ensure the preservation of landscape and biological diversity, create preconditions for rational use of nature and ecological improvement of the territory. The constituent elements of ecological networks are natural regions, natural corridors and buffer zones. At present, only natural regions - territories and objects of the nature reserve fund - have been identified in the area to some extent. Natural corridors and buffer zones are still waiting to be substantiated and designed.

Legislation in Ukraine clearly defines the possible spatial resources for the development of the ecological network. It is important that the inclusion of territories in the eco-network does not change the form of ownership and category of land. In addition, the owners and users of these areas can receive budget funding for wildlife conservation. The basis of the ecological network is, of course, protected areas, but in fact all areas where natural landscapes have been preserved to some extent can become elements of the ecological network. The natural ecological framework in the form of ecological networks should approximately cover a quarter of a third of the country's territory.

2 Results and discussion

The territory of the Odessa region is located mainly in the steppe, partly - forest-steppe natural zones. This determines its high agricultural production potential and rich recreational resources. This situation, however, determines the negative features of the nature of the region, in particular its low water and the region's small forest cover. This is especially felt in the coastal areas where a significant part of the nature reserve fund is concentrated, covering unique natural complexes and ecosystems of floodplains, estuaries, lakes, sea coasts, spit, sea shelf, which should be considered as a unique and invaluable regional and national resource. Its use requires strict regulation and regulation.

The coastal position of the region determines the most common problems of regional nature management. The territory is crossed by the rivers Danube, Dniester, Southern Bug, Dnieper, which remove pollutants from large water intakes and accumulate them in places of maximum concentration of population and economic activity. The contact zone land with sea, as well as the junction of river and estuarine waters with marine waters plays an important role in the regional and global biosphere barrier. It is in this strip that the most valuable objects of the country's nature reserve fund are located.

The lack of the most scarce water resources in the coastal strip is due to the distribution, primarily of surface runoff. The transit nature of rivers and the redistribution of water resources through main canals increase the potential of water resources, but do not radically solve the water problem in the region.

There is an acute shortage of forest stands, especially in the coastal areas of large cities and in recreational areas. High-value recreational resources of the region, especially the sea coasts, are limited in the use of significant pollution of sea and estuarine waters, their insufficiently clearly regulated economic operation. World experience shows that sea and estuarine coasts should be considered as a unique and valuable regional and national resource. Its use requires strict regulation and regulation.

The concentration of population, infrastructure, and economy in the coastal zone of Odessa region exceeds its level in the deep land part, which determines the characteristic type of territorial organization of the economy - coastal facade, the main economic axis of which is the coastline, estuaries and below large rivers along which are the main economic centres. The latter are represented by powerful port-industrial complexes, industrial hubs, resorts. At the same time, there is a high unevenness in the settlement of the population and the location of the economy. The 50-70% of the region's inhabitants, 80% of its industrial potential, most of the transport infrastructure, almost all recreational facilities are concentrated in the coastal strip.

The law provides for the creation of a framework of ecological security of the country as a single system of territories subject to special protection, which are composed of lands of nature reserves, health and recreational purposes, historical and cultural purposes, as well as lands of forest and water fund, areas of natural vegetation agricultural, residential and public buildings and other purposes. The law defines the formation of national and regional ecological networks as one of the most important prerequisites for sustainable socio-economic development of the country.

Ecological network (eco-network) is a single territorial structure of protected and undisturbed areas, which ensure the preservation of landscape and biological diversity, create preconditions for rational use of nature and ecological improvement of the territory. The constituent elements of ecological networks are natural regions, natural corridors and buffer zones. At present, only natural regions - territories and objects of the nature reserve fund - have been identified in the area to some extent. Natural corridors and buffer zones are still waiting to be substantiated and designed.

The main feature of the formation of the ecological network of the Ukrainian Black Sea region is that it is a part of the Pan-European (Eastern European) Ecological Network, as well as the national ecological network, and must be agreed with them [1, 2].

The regional ecological network, which forms the natural framework of ecological safety of Odesa region, includes elements of ecological networks of different levels: international - Pan-European ecological network; national - National Ecological Network of Ukraine; regional and local. Elements of the Pan-European Ecological Network are the natural region of the Lower Danube and two ecological corridors - Prut (Lower Danube-Carpathians) and Azov-Black Sea, which covers the coastal-maritime territorial and aquatic natural complexes of the Black Sea coast, and the Danube (Lower Danube valley) [3-5].

Ukraine takes part in international cooperation on the formation, conservation and use of the Pan-European Eenvironmental Network and the creation of its transboundary elements. Existing international agreements on the formation, preservation and use of the ecological network, approved by the Supreme Council of Ukraine, are part of the national legislation of Ukraine.

It is envisaged to ensure the connection of the regional ecological network of Odesa region with the ecological networks of neighboring regions within the framework of the unified national ecological network of Ukraine, as well as with certain elements of the Pan-European ecological network. These include the Danube Biosphere Reserve, the Lower Prut Biosphere Reserve of Moldova and the joint Lower Dniester National Nature Park with Moldova, as well as international natural corridors running through the Danube and Dniester valleys.

The decision of the Odesa Regional Council of 20.05.2011 № 136-IV approved the Regional scheme of formation of the ecological network of Odesa region, which was developed by the Southern Research Center of the National Academy of Sciences of Ukraine and the Ministry of Education and Science of Ukraine under the scientific guidance of Odesa I .I. Mechnykov National University in the framework of research work "Development of a regional scheme for the formation of an ecological network." Despite the fact that the task of forming ecological networks at the regional level as a whole can be solved, the most difficult remains the development on a large geographical scale of the composition and boundaries of ecological (natural) corridors.

The regional eco-network has cross-border elements and should be connected with the eco-networks of Moldova and Romania, Vinnytsia, Kirovohrad and Mykolayiv regions of Ukraine.

Transboundary elements of the regional ecological network of Odesa region: With Vinnytsia and Kirovohrad regions Odesaregion borders along the Galicia-Slobozhansky forest-steppe natural corridor, and with Mykolayiv - by Tiligul estuary, which in its lower part already has the status of regional natural parks.

Natural corridors that are part of the regional ecological network of Odesa (Lower Danube, Black Sea coastal, Khadzhyderskyi, Sasyk-Kohylnytskyi, Kyrhyzh-Kytaiskyi, Yalpuzkyi; national - Azov-Black Sea, Coastal-Dniester) gravitate to the coastal zones [6].

The core of the regional eco-network is the coastal strip, which we consider as a unique territorial-aqual natural system and which is designated as the Azov-Black Sea Ecological Corridor of the European Eco-network, where biosphere and nature reserves and national nature parks of Ukraine are concentrated. At the same time, the natural complexes of the coastal strip are very vulnerable to anthropogenic and man-caused loads: the mouths of the river collect pollution from huge watersheds every year; sea shores have complex and various geogdynamics (abrasion, landslides, siltation and accumulation of sediments, etc.); shallow shelf water has a small "environmental capacity" and low capacity for self-cleaning. In coastal areas, there is and is constantly growing socio-ecological contradiction between their extremely high attractiveness for settlement and many economic activities, on the one hand, and extremely high environmental vulnerability of sea coasts and adjacent waters - on the other.

World practice considers coastal strips (coastal zones) as a unique and valuable resource of regional, national and global importance, which requires a special legislative and regulatory framework for its use. In many countries it has been operating for a long time a special legal framework for nature usage and integrated management in coastal zones [7]. Ukraine has approved the Concept of Environmental Policy of Ukraine for the period up to 2020 (Decree of the Cabinet of Ministers of Ukraine of October 17, 2007 № 880-r), which emphasizes the problems of environmental protection of sea coasts and shelf strips.

Experience in the formation of ecological networks shows that for the formation of regional ecological networks it is necessary to review land use in the region, as well as to move at the regional and local levels to the agro-landscape organization of rural areas. The agro-landscape organization should take into account as much as possible the natural and geographical features of the territory during its agricultural use, in particular the cutting of land uses, massifs and fields of arable lands with their coordinated reference to the natural basis, to landscape areas and tracts. Note that in the context of radical agrarian and land reforms currently underway in Ukraine, there is a significant fragmentation of land tenure and land use, there is a significant departure from the large-scale organization of the territory and cutting large fields, which creates the necessary conditions for agro-landscape organization. This problem is especially relevant for forest-steppe and steppe landscape zones, in which the Odessa region is located.

The basis of the regional ecological network should be a nature reserve fund, also the share of regional ecological network should be increased at the expense of recreational areas and degraded and unproductive agricultural lands, which occupy large areas and require their return to the category of natural lands.

Degraded and unproductive agricultural lands are a significant reserve of land for the creation of ecological networks. Ukraine is a world record holder in terms of the level of plowing and agricultural development of the territory, and this is a sad and unnecessary record. Since 1995 to 1997, the State Land Cadastre was introduced in Ukraine, in the presence of which it became possible to actually withdraw from cultivation of unproductive and degraded agricultural land, preserve it and gradually return it to the state of natural lands. In most regions, at least 10-15% of agricultural land needs such a transformation.

Territories and objects of the nature reserve fund of the region form the basis of the regional ecological network of Odesa region and form most of its nodal elements. The development of the nature reserve fund of Odesa region is carried out in accordance with the regional Program of formation of the national environmental network in Odesa region, which envisages increasing the share of the nature reserve fund of the region to 5%. The latter component includes a list of protected areas that were reserved by the decision of the Odesa Regional Council of October 1, 1993 №496-XXI "On measures to preserve and develop the nature reserve fund of the region." Currently, the nature reserve fund of the region includes 125 territories and objects with a total area of 163523.05 hectares, of which 18 are of national importance and 107 local [8].

Territorial analysis of the structural elements of the ecological network of the Odessa region showed that the share of territories suitable for the formation of the ecological network is 30.5% of the area of the region. At the same time, there is a heterogeneity of this indicator in the administrative-territorial units of the region. The location of nature reserves in the region is uneven. In Lyubashivsky, Krasnooknyansky and Saratsky districts they are completely absent, and in other administrative units of the region their number is from the 1st to the 8th. Lands of the nature reserve fund occupy 4.9% of the territory of the Odessa region, this figure can be increased to 7-10% due to the expansion of existing NPF facilities and the creation of new nature protection areas.

The distribution between users and owners of land plots of territories and objects of the nature reserve fund is as follows: 43% of protected areas are used by forestry enterprises, 35% are communally owned by city and village councils, the rest are used by objects of the Ministry of Education and Science Ukraine, the Ministry of Health of Ukraine, the National Academy of Sciences of Ukraine, individual organizations of collective ownership.

The ecological network is defined as a natural framework of ecological safety of the territory and is considered as an obligatory component of planning of the territory. For the needs of planning of the regions territory we have developed the Concept of Geoplanning (Fig.1).

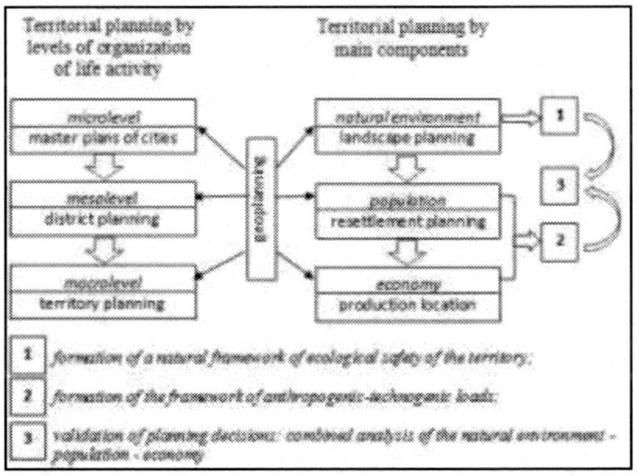

Fig.1. Conceptual scheme of "geoplanning" [12]

It is based on planning developments of the main components of the territory: the natural environment; population; economic activity. Let`s consider the methodological principles of the General scheme of planning of the territory of Ukraine on its main components - the environment, population and economy [9]. The basic characteristic of the natural environment is the landscape map and physical and geographical zoning of the territory. Schemes of agroclimatic, agrosoil and many other component zonings are widely used. Recently, dozens of applied natural resource and natural ecological maps of Ukraine have been developed. Maps of resistance of the natural environment to technogenic loading, natural and ecological potential of the territory, levels of ecological and economic balance of the territory are already made (V.A. Baranovsky & P.G. Shishchenko, 2005). Such a series of maps for the planning of the territory of Ukraine needs to be extended by project maps of national and regional ecological networks. And also maps of the ecological capacity of the territory to the resettlement of the population, maps of the different types of economic activity [10, 11]. Note that a sketch map of the national ecological network has already been developed [9].

Perhaps the most developed area of modern geography is the territorial organization of the population. With the help of thematic series of maps in the scientific and primary atlases, the composition and structure of the population, their dynamics, and population settlement are shown. For the needs of the planning of the territory, it is necessary to accentuate the loading levels of the anthropogenic and urban areas on the natural environment and to typify the territories according to the general nature of settlement and peculiarities of residential land use.

Geographers have to avoid of the strict linkage of the settlement systems to the administrative-territorial subdivision. First, settlement systems are historically and genetically more ingrained than any scheme of administrative-territorial organization. Secondly, a new administrative reform is underway in Ukraine. The new administrative-territorial subdivision has been solidified by explicit systems of settlement, and not by “squeezing” the settlement into new administrative units. Third, an in-depth analysis of the historical and geographical genesis of population settlement and its geospatial differences will make it possible to regionalize Ukraine fairly objectively and thus fulfill one of the prerequisites for its European integration.

A difficult methodological issue is the typification of populated areas for geoplanning purposes. In the General scheme of planning of the territory of Ukraine [9] the zone of urbanization which is divided on levels of industrial and town-planning development into three types is allocated: 1) critical; 2) high; 3) average. Typification of areas by the nature of settlement and development should be continued. Urbanized areas are the territories of individual cities and urban agglomerations, they are large cities with suburban areas, and they are urbanized areas of country houses and cottages (which do not have their own administrative status). Additionally, urbanized zones and areas can be differentiated by building density, population density, and infrastructure. Territories with rural settlement differ in genetic types of settlement, in population and density of settlements, in the density of the rural population.

The most complicated and least developed is the analysis of the territorial organization of economic activity. It is, in particular, about the systematics of types of economic use of territories - industrial, agricultural, forestry, transport, maritime, recreational, and environmental. Geographers traditionally develop maps of agricultural specialization. The General scheme shows eight functional types of agricultural use of the territory and one - forestry. But such a system for the needs of spatial planning should be further detailed. In particular, it is necessary to show different levels of intensity of agricultural use in the structure of land and land use, the complexity of existing production, etc.

The antithesis of the framework of ecological safety is the framework of anthropogenic and man-made loads, which integratedly represents the geospatial (territorial) distribution of environmental impacts of all socio-economic components - population, artificial material world created by man, economic and spiritual activities. Its structural elements are settlements, transport communications, lands of various economic purposes and uses, as well as socio-economic objects that pose an ecological danger. To date, as a rule, all of the above components are considered separately.

According to the concept of the anthropogenic and man-made loads framework, its target guidelines are "determining the nature of the territorial elements of the anthropogenic and man-made loads framework and identifying their regular spatial connections in order to identify ways to optimize land use in the regions and improve the overall structure of land use; systematization and typification of anthropogenic and man-made loads framework elements, optimization of the area, structure, condition of the frame elements in order to ensure favorable living conditions and environmental improvement of the territory, justification of measures to improve the process of anthropogenic and man-made loads frame elements in the regions and appropriate regulation of economic activity [13, p. 229]".

Determining and mapping the framework of anthropogenic and man-made loads in the study region is of great scientific and applied importance. It allows you to compare the natural and anthropogenic-technogenic frameworks of the region, to identify the most problematic areas. In the future, on the basis of an integrated map, we have the opportunity to develop measures to reduce the harmful effects of anthropogenic-technogenic load, to justify in the process of geoplanning strategy of rational territorial organization of nature-population-economy in order to achieve environmental safety and improve the quality of life.

Mandatory basis for the formation of eco-networks is land management. The main methodological problem: the categories of land that are part of the regional ecological network do not correspond (do not agree) to the structure of land, which is used by modern land accounting:

-not taken into account the land of the water fund (land design is required)

- non-certified (mostly) recreational and resort lands;

- disorderly accounting of degraded and unproductive agricultural lands, etc.

Experts are faced with the question - at the expense of which lands can be formed eco-networks? There are already the first concepts and programs for the formation of regional eco-networks (Odesa region, etc.), in which the share of eco-networks in the total area of the regions reaches 25-30% [6]. There are categories of lands that will be fully part of ecological networks. These are territories and objects of the nature reserve fund, forest lands, and recreational territories. But there are few such lands in most regions of the country. The main reserve for the formation of ecological networks should be the lands of the water fund. Currently, five categories of water fund lands are legally defined, and nature use regulations have been established for them.

In accordance with Part 1 of Art. 58 of the Land Code of Ukraine, the lands of the water fund include lands occupied by:

a) seas, rivers, lakes, reservoirs, other bodies of water, swamps and islands not occupied by forests;

b) coastal protection strips along seas, rivers and around reservoirs, except for lands occupied by forests (rivers less than 3 ha - 25 m; rivers and reservoirs more than 3 ha - 50 m; for large rivers and lakes - 100 m; along the sea, sea bays and estuaries not less than 2 km);

c) hydraulic, other water management structures and canals, as well as lands allocated for allotment strips for them; (they are designed primarily to protect facilities from pollution, littering, depletion and certain economic activities related to the use of water bodies. The peculiarity of these lands is that a special regime of use is established for them);

d) coastal strips of waterways;

e) artificially created land plots within the waters of seaports.

But these categories of land are not allocated on the ground and, therefore, have no real restrictions on economic use. Allocation of water fund lands in kind and strict regulation of their use is the main prerequisite for the formation of ecological networks of Ukraine and its regions.

Within the Odesa region, water fund lands occupy 210.73 thousand hectares or 6.3% of the region's territory, including natural rivers and streams - 15.3 thousand hectares, lakes and estuaries - 167.2 thousand hectares, ponds - 12.1 thousand hectares, artificial reservoirs - 7.6 thousand hectares, artificial drains (canals, collectors, ditches) - 8.4 thousand hectares. Open wetlands - these are mostly tidal or estuarine areas, occupying 72.5 thousand hectares. All these lands are suitable for the development of the ecological network.

As for other categories of water fund lands (coastal protection strips, etc.), at present their allocation on the ground has only just begun. Thus, water protection zones occupy 183.3 thousand hectares on 01.01.2020, coastal protection strips - 35.17 thousand hectares [13].

Due to its coastal location, the region has large areas of recreational and resort lands that could potentially be part of the eco-network. To date, the development of land design documentation for these areas is underway, and although compared to 2014 their area has increased slightly - from 2.7 thousand hectares on 01.01.2014 to 3.1 thousand hectares at the beginning of 2020. But even the last figure is a very small share in the structure of the regional potential of the eco-network.

A significant reserve for the formation of a regional ecological network are unproductive and degraded agricultural lands, which have lost their agricultural potential due to intensive and irrational use. These are eroded, swampy, wet, saline, contaminated with heavy metals, rocky, acidic agricultural lands. The ecological network, in particular its natural corridors and buffer zones, may also include agricultural lands of extensive use - extensive use - hayfields, which and pastures, which occupy 401.7 thousand hectares or 12.1% of the entire region.

The total area of degraded agricultural land in the region, which must be included in the ecological network, is 1242.3 thousand hectares. In the near future it is necessary to withdraw these unproductive lands from agricultural use and transfer them to the composition of natural lands - forests, shrubs, meadows, steppes.

When comparing the developed scheme of the regional ecological network [13] with the landscape map [14], it can be noted that the following landscapes are most covered by the ecological network in the Odesa region: floodplain forest, meadow-swamp, meadow steppe floodplains, floodplains, which are located in the valleys of the Danube, Dniester and their tributaries; they are the meridional ecological corridors of the national and regional level. Also, in the ecological network of the region are widely represented dry-steppe terraced and ancient delta hilly sandy landscapes with slightly humus-rich soils under the forests - the lower reaches of the Danube and Dniester rivers. The north-eastern part of the region, which is located in the Forest-Steppe zone of the South Podilliy Upland region, is also distinguished by its high share of territories that can be included in the ecological network. It is characterized by forest-steppe upland undulating forest landscapes with gray forest and dark gray podzolic soils, with hornbeam oaks and forest-steppe upland eroded landscapes with deep chernozems, hornbeam and oaks forests, ravines. It is through this territory the Galichansko-Slobozhansky latitudinal ecological corridor of the national level passes. Forest-steppe ecosystems are represented by a large number of nature reserve fund objects of both national and local significance.

Despite the fact that most of the region belongs to the Steppe zone, within which the Northern and Middle Steppe subzones are represented, the steppe landscapes of the subzones are insufficiently reflected in the regional ecological network due to extreme variability and fragmentation. To date, there is no scheme of the South Ukrainian latitudinal eco-corridor, which runs within the North Steppe subzone, only its natural cores (Kuchurgan-Andriyashivske and Kuyalnytske) are presented, which are quite difficult to unite with connecting corridors due to plowed lands. That is, steppe ecosystems in the middle part of the region, with the exception of protected areas, have been so altered by agricultural activities that at present it is not even possible to consider them as potentially suitable for the creation of an ecological network. Further more detailed design of the regional eco-network requires additional surveys and analysis of land use, in order to cover the eco-network of all subtypes of landscapes in the region.

Conclusions

Thereby the development of the ecological network as a mandatory element of the rational territorial organization of society allows to some extent to eliminate the imbalances between economic development and environmentally friendly nature management. Legislative definition of the basic theoretical, methodological, methodical questions of formation of ecological networks of various levels allows developing reasonably regional programs and schemes of formation of an ecological network, and such work is already actively carried out. It is clear that at the regional level at the initial stage of such research it is necessary to make some inventory of spatial resources that could potentially be used to build an eco-network. Execution of such tasks in the context of complex planning of territories of regions is an important direction of activity of modern social geographers.

So, the features of creation of the regional ecological network of Odesa region are caused by its coastal position. In coastal areas, there is a growing contradiction between the attractiveness for settlement and the main economic activities, on the one hand, and the exceptionally high environmental vulnerability of sea coasts and adjacent waters - on the other. The natural framework of ecological safety of Odesa region includes elements of the ecological network of different levels - international, national, regional and local. The main reserve for the formation of the regional ecological network of Odessa region is water fund lands, degraded and unproductive agricultural lands. Allocation of water fund lands in kind and their mapping on land use plans requires appropriate funding and design. Regional ecological networks must be coordinated with each other within a single national ecological network, as well as with micro-elements of the Trans-European Ecological Network. Therefore, it is important to develop cross-border coordination and connection of elements of ecological networks of neighboring regions and neighboring countries.

References

1. V.Oleshchenko, Ya. Movchan, H. Parchuk, *Normatyvno-pravovi pidstavy dlia rozbudovy ekomerezhi Ukrainy* (Regulatory and legal grounds for the development of the ecological network of Ukraine). (Kyiv, 1999)
2. *Formuvannia rehionalnykh skhem ekomerezhi (metodychni rekomendatsii)* (Formation of regional schemes of an ecological network (methodical recommendations)) ed by Ju.R.Sheljah-Sosonko. (Fitosociocentr, Kyiv, 2004)
3. O.H. Topchiiev, L.P. Platonova, A.M. Shashero, *Kontseptsiia karkasiv ekolohichnoi bezpeky krain i rehioniv yak suchasna paradyhma heohrafii* (The

concept of ecological security frameworks of countries and regions as a modern paradigm of geography). (Akademperyodyka, Kyiv, 2006)

4. O.V. Kohan, Ya.I. Movchan, L.D. Protsenko, V.A. Kostiushyn, V.D. Siokhin, *Ekomerezha Azovo-Chornomorskoho pryrodnoho korydoru* (Eco-network of the Azov-Black Sea natural corridor). (Khimdzhest, Kyiv, 2012)
5. *Ekomerezha stepovoi zony Ukrainy: pryntsypy stvorennia, struktura, elementy* (Eco-network of the steppe zone of Ukraine: principles of creation, structure, elements) ed by D.V. Dubina, Ya.I. Movchan. (LAT&K, Kyiv, 2013)
6. V.A. Sych, V.V. Yavorska, K.V. Kolomiyets, Features of the formation of the ecological network of the Ukrainian Black Sea region. Odesa National University Herald, Series Geography&Geology, **20**, 4 (2015)
7. V.I. Karamushka, *Prostorove planuvannia rozvytku pryberezhnoi smuhy moriv Ukrainy* (Spatial planning of the development of the coastal strip of the seas of Ukraine). (Print Bistro, Odesa, 2009)
8. Department of Ecology and Natural Resources of the Odessa Regional State Administration, https://oda.odessa.gov.ua. Accessed at 14 Jan. 2021
9. *Zakon Ukrajiny "Pro Heneralnu sxemu planuvannja terytoriji Ukrajiny"* (Law of Ukraine "On the General Scheme of Territorial Planning of Ukraine"). (Kyiv, 2002)
10. O.H. Topchiiev, *Suspilno-heohrafichni doslidzhennia: metodolohiia, metody, metodyky (navchalnyi posibnyk)* Socio-geographical research: methodology, methods, techniques (textbook) (Astroprint, Odesa, 2005)
11. V.A. Sych, A.M. Shashero, K.V. Kolomiyets, Framework of anthropogenic and technogenic loads on the territory of the Ukrainian Black Sea region. Odesa National University Herald, Series Geography&Geology, **24**, 1 (2019)
12. O.H. Topchiiev, D.S. Malchikova, I.O. Pylypenko, V.V. Yavorska, *Metodolohichni osnovy heohrafii: Landshaftna obolonka Zemli. Dovkillia: navch. posib.*(Methodological bases of geography: Landscape of the Earth. Environment: textbook. (Helvetyka, Kherson, 229, 2018)
13. *Rehionalna dopovid pro stan navkolyshnoho pryrodnoho seredovyshcha v Odeskii oblasti u 2019 rotsi* (Regional report on the state of the environment in Odessa region in 2019). (ODA, Odesa, 2020)
14. *Natsional'nyy atlas Ukrayiny*, (National Atlas of Ukraine) ed. by L.H. Rudenko. (DNVP Kartohrafiya, Kyiv, 2007)

Treatment of Mining and Thermoelectric Waste Through the Geopolymerization Process

Morales-Aranibar Carlos[1]*, *Linares* Nataniel[2], *Soto* Tolomeo[3], and *Morales-Aranibar* Luis[4]

[1]Metallurgist Engineer from the Jorge Basadre Grohmann National University
[2]Principal professor of the Professional School of Metallurgical and Materials Engineering of the Jorge Basadre Grohmann National University
[3] Rector of the Jorge Basadre Grohmann National University.
[4] Director of the Office of Innovation, Technology Transfer and Intellectual Property at the National Intercultural University of Quillabamba

Abstract. Currently, energy and extraction activities generate large amounts of highly polluting waste, so there is a need for sustainable and environmentally friendly technologies. The aim of the study is to treat mining tailings and fly ash through the process of geopolymerization. The samples studied were obtained from the Toquepala mine, Tacna and from the ENGIE-Moquegua hydroelectric plant (Peru). The methodology was based on two stages, the first characterization of Fly Ash (FA) and mine tailings (MT) by EDX chemical characterization, SEM morphology, the second was prepared mixtures of MT and FA in 10 M alkaline solution, cured in 35 days at room temperature and the characterization of the geopolymer by organoleptic analysis, SEM and TCLP. The first stage shows high aluminosilicate content 20.44% Al_2O_3 and 53.39 % SiO_2 for (MT); 22.11% Al_2O_3 and 51.76% SiO_2 for (FA), presents metal and pyrite content. In the second stage, the samples show health and environmental harmlessness, with the formation of tetragonal structures typical of the geopolymer, the samples show a significant reduction of Sr, Ca, Fe, Pb, Ba, Be, and Cu, demonstrating the effectiveness of treatment by means of geopolymerization opening a new field for the environmental passive treatment.

1 Introduction

Pollution from industrial mining and thermoelectric waste, which causes severe environmental alterations worldwide, presents great difficulty in its treatment because large amounts of mass are produced daily. This has motivated different researchers to develop technologies for the management and processing of mining and thermoelectric liabilities [1].

At a global level, the growing boom in technology has generated an important need for raw materials such as metal elements for the shaping of these technologies, leading to a mining boom, which has become a major socio-environmental problem for the population as the environment has been altered and sustainable methods are being sought to mitigate their impact [2]. On the other hand, the thermoelectric industries are of great concern, given that the transition to new green technologies will take a long time for Third World countries such as Peru [3], so it is of vital importance to develop new affordable and low-cost technologies for the treatment of mining and thermoelectric waste.

Industrial processes such as copper mining, where huge quantities of tailings are discharged into tailings ponds [4] and coal burning for electricity generation, generate large quantities of coal, leaving fly ash as a by-product [5], use techniques that move enormous quantities of materials to extract valuable metals and produce electricity, however, these contaminants possess metallic elements such as Cu, B, Be, Co, Mo, present in the tailings and fly ashes, these are harmful to soil and water as they are potentially reactive by surrounding water bodies which could generate acidic water [6] [7],

These wastes are environmental liabilities that need immediate recourse, not only to remediation or mitigation but also to seek new technologies that allow us to restore and give added value to these pollutants [8], where society, environment and economy are in balance, being these industries the ones that move the country economically [9].

Globally, these mining and thermoelectric wastes have an inherent composition of aluminosilicates present in mining tailings and fly ashes that are industrial passives that need to be recovered and treated [10] [11]. Encapsulation is a process involving tailings and ashes that, when agglomerated under alkaline conditions, achieve appropriate properties and chemical stability, forming geopolymers that are capable of stabilising these contaminants, in addition to giving them other sustainable uses [12]. Therefore, it is proposed to treat mining tailings and fly ash by means of the geopolymerization process.

2 Materials and methods

* Corresponding author: cgmoralesa@unjbg.edu.pe

2.1 Sample selection

The samples were taken in situ, the fly ash samples (FA) were provided by the ENGIE thermal power plant in the city of Ilo, and the mining tailings from Quebrada Honda of the Southern Peru Copper Corporation mining company in Tacna. The samples were studied at the mineralurgy laboratory of the Universidad Nacional Jorge Basadre Grohmann in Tacna

3 Methodology

The methodology for carrying out this research was divided into two stages, the first in which the MT and FA are physically, morphologically and chemically characterized; and a second stage where the physical, morphological and chemical properties of the geopolymer formed by MT and FA are characterized.

3.1 First stage: Primary characterization of the raw materials.

(a) Physical characterization: Tailings and fly ash samples were characterized by sieve analysis and specific gravity by the pycnometer method [13] [29].

(b) Morphological characterization: The samples were analyzed by scanning electron microscopy instrumental analysis [14] [30].

(c) Chemical characterization: The elements present were determined by instrumental analysis EDX - Energy Dispersive X-Ray Spectroscopy [15] [31].

3.2 Second stage: Formulation of the geopolymer and characterization.

Different mixtures composed of fly ash and mining tailings were developed, these were activated and agglomerated by means of a sodium hydroxide solution with a liquid solid ratio of 5 to 1 with a molarity of 10 M, these were cured in the open air for 35 days for their characterization [16] [32]

(a) Physical characterization: The geopolymer briquettes formed from the industrial wastes passed organoleptic tests.

(b) Morphological Characterization: The geopolymer samples were analysed by scanning electron microscopy [17] [33].

(c) Chemical Characterization: For which the TCLP (Toxicity characteristic leaching procedure) instrumental analysis was performed for the best geopolymer. [18] [19].

4 Results

4.1 First stage: primary characterization of raw materials.

a) Physical Characterization: Particle size analysis, for the tailings shows a fine particle size below the 106 μm opening, while for the fly ashes it shows a fine particle size below 60 μm denoted in Figure 1, these ones have a specific gravity of 2.86 for the tailings and 2.21 for the fly ashes.

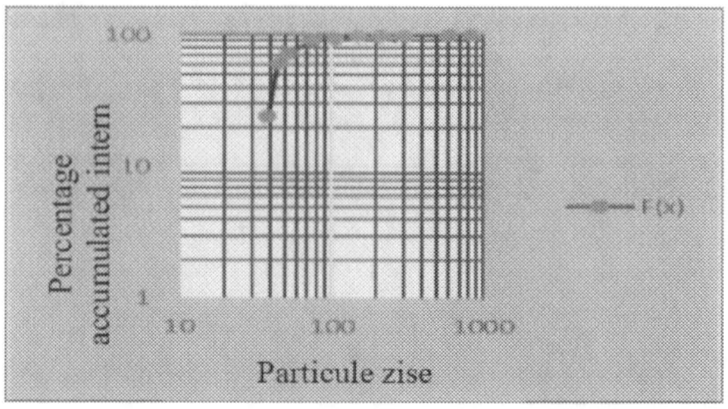

Fig. 1. Particle size distribution for fly ash.

b) Morphological Characterization: A heterogeneous particulate material with the presence of metallic content on the surface of the samples is observed at 100 magnification resolution the presence of pyrites and iron oxides, shown in figure 2 and 3.

Fig. 2. Microphotography of tailings at 100x resolution

Fig. 3. Microphotography of fly ash at 100x resolution

c) Chemical characterization: Quantitative chemical compositional analysis for tailings and fly ashes, showed a high percentage of elements that are important for geopolymer formation, such as silicates and oxides present in tailings and fly ashes and they are important for geopolymerization.

When these come into contact with the NaOH activating solution, they form the geopolymer. The presence of typical elements such as Sulphur oxides and iron oxides of the composition of tailings and fly ashes, which are highly contaminating due to their particle size and chemical reactivity, was also evident, as shown in Table 1 and Table 2.

Table 1. EDAX chemical composition analysis results for tailings

Formula	Composition % Relave
MgO	2,05
Al_2O_3	20,44
SiO_2	53,39
SO_3	8,81
K_2O	6,97
CaO	2,71
FeO	5,63

Table 2. EDAX chemical composition analysis results for fly ash.

Formula	Composition %. Fly ash
Na_2O	5,25
MgO	1,53
Al_2O_3	22,11
SiO_2	51,76
SO_3	2,58
K_2O	2,67
CaO	4,63
FeO	9,47

4.2 Second stage: formulation of the geopolymer and characterization.

a) **Physical characterization:** The formed briquettes showed particular properties according to their nature and mixture, it presents an average density of 2, 12 g/cm3, a significant reduction in the smell was evidenced, the briquettes could be easily manipulated without raising dust in the environment and health, the colour is maintained between lead shades evidenced in table 3.

b) **Morphological Characterization:** The geopolymer shows tetragonal structures typical of geopolymers, which shows the encapsulation and formation of the geopolymer, which provides high mechanical resistance and the insertion of the pollutants. This tetragonal structure allows the reduction of pollutants that are harmful to the environment, which can be seen in figure 4.

Table 3. Characterization of the organoleptic properties of the components and the geopolymer.

Type	Colour	Smell	Touch
Pure Tailings	They are in a leaden tone.	It has a mild sulphurous mineral odour and a rotten egg smell.	It is presented as a powder with a rough and fragile surface.
Pure Fly Ash	They have a dark lead colour.	Presents a mild smell of oxidised minerals	It is presented as a powder with a smooth and fragile surface.
Geopolymer	They vary with the presence of light to darker shades of lead fly ash	Odourless	Smooth and hard surface, in some cases the presence of silicon crystals similar to the common efflorescence is observed

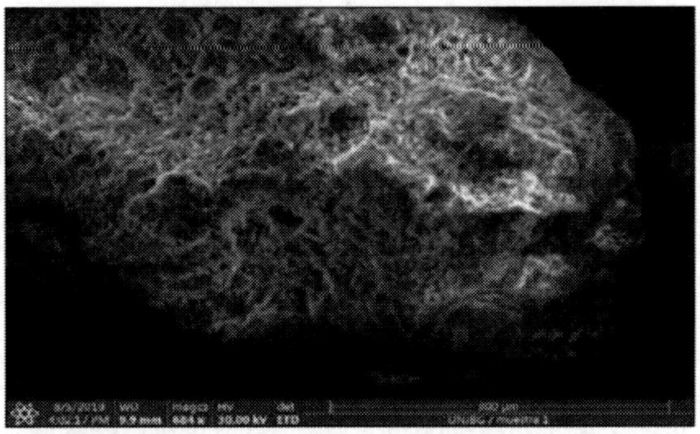

Fig 4. Micrograph of the geopolymer at 300 magnification.

c) **Chemical characterization:**

Results of the TCLP (Toxicity characteristic leaching procedure)

A large reduction in the levels of pollutants is observed in comparison with sample 1 pure binder mixture and the geopolymer formed in sample 2, a significant decrease in pollutants such as Pb, Ba, Be, and Cu, which are harmful to the environment, it is shown in (Table 4).

Table 4. TCLP Test Analysis

Type mg/L	Sample 1	Sample 2	Type mg/L	Sample 1	Sample 2
Ag(t)	<0.002	<0.002	Li(t)	<0.004	0.639
Al(t)	4.17	0.91	Mg(t)	16.21	9.79
Zn(t)	0.279	0.016	Mn(t)	1.277	0.012
B(t)	3.122	2.243	Mo(t)	<0.004	0.483
Ba(t)	0.307	0.188	Na(t)	2030.05	2618.02
Be(t)	0.0066	<0.0003	Ni(t)	<0.002	<0.002
Bi(t)	<0.02	<0.02	P(t)	<0.06	2.08
Ca(t)	493.42	228.77	Pb(t)	0.2	<0.01
Cd(t)	<0.001	<0.001	Sb(t)	<0.008	<0.008
Ce(t)	<0.02	<0.02	Se(t)	<0.02	1.76
Co(t)	<0.002	<0.002	Sn(t)	<0.007	<0.007
Cr(t)	<0.004	<0.004	Sr(t)	4.3157	2.3054
Cu(t)	0.469	0.005	Ti(t)	0.05	0.03
Fe(t)	1.12	0.33	V(t)	0.045	0.713
K(t)	9.89	68.5			

5 Discussion

The particle size of the samples that were studied, reveals a fine size; for mine tailings and fly ash this facilitates the mixing of the components and the geopolymerization reaction, due to the larger surface area of contact of the fine particles with the activating solution. The results are similar to [20] and [15]. The specific gravity for tailings and fly ashes shows that they are light materials, which is reflected in their density facilitating the transport of these encapsulated contaminants for better disposal.

The morphological characterization of the raw materials shows the presence of metallic elements on their surface, such as pyrites and other oxides, which are harmful to the environment and health [7], as well as the presence of environmental humidity typical of the Peruvian coast for Ilo and Tacna, which could generate acidic water, as demonstrated by [21], and contaminate underground tributaries if they are not treated appropriately. This demonstrates the importance of encapsulating industrial waste such as tailings and fly ashes in a chemical mesh such as a geopolymer, as explained by [22].

Due to the industrial origin of the samples, such as the Ilo thermal power plant and the tailings channels of Quebrada Honda, they show a high quantity of important elements for the geopolymerization, such as the aluminosilicates evidenced in the instrumental analyses EDAX with percentages of 20,44% Al_2O_3 and 53.39% SiO_2 for mine tailings and 22.11% Al_2O_3 51.76% SiO_2 for fly ash, these elements present in mining and thermoelectric waste demonstrate the formation of the geopolymer that was found by [19] and [12].

The activating solution in this case Sodium Hydroxide was very critical and it was demonstrated that the molarity was 10 M for the formation of the geopolymer, with important physical and chemical properties such as high hardness, high mechanical resistance and the formation of tetragonal structures of the geopolymer in which the contaminating elements are inserted for the encapsulation of hazardous waste which in a similar way are demonstrated by [23] and [24] The solid-liquid relationship shows the good workability of the mixture and its low water consumption for the formation of geopolymer briquettes, as well as the evidence of [25], which is a very important property for saving water consumption in desert areas such as the southern coast of Peru, since this is an environmentally sustainable material.

The organoleptic characterization for untreated fly ash and mine tailings in its pure state presents invasive properties to the environment such as bad smell due to the sulphides present and easy spreading in the environment due to the particle size and the form of dust in which it is presented with a fragile surface, which becomes very vulnerable to environmental climate conditions, as shown by [7], [8], this compared to the geopolymer which has a hard surface and greater resistance shows an ecological and environmental alternative to mitigate the impact of these industrial wastes due to the low cost and easy access to this technology.

The morphological characterization for the formed geopolymer shows tetragonal structures. This mesh is responsible for the special physical-chemical properties of the geopolymer, such as its structural resistance, low specific gravity and low chemical reactivity in the environment due to the encapsulation of the contaminating elements in its tetragonal network as evidenced by [26].

The TCLP analysis demonstrates a significant reduction in the levels of the contaminants Mr. Ca, Fe, Pb, Ba, Be, and Cu present in the raw materials after treatment by geopolymerization, demonstrates the chemical encapsulation of the contaminants, as well as the evidence of Cristelo et al. Elements such as Mo, Na, Ni, P, K, Li, Se and V, present difficulty to fix this kind of ions in the geopolymer matrix due to their high solubility in alkaline environments. This increases their presence in the formed geopolymer, as [27] show in their results, so these elements are scattered in the geopolymer matrix as [28], this predominant mechanism depends on the level of solubility of the heavy metal ions, i.e. the highly soluble elements are more spread throughout the matrix, while the less soluble elements form larger volume nuclei that are then involved by the gel matrix, which makes this geopolymer present a chemical and physical encapsulation for elements such as Sr, Ca, Fe, Pb, Ba, Be, and Cu and a physical encapsulation for the other elements Mo, Na, Ni, P, K, Li, Se and V, taking into account that the geopolymer formed is an aggregate with great mechanical properties, This marks an absolute difference between other treatments due to the properties obtained in this research work, which was treated under natural environmental conditions in the Tacna Peru region, which shows that it is possible to reduce the contaminating elements without the need to calcine the mixture of raw materials, however the curing time is 35 days to achieve optimal properties in the geopolymer formed, it is recommended to carry out more studies related to the encapsulation of heavy contaminating elements, in addition to studies of future applications of these materials for use in structures or roads.

6 Conclusion

The physical and chemical characteristics of mining tailings and fly ashes are ideal for the geopolymerization process at room temperature in alkaline solution, these untreated wastes represent a danger to the environment due to their physical and chemical characteristics such as their volatility and reactivity in the atmosphere, the geopolymer formed has optimum organoleptic properties for final disposal, it shows reduction of odour, fixes the contaminating particles in a tetragonal structure typical of the geopolymer and improves its mechanical properties, showing a significant reduction of the contaminating elements, such as Sr, Ca, Fe, Pb, Ba, Be, and Cu among others, which demonstrates the effectiveness of encapsulation of the contaminating residues, being innocuous with the environment, being able to consider its use for future applications in the treatment of mining and thermoelectric soils.

Aknowledgement

To the thesis "Immobilization of copper flotation tailings in Tacna through the use of geopolymers" financed with funds from the canon, sobrecanon and mining royalties by the Jorge Basadre Grohmann National University.

References

1. Segura-Beltran, F. y Sanchis-Ibor, C. (2018). Estrechamiento de cauces y cambio en cursos fluviales mediterráneos como consecuencia del cambio global en las últimas seis décadas: indicadores geomorfológicos. En: García, C., Gómez-Pujol L., Morán-Tejeda, E. y Batalla, R.J. (eds), Geomorfología del Antropoceno. Efectos del Cambio Global sobre los procesos geomorfológicos: 395-398. Universitat de les Illes Balears, Sociedad Española de Geomorfología, Palma.
2. Nieto, A. (2020), La Mina del Futuro: Las 5S para la Innovación en la Industria Minera, Revista Minería, Perú, May, Vol. 512, pp. 14-23. online: http://mineriaonline.com.pe/mineria/512/16/
3. Hughe, L. - De Jong, M Thorne, Z. (2020). (De)coupling and (De)carbonizing in the economies and energy systems of the G20, Environment Development and Sustainability. DOI: 10.1007/s10668-020-00834-7
4. Pinillos, J., (2015). Tecnologías ecológicamente racionales para el manejo de las aguas en los depósitos de relaves mineros. Lima, UNFV, p. 30.
5. Palacios M, Alonso MM, Vargas C, Puertas F (2019) Influence of the alkaline solution and temperature on the rheology and reactivity of alkali-activated fly ash pastes. Cem Concr Compos 95:277-284.DOI:10.1016/J. CEMCONCOMP.2018.08.010
6. Torrejón, J. A. V. (2015). Mecanismo de Producción más limpia: El Reúso de aguas residuales en la actividad minera. VOX JURIS, 30(2), pp. 263-278.
7. Wilson, D., Balkau, F. y Thurgood, M. (2005) "Solidificación y Estabilización". Manual de Formación en gestión de residuos peligrosos para países en vías de desarrollo. Disponible:www.uneptie.org/pc/hazardouswaste/menu.htm.
8. La Rotta, Á.,Torres, M. (2017). Mining and its health and environmental impacts. The case of Potosí in Bogotá https://doi.org/10.1590/0103-1104201711207
9. Instituto Peruano de Economía, (2017)." El valor agregado de la minería en el Perú", Primera edición. Hecho el Depósito Legal En La Biblioteca Nacional Del Perú N° 2017-06546, ISBN: 978-612-47458-0-5.
10. Giannopoulou, I. y Panias, D. (2006)." Development of geopolymeric materials from industrial solid wastes". "Conference: 2nd International Conference on "Advances in Mineral Resources Management and Environmental Geotechnology", AMIREG 2006Volume: pp. 69 – 73.
11. Waijarean N., Asavapisit, S., Sombatsompop, K., (2014). "Strength and microstructure of water treatment residue-based geopolymers containing heavy metals" Artículo publicado en Construction

and Building Materials 50:486-491. DOI: 10.1016 /j. conbuildmat.2013.08.047.

12. Barrie, E., Cappuyns, V., Vassilieva, E., Adriaens, R., Hollanders, S., Garcés, D., Paredes, C., Pontikes, Y., Elsen, J., Machiels, L. (2015). "Potential of inorganic polymers (geopolymers) made of halloysite and volcanic glass for the immobilisation of tailings from gold extraction in Ecuador", Applied Clay Science 109-110:95-106. DOI: 10.1016/j.clay.2015.02.025
13. Barbosa, V., MacKenzie, K., Thaumaturgo, C. (2000). "Synthesis and Characterization of Materials Based on Inorganic Polymers of Alumina and Silica: Sodium Polysialate Polymers." International Journal of Inorganic Materials 2(4): 309-317.

DOI: 10.1016 / S1466-6049 (00) 00041-6

14. Davidovits, J. (2011). Geopolymer Chemistry & Applications. 3rd Edition. Institut Géopolymère. Saint-Quentin, France. Publisher: Institut Géopolymère, Geopolymer Institute, Saint-Quentin, FranceEditor: Joseph Davidovits ISBN: 9782954453118 (5th ed.).
15. Davidovits, J. (1999) Chemistry of geopolymeric systems, terminology. In: Proceedings of 99 International Conference. France. 9-40. Davidovits, R. Davidovits & C. James, France.
16. Kastiukas, G.; Zhou, X.; Castro-gomes, J. (2017) "Preparation conditions for the synthesis of alkali-activated binders using tungsten mining waste". J. Mater. Civ. Eng. 2017, **29**, 1-9.DOI: 10.1061/(ASCE)MT.1943-5533.0002029
17. Mackenzie, K., Waijarean, N., Asavapisit, S., Guy, N., Jamesonv, L. (2017). "Synthesis and properties of geopolymers based on water treatment residue and their immobilization of some heavy metals", Journal of Materials Science 52(12). DOI: 10.1007/s10853-017-0970-4
18. Waijarean Naprarath, Suwimol Asavapisit, Kwannate Sombatsompop. (2014). "Strength and microstructure of water treatment residue-based geopolymers containing heavy metals" Artículo publicado en Construction and Building Materials 50:486-491. DOI: 10.1016/j.conbuildmat.2013.08.047
19. Komnitsas, K., Zaharaki, D. (2007). Geopolymerization: A review and prospects for the minerals industry. Minerals Engineering, pp.1261-1277. DOI: 10.1016/j.mineng.2007.07.011
20. Montes Valencia, Nancy. (2014). "Cementos Alcalinos: Materiales de Construcción Ecológicos" Revista CINTEX, Vol. **19**, pp.109-125.
21. Baquero J., Fernández R., Verdejo J., lorca D. (2008). Tratamiento de Aguas Ácidas. Prevención y Reducción de la Contaminación, revista de la Sociedad Española de Mineralogía, ISSN 1885-7264, N° **10**, 2008, págs. 44-47.
22. Davidovits, J. (2017) "Geopolymers: Ceramic-like inorganic polymers," J. Ceram. Sci. Technol., vol. **8**, no. 3, pp. 335-350, 2017. DOI: 10.4416/JCST2017-00038
23. Phair J., Deventer, V. (2001)." Effect of silicate activator pH on the leaching and material characteristics of waste-based inorganic polymers". DOI: 10.1016/S0892-6875(01)00002-4
24. Palomo, A., Blanco, M. (1999). Granizo M.L, Puertas F., Vázquez T. and Grutzeck M.W.; "Chemical stability of cementitious materials base on metakaolin". Cem. Concr. Res. Vol. **29**. DOI: 10.1016/S0008-8846(99)00074-5
25. Hardjito, D., Wallah S., Sumajouw, D., Rangan, B. (2004). "Properties of Geopolymer Concrete with Fly Ash as Source Material: Effect of Mixture Composition". Paper999 to the Seventh CANMET/ACI International Conference on Recent Advances in Concrete Technology. Las Vegas, USA.
26. Samarakoon, M.; Ranjith, P.; Rathnaweera, T.; Perera, M. (2019). "Recent advances in alkaline cement binders: A review". J. Clean. Prod. 2019, **227**, 70-87.4. DOI: 10.1016/j.jclepro.2019.04.103
27. Cristelo Nuno, Coelho João , Oliveira Mafalda, Consoli Nilo Cesar , Palomo Ángel Fernández-Jiménez Ana. (2020). Recycling and Application of Mine Tailings in Alkali-Activated Cements and Mortars-Strength Development and Environmental Assessment. Applied Sciences **10**(6):2084. DOI: 10.3390/app10062084
28. Lee, S.; van Riessen, A.; Chon, C.M.; Kang, N.H.; Jou, H.T.; Kim, Y.J. (2016). Impact of activator type on the immobilisation of lead in fly ash-based geopolymer. J. Hazard. Mater. **305**,59–66.DOI: 10.1016/j.jhazmat.2015.11.023.
29. I Capasso, S. Lirer, A. Flora, C. Ferone, R. Cioffi, D. Caputo, B. Liguori. (2019) Reuse of mining waste as aggregates in fly ash-based geopolymers. Journal of Cleaner Production 220. DOI: 10.1016/j.jclepro.2019.02.164.
30. Boca Santa R, Viana da Silva A, Padoin N, SoaresC, Gracher Riella H (2019) Novel porous geopolymeric formulation as green material applied to the recovery of contaminated effluent aiming environmental protection. Journal of Cleaner Production. DOI: 10.1016/j.jclepro.2019.05.213.
31. Kuenzela C, Ranjbarb N. (2019) Dissolution mechanism of fly ash to quantify the reactive aluminosilicates in Geopolymerisation. Journal Resources Conservation and Recycling. 10.1016/j.resconrec.2019.104421
32. Morales-Aranibar C, Linares N, Morales-Aranibar L (2021). Immobilization of copper sulfide flotation tailings through the use of geopolymers, Tacna – Perú. IOP Conference Series: Materials Science and Engineering, Volume 1065, 1st International Conference on Frontiers in

Engineering Science and Technology (ICFEST 2020) December 18th-19th 2020, Mangalore, India, doi:10.1088/1757-899X/1065/1/012009.

33. Yu Q, Li S, Li H, Chai X, Bi X , Liu J, Ohnuki T. (2019) Synthesis and characterization of Mn-slag based geopolymer for immobilization of Co. Journal of Cleaner Production. DOI:10.1016/j.jclepro.2019.06.149.

Green Infrastructure of Post-USSR Cities for Prevention of Noise Pollution

Nadiya Maksymenko[1], *Sergiy* Sonko[2], *Hanna* Skryhan[3], *Svitlana* Burchenko[1,4*], and *Anton* Gladkiy[1]

[1]Department of Environmental Monitoring and Nature Use, Karazin Institute of Environmental Sciences, V. N. Karazin Kharkiv National University, Svobody sq., 6, 61022, Kharkiv, Ukraine
[2]Department of ecology and safety of vital functions, Uman National University of Horticulture, Instytutska str., 1, Uman, 20300, Cherkasy region, Ukraine
[3]Department of Road Construction, Belarusian-Russian University, pr. Mira, 43, 212000, Mogilev, Belarus
[4]Ipris-Profil Ltd., Plytkova str., 12, Kharkiv, 61106, Ukraine

Abstract. The green infrastructure provides many different benefits for the environment and for human well-being. One of these benefits – protecting from noise pollution. The study of the deterioration of the ecological situation in cities is associated not only with the study of natural but also with physical environmental factors of natural and artificial origin. Increasingly, physical factors are defined as risk factors for the urban population and belong to the group of anthropogenic factors. Increased urbanization, mechanization of processes and the development of the transport network are intensifying influences for citizens. Territorial organization of the urban landscape in post-USSR cities, is quite complex and has signs of spontaneous formation, without protective function of green infrastructure from noise pollution. The aim of this research is to measure noise levels in Moskovsky district in Kharkiv, Ukraine. Kharkiv is a large industrial city and noise pollution is a considerable factor of influence on the environment. Based on these tasks, we have analyzed noise pollution in Moskovsky district in Kharkiv. Were identified 80 experimental points. Noise measurement was performed using an instrument "Digital Sound Level Meter". Was made the noise maps for working day and weekend in the mornings and in the evenings. For mapping was used QGIS instrument interpolation. The normal level of noise for this district is 60-70 according to the Ukrainian normative document. The WHO recommends the level of noise less than 20-30 decibels. According to this also was analyzed the change in the noise level in the zones of green spaces and was studied the objects of green infrastructure witch reduce noise in the cities.

1 Introduction

The level of noise is an important aspect of the environment and can reduce the quality of global ecosystems and human well-being. Sources of acoustic noise are different fluctuations in solid, liquid and gaseous environment. Noise pollution is no less dangerous than chemical pollution. Therefore, the creation a strategy for urban development, municipal development plan etc. it is necessary to take in account the possibility of reducing noise levels. One option for reduce the noise pollution is the development of green infrastructure in cities.

The definition of "green infrastructure" (GI) is provided by the European Commission – is a strategically planned network of natural and semi-natural areas with other environmental features designed and managed to deliver a wide range of ecosystem services such as water purification, air quality, space for recreation, and climate mitigation and adaptation [1].

[1]Corresponding author: s.burchenko@karazin.ua

To reduce excessive noise, preserve acoustically effective areas of the city, provide conditions for recreation and human health, it is necessary to develop special measures and implement technologies that reduce noise.

Traditional ways to reduce noise are legislative, construction and planning, organizational, technical and technological, design, and preventive measures. For reducing noise in the cities is noise control measures evaluated: 1) speed reduction, 2) pavement replacement, 3) decreased flow of heavy vehicles, 4) and all the measures together [2].

The acoustic load is the effect on a person of noise, is a set of sounds of different frequency and intensity that occur due to natural and technical factors [3].

The acoustic load is a parameter of noise pollution and allows forming ecological requirements to the construction of industrial infrastructure of the environment taking into account comfortable and safe stay of the person in it. The acoustic load is manifested in an increase in noise levels above the natural background, where natural sounds which have a beneficial effect on humans, are increasingly overlapped by anthropogenic, intermittent, with significant power noises that adversely affect human health and state of biotic and non-biotic components within the urban ecosystem.

Numbers of studies investigated the impact of noise on the environment and human-health: with questionnaire of people [4, 5]. The first step for regulation of level of noise is to compare experimental data (the measured noise) and legislative normative. Long-term studies of the effect of noise on human health allow us to speak about the positive role of green spaces [6, 7].

Territorial organization of the urban landscape in post-USSR cities, such as Kharkiv, is quite complex and has signs of spontaneous formation [8-10]. Green infrastructure had almost no protective function, with the exception of greenery in the sanitary protection zone of the plants. In our previous works [11] are investigated the shielding role of a green infrastructure in spreading of noise pollution.

Over the last decades, the level of acoustic pollution has been rapidly increasing due to the action of motor vehicles, along with the pollution of the environment with harmful toxic substances. This is due to the constant increase in the intensity of road traffic, the overall increase in engine power of cars, increasing speed and widening of roads. Noise is a form of physical pollution of the environment, and traffic noise is the excess of natural noise levels caused by the operation of engines, wheels, brakes and aerodynamic properties of the vehicle.

2 Materials and methods

2.1 Study area

As an experimental area, we have chosen the Moskovsky district in Kharkiv city. Moskovsky district occupies an area of 22,7 sq. km, the population is more than 300 thousand people (fig.1).

Fig. 1. Moskovsky district with the largest green zones

There are 35 large and medium-sized industrial enterprises in the district. There are also the main highways of the city. The territorial community of the city in the Moskovsky district has 1109 houses and 5064 private households. The total area of green zones is 4,62 sq. km.

2.2 Methods

To assess a noise in different countries, the sound level in decibels is the total sound pressure level measured by a noise meter on the correction frequency curve A, which determines the generally accepted frequency response of noise perception by the human ear.

The main criteria for providing acoustic comfort in residential areas and surrounding areas are the normative equivalent noise levels of the residential area in accordance to government buildings normative (Ukraine) – GBN B 1.1-31:2013 [12]. Based on the reconnaissance survey of the territory of the district, key points were identified, at which were carried out noise measurements and calculation of the number of passing vehicles.

The acoustical performance of green living systems includes two different aspects: the outdoor noise absorption and the insulation of indoor environments from outside noise [13]. The main requirement for the range of key areas is the uniformity of placement, coverage of all functional areas, and the possibility of parallel measurement of noise near its source and behind the "sound screen" (inside the residential area).

The methods of measurements: on weekends and working days, in the morning, and in the afternoon.. Noise measurement was performed using an instrument "Digital Sound Level Meter". Were identified 80 experimental points (table 1). Repeated measurements – three times at each point every 10 munites The total volume of the experimental data – is 1440 measurements.

To illustrate how noise pollution is distributed in the area of Moskovsky district we used software QGIS, and an instrument of interpolation. The noise map gives a visual representation of the noise regime of the study area. The main sources are plotted on the map and indicate their equivalent noise levels of decibels.

3 Results

Normative noise level on the territory next to residential buildings:

- from 7 to 23 hours - 55 (decibels);
- from 23 to 7 hours - 45 (decibels).

According to [12] equivalent and maximum noise levels generated by road, rail, and air transport measure in 2 m from the structures of the first echelon of residential buildings and at a height of 1,2 meters (fig. 2). Buildings that protect from noise, hotels and dormitories facing on the main streets of city, railways, sources of aviation noise, are allowed to take 10 decibels above.

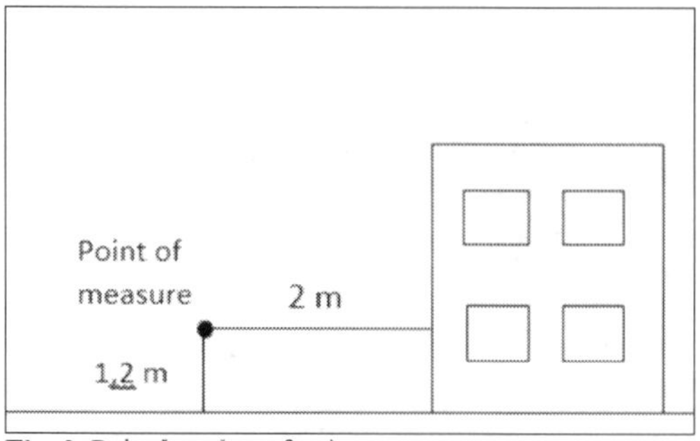

Fig. 2. Point location of noise measurement.

In addition, according to [12] in this area of urban development, is introduced an amendment +5 dB.

Thus, for this area, the standard noise level is:

- from 7 to 23 hours – 70 (decibels);
- from 23 to 7 hours – 60 (decibels).

The main source of external noise in this area are traffic flows consisting of cars and trucks, buses and trams. The initial noise parameter of the transport flow for various acoustic calculations is its noise characteristic.

As the noise characteristic is set an equivalent sound level generated by the flow at a distance of 7.5 m from the axis of the nearest lane and at a height of 1.5 m above the level of the roadway [14].

The initial data for calculating the noise characteristics of traffic flows are:

- traffic intensity during rush hours and the noisiest hour at night, units/hour;
- total share of freight and public transport in the flow, %.

All points of measurement are present in figure 3.

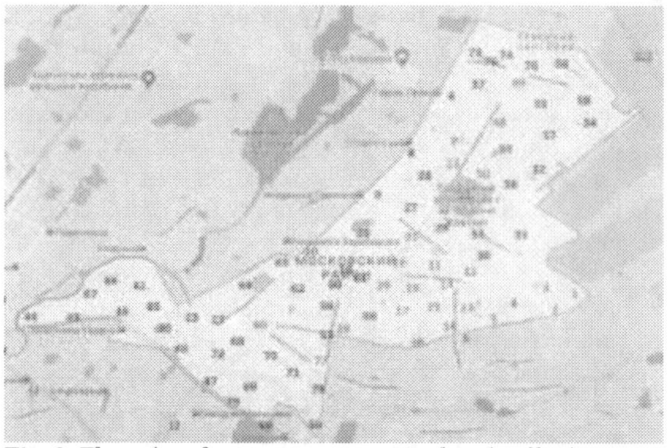

Fig. 3. The point of measurements at Moskovsky district

The requirement for the range of key areas is the uniformity of placement, coverage of all functional areas, and the possibility of parallel measurement of noise near its source and behind the "sound screen" was add. Common results of experimental data include 80 points of measurements (table 1). The mapping illustration of noise allocation is present in figures 4-7.

Table 1. The data of measurements of noise level in Moskovsky district.

№	Work day (morning) dB	Work day (evening) dB	Weekend (morning) dB	Weekend (evening)dB	№	Work day (morning) dB	Work day (evening) dB	Weekend (morning) dB	Weekend (evening)dB
1	65,8	68,5	69,1	67,6	**41**	75,1	75,3	70,7	71,9
2	68,9	70,6	72,5	72,6	**42**	79,4	80,3	77,4	76,7
3	66,5	66,2	67,7	64,8	**43**	79,7	82,6	77,6	77,4
4	66,0	64,9	65,0	61,8	**44**	79,2	81,2	77,8	80,6
5	68,2	69,8	64,9	62,7	**45**	80,4	82,0	78,4	79,2
6	73,0	72,9	74,4	76,0	**46**	83,1	81,2	76,7	77,6
7	70,1	70,2	71,3	67,4	**47**	81,6	81,4	79,0	79,5
8	71,2	71,8	72,6	67,0	**48**	81,2	81,2	81,6	82,6
9	71,6	72,8	74,0	73,7	**49**	70,0	69,4	67,7	66,3
10	76,0	77,6	75,1	73,3	**50**	77,1	75,9	72,6	70,5
11	64,7	70,5	67,6	64,3	**51**	65,6	67,8	60,9	60,3
12	65,7	70,6	70,5	65,5	**52**	57,2	55,0	51,9	52,6
13	65,2	71,0	69,5	65,5	**53**	75,7	77,2	73,7	74,3
14	74,6	71,2	72,5	73,2	**54**	57,6	55,4	52,3	51,9
15	68,1	70,5	69,2	72,6	**55**	65,2	64,8	65,9	66,2
16	69,2	68,2	70,8	70,5	**56**	73,3	74,8	73,0	73,6
17	68,5	69,5	68,0	70,0	**57**	66,8	65,1	63,6	66,7
18	70,6	72,8	72,5	71,6	**58**	66,5	62,0	65,1	64,1
19	70,2	72,7	71,3	72,6	**59**	69,4	69,0	69,0	69,6
20	66,4	66,2	68,9	69,8	**60**	70,3	68,0	71,2	72,6
21	70,6	73,3	65,1	64,5	**61**	70,7	71,4	69,4	69,2

22	74,6	75,0	72,5	62,8	**62**	59,2	57,3	46,3	45,9
23	71,3	71,2	75,7	71,2	**63**	54,5	52,2	47,4	48,2
24	61,0	59,7	62,6	68,4	**64**	52,0	49,7	47,6	45,0
25	62,0	54,2	59,3	69,7	**65**	51,7	51,9	47,9	47,8
26	63,4	66,1	57,2	54,8	**66**	51,3	51,0	51,4	51,2
27	60,3	61,1	52,1	56,3	**67**	53,0	52,7	49,8	49,5
28	52,7	56,9	52,7	54,5	**68**	53,1	44,4	47,4	48,0
29	53,2	49,5	54,1	55,4	**69**	51,0	49,5	50,1	43,7
30	53,8	55,3	48,6	45,6	**70**	54,2	54,8	48,1	46,5
31	53,4	51,5	46,4	47,0	**71**	50,3	46,9	45,1	45,7
32	51,7	46,8	51,7	54,6	**72**	50,1	50,8	46,4	44,0
33	53,1	45,7	51,9	49,9	**73**	48,9	47,3	43,5	44,3
34	52,8	52,0	48,0	46,9	**74**	48,1	43,8	44,4	41,9
35	51,4	48,9	55,4	46,6	**75**	47,9	48,4	45,8	44,1
36	51,7	52,1	45,8	47,1	**76**	46,7	47,5	46,4	44,8
37	54,6	54,2	44,2	45,0	**77**	70,7	70,6	71,1	71,4
38	68,9	69,4	66,0	66,9	**78**	79,7	82,6	77,6	77,4
39	67,4	66,9	62,3	64,7	**79**	79,2	81,2	77,8	80,6
40	76,1	76,7	73,8	72,8	**80**	69,4	69,0	69,0	69,6

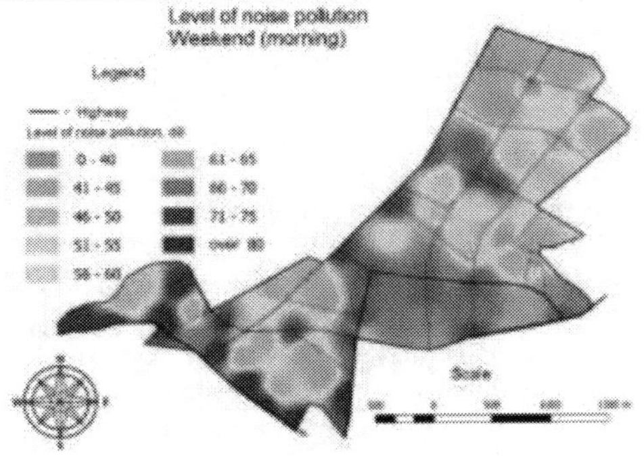

Fig. 4. Level of noise pollution at Moskovsky district (on weekend, morning).

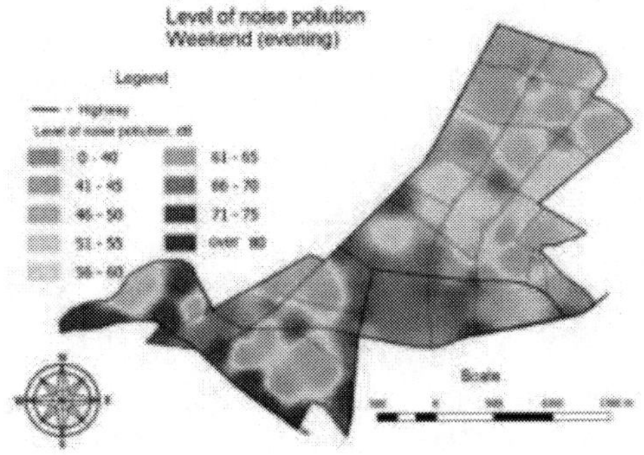

Fig. 5. Level of noise pollution at Moskovsky district (on weekend, evening).

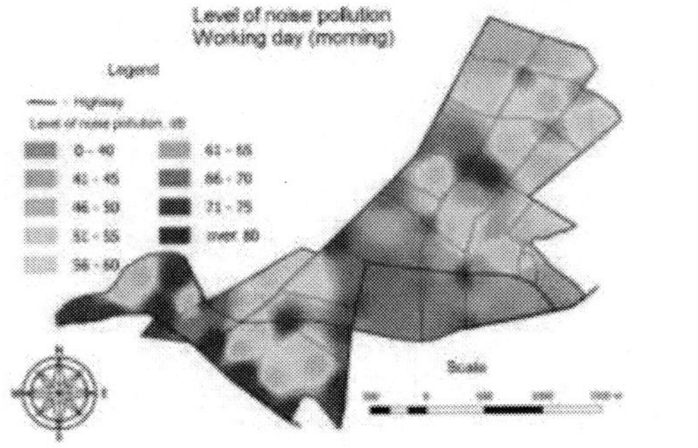

Fig. 6. Level of noise pollution at Moskovsky district (on working day, morning).

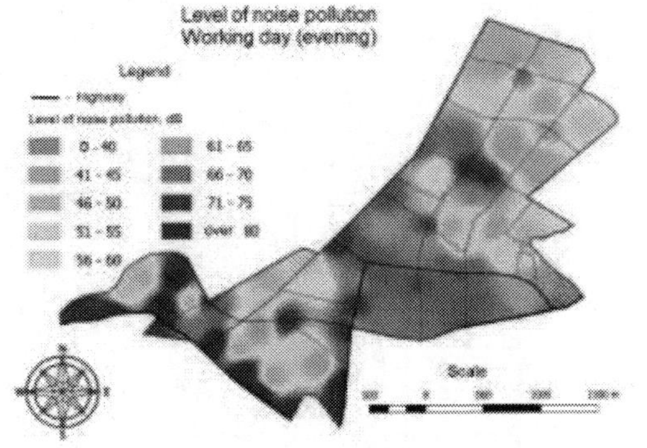

Fig. 7. Level of noise pollution at Moskovsky district (on working day, morning).

The highest noise levels in the Moscovsky region are observed in the south-western and central parts, both on weekends and on weekdays. This is because in these parts of the city there are industrial enterprises and large transport routes.

At the same time in these parts of the area, there are no means of protection against noise. The building and small private houses are facing on the main highways are most affected by noise. At the same time, noise from vehicles is often pulsed and intermittent.

This disproportion is typical because during the USSR times, realization of the protective function of green zones was not laid down in urban planning and urban development. Thus, the existing green infrastructure of the area are an isolated islands, there are only inside a decrease of the anthropogenic noise load.

Noise with a level of 50-60 dB creates a significant load on the nervous system. Noise above 70 dB causes physiological effects; noise above 80 dB can lead to hearing loss.

Areas with low noise levels (up to 45 dB) on weekdays and weekends are the greenest parts of the area.

In this case, the most interesting measurement is points next to green zones and inside there. The results of the measurement are present in table 2-4.

Table 2. Noise level next to Peremohy Park and inside

№	Work day (mor-ning)	Work day (eve-ning)	Weekend (mor-ning)	Weekend (eve-ning)	Stan-dard noise level
13	65,2	71	69,5	65,5	70
14	74,6	71,2	72,5	73,2	70
15 inside park	68,1	70,5	69,2	72,6	70
19	70,2	72,7	71,3	72,6	70
38	68,9	69,4	66	66,9	70

Based on the data of the table and maps, we can see that in the green zone – Peremohy Park the standards are also exceeded. This is due to the fact that in addition to the highway there are also Heat power station № 3, Saltovsky bakery plant, and a tram line near the park. Thus, we can say that the absence of a park in this area would significantly increase the noise pollution of this area.

Table 3. Noise level next to private gardens "50th anniversary of October" and inside

№	Work day (mor-ning), дБ	Work day (eve-ning), дБ	Week-end (mor-ning), дБ	Weekend (eve-ning), дБ	Stan-dard noise level
22	74,6	75	72,5	62,8	70
50	77,1	75,9	72,6	70,5	70
51 inside gar-den	65,6	67,8	60,9	60,3	70
55	65,2	64,8	65,9	66,2	70
56	73,3	74,8	73	73,6	70

Territories of private gardens "50th anniversary of October", and Nemyshlia river coastal zone have excesses over 70 dB, and are observed mainly on weekdays in areas of highways.

Table 4. Noise level next to tree lines of road

№	Work day (mor-ning)	Work day (eve-ning)	Weekend (mor-ning)	Weekend (eve-ning)
73	48,9	47,3	43,5	44,3
74	48,1	43,8	44,4	41,9
75	47,9	48,4	45,8	44,1
76	46,7	47,5	46,4	44,8

At these points, we see a decrease in the noise level due to the shielding of the forest. The average noise level in these points is 45,8 dB. Thus, linear plantings in combination with polygonal and point objects of green infrastructure can significantly reduce the level of noise load on residents of urbanized areas.

4 Discussion

In the general case, the methods of reducing traffic noise can be classified into the following three areas: reducing noise at the source of its occurrence, including the decommissioning of vehicles and changing the routes of their movement; noise reduction on the way of its distribution; the use of sound protection in the perception of sound.

Landscaping is an effective measure to combat noise in cities. Trees that are planted close to each other, surrounded by dense bushes, significantly reduce man-made noise and improve the urban environment. Plants are endowed with a special noise-absorbing ability. Plantations of maple, poplar, and linden absorb from 10 to 20 dB of sound signals. A thick hedge can reduce highway noise by 10 times.

Urban green areas are an important factor of the negative influence of noise in large cities. The harmful effects of noise on human health are generally recognized and manifested in a wide range: from subjective irritations to pathological changes in the hearing organs, central nervous and cardiovascular systems.

Have been identified two types of impact of green infrastructure on noise propagation:

- reducing its intensity deep into the green zone (appropriate only to those who are inside of green areas);
- retention of dissemination of noise from roads to residential buildings by forest belts between them.

In general, when adjusting the territorial organization of urban landscapes, the noise protection role of green infrastructure should be taken into account.

Greening the building roofs and facades are an nature-based solution in sustainable architecture that can restore losses of a natural environment in dense urban areas.

Types of GI, which can reduce noise pollution, and which we suggest to reduce noise are:

- parks near to coastal river zone, other parks, squares and gardens,
- green wall on new buildings (mainly houses),
- green roofs on commercial and entertainment centers,
- rain gardens in parks, municipal and private buildings,
- green corridors (rivers and canals, including their banks, road and railway corridors, bicycle routes, footpaths),
- green parking of commercial and entertainment centers' parking, municipal parking .

These zones can be classified by their functional use and accessibility for the population according to legislative (fig. 8-9).

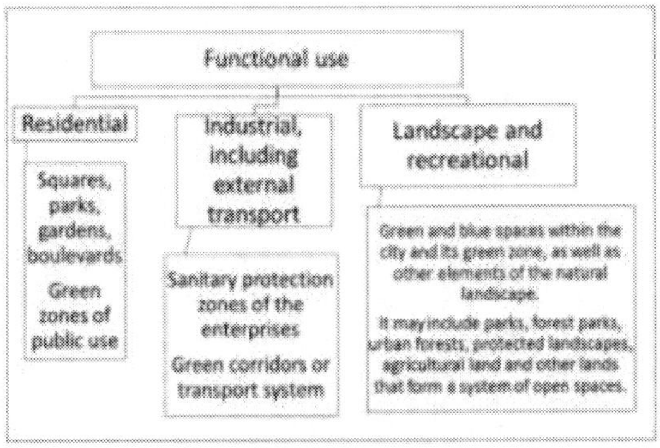

Fig. 8. Functional use of city's green areas

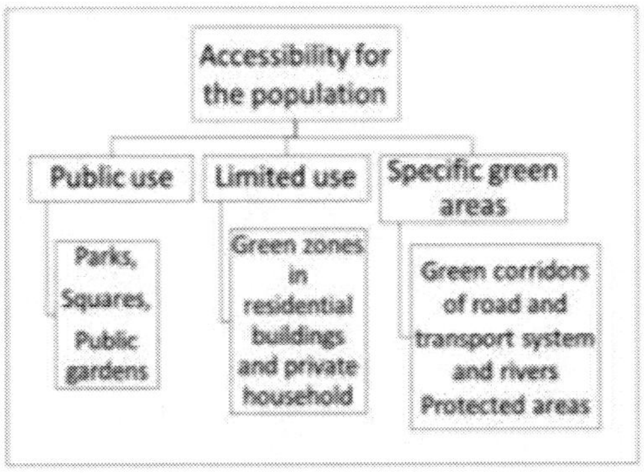

Fig 9. Accessibility of city's green areas for the population.

The important point is the absence of legislative and other normative documents of objects such as rain gardens, green roofs, walls and parking. Thus, such objects are mainly created on private territories and less often on urban land.

Parks, squares and gardens, regardless of their functional use and accessibility, occupy rather large areas, so the creation of new parks and polygonal objects may entail a conflict of interests of different land users.

Green roofs have the greatest potential for attenuating noise, and on certain roof shapes, may be able to reduce noise by up to 7,5 decibels. Also has other important environmental benefits, such as absorbing carbon dioxide, improving air quality, reducing the urban heat island effect, increasing urban biodiversity, and making streets and roofs look more attractive [5].

Green walls depending on the design and substrate can reduce as much as possible backward noise reflection and forward transmission from the wall [15].

Rain gardens is a garden of native shrubs, perennials, and flowers planted in a small depression, which is generally formed on a natural slope. It provide a temporarily hold and soak in rainwater runoff.

Green corridors in cities are linear zones of greenery. They are created not only along natural objects such as rivers but also along highways, railways etc. As shown in this research, it is linear plantings can reduce the noise level by about 20 decibels.

Green parking is a new way for cities to be environmentally more sustainable. They do not have a significant effect on the noise level in cities, but at the same time, they make the living space greener and not gray.

These areas and zones not only contribute to noise reduction but also provide ecosystem services such as:

- *regulative services:*

- regulation of surface runoff in the city – pollution control, groundwater recharging, flood control;
- air purification –trees or other plants also play an important role in regulating air quality by removing pollutants from the atmosphere;
- climate regulation – affect the climate both locally and globally. For example, on a local scale, changes in land cover (greening) can affect both temperature and precipitation;
- prevention of soil erosion – vegetation provides a vital regulatory service, preventing soil erosion;
- greenery creates a buffer against natural disasters, thus preventing possible damage, enhance the ability to adapt to the effects of climate change;
- carbon sequestration – regulate the global climate by storing and sequestering greenhouse gases. Trees and plants can remove carbon dioxide from the atmosphere and effectively lock it in their tissues.

- *Cultural services:*

- aesthetic values – many people find beauty or aesthetic value in various aspects of ecosystems, which is reflected in the support of parks and residential areas, the choice of places to live;
- social relations – affect the types of social relations that are established in a particular area;
- recreation and human well-being – people often choose where to spend their free time, based in part on the characteristics of natural or cultural landscapes in a particular area. Walking and playing sports in the green space is not only a good

form of exercise, but also allows people to relax. The role that green space plays in maintaining mental and physical health is increasingly recognized.

- *Provisioning services:*
- impact on the overall biochemical cycle;
- support and impact on the global hydrological cycle.
- *Supporting services:*
- habitat for species.

The creation of green infrastructure objects such as green roofs, green walls, green corridors, rain gardens and green parking does not require large territories and capital investments. However, they can significantly reduce noise pollution in the area, regulate surface runoff, improve human well-being and provide ecosystem services, such as regulation of building temperatures, reduced urban heat-island effects, and increased urban biodiversity.

References

1. European Commission. Communication from the commission to the European Parliament, The council, the European economic and social committee and the committee of the regions. Brussels, 2013.
2. F. Bunn, P. H. Trombetta Zannin, Noise Control Eng. J. **63**,1 (2015)
3. A. Tiasto, M. Kudymova, Young Sci, **10**, p. 45-48 (2015) *(in ukrainian)*
4. J. Jablonska, Polish Political Science Review, **8**(2), p. 101-109 (2020)
5. P.Ilić, Z. U. Rahman Farooqi, L. Stojanović, Indian Journal of Environmental Protection (2020) (to be published)
6. P. Morency, C. Plante, A.-S. Dubé, S. Goudreau1, C. Morency, P.-L. Bourbonnais, N. Eluru, L.-F. Tétreault1, M. Hatzopoulou,N. Chandra Iraganaboina, T. Bhowmik and A. Smargiassi, Int. J. of Environ. Res. Public Health, № **17**(14), p. 50-61 (2020)
7. M. Foraster, I.C.Eze, D. Vienneaua,M. Brink, C. Cajochend, S. Caviezel, H. Héritier, E. Schaffner, C. Schindler, M. Wannere, J.-M. Wunderlif, M. Röösli, N. Probst-Hensch, Environment International, № **91**, p. 341-349 (2016)
8. A. Klieshch, N. Maksymenko, J. of Geology, Geography and Geoecology, vol. **29** № 3, p. 539-549 (2020)
9. A. A. Klieshch, Man and environment. Issues of neoecology, № 3-4 (24), p. 28-31 (2015)
10. A. A.Klieshch, Visnyk of V.N. Karazin Kharkiv National University series "Ecology" № **17**, p. 63-71 (2017)
11. GBN V.1.1-31:2013, Protection of territories, buildings and structures from noise, Kyiv, Ministry of Regional Development, Construction and Housing of Ukraine, 2014, 48 p.
12. N. V. Maksymenko, I. I. Radionova, Visnyk of V.N. Karazin Kharkiv National University series "Ecology", №**16**, p. 113-118 (2017)
13. D. Dimitrijević, P. Živković, M. Dobrnjac, T. Latinović, *Noise pollution reduction and control provided by green living systems in urban areas*, in Scientific proceedings III international scientific-technical conference “Innovations”, p. 124-127 (2017)
14. GOST 20444-2014, Noise. Traffic flows. Methods for determining the noise characteristic. [Effective from 2015-07-01], 15 p. (2014)
15. A. Emmanuel, B. Dubus, T. Leblois, Cretin Building and Environment,187 (2021)

Assessment of Climate Change by Dendrochronological Methods in Polissya

Iryna Koval[1], *Valentyna* Meshkova[1], *Nadiya* Maksymenko[2], *Catalin-Constantin* Roibu [3] and *Iryna* Obolonik[1]

[1] Ukrainian Research Institute of Forestry and Forest Melioration named after G. M. Vysotsky, Pushkinska, 86, 61024 Kharkiv, Ukraine

[2] V.N. Karazin Kharkiv National University, sq Svoboda, 4, 61022, Kharkiv, Ukraine

[3] Stefan cel Mare University of Suceava, Str. Universitatii, nr. 13, 720229 Suceava , Romania Suceava, Romania

Abstract. Dendrochronological methods were used to study the response of Quercus robur L. to climate change in Polissya. Negative pointer years (1950, 1976, 1995, 1999, 2002, 2008, and 2011) resulted from the precipitation deficiency during the growing season, cold or extremely warm winters, and abnormal early-spring temperatures. Positive pointer years (1967, 1997, 2001, and 2007) had a favorable heat and moisture balance. Dendroclimatic analysis of oak regional tree-ring chronology showed that during the spring and summer terms of 1980 - 2013, as compared to the previous 1946–1979, there a decrease in the positive influence of temperatures on the oak radial growth. In the second period, the negative impact of precipitation during the cold period on radial growth increased. The adverse impact of April precipitation on radial growth for both periods was revealed as well as the positive influence of July precipitation in the first period and precipitation in June and July in the second one. The increased influence of temperatures and precipitation on the oak radial growth in 1980 - 2013, as compared to previous 1946-1979, indicates an increase in the tree sensitivity to climate change in the second period.

1 Introduction

English oak (*Quercus robur* L.) is one of the main forest-forming species in Ukraine. Oak forest ecosystems provide a wide range of ecological benefits [1, 2]. For the early detection of potential dangers, forest growth reactions must be more strongly connected to studies on adaptation relevant characteristics in future. Changes in climate, and weather extremes in particular, are radical selective factors that will change forest structure, including tree species composition, in the future in a regionally differentiated manner. Selectively effective extremes of weather will naturally shift competitive relationships and produce "losers", but also "winners" on all levels of the ecosystem (individual, population, ecosystem, forest community, landscape). Extreme climatic factors will lead to an increased dying off of individual trees within stocks. These trees or groups of trees must be detected early for different measures (timber use, forest protection, prevention of danger to ensure traffic safety etc.) [2].

Given the ecological and economic importance of forest and forestry, an analysis must be made of the potential threat from climate change and adaptation strategies must be developed. According to current knowledge, these observations can only be made for a few decades in advance, which barely represent half of a forest generation. This short time period is unsatisfactory in view of the temporal dimension of forest development [2].

Studying annual rings is essential to investigate English oak adaptation to climate change because they are an integral indicator of the environmental impact on the tree condition [2].

The assumption that the most common causes of forest degradation and large scale deforestation are global climate change and anthropogenic factors is an almost single hypothesis. Therefore discovering specific mechanisms of forest ecosystem development and revealing the causes of forest decline is extremely relevant.

The study aimed to identify the main climatic factors affecting the radial growth of English oak under climate change in Polissya.

2 Material and methods

The research was conducted in four sample plots in oak stands aged from 50 to 155 growing in fresh, wet, and moist fertile sites (Table 1, 2).

Table 1 Characteristics of the sample plots in Polissya.

SP	GEOGRAPHICAL LOCATION		A S. L., M	SITE CONDITION
	North latitude	East longitude		
1	50°35′00″	27°37′13″	218	Fresh fertile site
2	51°13′40″	27°38′53″	183	Wet relatively fertile site
3	51°13′40″	27°38′53″	183	Damp

1 Corresponding author: Koval_Iryna@ukr.net

				relatively fertile site
4	51°23′40	26°50′50″	150	Wet relatively fertile site

Notes: SP sample plot; a. s. l. – altitude; D – tree diameter; H–tree height

Table 2. Characteristics of oak stands in Polissya.

SP	AGE, YEARS	DIAMETER, CM	HEIGHT, M
1	155	44	28
2	>100	48	26
3	>100	45	27
4	>50	26	11

We used the data from the Zhytomyr meteorological station (50°15′53″ N, 28°40′36″ E, an altitude of 228 m a. s. l.).

Standard dendrochronological methods were used [3]. The cores were selected at a breast height (1.3 m above the ground). Tree-ring widths were measured using HENSON digital equipment with an accuracy of 0.01 mm. Cross-dating was performed including both counting the tree rings and comparing samples to recognize missing or false rings to identify the exact calendar year for each tree ring formation and reveal errors in the measurements.

The crossdating quality was checked using the COFECHA program [4]. The total signal strength in the series was determined through the intercorrelation coefficient (Rbar.). The coefficient was defined as the mean correlation between all tree-ring chronologies [5]. The expressed population signal (EPS), which states the reliability of measurements in chronology and is a function of Rbar and the number of tree-ring chronologies, was calculated using the formula:

$$EPS(t) = \frac{tRbar}{tRbar + (1 - Rbar)}, \quad (1)$$

Where: t is the number of tree-ring series;
R_{bar} is the mean correlation between tree-ring chronologies.

The suitability of the tree-ring series for further dendroclimatic analysis is evidenced by exceeding the EPS threshold of 0.85 [5].

Standardization of tree-ring chronologies was carried out using the ARSTAN program to remove the age trend. First, a linear approximation was used, and then autoregressive modeling was applied to eliminate the influence of previous years on the formation of the tree ring of the current year. Local chronologies were developed by averaging the values of individual chronologies, and regional chronology was built by averaging the values of local chronologies [3].

The standardization of ring width series allowed us to analyze the response of oak radial growth to climatic factors using the RESPO program from the DPL software package [6]. The response was analyzed for 1946–1979 and 1980–2013 to identify the oak radial growth response to climate variations during these periods. Average monthly temperatures and precipitation amounts from July of the previous year to August of the current year were used. We applied a stepwise regression analysis to select the most growth-affecting climatic factors, as well as a principal component analysis, to remove closely correlated climatic variables from the analysis. Moreover, we used multiple regression analysis to assess climate impact on radial growth [6].

3 Results and discussion

A regional tree-ring chronology of English oak for Polissya was built, consisting of 62 individual tree-ring series. The regional chronology was based on four local tree-ring chronologies, which contained 5,119 tree rings. Based on this chronology, the RESIDUAL index series was developed, from which the age trend was removed, making it suitable for dendroclimatic analysis (Fig. 1, Table 3, 4).

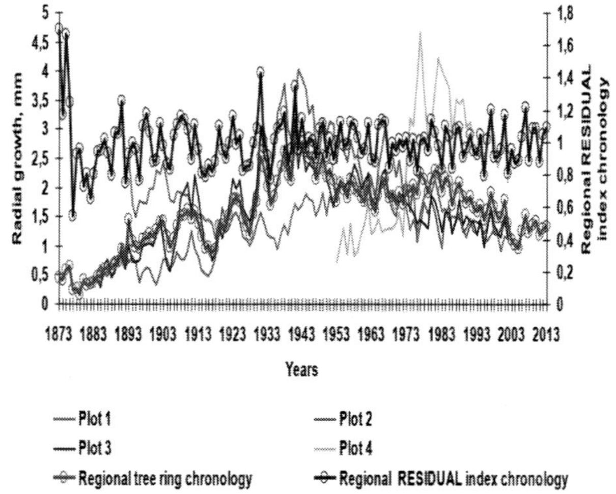

Fig. 1. Dynamics of English oak radial growth and regional RESIDUAL index chronology in stands within Polissya.

Interseries correlation coefficients range from 0.360 to 0.538, indicating high synchronicity and the possibility of using these tree-ring chronologies to identify the relationships between radial growth and climatic factors (Table 3, 4)

Table 3. Statistical characteristics of local and regional series of tree-ring chronologies of English oak in Polissya according to COFECHA and ARSTAN software.

SP	Period, years	Average, mm	Tree-ring series			
			$R_{bar.}$	$S_{td.dev}$	AC_1	MS_x
1	1890–2013	1.45	0.36	0.60	0.72	0.21
2	1897–2002	1.97	0.54	1.12	0.83	0.25
3	1873–2002	1.77	0.53	0.77	0.76	0.24
4	1953–2002	2.76	0.52	1.23	0.72	0.24

RT RS	1870–2013	1.86	0.40	0.87	0.76	0.24

Notes: SP sample plot $R_{bar.}$ – icorrelation coefficient; $S_{td.dev}$ – standard deviation; AC_1 – first-order autocorrelation; MS_x – mean sensitivity; RTRS – regional tree-ring series.

Table 4. Statistical characteristics of local and regional RESIDUAL series of tree-ring chronologies of English oak in Polissya according to COFECHA and ARSTAN programs.

SP	Residual series	
	MS_x	$S_{td.dev}$
1	0.11	0.12
2	0.21	0.19
3	0.23	0.21
4	0.17	0.17
RTRS	0.15	0.16

High values of first-order autocorrelation (0.716–0.832) indicate a high dependence of the current year's tree-ring width on that of the previous year. The EPS coefficient calculated for the regional RESIDUAL tree-ring chronology is 0.98. This also confirms the possibility of dendroclimatic analysis because it exceeds the threshold of 0.85, meaning the reliability of the chronology [5]. Sensitivity coefficients exceed 0.2, which allows this tree-ring chronology to be used in the dendroclimatic analysis (Table 2).

The use of the above factors to determine the reliability of the dendrochronological series has been proven by several studies [5, 7]. The stand growing in fresh relatively fertile site conditions (SP1) appeared to be the most resistant to the influence of environmental factors, while the stand growing in wet relatively fertile sites (SP 2) was the least resistant. This was confirmed by the values of standard deviation and mean sensitivity, which was the smallest for SP 1 and the largest for SP 2 (Tables 1, 2).

For the period 1873–2014, 23 pointer years in the regional RESIDUAL chronology were defined: 16 negative and 7 positive ones. They comprise about 11% of the total number of years analyzed. The pointer years of minimal growth (1950, 1976, 1995, 1999, 2002, 2008, and 2011) were mainly conditioned by the deficiency in precipitation during the growing season when rainfall deviation from the norm ranged from 33 to 77% (Fig. 1, table. 3).

In 1976, the precipitation deficiency, cold winter, and early spring temperatures caused a radial growth depression. The deviation from the norm at that time was 52% for winter temperatures, while March temperatures were almost four times below the norm (Table 5).

During the growing season in 1976, 1995, and 1999, temperatures in some months exceeded the norm by 37-56%. Due to the temperature increase during the winter period, the radial growth minimums were observed in 1995, 1999, 2002, and 2008 (deviations from the norm were 37–78%) (Table 5).

In March of 1995, 1999, 2002, 2008, and 2011, when xylogenesis started, temperatures increased 4-12 times compared to normal (norm was 0.33°C), and March temperatures ranged from 1.2 to 4.3°C, which could exceed the optimal limit for radial growth of trees (Table 5).

Table 5. Negative pointer years of the regional tree ring chronology of English oak in Polissya

Pointer years	Causes of radial growth depression
1877, 1926	The causes of radial growth depression are unknown due to the lack of meteorological data
1950	Radial growth was limited by precipitation in March-June. In 1950 the precipitation during this period was 90 mm, i.e. 57 % less than the norm, with was 207 mm
1976	Cold winter: average winter temperature was -5.9 °C; the rate was -3.9°C, i.e. the temperature was 52% lower than rate. In March the temperature was almost four times below the rate: -1.2 °C vs normal 17.1. Precipitation deficiency was observed in May-July; the precipitation was 40% less than norm: the amount was 132.8 mm during these months, while the norm was 221.
1995	Warm winter, during with temperatures were almost twice as high as the norm: average winter temperature was -3.9°C, while the norm was -1.6 °C. March temperature was eight times higher than normal: 2.6°C at a norm of 0.33 °C. In the summer months, precipitation was 40% below normal: 145 mm vs 242 mm, respectively.
1999	The extremely warm period from January to February, when the temperature was 55% above normal (-2.05 °C vs -4.6°C, respectively). March temperature was eight times higher than normal: 2.8 °C at a norm of 0.33°C. During May-July, precipitation was 37% below the norm: 136 mm at a norm of 221 mm.
2002	Extremely warm winter, when the temperature was 37% above normal -2.5 °C at a norm of -3.9°C. The precipitation deficiency in January, March, May, and July fluctuated within 40-53%, i.e. the precipitation were 19.6 mm vs 56.6 mm, and 4.4 mmvs 86.8 mm, respectively. March temperature s was 12 times higher than normal: 4.2°C vs 0.33°C, respectively. Dry growing season, when 210 mm of rain fell during April-August, i.e. precipitation was 20% below the norm, with was 253 mm. July was the driest when precipitation was 56% below normal.
2008	Warm winter, when temperature exceed the norm by 78%: 0.8% °C at a norm of -3.9°C. March temperature was almost four times normal: 1.2 °C vs normal 0.33 °C. Precipitation deficiency: precipitation during June -August was 95.4 mm at a norm of 242 mm, i.e. it was 78% below normal.
2011	March temperature was 13 times normal: 4.3 °C vs 0.33 °C, respectively. Precipitation deficiency: during January, March and August deviations from the norms ranged from 33 to 71%: 19.2 mm vs 32.7, 9.5 mm vs 32.7 mm, 32.2 mm vs 56.7, and 51,6 mm vs 77.7 mm, respectively

Thus, the pointer years of minimal growth (1950, 1976, 1995, 1999, 2002, 2008, and 2011) resulted from the precipitation deficiency during the growing season, cold or extremely warm winters, and early spring temperatures.

The favorable heat and moisture balance for tree growth was found during the positive pointer years: 1875, 1902, 1906, 1912, 1967, 1997, 2001, and 2007. In 1967, the precipitation in March-May exceeded the norm by 30%: 194.1 mm at the norm of 129.8 mm. In

https://doi.org/10.1051/shsconf/202110005005

1997, 2001, and 2007, the precipitation in March-September was almost twice the norm: 564.6 mm, 574.1 mm, and 548.5 mm, respectively, at a norm of 120.7 mm. At the same time, temperature deviations from the norm were insignificant.

The anomalous meteorological years of the Navacerrada (Spain) register have been analyzed in detail with the aim of establishing correlations with the extreme reductions of growth. Given that summer precipitation was found to be the most limiting factor for the radial growth of Pinus in the middle of the 20th century, it is also an explanatory factor in almost 50 % of the extreme growths. The negative pointer years 1962, 1963 and 1965 correspond with lack of precipitation during growing season in Navacerrada (1962, 1964) and were dry throughout Spain, whereas the1986 ring was formed after the dry year of 1985. Meanwhile, the positive 1959, 1973 and 1976 pointer years correspond with wet years. Nevertheless, other positive pointer years identified in the chronologies of the Sierrade Guadarrama – 1958, 1964, 1980 and 1994 – do not appear to have any clear climatic explanation, even when taking into account the anomalies during spring. This last-mentioned meteorological variable also to have a significant influence on the width of the rings in Central Spain, and could explain the positive tree radial increment of 1943 and 1945, and also the negative tree radial growth of 1986. Pointer year of 1996 was connected with extreme reduction in winter temperature. This year was anomalous to the normal development of the cambium, due to the branches breaking under the weight of snow. The period 1990–1995 is the driest period of the whole meteorological register. The decade of the 1990s was having one of the more extreme droughts that affected a major part of the Peninsula. 1995 is a pointer year related to this driest period [7].

The suitability of pointer years for determining extreme climatic events has been demonstrated by several investigations [7, 8, 9, 10, 11]. Climatic interpretations, such as hard winter frosts, severe summer droughts, or extremely cold springs, may explain most of the negative pointer years [7, 12]. Conversely, the most positive growth responses resulted from a local combination of favorable environmental factors [11].

Analysis of the response function showed that it is generally accepted that the tree-ring growth response to the climate is constant over time, covering the entire lifespan of trees [3]. However, the assumption that the relationship between tree radial growth and climatic factors is stable over time is doubtful, given the nature of the annual wood layer formation and the possibility that climate and other environmental factors may change over time [13, 14].

The oak radial growth response to climate variations was compared for two periods: 1946-1979 and 1980-2013. The average annual temperature of the first period was 6.9°C while that of the second one was 7.7·C; namely, there was an increase in temperature by 11%. For the first period, the average annual rainfall was 582 mm, and that for the second one amounted to 629 mm, i.e. the difference was 8% (Fig. 2).

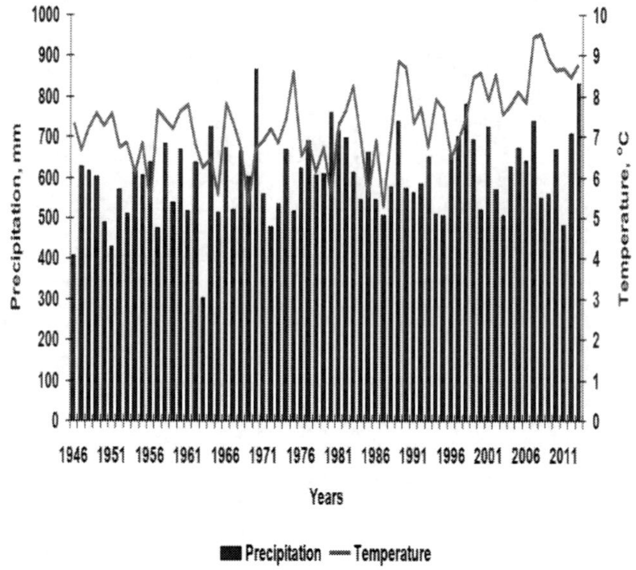

Fig. 2 Dynamics of temperature and precipitation according to the data of Zhytomyr meteorological station

We analyzed the response function for the periods 1946–1979 and 1980–2013. For the first period, the multiple correlation coefficient (R) for temperatures, precipitation and tree-ring indices RESIDUAL was 0.81 (F = 3.728, α = 0.004). Temperatures and precipitation determine the radial growth variability by 65.09%. For the second period, the multiple correlation coefficient was calculated: -0.91 (F = 7.568, α = 0.00004). The radial growth variability for oak was 83.10% dependable on temperatures and precipitation.

Thus, the dependence of radial growth on climatic factors increased in the second period.

Analysis of the response function when comparing the oak radial growth response for 1946–1979 and 1980–2013 showed a decrease in the positive effect of temperatures during the growing season in the second period. For example, for the first period, a significant positive effect of temperatures was recorded in May, July, and August, while for the second period, a positive effect was found only for May and July. The negative effect of temperatures on growth was registered in the first period in June, and in the second period two months earlier, in April (Fig. 3).

"The dendroclimatic analysis showed a similarly positive 'summer response' meaning the summer droughts had an inhibiting effect on the oaks. Water stress is equal to oaks in all vigour classes. Apparently, the sequential droughts during the preceding decade incited the dieback of the declining and the dead trees during 2005-2007. Thus, there appears to be a time lag between the mortality-inducing factors and actual death. Our results differ from the previous study in a lack of clear negative climate 'winter response'. On the seasonal scale, we can admit a favourable spring warmth effect on both early wood and total annual ring growth and a positive effect of mild winter on early wood formation" [18].

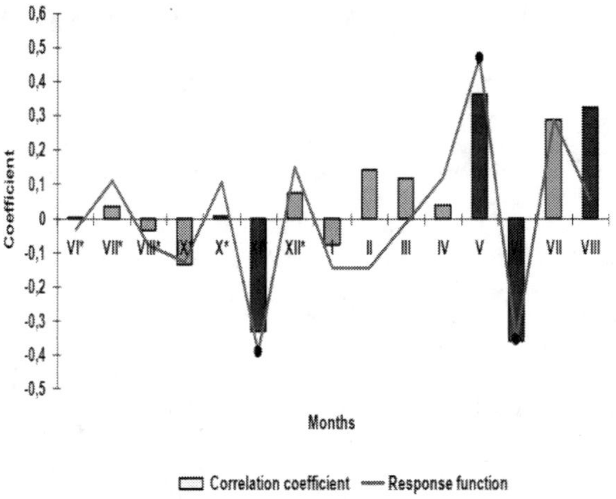

(A) 1946–1979

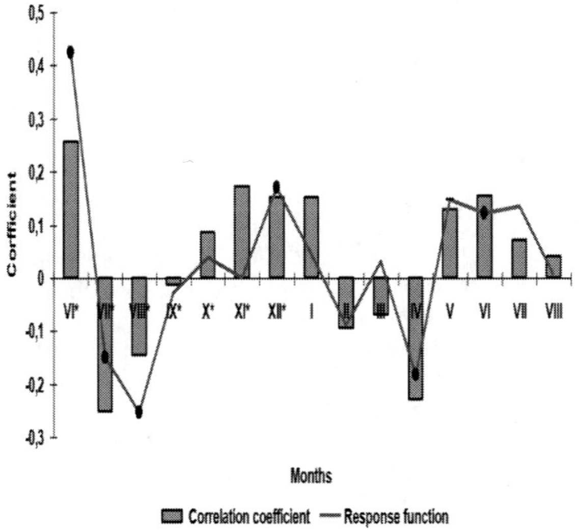

(B) 1980–2013

Fig. 3. Pearson correlation coefficients (bars) and response function coefficients (lines) between average monthly temperatures and regional tree-ring index chronology for Polissya (dark blue bars indicate significant correlation coefficients ($P < 0.05$); black circles show significant response function coefficients; asterisks (*) indicate the month of the previous year)

In the 1980-2013 (second period), the dependence of oak radial increment on precipitation increased. The impact of the previous year's precipitation on growth in the second period increased compared to the first one: negative effect on radial increment in June and November was found as well as positive in July and September, while for the first period there was only a significant negative impact of November precipitation for the previous year. In the spring, April precipitation had a negative influence on the radial growth in the first period, while in the second period there was a negative impact not only in April but also in February. The positive effect of temperatures was found for June and July in the second period and only for July in the first one (Fig. 4).

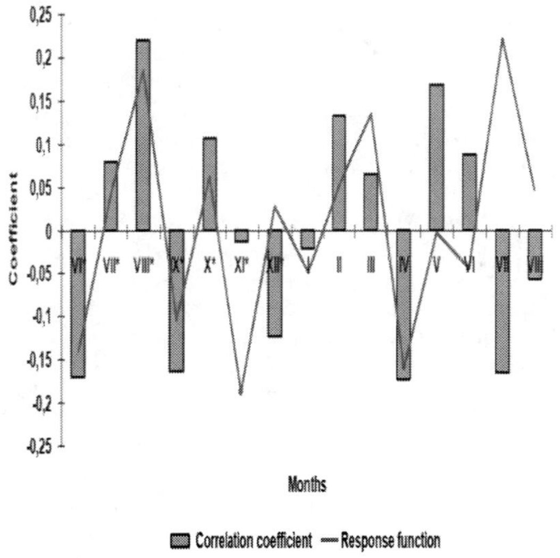

(A) 1946–1979

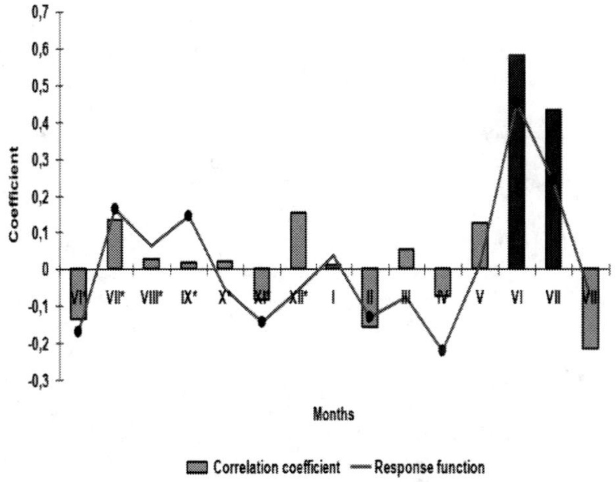

(B) 1980–2013

Fig. 4. Pearson correlation coefficients (bars) and response function coefficients (line) between monthly precipitation and regional tree-ring index chronology for Polissya (dark blue bars indicate significant correlation coefficients ($P < 0.05$); circles show significant response function coefficients; asterisks (*) indicate the month of the previous year)

Spring temperatures has weakened, and summer precipitation has more negative influence of the tree radial growth. The temperature - sensitivity of tree rings to spring to winter temperature weakened towards the end of the previous century. It was found that a positive influence of August precipitation and negative impact of July temperature in the previous year on tree ring width appeared. Climate change has led to lower sensitivity of oak tree rings to winter temperature. Radial growth of trees in stands growing in the eastern region of Latvia, can be limited by summer droughts [17].

It was investigated the decline of Quercus robur L. growing on shallow soil at the northern distributional limit of the species in southern Finland, using the dendroclimatic methods. Oak radial increment was positively related to the June and July precipitations. This was expressed more influenced by the warmer winter and spring. It was detected that the dead oaks and the declining oaks had caused growing during 2005-2007 after a decade-long summer drought series. The radial increment of the declining and the dead trees had dropped already since the 1990s, while the healthy oaks had better long-term growth and higher adaptive capacity to climate variation [18].

Hungarian researchers found a enough strong response of the oak tree rings to precipitation during the growing season, and negative correlations with late spring temperatures since the early 1960s [15], as in our study.

It was investigated that in central Italy drought in May-June and in prior-year summer and autumn was negatively influenced on the radial increment during 1974– 2006. November precipitation had strong signal in the tree ring series. Moving Correlation Functions (11 y windows) indicated that the May–June signal remained dominant until 1996, thereafter falling to non- significant values in parallel with the May–June water balance drying trend; the previous autumn correlations had significant values. Since 1994 there was a two-year lagged response to June water balance, suggesting that, when growth declined, loss of current-year climate signals was accompanied by the emergence of previous-year ones. Growth and productivity of deciduous oaks in Mediterranean conditions is linked to spring and summer precipitation; oak growth decline was associated with a delayed response to climate [16].

In Latvia it was found that changes in response of wood formation in English oak to climatic factors since 1900. It was hypothesised that the effect of winter and negative pointer years (1950, 1976, 1995, 1999, 2002, 2008, and 2011) resulted from the precipitation deficiency during the growing season, cold or extremely warm winters, and abnormal early-spring temperatures. Positive pointer years (1967, 1997, 2001, and 2007) had a favourable temperature and precipitation for growth of trees. It was found that the increased influence of temperatures and precipitation on the oak radial growth increased that tells about strengthening in the tree sensitivity to climate change [17].

It was investigated the decline of a oak (*Quercus robur* L.) forest growing on shallow soil at the northern distributional limit of the species in southern Finland, using the dendroclimatic methods. Radial increment of oak was positively related to the June and July precipitations. This was expressed more influenced by the warmer winter and spring. The analyses showed that the dead oaks and part of the declining oaks had ceased growing during 2005-2007 after a decade-long summer drought series. This indicates a time lag in the oak dieback. The radial growth of the declining and the dead oaks had dropped already since the 1990s, while the healthy oaks had better long-term growth and higher adaptive capacity to climate variation [18].

4 Conclusion.

A regional tree-ring chronology of English oak for Polissya was built, consisting of 62 individual tree-ring series. The regional chronology was based on four local tree-ring chronologies.

The response function for the periods 1946–1979 and 1980–2013 detected that for the first period, temperatures and precipitation determine the radial growth variability by 65.09%. For the second period, the radial growth variability for oak was 83.10% dependable on temperatures and precipitation.

Negative pointer years (1950, 1976, 1995, 1999, 2002, 2008, and 2011) resulted from the precipitation deficiency during the growing season, cold or extremely warm winters, and abnormal early-spring temperatures. Positive pointer years (1967, 1997, 2001, and 2007) had a favorable heat and moisture balance.

The increased influence of temperatures and precipitation on the oak radial growth in 1980–2013, as compared to previous 1946–1979, indicates an increase in the tree sensitivity to climate change in the second period.

References

1. Lakyda П. I., R.V. Sendzjuk, O.V. Morozjuk. Forests of Poltava region: bioproductivity and dynamics, **219** (2011)
2. K. Höppner, R. Kätzel. For Pol. A **53**, 9 (2011).
3. E.R. Cook, L.A. Kairiukstis. Methods of Dendrochronology. Applications in the Environmental Sciences, 394 (1990)
4. H.D. Grissino-Mayer. Tree-Ring Res. **57**, 16 (2001)
5. T. M. Wigley , K. R Briffa., P. D. Jones. J. Climate Appl. Meterol. **23**, 12 (1984).
6. R.L. Holmes. Dendrochronology program library users manual. Laboratory of Tree-Ring Research, 51 (1994)
7. M. Génova., Clim. Past. **8**, 13 (2011)
8. B. Neuwirth, F. H. Schweingruber, M. Winiger. Dendrochronologia. **24**, 10(2007)
9. F. Kienast, F. Schweingruber, O. Braker, E Schar. Can. J. Forest. Res. **17**, 13 (1987)
10. F.H. Schweingruber, D. Eckstein, F. Serre-Bachet, O.U. Braker. Dendrochronologia. **8**, 29 (1990)
11. C. Rolland, C. Desplanque, R. Michalet, F. H Schweingruber: Arct. Antarc. Alpine Res. **32**, 12 (2000).
12. I. Koval. Baltic Forestry. 19(2), 6 (2013)
13. V. Rozas Annals of Forest Science. **62**, 9 (2005).
14. H. Visser, J. Molenaar. Journal of Climate. **8**, 10 (1995)

15. M. Árvai , A. Morgós, Z. Kern. Biogeosciences and Forestry. **11** (2), 7 (2018)
16. Di Filippo, A. Alessandrini, F. Biondi, et al.. Ann. For. Sci. **67**, 706 (2010).
17. Ā Jansons, R. Matisons, L Puriņa., U Neimane., J.Jansons (2015). Silva Fen. **49,** 12 (2015)
18. K. Sohar, S.i Helama, A. Läänelaid, J.Raisio, and H. Tuomenvirta. Geochronometria. **41**(1), 11 (2014).

Ecological and Geographical Features of Ontogenesis of Holocene Soils of Kamianets-Podilskyi Fortress

Roman Malyk[1*], *Andriy* Kyrylchuk[1], *Zinovy* Pankiv[1], and *Ihor* Kasiyanyk[2]

[1]Ivan Franko National University of Lviv, 79000, University str. 1, Lviv, Ukraine
[2]Ivan Ogienko National University of Kamianets-Podilskyi, 32301, Ogienko str. 61, Kamianets-Podilskyi, Ukraine

Abstract. The article analyzes the degree of study of the genetic features of Holocene soils of beligerative complexes. A detailed description of natural and anthropogenic conditions and factors determining the geography, genesis and ecological condition of modern and buried soils of beligerative structures of Kamianets-Podilskyi Fortress is given. Considerable attention is paid to the problem of ontogenesis of the dominant natural and anthropogenic soils within the study area, represented by urborendzins and constructional soils. The morphogenetic features of these soils have been studied. The soils of the beligerative complex are represented by naturally anthropogenic soils – urborendzins and constructional soils. They are polygenetic with complex phylogeny and their genesis includes at least two onogenesis. For a long time soils have been evolving naturally in form of typical rendzin, but have undergone quantitative and qualitative changes due to intensive anthropogenic activity. The genetic profile of soils is characterized by significant variability of morphological traits, the presence of eluvium of soil-forming rocks in the entire soil layer. the studied Holocene soils (modern and buried) are an integral part of the historical and cultural lands and need protection, as they contain information about the evolution of the territory and ways of its use.

1 Introduction

Topicality. The study of the genesis of Holocene soils of beligerative complexes at the present stage of genetic-geographical research is important and arouses interest in the world soil science. This is mainly due to the presence of buried and fossil horizons in the genetic profile of these soils. This allows evolutionary, historical and paleo-soil science to solve the problems of time and evolutionary history by analyzing the genesis and geography of both modern and paleo soils within a single landscape complex with the possibility of accurate dating. The buried soils of beligerative complexes, which usually act as archeological monuments, are buried mainly due to anthropogenic influence. Consequently, they are well isolated from the digenetic action of modern soil-forming processes, in contrast to the relict features in the profile of modern soils. Holocene soils are better preserved because they were not subjected to pedometamorphic processes of the periglacial zone of the Pleistocene-Holocene, which distinguishes them from fossilized soils of the Pleistocene[1]. These genetic features make it possible to form fair presentation about the presence in the soil of signs of different age generations of all stages of its ontogenesis, from the parent breed to modern periods of soil life.
Soil evolution data contain information about the evolution of the territory of its formation and ways of its use. The genetic profile of the soil encodes information about the history of development and the course of landscape evolution in general, which in turn makes it possible to carry out reconstructions of paleolandscapes.
The purpose of the study is to study the genetic features of Holocene soils of the beligerative complex of Kamianets-Podilskyi Fortress. To achieve this goal it is necessary to solve the main objectives of the study: to analyze the degree of study of genetic features of Holocene soils of beligerative complexes, to characterize natural and anthropogenic conditions and factors influencing the geography, genesis and ontogenesis of modern and buried soils of investigated beligerative structures; to highlight the features of the genetics of the studied soils, to consider the complexity of the ontogenesis of natural and anthropogenic soils, represented by urborendzins and constructional soils within the study area.
The object of research is Holocene soils: urborendzins and constructional soils of the Old Castle of Kamianets-Podilskyi Fortress.
The subject of research is morphogenetic features of the studied Holocene soils at different stages of ontogenesis.
The following scientific methods were used while investigating the genetic features of Holocene soils of beligerative structures on the territory of Kamianets-Podilskyi Fortress: cartographic (when choosing the place for laying the sections, the following were used: the plan-scheme of the territory of the museum-reserve Kamyanets-Podilsky State Historical Museum-Reserve

* Corresponding author: derevokpkm1@gmail.com

(scale 1: 1000); large-scale topographic map (scale 1: 10000)), informational (analysis of materials on geomorphology, geological and tectonic structure, hydrogeology and historical and local lore archival materials of the museum-reserve), problem-chronological (study of soil evolution under the influence of the functioning of the beligerative complex), comparative-geographical (substantiation of conclusions about the genesis of soils and patterns of their location on the basis of comparison of soils and relevant factors of soil formation in their historical development and location), archeological with morphological-genetic, ecological-genetic and systemic approaches and evolutionary paradigm of soil formation.

2 Analysis of recent research and publications

Many research works have been done recently to study the genesis of Holocene soils of beligerative complexes. Most of the published monographs and articles deal with the genesis of buried and fossil soils of beligerative complexes in the context of paleopedological, geoarchaeological and paleobotanical research. Analyzing recent publications, it was established that the genetic features of Holocene soils of beligerative complexes are actively studied by scientists of evolutionary, evolutionary-genetic, historical soil science and anthropogenic landscape science. Genetic features of Holocene soils of beligerative complexes (archeological monuments) are considered thoroughly in the works of S. P. Pozniak, Y. M. Dmitruk, J. M. Matviishina, M. F. Veklych. O. M. Adamenko, O. L. Alexandrovskyi, G. I. Denisyk.

S. P. Pozniak and I. Y. Papish in their work "Soil-archaeological research of late Holocene black soils" define soil research and geochronological analysis of buried soils of archeological monuments as those that allow paleogeographic conditions reconstructing as well as reproducing of the evolution of soil cover at all stages of the Holocene [1]. In the monograph "Soils of Trojan shafts" by J. M. Matviishina and Y. M. Dmitruk genetic properties, problems and patterns of evolution of Holocene soils of beligerative complexes are considered in the context of evolutionary and ecological-genetic analysis [2]. A detailed analysis of the genesis of Holocene fossil soils using a geoarchaeological approach is presented in the works of S. P. Karmazinenko, S. P. Doroshkevych, A. S. Kushnir [3–4]. Timing and diversity of evolutionary development of Holocene soils in order to reconstruct the conditions of the past is considered in the works of O. G. Parkhomenko [5]. Buried soils of archeological monuments are considered in a number of works by foreign scientists S. Khamnueva-Wendt, H-R. Bork, F. Kurbanova, M. J. Storozuma, Z. Qin [6-8].

3 Research materials

The Holocene soils of the beligerative complexes of the Kamianets-Podilskyi Fortress are represented by naturally anthropogenic soils called urborendzins. Rendzins as soil-forming rocks contain a significant amount of calcium carbonates, which have a unique effect on the soil-forming process and, as a rule, give them properties that differ significantly from the properties of zonal climatic soils. Therefore, rendzins are considered and classified as typical intra zonal biolithogenic soils with mandatory presence of $CaCO_3$ in the soil-forming rock [9].

Depending on the composition and properties of soil-forming rocks, rendzins are divided into two groups: rendzins, i.e., soils formed on the weathering crust of dense carbonate rocks, such as chalk, limestone, marl; pararendzins, i.e. soils formed on loose carbonate rocks, such as loams, clays, moraines [10]. Thus, the ontogenesis of urborendzins occurs under the influence of eluvial and biogenic-accumulative processes (accumulation of litter, decalcification, decarbonation), but the dominant are anthropogenic and pedoturbation processes (erosion, mixing, burial) [11].

The geographical location of the fortress territory and its associate beligerative infrastructure is due to the tactical advantage of the combination of slope exposures at the confluence of the Smotrych River and its tributary – the Dibruha River. Morphological features of the relief and genetic combination of local landscapes are due to the peculiarities of split tectonics and lithological basis. They influenced the activity of erosion, the formation of the hydrological network, phytocolonization and the formation of natural soil cover.

The vegetation of the beligerative complex is represented by a transformed meadow-steppe formation. In geobotanical terms, the study area is located within the South Polish–Western Podolsk sub-province of deciduous forests and meadow steppes [12].

According to palynological studies, in particular spore-pollen analysis, the vegetation of this area during the Holocene till nowadays has not changed significantly its boundaries, but its floristic composition has undergone substantial changes [13]. The natural vegetation is preserved by fragments of meadow steppes and steppe meadows, the species composition of which in the anthropogenically altered formation is represented by only a small number of endemic species: *schivereckia podolica, cypripedium calceolus L., fritillaria montana and stipa pennata.*

The group is based mainly on neophytes: *elymus repens, amaranthus caudatus, phleum pratense, dáctylis glomeráta, medica gofalcata, brassica napus, amaranthu sretroflexus, malva sylvestris, bgerteroa incana, stellaria media, hyoscýamus niger* prevail. Ruderal forms are presented by *artemisia vulgaris, echinops sphaerocephalus, conium maculatum.*

The tectonic structure is characterized by a network of splits, being oriented sublatitudinally and submeridionally, determined the development of watercourse valleys. Block activity at the split intersection caused the formation of an omega-shaped meander within the historical part of Kamianets-Podilskyi and the emergence of an erosion remnant within the modern castle bridge. This remnant necessitated the strategic need to build fortifications to protect it as a direct route to the city. The relief here is due to the peculiarities of the lithological basis. The slopes and inner terraces of the canyon-like parts of the valleys are an erosive cut in the Silurian sediments. The supra-canyon part (fourth terrace) is an accumulative-erosion relic of the hydro network, which existed before the formation of Smotrych and was partially modified by it later.

The stratigraphic structure of the Silurian base has a monoclinic occurrence with a southwestern slope of the layers, which is noticeable even on visual inspection of rocky slopes. It is partly complicated by block microlifts within the old town. The composition of rocks is sedimentary carbonate sediments (Fig. 1).

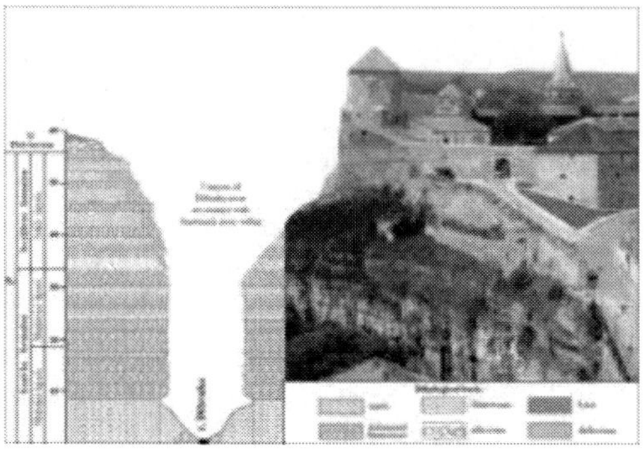

Fig.1. Lithological basis of the beligerative complex of Kamianets-Podilskyi fortress (developed by the authors)

They reflect the transgressive event of the beginning of the late Silurian with its natural changes. According to the local stratigraphic scale, they are united into the Malynovetskaya series, which is represented here by (formations): Konivska and Tsvyklivetska [14]. Konivska beds are composed of shallow coastal deposits. In the lower riverbed part there are bundles of domerites in form of lagoons with periodic drying, which turn gradually in the upward section direction into reef dolomitized limestones with numerous visible remains of favositides, tabulates and crinoids. These deposits are combined into Holoskivska subformation. There is a clear cut metabentonite horizon on its confining bed, which separates the Shutnivska subformation from deeper reef deposits. In the canyon this part forms a visually monolithic lower part with steep rocky slopes

The Tzvyklivets formation is represented by the alternation of powerful (up to 1,5 m) bundles of marls and domerites formed in deeper water environment in comparison with the Konivska formation conditions. They are united in the Sokil subformation and morphologically form a steep slope where accumulative deluvial areas alternate with indentations of native rocks. Accumulation areas are fixed by rock-steppe vegetation and demarcated by shrub strips. These deposits are the basis of the beligerative complex of the fortress.

On the opposite bank of the Dibruha River the thickness of the Sokil deposits reaches 20 m. They are covered with ancient alluvium of problematic origin. These are mainly transit silicate rocks, including sand and well-rolled jasperoids from the Carpathian region, weakly rolled fragments of Cretaceous spongilites from the Zbruch-Zhvanchytskyi watershed, as well as pebbles of local rocks. The structure of their occurrence does not coincide with the current configuration of the Smotrych river valley, reflecting the relict river network and the initial stages of the canyon part of the valley laying. The alluvium is covered with loess-like loams, the capacity of which increases with remoteness from the canyon parts.

The beligerative complex of the Kamianets-Podilskyi Fortress is located on the Sokil lithological basis. The eroded surface forms here a system of steps, bundles of tiled limestones. Their hypsometric levels decrease from the complex of the new fortress to the castle bridge. These rocks have a lumpy structure, with numerous organic remains. Due to the interaction with open air dark gray color of the fresh surface is covered with a light yellow oxidation layer up to 0,5 cm. This material is mostly used as a building material for fortress buildings owing to convenience of use (layers are broken by a system of cracks that form almost rectangular blocks that can be used directly for construction).

Lumpy limestones are separated by layers of marls with its thickness from 0,4 to 1,2 m. It is a fresh surface of light gray color with a blue tint (blue stone). The primary structure is a massive polydetrite structure that becomes shale (less) in a strict sense during one season of interaction with the environment. It leads to easy and rapid mechanical destruction of the rock to the clay fraction. In the structure of the slope, these layers correspond to areas of niches covered with deluvial-eluvial material and reflected by strips of rock-steppe phytocenoses. Blocks of such rocks are used in the reconstruction of fortress buildings (walls). They are visually identified by erosion and shedding without much effort.

The leveled areas of the root surface of the Sokil formation (the surface of limestone packs) are covered with carbonate eluvium and less-loams. The long stage of infrastructural development of the fortress complex has led to the accumulation of building material here, which is similar structurally to eluvium.

Taking into account naturally anthropogenically altered biolithogenic factors of urborendzins formation, influencing their genetic features within the studied beligerative complex, it should be noted that they are characterized by significant variability of soil profile capacity, morphological features, physicochemical features and the presence of buried horizons. The best general research of soil evolution is revealed in the studies of V. A. Kovda and B. G. Rozanov (Fig. 2). They indicate that the soil during ontogenesis undergoes changes due to the weakening or intensification of certain factors of soil formation, which do not allow its diagnosing on the basis of previously known properties. This indicates the evolutionary transition of soils to another stage of their development [10]. Due to the significant variability and violation of natural factors of soil formation, the case of ontogenesis of the studied soils is quite complex.

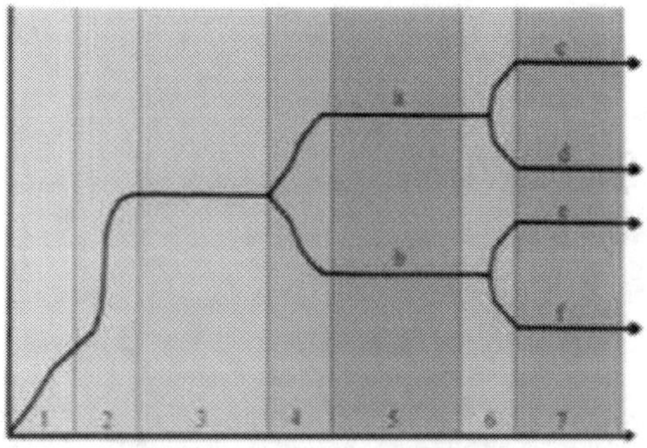

Fig. 2. General scheme of evolutionary sequence of stages and phases of soil formation (modified by the authors) [10]. 1 – initial stage; 2 – perfect stage; 3 – permanent stage; 4 – soil evolution in the "a or b" direction; 5 – permanent stage II (a or b); 6 – new evolution of soil in the c, d, e or f direction; 7 – permanent stage III (c, d, e or f)

They should be taken outside the dominant direction of evolution because they are based on the provisions of the evolutionary paradigm of soil formation. In fact, like most soils formed in the holocene, such soils are polygenetic, and their evolution, in general, should be called phylogeny. Polygenetic soils have several alternating ontogenesis (Fig. 3).

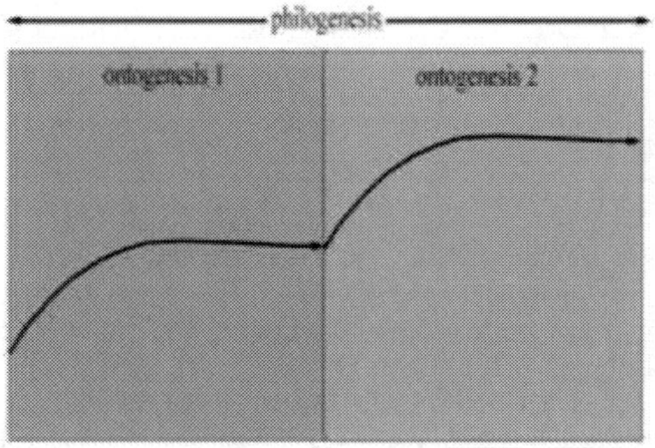

Fig. 3. The ratio scheme of ontogenesis and phylogeny of polygenetic soils (modified by the authors) [15]. Ontogenesis 1 – predominance of natural soil formation processes; ontogenesis 2 – predominance of anthropogenic influence and pedoturbation processes.

The initial stages of the evolution of the urborendzins of the beligerative complex of the Kamianets-Podilskyi fortress proceeded in the formation of typical rendzins. The initial stage is obviously characterized by the dominance of microprocesses of transformation and transfer of substances within the biological cycle with the participation of microbacterial and proactinomycete forms of radial fungi with the decomposition of primary and secondary minerals and the formation of chelates.

Perfect stage is accompanied primarily by the accumulation of organic matter with a decrease in the role of actinomycetes and their subsequent replacement by bacteria, fungi and lichens, which destroy carbonate rocks and primary minerals and thus cause the formation of primary fine soil with the onset of humus accumulation. Then, the development of the grass cover begins with the increase of the fine soil. Significant carbonate content of the parent rock (25,75 % – 40,21 %) together with grassy vegetation contributed to the intensive development of the sod process of soil formation, which was accompanied by a significant accumulation of humus.

Further evolutionary development is associated with the formation of a beligerative complex, anthropogenic transformation of natural soils, changes in their qualitative and quantitative characteristics, backfilling, burial, mixing and formation of constructional soils. As a result of anthropogenic transformation, intrasoil pedoturbation processes also take place. In order to study in detail the morphogenetic features of the current state of natural and anthropogenic soils, i.e. urborendzins and constructional soils, complex soil and geographical studies were conducted. As a result of research, four key areas of beligerative structures are laid down, which differ in functional purpose and period of formation. Soil sections (1–Dt, 3–Wc) were laid within the key areas and clearings of soils of vertical wall anthropogenic outcrops (2–Nb, 4–Ad) were made. When choosing the place for laying the sections, the following were used: the plan-scheme of the territory of the museum-reserve "Kamyanets-Podilsky State Historical Museum-Reserve" (scale 1: 1000); large-scale topographic map (scale 1: 10000); materials on geomorphology, geological and tectonic structure, hydrogeology and historical and local lore archival materials of the museum-reserve.To diagnose the objects of study and their nomenclature the classification of the FAO (WRB) [16] in combination with the substrate-functional classification of anthropogenic soils by O. B. Vovk and profile-genetic classification by M. M. Stroganova [17] was used.

The designation of genetic horizons is carried out according to the system of indices proposed by academician O.N. Sokolovskyi, with additions proposed by S. P. Pozniak and A. A. Kyrylchuk in combination with the classification developed by M. M. Stroganova [17–18]. The colors of genetic horizons were determined by the Mansell scale [19]. Genetic profiles at the present stage of evolution of Holocene soils of the beligerative complex have the following features: significant variability of morphological elements of the soil in the vertical and horizontal directions; the presence in the profile of the studied soils of day urborendzins, constructional soils, buried horizons of urborendzins of natural and anthropogenic origin, buried fragments of zonal soil types and buried soils of Chornoliska and Trypilska cultures. Within the entire thickness of the profile of the studied soils there is an eluvium of soil-forming rock which is represented by fragments of different shapes, sizes as well as fine carbonate material. Also earth cover contains a significant number of separates of different ages (inclusions), artifacts belonging to different cultural layers.

Constructional soils are characterized by a mosaic profile, spatial heterogeneity of structure and color. The nature and form of transitions between horizons indicate the repeated anthropogenic impact on them. The low content or absence of in the layers of constructional soil indicates their bulk nature and leaching of carbonates due to the washing regime (Fig. 4).

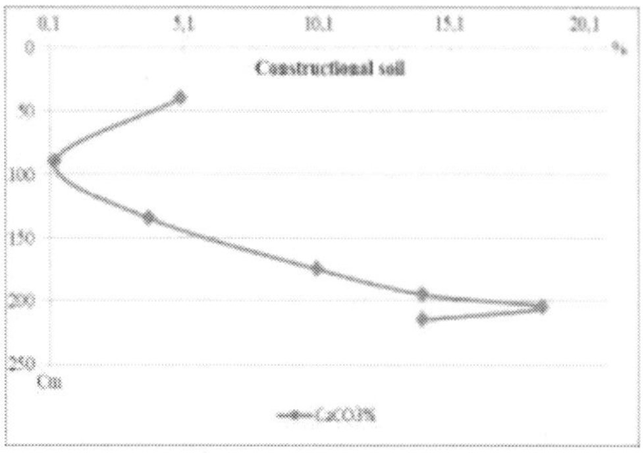

Fig. 4. Graph of $CaCO_3$ content and distribution in the constructional soil (Section 3–Wc is laid on the main shaft of the western curtain near its south-western wall (Geographic coordinates of 48 ° 40′27.03 ″ N w. and 26 ° 33′36.25 ″ east. e.))

Comparing the data on the content of absorbed CaCO3 urborendzin, we observe the process of decalcification with the removal of carbonates downward the profile in modern urborendzins ((Hcau) 0,79 % and (Hpcau/a1) 0,38 %) and, conversely, the accumulation of carbonates in buried urborendzins of the Trypillia period (Fig. 5.).

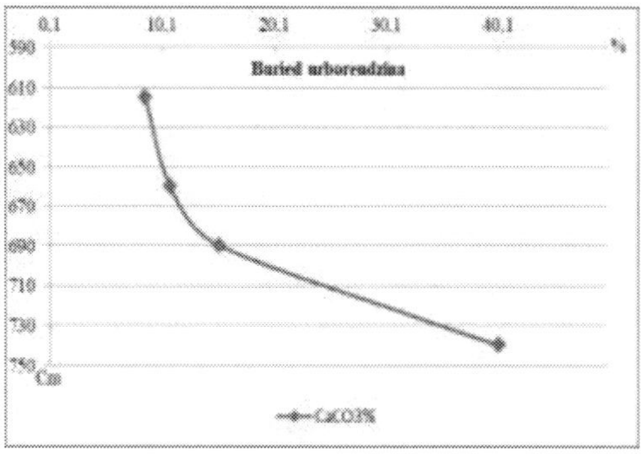

Fig. 5. Graph of $CaCO_3$ content and distribution in buried urborendzin (clearing of anthropogenic 2–Nb outcrop 770 cm high at a distance of 10 meters eastward from the casemated transition from Rozhanka tower to Pilna gate (Geographic coordinates of 48 ° 40′26.13 ″ N w. and 26 ° 33′45.62 ″ east. e.))

The humus-accumulative horizon of modern short-profile urborendzins (Hcau) has a lumpy-grained structure and a thickness from 25 cm. In the buried structure 50 – 60 cm thickness is oppressive, nut-prismatic (Fig. 6.). According to the results of laboratory analyzes, it was found that with the content of total humus from 3,41 % to 3,59 % in the horizon (Hcau) of diurnal and 1,09 % in the horizon [Hca / a1] of buried urborendzins soil belongs to a small (< 3 %) and medium humus (3 –5 %). The degree of humification of humus-accumulative horizons, both surface and buried, is mostly very high with fluctuations from 76,7% in the buried Trypillia soil to 34,4% in short-term modern urborendzins.

Fig. 6. Photo of the morphology of the soil section 1–Dt (located on the terrace of the upper open part of the "Day Tower", laid on the terrace of the stone floor in the upper part of the "Day Tower" at a distance of 1.5 meters from the inner edge of the wall in a northwesterly direction. Geographic coordinates of 48 ° 40′25.44 N w. and 26 ° 33′42′97 ″ east. e.)

The type of humus is mainly humate in the upper part of the profile and fulvate-humate in the transitional and buried horizons. The reaction of the soil solution of urborendzins, both buried and diurnal, is slightly alkali (pH H_2O 7,57 – 7,98), and alkali in the soil-forming rock (pH H_2O 8,48 – 8,65).

4 Conclution

Summing up the study, we note that the study of the genesis of Holocene soils of beligerative complexes is of current concern both in Ukraine and in the whole world.

Kamenets-Podolsky Fortress is a durable self-regulating military fortification beligerative landscape complex of slope type. Located in the canyon part of the valley of the river Smotrych. The peculiarity of this beligerative complex is that on a small area (≈ 4.5 hectares) there are various defensive structures from the XIII-XIV centuries to the time of the Commonwealth of the XVII-XVIII centuries and the Russian Empire of the XVIII-XX centuries. The surrounding area is protected for a long time and is not subject to active anthropogenic impact The Holocene soils of the Old Castle of Kamianets-Podilskyi are represented by naturally anthropogenic soils – urborendzins and constructional soils. Ecological and geographical conditions of their genesis (ontogenesis) are quite modified. The lithogenic basis of the beligerative complex is a dense sedimentary rock of sokil sediments. Vegetation is represented by a transformed meadow-steppe formation. The studied soils are polygenetic with a rather complex phylogeny, divided by the beginning of the construction of the beligerative complex into, at least, two successive ontogeneses. For a long time soils have been evolving naturally in form of typical rendzin, but have undergone quantitative and qualitative changes due to intensive anthropogenic activity. The genetic profile of soils is characterized by significant variability of morphological traits, the presence of eluvium of soil-forming rocks in the entire soil layer. The genetic (ontogenetic) features of the Holocene soils of the Kamianets-Podilskyi Fortress together with the elementary soil formation processes largely depend on their lithological composition, which directly affects the development of the soil profile. Therefore, to study the processes of formation, development and dynamics of the investigated soils, it is necessary to apply lithologically oriented research methods and trends, taking into account the determining role of the anthropogenic factor. The soils of Kamianets-Podilskyi Fortress (modern and buried) are an integral part of historical and cultural lands and should be protected, as they contain information about the evolution of the territory and ways of its use. Thus, soil profiles of beligerative complexes are heterogeneous (formed by different elementary soil processes that operate with different intensity and spatial localization), and heterochronous (different parts and features of the profile were formed at different times). Accordingly, some morphological features of the soil are the result of modern processes, while others are of ancient processes that corresponded to another combination of soil formation factors, and currently either absent or have much lower intensity, and their detection allows reconstruct paleogeographic conditions.

Reference

1. S. P. Poznyak, I. Y. Papish, Soil-archaeological research of chernozems of the late Holocene. Bul. Inst. Arch. **3**, 8–16 (2008)
2. Y. M. Dmytruk, Zh. M. Matviishyna, I. I. Slyusarchuk. *Soils of Trojan shafts: evolutionary and ecological-genetic analysis (*Ruta, Chernivtsi, 2008), p. 228
3. Zh. M. Matviishyna, S. P. Doroshkevych, A. C. Kushnir, Reconstruction of landscapes lifetime Trypillya culture based on paleopedological research. Bul. Lv. Univ. Ser. Geo. **48**, 107–115 (2014)
4. Zh. M.Matviishyna, S. P. Karmazynenko, S. P. Doroshkevych, O. V. Matsibora, Paleogeographical preconditions and factors of changes in human living conditions in the territory of Ukraine in Pleistocene and Holocene. Ukr. Geol journ. **1**, 19–30 (2017)
5. Zh. M. Matviishyna, O. G. Parkhomenko, Research of Holocene soils on the territory of the National Historical and Cultural Reserve "Hetman's Capital" in Baturyn in Chernihiv region. Sc. Bul. Chern. Univ. **824** (2020)
6. F. Kurbanova, T. Puzanova, O. Rudenko, G Starodubtsev, Dataset on the soils of Medieval archaeological monuments in the forest-steppe zone of the East European Plain. Dat. in Bri. **30** (2020)
7. M. J. Storozum, Z. Qin, Y. V. Wang, H. Liu, Buried soils as archives of paleo-pollution in the North China Plain. Anthropo. **31** (2020)
8. S. Khamnueva-Wendt, A. V. Mitusov, J. Wendt H. Bork, Classification of buried soils, cultural, and colluvial deposits in the Viking town Hedeby. Geoarchaeol. **35**, 313–337 2020
9. F. Dushofur. *Fundamentals of Soil Science: Soil Evolution (Experience in Studying the Dynamics of Soil Formation* (Progress, Moscow, 1970), p. 592
10. V. A. Kovda, B. G. Rozanov (ed.), *Soil Science. Soil and soil formation.* (Higher. Shk, Moscow, 1988), p. 400
11. A. A. Kyrylchuk, Agrogenic transformation of rendzins in the Western region of Ukraine in the conditions of intensive farming. Prob. Agrochem. Eco. **1**, 36–42 (2014)
12. Y. P. Didukh, Y. R. Shelyag-Sosonko, Geobotanic region of Ukraine and surrounding areas. Ukr. Bot. Jor.. **1**, 6–17 (2003)
13. V. V. Menderetsky, G. V. Chernyuk, I. P. Kasiyanyk, Natural rhythms of formation of Smotrytsya canyon according to palynological analysis of deposits of low terraces and floodplain of middle Dniester. Bul. Kam-Pod. Nat. Univ. **4** 85–97 (2019)
14. P. F. Gozhuk (ed.), *Stratigraphy of the Upper Proterozoic and Phanerozoic of Ukraine* (IGN NAS of Ukraine. Logos, Kyiv, 2013), p. 638
15. E. M. Samoilova, Y. S. Tolchelnikov, *Evolution of soils.* (MSU, Moscow, 1981), p 81
16. IUSS Working Group WRB, *World Reference Base for Soil Resources* 2015. (World Soil Resources Reports 106, FAO, Rome, 2015), 192 p.
17. M. N. Stroganova, M. G. Agarkova,. Urban soils: experience of study and taxonomy on the example of soils in the southwestern part of Moscow. Pochv. **7**, 16–24 (1992)
18. V. V. Harbar, S. P. Poznyak, Genesis and properties of rendzinas of the Podilski Tovtry. Pol. Jour. Soil. Sc. **48**, 229–240 (2015)
19. A. Munsell, *Munsell soil color charts.* (MD: Munsell Color Co, Baltimore, 1994), p. 16

Urban Green Space: Comparing the EU and Ukrainian Practice

Ihor Lishchynskyy[1]*, *Mariia* Lyzun[1], *Evangelos* Siskos[2], *Yevhen* Savelyev[1] and *Vitalina* Kuryliak[1]

[1]West Ukrainian National University, 46009 Ternopil, Ukraine
[2]University of Western Macedonia, 52100 Kastoria, Greece

Abstract. The goal of the paper is to compare the urban green space practice in the EU and Ukraine to find the advantages and drawbacks of both approaches. In order to reach the goal, following tasks has been raised: (1) to reveal the most common patterns of green space allocation in European cities; (2) to consider the urban green landscape in Ukrainian first-tier and second-tier cities; (3) to compare the trends and challenges of urban green space allocation in the EU and Ukraine. In Europe, a typical pattern of green space is a combination of greenbelts, regional green areas, green wedges, green middles, green infrastructure, etc. The main significant feature and advantage of green areas in European cities is the connectivity of the larger objects (parks and woods) and smaller ones (squares, boulevards, inter-quarter plantings, etc.). Urban territories in Ukraine involves quite large green areas (mostly by combining "communal parks" and "green infrastructure"). Transformation crisis leads to the vast decline of industrial zones, which were spontaneously transformed into de-facto green or brown spaces. After economic recovery green space is shrinking dramatically under the pressure of commercial and residential zones. For Ukraine, the challenge is the creation of small urban landscaping facilities and increasing functional content and quality of the existing green spaces.

1 Introduction

The pattern for spreading of cities in the outer green landscape has been an issue for more than a century. Depending on the design of the urban structure, green space could be considered both as the separating element between city and suburbia, and as the connecting link between people/districts/cities.

In terms of urban subject green space is rather diverse and could be classified from the perspectives of its size, purpose, vegetation cover, proximity to public transport and other facilities, etc [1, 2].

Public green space includes:

– parks, sporting fields, community gardens, reserves and nature conservation areas;

– greenways and trails, street trees, stream and riverbanks;

– other less conventional spaces such as green walls, cemeteries, green alleyways.

Private green space incudes:

– household backyards and communal grounds of apartment buildings;

– corporate campuses [3, 4].

However, the combination and spatial placement of the green areas varies depending from national traditions and geographical conditions of the city.

The **goal** of the paper is to compare the urban green space practice in the EU and Ukraine to find the advantages and drawbacks of both approaches. In order to reach the goal, following tasks has been raised: (1) to reveal the most common patterns of green space allocation in European cities; (2) to consider the urban green landscape in Ukrainian first-tier and second-tier cities; (3) to compare the trends and challenges of urban green space allocation in the EU and Ukraine.

2 Theoretical review of green space impact

There is a general consensus among the society that urban green space provides the ecological integrity of cities and has a positive effect on public health. Among the main proved advantages are:

– reduction of air pollution by absorbing certain airborne pollutants from the atmosphere [5];

– temperature moderation by providing shade and cooling an area [6];

– noise attenuation;

– groundwater replenishing [7].

Numerous investigations focused on linkages between green space and health or mortality [8-9]. There is a strong correlation between parks availability and local physical activity and respective health enhancement [10]. Parks and the obesity epidemic is also revealed [11]. Besides green space is associated with psychological well-being [12] and stress reduction [13].

However, some disadvantages should be also mentioned, including threats to vulnerable persons with allergies, the concentration of criminals and socially excluded persons, high costs of maintenance, etc. Besides some researches show that the irrigation, fertilizer, mowing and leaf blowing all add up, emitting more carbon dioxide and other greenhouse gases than the spaces absorb [14]. Sometimes park-adjacent neighbourhoods (especially in low-income communities) have higher pollution concentration [15]. Nevertheless,

* Corresponding author: lischynigor@ukr.net

 https://doi.org/10.1051/shsconf/202110005007

the abovementioned positive effects overwhelm these disadvantages.

The researchers highlight that green space is often uniquely distributed among society members. Access to green space could be stratified based on income, age, gender, (dis)ability, etc. Such a phenomenon is known as environmental injustice that is a subject of numerous recent investigations [16, 17].

Usually, low-income people reside in the urban core or inner-ring suburbs where areas are insufficiently landscaped or poorly maintained. More affluent residents live in areas of the suburban periphery with lots of greenery. Such areas are well-serviced and well-maintained [18].

This environmental injustice has been taken into account in urban planning, which has led to the implementation of park development programs and various strategies for landscaping poor areas. But increasing additional greenery in these areas can create an urban green space paradox. Due to the rise of green space, which is well-serviced and well-maintained, the attractiveness of these areas as well as the costs of housing are growing. The housing cost increasing can potentially lead to gentrification: the exclusion or displacement of the residents for whom the landscaping was carried out [19].

3 Green space pattern

Constant urban growth resulted in numerous environmental problems connected with air, water and noise pollution. Uncontrolled city expansion and chaotic blending of urban and rural areas lead also to the lack of adequate city infrastructure in the suburbs and respective traffic growth toward infrastructure-abundant districts. To bring control over the urban growth, different patterns of green space location are applied in the practice of city planning.

In Europe, the options of the common urban green landscape can be conditionally placed on a kind of scale between the two extreme scenarios:

- "greenbelts" (sometimes called "green belts"), which are aimed at curbing the spread of cities to the surrounding rural area;
- "green zones" / "green infrastructure", the task of which is to organize the urban space for its further development and expansion.

On other continents traditional urban green space pattern could be different. For instance, in East Asia historically the localities of urban and rural natures were situated next to each other. In such a case, the farms and wooded landscape provide the functions of visual amenity and cultural services even under the conditions of modern megapolises [20].

However, following the aim of the paper, we would emphasize more the Western type of urban green planning that is more typical to Ukrainian cities.

3.1 Greenbelts

Greenbelt mode of city planning became popular in Europe as a replacement of Medieval walls which were broken in most of the cities in the 18th-19th century [21]. The idea of greenbelt arises in popular Howard model of Garden Cities [22]. According to E. Howard, Garden Cities horizontal spreading should be strictly limited by surrounded Greenbelt – a space for agriculture and recreation.

The greenbelt concept is typically associated with the UK policy of post-World War II urban development known as "New Towns". During several waves of its implementation, new settlements have been created aiming population density reduction, first in London and Glasgow, and later in Liverpool, Birmingham, Manchester and other metropolises. Instead of the traditional peripheral extension of the urban zone along transport routes, new settlements were formed at a distance of about 50 km. from the megapolises. They were initially completely provided with infrastructure and housing and separated from the megapolises with a vast greenbelt (Fig. 1). In general, the success of New Town's policy is linked not so much to the ban on metropolitan expansion as to the more general trends towards decentralization and deconcentration that have coincided over time with this program. After the 2000s new towns within the UK program are being created under the principles of ecopolises.

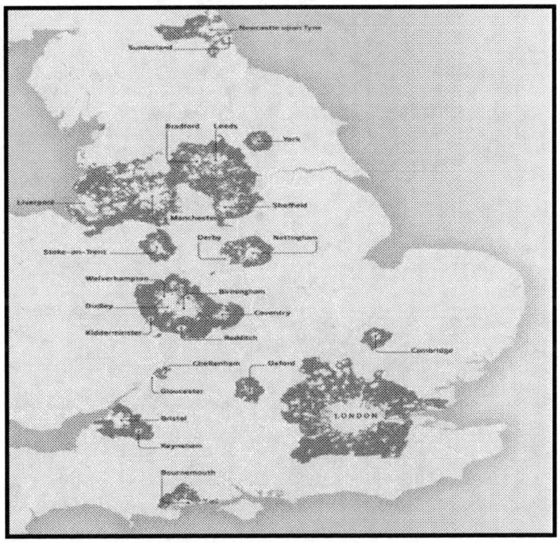

Fig. 1. Greenbelts in Great Britain (Source: theguardian.com).

The greenbelt approach was also used for some cities in France, Spain, Austria, Hungary.

Initially, the main role of greenbelt was considered in the field of controlling the urban growth, preventing conurbations (merge of agglomeration), separating of urban and rural functions. More recent arguments supporting the greenbelts are nature conservation and the issue of brownfield space [21].

3.2 Regional Parks and Green Wedges

The greenbelt approach has some long-standing disadvantages that make the city planners to look for the alternatives. Among the key drawbacks of greenbelts:

– rise of inequality between inhabitants of the area within greenbelt and outer residents;

– lack of accessibility for the general public or lack of scenical attractiveness;

– constrain of urban growth rises the average property prices and limits economic development;

– increasing the commuting distance for the residents beyond the greenbelt outer edges;

– decreasing the quality of the inner urban edges of the greenbelts due to the lack of land management.

To avoid the greenbelt drawbacks mentioned above a couple of solutions are possible. The first alternative is to "break" the circle of greenbelt to allow the controlled city overflows in the selected vectors, which are initially provided with appropriate commuting infrastructure for the residents outside the primal city ring.

A good example of such "breached" greenbelt is a system of Regional Parks around Berlin (Fig. 2). However, it should be noted that mentioned Regional Parks are not classical public parks or nature reserves. Being of a rather big size (144-964 km^2), they are perceived rather as a rural countryside. A review of the origin and structure of Regional Parks, as well as their comparison with classical greenbelts is provided by Kühn [21].

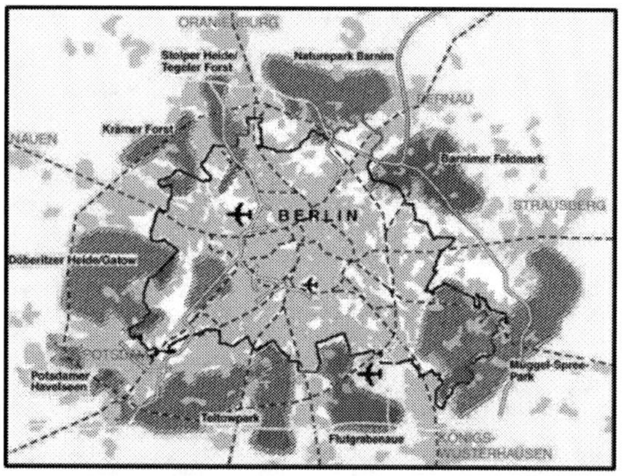

Fig. 2. Regional Parks in Berlin [21].

The second alternative to greenbelts is well described by Lindley under the name "green wedges" [23], which are specific green districts separating urban territories (Fig. 3). In the contrast to greenbelts, green wedges ensure continuing proximity of green space to the neighbourhood urban districts, providing the area for recreation, sport, education, air quality and water management, green commuting, nature conservation etc. An important feature of green wedges is their ability to be expanded alongside urban growth corridors, providing the sustainability of green space and lowering the real estate price pressure.

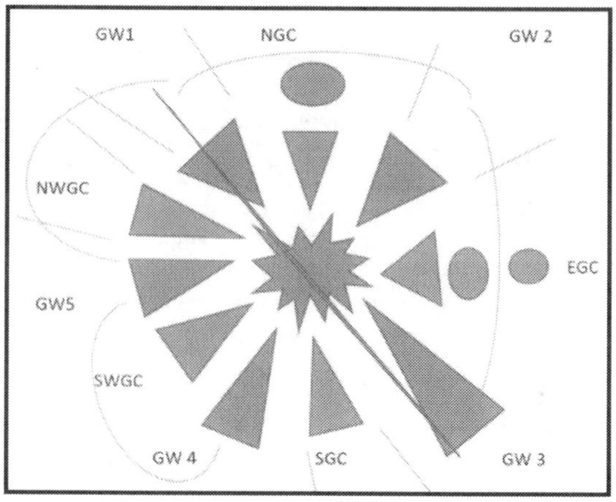

Fig. 3. Green wedges (GW) and growth corridors (GC) scheme [23].

The examples of green wedges implementation are "Finger Plan" developed for the post-WW II Copenhagen or 1988 Leicester city planning.

3.3 Green middle

Both previous cases were examples of green space as a stopper or separator of urban growth. However, green space could also fulfil the integrative function. Being an open space structure, it could connect different communities into a single unity.

A good example of the integrative green middle is Dutch city region Randstad – the conurbation of four major cities in the Netherlands: Amsterdam, Rotterdam, Utrecht and the Hague (Fig. 4). The Randstad is the case of the urbanized belt around the green core (the latter is known as "Green Heart").

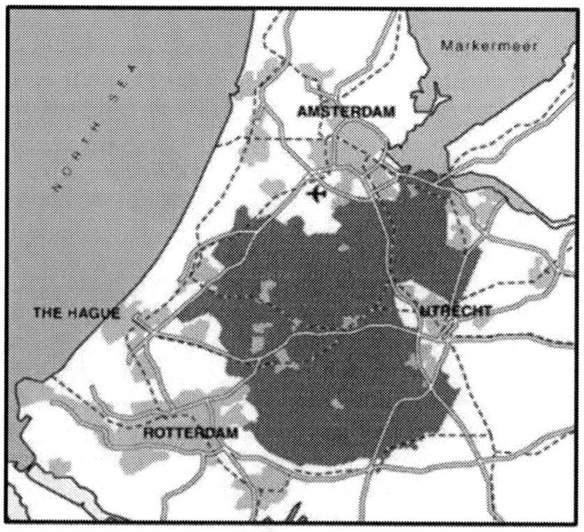

Fig. 4. Green Heart of the Randstad, the Netherlands [21]

The Green Heart is mainly the agricultural territory typically used for pastures. Its area covers around 1,500 km^2 and includes 70 local municipalities (though population growth in the area was an object of indirect limitation). The share of recreation and nature conserving areas are rather low. Thus, Randstad is the case of a regional city limited from the expanding to the middle. While the Green Hart is simultaneously the separator and integrator of urban growth [21].

3.4 Green infrastructure

In the previous sections, we emphasized mainly the vast green spaces influencing general urban-rural landscape. However, there are dozens of other green space types affecting city planning. Some of them were mentioned in the Introduction.

We could distinguish among the most influencing: green corridors and green infrastructure.

The green corridors form the linear natural network that connect different city districts or green/open spaces (example of Barcelona green corridors is provided in Fig. 5).

Fig. 5. Green corridors in Barcelona (Source: bloomberg.com)

Despite their relatively narrow size, green corridors can provide some functions of the larger green areas: protection from pollution and noise, enhancement of physical activity and public transport usage, increase the levels of social interactions, protection from high temperature and strong winds, provide urban biodiversity.

Green infrastructure issues are of crucial importance for the compact city planning mode. The main idea of the compact city is avoiding of classical urban zoning and provision the multifunctionality of land use to promote social cohesion and reduce motorization. Under such conditions, the minimum density of residential buildings should be 40-80 units of 2–4-storey buildings per hectare. The residential buildings are placed around the nodes, which also unite the structural elements of employment, education, commerce, recreation, etc. to maximize the share of local interaction.

However, compact city mode doesn't leave enough space for the green areas inside the city inner boundaries. Green infrastructure should be removed on the district outer fringes forming small green belts for the districts inside a city. In 2008–2014 compact city model was applied for 13 largest cities in Norway under urban revitalization strategy [24].

The case of green infrastructure is shown in the Fig. 6.

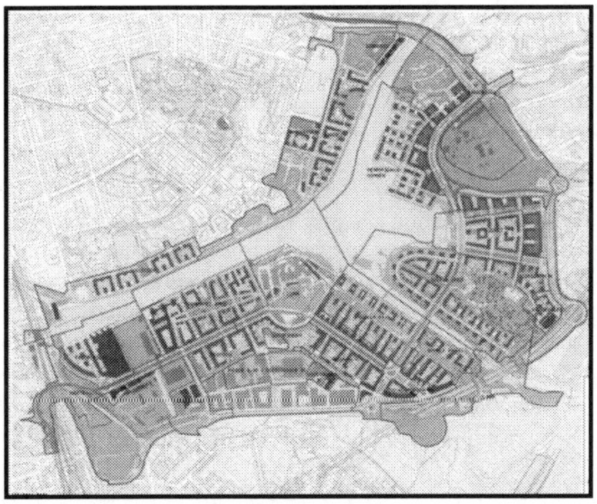

Fig. 6. Green infrastructure for the Hammarby Sjöstad district in Stockholm (Source: urbangreenbluegrids.com)

4 Urban green space in Ukraine

4.1 Urbanization in Ukraine

Urbanization is a global trend today. According to the UN, more than half of humanity already lives in urban areas. In 2018 this figure was 55%. Europe is considered as one of the most urbanized regions in the world, the average level of urbanization here is 74.5%, while in Ukraine this rate is 69%.

According to the State Statistics Service of Ukraine, at the beginning of 2019 in Ukraine, there were 461 cities (1344 urban settlements), where more than 69% of the population lived. The rate of urbanization is constantly growing despite the general depopulation. According to the State Statistics Service of Ukraine for 2015-2020, the growth rate of the urban population share was 0.35%. The largest share of the urban population is observed in Donbas, Kyiv, Dnipro and Kharkiv regions (Fig. 7.1a). However, high population density is also observed near the Carpathian Mountains, due to the high population density in rural and small urban settlements (Fig. 7.1b).

In recent years, 6 cities in Ukraine have implemented metropolitan functions. The city of Kyiv (3.0 mln inhabitants) is part of the world metropolitan network, the cities of Kharkiv (1.4 mln), Odesa (1.0 mln), Dnipro (1.0 mln), Lviv (0.8 mln) are regional metropolises of Ukrainian level. Donetsk has de-facto lost such function nowadays due to the occupation.

There are numerous typologies of cities with their limitations. In the paper we explore the concept of first-tier (capitals) and second-tier cities. We define the second-tier cities according to the methodology of ESPON/SGPTD 2012 Report as *"... those cities outside the capital whose economic and social performance is sufficiently important to affect the potential performance of the national economy"*. Thus, all abovementioned cities accept Kyiv we would consider as second-tier cities.

Simultaneously with the processes of population concentration in metropolitan agglomerations caused by migration, there is a significant decrease in the population of cities beyond first- and second-tier. The exception is satellite cities that form pendulum migration channels with their first-tier and second-tier vis-à-vis. Almost half a million people come to Kyiv every day from the outside region, which is about 30% of the latter's population. In turn, satellite cities are becoming centres of gravity for the rural population due to relatively cheap housing for permanent residence (compared to the centre of the agglomeration) and reasonable transport links. Rural peripheral settlements in most cases are experiencing accelerated socio-economic degradation. At the same time, rural settlements located at short distances from the 1st and 2nd tier cities are migratory stable or increase their population.

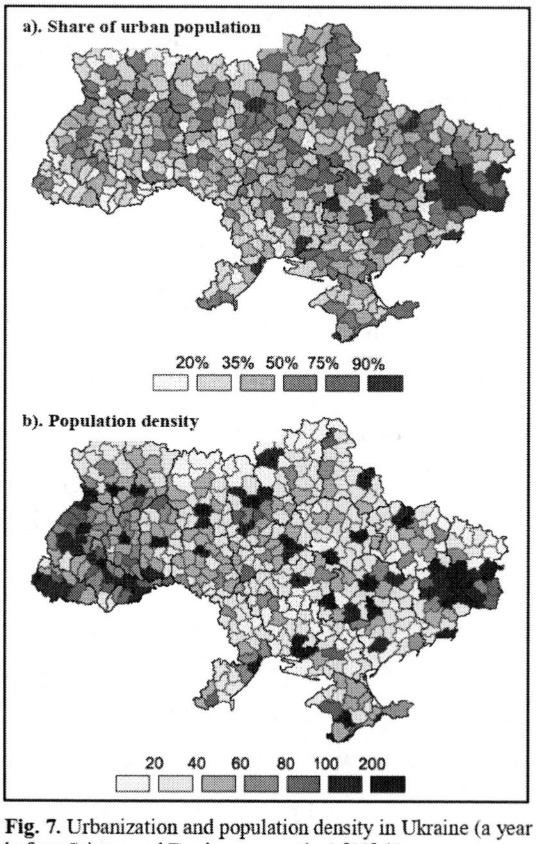

Fig. 7. Urbanization and population density in Ukraine (a year before Crimea and Donbas annexation) [25] (Source: ukrstat.gov.ua)

Besides, there is a clear pattern: the level of poverty increases with decreasing settlement size, and if small cities are close to rural areas in terms of poverty, then large ones, especially Kyiv, are sharply contrasted to small settlements. Thus, the poverty rate (according to the relative criterion of 75% of the median equivalent expenditure) in Kyiv is 14.1%, in other large cities - 17.8%, in small cities – 26.6%, in villages – 30.8%. Accordingly, a significant number of residents of villages and small towns migrate (want to migrate) to large cities to improve their financial situation. The above mentioned is making significant pressure on urban resident zones, which in their terms became a threat to the urban green scale.

4.2 Cases of green space in Ukrainian cities

Typical city pattern in Ukraine of the Soviet era has been of classical urban zoning type with clear central, industrial, and residential (so-called "sleeping") zones. Green space (communal park or preserved wooded area) in such a structure has often had the form of a specific zone that separates the districts of other types. Such structure is still observed in most first-tier and second-tier cities (see Fig. 8 for Ukrainian second-tier cities).

For instance, the green areas of Kharkiv cover an area of 15.4 thousand hectares. The indicator of the greenness of the city makes 50,4 per cent at the norm of 45 per cent (13.3 m^2 of green space per capita at the norm of 13 m^2). However, of these, only 7.5 thousand hectares are green areas for public use (parks, forests, meadows, water parks, gardens, squares, boulevards). The largest areas occupied by greenery are located along the western and northern borders of the city, going "wedges" deep into the built-up areas. The insufficient number of green areas are in the northeast, east, and southeast of the city, where most industrial zones are located. This negatively affects the sanitary and ecological conditions of residential districts in the neighbourhood. In general, the area of the green zone of Kharkiv is relatively sufficient. However, the bulk of green space is concentrated on the outskirts of the city. This indicates that, despite the high rate of green space, most of the city is inadequately provided with green spaces of a small type.

The geographical location of Odesa by the sea has led to the formation of a green space, which is not typical for other cities in Ukraine. The most vital in the urban landscape is the coastal park area (30 km length). This green area is of crucial importance for the general resort zone in the summer months, during the arrival of hundreds of thousands of vacationers. Besides, in every residential area of the city, there are parks, squares, boulevards. During the last decades due to the lack of vacant land within the city, the agricultural lands were allocated for construction. Often green space in these residential areas was not designed and created. The most significant factor limiting the increase in the efficiency of green plantations is the insufficient area of public green space. According to the General Development Plan of Odesa, the territory of green space in the city covers 5.4 thousand hectares, of which 1.0 thousand are for public use. Thus, the area of greenery is only 17 per cent of the city. In some city districts per capita indicator is lower than 20% of the national norm [26].

Lviv is one of the greenest cities in Ukraine (54 m^2 of green space per capita). Moreover, most of the green territory is situated within city borders. Due to the complicated relief of the city, the residential area doesn't threaten relatively the green zone. The historical centre of a city is surrounded by the first green ring that is rather wide. Besides, there are two other rings formed by parks and nature preserved areas. Among the disadvantages is the quality of some parks that are not adapted properly to the recreation function.

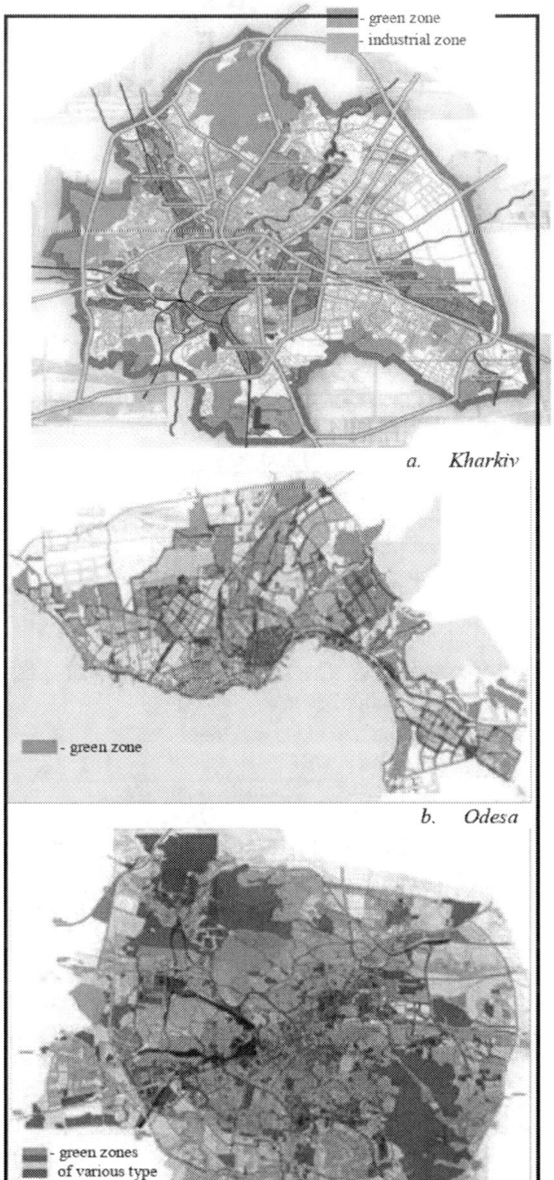

Fig. 8. Green (wooded) spaces in Ukrainian second-tier cities (Source: cities' General Plans)

On the other hand, the green pattern of Kyiv, the capital of Ukraine, is marked by a classical green wedge structure with powerful blue axe formed by Dnieper river (Fig. 9). In spite of constant growth of the metropolitan population, the latest revisions of the city General Plan doesn't prescribe the horizontal growth of residental zone. Though there are cases of hidden or semi-official transfer of green territory to fulfil the commercial and residential functions.

The type of green landscape is varied depending on the regions of Ukraine. Due to the climate, Northern and Western cities are more provided with wooden areas. While the outskirts of the Southern and Eastern cities are mainly cultivated plains with rare forests or bushes beside the roads.

In general, urban territories in Ukraine involves quite large green areas (mostly by combining "communal parks" and "green infrastructure"). Besides deindustrialization of the 1990s leads to the vast decline of industrial zones, which were spontaneously transformed into de-facto green or brown spaces of post-industrial character (abandoned buildings, a lot of garbage around). This process often closes the circles of greenbelts around a lot of 2-nd and 3-rd tier cities.

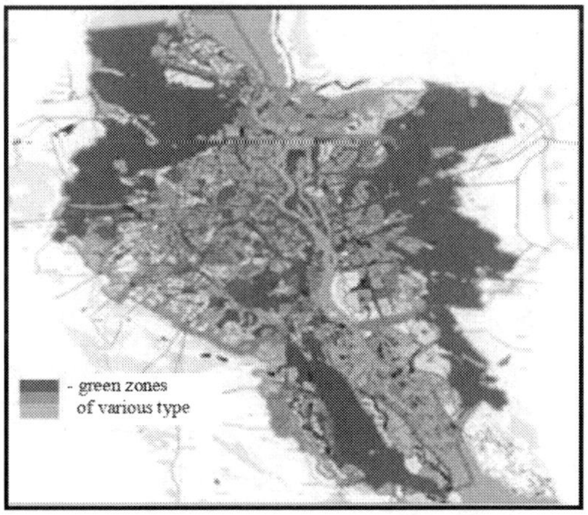

Fig. 9. Green (wooded) spaces in Kyiv (Source: city's General Plan)

The economic recovery of the 2000s reversed the tendency. Green space is shrinking dramatically under the pressure of commercial and residential (the corruption element is very high in the sphere of issuing the permits for activity in nature preserved areas). The suburban residential zones in a 15 km radius around the regional centres expand rapidly making additional pressure on the outer fringes of greenbelts.

Liberalization of automobile import taxation in 2014 increases the number of cars owned. City transport arteries in a lot of older cities are no longer able to maintain the traffic. To expand the roads green infrastructure is often removed. A typical related issue is a lack of space for the green landscape because of the redundant dwelling or commercial buildings. This problem requires alternative solutions or radical rearranging of city plans. However, the latter is often impossible due to the scarcity of city budgets.

5 Conclusions

Comparing the landscaping of European and Ukrainian cities, we see rather different approaches. In Europe, the typical pattern of green space is a combination of ring-radial, linear, wedge-shaped, spotted landscaping, etc. The main significant feature and advantage of green areas in European cities is the connectivity of the larger objects (parks and woods) and smaller ones (squares, boulevards, inter-quarter plantings, etc.). This allows forming a virtually indivisible ecosystem of the city green zone. The joint fragmentation of the green space is a surprisingly important characteristic, as it increases its resistance to external influences, and multiply the advantages of green territory.

Green space volume in Ukraine is not smaller than those in the EU. However, the indicator of green area per capita is not very relevant in our subject. More important is the quality of these landscapes. Given the scarcity of territory, which is an urgent problem in urban cities both in the EU and Ukraine, the special attention of city planners is paid to the more efficient use of the existing green areas. For Ukraine, the challenge is the creation of small urban landscaping facilities and increasing their functional content. Specialists in green space planning should pay attention to the wider introduction of such a widely used in Europe method as vertical landscaping, especially for its central part, where dense low-rise buildings predominate.

According to the national practice, the wooded area is considered mainly as a green space in Ukraine. Thus, European experience of green zone definition could be also applicable for better fulfilment of urban functions.

Acknowledgements

This work was supported by the Erasmus+ program of the European Union (Jean Monnet Module 612067-EPP-1-UA-EPPJMO-MODULE "European Regional Policy: Balance of Local, National and Pan-European Interests"). The European Commission support for the publication of this article does not constitute an endorsement of the contents, which reflects the views of the authors only, and the Commission cannot be held responsible for any use, which may be made of the information contained therein.

References

1. N. Dahmann, J. Wolch, P. Joassart-Marcelli, K. Reynolds, M. Jerrett, Health Place **16(3)**, 431–445 (2010)
2. C. Sister, J. Wolch, J. Wilson, GeoJournal **75(3)**, 229–248 (2010)
3. J. R. Wolch, J. Byrne, J. P. Newell, Landsc. Urban Plan. **125**, 234-244 (2014)
4. S. Roy, J. Byrne, C. Pickering, Urban for Urban Green **4(11)**, 351–363 (2012)
5. D. J. Nowak, D. E. Crane, J. C. Stevens, Urban for Urban Green **4**, 115–123 (2006)
6. S. K. Cummins, R. J. Jackson, Pediatr. Clin. N. Am. **48(5)**, 1241–1252 (2001)
7. F. J. Escobedo, T. Kroeger, J. E. Wagner, Environ. Pollut. **159(8)**, 2078–2087 (2011)
8. A. M. Dzhambov, M. H. Browning, I. Markevych, T. Hartig, P. Lercher, Environ. Res. **186**, 109613 (2020)
9. C. Coutts, M. Horner, T. Chapin, Geocarto Int. **25(6)**, 471–484 (2010)
10. J. F. Sallis, M. F. Floyd, D. Rodrıguez, B. E. Saelens, Circulation **125**, 729–737 (2012)
11. C. L. Ogden, M. D. Carroll, K. M. Flegal, JAMA **299(20)**, 2401–2405 (2008)
12. H. Ernstson, Landsc. Urban Plan. **109(1)**, 7–17 (2012)
13. J. Woo, N. Tang, E. Suen, J. Leung, M. Wong, Lancet, **373(9660)**, 299–300 (2009)
14. A. Townsend-Small, C. I. Czimczik, Geophys. Res. Lett. **37(2)** (2010)
15. J. Byrne, Geoforum **43(3)**, 595–611 (2012)
16. R. Bick, E. Halsey, C. C. Ekenga, Environ. Health **17(1)**, 1-4 (2018)
17. J. E. Givens, X. Huang, A. K. Jorgenson, Sociol. Compass **13(5)**, 12693 (2019)
18. N. Heynen, H. A. Perkins, P. Roy, Urban Aff. Rev. **42(1)**, 3–25 (2006)
19. R. Bentley, E. Baker, K. Mason, J Epidemiol Community Health **66**, 761–766 (2012)
20. M. Yokohari, K. Takeuchi, T. Watanabe, S. Yokota, Landsc. Urban Plan. **47**, 159–171 (2000)
21. M Kühn, Landsc. Urban Plan. **64**, 19–27 (2003)
22. E. Howard, Garden Cities of Tomorrow (Faber & Faber, London, 1946).
23. I. Lindley, Green belt vs. Green wedge (Landscape Institute, 2019, URL: https://www.landscapeinstitute.org/blog/green-belt-vs-green-wedge/)
24. H. Hofstad, Eur. J. Spat. Dev. 49, 1–23 (2012)
25. I. Lishchynskyy, Poliusy ta osi rozvytku v konteksti glokalizatsiinykh procesiv (TNEU, Ternopil, 2019)
26. I. Sholok, Bulletin of KhNU named after V. Karazin **1140** (2014)

Ecology of the City and the Role of Tourism in the Formation of Living Space (on the Example of Urban Ecosystems of the Cities of Kryvyi Rih and Uman)

Serhiy Sonko[1*], *Olena* Lakomova[2], *Daria* Shiyan[2], *Tetyana* Karpenko[2], and *Yulia* Kosenko[1]

[1]Uman National University of Horticulture, 1, Institutional Str., Uman, 20300, Ukraine
[2]Kryvyi Rih State Pedagogical University, 54, Gagarina Ave., Kryvyi Rih, 50086, Ukraine

Abstract. Despite the significant success of the human population in the development of the planet's resources, man will most likely never be able to completely get rid of dependence on the material and energy mechanisms of the biosphere. The clearest proof of this is the existence of urban ecosystems, within which human living space is formed. However, there is no consensus among researchers on the participation of social and natural components in the formation of urban ecosystems. Moreover, there is no clear answer to the question of how and through what mechanisms the space of human life is formed in modern cities. The article analyzes the ontological affiliation of the category "living space". Numerous aspects of the formation of modern living space have been studied on the specific examples of urban ecosystems of Kryvyi Rih and Uman. In particular, sample surveys of residents of these cities were conducted, based on the results of which conclusions are made about the quality of living space. Tourist activity and tourist resources are considered by the authors as a field of harmonization of human-environment relations in modern conditions of formation of living space.

1 Introduction

Nowadays, the study of human living space and the category of "living space" is becoming increasingly important from the standpoint of the conceptual and categorical apparatus of many sciences. In many ways, "fitting" for the space of life determines a person's self-identification, which is extremely important in today's stressful urban environment.

It is logical that the deep foundations of the idea of living space are rooted in the foundations of such basic sciences as ecology, sociology and psychology.

How far has man departed from other biological species in the perception of environmental factors and reactions to them? This is an eternal question of human ecology, projected on urban ecology and still remains unanswered.

What types of people relations with the space (more broadly, the environment) of life are most important and critical to their life?

These are the questions that need to be answered, especially against the background of the aggravation of such problems of human civilization as pandemics, terrorism, overpopulation and others.

Are there any real ways out of these crisis trends, or at least a reduction in their impact? Probably, these are the questions that will be very relevant in the near future.

Thus, from the point of view of science, the category of "living space" grows into an independent interdisciplinary scientific problem, the solution of which, according to V.I. Vernadsky, should be directed to the efforts of many sciences. The formalization of this problem through the methodological and conceptual-categorical apparatus of various sciences formulates the purpose of our study.

Trying to explore the multifaceted categories of living space, the authors set themselves the following tasks:

1. Determine the ontological affiliation of the category "living space".
2. Investigate the degree of influence of each dimension of this category on a person at different spatial levels.
3. Investigate various aspects of the use of living space in the urban ecosystems of Kryvyi Rih and Uman.
4. Investigate the role of the main branches of urban economy that form and use the living space of Kryvyi Rih and Uman.

The object of research is the urban environment of the cities of Kryvyi Rih and Uman.

The subject of research is the space of human life, which is formed under the influence of environmental, social and psychological factors.

Recently, this category has been interpreted mostly from the standpoint of racial theory, according to which a privileged nation objectively looks for expanding its living space. The founding father of the theory is Friedrich Ratzel, who set it out in "Political Geography" in 1897 [1]. Theory of the "continental bloc" (Berlin-Moscow-Tokyo axis) was eventually transformed by the Third Reich into the concept of "living space in the East." Our ideas about the modern space of life are

* Corresponding author: sp.sonko@gmail.com

incomparable and very far from the above meaning of the concept.

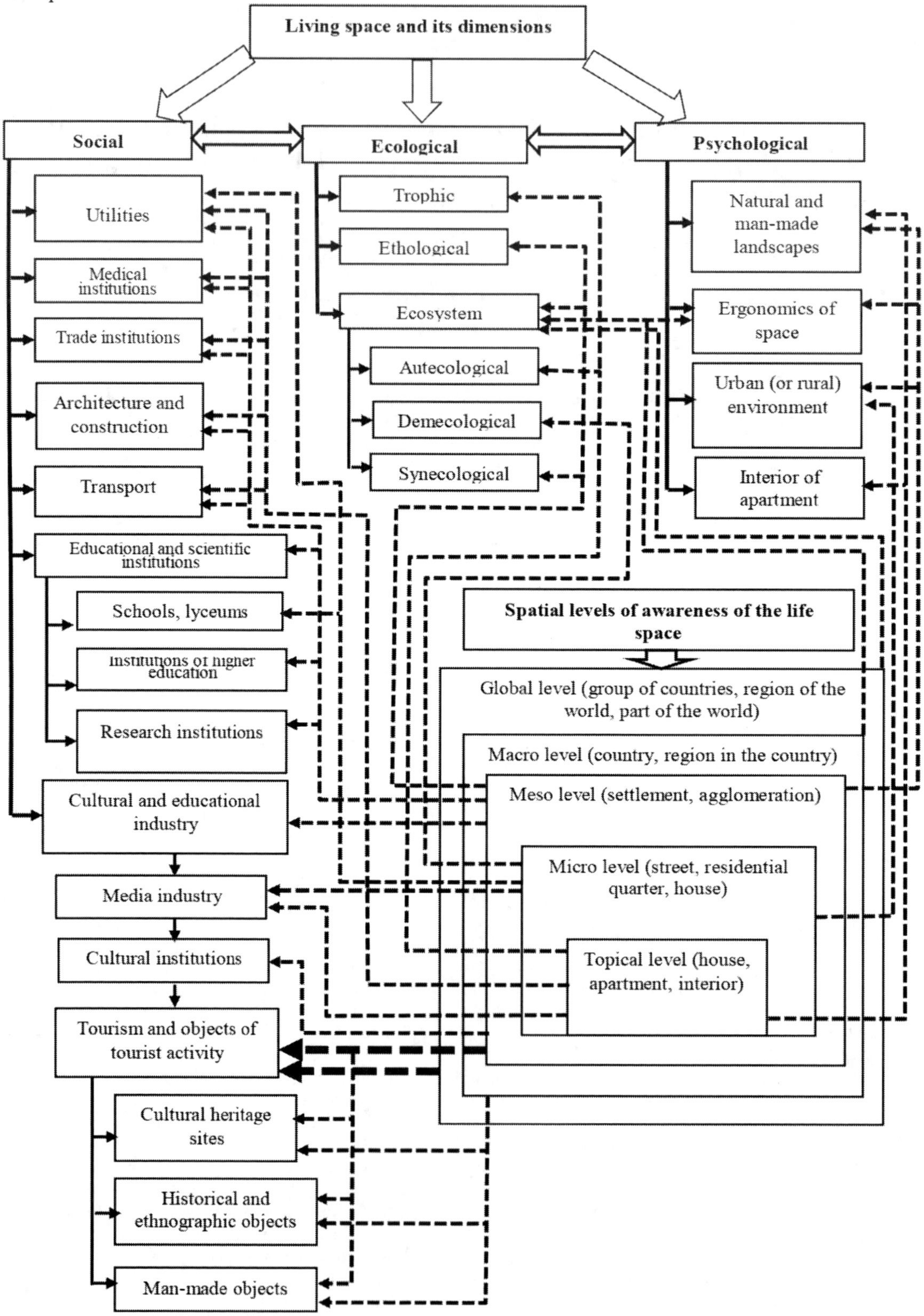

Fig. 1. Ontological scheme of the modern idea of the category "space of life".

Today's ideas about the space of life are formed above all through the requirements of modern civilization for comfort, convenience, and psychological state of man living and working in complex and stressful conditions of the human community, which due to various technological and cultural achievements "fits" for the environment.

Let's analyze in details the individual dimensions of the category "living space" (Fig. 1). Above all, the authors are closest to the ecological dimension, which is based on fundamental ideas about the biosphere "as a holistic, uniquely organized and constructed material object, which arose about 4 billion years ago and in which there is a qualitatively special form of matter - the biosphere." [2].

Based on this, the category of "living space" is primarily ecological precisely because it is based on the phenomenon of life, which through the life of the human population is realized in connection with the environment (Fig. 1). We have considered natural ecosystems, significantly transformed by man urboecosystems, it is necessary to consider their scientific content.

There are many definitions of an ecosystem Tansley A.G. in [3], P. Duvigne in [4], Eugene Odum in [5].

There are also different understandings of ecosystem boundaries [6].

However, more or less complete scope of the concept of "ecosystem" should include the following characteristics:

- being a natural or man-made functional system of the whole set of living beings, interconnected by trophic and other connections, and a certain relatively homogeneous physical (terrestrial, soil or aquatic) environment, which interact with each other… ";
- thermodynamically opened, interconnected with neighboring ecosystems, relatively resistant to external disturbances, stable over time, in its natural state, self-organized and self-regulating, living, non-entropic system;
- any size system, the structure and level of organization of which provide its long-term self-maintenance, material and energy transformation and biotic cycle [7].

It is important to understand what levels of living things are involved in formation of ecosystems. According to M.A. Golubets, "the infinite diversity of living beings on the planet belongs to one basic, universal, relatively independent, stable and primary level of organization - the *organismic*."

The second main level of organization of living things is considered to be *population*. It is a universal, integral, territorial, temporal, dynamic, independent structural and functional system. It has all without exception the biotic properties of the species. The third level is the *ecosystem*. It is on it that complex material-energy relations of man with the natural environment are formed, which, in fact, forms the basis of the category "space of life" (Fig. 1).

However, urban ecosystems, within which, the space of human life is formed, occupy a special place in the categorical apparatus of modern ecology. This is due to the fact that the development of urban ecosystems and urbanization in general in most urban scientists symbolizes the ability of human civilization to do in their lives without nature.

However, the deepening of urbanization creates a number of purely urban problems, including:

- extreme complication of transport problems;
- increase in price of engineering equipment;
- air pollution;
- distance of the population of large cities from nature;
- "suction" by large cities of productive forces from small and medium-sized cities [8].

In scientific works on the formation of urban ecosystems are dominated by either biologizing or sociologizing approaches. In the publications of urban planners, sociologists, futurists it is difficult to find the necessary information about the ecological specifics and impact of cities on the environment. In the ecological works of the city, along with terrestrial and aquatic, forest and agricultural ecosystems are characterized as ecological systems, and their main features are called material and energy metabolism, population dynamics, landscape transformation and more.

So, Y. Odum [5] considers the city as "a heterotrophic ecosystem that receives energy, food, fibrous materials, water and other substances from large areas outside it." To confirm this, the following indicators are given:

- the need for energy per square meter of the city is about 4000 kcal per day, which is 70 times higher than the need for the same area of oyster cans;
- in order to feed a city with a population of 1 million inhabitants need about 0.8 million hectares of land;
- to water and wash the necessary catchment, which would give 7.6 billion liters of water;
- more than 80 million kcal per inhabitant per year, while its annual physiological needs are only 1 million kcal.

Classifying ecosystems by energy performance, Odum [5] classifies cities as "industrial-urban fuel-driven ecosystems (fossil, organic, or nuclear)" and believes that they can be managed as functional ecosystems.

The general features of the city's ecosystem, according to modern ideas, can be summarized in several positions. Urban ecosystem is a spatially limited natural-technogenic system, a complex set of interconnected metabolism and energy of autonomous living organisms, abiotic elements, natural and man-made, forming an urban environment (space) of human life that meets its needs: biological, psychological, ethnic, labor, economic, social.

It consists of interconnected and interpenetrating subsystems (environments): quasi-natural (transformed geographical environment), landscape-architectural, socio-economic, socio-industrial [9]. The connection between them is so great that almost none of them can perform their functions separately, and at the same time the absence of one of the subsystems leads to the destruction of the ecosystem "City" as a whole (Fig. 1).

As you can see, the main system-forming factors (elements of the system) are man (he himself and all activities carried out within the urban area and the natural environment) (relief, geology, climate, water, etc.).

Thus, the urban ecosystem should be understood as a natural ecosystem in which during the evolutionary development of nature and society, the functions of social security associated with the removal of inert matter outside the body level of HomoSapiens were isolated and strengthened. At the same time, purposeful radical structural modification of the natural ecotope by man is obligatory. Ecological functions, such as participation in the food chains of ecosystems, have remained at the organismic level in the form of metabolism, but are provided not at the ecotope level, but at the level of ecological niche, which today covers the entire biosphere (Fig. 1).

All this determines the nature of a separate, relatively autonomous ecological system, which is based on the interaction of the social organism as a whole with the environment, which is undergoing changes on its part [10].

2 Presenting main material

Having determined the methodological guidelines of the category "living space", we justify the choice as objects of study of urban ecosystems of the cities of Kryvyi Rih and Uman. The main hypothesis of the study is the assumption that despite the significant differences between city-forming factors and mechanisms in the urban ecosystems of these cities, environmental and social factors play the same role in the formation of living space. To do this, we compare the main parameters that characterize the living space in these urban ecosystems (table 1). The principles of parameter selection were not accidental. The authors in the selection process were guided by the following requirements:

- commonality (and, consequently, the ability to compare) of the main characteristics;
- main spatial features (area, configuration, distance);
- natural and landscape environment;
- main ecological characteristics;
- saturation of the urban environment with objects of infrastructure, cultural, historical and industrial heritage.

The logical result of a comparative analysis of the selected parameters should be a general assessment of the quality of living space in urban ecosystems. It should be clarified that such an assessment is mostly subjective, despite the fact that we made it not only by analyzing the parameters listed in Table 1, but also by telephone surveys of the population of Uman and Kryvyi Rih. In accordance with the requirements of such surveys [11] we called up to 100 people in each city by random telephone numbers (70% by cellular numbers and 30% by Ukrtelecom). The questions asked required only an unambiguous answer - "yes" or "no". The questionnaires contained the following questions:

1. Do you like the urban lifestyle?
2. Do you consider Kryvyi Rih / Uman a big city?
3. Do you consider the urban environment of your city dangerous to health?
4. Do you consider the urban environment of your city safe for life?
5. Do you consider the urban environment of your city comfortable to live in?
6. In your opinion, can Kryvyi Rih / Uman be considered an ecologically dangerous city?
7. Is buying food in your city a big problem for you personally?
8. Is it a big problem for you personally to cover distances by public transport in your city?
9. Do you consider your personal home comfortable to live in?
10. Do you feel irritated when you meet individual tourists or tourist groups on the streets of your city?

The overall score was summed up in the last column on a 5-point scale in the dimension: 0 points - low quality of living space; 5 points - high quality of living space.

Emphasizing once again that the general assessment is mostly subjective, the authors suggest that even such an assessment makes it possible to assess in general such an ambiguous and non-mathematical category as "living space".

Given the ambiguity and a certain amount of subjectivity in these estimates, we consider it necessary to interpret them from the standpoint of the main objectives of our study.

- "Population" - gives a general idea of the living environment, as it largely determines the intensity and pace of life. By many estimates [12], highly urbanized metropolises due to the high pace of life cause the same high stress of the urban environment.
- "Average population density" - an average interpretation of the previous indicator, aimed at a correct comparison of different size and population of cities. However, the indicator may have an independent value, as to some extent characterizes the contagiousness of the urban environment, which in the context of disease epidemics becomes of paramount importance [13].
- "Configuration type" - an indicator that is actively used in the theory of placement [14, 15], architecture and urban planning [16] and in addition to purely geometric properties determines the ergonomic characteristics of the urban environment. In addition to verbal estimates of the configuration, mathematical models of such estimates are known, in particular, the Boyce index [17]. In our case, the elongated configuration of the territory of the city of Kryvyi Rih is far from optimal, both from a geometric and ergonomic point of view. However, this configuration creates the preconditions for the autonomy of certain areas (inner-city areas).
- "Spatial levels of awareness of the space of life" - a purely geographical indicator that characterizes the self-positioning of man in geographical space. In our case (Fig. 1) to the macro level (and below), inherent in both cities, in the city of Uman is added a global level through the annual pilgrimage of Hasids from around the world to the grave of his spiritual teacher - Tsadik Nachman.

Table 1. Comparison of development parameters that form the living space of urban ecosystems of Kryvyi Rih and Uman.

№	Parameters	Kryvyi Rih		Uman		∑Kr. Rig/∑Uman
1	Number of population	619.300 (January 1, 2020)		82 603 (January 1, 2020)		2/4
2	Average population density	1440.2 people / km²		2028.3 people / km²		4/3
3	Configuration type	Linear, elongated from south to north		Compact, close to a car		2/4
4	Spatial levels of awareness of the space of life	Macro level, meso level, macro level, topical level		Global level, macro level, meso level, macro level, top level		4/5
5	The level of contagiousness of the urban environment	Mostly average, due to the high autonomy of certain areas of the city and the reduced probability of contacts		Mostly high due to the significant compactness of the urban environment and high probability of contact.		4/2
6	The main type of natural landscape	Steppe and valley-river landscapes		Forest-steppe, valley-river and protected landscapes		3/5
7	The main type of anthropogenic landscape	Mining and settlement		Settlement		3/4
8	Intensity of trophic relations	High, due to the large population		Average due to the possibility of running a farm		2/4
9	The degree of rupture of trophic chains	Significant due to the long distances to the natural landscape		Insignificant due to proximity to natural landscapes		2/4
10	The predominant town-planning function	Industrial		Administrative and tourist		2/4
11	Saturation of the urban environment with natural and man-made objects, including	High		Insignificant		
- the predominant type of housing		Communal, multi-storey		Personal one-story		2/4
- social infrastructure		Well developed		Insufficiently developed		4/2
- Eco-network objects		Insufficiently represented		Presented in different types		2/4
- cultural and industrial heritage sites		Mostly industrial heritage		Mostly cultural heritage		3/4
12	Overall estimation of living space (according to the results of the survey)	satisfactory		good		3/4
Questions		Number of answers "yes"	Number of answers "no"	Number of answers "yes"	Number of answers "no"	∑Kr.Rig/∑Uman
1. Do you like the urban lifestyle?		83	17	52	48	5/3
2. Do you consider Kryvyi Rih / Uman a big city?		96	4	12	88	3/4
3. Do you consider the urban environment of your city dangerous to health?		68	32	23	77	4/2
4. Do you consider the urban environment of your city safe for life?		34	66	88	12	2/4
5. Do you consider the urban environment of your city comfortable to live in?		52	48	67	33	3/4
6. In your opinion, can Kryvyi Rih / Uman be considered an ecologically dangerous city?		87	13	8	92	1/4
7. Is buying food in your city a big problem for you personally?		23	77	9	91	4/5
8. Is it a big problem for you personally to cover distances by public transport in your city?		76	24	12	88	2/4
9. Do you consider your personal home comfortable to live in?		43	57	67	33	3/4
10. Do you feel irritated when you meet individual tourists or tourist groups on the streets of your city?		24	76	33	67	3/3
Total values of habitat quality indicators Kryvyi Rih / Uman		**72/94 (out of 110 points)**				

According to [18] in this case, Ukraine exports tourist impressions thanks to this attraction in the city of Uman. From the standpoint of the priorities of the post-industrial economy [19] export of raw materials (iron ore Kryvbas) is considered less important.

- "Level of contagiousness of the urban environment" - an indicator used in medical geography [20], and in human ecology [21] and characterizes the degree of probability and possible recurrence of physical contact between people. It was actively used by the Swedish geographer Thorsten Hegerstrand in developing the theory of the spread of innovations [22]. In our case, the more compact territory of the city of Uman and higher population density contribute to a higher level of contagiousness of the urban environment.- "The main type of natural landscape" - a geo-ecological indicator that determines the main direction of transformation of natural landscapes into man-made. In addition, its independent significance lies in the aesthetic perception of the environment by the individual. In particular, forest-steppe landscapes by M.D. Grodzinsky and G.I. Denisyk [23] due to their greater diversity (vegetation, relief, hydrography) carry a greater aesthetic load than the steppe.

- "Intensity of trophic relations" - a purely environmental indicator, projected on human life. Despite being in an urban environment, mostly detached from the natural landscape, man still can not completely exclude himself from the system of material and energy relations that have developed in natural ecosystems. According to Y. Odum [5] and M. A. Golubets [7] in urban ecosystems there is a certain proportion in the ratio of purely ecological (in our case trophic) functions, and social (Fig. 1). In our case, in large cities, "detached" by the artificial urban environment from natural landscapes, social functions far outweigh the ecological ones.

- "The degree of rupture of food chains" - a more detailed interpretation of the previous indicator. In our case, it means the presence of more intermediaries as analogues of trophic levels in the ecological pyramid. The ideal case is when the food comes directly from the producer to the consumer (plant - herbivore), or from the consumer of the second order (herbivore) to the consumer of the first order (predator). According to the important environmental law Lindeman [25] when moving from one trophic level to another, no more than 10% of energy is lost. In the urban environment, analogues of trophic levels can be elements of the food distribution system (wholesale and retail trade network, natural markets, subsidiary farms).

- "Predominant urban function" - the indicator does not require detailed comment, as the negative impact on the living space of any industrial functions compared to others is obvious.

- "Saturation of the urban environment with natural and man-made objects" - this indicator is a certain hybrid of ecological and social content of living space (Fig. 1)

- "the predominant type of housing" - according to [26] one-floor buildings and the location of the house against the background of unchanged natural landscape significantly improves human living space.

- "social infrastructure" - a large list of artificially created facilities that provide mostly social needs of people living in cities [27]. In our case, there is a direct relationship - the larger the city, the more numerous are the objects of social infrastructure.

- "Eco-network objects" - their presence either in the city (arboretum "Sofiyivka" of NASU), or near it (Kryvyi Rih Botanical Garden of NASU) is evidence of the remoteness or proximity of the urban ecosystem to natural analogues. In addition, from a scientific point of view, the objects of the ecological network can be reference sites, through the study of which the degree of transformation of natural ecosystems as a result of anthropogenic activity is determined.

- "objects of cultural and industrial heritage" - a meaningful list of anthropogenic objects that play a significant role in the aesthetic perception of living space. In our case, they are represented as objects of material culture, formed in the process of long-term mining and industrial development of Kryvbas, and objects of spiritual culture, formed in Uman region in different historical epochs (Trypillia culture, garden art, pilgrimage center).

- "General assessment of living space (according to the results of the survey)" - is an average assessment obtained from the mass survey of residents of Kryvyi Rih and Uman. Has the degree of objectivity inherent in such a survey [11].

- "Do you like the urban lifestyle?" - The answer to the question largely determines the preferences of city dwellers in relation to the actual urban living space.

- "Do you consider Kryvyi Rih / Uman a big city?" - a logical continuation of the previous question, however, the answer to which better allows to determine a person's self-assessment of belonging to the urban environment.

- "Do you consider the urban environment of your city dangerous for health?" - Probably the most pressing issue, especially against the background of the COVID-19 epidemic.

- "Do you consider the urban environment of your city safe for life?" - a question, the answer to which characterizes not only the criminogenic situation, but also the quality of work of municipal utilities and transport services.

- "Do you consider the urban environment of your city comfortable to live in?" - the antithesis of the previous question, the answer to which makes more clear the psychological dimension of the category "space of life" (Fig. 1)

- "In your opinion, can Kryvyi Rih / Uman be considered an ecologically dangerous city?" - Given a certain populism in the use of the phrase "environmentally dangerous", most respondents quite clearly understand what is being said, as evidenced by the number of positive answers (83 "yes" for residents of Kryvyi Rih and 8 "yes" for residents of Uman).

- "Is buying food in your city a big problem for you personally?" - The question has an age imprint, as most of the negative answers came from Ukrtelecom's city numbers, which is mainly due to the lack of mobile phones or smartphones among the elderly.

- "Is it a big problem for you personally to cover distances by public transport in your city?" - The answer to this question helps to clearly trace the relationship between the optimal configuration of the city and the need to move it. As in the previous question, mostly elderly people gave positive answers (76 "yes" in Kryvyi Rih).

- "Do you consider your personal home comfortable to live in?" To a large extent, the positive answer to this question testifies to the reaction of city residents to the latest trends in the arrangement of personal housing (Fig. 1), which is impossible without the presence of such elements of urban infrastructure as construction supermarkets and construction and repair specialists.

- "When you meet individual tourists or tourist groups on the streets of your city, do you feel annoyed?"

Almost the same number of positive and negative answers (76/67, respectively) to the last question in our opinion indicates the special role of tourism in the formation of positive content of the category "living space" in the city of Kryvyi Rih and Uman. Probably, this is the branch of human activity that should "sew" all the different characteristics of the living space for its holistic positive perception.

The authors have some experience in the study of tourism [28]. Today, the living space of the cities of Kryvyi Rih and Uman can be enhanced by positive content precisely due to its use in tourism. To do this, the authors have developed appropriate tourist routes and guides.

Participation of objects of tourism and tourist activity in the formation and use of the living space of the city of Kryvyi Rih.

1. Transport tourist route "My Kryvyi Rih". Length 60 km. Passes through the territory of the Metallurgical, Pokrovsky, Saksagansky areas of the city of Kryvyi Rih.

2. Transport tourist route "Journey to the world of plants (Kryvyi Rih Botanical Garden)". The length is 70 km. It passes through the territory of Metallurgical, Saksagan, Pokrovsky, Ternivsky districts.

3. Transport tourist route "Kryvbas - geological". The length is 60 km. It passes through the territory of the Central City, Saksagan and Pokrovsky districts.

4. Transport tourist route "They defended the Motherland". The length is 50 km. It passes through the territory of the Central City, Saksagan and Terniv districts of the city of Kryvyi Rih.

5. Transport tourist route "Thorny roads of Kryvyi Rih". Length 15 km. Passes through the territory of Metallurgical, Ingulets and Central City districts.

6. Transport tourist route "By the way of the liberated city". Length 4.1 km. Passes the territory of Central City and Saksagan districts.

7. Transport tourist route "Remember the past - for the future". Length 30 km. Passes through the territory of Ingulets district.

8. Transport tourist route "From Ingulets to Ingulets". Length 45 km. Passes the territory of the Central City District.

9. Transport tourist route "Northern Gate of Kryvbas". Length 21 km. Passes the territory of Terniv district.

10. Transport tourist route "Pokrovsky district - the pearl of Kryvyi Rih". Length 10 km. Passes through the territory of Pokrovsky district.

11. Transport tourist route "Living River of Time". Length 19 km. Passes through the territory of Pokrovsky and Saksagansky districts.

12. Transport tourist route "Cossack roads of Sicheslav region". Length 100 km. Passes through the territory of Metallurgical, Central City, Ternivsky districts.

13. Transport tourist route "My native Kryvyi Rih". Length 30 km. Passes through the territory of Metallurgical, Central City, Ternivsky districts.

14. Transport tourist route "Industrial Kryvbas". Length 30 km. Passes through the territory of Metallurgical, Ternivsky, transit through the Pokrovsky area.

15. Transport tourist route "Kryvyi Rih - the industrial heart of Ukraine". Length 30 km. Passes the territory of Ingulets and Metallurgical districts.

The analysis of these transport and transport-pedestrian routes showed that the largest transport and tourist load is experienced by Metallurgical and Central City districts (7 routes each), Ternivsky, Pokrovsky, Saksagansky districts (6 routes each), the least - Ingulets district (3 routes). However, the elongated configuration of the city "provides" each of its inner-city districts with participation in the implementation of tourist activities.

Due to the elongated configuration of the city of Kryvyi Rih in previous publications in order to optimize tourist routes, the authors used the tool "Poligons of Thyssen-Voronoi" in GIS Mapinfo.Prof. (Fig. 2, 3) [28]. The authors plan to carry out a similar optimization in future research for the city of Uman.

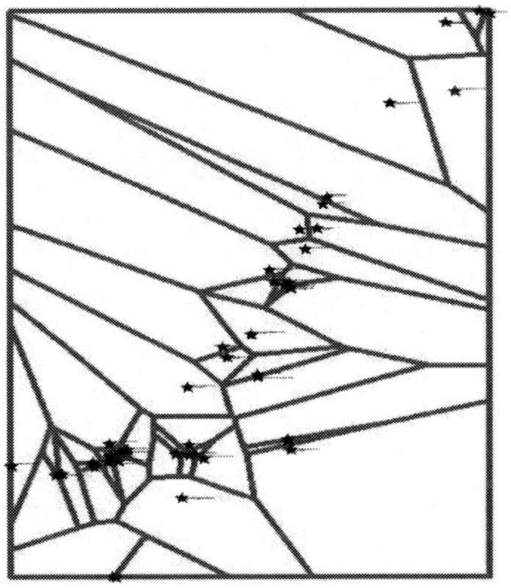

Fig. 2. General scheme of Voronoi polygons throughout the city of Kryvyi Rig.

The use of historical and cultural sites of the city of Uman in tourism and in the formation of the living space of the city.

The city of Uman is one of the oldest cities in Ukraine with a rich history, the chronicle of which is adorned with numerous bright historical events and prominent personalities.

The city is geographically located in the center of Ukraine since ancient times and today plays an important role as a highway. Cherkasy region is located in the heart of Ukraine. The public highway of state importance Kyiv-Odessa, M-05 passes through it. It is part of the European road route E-95 (St. Petersburg-Kiev-Odessa-Samsun-Merzifon). This highway connects many prominent cities, among which the dominant position is occupied by Uman, Cherkasy region.

Fig. 3. Fragment of the scheme of Voronoi polygons in inner-city areas with the highest density of tourist attractions.

The city is geographically located in the center of Ukraine since ancient times and today plays an important role as a highway. Cherkasy region is located in the heart of Ukraine. The public highway of state importance Kyiv-Odessa, M-05 passes through it. It is part of the European road route E-95 (St. Petersburg-Kiev-Odessa-Samsun-Merzifon). This highway connects many prominent cities, among which the dominant position is occupied by Uman, Cherkasy region.

Uman is a city of regional subordination and the second most populous city in the region. The city has many historical and cultural sites (monuments) of world importance, which in turn needs special attention and support from the state. There are about 40 monuments, memorials and other memorials in the city.

The city is divided into the following neighborhoods:
- Suburbs "Turok" - the oldest district of Uman with many architectural monuments. The Stara Uman State Historical and Architectural Reserve is located in the larger territory of the district.
- "Bald Mountain": an old area with one-story private estates.
- "Domansky": the smallest neighborhood in the east of the city with one-story private estates.
- "Babanske KP": an old area with one-story private estates in the south of the city with underdeveloped infrastructure.
- "Eastern": founded in the early 90's of last century. Streets parallel to the E-95 highway are named after the cities of the region (Cherkasy, Smilyanska, Shpolyanska, Korsun-Shevchenkivska and others), and between them - Twin Cities. Streets perpendicular to the route - "East 1, East 2… East 12".
- "Zvenigorod suburb": this name is almost never used. Located in the central part of the city with old one-story private estates and height differences.
- "Ivangorod suburb": a quiet neighborhood above Ostashivsky pond, developing, has many high-rise buildings, shops and cafes.
- "Suburbs of Bessarabka": also called "sleeping area", is located in the southwest of the city with a large number of high-rise buildings and popularly known as "railway station area".
- "Mischanka": an area of private estates with developed infrastructure, located in the north-western part of the city.
- "Slobidka": lies in close proximity to the park "Sofiyivka" in the northern part of the city.
- "Nova Uman": is located in the northeastern part of the park "Sofiyivka" and is directly adjacent to the center of the Hasidic pilgrimage "Tomb of Tsadik Nakhman", there are both private estates and five-story buildings.
- "DOS": officers' houses in the northern part of the city. More than ten five-storey houses of old buildings, several nine-storey buildings were built at the end of the last century and two are still under construction.

In addition, the following names of the following districts are used in everyday life:
- "Zarembova dam": an area of private estates above "Zarembovymstav" in the southern part of the city with underdeveloped infrastructure.
- "Oleksiyivka": the most remote area in the northern part of the city, consisting of one street - Karmelyuka (№ 145-205). It is located near the "Oleksiyivsky" pond - the river Palanka.
- "GRES microdistrict": is one of the most remote from the center. Five-storey buildings were built in the middle of the last century for the workers of the greenhouse plant.
- "Center": a densely populated area with different types of houses, many shops and financial institutions. The best infrastructure and proximity to the center has determined the popularity of the area.
- "Proletarskaya": north-western part, sleeping area, more than ten five-storey panel type.
- "Tyshchyka": five-storey and nine-storey buildings, shops, bazaars and more.
- "Breeding": several three-story buildings and private estates in the immediate vicinity of the "breeding station" (Uman Research and Breeding Station). Located in the northeastern part at the entrance to the city.
- "Avtostrada" (Belogrudovskaya Street): is the most remote area from the city center in the north of Uman. There are more than ten five-story buildings, shops, a hospital and the Belogrudovsky forest nearby.
- "Greek Forest": the newest neighborhood of the city. The project was developed by Umanavtodor. It is planned to build five- and fifteen-storey residential buildings, shopping, office, entertainment, cultural and

educational centers, a park area, and underground parking lots on the area of 22.9 hectares. That is, for the first time in the region it is planned to build a modern residential complex with a full infrastructure. The founders of the project plan to provide residents of the Grekiv Lis residential district with street lighting, playgrounds, round-the-clock water supply with multi-level water purification and filtration, digital telephone lines, dedicated Internet channel, silent high-speed elevators, security and video surveillance. The new complex is designed for 750 apartments, which can accommodate three thousand people from Uman.

In the city of Uman there are 178 km of roads, most of them with hard (asphalt) pavement; most intersections are equipped with traffic lights. Within the city there are 4 squares: Ivangorod Square, Gagarin Square, Victory Square and Sobornosti Square. The last new streets (about 20) were developed for the Vostok district in 1992. Every fifth square meter of housing in the region is being built in Uman

Longest streets:

- Stepana Banderi - 5.4 km;
- Energy - 3.6 km;
- Karmelyuk - 2.8 km;
- Kyiv - 2.5 km;
- International - 2.4 km;
- Independence - 2.3 km;
- European - 2.3 km;
- Mykhailivska - 2.2 km;
- Assumption - 2.2 km;
- Liberators - 1.8 km;
- Factory - 1.7 km;
- Ivan Vyhovsky - 1.7 km.

Given that in the formation of tourist routes in most cases they are based on the world-famous attractions "Sofiyivka" and "Tomb of Tsadik Nakhman", adding other cultural and historical heritage sites can help in the formation of a higher quality of life.

3 Conclusions

1. The article defines the ontological affiliation of the category "life space", which is realized in the field of such basic sciences as sociology, ecology, psychology, geography.
2. The degree of influence of each of the dimensions of the category "space of life" on a person at different spatial levels is studied.
3. Various aspects of the use of living space in the urban ecosystems of Kryvyi Rih and Uman have been studied.
4. The role of the main branches of municipal economy which form and use a living space of the cities of Kryvyi Rih and Uman in particular in tourist activity is investigated.

References

1. Mitch Rose, *The question of culture in cultural geography*: Latent legacies and potential futures. Prog. Human Geography, (2020). DOI:https://doi.org/10.1177/0309132520950464
2. Igor K. Liseev, Ecology as a Way to Combine Knowledge about the Natural and Social in Human Being. Epistemol. Philos. Sci. **57** (4), 133-137 (2020). DOI: https://doi.org/10.5840/eps202057466
3. Laura J. Cameron, *Sir Arthur Tansley*. (2017). DOI: https://doi.org/10.1093/OBO/9780199830060-0094
4. S.M. Dombrovska, V.V. Kovregin, A.L. Pomaza-Ponomarenko, O.M. Kolenov, *Derzhavne upravlinnya u sferi bezpeky sotsial'noekoloho-ekonomichnykh system: monohrafiya* (Public administration in the field of security of socio-ecological and economic systems: a monograph). (NUTSZU, Kharkiv, 2017), p. 244
5. N.V. Koneva, Ekologicheskiye imperativy v trudakh Gennadiya Rozenberga (Environmental imperatives in the works of Gennady Rosenberg). Samarskaya Luka: probl. regional glob. ecol. **28** (2), 127-151 (2019). DOI: https://doi.org/10.24411/2073-1035-2019-10219)
6. A. Klemmer, M. Loreau, Breaking Through Ecosystem Boundaries, Bull. Ecol. Soc. Am. **98**, 95-98 (2017). DOI: https://doi.org/10.1002/bes2.1313
7. Yu. Chernobay, *Muzeyne vidobrazhennya koevolyutsiynykh metamorfoz seredovyshcha i povedinky* (Museum representation of coevolutionary metamorphosis of the environment and behavior). Proc. of the State Nat. Hist. Mus. **36**, 3-14 (2020). DOI: https://doi.org/10.36885/nzdpm.2020.36.3-14
8. E.N. Pertsik, *Geourbanistika: uchebnik dlya akademicheskogo bakalavriata* (Geourbanistics: a textbook for an academic bachelor's degree). (Yurayt Publishing House, Moscow, 2018), p. 481
9. E.Sh. Akhmetova, *Sotsial'no-institutsional'nyye usloviya sovershenstvovaniya ekosistemy kak sostavlyayushchey kachestvo zhizni naseleniya monopromyshlennogo goroda* (Socio-institutional conditions for improving the ecosystem as a component of the quality of life of the population of a mono-industrial city) Abstract of a dissertation for the degree of candidate of sociological sciences. FSBEI HE "Penza State University" (2019), p. 22
10. Siân de Bell, H. Graham, Piran C.L. White, The role of managed natural spaces in connecting people with urban nature: a comparison of local user, researcher, and provider views. Urban Ecosyst. **21**, 875-886 (2018). DOI: https://doi.org/10.1007/s11252-018-0762-x
11. M.G. Shapiro, Klassifikatsiya metodov oprosa v sotsiologii (Classification of survey methods in sociology) Social Research. **2**, 51-59 (2017)
12. Richard Florida, Kryza urbanizmu. *Chomu mista roblyat' nas neshchasnymy* (The crisis of urbanism. Why cities make us unhappy). (Our format, Kyiv, 2019), p. 320
13. Ray M. Merrill, *Introduction to Epidemiology* (Jones & Bartlett Learning, USA, 2015), p. 340

14. L.E. Limonov, *Regional'naya ekonomika i prostranstvennoye razvitiye* (Regional economy and spatial development). (Yurayt, Moscow, 2015), pp. 71-73
15. A.V. Kuznetsov, *Sovremennyye teorii razmeshcheniya. Lektsiya dlya praktikantov geograficheskogo fakul'teta MGU im. Lomonosova* (Modern placement theories. Lecture for trainees of the Faculty of Geography, Moscow State University Lomonosov). (INION RAS, Moscow, 2020), p. 39
16. M.A. Sokolova, K. Yu. Aleksandrova, Gorodskiye pustoty kak komponent gumanizatsii arkhitekturnoy sredy (Urban Voices as a Component of Architectural Environment Humanization). Archit. Modern Inform. Technol. **4** (49), 262-280 (2019). DOI: https://doi.org/10.24411/1998-4839-2019-00018
17. Andre Dauphine. *Geographical Models with Mathematica* (ISTE Press - Elsevier, 2017), p. 314
18. V.F. Kifyak, G.Yu. Bondarenko, Inozemnyy turyzm yak holovna skladova turystychnoyi diyal'nosti v prykordonnomu rehioni (Foreign tourism as the main component of tourist activity in the border region). Bullet. Chernivtsi Trade and Econ. Inst. **3** (67), 72-83 (2017)
19. O.V. Taranych, Osoblyvosti formuvannya konkurentnykh perevah rozvynenykh krayin na svitovomu rynku (Peculiarities of formation of competitive advantages of developed countries on the world market). Eco.org.manage. **3**, 371-328, (2016)
20. D. Nikolaenko. Medical geography and infectious ecology of COVID-19 in the United States: approaches to a systematic GIS description. Environ. Epidem. J. **5,** 1-18 (2020). DOI: https://doi.org/10.13140/RG.2.2.28685.97768.
21. I.A. Vasilenko, I.M. Trus, O.A. Pivovarov, L.A. Frolova, *Ekolohiya lyudyny* (Human ecology). (Accent PP, Dnipro, 2017), p. 183
22. D.Audretsch, M.P. Feldman, Knowledge Spillovers and the Geography of Innovation. Editors. Handbook of Urban and Regional Economics: Cities and Geography. **4**, 2713-2739 (2004)
23. M.D. Grodzinsky, Landshaftna heohrafiya: stara nazva novoyi nauky chy vidrodzhennya mayzhe zabutoho? (Landscape Geography: The Old Name of a New Science or the Revival of the Almost Forgotten?). UGJ. **2**, 59-64 (2017). DOI: https://doi.org/10.15407/ugz
24. G.I. Denisyk, K.V. Mezentsev, E.A. Antipova, A.G. Kizyun, Heohrafiya povsyakdennya: prostorove riznomanittya povsyakdennoho zhyttya (Geography of everyday life: spatial diversity of everyday life). Bulletin of Kharkiv National University named after V.N. Karazin, series "Geology. Geography. Ecology" **52**, 63-70 (2015). DOI: https://doi.org/10.26565/2410-7360-2020-52-10
25. V. Khudoba, Y. Chikailo, *Ecology* (LDUFK, Lviv, 2016), p. 92
26. I.A. Saenko, L.R. Akhmetova, M.Yu. Belichenko, Izucheniye mneniya naseleniya o klassakh komfortnosti zhilykh domov (Study of the public opinion on the comfort classes of residential buildings). Young scientist. **10 (144)**, 88-92 (2017)
27. V.M. Novikov, N.M. Deeva, A.G. Gwelesiani, *Sotsial'na infrastruktura na shlyakhakh reformuvannya mistsevoho samovryaduvannya* (Social infrastructure on the way to local government reform). (Kyiv-Warsaw, 2018) p. 431
28. D. Shiyan, T. Kazakova, S. Sonko, Spatial organization of tourist attraction network of Kryvorizhzhia. J. Socio-Econ. Geog. **24** (1), 56-63 (2018). DOI: https://doi.org/10.26565/2076-1333-2018-24-06

https://doi.org/10.1051/shsconf/202110005009

Ensuring of Consumer Choice Ecologization on the Basis of Consumption Safety: International Experience for Ukraine

Kateryna Antoniuk[1*], *Anatoly* Mokiy[2], *Mariia* Pikh[3], *Lyudmila* Bukharina[4] and *Dmytro* Antoniuk[4]

[1]National University "Zaporizhzhia Politechnic", Department of International Economic Relationships, 69063 Zaporizhzhia, Ukraine
[2]SI "Institute of Regional Studies named after M.I. Dolishny NAS of Ukraine", Department of Regional Economic Policy, 79026 Lviv, Ukraine
[3]Lviv Professional College of Food Processing Industry of the National University of Food Technologies, 79000 Lviv, Ukraine
[4]Zaporizhzhya National University, Department of Entrepreneurship, Management and Logistics, 69600 Zaporizhzia, Ukraine

Abstract. The article is devoted to the disclosure of the results of the study of threats to the environment, sustainable development and consumption safety that arise in the process of consumption. The problems of formation of the environment which promotes growth of ecological demand in Ukraine in comparison with other states are analyzed. Factors of internal, external and situational influence on consumer choice at each stage of consumption (awareness of needs, search for information, evaluation of alternatives, purchasing, use and utilization) are investigated. The main of them such as values, culture of consumption, marketing (promotion), level of development of institutional base (legislation, social norms) are established and their optimal level in conditions of uncertainty is proved. Conclusions are made on the basis of using methods of office research, sociological survey, modeling on the basis of fuzzy logic with the use of expert assessments. The leading role of the factor of internal influence, which covers human values that can be improved by raising the educational and cultural level of the population, which is undoubtedly related to income is proved.

1 Introduction

In today's conditions, when there is a growing need to reduce the ecological footprint of human activity, the transition from consumption approaches to the principles of sustainable development, the task of ensuring consumption safety is relevant. At the current stage of the global economy development, it is important to alleviate the relationship between socio-economic growth and environmental degradation due to the internationalization of production and goods and services consumption, which in Ukraine, according to the UN Development Program, reaches 75% of total environmental impact and 60% consumer spending [1]. Problems of unsatisfactory health, obesity, diseases are caused by environmental conditions and unreasonable consumption. And these problems require solutions at all levels of the systemic hierarchy of management (global, megaregional, national, regional, of economic entities and end users) and that also depends on all stakeholders interests harmonization. Though there is the primacy of a man as an object of national interests protection, at the same time an individual as the end consumer is relative insecure in economic processes (due to lack of education, institutional imperfection, aggressive communications, etc. Thus it is important to create conditions for conscious, reasonable, sustainable, environmental choices with long-term consequences, here we mean focus on safety of consumption. In a broad sense, the consumption safety is interpreted as the ability of the state, society and business to create the preconditions for the conscious satisfaction of human needs for self-reproduction to protect the health of the nation and the environment as imperatives for future development. Ensuring consumption as a basis for sustainable consumption concept implementing and sustainable development providing to solving global problems of waste, irrational nature management, excessive consumption. The conceptual principle of sustainable consumption is the formation of rational consumer habits not only in saving money, but also in maintaining health, saving resources and limiting the negative impact on the environment, which remains an important component of quality and safety of human life. That is, within the outlined issues, it is necessary to take into account the factors of internal and external influence on the consumer behavior ecologization, which in today's conditions mostly have a negative context and need to change with a focus on long-term results.

2 Literature review

Summarizing the results of research in developed countries [2] suggests that scientists pay considerable attention to the problematic aspects of sustainable consumption, analysis of demand, perception, consumer behavior patterns, lifestyles, consumption culture, corporate social responsibility, brand ethics, storage and disposal, etc. There is a sufficient prevalence in developed countries of sustainable

* Corresponding author: ekaterinaia@ukr.net.

 https://doi.org/10.1051/shsconf/202110005009

aspects of consumption (Table 1), although a significant role is given to the state and scientific and technological progress. Instead, domestic consumers, also hoping for a political and technological solution, are characterized by a less progressive attitude to the environment and, consequently, to sustainable development (which can be explained by the low solvency of a large part of the population), although they try to take steps to change lifestyles in this direction.

Table 1. Attitudes of consumers to environmental protection, % of respondents.

	Australia	Canada	Chile	France	Israel	Japan	Korea	Netherland	Spain	Sweden	Switzerland	Ukraine
Transferring the solution of the problem to future generations	19	18	17	20	8	24	13	18	22	12	11	23
Willingness to accept the environmental decisions with regard to others	18	17	7	36	10	27	27	29	13	23	26	5
Underestimation of environmental consequences	46	38	20	35	23	47	38	42	39	37	38	10
Belief in solving environmental problems through the technological process	35	35	38	28	21	49	51	53	46	26	34	21
Unwillingness to pay their own money to solve environmental problems	72	68	61	67	72	58	64	54	78	60	60	22
Willingness to change the way of life in favor of the environment	79	86	93	87	87	56	98	78	91	81	87	50

Source [2, 3]

Prevention of environmental degradation, rejection of those types of goods and services that significantly worsen the environmental situation can be provided due by development of the consumption culture, which G. Shambrovsky understands as a socio-psychological characteristic, individual manner, peculiarities of human behavior in the process of consumption and use of material and spiritual property, and also beliefs about the usefulness of performance results. It is formed under the influence of internal and external psychological, social and economic factors. And the basis for its development is formed by cultural values, which are a set of social norms, rules, standards, that most people follow and which perform a regulatory function for their behavior [4]. Values are consideration mandatory in the marketing system, because the goods, services and social ideas that contradict them are perceived by people as hostile. However, in conditions of acute socio-economic crises cultural values can change significantly [5]. Among the obstacles to strengthening consumption safety in Ukraine the main are the lack of appropriate values, existing habits, as well as insufficient protection of consumer rights by the state and the presence of dangerous goods.

Correlated with national values is the fact that domestic consumers are primarily concerned about caring for the environment (35%), respect for human rights (32%) and support for domestic producers (31%). At the same time, we consider it unfavorable that none of these principles worries the majority of consumers, and 18% of respondents did not even thought about them.

According to the results of the author's research [3], the most important motive for choosing products is health safety, and this condition is the most important in terms of quality of goods and services (especially the quality of food, which, according to surveys [6; 7, p. 62]) domestic consumers consider the most pressing problem). This feature was mentioned by 88% of respondents, as opposed to 33% of respondents, for whom environmental safety is more important. Also, among the main features of consumption safety, environmental safety gives way to health safety and compliance with standards and norms. This correlates with the conclusion that consumers are primarily inclined to worry about short-term or more significant needs [8].

Sociological surveys of consumers [8, 9] show that most respondents consider it necessary to use organic products, but not all use it. The main attention is paid to the degree of usefulness and safety for health, and not to the elimination of negative impact on the environment [10, p. 13].

Along with the inability of a significant part of the population of Ukraine to buy ecological goods mainly due to lack of funds, there is a tendency to reduce the share of consumers willing to pay more for ecological goods. In this regard, there is a certain connection between the data of surveys in 2011 [8, p. 63-68] and 2019 (field author's sociological study [3]). Despite the fact that it is possible to indicate temporal, spatial (different regions of Ukraine), quantitative (sample size) differences, they rather reflect the relevance of the results and the dynamics of change.

Situational factors today are mostly characterized by a negative impact on consumption safety. It is worth noting the imperfection of the institutional basis of consumption safety, in particular the low level of consumer protection. This, along with the low level of development of corporate social responsibility and unsatisfactory socio-economic and environmental status of the country complicates the process of consumer decision-making on the choice of goods (services). Previous events, experiences and mistakes can also be influential in this case. Thus, the majority of respondents (56%) sometimes encountered unsafe consumption, and among the cases of misleading unfair advertising and promotions, concealment of dangerous ingredients and inaccuracy of information on the packaging were often called.

In the context of corporate social responsibility, it can be argued that Ukrainian consumers primarily consider it in the narrow aspect of consumption safety, they understand it as health care (88%), while only 45% of respondents are concerned about the impact on the environment. Comparison of the obtained results with the results of foreign research indicates a low level of demand of domestic consumers. Accordingly abroad consumers choose social responsibility and image (brand) as the main features for evaluating companies and identify interest in companies efforts to become more responsible [11, 12]).

Based on this, it is worth agreeing with the experts [13] that a true understanding of the social and institutional context of consumer action opens a creative view of innovation processes, the key task of which should be the transition from "reflection" to "development" models. For which it is necessary to determine the main factors influencing the ecologization of consumer choice and justify the direction of action aimed at creating conditions for its provision.

3 Methods

To determine the most significant factors influencing consumer behavior in the context of ensuring, including ecologization of consumption, the author a sociological study in the form of a questionnaire based on a simple unique quota sample with quotas by age, income, place of residence, education and gender is conducted.

Based on the analysis of the main factors of influence taking into account the peculiarities of consumption safety (uncertainty, hidden, fuzzy relationships between components, time delay), a model of probability of consumer decision on the choice of goods (services) based on consumption safety has been built [3]. The incompleteness or absence of statistical information on consumption safety parameters does not allow the use of formal models based on certain assumptions. That is why we consider the methods of fuzzy logic as a rational way to predict the behavior of individual consumers, which allows to formalize processes that are uncertain or fuzzy in nature, and to describe the algorithm of consumer behavior and assess its state based on modeling initial (factor) features [14; 15].

To describe the input (X_1, X_2, … X_n) and output (Y) variables we use linguistic evaluation according to the terms on the basis of which using a tree of logical conclusions (Figure 1) we formulate databases of expert evaluations (by group and individual survey). They are fuzzy carriers of information about available relationships between input and output variables. That is, decisions on the membership function of fuzzy subsets, which describe the consumption safety in conditions of uncertainty, are made at a certain point in time on the basis of factors selected by the expert method.

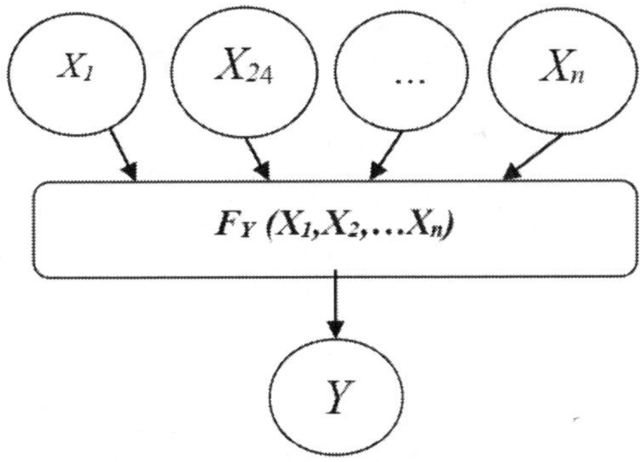

Fig. 1. The relationship between input and output variables of consumer behavior on the basis of consumption safety
Source: proposed by the authors

The obtained results are the basis for building a model of safe, including environmental, consumption, which takes into account the positive, neutral or negative impact of external, internal and situational factors at different stages of this process.

4 Results and discussion

To identify and counteract threats to consumption safety based on impact factors, we will consider in detail the process of consumption, taking into account the peculiarities of each of its stages: awareness of needs, search for information, evaluation of alternatives, purchasing, use and utilization. The stage of behavior after purchase should be investigated within the sub-stages that differ in needs and functional features: transportation and storage, use, disposal, perception. Note that although the perception may precede some other stages, in this case it is a question of forming a final, generalizing judgment, a conclusion according to which the consumer will plan his further actions. We will also remind you of the importance of feedback, especially at the current stage of development, when e-commerce is gaining in scale, and reviews of goods (services) are one of the main criteria for choosing products.

1. At the stage of awareness of the need the consumer feels dissatisfaction of interest under the influence of various factors. Thus, the real need is usually determined by internal factors, although its expression, in particular according to consumption safety, depends on values, motives, level of education, income and other psychological and individual indicators of previous experience, and lower or higher level of need and timing (present – future).

At the same time, external factors in the conditions of obsessive marketing or the corresponding environment (influence of a social group, subculture) can cause spontaneous needs, which will lead to waste. It is worth agreeing with O. Prokopenko [8] that an important and promising "purchasing power" in terms of consumption safety can be children who significantly influence parents' purchasing decisions and are future consumers and intermediaries between producers and parents. In the United States, more than a third of parents have changed their buying behavior because their children have convinced them of the importance of environmental issues. Students who are the most socially active group of the population are also active supporters of environmental organizations. Thus, 26% of freshmen in US colleges participate in environmental cleansing and conservation programs.

At this stage, the significant influence is made even by such factors of the global environment as political, legal, geopolitical, technological, which determine: confidence in the future, market demand (solvency, demanding, propensity to consume) and supply (innovation, safety, price, etc.).

The threat to the consumption safety is the ability (skills) to formulate correctly the need (taking into account the consequences for health, environment, society, self-development), to hypothesize the choice criteria (quality, naturalness, social responsibility of contractors (manufacturers, intermediaries) etc.), the required amount, etc.

A combination of rational and emotional motives is theoretically desirable. It should be borne in mind that the concepts and, consequently, the characteristics of the safety of goods (services) are narrower to the safety of consumption, and safety requirements may differ for goods (services). In addition, behaviors in which the usual choice criteria do not take into account the safety can be threatening. It is worth agreeing with the opinion [13] that determining the exact correlation of affective, moral, habitual and social motivations and situational conditions is not an easy task.

Various events and situations, such as emergencies, which increase the motivation of safety-oriented consumption, as well as accidental and predictable situations of interaction between consumers and products (primarily merchandising), which encourage a decrease or increase in safety-oriented consumption can be considered as situational factors influencing the safety of consumption [2].

The moderating effect of external situational factors on consumer intentions emphasizes the need to improve and alleviate conditions in a wide range of environmentally important situations. And, perhaps most notably, the integration of the individual into a social group indicates the vital influence of social and cultural context on consumer behavior.

Consumption culture determines the safety of consumption in stages of:

- awareness – affects the definition and formulation of safety requirements;
- search – influences the choice of information sources according to their reliability evaluation;
- choice - affects the definition of essential features and criteria;
- use - affects consumption without endangering health and the environment during transportation, storage, use and disposal.

The influence of the culture of consumption in the process of feedback in the conditions of satisfactory, and especially unsatisfactory experience of consumption (use) of the goods (service) is important.

2. At the stage of information search such threats as its completeness, reliability, accessibility, may be affected by: information search skills, level of awareness and education of the consumer, experience gained in previous use, perception; aggressiveness of the supply (level of competition, competitiveness, obsession of communications: frequency, channels, content of messages, etc.); institutional environment (regulatory mechanisms, protection tools), in particular information security, development of information and communication technologies; reference groups, their values, experience, feedback as a source of information about goods (services).

If the product (service) belongs to the category of daily demand, which is characterized by a model of normal behavior and choice, consumer orientation and situational factors (availability at the point of sale, means of sales promotion on the ground) are important. T. Jackson considers the need to understand and study the possibility of influencing the formation and change of habits in consumer actions the key factors for further study and promote behavior change that shape and constrain consumer choice [13].

O. Prokopenko notes that the bulk of consumers in the case of new product appearance on the market first gets acquainted with composition product with the information provided on the package, evaluates the environmental friendliness of the product [8]. In the future, consumers only compare the price of the product with the idea of environmental friendliness and so on. At the same time, middle-income people focus on environmental friendliness and taste characteristics of products.

In addition, the "cognitive trap" is reflected in routine behavior, even when it contradicts rational considerations and social norms and causes changes in best intentions. As stated in the report “Motivation for Sustainable Consumption” “… even if we can persuade people to change their views and beliefs in favor of actions to protect the environment, even if we can convince them of the need to behave prosocially through prohibitive or descriptive social norms, even if we can get people to adopt personal environmental standards, there is still no guarantee that they will behave accordingly in the world around them” [13]. Thus, behavioral change to protect the environment and ensure the safety of consumption is a daunting task.

3–4. In the context of consumption safety, the stages of evaluating alternatives and making a purchase decision can be considered together, as the consumer is influenced by the following factors:

- internal - values, motives, income level;
- external - marketing; advertising, in particular at points of sale, packaging; price; socio-cultural, reference groups;
- situational - availability, conditions, advice of the consultant, promotions.

In our opinion, the greatest threat at these stages is the unreasonable choice of criteria for comparing and evaluating goods (services) (emotional, affective; due to the environment, obsession).

At these stages, the threat may be:

 https://doi.org/10.1051/shsconf/202110005009

- lack of skills, lack of necessary information, complexity of conditions (limited time, space) for evaluating criteria;

- use first of all of compensatory assessment models where non-compensatory ones should be used: for hazardous components, doubts about shelf life, etc;

- disregard for tangible and intangible costs for storage, disposal after the purchase of goods (services).

5. Regarding consumer behavior after purchase, in particular in our country, it is advisable to pay attention to safety issues at the stage of direct use and disposal. This is due to greater regulation and institutional protection of the processes of purchase, transportation and storage of products (goods), which are more responsibly followed by most consumers. Although the conditions of use and disposal are also clearly defined, the problem of their compliance remains unresolved. Thus, as noted in [16, p. 29], in the future this will be demanded not only by politicians but also by consumers, but at present no country combines a high level of development with sustainable environmental consumption.

Our country is characterized by a situation where a significant part of the population (53%) does not sort garbage. That is largely due to the lack and imperfection of the relevant infrastructure. While the main reasons for significant volumes of consumption waste, respondents believe that too much packaging (63%).

The conditions of our country at the stage of goods (services) using are also threatened by ignorance, ignoring the requirements for storage and disposal, underdeveloped disposal infrastructure, inability of consumers due to lack of knowledge, skills, information, underdeveloped institutional environment to protect their rights and more. The latter can be mitigated by establishing negative feedback.

Dangerous, especially in the long run, is the inability or difficulty to assess the level of satisfaction of the need through imposed means of marketing communications (advertising, personal sales, public relations tools), feeling excessive or insufficient satisfaction when there is a desire to repeat the purchase.

Factors of internal and external influence as values, culture of consumption, marketing (promotion), level of development of institutional base (legislation, social norms) are chosen as input variables of consumers behavior modeling by methods of fuzzy logic (Table 2) evaluated by expert method. Regarding such a factor of external influence as marketing the unevenness of the intervals can be explained by its nature, which is to achieve the main goal - the persuasion of consumers to make a purchase and to be loyal. Thus, most marketing communications have a negative or relatively negative impact on consumption safety. Reports of positive impact are likely to be exceptions at the request of society and the state, or measures within the framework of corporate social responsibility, the imitation of which is more costly and insufficient in

Ukraine. The difference in the intervals for the factors of values and the institutional basis is also due to the imperfect level of their development and non-compliance with the requirements of consumption ensuring. The formation of appropriate consumption habits is a gradual process, as is the indicator by which we assess the safety of consumption.

The conducted modeling of consumption safety depending on the chosen factors of internal and external influence on the basis of methods of fuzzy logic allowed to optimize indicators of consumption safety level (Table 3, Figure 2). To achieve the optimal (not below average) level of consumption safety (0.4 and above) it is necessary to fulfill simultaneously the following prerequisites:

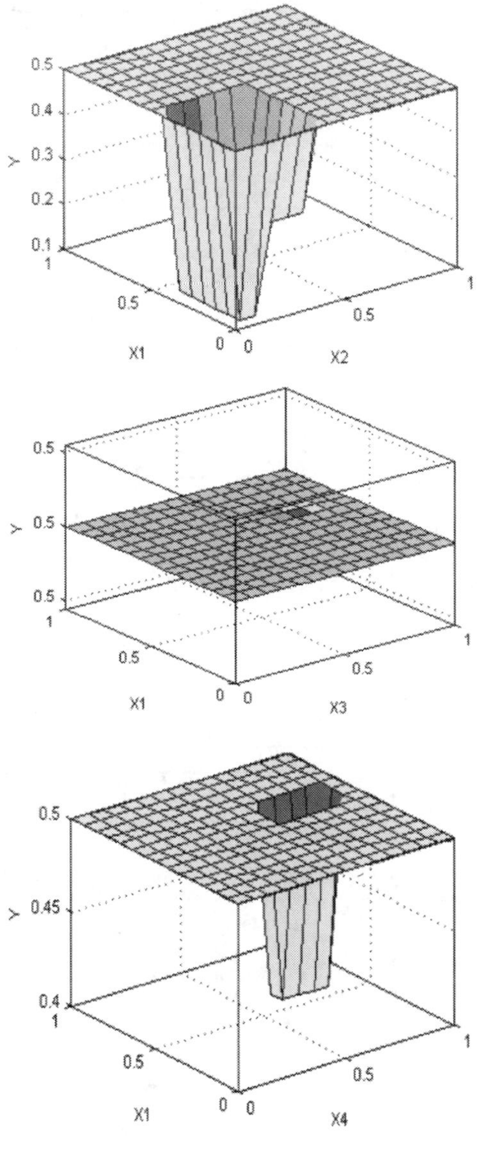

Fig. 2. Dependence of consumption safety level on indicators of internal and external influence, which include human values (X_1), consumption culture (X_2), marketing (promotion) (X_3) and the state of the institutional base in the country (X_4)
Source: built by the authors

Table 2. Linguistic evaluation and interpretation of variables in the process of modeling consumer behavior on the basis of consumption safety by fuzzy logic

Parameters	Type	Terms: Very low	Low	Average	High	Very high
X_1	Values	Negative for consumption safety (survival values, overconsumption, distrust to institutions)	Negative (the presence of one or more signs of negative values)	Neutral: in certain areas (product safety, health care, family - egocentric) or not always	Relatively positive: following of a significant number of positive values	Positive: values of self-development, care for the environment, well-being of future generations
		X_1 = 0–0.25	X_1 = 0,26–0.5	X_1 = 0.51–0.7	X_1 = 0.71–0.85	X_1 = 0.86–1
X_2	Consumption culture	Low level (consumpting everything without thinking about the usefulness and consequences)	Relatively low level (more often not thinking about the consequences for health or the environment)	Average (worrying about the consequences from time to time)	Relatively high (providing several high-level conditions)	High level (balanced approach to how much, when, how and what consumes, care about the consequences)
		X_2 = 0–0.2	X_2 = 0.21–0.4	X_2 = 0.41–0.6	X_2 = 0.61–0.8	X_2 = 0.81–1
X_3	Marketing (promotion)	Negative impact (presence of a significant amount of deceptive, intrusive, aggressive advertising, deceptive packaging, promotions, PR)	Relatively negative impact (a significant number of negative marketing factors, sometimes - an informative component)	Neutral influence (informative and information nature of advertising and other communication activities about the product or service)	Relatively positive impact (mostly informative comercial advertising, sometimes - popularization of sustainable development principles)	Positive impact (mostly social advertising, promotion of sustainable development principles and informative commercial advertising)
		X_3 = 0–0.3	X_3 = 0.31–0.6	X_3 = 0.61–0.8	X_3 = 0.81–0.9	X_3 = 0.91–1
X_4	Institutional environment (legislation, social norms)	Undeveloped (legislative and social norms of consumption are not carefully developed and vaguely formulated, there are no behavior ranks for the consumer)	Relatively underdeveloped (most conditions for the base development are not met)	Partially developed (consumer does not follow or partially follows) legislative and (or) social norms of consumption, or they are not thoughtfully developed and not clearly formulated)	Relatively developed (providing most of the conditions of development)	Developed (consumer follows legislative and social norms of consumption, which are carefully designed and clearly formulated)
		X_4 = 0–0.3	X_4 = 0.31–0.6	X_4 = 0.61–0.9	X_4 = 0.91–0.95	X_4 = 0.95–1
Y	Level of consumption safety	Lower limit	Below average	Average	Above average	High
		Y = 0–0.2	Y = 0.21–0.4	Y = 0.41–0.6	Y = 0.61–0.8	Y = 0.81–1

Source: authors' calculations.

Table 3. Perspective marks to ensure the optimal level of consumption safety

The value of the indicator of internal influence on the basis of human values, X_1	The value of the indicator of external influence, which covers the culture of consumption, X_2	The value of the external influence indicator, which covers marketing (promotion), X_3	The value of the indicator of external influence, which covers the state of the institutional base, X_4	Level of consumption safety, Y
0,598	0,845	0,902	0,931	0,5
0,114	0,315	0,477	0,469	0,3
0,386	0,315	0,447	0,469	0,2
0,386	0,1	0,705	0,777	0,3
0,144	0,0846	0,174	0,454	0,2
0,144	0,0846	0,174	0,777	0,1
0,144	0,0846	0,492	0,177	0,2
0,144	0,315	0,492	0,177	0,1
0,402	0,315	0,492	0,177	0,3
0,129	0,315	0,462	0,469	0,3

Source: authors' calculations.

- in the presence of negative for consumption safety values (X_1≥0.114) a safe level of consumption can be ensured only if the value of consumption culture is X_2≥0.315, the marketing impact is neutral (X_3≥0.477) and institutional base is undeveloped (X_4≥0.469);

- for a low level of consumption culture (X_2 ≥0.1) to ensure the consumption safety at level is possible by providing the value of the indicator, which covers human values X_1≥0.386, neutral influence of marketing (X_3≥0.705) and the average level of institutional development X_4≥0.777);

- the value of the indicator of external influence, which covers marketing (promotion) (X_3) should not be lower than 0.462, which corresponds to the range of relatively negative influence, ie the use by providers of a significant number of marketing factors of negative influence, and sometimes informative component;

- for a very low level of development of the institutional base (X_4≥0.177) a safe level of consumption can be provided when the value of the human values indicator X_1≥0.402, there is a relatively low level of consumption culture (X_2 ≥ 0.315) and relatively negative marketing impact (X_3 ≥ 0.492).

To achieve a high level of consumption safety, it is necessary to ensure certain levels of indicators of internal and external influence, and in the long run it is necessary to strengthen the factor of internal influence, which covers human values. This problem can be solved by raising the educational and cultural level of the population, which is undoubtedly related to income.

5 Conclusion

Consumer behavior reflects changes in demand for different types of goods and the level of welfare of the population, which must be taken into account when determining strategic priorities and modernizing mechanisms for regulating consumer market development, substantiation of state policy to protect consumer rights and promoting the consumer market commodity saturation of with quality domestic products.

The study of consumer behavior in the context of consumption safety ensuring identified the main factors of influence at all stages of purchasing decisions. They include the level of consumption culture, consumer habits, marketing and institutional basis, the composition of which determines the overall level of consumption safety. Achieving a high level of consumption safety is possible by providing defined on the basis of fuzzy logic levels of internal and external influence indicators. In the long run this requires an improvement in the internal impact indicator, which includes human values, through an increase in the educational and cultural level of the population, which is undoubtedly related to income. Thus, further research is needed to develop measures to improve the decision-making environment for safe, including green/ecological consumption, as well as the implementation of educational activities to form values and culture of environmental consumption.

References

1. Global Environment Outlook-5. UNEP. URL: http://web.unep.org/geo/resources/publications (22.08.2019).
2. N. Johnstone, Y. Serret, Greening Household Behaviour. Main Results of OECD Survey. URL: http://www.sviva.gov.il/English/env_topics/ InternationalCooperation/OnTheIntlFront/Documents/ GreeningHouseholdsBehavior-OECDSurveyResults-Nov2013.pdf (15.11.2019)
3. K. I. Antoniuk, Ensuring consumption safety in Ukraine in the process of European integration: theoretical and methodological foundations and applied aspects (NISS, 2020)
4. G. Shamborovsky, Welfare of the population of international integration associations: strategies for Ukraine (Ivan Franko Lviv National University, 2018)
5. The use of consumer psychology in modern marketing. URL: http://osvita.ua/vnz/ reports / psychology/10117/ (15.11.2019)
6. A. V. Rashchenko, Features of the formation of demand for agricultural products in the context of attitude, Agrosvit, 22 (2012)
7. N. V. Zinovchuk, A. V. Rashchenko, Ecological marketing (ZhSU Publishing House. I. Franko, 2015)
8. O. V. Prokopenko, Ecological marketing (Znannya, 2012)
9. A. Izvekov, Marketing research of the fruit and vegetable market and its development, Marketing, 50 (2020)
10. R. A. Perelet, Ya. P. Molchanova, E. V. Pashkov and others, Environmental certification and eco-labeling: the final technical report on the project "Harmonization of environmental standards (HPP II) - Russia", URL: http://air-q-gov.baes.by/user files/file/air_q_RU/CD/chapter_4/4-2/4-2-10.pdf (14.03.2018)
11. Responsabilidade Social das Empresas – Percepcao do Consumidor Brasileiro (Pesquisa, Instituto Ethos). URL: https://www.ethos.org.br/wp-content/uploads/ 2012/12/5Pesquisa-2001.pdf (18.11.2019)
12. The Millenium Poll on Corporate Social Responsibility. Global Public Opinion on the Changing Role of Companies. Environics International Ltda, 2000. URL: https://globescan. com/wp-content/uploads/2018/01/GlobeScan_ MillenniumPoll_1999_FullReport.pdf (18.11.2019)
13. T. Jackson, Motivating Sustainable Consumption – a review of evidence on consumer behaviour and behavioural change (University of Surrey, 2005)
14. A. Kofman, Introduction to the theory of fuzzy sets (Radio and communications, 1982)
15. M. S. Syavavko, O. M. Rybytska, Mathematical modeling under conditions of uncertainty (Ukrainian Technologies, 2000)
16. Consumption:Stakeholder Perspectives A Collection of Personal Essays by World Economic Forum Members and Constituents involved in the Forum's Sustainable Consumption Initiative 2008–2012 and associated Global Agenda Council on Sustainable

Consumption. (World Economic Forum, Geneva, Switzerland, 2013). URL: http://www3.weforum.org/docs/WEF_ENV_SustainableConsumption_Book_2013.pdf (18.16.2017).

Influence of Biologically Active Substances on Key Indicators of the Conditions of Winter Wheat Ecocenosis

Oleksandr Zabolotnyi[1]*, *Larysa* Rozborska[1], *Iryna* Leontiuk[1], *Ivan* Zhilyak[1], and *Anna* Datsenko[1]

[1]Uman National University of Horticulture, Instytutska Street, 1, 20300, Uman, Ukraine

Abstract. The article deals with the study of the effect of the application of the herbicide Granstar Gold 75, w.g. (water-soluble granulate) and plant growth regulator Regoplant on some ecological indicators of ecocenosis of winter wheat sowing (lipid peroxidation reactions in winter wheat plants by the activity of malonic dialdehyde content accumulation, enzymatic activity, total number of rhizosphere bacteria and anatomical structure of winter wheat leaves). The choice of research topic is due to the fact that currently obtaining high yields of winter wheat is closely connected to the widespread use of chemicals, in particular, herbicides, which by their nature are physiologically active substances that can affect both plants and soil microbiota. This, in turn, leads to the search for environmentally safe elements of technologies for growing winter wheat. One such element may be the use of herbicides together with plant growth regulators. The obtained experimental data testify to the protective ability of the growth regulator Regoplant against the winter wheat plants, as evidenced by a decrease in the activity of malonic dialdehyde accumulation, changes in the activity of the enzymatic defense system, decrease the number of epidermis cells and an increase in total rhizosphere bacteria in case of Regoplant use together with Granstar Gold 75, w.g. compared with the experiment variants, where the herbicide was applied without a growth regulator. That is, the use of growth regulator in a mixture with herbicide to some extent eliminates the toxic effect of xenobiotics, which has a positive effect on the state of ecobiosis of winter wheat sowing.

1 Introduction

The current stage of agriculture intensification is associated with the widespread use of various chemical compounds, which, along with increasing crop yields, significantly change the living conditions of both cultivated plants and soil biota. Therefore, research aimed at developing more environmentally friendly technologies for growing major crops are relevant and promising [1].

Also an urgent issue of modern agroecology is the development of scientifically sound criteria for assessing the environmental risks of herbicides in the technology of growing major cereals, in particular, winter wheat. After all, the peculiarity of agriculture in Ukraine is the extremely high variety of agri-environmental conditions, which cause certain environmental restrictions on the permissible range and conditions of application of herbicides.

In the practice of the chemical method of winter wheat crops protection from weeds, a promising measure is the integrated use of herbicides and plant growth stimulants in tank mixtures. This method of using preparations in terms of environmental feasibility contributes to leveling the negative effects of herbicides on winter wheat plants and ecocenoses in general, which in turn has a positive effect on improving the ecological state of the environment [2].

However, the ingress of herbicide active ingredients into the plant organism can cause stress, which contributes to the formation of reactive oxygen intermediates (ROI) such as hydrogen peroxide, singlet oxygen etc., which are produced both by enzymatic oxidation and by redox reactions. Subsequently, the enhanced ROI formation activates free radical oxidative processes [3, 4].

To reduce the level of ROI, which are toxic to the plant organism, is the activation of lipid peroxidation reactions (LP) [5]. Activation of LP is one of the rapid and nonspecific reactions of living organism cells to the stress factors [6].

Also, the LP activation can cause structural-functional and physiological-biochemical disorders in cell membranes [7].

It is established that the process of ROI (reactive oxygen intermediate) formation is natural and also occurs under optimal conditions of plant growth and development, but under the influence of external factors, including xenobiotics, it is greatly enhanced, so to eliminate ROI (hydrogen peroxide, singlet oxygen, etc.) in plant cells activate antioxidant systems protection, including enzymatic ones [8].

Adaptation of the plant organism to stressful conditions due to the toxic effects of xenobiotics occurs with the help of components of the enzymatic defense system, including enzymes of oxidoreductase (catalase,

*Corresponding author: aleks.zabolotnyi@gmail.com

peroxidase and polyphenol oxidase), because they play a significant role in plant protection reactions [9].

In general, the activity of lipid peroxidation (LP) and the activity of the enzymatic defense system components can be the environmental indicators of assessing the degree of xenobiotics influence.

Therefore, the LP activation is considered as a universal response of a living system to the action of stressors, in particular, herbicides. In the body of plants under normal conditions, the intensity of LP is maintained at a steady level, which is provided by a system of antioxidant protection, consisting of antioxidant enzymes and low molecular weight compounds.

Also, when using herbicides, there is a danger of exposure of this type of xecnobiotics on non-target objects, which are soil microorganisms. Herbicide studies have shown that xenobiotics and other biologically active substances can change the number of soil microbiota. In turn, microorganisms, due to their physiological properties, are quite sensitive to various environmental factors and anthropogenic influences and can be used in the diagnosis of environmental assessment of crop growing technologies [10, 11].

Thus, the studies by H.A. Valid Ibrahim [12] have shown that the application of the herbicide Stomp (5.0 l/ha) in sunflower and maize areas caused a decrease in the micromycetes number in the rhizosphere of plants, along with the use of Frontier in the rates of 1.7 and 1.5 l/ha contributed to the growth of their number.

The use of herbicides in winter wheat crops revealed a toxic effect of Lentipur, which inhibited the development of bacteria and actinomycetes in the crop rhizosphere by 1.5 and 1.3 times, respectively, and micromycetes – 1.8 times [13], while on maize areas the herbicide Bazys 75 at the rate of 20 g/ha did not have a pronounced negative effect on the number of major microbiota groups [14].

Other studies have shown that tank mixtures of herbicides with plant growth regulators contribute to its growth, which has a positive effect on the ecocenosis of crops by activating the rate of xenobiotics detoxification [15].

In general, most works covering the effect of xenobiotics on the soil microbiota contain data on both the inhibitory and stimulating effect of the latter on microorganisms and the course of microbiological processes in the soil [16, 17].

That is, today there is no single opinion about the impact of herbicide application on the soil microbiota and its number [18].

Also one of ecological indicators of an organism condition of winter wheat plants can be anatomic and morphological structure of epidermis of its leaves. Anatomical signs of a plant organism during its development under the influence of ecological factors, including anthropogenic character, constantly change. plants, the condition of which can characterize the depth of different factors influence on the plant organism.

The epidermis of plant leaves is a multifunctional tissue, the size and condition of the cells of which are able to characterize the water balance of the plant, the intensity of the processes of carbon dioxide absorption, which in turn has an impact on plant productivity. Histological features of primary integumentary cells depend on many biotic and environmental factors [19].

It is known from the practice of growing crops that the introduction of biologically active substances, including herbicides and plant growth regulators, significantly affects the formation of the structure of the epidermis of plant leaves [20].

Thus, the use of herbicides in crops of spring barley, winter wheat and other increases the number of epidermal cells of leaves by 5–30% and the formation of the number of stomata per unit of the leaf surface area compared to control [21].

It is known that the anatomical and morphological structure of a plant leaf significantly depends on changes in environmental factors, especially anthropogenic, which are herbicides [22].

Therefore, the anatomical and morphological structure of the vegetative organs of plants, in particular the leaves, is evidence of the plasticity of the plant organism [23].

In this regard, the change in the structure of the leaves epidermis is an indicator that characterizes the ecological condition and adaptation of the plant organism to environmental factors [24].

Modern toxicants such as herbicides have a systemic effect, which is manifested immediately by application of the toxicant on the leaves. From the leaf surface through the epithelial and cuticular formations of the active substances of products penetrate into the cells of the columnar and spongy parenchyma of the leaf mesophyll and from there - to other tissues and organs of plants.

Due to the change in the balance of endogenous phytohormones due to the influence of the active substances of herbicides, disturbances in the course of growth processes in the plant organism can occur, causing morphological and anatomical changes in tissues and organs. Growth regulators of plants of exogenous origin also have a significant impact on the anatomical structure of plant tissues and organs, in particular due to the growth of mitotic activity of meristematic tissues [25, 26].

Based on the above, one of the tasks of our research was to study the change in lipid peroxidation reactions, enzymatic activity, the number of rhizosphere microbiota and the anatomical structure of winter wheat leaves as environmental indicators of the herbicide Granstar Gold 75, w.g. and plant growth regulator Regoplant on the ecocenosis of the winter wheat sowing.

2 Material and Methods

Research on this topic was conducted in the field and laboratory conditions of the Department of Biology of Uman National University of Horticulture on the plantings of soft winter wheat Lazurna (originator – Institute of Plant Physiology and Genetics of the National Academy of Sciences of Ukraine) during 2018–2020. The herbicide Granstar Gold 75 v.g. (manufacturer

– American company DuPont, active substance – Tribenuron-methyl, 750 g/kg) in the rates of 15, 20, 25 and 30 g/ha and the growth regulator Regoplant (50 ml/ha) (manufacturer – CJSC «Vysokyi Vrozhai») before the emergence of the flag leaf of the crop.

The soil of the experiment was the low-humic heavy loam podzolized chernozem on loess. Its arable layer contains 3.5 % of humus, 88 and 132 mg/kg of phosphorus and potassium mobile compounds (by Chirikov method) respectively, 103 mg/kg of nitrogen easily hydrolyzed compounds (by Cornfield method), 6.2 of pH, 2.26 smol/kg of soil hydrolytic acidity [27].

The experiment was laid down by a systematic method with sequential placement of variants in four repetitions. A detailed scheme of the experiment is shown in figure 1.

The herbicide and the growth regulator were applied with the DS-3WF-3 battery-powered knapsack sprayer at the rate of 200 l/ha of the working mixture. The rates of herbicide application on the experimental plots were calculated based on the hectare rates of their application, taking into account the area of the plot and the rate of liquid consumption. The total area of one experimental plot was 40 m^2 and the accounting area – 20 m^2.

The intensity of the LP reactions in winter wheat leaves was determined by the method of malonic dialdehyde accumulation in the modification of V.V. Rogozhyn [28]. The activity of catalase, peroxidase and polyphenoloxidase in winter wheat leaves was determined by the methods described by V.V. Rogozhyn [28].

The number of microorganisms in the ecocenosis of winter wheat plantings in the crop flowering stage. The soil samples from winter wheat rhizosphere were taken in accordance with the generally accepted methods [29]. The total number of rhizosphere bacteria are used as a supply source mainly by the organic forms of nitrogen (ammonification group) – by sowing a soil suspension of appropriate dilutions on agar medium MPA (meat-and-peptone agar) [30]. The number of microorganisms was expressed in the colony forming units (CFU) and thousands of cells in 1 g of completely dry soil.

The anatomical structure of winter wheat leaves was studied according to the Hrytsaienko's method [31].

3 Results and Discussion

As indicated above, the introduction of xenobiotics, which are the anthropogenic toxicants, causes the activation of lipid peroxidation reactions in response to increasing levels of ROI.

Studies conducted during 2018–2020 found that the intensity of LP reactions, which was monitored by the level of malonic dialdehyde accumulation in winter wheat plants, was different and depended on the rate of xenobiotic use, both individually and in combination with the growth regulator Regoplant (fig. 1).

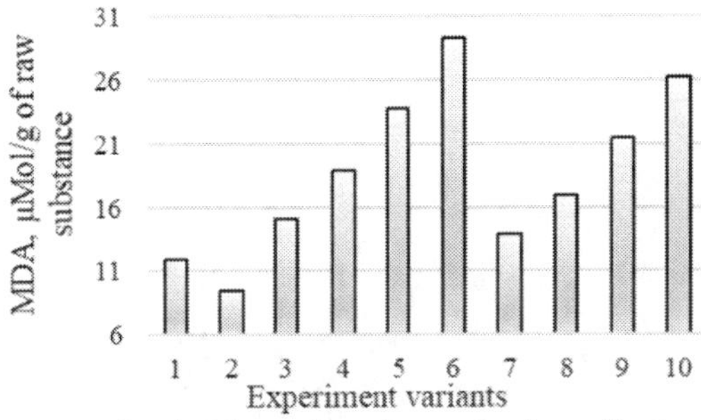

Fig 1. LP reactions in winter wheat plants under the application of the herbicide Granstar Gold 75, w.g. and growth regulator Regoplant (third day after application), the average for 2018–2020 (LSD_{05}=4.3).

1. Without preparation (control); 2. Regoplant 50 ml/ha; 3. Granstar Gold 75, w.g., 15 g/ha; 4. Granstar Gold 75, w.g., 20 g/ha; 5. Granstar Gold 75, w.g., 25 g/ha; 6. Granstar Gold 75, w.g., 30 g/ha; 7. Granstar Gold 75, w.g., 15 g/ha + Regoplant 50 ml/ha; 8. Granstar Gold 75, w.g., 20 g/ha + Regoplant 50 ml/ha; 9. Granstar Gold 75, w.g., 25 g/ha + Regoplant 50 ml/ha; 10. Granstar Gold 75, w.g., 30 g/ha + Regoplant 50 ml/ha.

In particular, on the third day after the application of Granstar Gold 75, w.g. in the rates of 15, 20, 25 and 30 g/ha, the intensity of LP reactions in winter wheat plants increased significantly compared to the control variant and averaged 15.1 over the years of research; 18.9; 23.8 and 29.3 µMol MDA/g of raw substance, respectively.

The increase in the intensity of LP reactions in comparison with the control variant indicates the active production of ROI by the plant body in response to the herbicide action.

In the variant of the experiment with the use of 50 ml/ha of the growth regulator Regoplant without herbicide, the intensity of LP reactions decreased compared to the control variant by 2.5 µMol MDA/g of raw substance.

Under the application of the herbicide Granstar Gold 75, w.g. in the studied rates together with the growth regulator Regoplant there was a slight decrease in the activity of the LP reactions compared to the variants of the experiment with the use of herbicide without a regulator, although this figure exceeded the control variant by 1.9; 5.0; 9.6 and 14.3 µMol MDA/g of raw substance. However, the obtained indicators of LP reactions decreased by 1.4; 2.0; 2.3 and 3.1 µMol MDA/g of raw substance, compared with the herbicide application without Regoplant.

Such indicators of LP reactions obtained with the combined use of drugs may indicate a positive environmental effect of the growth regulator on the winter wheat ecocenosis, because the slowdown in the accumulation of malonic dialdehyde indicates a decrease in the toxic effect of the herbicide on winter wheat plants.

The study of the intensity of LP reactions on the fifth day after application of the herbicide and plant growth regulator showed that it increased compared to the previous period, but there was a similar between the LP reaction and the application rates of Granstar Gold 75, w.g. both separately and in conjunction with Regoplant (fig. 2).

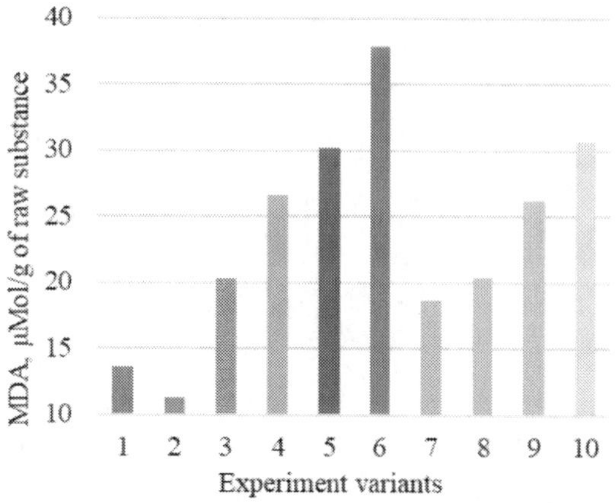

Fig 2. LP reactions in winter wheat plants under the application of the herbicide Granstar Gold 75, w.g. and growth regulator Regoplant (fifth day after application), the average for 2018–2020 (LSD_{05}=5.2).

1. Without preparation (control); 2. Regoplant 50 ml/ha; 3. Granstar Gold 75, w.g., 15 g/ha; 4. Granstar Gold 75, w.g., 20 g/ha; 5. Granstar Gold 75, w.g., 25 g/ha; 6. Granstar Gold 75, w.g., 30 g/ha; 7. Granstar Gold 75, w.g., 15 g/ha + Regoplant 50 ml/ha; 8. Granstar Gold 75, w.g., 20 g/ha + Regoplant 50 ml/ha; 9. Granstar Gold 75, w.g., 25 g/ha + Regoplant 50 ml/ha; 10. Granstar Gold 75, w.g., 30 g/ha + Regoplant 50 ml/ha.

Along with the study of the effect of the herbicide and plant growth regulator on the LP reaction to the malonic dialdehyde accumulation, the change in the activity of key enzymes of the oxidoreductase class (catalase, peroxidase and polyphenoloxidase) was also studied. These components of the enzymatic protection system are part of the formation of plant resistance to stressors, which in our case are herbicides.

It is established that on the average for the years of researches by the application of herbicide in the rates of 15, 20, 25 and 30 g/ha without plant growth regulator catalase activity increased compared to the control version of the experiment by 1.28; 1.35; 1.41 and 1.31 times; peroxidase activity – 1.19; 1.25; 1.28 and 1.22 times and polyphenol oxidase activity – 1.09; 1.15; 1.23 and 1.25 times in accordance with the standards of Granstar Gold 75, w.g. (fig. 3).

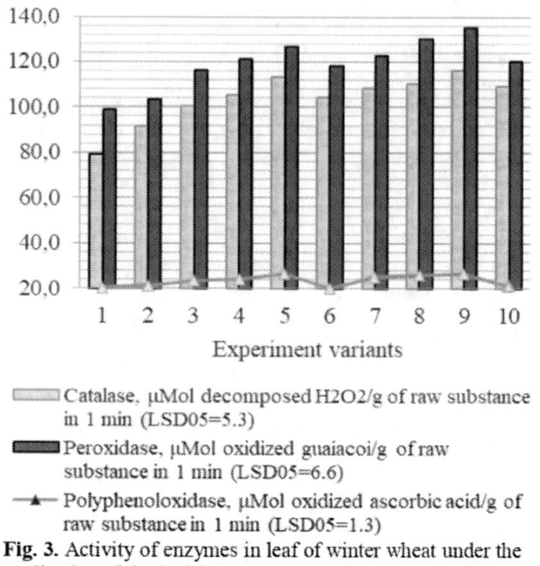

Fig. 3. Activity of enzymes in leaf of winter wheat under the application of the herbicide Granstar Gold 75, w.g. and growth regulator Regoplant (fifth day after application), the average for 2018–2020.

1. Without preparation (control); 2. Regoplant 50 ml/ha; 3. Granstar Gold 75, w.g., 15 g/ha; 4. Granstar Gold 75, w.g., 20 g/ha; 5. Granstar Gold 75, w.g., 25 g/ha; 6. Granstar Gold 75, w.g., 30 g/ha; 7. Granstar Gold 75, w.g., 15 g/ha + Regoplant 50 ml/ha; 8. Granstar Gold 75, w.g., 20 g/ha + Regoplant 50 ml/ha; 9. Granstar Gold 75, w.g., 25 g/ha + Regoplant 50 ml/ha; 10. Granstar Gold 75, w.g., 30 g/ha + Regoplant 50 ml/ha.

With the joint application of Granstar Gold 75, w.g. and Regoplant there is an increase in the activity of the studied enzymes compared with the use of herbicide without plant growth regulator.

In particular, for making 15, 20, 25 and 30 g/ha Granstar Gold 75, w.g. in combination with Regoplant, catalase activity against control increased by 1.37; 1.41; 1.46 and 1.38 times; peroxidase activity in 1.24; 1.32; 1.36 and 1.22 times and polyphenoloxidase activity in 1.18; 1.21; 1.24 and 1.29 times according to the rates of herbicide application.

Study of the soil microbiota of the rhizosphere of winter wheat in the flowering stage revealed that such an indicator of the state of ecocenosis of crops as the total number of rhizosphere bacteria has a certain tendency to change under the action of the herbicide Granstar Gold 75, w.g. introduced both separately and in conjunction with the growth regulator Regoplant (fig. 4).

It was found that in case of 50 g/ha Regoplant application, the total number of rhizosphere bacteria compared to the control version of the experiment on average over the years of research increased by 230 thousand CFU. In case of the Granstar Gold 75 w.g. herbicide application in the rates of 15, 20, 25 and 30 g/ha there was a certain inhibition of the rhizosphere bacteria development compared to the action of Regoplant, but their number exceeded the control variant by 180, 120, 50 and 10 thousand CFU, respectively.

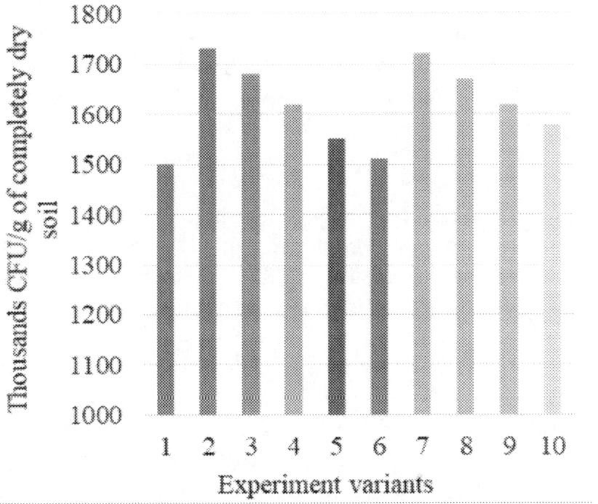

Fig 4. Number of rhizosphere bacteria of winter wheat plants under the application of the herbicide Granstar Gold 75, w.g. and growth regulator Regoplant (flowering stage), the average for 2018–2020 (LSD_{05}=55).

1. Without preparation (control); 2. Regoplant 50 ml/ha; 3. Granstar Gold 75, w.g., 15 g/ha; 4. Granstar Gold 75, w.g., 20 g/ha; 5. Granstar Gold 75, w.g., 25 g/ha; 6. Granstar Gold 75, w.g., 30 g/ha; 7. Granstar Gold 75, w.g., 15 g/ha + Regoplant 50 ml/ha; 8. Granstar Gold 75, w.g., 20 g/ha + Regoplant 50 ml/ha; 9. Granstar Gold 75, w.g., 25 g/ha + Regoplant 50 ml/ha; 10. Granstar Gold 75, w.g., 30 g/ha + Regoplant 50 ml/ha.

Decrease in the total number of rhizosphere bacteria under the action of Granstar Gold 75 w.g. while increasing the rate of xenobiotics use indicates the toxic effect of the chemical on the ecocenosis of winter wheat sowings.

With the joint application of herbicide and plant growth regulator, there was an increase in the total number of rhizosphere bacteria both against control and against the experiment variants with the use of Granstar Gold 75, w.g. without Regoplant, which indicates a positive effect of plant growth regulator on the ecocenosis of winter wheat plantings.

In particular, under the action of 15, 20, 25 and 30 g/ha of herbicide in a mixture with Regoplant, the total number of rhizosphere bacteria compared to the control variant increased by 220, 170, 120 and 80 thousand CFU, respectively. Compared with the use of herbicide without growth regulator, there was an increase in the number of rhizosphere bacteria by 40, 50, 70 and 70 thousand CFU in accordance with the rates of xenobiotics.

It was established that the studied preparations also had an effect on the anatomical structure of winter wheat leaves.

In particular, under the action of 15, 20, 25 and 30 g/ha of Granstar Gold 75, w.g. there is an increase in the number of epidermal cells from 134 to 178 pcs/mm^2. When the herbicide is co-administered with a growth regulator, there is a slight decrease in the number of epidermal cells compared to a single application of Granstar Gold 75, w.g., however, with increasing rate of the preparation in the mixture, the number of cells also to increase from 115 to 161 pcs/mm^2 (fig. 5).

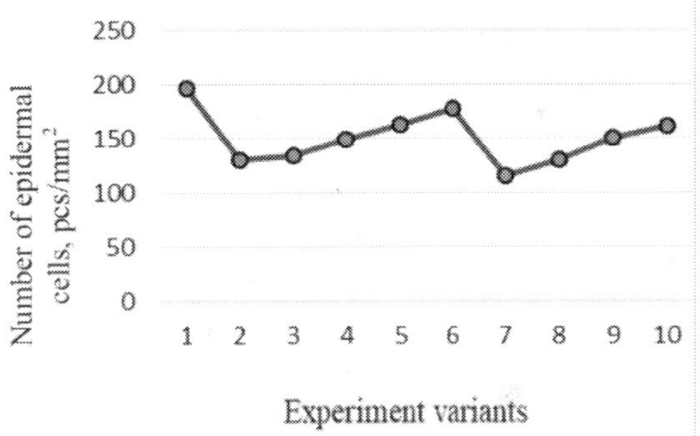

Fig 5. Number of epidermal cells of winter wheat leaves under the application of the herbicide Granstar Gold 75, w.g. and growth regulator Regoplant, the average for 2018–2020 (LSD_{05}=17).

1. Without preparation (control); 2. Regoplant 50 ml/ha; 3. Granstar Gold 75, w.g., 15 g/ha; 4. Granstar Gold 75, w.g., 20 g/ha; 5. Granstar Gold 75, w.g., 25 g/ha; 6. Granstar Gold 75, w.g., 30 g/ha; 7. Granstar Gold 75, w.g., 15 g/ha + Regoplant 50 ml/ha; 8. Granstar Gold 75, w.g., 20 g/ha + Regoplant 50 ml/ha; 9. Granstar Gold 75, w.g., 25 g/ha + Regoplant 50 ml/ha; 10. Granstar Gold 75, w.g., 30 g/ha + Regoplant 50 ml/ha.

The herbicide application without a growth regulator at the rate of 15 g/ha increased the area of one epidermal cell by 6 μm² against the control variant, while under 15 and 20 g/ha of herbicide the cell area exceeded the control by 10 and 18 μm². The application of the maximum rate of 30 g/ha had a pronounced toxic effect on the body of winter wheat plants, which was manifested in a decrease in cell area compared to previous rates, and this figure was almost at the level of the control variant (fig. 6).

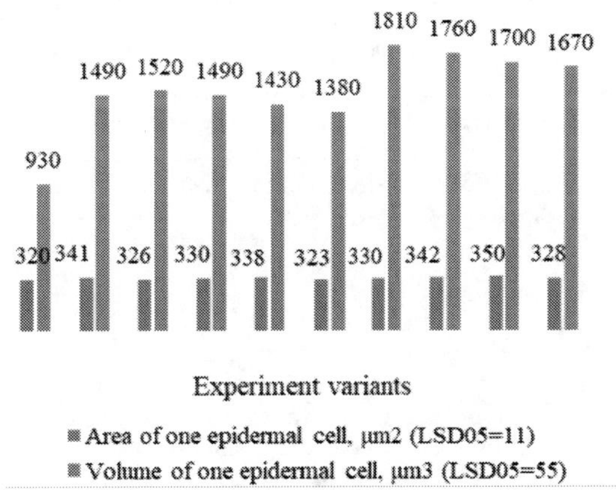

Fig 6. Size of one epidermal cells of winter wheat leaves under the application of the herbicide Granstar Gold 75, w.g. and growth regulator Regoplant, the average for 2018–2020.

1. Without preparation (control); 2. Regoplant 50 ml/ha; 3. Granstar Gold 75, w.g., 15 g/ha; 4. Granstar Gold 75, w.g., 20 g/ha; 5. Granstar Gold 75, w.g., 25 g/ha; 6. Granstar Gold 75, w.g., 30 g/ha; 7. Granstar Gold 75, w.g., 15 g/ha + Regoplant 50 ml/ha; 8. Granstar Gold 75, w.g., 20 g/ha + Regoplant 50 ml/ha; 9. Granstar Gold 75, w.g., 25 g/ha + Regoplant 50

ml/ha; 10. Granstar Gold 75, w.g., 30 g/ha + Regoplant 50 ml/ha.

In case of joint use of Granstar Gold 75, w.g. and Regoplant, the protective effect of the growth regulator on the organism of winter wheat dew was observed, which was manifested in an increase in the area of epidermal cells compared to the variants of the experiment, where the herbicide was applied alone. The largest cell area was observed under the action of 25 g/ha of herbicide in a mixture with Regoplant – 30 μm² more than the control and 12 μm² more than the variant where Granstar Gold 75, w.g. made without a growth regulator.

When determining the volume of one cell, there was a tendency to reduce this indicator in proportion to the increase in the rate of herbicide application. In particular, under the action of 15, 20, 25 and 30 g/ha of product without Regoplant there was a decrease in the volume of one cell from 1520 to 1380 μm³, but these figures exceeded the control value.

With the joint use of Granstar Gold 75, w.g. and Regoplant there was a certain increase in the volume of one cell compared to the variants of the experiment, where the herbicide was used without a growth regulator, but the tendency to decrease the volume of one cell in proportion to the increase in herbicide application in the mixture was maintained. Thus, for making 15, 20, 25 and 30 g/ha of Granstar Gold 75, w.g. with the growth regulator, the volume of one cell varied from 1810 to 1670 μm³ according to the herbicide standards.

The tendency to decrease the number of epidermis cells of winter wheat leaves with a simultaneous increase in their surface, which can be traced in the variants of the experiment with the combined use of toxicant and plant growth regulator, indicates the formation of mesomorphic type of leaf blade. This type of leaf blade begins to form under more favorable growth conditions for the plant, so we can assume that the combined effect of the studied products has a positive effect on the winter wheat ecocenosis as a whole. In contrast, a slightly larger number of cells in the control variat (with a high level of weeding) and an increase in the rate of herbicide application with a decrease in the area and volume of one cell indicates the formation of xenomorphic leaf type, which develops under less favorable living conditions.

4 Conclusion

Therefore, the obtained experimental data indicate that the use of herbicide Granstar Gold 75, w.g. and plant growth regulator Regoplant has a certain influence on such indicators of ecological stability of ecobiosis of winter wheat plantings as the intensity of lipid peroxidation reactions in the plants, the activity of enzymes and the total number of rhizosphere bacteria in the soil.

The application of Granstar Gold 75, w.g. without plant growth regulator is characterized by a certain toxic effect on both winter wheat plants and soil microbiota, which is manifested in increased activity of accumulation of malonic dialdehyde (indicating the activation of the lipid peroxidation reactions), changes in the antioxidant enzymes activity and a certain decrease in the total number of rhizosphere bacteria. The degree of xenobiotic influence on the specified indicators of a condition of ecobiocis of winter wheat plantings has a tendency to strengthening together with the increase of the preparation use rate. So, in particular, on the third day after the application of Granstar Gold 75, w.g. in the rates of 15, 20, 25 and 30 g/ha without Regoplant, there was an increase in the intensity of LP reactions in winter wheat plants against control from 3.2 to 17.4 μMol MDA/g of raw substance, depending on the rate of xenobiotic application. The activity of the enzymatic defense system increased in 1.18–1.46 times, depending on the rate of herbicide and type of enzyme. The total number of rhizosphere bacteria under the action of the herbicide without a growth regulator decreased from 1680 to 1510 thousand CFU by increasing the rate of the drug from 15 to 30 g/ha.

The combined use of preparations indicates the protective effect of the growth regulator on winter wheat plants and soil microbiota, which is manifested in a certain slowdown in the reactions of lipid peroxidation and an increase in the number of rhizosphere bacteria. In particular, on the fifth day after the application, the accumulation of malonic dialdehyde, although exceeding the control variant, but against the variants of the experiment with the use of herbicide without Regoplant decreased by 1.7–7.1 μMol MDA/g of raw substance according to the herbicide rates in a mixture with plant growth regulator. Under the joint application of preparations by 40–70 thousand CFU, the number of rhizosphere bacteria increased compared to the use of herbicide without plant growth regulator. Also under the using of preparationi was noted dependence to decrease the number of epidermis cells of winter wheat leaves with a simultaneous increase in their surface.

References

1. O.O. Tkachuk, Visnyk of Vinnytsia Polytechnical Institute. **3**, 41–44 (2014)
2. L.I. Moklyachuk, A.M. Lishchuk, G.D. Matusevich et al., Balanced nature using. **2,** 131–135 (2015)
3. Chupakhina GN, Maslennikov PV, Skrypnik PN (2011) Natural antioxidants (ecological aspect). BFU Publishing House. I. Kanta, Kaliningrad
4. L. Miteva et al, Reports of the Bulg. Academy of Sciences. **56(3)**, 53–58 (2003)
5. J.G. Scandalios, Trends Biochem. Sci. **27**, 483–486 (2002)
6. N.A. Zherebtsova, T.N. Popova, T.V. Zyablova, Applied Biochemistry and Microbiology. **37**(2),164–169 (2001)
7. G.S. Rossikhina, V.M. Gluboka, Visnyk of Dnipropetrovsk University. Biology, ecology. **15(1)**, 140–144 (2007)
8. G. Rossykhina-Galycha, Bulletin of Lviv University. **62**, 315–324 (2013)

9. A.V. Kartashov, N.L. Radyukina, Yu.V. Ivanov, Plant physiology. **55(4)**, 516–522 (2008)
10. O.S. Demyanyuk, L.Yu. Simochko, O.V. Tertychna, Modern methodological approaches to assessing the Problems of bioindications and ecology. **22(1)**, 55–68 (2017)
11. O.S. Demyanyuk, D.O. Shatsman, Agroecological journal. **3**, 93–99 (2019)
12. Walid Ibrahim Hussein Abu Ahmadih, Dissertation, Kyiv, 2003
13. I.M. Storchous, Plant protection and quarantine. **59**, 277–284 (2012)
14. O.I. Zabolotny, A.V. Zabolotna, Young scientist. **2(05)**, 16–20 (2014)
15. V.P. Karpenko, S.S. Shutko, Tavria Scientific Bulletin. **102**, 46–52 (2018)
16. M. Asad, U. Asad, M. Lavoie et al., Science of The Total Environment. **580**. 1287–1299 (2017)
17. D.B. Nguyen, M.T. Rose, T.J. Rose et al., Soil Biology and Biochemistry. **92**, 50–57 (2016)
18. G. Imfeld, S. Vuilleumier, European Journal of Soil Biology. **49**, 22–30 (2012)
19. S.V. Konstantinova, Dissertation, Moscow, 2007
20. Z.M. Hrytsaienko, V.P. Karpenko, Title of preprint (2021), https://nd.nubip.edu.ua/2011_2/11gzm.pdf. Accessed 01 Feb 2021
21. V.P. Karpenko, Title of preprint (2021), http://lib.udau.edu.ua/handle/123456789/1863. Accessed 01 Feb 2021
22. O.M. Nedukha, T.B. Kotenko, Modern Phytomorphology. **2**, 29–33 (2012)
23. E.I. Kovaleva, A.Yu. Pugacheva, Problems of ecology and nature protection of the technogenic region. **1(9)**, 51–56 (2009)
24. O. Paniuta, R. Palahecha. Bulletin of Taras Shevchenko Kyiv National University. **25–27**, 131–133 (2009)
25. Z.M. Hrytsaienko, Yu.I. Ivasiuk, Bulletin of Uman National University of Horticulture. **2**, 80–84 (2014)
26. V.P. Karpenko, S.S. Shutko, M.G. Gnatyuk, Collected Works of Uman National University of Horticulture. **94**, 264–274 (2018)
27. S.P. Poltoretskyi, Bulletin of Uman NUH. **1**, 59-64 (2017)
28. V.V. Rogozhin, T.V. Rogozhina, *Praktikum on physiology and biochemistry of plants* (Publishing house «Giord», SPb, 2013)
29. V.V. Volkohon (ed), *Experimental soil microbiology* (Agrarna nauka, Kyiv, 2010)
30. N.N. Tereshchenko, E.E. Akimova, O.M. Minaeva, *Modern methods for assessing soil microbiological properties and ecological status* (Publishing House of TSU, Tomsk, 2017)
31. Z.M. Hrytsaienko (ed), (2003). *Methods of biological and agrochemical researches of plants and soils*. (CJSC "NICHLAVA", Kyiv, 2003)

Eco-Controlling: Prospects for the Environmental Policy in Enterprise Management

Hanna Matukova[1], *Valentyna* Khrapkina[2], *Natalia* Bahashova[3]*, *Daria* Matukova-Yaryha[1] and *Alla* Levitskaia[4]

[1]National Medical University named after O.O. Bohomolets, 01601 Kyiv, Ukraine
[2]National University of Kyiv-Mohyla Academy, 04655 Kyiv, Ukraine
[3]State University Economics of Technology, 50005 Kryvyi Rih, Ukraine
[4]Comrat State University, «Innotsentr», MD-3801 Republic of Moldova Comrat

Abstract. The research defines eco-controlling as a new trend of the enterprise environmental policy in the field of environmental protection, its place in the enterprise control system under conditions of permanent changes of the economic environment and environmental challenges. There are outlined tasks of eco-controlling that covers a wide variety of issues associated with ensuring the enterprise's effective activity in the long run considering requirements of environmental protection. The authors substantiate efficiency of introducing eco-controlling as a set of tasks of planning, accounting, controlling, analyzing and auditing environmental aspects of industrial enterprises' activity.

1 Introduction

To avoid consequences of irresponsible, irrational or consciously criminal use of natural resources in various life activity spheres both locally and globally, it is reasonable to concentrate on search, identification, theoretical substantiation and practical introduction of managerial technologies of efficient natural resources use. It is required to consider opportunities, trends, ways and aims of using any natural resources and make thoroughly grounded, economically feasible and materially secure results. In other words, without implementing smart resource management, we will not be able to form economically efficient methods of social natural-production to achieve sustainability by activating cognitive processes [1].

Nowadays, smart management is the foundation of environmental culture that implies understanding, formation and support of such economic relations the evaluative prerogative of which is sustainable attitude of economic entities to the environment with simultaneous obtainment of stable economic advantages. Being aimed at sustainable development, culture of intellectually substantiated production of goods and services required for society will imply either complete refusal to manufacture necessary, yet environmentally unfriendly products (unrenewable elimination of natural resources, contamination of the environment with industrial and household wastes) or application of people's intellectual potential to avoid the need to eliminate or to exhaust natural resources in order to receive economic wealth the total value of which determines the economic boost level, yet fails to restore the resources lost.

Thus, global use of natural resources demands arrangement of global smart management. This arouses a specific interest in the concept of rationalism in natural resource use in which culture of global natural resource use is considered in the context of human socially, environmentally and economically intellectualized activity. Nowadays, illegal waste dumps are made spontaneously without any engineering and environmental assessment of the negative impact level for adjacent territories. During illegal waste storage, the measures for reducing the technogenic load are neglected, there is no control of the morphological composition of wastes transported, this not excluding possible storage of medical, toxic or other hazardous waste types coming. Organic wastes within solid household wastes cause creation of sources of rodent and insect reproduction, which can worsen the epidemiological situation. Solid household wastes decompose at waste dumps due to special combination of chemical, physical and biological processes. This results in solid, liquid and gaseous secondary substances and materials. Thus, organic acids, phenol, aldehydes, ammonia, nitrites, etc. are decomposition products of paper and food wastes. Carbon dioxide, methane, hydrogen sulphide, volatile organic acids are gaseous decomposition products [2].

Decomposition periods for different household wastes are the following:

- paper – from a few months to a few years depending on its quality; it decomposes to organic compounds; chalk overlay and coloured paper decompose longer; decomposition of chalk overlay and coloured paper takes longer:

*Corresponding author: bahashova_nv@kneu.dp.ua

- glass – several thousands of years, it decomposes to the state of sand;
- cans – from 10 to 30 years depending on quality; the cover of iron cans slows down their decomposition; they decompose to iron compounds;
- plastic bottles – about 400 years (they contain urethane, phenol, formaldehyde, styrol);
- leather footwear – from 25 to 40 years;
- batteries – 110 years, they contain heavy metals (zinc, manganese, copper, cadmium, mercury [3].

The authors suggest conducting rapid analysis of the enterprise's state by calculating and then analyzing the following coefficients:

1) financial stability (autonomy, times interest earned (TIE), net working capital, mobility);
2) liquidity (current/fast/absolute liquidity);
3) profitability/return (return on equity (ROE), return on sales (ROS));
4) dynamics of development (dynamics of sales, dynamics of assets, dynamics of net income);
5) turnover (days sales in inventory, time of debit debt turnover, time of credit debt turnover), and
6) conduction of eco-controlling.

The suggested list of indices for analysis can be disputable as to its completeness, yet it should be mentioned that it meets the requirements of clearness and simplicity of assessing the financial state and dynamics of an entity's development and is the most applicable to practice [3, 4].

2 Results

Environmental audit as a mechanism for environmental management was developed in the economically developed countries of the USA, Canada, Great Britain, Germany, etc. - in the 70s. Environmental audit began to develop as a branch of entrepreneurial activity in the environmental industry [5].

An innovation in the domestic legal system was the introduction of the so-called mandatory environmental audit (along with the traditional form of voluntary environmental audit). which should be exposed to objects that pose an increased environmental hazard, by order of the concerned executive authorities or local government bodies in relation to objects or activities that pose an increased environmental hazard (in accordance with the list approved by the Cabinet of Ministers of Ukraine), in cases provided for by law.

The Law "On Environmental Audit" has opened a new way to ensure environmental safety in Ukraine. Environmental auditing is designed to put a barrier to neglect of environmental requirements in the process of privatization, long-term lease or concession of objects that pose an increased environmental hazard. In certain cases of environmental auditing, accounting for the ecological state of an object, it is transferred or acquired into state or communal ownership, it can play a decisive role in the formation of its price. The application of the new legislation on environmental audit will contribute to the creation and implementation of environmental management systems in various sectors of the Ukrainian economy, will provide greater scientific justification for environmental insurance of activities and objects hazardous to the environment.

Environmental audit is understood as the entrepreneurial activity of environmental auditors or audit organizations to carry out independent non-departmental qualified analysis and assessment of economic activities, have an impact on the environment, and develop recommendations to reduce the negative impact on the environment and public health. The need to apply environmental audit procedures is determined by the reasons for the insufficient budgetary financing of environmental protection activities, which negatively affected the state of the environment, as well as the possibility of increasing competitiveness in the world market. Environmental audit should be performed in a certain sequence, it provides for three stages: preparation of environmental control at the enterprise; research and information gathering; completion of verification and development of recommendations. The stages are detailed with specific audit engagements and the performance of certain procedures [5].

So, at the first, preparatory, stage, the auditor must determine the goals and objectives of the environmental audit, a list of issues that need to be resolved; conduct a general familiarization with the environmental situation of the enterprise, its technical condition, the availability of the necessary licenses and permits; get acquainted with the environmental policy of the enterprise, carry out a survey of the management on environmental issues; examine internal documentation on the issues under study (orders, protocols, technical documentation, instructions, reports, management information and accounting data). The next step is to conclude an agreement for the implementation of an environmental audit and develop an audit plan and program indicating specific control procedures, deadlines and performers, and location. At the second stage, close contact of the auditor with the staff is established to provide and prepare the necessary information, inspect the territory and premises, observe production processes, conduct the necessary tests, analyzes, examinations, measurements, and the like. Areas of potential environmental risk are identified, the efficiency and economy of the enterprise's system of measures to reduce the environmental risk are assessed. At this stage, an assessment of the enterprise's environmental management system is carried out (a detailed study of the environmental policy, the qualifications of personnel, the formation and use of information on environmental issues, the delineation of personnel duties and responsibilities). At the third stage, the collected information is summarized, the results are summarized, the preliminary results are discussed with the management of the enterprise, the final report and conclusions are drawn up, recommendations are developed and the report is submitted to the audit client. Environmental audit of the agrarian sector is an organizational and economic mechanism of independent (internal and external) control and controlling of production, environmental and environmental activities, regardless of the forms of management [5]. So eco-audit is an organizational and economic mechanism that restrains the impact of production on the environment, my elements of planning (programming), motivation, organization and regulation; focused on assessing the compliance of environmental activities with existing environmental legislation, and also on the relationship of environmental activities with the final financial and economic results of production.

Social consequences of the ecological crisis in Ukraine: Life expectancy in Ukraine is on average about 71 years (in Sweden - 80, in Poland - 74). To a large extent, this is due to environmental pollution due to the production activities of enterprises of the mining, metallurgical, chemical industries and the fuel and energy complex. Medical and genetic studies have established that in connection with long-term environmental pollution in the population of the nation, the number of hereditary genetic anomalies is growing. As you know, if this figure reaches 30%, then, according to biological laws, the nation begins to disappear.

When the pollution of the atmosphere is 1.2–1.5 times higher than the sanitary and hygienic standards, diseases of the immune system begin. Today in Ukraine, there are about 1,700 dangerous sources of atmospheric pollution, of which 1,000 are especially dangerous chemical enterprises. The development of agglomeration processes is accompanied by an increase in the technogenic impact on the living environment. In cities, it manifests itself in various forms - changes in the microclimate, pollution of atmospheric air, water and soil, physical (electromagnetic, acoustic, dust) pollution, the formation of a significant amount of not only household, but also industrial waste hazardous to humans.

The Law of Ukraine "On Environmental Audit" provides for two types of environmental audit: voluntary and mandatory.

Voluntary – carried out on the initiative of the management of the audit object or in agreement with it, if the Audit Customer is a third interested party (potential buyer, financial institution, etc.).

Mandatory – carried out by order of interested government agencies for facilities or activities that pose an increased environmental hazard.

Mandatory environmental audit in Ukraine is carried out in the following cases:

- bankruptcy;
- privatization;
- transfer to concession of objects of state and municipal property;
- transfer or acquisition into state or communal ownership;
- transfer to long-term lease of objects of state or municipal property;
- creation of joint ventures on the basis of objects of state and municipal property;
- environmental insurance of objects;
- termination of the production sharing agreement in accordance with the law;
- in other cases provided by law.

It is advisable to carry out an environmental audit when you need:

- determine the compliance of the enterprise with the requirements of environmental legislation;
- check whether environmental payments or penalties of the environmental inspection are not overstated;
- to avoid excessive environmental payments and penalties;
- to reduce the costs of waste management, water supply, electricity, heat, gas supply;
- find out the availability of reserves of raw materials and energy resources at the enterprise;
- to reduce the risk of emergencies related to environmental pollution;
- to increase the investment attractiveness of the enterprise;
- to form the environmental strategy and policy of the enterprise;
- increase the competitiveness of the enterprise and increase customer loyalty through the production of products that meet all environmental, sanitary, and hygienic requirements;
- go international and strengthen the company's position among foreign partners;
- to certify the company's environmental management system for compliance with ISO 14001.

Environmental audit in Ukraine is carried out in order to ensure compliance with the legislation in the field of environmental protection in the course of economic and other activities.

An environmental audit is carried out according to the requirements of state bodies (in cases established by law) or at the initiative of an enterprise, if it is necessary to modernize production and/or technological processes, in order to obtain sound recommendations for rational use of natural resources and environmental protection, and increase efficiency to a level that meets the requirements of legislation.

Eco-controlling is a system of planning, accounting, controlling, analyzing and auditing, which facilitates alternative decision-making in operating and strategic management to achieve enterprises' environmental and economic efficiency [5].

Introduction of eco-controlling as a system of informational support of controlling enterprises' environmental activity will enable: planning, standardizing and calculating environmental expenses, controlling and analyzing data on an enterprise's natural resource use; reporting production data to support making effective, environmentally sustainable managerial decisions; defining aims and measures to reduce loads on the environment; creating conditions to compare operating, investment, financial and environmental activity, their assessment and encouragement; studying and analyzing financial consequences of the environmental activity and determining impacts of these expenses on an enterprise's profitability and product prices; facilitating solution of the problem of defining economic efficiency of environmental measures and assessment of economic losses caused by pollution assessing the importance of healthcare in an enterprise's activity, etc., defining places of emerging expenditures that reflect environmentally significant input of raw materials, materials and energy or output of raw material remainders.

In terms of eco-controlling, it is expedient to refer distribution of places of expenditure emergence (control over quality of products and services) [6] to the aims of environment protection only partially, the same referring to other expenditures not involving environment protection (e.g. trucks).

It is evident that in places of emergence of expenditures designated for environmental protection only, all the expenditures can be treated as environmental ones. In other places, environmental expenditures should be specified through corresponding expenditure types. Assessment of environmental impacts by making eco-balances is an important precondition of introducing efficient systems of eco-controlling at industrial enterprises. Assessment of technogenic load on the environment is a new methodological tool for realizing environmental aims and defining ways of their fulfillment. These standards provide for assessing the whole life-cycle of products, in other words, of all the stages starting from mining raw materials to utilization of products or its secondary inclusion into the material cycle. The assessment conducted in the context of eco-balance of products is an efficient tool of producers' and consumers' decision-making [6].

The enterprise's eco-balance can be viewed as a scheme of impacts on the environment (Fig. 1).

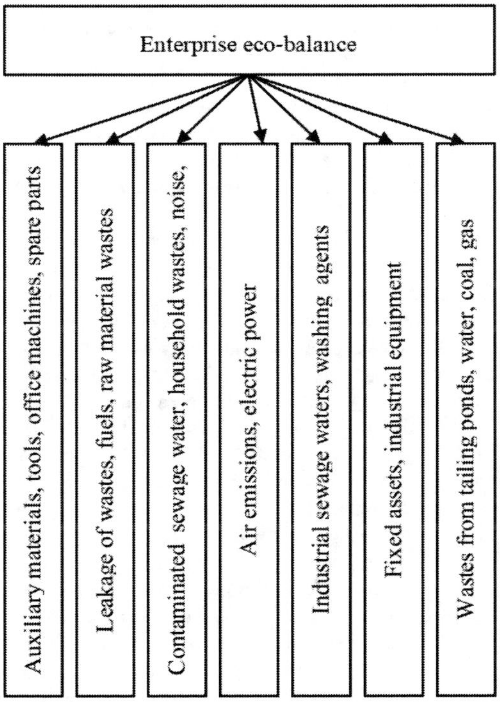

Fig. 1. The scheme of the enterprise's eco-balance.

Eco-balance is aimed at specifying the level of economic and environmental responsibility for its own style and method of economic activity. Eco-balance is grounded on voluntary, systematic, recurrent and objective control over impacts on the environment [7]. Eco-balance is a detailed reflection of environmentally significant resources spent on product manufacturing and resulted wastes as well as their impacts on the environment during a certain time period. Development of eco-balances enables the enterprise to identify environmental aspects and impacts on the environment caused by products during their life-cycle, revealing environmentally narrow spots in the industrial process; informing the community, personnel and consumers about products' impacts on the environment; applying eco-balance data to designing products. To receive a more complete reflection of various aspects of the enterprise's impacts on the environment, there are suggested eco-balances of raw materials, products, production, the enterprise and its location [8].

The enterprise's eco-balance contains information on the input and output of all the materials and energy including material and energy losses as well as the state of real estate, equipment and materials. With that, internal industrial processes are treated as 'a black box' and only flows of raw materials, basic and additional materials, end products and those based on a single production process are considered. Besides, impacts on the environment resulted from production as return air, heat, wastes and sewage waters are taken into account. Eco-controlling is noted for various managerial functions (normative, strategic, operating levels) aimed at integrating operating and strategic aims into concrete plans and actions in such activity areas as development of personnel, design of products, sales, supply, production, quality provision, logistics, distribution, marketing, logistics of products return and re-treatment) [9].

The mechanism of eco-controlling functioning can be presented as a combination of three components – analysis, control and regulation of processes aimed at reducing the enterprise's impact on the environment (Fig. 2).

Eco-controlling can be of two types – operating and strategic. In case of operating eco-controlling, the data on development in the limited period are processed and information on events or measures negatively impacting the environment is provided. This type of controlling is aimed at improving the environment in the short-term period. In the strategic eco-controlling, actual indices are compared with the scheduled ones and the potential of improving the results and risks of increasing expenses or deterioration of activity results is determined due to timely data on development trends [10]. Controlling is treated as a modern technology and the system of intellectualizing business processes, which, being managers' philosophy and a controlling tool of management systems, involves all the managerial, executive and supplying business-processes within a certain enterprise.

This enables forming a specific technology of total process controlling, which, unlike the current ones, enables including all the business-processes of the given enterprise [11].

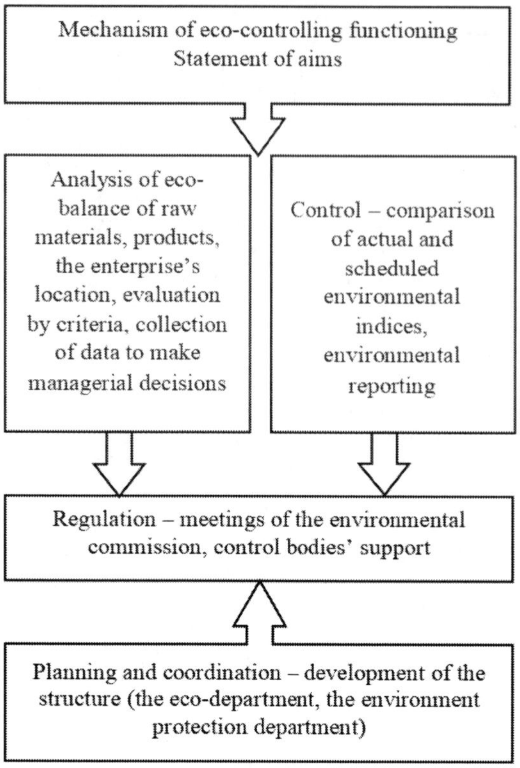

Fig. 2. The mechanism of eco-controlling functioning.

Industrial wastes are environmental pollutants. In Ukraine, the main source of waste generation is mining, chemical and metallurgical, machine building, fuel and energy, complexes.

Waste containing heavy metals, oil products, pesticides (pesticides) unsuitable for use, the bulk of which is generated in Donetsk and Dnipropetrovs'k regions.

There are almost 20,000 hectares of land under the toxic waste storage facilities. According to statistics, approximately 80% of all harmful emissions into the air are the result of energy processes (extraction, processing and use of energy resources).

Environmentalists analyzed the published non-financial reporting for the period 2015-2019. For the study, enterprises were selected, in particular, OJSC "EVRAZ Dneprovsky Metallurgical Plant", Dnepr, "Arcelor Mittal Kryvyi Rih". When analyzing the non-financial reporting of selected enterprises, it was found that there is a total advantage of using a passive communication strategy with the relative weakness of external stakeholders.

Non-financial reporting is a relatively new corporate practice in Ukraine; therefore, a unified approach to the definition of this concept has not yet been developed. In 2017-2019 At PJSC "Arcelor Mittal Kryvyi Rih" there is an increase in production volumes by 15.12% and 9%, respectively.

During this period, there has been a significant increase in the volume of discharges into the atmosphere and water resources by 6% and 8%, respectively. The enterprise is modernizing sinter plant - 2 (implementation period – 2015-2021), major overhaul of the 1st category with reconstruction of blast furnace - 9 (implementation period – 2016-2021), which will significantly reduce the use of expensive natural gas and coke used in smelting pig iron, improve energy efficiency and improve the environmental situation in the region.

The process of recycling waste in its own production needs to be optimized, which will increase the cost efficiency of waste management.

The increase in expenses for the protection of atmospheric air is also associated with the introduction of an integrated air monitoring system in the area of production impact.

To optimize environmental losses, enterprises are recommended to introduce: accelerated depreciation of fixed assets of environmental importance (treatment facilities, environmental quality monitoring systems); material and non-material incentives for personnel in the development of engineering solutions and rationalization proposals for the greening of production activities and the system of environmental management, energy efficiency and optimization of waste management; adherence to ISO and EMAS standards in the field of environmental protection, control over the effectiveness of the environmental responsibility system at enterprises.

It should be emphasized that in addition to optimizing the environmental strategy of the enterprise, these measures will create a positive image of the company, increase the competitiveness of products due to "green" advantages, which is especially important in European sales markets. At PJSC "Arcelor Mittal Kryvyi Rih", the indicator of the ecological loss rate of products is satisfactory, although its growth on average by 0.004% per year is mainly due to inflationary processes, which are reflected in the growth of income from sales of manufactured products.

An increase in the share of recovered environmental costs is observed at the enterprise, which negatively affects the financial results of the enterprise, because the unrecovered environmental costs are compensated from profit, which requires strengthening control in the field of tax planning and environmental monitoring. Because of strengthening control in this area, during 2016-2018. The enterprise did not pay fines or penalties on profit. The enterprise is characterized by an increase in the volume of environmental costs, on average, the increase occurs by 26% per year.

Regarding PJSC "EVRAZ DMZ", it is worth noting that today the enterprise is the only metallurgical plant in Ukraine, where, without the injection of pulverized coal, natural gas was removed using blast furnace technology. This made it possible to significantly reduce the energy consumption of production without significant impact on its volumes. During the study period (2015-2019), the enterprise has seen a steady increase in the production of marketable products. This trend also entails an increase in emissions into the environment.

The average growth rate of production volumes for the year was 5.18%, which in absolute terms as of 2018

https://doi.org/10.1051/shsconf/202110005011

amounted to 19,670,000 tons. The volume of annual emissions into the atmosphere on average for the year increased by 4.84%, the absolute value of 1% of the increase was 9,400,000 tons (as of 2018). The trend towards a decrease in the total annual volume of discharges into water bodies is positive. On average, discharges decreased by 6.57% over the year.

The enterprise is a water user of the Dnepr River and remains one of the main pollutants, which requires additional environmental protection measures. In the field of waste management, there is an annual increase in waste generation by an average of 1.43%, which is a natural trend with an increase in production volumes. The annual volume of used waste is increasing by an average of 7.2% per year.

This effect was achieved by increasing the use of waste in our own production.

Despite the increase in production volumes, on average, the coefficient of waste intensity in the category "air emissions" remains at a constant level. This is undoubtedly a positive trend, indicating the high efficiency of the measures introduced at the enterprise to protect the atmospheric air.

Thus, the coefficient of waste intensity in the category of “discharges into water bodies” in 2017 compared to 2015 increased by 4.86%, and on average, the annual growth was 1.09%. In the category of waste use, it should be noted that the coefficient increased annually by an average of 5.1%, which is due to the use of waste for construction projects of the enterprise.

Enterprises are participants in regional and local environmental programs, such as "Long-term program to solve environmental problems of Kryvbas and improve the state of the environment for 2011-2022", "Dnipropetrovs’k regional comprehensive program (strategy) of environmental safety and prevention of climate change for 2016-2025 years".

A high level of atmospheric air pollution is characteristic of the specifics of the activities of metallurgical enterprises in old industrial regions, which characterize the Dnipropetrovs’k region; the studied enterprises have certificates of the environmental management system and regularly undergo appropriate audits. Nevertheless, it should be emphasized the low level of assessments of the effectiveness of environmental management at "EVRAZ DMZ". Particular attention should be paid to issues of improving the effectiveness of measures in such an assessment category as the preparation and publication of non-financial reporting.

At the same time, the non-financial reporting of Arcelor Mittal Kryvyi Rih is prepared in accordance with the GRI system and is annually celebrated at the All-Ukrainian competitions for corporate social responsibility. As a result of the analysis of the assessment of the levels of significance of the proposed categories, the following conclusions can be drawn.

The experts identified the following as the most significant assessment categories that have the greatest impact on the level of the enterprise's environmental initiative: activities to compensate for general environmental damage, protection of atmospheric air and water bodies, the users of which are enterprises, the effectiveness of environmental management and environmental audit.

Among the recommendations for optimizing environmental protection costs, it is worth highlighting the following aspects: improving the non-financial reporting system, strengthening control in this area; conducting additional internal environmental audits to establish compliance with international environmental standards, carrying out a comprehensive assessment of the level of environmental responsibility of the enterprise with the subsequent development of recommendations to improve its efficiency.

The amount of compensation for losses for the excess emission of one ton of a pollutant into the air is calculated on the basis of the size of the minimum wage established on the date of detection of the violation, multiplied by a factor of 1.1, taking into account the regulatory factors and the relative hazard indicator of each pollutant substances. The amount of losses is calculated using the formula (1) [12]:

$$Ç = m_i \cdot 1{,}1\ddot{I} \cdot A_i \cdot K_t \cdot K_{ç^з} \quad (1),$$

where $З$ – amount of losses, UAH;
m_i – the mass of the excess emission of the i-th pollutant, t;
$1.1П$ – the size of the minimum wage (P) as of the date of detection of the violation for one ton of the conditional pollutant multiplied by the coefficient (1.1), UAH / t;
A_i - dimensionless indicator of the relative hazard of the i-th pollutant;
K_i – coefficient taking into account the territorial socio-ecological characteristics;
$K_{зi}$ – a coefficient depending on the level of atmospheric air pollution in the settlement with the i-th pollutant.

The total amount of compensation for damages is calculated as the sum of the amount of losses for the excess emission of each pollutant into the atmospheric air.

3 Conclusions

Under current economic conditions, the eco-controlling system is of particular importance as it is aimed at controlling current analysis of scheduled and actual indices in order to exclude errors and deviations of both the present-day and the future.

At the initial stage of its implementation, eco-controlling reveals itself as a system of informing of negative deviations in the enterprise’s activity for taking timely preventive measures.

In the enterprise’s activity, eco-controlling is an integral concept of enterprise management aimed at revealing chances and risks of receiving revenues.

In the latest years, enterprise managers have to face the problem of searching for and applying management methods, which are efficient and useful.

Within the system of anti-crisis management, eco-controlling ensures implementation of the enterprise's strategic potential and achievement of aims of liquidity and profitability. In our opinion, in the nearest future, eco-controlling should become a priority in the enterprise's management system.

The role of environmental audit in the environmental monitoring system is as follows:

- environmental audit is an element of these systems and the requirements for the presence of such an element are set by the DSTU ISO 14001 standard. So, environmental audit should be considered as an integral part of environmental management systems.

- environmental audit is a tool for the formation of the environmental policy of the object or territory subject to audit.

Of course, the scope and role of environmental audit is very wide and multifaceted. An environmental audit is considered as:

1) organizational and managerial tool for ensuring the environmental safety of the facility;

2) as an instrument of state ensuring the environmental safety of objects and territories (if it is considered on a national scale, when the "state system of environmental audit" has been introduced and operates;

3) as a tool to ensure the prevention and limitation of the consequences of accidents at the inspected facility (if we are talking about an environmental audit in order to determine environmental risks);

4) as an element of the environmental insurance system (if we are talking about an insurance environmental audit in order to determine the likelihood of emergencies and possible consequences. In this case, the audit results will affect the "economy" of the environmental insurance process, that is, the amount of insured amounts and tariffs);

5) as a new direction of activity in the market of environmental services (we are talking about the fact that a new direction for entrepreneurship appears on the market of environmental services, which is also very important in the context of state priorities for the development of small and medium-sized businesses).

We are quite optimistic about the prospects for introducing international standards in Ukraine. Measures taken at the state level in combination with market trends, raising the level of consciousness and business culture of managers will lead to the fact that Ukrainian enterprises will widely apply the principles of environmental controlling and auditing, which are a very important component that ensures the sustainable development of not only enterprises, but also ultimately the whole society.

Eco-audit should become an effective means of regulating the relationship between production and the environment. Based on the results of the audit, plans for agricultural and environmental activities are developed for both individual agricultural enterprises and for regions, the amount of subsidies, the legality of price increases and their value, as well as sanctions for offenders are determined.

References

1. I.S. Marekha, Resource saving as a form of rationalization of nature use in the conditions of modern economic relations, Global and national problems of economy, 2 (2014) http://www.global-national.in.ua/archive/2-2014/187.pdf
2. Some facts about garbage http://climbinduk.org/node/256
3. K. Walsh, Key financial indicators. Analysis and management of enterprise development (2001)
4. T. Lembert, Key problems of the head. 50 proven ways to solve problems: a guide (2001)
5. T.A. Kirsanova, Environmental controlling is an environmental management tool (2004)
6. N.V. Pakhomova, Environmental management (2003)
7. U. Stoltenberg BetrieblichesOekocontrolling: Leitfadenfuerdie Praxis (1996)
8. H. Matukova, (ed), Innovative Forms of Management for Companies Under Rehabilitation Advances in Economics, Business and Management Research III International Scientific Congress Society of Ambient Intelligence 2020 (ISC-SAI 2020) Advances in Economics, Business and Management Research **129** (2020) DOI: https://dx.doi.org/10.2991/aebmr.k.200318.013
9. L.I. Maximov, Organizational and methodological aspects of operational eco-controlling (2005)
10. I.A. Markina, Controlling for managers: textbook. way (2013)
11. E. Shvydky, Controlling as a technology of intellectualization of business process management in enterprises: (financial-accounting-analytical aspects) (2012)
12. Order of the Ministry of energy and environmental protection of Ukraine No. **277** (2020) April, 28 "On approval of the Methodology for calculating the amount of compensation for losses caused to the state as a result of excess emissions of pollutants into the atmospheric air" http://search.ligazakon.ua/l_doc2.nsf/link1/RE34697.html

The Intellectual Component of Labor Potential as a Factor of Sustainable Development of the Agricultural Sector

Olena Vasyl'yeva[1*] and *Andrii* Karpenko[1]

[1]National University "Zaporizhzhia Polytechnic", 69063 Zaporizhzhia, Ukraine

Abstract. In the article, the proposition is substantiated that the current development of the agricultural sector is possible on condition of harmonization of environmental, social and economic factors based on the principles of consistency, balance and social justice. The paper examines the impact of the intellectual component of labor potential on economic growth in the agricultural sector. The study of the main types of production functions for macroeconomic analysis is conducted. The advantages of using the Cobb-Douglas production function in determining the levers of agricultural production growth in the context of sustainable development are substantiated. The present-day priority of activities intellectualization in the agricultural sector is proved. It is proposed to particularly focus on the use of intellectual assets of human potential for agricultural production intensification. The accomplished modeling of the agricultural sector sustainable development gives grounds for an optimistic forecast as to increasing the resource potential of agricultural production due to a growth in the quantitative and qualitative indicators of labor potential.

1 Introduction

The main purpose of sustainable development is to provide the population of the world with food, therefore, the problem of sustainable development for the agricultural sector is utterly relevant and developed at the sectoral level. We believe that it is the agricultural sector that is concerned with environmental and food issues most, since it is a complex system of agricultural production, environment and rural population.

Not only does the sustainable development of the agricultural sector in Ukraine ensure food security, which is a constituent of the state's economic security, but it also provides for the development of the economy as a whole. A necessary condition for sustainable development is positive dynamics of the economic potential of the agricultural sector, which is a set of all available means, opportunities, productive forces, resources, reserves and competencies that can be used in production activities and may utilize market opportunities in order to achieve the goals of society's socio-economic development.

The content of sustainable development can be determined through formation of appropriate potentials of its components:

– the economic potential of sustainable development (a system of industrial relations that uses environmentally and socially dangerous means of production, safe resource-saving technologies; organization of production aimed at coordinating the economic, social and environmental outputs);

– the environmental potential ("the ability of an ecosystem to continuously restore itself in those qualitative parameters that meet the requirements of social and economic development", whose basis is ensuring the conditions for restoration of natural resources, economical use of non-renewable resources and preservation of biological diversity);

– the social potential of sustainable development (human potential, human and intellectual capital, social consciousness, social infrastructure, social policy). A special role is assigned to accumulation of human capital within the framework of the theory of endogenous growth by improving the levels of education and science, creating and using new technologies and innovations, and developing corporate culture. In modern conditions, it is an effective corporate culture that not only performs the function of favorable social environment formation, but also attracts and maintains talents, develops human capital [1]. Sustainable economic growth is based on knowledge, creativity (formulation of new ideas and their use to create cultural products and works of art, a capability of creating new technologies and making inventions) and access to information as components of the social potential of sustainable development [2].

We believe that the determining factor for achieving the goals of the agricultural sector's sustainable development is a system approach, according to which the components of sustainable development should develop in a comprehensive and coordinated manner, ensuring the efficient, secure production functioning, preservation and restoration of natural resources, quantitative and qualitative reproduction of labor potential.

In our opinion, the modern paradigm of the concept of the agricultural sector sustainable development fully corresponds to the new economic paradigm "The Economics of Happiness" [3], which grounds on the state of well-being and environment, ensuring of

* Corresponding author: olena.vasilyeva@gmail.com

material and spiritual satisfaction with life, provided that there is an optimal balance between the needs of society and limited resources. The production of high-quality food products within the reasonable use of natural resources is a necessary, but not sufficient, condition for improvement in the quality of life, since non-economic aspects of life determine its duration and quality as well as mental well-being of citizens. A human-centered economics of happiness changes the priorities of social existence and expands the interaction of the sustainable development components. It is the socio-economic model that comes to the fore and determines sustainable economic development: the alternative to the gross domestic product is the Human Development Index, which measures the people's capability of living a long, healthy, creative life. This model further develops the ideas of the Lukas and Romer [4-6] endogenous growth models, according to which the main factor of economic growth is an increase in investments in human capital that ensures the scientific and technological progress.

Informed management decision-making in the context of sustainable development requires use of modern methodological tools for obtaining the results, which are difficult to obtain by applying traditional methods. We assume that in order to analyze the levers for economic growth of the agricultural sector in the context of sustainable development, the most successful model to be applied is the Robert Solow neoclassical growth model with a stable path of balanced development, according to which it is the efficient use of resources and technological progress that are the imperatives of economic growth [7]. The product created in the economy is the result of the following production factors' interrelationship: labor *(L)*, capital *(K)*, land resources and technological progress factors *(N)*, whose impact on the volume of the total output can be described with the use of production function as one of the ways to forecast the development of complex economic systems.

2 Results and discussion

2.1 Sustainable development factors of the agricultural sector

Agriculture as a core element of the agrarian sector, which affects overcoming the global problem of hunger as well as maintaining food and environmental security of humanity, has the features that consist "in the use of production means of natural origin: plant and animal organisms, water resources alongside with climate-related resources – light, heat, precipitation, etc." [8]. The use of specific means of agricultural production, which are related to the environment and its natural laws in a complimentary way, limits the opportunities for economic growth. The ambiguous impact of external conditions on agricultural production can be compensated by internal factors, which include use of advanced technologies, availability of qualified personnel, efficiency of personnel use and their work motivation.

In our opinion, the manifestation of the components of the triune concept of sustainable development – economic, environmental and social, can be represented by the unity of three types of capital: physical (production), natural and human, which is formed, accumulated and realized in the agricultural production, agrarian landscape through labor potential.

The economic component of sustainable development of the agricultural sector provides for a constant increase in the volume of agricultural production and improvement of its quality, strengthening of the production potential, creation of a rational production structure, maintaining of an appropriate level of financial stability, increase in the labor productivity, establishment of an appropriate market infrastructure, improvement of the model for investment support and tax policy.

The environmental component of sustainability should ensure the integrity and restoration of natural capital, compliance with the agricultural and environmental production standards, ability of the ecosystem to self-regenerate, optimization of the agrarian landscape structure, improvement of the acreage and crop rotation structure, increase in soil fertility, reduction of the environmentally destructive impact of agricultural production on the agrarian landscape and environmental pollution, prevention of the negative impact of degraded land on the health and well-being of the population.

The social component is focused on the development of human capital, maintenance of the number of employed population and reduction of unemployment in agricultural production, stabilization of the demographic situation, improvement in the level and quality of life, development of the social infrastructure, rural areas and communities, improvements in the administrative and territorial structure, accessibility of healthcare in rural areas, increase in incomes of rural population.

Sustainable development of the agricultural sector is possible on condition that environmental, social and economic factors are harmonized on the basis of consistency, balance and social justice. The transition to sustainable development, to a new paradigm for society development, makes relevant the issue of rational use of all types of resources in the process of product creation in the economy. Agricultural produce occupy a significant share in the structure of generating gross value added, which gives the grounds to state about the prospects for increasing the sales volume of exported agricultural produce, provided that the interrelationship of production factors is optimal [9].

2.2 The main types of production functions in macroeconomic analysis

The impact of labor potential on the quantity of output goods can be described by the following production function equation:

$$Y = f(K, L, N) \quad (1)$$

The components in this equation (factors of economic growth) are represented in one form or another by: the total amount of production assets used K (the area of farmland, number of fixed and production assets in value form), the total cost of labor L (the number of workers employed in agriculture, cost of working time) and other predictors that take account of the technological progress N. In this form, the production function only characterizes a quantitative impact of one or more production factors on the output volume whereas a growth in production volumes due to extended use of all types of resources (extensive economic growth) is very limited. In the modern economic theory, the main role among the production factors belongs to such factors as labor, land, capital and entrepreneurial capabilities.

The production function can be represented in various forms, from a linear dependence on a single production factor to the complex systems described by recurrent equations that relate the system state at different time intervals.

The main types of production functions used in theoretical and applied macroeconomic analyses are as follows:

1) the linear function: $Y = a_1x_1 + a_2x_2 + ... + a_nx_n$. This model for estimating the parameters $a_1, a_2, ..a_n$ characterizes the marginal increase in the final value depending on the marginal increase in each resource per unit and in the units, in which they are presented in the input data (with unlimited elasticity of resource substitution);

2) the Cobb-Douglas aggregated function: $Y = a_0K^{a_1}L^{a_2}..X^{a_n}$. The model for estimating parameters $a_1, a_2, ..a_n$ characterizes the elasticity of production by resources, that is, the quantitative relationship between production volumes as provided by resources in relative (percentage) measurement (elasticity of resource substitution equals 1);

3) the Leontief function $Y = A\min(\frac{K}{K_0};\frac{L}{L_0})$ represents a production function with constant ratios of production factors (with zero elasticity of resource substitution);

4) the CES function: $Y = A(\sum_{i=1}^{m} a_ix_i^{-b})^{-h/b}$ represents a production function with constant resource substitution elasticity.

The most common among the production functions is the Cobb-Douglas production function, which models the relationship between the output (or another final indicator) and volumes of production factors as the product of the volumes of applied values with certain power coefficients. According to [10, 11], in the classical Solow growth model, it is advisable to apply the production function in the form of the Cobb-Douglas function, where the possibility and limitation of factor substitution are essential. R. Solow's contribution to the practice of using production functions consisted in modifying the Cobb-Douglas production function and extending it, in addition to labor (*L*) and capital (*K*), with the third factor – technological progress, which became the main reason for the development of the US economy in the first half of the XX century. According to R. Solow [12], only 12.5% of the growth in labor productivity during this period could be attributed to an increase in capital and the number of employees, whereas 87.5% was the result of the changes in the quality characteristics of labor potential and improvement of labor organization (professional development of employees, improvements in production organization, etc.).

In our opinion, today there is no practice of a widespread use of the Cobb-Douglas production function in management of labor potential; some elements of mathematical modeling are only applied, with the regression equation being used. It is the parameters of the Cobb-Douglas function equation that allow us to find answers to the questions "to what extent", "to what degree", "why", which, as we believe, furnishes grounds for development of adequate management models for production resources in the agricultural sector [13].

2.3 Justification of the possibility of the Cobb-Douglas production function use in determining the levers for agricultural production growth in the context of sustainable development

In addition to economic results, sustainable development implies an environmental effect as well as consideration of environmental impacts on economic development of the industry. We believe that conclusion as to the use of the environmental and economic balance, application of a production optimization model with consideration of efficient use of production resources and minimization of amounts of pollutant emissions deserves attention [14]. Therefore, it is quite appropriate to use an environmental and economic production function, which may include an environmental component as the external factor that affects output adversely. The methodology for forming and verifying a production function allows determination of the optimal combination of all resources for gaining maximum profits [15].

The advantages of the Cobb-Douglas production function are as follows:

– a relative simplicity of functional dependence, which makes it easy to determine the indicators of labor productivity and fund returns, elasticity of outputs in all parameters, margins for resource substitution;

– practical universality and adequacy: a possibility to determine the contribution of each factor to the total production output, intensive and extensive factors of economic growth, practical application in labor potential management;

 https://doi.org/10.1051/shsconf/202110005012

– macroeconomic orientation: based on the actual economic indices of official statistical reports and can be easily parameterized with the use of the correlation and regression analysis;

– feasibility: functional dependence of output on costs is non-linear and does not contain the disadvantages characteristic of linear production functions that describe the processes of an ideal economy.

But, despite the above advantages, the production function also has a number of disadvantages [13]:

– the production function with a constant scale effect may not adequately reflect the production process (under conditions of intensive growth of production factors, the scale effect is greater than 1);

– it is based on the assumption that production resources are fully substitutional;

– determination of the production function parameters is based on the marginal prices of factors that are equal to average prices and are calculated on the basis of market prices, which is possible in the context of perfect competition and market equilibrium and does not "work" in the actual economics;

– the principle of complementarity, which takes into account the capital structure, is ignored.

2.4 Intellectual assets of human potential as a factor of intensification of agricultural production

At the end of the XX century, the pace of economic development in the leading countries of the world was mainly stipulated by contribution of the intensive factors associated with the emergence of new technologies rather than by the number of people employed and increase in investments. American economists G. Mankiw, D. Romer and D. Weil proposed that the quality of labor force should be taken into account in the model of economic growth, introducing the factor of intellectual capital, which included the cost of education, publication of scientific literature, construction, material and technical resources of educational institutions, etc. [16]. In the theory of economic growth, P. Romer [17] distinguished two different types of production factors: tangible and intangible. The scientist believed that the emergence of new ideas and technologies (intangible resources) leads to the transformation of material resources into more valuable ones. S. Oliner and D. Sichel, based on the analysis of the reasons for the 1995-1999 acceleration of scientific and technological progress in the United States and on the assessment of the production function parameters, considered information technologies to be a separate significant factor of economic growth [18]. Modification of the production function and consideration of the qualitative parameters of labor potential have resulted from the emergence of the ideas of human capital accumulation at the end of the XX century, which became the basis for models of endogenous scientific and technological progress, according to which technology and knowledge are a public good, and technological progress is the result of knowledge accumulation.

Since human capital is defined as individuals' knowledge, education and competences in the implementation of national goals and objectives (mainly based on intangible resources and hidden opportunities), the model of its assessment should include qualitative and quantitative indicators. At the same time, the development potential is represented by intellectual capital, which is considered a key factor for success [19].

Not only have the large-scale and multi-vector changes of recent decades transformed a significant part of the world's socio-economic processes, but they have also expanded and deepened the existing risks. The rate of changes has accelerated significantly, and competitiveness has increasingly become determined by the intellectual grounds of capital. It is an economically active person who is the main basis for innovative changes and economic development.

Today, we observe intellectualization of the economy, with an objective process of expanding the conditions for using people's intellectual and creative abilities being underway, with scientific knowledge and specialized unique skills of their possessors becoming the main source and key factor in the development of material and non-material production, in ensuring sustainable economic development. It is determined that the educational, intellectual, and creative potentials of a person are not only a powerful factor of economic growth, but also a source of income at all levels: of a person, of an enterprise and of a state.

The results of studying various approaches to the definition and structuring of human potential allow us to state that the majority of them consider the intellectual component to be the most essential. The due regard for this component and its analysis enables distinguishing intellectual assets, that is, exactly that part of human potential that can be further capitalized and is capable of ensuring creation of added value which will determine the competitiveness level at both micro and macro levels in the future.

In our opinion, intellectual assets can be defined as an economic category that reflects the process of transformation of part of human potential capable of development and changes into other capital of economic units. Intellectual assets represent an integrated category that includes the intellectual capital objects already reflected in accounting (intangible assets), and those that should be included in the accounting system (customer relations, marketing, technology improvement, training and development of personnel, development of new products and services).

It is determined that intellectual assets are a universal resource with a unique character in terms of increasing the income and value of an enterprise. Intellectual assets act as a form of embodiment of possible future benefits and can make profits. The intellectual assets' capability of bringing economic benefits is one of the main conditions for their involvement into economic turnover. The possibility of alienating these assets to other entities by transferring different amounts and nature of rights enables their wider use in the process of commercialization through involvement of necessary economic entities. Thus, formation of intellectual assets and their effective use is an important strategic area of a company's development, which should be taken into account when drawing up financial budgets and disseminating information about the company's activities. Accounting for intellectual assets contributes to capitalization of the company, maximizing its market value. An increase in the company's value contributes to ensuring sustainability and further development through a greater opportunity to attract investment resources.

Intellectual assets of human potential can be defined both as individual elements of human potential and outcomes of this potential capitalization in activities. These are separate elements of human potential that gradually accumulate and transform starting from the very birth of a person during his/her upbringing, cultural development and involvement in a healthy lifestyle, while obtaining education and professional training, and also these are intellectual outcomes of human activities. Intellectual assets of human potential are considered as competencies (cognitive, emotional, creative), since they have an intangible nature, and as outcomes of intellectual activities (intangible and tangible forms) [20].

The intellectual assets of the human potential in Ukraine are estimated by the following indicators:

– the share of employees engaged in professional, scientific and technical activities to the total number of employees, %;

– the share of those employed in education to the total number of employees, %;

– the share of candidates and doctors of sciences in the total number of employees, %;

– the share of employees involved in scientific research and development in the total number of employees, %;

– the number of researchers per 1000 people of the employed population (aged 15-70), persons;

– the ratio of the average number of employees who used a computer with internet access to the average number of employees who used a computer, %);

– the number of people using broadband internet access, per 100 people of the population, persons;

– the specific weight of innovative products sold in the industrial production volume, %;

– the number of patents for inventions received per 1 million people, units;

– the number of utility patents received per 1 million people, units;

– the number of design patents received per 1 million people, units;

– the number of certificates for trademarks and service marks per 1 million people, units;

– the number of patents received in the United States per 1 million people, units.

The measurement of these indicators enabled determination of the integral coefficients for the intellectual assets of Ukraine's human potential [20].

2.5 Modeling of sustainable development of the agricultural sector based on labor potential

Given the complex structure of labor potential, which includes intellectual components that characterize the ability to innovate, entrepreneurial and creative abilities, educational level, professional experience [21], we believe that in the context of sustainable development of the agricultural sector, it is advisable to use a four-factor Cobb-Douglas function, which takes into account not only the quantity but also the quality of labor as well as the environmental factor of impact:

$$Y = a_0 K^{a_1} L^{a_2} I^{a_3} E^{a_4} \tag{2}$$

where Y stands for the efficiency indicator (output);

– K – the amount of fixed capital or the amount of fixed assets used (capital investments);

– L – labor costs (the number of people employed in agriculture);

– I – integral coefficient of intellectual assets;

– E – pollutant emissions;

– a_0 – a technological coefficient that characterizes production efficiency, takes into account complex influence of qualitative determinants of labor potential, impact of the factors which cannot be quantified (the indicator of technological progress);

– a_i – elasticity coefficients that characterize the effect of capital, labor and intellectual asset growth as well as pollutant emissions on Y output growth (i.e., a_i is a fraction of the factors) [22].

The sum of the elasticity coefficients $a_1 + a_2 + ... + a_n$ characterizes the effect of production scale: the effect is growing, if $a_1 + a_2 + ... + a_n > 1$ (intensive economic growth), the increase in production outgoes the growth of factor costs; the effect is constant, if $a_1 + a_2 + ... + a_n = 1$ (extensive economic growth), the increase in production volumes is not affected by production scale factors; the effect is receding, if $a_1 + a_2 + ... + a_n < 1$ (lack of economic growth), the increase in factor costs is accompanied by a slowdown in growth of production volumes.

When conducting an empirical study of the agricultural sector of the economy and forming the production function, we used the dependence that shows the relationship between the volumes of the fixed production resources use (labor, capital, intellectual assets and pollutant emissions) and the output. Since the main indicator of the industry's degree of development in a market economy is value added, which reflects a possibility of production expansion, the gross value added of agriculture was considered as the efficiency indicator.

The statistics shown in Table 1 were used to calculate the Cobb-Douglas production function with the gross value added as the efficiency indicator.

Having made the calculations with the use of the linear regression analysis based on the least squares method, the values of the Cobb-Douglas function coefficients were determined. The resulting production function is thus [23]:

$$Y = 422388 K^{0.34474} L^{2.63344} I^{1.60806} E^{-2.86638} \quad (3)$$

Table 1. The statistics for calculating the production function

T, year	Y(T), UAH million	K(T), UAH million	L(T), thousand persons	I(T)	E(T), thousand tons
2008	65148	16682	3322,1	0,5	7210,3
2009	65758	9295	3152,2	0,5	6442,9
2010	82948	11311	3115,6	0,6	6678,0
2011	109961	17981	3410,3	0,6	6877,3
2012	113245	18564	3506,7	0,6	6821,1
2013	132354	18175	3389	0,6	6719,8
2014	161145	18388	3091,4	0,6	5346,2
2015	239806	29310	2870,6	0,5	4521,3
2016	279701	49660	2866,5	0,5	4498,1
2017	303949	63401	2860,7	0,5	3879,1
2018	360757	65059	2937,6	0,4	3866,7

Calculated by the authors on the basis of [20, 24]

To evaluate the calculated production function (3), the regression analysis parameters were analyzed. According to the calculations, the multiple correlation coefficient is R=0.983, the standard approximation error is 0.148. The calculated value of the Fischer F-criterion equals 46.71 and is greater than the table value of the Fischer F-criterion (with 99% confidence probability, reliability), which is 8.45. Therefore, the resulting regression equation can be recognized as significant. This means that with a probability of 99%, the found Cobb-Douglas production function (3) corresponds to the input data of the problem.

The production function formed has reliable statistical characteristics. The value of the multiple correlation coefficient proves a closest relationship between the efficiency indicator and the selected factors; 98.3% of the variation in gross value added depends on the fluctuations of the factors included in the equation, and only 1.7 % could be attributed to the factors that are not taken into account. The value of the coefficient of determination R^2 (0.966) is quite close to 1, so the regression model is successful, and the relationship between the resulting indicator of the production function and the input factors is strong. The 96.6% variance in the output *Y(T)* is due to the regression of the selected levers of influence (*K, L, I, E*). This confirms that the model considers the most important factors.

Thus, the formed equation (3) meets all the requirements and can be used for economic analysis (Table 2).

Table 2. The results of the production function modeling

Parameters	Y(T) – gross value added
function	$Y = 422388 K^{0.34474} L^{2.63344} I^{1.60806} E^{-2.86638}$
a_0	422388
a_1	0,34474
a_2	2,63344
a_3	1,60806
a_4	-2,86638
$(a_1 + a_2 + a_3 + a_4)$	1,72

The analysis showed that the growth of quantitative and qualitative indicators of the agricultural sector's labor potential has a direct impact on the increase in the gross value added of agricultural produce, since there is a direct relationship between them. Respectively, sustainable development of the agricultural sector of the economy is possible on the basis of the endogenous technological progress, which is considered to be the result of purposeful human activities, whose driving force is the intellectual assets of human potential.

Conclusions

The econometric analysis using the Cobb-Douglas production function, that takes into account environmental, social and economic factors, showed that, first and foremost, the economic growth in the agricultural sector is associated with an improvement in the quantitative and qualitative characteristics of labor potential (the number of employed people and intellectual assets respectively); and secondly, it is associated with an increase in capital investments and a decrease in the amount of pollutants emitted.

The economic growth of agricultural production is labor-intensive, since ($a_2>a_1$) depends significantly on the quantitative indicators of the rural labor potential (a_2), which are gradually decreasing. It is caused by the negative demographic trends, which, in turn, are a consequence of the decline of rural areas and reduction in employment and income of the population in rural areas. In our opinion, the results obtained with regard to the growing effect of scale give grounds for an optimistic forecast as to increasing the resource potential of agricultural production due to a growth in the quantitative and qualitative indicators of labor potential.

Among the qualitative characteristics of labor potential, special attention should be paid to intellectual assets (cognitive, emotional and creative competencies), which gradually accumulate and transform starting from the very birth of a person during his/her upbringing, cultural development and involvement in a healthy lifestyle, while obtaining education and professional training, and also to the intellectual outcomes of human activities.

References

1. A. Karpenko, K. Basenko, Highly Effective Corporate Culture as an Instrument of Talents' Attracting and Retaining. Balt. J. of Econ. Stud. **3**(4), 101-106 (2017). doi.org/10.30525/2256-0742/2017-3-4-101-106.
2. V. Antonyuk, I. Pidorycheva, Strat. Pan. **2**, 116-123 (2017).
3. Iu.Yarmolenko, Theor. & Pract. Publ. Admin. **2**(53), 107-111 (2016).
4. P. M. Romer, J. Econ. Perspect. **8**(1), 3-22 (1994).
5. R. E. Lucas, J. Monet. Econ. **22**, 3-42 (1988).

6. C. I. Jones, Paul Romer: Ideas, Nonrivalry, and Endogenous Growth. Scand. J. of Econ. **121**(3), 859-883 (2019). doi: 10.1111/sjoe.12370.
7. R. M. Solow, Quart. J. Econ. **70**(1), 65-94 (1956).
8. O. Borodina, Imperatives of complementary development in agrosphere. Econ. & Forecast. **4**, 125-136 (2017). doi.org/10.15407/eip2017.04.125.
9. O. Vasylieva, Factors of Labor Productivity Growth in Agriculture of the Agrarian Region. Balt. J. of Econ. Stud. **3**(4), 1-6 (2017). doi.org/10.30525/2256-0742/2017-3-4-1-6.
10. R. M. Solow, Rev. Econ. Stat. (1961), https://msuweb.montclair.edu/~lebelp/ACMSCapLabSubREAS1961.pdf. Accessed 28 Sept 2020.
11. R. Barro, X. Sala-i-Martin, J. Pol. Econ. **100**, 223-251 (1992)
12. V. Gonda, BIATEC. **XIII**(11), 22-25 (2005).
13. O. M. Vilchynska, Y. M. Panochyshyn, T. O Kushnir, Her. Khmeln. Nat. Univ. Econ. Sc. **2**(1), 177-181 (2016).
14. I. N. Lyashenko, L. Z Khrushch, Econ. Her. Donbas. 16 184-188 (2012).
15. Y. Mundiak, R. Butzer, D. F. Larson, Heterogeneous technology and panel data: The case of the agricultural production function. J. Dev. Econ. **99**(1), 139-149 (2012). doi: 10.1016/j.jdeveco.2011.11.003.
16. G. Mankiv, D. Romer, D. Weil, Quart. J. Econ. **107**(2), 407-438 (1992).
17. P. Romer, Amer. Econ. Rev. **86**(2), 202-206 (1996).
18. D. Oliner, D. Sichel, J. Econ. Pers. **14**, 3-22 (2000).
19. M. Kozak, Intellectual Capital as a Key Factor of Socio-Economic Development of Regions and Countries. Procedia Economics and Finance. **6**, 288-295 (2013). doi: 10.1016/S2212-5671(13)00142-1.
20. A. V. Karpenko, *Rozvytok intelektual'nykh aktyviv lyuds'koho potentsialu: teoriya ta praktyka* (The Development of Intellectual Assets of Human Potential: theory and practice). (FOP V. V. Mokshanov, Zaporizhzhya, 2018)
21. L. V. Shaulska, *Strategiya rozvytku trudovogo potencialu Ukrayiny* (Development strategy of labor potential of Ukraine), (In-t ekonomiky promyslovosti NAN Ukrayiny, Doneczk, 2005).
22. L. I. Voynicha, Youn. Sc. **3**(55), 642-646 (2018).
23. O.Vasyl'yeva, in *Organizational problems of life quality management in the conditions of globalization*, ed. by M. Duczmal, T. Nestorenko (The Academy of Management and Administration in Opole, Opole, 2020), p.428.
24. Official site of (2020) the State Statistics Service of Ukraine. http://www.ukrstat.gov.ua. Accessed 28 Sept 2020.

Socio-Geographical Concept of Labor Market Monitoring and Regional Employment Policy Formation

Kateryna Butkaliuk[1], *Valentyna* Shchabelska[1], *Mariia* Bykova[1], and *Iuliia* Pologovska[1]

[1]National Pedagogical Dragomanov University, Kyiv, Ukraine

Abstract. The necessity of socio-geographical concept elaboration and implementation for labor market monitoring and regional employment policy formation is grounded from the socio-geographical point of view. It has been established that: 1) the versatility and complex nature of the labor market development specificated by natural and socio-economic factors of important geographical content has determined its socio-geographical essence; 2) the development of the concept requires different approaches, principles and research methods, including socio-geographical, taking into account various labor market factors; 3) the concept should take into account the influence of the laws of the production and human settlement territorial organisation, as well as the level of development and structure of the region's economy; 4) the concept should take into account the peculiarities of the labor market formation and development in the economic system agricultural sector as well as the impact of current risks and challenges on the social and labor sphere of public life; 5) the implementation of such a concept has to provide the balancing of labor supply and demand within the country and its regions; 6) regional employment policy should be aimed at rational use of the territory labor potential.

1 Introduction

In terms of the globalisation processes and the competitive environment development formation, the integration of Ukraine into the world community and membership in the World Trade Organisation (WTO), it becomes necessary to identify the priority areas of economic activity in combination with the effective employment policy formation. In this regard, it is important to establish and develop the concept of labour market. The implementation of such a concept has to provide the balance of labour supply and demand within the country and its regions. This will contribute to the efficient usage of labour resources, promote the employment, provide the adequate wages, stimulate motivation to work and its social prestige. The concept development in the context of complex social transformations requires the new approaches and research methods, with the focus on the geospatial approach. This will provide a holistic picture of the labour market functioning, balance supply and demand, identify the main trends in its development and justify the rational use of labour forces and increase the employment rate.

One of the basic modern principles of the geospatial approach to the labour market formation is the full use of the territory features as a concrete geographical space for the placement of economic activity objects and population. Having a certain capacity for the development and location of production facilities, the territory itself is an independent specific resource that affects the formation of the labour market. This is especially important in Ukraine, when the real distance between objects is significant due to the natural and economic factors. The mutual location and availability of production facilities are important characteristics in determining the interaction conditions of the related industries, the rational use of labour force [11].

2 Literature review

The analysis of the publications on this topic shows that except economists, socio-geographers more often analyse a number of issues related to the regional labour markets formation and development. First of all, should be mentioned the O. Levada's thesis, where he made an attempt to consider the labour market from a socio-geographical point of view [8], a number of articles by K. Butkaliuk [1, 2, 3, 4], N. Zapukhlyak [5], I. Melnyk [9], where the attention is focused on the essence of the labour market, factors and features of its formation, analysis of the regional labour market, research methods etc. However, in our opinion, the grounding of the socio-geographical concept of regional labor market development and monitoring, except the disclosing its essence, factors, methods, features, should take into account the impact of production and resettlement territorial organisation. These issues are given relatively little attention in scientific publications. The attempt of defining the influence of the labour territorial division on the regional labour market formation was made by K. Butkaliuk [3]. But researches of complex-proportional development of the region's economy, territorial concentration of production and population in

* Corresponding author: schabelska_v@ukr.net

areas with the most favourable natural and socio-economic conditions remain out of focus.

The aim of the study is to ground the socio-geographical concept of regional labour market monitoring.

3 The tasks of the study

The versatility and complex nature of the labour market development specificated by natural and socio-economic factors of important geographical content has determined its socio-geographical research essence and foundation of socio-geographical concept development and monitoring. Which are as follows:

– identifying the relationship between structural changes in the economy, reforming the social sphere and attracting the labour potential to the economy, taking into account socio-geographical factors and labour market formation global models;

– defining the main concepts, scientific category of the regional labour market as a socio-geographical unit with a rich geographical content;

– forming the methodological approaches to the labour market research, based on the interaction between the economy, social sphere, demographic processes and human labour distribution;

– analysis of the main socio-geographical formation, functioning and development factors, of the labour market, pointing out the structure and specialisation of the region's economy as a necessary condition for its evolution;

– development of the main directions of the regional labour market improvement, increase the employment rate.

4 The results

The labour market is a socio-territorial entity which is formed as a result of the interaction of the sphere of employment and the economically active part of the population within a certain region. The basis for the development and, accordingly, labour market monitoring, is the territorial differentiation of production, which is manifested in differences in its structure, specialisation, management forms and territorial concentration. Thus the sphere of labour force application and, accordingly, the demand for labour is being formed. At the same time, the important component of the labour market are the territorial communities of people, especially their economically active part with the necessary physical and spiritual abilities, professional knowledge and specific skills – it is the basis of labour supply. The labour market from this point of view is considered much more broadly than it is interpreted in economics. The study of the interaction between the economically active population and the field of employment is closely related to the analysis of the socio-geographical environment and socio-economic condition of the territory.

The formation of labour markets is significantly influenced by socio-geographical and socio-economic conditions and factors: natural conditions, availability of natural resources, demographic processes, their territorial differences as a basis for labour reproduction, formation of labour potential and professional composition of the population, development of material and technical base and transport communications.

The formation and development of the functional structure of labour potential occurs through the labour market, which is influenced by the territorial and sectoral economy structure. The transformational shifts in the country's economy, sectoral transformations, changes in the ownership structure determine the processes of labour resources reproduction and functioning.

The main indicators of the labour market are as following: the structure of labour resources by type of employment and industry, economic activity of population, balancing the jobs and labour potential in the economy key sectors, demand and employment of labour force, unemployment, underemployment, the workers layoffs by type of economic activity, workload on one workplace, features of the economically inactive population [12].

In 2019, the share of the working age population was 67.7% (of which the share of the main working age (25 - 54 years) – 51.5%, mature working age (55– 64 years – 14.1%)), younger than working age – 15.2%, which is 6% less than in 1991, and the older age group – 17.1%, which is 4.7% even more. The total number of working population is declining, which corresponds to the fourth stage of the demographic transition: the population aged 65–69 (2.38 million people) is not replaced by a group of 15-19 years (1.86 million people); and the group of 60-64 years (2.83 million people) – a group of 10–14 years (2.22 million people). The total demographic burden per 1,000 people of working age in 2019 amounted to 475 people, including persons aged 0-14 years – 228, 65 years and older – 247.

During the period from 2000 to 2019, the number of economically active population has decreased by 4.5 million people from 23.1 million to 18.6 million people. This trend is due to the development of negative demographic processes – natural population decline and aging, increasing mortality, entering the working age of a small amount of population that is unable to fully replace those who coming out of the economically active population etc.

According to the statistics of the State Statistics Service of Ukraine, there is a tendency of reducing the demand for labour force, as well as reducing the number of employed people. In 2019, the number of employed citizens of Ukraine aged 15–70 was 16.5 million, which is 4.7 million less than in 2000 and 8.9 million than in 1991. The most negative affects its dynamics had in Ternopil, Kherson and Chernivtsi regions.

The unemployment rate in Ukraine, despite some stabilisation, remains significant and in 2019 amounted 8.2%, which is 0.9% and 3.4% less than in 2015 and 2000, respectively. The highest unemployment rate is observed among the age groups 15-24 years and 15-29 years and is 11.9%. The regions with the highest unemployment rate are

https://doi.org/10.1051/shsconf/202110005013

Donetsk, Luhansk, Poltava, Kirovohrad, Ternopil, Zhytomyr regions, and the lowest – Kyiv, Odessa, Lviv, Kharkiv regions. However, in the regions with the lowest unemployment rates there is a tendency to its increase. In order to strengthen the sustainable integration of the unemployed into the labour market in the medium and long term, it is necessary to develop and implement active strategic measures and programs on the labour market. Such programs can provide temporary employment opportunities and eliminate the shortage of qualified personnel for integration into the labour market in the longer term perspective.

Significant changes also occurred in the structure of employment. During the years of independence, the share of the population employed in the traditional sectors of the economy decreased, while the share of services increased. In 1990 the share of people employed in industry was – 30.7%, agriculture, forestry and fisheries – 17.3%, construction – 9.4%, services – 42.6% and in 2019 – 14, 8%, 18.5%, 4.2%, 62.8%, respectively. It is clear that there is a growing trend in the number of people employed in the tertiary and quaternary sectors of the economy due to the growing role of business services. This can be clearly seen in the large cities of Ukraine, where the share of those employed in the service sector has reached a similar level in the large cities of Western Europe, which have undergone post-industrial transformations.

In the context of globalisation in Ukraine's economy, as in the most countries of the world, there is a restructuring of employment in accordance with the requirements of the market, which is the cause of new, non-traditional forms. Non-traditional or "non-standard" forms of employment include: part-time employment, employment on fixed-term employment contracts, temporary, non-permanent employment, including borrowed labour, secondary employment, remote employment, employment on the basis of civil employment contracts, non-employment, including self-employment, unregistered employment in the formal sector. They are distinguished from standard employment by the creativity and flexibility of their work schedule due to the use of scientific and technological progress and computer technology.

Employment in the informal sector of Ukraine's economy is quite common. In 2019, it amounted to 3.5 million people, or 21% of the total employed population aged 15-70 (according to the ILO methodology). Compared to 2000, this category of employees has increased in 1.2 times. Among the economic activities, the most common informal employment was in agriculture, forestry and fisheries (42%), wholesale and retail trade, repair of motor vehicles (18%), and construction (16%). The most common informal employment is among people aged 40 to 49 [17].

The development of the labour market occurs in certain historical, economic, social and natural conditions. Nowadays in Ukraine, in the context of market environment development the significant changes in the role of factors influencing the development of regional labour markets are happening. The role of macroeconomic conditions, economic-spatial and geopolitical location of the region, investment activity, scientific and technological progress and globalisation processes has significantly increased. It is a matter of wide implementation of world scientific and technical achievements in economy, provide the investment activity and innovation policy that is especially shown in the intellectual labour market formation.

Thus, the modern economy and society, generated by the fourth industrial revolution, are transforming, changing and emerging in a new format. The new economy (it is also digital, and network, and knowledge, and creative, etc.) opens the new opportunities and limitations of material, social, psychological, socio-cultural nature for the working person. Many assessments and forecasts state that with the development of a new (digital) economy and other today's realities, the global society should expects a collapse in the field of labour and employment. In particular, the well-known ratio of 20:80, at which economic activity will require no more than 20% of the economically active population, who can acquire the status of employed. The opposite assessments and forecasts, tell that in the new (digital) economy the mass disappearance of some jobs will be accompanied by the emergence of the new ones that can relieve the tension in the labour market [7].

According to the well-known modern economist R. Sharma, with the further transformation of jobs, people will probably replace the workplaces dedicated to robots and artificial intelligence with the new ones that we can't even imagine yet [16].

Conditions have been formed under which the symbiosis of new technologies, institutions of labour organisation, production and management, the new structure of resources and factors of acquiring undeniable competitive advantages and many other factors radically change all components of the social and labour sphere. [7]. The concept of labour market monitoring should take into account the nature of current phenomena and processes, which produce both the new opportunities and the new constraints and challenges in the field of social and labour development. Nowadays, the intellectual potential of Ukraine and its regions is ranked first among the factors of economy development. It is impossible to gain a significant advantage in today's globalised world only through the latest technologies. The human capital in the form of abilities, skills, professional knowledge is more powerful and promising. Along with purely professional knowledge of the system forming industries, in particular in the fields of agricultural production, food industry, serving industries, it is important to have market-oriented specialists. Investments in human capital can provide a powerful, stable and integrated economic and social effect.

The labour market has to take into account the country level of economic development, its position in the world community, the competitiveness of products, including intellectual, in domestic and world markets. The territorial structure of population production and settlement, the degree of territory urbanisation also play an important role. Regional differences in the labour market development significantly depend on the existence of industrialised regions, growth poles, centres that attract labour force, including highly qualified. In particular, from peripheral agricultural regions, where the opportunities for employment are significantly reduced, migration processes are intensifying to the urban systems with the developed industry and tertiary and quaternary service sectors. The labour market is in fact an indicator of socio-economic development of the country or region, the territorial organisation of society, demographic and migration processes, the living conditions of the population in the regions [10]. The formation of regional labour markets is closely linked with the laws of territorial production and settlement organisation of population, the complex-proportional development of the region's economy.

The formation of regional labour markets is based on the territorial division of labour, which determines the spatial differentiation of socially useful labour. Spatial forms of organisation of production and the population interact here through the exchange of goods, including "labour force". The effect of the regularity of the labour force territorial division is manifested in the production specialisation, it is concentrated in certain areas. As a result, certain combinations of productive forces are developing, which act as the core of local or regional economic entities. Respectively develop labour demand and supply, which is the basis of local or regional labour markets [3, p. 268].

Complex-proportional development of the Ukrainian regions is the result of intra-regional specialisation. It is achieved on the basis of coordination between industries, interrelations between the basic, auxiliary and service parts of production, between work application sphere and economically active population. It is one of the main preconditions for optimal labour market creation and maximal public involvement in specific activities. The displaying of regularity of the region complex development concerning the formation of the regional labour market consists in interconnected functioning and the most rational objects arrangement of economic activity and population settlement systems, balanced development of labour potential and spheres of labour application.

Territorial concentration of production and population in areas with the most favourable natural and socio-economic conditions is an important pattern of production territorial organisation, which also significantly affects the formation of the regional labour market. At the same time, the possibilities of production concentration and labour demand growth increase the level of employment. The forms of territorial production concentration such as specialised areas and local entities, on the basis of which regional specialised markets are developed (sugar, grain, fuel and energy, engineering), are emerged. Increasing the production concentration and the level of its territorial concentration, the development of agglomeration in the production and resettlement system, the formation of industrial and territorial complexes, the creation of megacities, technology parks, etc. are caused large-scale transformation in market conditions.

The territorial concentration allows to increase the rate of territory development and to form a complete industrial infrastructure, the most appropriate use the group forms of population resettlement. As the concentration of production increases labour-intensive industries become more dependent on the allocation of labour resources. At the same time, the concentration of production activities and the introduction of innovations lead to increased demands on the labour force, their educational and professional level. For example, in such regions as Dnieper, Donbass and within certain urban agglomerations due to the concentration of production, the supporting centres for the location of productive forces and the territorial organisation of the economy and population resettlement are formed. They perform a system-forming role in the development of regional labour markets. Thus, industrial centres and nodes, which arise on the basis of several small and medium-sized enterprises of the same type or one large and two or three related enterprises, create in these economic regions the zones of attraction for labour force, which gradually increases the impact on nearby settlements and strengthen the labour relations. The effectiveness of the regional labour market development mostly depends on the availability of such centres in each administrative region and level of their formation.

Thus, the supporting centres of productive forces location and territorial organisation of the economy and population settlement in each administrative region became the core of intra-regional labour markets, and their functional development determines the labour markets specialisation. It is important to take into account the hierarchy of the supporting centres, which is represented by four levels (local, district, inter-district and regional centres) in the administrative areas [13]. These centres determine the hierarchy of labour markets in the region. Local centres are the basis for the formation of local labour markets, district centres are the core of the labour markets in administrative districts. Inter-district centres are the basis for the development of inter-district labour markets. The regional centre determines the formation of the labour market of inter-district importance using of the territory of the adjacent administrative districts.

The development of labour markets and their territorial organisation in Ukraine is significantly influenced by the settlements structure and territorial composition of settlements which outlines the territorial boundaries of the labour market to some extent. During the interaction between production systems and population resettlement, the creation of transport arteries, a spatial compositional axis is formed around which the market process develops. Depending on the specialisation, the level of production concentration, the character of the economic centres location and the development of transport infrastructure, the economic system-forming framework is formed, which affects the capacity, size and market conditions.

The structure of the district's economy has a crucial influence on the regional labour market formation. The development of different types of economic activity, their sectoral distribution, intersectoral formations, integration of production and marketing are the basis of market segmentation and the allocation of its types. The economy structure mostly determines the social structure of the economically active population. There is a distribution of employees by areas of employment, taking into account the educational, professional and qualification composition of the working population. Eventually, the structure of the economy affects the level of employment in the region and the formation of reserve labour force.

As for the territorial structure of Ukraine's economy, it has undergone significant changes as a result of the transition to the market economy. Thus, the specialised sugar beet, grain, fruit and vegetable canning areas have lost their positions. It should be noted that currently the territorial labour market of agricultural regions with a high agro-industrial and environmental load on land resources is characterised by an imbalance in labour supply and demand.

The labour market formation is influenced by investment activities. This leads to the emergence of new economic centres, changes in interregional proportions, increasing the area of active economic development [6, p.129], which will positively affect the formation of demand in the regional labour market.

The formation and development of the labour market in the agricultural sector has its own characteristics. On the one hand, labour in agriculture is seasonal. Specialisation, concentration, technical level and structure of agricultural production are different from the industrial one. All this determines the territorial differentiation of agricultural production and employment, differences in conditions and amount of work in this sector. On the other hand, the problem of market development is complicated by the demographic situation in rural areas, unfavourable conditions for the economically active population development, low wealth and rural residents' quality of life.

The integral nature of the regional labour market and at the same time its diversity due to a combination of different elements determine the diversity of research, the use of different approaches involving a large volume of statistical, departmental, reference and cartographic information. The study process does not exclude the possibility of using the certain specific methods, techniques and principles of research.

With the development of a competitive environment, intensified attention is given to the study of special labour markets - industrial, agricultural, social orientation and so on. With the reform of the agricultural sector of the economy, the accession to the World Trade Organization, new aspects of socio-geographical research of the agricultural labour market were appeared. The revival of private land ownership and the creation of conditions for the development of management private forms, in particular in agriculture, a significant reduction the role of collective enterprises, equal functioning of different organisational and legal structures determine the necessity of the territorial differences identification in these processes taking into account regional features, their impact on the formation and development of the agricultural labour market.

Technological re-equipment of agro-industrial production, development of the innovative processes with the banks and foreign investments involvement form the competitive advantages of certain regions, their origin opportunities taking into account the natural resource potential, role in the labour force territorial division, socio-economic factors. Therefore, researches of the impact of these processes on the development of the agricultural labour market are of extreme importance.

The formation and development of the labour market as a socio-production-territorial system are based on the following principles:

– objectivity, which means that the regional labour market is the result of the interaction of the employment sphere and the economically active population in the territorial dimension;

– territorial differences in the location of production and population, their territorial differentiation as the main condition of the regional labour markets emergence;

– industrial-territorial relations, accessibility to the place of employment as one of the main criteria for determining the labour market boundaries;

– the relationship between the region's economic activity, the peculiarities of population settlement and the functional significance of the territory, which affect the sectoral and territorial distribution between the labour activity spheres, labour market segmentation and hierarchy;

– coordination of the labour markets development with the state administrative-territorial structure, the system of population settlement, taking into account the urbanisation of the territory;

– the leading role of urban settlement systems, in particular cities (large, medium, small) in the formation of regional labour markets, as they are centres of concentration of labour, economic, cultural and political life, which act as "cores" of the labour market.

Socio-geographical studies of the labour market should be aimed at solving the problems of Ukraine's productive employment and stimulate the motivation for legal employment, creating conditions for self-employment and entrepreneurship, targeted for improvement of region's labour resources, human capital reproduction, rural areas social revival, tracking regional differences in labour market development in order to develop an effective regional employment policy. These tasks are reflected in the Law of Ukraine 'On Employment' [14] and determine the main strategic directions of socio-geographical research of the labour market. They include identification of labour market formation patterns; grounded the socio-geographical principles of labour market monitoring creation taking into account tendencies of changes in the structure and territorial organisation of economic activity; study of the functioning mechanism of the regional labour market in the conditions of economy globalisation, integration into the EU and membership in the WTO; identification of the large-scale cooperative economic formations impact on the agricultural labour market; the development of the principles of state regional policy on employment etc.

Only the implementation of management activities on a program-targeted basis in the field of demography reproduction processes and labour use can provide the projected positive changes in the structure of the region's economy, which will lead to a balanced and less labour-intensive and redundant path of economic development. Socio-geographical studies of the economic territorial organisation, resettlement and migration flows are of greatest importance for the development of Ukraine's regional employment policy principles.

Regional employment policy is an organisational legal, economic, social, cultural, educational, environmental and administrative state activity, aimed for the rational use of labour resource potential of the territory. Its purpose is to increase the level of efficiency of labour resource potential in the region (socio-territorial complex) due to the competitive advantages (opportunities) of the territory, innovation and investment activities, reducing the level of labour capacity of economy. Its implementation taking, into account socio-geographical factors, requires the creation of geographic information systems (GIS). It is a system of collecting, accumulating, processing and analysing spatial information on labour supply and demand in order to increase the efficiency of the regional labour market.

In determining the regional employment policy priority trends, social, demographic, industrial and other problems are taken into account. The solution of these problems should contribute to the general socio-economic prosperity of the regions [15].

In the context of a global pandemic, the measures that were taken by national state institutions of different countries, including Ukraine, to limit the spread of COVID-19 and avoid the large-scale population losses, have at least threefold consequences for economic growth. First of all, it is the shock of the proposal. Legislative and executive authorities and local self-government have suspended economic and business activities in a number of economy sectors, including tourism, catering, transport, sports and entertainment. Secondly, it is the shock of demand. The point is that the incomes of the business entities and, consequently, the households in sectors of the economy whose activities are suspended, banned and restricted have fallen sharply and this process is ongoing. The vast majority of households limit their spending due to declining incomes, deteriorating consumer emotions and an unpredictable future. For a short period of time, private consumption has focused on basic necessities. Demand for other goods has declined sharply, which has a negative impact on all areas of economic activity. Thirdly, it is the shock of international trade. Restrictions on the free movement of goods, people, declining the migration flows, trips have led to disruption of global logistics chains, shortages of intermediate goods from countries that have suffered the greatest losses associated with the COVID-19 pandemic. [7, p. 16] The mentioned above makes significant changes in all components of global and national labour markets, while affecting the demand, supply and price of labour force services.

5 Conclusions

The employment, the formation and development of the labour market are the reflection of the interaction between the economy, social sphere, demographic processes and the distribution of human labour and require constant attention. The socio-geographical concept of regional market monitoring is based on the theoretical principals of the socio-geographical nature, principles and factors of labour market formation. They are based on the interaction of the employment sphere and the economically active population in their territorial dimension. Therefore, such research is multifaceted and involves the study of branches of tangible and intangible production as a field of labour application. In particular their territorial differentiation, structure, specialisation, forms of territorial concentration. Simultaneously, it is a study of territorial communities of people, their socially useful activities. This takes into account the population dynamics, the peculiarities of its natural and mechanical movement, gender and age structure, spiritual and physical abilities, level of education, professional knowledge, production skills etc.

It is important to take into account the laws of production and resettlement territorial organisation, as well as the economic system structure, which will develop promising areas of labour market, justify the forecast of the most appropriate combination of territorial communities and concentration of production and other economic activities. In order to increase the level of employment productive and balance the supply and demand in the labour market, it is important to provide their monitoring.

References

1. K.O. Butkaliuk. Socio-geographical factors of regional labor markets formation. *Economic and social geography*. **53**, 78-83 (Kyiv, 2002).
2. K.O. Butkaliuk. The essence and functions of the regional labor market. *Geography and Modernity*. **9,** 82-88 (Kyiv, 2003).
3. K.O. Butkaliuk. The influence of the labor territorial division on the formation of the regional labor market. *Socio-geographical problems of productive forces development of Ukraine*. 308 (Kyiv, 2004).
4. K.O. Butkaliuk. The methods of socio-geographical research of the regional labor markets. *Scientific notes of Vinnytsia State Pedagogical University named after M. Kotsyubynsky. Series: Geography*. **9** (Vinnytsia, 2005).
5. N.I. Zapukhlyak. Socio-geographical features of labor market formation. *Economic and social geography*. **53**, 83-88 (Kyiv, 2002).
6. Yu.D. Kachaev. The influence of investment activity on the region's territorial development. *Geography in the information society*. **4,** 129-131 (Kyiv, 2008).
7. A. Kolot, O. Gerasimenko, K. Yarmolyuk-Kryok. The labour sphere in the conditions of the global socio-economic reality 2020: challenges for

Ukraine, (2020), http://www.fes.kiev.ua Accessed 25 Feb 2021
8. O.M. Levada. Socio-geographical bases of formation and functioning of the countryside regional labor market (on an example of the Zaporozhye area). Thesis abstract. 18 (Odessa, 1999).
9. I.G. Melnyk. The labor market study as the direction of socio-geographical research. Actual problems of economic-geographical science and education development in Ukraine. 45-46 (Kyiv, 2003).
10. V.P. Nagirna, K.O. Butkaliuk. Socio-geographical principles of labor market research as an indicator of the efficiency of the region's economy. Geography in the information society. **2**, 114 (Kyiv, 2008).
11. V.P. Nagirna, K.O. Butkaliuk. Cluster approach to the labor market formation and the principles of its improvement (socio-geographical aspect). *Geography and Modernity.* **21,** 53-861 (Kyiv, 2009).
12. L.M. Nemets, Yu.Yu. Silchenko, K.A. Nemets, K.Yu. Segida. Labor potential of the Kharkiv region (socio-geographical aspect). 186 (Kharkiv, 2014)
13. E.I. Pityurenko. Settlement systems and territorial organisation of the national economy, 140 (Kyiv, 1983).
14. Law of Ukraine "On Employment of the Population". Information of the Verkhovna Rada, **24** (243), (2013), https://zakon.rada.gov.ua/laws/show/5067-17#Text Accessed 12 Jan 2021
15. V.S. Vasilchenko. State regulation of employment. 252 (Kyiv, 2005).
16. R. Sharma. Rise and Fall of States. The forces of changes in the post-crisis world. 84 (Moscow, 2018).
17. State Employment Center. Analytical and statistical information (2021), https://www.dcz.gov.ua/analitics/68 Accessed 25 Feb 2021

Assessment of the Ecological Condition of Soil Cover Based on Remote Sensing Data: Erosional Aspect

Andrii Achasov[1*], *Alla* Achasova[2], *Ganna* Titenko[1], *Oleg* Seliverstov[1] and *Vladimir* Krivtsov[3]

[1]V.N. Karazin KhNU, School of Ecology, 61022 Kharkiv, Ukraine
[2]NSC «O.N. Sokolovsky ISSAR», Soil Erosion Control Laboratory, 61024 Kharkiv, Ukraine
[3]Royal Botanic Garden Edinburgh, EH3 5LR Edinburgh, United Kingdom

Abstract. Soil erosion by water is the most important global environmental problem. A modern system for assessing and monitoring soil erosional degradation should be based on the use of remote sensing data. This raises the issue of correct data decoding. The article proposes a method for visual interpretation of eroded soils according to the Sentinel image obtained in the visible range. The authors give some combinations of decoding signs to determine the manifestations of linear and surface water erosion from images. The article shows possible errors in decoding the manifestations of water erosion and gives an example of assessing the erosion of the soil cover based on the results of decoding the Sentinel-2 satellite image. Moderately and heavily eroded soils are reliably distinguished, the area of which, according to the interpretation data, was 2.4% of the area of arable land in the studied territory. In the future, the obtained sample of spectral images of eroded soils can be used to develop an automated method of interpretation based on the principle of "computer vision".

1 Introduction

Soil erosion by water is a "shadow of world agriculture" according to Professor S.Yu.Bulygin's poetic statement [1]. After all, human agricultural activity has always caused more or less destructive erosion processes everywhere. Erosion is not only a huge environmental problem [2], but also an economic problem estimated at tens of billions of US dollars. Every year, the world loses 33.7 million tons of food due to erosion [3].

Unfortunately, Ukraine is no exception. According to experts, the area of land affected by water erosion in Ukraine is about 40%. Due to this and direct and indirect losses, the country's economy loses up to 10 billion of US dollars annually [1]. Direct negative consequences of soil erosion include the loss of crops, decrease in the harvest quality, and deterioration of infrastructure. However, soil erosion also results in a multitude of less obvious indirect effects, including e.g. disturbance of the local hydrological balance, pollution of local water bodies and smothering of aquatic organisms, among others. The lower lying areas with depositional environment may receive an excessive amount of sediments; that alters their geomorphological settings and decreases the hydraulic conductivity of the surface deposits thus making those areas more prone to flooding and inundation. At the same time, however, higher grounds situated on hillsides tend to experience insufficient water resources due to the increase in surface runoff. Furthermore, erosion processes also contribute to the climatic changes both through the increase in the mineralization of soil organic matter and consequent increased emissions of greenhouse gasses, and through alteration of the radiological balance due to the deterioration of the surface vegetation cover and the accompanying changes in albedo. All the processes listed above have negative implications for the economy and wellbeing of local residents, as well as for the global climate change.

The problem of erosion, like any other problem, requires, first of all, clear understanding of the condition of the country's soil cover and the pace of ongoing erosion processes.

Both issues require the use of modern research technologies, such as geographic information systems and remote sensing of the Earth surface. After all, it is impossible to monitor 32 million hectares of arable land in the country without space survey data [4]. Space sounding data is a reliable basis for monitoring soil resources due to the high detail and visibility of images. Being relatively inexpensive, they can be easily updated [5].

Fortunately, the current development of space observation systems helps to address this issue. At present there are, at least, two programs, providing free space survey data from Landsat and Sentinel satellites. These data, provided by the Sentinel-2A and Sentinel-2B spacecrafts, are particularly interesting. Spatial resolution of images in the visible range is 10 m. Two satellites take photos of the same area every 5 days, which is completely suitable for monitoring and mapping of eroded soils [6].

* Corresponding author: achasov@karazin.ua

 https://doi.org/10.1051/shsconf/202110005014

It should be noted that space monitoring involves continuous collection of huge amounts of data that need to be analyzed quickly and efficiently. The space image itself is a "semi-finished product". It becomes thematically informative only after instrumental processing by an expert, using special hardware and software, i.e – decoding. The most desirable and difficult is the option of automated decoding of remote sensing data. The computer selects and interprets objects in the image with little human intervention.

There are many examples of attempts to automatically decode the manifestations of water erosion on various scales in the scientific literature. Global forecasts of erosion spread according to satellite data of medium resolution are given in [7].

A previous study [8] presented the results of semi-automated decoding of gullies by integrating geographical object analysis of images with modern models of machine learning. For the analysis presented in this paper, we used Sentinel 2A multispectral survey data, a global digital terrain model obtained from ALOS-2 satellite data, and ground-based observation data. Karami and co-authors [9] estimated linear erosion from IRS-P6 satellite images, using pixel-by-pixel and object-oriented automated classification methods. In [8] there is an example of successful application of machine learning technology for the recognition of gully erosion. A variant of the method of determining the gully network according to high-resolution space sounding (on WorldView imagery) is presented by S. Uchida [10]. Using additional digital elevation data of ASTER / GDEM2, he obtained the recognition accuracy of the gully network of 63.4%. Some approaches to automatic decoding of erosion are also considered in [11]. However, the authors emphasize that "a researcher, analyzing satellite imagery, can often provide more accurate results in determining eroded soils".

It should be noted that most of the publications relate to the definition of linear (gully) erosion. Little attention is paid to the issue of surface water erosion, which is not surprising, given the complexity of decoding this phenomenon. Thus, Žížala, D. et al., 2019 [12] showed that even on a regional scale, the accuracy of automatic selection of eroded areas by uncontrolled classification of images according to Sentinel-2A reached 55.2%. In comparison, the accuracy of decoding was 86.9%, a remarkable increase considering the fact that only heavily eroded soils are separated from all others with the additional use of visual decoding of images to adjust the results.

Therefore, visual decoding of water erosion is a mandatory step in the development of methods for automatic decoding of images and verification of its results. At the same time, decoding manifestations of erosion according to the satellite imagery has many "pitfalls", conditioned by the difficulty of complete formalization of such features as the shape of the object, its structure, texture, etc.

The purpose of this article is to analyze the features of visual decoding of eroded soils and erosion processes in space images of the visible range.

Clear establishment of such features and creation of a correct method of visual decoding of eroded soils will allow us to move to the next stage - the automation of this process.

2 Materials and methods

2.1 Characteristics of research sites

The research was carried out in the territory of the forest-steppe physical-geographical zone and the north-steppe physical-geographical subzone of Ukraine at the junction of Kharkiv, Poltava and Dnipropetrovsk regions (Fig. 1) during 2019-2020. The study area was 12058.7 km^2, determined by the size of the standard site of the Sentinel-2a spacecraft.

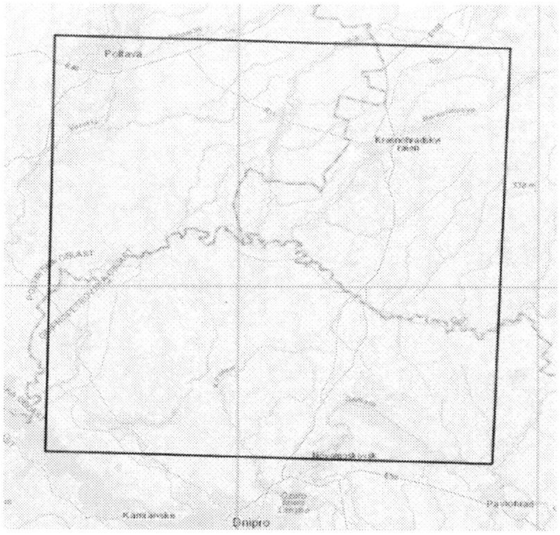

Fig. 1. Map of the research territory

The study area is characterized by a temperate continental climate. The average annual rainfall regularly decreases from northwest to south and southeast in the range of 550-460 mm per year. Its relief is characterized by the alternation of flat watersheds with wide river valleys. Most of the territory is actively used in agriculture. All this contributes to the significant development of water erosion of arable soils.

The soil cover of the territory is represented mainly by typical chernozems, podzolic chernozems and dark gray forest soils. All of them are characterized by a significant content of organic matter in the upper genetic horizon, determining their corresponding color (from gray to dark gray and black-brown) color and is an important feature in the visual decoding of space images.

According to the previous agrochemical studies, the average content of soil humus was quoted as 4,2% and 3,8% for Kharkov and Dniepropietrovsk regions respectively [13], whilst for Poltava region it was reported as 3,26% [14]. These levels of the soil humus content are considered as 'elevated' and 'high' [15], and soil horizons with such humus content are characterized by typical dark colours. Furthermore, humus-containing substances of the chernozem soils have a stronger colouring capacity compared to the podzolic soils [16], i.e. chernozem soils appear much darker than podzolic soils with the same organic carbon content.

2.2 Characteristics of the Remote Sensing Data Used in the study

A space image taken by the Sentinel-2a on April 27, 2019 (T36UXV_20190427T083601_TCI) was selected for the research. From three channels of the satellite (blue B2, green B3, red B4) the image corresponding to the color image made in the visible range of electromagnetic waves was synthesized.

The main characteristics of the image are: the level of pre-treatment 1C, spatial resolution of 10 m, cloudiness 0 points. The chosen shooting date allowed, on one hand, to observe the largest possible number of fields without cleared vegetation cover, and on the other - to avoid the influence of significant soil moisture after snow melting.

2.3 Methods of decoding space images

Visual decoding is the oldest type of decoding, which arose simultaneously with the first attempts to lift a person into the air. Its essence is to identify real-world objects in aerospace images by their characteristics without using any special hardware or software.

Visual decoding was performed according to the classical method, based on the use of direct (geometric, optical and structural characteristics of objects) and indirect (various relationships and interdependencies between the studied objects) decoding features.

"Recognition" of objects is the result of a complex logical-intuitive process that only the human brain is capable of. It is the complexity of formalizing this process that has led to the fact that, despite all modern devices and developed software, visual decoding remains a very effective tool for the remote sensing data analysis.

The image is decoded in GIS QGIS 3.16.1-Hannover. The algorithm for decoding planar erosion consists of the following main stages:

1. Pre-processing of a space image, including adjustments to brightness and contrast for the best visual perception of the image.
2. Selection of open ground fields visually or by numerical analysis of NDVI index.
3. Making sure that the soil at the time of survey was in an air-dry state. Checking whether there was any precipitation in the study area at least three or four days before the date of the survey according to the meteorological stations.
4. Taking into account direct decoding features, such as the color of the open ground and the structure of the image, and indirect features - the landscape location of the object, its proximity to river valleys and gullies.

Since the research methodology is the subject of this article, it will be described in more detail in Section 3.

3 Results and discussion

3.1 Decoding of typical erosional features

It is known that water erosion can be surface and linear. Linear erosion is the result of a significant concentration of water in natural depressions, resulting in the formation of linear soil erosion.

Such blurs are manifested in the image through a set of the following features (Fig. 2):

1) Dendroidal structure of the image. It is known that such a structure is a manifestation of natural processes of water movement in the landscape. At one time, it was the discovery of such structures on the surface of Mars that proved the presence of water currents on its surface in the past [17].

Fig. 2. View of plowed and unplowed linear erosion forms in the space image

2) Color. If linear erosion is of significant size and, therefore, has not been plowed for a long time, the process of its growth slows down, it eventually overgrows with natural vegetation. Accordingly, its color in the images will depend on the species composition and condition of the vegetation in the ravine or gully (Fig. 2). If the erosion has not reached a critical depth that interferes with processing, its color is usually characterized by a combination of light and dark parallel stripes. This is due to the uneven illumination of the opposite slopes of the basin: the light strip is eroded soils on the illuminated slope of linear erosion, and the dark - its shadow slope.

Surface erosion occurs in the case of relatively uniform soil erosion by small streams over the entire surface of the site. Its consequence is the formation of large areas of fairly evenly eroded soils without significant manifestations of linear erosion (Fig. 3).

Diagnosis of planar erosion from space images is also based on a combination of the two above features: color and structure.

1) Color. Spectral properties of soils affected by water erosion have much greater brightness compared

https://doi.org/10.1051/shsconf/202110005014

to unaffected soils. The reason for this is the loss of the upper humus layer and the exposure of the lower soil layers, which have a much lower content of organic matter.

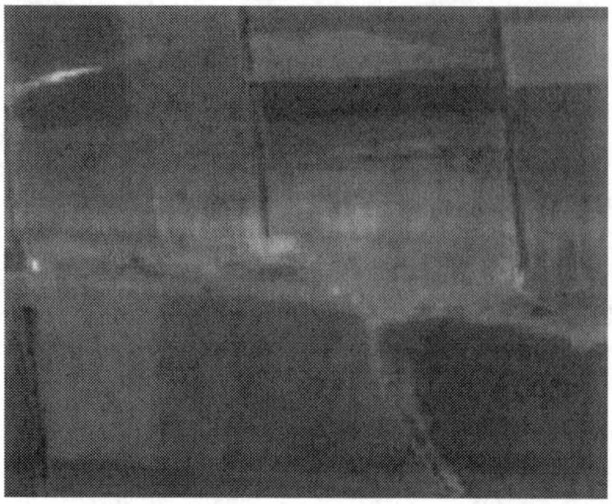

Fig. 3. Manifestations of planar erosion in space images

2) The structure of the image. Clearly expressed dendroidal structures are not typical for surface erosion. Rather, the structure of the image can be called "feathery". Small blurs form a pattern of densely arranged well-oriented, almost parallel light stripes. These stripes are like feathers adjacent to ravines and gullies.

The latter is very important, as the absence of such a topological connection may indicate a misdiagnosis. Therefore, the third decoding feature for surface erosion is topological interdependence.

3) Topological interdependence. This is an indirect decoding feature, connecting small erosions to the clearly defined elements of the ravine-gully network almost at right angles.

3.2 Decoding of artifact objects

It is worth emphasizing a very important methodological position – one can select manifestations of any type of water erosion only if there is a combination of these decoding features. Diagnosing soil as eroded by only one feature, such as color, can lead to error. This is especially important when it is impossible to inspect the whole field.

Fig. 4 clearly distinguishes a large area of open light colour ground. However, it would be incorrect to consider this area a manifestation of surface erosion.

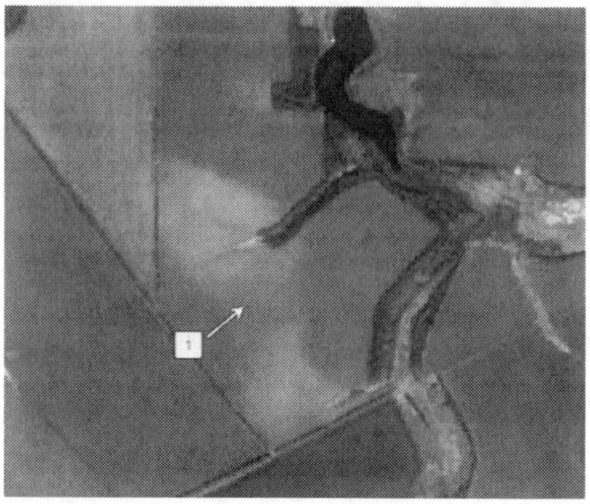

Fig. 4. Possible misdiagnosis of soil erosion (podzolic soils)

Although there are two notable manifestations of linear erosion, there is no feathery structure, characterizing surface erosion.

Looking at the shape of this plot, we can see that it is irregular with smooth twists. Thus, a contour delineating the area with uniform spectral properties is traced on two fields with different cultivation. Therefore, it cannot be considered a result of any agrotechnical measures, visually lightening the surface of the field (for example - leaving straw in the field).

The above-mentioned decoding features are typical for the areas of podzolic soils. These soils are formed under centuries-old forest stands. As a result, their surface horizon has a reduced content of organic matter compared to chernozems and the presence of fine-grained quartz (crystalline SiO_2), which has high reflectivity and gives the soil a characteristic light color. Hence, the picture might be revealing the selection of podzolic soils.

To test this assumption, an archival soil map of scale 1: 25000 was used [18] (Fig. 5).

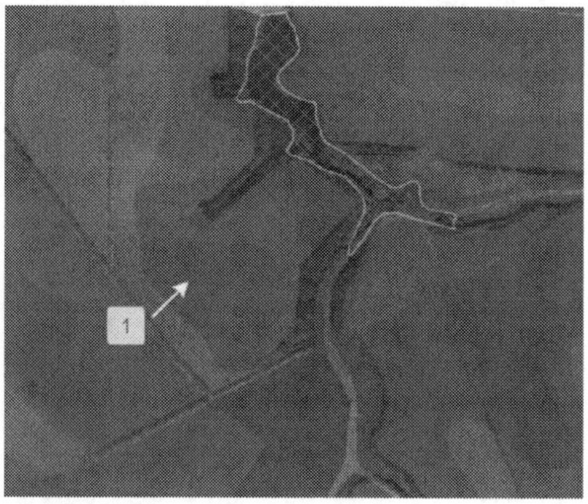

1 – dark-gray podzolic soil

Fig. 5. Overlaying of archival soil map on space image

The light contour, highlighted in the space image shown in Fig. 4, almost completely repeats the section, marked as "Dark gray podzolic and degraded soils and chernozems podzolic and degraded slightly washed heavy loam) on the soil map. Thus, in fact, these are not eroded, but podzolic soils bordering on typical chernozems.

An example of possible misdiagnosis of linear erosion is shown in Fig. 6. In this case, the light straight line that passes through the entire image is not an erosive form. First, erosions do not have such clear straight contours. Secondly, the image does not show the dendroidal structure, mandatory for the ravine-gully network.

This light line is an indicator of the gas pipeline, laid underground. The soil above the gas pipeline is lighter in color compared to the undisturbed soil. This is because the upper humus and lower, lighter soil horizons and rocks were mixed during the laying and backfilling of the gas pipelines.

Fig. 6. Anthropogenic linear shapes in space images. Gas pipeline

When decoding, even more difficult situations can arise. Fig. 7 shows a fragment of the image which can be identified as manifestations of linear and surface erosion, as well as the results of human activity, very similar to them.

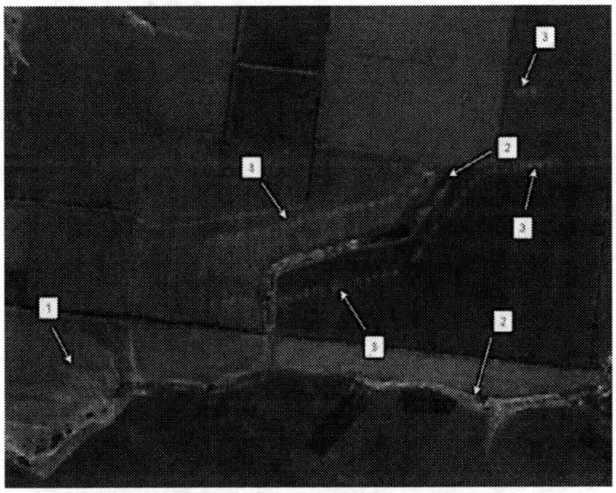

Fig. 7. A fragment of a space image with existing manifestations of erosion and anthropogenic artifacts (Sentinel image 2019)

Number 1 in Fig. 7 is a classic manifestation of surface soil erosion with all its inherent decoding features. The gully network (number 2) is also clearly visible in the picture. However, two light straight lines located on both sides of the ravine appear to be problematic (number 3).

Both lines are formed by separate light rounded spots. The diameter of each spot is 18-20 m. The distance of the lines from the ravine varies between 100-140 m.

Monitoring of this area, using the Internet service GoogleEarth, shows that this phenomenon has been traced unchanged in all available images since 2011 (Fig.8).

It is very difficult to assume that this phenomenon is natural because you can hardly find such clear structures in nature. This phenomenon might be of an anthropogenic origin. It may be related to some land reclamation activities in the area.

Fig. 8. A snapshot of GoogleEarth 2011. Results of the x visual decoding for a satellite image of the given territory

3.3 An example of visual decoding the space image

To improve the accuracy of diagnosis, visual decoding was performed only in fields where there was no vegetation and plant remains. The values of the NDVI coefficient of such fields ranged from 0 to 0.025. As a result, the total surveyed area was 2,055.3 km^2.

Both linear and surface erosions were identified in the area. In such cases, the soils were considered eroded only if the expert was 100% confident about it. About 10% of all cases were additionally verified, using the Internet service GoogleEarth (QGIS Send2GE module) and the site "Regulatory monetary valuation of agricultural land" [18], which contains archival maps of agricultural soil groups.

As a result of the decoding, 935 vector polygons were identified, characterizing the erosion processes on arable lands. The total allocated territories are 50.2 km^2. This is 2.4% of the surveyed area.

Here, we should note two important methodological aspects of the eroded soils allocation according to remote sensing. First, the content of organic matter (Corg) in the upper genetic horizon varies within 3-5% for the studied non-eroded soils. Studies show dependence of the spectral properties of chernozem soils on the content of organic carbon [19], this dependence being of the exponential decay type. The curve levels out at approximately the value of Corg = 3%. Further increase in Corg has very little effect on the soil color and its reflectivity. It follows that soils that differ markedly in the pictures in their color from the background contain considerably less than 3% organic matter. That is, they lost more than half of the upper humus horizon.

From the viewpoint of such soils classification according to the degree of erosion, they belong to the

 https://doi.org/10.1051/shsconf/202110005014

moderately -eroded or heavily eroded soils (FAO, 2006) [20]. Therefore, the figures obtained by us do not include weakly eroded soils, which do not differ in color from modal (non-eroded) soils.

The second aspect is the correctness of the spatial selection of eroded soils.

Both light and dark areas are in the zones of surface erosion manifestation (Fig. 3). It may seem that the overall erosion will be overestimated.

But this is not the case. Figures 9-10 give an example of a detailed field soil survey of one of the linear erosions. The survey was conducted near the village Vasylivske (Nova Vodolaga district, Kharkiv region) in the framework of another research project in 2010.

As part of the study, 18 soil sections were laid. The laying scheme was as follows: 1) 6 sections were laid along the bottom of the runoff basin at a distance of 50 m from each other, 2) two more sections were laid opposite each of these sections on opposite sides of the basin. The distance between them and the sections at the bottom of the basin was 20 m.

Figs. 9-10 show the depths of humus-rich horizon (H). As we can see, the depth of the horizon ranges from 20 to 100 cm. For comparison, it should be noted that the depth of this horizon for modal (non-eroded) soil in this field is 40 cm.

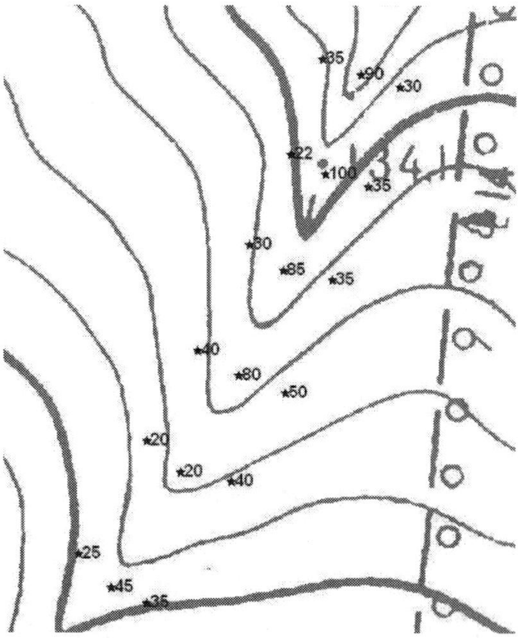

25 – the depth of the upper humus-accumulative genetic horizon

Fig. 9. A fragment of a topographic map showing the locations (marked with asterisks *) of the sampling points and the depths of the humus-rich horizon

Respectively, soils with H less than 40 cm according to the existing method of field diagnosis of eroded soils are considered eroded, and with H greater than 40 cm – uneroded; in sections placed along the talweg of a temporary water course such soils can be characterized as alluvial. It should be noted that in nature the rates of erosional and depositional processes exhibit both spatial and temporal variability. Consequently, eroded and alluvial soils alternate in some areas. All these soils can be called erosion-alluvial (i.e. washed or re-worked) soils.

As we can see (Fig. 10), the areas adjacent to most of the sampling points on both slopes of the basin do not differ in color from modal soils, although they are classified as eroded by the value of H.

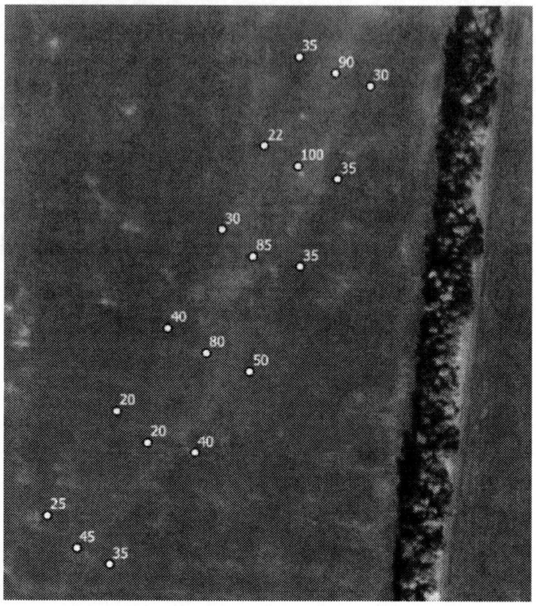

25 – the depth of the upper humus-accumulative genetic horizon

Fig. 10. A snapshot of GoogleEarth (11.10.2018)

Thus, the continuous allocation of areas characterized by a "feathery" image structure as superficially eroded, is justified. The spectral samples collection of eroded lands, created during the research, will be further used to test the method of "computer vision" [21] for automated decoding of space images.

4 Concluding remarks

Remote sensing is an uncontested method of real-time monitoring of eroded soils. Automatic interpretation of water erosion manifestations in satellite images is the most attractive option for operational monitoring of erosion. However, despite many attempts to automatically decode erosion all over the world, the technique has not yet been fully developed.

Creation of a technique for the correct automatic decoding is impossible without the development of an algorithm for delineation of areas covered by eroded soils according to a set of decoding signs providing clues for their identification. An important methodological point is also the elimination of decoding errors, which include recognition of various anthropogenic objects and natural heterogeneity of the soil cover as erosion manifestations.

In our studies, we established a combination of decoding signs to determine the manifestations of linear and surface water erosion features from images. Specific characteristics of the reflected signal in the areas covered by chernozem soils with high humus content makes it possible to correctly identify moderately and heavily eroded soils. Analysis of the Sentinel satellite image allowed us to assess the share of moderately and heavily eroded soils on arable lands within the studied territory at 2.4% of their total area.

Currently, soil erosion represents one of the most serious environmental problems, and our results are therefore relevant for researchers and practitioners dealing with negative consequences of soil erosion for agriculture, economy and society. The eroded sediment particles alter light penetration and biogeochemical balance of the receiving water courses [22]. Furthermore, suspended sediments adsorb large quantities of pollutants [22], which may then be transported for long distances affecting not only freshwater, but also marine ecosystems. Our study is, therefore, relevant to all these issues. In addition, the methodology presented here will also be of use for a wider range of applications relying on the use of remote sensing data.

Conflict of interest

The authors declare that there is no conflict of interests regarding the publication of this manuscript. In addition, the ethical issues, including plagiarism, informed consent, misconduct, data fabrication and/or falsification, double publication and/or submission, and redundancy have been completely observed by the authors.

References

1. S. Bulygin, The Ukrainian Farmer, **6,** 46-47 (2015)
2. Outcome document of the Global Symposium on Soil Erosion, FAO, 15-17 May 2019, Rome, Italy (2019)
3. M. Sartoria, G. Philippidis, T. Ferrari, P. Borrelli, E. Lugato, L. Montanarella, P. Panagos, Land Use Policy, **86**, 299-312 (2019)
4. M.I. Luleva, H. van de Werff, F. van der Meer, V. Jetten, Chemistry: Bulgarian Journal of Science Education, **21 (5)**, 748-764 (2012)
5. K.E. Seutloali, T. Dube, O. Mutanga, Phys Chem Earth, **100**, 296-304 (2017)
6. A.B. Achasov, A.O. Achasova, Visnyk of V. N. Karazin Kharkiv National University, ser. "Ecology" **944 (6)**, 20 - 27 (2011)
7. Vågen T-G, Winowiecki LA. Remote Sens. **11(15)**,1800 (2019)
8. H. Shahabi, B. Jarihani, S. Tavakkoli Piralilou, D. Chittleborough, M. Avand, O. Ghorbanzadeh, Sensors, **19 (22),** 4893 (2019)
9. A. Karami, A. Khoorani, A. Noohegar, S.R.F. Shamsi, V. Moosavi, Environmental and Engineering Geoscience. **21 (2),** 101-110 (2015)
10. S.Uchida, Procedia Environ. Sci. **24**, 29–34 (2015)
11. K. A. Lukyanchuk, I. P. Kovalchuk, O. M. Pidkova *Application of a remote sensing in monitoring of erosion processes*, in Proceedings of the Conference "Geoinformatics: Theoretical and Applied Aspects", 11-14 May 2020, Kiyv, Ukraine (2020)
12. D. Žížala, A. Juřicová, T. Zádorová, K. Zelenková, R. Minařík, Eur. J. Remote Sens. **52**, 108-122 (2019)
13. Periodical report on the condition of soils on agricultural land of Ukraine based on the results of the 9th round (2006–2010) of agrochemical survey of lands. (GA "Institute of Soil Protection of Ukraine", 2015)
14. V. Koval, V Natalochka, S. K. Tkachenko, Visnyk of Poltava State Agrarian academy, **3**, 84-88 (2013)
15. Soil quality. Methods for determination of organic matter: DSTU 4289:2004. (Derzhspozhyvstandart of Ukraine, 2006)
16. E. I. Karavanova Optical properties of soils and their nature. (2003).
17. M. H. Carr, Philos. Trans. R. Soc. Lond. **A 370**, 2193–2215 (2012)
18. Ukraine Normative Monetary Land Assessment Portal
19. A. Achasov, A. Achasova, 50North (2017)
20. Guidelines for soil description. Fourth edition. (FAO, 2006)
21. M. Zhu, Y. He, Q. He, International Journal of Geosciences, **10,** 1-11 (2019)
22. S. Arthur et al. Blue-green infrastructure-perspectives on water quality benefits (CIRIA, 2019)

Evaluation of the State of Ecological Sphere Vital Activity in the Medium-Size Cities

Mykola Klymenko[1], *Oleksandr* Klymenko[1], *Ludmyla* Klymenko[1], *Maria* Kopiy[2], and *Ilya* Tsinivskiy[1]

[1]National University of Water and Environmental Engineering, 33028, Ukraine
[2]Ukrainian National Forestry University,79057, Ukraine

Abstract. The analysis of the state of ecological sphere development in the medium-size cities of Ukraine certifies the presence of negative tendencies there, exactly, the decrease of population quantity, the growth of numbers of movable sources and the level of population morbidity. There has been suggested a scale to make qualitative and quantitative evaluation of the changes of ecological sphere state according to which indices ascertain the correspondence with their states: 1,0 – 0,75 as high, 0,749 – 0,50 as medium, 0,499 – 0,25 as less than medium, 0,249 –0, as low. It is revealed that high level of ecological development refers to Ivano-Frankivsk (0,777), which is the leader according to the level of planting greenery, the lowest accumulation of waste materials of the 1st – 3d level of danger, the lowest emission of pollutants into atmosphere from stationary and movable sources. Melitopol has medium level (0,57) and is the leader as to the indices of drinking water consumption and is characterized by good indices concerning the waste treatment, considerable expenses spent on nature-protecting activity and low amount of pollutant emission into atmosphere. The cities of Lutsk and Rivne are characterized by lower than medium level of the development of ecological sphere vital activity (0,482; 0,384 in the year 2017 correspondingly), because of a great amount of pollutants emission into atmosphere, low level of planting greenery. The city of Rivne, besides, has the highest indices as to the assumption of waste materials and the least expenses on nature-protecting measures. It is ascertained that the influence of destimulators and stimulators on the level of development of medium-size cities ecological sphere is described by dependencies which look like straight lines at the coefficients of determination R^2 from 0,381 to 0,721.

1 Introduction

The successful development of regions and cities based on sustainable development is possible only under the conditions of analytical studies of the dynamics of indicators that characterize the state of social, economic, and environmental spheres of its lives. Based on the results of research and SWOT-analysis for regions and cities, it is possible to formulate a strategic vision, mission, strategic and operational goals, as well as tasks for their development, which will contribute to their sustainability in the short and long term [1-6].

A number of scientists studied and assessed the state of the ecological,economic and social sphere of cities and regions, namely, Z.Gerasymchuk, I. Condius (2010),L. Shen, J.J.Ochoa, M.N.Shah, andX. Zhang (2011), K. Mori, A. Christodoulou (2012), U.Mörtberg, J.Haas, A.Zetterberg, J.P.Franklin, D.Jonsson (2013),M. Klymenko, A.Pryshchepa, N.Khomich (2014),T. Tolstykh, L.Gamidullaeva, N. Shmeleva and Y. Lapygin (2020) [5-10].

According to the recommendations of Z.V. Herasymchuk, OV. Sereda for the analysis of the state of development of the ecological sphere we have chosen medium-sized cities [1].

* Corresponding author: o.m.klymenko@nuwm.edu.ua

In recent years, developing Strategies for Sustainable Development of Cities of Ukraine, researchesprovide the evaluationof the state of the environmental sphere based on the results of surveys accompanied by standard and declarative decisions to improve the state of this sphere.

In our opinion, the assessment of the state of the ecological sphere of life of cities should be carried out using quantitative and qualitative indicators that characterize the state of this sphere in the past and in the future.

To achieve this goal, medium-sized cities of Ukraine with a population of 150 to 250 thousand people were selected for the research, which has common features in terms of levels of sustainable development and environmental safety.

Conducting the study of the ecological sphere of life of medium-sized cities is due to the fact that in recent decades, against the background of climate change and the growing number of mobile sources of air pollution, there was the negative impact on the functioning and development of their socio-economic and ecological system. According to the research many scientists have shown a growing trend in urban morbidity of respiratory diseases and circulatory systems, diseases of allergic origin, tumors, due to increasing emissions of pollutants into the atmosphere from stationary and mobile sources [6,11-14].

Accordingly, it is extremely important and urgent to solve problems related to the study of the functioning and development of the environmental sphere, assessment and forecasting of its changes in the nearest future.

Aim of the research was providing of a comparative estimation of the state of the environmental sphere of medium-sized cities in Ukraine.

Achieving the goal involved studying **the following tasks**:

– substantiation of the methodology for assessing the state of the ecological sphere of medium-sized cities;

– conducting a comparative evaluation of the state of the environmental sphere of medium-sized cities;

– developing of a mathematical model of improvement of the ecological sphere of cities.

The object of research is the processes taking place in the environmental sphere of medium-sized cities.

The subject of research is the ecological sphere of medium-sized cities, which is characterized by quantitative and qualitative indicators.

2 Methodology, research methods

It is recommended to estimate the state of the ecological sphere of medium-sized cities using quantitative indicators (primary data of statistical materials of the State Statistics Committee of Ukraine) and their ordering in the rating scale which varies from 0 to 1.0 according to the formulas:

$$X_1 = \frac{Ni - N(\min)}{N(\max) - N(\min)} \text{ for stimulators (1)}$$

$$X_2 = \frac{N(max) - Ni}{N(\max) - N(\min)} \text{ for destimulators (2)}$$

where Ni – actual values of the indicator of the ecological sphere;

N(max), N(min) –maximum and minimum values (for maximum stimulators and minimum destimulators we took quantitative values of indicators that correspond to the best values among medium-sized cities).

Vertical aggregation of indicators at all levels (basic (b_i) in aggregate (AI_i), aggregated in the integrated index of the state of the ecological sphere) is carried out using the arithmetic mean (Fig. 1).

We carried out the evaluation of the state of the ecological sphere of the city's vital activity according to developed unified measuring scale, and according to which the state and level of development of the sphere are assessed quantitatively and qualitatively: high level 1.0-0.75; average level 0.749-0.50; below the average level of 0.499-0.25; low level of 0.249-0.

This scale should also be used to assess the level of sustainable development of the medium-sized city on the basis of the integrated index of the level of sustainable development of the city (IILSDC) calculated as the geometric mean of the indices of social, economic, environmental spheres.

In addition, methods of analysis, comparison, correlation, and regression analysis using Microsoft Excel and Statistics were used.

3 Research results

To assess the state of the ecological sphere of life of the city, we used seven indicators, namely: emissions of pollutants from stationary sources; emissions of pollutants from mobile sources of pollution; waste generation of I-III hazard classes (destimulators); the level of freshwater use; capacity of treatment facilities; the total area of green areas and plantations; capital investment and running costs for environmental protection (stimulants). With vertical folding, seven indicators are aggregated into five criteria, namely: air emissions, water use, waste management, the level of urban landscaping, the cost of environmental activities.

The integrated index of development of the ecological sphere of life of the city is calculated as the arithmetic mean of seven indicators. The results of the study of the indicators of the four medium-sized cities are presented in the table1, 2.

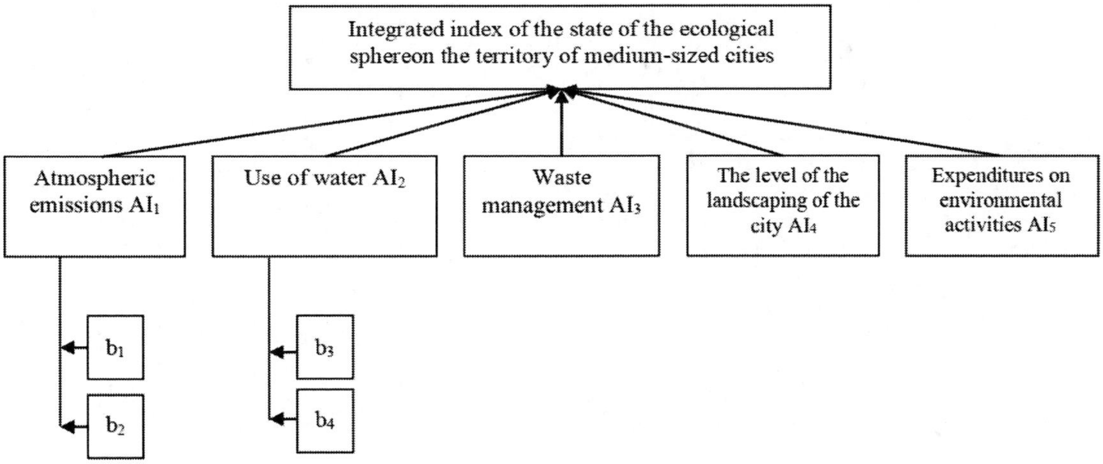

Fig.1. Scheme of calculation of the integrated index of the ecological sphere of the territory of medium-sized cities

Table 1. Theresults of the assessment of the level of sustainable development of the ecological sphere of medium-sized cities

Criteria	Indicators	Ivano-Frankivsk			Lutsk			Melitopol			Rivne		
		1	2	3	1	2	3	1	2	3	1	2	3
Atmospheric emissions,**AI₁**	Emissions of pollutants from stationary sources, kg/person	1.0	1.0	1.0	0.82	0.75	0.789	0.84	0.65	0.745	0.01	0.01	0.01
	Emissions of pollutants from mobile sources, kg/person	0.52	1.0	0.76	0.01	0.01	0.01	0.84	0.66	0.75	1.0	0.86	0.93
	AI₁	0.76	1.0		0.42	0.38		0.84	0.66		0.50	0.44	
Use of water, **AI₂**	Level of freshwater use, m^3 / person	0.01	0.01	0.01	0.16	0.16	0.16	1.0	1.0	1.0	0.82	0.79	0.805
	The capacity of treatment facilities, m^3 / person	0.40	0.46	0.43	0.34	0.32	0.33	0.01	0.01	0.01	1.0	1.0	1.0
	AI₂	0.21	0.23		0.25	0.24		0.50	0.50		0.91	0.90	
Waste management, **AI₃**	Waste generation of I-III hazard classes, kg/person	1.0	0.98	0.99	0.96	1.0	0.98	0.88	0.80	0.84	0.01	0.01	0.01
The level of urban landscaping, **AI₄**	The total area of greens and plantings, m^2 / person	1.0	1.0	1.0	0.23	0.24	0.235	0.01	0.01	0.01	0.01	0.01	0.01
Expenditures on environmental activities, **AI₅**	Capital investments and current expenditures on environmental protection, UAH / person	0.01	1.0	0.505	0.71	0.90	0.805	1.0	0.86	0.93	0.02	0.01	0.015
Integrated index of ecological sphere		0,563	0.777	0.671	0.461	0.482	0.473	0.654	0.57	0.612	0.41	0.384	0.396

Note: data for 1 - 2010; 2 - 2017; 3 - the average for 2010-2017.

Table 2. Analysis of indicators of the ecological sphere in terms of variability over time

Criteria	Indicators	Ivano-Frankivsk				Lutsk				Melitopol				Rivne			
		1	2	3	4	1	2	3	4	1	2	3	4	1	2	3	4
Atmospheric emissions	Emissions of pollutants from stationary sources, kg/person	2.07	1.65	1.8	↘	4.9	4.3	4.3	↘	4.7	5.3	5.3	↗	18.02	15.0	11.7	↘
	Emissions of pollutants from mobile sources, kg/person	52.7	50	43.5	↘	60.7	55.4	58.0	↘	47.7	48.4	48.4	→	45.2	48.4	45.5	↗
Use of water	Level of freshwater use, m3 / person	65.5	60.4	60.4	↘	71.3	68	68	↘	101.7	108	108	↗	95.4	97	97.6	↗
	The capacity of treatment facilities, m^3 / person	220.2	219	230.4	→	214	213	213	→	173.1	172	172	→	292	292	298	→
Waste management	Waste generation of I-III hazard classes, kg/person	1.85	1.4	1.4	↘	2.0	1.3	1.3	↘	2.4	2.4	2.4	→	6.7	6.7	6.9	→
The level of urban landscaping	The total area of greens and plantings, m^2 / person	47.8	47.9	47.9	→	28.6	28.8	28.8	→	23.2	23.2	23.2	→	23	23	22.9	→
Expenditures on environmental activities	Capital investments and current expenditures on environmental protection, UAH / person	8.9	224	224.2	↑	140.5	202	202	↑	195.5	194	194	→	12.5	-	12.5	→

Note: The data for: 1 - 2010; 2 - 2011; 3 - 2017; 4 - symbols: ↑ - rapidly growing; ↗ - increases; → - stable; ↘ - decreases; ↓ - decreases sharply.

From the data in the table. 1.2 shows that a high level of environmental development is characterized only by the regional center of the Western region of Ivano-Frankivsk (0.778), which occupies a leading position in the level of landscaping (47.9 m^2 / person), according to the criterion of waste management -1.4 kg/person), with the lowest emissions of pollutants into the atmospherefrom stationarysources (2.07-1.8 kg/person) and mobile sources (52.7-43.5 kg/person). The average level of ecological development is typical of the city of Melitopol (0.57), which leads in the level of freshwater use (101.7-108 m3 / person), has good indicators of waste management (2.4 kg/person), significant costs for environmental protection activities (195.5-194 UAH / person) and has low indicators of air pollutant emissions from stationary sources (4.7-5.3 kg/person) and from mobile (47.7-48.4 kg/person).

The cities of Lutsk and Rivne are characterized by a lower than average level of development of the ecological sphere of life of cities (0.482 and 0.384 in 2017, respectively). The reasons for the lower than average level of the ecological sphere of these cities are: for Lutsk significant volumes of emissions from mobile sources (60.7-55.4 kg/person), relatively low level of freshwater use (101.7-108 m3 / person)); and for the city of Rivne - high volumes of pollutant emissions into the atmosphere (18.02-11.7 kg/person), high levels of waste generation (6.7-6.9 kg/person), low level of landscaping (23- 22.9 m2 / person) and the use of small amounts of funds for environmental activities (12.5 UAH / person).

At the same time, it should be noted that for Rivne the positive factors that will improve the level of development of the ecological sphere of the city's life are: reduction of pollutant emissions from stationary sources, an increase of green areas, and environmental costs.

For the city of Lutsk, a competitor for the status of the regional center, promising factors for increasing the level of environmental development may be: reducing emissions of pollutants from stationary and mobile sources, increasing the area of greenery, and environmental costs. As can be seen from Table 2, the cities of Rivne and Lutsk have approximately the same chances, in the long run, to fight for the improvement of the ecological sphere. However, judging by the variability of factors over time, in Lutsk the number of factors whose values are slowly decreasing (disincentives) and increasing (stimulants) during 2010-2017, is much greater than in Rivne. Therefore, the city of Lutsk currently has significant advantages in the development of the environmental sphere compared to the city of Rivne.

Capital investments and current expenditures on environmental protection, UAH / person the sources are described by the equation of descending lines at the coefficients of determination of 0.629 and 0.542, respectively; the formation of waste I-III hazard classes by the equation of the descending line at R^2 = 0.527(Table 3).

Table 3. Results of correlation and regression analysis of the dependence of the index of development of the ecological sphere of cities on the values of indicators

№ п/п	The name of the indicator	Type of equation	R^2
1.	Emissions of pollutants from stationary sources, kg/person.	y_1=-0.017x_1+0.652	0.629
2.	Emissions of pollutants from mobile sources, kg/pers.	y_2=-0.00x_2+0.801	0.542
3.	Waste generation I-III hazard classes, kg/pers.	y_3=-0.04x_3+0.658	0.527
4.	The total area of green massifs and plantings, m2 / person.	y_4=0.007x_4+0.332	0.381
5.	Capital investments and current expenditures on environmental protection, UAH / person	y_5=0.00135x_5+0.35	0.721

Table 4. Calculation of matrix coefficients to establish the relationship between the index of the ecological sphere of cities and indicators (2017)

	Index of Environmental sphere	Emissions of pollutants from stationary sources of pollution	Emissions of pollutants from mobile sources of pollution	The level of freshwater use	The capacity of treatment facilities		Waste generation of I-III hazard classes	The total area of greens and plantations	Capital investment and current environmental costs
Environmental index	1.000								
Emissions of pollutants from stationary sources of pollution	-0.812	1.000							
Emissions of pollutants from mobile sources of	-0.216	-0.146	1.000						

pollution									
The level of freshwater use	-0.859	0.648	-0.173	1.000					
The capacity of treatment facilities	-0.147	0.687	-0.356	-0.049	1.000				
Waste generation of I-III hazard classes	-0.653	0.964	-0.388	0.573	0.790		1.000		
The total area of greens and plantations	0.988	-0.727	-0.352	-0.818	-0.034		-0.537	1.000	
Capital investment and current environmental costs	0.660	-0.974	0.275	-0.505	-0.826		-0.990	0.554	1.000

For stimulants, namely: the total area of green areas and plantations and capital investment and current expenditures on environmental protection, affecting the level of development of the environmental sphere, there are dependencies that have the form of increasing lines with coefficients of 0.381 and 0.721, respectively.

Other indicators, namely: the level of freshwater consumption per capita, the capacity of treatment plants are values that do not change significantly in cities and are close in value. This information should be taken into account when creating mathematical models for forecasting changes in time of the index of development of the ecological sphere of life of medium-sized cities of Ukraine.

At the same time, we analyzed the matrix coefficients that characterize the strength of the impact of indicators on the index of environmental development of medium-sized cities (Table 4).

As can be seen from the table 3 the highest correlation coefficients were found for destimulators: emissions of pollutants from stationary sources of pollution – 0.812; emissions of pollutants from mobile sources – 0.216; waste generation of I-III hazard classes – 0.653; for stimulants: total area of green areas and plantings 0.988; capital investment and current expenditures for environmental protection 0.66.

Based on the calculated matrix correlation coefficients, we select a system of indicators that together can most fully characterize the integrated indices of the ecological sphere of life of medium-sized cities in Ukraine (Table 4).

The algorithm for constructing a mathematical model, presented in Fig. 2, provides the following structural elements: regression analysis, evaluation of the results, verification of the adequacy of the mathematical model, forecasting the development of the ecological sphere of cities.

According to the calculations, the integrated index of the ecological sphere of life of medium-sized cities (y) depending on the value of indicators of multifactor linear regression at $R^2 = 0$, has the form:

$$y=3,3662-0,00409x_1-0,01475x_2-0,01301x_3-0,00752x_4+0,168403x_5+0,003603x_6+0,000751x_7 \quad (3)$$

where:

x_1 - emissions of pollutants from stationary sources of pollution;

x_2 - emissions of pollutants from mobile sources of pollution;

x_3 - the level of freshwater use;

x_4 - capacity of treatment facilities;

x_5 - waste generation of I-III hazard classes;

x_6 - total areas of green areas and plantations;

x_7 - capital investment and current costs for environmental protection.

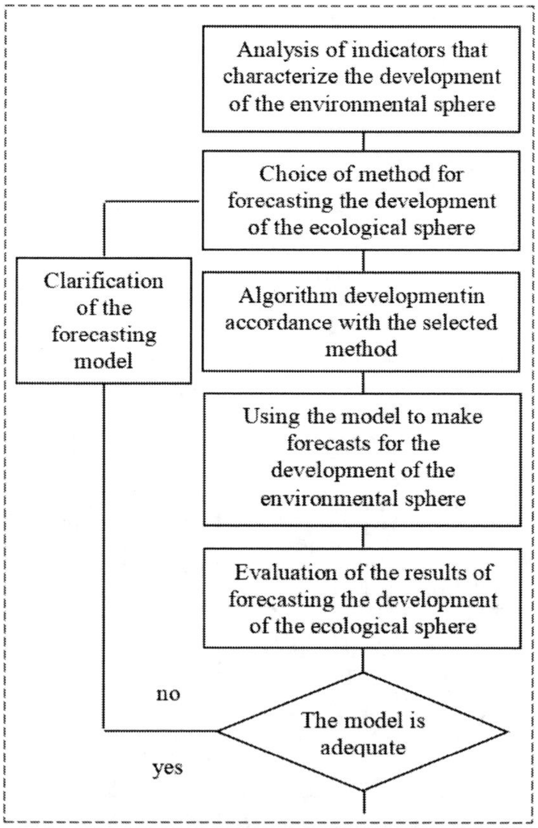

Fig.2.Block diagram of building a mathematical model of development of the ecological sphere of life of medium-sized cities of Ukraine

The assessment and verification of multifactor linear regression testify to its suitability for sufficiently high accuracy of forecasting the values of the integrated index of the ecological sphere of life (IIESL) of medium-sized cities of Ukraine. The deviation of the established IIESL from the calculated by multifactor linear regression IIESL does not exceed values + - 1% (table 5).

Table 5. Comparative analysis of integrated indices of the ecological sphere of life (IIESL) of medium-sized cities of Ukraine

Name of the city	2010			2014			2017		
	1	2	3	1	2	3	1	2	3
Ivano-Frankivsk	0.563	0.566	+0.53	0.66	0.658	-0.3	0.777	0.776	-0.13
Lutsk	0.461	0.46	-0.22	0.471	0.47	-0.21	0.482	0.481	-0.21
Melitopol	0.654	0.653	-0.15	0.614	0.616	+0.32	0.57	0.569	-0.18
Rivne	0.41	0.409	-0.24	0.392	0.39	-0.51	0.384	0.383	-0.26

Note: 1. IIES calculated by indicators.
2. IIES calculated by multifactor linear regression.
3. Deviation, %.

Modeling the development of ecological spheres of life of medium-sized cities in the near future using multifactor linear regression (1) provides for cities: Ivano-Frankivsk maintaining a high level of development of this area (0.828) by reducing emissions from mobile sources from 43.5 to 40 kg / person; Melitopol to ensure the average level of development (0.72) by reducing emissions of pollutants from stationary sources from 5.3 to 4.0 kg / person and mobile - from 48.4 to 40 kg / person; Lutsk achievement of the average level of development of the ecological sphere (0.61) by reducing emissions of pollutants from mobile sources from 58 to 50 kg / person and increasing the area of green areas and plantations from 28.8 to 30 m^2 / person; Equal achievement of the average level of development of the ecological sphere (0.517) by reducing emissions of pollutants from stationary sources from 11.68 to 5 kg / person, mobile sources from 45.5 to 40 kg / person and increasing the area of green areas and plantations from 22, 9 to 30 m^2 / person.

4 Conclusions

1. To assess the state of the ecological sphere of medium-sized cities, it is recommended to use a measuring scale, according to which the state of its development is assessed quantitatively and qualitatively: high 1.0 - 0.75; average 0.749 - 0.50; below the average 0.499 - 0.25; low 0.249 - 0.

2. It is established that the city of Ivano-Frankivsk has a high level of development of the ecological sphere (0.777), the city of Melitopol average (0.57), the city of Lutsk (0.482), and the city of Rivne (0.384) below average, with a tendency to improvement during 2010-2017 for Ivano-Frankivsk, Lutsk and to deterioration for Melitopol and Rivne.

3. Influence of dynamic indicators-stimulators: total areas of green plantings, capital investments and current expenses for environmental protection; as well as indicators-disincentives: emissions from stationary and mobile sources of pollutants, waste generation I-III hazard to the level of environmental development of medium-sized cities is described by dependencies that have the form of descending or increasing lines with coefficients from 0.381 to 0.721.

4. Modeling of the development of the ecological sphere activity of medium-sized cities of Ukraine should be carried out using a mathematical model in the form of the multifactor linear regression of seven indicators with a coefficient of determination R^2 = 0.912.

References

1. Z.V. Gerasymchuk, O.V. Sereda. Strategic management of sustainable development of the cities: theory, methodology, practice: monograph. Lutsk, (2014).
2. O. Kariy. Complex development of cities: theory and methodology of strategic planning: monograph. Lviv, (2011).
3. O.A. Karlova. The conception and strategy of socio-economic development of the city. Economic forum (scientific journal). Edition 3, Lutsk: LNTU, (2002). 185-191.
4. Z.V. Gerasymchuk, I.M. Yakhovich. Organizational and economic mechanism of formation and realisation of the strategy of the development of the region: monograph. Lutsk: LSTU, (2002).
5. Z.V. Gerasymchuk, I.S. Condius. Theoretical and applied principles of prognostication of sustainable development of the region: monograph. Lutsk: «Nadstyria», (2010)
6. M. Klymenko, A. Pryshchepa, N. Khomich. Evaluation of the state of Rivne as to the indices of ecological and social monitoring: monograph. Rivne: National University of Water Management and Natural Resources Use, (2014).
7. T.Tolstykh, L.Gamidullaeva, N.Shmeleva and Y.Lapygin.Regional Development in Russia: An Ecosystem Approach to Territorial Sustainability Assessment.Sustainability, 12, (2020).
8. L. Shen, J.J.Ochoa, M.N.Shah, and X. Zhang. The application of urban sustainability indicators – A comparison between various practices. Habitat Int., 35, (2011),17–29.
9. K. Mori, A. Christodoulou. Review of sustainability indices and indicators: Towards a new city sustainability index (CSI). Environ. Impact Assess. Rev., 32, (2012), 94–106.

10. U. Mörtberg, J. Haas, A. Zetterberg, J. P. Franklin, D. Jonsson, B. Deal. Urban ecosystems and sustainable urban development-analysing and assessing interacting systems in the Stockholm region. Urban Ecosyst., 16, (2013), 763–782.
11. Luko D. V., Kaskiv M. V. Effects of air pollution on the Rivne population morbidity rate. Man and the environment. Problems of neoecology. № 1-2, (2012), 35-40.
12. Kozlovska T.Ph. Medical and ecological risk as a way to estimate child's cancer depending on the level of air pollution. II All-Ukrainian congress of ecologists with international participation: collection of scientifics investigations. Vinnytsia. (2009), 416-419.
13. Stakhiv I.R. Influence of air pollution on health of population from 2001 to 2010. Theoretical and applied aspects of geoinformatics: collection of scientifics investigations. (2013), 126-132.
14. Shushpanov D. Determinants of health of the population of Ukraine: ecological dimension. Regional aspects of development of productive forces of Ukraine. Issue 22. (2017), 132-142.

Natural Resource-Based Socio-Economic System of Local Communities: Vulnerability and Adaptation to Climate Change

Sergii Kyryziuk[1]*, *Viktor* Yarovyi[1], *Igor* Prokopa[1], *Mariia* Lukyanova[1]

[1]Department of Economy and Policy of Agrarian Transformations, Institute for Economics and Forecasting of the National Academy of Sciences of Ukraine, Kyiv, 01011, Ukraine

Abstract. Local rural communities challenge due to regional disparities on conditions and features of economic, infrastructural, social development etc. Most rural communities are strengthened by natural and environmental bases for economic development compared to urban areas. But during the last two decades they face the global challenge – climate change, which takes a risk of socio-economic system (SES), based on natural resources use. Agriculture and natural-based activities are the most vulnerable systems to climate change. This study aims to test the possibilities for Ukrainian local rural communities to adapt their SES, which are based mainly on natural resources use, toward the building of more resilient to climate change systems. Here, we develop the framework for assessment of vulnerability of community' SES to climate change, using the integrated indicator-based approach. We estimate the Vulnerability Index for six rural communities, describing the integrated level and by subsystems level of vulnerability. Our assessment confirms the high vulnerability of agriculture and natural ecosystem to climate change. It is a crucial challenge for further socio-economic development of the rural SES with the monofunctional economy which is based on natural resources use. In addition, the problem is exacerbated by the high vulnerability of the social system and infrastructure for most studied communities. Understanding the vulnerable subsystems of community' SES ensure timely response and the development of concrete policy actions at the community and state levels. In this study we substantiate the hotlist of actions for the most vulnerable subsystems.

1 Introduction

We argue that the ecological component affects the formation of the living space of rural residents. There are at least two most important aspects of such an impact in rural areas as follows:

First, the general climate changes affect the characteristics of the rural living space due to extreme weather events, changes in temperature and patterns of precipitation and so on.

Second, it is an effect of the large corporate farming. There is a dual structure of Ukrainian agriculture represented by large corporate farms and small individual farms. Among the most distinguishing features of the corporate land use are the following: excessive land concentration, intensive use of local natural resources and infrastructure, focus on the most profitable crops without proper crop-rotations causing soil depletion and others. Usually, corporate farming is more unsustainable comparing with small individual farms in terms of balancing the major components of sustainable development as economic, social and ecological.

For a long time, the term 'sustainable development' meant, first of all, that natural systems are able to provide both natural resources for the economy and quality ecosystem services for the whole society. According to the modern understanding of the concept, using the term 'sustainable development' in this study, we emphasize not only the importance of the environmental sustainability, but also the necessity to balance environmental protection with the economic and social development.

There are lots of evidences that Ukrainian corporate farms, trying to maximize their own profit, ignore social and environmental issues negatively affecting local rural communities. For instance, intensive corporate land use and animal breeding leads to depletion and pollution of local water resources, including drinking water sources for local residents, excessive use of chemicals on the corporate fields causes increased incident of respiratory diseases, leads to disappearance of pollinating insects, decreasing harvest in individual gardens of local households. At the same time, corporate farms actively use the local communal infrastructure without any significant input to maintain or restore it (for instance, heavy tracks with grain broke local roads).

All together, the mentioned ecological factors strongly determine the characteristics of local living space in rural areas. Such an influence is especially significant at the territories and in communities with local economies based primarily on the use of natural resources because of increased competition for resources that has a pressure on living space.

Ukrainian administrative subdivision significantly determines the patterns of local economies. Ukrainian territorial reform resulted in establishing new administrative units of basic level – territorial communities, which are the principal units of the administrative divisions of Ukraine. As of beginning of 2021, 1438 new UTCs – so called "United Territorial Communities" (without the capital city Kyiv and communities in the occupied territories) are fully functional that means they have their own local financial budgets and relations with the state budget.

The territorial communities are very different in their characteristics, namely area, urban and rural population, economic structure, financial autonomy and others. Most of the rural territorial communities are subsidized from the state budget, and the share of state subsidies and grants constitutes 40-70% of all their budget revenues. Except state subsidies and grants, taxes are the most important source of local budget revenues, but it is limited because the tax base remains relatively narrow in many rural economies. It is a constraint for local development of such territorial communities.

Agriculture is the main economic activity in most of rural territorial communities, creating the basis for local autonomy and providing financial capacity to realize local social and economic development programs.

Rural communities strongly depend on agriculture that provide financial flows allowing to fund local social and economic development. Almost 3 million people are engaged in agriculture [1]. Ukrainian agriculture has a dual structure with two basic types of agricultural producers, namely the corporate (agricultural corporations) and individual or peasant (private and subsidiary farms) ones. The corporate farms focus on commercial and export-oriented crops, while individual farms target the domestic market.

Except agriculture, such sectors as trade and social infrastructure are also important for employment of rural population (¼-⅓ of formal employment in rural areas), but provide less for local budgets.

The increase in agricultural production capacity in Ukraine, which has taken place over the last decade, is largely due to the stock of unrealized agricultural potential [2]. In addition, the impact of global climate change also provided a positive effect to the increase of agricultural production by reducing the negative effects of certain weather and climatic factors (reducing freezing, increasing heat supply). However, the partly positive influence of climatic factors has a fairly pronounced regional character. In the south-eastern and southern regions, global climate change has become threatening in the last 2-3 years due to a sharp decline in moisture supply, increased moisture evaporation, which has increased the risks and scale of crop loss [3].

Climate risks pose a threat not only to individual farmers, but also to communities whose incomes depend on their efficiency. While the state is working on developing and implementing mechanisms to reduce production risks (resumption of partial financing for risk insurance, debt relief for climate-affected farms, etc.), local communities are left alone with the problem of budget revenues, unemployment and others.

The **research objective** of this study is to test the possibilities for Ukrainian administrative units of basic level to adapt their local development strategy, which are based mainly on natural resources use, toward adaptation to the climate changes and deal with other environmental challenges in order to mitigate their negative impact on the living space, especially in rural areas.

2 Materials and Methods

2.1 Data and analysis of natural resources availability

For the analysis, we used open local data and data provided by self-governments of six selected UTCs as follows: Buzhanska (hereinafter referred as Bzh), Vytvytska (Vt), Vovkovynetska (Vk), Losynivska (Ls), Novoslobidska (Ns) and Schumska (Sch). Selected UTCs represented different Ukrainian regions (Fig. 1) and three of the four natural and climatic zones, covered above 90% of the Ukraine' territory (the forest-steppe (Ns, Vk, Sch, Ls), steppe (Bzh) and the Ukrainian Carpathians (Vt). The last is quite important in terms of the objective of this study - assessing the impact of climate change, which have different effects depending on the zone [4].

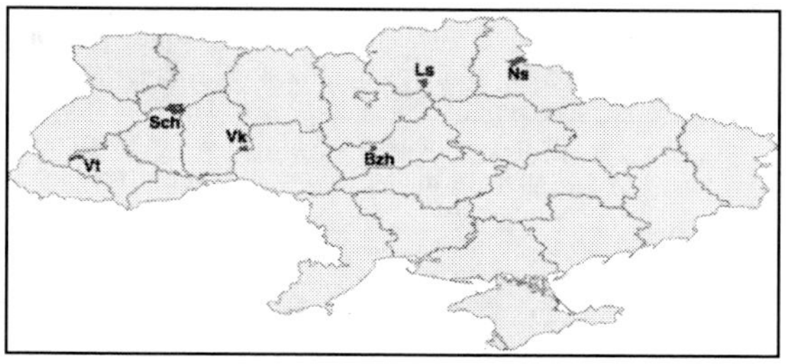

Fig. 1. Spatial allocation of the studied UTCs

For the analysis of local resources, we used data on on soils, natural reserves, mineral deposits from the State Land Cadaster, data on forests from State Forest Cadaster, data on licenses to use mineral deposits from State Service of Geology and Subsoil of Ukraine, data on state supervision on land use obtained from the State Service of Ukraine for Geodesy, Cartography and Cadastre, data on land covers obtained from satellite images for 2016-2018 and data on areas under agricultural crops in 2019 obtained from the State Land Cadaster [5].

Economic characteristics of agricultural lands are obtained from the official maps of natural and agricultural regions of Ukraine (prepared as a part of monetary assessment of agricultural land).

We used data on legal persons and individual entrepreneurs from the State Register to analyze economic activities and economic potential of UTCs [6]. We used also data on local budgets obtained from the official State online portal on local budgets [7].

We conducted two surveys in selected communities to get data on vulnerability of UTCs to climate changes: one investigates the opinion of local governments, involved in the development of local socio-economic strategy, programs and measures, on climate change issue; second investigates the opinion of local rural inhabitants, faced with the effects of climate change. We developed a questionnaire based on [8], which was distributed among the focus groups of respondents. Also, we gathered the assessment of climate change impact on SES in the different natural and climatic zones from the climate-related practitioners and scientists.

2.2 Climate change impact

For the assessment of UTCs vulnerability to climate change we used the framework consisted of the three components: factors of climate change impact; elements or subsystems of UTC' SES; vulnerability index (Fig. 2).

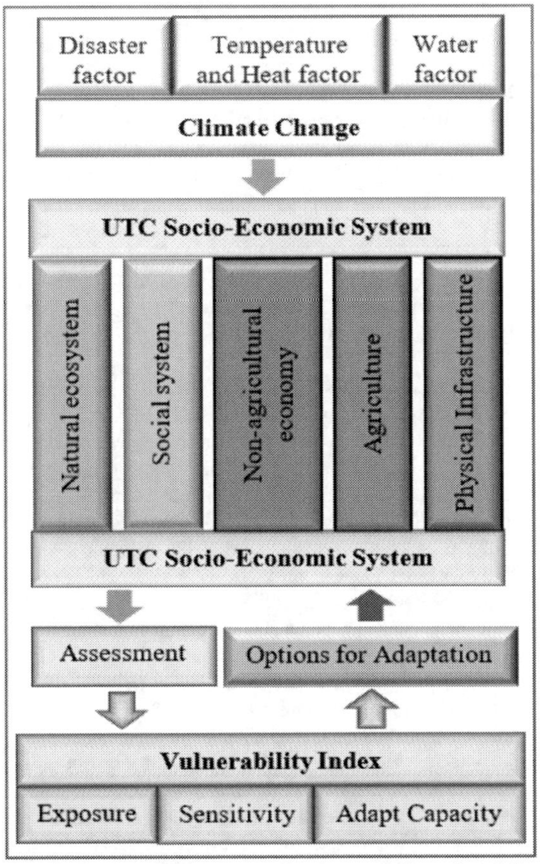

Fig. 2. Framework for assessment of community vulnerability to climate change

Methodologically this framework takes the IPCC concept on vulnerability [9] as a function of exposure, sensitivity and adaptive capacity and uses the indicator-based approach as a most common for the assessment [10, 11]. Disaster impact (in particular, hydrometeorological events), air temperature and heat stress, precipitation and water availability changes are the key three factors, described the exposure socio-economic system of UTCs to climate change. Also, each SES is tested for their sensitivity and adaptation capacity to climate change though the numerous of quantitative and qualitative indicators. The qualitative indicators were collected from the three groups of expert survey: UTC representatives, local inhabitants and scientists, and the quantitative indicators – from official national and local statistical sources, in particular, listed in the 2.1 chapter. Besides, we used the results of the Ukrainian hydrometeorological Institute [12] on climate change projections, in particular, precipitation, average and extreme air temperatures etc.

The qualitative indicators have the three-mark scale: low, medium and high level of vulnerability. The quantitative indicators were normalized also to the three-mark scale. Sum of marks for each of subsystems comparing with the maximum level represents the subsystem vulnerability based on the following scale: below 30 percent of maximum level – low vulnerability, from 30 to 60 percent of maximum level – moderate vulnerability, and more than 60 percent – high vulnerability. The SES vulnerability index applies an equal weight sub-systems vulnerability index.

3 Results and discussion

3.1 Availability and potentials of natural resources for local development

All six selected ATC have natural resources sufficient for sustainable development. At the same time, they are very different by the composition of natural resources, possibility and potential for their use. The ATC are situated in different natural climatic zones, have different areas and landscapes etc.

The main natural resources are the following: land, water, forests, minerals, soils and climate.

Being a constant and regular source of revenue for municipal budget, **land or territory** is a resource that is one of the most important for sustainable development of local communities (table 1). Economic potential of the land area depends very much on its functional structure. For instance, substantial areas of Vt UTC are covered by mountain forests and Ns UTC has a significant share of natural protected areas, which means low economic pressure on that lands. As opposed to such a low economic pressure on lands of Vt and Ns UTCs, land area of Sch OTG is densely populated and provides a high rate of economic output.

Table 1. Land resources of local communities (UTCs)

UTC	Territory, square km	Territory per 100 inhabitants, square km
Bzh	136,20	3,99
Vt	154,94	2,62
Vk	109,92	2,48
Ls	189,82	3,10
Ns	356,98	10,69
Sch	499,50	2,37

Agriculture is the main economic activity and way how the land is used in all UTCs except Vt UTC with its forestry, but even for population of Vt UTC the output from their agricultural land plots plays the most important role for their well-being. The combination of territorial characteristics and natural resources used in agriculture (particularly soils) is the basis of the local agricultural natural potential.

Comparing with other communities, Bzh UTC has the best soils providing the highest land rent. There are estimated land rent values officially calculated for different natural agricultural districts in Ukraine. This value for arable land of Bzh UTC is 43801 UAH per ha for 1/3 of its territory and 28378 UAH per ha for other 2/3 of the territory. For other UTCs the values are as follows: 32221 UAH for Ns UTC, 28213 UAH for Ls UTC, 22990 UAH for Sch UTC, 16218 UAH for Vk UTC, and 6893 UAH for Vt UTC.

The common feature for Bzh, Vk, Ls, Ns and the lowland part of Vt UTC is the excessive shares of agricultural land in their territory and arable land in their agricultural land. Such an economic pressure on land leads to violations of environmental standards, soil erosion, degradation of landscapes and other negative consequences.

Forest resources are very important of Vt UTC because 70.9% of its area are covered by managed forests. The share of area covered by forest is also higher than the average Ukrainian value (16%) in Sch UTC (34.3%), Ns UTC (22.6%) and Bzh UTC (22%). Forest covers 14.6% of all area of Vk UTC, and area under forest is very insignificant in Ls UTC.

Forest landscapes are valuable natural and recreational areas. Unfortunately, the UTCs use such their potential in inefficient and unsustainable way. It is true especially for Vt UTC, where there are processes of intensive cutting of woods leading to the deterioration of forest structure, soil degradation, floods and landslides.

The UTCs are also very different by their **water resources**.

The area of Bzh UTC belongs to the areas with poor water resources and water supply. The surface water outflow to small rivers and water levels in wells reduced in recent years. Vt UTC is situated in the Dnister River Basin, and the tributaries of the Dnister river contribute to the local tourism potential. Water resources are sufficient, but there is no communal water supply there. The community uses underground water for business and social needs. There is no communal central sewage system, so surface and groundwater pollution became a serious problem. Water resources of Vk UTC include small tributaries of the river Southern Bug and ponds. The increasing degradation of surface and groundwater water quality, loss of water quantity are the important problems for the community. Ls UTC has the similar problems. Its water resources are represented by small river system and underground waters. Large industrial corporate farms use the water resources in intensive way that leads to overexploitation of water and depletion of water resources. The situation with water resources in the Ns UTC is better than in other UTCs. Its water resources are represented by the Seym River and its tributaries. There are also ponds in all settlements. The economic pressure on water resources is also relatively low. In this community, their water resources are very important components of the local natural potential that is protected in reserves. The water resources of Sch UTC are represented by small tributaries of the Pripyat River Basin, ponds. Much of the area belongs to the protected riverbank territories. The main problems here are pollution of surface waters by local business, and pollution of local drinking water sources.

While come UTCs are situated on the areas with poor water resources, other UTCs have sufficient water resources but suffer from pollution of surface and underground waters. Water quality declined in all UTC during last decades. Recent climate changes also affected water quantity. All mentioned above are the factors to be taken into account for the local development strategies.

Minerals. All UTCs have local mineral deposits (peat, chalk, sand, clay, sapropel and others) on their territories, but not all deposits are being exploited because of different reasons (small reserves, danger to the environment etc.). Some UTCs have also mineral deposits that are of national interest (oil and natural gas).

The common feature in exploitation of local subsoils and mineral resources is the fact that usually all rights belongs to third parties, which registered outside communities. It means that UTCs have little control over and benefit minimally from the extractive projects. Another common feature – some mineral deposits, which were recognized impractical to exploit during Soviet era, could became an important factor of local economic development in the current context when local authorities have more power.

Natural reserves, recreational resources. According to the Ukrainian legislation, the share of natural reserves and other territories with the similar conservation status should be at least 10.4% of the total area. In some analyzed UTC such a share is even higher now. There are natural reserves and other territories with similar status of special natural value (like objects of Emerald Network) in Vt, Ls, and Ns UTCs. Such areas have significant recreational and educational potential, which is not yet used. There is also a common risk of negative economic impact on such natural areas in the UTCs.

At the same time, there are no even territories that potentially could pretend to be transferred to the status of protected in Bzh UTC and Vk UTC.

Another common problem for all communities in their resource use is the fact that not all natural resources have been inventoried and registered in state cadasters. It creates problems both for the planning of efficient local resource use and for the effective protection of unregistered resources, especially for the communal lands. For example, about a quarter of communal land still remained unregistered in the land cadaster in 2020. Inventory of local natural resources is one of the highest priorities for all UTCs.

3.2 Climate change: opportunities and challenges for local development

The investigated UTCs are located in the different agro-climatic zones of Ukraine: the forest-steppe (Bzh, Ns, Vk, Sch), the forest (Ls) zones and the Ukrainian Carpathians (Vt). During the last two decades the traditional climatic conditions were changed under the global warming: 0,8-1°C compare to the average for the

1961-1990 period, including 1-2°C – for winter season. And the traditional borders of the agro-climatic zones shift north by 200-300 km [4]. As a consequence, two UTCs belong to the new agroclimatic zones: Ls UTC – forest-steppe instead forest and Bzh UTC – steppe instead forest-steppe.

For further decade (till 2030) it is forecasting the significant increase of the yearly air temperature and slight growth of precipitation in Ukraine in general: up to 0,5°C and +5 percent comparing with 1991-2010 respectively. But considering the higher level of evaporation and the lower frequency of rainfall, there is a risk of the lower moisture supply and availability of water resources. The higher increase in precipitation occurs in spring and winter seasons. Instead, in the summer months, the amount of available precipitation reduces, as well as in the first autumn months, which challenges for winter sowing campaigns. But note that the forecasts, averaged within the different climate models, may have both arid (as the 2019/2020 season) and humid (as the 2018/2019 season) scenarios of global climate change.

In general, all of the investigated UTCs challenge by hydro-meteorological factor, in particular, hail, thunderstorms, squalls, strong gusts of wind. However, most UTCs face with problem of precipitation scarcity, except Vt UTC, which challenge by flooding.

Another climate factor, effected on UTCs vulnerability, is high air temperature and heat stress. In Ukraine, droughts, which cover up to 30% of the country, already occur once every 2-3 years, and in the future their frequency will double. After long-term drought there is high risk of extreme hydro-metrological disasters, causing great damage to the entire socio-economic UTCs' system.

The above-mentioned climatic factors impact on, first of all, agriculture, which leads to lower crop yields, especially, in the central, southern and eastern regions of Ukraine (in particular, in Bzh UTC). Besides, climate factors impact on energy production and supply due to higher demand for water distribution, air condition etc. On the other side, energy infrastructure stresses by extreme meteorological damages (storms, wind, freezing etc.). These are only the most visible and frequent negative impacts and vulnerable sectors to climate change.

In general, our assessment confirms the exceeding the threshold of the high vulnerability of the investigated UTCs SESs, except Ls UTC (but it is to close to threshold – 0.59). But there is significant variability of UTCs vulnerability between the subsystems (Fig. 3).

Agriculture is the among the most vulnerable UTCs socio-economic subsystem. The assessment reveals that agriculture of all UTCs (Bzh, Vk, Ns and Ls) are high vulnerable (agriculture vulnerability ranges from 0.64 to 0.80) to climate change due to (1) high exposure to drought and land erosion, (2) high sensitivity of agriculture to climate change due to high percentage of arable land and rain-fed crop area, high level of monocropping in agricultural practice), (3) low adaptive capacity of farming system due to the poor implementation of the modern agricultural practice (low- and no-till) and the innovative irrigation system, unbalanced fertilization of crop land, despite the high level of financial assets, gained from agricultural activities during the last decade.

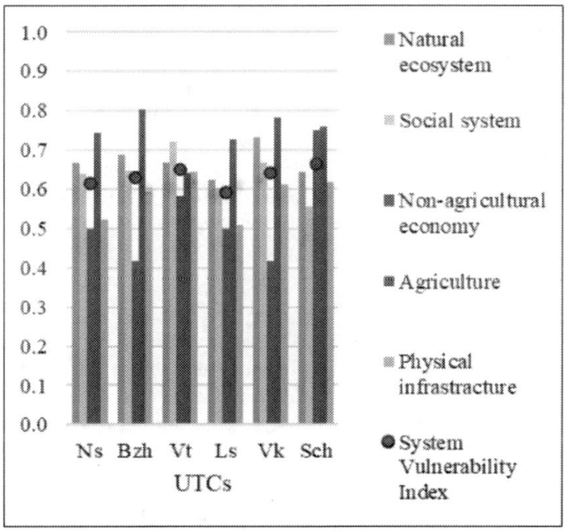

Fig. 3. Communities vulnerability to climate change

Natural ecosystem is the next most vulnerable subsystem with also high vulnerability index for all UTC (0.62-0.73). High vulnerability of natural ecosystem causes by high sensitivity of ecosystem to warmer climate, which are detected by the key indicators, like tree pests and diseases, lower river runoff, lower biodiversity etc.

Natural ecosystem can eliminate some of these negative effects thought self-reproduction or self-renewal. But this ability is limited by the intensive exploitation of natural resources and high anthropogenic pressure (for example, soils do not cope with the function of filtering groundwater, supersaturated with mineral fertilizers, etc.).

Hereby, high vulnerability of agriculture and natural ecosystem is crucial challenge for socio-economic development of the investigated UTCs with monofunctional economy which are based on natural resources use.

Moreover, UTCs SESs challenge by low incomes and high poverty rate of rural families (in particular, due to the high dependency of family income from farming household), unfavorable age structure of rural society, low level of awareness and activities of the rural inhabitants. Our study reveals the high vulnerability of the social subsystem for almost all investigated UTCs (except Sch UTC), which ranges from 0.60 to 0.72.

In the process of decentralization of power, most rural UTCs inherited the destroyed physical infrastructure. Although the government provides the support to the development of infrastructure through the different subventions and subsidies, it is not enough to overcome this problem. Therefore, communities should try to find internal assets to renovate rural physical infrastructure on their own. So, it is not surprising that our assessment of infrastructure vulnerability to climate change confirms this issue. And although there is slight exceeding of the threshold of the high vulnerability for the investigated UTCs, it does not provide any optimism concerning the UTCs resilience to climate change.

And finally, the relatively lower level of vulnerability of the non-agricultural economy (0.42-0.58) of the investigated UTCs is mainly due not to its lower sensitivity and higher adaptation capacity, but to its underdevelopment (except Sch and Ls UTCs with few food processing enterprises). On the one hand, it strengths the capacity (in particular, due to the low negative environmental impact), and on the other hand – it challenges due to the limiting financial resources for the implementation of adaptation policy measures.

4 Conclusions

Decentralization reform in Ukraine and transition of the power to the local level has exposed a number of problems in the field of local socio-economic development, especially rural development. Abandoned infrastructure, a monofunctional economy, and a weak level of human capital are just the main challenges to socio-economic development. At the same time, local communities have faced a global challenge – climate change. While local communities are already aware of the socio-economic challenges and are trying to take them into account in their development strategies, global climate change is fragmentary understood by local self-governments and communities. They don't take it into account in their long-term development strategies and plans.

UTCs, which are selected for this study, belong mainly to rural communities with a monofunctional economy based on the intensive use of natural resources, mainly agricultural land, which is very important resource for local communities. At the same time, under the climate change, the value of water resources increases exponentially, and it affects almost all elements of the local SESs.

The methodological tool we developed for assessing the vulnerability of the SES and its components to the climate change is adapted to use available national and local data sources. Thus, it could be used to analyze all Ukrainian UTCs, despite some limitations of the developed method. In particular, we understand some risk of subjective assessment of climate change effects on UTC' SES within the collected surveys. We gathered the professional assessment (scientists and practitioners in the field of climate change impact) to minimize this risk.

The obtained results confirmed the assumptions about the high vulnerability of local SESs to climate change. The most important elements of the local SESs are agriculture and and natural ecosystem, and they are the most vulnerable to climate change.

Possible scenarios of climate change indicate a further increase in temperature, which with slight increase in annual rainfall will result in further reduction of groundwater levels. This will negatively affect the availability of water resources from traditional sources (surface sources) and will increase the demand for the use of deep-water sources. It is a significant challenge for the monofunctional economies of the UTCs analyzed in this study.

Effective measures to reduce the negative effects of drought and reducing moisture supply are the following: implementing the moisture-saving tillage technologies; using rainwater for agriculture and other purposes; replacing traditional crops with drought-resistant ones (for example, replacing potatoes with chufa; traditional cereals (wheat, corn, barley) with millet, sorghum, chickpeas, etc.).

Additional effective measures to reduce the vulnerability of agriculture are the insuring production risks and increasing areas under perennial crops and pastures to reduce the share of arable land with a positive effect on the sensitivity of agricultural land use to climate change. Currently, the Ukrainian government is studying the possibility of introducing preferential risk insurance in agriculture, which is especially relevant after the large-scale drought in 2020.

The key measures to overcome the vulnerability of the local infrastructure are the following: enhancing the readiness and opportunities of local communities to eliminate the consequences of adverse hydro-meteorological phenomena (formation of mobile local teams of volunteers, purchase of special equipment and tools) and development of flooding prevention networks (support the working conditions of drainage, sewerage and water networks).

We argue and the practice confirms that, at the local level, small individual farms usually have higher potential for adaptation to climate change. In general, the potential for adaptation depends not so much on land size, but mainly on land use strategy. In this context, large corporate farms that renting their land from small landowners and small family farms have different motives for their strategies. Large corporate agro holdings have poorer connections with the local communities where they operated under the lease agreements with local land owners. They are not diversified and oriented mainly on production of wheat, maize, rapeseed and sunflower that are most commercially attracted agricultural crops. Such corporate farms are often registered and pay their taxes outside the local communities where they cultivate land rented from local peasants. Usually, large corporate farms don't really care about interests of local communities because their economic interest is the only driving force of their activities.

At the same time, owners of small family farms live close to their land plots. Such farmers and members of their families are integrated in the communities and shares the interests with other local rural residents. They are more dependent on the opinion of their neighbors. Small local farms are more likely to act in a responsible way that meet the local sustainable development goals. At the same time, they are more mobile to change their plans, diversify the crops, and face new environmental challenges.

The mentioned above means the excessive concentration of agricultural land use, which is forced by corporate farms, leads to declining local agricultural potential for adaptation to climate change. That is why it is the interest of local authorities and communities to state the priority and implement the support measures for family farming into their local development strategies.

References

1. Statistical Yearbook of Ukraine 2019. State Statistics Service of Ukraine. (2020) [On-line]. Available at: http://www.ukrstat.gov.ua/druk/publicat/kat_u/2020/zb/11/zb_yearbook_2019.pdf
2. Alexandratos, N.; Bruinsma, J. World Agriculture Towards 2030/2050: the 2012 revision. FAO, ESA Working paper No 12-03, (2012).
3. Ackermann A. Climate change may prevent Ukraine from becoming agricultural superpower [On-line]. Available at: https://www.atlanticcouncil.org/blogs/ukrainealert/climate-change-may-prevent-ukraine-from-becoming-an-agricultural-superpower/
4. Technology Need Assessment Report. Adaptation. Ministry of Ecology and Natural Resources of Ukraine (2019) [On-line]. Available at: https://tech-action.unepdtu.org/wp-content/uploads/sites/2/2019/09/final-ukraine-tna-adaptation-report.pdf
5. Public Cadastre Map. State Land Cadaster of Ukraine. [On-line]. Available at: https://map.land.gov.ua/?cc=3461340.1719504707,6177585.367221659&z=6.5&l=kadastr&bl=ortho10k_all
6. USR Register. [On-line]. Available at: https://irc.gov.ua/en/el_reg
7. Open budget. [On-line]. Available at: https://openbudget.gov.ua/local-budget
8. National Climate Vulnerability Assessment: Ukraine. Climate Forum East (CFE) and NGO Working Group on Climate Change (2014) [On-line]. Available at: https://climateforumeast.org/uploads/other/0/708.pdf
9. IPCC. Climate Change 2007: Impacts, Adaptation and Vulnerability; Contribution of Working Group II to the Fourth Assessment Report of the Intergovernmental Panel on Climate Change (IPCC); Cambridge University Press: Cambridge, UK (2007).
10. Neset, T.S.; Wiréhn, L.; Opach, T.; Glaas, E.; Linnér, B.O. Evaluation of indicators for agricultural vulnerability to climate change: The case of SwediSch agriculture. Ecol. Indic. 105, 571–580 (2019).
11. Wiréhn, L.; Danielsson, Å.; Neset, T.S.S. Assessment of composite index methods for agricultural vulnerability to climate change. J. Environ. Manag. 156, 70–80 (2015).
12. Results of scientific research projects. Ukrainian Hydrometeorological Institute. [On-line]. Available at: https://uhmi.org.ua/project/rvndr/

Development of Supplements Prevention System in Oxygen Converter Process in Order to Increase the Economic Efficiency of Steel Melting

Evhen Chuprinov[1*], *Vitalij* Lyalyuk[1], *Hanna* Andrushchenko[1], *Daria* Kassim[1], and *Natalia* Rad'ko[1]

[1]State University of Economics and Technology, 50006, Kryvyi Rih, Ukraine

Abstract. The article analyzes the work of the oxygen-converter shop of PJSC «ArcelorMittal Kryvyi Rih». It is shown that almost half of the total number of blowing supplements (~ 44%) is carried out by temperature. In order to improve the conditions of steel smelting and reduce resource costs, calculations were performed and new technological solutions were found in steel smelting to reduce the number of additives. The study of real smelters demonstrated the need for corrective operations in the smelting of steel using scrap metal of different types. Scrap smelts have a clear relationship between the amount of scrap in the charge and the temperature of the melt in the first dump. The addition of «goat» scrap to the charge has a number of uncontrolled effects on the process due to significant fluctuations in the chemical composition of this type of metal charge. An economic analysis with introducing the developed control mechanism over the course of steel smelting into the software system for conditions PSJC «ArcelorMittal Kryvyi Rih» was performed.

1 Introduction

The modern period of development of the metallurgical industry is characterized by intensification of production, which consists in continuous increase of specific and absolute productivity of metallurgical units, improving the range of manufactured metal products and its quality, which reduces the specific costs of metal, energy and labor costs. As a result, the scale of production of even metal-containing finished products increases in the absence of growth of steelmaking.

Along with the reduction of metal costs, there is a problem associated with production losses. One of the problems is supplements in steel production, which reduce the productivity of the unit and deteriorate the quality of steel. The main cause of blowing supplements is the unstable composition of the metal charge and various fluctuations in the temperature of the loaded charge [1].

The constructive solutions available in real practice to eliminate this problem are not possible for implementation in the conditions of the oxygen-converter shop of ArcelorMittal Kryvyi Rih, which requires a search for new approaches to reduce blowing supplements. In the current conditions, the most optimal direction will be work in the field of software engineering for operational control over the technological process.

The development of information methods of control and management of the technological process has recently opened up new opportunities to improve the processes in metallurgy at all production stages and steel-making production did not stand aside. Researchers obtained mathematical models, discovered new dependencies and refined already known ones [2].

Thus, the purpose of this work is the analysis of melting passports at PJSC "ArcelorMittal Kryvyi Rih" and the search for new technological solutions for steelmaking to reduce the number of additives based on the analysis.

2 Analysis of previous studies

A number of studies indicate that the available methods of carbon control do not provide steel of a given chemical composition from the first felling [2, 3]. A significant contribution to solving this problem was made in the works of V.S. Bogushevsky and co-authors [2-4], who developed a series of mathematical models, including static calculation, dynamic prediction and control mechanisms of melting in a closed mode. In [5], an improved method for calculating the material-thermal balance of oxygen-converter steelmaking was developed, which increases the flexibility of the steelmaking process by using the influence of more correct melting conditions. This will improve the technological process and significantly reduce the number of blowing supplements. However, all currently available developments do not allow to finally solve this problem, therefore, the search for additional mechanisms to control the converter process is an urgent problem.

In recent years, advanced methods have been developed to monitor and control processes in metallurgy. The introduction of powerful computing technology allowed technologists to accumulate and analyze large amounts of data in order to introduce

* Corresponding author: itchupa@gmail.com

advanced automation systems in metallurgy [6], and the development of programming languages allows introducing machine learning elements into metallurgy [7] to refine the developed models directly in the course of technological processes. Due to this, a lot of works have appeared devoted to the introduction into a real production process of models developed to control both the entire technology of steel production, and individual processes. For example, in the literature, there are works on improving the efficiency of using the conversion of CO_2 [8], which leads to an improvement in the energy efficiency of the converter process. Much has been done in the development of models for studying processes at the micro level. Thus, in [9], a model was created for macrosegregation of adhesion under the bloom surface, which takes into account the electromagnetic field, flow, heat and transfer of solutes based on the average volume method to study the effect of electromagnetic stirring in the form of M-EMS, and in the study [10], a model was developed dynamic behavior of the molten bath of the MLA-PBF.

Since 2017, the technical department of the PJSC «ArcelorMittal Kryvyi Rig» has begun a large-scale implementation of the software environment «Steelmaking process in the smart-online» mode [11]. The purpose of this project is to ensure the cyclical operation of the steelmaking process and constant correct informing of the manager and technological personnel involved in the process about the state of production online. The system covers all measuring instruments and sensors, starting from the control of the chemical and physical characteristics of the liquid iron leaving the blast furnaces (fig. 1) and scrap (fig. 2), ending with control over the processes in the continuous casting department.

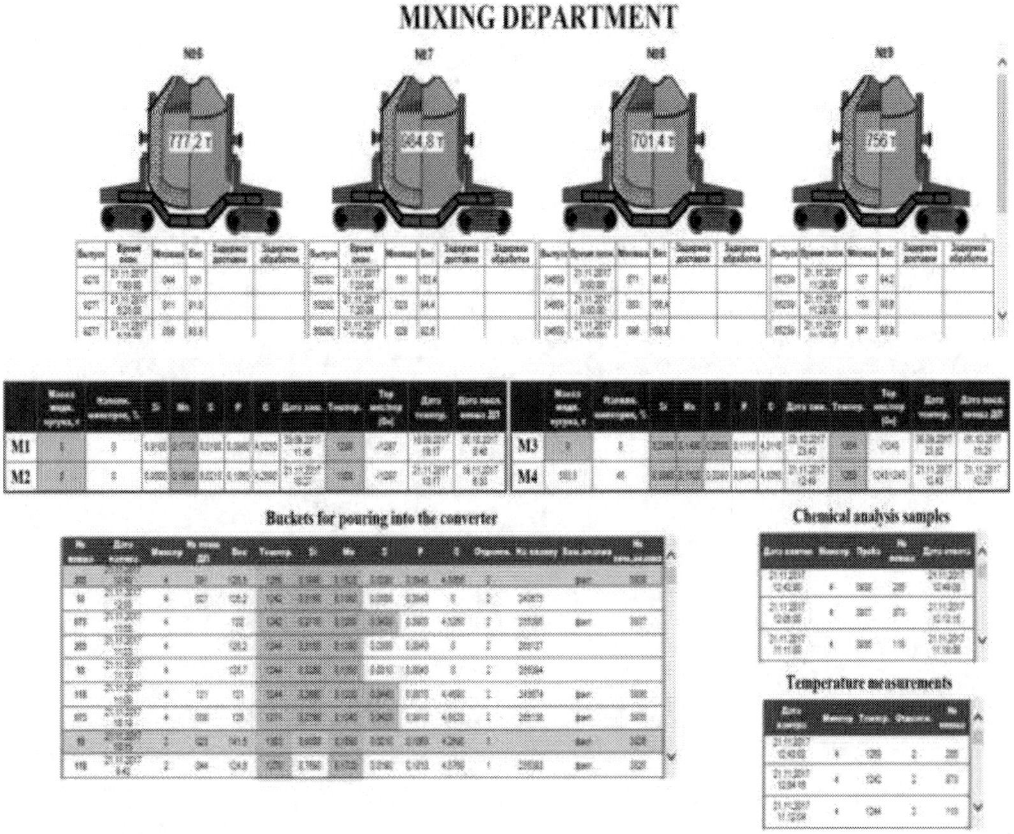

Fig. 1. Characteristics of the liquid iron

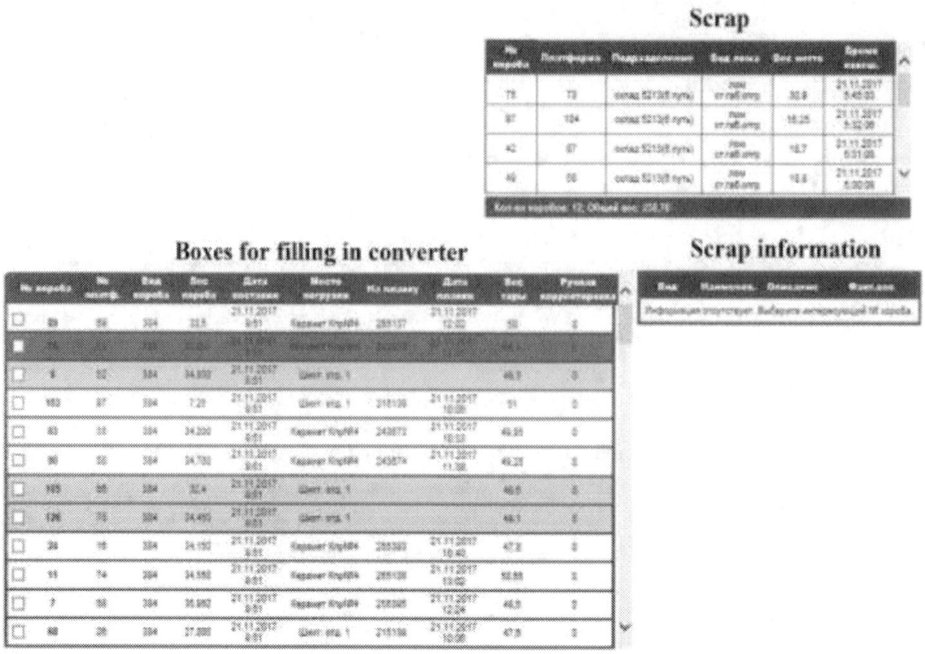

Fig. 2. Characteristics of the scrap

3 Materials, methods and results of the study

The use and implementation of this system made it possible to carry out a complete analysis of a number of technological processes, identify the relationship between various factors and identify problem areas in the adopted technology. The most important role for identifying problem points in the steel smelting process is played by the smelting passport and batching model (fig. 3), allowing to control all influences on the molten metal. As part of this work, the authors analyzed an array of more than 500 melting's passports.

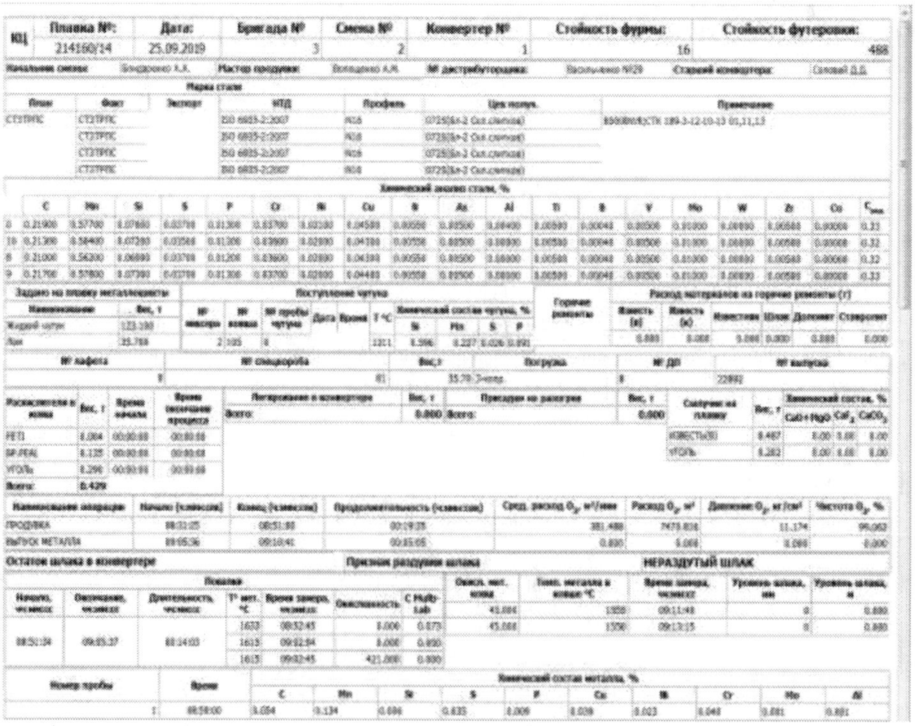

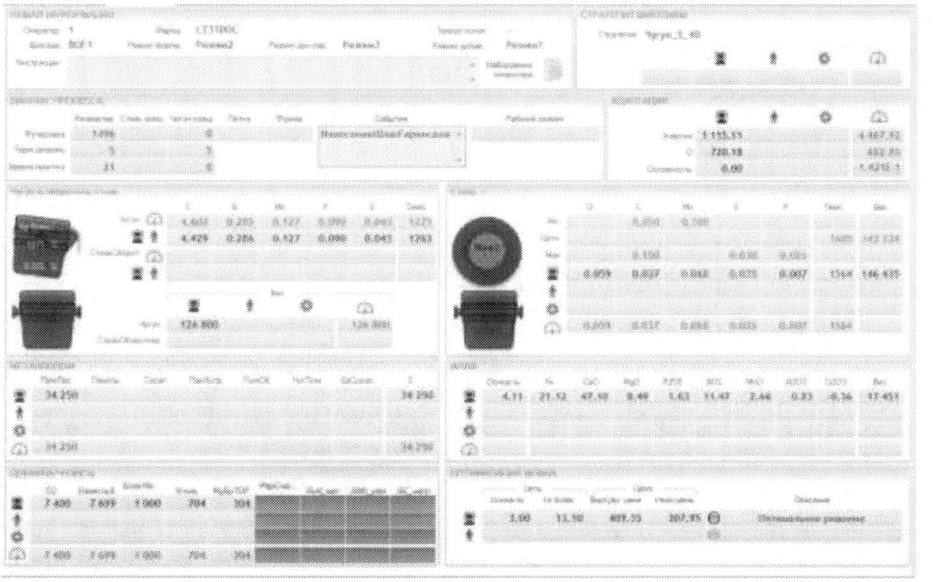

Fig. 3. Passport of the melting and batching model

According to the analysis of the oxygen-converter shop of PJSC «ArcelorMittal Kryvyi Rih» almost half of the total number of blowing supplements (~ 44%) is carried out by temperature. The physical heat of liquid cast iron contributes to the converter bath about 50-54% of the total heat, and an increase in the temperature of the cast iron, respectively, leads to an increase in the temperature of the bath.

Analyzing of the melting's passports, new relationships were obtained between the indicators of the oxygen-converter process, which are typical for the conditions of the steelmaking shop of PJSC «ArcelorMittal Kriviy Rih».

The dependence of the metal temperature on the first pile on the temperature of cast iron (fig. 4) is represented by the equation:

$$T_M = -0.0002 \cdot T^2_{c.i.} + 0.5444 \cdot T_{c.i.} + 1142.6,\ r^2=0.45, \quad (1)$$

where T_M is the temperature of the metal on the first dump, °C; $T_{c.i.}$ – temperature of cast iron, which is poured into the converter, °C.

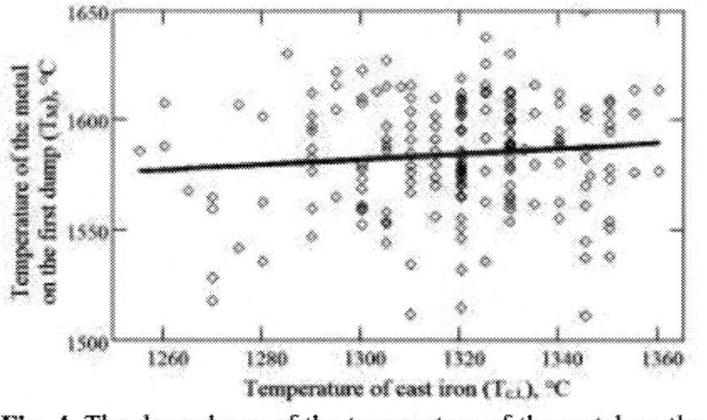

Fig. 4. The dependence of the temperature of the metal on the first dump on the temperature of cast iron

The influence of the amount of cast iron on the temperature of the metal on the first dump (fig. 5) is expressed by the equation:

$$T_M = 0.0109 \cdot Q^2 - 0.4442 \cdot Q + 1556.3,\ r^2=0.51, \quad (2)$$

where Q is the amount of cast iron, %.

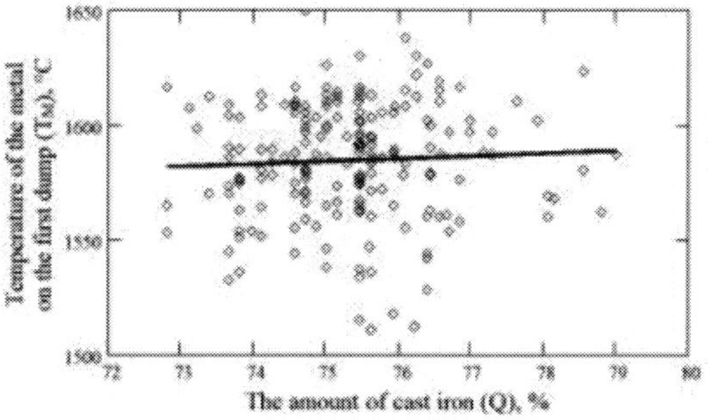

Fig. 5. The dependence of the temperature of the metal on the first dump on the amount of cast iron

The dependence of the metal temperature on the Mn content in the cast iron was also revealed – with the increase of the manganese content in the cast iron the metal temperature decreases (fig. 6):

$$T_M = 7.1733 \cdot [Mn]^2_{c.i.} - 44.9 \cdot [Mn]_{c.i.} + 1603.8,\ r^2=0.47, \quad (3)$$

where $[Mn]_{c.i.}$ – manganese content in cast iron, %.

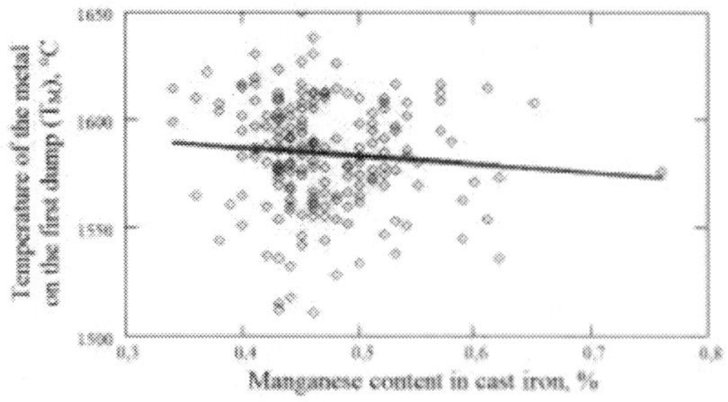

Fig. 6. Dependence of metal temperature on the first dump on the Mn content in cast iron

As the amount of scrap in the filling decreases, the temperature of the metal on the first dump increases (fig. 7). This is due to the fact that the scrap is a coolant, for its dissolution consumes a certain amount of heat. This dependence is expressed by the equation:

$$T_м = -0.0559\cdot P^2 + 1.3936\cdot P + 1584.2,\ r^2=0.56, \quad (4)$$

where P is the content of scrap in the metal charge, %.

Scrap enters to the landfill with a certain degree of slagging, its cooling capacity is greater than that of pure scrap.

All this contributes to the fact that with increasing amount of slag scrap in the filling there is a decrease in metal temperature (fig. 8). This dependence is expressed by the equation:

$$T_м = 0.006\cdot P_s^2 - 1.2823\cdot P_s + 1589.6,\ r^2=0.41, \quad (5)$$

where P_c is the content of slag scrap in the metal, %.

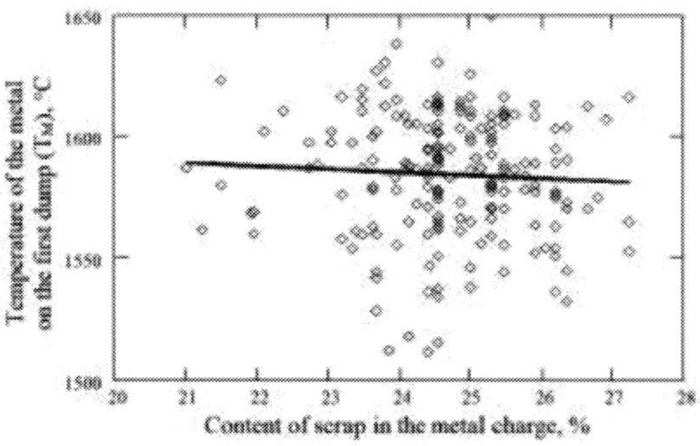

Fig. 7. The dependence of the temperature of the metal on the first dump on the amount of scrap in the charge

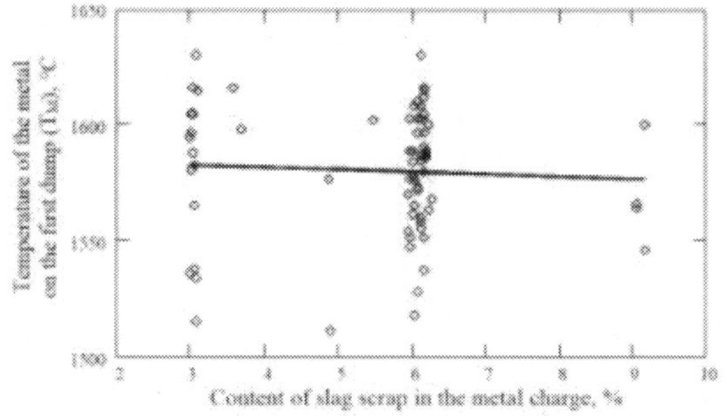

Fig. 8. The dependence of the temperature of the metal on the first dump on the amount of slag scrap in the charge

The impact of the so-called "goat" scrap has a slightly different character. With an increase in the amount of "goat" scrap loaded, there is an increase in metal temperature (fig. 9).

All these dependences confirm the fact that the "goat" scrap does not actually perform its function as a coolant and leads to an increase in the number of blowing supplements. When sampling unmelted scrap increases heat dissipation due to the cooling effect is 1400-1500 kJ/kg, which is equal to the same value of scrap metal. As a result, technologists are forced to perform blowing supplements operations on temperature with overheating of the melt [12].

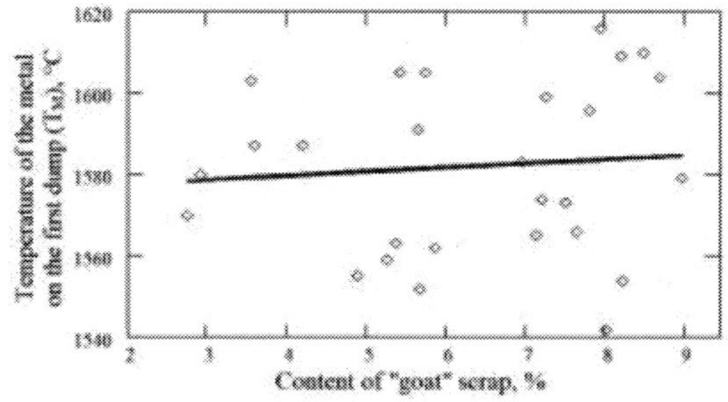

Fig. 9. The dependence of the temperature of the metal on the first dump on the amount of «goat» scrap in the charge

The expression of this dependence is shown by the following equation:

$$T_м = -0.0371\cdot P_g^2 - 0.619\cdot P_g + 1574.9,\ r^2=0.43, \quad (6)$$

where P_g is the content of «goat» scrap in the metal charge, %.

The use of the obtained mathematical models and their introduction into the smart matrix of the oxygen-converter process (fig. 10) made it possible to significantly improve the efficiency of the shop and improve the technology of steelmaking.

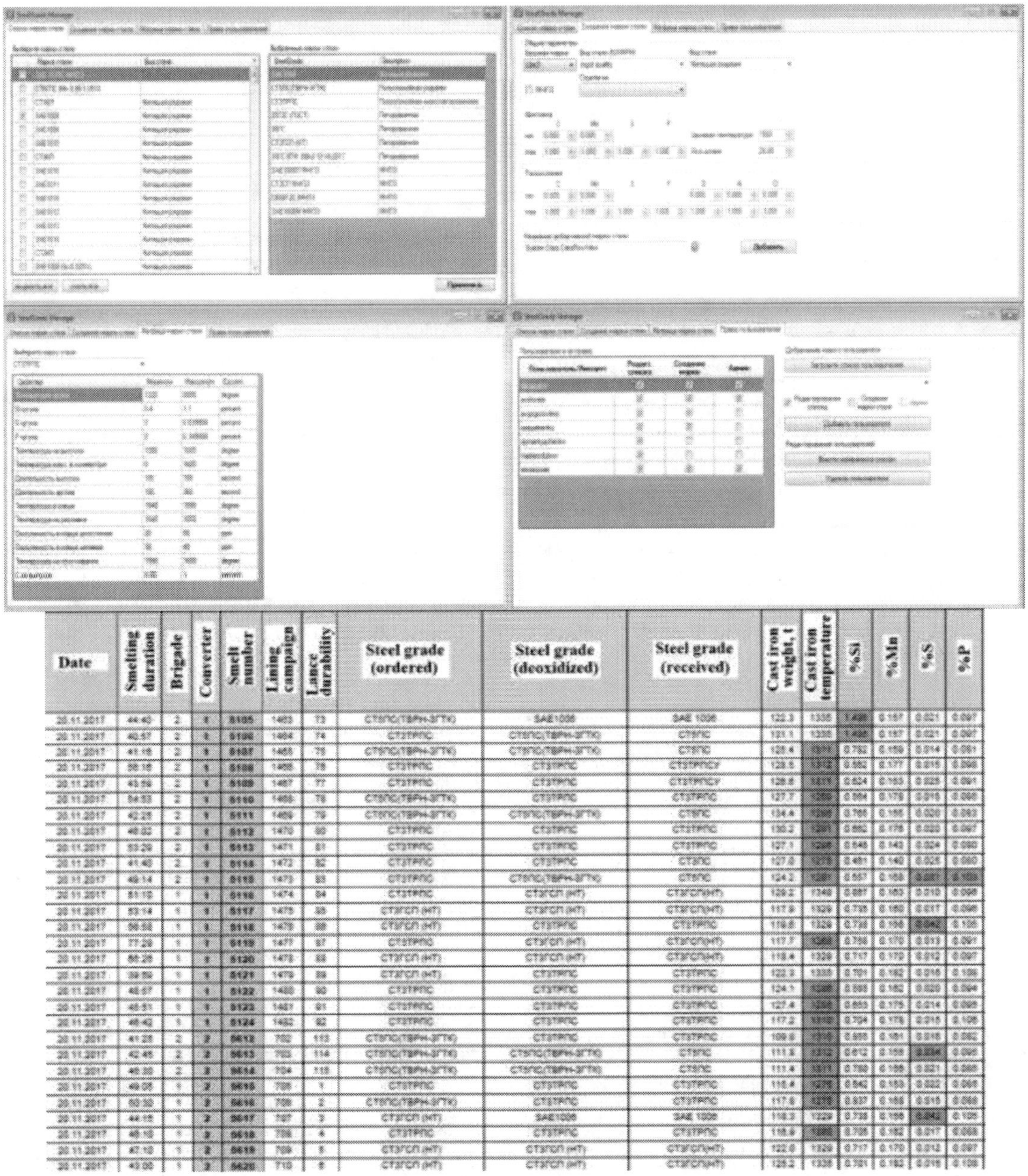

Fig. 10. Smart matrix of the oxygen-converter process

The corrective operations proposed by the authors of the article are based on calculations performed using the material-thermal balance of the oxygen-converter process.

The amount of heat required on average for converter melting is $285 \cdot 10^3$ MJ. Of these, the largest amount of heat (50-54%) gives the physical heat of cast iron, coming from the blast furnace shop. The share of chemical heat accounts for 34-35% of heat input (in the processing of pig iron, the amount of additional heat increases slightly). Blast furnace slag, which enters the converter from the mixer, contributes about 0.6 % of heat [13].

Thus, in the processing of pig iron weighing 10 tons, the same chemical composition as pig iron, is additionally made:

$$[C] = 1{,}10 - 0{,}20 = 0{,}90\ \% = 378\ \text{kg}\ [C]$$
$$[Si] = 0{,}23 - 0{,}08 = 0{,}15\ \% = 63\ \text{kg}\ [Si]$$
$$[Mn] = 0{,}46 - 0{,}46 = 0{,}00\ \% = 0\ \text{kg}\ [Mn]$$
$$[P] = 0{,}06 - 0{,}04 = 0{,}02\ \% = 8{,}4\ \text{kg}\ [P]$$

which gives due to chemical reactions the arrival of heat in the amount 17773 kJ.

That is, for 1 ton of pig iron enters the bath of the converter an additional 17773 kJ. The cooling effect of cast iron is slightly greater than in comparison with ordinary scrap metal, which is associated with a smaller particle size distribution and, accordingly, a larger area of interaction (table 1).

Table. 1. The amount of heat given off by the metal additive

Name	Scrap (1 t)	Pig iron (1 t)
Heat capacity	714 kJ	2256 kJ
Cooling effect	- 1.45 MJ	- 1.8 MJ
Total	- 0.736 MJ	0.456 MJ

As can be seen from table. 1, the difference in heat input is 1,192 MJ per 1 ton of pig iron. The most effective compensator for excess heat in terms of economic feasibility is limestone ($CaCO_3$), the cooling effect of which is $12 \cdot 10^3$ kJ per 1 ton of cast iron. The main disadvantage of limestone – low reactivity, which does not allow the optimal amount to react in the necessary time interval to ensure high productivity of the steelmaking unit. As a result, the technologist is forced to increase the amount of limestone consumption by 30-40 %, which in total is 130-140 kg of $CaCO_3$ per 1 ton of pig iron.

Along with pig iron, steel scrap is used, the degree of contamination of which ranges from 5 to 40 %, which negatively affects the arrival of heat. The heat capacity of scrap is 714 kJ/t. And from the heat of formation and the content of chemicals in the slag, from the steel scrap can calculate its heat capacity [13]:

FeO	= 23.32 kJ
Fe_2O_3	= 23.86 kJ
CaO	= 168.72 kJ
SiO_2	= 108.51 kJ
MgO	= 49.97 kJ
Al_2O_3	= 88.83 kJ
MnO	= 40.08 kJ
S	= 0.133 kJ
Total:	= 503 kJ

Analysis of the difference between the heat capacities of scrap and steel scrap (714 kJ/t and 503 kJ/t, respectively) gives an answer to the questions of technologists of PJSC «ArcelorMittal Kryvyi Rih» relatively large number of «cold» melting trunks that come with steel scrap. Given that the amount of steel scrap averages 10 tons per smelting, it becomes possible to calculate the amount of heat lost due to low heat capacity and a huge cooling effect. Thus, in scrap it varies between 1400-1500 kJ/t, and in steel scrap – 1600-1800 kJ/t. The amount of slag in the steel scrap is 20 %, and the heat input from 1 ton is 671.8 kJ [13]. The heat capacity of steel scrap is less than in scrap by 42.2 kJ/t. Along with this, there is a sharp increase in the cooling effect of scrap (table 2).

Table. 2. Cooling effect of scrap and steel scrap

Name	Scrap (1 t)	Steel scrap (1 t)
Heat capacity	714 kJ	671.8 kJ
Cooling effect	- 1.45 MJ	- 1.7 MJ
Total	- 0.736 MJ	- 1.028 MJ

As can be seen from table. 2, the difference between scrap and steel scrap is 0.292 MJ per 1 ton of scrap metal, which requires additional technological operations for its leveling. The most appropriate in this case is to add ground coke to the purge. Calculations performed according to the existing models of material-heat balance of steel smelting in an oxygen converter [14, 15] show that 104.65 kg of carbon or 120 kg of coke per ton of scrap are needed to generate this amount of heat.

A study of meltings passports shows that the share of "goat" scrap is 8 % of all meltings. In the analysis of these meltings passports the tendency is noticed that at increase in quantity of «goat» scrap the temperature on the first dump increases, which is mainly due to incomplete penetration of scrap. This is confirmed by the data on the average yield of the metal. Thus, in conventional batching the yield of metal is 146.9 tons, and in the processing of «goat» scrap – 141.6 tons.

In the process of dissolving the steel scrap is heat absorption with the same as in scrap, heat dissipation in the range of 1400-1500 kJ/kg, and its heat capacity is lower than that of scrap metal. Therefore, to avoid blowing supplements by temperature, it is necessary to overheat the bath of the converter by 20-30 °C, which affects on the oxidation of the metal and the amount of dissolved nitrogen in the metal.

4 Results, discussion and economic justification

The performed analysis of the oxygen-converter smelts showed that the average duration of one smelting on AMKR is 54 minutes, and the blowing supplements with duration of 4-5 minutes is carried out on each melt (fig. 11).

Experimental smelting operations carried out in the converter shop of PJSC «ArcelorMittal Kriviy Rig» with using mathematical models developed by the authors, showed the possibility of obtaining the specified physical and chemical properties of the metal at the outlet in 75-85 % of all cases.

An analysis of the shop's operation shows that one converter smelts 3724 tons of steel per day, with the duration of technological operations spent on one melt 54 minutes. If the developed and proposed adjustments are made, the duration of one melt will be reduced to 50 minutes, and the converter productivity will increase to 4032 tons of steel per day. Thus, the daily increase in the volume of steel will be 308 tons. Since not all smelting can be carried out without blowing supplements, and taking their quantity as 50 %, additional steel volume from one converter will be 154 tons. With the average price of 1 ton of steel at $ 225, the cost of additional production will amount to $ 38.500 per converter.

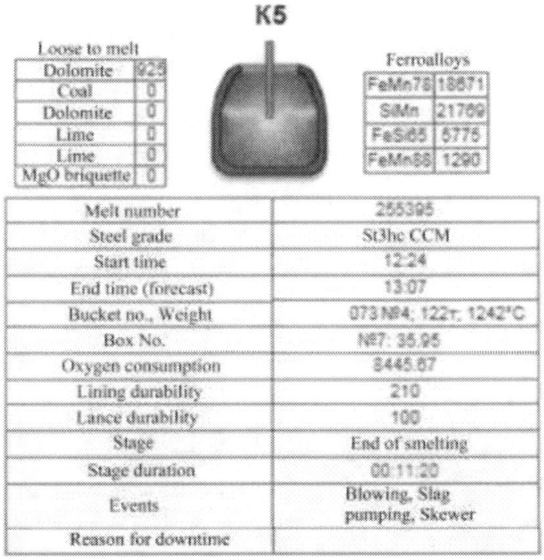

Fig. 11. Converter control screen

5 Conclusions

The use of modern software solutions for the automation of industrial processes is gradually becoming the most important attribute of large metallurgical enterprises. The results presented in the work clearly demonstrate that traditional steelmaking technologies have huge reserves for improvement.

The use of mathematical models, developed and implemented by the authors in production processes in PJSC «ArcelorMittal Kryvyi Rih» conditions, made it possible to obtain an additional yield of suitable metal worth $ 38.500 on a daily basis.

Comparing the results of the analysis of smelting passports and the obtained mathematical models through the prism of the possibility of introducing them into a real production process with the results of foreign studies [6-10], we can say that the proposals developed in the article are at a high technological level.

The value of the accuracy of the approximation of the developed computational models shows their sufficient accuracy for use in the modern oxygen-converter process. Despite the fact that the obtained equations, although they reflect the trends known to technologists, their introduction into the converter process control system through the use of modern computer technologies will allow improving and improving these models "underway", thereby increasing the efficiency of the shop with each melting.

Given the constant growth of the meltings pool passport database, the next stage of this work will be the development and implementation of a «smart» control system for the oxygen-converter process, which could take into account the accumulated «experience» to select the most optimal course of steelmaking.

References

1. S.V. Kolpakov, R.V. Starov, V.V. Smoktij and others, "Steel production technology in modern converter shops", M: Mashinostroenie, 1991, 463 p.
2. V.S. Bogushevskij, S.G. Melnik and S.V. Zhuk, "Steel decarburization as the main parameter of optimal control of oxygen-converter smelting", Metall i lityo Ukrainy, 2014, **2**, S. 14-16.
3. V.S. Bogushevskij and K.V. Yegorov, "Control of bath dynamics during purge as a component of converter melting control system", Naukovi visti NTUU «KPI», 2013, № 1, S. 51-56.
4. V.S. Bogushevskij, G.G. Grabovskij, V.M. Mihajlov and others, "Computer model for calculating the charge and purge converter smelting", Stal, 2006, **1**, S. 18-21.
5. A.V. Sushenko, E.S. Popov, R.S. Sidorchuk and others, "Improving the methodology and developing an application program for calculating the oxygen balance of the cast melting", Visnik Priazovskogo derzhavnogo tehnichnogo universitetu, 2013, Seriya: Tehnichni nauki, **26**, pp. 46-56.
6. Qi, Cc. Big data management in the mining industry. Int J Miner Metall Mater27, 131–139 (2020). https://doi.org/10.1007/s12613-019-1937-z
7. Holm, E.A., Cohn, R., Gao, N. et al. Overview: Computer Vision and Machine Learning for Microstructural Characterization and Analysis. Metall Mater Trans A 51, 5985–5999 (2020). https://doi.org/10.1007/s11661-020-06008-4
8. Zhu, R., Han, Bc., Dong, K. et al. A review of carbon dioxide disposal technology in the converter steelmaking process. Int J Miner Metall Mater 27, 1421–1429 (2020). https://doi.org/10.1007/s12613-020-2065-5
9. Huo, Yk., Zhao, Lh., An, Hh. et al. Model prediction of the effect of in-mold electromagnetic stirring on negative segregation under bloom surface. Int J Miner Metall Mater 27, 319–327 (2020). https://doi.org/10.1007/s12613-019-1906-6
10. Cao, L. Numerical Investigation on Molten Pool Dynamics During Multi-laser Array Powder Bed Fusion Process. Metall Mater Trans A 52, 211–227 (2021). https://doi.org/10.1007/s11661-020-06076-6
11. Technical department of the PJSC «ArcelorMittal Kryvyi Rig», «Steelmaking process in the smart-online» mode, VII Conference of the Steelmakers, 23-24 November, 2017, 15 p.
12. V.I. Baptizmanskij, B.M. Bojchenko, O.G. Velichko and others, "Steelmaking: Manual", K.: IZMN, 1996, 400 s.
13. T.G. Saberzyanov, "Subregular version of the general mathematical model of thermodynamics of metallurgical slag", Izvestiya vysshih uchebnyh zavedenij. Chernaya metallurgiya, 2001, **12**, pp. 53-55.
14. A.V. Sushenko and I.A. Fejerejzen, "Comparative analysis of methods for calculating the material and thermal balances of oxygen-converter smelting",

Visnik Priazovskogo derzhavnogo tehnichnogo universitetu, 2013, **27,** pp. 17-28.

15. V.N. Shinkin, "Mathematical model of technological parameters' calculation of flanging press and the formation criterion of corrugation defect of steel sheet's edge", CIS Iron and Steel Review, 2017, Vol. **13**, pp. 44–47.

Principle of Organization for Laboratory Stand of the Electric Drive with a Real Regulatory System No Time Scaling

Andrii Pirozhenko[1], *Yevhenii* Modlo[1*], *Ruslan* Shaida, *Viktor* Batarieiev[1], *Mykola* Zhukov[1], and *Mykhailo* Drukker[1]

[1]State university of economics and technology, 50006, Kryvyi Rih, Ukraine

Abstract. The existing practical training of specialists in the field of the electric drive is recognized as insufficient, not allowing them to conduct independently a complex of adjustment works or works on elimination of refusals. All known virtual methods of research of electric drives are scaled in time therefore at the trained skills on use of the measuring and registering devices during the work on real installation don't develop, and also skills on work with real knots of control systems for control of the set drive modes. We propose a stand consisting of a real system of regulation and the model of the power drive to work without time scale, thus significantly closer to the actual laboratory setting drive. The structure of the laboratory stand on the basis of the engine of a direct current of independent excitement with the reversible thyristor converter is in details considered. It is proved that an optimal algorithm for the simulation of three-phase bridge converter operating at anchor chain is an algorithm which rooms include a thyristor and of the remaining thyristors allows you to choose the design scheme and carry out the integration of equations describing electric processes in the circuit detect a change in state of the thyristors and the transition to other design scheme. Given the discrete model and considered design scheme of three-phase bridge Converter in a normal mode switching thyristors and emergency mode, if false turn on of the thyristor during commutation. Mathematical model of the device - the solution of differential equations by numerical integration of the Runge-Kutta. Proposed hardware mathematical model based on the family of microcomputer ARM CortexTM fourth generation.

1 Introduction

The problem and its connection with scientific and practical tasks. Training of specialists in the field of the electric drive in higher educational institutions was generally reduced the last years to theoretical part. Practical preparation was limited to performance of standard laboratory works, thus owing to known circumstances (the high cost of stands, restrictions connected with safety measures, restrictions on drive power, etc.) students independently can't conduct a complex of adjustment works or works on elimination of refusals. Besides, practical training of specialists has to provide variety of tasks, for example, ability to carry out adjustment of control systems for drives with engines of various power, various character of loading, etc. Thus it is necessary to provide the minimum capital and operational expenditure connected with creation of the laboratory stand.

2 Literature Review

The analysis of researches and publications. Now virtual methods of research of electric drives which provide variety of tasks, rather low capital expenditure are widely used and don't demand special events for safety of performance of laboratory works [1-6]. But, as known methods of research of electric drives are scaled in time, at the trained skills on use of the measuring and registering devices during the work on real installation, and also skills on work with real knots of control systems for control of the set operating modes of the drive don't develop. Thus, the existing approaches in creation of the laboratory stand for training of specialists in the field of the electric drive using the modeling installations working in time scale won't allow to solve objectives.

3 Statement of the problem

The creation of the laboratory stand for training in setting up and Troubleshooting control systems electric drive, consisting of a real regulatory system and model of the power part of the actuator operating without time scale.

4 Material statement and results

If we consider the laboratory plant consisting of a real control system and model of the power unit (Converter installations and motor) operating without time scale, we get a significant approximation of the laboratory setup to a real drive. Structural scheme of such installation is shown in Fig.1.

* Corresponding author: eugenemodlo@gmail.com

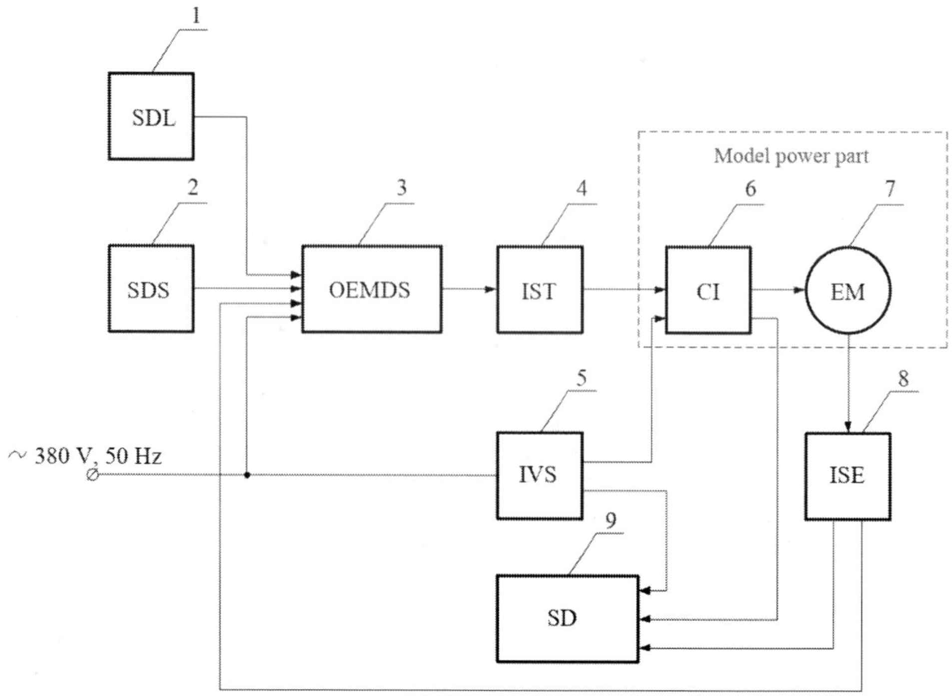

Fig. 1. Laboratory setting consisting of real control system and model of power part.

On fig. 1 shown: 1 – specifies the device load on the motor shaft (SDL), 2 – specifies the device speed of rotation of the motor shaft (shaft position)(SDS), 3 – an operating electric motor drive system (OEMDS), 4 – an interface of transformation of output signals of OEMDS into the discrete input signals of microcontroller (IST), 5 – an interface of transformation of three-phase supply voltage analog input signals of the microcontroller (IVS), 6 – a model of converter installation (CI), 7 – a model of electric motor (EM), 8 – digital signal converter EM in the relevant voltages and currents for organization of feedbacks of OEMDS and connect measuring devices (ISE), 9 - SD are devices for measuring of current (ammeters), voltage (voltmeters) or terminals for connecting of supervision devices.

As the article has limited size it was only considered the matter subject to developing the laboratory stand for educating the methods of tuning and repair of electric motor drive control systems based on separately excited direct-current motor with a reversible thyristor converter. In this case SDL and SDS are potentiometer, synchrotransmitter and the like reference-input elements of the set value of speed (load moment), or programmable devices for determination of diagrams to speed, position and loading. OEMDS is, for example, the analog control system of ETAL ETU 200, or the digital system "DC MASTER" by Siemens. The outputs of these and other control systems are impulse signals of turn-on of thyristors. That's why the ICT contains the elements of transformation of signal level and its galvanic isolation. CI is a model of reversible thyristor converter for the feed of motor armature and non-reversible converter for the feed of drive winding. EM is model of the electric motor of a direct current of independent excitement. All models should work in real time. ISE is a set of digital-to-analog converters of digit values of thyristors' currents and voltages, of electric motor windings and speed, of motor shaft position calculated by models, and also the converter of digital value of motor shaft position into pulse string similar to the signals of incremental encoder. IVC is a galvanic isolation with the voltage divider and quick-operating analog digital converters.

Model of the mechanical parts of the actuator advisable to execute on a separate microcontroller. On the speed (the provision of a shaft of the electric motor) and known mechanical characteristics of elements of the kinematic scheme, the loading moment on an engine shaft is defined, and its moment – on current of an anchor and a stream of excitement. The moment of the electric motor is the operating influence of model of mechanical part of the drive and the loading moment – revolting influence of model of electric part. This issue is well studied, and its implementation does not contain problems.

Electric part of model of electric motor is standard, without the account of reaction of anchor and at supposition, that the number of slots is infinitely large, i.e. the anchor chain of engine and chain of excitation is described linear differential equalization of the first order, taking into account nonlinear dependence anti - E.M.F. engine from the current of excitation. The model of anchor chain of engine, as a rule, is combined with the model of transformer.

Basic problem, arising up at creation of similar stands, it is the limited fast-acting of microcontroller for realization of model, working in real time. For the estimation of necessary fast-acting of microcontroller will consider the algorithm of work of models of transformers and engines of direct-current which is expedient for creation of the indicated stand. The models of electric drive of direct-current with a reversible thyristor transformer are known and widely used for planning of thyristor transformers [7,8] and for the virtual models of electric drive of direct-current. In first case they are a volume, because contain the models of semiconductor devices, in second case usually models simplified, as ignore the processes of commutation and the terms of disabling thyristor are unchecked.

In our case it is not necessary to consider transients in thyristor, it is sufficient to assume that:

the inclusion of the thyristor occurs instantly if the conditions of its inclusion (positive edge pulse on the control electrode of the thyristor and the positive voltage of the anode - cathode thyristor);

the inclusion of the thyristor occurs at a given, for this model of thyristor, the reverse current through it.

To obtain emergency modes of operation of the converter it is necessary to model the converter take into account switching processes when you enable (turn off) thyristor and contained a control voltage to the thyristors.

Two approaches are used when developing algorithms for simulation of three-phase bridge converters, working on the anchor chain of the engine:

the number of thyristor, which is fed with pulse on and the other thyristors (enabled/disabled) select a design scheme and executed the integration of the equations describing the electrical processes in the scheme, the following defines the moment of changing the state of the thyristor and switch to a different design scheme, and in the on state resistance of the thyristor is equal to zero, off – infinity;

calculated currents and voltages in all branches of the circuit, wherein the thyristor is replaced by a dipole whose parameters are changed depending on condition of the thyristor.

For example, if we consider the converter modes: a) normal mode rectification, b) the switching of the thyristors and c) emergency mode, if false turn on of the thyristor during commutation, we get three calculation schemes are shown in figure 2.

Sources of EMF in settlement schemes, the number of the controlled current to determine the moment of disconnection of the thyristors, the sign of the back EMF of the motor (zne), depend on the operating rooms and turn on the thyristors, as shown in table 1. The numbering of the thyristors in the table K = 1...6 corresponds to the sequence of their inclusion in time, the thyristor 1 is installed in the cathode group of phase A. In the design schemes

$$i = \begin{cases} i = k & if \quad (1 \le i) \cap (i \le 6) \\ i = k - 6 & if \quad i > 6 \end{cases}$$

i+2 included the thyristor; i+3,4,5 – falsely included thyristors.

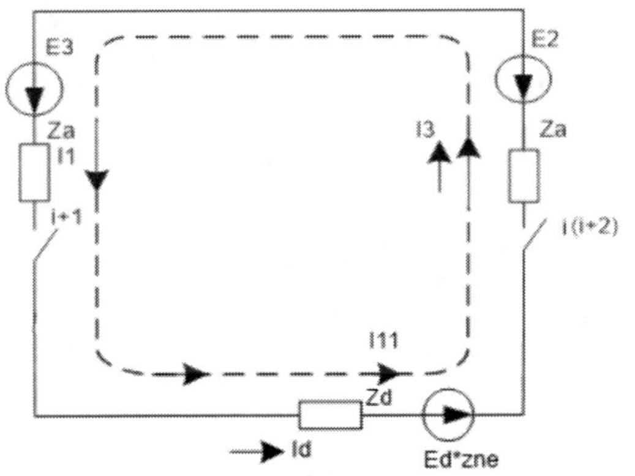

A)

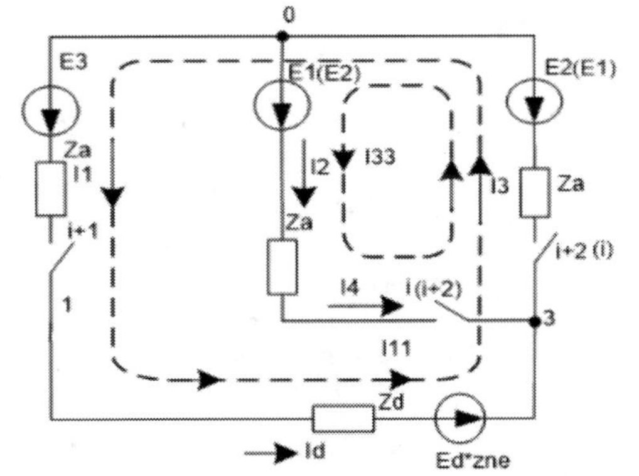

B)

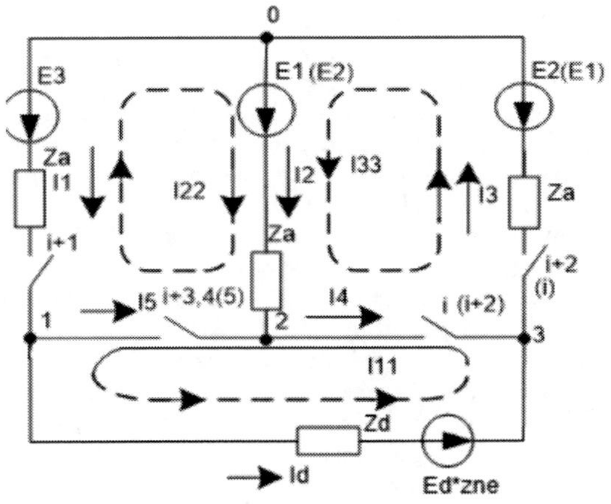

C)

Fig. 2. Calculation schemes/models of three-phase bridge converter

Table 1. Characteristics of thyristors 3-phase bridge converter

Commutation in group	Working thyristors	The included tiristors	Sources			i	i+1	i+2	Sign Ed zne
			E1	E2	E3				
cathode group	1,2	3	Ua	Ub	Uc	1	2	3	+
	3,4	5	Ub	Uc	Ua	3	4	5	+
	5,6	1	Uc	Ua	Ub	5	6	1	+
anode group	6,1	2	Ub	Uc	Ua	6	1	2	-
	2,3	4	Uc	Ua	Ub	4	3	4	-
	4,5	6	Ua	Ub	Uc	2	5	6	-

The order of switching of the thyristors is determined by a hardware-implemented control system of thyristor converter and electric drive in normal mode. Emergency operation of thyristor can forcibly be set by the program of management of model depending on an electric drive research problem. The program of model defines number of the settlement scheme then the subprogramme of the solution of the differential equations corresponding to this scheme joins.

The accuracy of the solution is determined step integration of the equations. Thus, the main impact on errors of a method of modelling will be had by time of the solution of the differential equations.
The simplified diagram of the algorithm of the model showing on the figure 3.

Most of the time the calculations are carried out according to the scheme of figure 2a, in the period of commutation calculations are carried out according to the scheme of figure 2b, and the time of switching for three-phase bridge converter is about 34% of the duration of the conversion interval.

Despite the last circumstance, it is necessary to choose an integration step proceeding from need of the solution of two differential equations which are worked out on a method of planimetric currents for electric part of the drive, the differential equation for mechanical part of the drive and definition on in advance set logical equation of values of EMF in the equations for an anchor chain of the engine.

The coefficients of the equations is a positive number, constant for the actuator, to define them in part of the preparations for the decision. When using a micro-computer that computes in the format with a fixed decimal point, are calculated according to the magnitude of currents and voltages. For modeling the converter with the engine in the second method we will use methods of calculation of electric circuits on the PC, was considered in [9]. For the calculation method of nodal potentials the discrete circuit model ra–La is given on the figure 4a, and the circuit rd– Ld on the figure 4b, the discrete model of the converter circuit non-reversing engine shown on the figure 4c. You can see on the figure that ga – anode circuit of the thyristors, gd – anchor chain of the engine, the gt is a two – terminal thyristor of substitution (in our case the conductivity).

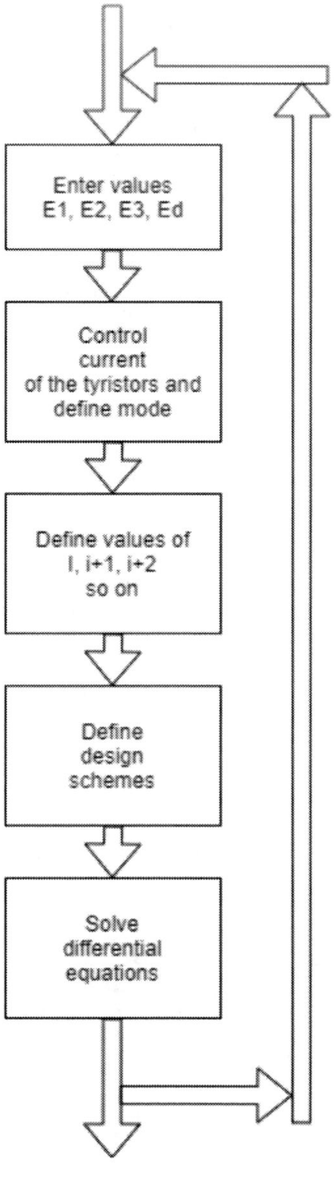

Fig. 3. A simplified diagram of the cycle algorithm within the time step of integration

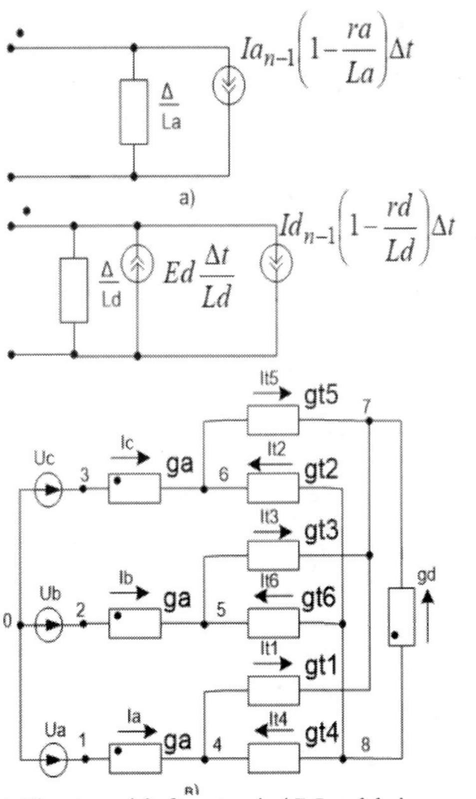

Fig. 4. Discrete model of two-terminal R-L and design scheme

Therefore, further we will stop on the first way of modelling for an assessment of level of sampling of process on time.

The mathematical apparatus of the model: the solution of differential equations by the method of numerical integration of Runge–Kutta methods providing sufficient accuracy of calculations. Calculations are performed on 64-bit numbers in floating point format. Supply voltage is entered using the built-in microcomputer ADC, and to ensure the safe voltage is fed through step-down transformers. In analog form are issued by the currents of the thyristors, anchor the motor current and the speed of rotation of the motor shaft.

Hardware mathematical model is a micro-computer family of ARM Cortex™ (A – application, R - real time, M - microcontroller) [10], as technology has widely spread lately. The company manufacturer of micro-computers selected from the calculation of the most fast-acting of the crystal, is presented in the line of this technology – ST Microelectronics. This micro-computer comprises peripheral means of analog-to-digital converter (ADC) with a resolution of 12 bits and the digital-to-analog converter (DAC) permission. The speed of the ADC of the micro-computer at a given precision of the calculation is 5 microseconds. The implemented algorithm of calculation of parameters of electrical machines by the method of Runge-Kutta methods, when issuing the sync pulse, showed a frequency of 19.6 KHz, which corresponds to 50 microseconds per clock cycle calculation (cycle computing). This result is not final, in view of the fact that the frequency of the generator of the kernel of the micro-computer corresponded to 48 MHz at 168 MHz is possible for this series. Thus, the sampling frequency of the processes under control in real time close to the sampling frequency processes when using the standard virtual models of engines.

Modelling of the excitation circuit does not cause problems due to large time constants and is executed by the first method without taking into account the processes of switching thyristors. To save time, the model of the circuit runs on a separate micro-computer working with a fixed decimal point.

In the system of the stand, it is advisable to include also the load node and set various options for the load-dependent rotational speed and position of the motor shaft, to simulate the elasticity of the connecting shafts, the errors of the gears, etc [11]. For this purpose you need an additional micro-computer, in connection with a sufficiently large amount of computational tasks. All micro-computers are linked by a synchronous data transfer system that will not cause an increase in the duration of the sampling interval.

Conclusions.

Training of specialists in the field of the electric drive in higher educational institutions was generally reduced the last years to theoretical part.

Now virtual methods of research of electric drives which provide variety of tasks, rather low capital expenditure are widely used and don't demand special events for safety of performance of laboratory works.

A significant approximation of the laboratory setup to a real drive is the laboratory plant consisting of a real control system and model of the power unit (Converter installations and motor) operating without time scale.

Basic problem, arising up at creation of similar stands, it is the limited fast-acting of microcontroller for realization of model, working in real time. For the estimation of necessary fast-acting of microcontroller we consider the algorithm of work of models of transformers and engines of direct-current which is expedient for creation of the stand.

Using the proposed principle of organization of laboratory work allows the student at low capital cost:
to learn how to work with real control systems, master the techniques of setting-up and definition of failure;
to explore a wide range of tasks for different power levels, different nature of loads;
to investigate systems of the interconnected electric drive.

Further research should be aimed at the approximation of the mechanical part of the drive to real conditions, in particular taking into consideration the gaps in it, a nonlinear transmitting devices and etc. Modeling of gaps and nonlinear transfer devices considerably increase the volume of calculations that will demand creation of multicontroller systems and development of modeling methods for them in real time.

It is also necessary to carefully select the element base, since microcontroller manufacturers are constantly improving their models. The main selection criterion should be the requirement to ensure sufficient productivity at a low cost.

References

1. Chornyy O. P., Lugovaya A.V., Rodkin D. Y., Sisuk V. Y., A. V. Sadovoy. Modeling of Electromechanical systems, textbook for universities, Kremenchug, (2001).
2. Esin E. K. Modeling of Electromechanical processes multimotor electric drives of mining machines. – Kemerovo: Kuzbass state technical.Univ., (1999).
3. Terekhin V. B. Simulation of electric drive systems in Simulink (MatLab 7.0.1). Tomsk: Publishing house of Tomsk Polytechnic University, (2008).
4. Shreiner R. T. Mathematical modeling of AC drives with solid state frequency converters DJVU DOC. Ekaterinburg: URO ran, (2000).
5. Patent Russian Federation N 2442995 G 01 R 31/34. The stand with the electromagnetic load module for research and test of electric drives. / A.I. Nekrasov, A.V. Yefimov, A.A. Nekrasov, N. N. Serykh, V. Z. Trubnikov, (publ.12.10.2012).
6. Patent Russian Federation N 24473921, G 01 R 31/34. The educational and laboratory stand for studying of electrical machines and electric drives. / Y.S. Roytburg, And, A. Sesin, Y.P. Dolgov, A.A. Prentsel, (publ. 27.01.2013).
7. D. N. Aminova, V. S. Mukhamedzhanov, V. P. Smirnov, M. R. Khalilova. A method of modeling and library of models of elements of schemes of valve converters for ATVK. The collection the Automated electric drive, Moscow, Energoatomizdat, (1983).
8. Kuzmin V.A., Moustapha G. M., Minosyan V. L. Modeling of transition process of inclusion of a tiristor on AVM. Electrotechnical industry. Converting tekhnika.(1978).
9. Nerreter V. Calculation of electric chains on the personal COMPUTER, Moscow, Energoatomizdat, (1991).
10. Y. Joseph. Kernel of Cortex-M3 of the ARM company. Full management. World Electronics series. Dodeka — XXI publishing house, Moscow, (2012).
11. Ivanushkin V.A. etc., Structural modeling of electromechanical systems and their elements/VA Ivanushkin, V. N. Sarapulov, P. Shymchak: Szczecin, (2000).

Current Aspects of Prevention of Coal Freezing by Means of Inorganic and Organic Reagents

Ekaterina Shmeltser[1], *Maryna* Kormer[1]*, and *Vitalij* Lyalyuk[1]

[1]State University of Economics and Technology, Kryvyi Rih, 50006, Ukraine

Abstract. In the cold season, irregular coal supplies to coke plants are aggravated by the need to heat the coal cars. Thawing of rail cars in garages (enclosures) is the least efficient and most expensive approach. Treatment of the coal concentrates with chemical additives reliably prevents freezing in winter during transit from suppliers to consumers. With a view to finding new reagents for preventing the freezing of coal in winter, the lime, the acetates and chlorides of alkaline-earth and alkaline metals are studied. Attention focuses on their physicochemical characteristics, methods of preparation and of introduction in coal concentrate, and their influence on freezing. The results of studies have shown that the use of organosilicon is more effective use of acetates and chlorides of metals. The high activity of organosilicon substances is explained by their elemental composition and structure of molecules.

1 INRODUCTION

Among the problems encountered by coke plants in the cold season is freezing of the coal concentrates on transportation, which hinders their discharge from the rail cars. The individual coal particles freeze together and also bond with the floor and walls of the car.

This is not a new problem. It has traditionally been addressed by heating the cars with coal in special garages (enclosures), which consumes considerable quantities of energy. It is much less expensive (by several orders of magnitude) to prevent freezing than to heat the frozen coal.

Freezing of coal is possible not only in northern and eastern Russia and in Kazakhstan, which provide coal concentrates for coke production at PAO ArcelorMittal Krivoi Rog, for example, but also in the relatively mild climate of the Ukraine. Recently, despite global warming, the air temperature in winter has been falling to –20°C or less, which interferes with coal supplies to coke plants on account of the longer time required for sorting of the wagons by coal rank before heating and discharging when the freezing of the coal reaches a depth from 0.25–0.30 to 0.6–1.0 m.

The freezing of coal is a complex thermophysical process accompanied by the migration of moisture and temperature variation. It depends primarily on the thermal conductivity and specific heat of the coal, the duration of the low ambient temperatures, the moisture content of the coal, its metamorphic stage, the composition of mineral impurities, the granulometric composition, and the hydrophilic properties of the surface. However, the key factor is the moisture content of the charge [1].

The water in the coal charge may be divided into free and bound types, depending on its bonds with the solid particles and mobility. The ratio between the quantity of free and bound water depends on the petrographic characteristics: the metamorphic stage, petrographic composition, and dispersity of the coal [2]. Free moisture, with purely physicomechanical bonds to the coal, is formed at the interface of three phases under the action of capillary forces. Its proper ties do not differ from those of ordinary water and it freezes at around 0°C. It evaporates at the same rate as pure water [3]. On account of its latent heat, bound water does not freeze even at – (50–70) °C. The con tent of free water increases with the dispersity of the load, which promotes freezing of the coal.

The moisture content of the coal such that it contains only unfrozen water is regarded as the safe moisture content, whose value depends on the metamorphic stage of the coal. The safe moisture content increases with increase in the yield of volatiles and with decrease in ash content of the fuel. That is because the safe moisture content is less for the coal's mineral compound than for the organic component. Increase in ash content of the coal (increase in the content of clay and other impurities able to form colloidal solutions) accelerates its freezing on transportation and storage [3].

The safe moisture content is highest for low metamorphic coal and declines when the coal reaches intermediate stages of metamorphism. That may be attributed to the different porous structure at the surface of the coal particles in different stages of metamorphism, which are associated with different total internal moisture content; by phase composition of the water at the particle surface; and by the strength of the ice that forms [4]. If the moisture content of the coal is at or below the safe level, freezing of the particles will not occur, and the coal will remain friable at negative temperatures [3].

*Corresponding author: maprina1955@gmail.com

On freezing, three materials interact in the railroad car: the coal, the material of the car's floor and walls, and the air. The main components of the freezing loads are coal particles, water (in three aggregate states), and the gases between the coal particles. At negative temperatures, the friable charge takes on new factors due to redistribution of the free moisture and its conversion to ice and to the formation of cooled structure and texture. The free moisture is mainly converted to ice layers. In the freezing coal, migration of free moisture to the floor of the car is sharply reduced, on account of ice barriers. The transfer of moisture mainly occurs under the action of the temperature gradient [2].

The influence of moisture on the freezing of coal may be assessed in terms of the effort required to separate the freezing samples at the same negative temperatures. The binding strength of the frozen coal particles increases with increase in moisture content and decrease in temperature.

Since the thermal conductivity of different coal samples fluctuates over a small range (0.63–1.26 kJ/hm °C), the freezing of the coal pieces to one another begins at the same time as their freezing to the walls of the car [5]. Then, depending on the ambient temperature and the residence time, the freezing process extends in all directions [6]. The binding strength on freezing depends not only on the coal's moisture content and the ambient temperature but also on the material from which the car is made. The force required to separate a coal layer frozen to walls of different materials at –15°C is as follows (kN/m^2), according to [3]:

Table 1. Forces of separation of frozen material from different surfaces

Material	Forces of separation, kN/m^2
Steel	78.4
Wood	71.5
Aluminum	32.3
Organic glass	27.4
Ebonite	18.6
PVC	12.7
Fluoroplastic	12.7

2 Analysis of previous studies

Protection against freezing depends on analysis of the responsible factors and study of the heat and mass transfer and the rate of freezing of coal on transportation. At present, we may identify two basic approaches to addressing the problem [5, 6]:

1) prevention of freezing;

2) restoration of the frozen coal's friability before or during discharge.

Methods of the first type include drying of the coal, mixing of wet and dry coal, refreezing of the coal, and the application of hydrophobic protective coatings to the coal and the walls of the rail cars.

Methods of the second type include heating of the cars with frozen coal in special garages (enclosures) and mechanical action by drilling, vibration, and agitation machines.

The most promising preventive method is to reduce the pour point of the moisture's active component and reduce the strength of the bonds in the frozen coal. The materials employed should be harmless to the operating staff and the environment, should not cause corrosion of metal components, should not impair coal quality, should not significantly reduce the capacity of the rail car, should not require special storage conditions, and should mix well with coal.

3 Results of the study

In laboratory research on the freezing of coal concentrate samples from the coal preparation shop in the coke plant at PAO ArcelorMittal Kryvyi Rig, we using the lime, the acetates and chlorides of alkaline-earth and alkaline metals are tested as reagents to prevent freezing of coals.

It is found that CaO may expediently be used to prevent freezing if its activity is no less than 85–88% and the ambient temperature is not below – (15–16) °C [7]. At lower temperatures, freezing is not prevented by unslaked lime even in quantities of 3–6% of the mass of coal. Unslaked lime only serves an auxiliary function; it may be used as a mechanical layer between the coal layers.

The positive effect of the use of acetates, formiates, and carbamides can be explained by the impact acetyl radicals on centers of the crystallization in the formation of ice, which ensure that it is loose and weak.

We consider acetates of sodium, potassium, calcium [8]. The acetates of potassium and sodium are mixed with coal in quantities of 1.5, 2.0, 3.0, 4.5, 5.5, and 6.0 wt %. The moisture content of the coal (≤3 mm class) is 12%, ash content of coal in the dry state – 11.7%, volatile matter in the dry ash-free state – 25,6%.

In Table 2, we show the results of studies and laboratory data of dependence freezing temperatures of coal from the mass fraction of potassium and sodium acetates added to the coal.

Table 2. Dependence of the freezing point of the coal on the added potassium and sodium acetates

Added acetate, wt %	Freezing point with added potassium acetate, °C	Freezing point with added sodium acetate, °C
1,5	-4,5	-4,3
2,0	-5,0	-5,2
2,5	-6,4	-6,8
3,0	-7,6	-8,0
3,5	-8,6	-8,5
4,8	-9,4	-8,6
5,5	-10,6	-8,7
6,0	-12,3	-8,7

Potassium acetate effectively reduces the freezing point of coal. There is a clear dependence of the salt content on the temperature. In contrast to potassium acetate, sodium acetate reduces the freezing point to – 8.5°C, at concentrations no greater than 3.0%. With further increase in its concentration, the freezing point of the coal remains unchanged.

These reagents are less corrosive and are characterized by lower environmental impact. Metal acetates are safe for most surfaces, including concrete, metal, and wood. They are nontoxic and biodegradable under the action of bacteria. Alkaline-earth metals, in contrast to alkaline metals, are present in fluxes and improve the blast-furnace process. Their addition in salt form to coal does not impair its technological properties. Therefore, we now focus on the use of calcium and magnesium acetates to prevent the freezing of coal [8].

Calcium and magnesium acetates were added to the coal raw materials in various ways:

1.The acetates are mixed with coal of 12% moisture content in quantities of 1.5, 2.0, 3.0, 4.5, 5.5, and 6.0 wt %;

2.Method of producing calcium and magnesium acetates within the coal mass. To that end, the coal is preliminarily mixed with calcium or magnesium oxide. Then the equivalent quantity of 60% acetic acid is carefully added to the mixture obtained, in accordance with the reaction

$$2CH_3COOH + CaO \rightarrow Ca(CH_3COO)_2 + H_2O \quad (1)$$

Since water is liberated by reaction with 60% acetic acid, we need to calculate the effect on the overall moisture content of the coal. It follows from calculation that, even with 6.0% salt, the moisture content of the coal is increased by no more than 0.6%.

3. Method of producing calcium and magnesium acetates based on the reaction of the carbonates of these metals with acetic acid by the reaction:

$$2CH_3COOH + CaCO_3 \rightarrow Ca(CH_3COO)_2 + H_2O + CO_2\uparrow \quad (2)$$

In contrast to the previous method, carbon dioxide is formed in this case. In the case of careful mixing with acetic acid, the freezing point of the coal remains precisely the same. However, if mixing is minimized so that the gas formed is largely retained within the coal, the freezing point falls. In this case, a weighed sample of coal is divided into five 100-g portions. Each portion is carefully mixed with calcium or magnesium carbonate, and then built up in layers within a vessel. Each layer is combined with the calculated quantity of acid, without mixing.

Comparison of results indicates that treatment with 5.0% salt by the proposed third method reduces the freezing point of the coal from to –14.7°C and not at –13.5°C (1 and 2 methods) [7].

We now turn to solutions of calcium chloride and magnesium chloride [7, 9].

For coal of moisture content up to 12%, the optimal addition of $CaCl_2$ solution is 1.5– 2.0%/t of coal. About 30% of the solution should be mixed with the lower layers of the coal, which freeze more severely; the remainder should be distributed uniformly through the rest of the coal. In Fig. 1, we show laboratory data for calcium chloride. The use of calcium chloride is not advised if the coal will subsequently be subjected to high temperature treatment.

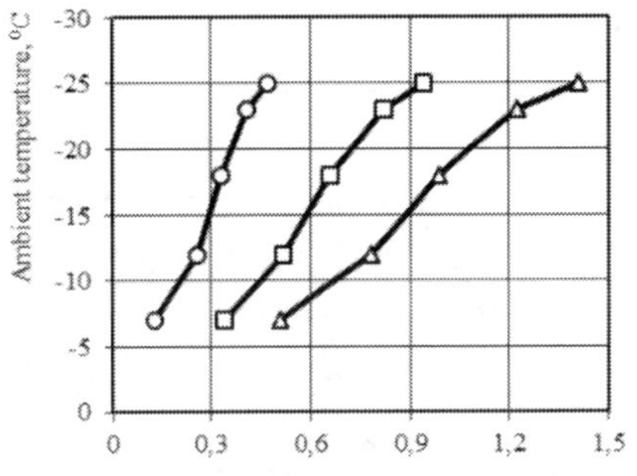

Fig. 1. Calcium chloride consumption to prevent the freezing of coal with 3% (1), 6% (2), and 9% (3) moisture content

Above 900°C, the calcium chloride added to the coal begins to break down, with the liberation of chlorine. This process is especially vigorous in the presence of oxides of silicon and aluminum, which act as catalysts. The presence of free chlorine and hot coke may lead to the formation of toxic dioxin. Other problems with calcium chloride are its high cost and corrosive properties.

These concerns do not apply to magnesium chloride, the main component of natural bischofite. It is mined, for example, in the Poltavsk and Chernigovsk deposits in Ukraine. This solution is safe and has medicinal properties. In contrast to calcium chloride, it is not corrosive; in fact, it can slow corrosion rates. In laboratory tests, the quantity of magnesium chloride added is 1–5% of the coal mass. The moisture content of the coal concentrate employed (≤3 mm class) is 12%. The table 3 presents the experimental results.

We see that adding 4–5% bischofite reduces the freezing temperature of the coal particles to –16–18 °C. Since the air temperature in the Ukraine rarely reaches –20°C in winter, adding 2.5–3% bischofite to coal transported within the country is best; the effectiveness of mixing must be no less than 96–98%. As shown by experiments, adding bischofite to coal does not change the packing density of the coal and raises the basicity index by only 0.5%.

In this approach, the moisture between the coal pieces is replaced by a solution with a low eutectic freezing point. The ice formed is characterized by a defective flake structure and consequently is fragile.

Table 3. Freezing temperature of coal treated with bischofite

Added bischofite, %	Calculated freezing temperature of bischofite solution, °C	Freezing temperature of coal treated with bischofite, °C
1,0	-7,3	-4
2,0	-11,6	-8
3,0	-14,6	-10
4,0	-16,3	-16
5,0	-19,3	-18

Another benefit of magnesium chloride over calcium chloride is that its decomposition at high temperatures does not release chlorine

$$MgCl_2 \cdot 6H_2O \xrightarrow{20-60\,^{\circ}C} MgCl_2 \cdot 4H_2O \xrightarrow{90\,^{\circ}C} MgCl_2 \cdot 2H_2O \xrightarrow{119{,}5\,^{\circ}C} MgCl_2 \cdot H_2O \xrightarrow{505\,^{\circ}C} MgO + 2HCl \quad (3)$$

Research shows that coal with low moisture content (no more than 11–12%) may expediently be treated with an aqueous solution of bischofite, since such treatment will raise the coal's moisture content.

Comparison of protective additives shows that sodium chloride is active to –10°C, calcium chloride to –10–16 °C, and magnesium chloride to –16–18 °C.

The chloride consumption depends on the coal's size class, as is clear from the dependence of the degree of freezing on the moisture content (Table 3).

With increase in size class, as we see in Table 3, the chloride consumption falls. Thus, the calcium chloride consumption declines from 9.2 to 5.3 g/kg of coal at –5°C, from 15.3 to 8.4 g/kg at –10°C, and from 18.3 to 12.3 g/kg at –15°C.

Analogous results are obtained for magnesium chloride. Its consumption declines from 6.2 to 3.8 g/kg of coal at –5°C, from 10.5 to 5.5 g/kg of coal at –10°C, and from 15.6 to10.8 g/kg of coal at –15°C.

Note also that the consumption of calcium chloride practically doubles with decrease in temperature from –5 to –15°C for all size classes: from 9.2 to 18.3 g/kg for the ≤3 mm class; from 7.4 to 14.2 g/kg for the 3–7 mm class; and from 5.3 to 12.3 g/kg for the 7–15 mm class. In the same conditions, the consumption of magnesium chloride increases from 6.2 to 15.6 g/kg for the ≤3 mm class; from 5.4 to 13.5 g/kg for the 3–7 mm class; and from 2.3 to 10.8 g/kg for the 7–15 mm class. The magnesium chloride consumption is less than the calcium chloride consumption [9].

Table 4. Consumption of calcium chloride and magnesium chloride to prevent freezing

Size class, mm	Chloride consumption, g/kg of coal					
	$CaCl_2$	$MgCl_2$	$CaCl_2$	$MgCl_2$	$CaCl_2$	$MgCl_2$
	at -5 °C		at -10 °C		at -15 °C	
0-3	9,2	6,2	15,3	10,5	18,3	15,6
3-7	7,4	5,4	10,7	8,6	14,2	13,5
7-15	5,3	3,8	8,4	5,5	12,3	10,8

As is evident from Table 4, treating only the ≤3 mm class increases the chloride consumption. Therefore, to prevent freezing, careful mixing of all the coal with reagent is expedient, without division into size classes

But it should be noted that it is advisable to treat with bischofite aqueous solution only coal with low humidity, because when processing coal with an aqueous solution of bischofite increases its humidity. The ultimate increase of coal humidity is 10-12%. Within this range, when the temperature drops below minus 15 °C, the coal will freeze. At lower air temperatures, deep freezing of coal should be expected. The search for the most effective chemical preventive agent to prevent the freezing of coal concentrates in winter at very low temperatures (below minus 18 °C) during transportation from the supplier (coal processing plants) to coke plants showed that these requirements are satisfied when using silicone polymer.

Table 5. Consumption of magnesium chloride in treating the ≤3 mm class [9]

Content of ≤3 mm class, %	$MgCl_2$ consumption, g/kg of coal		
	at -5 °C	at -10 °C	at -15 °C
50	8,2	11,4	16,8
70	12,4	17,4	21,6
80	16,2	22,1	26,6
90	18,7	26,5	30,5

Molecules of organosilicon compounds combine in part the structure of polymeric inorganic and organic molecules. The basis of their molecules is the siloxane skeleton - a chain of alternating atoms of silicon and oxygen. Other silicon bonds are compensated by organic radicals or groups of atoms [10].

The peculiarity of the structure is associated with the unusual properties of these polymers, which combine the stability of these compounds to the action of low temperatures. Organosilicone liquid polymers can be obtained with a freezing temperature of minus 130 ° C or even lower (fig. 2).

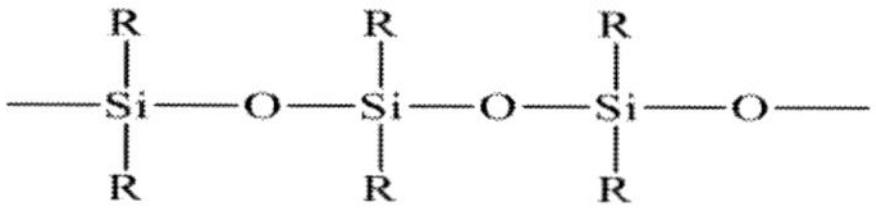

Fig. 2. The formula of organosilicone

A valuable property of silicone liquids is the weak dependence of their viscosity on temperature. Thus, as the temperature decreases, the viscosity of the organosilicon compounds increases to a lesser extent than for petroleum oils. This makes it easier to apply to the surface of a metal wagon and mix with coal [6].

For polysiloxanes, low freezing temperatures and chemical inertness to metals are typical. The flammability of organosilicon liquids is much lower than that of organic compounds. The final combustion products are carbon dioxide, water and silica as a fine powder. Polysiloxane fluids do not irritate the skin and eyes.

Thus, the most valuable technical properties of organosilicon liquids are determined by their physical and chemical properties: low pour point (minus 70 to minus 140 ° C), low viscosity dependence on temperature (multiplicity of viscosity change is 5-10 times in the interval temperature temps from - 50 to + 50 ° C) [6].

Emulsions of silicone polymers are stable structures of the type «oil in water». Due to the fact that it is practically more convenient to use aqueous emulsions of silicone liquids instead of their solutions in organic solvents, the industry produces 30-70% silicone emulsions. These emulsions are easily diluted with water and used in 5-10% concentration. In commercial form, the emulsions are a white mass of creamy consistency.

Molecules of organosilicon compounds contain an oxygen atom, so the electron density at oxygen atoms is localized. As a result, the reagents can specifically interact with the sorption-active centers of the coal surface bearing a positive charge (protonated hydrogen atoms of phenolic, carboxyl groups in macromolecules of organic mass of coal) according to the scheme (fig.3):

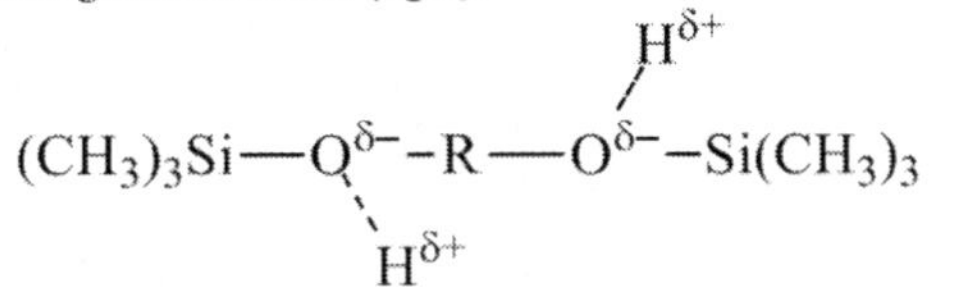

where R – alkyl radical

Fig. 3. Scheme the interactions of protonated hydrogen atoms of phenolic, carboxyl groups in macromolecules organic mass of coal

Molecules of oxygen-containing organosilicon compounds also contain hydrocarbon radicals that can interact with the apolar centers of the coal surface due to Van der Waals intermolecular forces. Therefore, when adsorbed on the coal surface, molecules of organosilicon exhibit both specific (hydrogen bonds) and universal nonspecific intermolecular forces of interaction with positive parts of the coal surface. This determines their high adsorption on the coal surface.

If the solution or emulsion is mixed with coal, these atoms or groups are reacted with the organic mass of the coal and water hydrate. Organosilicone bonds - Si - O - Si - O -, which provide physical and chemical interaction with the coal facing the surface, and organogenic groups framing the silicon atom in the opposite direction - outwards. The orientation of organosilicon bonds and hydrocarbon radicals when applied to the surface can be schematically represented Fig.2.

Organosilicon compounds, when applied to a coal surface, react with water approximately molecule to molecule. The films of organosilicon compounds are very thin. The thickness of the film, which is calculated by the material flow and film weight, assuming that its density is equal to one, is $0.1 \cdot 10^{-6}$-$2.5 \cdot 10^{-5}$. The film on the surface is invisible, does not wear away when rubbed, does not wash off with water, is stable over a wide temperature range (from - 200 to +300 °C). Polymeric silicones on the surface of coal, even in small quantities, provide a great water repellent effect [10].

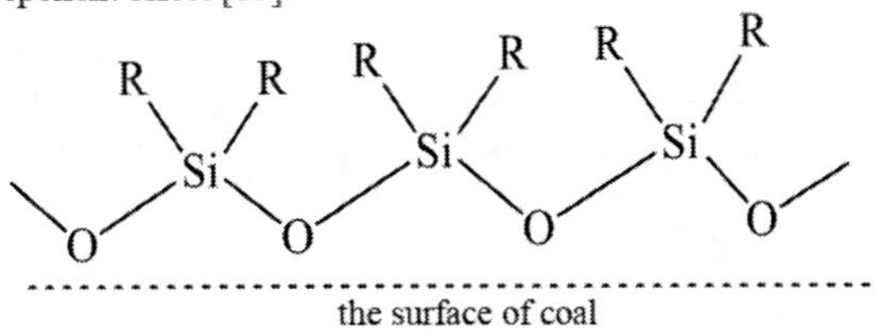

Fig. 4. Orientation of organosilicon bonds and hydrocarbon radicals on the coal surface

For the experiment, a sample of coal weighing 500 g was taken, the moisture content of the coal was 12%, the size class was 0-3 mm. The coals were carefully treated with an emulsion. Next, the processed coal was loaded into a metal tank with a diameter of 50 mm, height 50 mm, without bottom and cover. This metal container was placed in the freezer and kept at a predetermined temperature for 24 hours. Then the tank was turned over, and the coal that was poured was weighed. The degree of freezing was thus determined. The study used an organosilicon fluid GKZh-94, which is a colorless slightly yellow liquid. Density – 0,996-1,003 g/cm^3; freezing temperature -50 °C; pH not less than 6. In water it is insoluble, but well forms an emulsion, corrosion inactive, does not emit harmful vapors and gases, weatherproof [11].

The organosilicon additive was used as 5 and 10% aqueous emulsions. The emulsion was thoroughly stirred with charcoal for 3-5 minutes. The amount of emulsion was changed from 4 to 40 g per 1 kg of coal. The concentration of the emulsion was 5 and 10%. The results of the experiment are shown in Table 6.

Studies have also been conducted to identify consumption rates of known and claimed prophylactic agents to prevent freezing of coal with a moisture content of 12% (Table 7) [11].

The analysis of these tables shows that the organosilicon emulsions reduce the freezing point well, and when compared with salts it can be seen that the consumption of the organosilicon emulsion in comparison with the consumption of calcium chloride, magnesium and potassium acetate is lower.

Table 6. The degree of freezing of the treated coal GCZh-94 emulsion at minus 15 °C

Quantity of GCZh-94, g/kg	Mass of spilled coal, g	Variation of mass, g	Degree of freezing, %
5 % emulsion			
4	150	350	70
8	170	330	66
12	190	310	62
16	220	280	56
20	260	240	48
24	290	210	42
28	310	190	38
32	390	110	22
36	450	50	10
40	490	10	2
10 % emulsion			
4	200	300	60
8	220	280	56
12	250	250	50
16	270	230	46
20	290	210	42
24	360	140	28
28	400	100	20
32	420	80	16
36	480	20	4
40	500	0	0

Table 7. Dependence of the freezing point of the coal with a moisture content of 12% added of prophylactic agents

Freezing point of sample, °C	Added of prophylactic agents, wt %			
	$CaCl_2$	$MgCl_2$	CH_3COOK	10 % GCZh - 94
-15	5,0	4,6	4,0	3,6

The successful use of chemical reagents additives depends on the effectiveness of their mixing with coal, with must be at least 96–98%. At various times, coke plants have used plate, blade, beater, drum, screw, and disintegration units for batch mixing.

We propose [7] a system of rotor type for final mixing of the coal batch with any protective additive. The mixer (Figs. 5 and 6) is mounted above a conveyer belt, so that its rotors fit within the cross sections formed by the ribbed conveyer belt, with a gap between the plane of the belt and the rotor blades.

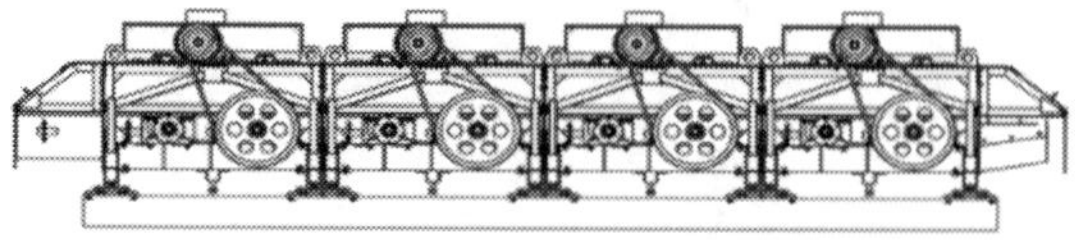

Fig. 5. Four-section rotary mixer: (*1*) mixer section; (*2*) support; (*3*) adjustable screws; (*4*) V-belt transmission; (*5*) electric motor; (*6, 8*) shielded input and output sections; (*7*) seal.

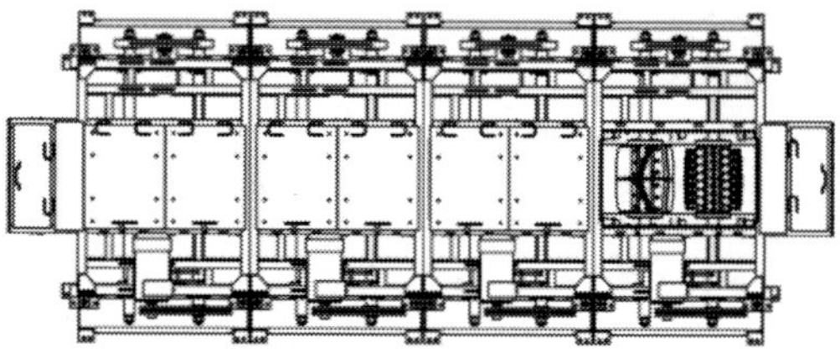

Fig. 6. Four-section rotary mixer (top view): (*1*) electric motor; (*2*) cable rotor; (*3*) blade rotor; (*4*) V-belt transmission; (*5*) tension roller

The rotors (Figs. 7 and 8) are responsible for the mixing. Each section of the mixer contains two types of rotors: one with regular blades and one with cable elements. The rotors turn in the same direction as the coal batch moves through the mixer.

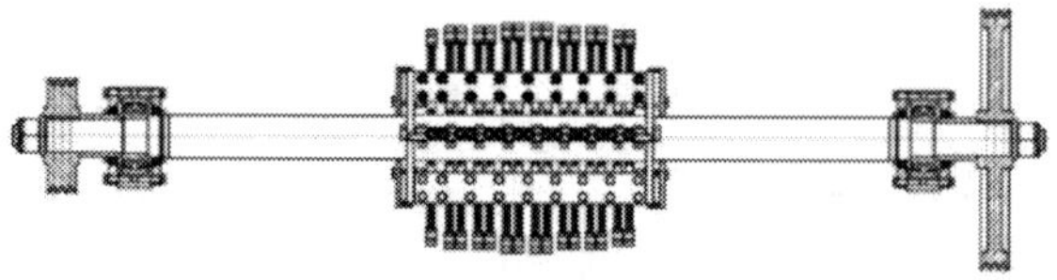

Fig. 7. Cable rotor of mixer: (*1*) shaft; (*2*) drum; (*3*) removable blades; (*4*) cable elements; (*5*) lock.

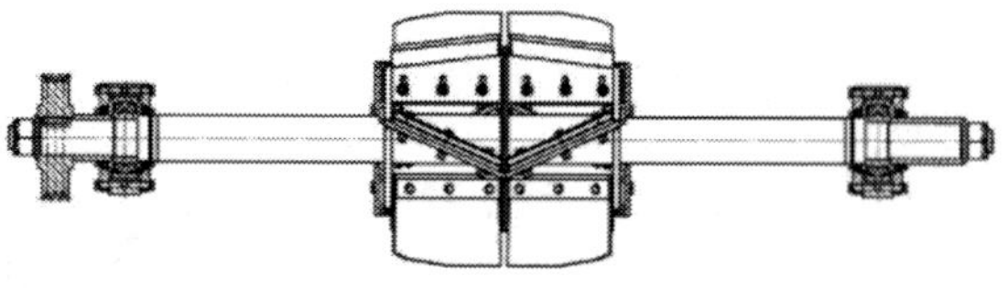

Fig. 8. Blade rotor of mixer: (*1*) shaft; (*2*) drum; (*3*) removable blades; (*4*) flexible blade elements; (*5*) lock

As the coal batch moves over the belt, it is acted on by the rotor blades and mixed with the protective additives. The characteristics of the SR2520×1.0 mixer are as follows:

Table 8. The characteristics of the SR2520×1.0 mixer

Parameter	Value
Width of conveyer belt, mm	1000
Number of sections	2
Rotor diameter, mm	520
Rotor speed, rpm	250
Length of rollers in horizontal support, mm	410
Conveyer slope, deg	0
Rated power, kW	15
Power of a single electric motor, kW	7.5
Number of motors	2

4 Conclusions

The treatment of coal concentrates with protective additives may reliably prevent freezing on transportation from suppliers to consumers in winter. The coal may easily be discharged from the rail cars, without large energy expenditures to heat the cars in special garages (enclosures). The increase in coal cost due to its treatment with protective additives by the supplier is more than compensated by the savings at the consumer, where no heating is required. Also, the coal does not need to be crushed by heavy duty machines, which usually results in excessive reduction in piece size and increase in leanness of the batch, especially in the case of valuable coal ranks, and sharply impairs coke quality.

Thus, the proposed method not only prevents the concentration of coal concentrates in winter during their transportation from the manufacturer to the coke plant, but also reduces the cost of coke by reducing the cost of defrosting (in greenhouses) and unloading coal, to prevent the corrosion of cars and equipment of coal-preparation shops of coke-chemical plants. Organosilicon compounds are less toxic. The use of organosilicon is more effective than the use of acetates and chlorides of metals. The high activity of organosilicon substances is explained by their elemental composition and structure of molecules.

References

1. S.F. Matasov, L.M. Kurtunov and A.S. Khorunzhii "Bor'ba so smerzaemost'yu pri perevozke po zheleznymdorogam" (Prevention of freezing in railroad transportation), Moscow: Metallurgiya, (1974)
2. M.I. Lepnev and E.P. Severinov. "Gruzy i moroz (Freezing of cargo), Moscow: Transport, (1988)
3. N.M. Mikhailov and A.T. Sharkov "Fizicheskie svoistva topliva i bor'ba s zatrudeniyami na toplivopodache elektrostantsii" (Physical properties of fuel and prevention o supply problems at power stations), Moscow: Energiya, (1974)
4. I.V. Eremin, V.V. Lebedev and D.A. Tsikarov "Petrografiya i fizicheskie svoistva uglei" (Petrography and physical properties of coal), Moscow: Nedra, (1980)
5. N.N. Kozhevnikov, V.I. Popov. "Prognozirovanie protsessov promerzaniya v spuchykh materialakh pri zheleznodorozhnykh perevozkakh" (Predicting

freezing in friable materials during rail transportation), Novosibirsk: Nauka, (1978)

6. V.E. Parunakyan, R.I. Sinyavskaya "Bor'ba s prilipaniem i primerzaniem gornoi massy k rabochim poverkhnostyam transportnogo oborudovaniya na kar'erakh" (Preventing the freezing of cargo to the working surfaces of vehicles in mines), Moscow: Nedra, (1975)
7. A.D. Uchitel, M.V. Kormer, V.P. Lyalyuk, E.O. Shmeltser "Transportation of Coal Concentrates at Negative Ambient Temperatures" Coke and Chemistry, **56**, № 5, 167-172 (2013)
8. E.O. Shmeltser, M.V. Kormer, V.P. Lyalyuk et al "Prevention of coal freezing by means of acetates" Coke and Chemistry, **59**, № 4, 132-136 (2016)
9. M.V. Kormer, V.P. Lyalyuk, E.O. Shmeltser, A.D. Uchitel et al "Dependens of coals freezing point on its granulometric composition" Coke and Chemistry, **58**, № 1, 9-14 (2015)
10. M.V. Sobalevskiy, O.A. Muzovskaya, G.S. Popeleva "Svoystva i primeneniye kremneorganicheskikh produktov", M.: Khimiya, (1975)
11. M.V. Kormer, V.P. Lyalyuk, E.O. Shmeltser, D.A. Kassim, I.A. Lyakhova. UA Patent 113500: MPK (2006.01) C09K 3/18, Byull. Izobret., № 2 (2017)

AUTHOR INDEX

AUTHOR INDEX

AUTHOR INDEX